With a Chapter on

HORMONE ASSAY AND ENDOCRINE FUNCTION

by

Reginald Hall, M.D., F.R.C.P.
Professor of Medicine
University of Newcastle upon Tyne

and

Gordon M. Besser, M.D., F.R.C.P.
Professor of Endocrinology
In the University of London
At St. Bartholomew's Hospital Medical School

Cantarow and Trumper

CLINICAL BIOCHEMISTRY

Seventh Edition

ALBERT L. LATNER,
   D.Sc., M.D., F.R.C.P., F.R.C.Path.

*Professor and Head of Clinical Biochemistry,
University of Newcastle upon Tyne;
Consultant, Newcastle University Hospitals.*

W. B. SAUNDERS COMPANY
Philadelphia, London, Toronto   1975

W. B. Saunders Company: West Washington Square
Philadelphia, Pa. 19105

12 Dyott Street
London, WC1A 1DB

833 Oxford Street
Toronto, Ontario M8Z 5T9, Canada

Library of Congress Cataloging in Publication Data

Cantarow, Abraham, 1901-
  Cantarow and Trumper, Clinical biochemistry.

  First ed. by M. Trumper and A. Cantarow published in 1932 under title: Biochemistry in internal medicine; 2d-6th ed. by A. Cantarow and M. Trumper published under title: Clinical biochemistry.

  Includes index.

  1. Chemistry, Medical and pharmaceutical.   2. Biological chemistry.   I. Latner, Albert L.   II. Trumper, Max 1892-1971 Biochemistry in internal medicine.   III. Title.   IV. Title: Clinical biochemistry.   [DNLM: 1. Biochemistry. QU4 C229b]   RB40.T7   1975   616.07   73-89933

ISBN 0-7216-5637-4

Cantarow & Trumper   ISBN 0-7216-5637-4
Clinical Biochemistry

© 1975 by W. B. Saunders Company. Copyright 1932, 1939, 1945, 1949, 1955, 1962 by W. B. Saunders Company. Copyright under the International Copyright Union. All rights reserved. This book is protected by copyright. No part of it may be reproduced, stored in a retrieval system, or transmitted in any form or by any means, electronic, mechanical, photocopying, recording, or otherwise, without written permission from the publisher. Made in the United States of America. Press of W. B. Saunders Company. Library of Congress catalog card number 73-89933.

Last digit is the print number: 9   8   7   6   5   4   3   2   1

This edition is dedicated to the memory of

                                        EARL J. KING

Preface to the Seventh Edition

More than a decade has passed since the last edition of this book first appeared, and this particular period has been especially productive in regard to new advances. This edition therefore includes a good deal of new material and most of the modern advances in clinical biochemistry. The author felt himself both honored and burdened in the invitation to re-think and re-do a basic, highly valued text in clinical biochemistry written with fine care and sound scholarship through six editions by Professor Cantarow and Dr. Trumper. The format has been deliberately kept to that of previous editions, since it seems to have proved especially popular. There has been no hesitation in regard to the inclusion of accounts of advances in human biochemistry and those in physiology which relate to disease from the biochemical point of view. Subjects such as nucleic acid metabolism, especially in relation to replication and the control of protein biosynthesis, are advancing so rapidly that there is no doubt that much new knowledge will have been obtained while this book is being prepared. Wherever necessary, short clinical descriptions are also included.

This edition contains a more complete account of many of the hereditary disorders, such as those affecting amino acids, hemoglobin, the urea cycle, porphyrins, uric acid, glycogen, and the mucopolysaccharides. In a book of this size, it is obviously impossible to include all the known variants of each main type. Tay-Sachs disease, for example, exists in more than one form, but only the primary one has been mentioned. Accounts are also given of recent advances in relation to liver and kidney disease, and there is a somewhat more detailed account of clinical enzymology. Other new, or more detailed, additions include disaccharidase deficiency, insulin, diabetic ketoacidosis, the lipoproteinemias, the immunoglobulins, nucleic acid metabolism, bone metabolism, water homeostasis, renal and urinary lithiasis, hemodialysis and kidney transplantation. The section related to jaundice has been considerably updated, and recent advances concerning bile salts are also included.

The chapter on endocrine function has been virtually rewritten by Professors R. Hall and G. M. Besser, who are both clinical endocrinologists of great distinction. Without their help, it would have been almost impossible to give a succinct account of the most up-to-date aspects of this rapidly advancing topic.

A new final chapter has been added. This deals briefly with the history and development of clinical biochemistry and includes an account of SI units.

I wish to acknowledge the great patience and many social sacrifices made by my wife during the years in which this edition was being prepared. Thanks are also due to the publishers for their unfailing co-operation.

A. L. LATNER

Preface to the First Edition

Modern advances in physiology and biochemistry have developed a need, not for another laboratory manual, but for a book designed to correlate established facts with problems encountered daily in internal medicine. The rapidity and magnitude of these developments have resulted in the growth of a highly specialized branch of laboratory medicine, namely, chemical pathology. The evolution of this specialty within a specialty has unfortunately tended to remove the clinician still further from a thorough understanding of those phases of internal medicine that require the assistance of the biochemical laboratory for their complete solution.

The remarkably fruitful researches of recent years in the fields of experimental physiology and pathology, by demonstrating the significance of biochemical observations in a constantly increasing number of abnormal states, have correspondingly increased their value to clinical medicine and surgery as well as to the laboratory. Modern practice demands the application of present knowledge regarding aberrations of endocrine, renal and hepatic function, abnormalities of organic and inorganic metabolism, nutritional defects, edema, dehydration, etc., in all branches of medicine as well as in pre- and postoperative treatment. The enormous increase in the use of chemicals in industry and in the treatment of disease and the growing appreciation of the possibly deleterious effects of such agents upon the organism have also increased the service that the biochemical laboratory may render to the clinician. The intelligent employment of these facilities will be of fundamental value in the increasingly important field of industrial toxicology.

The essential function of the laboratory is to supply the clinician with information which will complement that which he may obtain by other methods. Each patient presents a problem which cannot possibly be appreciated on the basis of dissociated laboratory studies. However, in order to take full advantage of the findings of the biochemist, the clinician must have a clear understanding of the significance and limitations of the results of laboratory investigation. This must be based upon an appreciation of the biochemical and physiologic factors involved in the preservation and alteration of organ and tissue function, for it may be stated, more truly than ever before, that physiology is the handmaid of medicine.

The progress made in the fields of biochemistry, metabolism, nutrition, colloidal and physical chemistry is based largely upon observations of a highly specialized and technical nature. This often renders the original literature unavailable to the majority of students and clinicians. As a result usually either they accept the brief interpretive statements made in most works on diagnosis by laboratory methods or they rely upon the chemical pathologist for an interpretation of his findings. The position of the latter is little better than that of the clinician who is required to explain the significance of an enlarged liver in an individual whom he has not seen and concerning whose clinical history and physical condition he knows nothing. Experience in the laboratory and in the clinic has impressed the authors with the difficulty which students and physicians experience in bridging the wide gap between abstract biochemistry or physiology and clinical medicine. Books and articles in abundance have been written for and by specialists, but only a few have attempted to interpret specialized knowledge for the physician. The undergraduate student of medicine and the progressive physician wish to be familiar with the applications of biochemistry to clinical medicine and surgery. They should be as well acquainted with the limitations as with the significance of biochemical findings in any given case. This volume constitutes an attempt to supply this information.

Haldane has stated that the aim of physiology is to consider how the internal environment of the body is kept constant in spite of continual alterations in the external environment. The aim of this treatise is to consider how the internal environment of the body is altered by certain specific changes in tissue and organ physiology. It is further intended to indicate the manner in which the physician may best avail himself of information which can be obtained by biochemical studies. To this end the subject of functional diagnosis by chemical methods has been discussed in considerable detail. With few exceptions, the technic of laboratory methods has not been discussed, being available in many admirable standard texts on that subject. The discussion has been restricted to those phases of biochemistry which are concerned with problems commonly encountered in clinical medicine and, therefore, purely abstract and theoretical considerations have been excluded.

A. C.
M. T.

Contents

Chapter 1
CARBOHYDRATE METABOLISM 1

Digestion and Absorption 1
    Dissaccharidase Deficiency 3
    Fate of Absorbed Carbohydrates 4
    Endogenous Sources of Glucose 4
Utilization of Glucose 5
    Storage 5
    Oxidation 5
    Conversion to Fat 5
    Conversion to Other Carbohydrates 6
    Conversion to Amino Acids 6
General Processes in Carbohydrate Metabolism 6
    Anaerobic Metabolism (Glycolysis) 7
    Aerobic Metabolism of Glucose 11
    Alternate Aerobic Pathway (Pentose Shunt) 12
Role of Liver in Carbohydrate Metabolism 13
    Glycogenesis and Glycogenolysis 13
    Gluconeogenesis 14
    Assimilation of CO_2 14
Muscle in Carbohydrate Metabolism 15
    Glycogenesis 15
    Glycolysis 15
    Muscle Contraction 16
Interrelation of Carbohydrate, Lipid and Protein Metabolism 17
Endocrine Influences in Carbohydrate Metabolism 18
    Insulin 18
    Adrenocortical Hormones 24
    Anterior Pituitary Factors 25
    Epinephrine 26
    Thyroid Hormone 26
    Glucagon 26

Normal Postabsorptive Blood Sugar	27
Glucose in Body Fluids Other than Blood	28
Sugars Other than Glucose in Body Fluids	29
Regulation of Blood Glucose Concentration	30
Rate of Supply of Glucose to Blood	30
Rate of Removal of Glucose from Blood	31
Fundamental Regulatory Mechanism	32
Experimental Diabetes Mellitus	33
Total Pancreatectomy	33
Subtotal Pancreatectomy	34
Alloxan Administration	34
Administration of Adrenocortical Hormones	35
Anterior Pituitary Extracts	35
Thyroid Administration	35
Carbohydrate Metabolism in Hypophysectomized-Depancreatized Animals	35
Normal Alimentary Reaction (Absorptive Response)	36
Sugar Tolerance	36
Intravenous Glucose Tolerance Test	38
Mechanism of Production of the Normal Glucose Tolerance Curve	39
Factors Influencing Absorptive Blood Sugar Response	40
The One-hour, Two-dose Glucose Tolerance Test	42
Effect of Other Sugars	43
Galactose Tolerance	44
Insulin Tolerance Test	44
Insulin-Glucose Tolerance Test	44
Cortisone and Prednisone Glucose Tolerance Tests	45
Cortisone	45
Prednisone	45
Tolbutamide Tolerance Test	45
Plasma Insulin Levels	46
Epinephrine Tolerance Test	46
Phenomena Associated with Normal Alimentary Glucose Reaction	46
Decreased Serum Phosphate	46
Decreased Serum Potassium	47
Increased Respiratory Quotient	47
Abnormalities of Postabsorptive Blood Sugar Level	47
Fasting Hyperglycemia	47
Diabetes Mellitus	47
Hyperthyroidism	49
Increased Secretion of Epinephrine	50
Adrenal Cortical Hyperfunction	51
Hyperpituitarism	51

Anesthesia, Asphyxia, Hypnotics	51
Acidemia	53
Hepatic Disease	53
Miscellaneous	53
Fasting Hypoglycemia	54
Hyperinsulinism	54
Hepatic Disease	56
Adrenal Cortical Insufficiency	57
Anterior Pituitary Insufficiency	57
Hypothyroidism	58
Nervous System Disorders	58
Miscellaneous	58
Abnormal Alimentary Response	58
Exaggerated Response—Diminished Glucose Tolerance	58
Diabetes Mellitus	60
Hepatic and Biliary Tract Disease	63
Hyperthyroidism	64
Hyperfunction of the Anterior Hypophysis and Adrenal Cortex	64
Pregnancy	64
Miscellaneous	65
Decreased Response—Increased Glucose Tolerance	66
Hyperinsulinism	66
Adrenocortical Insufficiency	67
Adenohypophyseal Hypofunction	68
Hypothyroidism	69
Miscellaneous	70
Abnormal Insulin Tolerance	70
Abnormal Cortisone-Glucose and Prednisone-Glucose Tolerance Tests	72
Abnormal Tolbutamide Tolerance Test	72
Abnormal Plasma Insulin Levels	72
Obesity	72
Diabetes Mellitus	73
Insulinoma	74
Other Pancreatic Disorders	74
Abnormal Epinephrine Tolerance Test	74
Abnormal Tolerance for Other Sugars	74
Abnormal Fructose Tolerance	74
Abnormal Galactose Tolerance	75
Blood Lactic Acid and Pyruvic Acid	76
Hyperlactatemia and Lactic Acidosis	77
Diabetic Ketoacidosis	78
Euglycemic Diabetic Ketoacidosis	79
Glycogen Storage Diseases	79
Type I Glycogenosis (von Gierke's Disease)	80

　　　　　　　　　　Type II Glycogenosis (Pompe's Disease) 80
　　　　　　　　　　Type III Glycogenosis (Limit Dextrinosis) 81
　　　　　　　　　　Type IV Glycogenosis (Amylopectinosis) 81
　　　　　　　　　　Type V Glycogenosis (McArdle's Disease) 81
　　　　　　　　　　Type VI Glycogenosis (Hers' Disease) 81
　　　　　　　　　　Type VII Glycogenosis 81
　　　　Glycogen Storage Deficiency 82
　　　　Excretion of Sugar in Urine 82
　　　　　　　　　　Mechanism of Glycosuria (Glucosuria) 82
　　　　　　　　　　Other Sugars in Urine 83
　　　　Abnormal Urine Sugar (Melituria) 83
　　　　　　Glycosuria 83
　　　　　　　　　　Nonhyperglycemic Glycosuria 83
　　　　　　　　　　Hyperglycemic Glycosuria 85
　　　　　　Fructosuria 87
　　　　　　Pentosuria 88
　　　　　　Lactosuria 88
　　　　　　Galactosuria; Galactosemia 89
　　　　　　Maltosuria 89
　　References 89

Chapter 2

LIPID METABOLISM 91

　　　　　　　　　　Fats (Triglycerides) 93
　　　　　　　　　　Phosphatidyl Derivatives (Phosphoglycerides) .. 94
　　　　　　　　　　Prostaglandins 95
　　　　　　　　　　Steroids 96
　　　　　　　　　　Lipoproteins 97
　　　　Digestion and Absorption 98
　　　　　　　　　　Triglycerides 98
　　　　　　　　　　Sterols 99
　　　　　　　　　　Phosphoglycerides 100
　　　　　　　　　　Fat-soluble Vitamins 100
　　　　　　　　　　Fecal Fat 100
　　　　Metabolism of Fat 101
　　　　　　　　　　Immediate Fate of Dietary Lipid 101
　　　　　　　　　　Anabolism and Catabolism of Fatty Acids 102
　　　　　　　　　　Anabolism and Catabolism of Lipids 105
　　　　　　　　　　Lipid Mobilization 105
　　Metabolic Interrelations of Lipids, Carbohydrates, and Proteins 106
　　Ketosis 109
　　Fat in Feces 110
　　Fat in Urine 111
　　Metabolism of Phospholipids 111

Turnover	112
Catabolism	112
Metabolism of Sterols and Bile Acids	112
Catabolism and Excretion of Cholesterol; Conversion to Bile Acids	113
Transport of Lipids	114
State of Lipids in Blood; Lipoproteins	115
Ultracentrifugation	115
Chylomicrons	115
Very Low-density Lipoproteins	116
Low-density Lipoproteins	116
High-density Lipoproteins	116
Very High-density Lipoproteins	116
Electrophoresis of Plasma Lipoproteins	116
Postabsorptive Plasma Lipid Concentration	117
Influence of Food and Nutrition	119
Influence of Age and Sex	121
Pregnancy	121
Deposition and Storage of Lipids	122
Role of Adipose Tissue in Lipid Metabolism	122
Role of Liver in Lipid Metabolism	124
Fatty Liver and Lipotropism	125
Lipidosis, Xanthomatosis	129
Abnormalities of Plasma Lipids	131
Plasma Lipoprotein Abnormalities	131
Hyperlipoproteinemia	134
Type I	134
Type II	134
Type III	135
Type IV	135
Type V	135
Hypolipoproteinemia	136
A-beta-lipoproteinemia	136
Hypobeta-lipoproteinemia	136
Tangier Disease	136
Lecithin: Cholesterol Acyltransferase Deficiency	137
Hypercholesterolemia	137
Diabetes Mellitus	137
Anesthesia	138
The Nephrotic Syndrome	138
Hepatic and Biliary Tract Disease	139
Hypothyroidism	140
Atherosclerosis	141
Xanthomatosis	141
Miscellaneous	141
Hypocholesterolemia	142

xvi   CONTENTS

        Anemia 142
        Hepatic Disease 142
        Infection 143
        Hyperthyroidism 143
        Inanition 144
        Terminal States 144
        Miscellaneous 145
References 145

Chapter 3

PROTEIN METABOLISM 147

Digestion and Absorption 148
    Dynamic State 150
    Metabolic Pool 151
Over-all Metabolism of Protein 152
    Nitrogen of the Food 152
    Nitrogen of the Body 152
    Excretion of Nitrogen 152
    Nitrogen Balance 153
    Essential Amino Acids 154
    Biological Value of Proteins 156
    Dietary Protein Requirements 156
Intermediary Metabolism 158
    Protein Turnover 158
    General Pathways of Protein Metabolism 158
Interrelations of the Metabolism of Proteins and Other Foodstuffs 162
Endocrine Influences in Protein Metabolism 164
    Growth Hormone 164
    Androgen 165
    Adrenal Glucocorticoids 165
    Insulin 165
    Thyroid Hormones 166
    Glucagon 166
Nitrogenous Constituents of the Blood 166
    The Plasma Proteins 166
        Identity and Properties 166
        Methods of Study 168
        Metabolism 173
        Functions 173
    Nonprotein Nitrogen 177
        Urea 178
        Uric Acid 178
        Creatinine 178

Amino Acids	178
Ammonia	179
Undetermined Nitrogen (Rest Nitrogen)	179
Nitrogen Excretion	179
Urea	180
Uric Acid	180
Creatine and Creatinine	180
α-Amino Acids	181
Ammonia	182
Plasma Protein Abnormalities	182
Total Plasma (Serum) Protein	183
Fibrinogen	184
Albumin	185
Prealbumins	189
Globulins	190
Serum Protein Abnormalities in Disease	200
Globulin Reactions	205
Zinc Sulfate Turbidity	205
Cephalin-Cholesterol Flocculation Test	206
Thymol Turbidity Test	206
Abnormal Blood Nonprotein Nitrogen	207
Urea Nitrogen	207
Uric Acid	209
Creatinine	209
Amino Acid Nitrogen	210
Ammonia	211
Total Nonprotein Nitrogen	211
Abnormal Urinary Nitrogen	212
Proteinuria	212
Benign Albuminuria	214
Organic Albuminuria	215
Quantity of Protein in Urine	218
Other Proteins in Urine	219
Urinary Nonprotein Nitrogen	220
Urinary Urea	220
Uric Acid in Urine	222
Creatine and Creatinine in Urine	222
Amino Acids in Urine	223
Ammonia in Urine	229
Negative Nitrogen Balance in Disease	229
Disorders of the Urea Cycle	230
Argininosuccinic Aciduria	230
Citrullinemia	231
Hyperargininemia	231
Hyperammonemia	231
Congo Red Test for Amyloidosis	231

Acid Mucopolysaccharides ... 232
    The Mucopolysaccharidoses ... 233
References ... 234

Chapter 4

NUCLEIC ACID METABOLISM ... 235

Biosynthesis ... 237
Digestion and Absorption ... 239
Interrelations of Metabolism of Nucleic Acids with Other Foodstuffs ... 239
Biological Significance of Nucleic Acids ... 240
    Viruses ... 240
    Chromosomes—Genes ... 241
    "Transforming Substance" of Bacteria ... 243
    Role of Nucleic Acids in Mutation and Carcinogenesis ... 243
    Role of Nucleic Acids in Protein Synthesis ... 244
Cancer Chemotherapy ... 249
    Antimetabolites ... 249
    Alkylating Agents ... 249
    Other Anticancer Substances ... 250
    Drug Combinations ... 250
Free Nucleotides of Biological Importance ... 250
    Nucleoside Triphosphates ... 250
    Coenzymes ... 250
    Cyclic Nucleoside Monophosphates ... 251
Intermediary Metabolism ... 251
    Purines and Pyramidines ... 252
    Uric Acid ... 253
    Orotic Aciduria ... 259
References ... 259

Chapter 5

METABOLISM OF HEMOGLOBIN AND PORPHYRINS ... 260

Biosynthesis of Porphyrins and Hemoglobin ... 262
    Catabolism of Hemoglobin ... 265
    Haptoglobins ... 265
    Varieties of Human Hemoglobin ... 266
    Abnormal Hemoglobins ... 267
    Detection of Hemoglobinopathy ... 270
Hemoglobinemia; Hemoglobinuria ... 270

Myoglobinuria ... 271
Abnormal Hemoglobin Derivatives ... 272
    Methemoglobin ... 272
    Sulfhemoglobin ... 273
    Carboxyhemoglobin ... 273
    Methemalbuminemia ... 273
Porphyria and Porphyrinuria ... 274
    Porphyria ... 275
    Erythropoietic Porphyrias ... 276
    Hepatic Porphyrias ... 276
    Acquired Porphyrinurias ... 278
References ... 278

Chapter 6

CALCIUM AND INORGANIC PHOSPHATE METABOLISM ... 279

Absorption ... 280
Deposition and Mobilization of Bone Minerals ... 281
Calcitonin ... 284
Interrelations of Citrate and Calcium Metabolism ... 285
Mechanism of Regulation of Serum Calcium Concentration ... 285
Blood Calcium and Phosphate ... 287
    Parathyroid Hormone ... 289
    Vitamin D ... 290
    Calcitonin ... 290
    Plasma Proteins ... 290
    Plasma Phosphate ... 291
    Acid-base Equilibrium ... 291
    Miscellaneous ... 291
Miscellaneous Factors ... 291
Excretion ... 292
Calcium and Phosphorus Requirement ... 293
Calcium Content of Other Body Fluids ... 294
Abnormal Serum Calcium ... 294
    Hypercalcemia ... 294
        Primary Hyperparathyroidism ... 294
        Secondary Hyperparathyroidism ... 295
        Familial Hyperparathyroidism ... 296
        Diagnostic Aids in Hyperparathyroidism ... 296
        Hypervitaminosis (Vitamin D) ... 297
        Multiple Myeloma ... 297
        Neoplastic Disease ... 298
        Immobilization ... 298
        Sarcoidosis ... 298

CONTENTS

- Idiopathic Infantile Hypercalcemia 299
- Milk and Alkali Intoxication 299
- Miscellaneous 300
- Hypocalcemia 300
  - Hypoparathyroidism 300
  - Pseudohypoparathyroidism 301
  - Pseudo-pseudohypothyroidism 301
  - Vitamin D Deficiency (Rickets and Osteomalacia) 302
  - Steatorrhea 302
  - Starvation 302
  - Nephrotic Syndrome 303
  - Chronic Renal Failure 303
  - Maternal Tetany 305
  - Neonatal Hypocalcemia 305
  - Acute Pancreatitis 305
  - Anticonvulsants 305
  - Magnesium, Oxalate and Citrate Tetany 305
  - Alkalemia 306
  - Miscellaneous 306
- Abnormal Urine Calcium 307
  - Increased Urinary Calcium 307
    - Hyperparathyroidism 307
    - Hyperthyroidism 307
    - Acidemia 307
    - Hypervitaminosis D 307
    - Idiopathic Hypercalciuria 308
    - Miscellaneous 308
  - Decreased Urinary Calcium 308
    - Hypoparathyroidism 308
    - Vitamin D Deficiency 308
    - Hypothyroidism 308
    - Miscellaneous 308
- Abnormal Fecal Calcium 309
- Hyperphosphatemia 309
  - Hypervitaminosis (Vitamin D) 309
  - Hypoparathyroidism 309
  - Renal Failure 309
  - Healing Fractures 310
  - Miscellaneous 310
- Hypophosphatemia 310
  - Rickets 310
  - Osteomalacia 310
  - Idiopathic Steatorrhea 311
  - Hyperparathyroidism 311
  - Fanconi Syndrome 311

        Renal Tubular Acidosis 311
        Vitamin D-resistant Rickets 311
        Miscellaneous 312
    Abnormal Urinary Phosphate 312
    Disturbances of Bone Formation and Mineralization 313
    References 315

Chapter 7

MAGNESIUM METABOLISM 316

    Absorption and Excretion 316
    Blood Magnesium 317
        Abnormal Serum Magnesium 317
    References 318

Chapter 8

IRON METABOLISM 319

    Absorption and Excretion 320
    Transport 322
    Utilization; Storage 323
    Requirement 325
    Abnormal Iron Metabolism 325
    References 326

Chapter 9

SULFUR METABOLISM 327

    Absorption 327
    Intermediary Metabolism 327
    Sulfur in Blood 328
    Excretion 328
    Reference 329

Chapter 10

IODINE METABOLISM 330

    Absorption and Excretion 330
    Blood Iodine 331
    Distribution and Intermediary Metabolism 331
    Abnormal Iodine Metabolism 333
    References 333

Chapter 11

TRACE ELEMENTS ... 334

Copper ... 334
    Absorption; Excretion; Metabolism ... 334
    Function; Deficiency Manifestations ... 335
Cobalt ... 337
Zinc ... 337
Fluorine ... 338
Other Essential Trace Elements ... 338
References ... 338

Chapter 12

WATER, SODIUM, CHLORIDE, POTASSIUM: PHYSIOLOGICAL CONSIDERATIONS ... 339

Water Intake ... 339
Water Output ... 340
    Feces ... 340
    Insensible Perspiration ... 341
    Perspiration ... 341
    Urine ... 341
    Equilibrium Requirements ... 341
Volume of Body Fluid Compartments ... 342
    Volume of Body Fluid Compartments ... 343
        Blood and Plasma Volume ... 343
        Total Extracellular Fluid Volume ... 344
        Total Body Water ... 345
    Composition of Body Fluid Compartments ... 346
        Milliequivalents (mEq.) ... 346
        Molar Concentration ... 347
        Osmolality ... 347
Extracellular Fluid ... 348
Intracellular Fluid ... 351
Exchanges Between Fluid Compartments ... 352
    Gibbs-Donnan Equilibrium ... 353
    Plasma: Interstitial Fluid Exchange ... 354
    Interstitial Fluid : Intracellular Fluid Exchange ... 355
Regulatory Mechanisms ... 357
    Renin-Angiotensin System ... 360
Sodium, Potassium, Chloride ... 360
    Absorption and Excretion ... 361
        Excretion in Urine ... 361
        Excretion by the Skin ... 363
        Excretion in Digestive Fluids ... 363

Normal Blood Na, Cl, and K	364
Distribution and Intermediary Metabolism	365

Chapter 13

WATER, SODIUM, CHLORIDE, POTASSIUM: ABNORMALITIES ... 368

Deficits of Water, Sodium, Chloride, Potassium	368
Dehydration	368
Routes of Loss	369
General Principles in Electrolyte Abnormalities	370
Biochemical Consequences of Dehydration	374
Gastrointestinal Disorders	375
Diabetes Mellitus	376
Kidney Disease	376
Excessive Sweating	378
Congestive Heart Failure	378
Cerebral Disorders	379
Pulmonary Neoplasms	379
Adrenocortical Insufficiency (Addison's Disease)	379
Excessive Therapy	379
Primary Water Deficit	380
Excesses of Water, Sodium, Chloride	380
Renal Disease	380
Congestive Heart Failure	381
Liver Disease	381
Toxemia of Pregnancy	381
Adrenocortical Hyperfunction	382
Postoperative Response	382
Miscellaneous Sodium and Chloride Abnormalities	382
Potassium Abnormalities	383
General Principles	383
Starvation; Malnutrition	385
Administration of NaCl and Glucose Solutions	385
Gastrointestinal Disorders	386
Diabetic Acidosis	387
Adrenocortical Hormones	387
Diuretic Drugs	387
Familial Hyperkalemic Paralysis	388
Potassium Therapy	388
Postoperative States	388
Renal Disease and Malfunction	389
Familial Periodic Paralysis	390
Miscellaneous	390

xxiv CONTENTS

Abnormal Blood Volume ... 391
    Increased Blood Volume ... 391
    Decreased Blood Volume ... 391
Transudates and Exudates ... 393
    Specific Gravity ... 393
    Protein ... 393
    Glucose ... 395
    Chloride ... 395
    Lipid ... 395
    Other Constituents ... 397
References ... 397

Chapter 14

HYDROGEN ION CONCENTRATION IN BODY FLUIDS ... 399

    Acids and Bases ... 400
Physiological Buffer Systems ... 402
    Buffer Action of Hemoglobin ... 402
    Buffer Systems for H_2CO_3 (CO_2) ... 403
    Buffer Systems for Nonvolatile Acids ... 405
    Net Effect of Buffer Mechanisms ... 406
    Bone ... 406
    Respiratory Regulation ... 406
    Renal Regulation ... 407
Intracellular-Extracellular pH Interrelations ... 412
Abnormalities of Hydrogen Ion Concentration ... 413
Acidosis ... 415
    Primary Excess of Acid ... 416
        Respiratory Acidosis ... 416
        Nonrespiratory Acidosis ... 417
    Primary Reduction of Base ... 421
        Nonrespiratory Acidosis ... 421
Alkalosis ... 421
    Primary Acid Deficit ... 422
        Respiratory Alkalosis ... 422
        Nonrespiratory Alkalosis ... 423
    Primary Excess of Base ... 424
        Nonrespiratory Alkalosis ... 424
Mixed Disturbances of Acid-Base Status ... 425
Methods of Studying Acid-Base Status ... 426
    CO_2 Content of Plasma ... 427
    Carbon Dioxide Combining Power of the Plasma ... 428
    Blood CO_2 Tension ... 429
    Astrup Microtechnique ... 429

        Determination of pH of Blood, Plasma, or Serum   429
        Other Methods of Investigation   430
  References   432

Chapter 15

RESPIRATORY EXCHANGE AND BASAL METABOLISM   433

  Chemical Control of Respiration   435
        Influence of CO_2 Tension and pH   436
        Influence of O_2 Tension   436
  Transport of Oxygen   436
        Dissociation of Oxyhemoglobin   437
  Transport of Carbon Dioxide   440
        CO_2 in Arterial Blood   440
        Entrance of CO_2 in Tissues   440
  Hypoxia   442
        Hypoxic Hypoxia   442
        Anemic Hypoxia   443
        Stagnant Hypoxia   443
        Histotoxic Hypoxia   444
        Relative Hypoxia   444
  Energy Metabolism   444
        Caloric Value of Foods   444
        Heat Production   445
        Respiratory Quotient   445
  Calorimetry   448
        Direct Calorimetry   449
        Indirect Calorimetry   449
  Basal Metabolism   450
        Physiological Variations in BMR   450
    Specific Dynamic Action (SDA) of Foods   451
    Total Metabolism (Caloric Requirement)   452
    Clinical Significance of BMR   453
  References   453

Chapter 16

RENAL FUNCTION   454

        Morphological Features of Function Importance   455
  Glomerular Filtration   457
  Tubular Function   459
        Reabsorption   459
        Tubular Excretion and Synthesis   465
        Competition for Transport Mechanisms   465

Water Homeostasis ... 466
Diuretics ... 468
Characteristics of Normal Urine ... 470
    Volume ... 470
    Specific Gravity ... 470
    Acidity ... 471
    Nonprotein Nitrogenous Constituents ... 472
    Allantoin ... 475
    Oxalic Acid ... 475
    Glucuronic Acid ... 476
    Hippuric Acid (Benzoylglycine) ... 476
    Citric Acid ... 476
    Other Organic Acids ... 477
    Ketone Bodies ... 477
    Sulfur-containing Compounds ... 477
    Phosphate ... 478
    Chloride ... 478
    Sodium; Potassium ... 479
    Calcium; Magnesium ... 479
    Carbohydrates ... 480
    Miscellaneous ... 480
Clinical Study of Renal Function ... 481
Clearance Tests ... 481
    Glomerular Filtration ... 482
    Renal Blood Flow ... 487
    Maximum Tubular Excretory Capacity ... 488
    Percentage Tubular Reabsorption ... 489
    Filtration Fraction ... 489
    Maximum Tubular Reabsorption Capacity ... 490
    Other Clearance Procedures ... 490
Elimination of Water ... 491
    Comparison of Fluid Intake and Fluid Output ... 491
    Oliguria; Anuria ... 492
    Water Function Test (Dilution Test) ... 494
Elimination of Solids—Urine Specific Gravity ... 495
    Concentration Test of Renal Function ... 496
        The Two-hour Specific Gravity Test ... 496
        The Urine Concentration Test ... 497
        Significance of Hyposthenuria ... 498
Elimination of Nonprotein Nitrogenous Substances ... 501
    Urinary Studies ... 501
    Simultaneous Study of Blood and Urine ... 501
        Blood Urea Clearance ... 501
        Endogenous Creatinine Clearance ... 506
    Blood Nitrogen Studies ... 507
        Renal Functional Impairment ... 507

Prerenal Deviation of Water	507
Excessive Protein Catabolism	508
Blood Nonprotein Nitrogen in Renal Disease	509
Urea in Other Body Fluids	514
Elimination of Foreign Substances	515
The Phenolsulfonphthalein Test	515
Localization of Renal Functional Defect	518
Acute Glomerulonephritis	518
Chronic Glomerulonephritis	520
Tubular Nephroses (Acute Renal Failure)	522
Pyelonephritis	522
Benign Nephrosclerosis (Essential Hypertension)	523
Malignant Nephrosclerosis (Necrotizing Arteriolitis)	523
The Nephrotic Syndrome	523
Toxemias of Pregnancy	524
Congestive Heart Failure	526
Other Biochemical Manifestations of Chronic Renal Insufficiency (Chronic Renal Failure)	526
Na and Cl Depletion; Dehydration	526
Serum K Abnormalities	528
Hypermagnesemia	528
Hypocalcemia; Hypercalciuria; Osteodystrophy	528
Acidosis	529
Hypoproteinemia	530
Plasma Insulin	531
Plasma Lipids	531
Phenol and Other Organic Substances	531
Anemia	533
Renal Tubular Functional Defects	533
Renal Glycosuria	534
Albuminuria	534
Renal Aminoaciduria	534
Cystinuria	535
Hartnup Syndrome	536
Prolinuria (Familial Iminoglycinuria)	536
Methionine Malabsorption (Oasthouse Syndrome)	536
Renal Glycinuria	536
Galactosemia	536
Hepatolenticular Degeneration (Wilson's Disease)	537
Fanconi Syndrome	537
Vitamin Deficiencies	537
Toxic Agents	537
Cystinosis	538
Lowe's Syndrome	538
Idiopathic Hypercalciuria	538

xxviii   CONTENTS

    Pseudohypoparathyroidism 538
    Vitamin D-resistant Rickets 538
    Renal Tubular Acidosis 538
    Salt-losing Nephritis 539
    Potassium-losing Nephritis 539
    Nephrogenic Diabetes Insipidus ("Water-losing Nephritis") 539
  Renal and Urinary Lithiasis 540
    Hypercalciuria 540
    Idiopathic Renal Lithiasis 541
    Hyperoxaluria 542
    Cystinuria 543
    Hyperuricuria 543
    Xanthinuria 543
  Hemodialysis 543
    Disequilibrium Syndrome 543
    Acute Renal Failure 544
    Chronic Dialysis 544
  Renal Homotransplantation 545
  References 546

Chapter 17

ENZYMES 547

    International Unit of Enzyme Activity 548
    Isoenzymes (Isozymes) 548
  Digestive Enzymes 549
    Serum Amylase (Diastase) 550
    Serum Lipase 551
    Uropepsin 551
  Serum Cholinesterase 552
  Aminotransferases (Transaminases) 553
    Liver and Biliary Tract Disease 553
    Heart Disease 553
    Miscellaneous 554
    Isoenzymes 554
  Enzymes of Carbohydrate Metabolism 554
    Lactate Dehydrogenase 555
    Malate Dehydrogenase 557
    Isocitrate Dehydrogenase (ICD) 558
    Aldolase 558
    Glucose Phosphate Isomerase 559
    Hexokinase 559
    Glucose-6-phosphatase 559
  Phosphatases 559

Alkaline Phosphatase ... 560
   Isoenzymes ... 561
  Skeletal Diseases ... 562
   Rickets ... 563
   Hyperparathyroidism ... 563
   Osteitis Deformans ... 563
   Hypophosphatasia ... 563
   Miscellaneous Bone Disorders ... 564
  Hepatic and Biliary Tract Disease ... 565
  Miscellaneous Disorders ... 566
Acid Phosphatase ... 566
  Leukocyte Acid Phosphatase ... 569
  Lysosomal Acid Phosphatase Deficiency ... 569
Creatine Kinase ... 569
  Skeletal Muscle Disease ... 569
  Heart Disease ... 570
  Miscellaneous Disorders ... 570
Gamma-Glutamyl Transpeptidase ... 570
Sorbitol Dehydrogenase ... 571
Leucine Aminopeptidases ... 571
  Leucine Aminopeptidase ... 571
  Oxytocinase (Cystine Aminopeptidase) ... 571
  Isoenzymes ... 572
Erythrocyte Enzyme Defects ... 572
Enzymes in Pleural and Peritoneal Effusions ... 573
Cerebrospinal Fluid Enzymes ... 574
Enzyme Abnormalities in Diagnosis ... 574
  Liver Disease ... 575
  Heart Disease ... 576
  Placental Insufficiency (Placental Dysfunction) ... 576
  Central Nervous System Disease ... 577
  Genitourinary Disease ... 577
  Malignant Disease ... 578
  Miscellaneous Disorders ... 578
  Inherited Disease ... 579
References ... 579

Chapter 18

CHEMICAL INVESTIGATION OF GASTRIC FUNCTION ... 580

  Psychic or Cephalic Phase ... 580
  Gastric Phase ... 581
  Intestinal Phase ... 581
  Other Factors ... 582
Normal Gastric Juice ... 582

- *Gastric Residuum* ... 582
- Investigation of Gastric Secretory Activity ... 584
  - *Insulin (Hypoglycemia) Stimulation* ... 585
  - *Serum Gastrin* ... 585
  - *Intrinsic Factor* ... 585
- Abnormal Response ... 585
  - *Achlorhydria* ... 585
  - *Hyposecretion* ... 586
  - *Hypersecretion* ... 586
  - *"Tubeless Gastric Analysis"* ... 587
- Peptic Activity ... 587
- Serum Gastrin ... 587
- References ... 588

Chapter 19

PANCREATIC FUNCTION ... 589

- Examination of Pancreatic Juice: Secretin Test ... 590
  - *Volume* ... 591
  - *Bicarbonate* ... 591
  - *Amylase* ... 592
  - *Trypsin* ... 592
  - *Lipase* ... 592
  - Abnormal Findings with the Secretin Test ... 592
- Lundh Test Meal ... 593
- Serum Enzymes ... 594
  - *Serum Amylase* ... 594
  - *Serum Amylase after Stimulation* ... 595
  - *Serum Lipase* ... 595
  - *Serum Trypsin* ... 596
  - *Amylase Isoenzymes* ... 596
- Amylase in Urine ... 596
- Miscellaneous Findings in Pancreatic Disease ... 597
- Examination of Feces ... 598
- References ... 599

Chapter 20

INTESTINAL MALABSORPTION SYNDROMES ... 600

- *Steatorrhea* ... 600
- *Protein Abnormalities* ... 601
- *Vitamin Deficiencies* ... 602
- *Anemia* ... 602

Skeletal Abnormalities	604
Miscellaneous Abnormalities	604
Clinical Methods of Studying Absorption	604
Lipid Absorption	605
Carbohydrate Absorption	606
Protein Absorption	607
Vitamin B_{12} Absorption	607
Miscellaneous Procedures	608
Biopsy of Small Intestine Mucosa	608
References	609

Chapter 21

HEPATIC FUNCTION — 610

Carbohydrate Metabolism	613
Fasting Blood Sugar Level	614
Glucose Tolerance	615
Epinephrine Hyperglycemia	616
Lactate and Pyruvate Levels	617
Protein Metabolism	617
Amino Acids	617
Urea	618
Uric Acid	619
Ammonium Ion	619
Plasma Proteins	620
Thymol Turbidity and Flocculation	626
Cephalin-cholesterol Flocculation	627
Zinc Sulfate Turbidity	628
Plasma Prothrombin	628
Lipid Metabolism	630
Fat in Feces	631
Plasma Cholesterol	631
Other Plasma Lipids	633
Pigment Metabolism—Jaundice	633
Serum Bilirubin	636
The van den Bergh Reaction	636
Serum Bilirubin Concentration	636
Pathological Hyperbilirubinemia	638
Total Serum Bilirubin	638
Pathogenesis of Jaundice	640
Predominant Increase of Unconjugated Serum Bilirubin	641
Predominant Increase of Conjugated Bilirubin in Serum	644

Jaundice Due to Drugs	645
Pathological Neonatal Jaundice	646
Kernicterus (Bilirubin Encephalopathy)	648
Bilirubinuria	649
Bilins and Bilinogens in the Urine and Feces	650
Urobilinogen in Feces	651
Urobilinogen in Urine	652
Bile Pigments in Feces	655
Porphyrin in Urine and Feces	655
Bile Acid Metabolism	656
Steatorrhea	659
Gallstones	660
Detoxification-Conjugation	661
Hippuric Acid Synthesis	662
The Liver and Hormone Metabolism	663
Androgens	664
Estrogens	664
Adrenocortical Hormones	665
Elimination of Dyes	666
Bromsulphalein (BSP) Excretion	666
Significance of Abnormal BSP Retention	668
Serum Enzymes	670
Serum Iron	671
The Liver in Water and Salt Metabolism	672
Hepatic Coma	673
Drug Induction of Liver Enzymes	675
Liver Function Studies in Differential Diagnosis	676
Selection of Tests	679
References	683

Chapter 22

ENDOCRINE FUNCTION 685

By R. Hall and G. M. Besser

Cyclic Adenosine Monophosphate	686
Pituitary Hormones	687
Hypothalamic Regulation of Anterior Lobe Hormones	687
Thyrotrophin-releasing Hormone (TRH)	688
Luteinizing Hormone/Follicle-stimulating Hormone and Releasing Hormone (LH/FSH-RH)	689
Corticotrophin-releasing Hormone (CRH)	689
Melanocyte-stimulating Hormone-releasing Hormone (MRH)	689

Prolactin-releasing Hormone (PRH)	690
Growth Hormone-releasing Hormone (GRH)	690
Growth Hormone-Release Inhibiting Hormone (GH-RIH)	690
Prolactin-Release Inhibiting Hormone (P-RIH)	690
Melanocyte-stimulating Hormone-Release Inhibiting Hormone (M-RIH)	690
Anterior Lobe (Adenohypophyseal) Hormones	690
Gonadotrophic Hormones	691
Follicle-stimulating Hormone (FSH)	693
Luteinizing Hormone (LH, ICSH)	693
Human Menopausal Gonadotrophin	693
Human Chorionic Gonadotrophin (HCG)	693
Physiological Considerations	693
Normal Gonadotrophin Values	695
Abnormal Gonadotrophin Secretion	695
Thyroid-stimulating Hormone (Thyrotrophin)	699
Assay	699
Actions	700
Secretion and Metabolism	700
Abnormal Thyrotrophin (TSH) Secretion	700
Prolactin	700
Assay	701
Secretion and Action	701
Overproduction (Galactorrhea)	701
Underproduction	702
Adrenocorticotrophin (ACTH) and Melanocyte-Stimulating Hormones (α- and β-MSH)	702
Secretion and Transport	703
Metabolic Effects	703
Assays for ACTH	704
Abnormal Adrenocorticotrophin (ACTH) Secretion	704
Growth Hormone (Somatotropin)	705
Secretion and Transport	705
Metabolic Effects	706
Excessive Growth Hormone Secretion	708
Decreased Growth Hormone Secretion	708
Posterior Lobe (Neurohypophyseal) Hormones	709
Vasopressin	710
Abnormal Vasopressin Secretion	711
Oxytocin	712
Uterine Actions	712
Breast Actions	712
Steroid Hormones	713
Adrenal Gland	713

Anatomy and Embryology	713
Hormones of the Adrenal Cortex and Their Measurement	713
Structure of Steroid Hormones	714
Biosynthesis of Adrenocortical Hormones	714
Transport and Metabolism of Corticosteroids	717
Determination of the Rate of Secretion of Corticosteroids	719
Cortisol and Its Metabolites	720
Aldosterone and Its Metabolites	721
Androgens and Their Metabolites	721
Metabolic Effects	722
Anti-inflammatory Action	726
Effect on Immune Reactions	726
Miscellaneous Effects	726
Physiology of Cortisol	727
The "Alarm Reaction"; The Adrenal Cortex Under Stress	727
Control of Aldosterone Secretion	729
Adrenocortical Hyperfunction	730
Cushing's Syndrome	730
Hyperaldosteronism	734
Adrenogenital Syndrome	735
Adrenocortical Hypofunction	737
Clinical Features	738
Hormone Studies	738
Carbohydrate Metabolism	738
Electrolyte and Water Metabolism	739
Adrenal Medullary Hormones	740
Storage of Catecholamines	740
Release of Catecholamines	741
Biosynthesis and Metabolism	741
Underproduction of Adrenal Medullary Hormones	744
Overproduction of Adrenal Medullary Hormones	744
Sympathoblastoma	744
Neuroblastoma	744
Ganglioneuroma	745
Pheochromocytoma	745
Thyroid Hormones	746
Chemistry	746
Biosynthesis and Secretion	747
Circulating Thyroid Hormones	750
Metabolism of Thyroid Hormones	750
Metabolic Effects	751
Mechanism of Action	754

  Agents Interfering with Synthesis of Thyroid Hormone   754
   Thyroid Hormone   754
   Iodine   754
   Thiocyanate   755
   Antithyroid Agents   755
  Laboratory Diagnosis of Thyroid Disease   755
   a. Thyroid Iodide Trap Tests   756
   b. Thyroid Hormone Release Tests   756
   c. Peripheral Tissue Response Tests   760
   Basal Metabolic Rate   760
   Tyrosine Tolerance   760
   Red Cell Sodium   760
  Thyroid-stimulating Hormone Immunoassay   760
   Hypothyroidism   760
   Hyperthyroidism   761
   Nontoxic Goiter   761
  Thyrotrophin-releasing Hormone Test   761
   Normal Subjects   761
   Hypothyroid Subjects   762
   Hyperthyroid Subjects   762
   Triiodothyronine (T_3) Suppression Test   762
   TSH Stimulation Test   763
   Perchlorate Discharge Test   763
   Thyroid Antibody Tests   763
   Application of Tests to Determine the Level of Thyroid Function   764
Nontoxic Goiter and Hypothyroidism   764
  Hypothyroidism   765
  Laboratory Diagnosis of Hypothyroidism   765
   Overt Hypothyroidism   765
   Mild Hypothyroidism   766
   Compensated Hypothyroidism   766
Hyperthyroidism   766
  T_3 Toxicosis   766
  Diagnosis of Hyperthyroidism   767
Calcitonin   768
  Structure of Human Calcitonin   769
  Factors Affecting Secretion of Calcitonin   769
  Sites and Mechanisms of Action and Distribution of Calcitonin   769
  Assay of Calcitonin   769
  Physiological Significance of Calcitonin   769
  Pathological Significance of Calcitonin   770
  Clinical Applications of Calcitonin   770
Gastrointestinal Hormones   771

Gastrin ... 771
    Assay of Gastrin ... 771
    Physiology and Regulation of Secretion ... 772
    Syndromes of Gastrin Excess ... 772
Cholecystokinin ... 773
    Assay of CCK ... 773
    Physiology and Regulation of Secretion ... 773
    Syndromes of Excess and Deficiency ... 773
Secretin ... 773
    Assay of Secretin ... 774
    Physiology and Regulation of Secretin ... 774
    Syndromes of Excess and Deficiency ... 774

Ovarian Hormones ... 774
  Estrogens ... 774
    Actions of Estrogens ... 775
    Assay of Estrogens ... 776
  Clinical Applications of Estrogen Assays ... 776
    Increased Estrogen Values ... 776
    Decreased Estrogen Values ... 777
    Indications for Estrogen Therapy ... 778
    Estrogen Preparations ... 778
  Progesterone ... 779
    Actions of Progesterone ... 779
    Assay of Progesterone ... 779
  Clinical Applications of Progesterone Assays ... 779
    Increased Progesterone Values ... 779
    Diminished Progesterone Values ... 779
    Indications for Progesterone Therapy ... 780
    Progesterone Preparations ... 780
  Androgens ... 780
  The Hormonal Control of the Menstrual Cycle ... 780
Puberty in the Female ... 781
The Menopause ... 781
    Hormonal Changes ... 781
Hypogonadism in the Female ... 782
Oral Contraceptives ... 782
  Choice of Preparation ... 784
    Combined (Recommended Dose Containing 0.03–0.05 mg. Estrogen) ... 784
    Combined (Containing More Than 0.05 mg. Estrogen) ... 784
    Sequential ... 784
  Metabolic Effects of Oral Contraceptives ... 785
  Endocrine Effects of Oral Contraceptives ... 785
    Hypertension ... 785

Amenorrhea	786
Pigmentation	786
Thyroid and Adrenal Function Tests	786
Thromboembolic Effects of Oral Contraceptives	786
Gynecological Effects of Oral Contraceptives	787
Cervical Changes	787
Breast Changes	787
Vaginal Candidiasis	787
Placental Hormones and Pregnancy	787
Estrogens	788
Progesterone	789
Human Chorionic Gonadotrophin (HCG)	790
Human Placental Lactogen (HPL) (Human Chorionic Somatomammotrophin, HCS)	790
Human Chorionic Thyrotrophin (HCT) and Human Molar Thyrotrophin (HMT)	791
Prostaglandins (PG)	791
The Adrenal Gland and Pregnancy	792
The Thyroid Gland and Pregnancy	792
The Pituitary Gland and Pregnancy	793
Androgens and the Testis	793
Testosterone Synthesis and Metabolism	793
Metabolic Actions	796
Abnormal Androgen Values	797
Ectopic Hormone Production by Nonendocrine Tumors	797
Significance of Ectopic Hormone Production	799
Ectopic ACTH Syndrome	800
Hypoglycemia	800
Hypercalcemia	800
Inappropriate Secretion of Vasopressin	801
Hyperthyroidism	801
Ectopic Growth Hormone Secretion	802
Ectopic Gonadotrophin Secretion	802
References	802

Chapter 23

VITAMINS — 804

Vitamin A	805
Deficiency in Man	808
Detection of Deficiency in Man	808
Effects of Excess of Vitamin A	809
Hypercarotenemia	810
Thiamine	810

Metabolism of Thiamine	811
Demonstration of Deficiency	811
Thiamine Deficiency	812
Riboflavin	813
Demonstration of Deficiency	814
Niacin (Nicotinic Acid)	814
Demonstration of Deficiency	815
The Vitamin B_6 Group	815
Folic Acid	817
Vitamin B_{12}	818
Congenital Methylmalonic Aciduria	821
Ascorbic Acid	821
Detection of Deficiency in Man	822
Vitamin D	825
Deficiency Manifestations	827
Effects of Excess Vitamin D	829
Vitamin E (Tocopherols)	829
Vitamin K	830
Deficiency	832
Demonstration of Vitamin K Deficiency	833
Other Vitamins	834
References	834

Chapter 24

CEREBROSPINAL FLUID 835

Protein	835
Meningitis	836
Serous Meningitis	836
Aseptic Meningitis	837
Convulsive States	837
Organic Disease of Brain and Cord	837
Myxedema	837
Myelography	838
Glucose	838
Hyperglycorrhachia	839
Hypoglycorrhachia	839
Nonprotein Nitrogenous Constituents	840
Chloride	841
Inorganic Phosphate	842
Cholesterol	842
Lactate	842
Hydrogen Ion Concentration	843
Sodium, Potassium, Calcium, and Magnesium	843
Xanthochromia	844

Enzymes 845
References 846

Chapter 25
HISTORY AND DEVELOPMENT OF CLINICAL BIOCHEMISTRY 847

The "Normal Range" 852
SI Units 854
Derived Coherent Units 854
Derived Noncoherent Units 855
References 857

INDEX 859

Chapter 1

CARBOHYDRATE METABOLISM

The chief function of carbohydrate in the animal organism is that of a fuel, the degradation of which to carbon dioxide and water represents a major source of energy. Carbohydrate can also be used as a starting material for the biological synthesis of other types of compounds in the body, such as fatty acids and certain amino acids. In addition, it has a role in the structure of certain biologically important compounds, such as glycolipids, glycoproteins, heparin, nucleic acids and other substances.

DIGESTION AND ABSORPTION

Ingested disaccharides (mainly sucrose and milk lactose) and polysaccharides (starch and glycogen) are converted into monosaccharides by the action of enzymes present in saliva (amylase), pancreatic secretion (amylase) and the mucosa of the small intestine (disaccharidases).

The polysaccharides are made up of glucose and molecules combined together by two types of linkage, the so-called 1→4 and the 1→6. The former forms straight chains and the latter is the linkage at the point where one straight chain branches from another. In this way highly branched structures are formed. The amylases of saliva (to a limited extent) and pancreatic juice can attack only the 1→4 linkages. In this process molecules of glucose and maltose are split off as well as isomaltose, which represents the original 1→6 linkage. Glucose produced in this way, and some from the diet itself, represents the only polysaccharide-derived monosaccharide material presented to the intestinal mucosa for absorption.

The remainder, in fact, the large bulk of carbohydrate is presented to the mucosa in the form of disaccharides (sucrose and lactose in the diet and maltose and isomaltose derived from polysaccharide digestion). Enzymes capable of splitting disaccharides into monosaccharides are known as glycosidases or disaccharidases. It is now well recognized that these enzymes occur in the plasma membranes covering the microvilli at the brush borders of the intestinal mucosal cells. There appear to be at least seven such enzymes in human material. Five are glucosidases and two are galactosidases. The former have maltase, sucrase, isomaltase and probably some amylase activity. It seems likely that one of the galactosidases is derived from lysosomes. The glucosidases also have trehalase activity—in other words they can break down trehalose, a disaccharide present in insect hemolymph. This is possibly fortuitous, unless nature intended us to eat chocolate-coated ants. There are also other enzymes present in the brush border membrane, notably alkaline phosphatase, which may have some part to play in carbohydrate absorption.

As a result of the disaccharidase activities, the disaccharides within the plasma membrane are broken down to the monosaccharides: glucose, fructose and galactose. It seems likely also that the disaccharidase activity is in some way linked to absorption into the mucosal cell, and thence into the portal blood stream. Other monosaccharides presented to the intestinal mucosa are mannose derived from glycoproteins and pentoses derived from digestion of nucleic acids.

The comparative rates of absorption of the monosaccharides may be indicated as follows: galactose > glucose > fructose > mannose > pentoses. These striking differences, in conjunction with the fact that they are all absorbed from the peritoneal space at the same low rate, i.e., that of pentoses, indicate that at least galactose, glucose and fructose are transferred across the intestinal mucosa by some active process, and not entirely by diffusion. Moreover, the rate of absorption of certain sugars is modified by simultaneous administrations of another; e.g., that of galactose is diminished when it is given with glucose. Two mechanisms are apparently involved: (1) simple diffusion, dependent on the sugar concentration gradients between the intestinal lumen, mucosal cells and blood plasma; (2) active transport, independent of concentration gradient, and dependent upon the provision of energy by metabolic processes in the intestinal tissue. Although the transport mechanisms for different sugars possess certain features in common, they also differ in certain respects.

It appears that the transport processes for monosaccharide absorption are also located within the brush border limiting plasma membrane. One active transport process, which is sodium dependent, is shared by glucose and galactose. The mechanism for absorption of fructose is a different one. It does not seem to show the characteristics of active absorption, but is nevertheless more rapid than simple diffusion.

Active transport of sugars is depressed by agents that inhibit respiration (cyanide, azide, malonate, fluoroacetate), and by dinitrophenol, which uncouples oxidation from phosphorylation. Phloridzin, which apparently interacts with the membrane site at which the sugar enters, also inhibits intestinal absorption of glucose and galactose. However, whereas dinitrophenol interferes with conversion of fructose to glucose, phloridzin does not. The hexokinase reaction is not critically involved in the control of the rate of transport of sugars; e.g., mannose, although phosphorylated by a hexokinase, is not actively transported, whereas other sugars (e.g., deoxyglucose) are actively transported but are not phosphorylated by hexokinases.

In certain species, e.g., the guinea pig, a considerable fraction of absorbed fructose is converted to glucose in the intestinal mucosa via the fructokinase reaction (p. 43). A portion of glucose and fructose also undergoes glycolysis to lactic acid.*

Absorption of glucose from the intestine is influenced by the general condition of the organism, being interfered with in infections, various intoxications, prolonged undernutrition and vitamin deficiencies, especially of thiamine, pantothenic acid and pyridoxine. The period of contact of the sugar with the absorptive surface is of importance in this connection, the amount absorbed being diminished with increased intestinal motility, as in diarrheal conditions and gastrocolic fistula. Absorption may be decreased in the presence of abnormalities, structural or functional, of the mucous membrane, as in inflammation (enteritis), edema (congestive heart failure) and celiac disease. It is retarded in hypothyroidism and accelerated in hyperthyroidism. These facts must be borne in mind in interpreting results of the oral glucose tolerance test (p. 36). In experimental animals, active absorption of glucose is influenced strikingly by alterations in the concentrations of cations, especially Na and probably K. Diminished absorption in adrenal cortical insufficiency is dependent upon the decreased concentration of Na in the body fluids, and probably not upon a specific influence of the adrenal hormones on the absorptive process.

Disaccharidase Deficiency. It is now well recognized that it is possible for symptoms such as abdominal cramps, bloating, distention and diarrhea to be associated with congenital deficiency of intestinal lactase activity. The symptoms follow the ingestion of milk. Some individuals suffer from severe diarrhea and malnutrition from the first days of life. There is a severe form of disaccharidase deficiency, in which lactose is excreted in the urine and there is liver and kidney damage with aminoaciduria.

Lactose intolerance of congenital origin may not become manifest, in other individuals, until puberty or late adolescence. There is a tendency in most mammals to lose their intestinal lactase early in life. In the human this does not seem to occur in most western populations but appears to be quite common in American Negroes, Bantus and Orientals. The latter group of peoples tends to show symptoms of lactase deficiency quite commonly.

Acquired lactase deficiency is often seen as a result of certain gastrointestinal diseases, e.g., sprue, regional enteritis, bacterial infections.

The symptoms are due partly to the osmotic effect of unhydrolyzed lactose and also to its fermentation by intestinal bacteria to produce lactic acid and short chain fatty acids, with resultant bowel irritation. The disease is best diagnosed by peroral intestinal biopsy and lactase determination

---

* Organic acids are frequently described as if occurring as such in the tissues. This convention has been adopted in this book since it is the one normally met with in clinical practice. In actual fact, organic acids usually occur mainly in the form of their salts.

on the specimen. There is also an oral lactose tolerance test, which is not entirely reliable.

Another recognized congenital disaccharidase deficiency is that of sucrase and isomaltase in combination. Diarrhea starts when sucrose or starch is fed, and consequently the disease is not manifest from birth.

Fate of Absorbed Carbohydrate. After absorption into the portal blood, carbohydrates must pass through the liver before entering the systemic circulation, a fact of considerable physiological significance. Certain hepatic mechanisms contribute to the withdrawal of carbohydrates from the blood: (1) uptake of hexoses such as fructose and galactose for conversion to glucose by the liver cells; (2) conversion of glucose to glycogen for storage in the liver (glycogenesis); (3) utilization of glucose, by oxidation, for energy production; (4) utilization of glucose for the synthesis of other compounds, such as fatty acids and certain amino acids. Opposed to these mechanisms are others which lead to the release of glucose by the liver to the blood: (1) formation of blood sugar from hexoses other than glucose by the liver; (2) conversion of liver glycogen to blood glucose (glycogenolysis*); (3) synthesis of blood glucose by the liver and kidney from noncarbohydrate sources, such as certain amino acids (gluconeogenesis). The amount of glucose reaching the systemic circulation at any instant is consequently the resultant of the operations of these two groups of opposing processes.

Once it is in the systemic circulation, blood glucose becomes available for utilization by the extrahepatic tissues. From what has been said about portal absorption, it is evident that these tissues are presented with carbohydrate which has already been "picked over" by the liver in a selective manner. It is to be expected, therefore, that the functional state of the liver will have a profound influence on the carbohydrate metabolism of the entire organism, an expectation abundantly fulfilled in practice.

Endogenous Sources of Glucose. Among the minor carbohydrate sources of body glucose may be listed the small quantities of endogenous galactose, mannose and, possibly, pentoses which may be converted to glucose under certain circumstances. The major endogenous carbohydrate source is liver glycogen. The glycogen of muscle is not directly convertible to blood glucose; however, the lactic acid formed during glycolysis of muscle glycogen can be converted to glucose and glycogen in the liver.

Certain noncarbohydrate sources are available to the organism for production of glucose. Although it is generally believed that the fatty acids cannot undergo net conversion to carbohydrate, their oxidation provides the energy for gluconeogenesis. Also, the glycerol moiety of lipids can serve as a source of glucose (from which it can also be derived). However, amino acids are the major raw material for the synthesis of carbohydrate from noncarbohydrate sources (gluconeogenesis). It has been estimated that over

---

* The term "glycogenolysis" has been used by some authors to mean any breakdown of glycogen, whether it leads to blood glucose (one of the two available pathways in liver), or down the glycolytic sequence of reactions to pyruvic and lactic acids (the second of the two pathways in liver, and the probable sole pathway in muscle).

half of the average animal protein is potentially capable of conversion to carbohydrate. Those amino acids which are carbohydrate-formers are called "glucogenic." The major site of gluconeogenesis is the liver.

UTILIZATION OF GLUCOSE

Storage. When a molecule of glucose is introduced into the body or synthesized *de novo* therein, several metabolic pathways are open to it. In the absence of urgent physiological demands for oxidative energy or conversion to special products, excess glucose may be deposited as glycogen in the liver, muscles and other tissues (glycogenesis). Inasmuch as the amount of glycogen which can be stored in the body is limited, quantities of glucose in excess of this upper limit are converted to fatty acids and stored as triglycerides in the fat depots. There seems to be no fixed limit to this process, as may be seen from everyday observations of human beings.

Oxidation. In response to physiological demands for energy, glucose may be completely oxidized to CO_2 and H_2O, a process which occurs in all tissues. At certain times, special circumstances in muscle may result in only partial degradation of glucose (glycolysis). The product, lactic acid, is then largely disposed of by other tissues, notably liver.

Conversion to Fat. As indicated above, a net conversion of glucose to fatty acids occurs when accommodations for storage as glycogen are exceeded. However, since one of the major metabolites of glucose is in rapid equilibrium with fatty acids (p. 18), a constant interconversion occurs between certain molecules common to the carbohydrate and fatty acid pathways. This interconversion is very rapid, and investigations with isotopes have led to the conclusion that a large proportion of glucose that is oxidized actually forms fatty acids before final degradation to CO_2 and H_2O. It must be noted that the over-all conversion of glucose to fatty acids is irreversible, for reasons which will be pointed out subsequently (p. 12). In contrast, the transformation of glucose to the glycerol moiety of the lipids is readily reversible.

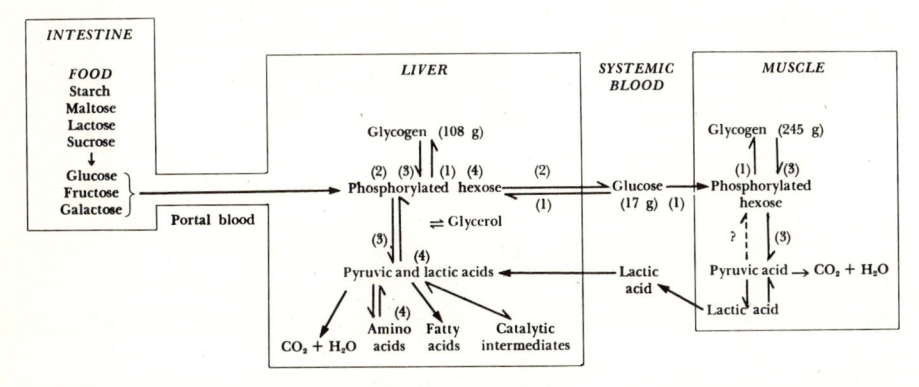

(1) Glycogenesis; (2) Glycogenolysis; (3) Glycolysis; (4) Gluconeogenesis

Figure 1-1. Sources and routes of utilization of carbohydrate. (Numerical data for 70 kg. man.)

Conversion to other Carbohydrates. Small amounts of glucose are used, directly or indirectly, in the synthesis of certain other carbohydrates which play important roles in the economy of the organism. These include (1) ribose and deoxyribose, required for the synthesis of the nucleic acids (p. 251), (2) mannose, glucosamine and galactosamine, which form parts of the mucopolysaccharides and glycoproteins, (3) glucuronic acid, also involved in the mucopolysaccharides and in "detoxication" reactions, and (4) galactose, which is a component of the glycolipids, as well as of the disaccharide, lactose, secreted in milk (p. 30).

Conversion to Amino Acids. Certain amino acids are not required in the diet, although they occur in the tissue proteins. It may be concluded, therefore, that they are synthesized by the body. Evidence at hand indicates that this group of amino acids, commonly designated "dispensable" or "nonessential" (p. 153), derives its carbon skeletons from glucose or its metabolites. Although all amino acids which are formed from glucose are also "glucogenic," the converse generalization does not hold.

GENERAL PROCESSES IN CARBOHYDRATE METABOLISM

The major sequence of catabolic reactions whereby glucose is degraded to CO_2 and H_2O may be divided, for didactic convenience, into anaerobic and aerobic phases. It is currently believed that these phases do not involve separate pathways, but that an initial, anaerobic series of reactions takes place, continuing directly in an aerobic series in the presence of oxygen. The anaerobic phase of glucose metabolism occurs whether oxygen is present or not, its anaerobic character deriving from the fact that participation of oxygen is not required in any of its reactions. "Glycolysis" is the term commonly applied to the production of lactic acid from glucose or glycogen, a phenomenon which is rare under physiological conditions, being largely restricted to circumstances of muscle contraction (p. 16) in which the rate of metabolism of carbohydrate outstrips the oxygen supply to the tissue (relative anaerobiosis).

Under the usual aerobic conditions, lactic acid is readily oxidized, *in situ,* to pyruvic acid. In any event, the aerobic phase of glucose metabolism begins with the end-products of the anaerobic phase, lactic or pyruvic acids, which, by means of an ingenious cyclic mechanism (p. 11), are degraded stepwise to CO_2 and H_2O. The efficiencies of the anaerobic and aerobic phases are similar, insofar as recovery of liberated free energy (p. 11) is concerned. However, if calculated in terms of energy obtained per molecule of glucose degraded, the aerobic is superior to the anaerobic by a factor of about 15 to 20. The metabolism of pentoses is not understood so completely as that of hexoses, but it appears that pentoses arise mainly from hexoses or products of hexose metabolism. The pathway involved can by-pass some of the glycolytic sequence of reactions (pentose shunt; oxidative shunt; hexose monophosphate shunt).

Anaerobic Metabolism (Glycolysis). Glycolysis can be subdivided into the phases of (*a*) initial phosphorylation, (*b*) conversion to trioses (cleavage), (*c*) the oxidative step, and (*d*) formation of lactic acid. These are outlined in Figure 1–2, which also shows the synthesis of glycogen.

Phosphorylation is an obligatory initial step in the metabolism of glucose (and other sugars). This reaction is catalyzed by a class of enzymes termed "kinases," that concerned with glucose being designated "glucokinase" ("hexokinase"). In the presence of this enzyme and adenosinetriphosphate (ATP), glucose-6-phosphate is formed ("hexokinase reaction"). As indicated in Figure 1-2, fructose and galactose are brought into the pathway of glucose metabolism at or beyond this point, which means that they are potential sources of blood glucose and liver glycogen.

The "hexokinase reaction" is inhibited by growth hormone (adenohypophysis, p. 25) and adrenocortical hormones (p. 24). These hormones, therefore, exert a fundamental, but not necessarily direct, influence upon the first, obligatory step in the metabolism of glucose, which is one reason for the disturbances in carbohydrate metabolism which occur in the presence of abnormal supply of these factors.

Depending on circumstances, the subsequent anaerobic metabolism of glucose may take the direction of (a) glycogen formation (glycogenesis) (liver and muscle, mainly) or (b) degradation to pyruvic acid and, in muscle, to lactic acid (Fig. 1-2), with the liberation of energy (p. 16). Glycogen is a highly branched polysaccharide of varying molecular weight (Fig. 1-3) con-

```
                              Glycogen
                                 ↓
              Glucose-1-phosphate ⇌ Galactose-1-phosphate ⇌ Galactose
                                 ↓
         (hexokinase)
Glucose  ⇌⇌⇌⇌⇌⇌⇌⇌⇌  Glucose-6-phosphate
         (phosphatase)
         (in liver)              ↓
                       Fructose-6-phosphate ⇌ Fructose
                                 ↓              ↑↓
                     Fructose-1,6-diphosphate ⇌ Fructose-1-phosphate
                                 ↓
                     Glyceraldehyde-3-phosphate
                                 ↓
                       Phosphoglyceric acid
                                 ↓
                       Phosphopyruvic acid
                                 ↓
                         Pyruvic acid ⟵⟶ Lactic acid
```

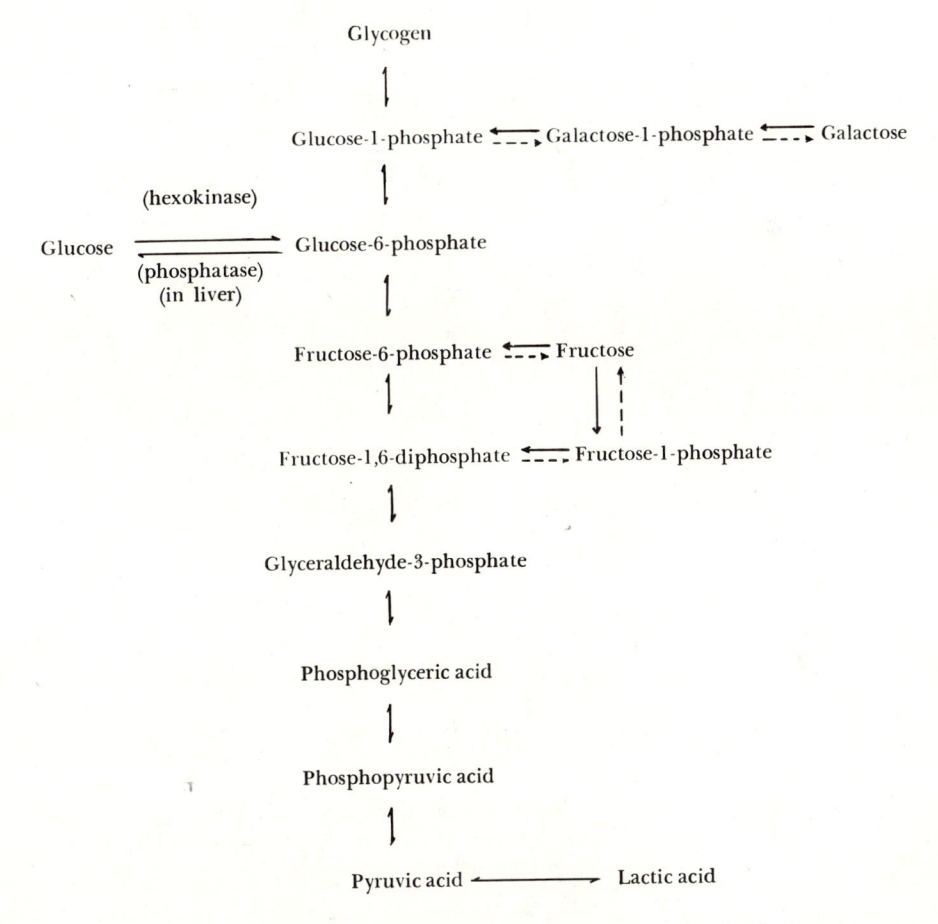

Figure 1-2. Anaerobic metabolism (glycolysis) of glucose and related hexoses.

8   CARBOHYDRATE METABOLISM

Glycogen contains linear amylose chains, made by linking carbons 1 and 4 of glucose through oxygen.

Branches occur in the amylose chains where carbon 6 of a residue is also linked through oxygen to the C-1 terminal of another chain segment.

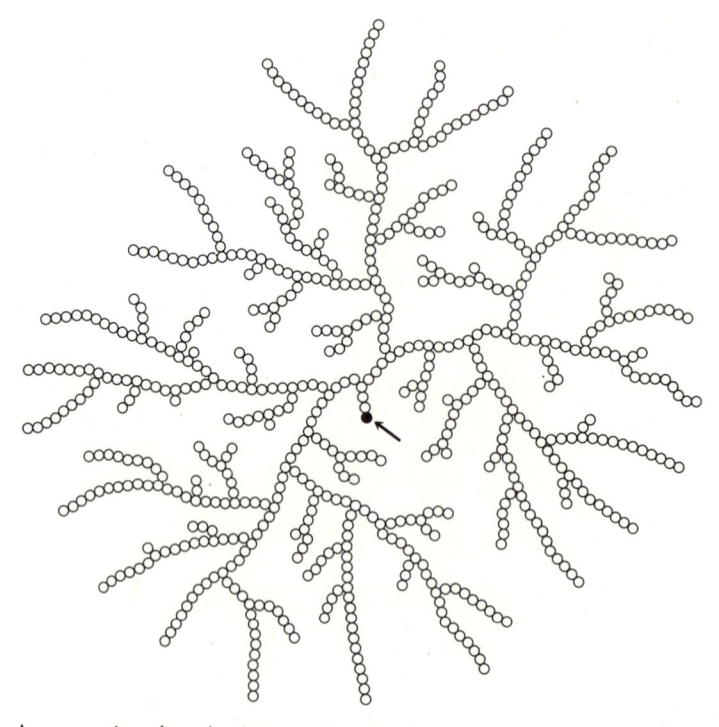

A cross-section through glycogen showing the tree-like structure created by branched amylose chains. The short inner segments are actually part of branches extending above and below the cross-section. The circles represent glucosyl residues; only one residue (arrow) doesn't have C-1 attached to an ether linkage so that it may assume an open-chain form and mutarotate.

Figure 1-3. The structure of glycogen. (From McGilvery: Biochemistry: A Functional Approach, W. B. Saunders Company, 1970.)

sisting of glucose molecules in linear 1→4 linkages, and 1→6 linkages at the branching points. The linear chains are formed by the action of a transferase enzyme system, so-called glycogen synthetase (UDPG: glycogen α-4-glucosyltransferase), formation of the branching-point linkages being catalyzed by a "branching enzyme" (α-1,4-glucan; α-1,4-glucan 6-glycosyltransferase). The glycogen synthetase actually adds glucosyl residues to the ends of branches, thereby increasing their length. It therefore requires the presence of a so-called primer substance, which can be a glycogen, a large molecule, which is to be made even larger, or a smaller molecule such as a maltodextrin, i.e., a member of a group of substances consisting of a relatively short chain of glucosyl residues, which are 1→4 linked. Of these two primer groups, glycogens are much more effective. Glycogen synthetase transfers glucosyl residues obtained from uridine diphosphate glucose (UDPG), which in turn is synthesized from uridine triphosphate (UTP) and glucose 1-phosphate by the action of an enzyme, which is a uridyltransferase. The glucose 1-phosphate is derived from glycogenolysis or from glucose 6-phosphate by the action of the enzyme phosphoglucomutase.

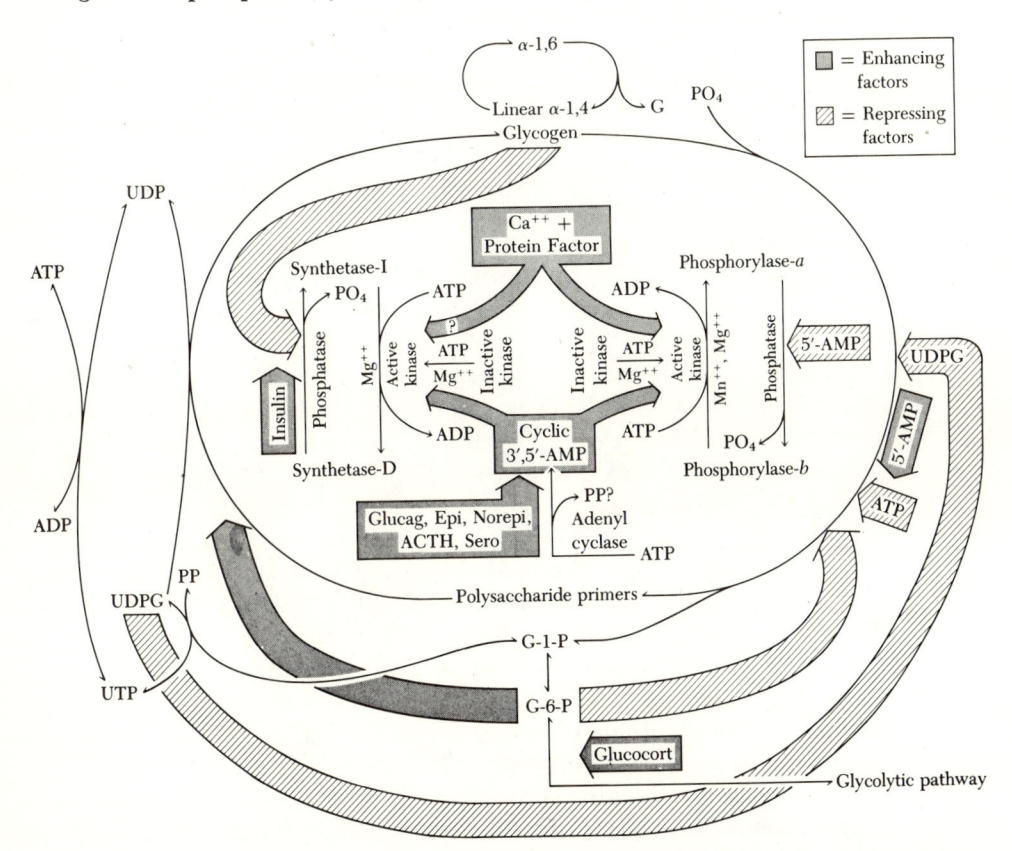

Figure 1-4. Pathways and regulatory factors in glycogen metabolism. *ACTH,* Adrenocorticotropic hormone; *Epi,* epinephrine; *G,* glucose; *G-1-P* and *G-6-P,* glucose-1- and glucose-6-phosphate; *Glucag,* glucagon; *Glucocort,* glucocorticoid hormone; *Norepi,* norepinephrine; PO_4, phosphate; *PP,* pyrophosphate; *Sero,* serotonin; *UDP* and *UTP,* uridine di- and triphosphate. (From Cantarow and Schepartz: Biochemistry, 4th edition. W. B. Saunders Company, 1967.)

The enzyme glycogen synthetase kinase converts the synthetase into a phosphorylated form (synthetase D), which is dependent on glucose-6-phosphate for its activation. This substance would accumulate in the absence of glucose-6-phosphatase, and hence stimulate the formation of glycogen. Such a mechanism could possibly account for the accumulation of glycogen in Type I glycogenosis (p. 80).

The kinase exists in an active and inactive form. The latter is activated by cyclic AMP (cyclic adenosine monophosphate) produced from ATP (adenosine triphosphate) by adenylcyclase, which in turn is activated by epinephrine. In stress situations, therefore, these various enzyme interactions result in the formation of phosphorylated glycogen synthetase, and glycogen synthesis becomes dependent on the availability of glucose-6-phosphate.

There is also an enzyme, glycogen synthetase phosphatase, which produces glycogen synthetase (synthetase I) from its phosphorylated form and releases the enzyme system from dependence on glucose-6-phosphate. This enzyme is inhibited by glycogen.

The reverse reaction, glycogen to glucose-1-phosphate, is catalyzed by phosphorylase. This exists in a relatively inactive form, phosphorylase b, which is converted into an active form, phosphorylase a, by phosphorylation by ATP brought about by another enzyme, known as phosphorylase kinase. This is active only in the presence of $3',5'$-cyclic AMP ($3',5'$-cyclic adenosine monophosphate), which is formed from ATP by the action of the enzyme adenylcyclase. The latter is, in turn, activated by glucagon in the liver and by epinephrine in liver and muscle. These hormones can, therefore, stimulate glycogen breakdown by starting off the whole chain of events. The increased production of cyclic AMP also results in phosphorylation of the most active glycogen synthetase, rendering it considerably less active. Glucagon and epinephrine therefore diminish glycogen synthesis as well as stimulate its breakdown. On the other hand, insulin inhibits adenylcyclase in the liver and so prevents glycogen breakdown. Insulin also causes an increased production of glycogen synthetase, which then gives rise to an increase in liver glycogen.

Phosphorylase can only break down the linear $1 \to 4$ linkages of glycogen. The $1 \to 6$ linkages at the branching points are dealt with by a debranching enzyme system, which manifests two different enzymic activities, so-called oligo-1,4 $\to$ 1,4-glucantransferase and amylo-1,6-glucosidase.

The various reactions involved in glycogen synthesis and breakdown are shown in Figure 1-4. Congenital absence of certain of the enzymes concerned with glycogenesis and glycogenolysis results in the development of glycogen storage diseases (p. 80).

Under normal conditions of oxygen supply in most tissues, the intermediary metabolism of glucose follows the path through pyruvic acid and the tricarboxylic acid (Krebs; citric acid) cycle (Fig. 1-5), with the ultimate production of CO_2 and H_2O. However, lactic acid is produced physiologically during the anaerobic phase of muscle contraction and in excess during strenuous physical exercise, passing into the blood stream, from which it is removed and converted to glucose or glycogen by the liver. Lactic acid is utilized for energy production (oxidation to CO_2 and H_2O) by heart muscle, but not to a significant extent by other muscles, under physiological conditions.

Aerobic Metabolism of Glucose (Fig. 1-5). The major route of aerobic metabolism of glucose is identical with the anaerobic through the stage of pyruvic acid. In the presence of adequate amounts of oxygen, the latter is oxidized (and decarboxylated) to CO_2 and H_2O, instead of being reduced to lactic acid. The several steps involved are collectively designated the tricarboxylic or citric acid cycle (Krebs cycle), and are responsible for the production of over 90 per cent of the energy derived from the metabolism of carbohydrate. The over-all oxidation of pyruvic acid may be represented as follows:

$$CH_3 \, CO \, COOH + 5 \, O \rightarrow 2 \, H_2O + 3 \, CO_2$$

Certain features of this process should be pointed out because they have an important bearing on the metabolic interrelations of carbohydrates, fatty acids (fats) and amino acids (proteins).

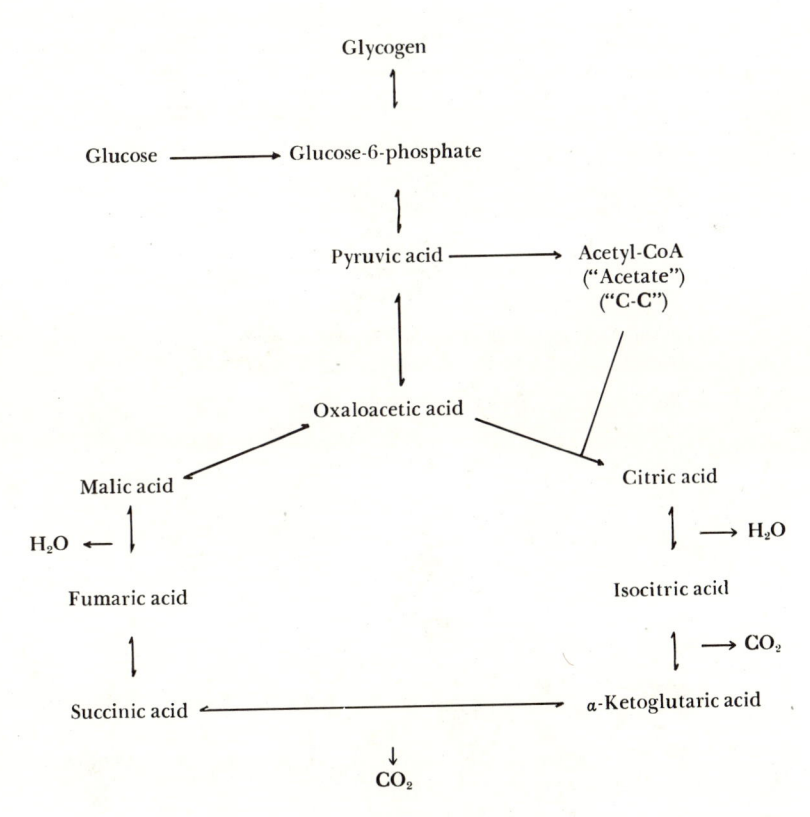

Over-all reaction: Pyruvic acid ($CH_3COCOOH$) + 5 (O) $\longrightarrow$ $2H_2O + 3CO_2$

Figure 1-5. Aerobic metabolism of carbohydrate. Tricarboxylic acid (citric acid) (Krebs) cycle.

The major portion of pyruvic acid undergoes oxidative decarboxylation to a 2-carbon (acetyl) fragment which may be regarded as acetate, although it is in reality "active acetate," or acetyl-CoA, a compound of acetic acid and coenzyme A, which contains pantothenic acid (one of the B vitamins). This initial step, i.e., oxidative decarboxylation of pyruvic acid, requires other B vitamins, i.e., thiamine pyrophosphate (p. 811) and lipoic acid, as coenzymes. Consequently, thiamine deficiency (as in beriberi) results in accumulation of excessive amounts of pyruvic (and lactic) acid in the body fluids (p. 77).

Acetyl-CoA is formed also from fatty acids and certain amino acids, and represents one of the major junctions in the integration of the metabolism of carbohydrate, fat and protein (Fig. 1-9). This fragment condenses with oxaloacetate (from pyruvate and amino acids), forming citrate. After a series of reactions involving oxidation and decarboxylation, a molecule of oxaloacetate remains, and the effect is as if the 2-carbon fragment with which it had originally condensed had been transformed to CO_2 and H_2O in its passage through the cycle.

Although fatty acids are metabolized via acetyl fragments similar to those from carbohydrate, they cannot increase the quantity of the latter material in the body since the reaction, pyruvate to acetate, is irreversible in man. Certain amino acids, such as alanine, aspartic acid and glutamic acid, are directly or indirectly convertible (oxidative deamination, transamination) to Krebs cycle intermediates (Fig. 1-9). Inasmuch as the reactions within the cycle and those of glycolysis are reversible (in the liver), such amino acids can produce new molecules of glucose or glycogen, and are, therefore, "glucogenic." These reactions form the basis for the phenomenon of gluconeogenesis (p. 14).

It is probable that carbohydrate unavailability, in addition to causing acceleration of catabolism of fatty acids and amino acids for energy production, leads to a further breakdown of amino acids to form members of the tricarboxylic acid cycle. A major part of the metabolism of the body depends upon the proper functioning of this cycle.

Alternate Aerobic Pathway (Pentose Shunt). Although the combined anaerobic and aerobic pathways outlined above probably represent the major routes of carbohydrate metabolism in the body, alternate aerobic pathways serve important functions, at least in certain tissues. The best established is referred to as the "oxidative, pentose, or hexose monophosphate shunt," a pathway of particular significance in liver and adipose tissue.

The sequence begins with glucose-6-phosphate, which is oxidized to 6-phosphogluconic acid under the influence of glucose-6-phosphate dehydrogenase in the presence of NADP. This is a key reaction in this sequence, among other things being one of the few that provide NADPH for certain important metabolic reactions, e.g., fatty acid synthesis. Genetically determined deficiency of glucose-6-phosphate dehydrogenase in erythrocytes is associated with a tendency toward hemolysis by primaquine, sulfonamides, and fava beans (p. 554).

```
                    Glucose
                       ↓           NADP
        ┌──→ Glucose-6-phosphate ──────→ Gluconate-6-phosphate
        │                    CO₂ ↙  NADP
        │
        │        Ribulose-5-phosphate ←──→ Ribose-5-phosphate
        │              ↑↓
        │        Xylulose-5-phosphate ────────────┐
        │                                         │
        └── Fructose-6-phosphate ←──→ Sedoheptulose-7-phosphate
                                                  │
                                                  ↓
             Krebs cycle ←────── Glyceraldehyde-3-phosphate
```

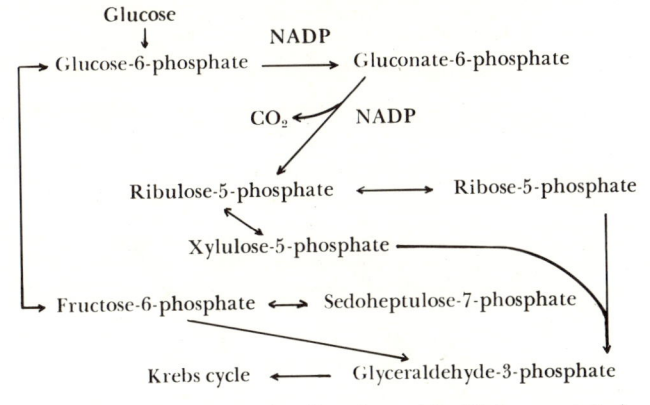

Figure 1-6. "Pentose-shunt" pathway (simplified presentation).

6-Phosphogluconic acid then undergoes oxidative decarboxylation (producing an additional molecule of NADPH) to form, successively, ribulose-5-phosphate and ribose-5-phosphate. The latter are convertible (1) to nucleotides and (2) to fructose-6-phosphate and glyceraldehyde-3-phosphate. Through the latter, re-entry is made into the glycolytic reaction sequence (Fig. 1-2, p. 7).

Since fructose-6-phosphate is readily converted to glucose-6-phosphate, it is evident that the elements of a cyclic mechanism are provided by this pathway. The "oxidative shunt" therefore has a dual significance: (1) it provides a mechanism for synthesis of ribose, which in turn is an important component of free nucleotides and nucleic acids; (2) it furnishes certain tissues with an aerobic by-pass around the glycolytic sequence and tricarboxylic acid cycle, a pathway which provides for the oxidation to CO_2 of one carbon atom of glucose at a time. In addition, because of its NADP requirement, it provides a supply of NADPH essential for certain important metabolic reactions.

ROLE OF LIVER IN CARBOHYDRATE METABOLISM

Glycogenesis and Glycogenolysis. In addition to the general reactions of carbohydrate metabolism shown in Figure 1-5, the liver is able to perform certain transformations peculiar to itself. One of these is the glucose-6-phosphatase reaction. As a result of the presence of this specific phosphatase in liver, the equilibria obtained are shown in Figure 1-7.

Glycogenesis in liver can occur from blood glucose or from any substance capable of giving rise to pyruvate. In the latter category are glycerol, blood lactate (from muscle, p. 16) and the glucogenic amino acids (p. 16).

Owing to the glucose-6-phosphatase, liver glycogen can contribute directly to blood sugar (glycogenolysis, p. 30) by way of glucose-1-phosphate and glucose-6-phosphate. In certain tissues, nonspecific phosphatases may substitute for glucose-6-phosphatase. It should be pointed out that, accord-

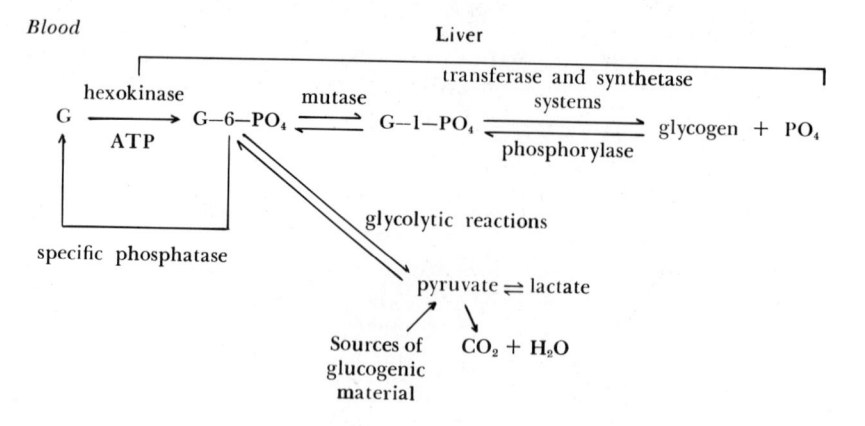

```
Blood                         Liver
┌──────────────────────────────────────────────────────────────┐
                       transferase and synthetase
   hexokinase    mutase       systems
G ──────────→ G–6–PO₄ ⇌ G–1–PO₄ ⇌───────────────→ glycogen + PO₄
↑    ATP                      phosphorylase
│            ╲
│             ╲ glycolytic reactions
└──────        ╲
specific phosphatase
                  pyruvate ⇌ lactate
                 ╱        ╲
          Sources of    CO₂ + H₂O
          glucogenic
          material
```

Figure 1-7. Equilibria between blood glucose (G), liver glycogen and glycolytic reactions.

ing to the above scheme, glucogenic substances in liver can form blood sugar without necessarily passing through the stage of liver glycogen.

The concentration of glycogen in the liver at any time is the resultant of glycogenolysis and the various routes of glycogenesis, the influx of dietary carbohydrate being one of the most important of the latter. Human liver contains, on the average, some 5 to 6 per cent glycogen. In dogs, the concentration may rise to 10 to 15 per cent after heavy carbohydrate feeding, and falls to less than 1 per cent on fasting. In certain cases of glycogen storage disease in human beings (von Gierke's disease), a deficiency of glucose-6-phosphatase has been found in the liver. Inability to carry out glycogenolysis, in the face of undiminished glycogenesis from blood sugar, probably accounts for the abnormal accumulation of liver glycogen in these subjects. Other types of glycogen storage disease (glycogenosis) are due to the absence of other enzymes (p. 79). The level of the blood sugar itself is a major factor in determining the relative rates of glycogenesis and glycogenolysis (p. 30). Endocrine influence on the several reactions involved is discussed elsewhere (p. 18).

Gluconeogenesis. The major conversions of other types of foodstuffs into carbohydrate occur in the liver. The pathway whereby glycerol may be converted to glucose was indicated previously. This, however, is a relatively minor source of carbohydrate; the chief gluconeogenetic substances are the amino acids. The nitrogen-free carbon skeletons of certain amino acids, as will be shown elsewhere (p. 17), are capable of conversion to pyruvic acid, from which point glucose or glycogen can be formed by reversal of the glycolytic reactions. This process is accelerated under conditions of carbohydrate deprivation. Gluconeogenesis from amino acids is under the control of adrenal cortical hormones (p. 25).

Assimilation of CO_2. Although carbon dioxide is usually considered a waste product of metabolism, evidence has accumulated that this substance is an important building-block in anabolic reactions. The reactions which appear to account for much of the CO_2 assimilatory activity of the animal body occur largely in liver and may be regarded as adjuncts to the reactions of the tricarboxylic acid cycle. One such reaction is the conversion of pyruvic to oxaloacetic acid by the mitochondrial enzyme pyruvic carboxylase. The CO_2 is "activated" by biotin.

The physiological significance of CO_2 incorporation reactions seems to be that the tissues in which such reactions are well developed can readily synthesize their supply of catalytic acids for the cycle (e.g., oxaloacetate), as long as a source of pyruvate is available. (CO_2, of course, is always in plentiful supply in any living tissue.) During the metabolism of carbohydrate, pyruvic acid is formed as a normal intermediate, hence the aerobic catabolism of carbohydrate is never retarded as a consequence of lack of oxaloacetate.

MUSCLE IN CARBOHYDRATE METABOLISM

The ensuing discussion is concerned chiefly with skeletal muscle. Certain peculiarities of other muscles should be noted, however. Cardiac muscle differs from skeletal in its capacity to oxidize blood lactic acid for energy production. Diaphragm, in contrast to skeletal muscle, is said to be better able to reverse the glycolytic sequence of reactions and synthesize glycogen from pyruvate or lactate.

Glycogenesis. The formation of muscle glycogen from blood glucose follows the same path as in liver. The absence of glucose-6-phosphatase from muscle, however, makes it impossible for muscle glycogen to contribute directly to blood sugar; consequently glycogenolysis, as ordinarily defined, does not occur in muscle. It should be noted that nonspecific phosphatases, which might be expected to catalyze the same reaction, are also virtually absent from muscle.

The level of muscle glycogen varies with the dietary intake of carbohydrate and its precursors but does not fluctuate so widely as liver glycogen. The concentration in human muscle ranges from 0.8 to 3.9 per cent. In general, muscle glycogen is subject to the same hormonal controls as liver glycogen, except in the case of glucagon (p. 26). Muscle glycogen is probably not formed to any extent by gluconeogenesis; its major source seems to be blood glucose.

Glycolysis. The glycolytic reactions in muscle follow the general course outlined previously. In muscle, however, the process seems to be irreversible, since experiments have shown that isotopically labeled lactic acid is not converted to labeled glycogen in the perfused muscle, although it is readily oxidized to CO_2. The reason for the irreversibility has not been definitely established. Studies with rabbits and chickens have indicated that a number of enzymes of the glycolytic pathway apparently exist in two forms. One is predominant in muscle and the other in liver. Those in muscle seem to be geared toward glucose breakdown, whereas those in liver are apparently geared toward gluconeogenesis. Another consideration

16 CARBOHYDRATE METABOLISM

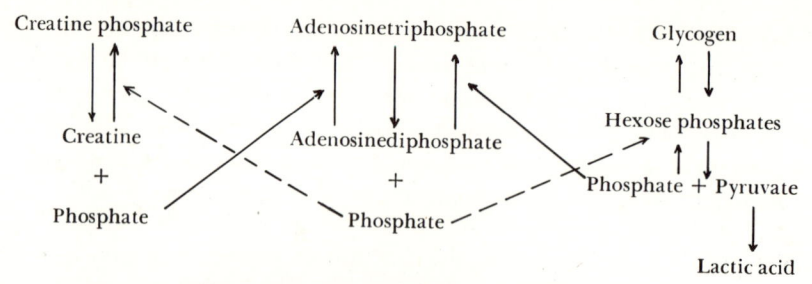

Figure 1-8. Chemistry of muscle contraction.

is the great permeability of muscle to lactic acid; any that is formed (under anaerobic conditions) is readily lost to the blood. This lactic acid is taken up by the liver, and may be converted to glycogen or blood glucose, the complete route of: blood glucose → muscle glycogen → lactate → liver glucose-6-phosphate → blood glucose, being known as the Cori cycle.

Muscle Contraction. As a form of mechanical work, it is to be expected that muscle contraction would use as its source of energy that common currency of bioenergetics, adenosinetriphosphate (ATP) (p. 250). The contractile phenomenon involves a reaction between ATP and myosin, a muscle protein, resulting in the formation of ADP and inorganic phosphate, the energy of hydrolysis of the terminal phosphate linkage of ATP being transformed to the mechanical work of contraction (Fig. 1-8). This is brought about by a sliding toward each other of the thin longitudinal filaments (actin) between the thick longitudinal filaments (myosin) of each muscle fibril. In this way, each fibril becomes shorter and thicker, which results in a shortening and thickening of the whole muscle. Although the ATP of resting muscle may be derived from other energy-producing reactions, such as oxidation of fatty acids and ketone bodies, it is probable that oxidation of carbohydrate is the main source of supply in muscle during active contraction.

Concomitantly with the utilization of ATP in the contractile act, a series of recovery processes occurs. Phosphocreatine, the most important storage form of high-energy phosphate in muscle (p. 180), reacts with ADP. This is accompanied or shortly followed by glycolysis (anaerobic) of stored muscle glycogen or glucose that has entered from the blood, the end product (under relatively anaerobic conditions) being lactic acid. This largely passes into the blood for utilization elsewhere, e.g., reconversion to glycogen or glucose in the liver (p. 13). However, the comparatively low rate of regeneration of ATP during glycolysis (p. 7) would not permit continuous muscular exertion for prolonged periods. This requirement is met by aerobic recovery processes, actually practically synonymous with the operation of the tricarboxylic acid cycle, which provides for rapid regeneration of ATP. When the muscle is at rest, the usual oxidative catabolism of carbohydrate (and probably other substances) produces a surplus of ATP, the excess of high-energy phosphate being stored as phosphocreatine (creatine + ATP → phosphocreatine + ADP), thus providing a ready reservoir of potential ATP for future contractions.

INTERRELATION OF CARBOHYDRATE, LIPID AND PROTEIN METABOLISM

One of the great contributions of isotopic studies has been the demonstration of the intimate interrelation of carbohydrate, lipid and protein metabolism. This has led to a better understanding of many abnormalities that occur in disease states.

The metabolism of the three major foodstuffs exhibits many points of interaction (Fig. 1-9). Metabolites of carbohydrate can form fatty acids and the carbon skeletons of certain amino acids, and many amino acids can be converted to carbohydrate. Fragments from all classes of foodstuffs are eventually channeled into a common pathway of aerobic catabolism, the tricarboxylic acid cycle, for ultimate conversion to CO_2 and H_2O, and liberation of energy.

The most important common junction in the metabolism of fatty acids and carbohydrate is the 2-carbon fragment ("active" acetate; acetyl-CoA), which is formed (a) by oxidative decarboxylation of pyruvate (which is formed from carbohydrates and certain amino acids (glucogenic)), (b) by oxidation of fatty acids, and (c) directly from certain amino acids (ketogenic amino acids). Moreover, the majority of amino acids are directly or indirectly convertible to carbohydrate intermediates (p. 162), and since, in the liver, the main reaction sequences of carbohydrate metabolism are reversible, these amino acids can form new molecules of glycogen or glucose (glucogenic amino acids). Likewise, by reversal of these processes, carbohydrate is convertible to (a) fatty acids (via "active acetate"), and, therefore, into neutral fat (fatty acids plus glycerol), which itself can be formed from carbohydrate (p. 102), and (b) nonessential amino acids, by amination of α-keto acid intermediates (e.g., pyruvate $\rightarrow$ alanine; oxaloacetate $\rightarrow$ aspartic acid; α-ketoglutaric acid $\rightarrow$ glutamic acid).

In the liver, each "active acetate" fragment can take one of a variety of metabolic pathways including: (a) condensation with oxaloacetate, forming citrate, and thus entering the tricarboxylic acid cycle for conversion to CO_2 and H_2O; (b) synthesis of fatty acids; (c) synthesis of cholesterol; (d) condensation with an "active acetoacetate" molecule (acetoacetyl-CoA) to form acetoacetate (ketogenesis). The last indicated pathway (ketogenesis) is available in both liver and kidney.

Certain hormonal mechanisms are integrated to a common purpose in this connection. The system fundamentally involved is the balance between insulin, on the one hand, and the adrenal 11-hydroxysteroids and pituitary growth hormone on the other. The pertinent actions of these hormones are outlined in detail elsewhere (pp. 11, 24, and 25). It is necessary here merely to indicate the over-all metabolic consequences of imbalance of these factors. Insulin deficiency, absolute (i.e., pancreatic islet insufficiency) or relative (i.e., adrenocortical or anterior pituitary excess), results in: (a) depression of carbohydrate utilization; (b) stimulation of mobilization of fat from depot stores (pituitary, adrenal and glucagon action), with hyperlipemia and accumulation of free fatty acids, which inhibit glycolysis and encourage

18 CARBOHYDRATE METABOLISM

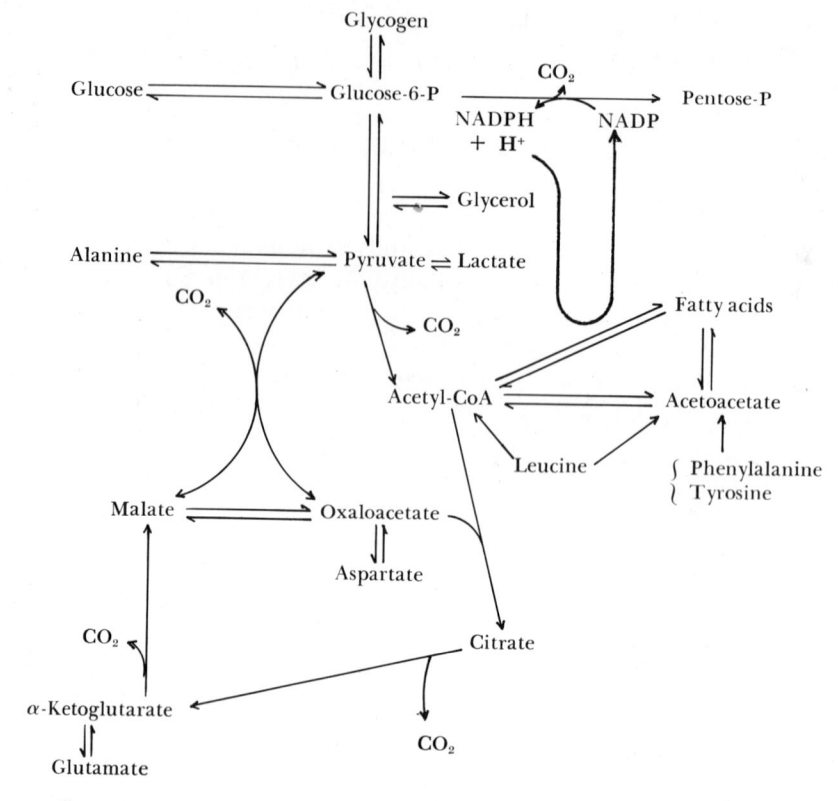

Figure 1-9. Interrelation of carbohydrate, protein, and fatty acid metabolism.

gluconeogenesis; (c) excessive formation of "active acetate" from fatty acids; (d) excessive diversion of "active acetate" into ketogenesis (in liver and kidney); (e) excessive cholesterol synthesis (in liver); and (f) excessive protein catabolism (adrenal action), with increased gluconeogenesis and negative nitrogen balance. Similar phenomena follow undue restriction of carbohydrate intake.

The hormone glucagon stimulates lipase activity in the liver. The resultant breakdown of triglycerides liberates free fatty acids which, as already described, affect glycolysis and gluconeogenesis.

ENDOCRINE INFLUENCES IN CARBOHYDRATE METABOLISM

The several processes concerned with the supply of glucose to the blood by the liver and with its utilization in the tissues must be regulated and integrated, not only with one another, but also with certain phases of the metabolism of proteins and lipids. Endocrine organs play an important role in this homeostatic mechanism.

Insulin. The over-all effects of insulin on the organism are illustrated by certain observable metabolic consequences of (1) pancreatectomy and (2) administration of insulin to normal animals.

1. *Pancreatectomy* is followed by:
    a. Decreased glucose tolerance.
    b. Hyperglycemia and glycosuria.
    c. Depletion of glycogen in liver and muscle.
    d. Decreased rate of oxidation of glucose (decreased R.Q.).
    e. Increased gluconeogenesis.
    f. Negative nitrogen balance.
    g. Increased mobilization of depot fat; increased ketogenesis.
    h. Increase in cholesterol and other lipids in the plasma.
2. *Insulin administration* is followed by:
    a. Decrease in blood sugar.
    b. Increased oxidation of glucose (increased R.Q.).
    c. Increased muscle glycogen.
    d. Decreased gluconeogenesis.
    e. Decreased ketogenesis.

The most important known actions of insulin are as follows:

1. Effect on glucose transport across cell membranes.
2. Increased glycogen formation.
3. Acceleration of conversion of glucose to fat—an indirect effect.
4. Stimulation of protein synthesis.
5. Inhibition of ketogenesis.
6. Inhibition of gluconeogenesis.
7. Increase in glucose breakdown.
8. Effect on muscle ion balance.

On the basis of these over-all effects, insulin may be considered to promote the utilization of glucose in every direction, including its oxidation, glycogenesis, and lipogenesis; it also promotes protein synthesis, in a manner as yet unknown; it has been suggested that it acts on ribosomes. The mechanism of action of insulin in the production of these effects has not been established with certainty, much of the evidence being of an indirect nature and still controversial. Liver slices from insulin-deficient animals exhibit the following phenomena, all of which are corrected by supplying insulin:

(a) Decreased phosphorylation of glucose. This would result in its decreased utilization in every known direction.

(b) Increase in glucose-6-phosphatase activity. This would result in increased output of glucose from the liver, contributing to hyperglycemia.

(c) Decreased glycogenesis.

(d) Decrease in glucose-6-phosphate dehydrogenase activity. This would result in depression of oxidation by the direct oxidation pathway (pentose shunt).

The glycogen of the brain and myocardium differs from that of the liver and skeletal muscle with reference to insulin action. This hormone has little or no effect upon glycogen in nervous tissue, which is also influenced but slightly, if at all, by the level of blood sugar. Although, under certain experimental conditions, administration of insulin may result in an increase in myocardial glycogen, the latter may be maintained in depancreatized animals. It would appear to be dependent more on the level of blood sugar than on the presence of insulin. However, high values for cardiac muscle glycogen have been obtained in phloridzinized animals with low blood sugar concentrations. There is evidence that the pituitary growth hormone plays an important role in the deposition of myocardial glycogen.

Substantial evidence supports the view that one of the fundamental actions of insulin is to promote the passage of glucose from the extracellular fluid into the cells of muscle and adipose tissue but not those of liver or the nervous system. This is, of course, a prerequisite for its metabolic utilization, including the initial phosphorylation (hexokinase reaction). There is evidence that insulin accelerates the entry into cells, not only of glucose, but also of nonutilizable sugars with structural configurations resembling that of glucose. These observations have led to the following conclusions: (1) that the cell "wall" presents a barrier to the entry of glucose (and related sugars), which limits the rate of their passage from extracellular fluids into the cell; (2) that insulin acts upon a sugar transfer mechanism to accelerate the entry of certain sugars into the cell; (3) it is believed by some that this transfer mechanism may not be enzymatic, inasmuch as the transfer is accelerated in the absence of enzymes capable of acting upon this sugar. It seems to be the consensus at present that a "sugar transfer mechanism" of as yet undetermined nature is one, but not necessarily the only, site of action of insulin. There is evidence that insulin and other hormones combine with muscle chemically and in some instances competitively (e.g., insulin and an anterior pituitary hormone). It has been suggested that such competitive interaction between hormones with antagonistic metabolic effects for sites of chemical combination with the tissues of specific target organs may constitute the basis for the action of hormonal regulatory mechanisms.

The hypoglycemia induced by insulin could conceivably result from one or both of two processes: (1) decreased rate of entrance of glucose into the blood stream from the liver; (2) acceleration of removal of glucose from the blood and the interstitial fluid, that is, increased rate of entry into and utilization in the cells. According to one view, the initial effect of insulin is to reduce the output of glucose by the liver; another maintains that the initial effect is to increase the utilization of glucose. There is general agreement, however, that both of these phenomena occur as later manifestations of insulin action.

Increase in muscle glycogen is usually readily demonstrable following administration of insulin. However, increased hepatic glycogenesis may be masked by the effects of the existing hypoglycemia, which itself accelerates hepatic glycogenolysis (p. 30), e.g., via increased secretion of epinephrine.

One of the functions of insulin is to promote conversion of carbohydrate to fat (i.e., long-chain fatty acids) in liver and adipose tissue. In the livers of diabetic animals, the conversion of glucose to fatty acids is depressed. Moreover, production of fatty acids from both acetate and fructose is also depressed, although their oxidation proceeds normally. This effect on fatty acid synthesis is now believed to be an indirect phenomenon. In the first place, insulin stimulates the pentose phosphate cycle, which provides additional NADPH for fatty acid synthesis. In the second place, insulin promotes the combination of glycerol and fatty acids to form triglycerides in adipose tissue. It also inhibits their enzymic hydrolysis. Both these effects prevent the accumulation of free fatty acids (FFA), which are known to inhibit the conversion of acetyl-CoA to malonyl-CoA, a rate-limiting reaction in their own synthesis. By lowering the level of FFA, insulin therefore causes an increase in their synthesis. On the other hand, insulin is not required for the conversion of acetate to cholesterol.

There is abundant evidence that insulin promotes protein synthesis. Depancreatized or alloxanized animals are in a state of negative nitrogen balance. An adequate supply of insulin is not required for the protein anabolic activity of pituitary growth hormone (p. 165). Insulin increases the incorporation of ^{14}C-labeled amino acids into the proteins of the isolated rat diaphragm and in isolated adipose tissue. It also increases the transport of the amino acids into the cells. This increase, unlike that of sugars, can occur against a concentration gradient. It involves an active process that depends on sodium ions. The transport system of amino acids is different from that for glucose, and it is not dependent upon it. This effect of insulin on protein metabolism is highly important and is regarded by some biochemists as its major effect. The effect of insulin on carbohydrate metabolism is largely dependent on its ability to induce the biosynthesis of certain enzymes, although it can inhibit that of others.

In diabetes (insulin deficiency, absolute or relative), there is an increase in liver fat, excessive ketogenesis, hyperlipemia and hypercholesterolemia. The increase in fat in the blood and liver is due to excessive mobilization of depot fat, due to lack of insulin, and also in consequence of the state of hormonal imbalance in the direction of relative excess of adrenocortical and anterior pituitary hormones, which promote lipolysis. In the absence of an adequate supply of insulin, the increased catabolism of fatty acids and certain amino acids results in the production of excessive amounts of acetoacetyl-CoA and acetyl-CoA which cannot be utilized adequately for synthesis of long-chain fatty acids, and which are present in amounts exceeding the limited capacity for oxidation. They are diverted, therefore, in increased amounts to formation of acetoacetate (excessive ketogenesis) and cholesterol (hypercholesterolemia). The metabolic pathways involved are shown in Figure 1-10. It is important to note that most acetoacetate is not formed, as was previously thought, from the condensation of two molecules of acetyl coenzyme A. The liberated acetoacetate is partially reduced to hydroxybutyrate and partially decarboxylated to form acetone. Acetoacetate, hydroxybutyrate and acetone are, of course, the "ketone bodies" of clinical practice.

This ketogenic pathway occurs in the liver, mostly in the mitochondria.

22 CARBOHYDRATE METABOLISM

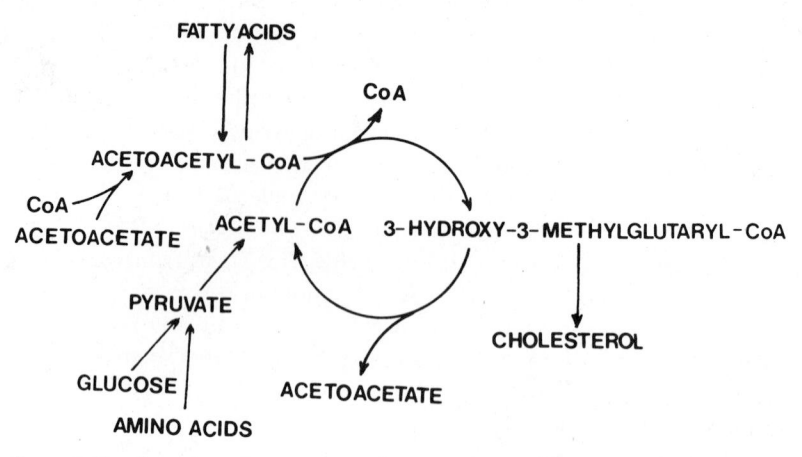

Figure 1-10. Formation of acetoacetate from acetyl-CoA and acetoacetyl-CoA.

The hydroxybutyrate diffuses into the blood and thence into both skeletal and cardiac muscle. It is now converted back to acetoacetate, which reacts with succinyl-CoA to form acetoacetyl-CoA and free succinate. The former is now converted into two molecules of acetyl-CoA and the "active acetate" enters the citric acid cycle. Although these reactions are reversible in the test tube, under physiological conditions they proceed in one direction and provide a mechanism for supplying energy to muscle from the four carbon atom compounds produced in the liver.

The hormonal imbalance referred to above (adrenocortical preponderance) results also in stimulation of increased glucose production from amino acids (excessive gluconeogenesis) with associated increased protein catabolism. This increased production of glucose, together with its decreased utilization, contributes to the hyperglycemia associated with insulin deficiency. Insulin promotes the biosynthesis of certain key enzymes of glycolysis, lipogenesis and the pentose phosphate pathway. The result is a stimulation of glucose breakdown. The lipogenic effect is important, since FFA (free fatty acids) inhibit the key enzymes of glycolysis and so their removal also helps to promote glucose breakdown. A fall in FFA also results in a corresponding fall in acetyl-CoA, which will lower the activity of pyruvate carboxylase. This is a key enzyme in gluconeogenesis, and so glucose synthesis is inhibited. Insulin also inhibits gluconeogenesis by acting as a suppressor of the biosynthesis of its key enzymes.

Stimulation of glucose utilization, by administration of either glucose or insulin, is accompanied by a decrease in the concentration of potassium and inorganic phosphate in the blood. These electrolyte changes are apparently associated with two phenomena: (1) acceleration of entrance of glucose (or other hexoses) into the tissue cells, and (2) deposition of K and PO_4 with glycogen in liver and muscles. It is known that increase of intracellular K and PO_4 does not depend on glucose uptake. The closely parallel changes in blood glucose, K, and PO_4 that follow administration of insulin, however, could be dependent upon common cellular metabolic mechanisms, probably involving phosphorylation processes.

Insulin Biosynthesis and Release. Insulin arises in the granules of the beta cells of the pancreatic islets from a single peptide *proinsulin*. A portion of the molecule, known as the connecting peptide, is then removed, leaving two peptide chains, A and B, connected by disulphide bridges (Fig. 1-11). The insulin is released by migration of the granules to the cell membrane, with which they fuse and then discharge their contents outside the cell. This process is known as emiocytosis.

The release of insulin is primarily a response to an increase in blood glucose levels. It is, however, affected by a number of other factors. It is of some interest that the sugar mannoheptulose, which occurs in the avocado pear, inhibits the glucose effect and that 2-deoxyglucose itself has no effect on insulin release, but can block the effect of glucose. Ingestion of protein or infusion of amino acids leads to a marked increase of plasma insulin. In certain sensitive individuals and in people with no such sensitivity who have been given chlorpropamide, leucine stimulates insulin release to the extent that it produces hypoglycemia. The ketone bodies also increase plasma insulin. Some hormones have a profound effect. Both epinephrine and norepinephrine are potent inhibitors, whereas glucagon produced in the pancreas and the intestine stimulates insulin release. Secretin, the chemical

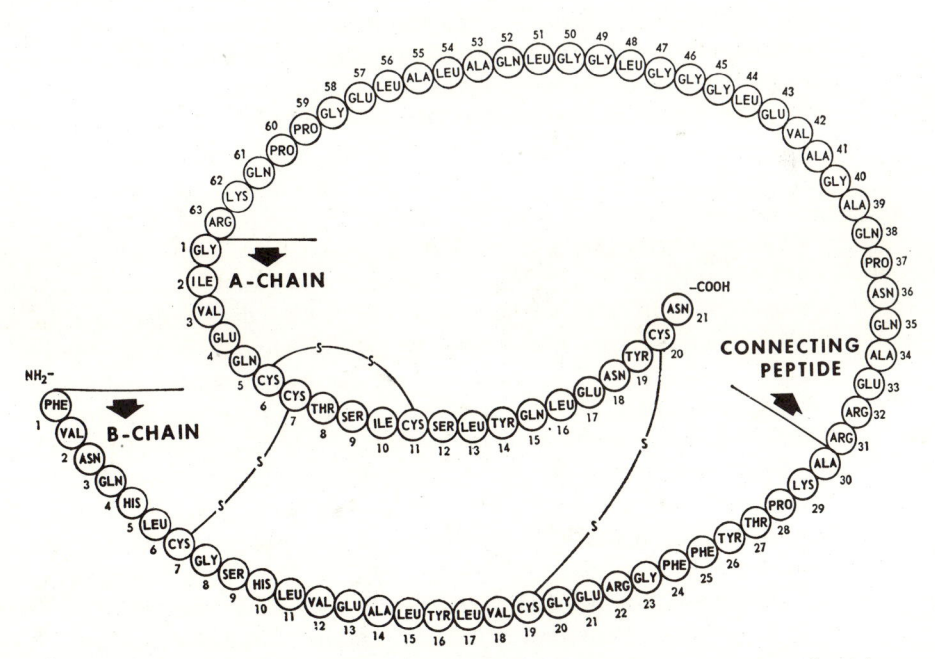

Figure 1-11. The primary structure of proinsulin from pigs. The molecule is synthesized as a single peptide chain. Three disulfide bridges are formed within the molecule from cysteinyl residues. The chain is then cleaved at two points, but the terminal segments are held together by the disulfide bridges and constitute the active insulin molecule, which now has two separate peptide chains linked by the bridges. (Reproduced from Chance, Ellis, and Bromer, Science, 161:166. Copyright 1968 by the American Association for the Advancement of Science.)

structure of which resembles glucagon, acts in a similar fashion to it. For stimulation by glucose, Ca^{2+} ions are required. A very high level of K^+ ions (8 mmol./l.) stimulates release independently of glucose.

The autonomic nervous system plays a part in insulin release. Stimulation of both vagus nerves simultaneously releases insulin and the effect can be blocked by atropine, Isoproterenol, which stimulates β-adrenergic nerves, produces insulin secretion. This effect can be blocked by propranolol and is, of course, opposite to the effect produced by epinephrine, which is on an α-receptor.

The sulfonylureas—for example, tolbutamide—are potent stimulators of insulin release. The benzothiadiazines suppress it. Diazoxide, which is the most active of these substances, can reduce the plasma insulin concentration in patients with insulinomas.

Metabolism of Insulin. The most reliable studies of insulin metabolism have been carried out with the technique of radio-immunoassay, although some information has been obtained from *in vitro* bioassays using the epididymal fat pad or the diaphragm of rats. The hormone leaves the pancreas both by the pancreatic vein and the lymphatic drainage. A good deal of activity is present in thoracic duct lymph. The hormone can also be found, at lower concentrations, in parotid saliva and cerebrospinal fluid. It is weakly associated with an α_1-globulin. Insulin is rapidly removed from the blood by the liver and most of it is destroyed by insulinases such as glutathione-insulin transhydrogenase, which separates the A and B chains, both of which are present in blood. Peptidases also play a part in its destruction. In the kidney, the hormone is filtered through the glomerulus and destroyed in the cells of the proximal tubule. The latter also remove some insulin from the blood. Some insulin is present in urine, even when the patient is fasting. This increases approximately fourfold after an oral glucose load.

Adrenocortical Hormones. The fasted, untreated, adrenalectomized animal (or patient with Addison's disease) exhibits the following changes in carbohydrate metabolism: (1) striking decrease in liver glycogen; (2) less marked decrease in muscle glycogen; (3) hypoglycemia; (4) decreased intestinal absorption of glucose. These changes are prevented or corrected by administration of 11-oxygenated adrenal hormones (glucocorticoids). Cortisol is three to five times as potent in this respect as corticosterone or 11-dehydrocorticosterone (Compound A), and 11-dehydro-17-hydroxycorticosterone (Compound E; cortisone), two to three times as potent. There are other indications of the influence of adrenocortical hormones on carbohydrate metabolism.

1. Adrenalectomized animals exhibit increased sensitivity to insulin.

2. Adrenalectomy results in amelioration of diabetes produced by pancreatectomy or alloxan. Administration of 11-oxygenated adrenal hormones or cortical extracts aggravates the diabetes.

3. Administration of cortical extract or 11-oxygenated adrenal hormones causes a rise in blood sugar, liver glycogen and total body carbo-

hydrate, glycogen formation being accelerated and its breakdown to glucose retarded.

The adrenocortical deficiency manifestations are related mainly to two fundamental phenomena: (1) decreased gluconeogenesis from body protein (p. 18), and (2) decreased output of glucose by the liver. Conversely, cortical hormones active in this sphere increase gluconeogenesis from body protein. The livers of animals given large amounts of cortisol show selectively increased activity of glucose-6-phosphatase, fructose-1,6-diphosphatase, phosphoenolpyruvate carboxylase and pyruvate carboxylase. Inasmuch as these enzymes occupy key positions in gluconeogenesis in the liver, such changes could contribute to the hyperglycemic action of the glucocorticoids.

The glucocorticoid hormones release gluconeogenic precursors, such as amino acids, from muscle. These pass to the liver and their increased concentration is one factor in stimulating gluconeogenesis in that organ. The hormones also release FFA from adipose tissue, resulting in inhibition of hepatic glycolysis, as previously discussed. This shifts toward glucose production the equilibrium between glucose breakdown and its synthesis. The increase in FFA in the liver also results in an increase of acetyl-CoA, which activates pyruvate carboxylase, and so stimulates gluconeogenesis. Glucocorticoid, as has already been indicated above, also results in increased biosynthesis of the key enzymes of gluconeogenesis in the liver. This does not apparently happen in the kidney cortex. Another factor in tipping the balance in favor of gluconeogenesis in the liver is the inhibition by glucocorticoids of glycolytic enzyme induction by insulin.

Anterior Pituitary Factors. Hypophysectomy leads to the following changes in carbohydrate metabolism: (1) tendency to hypoglycemia on fasting; (2) decrease in liver and muscle glycogen on fasting; (3) increased sensitivity to insulin; (4) increased utilization of carbohydrate; (5) amelioration of diabetes in depancreatized or alloxanized animals. Certain of these manifestations are due in part to adrenocortical hypofunction resulting from loss of ACTH. However, they are completely reversed only by administration of growth hormone or crude pituitary extracts in addition to adrenocortical hormones or ACTH.

Conversely, administration of growth hormone produces the following changes in carbohydrate metabolism, the effects varying in different species: (1) hyperglycemia and aggravation of diabetes in depancreatized-hypophysectomized animals ("diabetogenic" effect); (2) inhibition of insulin action ("anti-insulin" effect), with decreased utilization of carbohydrate and lowering of the respiratory quotient; (3) increase in muscle glycogen in hypophysectomized animals ("glycostatic" effect). The permanent diabetes produced in certain species (e.g., dog) by growth hormone is due to destruction of the islands of Langerhans. This effect is probably due to the primarily induced hyperglycemia, which in turn causes hyperactivity and hyperplasia of the islet beta cells and, subsequently, functional exhaustion and atrophy.

Not only does growth hormone decrease the breakdown of muscle glycogen, it also decreases the breakdown of muscle protein. Both these effects resemble those of insulin. The effect of growth hormone on protein actually requires the presence of insulin. On the other hand, it decreases the transport action of insulin in relation to glucose. It also decreases the rate of deamination of amino acids. These effects mean that the only energy source growth hormone permits is fat. It mobilizes the fat in the form of FFA from adipose tissue by increasing lipase activity and promotes its absorption into muscle. This is similar to the fasting state, in which, as a matter of fact, plasma growth hormone levels are elevated. The increase of FFA is believed to diminish glucose utilization by muscle and, as has already been discussed, it inhibits the action of key glycolytic enzymes in the liver. These phenomena would also result in the tendency to hyperglycemia. The increase of FFA, and its use as the source of energy, also results in the excessive production of ketone bodies.

Growth hormone secretion is increased in hypoglycemic states, as an attempt to restore the glucose level toward normal. Secretion is thus increased in the fasting state and decreased after glucose administration.

Epinephrine. Epinephrine produces an increase in blood sugar and in blood lactic acid. These are due to acceleration of glycogenolysis in the liver and muscles, respectively, and are accompanied by a decrease in the glycogen content of these structures. The mechanism involved is stimulation of phosphorylase activity (reactivation of inactivated phosphorylase by a mechanism similar to glucagon and described under that heading). There is some evidence, too, that epinephrine may diminish the uptake of glucose by tissue cells, thus interfering with its utilization. There is, however, some doubt about this, since the diminished arteriovenous glucose difference after epinephrine administration may be due to increased blood flow to muscle. This hormone also causes diminution in the amount of insulin released from the pancreas.

Thyroid Hormone. Thyroxine accelerates hepatic glycogenolysis, with consequent rise in blood sugar. This may be due in part to increased sensitivity to epinephrine. This effect is partially offset by simultaneously increased utilization of glucose incident to the acceleration of metabolism and blood flow. Thyroxine also increases the rate of absorption of hexoses from the intestine.

Glucagon. A factor is present in extracts of the pancreas, as well as in most commercial insulin preparations, which causes an increase in blood sugar when injected intravenously. It has been designated the "hyperglycemic factor" of the pancreas, or glucagon. It exhibits protein characteristics, is formed in the alpha cells of the islets, and is classified as a hormone. It is now known to be a polypeptide containing 29 amino acid residues, the sequence of which has been determined. A similar, but not identical, substance is derived from the mucosa of the upper intestine. This cross reacts immunologically with the pancreatic entity. The production of the latter is stimulated by intravenous glucose but inhibited by glucose given orally, which, however, stimulates the release of intestinal glucagon (enteroglucagon). Both forms stimulate the release of insulin from the pancreas. The glycogenolytic activity of pancreatic glucagon is due to stimulation of liver adenylcyclase and so of phosphorylase because of the increased formation of cyclic AMP (cyclic adenosine monophosphate). This also stimulates gluconeogenesis by increasing the activity of the metabolic pathway whereby pyruvate is converted to phospho-enol-pyruvate, with the formation of oxaloacetate as an intermediary. Intestinal glucagon does not apparently have glycogenolytic activity.

In normal subjects, after an overnight fast, the plasma levels, as determined by immunoassay, are less than 1 ng./ml. This figure includes both pancreatic and intestinal forms. Methods are now available for determining each form separately. After orally administered glucose, there is a rise of plasma enteroglucagon and a fall in that derived from the pancreas. The latter increases after intravenous amino acids or after pancreozymin is administered by the same route. In man, unlike dogs, fasting or hypoglycemia does not seem to affect the level of plasma glucagon. There has been a suggestion that the hormone has some effect on lipolysis. When administered to human beings, glucagon produced increased secretion of nitrogen compounds in the urine. This coincides with animal *in vitro* experiments, in which it was shown that glucagon reduced the incorporation of amino acids into muscle protein.

The sulfonylureas decrease the release of pancreatic glucagon.

When glucagon is injected intravenously, the blood sugar begins to rise immediately, reaches a maximum in about 30 minutes, and falls to the original level in about an hour. Liver slices incubated with this factor show a decrease in glycogen and an increase in glucose-1-phosphate as a result of increased phosphorylase activity.

Liver phosphorylase undergoes continual enzymatic inactivation and reactivation; glucagon (and also epinephrine) accelerates its reactivation.

NORMAL POSTABSORPTIVE BLOOD SUGAR

In the resting, postabsorptive state (overnight fast), glucose is present in systemic venous blood in the following concentrations: whole blood, 60–100 mg./100 ml.; plasma, 65–110 mg./100 ml. Except for low values (30 mg./100 ml.) in the newborn and even lower values in premature infants, there is no clinically significant sex or age difference, although average values are somewhat higher in elderly subjects. Certain analytical procedures once in common use yielded higher values, i.e., 80–120 mg./100 ml. whole blood, because they included also nonsugar reducing substances, e.g., ergothioneine and glutathione, present chiefly in the erythrocytes. In the fasting state, the concentration of glucose in arterial (or capillary) blood is slightly higher (2–10 mg./100 ml.) than in venous blood, owing to its continual passage into and utilization by tissue cells. This "arterial-venous blood sugar difference" is increased under conditions of decreased blood flow or increased glucose utilization, e.g., following administration of carbohydrate (with rise in blood sugar) or of insulin. In man, glucose is distributed equally in the water of the plasma and erythrocytes. The lower concentration in the cells as compared to plasma, per unit volume, is due to their lower water content (because of large amount of hemoglobin). This uniform distribution of glucose between erythrocytes and plasma does not occur in other mammalian species (except anthropoids).

Owing to the operation of an efficient regulatory mechanism, the postabsorptive blood sugar concentration is maintained within the limits indicated in normal subjects except under unusual circumstances. What deviations do occur are usually relatively slight and transitory. Pain and emotional excitement (especially apprehension, anxiety, fear, anger) produce a brief rise (10–20 mg./100 ml.), due to acceleration of hepatic glycogenolysis resulting from increased epinephrine secretion. As the blood sugar rises, insulin secretion is stimulated, which, together with the accelerated blood flow induced by the epinephrine, results in increased utilization (i.e., removal) of glucose by the tissue cells and, consequently, prompt fall in blood sugar. The same may occur with brief, strenuous exercise (increased hepatic glycogenolysis), but may here, too, be dependent largely on the accompanying emotional excitement. Protracted strenuous exercise (e.g., marathon race) may at times result in a fall in blood sugar as a result of depletion of hepatic glycogen. During a prolonged fast in subjects whose activity is maintained, the blood sugar falls somewhat after about 2 days, usually reaching a minimum at 4 to 6 days (average drop 15–30 mg./100 ml.), occasionally falling as low as 40–50 mg./100 ml. The fall is of greater degree and occurs more promptly in infants than in adults (greater utilization of glucose and faster depletion of hepatic glycogen). With continued fasting, the blood sugar rises gradually during the second week to approximately the original level. The initial drop is due to depletion of preformed hepatic glycogen, which is necessarily maintained subsequently solely by gluconeogenesis from tissue protein (small amount from lactic acid). The gradual return to normal is effected by the homeostatic mechanism; important adjustments include decreased secretion of insulin (due to hypoglycemia) and thyroxine (due to starvation), with consequent decrease in utilization of glucose, and increased adrenocortical activity (due to stress of starvation), with consequent increased gluconeogenesis.

Glucose in Body Fluids Other than Blood. Glucose diffuses readily across blood and lymph capillary walls and, therefore, in the postabsorptive state, is equally distributed in the water of the blood plasma and interstitial fluid (also synovial fluid). Its concentration in cerebrospinal fluid removed from the ventricles is about 10–20 mg./100 ml. lower, and that in the lumbar fluid about 20–30 mg./100 ml. lower than in the blood plasma obtained simultaneously (postabsorptive state). The glucose content of the aqueous humor of the eye is also somewhat lower than that of blood plasma, per unit of water. Increase in the blood sugar concentration, e.g., after ingestion of glucose, is reflected in the extracellular fluids, the increase in the latter lagging somewhat behind that in the former.

The digestive secretions, with the exception of bile, contain virtually no glucose, even at high blood sugar concentrations. The same is true of milk and of secretions of the prostate and seminal vesicles. Unless the blood sugar is raised to unphysiological levels, viz., by intravenous injection, glucose does not appear in the urine of normal subjects except in negligible quantities. There are occasional exceptions to this statement, e.g., in pregnancy. Although glucose diffuses perfectly through the glomerular capillaries, it undergoes practically complete reabsorption in the renal tubules under normal circumstances, phosphorylation being probably involved in this process, as in its absorption from the intestine.

The concentration of glucose in intracellular water varies considerably in different tissues but is uniformly considerably lower than in the plasma, except, occasionally, in the liver. The hepatic cells differ from other cells in this connection in that they form glucose (from glycogen) in addition to receiving it from the extracellular fluid. This sugar cannot be stored, as such, to any significant extent in the cells and, consequently, is utilized at least as rapidly as it enters. Indeed, there is evidence that free glucose does not pass readily across the tissue cell membranes by passive diffusion, but is largely carried in actively by a "transfer mechanism" (p. 20), which may or may not involve phosphorylation. Variations in the concentration of glucose in the blood, e.g., after administration of glucose, epinephrine or insulin, are reflected in that in the tissue cells, more closely in liver and kidney, less so in muscle and brain, in which the concentration of free glucose is relatively low.

Sugars Other than Glucose in Body Fluids.

Sugars other than glucose are not ordinarily detectable in the blood or interstitial fluid under normal conditions in the postabsorptive state. In a considerable proportion of normal lactating women, a small portion of the lactose which is formed from glucose in the breast and is excreted in the milk passes into the blood stream and, not being utilizable by the tissues, is excreted in the urine. However, its concentration in the blood is rarely sufficiently high to permit its detection by ordinary methods. Fructose, galactose and pentoses are readily absorbed in the intestine from dietary sources. However, under ordinary conditions of intake their removal from the circulation and their utilization proceed so rapidly that they are not easily detectable in the blood of normal subjects. Using special techniques, it has been possible to demonstrate the presence of fructose, lactose and even sucrose, but not pentose. The same is true in subjects with essential fructosuria and essential pentosuria, in which conditions the excretion of these sugars in the urine presupposes their presence in the blood. Fructose phosphates are continually formed in all cells during the process of glycolysis but apparently do not leave the cells. Pentoses and galactose also are formed in the organism from glucose, the former being incorporated in nucleotides and nucleic acids, the latter in galactolipids (and lactose, in lactating breast). However, they, too, are either utilized completely *in situ* or enter the blood stream in amounts too small to permit their detection.

After ingestion of supertolerance quantities of fructose or galactose (e.g., 40 g.) by normal subjects, a sufficient quantity may escape prompt conversion to glycogen in the liver to cause their transitory appearance in the systemic circulation (30 to 75 minutes) in demonstrable amounts. They are also excreted in the urine under such circumstances inasmuch as, in contrast to glucose, they do not undergo extensive reabsorption from the glomerular filtrate in the uriniferous tubules. This may occur, too, in subjects with liver damage (i.e., impaired glycogenesis) after ingestion of small amounts of fructose or galactose. An alimentary form of pentosuria is also encountered occasionally, usually without appreciable amounts in the blood.

Lactose is present in milk, being synthesized in abundance by lactating mammary tissues from glucose and galactose, both of which originate in the blood glucose, the galactose, too, being formed *in situ*. Insulin is apparently not involved in this process. Fructose is present in high concentration (300–700 mg./100 ml.) in seminal fluid. It is apparently formed by the seminal vesicular epithelium, this function, in common with others of that structure, being stimulated by androgens (e.g., testosterone) and decreased by castration or hypophysectomy.

REGULATION OF BLOOD GLUCOSE CONCENTRATION

The concentration of glucose in the blood is the resultant of two general factors: (1) the rate of its entrance into and (2) the rate of its removal from the blood stream. In the normal postabsorptive state (no glucose in the intestine), it is probably contributed to the blood solely by the liver (p. 14), entering the hepatic vein at the rate of approximately 115 mg. ($\pm$ 10%)/m^2/min. Inasmuch as its concentration in the blood remains fairly constant under these conditions (resting, fasting), it must leave the blood, in the capillaries, at approximately the same rate. These two general processes are influenced by a number of factors under physiological conditions.

Rate of Supply of Glucose to Blood. Except for a possible minor contribution by the kidney, the blood glucose may be derived directly from the following sources (Fig. 1-1, p. 5): (1) hepatic glycogen; (2) amino acids (gluconeogenesis, in liver); (3) other carbohydrates, e.g., fructose, galactose, pentose, and also from lactate and glycerol; (4) glucose absorbed from the intestine, which may be incompletely removed from the portal blood by the liver.

Hepatic Glycogenesis and Glycogenolysis. In the postabsorptive state in a well-nourished person, the most important of the above factors is the liver glycogen. Under physiological conditions the following factors influence the rate of hepatic glycogenesis and glycogenolysis:

(a) Blood Glucose Concentration. A rise in blood sugar increases, and a fall in blood sugar decreases, the rate of hepatic glycogenesis. This results in part perhaps from a "mass action" effect (glucose → glycogen) and also from the incident changes in secretion of insulin by the pancreatic islet cells. A fall in blood sugar to hypoglycemic levels stimulates secretion of epinephrine, with consequent increased hepatic glycogenolysis.

(b) Insulin. Increase in the amount of insulin reaching the liver increases the rate of hepatic glycogenesis and decreases that of hepatic glycogenolysis (decreased output of glucose from liver). It is particularly interesting in this connection that the blood glucose concentration is the most important but not the only factor governing secretion of insulin, which increases as the blood sugar rises and decreases as it falls.

(c) **Epinephrine.** Increase in the amount of epinephrine reaching the liver increases the rate of hepatic glycogenolysis. Secretion of epinephrine is stimulated by hypoglycemia, however induced, and by a variety of emotional and other factors, including fear, anxiety, apprehension, anger, pain, unusual exertion, etc.

(d) **Thyroxine.** Increase in the amount of thyroxine reaching the liver increases the rate of hepatic glycogenolysis. This may occur during exposure to cold.

(e) **Muscular Activity.** Vigorous exercise is accompanied by an increase in the rate of hepatic glycogenolysis (epinephrine?).

Absorption from Intestine. A variable portion of the glucose absorbed from the intestine is removed from the portal blood by the liver (glycogenesis, lipogenesis, oxidation). The remainder passes directly into the systemic circulation. The relative proportions taking these routes depend largely upon (1) the capacity of the liver, at the moment, for further storage of glycogen, and (2) the rate of absorption of glucose from the bowel. The latter is increased by thyroid hormone and is decreased in the presence of increased intestinal motility, the glucose being hurried through the upper intestine, the region of maximal absorption.

Gluconeogenesis. If provided in adequate amount, dietary carbohydrate is the main source of liver glycogen. In carbohydrate restriction, increasing amounts, and in starvation virtually all of the liver glycogen is derived from amino acids. This process of gluconeogenesis (p. 14) is accelerated by the 11-oxygenated adrenal hormones and is retarded by insulin. If adequate amounts of glycogen are present in the liver, glucose may be formed directly by gluconeogenesis and pass into the blood stream.

Rate of Removal of Glucose from Blood.

Glucose leaves the blood in all tissues, being utilized by all cells for the production of energy (oxidation), and by certain cells for more specialized purposes, e.g., glycogenesis, lipogenesis, etc. (p. 5). Inasmuch as free glucose cannot be stored as such in cells in significant amounts, the rate of utilization of glucose in the tissues will determine the rate at which it is removed from the blood. Under physiological conditions, this is influenced by the following factors:

(a) **Blood Glucose Concentration.** A rise in blood sugar increases and a fall in blood sugar decreases the rate of glucose utilization. As in the case of hepatic glycogenesis, this may be due in part to a "mass action" effect (glucose → end-products), and also to the incident changes in insulin secretion.

(b) **Insulin.** Increase in the amount of insulin reaching the tissues increases the rate of utilization of glucose for all purposes (oxidation, glycogenesis, lipogenesis). Again, attention is directed to the fact that secretion of insulin increases as the blood sugar rises and decreases as it falls. One of the actions of insulin is to accelerate transfer of glucose across certain cell membranes, i.e., from intercellular to intracellular fluids.

(c) **Adrenocortical and Pituitary Hormones.** Utilization of glucose is depressed by growth hormone and by the 11-oxygenated adrenocortical hormones. Increase in the latter occurs, under physiological conditions, in response to a great variety of "alarming" stimuli (alarm reaction), which include practically all unpleasant emotions (anger, fear, anxiety, etc.) and pain, as well as other types of stress (exposure to cold, physical and mental tension, etc.).

Fundamental Regulatory Mechanism. The processes of hepatic glycogenesis and tissue utilization of glucose are sensitive to relatively slight deviations from the normal blood sugar concentration. As the latter rises, glycogenesis is accelerated and utilization increased, with consequent fall in blood sugar. The reverse occurs as the blood sugar concentration falls. The normal balance between production and utilization of blood sugar at a mean level of circulating (plasma) glucose of approximately 80 mg./100 ml. (65 to 110 mg.) is, therefore, dependent upon the sensitivity of these processes to variations above and below this concentration. This level of sensitivity is determined to a considerable extent by the balance between insulin, on the one hand, and hormones of the adrenal cortex and anterior pituitary on the other.

The over-all effect of insulin is to lower the blood sugar, that of the adrenocortical and growth hormone to raise it. Inasmuch as these two sets of factors are mutually antagonistic in this respect and, indeed, in several aspects of their metabolic actions, it is the ratio between them rather than their absolute amounts that is of primary importance in this connection. With minor differences, decrease in insulin produces much the same effects on carbohydrate metabolism as does increase in adrenocortical and growth hormones. Similarly, the effects of increase in insulin resemble closely those of decrease in either of the other factors. The fundamental importance of this balance is demonstrated strikingly by the fact that diabetes induced by removal of the pancreas is ameliorated considerably by subsequent hypophysectomy (Houssay preparation) or adrenalectomy.

The processes of hepatic glycogenesis and glycogenolysis and glucose utilization, and also the blood sugar concentration, are continually exposed to disturbing influences under physiological conditions. These include absorption of glucose from the intestine, physical and mental activity, emotional states, etc. The primary effect of the majority of these is a rise in blood sugar. As indicated previously, this automatically results in a net decrease in the delivery of glucose by the liver and acceleration of its utilization by the tissues. There is a simultaneous increase in secretion of insulin, stimulated by the elevated blood sugar concentration, with consequent increase in the ratio of insulin to adrenocortical and anterior pituitary hormones. This change in hormonal balance results in increased hepatic glycogenesis, decreased gluconeogenesis, decreased output of glucose by the liver, and increased utilization of glucose. The blood sugar falls accordingly.

A drop in blood sugar below the normal resting level causes decreased secretion of insulin, decrease in the ratio of insulin to the antagonistic adrenocortical and pituitary hormones, increased production of blood sugar, and decreased glucose utilization. Accordingly, the blood sugar rises. If the blood sugar falls to hypoglycemic levels, an additional emergency mechanism comes into operation, i.e., stimulation of epinephrine secretion (by hypoglycemia), resulting in acceleration of hepatic glycogenolysis and rise in blood sugar. The increase in epinephrine may also stimulate production of ACTH and, therefore, adrenocortical hormones, causing increased gluconeogenesis.

In the ultimate analysis, the blood sugar concentration regulates itself. Efficient operation of this autoregulation at physiological levels, however, requires a normal balance between insulin and the carbohydrate-active adrenocortical and anterior pituitary hormones, and also normal responsiveness of the pancreatic islet cells to variation in blood sugar. This constitutes the central regulatory mechanism. If it is intact, the blood sugar tends to remain within rather narrow limits in the face of periodic disturbances which may cause temporary fluctuations. On this basis, it can be readily understood why conditions of abnormal function of the islands of Langerhans, the adrenal cortex or the adenohypophysis are accompanied by serious and fundamental disturbances in carbohydrate metabolism and blood sugar regulation. Changes in blood sugar induced by other factors, e.g., epinephrine, thyroxine, exercise, glucose absorption from bowel, etc., are promptly and effectively counteracted by this control mechanism and are, therefore, usually limited in extent and of rather brief duration.

Experimental Diabetes Mellitus

Metabolic counterparts of clinical diabetes mellitus can be produced experimentally by disturbing the balance between insulin, on the one hand, and adrenocortical and anterior pituitary hormones on the other (p. 18 ff.). This induced imbalance is in the direction of a decrease in the amount of insulin in relation to that of the other two types of hormones. The importance of the balance between these fundamental regulatory factors is reflected in the following facts: (1) the typical diabetic state, with minor differences, results from either (*a*) absolute deficiency in insulin or (*b*) absolute excess of adenohypophyseal or adrenocortical hormones (i.e., relative deficiency in insulin); (2) the severe diabetes produced by total pancreatectomy is alleviated strikingly by subsequent removal of either the anterior pituitary or the adrenals.

Total Pancreatectomy. Surgical removal of all pancreatic tissue is followed by characteristic manifestations of diabetes mellitus, the severity of which varies considerably in different species (e.g., severe in dog and cat; moderately severe in man; mild in monkey, goat, pig; none in duck). Removal of the sources of insulin (islet β-cells) (in susceptible species) results in the following metabolic phenomena:

(1) Decreased utilization of glucose indicated by: (a) decreased glucose tolerance (blood sugar curve) (p. 60); decreased R.Q. (p. 19); fasting hyperglycemia (p. 47); glycosuria (p. 85).

(2) Increased gluconeogenesis (from protein), indicated by negative nitrogen balance and increasing urinary glucose: nitrogen (G:N) ratio. This is due to relative preponderance of protein catabolic adrenocortical hormones (p. 165).

(3) Increased mobilization of body lipids due to relative preponderance of pituitary and adrenocortical hormones (p. 105). This is reflected in loss of body fat and increase in blood plasma lipids.

(4) Increased oxidation (in liver) of the fatty acids, mobilized in excess, to active two-carbon and four-carbon fragments (p. 102). These cannot be utilized adequately for lipogenesis in the absence of insulin and are present in amounts greater than can be oxidized in the liver (tricarboxylic acid cycle). They are, therefore, diverted in increased amounts to the formation of acetoacetate and cholesterol. This results in ketosis, ketonuria and hypercholesterolemia.

(5) The excessive production of acetoacetic and β-hydroxybutyric acids, resulting in acidemia or acidosis, due to primary reduction in bicarbonate (p. 417).

(6) Dehydration, initiated by the polyuria incident to the glycosuria. Excessive amounts of Na and K are lost from the organism, in addition to water. In advanced stages, e.g., in diabetic coma, the plasma volume decreases, and shock and renal failure develop.

In man and the dog, the diabetes produced by removal of 90 to 95 per cent of the pancreas is more severe than that which follows total pancreatectomy. This has been attributed to the aggravating influence, in the former case, of the hyperglycemic factor, presumably secreted by the α-cells of the pancreatic islets. The residual β-cells undergo degeneration as a result of the prolonged hyperglycemia, as indicated below.

Subtotal Pancreatectomy. In certain species if an amount of pancreatic tissue inadequate to produce diabetes is removed and the blood sugar is maintained subsequently at a high level by various means, diabetes ensues. At first this is mild and reversible; later it is severe and irreversible. The hyperglycemia may be induced by intraperitoneal administration of glucose, administration of a high caloric diet, or anterior pituitary extracts (p. 25).

The elevated blood sugar constitutes a stimulus to the remaining β-cells which, if the hyperglycemia continues, presumably pass through phases of hyperfunction, functional exhaustion, vacuolization, and other degenerative changes eventuating in hyalinization or fibrosis, i.e., permanent destruction. The importance of hyperglycemia in inducing morphological damage of the β-cells is indicated by the fact that these lesions do not develop if the elevation of blood sugar is prevented by simultaneous administration of insulin or phloridzin.

Alloxan Administration. Injection of alloxan, a substance related to the pyrimidine bases and to uric acid, causes permanent diabetes, after brief transitory phases of hyper- and hypoglycemia. The diabetes is a result of degeneration and resorption of the β-cells of the pancreatic islets, the α-cells and acinar tissue being unaffected. The alloxan acts directly, promptly and specifically on the β-cells, and its effect can be prevented by administration of cysteine, glutathione, BAL (dimercaptopropanol), or thioglycollic acid immediately before or within a few minutes after injection of the alloxan. This protective action is due apparently to the $-SH$ (sulfhydryl) content of these compounds, the alloxan being perhaps reduced to an inactive substance.

There are interesting differences between the diabetes that follows pancreatectomy and that produced by alloxan. In alloxan diabetes, hyperglycemia and glycosuria are more severe and the insulin requirement is higher, but the animals survive longer without insulin than do depancreatized animals, show little ketonuria, and do not go into coma. If alloxan diabetic dogs are subsequently depancreatized, glycosuria and the insulin requirement diminish, but ketosis and coma supervene rapidly if insulin is withheld. It has been suggested that the difference between the two diabetic states are dependent upon a hormone elaborated by α-cells (glucagon), present in the alloxanized but not, of course, in the depancreatized animal.

Administration of Adrenocortical Hormones. The effects of the 11-oxygenated adrenal hormones on carbohydrate metabolism are discussed elsewhere (p. 24). Diabetes develops in subjects with certain types of adrenocortical hyperfunction and can be induced in rats by administration of these hormones together with a high carbohydrate diet. There is a considerable species difference in this regard. A temporary diabetic state occurs at times, but not invariably, in man following prolonged administration of adrenal glucocorticoids or ACTH. "Adrenal diabetes" differs from "pancreatic diabetes" chiefly in the associated increased resistance to insulin and the rather consistent increase in liver glycogen.

Anterior Pituitary Extracts. The influence of hormones of the anterior hypophysis on carbohydrate metabolism is described elsewhere (p. 25). Injection of crude saline anterior pituitary extracts (or growth hormone) causes hyperglycemia and, if repeated over periods of 1½ to 4 weeks, permanent diabetes in certain species, e.g., adult dog and cat. The rabbit is moderately susceptible, but the rat and mouse are highly resistant to this influence. The permanent diabetes so induced results from destruction of the islet β-cells induced by the prolonged hyperglycemia. It does not develop if the latter is prevented by simultaneous administration of insulin.

Thyroid Administration. Diabetes ("metathyroid diabetes") can be produced in partially pancreatectomized dogs in which hyperglycemia is maintained by administration of thyroid hormone. The islet β-cells undergo degeneration, as described above with hyperglycemia induced by other means.

Carbohydrate Metabolism in Hypophysectomized-Depancreatized Animals. Houssay demonstrated that manifestations of diabetes in depancreatized animals are prevented or alleviated by hypophysectomy. Such animals ("Houssay animals") differ from those merely depancreatized as follows:

(1) Glycosuria and polyuria are diminished and at times absent.

(2) They survive for longer periods, with less weight loss, are less susceptible to infection, and their wounds tend to heal more rapidly.

(3) The blood sugar is maintained at lower levels and is occasionally subnormal.

(4) The Houssay animal is extremely sensitive to insulin, as is the simply hypophysectomized animal.

(5) There is less ketosis, and the plasma bicarbonate remains within normal limits.

(6) Hepatic and muscle glycogen may be normal and the R.Q. may exhibit an almost normal increase following ingestion of glucose.

(7) The glucose tolerance curve in some cases is less distinctly abnormal, although it usually remains of the "diabetic" type.

(8) The nitrogen balance becomes less markedly negative.

These animals are, however, far from normal. They are rather precariously balanced between hypoglycemia and hyperglycemia. The blood sugar exhibits wide fluctuations, responding unduly to abstinence from (hypoglycemia) and administration of (hyperglycemia) carbohydrate. The ameliorating influence of hypophysectomy on pancreatic diabetes is apparently a result of diminished gluconeogenesis and ketogenesis and improved utilization of glucose at a more normal blood sugar concentration. Similar improvement occurs in depancreatized animals if the adrenals are removed and not the hypophysis. These observations emphasize the physiological importance in this connection of the balance between insulin, on the one hand, and hormones of the adenohypophysis and adrenal cortex, on the other.

NORMAL ALIMENTARY REACTION (ABSORPTIVE RESPONSE)

Sugar Tolerance

Ingestion of glucose, starch and, to a lesser degree, other carbohydrates (fructose, galactose) or protein (but not fat) is followed, in normal subjects, by a rise in blood sugar. The degree of elevation depends somewhat upon the amount of glucose contained in or produced from the ingested material, but this relationship is by no means quantitative. This "alimentary reaction" forms the basis for the carbohydrate tolerance tests commonly employed for the detection of disturbances of carbohydrate metabolism. The characteristics of a normal response may be illustrated by describing the changes which follow ingestion of 100 g. of glucose (or 1.75 g./kg.) (oral glucose tolerance test). Fifty g. of glucose in a 15 per cent solution with orange or lemon flavoring and ingested over 5 to 10 minutes is far less nauseating and produces very similar results. Samples of blood are

NORMAL ALIMENTARY REACTION (ABSORPTIVE RESPONSE) 37

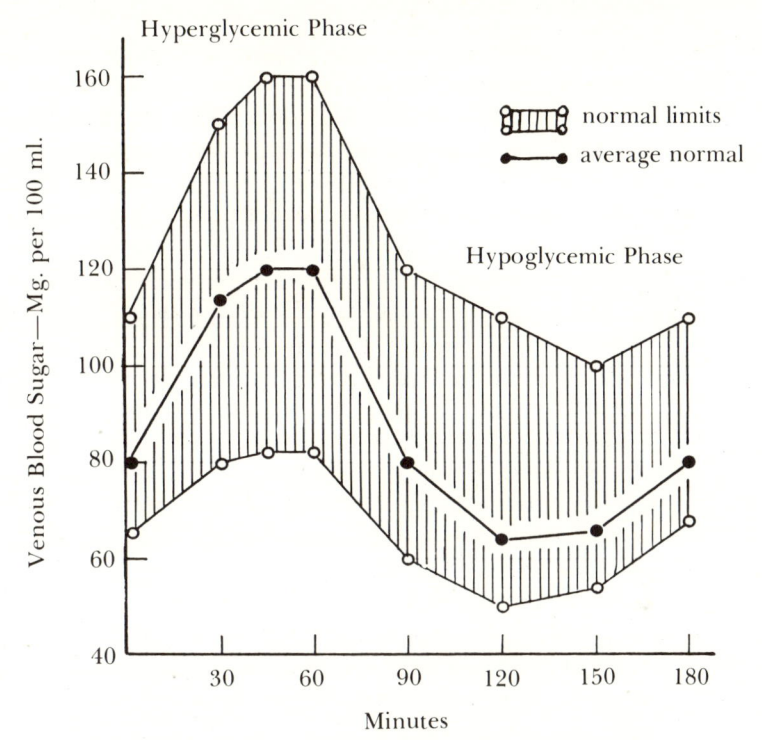

Figure 1-12. Venous blood sugar curve after ingestion of 100 g. glucose.

obtained before, and at intervals (Fig. 1-12) after, administration of the glucose. The upper limit for the fasting level shown in the figure is perhaps a little higher than normal. It is wise, however, not to diagnose abnormality until the fasting level is above 110 mg./100 ml.

The characteristic changes in venous blood are as follows: (a) A sharp rise to a peak, averaging about 50 per cent above the fasting level, within 30 to 60 minutes. The extent of the rise varies considerably, but the maximum should not exceed 170 mg./100 ml. in normal subjects, regardless of the initial level. This rise is due to the glucose absorbed from the intestine, which temporarily exceeds the capacity of the liver and tissues for removing it, and to glucose released by the liver. As the blood sugar concentration increases, however, the regulatory mechanism (p. 32) comes into play; hepatic glycogenesis is accelerated and glycogenolysis decreased, and glucose utilization increases, principally because of stimulation of insulin secretion. The peak of the curve is reached when utilization is accelerated to the point where the glucose is removed from the blood stream as rapidly as it enters. This point is usually reached while glucose is still being absorbed from the intestine.

(b) A sharp fall to approximately the fasting level at the end of 1½ to 2 hours. Glucose is now leaving the blood faster than it is entering. This is due to continuing stimulation of the mechanisms indicated above (increased utilization and hepatic glycogenesis) and to slowing or completion of glucose absorption from the bowel.

(c) Continued fall to a slightly subfasting (10 to 15 mg. lower) concentration (hypoglycemic phase) and subsequent rise to the fasting level at two and one-half to three hours. This hypoglycemic phase of the curve is caused by the inertia of the regulatory mechanism. The decreased output of glucose by the liver and the increased utilization, induced by the rising blood sugar, are not reversed so rapidly as the blood sugar falls. The higher the initial rise, e.g., after intravenous injection of glucose, the more pronounced the hypoglycemic phase.

The changes in arterial (capillary) blood differ from those in venous blood characteristically as follows:

(a) The rise begins somewhat earlier. When the blood sugar has increased about 20 mg./100 ml., the rise in venous blood begins to lag behind that in arterial blood.

(b) At the peak, usually reached at thirty to forty-five minutes, the level in arterial blood (150–200 mg./100 ml.) may be 20 to 70 mg. (average 30) higher than in venous blood. This arterial-venous blood sugar difference is an expression of the extent of removal of glucose from arterial blood by the tissues, i.e., of glucose utilization.

(c) The return to the fasting level (at one and one-half to three hours) is not so rapid as in the case of venous blood, indicating that active removal from the blood, i.e., increased utilization, continues beyond the period of hyperglycemia. The two curves converge at the resting level or slightly below in two and one-half to three hours.

With modern techniques for estimating blood glucose, e.g., glucose oxidase methods or the AutoAnalyzer procedure, Figure 1-12 would serve to show the range in capillary blood, rather than venous blood.

Intravenous Glucose Tolerance Test. The possibility of abnormalities in the glucose tolerance curve due to anormalities of absorption from the intestine may be obviated by injecting the glucose intravenously. In normal subjects the nature of the curve depends upon the amount of glucose injected and the duration of the period of injection. However, when relatively small amounts are injected continuously at a constant rate, the blood sugar rises to a peak and then falls steadily during the period of administration, even at times to hypoglycemic levels. A satisfactory procedure consists in the intravenous injection of 0.5 g. of glucose per kilogram of body weight in 20 per cent solution in distilled water by constant infusion over a period of thirty minutes. Blood is withdrawn for glucose determination before the injection is begun (control specimen), at the termination of the injection (30-minute specimen), thirty minutes later (1-hour specimen) and at subsequent hourly intervals for periods of 2 to 6 hours, as desired.

The blood sugar rises in normal subjects to a maximum of about 200 to 250 mg. per 100 ml. at the end of the injection, falls steadily to a slightly subresting level at two hours and returns to the pre-injection level at three to four hours, as in the case of the oral test (Fig. 1-13). The hypoglycemic phase of the curve is obtained more consistently after intravenous than after oral administration of glucose, rendering this procedure of particular value

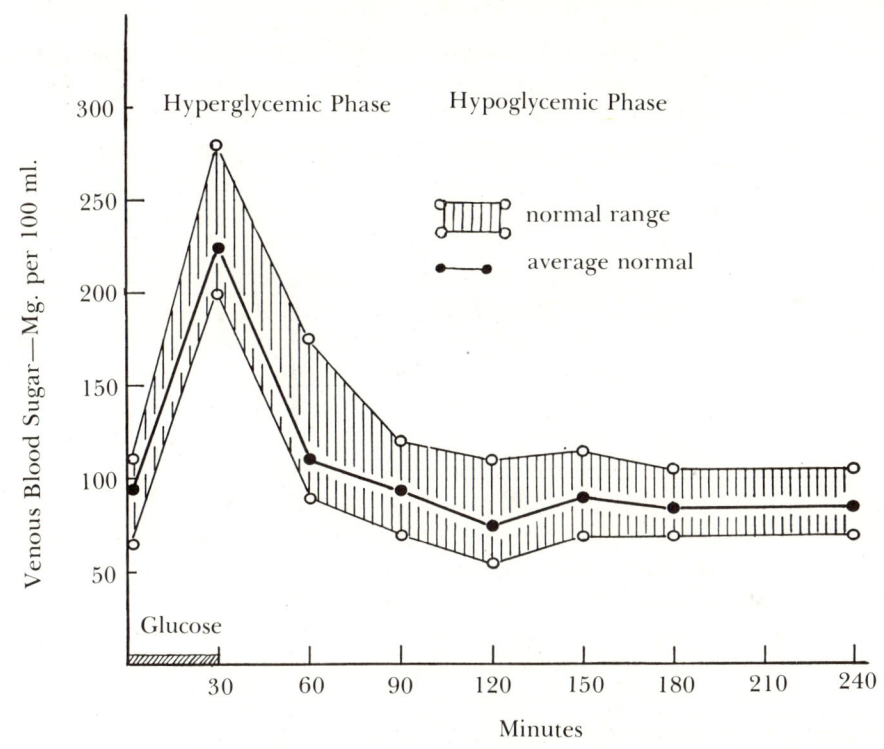

Figure 1-13. Intravenous glucose tolerance test.

in the study of conditions accompanied by hypoglycemia or increased glucose tolerance (p. 66).

A much more rapid and more commonly used technique is to inject intravenously at a steady rate, over a period of 2 to 4 minutes, 50 ml. of a 50 per cent glucose solution in distilled water. Blood samples are obtained before and at 5- or 10-minute intervals after the midpoint of the injection, over a period of one hour. After 15 minutes, during which the blood levels rise and the glucose is distributed throughout the extracellular space, the levels fall exponentially. Injected glucose disappears more slowly than the normal rate (3.71 ± 0.40%/min) in individuals with impaired tolerance. In the diabetic the disappearance is about half this rate.

Mechanism of Production of the Normal Glucose Tolerance Curve

Rise of Blood Sugar. The initial rise in both arterial and venous blood sugar in the oral test is due in large measure directly to the glucose absorbed from the intestine, which exceeds the capacity of the liver for removing it from the portal blood and passes into the systemic circulation.

Arterial-Venous Difference. This is an expression of the rate of removal of glucose from arterial blood by the tissues, particularly the muscles, for the formation of glycogen and for oxidation. Insulin plays an important part in these processes.

Fall in Blood Glucose. This appears to be due largely to three factors:

(a) Removal of Glucose by the Liver to Form Glycogen.

(b) **Removal of Glucose by the Extrahepatic Tissues for Glycogen Formation and for Oxidation.** It has been generally believed that the hyperglycemia caused by the administration of glucose stimulates the pancreas to secrete more insulin, the resulting increased storage and utilization of glucose causing the fall in blood sugar following the primary rise. The validity of this hypothesis has been questioned in recent years and evidence now available suggests that the homeostatic mechanism in the liver is probably of greater fundamental importance than the pancreas in this connection.

(c) **Diminished Hepatic Glycogenolysis and Gluconeogenesis.** The evidence referred to indicates that under normal conditions the liver responds to an increased blood sugar level by decreasing the output of blood sugar which it has been supplying from its own sources. An apt analogy has been drawn between this regulatory mechanism and a thermostatically controlled heating system. The liver corresponds to the furnace, the blood stream to the room, the tissues to the walls and the blood sugar concentration to the room temperature; the fuel is liver glycogen and the thermostatic regulation (i.e., of hepatic glycogenolysis) is represented by the interaction of hormones of the pancreatic islets (insulin), adrenal cortex, anterior hypophysis and, under certain circumstances, also the thyroid, adrenal medulla (epinephrine) and perhaps the gonads; radiation of heat from the room is represented by the removal of glucose from the blood for oxidation and storage as muscle glycogen. Under normal conditions, as the blood sugar (i.e., room temperature) increases above the normal resting level (e.g., 100 mg./100 ml.), hepatic glycogenolysis and gluconeogenesis (i.e., heat production by the furnace) automatically diminish, and removal of sugar for storage and oxidation in the tissues (i.e., radiation of heat) is accelerated. Decrease in the blood sugar is automatically followed by increased glucose production in the liver and decreased rate of removal by the tissues. As stated by Soskin, the glucose tolerance curve, with its hyperglycemic and hypoglycemic phases, resembles the fluctuations in temperature above and below the threshold of regulation when extra heat is introduced into the system. The nature of the curve depends upon (1) the quantity of heat (amount of sugar) introduced, (2) the sensitivity and setting of the thermostat (endocrine balance), and (3) the capacity of the furnace and the fuel supply (ability of the liver to produce sugar and the glycogen store in the liver).

Factors Influencing Absorptive Blood Sugar Response. In normal subjects, the height to which the blood sugar rises increases with increasing doses of glucose up to about 50 g. which produces a curve approximating that described above. The blood sugar rises no higher (in normal subjects) following ingestion of increased amounts of glucose (to 200 g. or more), but the fall may be somewhat delayed. This is due to the fact that absorption of glucose from the intestine proceeds at a relatively constant rate regardless, within wide limits, of the quantity present. As soon as the increasing rate of utilization matches the increased rate of entrance of glucose into the blood stream, no further increment occurs in blood sugar concentration regardless of continuing absorption from the bowel. This is not the case in the presence of defects in the blood sugar regulating mechanism, e.g., in diabetes mellitus, in which the degree of elevation of blood sugar increases with increasing doses of glucose.

The glucose tolerance tends to be somewhat lower (i.e., the curve of blood sugar is higher and more prolonged) in old than in young subjects. Emotional disturbances, which tend to raise the blood sugar, may cause undue and abnormally prolonged hyperglycemia. Strenuous exercise before the ingestion of glucose may cause the blood sugar to rise to an abnormally high level, whereas strenuous exercise after ingesting the glucose may cause an abnormally marked and prolonged hypoglycemic phase in some subjects. Because of the several factors, physical, emotional, dietary and gastrointestinal, that may influence the curve, the conditions under which the test is performed should be rigidly standardized. It should be performed in the morning at the time of the usual breakfast; the previous diet should be unrestricted (except in frank diabetes, when the test is usually superfluous); and, except when interest is centered particularly upon the hypoglycemic phase of the curve, physical exertion and emotional excitement should be avoided during the test period and for at least 15 minutes before it. Glucose tolerance tests must often be repeated before apparently abnormal results can be interpreted satisfactorily. Moreover, false normal or borderline normal curves are obtained at times as a result of disturbances of gastric emptying and intestinal motility and absorption. In such cases the intravenous test should be employed.

The state of carbohydrate nutrition influences the alimentary glucose response of normal individuals. An adequate deposit of glycogen in the liver and other tissues is essential to the production of a normal response as above described. If the subject is in a state of relative carbohydrate starvation, the rise in blood sugar following the ingestion of glucose will be more pronounced and its fall more delayed than under normal conditions. Conversely, if the organism is in a state of carbohydrate saturation,

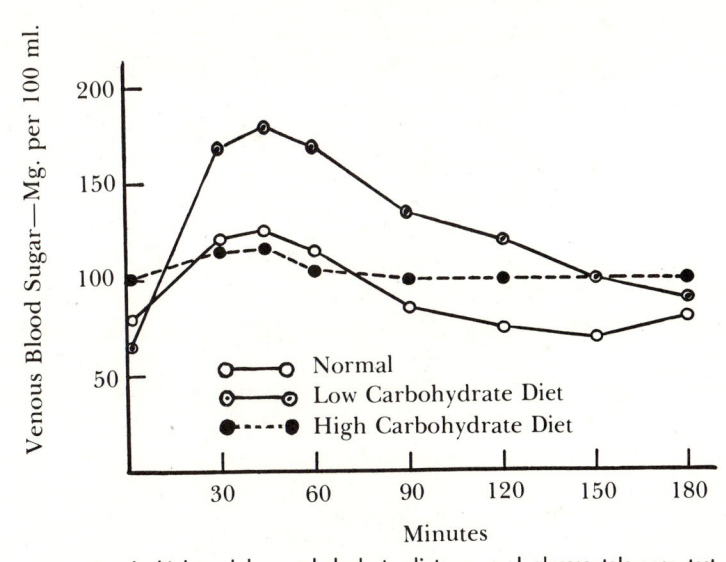

Figure 1-14. Effect of previously high and low carbohydrate diets on oral glucose tolerance test.

i.e., if a high carbohydrate meal has been taken within two to three hours of the performance of the test, the rise in blood sugar will be distinctly less marked than that described above.

It is therefore important that the subject shall have partaken of a well-rounded diet for some days prior to the performance of the test (Fig. 1-14). A diet high in fat and low in carbohydrate may result in a curve typical of a state of carbohydrate starvation. It is also well known that the repeated administration of glucose to normal persons at intervals of a few hours is followed by progressive lowering of the curve of alimentary hyperglycemia, i.e., a progressive increase in tolerance. These observations have been generally interpreted as indicating that the rather continuous absorption of adequate amounts of glucose from the intestine maintains active secretion of insulin by the pancreas and renders the islet cells sensitive to the stimulus to increased secretion furnished by the administration of glucose in the performance of the glucose tolerance test. When carbohydrates are withdrawn from the diet, insulin secretion is diminished, the machine idles, so to speak, and the sensitivity of the islet cells to sudden stimulation is diminished. When glucose is administered subsequently, an adequate amount of insulin is not secreted promptly and removal of the absorbed glucose from the blood stream is delayed. The relatively slight hyperglycemic effect of repeated doses of glucose is generally attributed to a state of increased responsiveness on the part of the pancreas, resulting from the previous administration of glucose. This influence of previous sugar administration on subsequent dextrose tolerance curves is known as the Staub-Traugott or Hamman-Hirschman effect.

It has also been suggested that the decreased blood sugar response that follows the repeated administration of glucose is a result of a decreased output of glucose from the liver (decreased hepatic glycogenolysis) and probably not to increased secretion of insulin, although these two phenomena are not necessarily mutually exclusive (p. 20).

The One-hour, Two-dose Glucose Tolerance Test. The Staub-Traugott phenomenon, referred to above, may be described briefly as follows: Normal human subjects react to repeated doses of glucose with either diminution or little or no change in glycemia, while diabetics react with definite hyperglycemia. This phenomenon has been utilized as the basis for several modifications of the glucose tolerance test. One of the most widely employed of these is the one-hour, two-dose test proposed by Exton. The test is performed as follows: Dissolve 100 g. of glucose in about 650 ml. of water. This solution is then flavored with lemon juice and divided into two equal doses, each containing 50 g. of glucose in about 15 per cent solution. In the morning after a twelve-hour fast collect blood and urine samples and give the first dose of glucose, allowing one-half minute for its ingestion. Thirty minutes later collect a second blood sample and give the second dose of glucose. Thirty minutes after the ingestion of the second dose of glucose collect blood and urine samples.

Normal criteria are: (1) Normal fasting blood sugar; (2) a half-hour level less than 50 mg. above the fasting value; (3) a one-hour value less than 30 mg. above the half-hour value. In our experience, however, the following criteria are the most acceptable: (1) normal fasting blood sugar; (2) one-hour blood sugar less than 160 mg. per 100 ml.; (3) all urine specimens sugar-free.

The use of this procedure is practically restricted to the diagnosis of diabetes mellitus and, in our experience, has not been so satisfactory for routine investigation of glucose tolerance as the longer oral and intravenous tests.

EFFECT OF OTHER SUGARS

The metabolism of fructose differs from that of glucose. The difference between the behavior of these two sugars may be illustrated by certain experimental observations.

(*a*) Liver and muscle contain separate hexokinases for glucose and fructose. Insulin (in the presence of growth hormone and adrenocortical hormones) is necessary for normal activity of the glucokinase ("hexokinase") (p. 20) but not the fructokinase reaction. Consequently, in the absence of insulin, phosphorylation of and, therefore, glycogenesis from fructose may be uninhibited, whereas phosphorylation of and glycogenesis from glucose is impaired considerably.

(*b*) Insulin hypoglycemia is relieved by glucose more effectively than by fructose.

(*c*) Fructose produces the same rise in blood pyruvic acid and fall in inorganic phosphate in diabetic as in normal subjects, whereas glucose does not.

(*d*) Following the ingestion of fructose there is comparatively little rise in blood sugar and practically no arterial-venous difference, indicating the minor role of the muscles in removing this sugar from the blood.

(*e*) Inasmuch as insulin is not necessary for essentially normal glycogenesis from and oxidation of fructose, this sugar has frequently been advocated as a substitute for glucose in the treatment of diabetes mellitus. However, a portion of the fructose is converted to glucose in the liver; the subsequent utilization of this fraction depends upon the availability of insulin. Moreover, this hormone is necessary for normal lipogenesis from fructose, as from glucose. In view of the fact that an appreciable proportion of the daily carbohydrate intake is converted to fat, impairment of this major pathway of utilization probably constitutes an additional reason why fructose has little if any practical advantage over glucose in the management of clinical diabetes.

(*f*) Cori has found that although fructose is absorbed from the intestine much more slowly than glucose, at the end of four hours the amount of glycogen formed in the liver by the former was 39 per cent and by the latter 17 per cent. In the absence of hepatic disease or functional insufficiency the ingestion of fructose, in tolerance doses, is followed by comparatively little elevation of blood sugar since, after absorption, it is effectively removed from circulation by the liver where it is stored as glycogen. The fructose tolerance test has, therefore, been utilized as a test of the integrity of liver function. It is no longer employed.

Galactose Tolerance. Galactose, like fructose, is metabolized chiefly by the liver. The metabolism of galactose differs from that of glucose chiefly in that (a) it is absorbed from the intestine somewhat more rapidly than the latter, (b) the magnitude of its increase in the blood after ingestion is much smaller than in the case of glucose, (c) the "renal threshold" for excretion of galactose is very low and (d) its intermediary metabolism is apparently influenced much less than that of glucose by the function of organs other than the liver.

After ingestion of 40 g. of galactose, the blood galactose concentration normally reaches a maximum of 15 to 35 mg. per 100 ml. in thirty to sixty minutes, falling to zero at the end of 2 hours. Normal subjects may excrete up to 3 g. of galactose (or galactose plus glucose) in the urine during the five hours immediately following ingestion of 40 g. of this sugar. Intravenous injection of 1 ml. of a 50 per cent solution of galactose per kilogram of body weight is followed by its rapid disappearance from the blood stream in normal subjects, none remaining at the end of seventy-five minutes.

OTHER TOLERANCE TESTS

Insulin Tolerance Test

The purpose of the insulin tolerance test is to determine (a) the sensitivity of the organism to insulin and (b) its responsiveness to insulin-induced hypoglycemia. Under normal conditions, the blood sugar falls to about 50 per cent of the fasting level twenty to thirty minutes after intravenous injection of 0.1 unit of insulin per kilogram of body weight. It then rises rapidly, reaching the pre-injection level in ninety to one hundred and twenty minutes (Fig. 1-18, p. 71). The duration of the hypoglycemia is of greater significance than its degree. The precautions that must be observed in performing this test are the same as those described in connection with the glucose tolerance test, since the response in both instances is influenced by the same factors. In cases in which an exaggerated or prolonged hypoglycemic response is anticipated (p. 66), it is advisable to provide for prompt administration of glucose as soon as clinical manifestations of hypoglycemia develop and, at times, to use one-half to one-third the standard dose of insulin. In such cases it is also advisable to administer carbohydrate about three hours after the test is completed.

Insulin-glucose Tolerance Test

This procedure was devised as a substitute for the conventional insulin tolerance test. It involves less risk than the latter and is regarded as a more sensitive means of detecting hypoglycemia unresponsiveness.

After an overnight fast, a sample of blood is taken and 0.1 unit of regular insulin per kg. is injected intravenously. After thirty minutes, or when the first significant manifestation of hypoglycemia develops (whichever is first), 0.8 g. of glucose per kg. is given by mouth. Samples of blood are taken at thirty, sixty, ninety, one hundred and twenty, and one hundred and eighty minutes after the injection of insulin.

In normal subjects, the rise in blood sugar following ingestion of glucose in the hypoglycemic state exceeds that which follows administration of glucose without insulin. Representative normal values for blood sugar with this procedure are: thirty minutes, 30 to 55 mg. (40); sixty minutes, 100 to 135 mg. (115); ninety minutes, 125 to 175 mg. (150); one hundred and twenty minutes, 105 to 165 mg. (140); one hundred and eighty minutes, 75 to 115 mg. (95).

Cortisone—and Prednisone Glucose Tolerance Tests

Cortisone. At both 8½ hours and 2 hours before oral administration of 1.75 g. glucose/kg. ideal weight, cortisone acetate is administered. Each dose is 50 mg. for subjects weighing less than 160 lbs. and 72.6 mg. for heavier individuals. The patient is in the postabsorptive state when the glucose is given. The normal response is a level below 140 mg./100 ml. of venous blood, collected 2 hours after the administration of glucose.

Prednisone. The rate of glucose excretion by the kidneys during sleep is normally quite constant and very low. In this test, steroid-induced hyperglycemia is determined by measuring the glucose excretion rate overnight. A 300 g. carbohydrate diet is taken during the day and no food is taken after 6:00 P.M. At midday, 4:00 P.M. and 8:00 P.M., 20 mg. prednisone is administered orally. Urine is then collected from 10:00 P.M. to 6:00 A.M. (test urine), and capillary blood samples taken at midnight and 1:00 A.M.

In normal individuals, the test urine shows a glucose excretion of less than 60 mg./hr. Patients with renal glycosuria can have a raised level of excretion, but can easily be excluded. The mean glucose level of the midnight and 1:00 A.M. capillary blood samples normally does not exceed 128 mg./100 ml.

Tolbutamide Tolerance Test

The effect of tolbutamide on insulin release, enables it to be used in an intravenous tolerance test for two purposes, viz., the diagnosis of diabetes mellitus and the recognition of hypoglycemia resulting from an insulinoma. For both purposes, 20 ml. of a 5 per cent solution of tolbutamide in distilled water (1 g. tolbutamide) is injected at a constant rate intravenously over a 2-minute period.

In relation to diabetes mellitus, the blood glucose concentration is determined at exactly 20 minutes and exactly 30 minutes after the midpoint of the injection, which is given to the patient in the fasting state and after a fasting specimen of blood has been collected. Although there is a normal range (see Fig. 1-19), the glucose level of the 20-minute specimen is usually lower than 75 per cent of that of the fasting specimen and almost all normal individuals have a 30-minute level, which is less than 77 per cent of that of the fasting specimen. Breakfast should be given immediately after the 30-minute specimen has been collected.

For the diagnosis of an insulinoma, in addition to the fasting specimen, samples of blood are taken every 15 minutes for the first hour and every 30 minutes for the next two hours. In a normal individual, the lowest level reached is approximately 30 mg./100 ml. The highest concentration reached during the period of 90 to 180 minutes is always above 68 mg./100 ml. (See Fig. 1-19.)

Plasma Insulin Levels

Plasma insulin levels can be measured by a variety of ways, viz., by measuring the glucose uptake of the rat diaphragm or epididymal fat pad, and by radio-immunological assay. The latter is regarded as the most reliable, but it also includes proinsulin, the insulin B chain and a substance of molecular weight of approximately 9000, known as "big insulin," probably related to or even identical with proinsulin. These facts must be borne in mind in relation to the finding that the immune-reacting insulin in plasma of normal fasting adult human beings usually lies between 10 and 20 μU./ml. and values above 30 μU./ml. are very rarely found; in children under 5 years of age, it is somewhat lower. After administration of glucose by mouth, the plasma insulin of adults rises to a value between 100 and 140 μU./ml. In normal pregnancy, raised levels occur.

Epinephrine Tolerance Test

The increase in blood sugar which follows administration of epinephrine has been utilized as an index of the quantity and availability of glycogen in the liver. After intramuscular injection of 1 ml. of a 1:1000 solution of epinephrine hydrochloride, the blood sugar concentration normally increases 35 to 45 mg. per 100 ml. in forty to sixty minutes, returning to the resting level in one and three-fourths to two hours.

Phenomena Associated with Normal Alimentary Glucose Reaction

The increased utilization of glucose which follows its absorption from the intestine in normal subjects is accompanied by other biochemical changes.

Decreased Serum Phosphate.
The inorganic phosphate of the blood plasma appears to be intimately related to the intermediary metabolism of glucose. Formation of hexose phosphates occurs in the process of utilization of glucose. Consequently, during this process inorganic phosphate is withdrawn from the plasma; the phosphate content of the tissues, particularly muscle, increases and phosphate excretion decreases. These changes in phosphate occur independently of the level of blood

sugar; e.g., they continue beyond the period of hyperglycemia induced by glucose administration and occur also during the hypoglycemia induced by administration of insulin. The hypophosphatemia is therefore, at least in part, a reflection of the increased utilization of glucose.

Decreased Serum Potassium. Stimulation of glucose utilization is accompanied by a shift of potassium from extra- to intracellular fluids. This occurs most strikingly when glucose and insulin are administered after a period of carbohydrate restriction or to subjects with uncontrolled diabetes mellitus. This shift may be sufficiently pronounced to produce a fall in the serum potassium concentration. The basis for this phenomenon is not clear, but it may be related to the passage of phosphate ion in the same direction, i.e., to the increased utilization of glucose.

Increased Respiratory Quotient. An increase in R.Q. above the normal resting level (0.82) is usually regarded as an indication of increased carbohydrate utilization (oxidation, lipogenesis). About one to one and one-half hours after oral administration of glucose (100 g.) the R.Q. rises from 0.82 to 0.88 to 0.90, reaching 0.95 or 0.96 in about two hours and then gradually falling to the resting level in about four hours. The period of this increase coincides approximately with that of hypophosphatemia, discussed previously.

ABNORMALITIES OF POSTABSORPTIVE BLOOD SUGAR LEVEL

Fasting Hyperglycemia

Fasting hyperglycemia may be due to either an increased rate of entrance of glucose into the blood from the liver (increased hepatic glycogenolysis or gluconeogenesis) or a decreased rate of removal of glucose from the blood by the tissues (decreased storage and utilization). In many clinical disorders both of these mechanisms are operative, and it seems probable that persistent hyperglycemia may be dependent fundamentally upon elevation of the threshold of sensitivity of the hepatic glycogenolytic mechanism to the blood sugar, i.e., a disturbance of the homeostatic mechanism in the liver (p. 30).

Diabetes Mellitus. The fundamental fault in diabetes mellitus is absolute or relative (p. 33) deficiency in pancreatic islet secretion (insulin). Relative insulin deficiency may be due to excess of anterior pituitary or adrenocortical hormones or to a state of insulin resistance. In the absence of an adequate amount of insulin, hepatic glycogenolysis and gluconeogenesis are increased and the rate of removal of glucose from the blood for glycogenesis in the extrahepatic tissues is decreased. The obvious consequence is an increase in the concentration of glucose in the blood. The utilization of glucose at ordinary levels of blood sugar is probably impaired, but there is evidence that at very high blood sugar levels glucose can be utilized by the tissues in the presence of insulin deficiency and even in the absence of insulin if the anterior hypophysis or adrenal cortex has been removed. These facts suggest that the fundamental mechanism underlying the hyperglycemia of diabetes mellitus involves overproduction (excessive hepatic glycogenolysis and gluconeogenesis) as well as underutilization of glucose. On the basis of the homeostatic mechanism in the liver, one may assume that because of the disturbed endocrine balance (diminished insulin, with excess of anterior pituitary and adrenal cortex hormones), the "thermostat" regulating the mechanism has been set at a higher level, allowing the blood sugar to rise to a hyperglycemic level before hepatic glycogenolysis and gluconeogenesis are depressed.

In large series of cases of diabetes mellitus the postabsorptive blood sugar has ranged from 70 to 1850 mg. per 100 ml. of blood. Complicating conditions such as acidosis, which should be considered a part of the disease, and hyperthyroidism tend to maintain the blood sugar at a higher level than that due to the insulin deficiency per se. Although it is true that the level of the fasting blood sugar usually parallels the severity of the condition, such a statement cannot be made unequivocally. In early cases the fasting blood sugar may be well within normal limits; as the condition becomes more advanced values of 180 to 300 mg. may be obtained; in advanced cases values of 400 mg. are not uncommon and, if acidosis is marked and the patient in coma, the degree of hyperglycemia may be extreme (700 mg. or over). Figures above 600 mg. are, however, rarely observed.

There is a remarkably wide range of variability in blood sugar values in patients in diabetic coma. This condition has been present with blood sugar concentrations as low as 130 mg./100 ml. and has been absent with values as high as 1500 mg./100 ml. Coma is most likely to develop in the presence of relatively low blood sugar values in patients below the age of 12 and above the age of 50 years, and in those with acute and overwhelming infection. Moreover, there appears to be no consistent relationship between the blood sugar concentration and the occurrence of death or recovery in patients with diabetic coma. Death may occur with values below 300 mg./100 ml., and recovery has been reported in patients with blood sugar concentrations as high as 1800 mg./100 ml. Obviously, the blood sugar concentration alone is not a reliable criterion for judging the severity of diabetic coma in any individual case.

In some instances the postabsorptive blood sugar concentration does not afford a true index of the presence or severity of an existing diabetes. As has been stated, in mild cases the fasting blood sugar may be within normal limits and further studies, such as the glucose tolerance test, must be resorted to in order to establish the true nature of the condition. This will be discussed later in greater detail (p. 60). Fasting hyperglycemia (above 110 mg. per 100 ml.) is highly suggestive of diabetes mellitus, but other conditions must be considered. In the presence of complicating factors, particularly hyperthyroidism, the diagnostic standard must be raised, according to some authorities, to a blood sugar of 150 mg. per cent fasting or 200 mg. per cent or more after meals, in addition to glycosuria.

At any given level of severity of diabetes, the fasting blood sugar concentration tends to be higher in obese than in undernourished subjects. Moreover, in diabetics, in contrast to normal subjects, the height and duration of the rise in blood sugar following meals are proportional to the amount of carbohydrate ingested. Consequently, the fasting blood sugar concentration reflects the severity of the disease only when a number of other factors are taken into consideration, including the presence of complications, the state of nutrition of the subject, and the diet.

The blood glucose is derived from hepatic glycogen in the absence of an exogenous supply. Removal of the liver in complete diabetic (depancreatized) animals results in the disappearance of sugar from the blood as in the case of nondiabetic animals. Advanced acute hepatic disease with serious disturbance of the glycogenic function of the liver, occurring in a patient with diabetes, may tend to diminish the blood sugar concentration (phosphorus or arsenic poisoning, toxemias of pregnancy, acute diffuse necrosis of liver, cirrhosis, and occasionally acute infectious jaundice). It is interesting to note in this connection that several observers have called attention to the occurrence of hypoglycemic reactions in patients with hemochromatosis receiving insulin, and to the marked fluctuation in glucose tolerance in that condition. This peculiar state is somewhat comparable to that of the hypophysectomized-depancreatized dog, in which the blood sugar rises higher after glucose than in the pancreatectomized dog, and hypoglycemic manifestations develop more rapidly and at a higher blood sugar level than in the hepatectomized animal. It must be emphasized that whereas the superimposition of hepatic disease occasionally is associated with a lowering of the fasting blood sugar concentration in diabetics, it usually further diminishes the already lowered glucose tolerance of these individuals and increases their resistance to insulin. A tendency toward lowering of the fasting blood sugar level occurs in undernourished individuals whose carbohydrate and protein intake has been so restricted that the available glycogen store in the liver has been depleted. Under these circumstances the postabsorptive blood sugar level may mask the severity of the diabetic condition. On the other hand, as has been indicated, hyperthyroidism, hypertension, nephritis, acidosis, and acute infections occurring in a patient with diabetes mellitus may, by their independent hyperglycemic effect, exaggerate the apparent severity of the condition. It should be realized, however, that the metabolic error in diabetes is seriously aggravated by factors which increase hepatic glycogenolysis, such as hyperthyroidism, infection and acidosis, and that their presence is of definitely adverse prognostic import.

Hyperthyroidism. Hyperthyroidism, whether due to thyroid disease (exophthalmic goiter, toxic adenoma) or temporarily induced by the administration of thyroid gland substance or thyroxine, results in a state of hypersensitiveness on the part of the liver for the conversion of glycogen into glucose. This glycogenolytic action of thyroxine may be excited directly, but it is more probable that the effect is produced by rendering the liver hypersensitive to epinephrine. Increased glycogenolysis and also gluconeogenesis perhaps result also from the increased rate of tissue metabolism. In the presence of an adequate supply of glycogen in the liver there is a tendency toward a state of fasting hyperglycemia, the tissues being constantly subjected to the necessity of handling a superabundance of glucose. This, unless excessive, is compensated by increased combustion of glucose in the tissues. Unless carbohydrates are supplied in abundance, the store of hepatic glycogen will be depleted and the blood sugar will automatically fall and may become subnormal. This perhaps accounts for the relatively infrequent incidence of fasting hyperglycemia in patients with hyperthyroidism, which is rarely severe and which has been reported as occurring in 0.5 to 8.5 per cent of large series of cases.

The mistake should not be made of interpreting a disturbance of carbohydrate metabolism dependent upon hyperthyroidism as due to diabetes mellitus (pp. 48 and 60). Patients with hyperthyroidism are undoubtedly more prone to develop diabetes than other individuals, the combination being encountered in about 1 to 3 per cent of patients with hyperthyroidism.

Increased Secretion of Epinephrine. The hepatic glycogenolytic action of epinephrine has been referred to previously. If the glycogen content of the liver is adequate, the injection of epinephrine is followed by a rise in blood sugar. This fact has been utilized as a test of the glycogen storing capacity of the liver. If 0.62 ml. of a 1:1000 solution of epinephrine hydrochloride is injected intramuscularly, the blood sugar rises 35 to 45 mg. per 100 ml. in three-fourths to one hour and returns to the resting level in one and three-fourths to two hours.

This effect of epinephrine is probably involved in several physiological and pathological states.

(a) As indicated previously, it appears to be a factor in the production of the hyperglycemic response to increased thyroid activity.

(b) Hypoglycemia, produced characteristically by the administration of insulin, results in increased epinephrine secretion with consequent mobilization of sugar from the liver. This arrangement represents another example of automatic adjustment when a disturbance threatens the equilibrium of the organism.

(c) Various emotional states such as anger, anxiety and fear are associated with an increased quantity of epinephrine in the blood. These states, as well as conditions of mental stress, excitement and excessive cold, have been found to be accompanied by an increase in blood sugar, apparently the result of the mobilization of liver glycogen by epinephrine. Pain and discomfort perhaps act in a similar manner. These conditions constitute what may be termed psychic hyperglycemia.

(d) Increased epinephrine secretion is a factor in the production of hyperglycemia in the so-called "diabetic piqûre" and in intracranial disorders (concussion, brain tumor, fracture of the skull and intracranial hemorrhage, with increased intracranial pressure) which cause the transmission of impulses through the splanchnic nerves to the adrenals and liver. It has also been observed in association with lesions in the region of the fourth ventricle and the hypothalamus, particularly the pons. Hyperglycemia may occur in a variety of convulsive states, including idiopathic epilepsy, Jacksonian epilepsy, tetanus, tetany, eclampsia and hypertensive encephalopathy, if the hepatic glycogen stores are adequate. If the latter are depleted, the blood sugar does not rise and indeed may fall if the convulsive state is prolonged.

(e) Fasting hyperglycemia, either constant or temporary and occurring periodically, has been observed in association with certain tumors of the adrenal medulla (pheochromocytomas), dependent upon the periodic outpouring of excessive quantities of epinephrine into the blood stream. There is a rather characteristic clinical picture of paroxysmal hypertension with simultaneous paroxysms of hyperglycemia. The blood sugar concentration in the intervening periods may be normal and the hyperglycemia (and hypertension) tends to be inhibited by ergotamine and benzodioxane compounds, which counteract the effect of epinephrine. In certain cases, hyperglycemia and hypertension are persistent rather than paroxysmal. Others have hypotensive attacks.

Adrenal Cortical Hyperfunction. This condition, resulting from hyperplasia or tumor of the adrenal cortex, may result in (a) the adrenogenital syndrome or (b) Cushing's syndrome. The former is seldom accompanied by abnormality in carbohydrate metabolism except in a congenital type of adrenal hyperplasia, in which hypoglycemia may occur; the latter is usually characterized by the gradual development of diminished glucose tolerance, followed by progressive elevation of the fasting blood sugar concentration, which may, however, remain normal for a long time. The hyperglycemia, when it develops, is due to increased hepatic gluconeogenesis and decreased glucose utilization (p. 25). This so-called "steroid diabetes" is characterized by resistance to insulin, mildness during fasting, and negative nitrogen balance in the absence of marked glycosuria.

Hyperpituitarism. The hyperglycemic effect of anterior pituitary extracts has been considered previously (p. 25). Two clinical hyperpituitary states may at times be accompanied by hyperglycemia, i.e., acromegaly (eosinophilic adenoma) and the type of Cushing's syndrome with pituitary basophilism or basophilic adenoma. The fasting blood sugar is usually normal in patients with acromegaly, particularly in the early stages of the disease, before the tumor has encroached extensively upon the remainder of the gland. In a large series of cases, about 10 per cent have moderate and 10 to 15 per cent mild fasting hyperglycemia but many more have a diminished glucose tolerance. Diabetes mellitus occurs in about 15 to 20 per cent of acromegalics, the former condition developing within fifteen years of the onset of acromegaly in the majority of instances (average 9.5 years). In several instances we have observed a fall in a previously elevated blood sugar, with improvement in glucose tolerance as the pituitary tumor increased in size.

Cushing's syndrome with pituitary basophilism may be accompanied by manifestations of disturbed carbohydrate metabolism described in connection with adrenal cortical hyperfunction.

Anesthesia, Asphyxia, Hypnotics. The hydrogen ion concentration of the blood increases rather suddenly during the induction of ether anesthesia and then rises gradually during its maintenance. Although ether and chloroform may exert a direct glycogenolytic influence upon liver glycogen, anoxia may play an important role in the production of this effect. A remarkable rise in blood sugar may occur during anesthesia. The degree of hyperglycemia is dependent upon the amount of glycogen in the liver, the quantity of anesthetic administered and upon the adequacy of the mechanism of carbohydrate utilization.

In the absence of any other factor which tends to produce hyperglycemia, such as hyperthyroidism or diabetes mellitus, the blood sugar has been found to rise 7 to 8 mg. per 100 ml. per ounce of ether administered. These figures may be greatly increased if the anesthetic is not skillfully administered, by the introduction of complicating factors such as excessive muscular effort, excitement and asphyxia. In diabetes mellitus the degree of hyperglycemia is much more marked as the excessive amount of glucose entering the blood stream cannot be stored or utilized adequately by the tissues. The duration of anesthesia hyperglycemia, usually four to twelve hours after operation, depends upon the quantity of anesthetic used and the duration of the state of anesthesia. The rise appears to be more pronounced in laparotomies than in extra-abdominal operations but is otherwise independent of the nature or severity of the operative procedure. Values as high as 400 mg. per 100 ml. have been observed in individuals with normal carbohydrate metabolism.

If the quantity of glycogen stored in the liver is deficient, as in severe hepatocellular damage, the blood sugar cannot rise as in normal individuals and there is great danger of depletion of the hepatic glycogen reserve with consequent serious damage to the liver by the ether or chloroform, which have a profound toxic effect upon the liver parenchyma. It has also been found that practically all of the barbituric acid series of hypnotics cause a rise in blood sugar and a simultaneous fall in the glycogen content of the liver. The effects of these various anesthetic agents may be attributable in part to asphyxia, to a direct action of the anesthetic on the liver, or to liberation of epinephrine from the adrenal glands.

Hypoxia, whether due to mechanical causes or occurring in carbon monoxide poisoning or heart failure or during anesthesia, particularly with nitrous oxide and ethylene, is accompanied by an increase in blood sugar. The two anesthetic agents mentioned do not, in the absence of asphyxia, have any appreciable effect upon the blood sugar concentration. Hypoxia produces this effect probably by increasing mobilization of hepatic glycogen. The blood sugar may rise 20 to 40 mg. per 100 ml. after the administration of morphine, the mechanism probably being similar to that involved in asphyxia. Transient hyperglycemia and glycosuria have been reported in patients with acute coronary artery occlusion. The mechanism operating here may be due to sudden circulatory inefficiency but may be contributed to by the pain and anxiety incident to the acute attack.

Initially, severe hemorrhage and shock are accompanied frequently by a sharp rise of blood sugar due to increased hepatic glycogenolysis. A fall, often to hypoglycemic levels, may occur in terminal stages, due to hepatic functional insufficiency.

Acidemia. Acidemia, or consequent acidosis, diminishes the blood sugar lowering action of insulin. Although the mechanism involved is not completely understood, it appears that in diabetic acidemia there is interference with the phosphorylation of glucose. This increased resistance to insulin may be due to the development of an insulin antagonist in the serum α-globulin fraction. It does not occur in uremic acidemia.

This factor is of particular importance in diabetes mellitus because it aggravates a pre-existing hyperglycemic mechanism, in this case hypoinsulinism. In the evaluation of the significance of the fasting blood sugar concentration in this disorder it is important to attempt to determine what proportion of the elevation is dependent upon this complicating state. Acidemia or acidosis due to other causes, as fever, nephritis and dehydration, may result in an increase in the level of blood sugar. The rise in such cases is not marked, but, occurring in association with other factors, its influence may be significant.

Hepatic Disease. Although diminished glucose tolerance (p. 63) and a normal or slightly subnormal fasting blood sugar level (p. 56) occur more commonly than fasting hyperglycemia in patients with hepatocellular damage, the latter is observed occasionally. Fasting values as high as 185 mg. per 100 ml. have been reported in patients with acute hepatitis in the absence of true diabetes mellitus. This is apparently the result of excessive glycogenolysis. In some cases, hyperglycemia in patients with hepatic or biliary tract disease may be due to associated pancreatitis, but this in itself seldom causes hyperglycemia. Fasting hyperglycemia, often marked, is a common but not invariable feature of hemochromatosis (p. 49), being due fundamentally to involvement of the islands of Langerhans in the fibrotic and degenerative process in the pancreas.

Miscellaneous. Strenuous muscular exercise causes liberation of glucose from the liver with resulting hyperglycemia. It must be realized that, as in all conditions acting through the agency of increased hepatic glycogenolysis, hyperglycemia can only be produced if the reserve supply of liver glycogen is abundant. When this is depleted, the blood sugar concentration decreases and, eventually, hypoglycemia may result. This fact is perhaps responsible for the contradictory reports of the blood sugar level in eclampsia and other convulsive states. Similarly, strenuous muscular exertion, in a normal individual, is at first associated with an increase and later, if continued, with a decrease in blood sugar, the source of supply having been virtually exhausted.

Fasting hyperglycemia may occur in many acute and chronic infections although usually the fasting blood sugar is within normal limits and the tolerance for glucose is lowered (p. 65). This is believed to be due to increased and prolonged hepatic glycogenolysis and gluconeogenesis, which may seriously aggravate a condition of latent or frank diabetes mellitus. A mild degree of hyperglycemia has been observed in about 5 to 40 per cent of cases of essential hypertension and occasionally in nephritic hypertension. Its incidence in the former condition is highest in patients with obesity, a condition that is frequently accompanied by a lowered tolerance for glucose (p. 65). Such patients (obese and hypertensive) may in reality be potential diabetics.

Moderate elevations of blood sugar may occur in acute and chronic pancreatitis, but this is by no means a constant finding. An increase occurs as a result of smoking, due probably to the epinephrine-stimulating effect of nicotine. It may occur also after administration of caffeine, quinine, pituitrin and amphetamine, and has been reported in methyl salicylate poisoning.

Fasting Hypoglycemia

The factors which tend to lower blood sugar are the opposite of those which have been dealt with in the discussion of hyperglycemia: (a) decreased rate of entrance of glucose into the blood (decreased hepatic glycogenolysis and gluconeogenesis); (b) increased rate of removal of glucose from the blood (increased storage or utilization in the extrahepatic tissues). The latter seldom produces fasting hypoglycemia by itself, except under unusual conditions of severe and prolonged muscular activity (marathon runners), but a tendency toward hypoglycemia due to other causes may be aggravated by exercise.

Hyperinsulinism. From the standpoint of etiology, hyperinsulinism may be classified as (1) induced, (2) organic, (3) functional, (4) undetermined.

Induced Hyperinsulinism. Following injection of insulin the blood sugar concentration falls, usually reaching a minimum in 2 to 4 hours, the degree of reduction and duration of hypoglycemia depending upon the amount of insulin administered. After protamine zinc insulin, the blood sugar usually reaches a minimum in twelve to sixteen hours, the effect lasting for about twenty-four hours. Symptoms of hypoglycemia commonly occur when the blood sugar level reaches 30 to 40 mg. per 100 ml., although some patients have an extraordinary tolerance for low levels of blood sugar. Cases of diabetes have been reported with blood sugar values of 20 to 30 mg. per 100 ml. over a period of 6 hours without symptomatic manifestations.

Organic Hyperinsulinism. Spontaneous hyperinsulinism may occur as a result of adenoma and, rarely, carcinoma of the islands of Langerhans, chronic pancreatitis, and perhaps islet hyperplasia due to other causes. The fasting blood sugar may be as low as 20 to 50 mg. per 100 ml., but is not necessarily consistently subnormal. In such cases, hypoglycemia may be induced or accentuated by (a) moderate exercise after a prolonged fasting period or (b) performing an oral or, preferably, an intravenous glucose tolerance test, the hypoglycemic phase of the blood sugar curve being characteristically exaggerated and prolonged (p. 66). In establishing the diagnosis, significant hypoglycemia must be shown to be present at the time of occurrence of the characteristic subjective and objective manifestations, which must also be largely relieved promptly by intravenous injection of glucose. However, the occurrence or severity of symptoms bears no constant relation, even in the same subject, to the actual level of the blood sugar. The rapidity as well as the extent of the fall may be important in this connection. In cases of latent or mild hyperinsulinism, the condition may be evidenced only by an exaggerated response to insulin administration (insulin tolerance test, p. 70). One of the most reliable diagnostic procedures for hyperinsulinism is a prolonged fast (48 to 72 hours), which usually produces hypoglycemia and its manifestations; intravenous glucose may be necessary at this stage to prevent a fall to dangerous levels.

Functional Hyperinsulinism. This designation should probably be applied to conditions in which increased secretion of insulin is not due to morphologic abnormality of the islands of Langerhans. In some cases, islet hyperplasia may result from continued overstimulation. However, this term is frequently applied to conditions of mild hypoglycemia for which no other cause is apparent. Hypoglycemia of this type occurs at times in the newborn of mothers with uncontrolled diabetes. In some fatal cases no morphologic abnormalities have been observed in the pancreas, whereas in others there has been hyperplasia or hypertrophy of the islands of Langerhans. It is believed that during intra-uterine life prolonged exposure of the fetus to an abnormally high blood sugar concentration results in excessive insulin secretion, with or without consequent hyperplasia of the islets. Following delivery, this increased insulin production often continues and exceeds the requirement of the infant, and hypoglycemia develops, frequently with serious consequences unless it is anticipated and treated promptly and adequately. In this connection it must be recalled that a few hours after birth, unless sugar is administered, the blood sugar concentration is considerably lower than in older children or adults, at times as low as 30 mg./100 ml. This "physiologic" neonatal hypoglycemia is seldom accompanied by symptoms.

Some patients with partial gastrectomy develop moderate hypoglycemia, with symptoms, one and one-half to three hours after a high carbohydrate meal. In the absence of the pyloric mechanism, a relatively large amount of carbohydrate enters and is absorbed from the intestine more rapidly than normally; the postprandial rise in blood sugar is exaggerated, as is the subsequent hypoglycemic phase (increased insulin secretion, p. 39). This may occur also following gastroenterostomy and in patients receiving carbohydrate by duodenal instillation. In this type of functional hyperinsulinism, which occurs as a reaction to hyperglycemia, the fasting blood sugar concentration is invariably normal, as are the results of intravenous glucose tolerance and insulin tolerance tests.

A peculiar form of hypoglycemia occurs in certain children following administration of a high protein meal, or casein hydrolysates, due apparently to sensitivity to L-leucine or its metabolites (e.g., isovaleric acid). In susceptible subjects, these substances produce a marked fall in blood sugar almost identical with that produced by insulin, the maximum depression occurring within thirty to forty-five minutes, with a return to normal values at sixty to ninety minutes. In certain instances there is also increased sensitivity to insulin.

The most common form of spontaneous hypoglycemia occurs in subjects with emotional and autonomic instability. The nervous system manifestations are out of proportion to the degree of hypoglycemia, which is invariably mild. The attacks occur about 2 hours after meals, not in the fasting state, and are not induced by exercise. Recovery usually occurs spontaneously but may be accelerated by taking carbohydrate. The underlying mechanism is not clear.

Hepatic Disease. In the absence of exogenous carbohydrate, liver glycogen is the only important immediate source of blood sugar. It would be anticipated, therefore, that fasting hypoglycemia might result from depletion or inadequate depositions of liver glycogen or from inadequate hepatic glycogenolysis. The liver is endowed with such an extensive functional reserve capacity and with such remarkable regenerative powers that in chronic diseases such as cirrhosis its glycogenic function is seriously impaired only in the late stages.

Blood sugar values of 50 to 60 mg. per 100 ml. may be observed, rarely, in the terminal stages of cirrhotic processes, particularly in obstructive biliary cirrhosis but at times also in the portal type. Hypoglycemia may be a terminal event after operations for biliary tract disease under general anesthesia which may exhaust a glycogen reserve already depleted by associated hepatic disease (hepatitis).

It is in the acute, rapidly progressive and extensive forms of hepatic disease that hypoglycemia is most commonly observed. Values as low as 25 to 40 mg. per 100 ml. have been reported in cases of phosphorus, benzol, chloroform, and carbon tetrachloride poisoning and following administration of arsphenamine and sulfonamides, usually, however, in terminal stages. Severe acute infections, such as diphtheria and scarlet fever, may produce similar grades of hypoglycemia, perhaps in the same manner, although adrenal insufficiency may be a factor in such cases. Very low values have been observed in acute necrosis of the liver (15 to 50 mg. per 100 ml.), acute and chronic infectious hepatitis and cholangiolitis, and in fatty liver. Marked hypoglycemia (25 mg. per 100 ml.) has been reported with primary liver cell carcinoma replacing 70 to 80 per cent of the liver substance. The presence of serious hypoglycemia may be unsuspected in the presence of hepatic coma in patients with cirrhosis or hepatitis.

The state of the glycogen reserve is of the utmost importance in surgical disorders of the biliary tract since the administration of ether in such cases, by depleting this reserve, may result in serious consequences (post-anesthetic hypoglycemia). Excessive, continued muscular exertion may be associated with hypoglycemia following the primary period of hyperglycemia, due likewise to secondary exhaustion of hepatic glycogen; this is also true of convulsive disorders such as occur in strychnine poisoning, eclampsia, uremia, and tetanus.

Fasting hypoglycemia, usually mild but occasionally below 30 mg. per 100 ml., may occur in glycogen storage disease (glycogenosis Types I, III, IV and rarely VI). The pathogenesis of the various types of glycogen storage disease is discussed elsewhere (p. 79).

Reference has been made (p. 49) to the occasional occurrence of hypoglycemia after administration of insulin to patients with hemochromatosis. The situation may be analogous to that of the hepatectomized-depancreatized dog, in which the blood sugar rises higher after glucose than in the depancreatized dog and hypoglycemic manifestations develop more rapidly than in hepatectomized animals. This phenomenon does not always occur, certain patients with hemochromatosis exhibiting an extreme degree of insulin resistance.

Adrenal Cortical Insufficiency. In Addison's disease, usually associated with atrophy or tuberculosis of the adrenal glands, the fasting blood sugar is frequently low normal and occasionally subnormal. The average value in a large series of collected cases was 75 mg. per 100 ml. In a few instances figures of 30 to 40 mg. have been reported, usually in fatal cases shortly before death. Although essentially normal values may be maintained by proper carbohydrate feeding, the blood sugar falls progressively during periods of carbohydrate deprivation. This hypoglycemic tendency is due to diminished gluconeogenesis, increased utilization of glucose, and perhaps also to diminished absorption of glucose from the intestine, resulting from adrenal cortical insufficiency. The defect in carbohydrate metabolism may be demonstrated most readily by performing an intravenous glucose tolerance test, the characteristic finding being exaggeration and prolongation of the hypoglycemic phase of the blood sugar curve (p. 67). Insulin sensitivity is increased (p. 70).

Hypoglycemia may occur following abrupt cessation of prolonged corticosteroid therapy, due to the depression of adrenocortical function induced by the exogenous hormone. A similar situation may develop following removal of a hyperfunctioning adrenal tumor, due to functional depression of the residual adrenocortical tissue. Hypoglycemia may occur in a type of congenital adrenal hyperplasia characterized by overproduction of adrenal sex steroids (virilism in girls, precocious development in boys) and underproduction of cortisol and corticosterone. As in other conditions of adrenal insufficiency, the attacks of hypoglycemia may be induced or aggravated by infection and other varieties of stress.

Hypoglycemia (0 to 30 mg./per 100 ml.), as a terminal event in extensive burns, diphtheria, and scarlet fever (adrenal apoplexy or Waterhouse-Friderichsen syndrome), has been attributed to adrenal failure.

Anterior Pituitary Insufficiency. The disturbance in carbohydrate metabolism is similar to that in adrenal cortical insufficiency. It is due in large measure to the same mechanism, being accompanied by atrophy of the adrenal cortex resulting from absence of the adrenocorticotrophic hormone. In addition, there is atrophy also of the thyroid (absence of thyrotrophic hormone), which further increases the tendency toward hypoglycemia and increases the glucose tolerance and the insulin sensitivity (pp. 68 and 70). Spontaneous hypoglycemia may occur, with convulsions and coma. Fasting blood sugar values of 60 mg. per 100 ml. or lower have been reported in about 50 per cent of verified cases of Simmonds' disease (pituitary cachexia) (range, 24 to 114 mg. per 100 ml.). A similar finding has been obtained in about 20 per cent of cases of anorexia nervosa (range, 26 to 134 mg. per 100 ml.), rendering the occurrence of spontaneous hypoglycemia of limited value in the differential diagnosis of these conditions.

A blood sugar of practically zero has been observed in a patient with diabetes mellitus following infarction of the hypophysis with destruction of the anterior lobe. Hypoglycemia has resulted from extensive destruction of the anterior hypophysis by chromophobe adenoma, craniopharyngioma and, rarely, metastatic tumor.

Hypothyroidism. In myxedema and cretinism the blood sugar concentration may be low, values of 50 to 60 mg. per 100 ml. having been reported. Values of 60 to 70 mg. are commonly observed. A drop in blood sugar (50 to 60 mg. per 100 ml.) may occur at times immediately following thyroidectomy in individuals who have not received adequate amounts of carbohydrate before operation.

Nervous System Disorders. Fasting hypoglycemia, usually of mild degree, has been observed occasionally in vagotonia, "neurocirculatory asthenia," various psychoses, paresis, subarachnoid hemorrhage, the postencephalitic syndrome, and certain brain stem (hypothalamic) lesions. Increased glucose tolerance (p. 70) occurs much more commonly than actual hypoglycemia in such cases. The mechanism underlying the abnormality in carbohydrate metabolism is believed to be disturbed nervous control of hepatic glycogenolysis, a similar condition having been produced experimentally by injury to the paraventricular nucleus in rabbits and the hypothalamic region in cats and dogs.

Miscellaneous. Severe spontaneous hypoglycemia has been observed in association with large mesodermal thoracic and abdominal tumors (fibromas, sarcomas). The mechanism is not known; it has been suggested that these tumors may produce a substance with insulin activity.

The term "idiopathic spontaneous hypoglycemia of infancy" is applied to a type of hypoglycemia in infants, of variable severity, frequently familial in nature, for which no cause has been demonstrated.

Marked fall in blood glucose has been reported following ingestion of fructose in subjects with a rare form of congenital fructosuria (p. 88). A severe, and even fatal, form of hypoglycemia can occur in malnourished chronic alcoholics after ingestion of alcohol. It has also been reported after excessive intake in some normal individuals, especially children. Poisoning by Akee nuts has been an important cause of hypoglycemia in Jamaica. Mild hypoglycemia has been reported in cases of status thymicolymphaticus, progressive muscular dystrophy, pregnancy, lactation, renal glycosuria, sensitivity to nicotine (tobacco), and in anorexia nervosa and other conditions accompanied by severe grades of undernutrition.

ABNORMAL ALIMENTARY RESPONSE

Exaggerated Response—Diminished Glucose Tolerance

By diminished glucose tolerance is meant inability of the organism to handle ingested glucose as efficiently as a normal organism; it indicates inefficiency of one or more of several factors involved in the metabolism of carbohydrate:

(1) Inadequate glycogen storage capacity of the liver, so that glucose reaching the liver in the portal blood is not adequately removed, and, as a result, enters the systemic circulation in abnormal amounts. The chief clinical conditions in which this factor plays an important part are those in which there is extensive and rapidly progressive hepatic disease.

(2) Increased hepatic glycogenolysis and gluconeogenesis. In this group are included hyperthyroidism, hyperadrenalinism, acidemia or acidosis and toxemia due to acute infections, and adrenal cortical hyperfunction. These phenomena probably contribute also to the decreased glucose tolerance of diabetes mellitus. In the past, the generally accepted view has been that infection or toxemia affects the pancreas in such a way as to diminish insulin production or that they interfere with the action of insulin, whether of endogenous or exogenous origin. More recent studies have led to the hypothesis that the diminished glucose tolerance and relatively high insulin resistance associated with toxemia and infection are dependent, in part at least, upon acceleration of hepatic glycogenolysis. Some believe this to be due to increased epinephrine secretion, but it may be due to interference with the mechanism whereby the liver normally diminishes its supply of glucose to the blood in response to an influx of exogenous sugar (p. 30). A similar mechanism may also operate in the presence of hepatic functional impairment associated with various types of biliary tract disorder and liver disease included in the preceding group.

(3) Decreased tissue utilization. If the ability of the extrahepatic tissues to form glycogen and to utilize glucose is diminished, an excess remains in the blood. The most important clinical condition in this group is diabetes mellitus.

(4) Increased rate of absorption. An abnormally high rise in the blood sugar level may result from marked increase in the rate of absorption of glucose from the intestine, believed by some to occur in hyperthyroidism.

The chief features of the blood sugar curve following the ingestion of glucose (Fig. 1-15) which characterize a diminished glucose tolerance are:

(a) An abnormally high rise in the venous and capillary blood sugar concentration (above 150 and 160 mg. per 100 ml. respectively).

(b) The maximum concentration is reached later than in the normal individual (after one hour).

(c) The period of hyperglycemia is longer than normal.

(d) The return to the postabsorptive level is delayed (more than two hours).

In conditions in which the dominant factors are abnormally rapid absorption (e.g., hyperthyroidism), inadequate storage in the liver (e.g., hepatic disease) or excessive hepatic glycogenolysis (e.g., hyperadrenalinism, acidosis, toxemia), the most marked abnormality occurs in the first phase of the glucose tolerance curve; i.e., there is an excessively high rise in the blood sugar, but the return to the control level is not so long delayed as in conditions in which there is believed to be impaired utilization of glucose.

60   CARBOHYDRATE METABOLISM

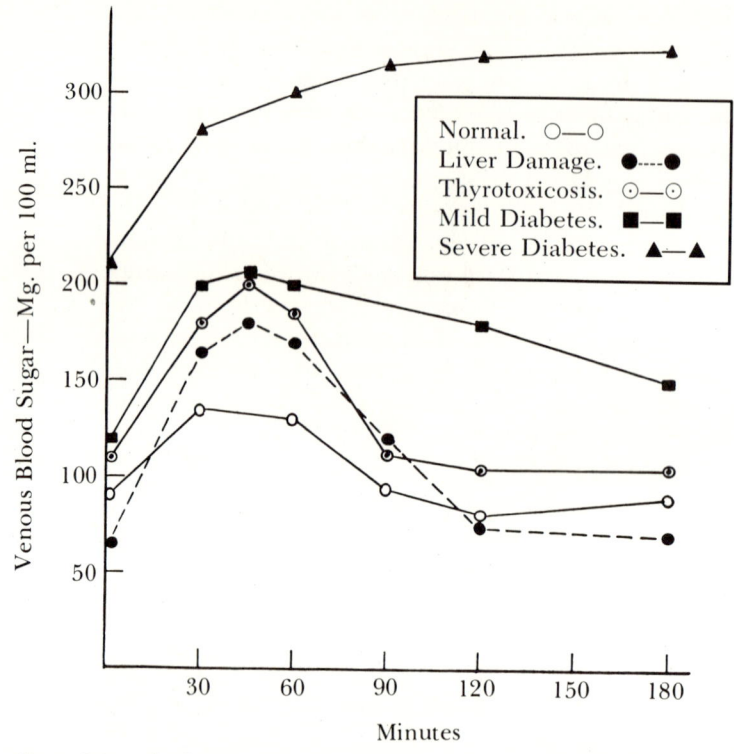

Figure 1-15. Blood sugar curves characteristic of decreased glucose tolerance (100 g. glucose by mouth).

The latter condition is observed characteristically in diabetes mellitus, the blood sugar curve being abnormally high and markedly prolonged.

Diabetes Mellitus (p. 47). Diminished glucose tolerance is the most characteristic metabolic feature of this condition, occurring regardless of the postabsorptive blood sugar concentration. It is probably dependent in part upon disturbance of the homeostatic mechanism in the liver (p. 30), resulting in continuation of hepatic glycogenolysis and gluconeogenesis at abnormally high blood sugar levels, and also upon diminished capacity of the tissues for utilizing glucose at normal blood sugar levels.

The sugar tolerance curve in diabetes mellitus has the following characteristics:

(a) Fasting hyperglycemia is present in all but very mild cases.

(b) A gradual rise to an excessively high level (above 160 mg. per 100 ml.), the degree of elevation being approximately proportional to the severity of the condition. It is, perhaps, not wise to make a diagnosis of diabetes mellitus until the level is above 200 mg. per 100 ml., unless there is undoubted delay in return to the fasting level as well as other clinical indications.

(c) The maximum concentration is reached after a variable interval following the ingestion of glucose, but practically always more than 1 hour.

In general, the greater the rise, the longer is the time elapsing before the highest level is attained. In severe cases the peak may not be reached for 3 or more hours.

(d) The most characteristic feature is the delayed return to the postabsorptive level. A failure to return to normal at the end of three hours is usually indicative of diabetes mellitus. After reaching a maximum, the blood sugar concentration remains at a high level for a variable period, decreasing slowly to the fasting level. The greater the elevation the slower is the rate of decrease. Characteristically the curve is of the high plateau rather than the peaked type. In contrast to the normal subject, the height and duration of the rise of blood sugar in diabetes is proportional to the amount of glucose ingested.

(e) The arterial-venous difference is often diminished.

Other phenomena may occur which suggest impaired glucose utilization. The respiratory quotient may remain at the postabsorptive level (p. 47) instead of exhibiting the normal increase which attends active carbohydrate storage and combustion. However, this may also be interpreted as due to increased gluconeogenesis from protein and fat. The serum phosphate concentration characteristically remains unchanged instead of decreasing during the period of hyperglycemia; in some cases it may increase slightly. Serum phosphate levels are of little value in the clinical situation.

The oral glucose tolerance test is of particular value in cases in which the fasting blood sugar is normal or only slightly elevated. The intravenous test is practically never indicated in cases in which diabetes mellitus is suspected and, since interest is centered chiefly in the first phase of the curve, the test may usually be terminated at the end of two or three hours for all practical purposes in such cases. Although this procedure is of great diagnostic value, the curves obtained in diabetes do not necessarily reflect the severity of the disease. Improper antecedent diets or complicating conditions may produce markedly abnormal curves in subjects with mild diabetes. Obesity and pregnancy also aggravate the abnormality in glucose tolerance in diabetics.

Screening for Diabetes. Attempts have been made to determine the incidence of diabetes mellitus in the general population. In one such survey, 70 per cent of the population of the town of Bedford, U.K., submitted a postprandial urine sample for testing by Clinistix for the presence of glucose. The 4 per cent who proved positive were given a modified glucose tolerance test, namely, the determination of the blood glucose level by AutoAnalyzer, 2 hours after 50 g. oral glucose. In addition, 542 randomly chosen members of the whole population were subjected to an age-stratified 2½-hour glucose tolerance test. Normal individuals showed a 2-hour single blood glucose level of less than 120 mg./100 ml., border-line individuals were between this figure and 199 mg./100 ml., and those with a figure of 200 mg./100 ml. and above were regarded as overt diabetics.

Similar and often much more extensive investigations have been carried out in the United States (Sudbury), Japan (Hiroshima) and elsewhere. The results indicated that much depended on the manner of performance of the tolerance test and the criteria of abnormality. Variations in these gave incidence rates varying from 0.8 to 6.8 per cent in the American series and dependence on a single blood value after a glucose load increased the incidence to as high as 13.4 per cent. The impression given, however, is that the incidence of the disease increases somewhat with age. Most studies have shown no statistically significant differences related to sex.

There are undoubtedly great difficulties besetting this type of investigation, especially since it is now also realized that the blood glucose abnormalities can be regarded as relatively late manifestations of a condition possibly present from birth. There is, in fact, a well recognized prediabetic stage. This can be manifested clinically by the occurrence of such phenomena as obstetric postmaturity or the development of abnormal blood glucose levels, either fasting or glucose tolerance, during pregnancy.

Sometimes predisposed individuals can be detected by the means of a cortisol-glucose tolerance test. When this is the case they are recognized as belonging to the group of subclinical diabetes. The stages in evolution of the disease are referred to as:

(a) Prediabetes.
(b) Subclinical diabetes.
(c) Latent diabetes (abnormal glucose tolerance, or fasting hyperglycemia, with no other signs or symptoms except possibly diabetic angiopathy).
(d) Overt diabetes.

No completely satisfactory account of the cause of the disease has yet been given. The old concept of islet cell exhaustion is no longer accepted. Etiology has been explained in terms of autoimmunity, an insulin antagonist synalbumin (believed to be excessive amounts of the insulin B chain bound to albumin) and to abnormal binding of insulin to protein. There has been much criticism of each of these suggestions.

Abnormal One-Hour, Two-Dose Test. Exton found that in the one-hour, two-dose test in mild diabetes, the first part of the curve is similar in trend to that obtained in the ordinary one-dose glucose tolerance test. After the second dose of glucose, however, the blood sugar rises sharply, this feature distinguishing diabetes from all other conditions, according to this author. In more severe diabetes the responses to the one-hour, two-dose test are in general the same as the responses in mild diabetes, except that the sugar values are higher in all the blood and urine samples in the more severe cases. The criteria for determining diabetes in the one-hour, two-dose test are (1) a more or less steep rise of not less than 10 mg. of blood sugar in the sixty-minute sample, following the second dose of glucose, and (2) the relation of blood and urine sugar values to the severity of the disease. The criteria of alimentary glycosuria are (1) a sugar-free urine after fasting, with sugar in the final urine specimen, and (2) blood sugars that follow the normal curve even when the level is higher than normal. The criteria for renal glycosuria are (1) blood sugars which follow the normal course, or in any event which never reach the diabetic level, and (2) sugar in both urine specimens.

According to some, this procedure is more valuable than the ordinary glucose tolerance test in that it frequently yields normal findings in non-diabetic subjects in whom abnormal or questionable results are obtained by the latter method. In our experience with hospital patients suffering from a variety of disorders, this procedure has not proven superior to the standard test. Distinctly abnormal curves were obtained in many patients in whom no evidence of diabetes could be demonstrated and in whom essentially normal curves were obtained by the single dose procedure.

Hepatic and Biliary Tract Disease. Since the sugar which enters the blood after the ingestion of glucose is removed both by the tissues and by the liver, one should theoretically expect abnormalities in the curve of alimentary glycemia in persons with liver disease. Because such individuals partially lack one of the mechanisms that reduce hyperglycemia, hepatic glycogen formation, the curve would presumably be excessively high or prolonged. Unfortunately, however, from a diagnostic standpoint, this theoretical observation is not always borne out in fact. A normal alimentary response may be obtained in the presence of advanced hepatic disease. This is particularly true of chronic disorders such as various types of cirrhosis, passive congestion, carcinoma, and lues of the liver. This is, in all probability, due to the great functional reserve and remarkable regenerative power of that organ. In acute diffuse forms of hepatic disease significant alterations in the blood sugar tolerance curve occur much more frequently. Such abnormalities have been described in acute and subacute hepatic necrosis, various forms of hepatitis, fatty liver, cholangiolitis, obstructive jaundice, acute alcoholism, portal cirrhosis, and poisoning with phosphorus, chloroform, carbon tetrachloride, sulfonamides, and arsphenamines. An abnormally high rise, occasionally prolonged, with an exaggerated hypoglycemic phase, occurs occasionally in glycogen storage disease (p. 80).

The type of curve obtained in typical cases of insufficiency of hepatic glycogenic function is characterized by these factors:

(a) Abnormally high rise in blood sugar concentration.

(b) The maximum concentration is attained at the end of three-fourths to one and one-half hours, and in most instances within one hour.

(c) The blood sugar level usually falls rather rapidly, returning to normal within two to three hours. Only in cases of extreme grades of hepatic insufficiency is hyperglycemia of long duration.

(d) The arterial-venous difference is usually normal or, in some instances, increased.

(e) The combination of a low normal or subnormal fasting blood sugar level with a tolerance curve having the characteristics described above is highly suggestive of diminished hepatic glycogenesis.

If determinations of the respiratory quotient and serum phosphate are made at frequent intervals after the administration of glucose, they will be found to change to an essentially normal manner; i.e., the R.Q. rises and the serum phosphate falls during the period of increased glucose supply and utilization (see p. 47).

Hyperthyroidism. Decreased tolerance for glucose occurs in 50 to 80 per cent of patients with hyperthyroidism but does not necessarily parallel the severity of the condition. It is probably due in part to an increased rate of absorption from the intestine and in part to increased hepatic glycogenolysis (p. 26). There is some doubt whether the thyroid plays any major part in intestinal absorption of glucose. The blood sugar curve is not so characteristically plateau in type as that in diabetes mellitus, usually rising to a high peak earlier and descending earlier than in the latter condition.

Hyperfunction of the Anterior Hypophysis and Adrenal Cortex. Cushing's syndrome, whether due to primary adrenal cortical tumor or hyperplasia, is characteristically accompanied by a state of decreased glucose tolerance. This is not present in all cases in early stages of the condition, but usually develops eventually. It is dependent mainly upon excessive gluconeogenesis and decreased glucose utilization (pp. 25 and 35). The curve in mild cases may resemble that seen in hyperthyroidism, acute infections, "toxemia," etc., but in more severe cases it is usually of the characteristic diabetic type. This so-called "steroid diabetes" is characterized by relative insensitivity to insulin, mildness during fasting, and negative nitrogen balance in the absence of marked glycosuria.

The glucose tolerance varies considerably in acromegaly but is usually decreased at some phase of the development of the condition. In advanced cases, with destruction of the remainder of the gland by the eosinophilic tumor, normal curves may be obtained. The variable findings may also be due to opposing influences in this connection of factors stimulating gluconeogenesis from protein and of the factor or factors stimulating growth and thus tending to conserve protein.

Glucose tolerance is usually normal in the adrenogenital syndrome, characterized by precocious puberty with either isosexuality in children, masculinization in the female, or feminization in the adult male. In some instances, however, a diabetic type of curve may be obtained. Patients with adrenal medullary hyperfunction (pheochromocytoma) exhibit decreased tolerance for glucose during, and at times between, the characteristic attacks of paroxysmal hypertension and hyperglycemia (p. 51).

Pregnancy. Mild to moderate decrease in glucose tolerance, with a normal fasting blood sugar, occurs rather commonly in the latter half of apparently normal pregnancy. This is accompanied by increased resistance to exogenous insulin. Diabetic patients exhibit an increase in insulin requirement during pregnancy and active disease may be precipitated in prediabetic subjects.

This has been attributed generally to increased production, in normal pregnancy, of hormones that exert effects in carbohydrate metabolism antagonistic to those of insulin, e.g., probably adrenocortical hormones (produced also by the placenta), thyroid hormones, possibly growth hormone. However, no consistent relationship has been established between the decreased glucose tolerance of normal pregnancy and demonstrable changes in production of these hormones. It has been suggested that the change in carbohydrate metabolism may be due in part to active destruction of insulin by proteolytic mechanisms in the placenta.

Decision as to whether the decreased glucose tolerance in a given case is physiological or whether it reflects a condition of latent diabetes is a matter of considerable importance. This is not always easy to decide in the absence of previous observations in the nonpregnant state.

In general, the earlier the appearance of decreased glucose tolerance (e.g., in the first or second trimesters) and the more severe the abnormality, particularly if there is an associated elevation of the fasting blood sugar concentration, the more justified one is in regarding the condition as diabetes mellitus and in instituting treatment. There is experimental basis for the belief that early correction of the periodic hyperglycemia by insulin may prevent the development of frank diabetes mellitus in the mother and reduce the fetal mortality.

It is firmly believed by some authorities that the glucose tolerance of normal pregnant women does not differ from other normal adults. A diminished tolerance is regarded as a probable indication of prediabetes.

Miscellaneous. Diminished tolerance to glucose is commonly observed in patients with severe anemia, including pernicious anemia. Fasting hyperglycemia and diminished glucose tolerance are at times observed in patients with glomerulonephritis and nephrosclerosis, particularly in the presence of renal failure. There is no associated disturbance of glucose utilization, the respiratory quotient rising in a normal manner following the ingestion of glucose. The abnormality apparently lies in a retarded transfer of glucose from blood to tissues rather than in retarded combustion of glucose in the tissues.

Decreased glucose tolerance may occur in certain central nervous system disorders, including head injuries or other causes of increased intracranial pressure (brain tumor, abscess, hemorrhage), hypothalamic lesions, emotional instability and, at times, dementia praecox and manic-depressive insanity.

It also occurs in about 75 per cent of cases of hemochromatosis (fibrosis and degeneration of the islands of Langerhans) and occasionally in acute and chronic pancreatitis, due to involvement of the islands of Langerhans and to associated hepatic and biliary tract disease.

Various infectious and toxic states often depress glucose tolerance. This is particularly true of pyogenic infections, especially of the skin (furunculosis, carbuncle), but diminished tolerance may also occur in such conditions as rheumatoid arthritis, diphtheria, scarlet fever, pneumonia, tuberculosis, and others. The influence of infections is perhaps exerted through the medium of the liver, i.e., by decreased responsiveness of the mechanism of hepatic glycogenolysis and gluconeogenesis to an increase in the blood sugar concentration (p. 30).

Curves indicative of diminished tolerance may be obtained in subjects with simple obesity (improving after weight reduction), essential hypertension, arteriosclerosis, and Paget's disease (osteitis deformans), in about 50 per cent of cases of advanced malignancy and certain other conditions accompanied by severe undernutrition, in infants with extreme dehydration and in subjects in whom the carbohydrate intake has been restricted prior to the performance of the test.

Glucose tolerance is frequently reduced temporarily by severe trauma (including surgical). Occasionally, latent diabetes is activated in this way.

Decreased Response—Increased Glucose Tolerance

By increased glucose tolerance is meant increased ability of the organism to handle glucose. The alimentary blood sugar response may be affected in one or both of two ways: (a) the initial blood sugar rise (hyperglycemic phase) is abnormally low or, at times, entirely absent, and/or (b) the hypoglycemic phase is abnormally pronounced and prolonged. Spontaneous hypoglycemia may or may not be present. Subjects with increased tolerance for glucose almost invariably also exhibit increased sensitivity to insulin.

Hyperinsulinism. As indicated elsewhere (p. 54), this condition may occur as a result of hyperfunction, hyperplasia, adenoma, or carcinoma of the islands of Langerhans. The metabolic phenomena result from a constant and excessive supply of insulin instead of a periodic secretion in response to the requirements of the organism. The characteristic blood sugar curve following administration of glucose should theoretically resemble that which follows the simultaneous administration of glucose and a large dose or repeated doses of insulin.

(a) Fasting hypoglycemia.

(b) Little or no elevation of blood sugar following glucose ingestion, resulting in a flat type of curve.

(c) Exaggeration of the hypoglycemic phase.

(d) Decrease in serum phosphate.

(e) Increase in the respiratory quotient.

However, the glucose tolerance curves in many cases of hyperinsulinism do not necessarily conform to this description. Sometimes the blood sugar concentration may rise to excessive heights (> 250 mg. per 100 ml.) This observation, suggestive of diminished sugar tolerance, has been interpreted as due to saturation of the liver with glycogen. The glucose tolerance test of three or four hours, as ordinarily carried out, is usually of little value in establishing a diagnosis of hyperinsulinism in such cases. Although the low blood sugar values in the fasting state are significant and should arouse suspicion, the normal or even excessive rise during the first hour and the subsequent fall to a relatively normal level within three hours may be misleading. However, if one continues to make observations of the blood sugar concentration over a period of six hours, much more characteristic findings are usually obtained. After reaching a normal level, the blood sugar continues to fall and may be extremely low at the end of five to six hours. This procedure should be carried out in all cases of suspected hyperinsulinism. As has already been indicated, a prolonged fast is even more effective in this respect, but not without danger.

Abnormal prolongation of the hypoglycemic phase of the curve is of greater significance than is the degree of hypoglycemia. The shape of the curve is influenced considerably by the nature of the antecedent diet, a high carbohydrate intake for a few days before the test favoring the production of a flat type of curve and a previously low carbohydrate intake often causing a normal or supernormal initial hyperglycemic phase. In some instances, removal of an adenoma of the pancreatic islet cells is followed by a temporary state of hypoinsulinism, with diminished glucose tolerance, due probably to suppressed function of the islets of Langerhans.

In certain cases exhibiting manifestations similar to those indicated above, no morphologic abnormality of the pancreatic islet cells can be demonstrated. These have been designated "functional hyperinsulinism," but the justification for this appellation is questionable. In this condition, the hypoglycemic attacks, although occurring usually two to four hours after meals, appear to be related more directly to emotional states or physical stress, are more transient and less inclined to progress in severity and frequency than in organic hyperinsulinism. The hypoglycemic phase of the glucose tolerance curve is less prolonged, as is the hypoglycemia induced by insulin or by insulin plus glucose. Autonomic imbalance in the direction of vagotonia may play a part in the pathogenesis of this type of hypoglycemia.

Adrenocortical Insufficiency. This condition is characteristically associated with (a) increased tolerance for glucose, (b) increased sensitivity to insulin and (c) a decreased glycemic response to epinephrine. Some or all of the following abnormalities of carbohydrate metabolism have been observed in a large proportion of cases of Addison's disease.

(1) Low-normal or subnormal fasting blood sugar concentration (p. 57).

(2) Flat type of oral glucose tolerance curve, i.e., diminished glycemic response after ingestion of glucose. This is due, in part at least, to a diminished rate of absorption of glucose from the intestine, which is the only abnormality of carbohydrate metabolism that is corrected by administration of deoxycorticosterone acetate. There is some doubt whether the adrenal cortex plays any major part in the intestinal absorptiton of glucose. Intravenous administration of glucose under standardized conditions (p. 38) is usually followed by a rise in blood sugar to an essentially normal level, i.e., the hyperglycemic phase of the curve is essentially normal.

(3) Exaggeration of the hypoglycemic phase of the curve after either oral or, more strikingly, intravenous administration of glucose. The blood sugar falls to an abnormally low level at two to three hours and may remain at this level for several hours instead of returning promptly to the control level, as in normal subjects. During this period, severe subjective and objective manifestations of hypoglycemia often develop at a higher blood sugar concentration than in the case of normal subjects.

68 CARBOHYDRATE METABOLISM

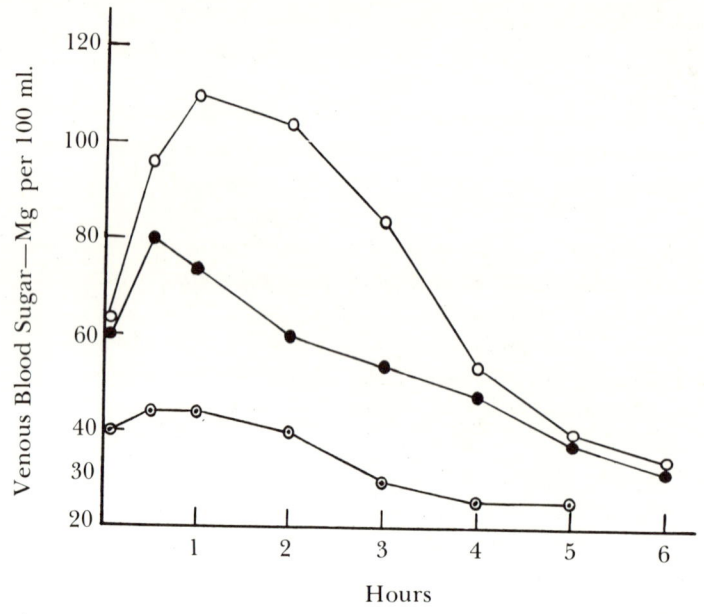

Figure 1-16. Three types of blood sugar response to ingestion of 100 g. glucose in cases of hyperinsulinism.

(4) Marked prolongation of the hypoglycemic response to insulin (p. 70), i.e., "hypoglycemia unresponsiveness."

(5) Slightly decreased glycemic response to a standard dose of epinephrine.

(6) Striking hypoglycemia (a) during fever or infections, (b) on a diet high in fat and low in carbohydrate, and (c) following a twenty-four hour fast.

(7) High standard R.Q. (p. 47), which increases more than normally after administration of glucose.

These defects in carbohydrate metabolism are due to (a) possibly diminished absorption of glucose from the intestine and (b) decreased gluconeogenesis from protein (amino acids) (p. 18). The ability to form glucose from lactic acid and pyruvate as well as from glucogenic amino acids is apparently impaired. The diminished absorption has been stated to be corrected by administration of an adequate amount of sodium chloride or deoxycorticosterone acetate. This, as well as the more fundamental defects in gluconeogenesis, glucose utilization, and glycogenesis, can be corrected by adequate amounts of adrenal cortical extract or 11-oxygenated adrenal hormones (p. 24).

Because of the ease with which hypoglycemic manifestations are produced in subjects with adrenocortical insufficiency, provision must be made for prompt intravenous administration of glucose when either glucose or insulin tolerance tests are performed. The combined insulin-glucose tolerance test (p. 44) is a safer procedure to employ in such cases.

Adenohypophyseal Hypofunction. Increased tolerance for

glucose (flat blood sugar curve) occurred in about 50 per cent of twenty-one verified cases of pituitary cachexia (Simmonds' disease) due to destructive lesions, such as atrophy, hemorrhage, chromophobe tumor, and cyst. It is difficult in such cases to evaluate the influence of severe undernutrition in this connection, as evidenced by the occurrence of a flat type of curve in about 30 per cent of cases of anorexia nervosa. However, in Simmonds' disease, there is characteristically a tendency toward spontaneous hypoglycemia similar to that in Addison's disease, and increased sensitivity to insulin.

The glucose tolerance is not so frequently altered in other types of pituitary hypofunction, such as pituitary infantilism or dwarfism. Although the glucose tolerance may be increased, fasting hyperglycemia and decreased tolerance also have been observed in these conditions.

Hypothyroidism. A flat type of glucose tolerance curve is commonly obtained in cretinism and myxedema. This is possibly due in part to a diminished rate of absorption from the intestine, for the intravenous glucose tolerance test yields essentially normal findings in many such cases.

The fasting blood sugar concentration has been found to be normal in nondiabetic patients with hypothyroidism induced by total thyroidectomy. However, the postoperative sugar values tend to be slightly lower than those obtained before operation. Similarly, although the hyperglycemia produced by glucose ingestion is usually slightly less after total thyroidectomy than previously, the glucose tolerance curve is generally within normal limits in nondiabetic patients with induced hypothyroidism.

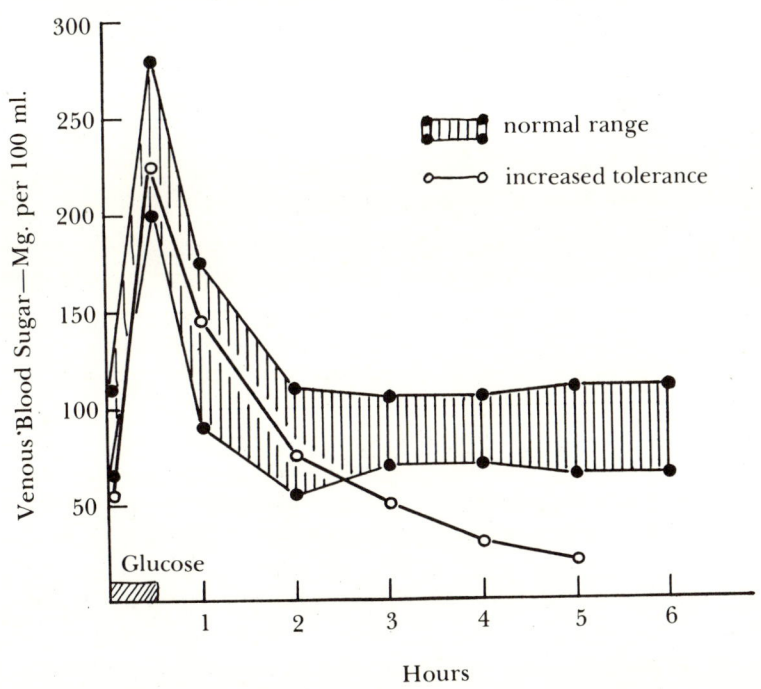

Figure 1-17. Intravenous glucose tolerance curves. Curve characteristic of increased tolerance is obtained in typical cases of Addison's disease and hyperinsulinism.

Carbohydrate metabolism is apparently not significantly influenced by hypothyroidism induced by total thyroidectomy except when a derangement of carbohydrate metabolism is evidenced prior to operation. Marked improvement in glucose tolerance has been observed in diabetic patients following removal of the thyroid gland, and also in patients with hyperthyroidism who present a diabetic type of curve before operation.

Miscellaneous. A flat type of glucose tolerance curve may be obtained in renal glycosuria, in hypothalamic lesions, in some, but not all cases of glycogen storage disease (p. 79), due apparently to increased stability of the hepatic glycogen, with diminished glycogenolysis, and in celiac disease and sprue, probably as a result of impaired absorption of glucose from the intestine. Curves of this nature may be produced by conditions associated with increased intestinal motility, such as vitamin B deficiency, "gastrointestinal neuroses," achlorhydria, tuberculous enteritis, and chronic ulcerative or mucous colitis. In such cases, a normal response is obtained with the intravenous glucose tolerance test. Curves indicative of increased tolerance have been reported in marasmus in infants and in anorexia nervosa, and may be produced in normal subjects by excessive carbohydrate feeding prior to the performance of the test. Hypoglycemia, occurring about two hours after meals, may occur in partially gastrectomized subjects (p. 55). Oral glucose tolerance tests in such individuals reveal an excessively high rise and exaggeration of the subsequent hypoglycemic phase of the curve. This has been attributed to the rapid passage of the glucose into the intestine, its rapid absorption and excessive stimulation of insulin secretion due to the abnormally high rise in blood sugar. The intravenous glucose tolerance test is normal.

Abnormal Insulin Tolerance

The purpose of the insulin tolerance test is to determine (a) the sensitivity of the organism to insulin and (b) its responsiveness to insulin-induced hypoglycemia. The normal blood sugar response to insulin is discussed elsewhere (p. 44). Abnormalities may occur in one or both of two phases of the curve, i.e., in the extent and duration of the hypoglycemia and in the subsequent rise toward the pre-injection level. There are two important types of abnormality: (1) "insulin resistance," characterized by a relatively slight (less than 50 per cent of the control level) or delayed (forty-five minutes or longer) fall in blood sugar; (2) "hypoglycemia unresponsiveness," characterized by undue prolongation of the period of hypoglycemia (absence of or marked delay in the subsequent rise in blood sugar).

Hypoglycemia unresponsiveness is observed characteristically in hyperinsulinism, Addison's disease (adrenocortical hypofunction), anterior pituitary hypofunction (Simmonds' disease and, at times, pituitary infantilism or dwarfism), and in some cases of glycogen disease. When performing this test in suspected cases of these disorders, provision must be made for prompt intravenous administration of glucose if serious manifestations of

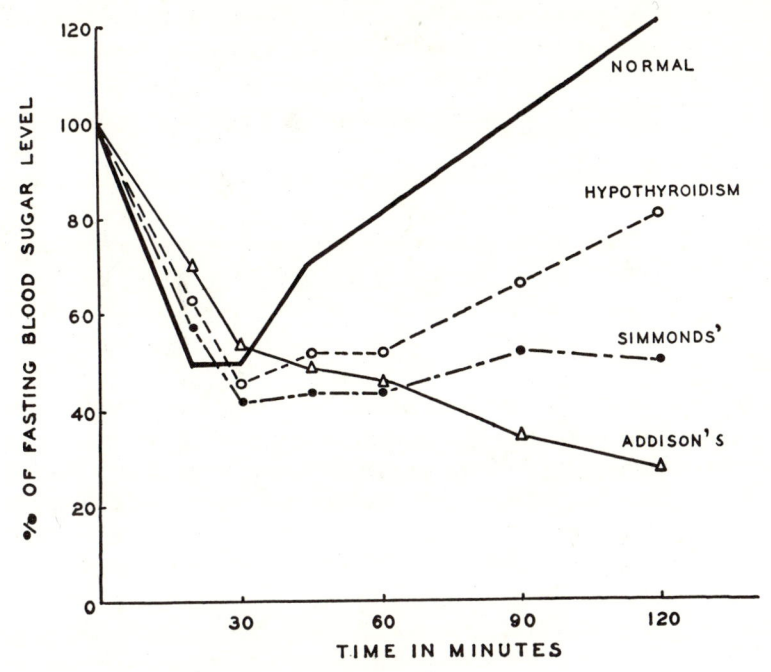

Figure 1-18. Insulin tolerance test.

hypoglycemia develop, or to terminate the prolonged period of hypoglycemia. This procedure may be useful in differentiating anterior pituitary hypofunction, as in Simmonds' disease, from anorexia nervosa and, at times, from primary hypothyroidism.

Insulin resistance may be observed in primary adrenocortical hyperfunction (Cushing's syndrome), anterior pituitary hyperfunction (acromegaly), and in some cases of diabetes mellitus. Subjects with diabetes mellitus have been classified as "insulin-resistant" and "insulin-sensitive" partly on the basis of this test and partly on the basis of the glucose-insulin tolerance test. Excluding "steroid diabetes" (adrenocortical or pituitary hyperfunction), the validity of this classification is questionable. There are, of course, instances of true insulin resistance, but this phenomenon is not related to the pathogenesis of the diabetic state in such cases.

Hypoglycemia unresponsiveness can usually be demonstrated by means of the combined insulin-glucose tolerance test (p. 44), with less danger of undesirable reactions than with the conventional insulin tolerance test. In normal subjects, hypoglycemia induces a homeostatic response which results in prompt rise in blood sugar to a normal level (i.e., decrease in insulin secretion, increase in hepatic glycogenolysis). If glucose is administered in the hypoglycemic state, the subsequent rise in blood sugar, in normal subjects, exceeds that which occurs if glucose is given at normal blood sugar levels. In hypoglycemia unresponsiveness, this phenomenon is not observed, characteristically, the response to glucose being usually less than that seen normally in the conventional glucose tolerance test.

Equivocal results have been reported with this procedure in occasional cases of anorexia nervosa and psychogenic vomiting.

Abnormal Cortisone-Glucose—and Prednisone-Glucose Tolerance Tests

These tests are, of course, stress tests used for patients with normal or suspicious reactions to the ordinary glucose tolerance test. They have been applied in the hope of revealing cases of latent diabetes. The cortisone test is positive if the two-hour blood glucose level is above 140 mg./100 ml. It has been shown in one group that 26 per cent of individuals showing this abnormality subsequently developed diabetes, although at the time of the initial test the ordinary glucose tolerance test proved normal. There is increasing incidence of abnormality with increasing age. One third of pregnant women also give an abnormally high two-hour blood glucose level. The test is more commonly positive in relatives of diabetics than in the general population. Very similar findings have been obtained with the prednisone test, which is, perhaps, somewhat more sensitive.

Abnormal Tolbutamide Tolerance Test

In patients with diabetes and fasting hyperglycemia, the percentage figures obtained at 20 minutes and 30 minutes are undoubtedly higher than those obtained from individuals with normal glucose tolerance tests (Fig. 1-19). Discrimination is, therefore, good in relation to overt diabetes. Unfortunately, diabetics with normal fasting levels of blood glucose show overlap with the normal range. The test can therefore give a goodly proportion of false negative results.

In the form of the test used for the diagnosis of insulinoma, it has been found that patients with an insulinoma show a minimum glucose concentration of less than 25 mg./100 ml., and a figure less than 40 mg./100 ml. is the highest concentration during the period 90 to 180 minutes after intravenous tolbutamide (Fig. 1-19). Patients with hypoglycemia due to liver disease, adrenocortical insufficiency or functional causes tend to react as normal individuals. In the first two conditions, there is a tendency for the maximum level between 90 and 180 minutes to be somewhat lowered but seldom below 50 mg./100 ml.

Abnormal Plasma Insulin Levels

Obesity. The fasting plasma insulin levels tend to be higher in obese than in normal weight individuals. The increase in level, after oral or intravenous glucose, is also higher than normal. This is in part due to the fact that obese individuals develop some resistance to insulin and so their pancreata have to secrete more insulin in order to maintain their insulin-controlled metabolic pathways. Abnormal glucose tolerance curves are more commonly found in fat people. It could be argued that a higher blood glucose is necessary in order to call forth the larger amount of insu-

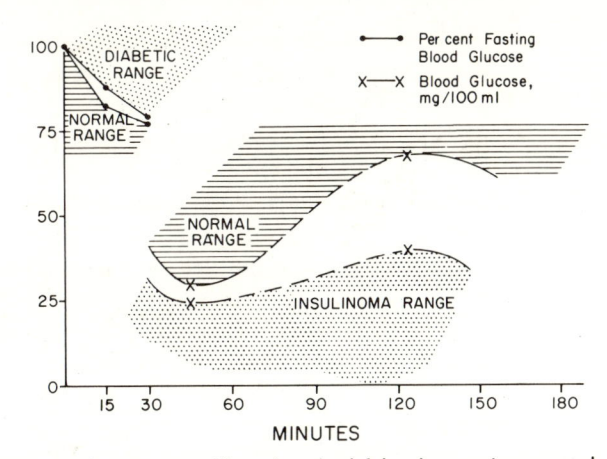

Figure 1-19. Intravenous tolbutamide tolerance tests. Note that the left-hand curve is expressed as per cent of fasting blood glucose concentration, whereas the right-hand portion of the graph refers to blood glucose concentrations in mg. per 100 ml. (From Bondy: *Duncan's Diseases of Metabolism*, 6th edition. W. B. Saunders Company, 1969.)

lin required because of the resistance. Animals made obese under experimental conditions show similar phenomena, which are reversible when weight is lost. This is also apparently the case with humans.

Diabetes Mellitus. There has been considerable confusion in relation to the plasma insulin found in those with diabetes mellitus. As a result of bio-assay techniques, it was originally suggested that patients with this condition could be divided into two groups. One group had very low insulin levels and the other, the obese diabetics, had high levels. This, not unnaturally, led to a good deal of speculation in relation to antagonists to insulin, and there was discussion of such things as relative insulin deficiency to explain the occurrence of diabetes in patients with high levels of plasma insulin. The studies on obesity have somewhat altered this point of view. When matched, weight for weight, against nondiabetic controls, it has become apparent that in patients with diabetes mellitus there is a diminution both in fasting levels of plasma insulin and in the response to glucose. The latter, on the whole, is inversely related to the severity of abnormality of the glucose tolerance test. In other words, the lower the glucose tolerance the lower and more prolonged is the response of the plasma insulin level to administered glucose. It is also now realized that much depends on the route of glucose administration, since oral glucose probably increases the production of intestinal glucagon, which in turn stimulates insulin production by the pancreas.

The fact still remains that obese diabetics may be producing less insulin than controls of equal weight but they certainly have hyperinsulinemia when compared with nondiabetic individuals of normal weight. They are diabetic, however, because of the increased resistance to endogenous insulin in obesity. It can therefore be suspected that obesity, which brings about insulin resistance, also leads to the production of diabetes mellitus in persons whose capacity to secrete insulin cannot match the increase in insulin resistance.

It is interesting that there has been no demonstrable alteration of insulin secretory capacity with advancing age. This has led to the suggestion that, since diabetes mellitus tends to be found more frequently in the elderly, with advancing age there is a diminution in the rate at which insulin leaves the circulation, possibly due to vascular changes. In non-diabetic individuals with a strong family history of diabetes mellitus, a lowered plasma insulin response to glucose is found quite often.

Evidence is accumulating that cyclic AMP production in the beta cells of the islets is necessary for insulin release. It could be that deficiency in AMP cyclase activity occurs in diabetes mellitus. It may or may not be causally related. Glucagon, which stimulates the activity of this enzyme, also increases, in diabetics, the early part of the plasma insulin response to glucose.

Insulinoma. Raised levels of fasting plasma insulin can be demonstrated in these patients. In normal individuals, proinsulin accounts for up to 20 per cent of immunologically determined insulin levels. In patients with insulinoma, the relative amount of proinsulin is considerably higher.

Other Pancreatic Disorders. A very flat plasma insulin response curve is sometimes found in pancreatitis and in carcinoma of the pancreas.

Abnormal Epinephrine Tolerance Test

Diminished glycemic response to epinephrine (rise of less than 35 mg. per 100 ml. in forty to sixty minutes) may occur in conditions in which the hepatic glycogen stores are depleted (in the absence of other disturbances of carbohydrate metabolism), particularly in advanced grades of hepatocellular damage (acute and subacute hepatic necrosis, various forms of hepatitis, fatty liver, cirrhosis, and so on). It is also observed in patients with certain types of glycogen storage disease (von Gierke), in which condition the liver contains an excessive quantity of glycogen which is resistant to glycogenolytic stimuli. A subnormal response to epinephrine also occurs commonly in conditions in which a state of "hypoglycemia unresponsiveness" is demonstrated by means of the insulin tolerance test, viz., hyperinsulinism, Addison's disease, and pituitary cachexia.

Abnormal Tolerance for Other Sugars

Abnormal Fructose Tolerance. The metabolism of fructose differs in several essential details from that of glucose (p. 43). Although fructose may be utilized without preliminary transformation to glucose or glycogen, it is probably largely removed from the blood and transformed into glycogen before contributing to the blood glucose or possibly before undergoing oxidation. The liver appears to be the chief site of transformation of fructose to glycogen. It appears that fructose can undergo this transformation in the absence of insulin. Because of the specific importance of the liver in the intermediary metabolism of this sugar, the determination of the tolerance of the organism for fructose has been suggested as a means of detecting impairment of hepatocellular function (p. 44). Following the ingestion of 50 g. of fructose, diminished capacity of the liver for transforming this sugar to glycogen or glucose is evidenced by the following phenomena:

(a) A rise in blood sugar of more than 35 mg. per cent, a concentration of 135 mg. or more being reached at some period during the performance of the test.

(b) A delayed return to the postabsorptive level (beyond two hours).

The demonstration of diminished fructose tolerance in the absence of evidence pointing to any disturbance of general carbohydrate metabolism is suggestive of impairment of liver function. However, negative results are commonly obtained in the presence of advanced hepatic disease, especially if chronic in nature (cirrhosis, lues, and malignancy of the liver, etc.). Positive results may be obtained in more acute forms of liver disease such as acute necrosis or hepatitis. As a rule, this test is of little practical clinical value in the estimation of liver functional efficiency since in most instances other evidence of functional impairment is present long before positive results are obtained. Of still less value is the occurrence of fructosuria following the ingestion of 100 g. of fructose. At least 10 per cent of normal individuals respond by eliminating some fructose in the urine, and many patients with hepatic disease yield findings within the limits regarded as normal.

More satisfactory results have been reported when the analytical procedure is modified so that the blood fructose alone is determined, rather than fructose plus glucose. In the absence of hepatocellular damage, after ingestion of 50 g. of fructose, the blood fructose concentration, usually 0 to 8 mg. per 100 ml. in the fasting state, increases not more than 15 mg. per 100 ml., usually within the first hour, and falls to 0 to 10 mg. at the end of two hours. Increases of 16 to 30 mg., with a delayed fall to the resting level, have been observed in patients with hepatocellular damage. The investigation has proved to be of no real clinical value. Similar findings have been obtained at times in patients with arteriosclerosis.

Abnormal Galactose Tolerance. When galactose is injected intravenously in the normal dog it disappears from the blood in about two hours, 10 to 30 per cent appearing in the urine. Removal of 50 to 70 per cent of the liver has little effect upon the tolerance for galactose. In the absence of the liver, 50 to 60 per cent of the amount injected is recovered in the urine. Similarly, increased excretion has been observed in the presence of acute degenerative lesions of the liver produced by hepatotoxic agents.

The usefulness of this procedure as a measure of hepatocellular function and as a means of differentiating between obstructive and hepatocellular jaundice is discussed in detail elsewhere (p. 616). Suffice it to state here that, with the oral test, the excretion by a jaundiced patient of less than 3 g. of galactose in five hours suggests that the jaundice is not of hepatocellular origin, while the urinary excretion of more than 4 to 5 g.

suggests that it is. With the intravenous test, the presence of more than 20 mg. of galactose per 100 ml. of blood at the end of seventy-five minutes, in a patient with jaundice, suggests that the latter is of hepatocellular origin, while less than 20 mg. per 100 ml. at this time points toward obstructive jaundice.

Determination of the increase in blood galactose after ingestion of 40 g. of this sugar has been proposed as a means of studying thyroid function. It has been found that in hyperthyroidism the peak of blood galactose concentration ranges from 25 to 150 mg. per 100 ml., being above 40 mg. in the great majority of cases (normal maximum 15 to 35 mg. per 100 ml.). This abnormally high rise is attributed to increased rapidity of absorption of the sugar from the intestine in patients with hyperthyroidism. Essentially normal results are obtained in hyperthyroidism when the galactose is given intravenously, indicating that the above findings are not due to abnormality of intermediary metabolism. This procedure may be of value in differentiating hyperthyroidism from other conditions accompanied by increase in the basal metabolic rate that may simulate it clinically, especially congestive heart failure of nonhyperthyroid origin.

Decreased tolerance for galactose has been observed in hyperpituitarism, certain pluriglandular disturbances involving gonadal dysfunction, and occasionally during menstruation and after the menopause. It occurs also in galactosemia (p. 89).

BLOOD LACTIC ACID AND PYRUVIC ACID

Under resting conditions there are small amounts of lactate and pyruvate in the blood. Whole venous blood, at bed rest and fasting, contains 5 to 15 mg. of lactate and 0.12 to 1.00 mg. of pyruvate in 100 ml. (0.55 to 1.70 mmol./l. and 0.01 to 0.11 mmol./l., respectively). Arterial blood under these conditions contains 3.1 to 7.0 mg. of lactate in 100 ml. (0.34 to 0.78 mmol./l.). For these determinations, a tourniquet must not be used for blood collection and plasma measurements are unreliable.

These intermediates in carbohydrate metabolism are derived mainly from glycolytic reactions (p. 7) in the muscles and erythrocytes. Lactate increases (up to 100 mg.), as does pyruvate, to a lesser degree (up to 5 mg.), in strenuous muscular exercise, when the requirement for energy outstrips the supply of oxygen to the muscles, in which the metabolic emphasis is therefore on anaerobic glycolysis, with excessive production of lactic acid; this passes into the blood stream, destined for reconversion to glycogen and glucose in the liver (p. 10).

Relatively slight exertion produces an abnormally high level of blood lactate and pyruvate in conditions associated with anoxia, e.g., congestive heart disease, valvular or hypertensive, and coronary artery disease. This is of little practical consequence, although it has been proposed as an objective means of evaluating circulatory efficiency or myocardial reserve.

Increase in blood lactate in patients in shock is of serious prognostic significance, indicating the onset of the irreversible stage of shock.

Thiamine pyrophosphate is an essential coenzyme (cocarboxylase) in the transformation of pyruvate to acetate (p. 12). In thiamine deficiency, therefore, pyruvate, and its reduction product, lactate, accumulate in excess and their concentrations in the blood increase. Determination of blood pyruvate has been employed for the detection of thiamine deficiency, but is obviously not specific for this purpose.

Although the liver is the organ chiefly concerned in the removal of lactate from the blood, the plasma lactate concentration is seldom increased in hepatic disease except, occasionally, in severe, acute hepatic necrosis or in terminal stages of decompensated cirrhosis, and in glycogen storage disease. However, injected lactate may not be removed from the blood at normal rate in patients with liver damage; consequently, a lactate tolerance test has been proposed as a test of liver function. It has not been widely employed and has no advantage over several simpler procedures.

In normal subjects, administration of glucose is followed by a significant increase in blood lactate and pyruvate within one hour, a manifestation of increased glucose utilization. This also follows injection of insulin. In diabetes mellitus this rise is delayed or absent and, therefore, determination of blood lactate and pyruvate in their relation to changes in blood sugar in the course of the glucose tolerance test has been proposed as a means of distinguishing between pancreatic diabetes and other conditions exhibiting decreased glucose tolerance, e.g., "steroid diabetes" arteriosclerosis, hepatic disease, "alimentary" glycosuria.

Increase in blood pyruvate has been reported in certain cases of decompensated hepatic cirrhosis and hepatic coma. It has been suggested that this may be due to failure of the liver to assimilate pyruvate efficiently into the Krebs cycle sequence of reactions.

Hyperlactatemia and Lactic Acidosis. There are a number of conditions, e.g., vitamin B_1 deficiency, which can lead to an increase in the blood lactate level without any marked effect on plasma pH. In other words, it is possible to have a *hyperlactatemia* either without change (hyperventilation, obesity, glucocorticoid therapy or oral contraceptives) or with a fall in plasma pH. When a fall occurs, the condition is then described as *lactic acidosis,* which is a metabolic acidosis. In this condition, the [lactate] : [pyruvate] ratio is raised (normal is approximately 10), which can be an indication of tissue anoxia.

Lactic acidosis can be divided into two groups. In the first group, there is accelerated glycolysis, which is secondary to hypoxia, shock or circulatory failure. If the defect is corrected, the blood lactate concentration falls toward normal and the prognosis is good. In the second group, there is no obvious explanation for the raised lactate level, which can be as high as 14 to 26 mmol./liter in venous blood. There is often associated severe disease such as uremia, bacterial infection, alcoholism, arteriosclerotic heart disease, liver disease, leukemia, severe anemia, starvation and diabetes mellitus. Individuals in this second group often do not survive. It is interesting to note that the blood lactate level is frequently raised in the agonal phase of disease.

So far, we have been considering acute lactic acidosis. There is also a chronic form, which occurs in children with mental retardation (mongolism) or other neurological disease. Chronic lactic acidosis, possibly familial, occurs in adults, too, and is manifested by an undue rise in blood lactate after moderate exercise or intake of alcohol. Chronic lactic acidosis also occurs in Type I glycogen storage disease.

The drug phenformin, used in the treatment of diabetes, can cause lactic acidosis, which may be severe. The same can be caused by ethanol.

Apart from the usual presentations of any severe metabolic acidosis, lactic acidosis tends to be associated with marked lethargy leading to coma.

DIABETIC KETOACIDOSIS

Untreated diabetes may eventually lead to so-called diabetic ketoacidosis, which then leads to diabetic coma and death. Even in the treated diabetic, a severe infection or other serious stress can set off the same chain of events.

There is at first an excessive production of ketone bodies due to lack of insulin or possibly increased resistance to its action. In badly controlled patients, ketosis may continue for weeks. The accompanying hyperglycemia leads to a large excess of glucose in the urine, which produces a prolonged osmotic diuresis with loss of water and electrolytes. Severe ketosis produces nausea and vomiting with further loss of fluid and electrolytes. These water losses rapidly lead to dehydration. There is a fall in blood volume, as well as in blood pressure, and a diminution in renal circulation, with the result that kidney function is greatly impaired and the level of blood urea rises. Thus far, the tendency for increase in the hydrogen ion concentration of the blood, due to excessive anion production (acetoacetate and hydroxybutyrate), has been compensated by renal mechanisms. This compensation involves loss of cations, particularly sodium, and some potassium, as well as nitrogen in the form of ammonium ions. With increasing renal failure, the hydrogen ions are retained in the blood stream and the pH falls to levels well below 7.35. In other words, an acidosis has become an acidemia. The hyperosmolality of the extracellular fluid causes intracellular dehydration with further loss of water, and particularly potassium and magnesium. This loss is no doubt aggravated by the apparent necessity of insulin for the maintenance of the integrity of the cell membranes, especially those of muscle. The continued vomiting leads to a severe loss of chloride ions, which have also been lost during the osmotic diuresis. The increase in hydrogen ion concentration of the blood results in a severe loss of plasma bicarbonate. The patient now has hyponatremia, hyper- or occasionally hypopotassemia (severe loss of body K), hypomagnesemia, hypochloremia and greatly lowered plasma bicarbonate with an increased level of blood urea.

Insulin resistance has also developed. This is due in part to the acidemia and in part to accumulation of an α_1-globulin, which combines with insulin and is thus rendered physiologically inactive. This grave lack of insulin activity encourages gluconeogenesis from amino acids and causes severe diminution in protein synthesis and great loss of nitrogen compounds in the urine. The insulin lack also leads to hyperlipemia, for reasons already discussed (p. 21). A fatty liver develops and this organ is also depleted of glycogen, partly due to insulin lack stimulating phosphorylase activity but largely due to the state of starvation resulting from the severe nausea associated with ketoacidemia. Cerebral metabolism is severely affected and there is decreased oxygen uptake by the brain, even though both arterial blood flow and oxygen tension may be normal. Consciousness is diminished and eventually coma sets in. These effects on brain metabolism are by no means understood. They seem to be due, in part at least, to the fall in pH of the body fluids. Since other equally severe acidemias do not produce the same degree of central nervous system depression, it is believed that acetone and acetoacetate also play a part. They are known to be cerebral depressants. The high plasma osmolality due to great elevation of glucose may also be important. In fact, this seems to be the main factor in one form of coma associated with diabetes, with very high glucose levels and little or no ketoacidosis. The condition has a high mortality rate.

Even though the serum potassium level may have been raised, it rapidly falls to low levels as a result of treatment with insulin. It is important, therefore, to administer potassium salts intravenously as part of the therapy of diabetic ketoacidosis. Because the initial blood level frequently is raised, potassium is not usually administered immediately.

Euglycemic Diabetic Ketoacidosis. Recently, a type of diabetic ketoacidosis occurring in young people has been described, in which blood glucose levels may be normal. Although usually raised, the level does not exceed 300 mg./100 ml. blood. The patients may be drowsy but can be fully alert. This seems to discount the concept that ketone bodies are responsible for diabetic coma.

GLYCOGEN STORAGE DISEASES

This designation is applied to a group of congenital disorders, frequently with a familial incidence.

Abnormal amounts of glycogen, or related compounds, are deposited in one or more of the following organs: liver, kidney, heart and muscle. In most cases, this is due to a known enzyme deficiency. Dependent on this and the main organ or organs, in which the material is deposited, seven main types have been recognized. There are also a number of subtypes.

The deficient enzyme is, of course, one of those known to be involved in the synthesis or breakdown of glycogen. It will be remembered (p. 9) that starting with UDP-glucose and a primer molecule, the enzyme glycogen synthetase can increase the length of glucosyl chains made up of glucosyl residues strung together in 1→4 linkages. Glycogen, however, is a very branching molecule and, at the commencement of each branch, two glucosyl residues are connected by 1→6 linkage. This is brought about by a brancher enzyme, which removes segments of outer glycogen chains, by splitting 1→4 bonds, and then deposits these segments on inner glucosyl residues by means of 1→6 linkages. The resulting glycogen molecule is then broken down by activated phosphorylase, which breaks the 1→4 linkages of each straight chain of glucosyl residues, yielding glucose 1-phosphate. This breakdown cannot proceed further than the fourth residue from the branching point, leaving a so-called limit dextrin molecule. There is, therefore, a debranching enzyme system, which removes three quarters of the four residue chain and deposits it on the end of any straight chain in 1→4 linkage, which can then be broken down by phosphorylase. The debranching enzyme system then attacks the remaining quarter of the four residue chain, which, of course, is one glucosyl residue in 1→6 linkage. This linkage is destroyed and a free glucose molecule is liberated for each branch attacked. The debranching system therefore consists of two different activities.

Type I Glycogenosis (von Gierke's Disease). Glycogen is deposited in the liver and kidneys. In one series reported, the content in the former organ ranged from 4.0 to 18.5 per cent. This accounts for the occurrence of hepatomegaly. The disease is due to a deficiency of glucose-6-phosphatase, which is also absent from the intestinal mucosa. The deficiency means that glycogen cannot be broken down to liberate glucose, and glucose-6-phosphate is maintained at a level which promotes glycogen synthesis. Since very little glucose is derived from the liver, children with this disease tend to develop hypoglycemia. They use fat mostly as an energy source and this leads to lipemia, acidemia and ketosis, as well as to an increase in plasma cholesterol sufficient to produce xanthoma. There is fatty infiltration of the liver. The hypoglycemia inhibits insulin secretion, which in turn also inhibits protein synthesis, and growth is stunted. Hypoglycemia stimulates epinephrine production, which causes the muscles to break down their glycogen with the liberation of large amounts of lactate, and thus gives rise to an increased level of blood lactate. This competes with urate for excretion by the kidney and so an increased blood urate level occurs.

There is also some evidence of increase in uric acid synthesis in those children who develop symptoms of gout. Although many of these children die young, a number have survived to adolescence, when for some unknown reason, much improvement can occur.

Type II Glycogenosis (Pompe's Disease). This disease is characterized by the fact that nearly all tissues contain excessive amounts of glycogen. The main clinical symptomatology arises from the effects on heart and skeletal muscle. The former is greatly enlarged and there is extreme muscle weakness. Death occurs usually before the ninth month of life. A few cases of a somewhat milder form of Pompe's disease have been described, in which the children have survived up to two and one-half years. The disease is due to deficiency of a lysosomal enzyme, acid maltase (α-1,4-glucosidase), whose function is to help destroy glycogen which enters lysosomes. The blood changes associated with Type I glycogenosis do not occur.

Type III Glycogenosis (Limit Dextrinosis). In this condition, there is an accumulation of limit dextrin in liver and muscles. This substance, it will be remembered, is produced by the action of phosphorylase on glycogen. Further breakdown requires a debrancher enzyme system, which is deficient in this disease. The accumulated material can reach as high as 18 per cent of wet weight in the liver and about 6 per cent in muscle. The enzyme deficiency can also be demonstrated in leukocytes, which makes a diagnostic test available and does not require biopsy. Sometimes only the transferase activity of the enzyme system is defective. The clinical manifestations are somewhat milder but very similar to those of Type I glycogenosis. There can be quite severe myopathy, which may present in adult life. Patients with Type III glycogenosis are known to survive well into adult life.

Type IV Glycogenosis (Amylopectinosis). In this condition, which mainly affects the liver, there is accumulation of a glycogen type of polysaccharide, which has relatively few branches and long outer chains—in other words, an amylopectin. A deficiency of the branching enzyme has been demonstrated in the liver and leukocytes in one case. The disease is fatal, the longest reported survival being four years. All the reported cases showed hepatomegaly and cirrhosis and a moderate degree of fasting hypoglycemia. Amylopectins have been found in liver, heart, kidneys, muscle and spleen.

Type V Glycogenosis (McArdle's Disease). In this condition, which affects children and adults, there is a defect of muscle phosphorylase. Cramp occurs in muscle groups after only moderate exercise. Recovery is attained during rest. This would be expected, since in the latter state muscles derive most of their energy from fatty acids. They require glucose, however, when exercised. In this disease, the exercised muscles do not release lactate to the blood stream, since they are not metabolizing glucose. They can, however, metabolize fructose.

Type VI Glycogenosis (Hers' Disease). This is a somewhat ill-defined group of diseases in which liver glycogen concentration is usually increased and there is sometimes, but by no means always, a deficiency of liver phosphorylase. The condition has also been reported to occur in association with Fanconi syndrome. As a rule the disease presents clinically rather like a mild case of Type I glycogenosis.

Type VII Glycogenosis. In this condition, there is a moderate accumulation of glycogen in skeletal muscle. There is also a marked accumulation of glucose 6-phosphate and fructose 6-phosphate and a marked diminution of fructose 1,6 diphosphate. This is due to a demonstrable deficiency of the enzyme phosphofructokinase. The symptoms are similar to those of Type V glycogenosis.

The four most commonly occurring glycogen storage diseases are Types

I, II, III and VI; 26 per cent, 17 per cent, 22 per cent and 31 per cent, respectively, were found in a group of 155 patients studied at Saint Louis. In those types, which are associated with hepatomegaly, there is frequently hypoglycemia, which is most severe in Type I, showing a nil or minimal response to glucagon or epinephrine. Some cases of Type I glycogenosis have shown a moderate response to glucagon with a nil response to epinephrine. The milder types tend to have a lesser degree of hypoglycemia and a somewhat better response to glucagon or epinephrine.

GLYCOGEN STORAGE DEFICIENCY

A case has been reported of glycogen synthetase deficiency in the liver of a child, which led to a very low level of liver glycogen. Another case showed deficiency of both glycogen synthetase and phosphorylase in liver and muscle. The clinical presentation was very similar to that of Type I glycogenosis.

EXCRETION OF SUGAR IN URINE

Under ordinary dietary conditions, in normal subjects, glucose is the only sugar present in the free state in the blood plasma in significant amounts. Under certain circumstances glucose, in abnormal amount, or other sugars may be excreted in the urine. This condition may be designated "melituria," the terms glycosuria, fructosuria, galactosuria, lactosuria, sucrosuria and pentosuria being applied specifically to the urinary excretion of glucose, fructose, galactose, lactose, sucrose and pentose, respectively. Some employ the term "glycosuria" to indicate the presence of any sugar in the urine, and the term "glucosuria" to indicate the urinary excretion of glucose.

Mechanism of Glycosuria (Glucosuria). Glucose is present in the glomerular filtrate in the same concentration as in the water of the blood plasma (arterial). Under normal conditions it undergoes practically complete reabsorption by the renal tubular epithelial cells and is returned to the blood stream. In normal subjects a very small amount may escape reabsorption and be excreted in the urine (p. 480). Under controlled conditions, normal kidneys are capable of reabsorbing 250 to 350 mg., and perhaps as much as 450 mg., of glucose per minute. If one assumes an average glomerular filtration rate of 125 ml./min. and a maximum arterial blood sugar concentration of about 200 mg./100 ml. (after glucose ingestion), the quantity of glucose delivered to the tubules seldom exceeds 250 mg./min. in normal subjects.

Excretion of abnormal amounts of glucose in the urine may be due to two types of abnormality: (1) increase in the amount entering the tubules per minute; (2) decrease in the glucose-reabsorptive capacity of the tubular epithelium (Tm for glucose or TmG).

(1) The quantity of glucose entering the tubules is the product of (*a*) the minute volume of glomerular filtrate and (*b*) the concentration of glucose in the filtrate, i.e., in the arterial blood plasma. Inasmuch as glomerular filtration is rarely increased markedly, glycosuria of this type is due almost invariably to an increase in the blood sugar concentration above the so-called "threshold" level, i.e., 160–170 mg./100 ml. (venous blood). This may be designated "hyperglycemic glycosuria." It should be pointed out that the amount of glucose presented to the tubules may remain within normal limits in the face of considerably higher blood sugar concentrations if the volume of glomerular filtrate is reduced simultaneously (e.g., glomerular damage complicating diabetes; dehydration; shock).

(2) Reabsorption of glucose by the renal tubular epithelium is accomplished mainly by an active process. The capacity for reabsorption may be diminished by induced (phloridzin), hereditary ("renal" glycosuria), or acquired (certain types of kidney disease) defects in this transport mechanism, with consequent glycosuria in the presence of normal or subnormal blood glucose concentrations (non-hyperglycemic glycosuria). These are forms of so-called "renal glycosuria" ("lowered renal threshold" for glucose).

Other Sugars in Urine. Sugars other than glucose are reabsorbed by the renal tubular epithelial cells to only a relatively slight extent. Consequently, when they appear in the blood (and, therefore, in the glomerular filtrate) in demonstrable concentrations, they are eliminated in the urine (p. 87 ff.).

ABNORMAL URINE SUGAR (MELITURIA)

Glycosuria

The term "glycosuria" signifies the presence of abnormal amounts of glucose in the urine. The normal daily excretion is 32 to 93 mg., but glycosuria usually means a daily excretion above 150 mg.

Nonhyperglycemic Glycosuria. Glucose may appear in the presence of a normal concentration of sugar in the blood. This condition may be produced experimentally by the administration of the glucoside phloridzin. It occurs clinically as a result of congenital or acquired defects in the renal tubular transport mechanism for glucose.

Phloridzin Glycosuria. The administration of phloridzin, orally, or, better, subcutaneously, is followed by glycosuria associated with a normal, and indeed, in many instances, a subnormal blood sugar concentration. The precise mechanism is not known. Interference with renal tubular transport of glucose was formerly attributed to inhibition of phosphorylation, but this is apparently not the fundamental defect.

Renal Glycosuria (Renal Diabetes). This condition is also known as "benign" glycosuria and "diabetes innocens." When rigid diagnostic criteria are applied, the reported incidence is usually less than 1 per cent of cases of glycosuria, although this condition has been reported as comprising 9 to 26 per cent of all cases of glycosuria in army recruits. The following standards have been set for the diagnosis of true renal glycosuria:

(1) Fasting blood sugar within normal limits and a normal or supernormal flat glucose tolerance curve.

(2) Glucose should be present in every specimen of urine, whether voided in the fasting state or after a meal. The quantity of sugar in the urine should be largely independent of the diet, although it may vary somewhat depending on the amount of carbohydrate ingested.

(3) Carbohydrate utilization should be normal, as evidenced by determinations of respiratory quotient and serum inorganic phosphate after glucose ingestion.

(4) There should be no disturbance of fat metabolism, ketosis being more likely to develop when the patient fasts than when he overeats.

(5) Moderate doses of insulin should have little or no effect upon the glycosuria.

This condition is genetically determined and frequently exhibits a familial incidence; there is a fundamental defect in the renal tubular transport mechanism for glucose. It may occur as an isolated phenomenon or may be associated with other tubular transport defects, particularly certain types of aminoaciduria. The importance of its recognition depends upon its apparent harmlessness; so far as can be determined, it seldom, if ever, results in diabetes mellitus or in any metabolic derangement whatsoever. Practical danger in making the diagnosis of renal glycosuria lies in the confusion of this condition with potential or mild diabetes mellitus. The patient must be observed carefully over a prolonged period before the diagnosis of renal glycosuria can be established definitely.

The essential cause is unknown. Whatever it may be, inefficiency of the mechanism for tubular reabsorption of glucose is the fundamental defect that determines the occurrence of glycosuria in this condition. From this standpoint it resembles phloridzin glycosuria. No metabolic disturbance occurs in subjects with renal glycosuria as long as the carbohydrate intake is adequate to compensate for the amount lost in the urine. Deprivation of carbohydrate may cause hypoglycemia, increased sensitivity to insulin and ketosis more readily than in normal subjects.

Fanconi Syndrome (p. 311). This term is applied to a more extensive type of congenital renal tubular dysfunction in which there is defective reabsorption of amino acids, other organic acids and phosphate, as well as glucose. There is an associated developmental defect of the tubules. Other types of renal glycosuria, associated with renal aminoaciduria, occur occasionally in hepatolenticular degeneration (Wilson's disease), and in poisoning by certain heavy metals (lead, uranium, cadmium), and in tubular damage produced by Lysol and nitrobenzene.

"Alimentary" Glycosuria. Opinion is divided regarding the metabolic status of so-called "alimentary" glycosuria. The term is employed to designate the urinary excretion of glucose by certain apparently normal individuals after the ingestion of excessive amounts of cane sugar, glucose or, at times, starch. The occurrence of glycosuria under such circumstances may be due either to a "lowering of the renal threshold" for glucose or to the absorption of glucose from the intestine at a rate too rapid to allow of its adequate removal from the circulation by the liver (e.g., in hyperthyroidism). As in the case of the glycosuria of pregnancy, the possibility of decreased glucose tolerance must be investigated, and such conditions as latent diabetes, thyrotoxicosis, and hepatic disease carefully excluded.

Glucose-Galactose Intolerance. Glycosuria occurs in this condition, in which it is associated with intestinal malabsorption of glucose and galactose. This begins in the first few days of life and is associated with severe diarrhea.

Glycosuria in Glomerulonephritis and Nephrosis. Glycosuria occurs at times in a considerable proportion of patients with glomerulonephritis, nephrosclerosis, and nephrosis. In some with glomerulonephritis and nephrosclerosis, glycosuria is associated with fasting hyperglycemia or diminished glucose tolerance or both (pp. 53 and 65). However, in many such cases and in individuals with nephrosis, the glycosuria appears to be dependent upon failure of the renal tubules to reabsorb glucose completely from the glomerular filtrate as a result of the degenerative changes in the renal tubular epithelium. In such cases the urine may contain more than 1 per cent of glucose after the ingestion of carbohydrate-rich meal (p. 83). The probable reason for the inconsistent occurrence of glycosuria in glomerulonephritis is the fact that the glucose load delivered to the tubules, i.e., the amount filtering through the glomeruli per minute, is decreased as a result of glomerular damage and may, therefore, remain within the reabsorptive capacity of the damaged tubular epithelium.

Hyperglycemic Glycosuria. The occurrence of glycosuria in association with hyperglycemia is readily understood. If one accepts the normal "renal threshold" value as being about 160 mg. of glucose per 100 ml. of blood (venous), the elimination of glucose in the urine may be expected in the presence of higher blood sugar levels. The fact must be kept in mind, however, that the renal threshold may exhibit rather wide variations in different individuals and under different conditions. Obviously, the causes of hyperglycemia are potential causes of glycosuria. These include the following:

Diabetes Mellitus. Diabetes mellitus is the most frequently observed individual cause of glycosuria dependent upon hyperglycemia. The view that glycosuria always indicates diabetes mellitus is, however, erroneous, even in the presence of hyperglycemia. Studies of large series of patients with glycosuria have revealed diabetes to be the cause in about 40 per cent of cases, the incidence falling to about 30 per cent in glycosuric subjects under 20 years of age. As has been mentioned previously, glycosuria is not uncommon in the later months of pregnancy. Its occurrence in the early months should lead to a rigid investigation of the possibility of the existence of diabetes mellitus.

Glucose is excreted in the urine when the blood sugar level rises above the "renal threshold" for glucose which, under normal conditions, ranges from 160–180 mg./100 ml. Since in mild diabetes the fasting blood sugar level may be below this point, glycosuria may occur intermittently, chiefly following ingestion of carbohydrate-rich meals. It is therefore obvious that single specimens of urine may not contain glucose, and it is important that the examination be conducted on twenty-four-hour specimens or on all voided specimens. In more severe cases, the blood sugar being maintained at a relatively high level, glycosuria of varying degree occurs continually. In mild cases the urine (twenty-four-hour sample) may contain from a trace to 1 per cent glucose; moderately severe cases, 1 to 3 per cent; severe cases, 3 to 8 per cent. Concentrations above 8 per cent are rarely observed. The kidney responds to the necessity for elimination of larger quantities of glucose by the simultaneous excretion of increased quantities of water. Polyuria is almost invariably observed as the concentration of glucose in the urine rises above 4 per cent. Under such circumstances, if water is supplied in abundance, extremely large volumes of urine may be passed daily, occasionally as much as 2 or 3 gallons. Elimination of such large quantities of water frequently results in extreme grades of dehydration. One characteristic feature of the polyuria of diabetes mellitus is the high specific gravity of the urine, frequently 1.030 or more, which is due to its high sugar content. This feature distinguishes it from polyuria resulting from other conditions, in which the specific gravity of the urine is usually in inverse relation to the urine volume.

In some cases glycosuria may be absent although the blood sugar concentration is above the normal "renal threshold" level (p. 83) and, at times, may be extremely high. The occurrence of hyperglycemia without glycosuria is ascribed to an increase in the "renal threshold" for sugar and is observed most commonly in association with arteriosclerosis or glomerulosclerosis in elderly individuals with diabetes of long standing. Sudden, marked variations in the "renal threshold" for sugar may occur in such individuals.

Hyperthyroidism. Glycosuria occurs in 25 to 35 per cent of patients with hyperthyroidism. Its incidence is somewhat higher in cases of primary (exophthalmic goiter) than of secondary hyperthyroidism (toxic adenoma). Associated with manifestations of hyperthyroidism, it is usually readily distinguishable from other causes of glycosuria and hyperglycemia. It must be recognized that diabetes and hyperthyroidism may coexist and that hyperthyroidism is likely to act as a predisposing cause of diabetes. Glycosuria in this condition may be due in part to an increased rate of absorption of glucose from the intestine (alimentary glycosuria) as well as to lowered tolerance.

Hyperpituitarism. Glycosuria occurs in 25 to 40 per cent of patients with acromegaly at some time during the inception of the disease.

Adrenal Hyperfunction. Glycosuria may occur during periods of paroxysmal hypertension and hyperglycemia in patients with adrenal medullary tumor (pheochromocytoma), due to increased epinephrine secretion. It is also a feature of the clinical picture of Cushing's syndrome due to adrenal cortical hyperplasia or tumor.

Emotional Glycosuria. Glycosuria may occur during periods of excessive nervous strain and emotional excitement. This condition has been termed "emotional" or "psychic" glycosuria and has been attributed to increased secretion of epinephrine (p. 50). It has been found in normal subjects exposed to the emotional excitement and worry of athletic contest, anticipated operative procedure, and college and insurance examinations. Its incidence in college students under the stress of final examinations has been reported as 1.6 per cent.

Glycosuria of Pregnancy. Glycosuria may occur in 15 to 25 per cent of all apparently normal pregnant women, particularly in the later months, and more frequently in primigravidae than multigravidae. Many have regarded this as a form of renal glycosuria, since the fasting blood sugar is usually not elevated. However, more extensive use of glucose tolerance tests suggests that decreased glucose tolerance, perhaps physiological, may form the basis for the glycosuria of normal pregnancy (p. 64). Care must always be exercised to exclude the possibility of latent diabetes mellitus. It must also be borne in mind that lactose, which gives a positive reduction test, may be present in the urine late in pregnancy, although usually in small amounts, and should not be mistaken for glucose (p. 88).

Miscellaneous. Glycosuria may occur in hemochromatosis, advanced pancreatitis, and severe hepatocellular damage, in the latter usually with a normal fasting blood sugar concentration and only after a high carbohydrate intake. It is seen at times with head injuries or other causes of increased intracranial pressure (brain tumor, abscess, hemorrhage), hypothalamic lesions, ether or chloroform anesthesia, narcosis by morphine or barbiturates, asphyxia, acidosis, acute and chronic infections, coronary artery occlusion, advanced malignancy, simple obesity and essential hypertension. It has also been observed after administration of caffeine, Diuretin, strychnine, bichloride of mercury, and chromates. In these conditions, the persistence and severity of glycosuria are related usually to the persistence and severity of the accompanying hyperglycemia (pp. 51–54), except in the presence of renal tubular damage. Consequently, it is usually transitory and of mild degree (less than 1 per cent) in conditions other than diabetes mellitus, hemochromatosis, or Cushing's syndrome.

Fructosuria

Fructose may appear in the urine under the following circumstances:

(*a*) Alimentary fructosuria, following the ingestion of large quantities of fructose, particularly in patients with hepatic insufficiency. This has been utilized as a test of hepatic function but is unsatisfactory; approximately 10 per cent of normal individuals eliminate fructose in the urine following the ingestion of 100 g. of fructose.

(*b*) Essential fructosuria is a rare congenital disorder characterized by inability to utilize fructose completely. A few cases have been reported in which there was a total absence of tolerance for fructose, the sugar being eliminated if any whatsoever was ingested. It is possible that some fructose is of endogenous origin. Insulin has no influence upon this condition; no rise in the respiratory quotient or blood lactic acid follows administration of fructose to such individuals, indicating failure of utilization of that sugar.

At least two varieties of this condition have been described. In one, the metabolism of other carbohydrates is undisturbed, there are no clinical symptoms, and the condition is innocuous, its only importance being that the presence of a reducing sugar in the urine may lead to the erroneous diagnosis of diabetes mellitus. The enzyme defect in the liver is a deficiency of fructokinase, which is concerned with the conversion of fructose into fructose 1-phosphate.

In another form, the rise in blood fructose that follows ingestion of this sugar is accompanied by a sharp drop in the blood glucose concentration, with severe symptoms of hypoglycemia. This is believed to be due to deficiency in phosphofructoaldolase in the liver, resulting in accumulation of fructose-1-phosphate, which blocks important pathways of utilization of fructose.

Pentosuria

Pentose may appear in the urine under the following circumstances:

(*a*) Alimentary pentosuria, occurring in normal individuals after the ingestion of large quantities of fruits which have a high pentose content (prunes, cherries, grapes, plums). It is of no clinical significance apart from the fact that it may be mistaken for glycosuria because of a positive copper-reduction test. The urinary pentose in this condition is L-arabinose.

(*b*) Essential pentosuria (chronic pentosuria).

This is one of the originally described "inborn errors of metabolism." It is genetically determined, has a familial incidence and occurs practically exclusively in Jewish subjects, predominantly in males. It is asymptomatic and innocuous, the utilization of other carbohydrates being unimpaired. Its chief importance lies in the possibility of mistaking it for glycosuria.

The urine sugar in this condition is L-xyloketose (xylulose), the quantity excreted bearing no relation to the amount of pentose ingested. Although the exact nature of the defect has not been established, it is believed to be in the transformation of L-xylulose to L-xylitol, which occurs normally in the liver.

Lactosuria

Lactosuria occurs in a considerable proportion of women during the period of lactation. Small to moderate amounts of lactose may be found in the urine in about 80 per cent of all pregnant women, the incidence increasing as pregnancy advances. It appears more frequently in the afternoon and in primigravidae.

Galactosuria; Galactosemia

Galactose may appear in the urine under the following circumstances:

(a) *Alimentary galactosuria*, following ingestion of supertolerance doses of galactose, particularly in patients with hepatic functional impairment (p. 44).

(b) *Galactosemia*. This is a serious hereditary disorder characterized by a genetically determined defect in conversion of galactose to glucose. It exhibits a familial incidence, and impaired galactose tolerance can usually be demonstrated in one or both parents. Infants with this condition exhibit the following characteristics: (a) malnutrition and wasting, with insatiable hunger; (b) hepatomegaly; cirrhosis may develop later; (c) lamellar cataracts; (d) frequently, mental retardation; (f) frequently, aminoaciduria; (g) galactosuria, which is related to the ingestion of milk and milk products. If the condition is recognized early, all signs and symptoms subside when milk is removed from the diet.

The fundamental defect is decrease or absence of galactose-1-phosphate uridyltransferase, the enzyme which catalyzes the transformation of galactose-1-phosphate to glucose-1-phosphate. This results in accumulation of galactose and galactose-1-phosphate in the tissues and body fluids, resulting in the characteristic morphologic changes. Increase in galactose may depress the blood glucose concentration. Renal tubular injury is reflected in the rather constant occurrence of proteinuria and aminoaciduria. Hyperchloremic acidosis and hypopotassemia may occur as the condition progresses, as well as hyperbilirubinemia and other manifestations of liver functional impairment. If the condition is recognized early and treated promptly, small amounts of milk may be tolerated as the child grows older, probably because of the development of accessory pathways of conversion of galactose to glucose in the liver.

A somewhat milder form of galactosemia results from a cellular deficiency of galactokinase.

Maltosuria

This is a rare condition, of no known clinical significance.

References

Ashmore, J., Cahill, G. F., Jr., and Hastings, A. B.: Effect of hormones on alternate pathways of glucose utilization in isolated tissues. *In* Recent Progress in Hormone Research, New York, Academic Press, Inc., *16*:547, 1960.

Behrens, O. K., and Bromer, W. W.: Glucagon, Vitamins and Hormones, *16*:263, 1958.

Bondy, P. K., ed.: Duncan's Diseases of Metabolism. Ed. 6. Philadelphia, W. B. Saunders Co., 1969.

de Bodo, R. C.: Insulin hypersensitivity and physiological insulin antagonists. Physiol. Rev., *38*:389, 1958.

de Bodo, R. C., and Altszuler, N.: The metabolic effects of growth hormone and their physiological significance. Vitamins and Hormones, *15*:206, 1957.

de Bodo, R. C., and Sincoff, M. W.: Anterior pituitary and adrenal hormones in the regulation of carbohydrate metabolism. *In* Recent Progress in Hormone Research. New York, Academic Press, Inc., *8*:511, 1953.

Cantarow, A., and Schepartz, B.: Textbook of Biochemistry. Ed. 4. Philadelphia, W. B. Saunders Co., 1967.

Dickens, F., Randle, P. J., and Whelan, W. J., eds.: Carbohydrate Metabolism and Its Disorders. London and New York, Academic Press, 1968.

Editorial: Lancet, *1*:1211, 1970.

Goldner, M. G., ed.: Chlorpropamide and diabetes mellitus. Ann. N. Y. Acad. Sci., *74:* 407, 1959.

Guest, G. M.: Diabetic coma; metabolic derangements and principles for corrective therapy. Am. J. Med., *7*:630, 1949.

Hsia, D. Y. Y.: Inborn Errors of Metabolism. Chicago, Year Book Publishers, Inc., 1959.

Lukens, F. D. W.: The pancreas: insulin and glucagon. Ann. Rev. Physiol., *21*:445, 1959.

McGilvery, R. W.: Biochemistry: a functional approach. Philadelphia, W. B. Saunders Co., 1970.

Mirsky, I. A.: Insulinase, insulinase inhibitors, and diabetes mellitus. *In* Recent Progress in Hormone Research. New York, Academic Press, Inc., *13*:429, 1957.

Paschkis, K. E., Rakoff, A., and Cantarow, A.: Clinical Endocrinology. Ed. 2. New York, Paul B. Hoeber, Inc., 1958.

Smith, M.: The diagnosis of diabetes mellitus. Med. Clin. of North America, *43*:579, 1959.

Sprague, R. G., et al.: "Steroid diabetes" and alkalosis associated with Cushing's syndrome. J. Clin. Endocrinol., *10*:289, 1950.

Symposium on Biochemical Lesions of Carbohydrate Metabolism. Federation Proc., *19:* 971, 1960.

Symposium on Insulin Glucagon, and the Oral Hypoglycemic Sulfonylureas. Diabetes, *6*:221, 1957.

Thorn, G. W., et al.: The adrenal and diabetes. Diabetes, *8*:337, 1959.

White, A., Handler, P., and Smith, E. L.: Principles of Biochemistry. Ed. 5. New York, McGraw-Hill Book Co., 1973.

Williams, R. H., ed.: Diabetes. New York, Paul B. Hoeber, Inc., 1960.

Yalow, R. S., and Berson, S. A.: Immunoassay of endogenous plasma insulin in man. J. Clin. Invest., *39*:1157, 1960.

Chapter 2

LIPID METABOLISM

Lipids are defined as substances consisting of the higher fatty acids, their compounds and the substances found in natural chemical association with them. One important role of the more simple lipids (fatty acids and triglycerides) is that of fuel. In some respects, this type of lipid is even superior to carbohydrate as a raw material for combustion, since it yields more heat per gram when burned and furthermore can be stored by the body in quite considerable amount in an almost anhydrous state. Some deposits of lipid may exert an insulating effect in the body, while others may provide padding to protect the internal organs. The nervous system is particularly rich in lipids, especially certain types which do not normally occur in such high concentrations elsewhere in the organism. Unfortunately, we know little about the functions of these particular substances in the nervous system or other tissues. Some compounds derived from the lipids are important building blocks of biologically active materials; e.g., the acetyl group (a normal breakdown product of the fatty acids, as will be seen later) can be used by the body to synthesize the rather complex compound, cholesterol, which, in turn, can give rise to certain hormones. One of the important functions of dietary lipid is that of supplying so-called "essential fatty acids." These are compounds which cannot be synthesized by the animal organism. Compounds of lipids with proteins, lipoproteins, are important constituents of many natural membranes and are involved in a number of other phenomena of possible clinical significance.

The lipids constitute a heterogeneous group of chemical types. First are the fatty acid esters of glycerol (triglycerides, neutral fats) and of other alcohols (waxes). Then, there are the phosphatidyl compounds which contain fatty acids (as esters), phosphoric acid and a base (ethanolamine, choline, serine) or inositol in place of the base. The phosphatidyl compounds (plasmalogens) are very similar, but one of the hydroxyl groups of the glycerol forms an α,β-unsaturated ether linkage with a long aliphatic chain. The sphingolipids contain the unusual basic alcohol sphingosine, $CH_3(CH_2)_{12}CH = CH.CHOH.CHNH_2.CH_2OH$(4-sphingenine), in amide linkage with a long chain fatty acid. This combination is known as a ceramide. There are a number of compounds closely related to sphingosine and these too, can be found in the ceramides. The different classes of sphingolipid are formed by combination of ceramide with different components, in ester linkage. Thus, the sphingomyelins are made up of ceramide and phosphorylcholine, the neutral glycosphingolipids of ceramide and a mono- or oligosaccharide, the sulfatides of ceramide and galactose-3-sulfate, the gangliosides of ceramide and an oligosaccharide containing sialic acid.

The neutral glycosphingolipids are derived from glucosylceramide (glucocerebroside) or from galactosylceramide (galactocerebroside) by the addition of galactosyl and N-acetylgalactosaminyl residues. They are subdivided into Types A, B, C and D. Only Type B is derived from galactosyl ceramide; the others are derived from glucosyl ceramide. Type D also contains glucosaminyl residues. The classification depends on the order in which the residues are arranged. Type A is that normally present in greatest amount in blood and tissues, with the exception of the nervous system, in which the major neutral glycosphingolipid is galactosyl ceramide. This is also present in kidney tissue. Type C is identical with the gangliosides after their sialic acid has been removed. This type is present in tissues only in small amounts under normal conditions but can accumulate in glycosphingolipidosis (Tay-Sachs disease). Type D belongs to some of the blood group active substances.

The gangliosides have been classified by Svennerholm into types G_{M3} (N-acetylneuraminidogalactosyl-glucosyl ceramide, or expressed symbolically NANAGal-Glc-ceramide), G_{M2} (GalNac-NANAGal-Glc-ceramide; GalNac represents N-acetylgalactosamine), G_{M1} (Gal-GalNac-NANAGal-Glc-ceramide), G_{D1a}, G_{D1b} (both are G_{M1} with an additional NANA) and G_{T1} (G_{M1} with two additional NANA residues). The capital letter in subscript indicates the number of sialic acid residues in the molecule (M = 1; D = 2; T = 3). Tissue catabolism produces the breakdown sequence $G_{M1} \rightarrow G_{M2} \rightarrow G_{M3} \rightarrow$ galactosyl-glucosyl-ceramide (lactosyl-ceramide) $\rightarrow$ glucosyl ceramide $\rightarrow$ ceramide. Each of the monosialo-gangliosides can lose its sialic acid to yield a glycosphingolipid. Biosynthesis of gangliosides occurs in the reverse order to the catabolic processes. The sugar residues are derived from their uridine diphosphate (UDP) derivatives and sialic acid is added from its cytidine monophosphate (CMP) derivative.

Finally, substances must be mentioned that are included as lipids, partly by metabolic connections, largely by common solubilities, e.g., the sterols and bile acids, other steroids, and the fat-soluble vitamins and provitamins.

In considering the metabolism of the individual constituents of the lipids, it is obvious that most of them are only remotely connected. The carbohydrates of sphingolipids and the glycerol moiety of the fats and phospholipids are synthesized and degraded via pathways common to the general metabolism of carbohydrates. Serine, ethanolamine, and choline are more directly related to amino acids than to lipids in their anabolism and catabolism. In fact, the only constituents of lipids that have a completely "lipoid" metabolism are the fatty acids, most of the other components apart from sphingosine entering into this area of metabolism only at such points as they are incorporated into the lipid molecule in the process of its formation.

As a consequence of the diversity of compounds encompassed by the class of lipids, the treatment of the metabolism of these substances necessarily differs from that applicable to the carbohydrates and proteins. The individual carbohydrates are connected, more or less directly, with the glycolytic sequence of reactions, and a common thread of metabolism runs through the reactions of the proteins and even of the individual amino acids, with some exceptions. On the other hand, the metabolism of the individual types of lipids and their constituents must be considered independently, for practically no metabolic interrelations are known among these substances of a type characteristic of the other major foodstuffs.

Another unusual feature of lipid metabolism is the importance of certain phenomena of transport and deposition. Although mobilization of carbohydrate and protein from one site in the body to another certainly occurs, as does deposition of stores of surplus (at least in the case of carbohydrate), these phenomena do not occupy the prominent position or display such aberrations under abnormal conditions as in the case of lipids.

Most of the lipid material occurring in nature differs from the carbohydrates and the proteins in being insoluble in water and soluble in certain organic solvents (ether, benzene, chloroform, etc.). The fat depots of animals contain mainly neutral fat, by which is meant esters of glycerol with three fatty acid molecules. In contrast, most cells other than adipose tissue contain much less fat; their lipids consist largely of phospholipids and cholesterol. The brain is particularly rich in cholesterol (largely in the free state), phospholipids, and sphingolipids. The lipoproteins have been increasingly investigated as a group, since their importance is given by researches which relate the occurrence of certain lipoproteins in the plasma to atherosclerosis.

Attention will be directed mainly to those classes of lipids that are involved in abnormalities of clinical significance.

Fats (Triglycerides). The storage lipid of animals and the chief lipid constituent of the diet is neutral fat (triglycerides). These compounds consist of three molecules of fatty acid esterified to a molecule of glycerol. They differ from one another in the nature of their constituent fatty acid, i.e., variety, chain length, saturation or unsaturation. Hydrolysis of the glycerides proceeds by the addition of the elements of water across the ester linkages, producing, successively, diglycerides (and one fatty acid), monoglycerides (and two fatty acids), and glycerol (and three fatty acids).

Glycerol is closely allied, chemically and metabolically, to carbohydrate, with which it is metabolically interconvertible. The fatty acids differ in chain length (molecular size) and with respect to the presence of double bonds between certain carbon atoms (unsaturation). The polyunsaturated fatty acids, viz., linoleic (two double bonds), linolenic (three double bonds), and arachidonic (four double bonds), are termed "essential" fatty acids, since there is evidence that they cannot be synthesized in the body, are essential for normal nutrition and must, therefore, be provided in the diet.

Neutral fat is present normally in the liver ($<$ 1 to 2 per cent) and, especially after a lipid-containing meal, in the blood plasma (p. 117). In extrahepatic tissues, it is confined largely to the fat depots (adipose tissue). Relatively large quantities of unsaturated fatty acids are found in the liver, as compared with other tissues.

The body uses fats mainly as fuel. On a weight basis, the calorific value of fat (9.3 Cal. per g.) is more than twice that of carbohydrate or protein (4.1 Cal. per g.), it can be stored in practically anhydrous condition and in almost unlimited amounts, and it is metabolically sufficiently labile to be readily mobilized from the fat deposits when needed. Nevertheless, carbohydrate, and not fat, is the preferred fuel of the body, and any attempt to utilize excessive quantities of fat can lead to serious consequences, e.g., ketosis, hypercholesterolemia, hyperlipemia, fatty liver (p. 126).

Phosphatidyl Derivatives (Phosphoglycerides). These are so designated because they contain phosphatidic acid (di-fatty-acyl glycerol phosphoric acid). The most important members of the group are *choline phosphoglycerides* (phosphatidyl choline; previously called lecithin), *ethanolamine phosphoglycerides* (phosphatidyl ethanolamine; previously called cephalin), *serine phosphoglycerides* (phosphatidyl serine) *inositol phosphoglycerides* (phosphatidyl inositol) and *cardiolipins* (bisphosphatidyl glycerol). The latter are constituents of mitochondrial membranes.

The phosphatidyl compounds (plasmalogens) are especially found in the membranes of nerve and muscle cells.

Phosphoglycerides are the dominant type of lipid in tissues other than adipose. They exist mainly in membranes.

Although the phospholipids can scarcely be considered metabolically inert, this fact in itself does not prevent these compounds from serving a structural function. Indeed, phospholipids participate in the lipoprotein complexes which constitute part of the matrix of cell membranes and of such structures as mitochondria and microsomes. In this role they impart certain physical characteristics to these structures that are of functional significance, viz., high permeability toward certain hydrophobic molecules, and lysis by surface-active agents such as bile salts and detergents.

Certain enzymes appear to require tightly bound phospholipid for their action. Whether or not the phospholipid deserves the status of a coenzyme, the fact remains that its removal from the enzyme protein irreversibly inactivates the latter.

Substances having the properties of phospholipids (ethanolamine phosphoglycerides) are believed to function as parts of both "thromboplastin" and "antithromboplastin" in the mechanism of blood coagulation. The specific compounds involved have not been identified.

That phospholipids play some role in the metabolism of fatty acids, at least in the liver, seems probable from observations that the turnover of phospholipids is correlated with the intensity of fat metabolism (fat feeding and diabetes).

Prostaglandins. These are a group of naturally occurring substances having in common a structure based on prostanoic acid, which contains 20 carbon atoms. They are separated into four groups, **A, B, E** and **F**, in accordance with differences in structure of the 5-carbon ring. **A, B** and **E** have an oxo-grouping at position 9, whereas **F** has a hydroxyl- group in this position. **A** has a double bond between positions 10 and 11, whereas **B** has a double bond between positions 8 and 12. **E** and **F** do not have a double bond in the ring but possess a hydroxyl- group at position 11.

Prostanoic acid

All active prostaglandins have at least one double bond between positions 13 and 14. Some have two double bonds, the second being between positions 5 and 6, and finally some prostaglandins have three double bonds, the additional bond being between positions 17 and 18. The common single double bond has the *trans-* configuration, whereas the other double bonds have the *cis-* configuration. All prostaglandins have a hydroxyl- group at position 15 and some have another hydroxyl- group at position 19. When the 5-carbon ring has two hydroxyl groups, their positions in space give rise to the possibility of α or β isomers.

Fourteen prostaglandins (P) have been isolated. These are PGA_1, PGA_2, PGB_1, PGB_2, PGE_1, PGE_2, PGE_3, $PGF_{1\alpha}$, $PGF_{2\alpha}$, $PGF_{3\alpha}$, 19-OH PGA_1, 19-OH PGA_2, 19-OH PGB_1 and 19-OH PGB_2. (The number in subscript indicates the number of double bonds and the Greek letter the isomeric form.)

The prostaglandins were first found in seminal fluid and vesicular glands. They are now known to be of very wide distribution in mammalian tissues, e.g., lung, kidney, thyroid, spleen, endometrium, gastrointestinal mucosa, amniotic and menstrual fluids. Biosynthesis occurs from the "essential" (unsaturated) fatty acids containing 20 carbon atoms. It is believed that PGE_1, PGE_2, PGE_3, $PGF_{1\alpha}$, $PGF_{2\alpha}$, and $PGF_{3\alpha}$ are primarily prostaglandins and that the others are derived from them.

The prostaglandins have been shown to have marked biological activity. The effect depends on the particular prostaglandin which is used. Because of the stimulatory effect of prostaglandins on contractions of the human uterus, PGE_1, PGE_2 and $PGF_{2\alpha}$, given intravenously, have been used to induce labor. PGE_2 and $PGF_{2\alpha}$ have been shown to be effective orally. These substances have also been administered intravenously or intravaginally, as well as into the amniotic sac, to induce therapeutic abortion. They have possibilities as contraceptives. PGA_1 infused intravenously is a natriuretic and vasodilator and lowers the blood pressure. PGE_1, by inhalation, has produced some improvement in asthmatic patients and has been shown to be an inhibitor of gastric acidity when administered intravenously. PGA_1, PGE_1, PGE_2, and $PGE_{1\alpha}$ have a vasoconstrictor action on the blood vessels of the nasal mucosa, and PGE_1, applied by atomizer, proved to be an effective nasal decongestant.

Steroids. This designation is applied to a large group of compounds having in common a structure based on the perhydrocyclopentano phenanthrene nucleus. They have greatly diversified physiological properties and include such substances as: (a) provitamin D; (b) cholesterol; (c) bile acids; (d) important hormones, viz., adrenocortical, androgens, estrogens, progesterone; and (e) certain potent carcinogens. Certain of the steroids occur in nature in association with lipids, either by reason of mutual solubility, or owing to some actual metabolic relationship.

Perhydrocyclopentanophenanthrene   *Cholesterol*

Cholesterol is a steroid alcohol and, in tissues (including blood plasma) outside of the nervous system, exists partly in the form of esters with fatty acids, although, with the exception of the liver, plasma, and adrenal cortex, it is present largely in the free state. The amount of cholesterol in muscles is apparently related to the degree of automatic activity; e.g., in man, smooth muscle invariably contains more than skeletal muscle, heart muscle occupying an intermediate position. Inasmuch as their phospholipid contents are approximately the same, the phospholipid : cholesterol ratio is lower in smooth (5 : 1) than in skeletal (10–16 : 1) muscle. The significance of these relationships is not known.

It is well established that cholesterol is the parent substance of the cholic acid component of the bile acids, this transformation occurring in the liver. It is also convertible into progesterone, testosterone, and the adrenocortical hormones. Administration of ACTH results in simultaneous increase in adrenocortical hormone formation and decrease in adrenal cholesterol content.

In the biosynthesis of cholesterol, acetyl-CoA and acetoacetyl-CoA interact to form 3-hydroxy-3-methylglutaryl CoA. This gives rise to mevalonic acid and eventually to squalene, which gives lanosterol. This, in turn, loses three methyl bonds, a double bond in the side chain and undergoes a shift of the double bond in the steroid nucleus to give cholesterol.

Lipoproteins. It is probable that many of the substances described in this chapter do not exist as such in their natural state and represent what the purists would call artifacts of the laboratory. This is a reflection of the primitive state of intracellular biochemistry (and unfortunately the situation is not confined to the lipids). From the available evidence, it appears that many of the lipids are combined with proteins in the tissues. These combinations are called "lipoproteins." The nature of the linkage between the lipid and protein moieties is not known; it is usually stable toward nonpolar solvents, such as ether, but can frequently be disrupted by more highly polar substances, such as alcohol. Despite the evidence for the widespread occurrence of lipoproteins, the technical difficulties in handling these materials have been such that, until the very recent introduction of newer physiochemical methods, few members of this group were characterized as chemical entities.

It is widely assumed that the cell membrane is a lipoprotein structure of some type and, although there is not much direct evidence available, the general properties of cell membranes are at least consistent with this view. Certain of the internal structures of the cell also contain lipids, apparently in combination with proteins. This is the case with visible entities, such as the nucleus and mitochondria, as well as the submicroscopic particles (microsomes).

The thromboplastic protein isolated from lung (p. 94) has been characterized as a lipoprotein. The lipid portion, as in the case of many lipoproteins, consists of more than one type of lipid. Egg yolk contains a lipoprotein known as lipovitellin, the lipid fraction consisting largely of phospholipids. A number of bacteria have been shown to contain firmly bound lipids, although it is not certain in some cases whether the lipids are bound to protein or to polysaccharides. Certain animal viruses appear to contain firmly bound lipids also, but this is not true of the viruses thus far isolated from plant sources. The film which stabilizes the fat droplets in milk appears to be a combination of protein and phospholipid. Most, if not all, of the lipid in blood serum is carried in the form of lipoprotein complexes. The statement also holds for the lipid-soluble vitamins, the carotenoids, and the steroids. Since the presence of lipid causes the lipoprotein molecule to have a lower density than that of ordinary proteins, it is possible, by proper adjustment of the experimental conditions, to force lipoproteins to travel toward the axis instead of toward the periphery of an ultracentrifuge.

The flotation (negative sedimentation) determinations are made in a medium of known density (sp.gr. 1.063). The rate of upward migration of a molecule under these conditions is expressed as "Svedberg flotation units" (S_f). Molecules exhibiting a flotation rate of 10 S_f units are referred to as S_f 10 molecules. Blood plasma has been shown to contain a wide range of S_f classes of molecules in various physiological and pathological states, the size of the lipoprotein molecule generally increasing with increase in the S_f number. The higher numbered classes contain particles that fall into the category known as "chylomicrons" (p. 115).

DIGESTION AND ABSORPTION

Triglycerides. During the process of digestion, ingested triglycerides first are emulsified, and then undergo enzymatic lipolysis. The resulting mixture is converted into aggregates known as micelles, which are "solubilized" by a kind of detergent action. Emulsion formation is initiated by the process of chewing and then by the contractions of the stomach. The gastric contents contain some free fatty acids, possibly due to the action of a gastric lipase. Further emulsification occurs in the duodenum and small intestine. This is brought about by peristalsis, which is greatly aided in this respect by the presence of bile salts, monoglycerides (lipolysis), phospholipid and the lysolecithin arising from it by the action of the enzyme phospholipase A in pancreatic juice. All these are emulsifying agents.

Pancreatic lipase hydrolyzes the 1- and 3-positions of the triglycerides, leaving, as a result of its action, a mixture of 2-monoglycerides and 1,2- and 2,3-diglycerides as well as the soaps of the free fatty acids. Because of the presence of bile salts, the lipase has a pH optimum that has been lowered from between 8.2 to 9.2 down to 6 to 8, which corresponds with that of the contents of the small intestine. About 75 per cent of the triglyceride fatty acid is liberated by lipolysis. There is now an aggregation of the various fatty substances into so-called micelles, which are polymolecular aggregates of size 5 nm. These are then "solubilized" by the detergent action of bile salts, which become part of each micelle. There is evidence that the micelles do not contain the residual triglycerides and diglycerides, which are not "solubilized" but remain in the form of an emulsion. Unless they are broken down further and become incorporated into "soluble" micelles, there is also evidence that the tri- and diglycerides of the emulsion are not absorbed into the mucosa of the small intestine. Passive diffusion seems to be the mechanism by which the micelles enter the brush border of the absorbing cells of the villi of the small intestine. There is no real evidence that pinocytosis occurs in man, and hence little or no triglyceride is absorbed in emulsion form. Having crossed the inner boundary of the brush border, the fatty acids and monoglycerides are re-esterified to triglycerides, by an unknown mechanism. They are then covered with a coating of protein, phospholipids and cholesterol to form tiny droplets known as chylomicrons, which are transported from the intestine via the lymphatic system to the blood stream. The rate of their removal from the blood is slower than their rate of entry postprandially. There is therefore an accumulation of chylomicrons (physiological hyperlipemia) with a peak between 2 and 4 hours after meals containing fat. They are all removed, however, at about 6 to 10 hours after the meal. In an older age group, removal took up to 24 hours. The chylomicrons are taken up by the liver, by adipose tissue and by the reticuloendothelial system—possibly by muscle. This requires the enzyme lipoprotein lipase for hydrolysis of the contained glycerides. It is a matter of some interest that chylomicrons are not necessarily obtained solely from the absorption of dietary fat, since they can be formed on a fat-free but high carbohydrate diet. The pre-beta lipoproteins (p. 116) act as carriers of dietary fat when the fat intake is low (less than 50 g./day).

Some free glycerol is liberated in the small intestine. This is soluble and absorbed by a process different from that of fat. It can be used as an energy source by the intestinal cell, which can also synthesize glycerol for combination with fatty acids via α-glycerol phosphate.

The shorter chain fatty acids (less than 12 carbon atoms) can apparently be absorbed directly into the blood stream as complexes with albumin.

Sterols. A high degree of specificity is exhibited in the intestinal absorption of steroids. Cholesterol is normally absorbed readily, but certain of its isomers and reduction products are not (e.g., coprosterol, cholestanol p. 114). Although most phytosterols (plant sterols) are poorly absorbed, if at all (e.g., ergosterol and sitosterol), certain closely related substances, such as vitamin D and digitalis glycosides, are absorbed quite readily.

Cholesterol enters the intestine in animal foodstuffs and in the intestinal secretions and bile. Cholesterol esters in the digestive mixture are probably all hydrolyzed by cholesterol esterase (pancreatic juice), this reaction being enhanced considerably by bile acids, perhaps in part at least through their action in forming water-soluble complexes with cholesterol. Absorption is poor in the absence of bile, but may be improved by other dispersing agents, including mono- and diglycerides. In the mucosal cell, the cholesterol is incorporated into the chylomicrons and about half of it is esterified, mainly with oleic acid.

Cholesterol is absorbed into the lymph, its concentration in the fasting state remaining quite constant at about one-third the concentration in the blood plasma, with approximately 70 per cent in ester form. This is derived in part from reabsorption of endogenous cholesterol from the lumen, but also from cholesterol synthesized in the intestinal mucosa. In the absence of bile, the fasting lymph cholesterol decreases, as does the percentage in ester form. This indicates that bile acids are important not only in hydrolysis of cholesterol esters in the lumen but also in the esterification of free cholesterol by cholesterol esterase. There is evidence, too (in liver), that bile acids influence the rate of catabolism of cholesterol.

Following administration of cholesterol, there is an increase in the lymph not only of cholesterol, esterified and free, but also of triglyceride, phospholipid and protein, i.e., of lipoproteins and chylomicrons. Conversely, during absorption of fat, there is an increase also in lymph cholesterol. This indicates that intestinal synthesis and absorption of cholesterol are intimately related to absorption and transport of other lipids. As in the case of fatty acids, the nature of the intracellular mechanisms involved in the transport of cholesterol across the intestine is not clear. Protein complexes may be formed and exchanges may occur between various lipoproteins and the pool of free cholesterol in the mucosa. Dihydrocholesterol and sitosterol, which are absorbed to a relatively limited extent, interfere with absorption of cholesterol; this may be due to saturation of specific lipoproteins with the foreign sterols, thereby reducing their capacity for transporting cholesterol.

Substitution in the diet of fats rich in unsaturated fatty acids (linoleic) for those rich in saturated acids (stearic, palmitic) frequently results in lowering of the plasma cholesterol concentration. This effect is probably due to an influence on synthesis of plasma cholesterol rather than on intestinal absorption (p. 120). It is of interest in this connection that cholesterol esters contain a particularly high concentration of polyunsaturated fatty acids.

Phosphoglycerides. Phosphatidyl derivatives, entering the intestine in the food or in the digestive secretions (bile, intestinal juice), may be absorbed as such and may also be hydrolyzed by lecithinases (pancreatic juice, p. 589). The end-products of digestion by these enzymes are free fatty acids, removed from the 2-position and the remainder of the molecules, known as lysolecithins and lysocephalins. The hydrophilic properties of phosphoglycerides are important in several aspects of the digestion, absorption and transport of triglycerides and cholesterol. They, too, appear in the lymph and blood as components of lipoprotein complexes. Their relation to the absorption of other lipids is indicated in the discussions of intestinal absorption of triglycerides and cholesterol.

Fat-soluble Vitamins. Both preformed vitamin A and carotene may be absorbed from the intestine in the absence of bile acids, but the amounts absorbed are influenced by factors which affect lipid absorption, as would be anticipated in view of the lipid solubility of these vitamins. Their absorption is diminished in various types of steatorrhea and is favored by conditions which promote emulsification in the intestinal lumen (p. 98).

Bile salts are required for optimal absorption of vitamin D, which is also promoted by fat absorption. Vitamin D deficiency due to inadequate absorption may occur in various forms of steatorrhea, contributing to the defective skeletal mineralization in these conditions, particularly in children (p. 302).

Naturally occurring forms of vitamin K, being fat-soluble, require bile salts for optimal absorption. Vitamin K deficiency may, therefore, occur in obstruction of the common duct and in external bile fistula, resulting in defective prothrombin formation and a hemorrhagic tendency (p. 629).

Fecal Fat. Under ordinary conditions of dietary fat intake and normal fat digestion and absorption, the lipid content of the feces remains fairly constant, usually less than 5 g. of fatty acids being excreted daily in the adult. In adult males, at about 60 years of age, the figure increases to 6 g. daily. In children, up to the age of 6 years, the amount is less than 2.0 g. This gradually increases until older children have a fecal fat excretion similar to adults. These figures depend on an average fat intake (50-160 g. daily) and are usually determined on a sample of feces collected over 72 hours, with special precautions to prevent fat breakdown.

This is derived in part from the dietary fat but mainly, apparently, from endogenous sources (intestinal mucosal cells, which have been shed) and by bacterial synthesis in the lumen of the intestine (p. 600).

Increase in fecal fat (steatorrhea) may result from defective emulsification (gastrectomy, defects of gastric motility and duodenal bypass), lipolysis, micelle formation, or absorption. These may occur due to absence of bile (obstructive jaundice), absence or defective formation of pancreatic juice (pancreatic duct obstruction, acute or chronic pancreatitis and fibrocystic disease) or loss of extensive areas of absorptive surface (extensive small bowel resection). It is a matter of interest that resection of the lower portion of the small intestine can produce severe steatorrhea. This seems strange, since fat absorption extends from the duodenum to the jejunum. However, bile salts are largely reabsorbed in the ileum and so removal of large parts of it interfere with the enterohepatic circulation of these substances, which under normal conditions are used over and over again in relation to fat emulsification and micelle formation. The operation therefore creates a relative bile salt deficiency in the intestinal lumen, and steatorrhea occurs. When much of the jejunum is removed, however, the ileum can effectively take over the absorption of lipid, but there is often a lower grade of steatorrhea. Diseases of the lower bowel such as regional ileitis, ulceration from various causes, enteritis due to radiation and infiltrations such as amyloidosis and Whipple's disease also cause steatorrhea secondary to malabsorption of bile salts. Cholestyramine, which binds bile salts and so interferes with their absorption, can in appropriate dosage lead to a moderate degree of steatorrhea.

In adult celiac disease (nontropical sprue), which is probably the same disease as the childhood form, the breakdown products of a wheat protein (gluten) are deleterious to the mucosa of the small bowel. This is probably, but not certainly, due to local hypersensitivity. Interference with mucosal function then leads to steatorrhea, which responds to a gluten-free diet.

Interference with chylomicron formation can also result in malabsorption of fat. This occurs in a-beta-lipoproteinemia (congenital) or in hypobeta-lipoproteinemia (congenital or acquired), since beta-lipoprotein is necessary for the formation of the chylomicron envelope.

Tropical sprue appears to be of mixed etiology, resulting from bacterial infection.

Experimentally, absorption of lipids may be decreased by administration of phloridzin, which also results in impaired absorption of carbohydrate (p. 2). This may be related to depression of metabolic activity in the intestinal mucosa.

METABOLISM OF FAT

Immediate Fate of Dietary Lipid. Newly absorbed lipid reaches the plasma mainly by way of the lymphatics (thoracic duct to subclavian vein), manifesting itself largely in the neutral fat fraction of the plasma lipids, although increases occur also in other lipids during the absorptive period. Neutral fat is taken up rapidly by fat cells (adipose tissue) and by the liver, the proportion going to each of these sites being determined by the physiological state of the organism. The comparatively small dietary contributions to other plasma lipids are probably removed chiefly by the liver, since they are taken up by extrahepatic tissues relatively slowly.

Anabolism and Catabolism of Fatty Acids. The fatty acid moieties of the lipids of a given tissue are derived from (1) the diet, (2) transfer from other tissues, or (3) synthesis *in situ* from carbohydrate or protein, chiefly in adipose tissue and the liver. Net formation of fatty acids occurs as a result of an intake of carbohydrate in excess of immediate requirements, but a "steady-state" conversion proceeds continually, the catabolic reactions keeping pace with the anabolic. Limitations to the synthesis of fatty acids are indicated by the existence of certain highly unsaturated fatty acids which are essential dietary factors.

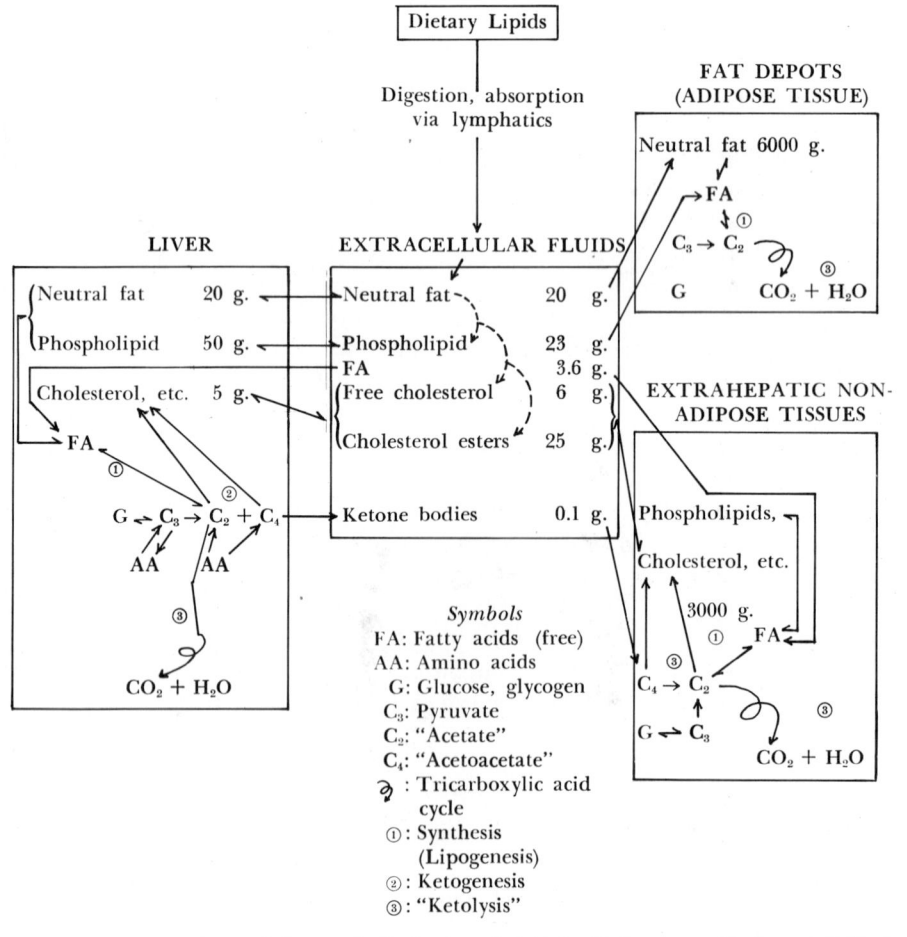

Figure 2-1. Overall metabolism and distribution of lipid in 70 kg. man. Binding of lipids in lipoproteins is not indicated.

Carbohydrate is the major raw material for the synthesis of fatty acids. Pyruvic acid forms "active acetate" (acetyl-CoA), which can take a variety of metabolic paths, one of which is fatty acid formation (p. 18). This is accomplished in a so-called fatty acid synthetase complex, which is made up of a number of enzymes and an acyl carrier protein. These complexes are situated in the cytoplasm and are extramitochondrial. Acetyl-CoA ($CH_3CO.CoA$) is converted to malonyl-CoA ($COOH.CH_2CO.CoA$) by the addition of CO_2 which has been "activated" by an enzyme system containing biotin. One molecule of acetyl-CoA and one molecule of malonyl-CoA transfer their acetyl and malonyl groups, respectively, to the acyl carrier protein under the influence of appropriate enzyme systems. Another enzyme then causes the two radicals to condense together with the loss of CO_2 to form an acetoacetyl ($CH_3CO.CH_2CO-$) group which remains attached to the carrier protein. By a series of enzyme actions the acetoacetyl group is first converted to a β-hydroxybutyryl ($CH_3.CHOH.CH_2CO-$) group by the addition of hydrogen, then to a crotonyl ($CH_3CH=CHCO-$) group by the loss of water, and this is then converted to a butyryl ($CH_3CH_2CH_2CO-$) group by the addition of hydrogens across the double bond. We thus have a 4 carbon chain fatty acid radicle attached to the carrier protein. Another malonyl group from malonyl-CoA is now attached to the carrier protein and then condensed with the butyryl group already there, by the enzymatic process involving the loss of CO_2. Thus, the chain length is now converted to 6 carbons. By the series of enzyme reactions described in relation to the 4-carbon chain, the 6-carbon chain is converted to a fatty acid radicle still attached to the carrier protein. The whole condensation process is repeated over and over again until it has occurred seven times in all. This results in a palmityl (16 carbon) fatty acid radicle attached to the carrier protein. This radicle can be set free as palmitic acid or transferred to CoA to form palmityl-CoA, which can react with α-glycerol phosphate to form a glyceride. The palmityl chain length can be increased by direct condensation with acetyl-CoA. This occurs inside mitochondria. The process by which the palmitic acid gets into the mitochondrion is not well understood. Some believe that carnitine is involved. In microsomes, the palmityl-CoA can be condensed with malonyl-CoA with the loss of CO_2. Here again, the chain length has increased by 2 carbon atoms. Microsomes also contain enzyme systems, which can to a limited extent introduce double bonds and form unsaturated fatty acid derivatives of CoA. Unsaturated fatty acids such as linolenic acid cannot be formed. They are nevertheless vital to the body and so must be taken in with the diet and are referred to as essential fatty acids. The various acyl-CoA radicles of varying chain length and degree of unsaturation can react enzymatically with α-glycerol phosphate to form triglycerides.

It will be noted that fatty acid synthesis requires a ready supply of hydrogen, by the mechanism of electron transfer. This comes from the NADPH formed largely in the pentose phosphate pathway (pentose shunt) described in Chapter 1. NADPH can also be formed by the transfer of an electron from NADH to NADP, i.e., NADH + NADP $\rightleftharpoons$ NAD + NADPH.

It is known that the rate of synthesis of fatty acids is dependent on the availability of malonyl-CoA. The rate-limiting reaction is therefore the addition of "active" CO_2 to acetyl-CoA. It is known that this reaction is inhibited by accumulation of free fatty acids—in other words, excessive triglyceride breakdown from any cause. On the other hand, the reaction is stimulated by citric and isocitric acids. These acids become easily available when glucose is being readily metabolized. This is one of the ways in which insulin stimulates fatty acid synthesis. The hormone also stimulates the pentose phosphate cycle and so makes more NADPH available for fatty acid synthesis. Moreover, in adipose tissue insulin is required for the synthesis of triglycerides. These can, however, be formed in the liver in the absence of insulin. Thus, deficiency of the hormone leads to breakdown of adipose tissue triglycerides, with the liberation of free fatty acids, which are carried to the liver, which then converts them into triglycerides and phospholipids. In other words, insulin deprivation leads to a "fatty liver." There are also metabolic pathways available for the synthesis of fatty acids from amino acids. The glucogenic amino acids (p. 161) are convertible to pyruvic acid; the ketogenic amino acids (leucine, tyrosine, phenylalanine) form acetate or acetoacetate directly, both of which are lipogenic.

Insulin appears to be essential for adequate formation of long-chain fatty acids from acetate, i.e., for lipogenesis from carbohydrate. Under normal conditions, in a well-nourished animal (rat), about 30 per cent of administered glucose takes this metabolic path. In the absence of insulin, this process of lipogenesis from carbohydrate suffers quantitatively more than does the oxidation of glucose to CO_2 and H_2O. This indicates an important direct disturbance of fat metabolism in diabetes mellitus, in addition to the indirect consequences of impaired utilization of carbohydrate.

Fatty acids are catabolized primarily by oxidation, the extent and rate varying with the physiological state of the organism. The breakdown occurs by the removal of consecutive 2-carbon units. The fatty acids must first be activated by combination with coenzyme A; each of these compounds is known as a fatty acyl-CoA ($RCH_2.CH_2COOH + CoA \rightarrow RCH_2.CH_2CO.CoA + H_2O$). One hydrogen atom on either side of the 2-3 carbon-carbon bond is first removed to give an unsaturated derivative ($RCH = CHCO.COA$), which is now converted into a hydroxy-derivative ($RCHOH.CH_2CO.CoA$) and then into a ketone ($RCO.CH_2CO.CoA$). This involves two oxidative steps, which are linked with the formation of ATP (adenosine triphosphate) and hence the storage of readily available energy. The keto-acid is now broken down to give free CoA and an acyl-CoA, which is two carbon atoms shorter than initially. In this process acetyl-CoA is formed ($RCO.CH_2CO.CoA + CoA \rightarrow RCO.CoA + CH_3CO.CoA$). The process is repeated. In this way, fatty acids containing an even number of carbon atoms are eventually broken down to acetyl-CoA molecules. The penultimate stage is acetoacetyl-CoA, which can be either broken down further or utilized with acetyl-CoA for the biosynthesis of cholesterol and the liberation of acetoacetate (Fig. 1-10), which can then be metabolized by muscle and other tissues. Fatty acids with an odd number of carbon atoms are broken down to propionyl-CoA. This is then dealt with by another mechanism, involving the production of methylmalonyl-CoA which becomes succinyl-CoA, and then enters the tricarboxylic acid oxidative pathway. Unsaturated fatty acids are broken down in much the same way as the saturated entities, but special enzyme mechanisms are necessary when the double bonds are approached. When the rate of formation of acetyl-CoA outstrips the capacity of the hepatic tricarboxylic acid cycle, condensation occurs between molecules of acetyl-CoA to form acetoacetyl-CoA. The latter can react with the former (Fig. 1-10) to produce free acetoacetate which leaves the liver and can be metabolized by muscle and other tissues. When, under various circumstances, acetoacetate is produced in large amounts, it gives rise to excessive ketonemia ("ketosis"), as previously described (p. 102), and then to ketonuria.

Anabolism and Catabolism of Lipids. Within each tissue, fatty acids and other constituents of lipids are involved in a constant interchange between free and bound molecules. The rate of "turnover" of each type of lipid, as well as the general direction of net movement (synthesis or degradation), is determined by the chemical character of the lipid, its location in the body in general and in the tissue in particular, as well as by the physiological exigencies of the moment.

Lipid Mobilization. Lipids synthesized or stored in one tissue may, in response to physiological demands, be transferred to another site. The mechanism of transport in the blood plasma from liver to fat depots involves lipoproteins (p. 115). Transport of lipid from the depots to other tissues also occurs in the form of unesterified fatty acids bound to plasma albumin. The level of lipid in the blood is a resultant, on the one hand, of the rate of influx from the diet, synthesis in such tissues as the liver and depots, and mobilization from the fat stores of the depots and, on the other hand, the rate of oxidation in the various tissues and deposition in the liver and depots. Excessive levels of blood lipid ("hyperlipemia") may result from an imbalance between these sets of factors.

Endocrine influences play an important role in the mobilization of lipid from the adipose tissue stores. Epinephrine, growth hormone, glucagon, thyroxine and cortisone stimulate lipolysis in fat cells, increasing their output of unesterified fatty acids into the blood stream. Insulin exerts the opposite effect, indirectly, through stimulation of glucose utilization and lipogenesis.

In adipose tissue, triglycerides are first broken down to diglycerides by a lipase, and these are then broken down to glycerol and fatty acids by another lipase, which is capable of acting upon diglycerides and monoglycerides. The conversion of triglycerides to diglycerides is relatively slow and is the rate-limiting reaction for fat mobilization as fatty acids. It is therefore not surprising that the lipase concerned is the hormone-dependent entity. It is activated by cyclic AMP (cyclic adenosine monophosphate) and so is affected by the hormones, which play a part in the production of this substance. These are essentially epinephrine, norepinephrine and glucagon. Growth hormone apparently acts by increasing the effect of these other hormones and by stimulating the biosynthesis of lipases. On the other hand, insulin decreases the level of cyclic AMP and consequently inhibits lipase activity. By increasing glucose metabolism, it also makes more α-glycerol-phosphate available for triglyceride synthesis rather than breakdown.

The roles played by adrenocortical hormones and by thyroxine are indicated by the diminished action of epinephrine and norepinephrine in adrenalectomized or thyroidectomized animals.

In periods of nutritional plenty, dietary lipids, as well as those synthesized from carbohydrate in the liver, are sent to the depots for storage. Adipose tissue itself is responsible for much of the synthesis of lipid from carbohydrate. In "lean" times, these stored materials are mobilized from the depots to the liver (and other tissues) for oxidation. The concentration of lipid in the liver, then, is a resultant of the rates and directions of mobilization of lipid between liver and depots, as well as of the rates of synthesis and utilization within the liver. Imbalances may result in the development of "fatty liver" (p. 125). Agents which promote the clearance of lipid from the liver are said to be "lipotropic."

METABOLIC INTERRELATIONS OF LIPIDS, CARBOHYDRATES, AND PROTEINS

Figure 2-2 illustrates the metabolic interrelations between the major types of lipids and other foodstuffs.

The lipids do not contribute significant quantities of material to the synthesis of amino acids. Glycerol is the only major raw material which may be provided for the synthesis of carbohydrate from lipid sources, although it is probable that the usual net flow of this substance is in the reverse direction. The characteristic relation of lipids to other foodstuffs involves either synthesis of the former from the latter, or some type of coupling between the rates of synthesis and degradation of the two families of compounds.

Carbohydrate is converted to lipid in various ways. Small amounts of galactose (and in some cases glucose) are undoubtedly provided for the synthesis of cerebrosides and gangliosides. The glycerol moiety of the triglycerides and glycerophosphatides is formed readily from the intermediates of glycolysis. However, the major conversion in this direction, and the one usually meant by the phrase, "synthesis of fat from carbohydrate," is the formation of fatty acids from acetyl units, which are in turn derived from the oxidative decarboxylation of pyruvic acid. The irreversibility of this decarboxylation, incidentally, forms the basis of the current view that fatty acids cannot be converted into carbohydrate.

The concluding statement of the previous paragraph seems to be belied by many experiments with isotopes, in which labeled carbon atoms from fatty acids have been shown to be incorporated into glucose or glyco-

LIPIDS, CARBOHYDRATES, AND PROTEINS

Figure 2-2. Metabolic interrelations of lipids with other foodstuffs. (From Tietz: *Fundamentals of Clinical Chemistry.* W. B. Saunders Company, 1970.)

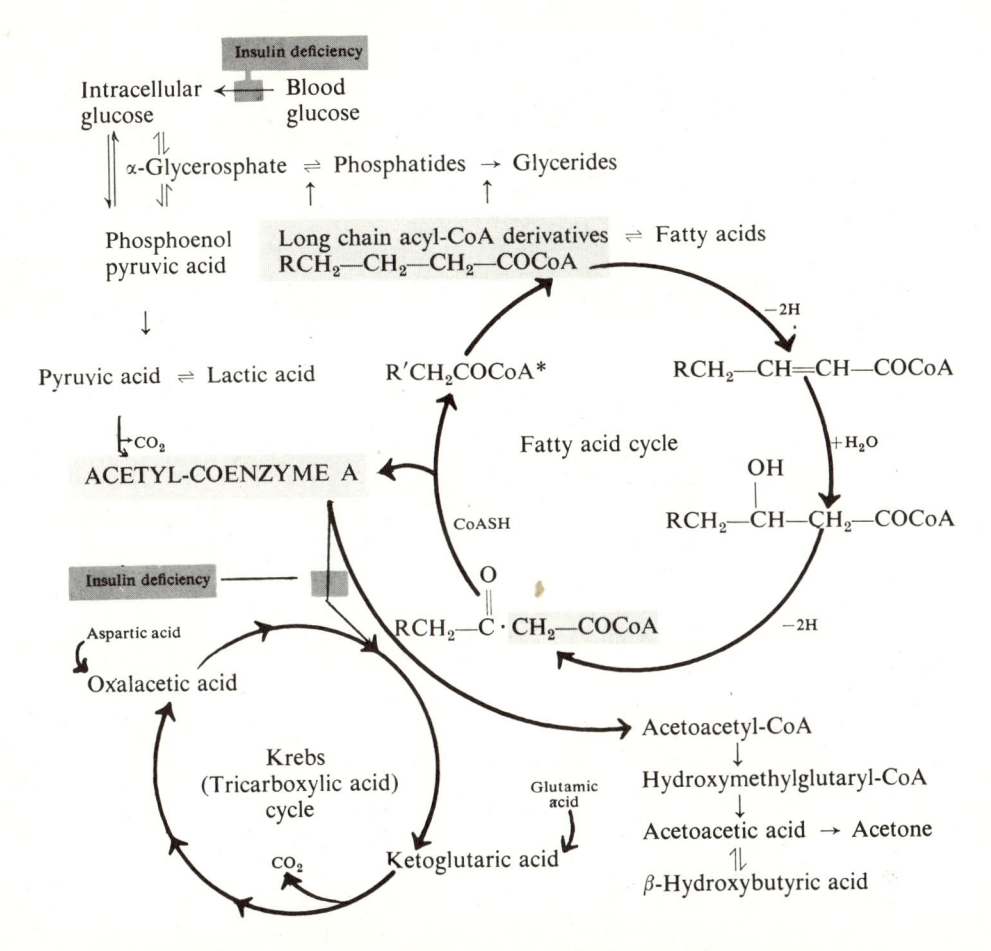

*R′ contains 2 carbons less than R, CoA represents Coenzyme A combined with the carboxyl end of the acyl chain.

gen. The contradiction is more apparent than real, however. Inspection of the reactions of the tricarboxylic acid cycle (p. 11) subsequent to the condensation of acetyl-CoA and oxaloacetic acid reveals that the two carbon atoms lost as CO_2 from the cycle are not the same atoms which entered the cycle as acetate. The carbon atoms of the acetate finally appear in the regenerated oxaloacetic acid, which can form phosphoenolpyruvic acid by decarboxylation and phosphorylation. Hence a labeled atom from acetate or a fatty acid can be incorporated into glucose or glycogen by subsequent gluconeogenesis. Nevertheless, the "material balance" shows that two carbon atoms have reacted with a molecule of oxaloacetic acid to form two molecules of CO_2 and regenerate the oxaloacetic acid. Since no additional molecules of oxaloacetic acid are formed, no net synthesis of pyruvic acid or carbohydrate from fatty acids has occurred, despite the transfer of carbon atoms from one to the other. Since fatty acids with an odd number of carbon atoms are metabolized to propionyl-CoA, which is then converted to succinyl-CoA, four of their carbon atoms enter the tricarboxylic acid cycle. This means that they make a definite contribution to oxaloacetate, which can be converted to phosphoenolpyruvate and so to glucose. In other words, odd-chain fatty acids, in contrast to even-chain fatty acids, can make a small contribution to carbohydrate.

Amino acids contribute to the synthesis of lipids. Ethanolamine, choline, serine, and sphingosine, for example, components of the phospholipids and glycolipids, are derived from the pathways of protein metabolism. The ketogenic amino acids form acetyl- or acetoacetyl-CoA from which fatty acids and sterols can be derived. The glucogenic amino acids, directly or indirectly, form pyruvic acid, which can be used for the synthesis of either carbohydrate or fatty acids.

An intimate quantitative relationship exists between the metabolism of fatty acids and of carbohydrate. An increase in glucose metabolism or supply causes a rise in plasma insulin levels, which, as already described, increases the synthesis of free fatty acids and diminishes the activity of the lipase of adipose tissue. Moreover, the increased catabolism of glucose provides the α-glycerol-phosphate required for combination with fatty acids to form triglycerides to be stored in the fat depots. This is important because these substances in this situation are being continuously broken down and resynthesized. Adipose tissue cannot phosphorylate the glycerol liberated by the hydrolytic breakdown, and so glucose catabolism becomes the source of the glycerol-phosphate required for resynthesis of the triglycerides.

When the rate of oxidation of carbohydrate, the preferred fuel of the body, falters, fatty acids are mobilized from the depots to the liver, probably by a mechanism involving various hormones (p. 105). The rate of degradation of fatty acids in the liver is then increased. Simultaneously, the rate of utilization of acetyl-CoA fragments for the synthesis of fatty acids is decreased, due to diminished supply of NADPH from the pentose shunt pathway, accompanied by an actual depression of activity of enzymes involved in lipogenesis due to decreased plasma insulin and inhibition of the formation of malonyl-CoA by the free fatty acids, which in turn inhibit glycolysis. The resulting increase in ketogenesis and its consequences are described elsewhere (below).

Although this discussion of the metabolism of fatty acids and carbohydrates has stressed the net flow of metabolites in one direction or another, it may be well to conclude with the reminder that, regardless of net synthesis or catabolism or steady states, all groups of molecules connected metabolically interchange material continually, as indicated by isotopic labeling. Whether the physiological situation of the moment calls for oxidation predominantly of carbohydrate or of fat, a certain amount of conversion of carbohydrate to fat always takes place. These and other aspects of the "dynamic state of the body constituents" are considered in greater detail in connection with the metabolism of protein (p. 150).

KETOSIS

Ketosis is the term applied to the accumulation of excessive quantities of "ketone bodies," i.e., acetoacetic acid, β-hydroxybutyric acid, and acetone, in the body. Ketonuria refers to their excretion in excess in the urine. These substances are formed in the liver and kidneys (p. 105), normally in relatively small amounts, and pass in the systemic circulation (0.5-3.0 mg./100 ml.) to the extrahepatic tissues, where they are oxidized to CO_2 and H_2O. On an average mixed diet, up to 125 mg., expressed as β-hydroxybutyric acid, may be excreted in the urine of normal subjects daily, if sufficient carbohydrate is ingested.

The extrahepatic tissues, especially skeletal and cardiac muscle, have a great capacity for utilizing these ketone bodies (ketolysis). Acetoacetic acid is "activated" and split into two molecules of acetyl-CoA ("active acetate"), which then enters the tricarboxylic acid cycle and is oxidized to CO_2 and H_2O. The liver, which is the site of formation of ketone bodies, apparently does not utilize these substances.

Even under conditions of severe carbohydrate restriction or impaired carbohydrate utilization (e.g., diabetes mellitus), with marked ketosis, the utilization of ketone bodies in extrahepatic tissues proceeds essentially normally. Consequently, the cause of ketosis is excessive ketogenesis, not inadequate ketolysis, and the liver is, therefore, the chief organ involved in this phenomenon.

When the capacity of the tissues for utilizing the ketone bodies is exceeded, the excess is excreted in the urine (ketonuria). When the capacity of the kidneys for excreting these substances is exceeded, they accumulate in the plasma (hyperketonemia or ketosis). Severe ketosis is usually accompanied by acidemia or acidosis, bicarbonate being displaced by the acid anions, acetoacetate, and β-hydroxybutyrate (primary bicarbonate deficit, p. 419). The fact should be emphasized, however, that there is no obligatory causal relationship between acidemia or acidosis and ketosis. Ketosis does not occur in such conditions unless there is an associated state of carbohydrate deprivation, (e.g., vomiting) or inadequate utilization. In fact, ketonuria and ketosis occur commonly in alkalemia or alkalosis due to persistent vomiting, unless glucose is administered parenterally.

Ketonuria and ketosis may occur in a variety of clinical conditions which have as a common denominator inadequate supply or utilization of carbohydrate:

(1) Diabetes mellitus, increasing with increasing severity of the condition (p. 34).

(2) Carbohydrate starvation, as seen in vomiting, due to any cause, severe anorexia and starvation due to other causes, and postoperatively, if glucose is not given parenterally.

(3) At times in protracted hyperinsulinism, with depletion of liver glycogen, due to hypoglycemia, and consequent unavailability of carbohydrate in adequate amounts.

(4) The hepatic forms of glycogen storage disease (p. 79).

(5) Acute increase in metabolism, with increased requirement for carbohydrate, as in thyrotoxicosis and fever.

(6) The Fanconi syndrome, as a reflection of both decreased renal tubular reabsorption of these organic acids and depletion of hepatic glycogen reserves incident to the loss of glucose in the urine in this condition (p. 84). Ketonuria has also been observed, although rarely, in other types of renal glycosuria. It may occur in severe form in phloridzinized animals (p. 83).

(7) Administration of growth hormone and 11-oxygenated adrenocortical hormones and occurs, occasionally, in the corresponding clinical conditions, i.e., acromegaly and Cushing's syndrome (decreased utilization of carbohydrate and increased fat mobilization) (p. 105).

(8) A high fat diet.

FAT IN FECES

Fat is normally present in the feces in three forms, soap fat (combined fatty acids), free fatty acids and neutral fat, the relative proportion of each being dependent somewhat upon the efficiency of fat digestion and absorption.

Normal values, based upon the examination of dried feces, are as follows:

	Per cent of dry weight
Total fat	15–25
Neutral fat	10–15
Free fatty acids	9–13
Combined fatty acids (soaps)	10–15

Normal values, employing the wet method, are as follows:

	Per cent of total dry matter
Total lipids	19.3
Neutral fat	7.3
Free fatty acids	5.6
Combined fatty acids (soaps)	4.6
Sterols	1.8

Fowweather laid down the following criteria for abnormal fat excretion in adults and children over two years of age:

(a) Any specimen in which the total fat exceeds 25 per cent of the total dry matter of the feces is probably abnormal. (In fact, up to 30 per cent is probably normal.)

(b) Deficient fat digestion is suggested in any specimen in which the neutral fat exceeds 11 per cent of the total dry matter or 55 per cent of the total fat.

(c) Deficient fat absorption is suggested in any specimen in which the total fatty acid (free fatty acid plus soap) exceeds 16 per cent of the total dry matter and 75 per cent of the total fat.

A daily fecal fat excretion in excess of the figures already mentioned (p. 110) also indicates deficient fat absorption. This is the investigation most commonly used for detection of steatorrhea.

The fecal fat may consist of either unabsorbed fat, fat that has been "excreted" into the bowel, or both, as well as that derived from bacteria. The amount excreted is quite constant under normal conditions and is independent of the diet, often remaining unchanged during periods of starvation; this is due to the fact that virtually all dietary fat in an average diet is absorbed under normal conditions. In the past, considerable diagnostic significance was attached to studies of the partition of fecal fat in relation to disturbances of fat absorption and digestion. This subject is discussed in greater detail in the section on the intestinal malabsorption syndrome (p. 600).

FAT IN URINE

Fat does not appear in the urine of normal individuals. It does occur in the tiger. Lipuria may occur in association with the lipemia of diabetes mellitus and the nephrotic syndrome, following fracture of long bones with injury to the bone marrow, in eclampsia, and also following extensive superficial injuries with crushing of the subcutaneous fat. It has been reported in subjects suffering from poisoning with such agents as phosphorus and alcohol. Fat in the urine may also be derived from fatty degeneration of leukocytes and epithelial cells, particularly in such conditions as pyelonephritis and the nephrotic syndrome.

The terms "chyluria" and "lymphuria" are applied to conditions in which there is some obstruction to the flow of lymph in the thoracic duct, with consequent distention and rupture of the lymph vessels of the kidney or bladder, the fat content of the urine in such cases varying with the quantity of fat ingested. The urine in chyluria is definitely milky in appearance, whereas in lipuria it is rather opalescent. It must be remembered that fat appearing in the urine after catheterization may be due to contamination with the lubricant used in passing the catheter.

METABOLISM OF PHOSPHOLIPIDS

It seems probable that all tissues synthesize phospholipids, but at widely varying rates. In all tissues but one these compounds are apparently synthesized, used, and degraded *in situ*. Liver is exceptional, in that a proportion of the phospholipids synthesized in this organ is liberated into the plasma. As a matter of fact, liver is practically the sole source of plasma phospholipids and is also the chief site of their further metabolism or degradation. Although there is every indication that the phospholipids play important roles in intermediary metabolism, the specific nature of these roles can scarcely be conjectured at this time. It is known that one of their fatty acids, usually unsaturated, can be transferred to cholesterol to form cholesterol esters. A familial deficiency of the appropriate enzyme has been described.

Turnover. By "turnover" is meant the rate at which a given type of molecule is renewed in the organism. Turnover may be a measure of the rate of total synthesis or degradation of the molecule in question or, as is frequently the case in isotopic-labeling experiments, it may merely indicate the rate at which one segment of the molecule "exchanges" with its free brethren in the "metabolic pool."

Whether measured by labeling the phosphate, fatty acids or bases, the turnover of phospholipids exhibits the same relative rates in the various tissues; liver, intestine and kidney comprise the most active group, such organs as pancreas, adrenals, and lungs having intermediate activity, and muscle and brain being in the slowest category.

Data obtained in experimental animals using labeled acetate indicate that, in liver and plasma, neutral fat turns over more rapidly than phospholipid, whereas the converse holds true in most other organs (including mesenteric lipid). Turnover rates are equal for fat and phospholipid in the carcass. The half-life of liver and plasma phospholipid is a matter of hours rather than days.

As might be anticipated, the turnover of phospholipids is markedly affected by dietary or metabolic influences. Fat feeding increases the turnover in liver and intestine, decreases that of the kidney, and has no effect on other organs. Experimental diabetes results in markedly increased turnover in liver and plasma, whereas other tissues are affected slightly or not at all. Administration of choline increases the rate of turnover of the lecithin fraction of liver.

Catabolism. The initial reactions of breakdown of phospholipids in the tissues are presumably hydrolytic, similar to those taking place in the digestive tract. Various "lecithinases" (p. 100), which are not specific for lecithin, are found in animal tissues.

METABOLISM OF STEROLS AND BILE ACIDS

Of the many sterols in nature, cholesterol is the only one which is of any great significance in animal metabolism. (The products of ultraviolet irradiation of the plant sterol, ergosterol, are absorbed through the intestine and certain of them have vitamin D activity in animals.) It has been known for some time that the animal organism can synthesize cholesterol. With the aid of isotopic labeling, small fragments such as acetic acid and acetoacetic acid have been shown to be precursors of the sterol molecule. Their acyl radicles in combination with CoA form 3 hydroxy- 3 methyl-glutaryl-CoA, and then a long biosynthetic process produces cholesterol from that substance.

In terms of total output, the liver and intestinal mucosa are probably the major sites of synthesis of cholesterol. Active synthesis occurs also in the adrenals, spleen, red cell, bone, heart, omentum, muscle, skin, lungs, kidney, gonads, and perhaps in all tissues. It is of interest in connection with atherosclerosis that cholesterol can be synthesized by the arterial wall. Brain, which has a very high concentration of cholesterol, synthesizes the sterol at a very slow rate in the adult. From isotopic labeling studies, it has been calculated that the "half-life" of serum cholesterol in man is eight days.

The rate of synthesis of cholesterol by the liver is influenced by a number of factors. It is, for example, inversely related to the supply of dietary cholesterol, which may be regarded as a homeostatic mechanism. In man, however, intestinal absorption of cholesterol is poor and a large proportion is excreted in the feces. As mentioned above the intestinal mucosa synthesizes the compound. The rate is depressed in the hypophysectomized animal, in pregnancy, and following estrogen administration and is frequently increased in the diabetic, matters which are discussed elsewhere (p. 137). The level of cholesterol in the plasma is influenced by the functional status of the thyroid (p. 143). Introduction into the blood stream of a number of surface-active agents, such as synthetic detergents as well as bile salts, results in greatly increased synthesis of cholesterol in the liver and mobilization of cholesterol as well as other liver lipids into the plasma.

The relationships between the cholesterol contents of liver, plasma and extra-hepatic tissues resemble to a considerable degree the situation which obtains in the case of the phospholipids (p. 111). Liver and intestinal mucosa are practically the sole sources of plasma cholesterol, and liver is the main depository of cholesterol already present in the blood. Other tissues synthesize their own supplies of cholesterol, liberate practically none into the blood, and draw upon the latter source to only a very slight extent; in other words, tissue and plasma cholesterol exchange is very slow.

Esterification of cholesterol occurs in the intestinal wall, liver, and in certain other tissues (e.g., the adrenal cortex). The liver is a chief source of free cholesterol of the blood plasma, a fact of importance in connection with changes in plasma cholesterol in hepatic disease (p. 139). Esterification of cholesterol in the plasma occurs in the circulation and is brought about by the enzyme lecithin-cholesterol acyltransferase.

It is now believed that free cholesterol liberated into the plasma forms part of the high density lipoprotein (p. 115). It is now esterified by the actyltransferase and then transferred to the low-density lipoprotein moiety. This acts as a vehicle for carrying cholesterol ester to the liver, in which organ a lipase liberates the cholesterol. The free cholesterol can be converted to bile acids, or some can return to the plasma to become part of the high-density lipoprotein.

Catabolism and Excretion of Cholesterol; Conversion to Bile Acids. The general pathways of catabolism of cholesterol and the routes of excretion of the waste products are indicated in Figure 2-3. For convenience, the catabolism of cholesterol may be divided into two compartments, the excretory route via neutral sterols, and that involving the bile acids.

(a) Neutral Sterols. In the liver and other tissues, cholesterol is ac-

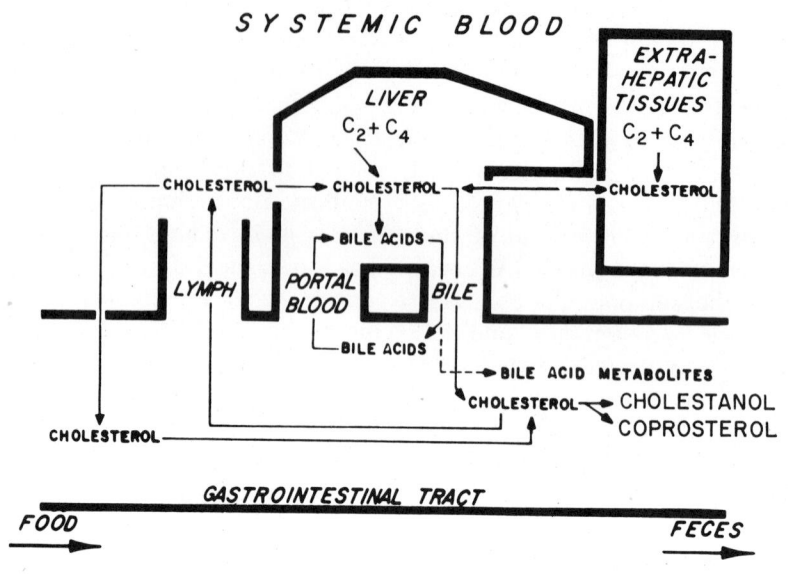

Figure 2-3. Overall metabolism of cholesterol.

companied by small amounts of cholestanol (dihydrocholesterol). A certain amount of cholesterol and cholestanol is excreted through the intestinal wall. Cholesterol and cholestanol are excreted also by the liver, via the bile, into the small intestine. Biliary cholesterol is largely formed in the liver. The cholesterol thus excreted mixes with dietary cholesterol and may be partially reabsorbed by way of the lymphatics, mainly after preliminary esterification in the cells of the intestinal mucosa. Unabsorbed cholesterol and cholestanol (which cannot be absorbed) are found in the feces. A third neutral sterol excreted in the feces is coprostanol (coprosterol). This compound is formed by intestinal bacteria (reduction).

(b) Bile Acids. As determined by isotopic tracers, 10 to 20 per cent of exogenous cholesterol is accounted for in the neutral sterol fraction of bile and feces. The remaining 80 to 90 per cent is converted to bile acids (cholic acid and derivatives) in the liver and excreted via the bile into the intestine, from which the major fraction is reabsorbed (p. 657). Unabsorbed bile acids are converted by intestinal bacteria into metabolites which are excreted in the feces. The loss of some bile acids, in this fashion, seems to be an important factor in plasma cholesterol homeostasis.

TRANSPORT OF LIPIDS

Phospholipids (chiefly lecithin), cholesterol and neutral fat are the most abundant lipids in the plasma. There are small amounts of unesterified fatty acids, glycolipids, and minute quantities of certain important hormones and vitamins of lipid nature. Normal values for the major plasma lipids are indicated in Table 2-2. Because of the heterogeneity of this group of substances and because different components have different metabolic origins, fates and significance, it is necessary to consider them individually from these standpoints. Before doing so, however, certain general features of their state in the plasma should be reviewed. These have to do largely with observations concerning the intimate relationship between the plasma lipids and certain fractions of the plasma proteins.

State of Lipids in Blood: Lipoproteins. Practically all of the lipids of plasma are present as lipoprotein complexes. Ultracentrifugal analysis of serum by the flotation technique has led to the detection of a hierarchy of lipoproteins of varying particle weight, subdivided into high and low density groups, corresponding generally, but not entirely, to the alpha- and beta-classifications determined by electrophoresis.

The plasma lipoproteins contain cholesterol (free and esterified), phospholipids, neutral fat, unesterified fatty acids (UFA), and traces of the lipid-soluble vitamins and the steroid hormones. The protein moieties are known as apoproteins. There are two main types, the A protein, or α apoprotein, and the B protein, or β apoprotein. They are synthesized in the liver and intestinal mucosa and are both glycoproteins containing appreciable amounts of hexose, hexosamine, sialic acid and fucose. Small amounts of a third apoprotein (C protein) have been found associated with the very low-density lipoproteins.

Ultracentrifugation. In the ultracentrifuge the components which undergo flotation in a medium of specific gravity < 1.006 are referred to as *very low-density lipoproteins*. Flotation between s.g. 1.006 to 1.063 produces the so-called *low-density lipoproteins*, that between s.g. 1.063 to 1.21 the *high-density lipoproteins*. After a fatty meal, chylomicrons separate out with the very low-density lipoproteins. The approximate average composition of these various groups is shown in Table 2-1.

There is also a lipoprotein fraction, the *very high-density lipoproteins*, which undergoes flotation in media of s.g. > 1.21. This is made up of two components, partially delipidized apoprotein, which still contains some phospholipid, and complexes of free fatty acid with albumin.

Chylomicrons. As has been previously indicated, the fatty com-

TABLE 2-1 CHEMICAL COMPOSITION OF MAJOR LIPOPROTEIN CLASSES*

CLASS	TRI-GLYCERIDE	PHOSPHO-LIPID	CHOLESTEROL ESTERS	FREE CHO-LESTEROL	UNESTERIFIED FATTY ACIDS	APOPROTEIN % AND TYPE
Chylomicrons	81	7	4	6	<1	2 A, B & C
VLD	55	22	8	5	<1	10 A & B
LD	11	22	37	8	1 or less	21 B
HD	6	26	15	3	—	50 A

*Tabulated from data cited by Scheig, R. *In* Bondy, P. K. (ed.): Duncan's Diseases of Metabolism. Ed. 6. Philadelphia, W. B. Saunders Co.

ponents of the chylomicrons are derived mainly from exogenous sources, that is, by way of small intestinal absorption.

Very Low-Density Lipoproteins. These have fatty components, which are almost entirely endogenous. They are synthesized in the liver. Diets high in carbohydrate cause an increase in this fraction and hence of plasma triglycerides. Ethyl alcohol and medium chain triglycerides have a similar effect.

Low-Density Lipoproteins. These contain about three-quarters of the cholesterol in serum. They are synthesized by the liver. The cholesterol and phospholipid are freely exchangeable with the high-density lipoprotein fraction.

High-Density Lipoproteins. Like the low-density lipoproteins, these contain a low percentage of triglycerides. They contain about the same amount of phospholipid but somewhat less total cholesterol than the low-density lipoproteins. They do, however, contain considerably more aproprotein.

Very High-Density Lipoproteins. This fraction represents the greatest amount of lipid transported in the blood, i.e., unesterified fatty acids bound to albumin. As much as 25 g. of the former can be transported during each hour. These fatty acids are mainly derived from adipose tissue stores.

Electrophoresis of Plasma Lipoproteins. If plasma is examined by paper, or agarose, electrophoresis at pH 8.6, preferably using buffers containing 1 per cent albumin, it is possible to demonstrate, by means of fat stains, the existence of four bands. Three of these have moved toward the anode and one remains at the origin, that is, the line of application of the serum. The fastest moving band occurs approximately in the same position as α_1-globulin. It is known as the α-lipoprotein and corresponds to the high-density fraction demonstrated by ultracentrifugal flotation. The next fastest band appears between the positions of α_2-globulin and β-globulin. It is known as prebeta-lipoprotein and corresponds to the very low-density lipoprotein of the ultracentrifuge technique. The slowest moving band appears at approximately the position of β-globulin and is known as the beta-lipoprotein, which corresponds to the low-density fraction in the ultracentrifuge. The band, which remains stationary at the origin consists of the chylomicrons. Since chylomicrons are not normally present in the postoperative state, this last band does not then appear in normal individuals. For clinical investigation only postabsorptive plasma samples are used. The very high-density fraction of the ultracentrifugal technique is not demonstrated by the electrophoresis methods. To summarize, in normal postabsorptive plasma there are two major bands, the α_1- and β-lipoproteins. A third band, the prebeta-lipoprotein, is just discernible.

The β-lipoproteins differ from the α-group in transporting more of the total plasma cholesterol (about two-thirds), in containing a higher concentration of both free and esterified cholesterol and phospholipid, and in showing a much higher cholesterol : phospholipid ratio.

It is probable that the abnormally high cholesterol : phospholipid ratio observed clinically in certain hyperlipemic serums may be a reflection of elevated concentration of the β-lipoproteins relative to that of the α- group, which do not vary greatly in concentration. The β-lipoproteins are of particular clinical interest, since it is claimed that the occurrence of elevated concentrations of these substances in human serum may be correlated with the incidence of atherosclerosis.

Injection of heparin results in marked clearing of the plasma in postalimentary hyperlipemia. Plasma thus "activated" in vivo can exert clearing action on another hyperlipemic plasma in vitro. Clearing action is due to an enzyme, lipoprotein lipase, which is either liberated or activated by heparin. Its inhibition by protamine (a basic protein) and heparinase suggests that the enzyme may contain as prosthetic group a heparin-like acidic mucopolysaccharide. The enzyme, which is found in adipose tissue, heart, lung and skeletal muscle, as well as plasma, acts specifically on protein-bound lipid in much the same way as ordinary lipase acts on free triglyceride, producing di- and monoglycerides, as well as some free glycerol. Proper functioning of the enzyme requires the presence of an acceptor for the fatty acids which are liberated; this role is performed by plasma albumin. The enzyme is activated by certain cations, calcium probably being the natural activator. Intravascular clearing is probably a minor metabolic pathway. The major part of this process appears to be accomplished by the liver, possibly involving phagocytosis by reticuloendothelial cells.

Postabsorptive Plasma Lipid Concentration. The range

TABLE 2-2 LIPIDS AND KETONE BODIES IN BLOOD PLASMA (FASTING)

SUBSTANCE	CONCENTRATION (mg./100 ml.)
Total lipid	385–675 (530)
Neutral fat	0–260 (140)
Unesterified fatty acids	8–31 (26)
Phospholipids and sphingolipids	110–250 (165)
Lecithin	80–200 (110)
Cephalin	0–30 (10)
Sphingomyelin	10–50 (30)
Cholesterol	140–260 (200)
Ester	90–200 (145)
Free	40–70 (55)

Total fatty acids
   Neutral fat × 0.95
   Phospholipid × 0.65
   Cholesterol ester × 0.43
	110–485 (300)

Ketone bodies (as acetone)	0.2–0.9

of normal variation in the concentration of plasma lipids is wide, even in the postabsorptive state (Table 2-2). Moreover, although the three major components, i.e., triglycerides (neutral fat), phosphatides (phospholipids), and cholesterol, frequently vary in the same direction, this is not always the case. Total fatty acids and neutral fat are particularly variable, even in the same individual. The cholesterol:phospholipid ratio is more stable and uniform, although its significance is not apparent. Little is known concerning the intrinsic mechanisms which regulate the equilibrium levels of the blood lipids. It is conceivable that the cholesterol:phospholipid ratio may be influenced, if not fixed, by the hepatic mechanisms for synthesis of the lipoproteins in which these molecules are found. The possibility has been mentioned previously of a "carrier" low density lipoprotein, of fixed composition, to which may be added increasing quantities of lipid, in particular triglycerides, as requirements for transport of these substances from liver to adipose tissue are increased.

Cholesterol. Cholesterol exists in the plasma in two forms, (1) free and (2) esterified (combined with fatty acids), both of which are incorporated in lipoprotein molecules (p. 115). Free cholesterol comprises about 20 to 40 per cent, and ester cholesterol about 60 to 80 per cent of the total of 140–260 mg./100 ml. This ratio is usually preserved with remarkable constancy even in the presence of wide variations in the total due to disease states that are not accompanied by disturbance of liver cell function. Although cholesterol is undoubtedly synthesized by a number of tissues, the liver is the main source of plasma cholesterol (also intestinal mucosa, p. 99). It plays an important role, too, in the removal of this substance from the blood and in its subsequent metabolism, as indicated by the fact that the bulk of exogenous cholesterol is recoverable as bile acids in the bile (p. 114).

Phospholipids. The range of normal variation in plasma phospholipids is indicated in Table 2-2. Lecithin is the main component of this fraction, which, like other lipids, occurs in the blood plasma in association with certain globulins, as lipoprotein complexes (p. 115). It has been shown, by the use of isotope-labeling techniques, that the plasma phospholipids, similarly to cholesterol, not only originate in the liver but are also removed from the blood and metabolized by that organ. It appears, therefore, that plasma phospholipids undergo most of their metabolic cycle within the liver, with a temporary sojourn in the blood stream, the purpose of which is not readily apparent. The constancy of the cholesterol : phospholipid ratio suggests that this (ratio) is of physiological significance.

Triglycerides (Neutral Fats). The variable concentration of neutral fat in the plasma in the postabsorptive state is indicated in Table 2-2. The components of this fraction, too, exist as lipoprotein complexes (p. 115). During the period of absorption of fats from the intestine, they occur temporarily in the almost purely triglyceride form of chylomicrons (p. 115). In the postabsorptive state they represent mainly triglycerides from the liver, en route to the fat depots.

Unesterified Fatty Acids. The normal concentration of unesterified fatty acids is indicated in Table 2-2. In addition to "unesterified fatty acids" (UFA), this fraction of the plasma lipids has been referred to as "non-esterified fatty acids" (NEFA) or "free fatty acids" (FFA). However, these substances are not truly free, since two-thirds of the total amount is attached to albumin and one-third to the lipoproteins (chiefly the high-density fractions). Plasma albumin, which has considerable affinity for many anions, binds 7 moles of fatty acids tightly per mole albumin, and holds perhaps 20 more in looser combination.

Unesterified fatty acids are the form in which stored lipid is mobilized from the fat depots to liver and other tissues for oxidation. Doubts concerning their importance in this regard, based on the relatively low concentration of unesterified fatty acids in plasma, have been allayed by the discovery of their extremely brief half-life (one to three minutes in man). This rate of turnover, even allowing for considerable recycling of fatty acids leaving the plasma, still is believed to represent a significant fraction of the daily caloric output.

The concentration of unesterified fatty acids in plasma is decreased by administration of glucose or insulin, and increased in diabetes mellitus, by fasting, or by administration of growth hormone, epinephrine and norepinephrine, with cortisone exhibiting "permissive" action in the last three instances. Thyroid hormones also play a "permissive" role with respect to a number of factors which tend to elevate the concentration of unesterified fatty acids in plasma. Adrenocorticotrophic hormone (ACTH) may exert an influence in the same direction, although results in the intact organism have not been so clear-cut as in vitro experiments with adipose tissue (p. 105).

The detailed mechanisms by which the above-mentioned factors influence mobilization into the plasma of unesterified fatty acids from the fat depots will be discussed in connection with the metabolism of adipose tissue. It may be noted at this point, however, that unesterified fatty acids thus liberated are converted by the liver into low density lipoproteins which may be returned to the plasma, resulting in hyperlipemia. It is apparent that unesterified fatty acids enter the blood from the fat depots at an accelerated rate under a variety of conditions in which glucose either is not derived in adequate amounts from the diet or cannot be utilized adequately because of hormonal or other abnormalities.

Influence of Food and Nutrition. After a mixed meal containing an abundance of fat, the plasma lipids begin to rise within two hours, reaching a maximum concentration in approximately four to five hours, subsequently returning to the basal level. The latter may not be reached within ten hours if a large amount of fat has been ingested. This postprandial hyperlipemia is due to the fact that the major portion of the dietary lipid is absorbed by the intestinal lymphatics and enters the systemic blood stream directly, via the thoracic duct (p. 98). The bulk of the increment is in neutral fat (30 to 150 per cent increase), cholesterol and phospholipids making relatively minor contributions, increasing up to about 10 and 20 per cent, respectively, even in their absence from the diet.

There is considerable individual variation in the lipemic response to meals. In certain normal subjects, the continual ingestion of a high-fat diet results in a cumulative effect, i.e., superimposition of the individual postprandial curves, with consequent elevation of the concentrations of total lipids and cholesterol in the plasma.

Under the conditions mentioned above, the plasma frequently exhibits turbidity, due mainly to the increase in neutral fat (chylomicrons). In abnormal states, e.g., in biliary cirrhosis, large independent increases in phospholipid (over 2000 mg./100 ml.) and cholesterol (over 1000 mg./100 ml.) may occur without turbidity of the plasma.

Because of the suggested relationship between hypercholesterolemia and atherosclerosis (p. 141), the influence of diet on the plasma cholesterol level has received considerable attention. It seems to have been established that, at least in normal subjects, the plasma cholesterol concentration is not influenced significantly by variations in the dietary intake of this substance over a rather wide range (0 to 700 mg.). However, a significant decrease may be produced in subjects with certain types of hypercholesterolemia by drastically reducing the total lipid intake, i.e., a diet of extremely low fat content. These facts are to be anticipated because of the demonstrated large capacity (1.5 to 2 g. daily) of normal subjects for synthesizing cholesterol from acetate, that common denominator of organic metabolism. Morever, it has been shown that cholesterol administration causes a decrease in the amount synthesized, so that the total provision of this substance varies but little in the presence of wide variations in exogenous supply.

The serum cholesterol concentration is influenced by the type as well as by the absolute amount of dietary fat. Although there is no complete agreement on all points, the following statements seem warranted on the basis of available evidence: (1) other factors remaining constant, addition of saturated fats (i.e., containing saturated fatty acids) to the diet increases the serum cholesterol concentration, and reduction in dietary saturated fats lowers the cholesterol concentration; (2) at any given level of total dietary fat, isocaloric substitution of unsaturated for saturated fats results in decrease in serum cholesterol. The cholesterol-lowering effect is generally attributed mainly to the polyunsaturated fatty acids (e.g., linoleic), but some believe it to be related principally to the degree of total unsaturation of the fat, including also oleic acid (mono-unsaturated fatty acid). The more highly unsaturated natural fats are of vegetable origin; in practice, therefore, diets recommended for reduction of the serum cholesterol concentration contain fats largely in the form of corn, cottonseed, peanut or soybean oils.

Racial differences in serum cholesterol concentration, which are considerable, have been attributed to differences in dietary fat. The values tend to be low in Orientals (100 to 140 mg./100 ml.) and in African natives (140 to 180 mg.), their diets containing relatively little animal fat (i.e., saturated) and relatively more vegetable fat.

There is no general correlation between the blood lipid levels and the amount of body fat, but chronic malnutrition and wasting diseases are often accompanied by subnormal levels of phospholipids and cholesterol. In the initial stages of starvation, or of carbohydrate deprivation, after depletion of the preformed glycogen stores (about 36 hours), hyperlipemia occurs as a result of increased mobilization of depot unesterified fatty acids and consequent increased synthesis of low density lipoproteins by the liver. This may be accompanied by fatty liver (p. 126). If the fasting period is prolonged to the point of depletion of available fat stores, the concentration of plasma lipids may fall.

Diets deficient in protein result in subnormal rates of synthesis of serum lipoproteins. With the exception of outright deprivation, the level of plasma triglycerides and S_f 12–400 lipoproteins varies generally inversely with the carbohydrate intake.

Influence of Age and Sex. The concentration of low-density serum lipoproteins, low in early childhood, increases steadily in the male to a maximum at 50 to 60 years of age, subsequently tending to diminish. In the female, the values are significantly lower than in the male up to 50 or 60 years of age. Subsequently, they continue to rise in women, at times to levels above those in men. The possible importance of this sex difference, corresponding to the difference in incidence of atherosclerosis, particularly early in life, is suggested by observations that the low-density lipoproteins decrease following administration of estrogens and increase following administration of androgens.

The plasma phospholipids are relatively low in preadolescence, rising to levels of 110 to 250 mg./100 ml. (lipid P = 4 to 10 mg.) in early adult life. The plasma cholesterol is quite low at birth, about 50 mg./100 ml., increasing rapidly within the first few days to about 20 to 25 mg. below the average normal adult level, where it remains until adolescence.

In the male, the concentration tends to increase steadily to about 50 years of age, tending to fall somewhat after 60 years. In the premenopausal female, the average concentration is about 20 mg./100 ml. less than in males, rising above the latter after 50 years of age.

The significance of much of the reported data on sex and age differences in plasma cholesterol is difficult to evaluate because of differences in analytical techniques and the wide range of normal variation. The following conclusions seem to be warranted on the basis of recent critical statistical analyses: (1) Between 17 and 40 years of age, the plasma cholesterol concentration shows an average increase of about 2 mg./100 ml. per year, being somewhat higher in males (20 mg.). (2) It continues to increase in men, reaching a maximum in the sixth decade. After the menopause, values in women approach those in men. (3) Values above 300 mg./100 ml. occur in 1 per cent of normal men over 35 years of age, above 300 mg. in 5 per cent of those 45 to 60 years old, and above 320 mg. in 1 per cent of those 45 to 70 years old.

Pregnancy. The concentrations of cholesterol, phospholipids and neutral fat increase during pregnancy. The maximum increase in cholesterol is reached at about the thirtieth week, the average increment in the free fraction being about 25 per cent and in the ester fraction about 5 per cent. The normal level is restored about eight weeks postpartum. The phospholipids increase to about the same extent as cholesterol, with an occasional additional rise during lactation.

The cause of the increased lipidemia in pregnancy is not known. Endocrine factors may be involved. The blood lipids, including cholesterol, increase in female birds during puberty and the period of egg production. This phenomenon can be produced by administration of estrogen to male and nonlaying female birds. However, attention should be directed to the observation that estrogen causes a decrease in the plasma cholesterol in atherosclerotic subjects, due mainly to decrease in the β-lipoprotein fraction (p. 116). Effects on blood lipids, similar to those of pregnancy, are produced by contraceptive pills.

DEPOSITION AND STORAGE OF LIPIDS

Role of Adipose Tissue in Lipid Metabolism.
Although adipose tissue has long been considered a rather inert storage depot for fat, it is apparent from more recent investigations that this tissue is by no means static. On the contrary, it carries on all the metabolic processes characteristic of any active tissue, and in addition performs certain specialized functions of paramount importance in the metabolism of lipids. The tissue is under nervous control, since conditions which cause sympathetic discharge result in the liberation of unesterified fatty acids and thus a loss of fat, whereas mobilization is inhibited by denervation, although fat deposition still may occur in this condition. Different types of adipose tissue (e.g., brown and white) and adipose tissues located in various areas of the body exhibit differences in metabolism.

The respiratory quotient (p. 445) of adipose tissue from normally fed animals is 1.0, suggesting mainly carbohydrate metabolism, whereas that from fasting animals may approach a value of 0.71, characteristic of fat metabolism. Respiratory quotients from 1.27 to perhaps 1.6 are obtained from adipose tissue under conditions in which carbohydrate is being converted into fat. Human adipose tissue has lower metabolic activity, in all respects that have been studied, including relative lack of response to hormones, which inhibit (insulin) or stimulate (epinephrine) lipolysis.

Figure 2-4 contains a simplified scheme of those aspects of carbohydrate and fat metabolism which are relevant to our discussion of adipose tissue, accompanied by indications of certain interchanges between adipose tissue and plasma.

Adipose tissue appears to contain the usual complement of metabolic pathways for the utilization of carbohydrate. Under conditions of normal carbohydrate metabolism, glycogen is deposited, a usual prelude to the synthesis of fat in adipose tissue. Both the glycolytic sequence of reactions and the pentose-shunt pathway are operative in adipose tissue. With respect to fat synthesis, the former pathway has the task of providing the raw materials, e.g., the glycerophosphate (derived from dihydroxyacetone-

Figure 2-4. Factors involved in lipid transport in the body. The liver is central in the formation of lipid-rich lipoproteins, which carry triglycerides to the adipose tissue for storage. Nonesterified fatty acids (NEFA) are released as needed to furnish available energy elsewhere. The lymphatic system cooperates in the absorption process. (From Tietz: *Fundamentals of Clinical Chemistry.* W. B. Saunders Company, 1970.)

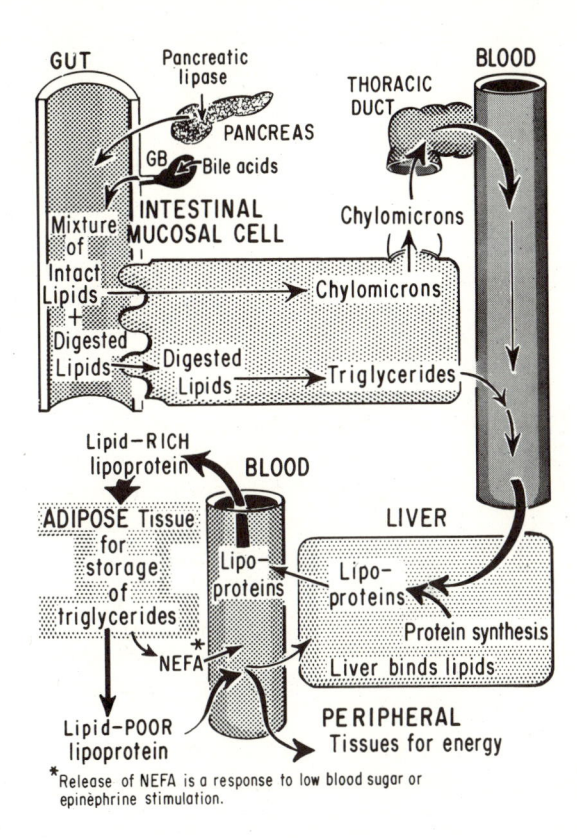

*Release of NEFA is a response to low blood sugar or epinephrine stimulation.

phosphate), required for the glycerol backbone of the triglycerides, and the acetyl-CoA fragments utilized in the synthesis of the fatty acids. The latter pathway provides the NADPH required for the reductive steps in lipogenesis.

A number of hormones influence directly the metabolism of carbohydrate in adipose tissue. As in most other tissues, the uptake and/or utilization of glucose are dependent upon an adequate supply of insulin. Epinephrine and ACTH increase the utilization of carbohydrate. However, the major effects which these hormones have upon adipose tissue are exerted in the area of fat metabolism and transport and are discussed below.

The role of carbohydrate in the synthesis of fat has been mentioned already. The glycerophosphate and fatty acyl-CoA's provided by carbohydrate are combined to form triglyceride, which collects in fat droplets. Adipose tissue contains lipases which hydrolyze this stored fat to free fatty acids and glycerol. The fatty acids may re-enter the pathways of fat synthesis. The glycerol, however, is not reutilized, due to the absence of a glycerokinase, and hence is liberated into the plasma and metabolized by other tissues. In addition to traversing the intracellular pathways of metabolism, the free or unesterified fatty acids also may combine with plasma albumin and be transported to liver and other tissues for utilization. Conversely, the intracellular fatty acid pool of adipose tissue may be augmented from outside sources, namely the chylomicrons, which represent fat from the diet, and the low density lipoproteins, which represent fat transported from the liver. Fatty acids are liberated from both by a lipoprotein lipase. It seems likely that, in the human being, this augmentation from outside sources is quantitatively greater than fatty acid synthesis in adipose tissue itself. The liver largely supplies these acids in the form of the very low-density lipoproteins carried in the circulation.

The influence of carbohydrate metabolism upon the fat stores of adipose tissue is profound. In conditions of carbohydrate deprivation, such as starvation or diabetes mellitus, fatty acids and triglycerides are not synthesized. The lipolytic process goes on, however, converting the stored fat to unesterified fatty acids, which are transported as albumin complexes to liver and other tissues. In these circumstances, adipose tissue is less able to take up chylomicrons and low density lipoproteins. The heavy influx of fatty acids results in greatly increased synthesis of low density lipoproteins by the liver. Their liberation into the plasma, in the face of decreased uptake by adipose tissue, contributes to the resultant hyperlipemia. In contrast, adequate carbohydrate utilization by adipose tissue, although it results in increased synthesis of fatty acids, nevertheless provides glycerol phosphate "backbones" to which fatty acids may be attached for storage as triglycerides. The consequences are a diminution of the fatty acid pool, decrease of the net output of unesterified fatty acids attached to albumin, and facilitation of the uptake of lipid from chylomicrons and low-density lipoproteins.

Insulin decreases the output of unesterified fatty acids from adipose tissue since it inhibits the formation of cyclic AMP necessary for the activation of hormone-sensitive lipase. Prostaglandins, which are formed from fatty acids as precursors, also decrease lipolysis, probably by a similar mechanism. This introduces a type of feed-back control; lipolysis increases the amount of free fatty acids, which then stimulates the production of prostaglandins, thus inhibiting lipolysis.

On the other hand, hormones which stimulate the release of unesterified fatty acids, such as epinephrine, norepinephrine, growth hormone and ACTH, act directly on the lipolytic mechanism.

All but growth hormone act by stimulating the production of cyclic AMP and hence the activation of hormone-sensitive lipase. Growth hormone augments the action of the others and also causes an increase in biosynthesis of the lipases.

The "permissive" actions of cortisone and thyroid hormones have been referred to previously.

Role of Liver in Lipid Metabolism. Although incidental mention has been made of the role of the liver in various aspects of lipid metabolism, it seems desirable to summarize here the pertinent information on this point:

(a) The liver is probably the most important site of synthesis of fatty acids from carbohydrate, and of cholesterol from acetate fragments.

(b) It is probably the sole source of bile acids.

(c) The phospholipids and cholesterol of the plasma, and the lipoproteins in which they are incorporated, are synthesized by the liver. In man, the phospholipids are an important source of fatty acids for triglyceride synthesis in adipose tissue.

(d) In the other direction, the liver is the organ chiefly concerned in removal of phospholipids, cholesterol, and probably also certain species of lipoproteins from the plasma.

(e) It is the major site of degradation of fatty acids of dietary or depot origin, when the physiological state of the body calls on fat for the major provision of energy.

(f) It is the only physiologically significant site of formation of ketone bodies.

(g) The liver is one of the poles of the "liver-depot axis," along which fatty acids or fat are transported in one direction or the other in response to physiological needs. In this connection, the level of fat in the liver at a given time reflects the current status and net direction of the mobilization mechanisms.

(h) The carbohydrate metabolism of the liver is also important in relation to fatty acid synthesis. Hepatic glucokinase increases with ready availability of carbohydrate in the diet. This increases glucose incorporation into the liver, and hence glycolysis and fatty acid synthesis. The acids are carried to the adipose tissue as triglycerides in very low-density lipoprotein. Starvation diminishes glucokinase, and so leads to diminished fatty acid synthesis in the liver. The hypoglycemia stimulates growth hormone production, which in turn stimulates lipolysis. The free fatty acids so liberated from adipose tissue can influence carbohydrate metabolism in the liver, where they are broken down with the formation of acetyl-CoA. This activates pyruvate kinase, which is a key enzyme in gluconeogenesis. It is also interesting to note that the lipolytic effect of growth hormone is also necessary for its protein anabolic action.

Fatty Liver and Lipotropism. The amount of lipid in the liver at any given time is the resultant of several influences, some acting in conjunction with, some in opposition to, others. (Normal lipid totals about 4 per cent, three-fourths of which is phospholipid, one-fourth neutral fat.) This situation is illustrated in Fig. 2-4. Factors that tend to increase the fat content of the liver are: (1) the synthesis of fatty acids in that organ from carbohydrate and protein, (2) influx of dietary lipid, and (3) mobilization of fatty acids from the depots to the liver. Decrease in liver fat results from: (1) mobilization of fat to the depots from the liver, (2) passage of cholesterol esters and phospholipids into the blood, (3) degradation of the fatty acids within the liver itself. Normal levels of lipid in the liver are the result of maintenance of a proper balance between these factors. A relative increase or decrease in the rate of one or other of these processes can result in the accumulation of abnormal quantities of lipid in the liver, so-called "fatty-liver." On this basis, five types of fatty liver can be distinguished, in theory at least, due to the following causes: (1) overfeeding of fat; (2) oversynthesis of fat from carbohydrate; (3) overmobilization from depots to liver; (4) undermobilization from liver to depots; (5) underutilization in the liver. Although it is doubtful whether all of these types have been observed in uncomplicated form, a summary of the probable general features characteristic of each is presented in Table 2-3.

Type 1. Overfeeding of fat results in the appearance of chylomicrons in the plasma. Although some are taken up by adipose and other tissues, and a small fraction may be cleared by the lipoprotein lipase of plasma, the major factor in removal of chylomicrons from the plasma is the liver, the fat content of which increases. In response to this increase, the liver synthesizes larger quantities of low-density lipoproteins which it liberates into the plasma for transport to adipose tissue, in which the stores of fat also increase. Depending upon the severity of the cause, the postabsorptive level of blood lipids may show no change or an increase in triglycerides and low-density lipoproteins. Although fatty livers of pure Type 1 can be produced in experimental animals, they are frequently deliberately aggravated by the imposition of conditions conducive to the development of Types 2 or 4. The lipids deposited in Type 1 reflect the composition of

TABLE 2-3 CLASSIFICATION OF FATTY LIVER

TYPE	EFFECT ON BLOOD LIPIDS (POSTABSORPTIVE)	EFFECT ON LIVER LIPIDS	EFFECT ON DEPOT LIPIDS	IMMEDIATE CAUSE	CURATIVE AGENTS* (LIPOTROPIC FACTORS)
1. Overfeeding	None or increase	Increase	Increase	Excessive fat in diet	Choline and precursors or substitutes
2. Oversynthesis	None or increase	Increase	Increase	Excessive carbohydrate, cystine, B vitamins in diet Deficiency of threonine	Choline and precursors or substitutes
3. Overmobilization to liver	Increase, normal pattern Increase in unesterified fatty acids.	Increase, normal pattern	Decrease	Carbohydrate deprivation (dietary, hormonal)	
4. Undermobilization from liver	Decrease, especially in phospholipid and cholesterol	Increase in fat and cholesterol; decrease in lecithin	Decrease	Deficiency of: Essential fatty acids Pyridoxine Pantothenic acid Choline (direct or indirect) Excess of: Diet cholesterol Biotin	Inositol, choline Choline and precursors or substitutes Choline and precursors or substitutes, lipocaic Choline and precursors or substitutes, plus inositol or lipocaic Inositol, lipocaic, choline
5. Underutilization	Increase in phospholipid and cholesterol Decrease later, if severe	Increase	Increase; decrease ?	Deficiency of pantothenic acid; hepatotoxic agents	Choline and precursors or substitutes

* Other than direct removal of cause.

the dietary lipid, as would be expected. The "cholesterol fatty liver" may belong to this category, although it is usually considered in another context (Type 4).

Type 2. Ingestion of quantities of carbohydrate greatly in excess of the caloric requirement soon overloads the capacity of the cells which normally store glycogen. The surplus carbohydrate is channeled into synthesis of fatty acids and occurs very readily in liver and adipose tissue. The liver responds to the excessive amount of endogenous fat just as it does to that from the diet; increased quantities of low-density lipoproteins are synthesized and transported to the fat depots. Consequently, an increase in plasma triglyceride and low-density lipoproteins may be observed, depending upon the severity of the condition. Oversynthesis from carbohydrate can result from forced feeding; it is more generally produced experimentally by the administration of excessive amounts of certain B vitamins (thiamine, biotin, riboflavin) or cystine, which seem to stimulate the appetite. Increased consumption of fat as well as carbohydrate lends to this type of fatty liver certain of the characteristics of Type 1. The situation is even more complex, however. The increase in general metabolic activity occasioned by the inclusion of large quantities of B vitamins or cystine in the diet results in a greater demand for certain factors, such as inositol and choline. The consequent relative deficit in these factors actually produces fatty livers of Type 4.

Type 3. Fatty liver of this type is referred to by some as "physiological" fatty liver, because it represents merely an exaggeration of a normal process, the mobilization of fatty acids from the depots to the liver. The liver responds in characteristic fashion; larger quantities of low-density lipoproteins are synthesized and liberated into the plasma. The normal proportions among the various types of lipids in blood and liver are maintained: all are equally elevated in concentration. Fatty livers of this type develop in conditions involving greatly increased utilization of fat as fuel, which is the physiological equivalent of saying all conditions in which there is interference with the oxidation of carbohydrate. Such conditions exist in: (1) diabetes, human or experimental, of the hypoinsulin, hyperpituitary or hyperadrenocortical types; (2) "pseudodiabetes" induced by phloridzin; (3) starvation; (4) carbohydrate deprivation. The fat which accumulates in the liver is derived from the fatty acids of the depots, the fat content of which decreases accordingly. (Owing to the nonutilization of carbohydrate, adipose tissue is unable to take up low-density lipoproteins, thus aggravating the hyperlipemia.) In addition to fatty liver and hyperlipemia, this condition is characterized also by ketosis and, in advanced cases, by acidosis.

Type 4. Fatty liver of this type has been differentiated from the preceding type by being designated "pathological." It is accompanied by a decrease in plasma lipids (hypolipemia) which affects mainly the phospholipids and cholesterol esters. The pattern of liver lipids is also abnormal; fat and cholesterol esters are especially increased, whereas lecithin is present in less than the usual proportion. Fatty livers of this type may eventuate in cirrhosis of the liver and there may be associated hemorrhagic lesions in the kidneys. They appear to be caused by agents or conditions which produce either an absolute or a relative deficiency in certain of the ingredients used by the liver for synthesis of phospholipids, viz., choline, inositol and the polyunsaturated fatty acids. The resulting condition appears to reflect some sort of blockage of metabolism of fat in, or its movement through, the liver. According to certain investigators, the degradation of fatty acids is impaired; the majority opinion holds that the integrity of phospholipid metabolism in the liver is necessary for the mobilization of fat from that organ.

Agents (e.g., choline, methionine, betaine and inositol) which have the apparent effect of facilitating the removal of fat from the liver are said to be "lipotropic," the phenomenon itself being called "lipotropism." Antagonistic agents and the converse condition are "antilipotropic" and "antilipotropism," respectively.

Fatty livers of Type 4 may be roughly classified according to the causative agent or phenomenon:

(a) **Deficiency of Essential Fatty Acids.** The fatty livers due to deficiency of the essential fatty acids are cured only by the reintroduction of these substances into the diet. Since the phospholipids of the liver are characterized by a relatively high content of polyunsaturated fatty acids, it is surmised that a shortage of the latter substances results in impairment of synthesis or turnover of the former, thus interfering with synthesis of the basic structure of the low-density lipoproteins.

(b) **Imbalance of B Vitamins.** Deficiencies of pyridoxine and pantothenic acid, and excessive amounts of biotin in the diet, give rise to fatty livers, the etiologic factor of which is in dispute. It is possible that the fatty liver of pantothenic acid deficiency properly belongs to Type 5 rather than 4, since a shortage of CoA (of which pantothenic acid is a part) might be expected to result in impairment of the degradative mechanisms for fatty acids. Lack of pyridoxine and a surplus of biotin are said to elicit a greater demand upon the supply of inositol, thus supposedly interfering with the synthesis of inositol-containing phosphatides. Although the fatty livers in these conditions respond well to inositol, this explanation leaves much to be desired.

(c) **"Cholesterol" Fatty Livers.** Fatty livers resulting from administration of excessive amounts of cholesterol are sometimes regarded as belonging to Type 4, although they may perhaps as readily fit the requirements of Type 1. One explanation advanced for the former classification is that cholesterol may compete with phospholipids for the polyunsaturated fatty acids, for the acids found in cholesterol esters are highly unsaturated. In support of this point of view, it may be noted that elevation of the level of dietary cholesterol increases the requirement for essential fatty acids; also, in cholesterol fatty livers, phospholipid turnover is depressed.

(d) **Choline Deficiency.** The remaining fatty livers of Type 4 are due, more or less directly, to a deficiency of choline. Since choline is synthesized by the successive methylation of ethanolamine, it is true that induced deficiencies of the methyl group produce fatty livers, solely, however, because of the resulting shortage of choline. An indication of the probable importance of choline and other components of phospholipids for the synthesis of lipoproteins is furnished by the observation that, in experimental animals, choline deficiency results in lowering of the high-density lipoproteins and virtual disappearance of the low-density lipoproteins. In man, lowering the dietary protein and choline, without changing dietary fat or calories, results in decreases in blood cholesterol and low-density lipoproteins.

Animals maintained on diets deficient in the usual sources of preformed methyl groups (choline, methionine, betaine), and in the vitamins necessary for synthesis of methyl groups from appropriate precursors (folic-folinic acid group, B_{12}), develop fatty livers of Type 4. These fatty livers are cured by the methyl donors mentioned above, by choline itself or its proper analogs, and are at least alleviated by folic acid and vitamin B_{12}.

A deficiency of choline can be induced also by inclusion in the diet of compounds which will compete with ethanolamine for available methyl groups. Nicotinic acid or amide and guanidoacetic acid, for example, are methylated in the body of N^1-methylnicotinamide and creatine, respectively. Administration of greatly excessive quantities of these compounds depletes the supply of methyl groups available for the synthesis of choline, and fatty liver results.

Type 5. As suggested previously, it is possible that the fatty livers of pantothenic acid deficiency are of this type, i.e., underutilization. Those due to hepatotoxic agents (chloroform, carbon tetrachloride, phosphorus) may also belong to this group, although they present in addition certain of the features of Type 4. The initial elevation of the serum lipids (during which time the depots may have an increased content of fat) may be due to impairment of the mechanisms for the degradation of fatty acids, but not of the mobilization of fat from the liver to the depots. The second, or hypolipemic phase (during which time the depots are probably somewhat depleted), may reflect interference with the mobilizing mechanisms of the liver. In support of the probably composite nature of the pathogenesis of this type of fatty liver, it may be mentioned that it is possible to counteract or mitigate with choline or methionine the fatty infiltration due to the hepatotoxic agents, if the therapeutic agent is given promptly after administration of the toxic agent, or better, prophylactically.

LIPIDOSIS, XANTHOMATOSIS

According to Fredrickson, the term lipidosis refers to abnormal concentrations of lipoproteins in blood or of *specific lipids* in tissues. The latter may be sphingolipids (Tay-Sachs disease, Gaucher's disease, Fabry's disease, metachromatic leukodystrophy, Niemann-Pick disease) or *other lipids* (Refsum's disease, Wolman's disease, hepatic cholesterol ester storage disease, ceroid storage disease). Each of these disorders may involve the nervous system. They are genetically determined enzyme deficiencies

TABLE 2-4 THE LIPID STORAGE DISEASES

STORAGE DISORDER	LIPID STORED	METABOLIC ABNORMALITY
Pseudo-Hurler's (generalized gangliosidosis Type I) and Juvenile G_{M1} gangliosidosis	Galactosyl-galactosaminyl-(N-acetylneuraminido) galactosyl-glucosyl-ceramide (ganglioside G_{M1})	β-galactosidase deficiency (G_{M1} galactosidase)
Tay-Sachs	N-acetyl-galactosaminyl-(N-acetylneuraminido) galactosyl-glucosyl-ceramide (ganglioside G_{M2})	Deficiency of hexosaminidase-A in blood and tissues including amniotic fluid
Fabry's	Galactosyl-galactosyl-glucosyl-ceramide	Deficiency of α-galactosyl hydrolase
Gaucher's	Glucosyl-ceramide	Deficiency of glucosyl-ceramide hydrolase
Metachromatic leukodystrophy	Sulfonylgalactosyl-ceramide	Deficiency of cerebroside sulfatase
Krabbe's	Galactosyl-ceramide	Galactocerebroside β-galactosidase deficiency
Niemann-Pick	Sphingomyelin and cholesterol	Deficiency of sphingomyelinase in some types
Refsum's	Phytanic acid (3, 7, 11, 15-tetramethyl hexadecanoic acid)	Deficiency of phytanic acid oxidase
Wolman's	Triglycerides and cholesterol	Acid lipase
Cholesteryl ester storage disease	Cholesteryl esters	—
Cerebrotendinous xanthomatosis	Cholesteryl esters and cholestanol	—

(Table 2-4). Tay-Sachs disease can be diagnosed by demonstrating a deficiency of hexosaminidase A in plasma, leukocytes or cultures of fibroblasts from skin.

It is now possible to diagnose a number of these diseases (Tay-Sachs disease, G_{M1} gangliosidoses, metachromatic leukodystrophy, and possibly Krabbe's disease) by examination of the cells in amniotic fluid obtained by paracentesis. These cells can be shown to have the appropriate enzyme deficiency. Prenatal diagnosis of this type is becoming much more common and the number of disorders diagnosed in this way is increasing. Similarly, it is possible to demonstrate hereditary enzyme deficiency disorders of various types by means of tissue culture.

According to Fredrickson, xanthomatosis means lipid accumulation in tissues in association with large "foam cells" (Hand-Schüller-Christian disease, Letterer-Siwe disease, and related conditions). The lipid is mostly cholesterol.

ABNORMALITIES OF PLASMA LIPIDS

Although variations occur in the plasma fat and phospholipid concentrations in certain abnormal states, determinations of these factors have only quite recently been made commonly, clinically, due mainly to technical difficulties and to the fact that alterations in cholesterol, which usually occur simultaneously, could be demonstrated much more readily.

Attention will be directed here mainly to observations on the plasma lipoproteins and cholesterol. Clinically significant abnormalities in fat and phospholipids will be mentioned incidentally. In general, with few exceptions, changes in plasma fat, phospholipids, and cholesterol concentrations in disease usually occur in the same direction, although not necessarily to the same degree. It should be understood that the several lipids do not exist in serum as homogeneous entities, but rather as components of large lipoprotein molecules.

PLASMA LIPOPROTEIN ABNORMALITIES

Reference is made elsewhere (p. 115) to the fact that the plasma lipids, including cholesterol and phospholipids, are combined with proteins to form lipoprotein complexes of high molecular weight. Studies in this field have been conducted intensively, with particular reference to the possible role of plasma lipoprotein abnormalities in the pathogenesis of atherosclerosis. Clinical interest in this subject necessitates a brief review of the current status of this important problem.

For many years there have been attempts to relate atherosclerosis to some abnormality in plasma lipids, particularly to hypercholesterolemia. Belief in the possibility of such a relationship arose in part from the observation that the lipid composition of the intima of arteries resembles that of the blood plasma, suggesting that the lipids may pass from the latter to the former. Moreover, it was shown that atherosclerosis develops in rabbits in which hypercholesterolemia had been induced by feeding cholesterol. The applicability of this observation to atherosclerosis in man was regarded as questionable, inasmuch as the rabbit is a herbivorous animal and normally ingests no cholesterol. However, this disease has now been observed also in chicks, monkeys and dogs with experimentally induced hypercholesterolemia. Moreover, the incidence of atherosclerosis is unusually high in certain conditions in which the plasma cholesterol concentration tends to be elevated for prolonged periods, e.g., diabetes mellitus and familial hypercholesterolemias (lipemias). There is also a statistical correlation between racial differences in serum cholesterol concentration and the incidence of atherosclerosis in these populations. On the other hand, there have been observations that suggest the lack of relationship of cause and effect between hypercholesterolemia and clinical atherosclerosis.

Improved understanding of the state in which lipids exist in the blood, i.e., as lipoproteins, has opened new approaches to the study of possible metabolic abnormalities that may consist in alterations in the plasma lipid distribution pattern rather than in the absolute concentration of individual lipid components. The influence of race, age, sex, pregnancy, diet, and nutrition has been reviewed elsewhere (p. 121).

Interest is centered particularly in lipoprotein abnormalities in diseases accompanied by hyperlipemia and/or hypercholesterolemia, and in their possible relation to the development of atherosclerosis. Certain generalizations may be made that have practical significance:

(1) Almost invariably, in cases of abnormal hyperlipemia, the most significant and consistent serum lipoprotein abnormalities are increased concentration of the low-density and/or the very low-density group. The high-density lipoproteins may be simultaneously decreased, increased, or normal.

(2) In many such cases there is also hypercholesterolemia.

(3) In cases of hyperlipemia in which the serum is lactescent and the cholesterol only moderately elevated, the increase is predominantly in the very low-density lipoproteins (S_f 20–400), including chylomicrons (> 400).

(4) In cases of hyperlipemia with marked hypercholesterolemia but no lactescence of the serum, the increase is predominantly in low-density lipoproteins of the class S_f 0–20.

Increase of serum low-density lipoproteins has been observed in virtually all conditions accompanied by hypercholesterolemia (p. 137). These include diabetes mellitus, the nephrotic syndrome, hypothyroidism, obstructive jaundice, familial and idiopathic hyperlipemia, the hypercholesterolemic type of xanthomatosis, and glycogen storage diseases. Its chief clinical significance in these disorders lies in its possible relation to their predisposition to the development of atherosclerosis, in the pathogenesis of which some believe this lipoprotein abnormality to be implicated. When large groups of subjects are studied statistically, the concentration of serum low-density lipoproteins, and also of total serum cholesterol, is significantly higher in subjects with clinically evident atherosclerosis than in "normal" subjects. Moreover, cases have been studied in which an increase in low-density lipoproteins antedated the development of clinical manifestations of atherosclerosis, e.g., coronary occlusion. However, there is considerable overlap between values in atherosclerotic and "normal" subjects, many of the former having serum lipid concentrations well within the generally accepted normal range. One problem of importance in this connection is the difficulty of diagnosing atherosclerosis except of advanced degree or involving the coronary or cerebral arteries. Because of this fact, probably many subjects regarded as normal, particularly in the older age groups, belong properly in the atherosclerotic group. Their exclusion might conceivably result in lowering of the accepted upper limit of normal values for serum lipids. It is perhaps significant in this connection that there is less overlap in the younger age groups, with lower normal values, than in older individuals. Under existing circumstances, however, studies of serum lipoproteins, although of considerable theoretical importance, are still of limited clinical value, e.g., in predicting the development of such complications as myocardial infarction, cerebral thrombosis, etc.

Studies related to increased susceptibility to ischemic heart disease have indicated that not only the low-density lipoproteins (cholesterol) but also the very low-density lipoproteins (triglycerides) are possible risk factors. In a Swedish study, it appeared that a plasma triglyceride increase is a better indication of risk, particularly in young men, than is increased plasma cholesterol. In fact, a significant increase in risk has been demonstrated when the plasma triglycerides are raised and the cholesterol is normal. From the Framingham, Massachusetts, studies, it has become obvious that hyperlipidemia is only one of a number of factors associated with increased susceptibility to premature atherosclerosis as indicated by ischemic heart disease. These are age, hyperlipidemia, cigarette smoking, hypertension, obesity, diabetes mellitus, physical inactivity, hyperuricemia, family history and abnormalities in the electrocardiogram. These factors vary in their relative importance with age and there are, no doubt, many other factors to be considered. The Framingham studies indicate that men with hypercholesterolemia, hypertension and cigarette smoking as factors have eight times the risk of ischemic heart disease than a group in whom these factors are absent. The risk, however, is diminished fivefold when any one of these three factors is present in the absence of the other two. In other words, hypercholesterolemia alone would increase risk by approximately one and one-half times. This puts the matter into its proper perspective. It is still quite possible that, in man, raised plasma cholesterol is merely an epiphenomenon and not necessarily a causative factor. Nevertheless, it appears that, in individuals under the age of 50, increased plasma levels of cholesterol and/or triglycerides are definite indications of higher risk, even when considered on their own.

Reference is made elsewhere to the "lipemia-clearing" action of heparin (p. 117). This effect in decreasing the turbidity of lipemic plasma is accompanied by a shift in the lipid distribution pattern toward molecules of the lower S_f classes. The abnormal plasma lipoprotein pattern in atherosclerotic subjects is shifted toward normal. A similar shift from β-lipoproteins to α-lipoproteins, i.e., toward a more normal pattern, has been observed in subjects with other types of hyperlipemia after administration of heparin. It has been suggested that regulation of the plasma lipoprotein pattern may be an important function of heparin, and that deficiency in this factor, or in some agent with similar properties, may underlie the presumed lipid metabolic (or transport) defect in the conditions under consideration.

The possible importance of this action of heparin is indicated by reports that it protects against the development of atherosclerosis in cholesterol-fed rabbits and relieves anginal pain in patients with coronary artery disease. It is pertinent in this connection to mention the observation that hyperlipemia, i.e., turbidity, and increase in the higher S_f classes, are accompanied by significant slowing of the circulation in various capillary areas. This abnormality, too, is corrected by administration of heparin.

HYPERLIPOPROTEINEMIA

Based on the lipoprotein patterns obtained by electrophoresis and the estimation of plasma cholesterol and triglyceride content, Fredrickson has defined five types of hyperlipoproteinemia. Each of these can be familial or acquired.

Type I. In Type I hyperlipoproteinemia, electrophoresis demonstrates the presence of a definite chylomicron band in plasma obtained in the postabsorptive state (this band is absent in normal postabsorptive plasma). There is a decrease in all other lipoproteins. The triglyceride concentration ranges between 1500 and 15,000 mg./100 ml. The cholesterol is abnormally high when the triglyceride figure is above 1000. The ratio of cholesterol to triglyceride concentration is less than 0.15, which is very low. The plasma appears milky. The clinical picture is one of eruptive xanthomas, hepatosplenomegaly and bouts of abdominal pain, usually after a period of high fat intake. Fever may be present and pancreatitis can occur.

The familial type of Type I hyperlipoproteinemia, usually detected in the first decade of life, is due to an inherited defect of chylomicron removal by the enzyme lipoprotein lipase, which is in some way dependent on heparin. There is very low lipolytic activity when heparin is added to the plasma, and low levels of lipoprotein lipase have been demonstrated in the adipose tissue.

The acquired type has been found in association with diseases which manifest abnormal proteins in the blood stream, e.g., myelomatosis, macroglobulinemia, systemic lupus erythematosus, lymphomas, etc. These show heparin resistance, possibly due to binding of the heparin by the abnormal protein.

Type II. In Type II, electrophoresis demonstrates an increased concentration of the beta-lipoprotein band in postabsorptive plasma. There are now considered to be two subgroups, Types II_a and II_b. The former, on examination in the ultracentrifuge shows an increase only in low-density lipoproteins, whereas the latter also shows an increase in very low-density lipoproteins (β and pre-β).

In the familial type (familial hypercholesterolemia), the abnormal pattern can be detected in heterozygotes at the age of one year. In the homozygote, the plasma cholesterol concentration ranges between about 600 and 1000 mg./100 ml.; the heterozygote shows values about half of these figures. The plasma is usually clear but is occasionally turbid when there are accompanying moderate increases in the prebeta-lipoproteins, but the plasma triglyceride level, which is raised, is usually not above 500 mg./100 ml. The cholesterol to triglyceride ratio is usually above 1 and sometimes markedly so. Clinically, the disease is manifested by the occurrence of xanthomas in tendons and around certain bony tuberosities (tibia, elbow). Xanthelasma and arcus cornea are common occurrences. There is a marked tendency for the development of atherosclerosis, ischemic heart disease and even aortic stenosis.

The acquired type occurs in association with diabetes mellitus, hypothyroidism, the nephrotic syndrome and various dysproteinemias. The plasma cholesterol is usually below 300 mg./100 ml. These patients may have xanthelasma but usually do not have xanthomas of tendons or bony tuberosities. Probably the most common cause of the acquired abnormal lipoprotein pattern is dietary excess.

Type III. Electrophoresis of postabsorptive plasma demonstrates a broad beta-lipoprotein band in Type III hyperlipoproteinemia. There is usually some increase in the prebeta-lipoprotein and a chylomicron band becomes evident. Both the cholesterol and triglyceride concentrations lie between 200 and 1000 mg./100 ml.

The familial type of this condition is usually detected after the age of 20 years. It is manifested by the occurrence of yellow plane xanthomas in the palmar creases or fingers and reddish yellow on the elbows. There may be pedunculated xanthomas on the buttocks and occasionally tendon xanthomas. Atherosclerosis is manifested by intermittent claudication, cerebrovascular and ischemic heart disease. There is marked diminution in glucose tolerance. High carbohydrate diets markedly aggravate the condition. There is sometimes an increase in the level of plasma urate. The mechanism of disease production is not known.

The acquired type of abnormal lipoprotein pattern has been seen during the recovery stage of diabetic coma.

Type IV. In this type of hyperlipoproteinemia, electrophoresis of postabsorptive plasma demonstrates a marked increase in the prebeta-lipoprotein band. The plasma triglyceride levels are increased. When the level is above 300 mg./100 ml., there is also an increase in plasma cholesterol and the development of turbidity.

The familial type shows a marked tendency to ischemic heart disease and diabetes mellitus can occur. The plasma clearing effect of heparin is normal. The disease is due to excessive production of triglycerides from carbohydrates. It is the most frequently seen form of familial hyperlipoproteinemia.

The acquired type of abnormal pattern has been seen in association with hypothyroidism, the nephrotic syndrome, myelomatosis, macroglobulinemia, glycogen storage disease, idiopathic hypercalcemia, gout and Werner's syndrome. It can be induced by contraceptive pills containing progestational hormones.

Beta-lipoprotein is polymorphic and a common variant can move somewhat faster than the normal β-lipoprotein during electrophoresis. Since it has pre-beta mobility ("sinking pre-beta"), the picture can be confused with Type IV hyperlipoproteinemia. The triglyceride content of the plasma is, however, within the normal range. This emphasizes the point that it is incorrect to rely on the electrophoresis picture alone in relation to the diagnosis of the hyperlipoproteinemias.

Type V. In Type V, electrophoresis of postabsorptive plasma demonstrates the presence of a chylomicron band and an increase in the prebeta-lipoprotein band. The plasma cholesterol level is usually somewhere between 200 and 1000 mg./100 ml. and the triglycerides between 500 and 5000 mg./100 ml. The cholesterol to triglyceride ratio is usually between 0.2 and 0.4. The plasma is turbid. The abnormal pattern is due to defective metabolism of triglycerides (exogenous or endogenous).

The familial type often presents as bouts of abdominal pain with transient increases in plasma amylase. The latter is due to pancreatitis. There may be hepatosplenomegaly, eruptive xanthomas and lipemia retinalis as well as painful extremities or paresthesias. Hyperuricemia is quite common. There may be a tendency to diabetes as well as atherosclerosis.

The acquired form can occur in association with diabetes mellitus, acute pancreatitis, dysproteinemias and a variety of other diseases associated with hyperlipemia.

Alpha-lipoproteinemia. This can be demonstrated by electrophoresis in plasma from patients with obstructive liver disease, in which there is hypercholesterolemia and a great increase in nonesterified plasma cholesterol. High dosage with estrogens can also increase the alpha-lipoprotein.

HYPOLIPOPROTEINEMIA

A number of conditions exist in which there is a deficiency in the amount of plasma lipoprotein.

A-beta-lipoproteinemia. This is a familial condition. Within the first year or two of life, there is failure to thrive, steatorrhea, diarrhea and abdominal distention (cf. celiac disease). At or after the age of five, a neurological picture develops, which is similar to Friedreich's ataxia. There may be retinitis pigmentosa. Acanthocytes (crenated red cells) are present in the blood. The plasma contains only alpha-lipoproteins. The beta- and prebeta-entities, as well as chylomicrons, are absent. The plasma cholesterol and phospholipid levels are markedly decreased. The genetic defect is believed by some authorities to be an inability to synthesize apoprotein B. This could also explain the existence of steatorrhea. The intestinal mucosa is infiltrated with fat, even after an overnight fast. The B protein seems to be required for transport of triglycerides, as well as for the formation of chylomicrons and so normal fat absorption is greatly impaired.

Hypobeta-lipoproteinemia. In this condition, there is a decrease in the low-density (beta) and very low-density (prebeta) lipoproteins. The plasma cholesterol and triglyceride levels are low. This picture is sometimes genetically determined but is more commonly acquired as a result of severe debilitating diseases or malabsorption syndromes.

Tangier Disease. This genetically determined condition is characterized by a marked decrease in the high-density (alpha) lipoproteins. The small amount, detectable only by ultracentrifugation, is mostly different antigenically from normal alpha-lipoprotein. There is also a marked lowering of plasma cholesterol and triglyceride. The tonsils are enlarged and have a peculiar orange or yellowish-gray discoloration (cholesterol oleate). There is also lymphadenopathy and hepatosplenomegaly. A peripheral neuropathy may occur.

Lecithin-Cholesterol Acyltransferase Deficiency.
This is a rare familial disorder which was discovered in Norway. The plasma alpha-lipoprotein is markedly diminished but the total cholesterol and triglyceride levels are either normal or raised. Cholesterol esters are absent from the plasma, which also shows a deficiency of the enzyme, which gives the disease its name. Serum albumin levels are low and serum acid phosphatase high. Proteinuria, anemia and arcus cornea are present.

HYPERCHOLESTEROLEMIA

Diabetes Mellitus.
The well-recognized occurrence of hyperlipemia in diabetes mellitus has attracted considerable attention (p. 21). Despite its chemical individuality, it has been shown that the concentration of plasma cholesterol in this condition tends to parallel approximately that of the total fatty acids of the blood, changes in the latter being reflected more or less accurately in the former. Early investigators of this problem came to the conclusion that diabetes mellitus is accompanied by a marked increase in plasma lipids, including cholesterol, and that this increase is progressive with the seriousness of the condition. Many have emphasized the importance of the plasma cholesterol concentration as an index of lipid metabolism and of prognosis in patients with diabetes. Patients with normal blood sugar values following treatment, but with persistent hypercholesterolemia, have been found to be more susceptible to relapse following slight dietary indiscretion or intercurrent illness, to severe complications (neuritis and gangrene) and to insulin refractoriness in the presence of complicating infectious processes. Changes in plasma cholesterol appear to be less marked in diabetes in children. Hypercholesterolemia is observed rather consistently in the presence of diabetic coma, and cases of progressive diabetes present values that are somewhat higher than those of decreasing severity. However, normal cholesterol values are usually obtained in uncomplicated cases in children and appear to bear no relation to the incidence of complications.

Occasionally, subnormal values for plasma cholesterol may be obtained in patients with advanced grades of diabetes. This finding, which is considered elsewhere (p. 144), is of serious prognostic significance.

In the great majority of cases the hypercholesterolemia of diabetes mellitus is due to an approximately proportionate increase in both free and ester fractions. The normal ratio usually tends to be maintained even though the total cholesterol concentration may be increased 400 to 500 per cent. The underlying mechanism is the same as that operating to produce excessive ketogenesis (p. 109). There is excessive production of acetyl-CoA, and of acetoacetyl-CoA. These combine to form 3-hydroxy-3-methyl-glutaryl-CoA in large amounts. This not only gives rise to acetoacetate but also provides the source of the metabolites involved in the increased biosynthesis of cholesterol by the liver.

There is evidence that hemoconcentration may contribute to the increase in the concentration of cholesterol, phospholipids and neutral fat in severe diabetic acidosis. Dehydration in this condition may be accompanied by an increase in the plasma protein concentration, which falls, rapidly, with the cholesterol during recovery.

There is no strict parallelism between hypercholesterolemia and hyperglycemia, glycosuria, ketonuria or acidosis in diabetes. Extremely high figures have been reported, a few being above 1000 mg. per 100 ml. of plasma, several above 500 mg. and many above 300 mg. per 100 ml. Administration of insulin is usually followed by diminution in plasma cholesterol. The determination of plasma cholesterol may serve as a useful measure of the adequacy of control of diabetes.

Anesthesia. An increase in plasma cholesterol ocurs rather consistently during and following ether and, less constantly, chloroform narcosis. Ether anesthesia is accompanied by a state of hyperglycemia which frequently tends to be proportional to the degree of hypercholesterolemia. It has been found that both of these phenomena may be prevented by the administration of insulin, which suggests that the simultaneous disturbances in cholesterol and carbohydrate metabolism incident to ether anesthesia are fundamentally interrelated.

The Nephrotic Syndrome. Hypercholesterolemia is a rather constant manifestation of the nephrotic syndrome (amyloid nephrosis; chronic active glomerulonephritis, p. 521). This increase is associated with retention of other lipids in the blood, chiefly neutral fat, but also phospholipids. Values for plasma cholesterol as high as 2200 mg. per 100 ml. have been reported and figures of 500 to 700 mg. are not unusual. The total lipids often exceed 2 g./100 ml. Although some investigators of this problem have reported an increase in the proportion of cholesterol esters (80 to 90 per cent of the total) in this condition, the majority have failed to find any alteration in the distribution of cholesterol in the plasma. This hypercholesterolemia is associated with increased amounts of cholesterol in the urine and with its deposition in the renal tubular epithelium. An increase in plasma cholesterol is also found occasionally in cases of acute glomerulonephritis. The major lipid increment occurs in the low-density lipoproteins.

The significance of this phenomenon and the factors responsible for its development are not clearly understood. Its association with a diminished concentration of serum albumin suggests the possibility that both phenomena may be concomitant if not related features of some underlying metabolic disorder. Although edema, hypoproteinemia and hypercholesterolemia are usually present concomitantly in this condition, there is no consistent quantitative relationship between the degree of cholesterolemia and the other two phenomena. A lack of fundamental dependence of the hyperlipemia on the hypoalbuminemia is suggested by the fact that the former does not occur in the presence of low plasma albumin levels due to nutritional causes (e.g., starvation, hepatic disease, celiac disease), or induced in animals by plasmapheresis. In fact, the plasma cholesterol may be decreased under such circumstances. It is of interest in this connection that in nephrotic patients responding favorably to administration of adrenal steroids a fall in serum cholesterol concentration usually, but not invariably, occurs some time after the increase in serum albumin. However, in experimentally induced nephrosis, hyperlipemia has been observed to precede the hypoalbuminemia.

It has been suggested that the increase in serum lipids may be related, in part at least, to the function of albumin as an essential acceptor of fatty acids released by lipoprotein lipase (p. 118). On this basis, hypoalbuminemia might conceivably result in an aberration of lipoprotein metabolism and transport. This would result in an excessive concentration of fatty acids in the liver, which would in turn inhibit their synthesis and also make acetyl-CoA available for the synthesis of cholesterol. In spite of contradictory observations and opinions, it seems likely that hypoalbuminemia plays an important role in the pathogenesis of the hyperlipemia of the nephrotic syndrome.

Some have attributed the hypercholesterolemia of the nephrotic syndrome to defect in the mechanism which ordinarily removes cholesterol from the blood or to a disturbance which causes excessive mobilization of lipids from the fat depots. Although an increase in other plasma lipids is observed commonly in all types of chronic glomerulonephritis, hypercholesterolemia occurs usually only in the presence of the nephrotic syndrome. In its absence the plasma cholesterol concentration is usually within normal limits in patients with chronic glomerulonephritis. Subnormal values may be obtained in patients in the terminal states of this condition (p. 144).

Hepatic and Biliary Tract Disease. The plasma cholesterol concentration is usually increased in jaundice due to uncomplicated common duct obstruction, the degree of hypercholesterolemia increasing with that of hyperbilirubinemia and returning to normal with release of the obstruction. Changes in phospholipid concentration roughly parallel those in free cholesterol. In some cases the cholesterol esters increase concomitantly with the total cholesterol, the normal ratio between the free and the ester fractions being maintained; in others, the increase may occur mainly in the free fraction. In some cases, particularly with partial obstruction and mild hyperbilirubinemia, there is an increment in both free and ester fractions, the normal ratio, although somewhat altered, tending to be preserved. Usually, however, with increasing bilirubinemia, free cholesterol contributes increasingly to the hypercholesterolemia. Accordingly, the proportion of cholesterol esters decreases progressively; in some cases the absolute amount of this fraction may be decreased. The mechanism of production of these changes is not clear, since cholesterol esters are formed in the plasma (lecithin-cholesterol acyltransferase), but in the presence of obstruction, decrease of cholesterol esters cannot be reliably interpreted as indicative of superimposed hepatocellular damage.

Hypercholesterolemia may occur also in biliary cirrhosis, often with an increase in phospholipids. In some instances, apparently, the increase occurs mainly in the beta-lipoprotein fraction, in others in the alpha.

The mechanism of production of the increase in plasma cholesterol in obstructive jaundice is not known. The view that it is a retention phenomenon is generally regarded as untenable. However, this opinion may require modification in the light of present understanding that the cholesterol of both the bile and the blood plasma originates in the liver cells, in contrast to the former view that the hepatic cells remove cholesterol from the plasma and excrete it in the bile. As indicated elsewhere (p. 114), cholesterol removed from the plasma by the liver is excreted in the bile largely as bile acids, not as cholesterol. This conversion to bile acids (cholic acid) occurs in the hepatic cells. Evidence has been presented which suggests that the hypercholesterolemia of biliary obstruction is dependent, in some manner, on the associated increase in bile acids in the plasma.

Patients with jaundice of extrahepatic obstructive origin not infrequently present normal or even subnormal plasma cholesterol values. Such findings are usually dependent upon the presence of some complicating factor, among the most common of which are cachexia, infection, superimposed hepatocellular damage or terminal cholemia. The occurrence of hypocholesterolemia in these conditions is discussed elsewhere (p. 142 ff.). When this phenomenon is observed in patients with common duct obstruction it should be regarded as of serious prognostic significance, since it usually indicates the presence of one or more of the complicating factors enumerated above.

Hypercholesterolemia occurs at times in cases of mild hepatocellular jaundice (hepatitis), but much less frequently than in obstructive jaundice. When it does occur, it is usually dependent upon an increase in the free cholesterol fraction, the ester cholesterol-free cholesterol ratio being diminished. Except in the terminal stages, plasma cholesterol concentration and partition are normal in portal cirrhosis and are either normal or insignificantly elevated in cholecystitis or cholelithiasis without biliary obstruction. The hypothesis of a relationship of cause and effect between hypercholesterolemia and cholelithiasis has not been substantiated. There is no constant relationship between the concentration of cholesterol in the bile and in the blood plasma. When the plasma cholesterol concentration is raised by cholesterol administration in experimental animals, the major portion of the increment is transformed, in the liver, to bile acids and is excreted as such in the bile and not as cholesterol. It would appear that the formation of gallstones is probably dependent largely upon factors operating within the extrahepatic bile passages.

Hypothyroidism. Untreated hypothyroidism is rather consistently associated with an increase in blood lipids, the degree of hypercholesterolemia being roughly proportional to the diminution in the basal metabolic rate. Values above 600 mg. per 100 ml. have been reported in both adults and children. The increase occurs usually in both ester and free fractions.

The plasma phospholipids are also increased. A return to normal usually follows the administration of thyroid extract. A rise in plasma cholesterol is quite often an early sign of hypothyroidism.

Estimation of the plasma cholesterol concentration may be of value in the diagnosis and regulation of thyroid dosage in the treatment of children with hypothyroidism in whom the determination of the basal metabolic rate may not be practicable. However, although it tends to be elevated, at times markedly (above 500 mg. per 100 ml.), the cholesterol concentration in hypothyroid children has been found to fluctuate within wide limits upon repeated determination and to be within normal limits in some cases of severe hypothyroidism. Decrease in plasma cholesterol during thyroid hormone therapy and a marked increase following cessation of therapy have been found to be of considerable diagnostic significance in children.

Atherosclerosis. The relation of plasma lipoprotein abnormalities to the development or incidence of atherosclerosis is discussed elsewhere (p. 132).

Xanthomatosis. Xanthomatosis is the designation applied to the cholesterol variety of lipidosis (p. 130). In certain forms of xanthomatosis, the plasma cholesterol is within normal limits. This category includes the great majority of cases of the Hand-Schüller-Christian syndrome. Hypercholesterolemia is present in a small percentage; according to some, it occurs during active stages of the disease, normal values being obtained during periods of apparent quiescence. Hypercholesterolemia occurs also in xanthomatous biliary cirrhosis and in other (hypercholesterolemic) types of cutaneous and visceral xanthomatosis.

Xanthomata frequently develop in conditions characterized by protracted hypercholesterolemia. These conditions include hyperlipoproteinemias, chronic relapsing pancreatitis, diabetes mellitus, biliary cirrhosis, hypothyroidism, glycogen storage disease, etc. The plasma phospholipids are also increased. With the exception of occasional cases of biliary cirrhosis, the increment occurs largely if not exclusively in the β-lipoprotein fraction.

Miscellaneous. Whereas the plasma cholesterol is often low in patients with other forms of anemia (p. 142), high values may be obtained following acute hemorrhage. This increase may be pronounced within forty-eight hours after the acute blood loss and may persist for several days. The cause is not known. It may be due to hypoxia and excessive mobilization of lipids from the fat depots.

Hypercholesterolemia has been observed occasionally in patients with hypertrophic osteoarthritis, senile cataract, psoriasis, and dermatitis. Its occurrence in these conditions is probably fortuitous.

There have been claims of an increase in the plasma phospholipid : cholesterol ratio in eclampsia, due to an independent increase in phospholipids, with no significant change in cholesterol. This may be related to an increase in the concentration of beta- at the expense of alpha-lipoproteins, which has been reported to occur in certain types of kidney disease, e.g., glomerulonephritis, intercapillary glomerulosclerosis.

HYPOCHOLESTEROLEMIA

Anemia. The plasma cholesterol is uniformly low in pernicious anemia during relapse and in hemolytic jaundice, values as low as 50 mg. per 100 ml. having been reported. Hypocholesterolemia is likewise observed in practically all patients with severe hypochromic anemia with the exception of severe aplastic anemia and anemia following acute hemorrhage. In severe hypochromic anemias and also in pernicious anemia and in hemolytic jaundice the diminution in plasma cholesterol is associated with reduction in plasma phosphatide and an increase in fat, these changes occurring typically with hemoglobin values below 50 per cent but bearing no direct relation to the degree of anemia. In typical cases of pernicious anemia in relapse the cholesterol rises suddenly at the onset of remission, whether spontaneous or induced by treatment. This occurs simultaneously with the reticulocyte response and occurs usually before any increase in hemoglobin or red blood cells. Consequently, normal or even elevated plasma cholesterol values may be present in patients with severe anemia, provided that remission has begun. This may account for certain of the discrepancies in reported observations of plasma cholesterol concentration in this condition. It must be remembered that pernicious anemia is now recognized as an "auto-immune disease." This would explain its association with hypothyroidism, which can also show manifestations of auto-immunity. That pernicious anemia can coexist with hypothyroidism is well recognized clinically. This could account for a raised plasma cholesterol level.

Hepatic Disease. Hepatocellular damage is frequently accompanied by a diminished proportion of cholesterol esters in the blood plasma (normally 60 to 80 per cent of the total). This may not be, but usually is, associated with a diminution in total cholesterol concentration. The chief practical significance of this observation lies in whether it may aid in differentiating hepatocellular jaundice from simple obstructive jaundice, in which hypercholesterolemia is the rule (pp. 139 and 631). Its clinical application has proved disappointing. Hypocholesterolemia, with a diminished proportion of cholesterol esters, has been observed in cases of hepatocellular damage due to the following causes: (a) drugs, such as cinchophen, phenobarbital, chloroform, carbon tetrachloride, and phosphorus; (b) pneumonia, myocardial failure, yellow fever, and spirochetal jaundice; (c) infectious (viral) and toxic hepatitis, and acute, subacute, and chronic diffuse necrosis of the liver. Similar findings have been obtained in experimental animals.

A pronounced discrepancy between the degree of bilirubinemia and of cholesterolemia is usually observed in hepatocellular damage; the more severe the hepatic damage the greater the tendency toward hypocholesterolemia. This discrepancy between the hyperbilirubinemia and cholesterolemia may aid in differentiating between purely mechanical obstructive lesions and superimposed or primary degenerative lesions of the hepatic parenchyma. In parenchymatous disease of the liver, a drop in the ester fraction often parallels the severity of the hepatic damage, and in rapidly fatal cases this fraction may be very low throughout the course of the disease. In the event of improvement, the initially low ester values eventually increase. Low values may also be obtained in terminal stages of portal cirrhosis. Decrease of the ester fraction occurs commonly in association with hypercholesterolemia in subjects with uncomplicated common duct obstruction, particularly with marked hyperbilirubinemia. However, in cases of known obstruction, with persistent or increasing hyperbilirubinemia, simultaneously decreasing cholesterolemia and ester fraction could be regarded as of adverse prognostic significance in relation to possible superimposed hepatocellular damage.

The occurrence of hypocholesterolemia in these conditions is explicable on the basis of present knowledge concerning the role of the liver in the synthesis of the plasma cholesterol. Experimental evidence suggests that the hepatic cell is probably the most important site of formation of the plasma cholesterol (p. 112). Impairment of liver cell function would therefore be expected to result in a decrease in total cholesterol levels. The usefulness of this determination in the differential diagnosis of obstructive and hepatocellular jaundice is discussed elsewhere (p. 631).

Infection. Hypocholesterolemia and a diminution in the concentration of other blood lipids are observed frequently during the course of acute infectious diseases. The most marked decrease usually occurs in the ester fraction which, together with the total cholesterol concentration, rises to normal and sometimes supernormal levels during convalescence. Similar changes occur simultaneously in fatty acids and phospholipids. There appears to be no constant relationship between the degree of hypocholesterolemia and the severity of the clinical manifestations. Such changes cannot be produced by the induction of artificial fever. According to some observers, hypocholesterolemia is of prognostic significance in tuberculosis, in which condition it appears to be related to the severity of the process. The cause of these changes in plasma cholesterol in infectious diseases is not understood.

Hyperthyroidism. In thyroid disease, there appears to be a roughly reciprocal relationship between basal metabolic rate and plasma cholesterol concentration. Low values (60 to 100 mg. per 100 ml.) may be obtained in patients in or near thyroid crisis. The average values in exophthalmic goiter are somewhat lower than in toxic nodular goiter. The apparent absence of such changes with increased basal metabolic rates due to administration of dinitrophenol suggests that the changes in plasma cholesterol in thyroid disease are not directly related to the metabolic rate but to other actions of thyroid hormone. The acetic acid analogs of thyroxine and triiodothyronine lower the serum cholesterol without causing significant changes in BMR. Diminution in the basal metabolic rate in patients with hyperthyroidism following administration of iodine or antithyroid drugs or subtotal thyroidectomy is usually accompanied by an increase in plasma cholesterol concentration.

Determination of plasma cholesterol is of little clinical value in hyperthyroidism. The nutritional status exerts a considerable influence and, whereas the plasma lipid level may be related to the degree of thyroid activity in uncomplicated cases, the problems of diagnosis and prognosis in such cases are relatively simple and are not usually further clarified by investigation of the plasma cholesterol concentration. Moreover, the latter has been found to be no criterion of the probable response of the patient to thyroidectomy or other specific therapy.

Inanition. A decrease in plasma cholesterol is commonly observed in conditions accompanied by wasting and cachexia. Its significance in these conditions is not clear. In many cases there is an associated decrease in plasma albumin, both phenomena apparently being directly related to the state of malnutrition.

Terminal States. The frequent occurrence of hypercholesterolemia in the nephrotic syndrome has been referred to elsewhere (p. 138). In non-nephrotic forms of glomerulonephritis, however, although the total fat, fatty acid, and phospholipid concentrations in the plasma are commonly increased, plasma cholesterol is frequently normal or subnormal, this dissociation suggesting a functional differentiation between these lipid fractions. The variability of the plasma cholesterol in chronic glomerulonephritis has attracted considerable attention, practically all investigators of this problem concurring in the opinion that a fall from a previously elevated or normal level is of serious prognostic significance, particularly if associated with increasing nitrogen retention. There usually appears to be a roughly reciprocal relationship between these two factors in the terminal stages of this disease, although there is no constant quantitative relationship between the degree of cholesterolemia and of nitrogen retention.

Low plasma cholesterol values (50 to 100 mg. per 100 ml.) have also been obtained in patients in the terminal stages of a variety of diseases without associated nitrogen retention Among these are congestive heart failure, carcinoma, acute pancreatitis, pulmonary tuberculosis, bacterial endocarditis and other forms of bacteremia, coronary artery occlusion, and diabetes mellitus.

No definite statement can be made at the present time regarding the cause of hypocholesterolemia in terminal states. Some believe that it is due to anemia and cachexia, which are frequently present in such patients. However, these factors are frequently absent in those dying of peritonitis, pneumonia, coronary artery occlusion, congestive and left-sided heart failure, and in certain patients with urinary obstruction due to prostatic enlargement. If one is to assume that the mechanism underlying the production of hypocholesterolemia in various terminal states is essentially the same in each case, it would appear that some other explanation must be sought. There is some evidence that this mechanism involves an increased rate of removal of cholesterol from the blood as a result of a state of increased reticuloendothelial cell activity. During periods of stress, both physical and mental, the adrenal cortex is stimulated to increased activity, an important feature of the "alarm reaction" or adaptation syndrome. As indicated elsewhere, such stimulation is accompanied by a decrease in adrenal cholesterol (p. 704), continuing to marked depletion in severe conditions of stress. It is possible that, under such circumstances, the cholesterol stores of the body may be depleted, leading to hypocholesterolemia. A similar mechanism may operate in severe infections (p. 143). It should be recalled, however, that experimental evidence indicates that, under physiological conditions at least, relatively little cholesterol is removed from the plasma by tissues other than the liver. Whatever may be the cause, the development of hypocholesterolemia under such circumstances is of serious prognostic significance, especially in diabetes mellitus and chronic glomerulonephritis, in which the plasma cholesterol may previously have been elevated.

Miscellaneous. Hypocholesterolemia has been observed in patients with prostatic obstruction and with intestinal obstruction, its occurrence in these conditions being of rather serious prognostic import. Low plasma cholesterol values have also been reported during epileptic seizures and in arthritis, sprue, and celiac disease.

References

Adlersberg, D., ed.: The Malabsorption Syndrome. New York, Grune & Stratton, Inc., 1957.
Bloch, K., ed.: Lipide Metabolism. New York, John Wiley & Sons, Inc., 1960.
Cantarow, A., and Schepartz, B.: Biochemistry. Ed. 4. Philadelphia, W. B. Saunders Company, 1967.
Deuel, H. J., Jr.: The Lipids. New York, Interscience Publishers, Inc., Vol. II, 1955; Vol. III, 1957.
Drury, A., and Treadwell, C. R., eds.: The influence of hormones on lipid metabolism in relation to atherosclerosis. Ann. New York Acad. Sci., 72:787, 1959.
Eder, H. A.: Lipoproteins of human serum. Am. J. Med., 23:269, 1957.
Eder, H. A.: The effect of hormones on human serum lipoproteins. Recent Progress in Hormone Research, 14:405, 1958.
Frazer, A. C.: Fat absorption and its disorders. Brit. Med. Bull., 14:212, 1958.
Fredrickson, D. S.: Disorders of lipid metabolism and xanthomatosis. In Harrison's Principles of Internal Medicine. Ed. 6. New York, McGraw-Hill Inc., 1970, p. 629.
Fredrickson, D. S., and Gordon, R. S., Jr.: Transport of fatty acids. Physiol. Rev., 38: 585, 1958.
Glover, J., and Morton, R. A.: The absorption and metabolism of sterols. Brit. Med. Bull., 14:226, 1958.
Gofman, J. W., et al.: Lipoproteins, coronary heart disease, and atherosclerosis. Physiol. Rev., 34:589, 1954.
Kannel, W. B., Castelli, W. P., and McNamara, P. M.: The coronary profile: 12-year follow-up in the Framingham study. J. Occup. Med., 9:611, 1967.
King, H. K.: The Chemistry of Lipids in Health and Disease. Springfield, Ill., Charles C Thomas, 1960.

Krebs, H.: Biochemical aspects of ketosis. Proc. Roy. Soc. Med., 53:71, 1960.
Kritchevsky, D.: Cholesterol. New York, John Wiley & Sons, Inc., 1958.
McGilvery, R. W.: Biochemistry: A Functional Approach. Philadelphia, W. B. Saunders Company, 1970.
Olson, R. E.: Nutrition-endocrine interrelationships in the control of fat transport in man. Physiol. Rev., 40:677, 1960.
Portman, O. W.: Dietary regulation of serum cholesterol levels. Physiol. Rev., 39:407, 1959.
Scheig, R.: Diseases of lipid metabolism. In Bondy, P. K., ed.: Duncan's Diseases of Metabolism. Ed. 6. Philadelphia, W. B. Saunders Company, 1969, p. 295.
Schwartz, K., et al.: Symposium on nutritional factors and liver diseases. Ann. New York Acad. Sci., 57:615, 1954 (fatty liver).
Symposium on Lipids in Health and Disease. Federation Proc., 20:115, 1961.
Tietz, N. W., ed.: Fundamentals of Clinical Chemistry. Philadelphia, W. B. Saunders Company, 1970.
Wertheimer, E., and Shafrir, E.: Influence of hormones on adipose tissue as a center of fat metabolism. Recent Progress in Hormone Research, 16:467, 1960.

Chapter 3

PROTEIN METABOLISM

Proteins may be defined as compounds of high molecular weight, consisting largely (or entirely) of chains of α-amino acids united in peptide linkage. The constituent amino acids can be obtained by hydrolysis of the proteins. While the individual amino acids may be considered analogous to the monosaccharides of carbohydrate chemistry, and the true proteins analogous to the polysaccharides, an important difference must be pointed out. The polysaccharides are polymers of a single sub-unit, or of a few types of sub-units at the most, while the proteins contain, in general, some twenty-odd individual amino acids, present in characteristic proportions and linked in a specific sequence in each protein. Hence, other things being equal, it is possible for nature to concoct many more different proteins than polysaccharides, a fact which lends fascination as well as difficulty to the problems confronting the protein chemist.

The molecular weights of the proteins vary from about 5000 to many millions. Most proteins, because of their molecular size, are not diffusible through membranes such as cellophane and, like the polysaccharides, are actually of colloidal dimensions and exhibit the properties associated with the colloidal state of matter. Some are soluble in pure water; some require the presence of salts or small amounts of acid or base to dissolve. One group is soluble in certain concentrations of alcohol, although proteins are generally insoluble in organic solvents. The structural proteins known as scleroproteins are dissolved only by reagents which cause considerable alterations in their structure.

The average protein is a very sensitive individual. In reprisal for its exposure to heat, extremes of pH, surface action, or various reagents, the native protein undergoes a series of changes known as denaturation, resulting in alterations in a number of its properties. Heat denaturation, under the proper circumstances, proceeds to a further state, coagulation.

The biochemical significance of proteins cannot be overemphasized. If carbohydrates and lipids, generally speaking, can be considered the fuels of the metabolic furnace, proteins may be regarded as forming not only the structural framework, but also the gears and levers of the operating machinery. Indeed, at the risk of pushing the analogy to extremes, we may regard the protein hormones (which acts as regulators of metabolism) as the policy-forming top management of the enterprise.

Such structures as cell walls and various membranes are protein coupled or associated with lipids. Connective tissue is composed of chemically rather inert types of proteins, in some cases combined with carbohydrates. The contractile elements of muscle are proteins. Aside from such obviously structural substances as those mentioned, the catalysts which permit the reactions of intermediary metabolism to proceed at a reasonable rate under mild conditions, the enzymes, are known to be protein in nature. Finally, as already pointed out, certain proteins in the body act as hormones, or regulators of metabolism.

Protoplasm may be regarded as a number of organized colloidal systems of proteins, together with some lipids and carbohydrates as well as metabolites. With some exceptions, the supply of carbohydrates and lipids may rise or fall, and the cell may flourish or do poorly as a result, but at any rate it survives for some time. If something interferes with the supply of protein, however, the cell inevitably sickens and dies.

DIGESTION AND ABSORPTION

Ingested proteins undergo digestion in the stomach and small intestine. This process consists essentially in hydrolysis into their constituent amino acids and some small peptides by means of enzymes, which are either *endopeptidases* or *exopeptidases*. The former break down peptide bonds of the inner portion of the amino acid chain and the latter attack only the terminal bonds. They are known as aminopeptidases when the N-terminal bond is involved and carboxypeptidases when the C-terminal bond is attacked. It is obvious that the enzymes, if active, could attack the tissue in which they were produced. They are, however, synthesized in inactive forms, known as proenzymes. These have an additional length of peptide chain, which must be removed for conversion to the active enzyme.

A typical proenzyme, pepsinogen, is secreted by the chief cells of the gastric mucosa. A portion of it is activated by the hydrochloric acid produced by the parietal cells. Once formed, pepsin the active enzyme, an endopeptidase, can activate more pepsinogen by so-called autocatalysis. There are actually a number of different pepsins. These initiate the hydrolysis of food proteins, either in their native state or denatured by cooking or hydrochloric acid. The enzymes show a preference for peptide bonds involving an aromatic amino acid (phenylalanine, tyrosine and tryptophan). Pepsins also cause milk to clot. Even though pepsin may be absent from the gastric juice (e.g., achylia gastrica), this does not have any obvious effect on the eventual breakdown of protein in the intestinal tract.

The mixture of partially broken down proteins and acid is discharged into the small intestine and the pH made slightly alkaline by the bicarbonate present in pancreatic juice. The latter also contains three endopeptidases, trypsin, chymotrypsin and elastase, and two exopeptidases, carboxypeptidases A and B. All these enzymes are also initially secreted as inactive proenzymes. Trypsinogen is first activated by a proteolytic enzyme, enterokinase, formed in the small intestine. The active trypsin then autocatalytically activates the remaining trypsinogen and plays a part in activating the other proenzymes. Trypsin splits peptide bonds involving the basic amino acids, chymotrypsin the aromatic amino acids, and elastase the nonpolar amino acids. Carboxypeptidase A liberates carboxy- terminal amino acids and carboxypeptidase B only carboxy- terminal basic amino acids. A small protein, which is a trypsin inhibitor is also secreted by the pancreas. This could be a form of insurance against the production of any active trypsin within the gland. This inhibitor is destroyed when sufficient trypsin has been produced in the intestine.

By virtue of these various enzyme actions, a mixture of amino acids and small peptides (not more than 8 amino acid residues) are presented for absorption by the cells of the intestinal mucosa. Within these cells there are a group of aminopeptidases, e.g., leucine aminopeptidase, and a number of dipeptidases, which act specifically on certain dipeptides. The amino acids and a small amount of the small peptides are now absorbed into the portal blood.

Amino acid transport is carried out by at least three different active processes. One process involves cystine and the basic amino acids, another the amino acids proline and hydroxyproline, and a third the neutral amino acids. The latter include the amides (asparagine and glutamine) of the dicarboxylic acids (aspartic and glutamic). Little seems to be known about the mechanism of absorption of these two acidic amino acids. Amino acid transport seems to require sodium ions and, possibly, pyridoxine. It is very likely that the various transport mechanisms closely resemble or are identical with those present in the renal tubules.

It appears that occasionally whole protein can be absorbed into the blood. This could account for food allergies. In the young mammal, the permeability of the mucosa, in this respect, is greater than that in the adult.

Food proteins are generally readily digested (90 to 97 per cent) under normal conditions, very little escaping in the feces. The only important exception is the insoluble fibrous protein, keratin, which is not hydrolyzed by enzymes of the human digestive tract. Most proteins are profoundly altered by many procedures commonly used in the preparation of foods. Heating to coagulation temperatures causes polymerization; superheated steam, excluding air (pressure cookers), causes hydrolysis; dry heat may cause oxidation. In the majority of instances, the digestibility and biological value (p. 156) of these proteins are not affected by such procedures. However, proper cooking may facilitate digestion and utilization, e.g., cooked egg albumin is digested more readily than raw; heating soybeans increases their biological value by inactivating a component with antitryptic activity. The nutritional value of cereal proteins is lowered by overheating or toasting (e.g., certain breakfast cereals).

With the few exceptions indicated above (i.e., small peptides, occasionally native protein), food proteins enter the organism in the form of their constituent amino acids. Certain other aspects of protein digestion may have an important bearing on their nutritional value, e.g., variation in the rates of liberation of amino acids. The amino acids are readily soluble in water and are promptly absorbed from the small intestine, mainly into the portal circulation (to the liver), and to a much smaller extent, via the lacteals, into the thoracic duct and thence directly into the systemic circulation. Only small amounts of free amino acid are found in the intestinal contents during the process of digestion, indicating the rapidity of their absorption. This is reflected also in the rather prompt postprandial increase in blood amino acid concentration (p. 160).

Dynamic State. It was realized even for some time prior to the use of isotopes, but particularly clearly since the introduction of labeling techniques, that the constituents of the body are in a constant state of flux. All body proteins, including plasma proteins, hemoglobin, and intracellular proteins, continually undergo degradation and synthesis, more than half of the protein of the liver and intestinal mucosa being broken down and resynthesized in ten days; the turnover is slower in muscles and erythrocytes. Active resynthesis occurs even during periods of starvation, and active breakdown during periods of nitrogen equilibrium; degradation of protein in one tissue may be accompanied by synthesis in others. Antibody protein (γ-globulins), induced by active immunization, also undergo continual breakdown and synthesis, the half-life of these and other plasma proteins being about two weeks.

Protein in one compartment of the organism may be drawn upon to supply a deficiency in another compartment. For example, either plasma protein or hemoglobin (in the dog), given intravenously, may supply the protein requirements during prolonged fasting periods.

The apparent stability of the adult organism is the result of a balance between the rates of synthesis and degradation of its constituents. In a growing organism, the rates of synthesis of many of its constituents must exceed the rates of breakdown in order that new tissue may be constructed. Wasting diseases, starvation, and related states are characterized by rates of catabolism which are greater than the rates of anabolism.

In a given physiological (or pathological) state, the rate of synthesis or degradation of a specific compound will be characteristic of (1) the chemical nature of the compound in question, (2) its location in a particular tissue or organ, and (3) its intracellular location.

Several methods of expression are currently used to describe the rate of synthesis or breakdown of a body constituent. Most of these have been developed in connection with isotopic-labeling techniques. The "half-life" is the time required for replacement of one-half of the molecules of the compound in question, the definition being analogous to that used in describing the unstable isotopes. The "average life" of a given compound is the time required to replace the molecule of "average stability." Sometimes the rate of a synthetic or degradative process is expressed directly in terms of mass of compound transformed per unit time per unit weight of tissue or animal, e.g., "moles of urea synthesized per day per kilo body weight." On other occasions, the rate of turnover of a compound is stated as the percentage of the amount present which is metabolized per unit time.

Metabolic Pool. Since the "endogenous" molecules are in a state of flux, they and their metabolites must be constantly mixing with those derived from the diet ("exogenous"). This mixture of originally endogenous and exogenous materials, which is drawn upon for both anabolic and catabolic reactions, constitutes a reservoir or "metabolic pool" of the compounds in question. For example, the alanine derived from the continuous breakdown of body protein and that obtained from dietary protein mix to form an "alanine pool." The amino group of the alanine, together with the amino groups of all other amino acids which are able to participate in nitrogen exchange reactions (e.g., transamination), form a pool of metabolically labile nitrogen. (Lysine and threonine are exceptional, in that they contribute their amino nitrogen to the general pool but do not accept nitrogen from this source.) This pool is contributed to by substances derived from catabolic processes in the tissues and substances absorbed from the intestine. Catabolism of proteins is stimulated by 11-oxygenated adrenocortical hormones and excessive amounts of thyroid hormone. Protein anabolism is stimulated by pituitary growth hormone, insulin, testosterone, and, at least in the growing organism, by physiological amounts of thyroid hormone.

Through certain compounds such as pyruvate, acetate, oxaloacetate, α-ketoglutarate, etc. (pp. 18 and 162), the metabolism of protein is integrated with that of carbohydrate and fat, members of one pool being thus metabolically equilibrated with other pools. Alanine, for example, is reversibly converted by deamination to pyruvic acid; hence the carbon skeleton of this amino acid is involved in the pyruvate pool, which is directly connected to the carbohydrate pool.

The size of a pool is the quantity of constituent instantaneously present and available for all of the reactions leading into and from that particular pool. The pool of metabolically labile amino groups in the human being, for example, has been estimated at 2 g. of nitrogen (70-kg. man), i.e., of the same order of magnitude as the free amino acids of the tissues.

The concepts of dynamic state and metabolic pool have been invaluable in the interpretation of data resulting from experiments with isotopic labeling. For example, consider the case of an adult experimental animal in nitrogen balance being fed a dose of glycine labeled with ^{15}N. According to the formerly held theories of biochemistry, the metabolic paths of endogenous and exogenous compounds are compartmentalized; hence exogenous glycine, since it is not required for tissue construction (adult animal, nitrogen equilibrium), should be degraded directly and the nitrogen excreted as urea within approximately twenty-four hours. What is found experimentally is that a quantity of urea is excreted equivalent to the glycine administered. However, only a small fraction of the ^{15}N is excreted in this urea; several days are required for the recovery of an appreciable fraction. Hence, the nitrogen output is equivalent to the intake, but the atoms excreted are mainly not those ingested. To begin with, the ingested glycine enters the glycine pool, so that the concentration of ^{15}N in the average glycine molecule is lowered by dilution with all of the ^{14}N-glycine molecules. Secondly, the amino group of glycine is readily equilibrated with that of most other amino acids (transamination, p. 159), resulting in further dilution or spread of ^{15}N throughout most of the amino acids of the body. The nitrogen pool which is drawn upon for the synthesis of urea, therefore, contains a much lower concentration of molecules labeled with ^{15}N than did the sample of glycine administered. It is obvious that the concept of independent exogenous and endogenous metabolism had to be abandoned.

OVER-ALL METABOLISM OF PROTEIN

Much of the information on the metabolism of protein can be expressed (and is frequently measured) in terms of nitrogen, the reasons for which are readily apparent. Carbon and hydrogen atoms of the proteins can form CO_2 and H_2O as end-products of metabolism, indistinguishable from the end-products of the metabolism of carbohydrates or fatty acids. Although certain proteins contain phosphoric acid, this constituent is more characteristic of the nucleic acids, phospholipids, and certain intermediates in carbohydrate metabolism. Sulfur is an almost invariable constituent of proteins, but its metabolism reflects mainly the reactions of cystine and methionine. In contrast to these elements, nitrogen is uniquely characteristic of proteins, from the standpoints of both its high concentration in the protein molecule (average, 16 per cent) and its specialized excretory products (ammonia, urea, etc.).

Nitrogen of the Food. Protein nitrogen outweighs all other forms of nitrogen in the diet. Traces of inorganic nitrogen are ingested in the form of nitrates and nitrites. Small amounts of organic nonprotein nitrogen (NPN) are also present in the food, including nucleic acids and their derivatives, and amino acids and peptides.

Nitrogen of the Body. Many nitrogenous compounds are found in the tissues and fluids of the body. Protein itself averages about 20 per cent of the wet weight of most tissues. In the blood, in addition to protein, several types of NPN are found. Urea, the major waste product of protein catabolism, formed in the liver, passes into the blood stream to the kidneys for excretion. Blood creatine may be en route to muscle for the synthesis of phosphocreatine, or may have leaked out of muscle. Creatinine is a waste product, formed from phosphocreatine and creatine. Uric acid is the end-product of purine catabolism. Free amino acids are also found in the blood; they are in transit from one organ to another, for purposes of synthesis or breakdown. Other components of the blood NPN include polypeptides, gluthathione, purines and pyrimidines, ATP, and ergothioneine.

Excretion of Nitrogen. Fecal nitrogen appears not to be related to the nitrogen of the ingested protein. Its quantity varies with the bulk of the diet and does not normally represent unabsorbed dietary protein. In the adult human, it amounts to 1 to 2 g. N/day. Perspiration, unless sweating is excessive, accounts for a loss of only 0.3 g. N/day or less.

The urine is the major route of excretion of nitrogen (p. 179). In the average, normal, human adult, the total nitrogen of the urine is about 13 g./day, and this is normally all NPN (no significant quantities of protein are found in the urine normally). Urea (ca. 85 per cent of the total) and ammonia (ca. 3 per cent) vary directly with the level of protein in the diet. Creatinine (5 per cent of the total) excretion is related to the muscle mass and is almost constant for an individual. Creatine excretion is more commonly found in young children than in adults, and women may possibly excrete more than men. In either case, the amounts are small. Excretion, however, is raised in pregnancy as well as in the early postpartum period. Only traces of amino acids are found in the urine. Uric acid (1 per cent of total) output fluctuates with the level of dietary purines. About 5 per cent of the total is composed of compounds not ordinarily determined. The factors which influence the urinary excretion of the various nitrogenous substances are discussed elsewhere (p. 179).

Nitrogen Balance. Since most of the nitrogen of the diet represents protein, and most of the nitrogenous excretory products are derived from protein catabolism, it is apparent that the balance between the two will reveal significant features of protein metabolism. Nitrogen balance is defined as the quantitative difference between the nitrogen intake and the nitrogen output, both expressed in the same units (such as grams N/day). By intake is meant the nitrogen of the food. Included in output are such routes of excretion as urine, feces, milk, sweat, expectoration, vomitus, desquamation of skin, menstrual fluid, and loss of hair; in practice, however, only the first two are taken into account, except under unusual circumstances.

Positive nitrogen balance exists when intake exceeds output. This condition obtains whenever new tissue is being synthesized, such as during growth of the young, in pregnancy, and in convalescence from states of negative nitrogen balance.

In negative nitrogen balance, the output exceeds the intake. This condition, which obviously cannot continue indefinitely, occurs on inadequate intake of protein (fasting, diseases of the gastrointestinal tract), in states of accelerated catabolism of tissue protein (fevers, infections, wasting diseases, trauma, including postoperative), and when the loss of protein from the body is accelerated in some way (lactation on inadequate diets, albuminuria, protein-losing gastroenteropathy). Negative nitrogen balance occurs also in the absence from the diet of an "essential" amino acid.

From the preceding discussion, it is evident that nitrogen equilibrium (intake = output) can be achieved only in the adult organism, and only in the absence of the abnormal conditions enumerated above. An "adequate diet" must also be specified, which means that all requirements for minerals and vitamins are fulfilled, that the protein of the diet is of high "biological value" and is administered at a sufficiently high level, and that the caloric needs of the body are met satisfactorily by the carbohydrate and fat of the diet. A certain minimum amount of dietary carbohydrate is required for the maintenance of nitrogen equilibrium (5 g./100 Calories), independently of the adequacy of the energy provision from fat and protein. This is referred to as the "protein-sparing action" of carbohydrate (pp. 157 and 164). In these circumstances, the nitrogen output equals the intake, and an increase or decrease in the intake is followed, within a day or two, by a corresponding adjustment in the output, so that nitrogen equilibrium is established at a new level. There appears to be no marked tendency toward storage of surplus nitrogen, as there is in the case of fats and carbohydrates.

Essential Amino Acids. It is possible to replace the protein of the diet completely with pure amino acids. The so-called "dietary requirement" for protein is, therefore, really a requirement for amino acids and is expressed in terms of protein only because the amino acids of the available foodstuffs are found in that form.

By the elimination of single amino acids from an otherwise complete diet, it has been found that the organism can successfully dispense with certain of these compounds, but not with others. Evidently all amino acids are not synthesized with equal facility by the body. An "essential," or "indispensable," amino acid is defined as one which cannot be synthesized by the organism, from substances ordinarily present in the diet, at a rate commensurate with certain physiological requirements. The definition includes the phrase, "cannot be synthesized by the organism, from substances ordinarily present in the diet . . . ," because certain of these amino acids may be replaced by the corresponding α-keto or α-hydroxy acids. The requirements specified in the most extensive series of investigations are: (1) optimal growth of the young, and (2) maintenance of nitrogen equilibrium in the adult.

Although the amino acid requirement for growth has been investigated in several species, the most thorough experiments have been performed in the white rat. The ten amino acids required by this animal for optimal growth (and, in the absence of data to the contrary, the same list is assumed to apply to other mammals) are: isoleucine, leucine, lysine, methionine, phenylalanine, threonine, tryptophan, valine, arginine and histidine. In the case of arginine, the requirement is not absolute; the animals grow, but suboptimally, in its absence, indicating a slow and (for this physiological requirement) inadequate rate of synthesis of this amino acid.

The requirement for nitrogen equilibrium in the adult appears to be less stringent. In the human, only eight amino acids are required for maintenance of nitrogen equilibrium, the list differing from the preceding in the absence of histidine and arginine. These are, however, probably essential for normal growth in childhood, although arginine does not appear to be essential for the human infant.

The requirement for certain essential amino acids may be spared by the inclusion of specific nonessential amino acids in the diet. In each case, the sparing action appears to be explicable on the basis that a significant fraction of the essential amino acid is required for synthesis of a related nonessential amino acid, a burden which may be removed by inclusion of the latter in the diet. Thus, tyrosine spares phenylalanine, cystine spares methionine, and glutamic acid and proline spare arginine (in the growing animal). In phenylketonuric individuals (p. 226), who are unable to convert phenylalanine into tyrosine, the latter becomes an essential amino acid.

A curious fact concerning the essential amino acids is that the complete group must be administered to the organism simultaneously. If a single essential amino acid is omitted from the group and fed separately several hours later, the nutritional effectiveness of the entire group is impaired.

It might logically be expected that the optimal ratio of essential amino acids in the diet should approximate that found in the carcass of the animal concerned. This appears to be true. In fact, significant deviations from this ratio result in a number of adverse effects:

1. Certain amino acids are toxic to the experimental animal when fed at a high level. In some instances, the toxicity is counteracted by increasing the concentration of a structurally related amino acid in the diet. Such cases are probably genuine examples of metabolic antagonism.

2. Vitamin deficiencies can be aggravated by addition to the diet of increased quantities of a single amino acid (or by improvement of the dietary intake of protein in general). These effects may be due to an increase in the rate of amino acid metabolism, leading to increased utilization and breakdown of vitamins.

3. True imbalance of amino acids can occur when the diet is marginal or suboptimal in one essential amino acid, whereupon increasing the dietary level of another essential amino acid sets up an imbalance and growth is decreased. This condition may be corrected by supplementation with small amounts of the limiting amino acid. The explanation appears to be that the body uses amino acids for synthesis of protein in ratios required by the carcass composition, in quantities determined by the limiting amino acid. The excess is catabolized. Furthermore, the process of catabolizing the excess may increase the rate of general amino acid catabolism, including that of the limiting amino acid, thus aggravating the condition.

From a general point of view, the negative nitrogen balance or decrease of growth resulting from omission of an essential amino acid from the diet, or its separate administration at a time appreciably different from the rest of the diet, may be considered special cases of amino acid imbalance.

The lists of essential amino acids must be taken provisionally, particularly in their application to human nutrition. In the case of the ten amino acids required for growth (rat), there is a certain risk in transferring nutritional requirements from one species to another. Although the nitrogen equilibrium experiments in human adults seem satisfactory, they were run for periods of time which represent only small fractions of the lifetime of the individual. It is possible that additional requirements, quantitative and qualitative, exist if longer periods of time are considered, or if the individual is subjected to stress.

Biological Value of Proteins. In order that a food protein may have value to the organism, it must be digested and absorbed. Keratins, for example, are practically refractory to the action of digestive enzymes and, consequently, are of no value in the diet. Even if digestible, a protein may be nutritionally inadequate owing to the absence (or presence in inadequate amount) of an essential amino acid. Gelatin (no tryptophan) and zein (no lysine, low tryptophan) are in this category. In general, animal proteins are nutritionally superior to plant proteins. The latter may be satisfactory if the food contains a sufficiently wide variety, but many individual plant proteins are seriously deficient in one or more of the essential amino acids.

Dietary Protein Requirements. *(a) General Nutrition.* If it is assumed that the protein of the diet is biologically adequate, the nutritional requirements can be discussed in terms of quantities of protein, rather than of amino acids. As a further assumption, sufficient fat and carbohydrate must be included in the diet to cover the caloric requirements of the individual. Increased requirements occur in certain situations, viz., pregnancy, lactation, and convalescence.

In connection with protein nutrition, the subject of "storage protein" is of interest. From the relatively rapid adjustment of nitrogen excretion to variations in intake in adult organisms, it appears that no significant quantity of protein is stored. It is nevertheless true that, in states of protein deprivation, certain tissues are able to contribute significant quantities of their protein for the preservation of certain vital functions. Although there is no chemically or morphologically distinct storage form of protein, corresponding to glycogen or depot fat, the more metabolically labile proteins of the tissues are available in case of need. These include the plasma proteins (however, other tissue proteins are usually sacrificed to provide plasma protein), and the proteins of the liver, gastrointestinal tract, and kidney. Only a small percentage of muscle protein appears to be labile, but the size of the muscle mass is such that it provides the greatest absolute amount of protein in conditions of deprivation.

(b) Synthesis of Special Proteins. If some of the blood is removed from an animal, centrifuged, and the cells suspended in saline and reinjected into the same animal, it is possible to lower the plasma protein level considerably ("plasmapheresis" technique). Initial bleedings of this type in dogs result in rapid regeneration of the plasma protein (from the labile protein of the tissues). Repeated plasmapheresis of dogs kept on appropriate diets eventually depresses the plasma proteins to a low, constant level. The effect can then be tested of various proteins or amino acids in the diet on the synthesis of plasma proteins.

Similar experiments in which the red cells are removed and the plasma replaced lead to an anemia-type of depletion. Iron, of course, is very important in the regeneration of hemoglobin. It is of interest that, in animals suffering from a double deficiency (plasma proteins and hemoglobin), the process of regeneration favors hemoglobin over the plasma proteins.

As might be expected, dietary proteins effective in the repletion of the protein of one depleted tissue often are ineffective in the "nutrition" of another. Restoration of the protein content of the livers of animals which have been deprived of proteins for various periods of time is accomplished effectively by certain dietary proteins which are considerably less efficient in the regeneration of plasma proteins.

(c) Protein-Sparing Effects. It has been observed that the dietary requirement for protein is influenced markedly by the level in the diet of fat and carbohydrate, these latter foodstuffs appearing to have a "protein-sparing" effect. This phenomenon is best understood by consideration of the physiological functions of the major foodstuffs. Fat, although it has certain structural uses, functions, primarily as a fuel (9 Cal./g.). Carbohydrate also serves as a fuel (4 Cal./g.), but, in addition, is required for the synthesis of certain catalytic compounds of metabolic cycles (e.g., oxaloacetic acid in the tricarboxylic acid cycle), and provides the carbon skeletons for the synthesis of the nonessential amino acids. Regardless of the adequacy of caloric intake in the form of proteins and fat, at least 5 g. of carbohydrate per 100 Calories must be supplied if nitrogen equilibrium is to be maintained (p. 154). Protein has an obvious structural role, but even more important from the dynamic point of view is its catalytic function, the enzyme proteins forming a major part of the actual working machinery of the body.

In the absence of fat and carbohydrate from the diet, protein must be degraded to provide fuel (4 Cal./g.) and catalytic compounds of metabolic cycles. Essential amino acids may be broken down to supply the raw materials for the synthesis of the nonessential amino acids. To provide for these increased burdens, the protein intake consequently must be increased. Conversely, addition of fat to the diet will take care of the caloric needs, a relatively small amount of carbohydrate will furnish the catalytic compounds of the cycles and carbon skeletons of the nonessential amino acids, and the protein requirement will be decreased as a result.

(d) Protein Deprivation. As stated previously, protein forms the structural framework of the body and the active machinery as well. Liver, intestinal mucosa, and kidney are sites of very active protein metabolism. The turnover rate of protein is very high in these tissues, and they also happen to discharge functions in protein digestion or metabolism which are important to the body as a whole. Protein deprivation, therefore, is particularly harmful to the activity of these tissues. Impairment of their metabolic efficiency results in further disturbances in protein digestion and metabolism, thus aggravating the state of deprivation. A "cycle of degeneration" sets in, reversal of which becomes increasingly difficult. It is obvious, therefore, why so much emphasis has been placed in recent years on the state of protein nutrition of the patient.

INTERMEDIARY METABOLISM

Protein Turnover. As indicated elsewhere (p. 150), proteins in the body are constantly breaking down to their constituent amino acids and being resynthesized (dynamic state), the rate of this turnover varying for different proteins in different tissues. The amino acids from the body proteins and those derived from the food form a metabolic pool, the members of which are drawn upon for reactions in either the anabolic or catabolic direction.

The metabolic lability of the proteins in the various tissues can be investigated by noting the rapidity with which each tissue loses protein when the organism is depleted (i.e., on a protein-free diet) and regains protein on repletion. A more recent technique measures the rate of uptake into the proteins of each tissue of an isotopically labeled amino acid or, in certain cases, the rate of loss of an isotope already incorporated. Whatever the method, high rates of turnover have been shown in plasma proteins, intestinal mucosa, pancreas, liver and kidney, and low rates in muscle, skin and brain. In man, the half-life of the plasma and liver proteins has been found to be about ten days, that of muscle about one hundred and eighty days. Tissue proteins are degraded by intracellular proteolytic enzymes ("cathepsins"), the intracellular and extracellular amino acids tending to become equilibrated in a common pool, although absolute uniformity of distribution obviously cannot be attained because of the differences in protein composition of various tissues. The concentration of free amino acids within the cells is many times that in the blood plasma.

General Pathways of Protein Metabolism. The most important of the general metabolic pathways of amino acids are illustrated in Figure 3-1. Some follow special metabolic routes; among these are reactions leading to the final excretion of sulfate (from methionine and cystine) and creatinine (from glycine and arginine). General features of the intermediary metabolism of amino acids of most importance from the clinical standpoint will be outlined here. Further details concerning important metabolites in the blood and urine are discussed elsewhere (pp. 177 and 179).

Circulating Amino Acids (p. 178). Amino acids absorbed from the intestine and escaping primary changes in the liver, and those produced by proteolysis in the tissues and not reutilized in situ for synthetic purposes, pass into the blood plasma and are distributed throughout the body.

Urinary Excretion. Comparatively little free amino acid is excreted in the urine (p. 188) under normal conditions. These substances pass freely into the glomerular filtrate but are very efficiently reabsorbed by the renal tubular epithelium. Even when the plasma level is raised by intravenous injection of amino acid mixtures, usually not more than 5 per cent of the quantity administered escapes in the urine.

Synthesis of Nitrogenous Substances. The amino acid pool is drawn upon for formation (i.e., replacement) of a large number of essential substances: e.g., enzymes, other cellular proteins; plasma proteins; hemoglobin;

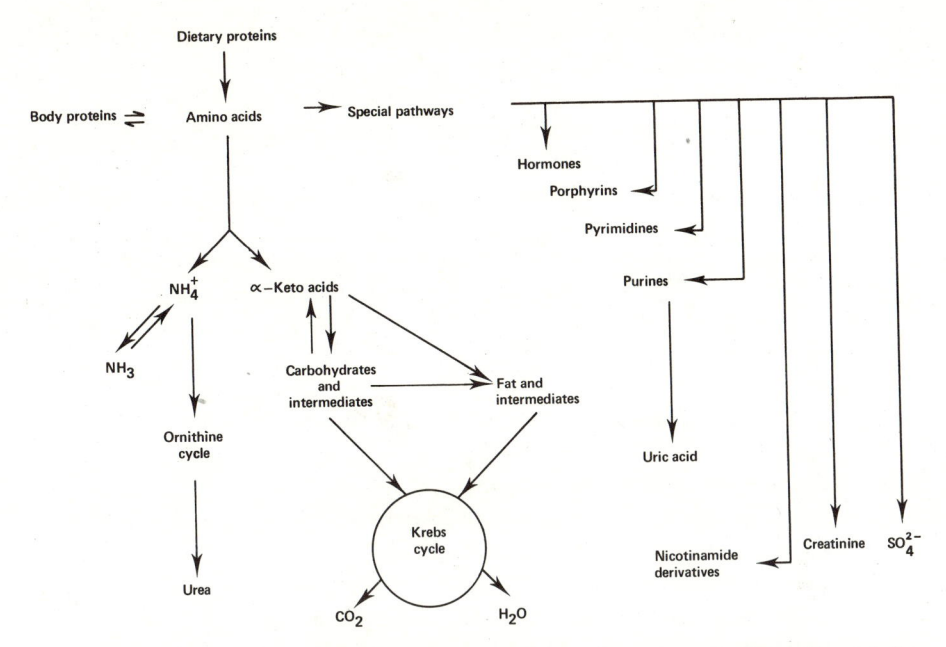

Figure 3-1. General pathways of protein and amino acid metabolism.

purines and pyrimidines (i.e., nucleotides and nucleic acids); certain hormones (e.g., anterior pituitary hormones, insulin, parathyroid hormone, thyroxine, epinephrine); glutathione; creatine.

Transamination and Deamination. With few exceptions, catabolism of amino acids begins with separation of the amino (NH_2) group from the carbon skeleton, which then becomes an α-keto acid. These reactions occur mainly in the liver. For example, alanine → pyruvic acid + ammonia; glutamic acid → α-ketoglutaric acid + ammonia; aspartic acid → oxaloacetic acid + ammonia. At physiological pH levels, virtually all the ammonia becomes NH_4^+ and the acids form anions. These reactions are reversible, and, consequently, amino groups of two amino acids may be exchanged, or an amino group may be transferred from an amino acid to an α-keto acid derived, for example, from the metabolism of carbohydrate, e.g., pyruvic, oxaloacetic, or α-ketoglutaric acids, with the formation of new amino acids. Pyridoxal phosphate, a pyridoxine derivative (vitamin B_6), is the coenzyme for transamination reactions. These interactions constitute an important mechanism for the integration of protein, carbohydrate and fat metabolism (p. 162).

Disposal of the Nitrogen. (a) Synthesis. Ammonia liberated from amino acids may be utilized, as indicated above, for amination of α-keto acids to form new amino acids. It also participates in the synthesis of purines and pyrimidines (i.e., nucleotides and nucleic acids), and also porphyrins.

Ammonia is a toxic substance, and large concentrations do not accumulate in cells or extracellular fluids. One of the apparently important mechanisms of its "detoxication" consists in the synthesis of glutamine, an amide of glutamic acid. This compound can thus serve as a carrier of ammonia, which can subsequently be liberated from it by the enzyme, glutaminase.

Certain amino acids enter into the synthesis of other nitrogen-containing compounds without undergoing preliminary deamination: e.g., glycine, in formation of porphyrins (hemoglobin), glutathione, creatine, glycocholic acid; serine, in formation of choline; arginine, in formation of creatine; cystine, in formation of taurocholic acid; tyrosine, in formation of thyroxine and epinephrine; tryptophan, in formation of nicotinic acid.

(b) **Urinary Ammonium Ion.** In the cells of the distal portion of the uriniferous tubules, ammonia is liberated from glutamine (60 per cent of total urine NH_4^+) and other amino acids (about 12 per cent) and passes into the lumen of the tubules. This phenomenon, the extent of which is determined, under normal conditions, by the urine pH, plays an important role in Na conservation by the kidney especially in acidemia or acidosis (p. 411). About one-quarter of the urinary ammonium ion is probably the result of excretion from arterial blood.

(c) **Urea Formation.** NH_4^+, derived from amino acids by oxidative deamination, and metabolic CO_2 (mainly Krebs cycle) combine in the presence of ATP and the appropriate enzyme to form the carbamyl group (NH_2CO-) of carbamyl phosphate. This group is transferred to the amino acid ornithine, converting it to citrulline, which receives an amino group from aspartic acid to form an intermediate compound (argininosuccinate), which yields arginine. This is split by the enzyme arginase into urea and ornithine, which can then begin again the so-called ornithine-citrulline-arginine cycle. The urea enters the systemic circulation and is excreted from the body, mainly in the urine. Apparently it is not generally realized that urea formation gives rise to hydrogen ions. Consequently, in acidemia or acidosis, the body tends to use more of the other mechanism for excretion of nitrogen, and so ammonium ion concentration increases in the urine.

In the normal animal, an increase in dietary protein is followed by an increase in the concentration of amino acids in the blood (which are taken up largely by the liver), and by an increase in urinary urea. When liver function is severely impaired, e.g., in acute hepatic necrosis, there is a rise in the blood amino acids and a fall in the urea concentration of blood and urine occurs not uncommonly, but its level in the blood depends on renal function. Similarly, removal of the liver in an experimental animal is followed by a rise in the concentration of amino acids and ammonia in the blood and a fall in blood and urinary urea. If the kidneys are removed and the liver left intact, the blood urea rises. Removal of both liver and kidneys results in a constant level of blood urea. From all of these data, it may be concluded that the liver is the chief, if not the sole, site of synthesis of urea, the precursors of which are amino acids and ammonium ion, largely derived from bacterial action in the large intestine. The bulk of the amino acids (NH_2 groups) not utilized in the organism is thus transformed to urea, which is excreted in the urine as a terminal (waste) metabolite.

Disposal of the Non-nitrogenous Residue. The α-keto acids resulting from deamination may be reaminated, as indicated above, re-forming the original amino acids. Certain of the carbon skeletons are used for special synthetic reactions. The remainder, not required for these replacement purposes, take the pathways of glucogenesis or ketogenesis.

(a) Glucogenesis. Most of the amino acids are convertible to carbohydrate ("gluconeogenesis from protein"). The routes vary with the compound concerned, but all converge ultimately at phosphoenolpyruvic acid. The pathways of three glucogenic amino acids to phosphoenolpyruvic acid are particularly direct:

Glucose　　　　　　　　Pyruvic acid ⇌ Alanine
⇅　　　　　　　　　　⇅
Phosphoenolpyruvate ← Oxaloacetic acid ⇌ Aspartic acid
　　　　　　　　　　　⇅
　　　　　　　α-Ketoglutaric acid ⇌ Glutamic acid

α-Ketoglutaric acid and oxaloacetic acid are interconvertible by means of the tricarboxylic acid cycle (p. 8). (In the central nervous system, glutamic acid is converted to succinic acid via γ-aminobutyric acid). Phosphoenolpyruvic acid and glucose are connected by the glycolytic series of reactions (p. 7).

It will be noted that the reactions listed above are reversible. This is generally true for the nonessential, glucogenic amino acids, which can be synthesized by the body by transamination of the α-keto acids provided by carbohydrate. Many amino acids in addition to those shown are glucogenic. Their pathways to pyruvic acid are more devious, however. Transamination plays a major role in gluconeogenesis from amino acids. The process requires pyridoxal phosphate as a cofactor and is facilitated by adrenal oxysteroids.

It has been estimated that over half of the amino acid constituents of the average animal protein are glucogenic. If a state of total diabetes or phloridzin-diabetes is induced in an experimental animal, it becomes virtually unable to utilize glucose. Preformed carbohydrate or any which may be synthesized from protein is lost practically quantitatively in the urine. If such an animal is fasted, its preformed glycogen is consumed in short order, followed by depletion of its fat stores, at which time the body attempts to support itself by the degradation of tissue protein. All of the protein which is degraded gives rise to urinary nitrogen (urea, NH_4^+, creatinine). Since the animal is in a diabetic state, the glucose derived from the glucogenic amino acids of the degraded protein is excreted, unused, in the urine also. The ratio of glucose to nitrogen in the urine is known as the "G/N ratio" (sometimes D/N, where D = dextrose). This varies with the severity of the diabetes, reaching a maximum value of about 3.65 in the phloridzinized animals. This means that each gram of nitrogen resulting from the degradation of tissue protein has been accompanied by the formation of 3.65 g. of glucose. Since nitrogen constitutes 16 per cent of the average animal protein (protein = N × 1/0.16), each gram of protein has produced 3.65 × 0.16 = 0.58 g. of glucose. Hence, 58 per cent by weight of the protein is glucogenic. Actually, there are a number of objections to this simple interpretation of the G/N ratio, and it is no longer regarded as an accurate index of the severity of the diabetes or the degree of gluconeogenesis from amino acids or proteins. However, other techniques have confirmed the general finding that a large proportion of the amino acids in the average protein is glucogenic.

This phenomenon of gluconeogenesis, which is the only source of carbohydrate in the fasted animal after the pre-existing glycogen reserves have been utilized, takes place in the liver. The 11-oxygenated adrenocortical hormones are required for, and increase the rate of, this process (p. 25); in their absence, e.g., in Addison's disease, the blood sugar falls rapidly if exogenous carbohydrate is not supplied (p. 57). Insulin decreases the rate of gluconeogenesis (p. 19); in its absence, e.g., in diabetes mellitus, gluconeogenesis is accelerated, protein catabolism being stimulated, with the production of negative nitrogen balance (p. 34).

(b) Ketogenic Pathway. The α-keto acids derived from a few amino acids are more closely allied to fats than to carbohydrates. In contrast to the glucogenic amino acids, the ketogenic group produce ketone bodies instead of glucose. In comparison to the glucogenic group, the ketogenic amino acids are in a minority, comprising only phenylalanine, tyrosine, leucine, and isoleucine. Furthermore, certain of these amino acids are metabolized along pathways which are glucogenic as well as ketogenic.

INTERRELATIONS OF THE METABOLISM OF PROTEINS AND OTHER FOODSTUFFS

The major interrelations between proteins, fats and carbohydrates are outlined in Figure 3-2. The importance of proteins in the metabolism of all types of compounds, inasmuch as enzymes and many hormones are proteins (or amino acid derivatives) in nature, is too obvious to require elaboration. Synthesis of specialized products, such as purines, pyrimidines, and porphyrins, is likewise an important function of proteins and amino acids, as has been discussed previously. The subject of interest in this section is the relationship between the metabolism of proteins and that of the two other major foodstuffs, as exemplified by direct interconversion and by caloric substitution.

The ketogenic amino acids and certain fragments derived from the metabolism of nonketogenic amino acids form acetate or acetoacetate, thus contributing directly to the synthesis of fatty acids. The glucogenic amino acids can function similarly, but subsequent to initial conversions to pyruvate or keto acids yielding pyruvate. Certain specialized lipid moieties are also derived from amino acids, viz., serine, ethanolamine, sphingosine, and choline. The pathway from the carbon skeletons of the ketogenic or glucogenic amino acids to the fatty acids is irreversible.

The majority of amino acids can form carbohydrate. In certain cases,

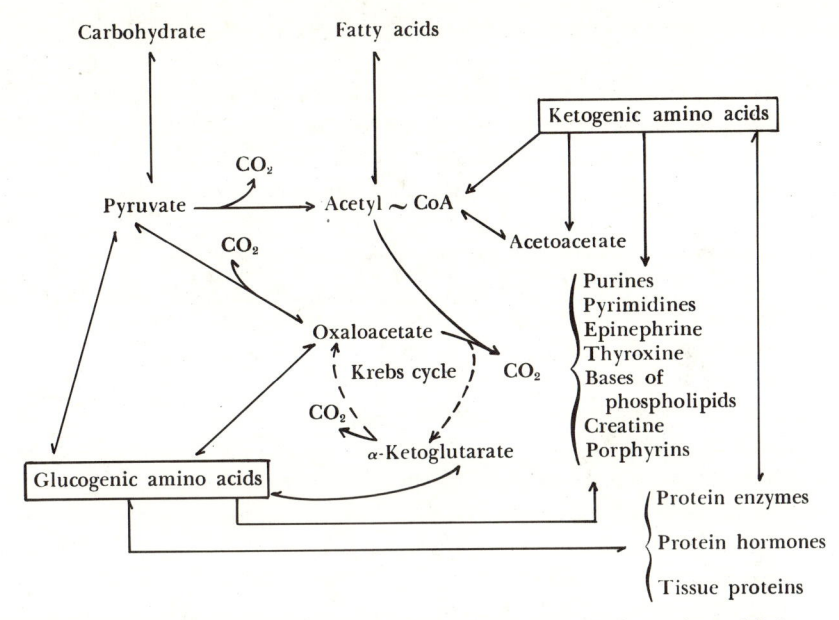

Figure 3-2. Metabolic interrelations of proteins with other foodstuffs (much simplified).

viz., alanine, aspartate, and glutamate, the conversion to precursors of carbohydrate is direct; in others the pathway may be quite devious. The conversion is reversible for most of the nonessential, glucogenic amino acids. It must be emphasized in connection with glucogenicity that the α-keto acids produced from the amino acids (pyruvate, oxaloacetate, α-ketoglutarate) function not only as raw materials for the synthesis of glucose, but also as catalysts in the channeling of acetyl units from all classes of foodstuffs through the tricarboxylic acid cycle (α-ketoglutarate and pyruvate being easily converted to oxaloacetate, the actual catalytic compounds of the cycle) for the provision of energy.

The energy content of the diet influences protein metabolism as reflected in the nitrogen balance. Nitrogen excretion is increased when the caloric intake is lowered to submaintenance levels. Conversely, increasing the caloric intake causes a decrease in urinary N excretion. At any given adequate level of dietary protein, the nitrogen balance is determined, in part at least, by the caloric intake. This relationship can be demonstrated in another manner. When the caloric intake is adequate, increasing the dietary protein from submaintenance levels leads to improvement in N balance. However, if the energy intake is below about 25 per cent of that required, this response to an increased protein intake is not obtained. Under these restricted conditions, the caloric intake is apparently a limiting factor in the utilization of dietary proteins.

On diets grossly inadequate in caloric value, nitrogen equilibrium may still be maintained, especially in obese subjects and if adequate amounts of protein and a small amount of carbohydrate are provided. During the initial days of such a regimen (or of a fast), the N excretion is lower than on subsequent days, owing to utilization of glycogen stores. Later, body fat is utilized for energy, exerting the beneficial effect on N balance indicated above for dietary fat.

These effects of the energy content of the diet are apparently largely independent of the nature of the energy-providing materials, being essentially identical, except temporarily (see below), with equicaloric amounts of carbohydrate, fat or alcohol. They are not influenced by the time of feeding of the latter in relation to that of the protein. The mechanism of production of these effects is not known.

Dietary carbohydrate exerts an influence on protein metabolism not shared by other substances. This is independent of its caloric effect, which is shared equally by fat and alcohol; it is referred to as the specific "protein-sparing" action of carbohydrate. It manifests itself in several ways:

1. Isocaloric substitution of fat for carbohydrate in the diet is followed by an increase in nitrogen excretion. This is of relatively brief duration, however, the original level of N balance being restored after several days. This effect is observed whether or not the energy and protein contents of the diet are adequate.

2. If the protein and carbohydrate components of an adequate diet are ingested separately, at wide time intervals, there is a transitory increase in N excretion. No such effect is exhibited by fat.

3. On an exclusively fat diet, the nitrogen output is the same as during starvation, whereas administration of carbohydrate reduces the nitrogen output.

The mechanism underlying this phenomenon is not known for certain. It is apparently not mediated by hormonal influences (pituitary, thyroid, insulin) nor by improved synthesis of amino acids or of protein. It is probably a reflection of the fact that carbohydrate, in addition to its function as a fuel, also provides the major source of oxaloacetate for the tricarboxylic acid cycle and the carbon skeletons of nonessential amino acids. In the absence of carbohydrate, protein is called upon specifically to assume these functions, since fat cannot. The increased degradation of protein and loss of nitrogen are necessarily greater than would occur if fat were omitted from the diet instead of carbohydrate.

ENDOCRINE INFLUENCES IN PROTEIN METABOLISM

Anabolism (synthesis) and catabolism (degradation) of proteins are influenced, if not indeed regulated, by certain hormones, which also influence the interrelationships between the metabolism of protein and of carbohydrate. The most important of these are the growth hormone (adenohypophysis), the adrenal 11-oxysteroids, the thyroid hormones, androgens, and insulin.

Growth Hormone. This exerts a protein anabolic effect. Its administration is followed by diminished urinary excretion of nitrogen (positive N balance), with a decrease in plasma urea and amino acids and increased deposition of protein in the tissues. Somatic growth is induced in prepubertal children. Not only does the hormone stimulate transcription of messenger RNA but it also stimulates translation in the ribosomal mechanism. The resulting increase in protein synthesis is made possible by an increased uptake of amino acids across cell membranes. It used to be believed that insulin was necessary for growth hormone protein stimulation. This is no longer regarded as correct, since it has been found that growth hormone stimulates protein synthesis in the isolated rat liver.

Androgen. Testosterone, too, exerts a protein anabolic effect, as do androstenediol and androstenedione, but androsterone and dehydroepiandrosterone exhibit little such action. Administration of testosterone is followed by decreased urinary excretion of nitrogen, positive N balance, stimulation of somatic growth in prepubertal children (adolescent growth spurt), and increased weight (i.e., protein deposition) not only of specific target tissues (e.g., prostate, seminal vesicles), but also of other organs, particularly the kidneys (renotropic action). Castration is followed by increased urinary excretion of creatine; this is reversed by testosterone. Methyltestosterone, a synthetic androgen which also exerts a protein anabolic effect, increases urinary excretion of creatine. It is believed to stimulate synthesis as well as storage of creatine, whereas testosterone stimulates only its storage.

Adrenal Glucocorticoids. Under usual experimental and clinical conditions, these 11-oxygenated adrenocortical hormones apparently exert a protein catabolic effect, i.e., they cause increased urinary excretion of nitrogen and a negative N balance. This effect is intimately related to their action in increasing gluconeogenesis (p. 25), which probably involves increased deamination of amino acids and increased conversion of the non-nitrogenous residues to glucose (or glycogen). This stimulation of gluconeogenesis is antagonized by insulin (p. 19). It seems that cortisol acts by stimulating the production of certain enzymes required for gluconeogenesis and for converting amino acids into appropriate residues for this purpose, e.g., by stimulating aminotransferase production or the production of the enzyme for breaking the tryptophan ring (tryptophan pyrrolase), and so on. It has been shown that cortisol increases cellular uptake of amino acids and increases protein synthesis, but probably only in relation to specific protein enzymes concerned with amino acid breakdown and gluconeogenesis. In other words, it stimulates the activity of certain messenger ribonucleic acids. It has been suggested that this type of mechanism accounts at least in part for the physiological effects of a number of different hormones.

Glucocorticoids also cause increased urinary excretion of uric acid, apparently a reflection of increased catabolism of purines (i.e., nucleic acids).

Insulin. Insulin antagonizes the gluconeogenic action of adrenocortical hormones. It increases the transport of amino acids into cells and stimulates protein synthesis, including that of certain enzymes involved in glycolysis, the pentose phosphate shunt and fat synthesis. It has been suggested that it stimulates the production of a messenger RNA for a protein, which stimulates ribosomal activity. Insulin probably stimulates the biosynthesis of the specific enzymes already mentioned by stimulating the production and action of specific types of messenger RNA.

Thyroid Hormones. In physiological doses, thyroid hormones exert a protein anabolic effect, as indicated by a positive N balance, especially in the growing organism. This action would be inferred from the fact that removal of the thyroid in early life results in retardation of growth, which can be corrected by administration of thyroid hormone. Thyroidectomy (and hypothyroidism) is followed by accumulation in the extracellular fluids of a mucoprotein, rich in hyaluronic acid and similar to fetal mucin in composition (myxedema).

In unphysiological (large) doses, thyroid hormones exert a protein catabolic effect, with increased urinary excretion of creatine (creatinuria), resulting from excessive catabolism of muscle tissue and also, perhaps, from a decreased capacity for prompt resynthesis of phosphocreatine during the recovery stage of muscle contraction.

Glucagon. Glucagon decreases the incorporation of amino acids into protein of isolated skeletal muscle. When administered to man in prolonged and high dosage, it results in an increased nitrogen excretion.

NITROGENOUS CONSTITUENTS OF THE BLOOD

The blood plasma normally contains a number of nitrogenous substances. These may be classified as (1) proteins and (2) nonprotein nitrogenous substances.

The Plasma Proteins

Segregation of discussion of the plasma proteins is justified by two considerations. First, because of their intimate relation to protein metabolism in the liver, as well as their interaction with other tissues throughout the body, the plasma proteins occupy a central position in the metabolism of protein. The second consideration is entirely practical: the plasma proteins happen to be the most conveniently obtainable sample of protein available in the body. Because of the interrelations between the proteins of plasma and the tissues, a great deal can be learned concerning the general status of protein metabolism from examination of the plasma proteins.

Identity and Properties. The heterogeneity of the plasma proteins is readily demonstrated in the ultracentrifuge. In addition to fibrinogen, albumin, and a group of globulins sedimenting at a rate between these two, the ultracentrifugal analysis of plasma reveals a group of very rapidly sedimenting globulins and the nonsedimenting lipoproteins, the flotation of which varies markedly with the density of the medium. This heterogeneity is reflected in the wide range of molecular weights, i.e., sizes, of the various plasma proteins. Albumin, the smallest, has a molecular weight of about 69,000, α-globulins 200,000 to 300,000, β-globulins 150,000 to 1,300,000, γ-globulins 150,000, and fibrinogen 400,000. The macroglobulins, which include α- and γ-globulins, have a molecular weight of 900,000 or more. Inasmuch as osmotic pressure is a function of the number of active particles (e.g., molecules) in solution, gram for gram, albumin has considerably more influence on osmotic pressure than do the other plasma proteins. Moreover, being the smallest of these molecules, it is the first to escape from the blood stream when the permeability of the capillary walls is increased, e.g., in inflammation.

The degree of resolution of the plasma proteins in the ultracentrifuge is inferior to that obtainable with electrophoresis. An electrophoretic diagram of normal plasma is shown in Figure 3-3. In addition to albumin and fibrinogen, this technique resolves the globulins into four main groups (α_1, α_2, β, and γ).

Before the introduction of paper electrophoresis, clinical data on the plasma proteins were obtained chiefly by the method of salting-out. Precipitates or supernatant solutions produced by the addition to the plasma of certain concentrations of salts were analyzed for protein content. Before the advent of the electrophoretic technique, clinical analyses of plasma by salting-out were performed according to the method of Howe. The Howe fractions disagree markedly with the results of electrophoresis. A comparison is shown in Table 3-1 of the Howe method and a more modern adaptation (Milne) which yields fractions more closely related to those obtained electrophoretically. It would appear that, with the Howe procedure, which was widely used, the "albumin" fraction contains about 1 gram of globulin. With the Milne procedure, pseudoglobulin corresponds rather closely to the α-globulins, and euglobulin to the β- and γ-globulins.

The fractions of the plasma proteins obtained by ultracentrifugation, electrophoresis or salting-out are chemically heterogeneous. For scientific and other purposes, methods have been elaborated which permit fractionation of the plasma proteins to the point of isolation of relatively homo-

Figure 3-3. Electrophoretic pattern of normal plasma.

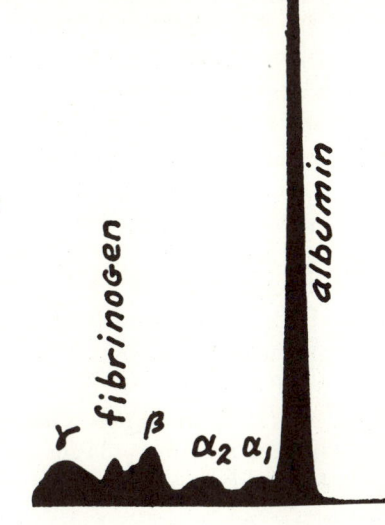

168   PROTEIN METABOLISM

TABLE 3-1   SERUM PROTEIN CONCENTRATIONS

PROTEIN	SALTING OUT HOWE (grams %)	MILNE (modified) (grams %)	ELECTROPHORESIS (grams %)	(% of total)
Total	6–8 (7)	6–8 (7)	6–8 (7)	100
Albumin	4.7–5.7 (5.2)	3.6–4.5 (4.0)	3.8–5.0 (4.4)	52–68 (60)
Globulin	1.3–2.5 (2.0)	2.1–4.2 (3.2)	2–4 (3)	32–48 (40)
Pseudo-		0.8–1.9 (1.3)	α 0.5–1.3 (0.9)	7–17 (12)
Eu-		1.3–2.5 (1.9)	β 0.6–1.2 (0.9)	8–16 (12)
			γ 0.8–1.6 (1.2)	10–22 (16)
Fibrinogen (plasma)	0.2–0.4 (0.3)	0.2–0.4 (0.3)	(0.3)	

geneous chemical constituents. These methods are of interest not only in that they permit a closer study of the physicochemical and physiological properties of the individual proteins of the plasma, but also because they have made possible the large-scale production of plasma protein products for therapeutic use, viz., fibrin film and foam (surgery), purified albumin (shock), γ-globulin (measles, hepatitis, poliomyelitis), and antihemophilic globulin, and other proteins for diagnostic and research purposes.

These procedures are based upon fractional precipitation of the plasma proteins with ethanol at low temperatures and low salt concentration. Proper adjustment of the concentrations of protein, salts and ethanol, temperature, and pH permits the isolation of six fractions ("Harvard" or "Cohn" fractions), each of which can be subfractionated by appropriate procedures into relatively pure components. Certain characteristics of the plasma fractions so obtained are indicated in Table 3-2.

Two lipoprotein fractions, an α_1- and a β_1-globulin, have been isolated by the methods described above (p. 167). However, there is evidence that a larger number of such proteins are present in plasma. By the addition of salt to plasma, its density may be elevated to the point where the lipoproteins (having a lower density than ordinary proteins due to their lipid content) will migrate toward the axis instead of toward the periphery in the ultracentrifuge. Whereas the rates of sedimentation of particles in conventional ultracentrifugation are expressed in Svedberg units (S = 1 × 10^{-13} sec.), those of the lipoproteins observed in the flotation method are expressed in "negative" S units, or "flotation" S units, S_f. In these terms, particles have been found distributed over a range of S_f, viz. S_f 0-20 (beta, or low density), S_f 20-400 (prebeta, or very low density) and S_f > 400 (chylomicrons).

Methods of Study. Appreciation of the clinical significance of abnormalities of serum protein concentration has been furthered as a result of the increasing availability of analytical techniques. In view of the widespread use of a variety of procedures, it is desirable to review briefly the types of information they may provide.

(1) Quantitative Chemical Methods.   (a) The total protein concentration of either plasma or serum may be determined readily by micro-Kjeldahl or biuret procedures. Although the status of this factor does not

TABLE 3-2 PROPERTIES OF MAJOR PLASMA PROTEINS*

ELECTROPHORETIC CLASS	ULTRACENTRIFUGAL CLASS, SVEDBERG UNITS ($s_{20,w}$)	pI	MOLECULAR WEIGHT	DIMENSIONS, Å	SPECIAL PROPERTIES
Albumin	4.6	4.9	69,000	150 × 38	Osmotic pressure, fatty acid transport
Prealbumin	4.2	4.7	61,000		Thyroxine binding
α_1-Globulins					
Orosomucoid	3.1–3.5	2.7	44,100		41% carbohydrate
α_1-Acid glycoprotein (Schmid)	3.5		54,000		14% carbohydrate
α_1-Antitrypsin	3.4	4.0	45,000		Emphysema
α_1-Lipoprotein* (HDL$_3$)	5.0	5.2	195,000		43% lipid
α_1-Lipoprotein* (HDL$_2$)	5.5		435,000		67% lipid
Thyroxine-binding globulin	3.3	4.0	40–50,000		
Transcortin	3.0 (S_{20} 1.2%)				Cortisol binding
α_2-Globulins					
Haptoglobin, Type 1-1	4.4	4.1	100,000		Binds hemoglobin, different genetic types, 23% carbohydrate
Type 2-1	4.3–6.5				
Type 2-2	7.5				
Prothrombin	4.85	4.2	62,700		Blood coagulation, 11% carbohydrate
Ceruloplasmin	7.1	4.4	160,000		Contains Cu
Cholinesterase	12	3.0	300,000		Various phenotypes
α_2-Macroglobulin	19.6	5.4	820,000		8% carbohydrate
β-Globulins					
β-Lipoproteins,* d, 0.98–1.002	S_f 20–100		5–20 × 10^6		91% lipid ⎫ lipid
β-Lipoproteins,* d. 1.03	S_f 10–12		5 × 10^6		79% lipid ⎭ transport
Transferrin	5.5	5.9	90,000	190 × 37	Transports Fe, 5.5% carbohydrate
Plasminogen	4.28	5.6	143,000		Precursor of plasmin, a fibrinolysin
Hemopexin	4.8		80,000		Heme-binding
Fibrinogen	7.63	5.8	341,000	475 × 40	Blood coagulation, 5% carbohydrate
γ-Globulins					
γM-Globulins	19	5.8	ca. 1 × 10^6		10% carbohydrate ⎫
γG-Globulins	6.6–7.2	5.8–7.3	160,000	235 × 44	3% carbohydrate ⎬ antibodies agglutinins
γA-Globulins	7.0		170,000		⎭
γE-Globulins	8.0		190,000		
γD-Globulins	7.0		160,000		

* Lipoproteins are presented also in Table 2-1, p. 115.

necessarily of itself indicate either the presence or nature of an abnormality of the plasma proteins, knowledge of the total protein concentration is essential for complete understanding of the entire plasma protein picture.

(b) **Salt Fractionation of Albumin and Globulins.** Procedures in this category are based on the premise that globulins are precipitated from solution by half-saturation with certain salts (e.g., ammonium sulfate, sodium sulfate), whereas albumin is not. As stated previously, by increasing the concentration of sodium sulfate (Milne) for separation of albumin and globulin, values may be obtained that correspond more closely to those obtained with electrophoresis (Table 3-1). This has been entirely superseded by more efficient quantitative procedures, such as electrophoresis. Albumin can be determined by specific dye-binding techniques.

(c) **Ethanol Precipitation at Low Temperatures.** This fractionation procedure (Cohn) is useful particularly for preparative purposes and does not lend itself readily to routine use clinically as an analytical procedure.

(d) **Fibrinogen.** This may be determined by chemical analysis (Kjeldahl) of the fibrin clot produced by recalcification of oxalated plasma.

(e) **Mucoproteins.** This designation is applied to a category of glycoproteins particularly rich in carbohydrate, that are not easily precipitated by the usual strong acid protein-precipitating agents, e.g., perchloric, trichloracetic, sulfosalicylic acids. They are, however, precipitated by phosphotungstic acid. These features form the basis for their quantitative estimation in the serum, in which they normally comprise somewhat less than 1 per cent of the total protein, migrating electrophoretically mainly, but not entirely, in the $alpha_2$-globulin component.

(2) **Electrophoresis.** Paper and cellulose acetate electrophoresis have largely replaced salt fractionation as the procedure of choice for quantitation of the various serum proteins. Since proteins differ in their isoelectric points, they will generally bear net charges of different magnitudes at any given pH. Consequently, individual components of a mixture of proteins will migrate at different velocities in an electrical field. If the pH is adjusted, by addition of a proper buffer, to a value alkaline to the isoelectric points of all the serum proteins, the proteins will all carry negative charges, but of different magnitudes. Passage of an electric current through the solution will then cause the proteins to migrate toward the positively charged electrode (anode) at characteristically different rates. Originally (Tiselius), this moving boundary of proteins was followed by recording changes in the refractive index of the solution (boundary electrophoresis). This expensive and time-consuming procedure has been almost completely replaced for clinical purposes by paper or cellulose acetate electrophoresis, in which similar migration of proteins is followed in an electrical field set up across a strip of filter paper or cellulose acetate saturated with buffer. Because of complications incident to the addition of an anticoagulant, serum is commonly employed in these studies except in occasional instances in which interest is focused upon fibrinogen.

The separated components are located usually by the application of a dye with protein affinity, and the relative amount of each is measured photometrically. The absolute concentration of each component can then be calculated from the known total protein concentration. Because of fundamental differences in the nature of the protein characteristic that forms the basis of these measurements, values obtained with boundary electrophoresis (refractive index increment) may not coincide with those obtained with paper (zone) electrophoresis (dye-binding capacity). Moreover, these do not correlate consistently with values obtained by salting-out and chemical procedures. Despite these inconsistencies, the combination of paper or cellulose acetate electrophoresis and total protein determination (Kjeldahl, biuret) is most useful for clinical purposes. This not only provides a rather precise indication of the concentrations of the normal serum protein components separable by electrophoresis, but also indicates the presence and amounts of certain abnormal components with different electrophoretic mobilities, e.g., myeloma protein. Paper or agarose electrophoresis, employing both lipid- and protein-staining reactions, permits qualitative examination of serum lipoproteins (p. 116).

(3) Ultracentrifugation. Although procedures of this type are seldom used clinically, they have been employed for certain studies of serum lipoproteins (p. 115). Moreover, the presence of macroglobulins (p. 198), which may be suspected on the basis of the electrophoretic pattern, can be established with certainty only by molecular-weight determinations in the ultracentrifuge.

(4) Immunological Techniques. It is now possible to obtain antisera to specific proteins or specific protein groups (e.g., IgG, IgA, IgM and IgE) and to determine quantitatively the amount of protein antigen present

TABLE 3-3 METHODS OF STUDY OF PLASMA PROTEINS

Quantitative Chemical:	Total Protein
	Albumin (also by dye-binding)
	Globulin
	Pseudoglobulin (alpha, beta)
	Euglobulin (gamma, beta)
	Mucoprotein (alpha)
	Fibrinogen
Electrophoresis:	Albumin
	Alpha$_1$-globulin
	Alpha$_2$-globulin
	Beta-globulin (iron-binding capacity)
	Gamma-globulin
	Fibrinogen
	Abnormal globulins
Temperature-sensitive Globulins:	
	Cryoglobulins
	Pyroglobulins
Test-tube Globulin Reactions (Flocculation and Turbidity):	
	Zinc sulfate turbidity
	Thymol turbidity and flocculation
	Cephalin-cholesterol flocculation
Ultracentrifugation	
Low-Temperature–Ethanol Fractionation	
Immunological	

by recognized and relatively simple immunological techniques. Alternatively, serum can be subjected to electrophoresis in agar and after electrophoresis is completed an antiserum to human proteins is placed in a long narrow slot parallel with the direction of protein migration. The antiserum diffuses inward toward the separated proteins, which are in turn diffusing outward. This results in the formation of a number of precipitin arcs, which can almost all be identified by virtue of their relative positions. This is known as immunoelectrophoresis and can be rendered semiquantitative. It is possible after electrophoresis is completed to carry out a second electrophoresis at right angles to the first and into a medium containing the antiserum to human serum proteins. This results in a number of overlapping, gaussian-shaped, precipitin curves. Each corresponds to a specific protein or protein group. The area under each curve is proportional to protein concentration. This provides an elegant method for the estimation of more than 40 recognizable entities.

Radioimmunoassay can be used in the estimation of proteins or peptides present in small quantity. It is, of course, particularly applicable to estimation of protein or polypeptide hormones, as well as to the study of the releasing factors for the anterior pituitary hormones.

(5) Miscellaneous Studies. Various qualitative or semi-quantitative procedures are frequently employed, usually for demonstration of some abnormality of serum globulins. These include tests for abnormal precipitability at elevated (pyroglobulins, p. 198) or lowered (cryoglobulins, p. 198) temperatures, and the cephalin-cholesterol flocculation, thymol turbidity, and zinc sulfate turbidity tests (p. 205). Immunological procedures are also available, some of a general and qualitative nature, e.g., C-reactive proteins (p. 198), others quantitative, measuring rather precisely the concentrations of specific antibodies, e.g., diphtheria antitoxin. Albumin can be estimated quantitatively by its ability to bind certain dyes under carefully defined conditions. The color of the dye complex is different from that of the free dye and so can be estimated by photometric procedures.

Recently, it has been shown that protein mixtures can be separated if placed in a stabilized pH gradient and subjected to a potential difference along the gradient. Each protein migrates to the zone of pH corresponding to its isoelectric point. In other words, there is *isoelectric focusing*. The potential gradient is produced by a mixture of ampholytes (poly-aminopolycarboxylic acids), which migrate to their own isoelectric region and produce a pH gradient, which then influences the much more slowly moving proteins. The whole procedure can be carried out in tiny acrylamide gel columns enclosed in glass tubes. The acrylamide cylinder can be removed and the protein bands fixed and stained by appropriate dye-containing solutions. When applied to serum, numerous bands can be demonstrated, which are somewhat difficult to identify. It is possible, however, after isoelectric focusing to place the unstained gel cylinder on top of a sheet of acrylamide containing buffer. Ordinary electrophoresis can now be performed into this sheet and at right angles to the axis of the original cylinder. Staining will demonstrate spots or shaped zones, many of which have been identified. It is possible even to identify IgG and IgA and so to classify monoclonal (myelomatosis, etc.) and polyclonal gammopathies without the use of specific antisera.

Metabolism. The liver is the sole site of formation of albumin, fibrinogen, prothrombin and the main site of formation of α- and β-globulins. The γ-globulins are synthesized in reticuloendothelial cells.

The nutritional status of the individual with respect to protein has a profound effect on the synthesis of plasma proteins, both directly, in terms of provision of raw materials for synthesis, and indirectly, due to the deleterious effect of protein deprivation on the liver. It is of interest that protein deprivation has its most marked effect on the levels of plasma albumin and γ-globulins, a decrease in the former of sufficient severity leading to edema (p. 185), a decrease in the latter fraction resulting in impaired resistance to infections.

It has been estimated, from studies with isotopes, that a 70 kg. man synthesizes and degrades approximately 15 to 20 g. of plasma protein per day. Not all of the fractions turn over at the same rate, however. Albumin and γ-globulins are synthesized at somewhat slower rates than the α- and β-globulins. Under optimum conditions, i.e., adequate protein supply and marked stimulus to synthesis, much more can undoubtedly be formed. It has been found that normal dogs, depleted of plasma proteins (plasmapheresis), are capable, under these conditions, of regenerating about 90 per cent of their total plasma proteins weekly. For a 70 kg. man, this would amount to 25 to 30 g. daily. The half-life of human serum proteins is, in fact, approximately ten days. It does, however, vary from protein to protein.

The plasma proteins, among their other functions (discussed below), serve as a source of protein nutrition for the tissues. Indeed, a dynamic equilibrium exists between the proteins of the tissues and those of the plasma, each group sustaining the other when the need arises. In conditions of protein deprivation, however, the normal level of the plasma proteins is apparently guarded more zealously than that of the tissue proteins, since the latter are degraded to provide for the former. As a result, considerable loss of tissue protein may occur with minimal decrease in the concentrations of plasma proteins. It has been calculated that a decrease in total circulating plasma protein of 1 g., in hypoproteinemia due solely to protein deprivation, represents a concomitant loss of 30 g. of tissue protein.

Observations in experimental animals indicate that the plasma proteins circulate within a much larger volume than that of the blood plasma, passing continually from the plasma and other extracellular fluids into the lymph and back again. In the dog, about 50 per cent of the total traverses the thoracic duct daily. It appears, also, that the intravascular plasma protein is in dynamic equilibrium with the extravascular plasma protein, a decrease in one compartment resulting in movement in that direction from other compartments.

Functions. *(1) Nutritive.* The nutritive functions of the plasma proteins are probably attributable largely to the albumin fraction, due to its quantitative dominance. Intravenously administered albumin has been shown in fact to be efficiently, although slowly, utilized in humans.

(2) *Water Distribution*. Because of the relative impermeability of the capillary walls to colloids, the colloid osmotic pressure of the plasma proteins, which normally ranges from 24 to 30 mm. Hg, plays an important part in the maintenance of the normal distribution of water between the blood and the tissues. The osmotic pressure of blood plasma is approximately 7 atmospheres, being dependent upon its crystalloid and colloid content. Because of the fairly high degree of diffusibility of the crystalloid constituents, the concentrations of these substances in plasma and interstitial fluids are practically identical; minor differences exist because of the effect of a Donnan equilibrium. It may therefore be considered that, in the resting state, the crystalloid osmotic pressure of the plasma is balanced by that of the interstitial fluids.

There is considerable disparity, however, between the protein content of plasma and that of interstitial fluid, the latter varying in different situations due to varying degrees of capillary permeability, as demonstrated by Starling. Thus, the lymph flowing from the liver contains 6 to 7 g. per cent of protein, that from the extremities only 2 to 3 g. per cent. Because of this difference, the colloid osmotic pressure of the plasma (24 to 30 mm. Hg) exerts a force which tends to direct the flow of water from the tissues into the blood. This force, the colloid osmotic pressure or oncotic pressure of the plasma, is a function largely of the molecular size of the various proteins. It is influenced consequently much more by albumin than by globulin, 2.4 to 3.9 times as much, according to different observers, because of the smaller size of the albumin molecule (molecular weight of serum albumin about 69,000; serum globulins, 150,000 to 1,300,000). Thus, albumin, comprising about 60 per cent of the total proteins, is responsible for about 80 per cent of the oncotic pressure of human plasma. By direct measurement, the colloid osmotic pressure of normal plasma or serum has been found to vary from 280 to 470 mm. H_2O. Under normal conditions, this force, which tends to direct the flow of water from the tissues into the blood, is counterbalanced by the opposing force of the capillary blood pressure (13 to 35 mm. Hg). Because of the drop in blood pressure from the arteriolar to the venous end of the capillary, fluid tends to pass from the blood to the tissues on the arteriolar side of the capillaries and from the tissue spaces to the blood at the venous end. Passage of fluid in the latter direction is also favored by the increased colloid osmotic pressure of the blood plasma at the venous end of the capillary, which results from the increased concentration of plasma proteins incident to the loss of fluid in the proximal end of the capillaries. Under conditions of active tissue metabolism the osmotic pressure of the tissue fluids is raised by the decomposition of carbohydrates, proteins, and fats into relatively much smaller molecules, thus temporarily aiding in the abstraction of water from the blood plasma.

(3) *Hydrogen Ion Regulation*. From a quantitative standpoint, the buffering capacity of the plasma proteins is not very great. In whole blood, however, the combined buffering action of the hemoglobin and plasma proteins is as important as that due to the bicarbonate and other inorganic buffer systems of the blood (p. 402).

(4) Transport. The lipid-transporting functions of the plasma proteins are discussed in detail elsewhere (p. 115). In addition to the lipids the lipoproteins transport certain lipid-soluble compounds, such as the fat-soluble vitamins (e.g., A, D and E). Bilirubin is associated mainly with albumin, also with fractions of the α-globulins. β_1-Metal combining globulin (siderophilin; transferrin) is responsible for the transport of iron in the plasma. Approximately half of the calcium of plasma is bound to protein in a linkage which is relatively weak in comparison to the metalloproteins just mentioned. Thyroxine appears to be transported in association with an α-globulin (thyroxine-binding protein, TBP) and cortisol by a mucoprotein (transcortin). Hemoglobin liberated intravascularly is carried to the reticuloendothelial system by complexing with the haptoglobins. In addition to the more or less specific transport functions already mentioned, due in most cases to the various globulins, plasma albumin has been found to exhibit considerable affinity for anions and cations having a wide variety of structures. Many drugs and dyes are transported in the plasma in combination with albumin.

(5) Coagulation. In addition to fibrinogen and prothrombin, plasma contains a number of other components which participate in the process of blood coagulation, viz., Ac-globulin (Factor V), serum prothrombin conversion accelerator (Factor VII), antihemophilic globulin (Factor VIII), Christmas factor (Factor IX), Stuart-Prower factor (Factor X), plasma thromboplastin antecedent (Factor XI) and Hageman factor (Factor XII). According to the "cascade" hypothesis of blood coagulation, each factor exists in an inactive form, which becomes active in a sequence. Surface contact produces the active form of Factor XII, which activates Factor XI, which activates Factor IX, which activates Factor VIII, which activates Factor X, which activates Factor V, which converts prothrombin (Factor II) to thrombin, which acts on fibrinogen (Factor I) to produce fibrin. Calcium ions are required for all stages, except the first and the last. It is postulated that each stage involves an amplification process, so that the effect at the final stage is greatly magnified, and so enables sufficient fibrin to be formed.

Plasma also contains plasminogen, the precursor of the enzyme plasmin, which is fibrinolytic. There is also a clot-stabilizing system which is concerned with fibrin cross-linking. This involves so-called Factor XIII and transglutaminase activity, which exists in the plasma in the form of a precursor.

(6) Immunity. Reference has been made to the immunological functions of the plasma proteins. The γ-globulins contain a large number of antibodies, among which may be mentioned those against influenza, mumps, poliomyelitis, measles, virus hepatitis, typhoid, whooping cough, and diphtheria. At least 11 of the components of *complement,* an activity necessary in certain types of immunological reactions, are proteins of the globulin type. Complement consists of 9 separate components but the first easily breaks into 3 separate fractions, in the absence of calcium ions, giving 11 proteins in all. The main components are designated C1 (M.W. c. 10^6), C2 (M.W. 117,000), C3 (M.W. 185,000), C4 (M.W. 230,000), C5, C6, C7, C8 and C9 (M.W. 75,000). The separate portions of C1 are designated C1q (M.W. 400,000), C1r and C1s (M.W. 110,000). C1 first becomes attached to the tail portions of antibody, which has already formed a complex with antigen. The antibody is either IgM or two IgG molecules in close juxtaposition. Much of the work with complement has involved systems containing sheep erythrocytes. Therefore, if one uses the symbol E for antigen and A for antibody, the antigen-antibody complex becomes EA and when C1 is attached (by its C1q portion) the complex EAC1 is formed. The C1 is converted to an active form $\overline{C1}$, which is an esterase (active site C1s) whose substrate is C4. The latter is broken into two fragments, the larger of which forms a complex with $EAC\overline{1}$, designated $EAC\overline{1}4$. This reacts on its substrate, which is C2. The molecule of the latter is split, but again a complex is formed by the attachment of a major portion of the C2 molecule. This complex is designated $EAC\overline{1}42$. This reacts with C3 to give C3 fragments, the larger one forming the complex $EAC\overline{1}423$. In a similar manner, $EAC\overline{1}4235$ is formed, which reacts with C6 and C7 to form fragments, which give $EAC\overline{1}423567$. Next $EAC\overline{1}4235678$ is formed and finally $EAC\overline{1}42356789$, which possesses the ability to rutpure membranes and so cause lysis of cells.

One of the fragments of C2 acts like a kinin. The small fragment released from C3 is anaphylatoxin, which produces a weal when injected subcutaneously and causes mast cells to release histamine. The attachment of the major fragment of C3 (C3b) in $EAC\overline{1}423$ brings about immune adherence, i.e., the attachment of the antigenic cell to specific sites on erythrocytes, platelets and polymorphonuclear leukocytes. The latter attachment enhances phagocytosis by the polymorphonuclear leukocyte. The fragments of C5 have two known biological activities. One functions as an anaphylatoxin, and the other induces chemotaxis for polymorphonuclear leukocytes. A high molecular weight complex of C5, 6 and 7 is also a chemotactic factor. Once the complex $EAC\overline{1}423567$ is formed, the cell to which it is attached becomes susceptible to lymphocytotoxicity.

There are also found in serum one inhibitor and two inactivator proteins. The first protein, designated $\overline{C1}$ INH, inhibits the esterase action of activated C1 ($\overline{C1}$) and so prevents the formation of $EAC\overline{1}4$ and subsequent stages toward the final complex. The second protein, designated C3 INA, inactivates cell-bound C3. The third inactivates cell-bound C6 and is designated C6 INA.

The disease *hereditary angioedema* is now believed to be due to an inherited deficiency of $\overline{C1}$ INH. This leads to excessive activity of activated $\overline{C1}$, with resultant increased breakdown of C4 and greater production of the fragment with kinin activity. By increasing capillary vascular permeability, this gives rise to transient circumscribed single areas of edema on the face or extremities. There is also involvement of the gastrointestinal tract, with vomiting and abdominal pain, as well as involvement of the upper respiratory tract, which may lead to fatal edema of the larynx.

Polymorphism as well as inherited deficiencies of complement components have also been described.

More recent studies have added considerably to our knowledge of complement. It has now been established that the C1 system is made up of one C1q, two C1r and four C1s polypeptides. There has also been further study of the way in which the final complement complex is formed. It is now realized that at least part of its structure is dependent upon some of its components being attached to adjacent sites on the cell membrane. There is also further knowledge with regard to the actual sequences involved. As soon as $EA\overline{C1}$ is formed, the active site $(\overline{C1s})$ splits C4 into C4a and C4b. The latter is the larger fragment. In the presence of magnesium ions, some C4b combines with a receptor on the cell surface and adsorbs C2, which is then split by a neighboring $\overline{C1s}$ into C2a and C2b, which are two large fragments. The former remains attached to give $\overline{C4b2a}$, which is an enzyme reacting with C3 to give C3b. The latter becomes bound to the cell membrane in the immediate vicinity of the C4b2a and thereby creates a complex $\overline{C4b2a3b}$ with enzymic activity for C5, which is thus split to give C5a (M.W. 15,000) and C5b (M.W. 170,000). $\overline{C4b2a3b5b}$ is now formed, and reacts with C6 and C7 to give rise to C4b2a3b67. The latter then splits into $\overline{C4b2a3b}$ (which may possibly activate another molecule of C5) and C5b67, which it is believed becomes attached to a different site on the cell membrane. C8 and C9 are now attached sequentially to this complex, the former to the C5b subunit, the latter to the C8. The group of proteins so formed (C985b67) gives rise to a structure with a hole in the middle, rather like a doughnut. The hole pierces the cell membrane and allows the passage of water and ions into the interior, which causes the cell to burst.

The Properdin System. This system together with complement participates in phagocytosis and inflammatory reactions. Incubation of serum with bacteria or certain polysaccharides gives rise to two enzymes, Factor B and Factor D, which bring about activation of C3 and C5. Factor B corresponds in function to the complement complex $\overline{C4b2a}$ and requires Factor D for its activation. Factor D corresponds to $\overline{C4b2a3b}$. It activates C5 and probably sets in train a series of events which correspond to those of the classic complement system. Once C3b is attached, phagocytosis is stimulated. Antibody may not be required as the initiator of the properdin system. It would appear, however, that the presence of C3b is essential for initiation, but its source is not clear.

(7) Enzymes. The protein constituents of the plasma include a number of enzymes, some of which are of clinical diagnostic importance, e.g., amylase (p. 549), lipase (p. 551), phosphatases (pp. 559, 560), aminotransferases (p. 553), glycolytic enzymes (p. 554), et al.

Nonprotein Nitrogen

The nonprotein nitrogen (NPN) of the blood, that portion of the nitrogenous substances not precipitated by the usual protein precipitants, includes urea, uric acid, amino acids, creatine, creatinine, ammonia, polypeptides, nucleotides, various bases and a number of other compounds. From a metabolic standpoint, the nonprotein nitrogenous constituents of the blood are usually of greater clinical interest than the plasma proteins, since they represent products of the intermediary metabolism of ingested and tissue protein. The normal NPN of serum is usually 15 to 35 mg. per 100 ml., figures as high as 40 mg. occurring occasionally.

Urea. The blood urea normally ranges from about 20 to 45 mg. per 100 ml., the urea N being 9 to 20 mg. (46.6 per cent of the total urea molecule). The extreme normal limits under conditions of very low to very high protein intake are 5 to 23 mg. urea N per 100 ml. Values of 5 to 12 mg. are common in the last months of normal pregnancy. Urea is the chief end-product of protein metabolism and is formed normally largely, if not entirely, in the liver by a mechanism described elsewhere (p. 160).

The blood urea nitrogen may rise to about 35 mg. per 100 ml. following a meal very high in protein and tends to maintain a relatively higher level in individuals on a high protein diet than in those on a low protein intake. It is essential, therefore, in making comparisons between blood urea values, that a standard diet be administered and that the specimen be taken in the fasting state.

Urea is an extremely diffusible substance and, as such, exists in all body fluids in practically the same concentration. Thus, it is present in the spinal fluid, saliva, exudates, and transudates in approximately the same amount as in blood. It is eliminated chiefly in the urine, but considerable amounts may be lost through the skin if perspiration is active.

Uric Acid. The uric acid content of plasma or serum determined by enzyme assay is 3.8 to 7.1 mg./100 ml. in males and 2.6 to 5.6 mg./100 ml. in females. Children show a range of 2.0 to 5.5 mg./100 ml. Uric acid is, in man, the chief end-product of purine metabolism, and is discussed in connection with the metabolism of nucleic acids (p. 253).

Creatinine. The creatinine content of normal blood ranges from 1 to 1.5 mg. per 100 ml., males having slightly higher levels than nonpregnant females. It is the anhydride of creatine, which is present in the muscles. Creatinine is readily diffusible and is excreted largely by the kidneys.

The methods usually employed for the determination of blood creatinine are nonspecific in nature, although specific methods are available. However, whatever may be the exact nature of the substances in the blood which respond to the reaction for creatinine, so remarkable is their constancy in health and so specific are variations in their concentrations in disease that the question as to whether or not this nitrogen fraction is in reality creatinine is of more academic than practical clinical interest.

Amino Acids. The alpha-amino acid nitrogen content of normal plasma is about 3 to 6 mg. per 100 ml. It is derived from that portion of the amino acids absorbed from the intestine which has escaped deamination and synthesis into protein in the liver, and from amino acids resulting from breakdown of tissue protein which have not as yet undergone deamination or synthesis. A rise in the amino acid concentration of the blood occurs following the ingestion of proteins. A decrease follows administration of insulin, growth hormone, or testosterone in normal subjects, due presumably to stimulation of protein synthesis. A similar effect is apparently produced by epinephrine and by administration of glucose (insulin increase?). For clinical purposes, individual amino acid levels are more useful than the total figure.

Ammonia. Because of technical considerations applicable to all available analytical procedures for blood ammonia determinations, there has been considerable uncertainty as to how much ammonium ion is normally present in the systemic circulation. The normal values obtained by the best resin methods, with adequate refrigeration of the specimen, should not exceed 20 μg. NH_3 nitrogen/100 ml. of plasma. The figure is raised by muscle contraction.

The toxicity of ammonia precludes the possibility of its accumulation in significant amount as an intracellular metabolite in deamination reactions. It is effectively disposed of generally by formation of glutamine and, in the liver, by urea formation (p. 160). Although relatively minute contributions to the plasma "ammonia" may be made by deamination processes in the liver and, probably, the kidneys, free ammonia is believed to enter the blood stream mainly from the intestine, where it is formed as a result of bacterial action on amino acids. Under normal conditions, this is removed from the portal blood by the liver cells and converted to urea. A small fraction may, however, normally escape such conversion and enter the systemic circulation in the hepatic vein blood.

Undetermined Nitrogen (Rest Nitrogen). After all of the known nonprotein nitrogenous constituents of the blood have been determined, including those mentioned above and others, such as creatine (0.2 to 0.9 mg. per 100 ml. plasma), there remains a considerable amount of nonprotein nitrogen of undetermined composition termed "rest nitrogen" or "undetermined nitrogen." This may constitute as much as 45 per cent of the total nonprotein nitrogen, ranging normally from 5 to 18 mg. per 100 ml. and residing chiefly in the corpuscles. Since its nature is unknown, its metabolic significance and its derivation are conjecturable.

NITROGEN EXCRETION

Under normal conditions the urine contains an extremely small amount of albumin which usually cannot be detected by ordinary qualitative methods. The excretion of albumin by essentially normal individuals (functional, orthostatic or adolescent albuminuria) will be considered in the discussion of albuminuria (p. 214). Therefore, from a practical viewpoint, with certain reservations, it may be stated that protein is not normally present in the urine.

Practically all of the nonprotein nitrogenous constituents of the blood are eliminated by the kidneys.

Urea. The quantity of urea excreted in the urine depends largely upon the protein intake. The percentage of the total urinary nitrogen represented by urea likewise varies directly with the amount of protein ingested. The lower the protein intake the greater is the relative proportion of products of endogenous metabolism such as creatinine and uric acid, although the total amount of the former is unaltered and the latter decreased. The relative proportion of ammonia nitrogen is likewise greatly increased under such circumstances. Upon a high protein intake (120 g.), urea may constitute approximately 90 per cent of the total urine nitrogen; upon an intake of 6 g. of protein this figure is reduced to about 60 per cent (Folin). With an average diet, approximately 30 g. of urea are eliminated in twenty-four hours, constituting about 50 per cent of the total urinary solids.

It is evident that the total urinary nitrogen, as well as the urea nitrogen, depends upon the nitrogen intake and, to be of any clinical significance, must be regarded in its relation to the latter. Furthermore, since nitrogen balance may be established with varying levels of nonprotein nitrogen in the blood, the isolated determination of total urinary nitrogen or urea nitrogen is of little clinical value under ordinary circumstances.

The mechanism of renal excretion of urea and the clinical significance of abnormalities in this function are considered in detail in the discussion of renal function (p. 501).

Uric Acid. The quantity of uric acid present in the urine of normal individuals depends upon the amount of nucleoprotein ingested (exogenous uric acid) and upon the amount formed as a result of the metabolism of tissue nucleoprotein (endogenous uric acid). This subject is discussed elsewhere (nucleic acid metabolism, p. 253).

Creatine and Creatinine. Creatine, derived from glycine, arginine (amidine group), and methionine (methyl group), is not a waste product, but is of fundamental importance in the phenomenon of muscle contraction (as phosphocreatine). The energy for muscle contraction is derived from ATP breakdown. Phosphocreatine is necessary for its resynthesis.

Creatine is present in the urine of prepubertal children (4.2 mg. per kilogram daily), during pregnancy and the puerperium (two to three weeks), and occasionally in normal nonpregnant women (0 to 100 mg./24 hr.) but is often absent from the urine of normal males on a balanced diet. It may appear in the urine of normal subjects maintained on a very low carbohydrate intake and during fasting periods (up to 100 mg. creatine daily), and is believed to be derived from excessive catabolism of muscle tissue. Creatine is present in the blood almost entirely in the erythrocytes (2 to 7 mg. per cent); when its concentration in the plasma exceeds 0.6 mg. per 100 ml., it almost invariably appears in the urine. This suggests that it is completely reabsorbed in the renal tubules at lower concentrations.

When 1.32 or 2.64 g. of creatine hydrate (equivalent to 1 and 2 g. of creatine) are ingested by normal adults, about 80 per cent is retained by males and about 70 per cent by females, about 20 per cent and 30 per cent, respectively, being excreted in the urine during the next twenty-four hours. This forms the basis of the creatine tolerance test, which is of value clinically in demonstrating excessive excretion of creatine in the absence of spontaneous creatinuria under ordinary dietary conditions. It is also of limited value in affording a quantitative index of the functional state of the muscles in various myopathies and other conditions (p. 222).

Creatinine (creatine anhydride) is a waste product derived from creatine, and is, therefore, usually of endogenous origin. Normal adult males excrete 1 to 2 g. and females 0.8 to 1.8 g. daily, the amount eliminated being uninfluenced by the urine volume or the character of the diet if the latter is adequate and contains no creatinine or creatine. Folin found that upon a diet containing 118 g. of protein and 2786 calories, 0.58 g. of creatinine was eliminated, constituting 3.6 per cent of the total urinary nitrogen; upon a diet containing 6 g. of protein and 2153 calories, 0.60 g. of creatinine was eliminated, constituting 17.2 per cent of the total urinary nitrogen. Because of this relative constancy of elimination, the determination of the total creatinine excretion has been employed as a means of checking the accuracy of twenty-four-hour urine collections. If marked variations in urinary creatinine are observed that do not depend upon corresponding changes in weight or surface area, one may usually assume that there has been some error in the collection of the urine specimens. Because of its origin from creatine, the quantity of creatinine excreted in the urine is an index of the level at which the body maintains its phosphocreatine and not of the rate of endogenous protein metabolism. The amount of creatinine excreted in the urine in twenty-four hours corresponds to about 2 per cent of the creatine in the body.

The term "creatinine coefficient" is used to indicate the number of milligrams of creatinine (plus creatine) nitrogen excreted per kilogram of body weight in twenty-four hours. Normal values range from 7.5 to 10 (20 to 26 mg. creatinine) for men and 5 to 8 (14 to 22 mg. creatinine) for women. The creatinine coefficient may be assumed to be an index of the amount of active muscle tissue in the body. It is believed by some that muscular exercise has no influence in this connection, but others believe that although the daily elimination is constant, the urinary creatinine is increased during of muscular activity, being derived from creatine liberated during this period.

α-Amino Acids. Amino acids are eliminated in both the combined and the free state. In normal adults the total urinary amino acid nitrogen varies from 0.5 to 1.0 g. daily. The free amino acid nitrogen ranges from 0.05 to 0.2 g. daily. Under normal conditions, amino acids pass freely from the blood plasma into the glomerular filtrate, but are very efficiently reabsorbed by the renal tubular epithelium and returned to the blood stream. Even when the plasma level is elevated by intravenous injection of amino acids, comparatively little escapes in the urine, usually not over 5 per cent of the amount administered. There is considerable variation in the extent to which individual amino acids are reabsorbed in the renal tubules. The fact that isolated reabsorption defects may occur, resulting in excessive excretion of certain amino acids, but not others, suggests the existence of separate tubular transport mechanisms for different groups of amino acids (p. 543).

Histidine, an amino acid not ordinarily excreted in significant amounts, appears in the urine at about the fifth week of pregnancy and disappears from the urine within a few days after delivery. There is evidence that the activity of liver histidinase is inhibited during this period, failure of the enzyme to catalyze the catabolism of histidine accounting for the presence of this amino acid in the urine. The amount excreted is increased by a high intake of protein.

Ammonia. The daily output of ammonia nitrogen in the urine varies from 0.5 to 1.2 g. As stated elsewhere (p. 160), urinary ammonium ion is formed in the kidneys, being derived from amino acids and glutamine through the activity of the epithelial cells of the renal tubules (p. 463). This subject is discussed in detail elsewhere (hydrogen ion regulation p. 411).

Urinary ammonia varies in accordance with the body's needs for excretion of acid with, at the same time, the maximum conservation of fixed base. Exceptions to this rule include acidemia accompanying renal disease, in which the ammonia-forming function of the renal epithelium is impaired (p. 419). Urinary ammonium ion is increased in normal subjects following ingestion of acid-forming foods, mineral acids or their ammonium, calcium or magnesium salts, by a low carbohydrate intake (ketosis), and in normal pregnancy. It is decreased following ingestion of alkalis or base-forming foods and during the "alkaline tide" accompanying gastric secretion of free HCl (p. 471).

PLASMA PROTEIN ABNORMALITIES

Several procedures are available for demonstration of quantitative and, in some instances, qualitative abnormalities of plasma proteins (Table 3-3). The total protein concentration is usually determined chemically (Kjeldahl or biuret). Salt fractionation may be employed for separation of albumin and globulin, the former being determined in the filtrate (Kjeldahl, biuret) and the latter by difference from the total protein. These procedures generally give falsely high values for albumin because of incomplete precipitation of globulins. By electrophoretic analysis one obtains the relative concentrations of the individual protein fractions, usually expressed directly as percentages of the total. Normal serum usually yields 5 major fractions, albumin, α_1-globulin, α_2-globulin, β-globulin, and γ-globulin; plasma, in addition, contains fibrinogen. However, more globulin bands may be obtained by electrophoretic procedures carried out in starch or acrylamide gels. The fact should be appreciated that values obtained with boundary electrophoresis (measured by increment in specific refractive index) are not identical with those obtained with paper or cellulose acetate electrophoresis (measured by dye-binding capacity); the latter method is commonly employed clinically. Moreover, no chemical fractionation procedure yield values that correspond satisfactorily to those for α- and β-globulins as determined by electrophoresis. Despite these drawbacks, the combination of total protein determination (Kjeldahl, biuret) and cellulose acetate electrophoresis is most useful.

Ultracentrifugation and low temperature-ethanol fractionation are seldom employed clinically; immunological procedures are now much more commonly used. The presence of certain qualitatively abnormal proteins may be demonstrated by electrophoresis (e.g., myeloma globulin) or by certain peculiar temperature reactions (e.g., cryoglobulins, pyroglobulins). Other abnormalities, quantitative and qualitative, may be reflected in abnormalities of certain relatively simple turbidity and flocculation reactions; these are particularly useful in the study of patients with hepatic and biliary tract disease (p. 625).

Total Plasma (Serum) Protein. Certain generalizations may be made concerning the trends of abnormalities in disease. These are most often in the direction of a decrease in total protein concentration, due to decrease in albumin. Although the globulin fraction may occasionally also be low, it generally exhibits a simultaneous increase. The causes and significance of these changes in individual protein components will be discussed subsequently.

The degree of hypoalbuminemia usually exceeds that of the frequently coexisting hyperglobulinemia, the total protein concentration being, therefore, subnormal. However, in certain cases, particularly of multiple myeloma, viral and protozoal diseases and, occasionally, hepatic cirrhosis, the converse may occur, with consequent increase in total protein concentration, e.g., as high as 16 g./100 ml. in multiple myeloma. Regardless of their magnitude, the changes in these two fractions in opposite directions tend to offset each other and, obviously therefore, the total protein concentration may not reflect either the nature or the extent of an existing abnormality and may, indeed, even fail to indicate its presence. Some type of fractionation procedure (p. 167) is required for these purposes.

There are other pitfalls in the interpretation of plasma protein values. On standing erect for one-half hour, the concentration may increase about 1 g., i.e., from 6.7 g. to 7.5 g./100 ml., due to hemoconcentration incident to increased transudation from capillaries in dependent portions (increased venous pressure). It must be borne in mind, too, that in many disease states (e.g., diabetes mellitus, glomerulonephritis, protracted diarrhea or vomiting), hemoconcentration due to dehydration may mask the presence or extent of an existing serious plasma protein deficit. The true situation is revealed only by determining the total circulating plasma protein, i.e., plasma volume $\times$ protein concentration. For example, during a period of protein depletion in dogs, the plasma protein concentration fell only from 7 g. to 5.9 g./100 ml. (about 15 per cent) although the total amount of circulating plasma protein fell from 16.5 g. to 6.5 g.

(about 60 per cent). Similar observations have been made in inmates of prison camps. However, measurement of plasma volume is not always feasible, and is certainly not necessary in every case.

Fibrinogen. The plasma fibrinogen (normal 200 to 400 mg. per 100 ml.) is not significantly altered in disturbances of protein metabolism per se.

Increased Plasma Fibrinogen. Rouleaux formation and erythrocyte sedimentation in vitro are influenced strikingly by the plasma fibrinogen concentration. Increase in this factor is a most important cause of the increased sedimentation rate in infections and other diseases. The alpha- and beta-globulins are next in order of activity. Although the sedimentation-accelerating effect of gamma-globulins is less marked, they usually contribute more to the production of this phenomenon in disease because the increase in their concentration generally far exceeds that of the other globulins. Albumin exerts an opposing effect, tending to diminish erythrocyte sedimentation.

(1) Conditions causing slight hepatic injury (hepatitis) and tissue destruction (inflammatory) may be associated with moderate increases in plasma fibrinogen.

(2) Most acute infections (except typhoid fever), septicemias, bacteremias, and particularly pneumococcic pneumonia, in which condition values up to 1000 mg. per 100 ml. may be observed.

(3) Pregnancy and menstruation (400 to 500 mg. per 100 ml.).

(4) Following x-ray irradiation.

(5) Nephrotic syndrome. In this condition the plasma fibrinogen may rise to 1000 mg. per 100 ml., the mechanism being unknown.

(6) Increased fibrinogen values have been reported in multiple myeloma, lymphopathia venereum, and certain cases of cirrhosis.

Decreased Plasma Fibrinogen. (1) **Congenital Afibrinogenemia.** This is a rare condition, genetically determined, characterized by extreme grades of fibrinogen deficiency, due to defective synthesis. Near relatives may show decreases of plasma fibrinogen to levels not adequate to produce bleeding.

(2) **Hepatic Insufficiency.** Since fibrinogen is formed in the liver, decreased concentrations may occur in severe acute liver failure, as seen occasionally in acute hepatic necrosis (e.g., chloroform, carbon tetrachloride poisoning). It is seldom observed in infectious hepatitis or cirrhosis.

(3) **Obstetric Complications.** These constitute the most common and most important causes of decrease of fibrinogen. It may occur in: (*a*) premature separation of the placenta; (*b*) delivery of a hydatidiform mole; (*c*) missed abortion of late pregnancy; (*d*) retention (several weeks) of a dead fetus; (*e*) tumultous labor, as in eclampsia or induced by Pitocin.

This condition may be due to one or both of two causes: (*a*) liberation of tissue thromboplastin into the systemic maternal circulation, with consequent fibrination and depletion of the circulating fibrinogen; (*b*) local, i.e., intrauterine defibrination of the maternal blood.

(4) **Acute Hypofibrinogenemia.** This may occur in shock, severe burns, extensive cancer of the prostate, bladder, stomach, or pancreas, and as a

complication of major abdominal or pulmonary operations. It may be associated with decrease in other clotting factors (e.g., prothrombin and proaccelerin) and with increased fibrinolytic activity of the blood. According to one view, the decrease in fibrinogen is due primarily to its increased destruction by fibrinolysin, produced in increased amount by activation of its precursor, plasminogen, by an activator (possibly Factor XII is involved) entering from the damaged tissue. Another theory holds that tissue factors entering the circulation act primarily as thromboplastins, initiating intravascular fibrination, as in the case of pregnancy complications. Increased fibrinolysis may also occur as a secondary event.

Albumin. Increase in the concentration of albumin in the plasma is seldom encountered clinically. It may occur in dehydration with diminution in the water content of the plasma and consequent increased concentration of all nondiffusible components. Although theoretically possible, this condition is actually rarely observed, for in most instances in which dehydration is a prominent feature, such complicating factors as malnutrition, diarrhea, vomiting, etc., exert an influence which tends to decrease the plasma albumin concentration. However, total protein values of 10–12 g./100 ml. due to this cause have been reported in patients with cholera, other forms of severe diarrhea, particularly in children, shock, burns, diabetic acidemia, Addison's disease, intestinal obstruction, intestinal fistula, pyloric obstruction, rigid restriction of fluid intake, heat exhaustion, and certain fulminating infections. In such cases, as indicated previously, the true nature of the situation can be revealed only by determination of the circulating plasma volume. The existence of an actual plasma albumin deficit is revealed at times by rapid restoration of extracellular fluid volume by therapeutic administration of salt solution (p. 189).

Decrease in plasma albumin (normal 3.6–4.5 g./100 ml.) is of practical significance in several groups of clinical disorders, which are outlined below. One of the most important functions of albumin is in the maintenance of the colloid osmotic pressure of the blood plasma (p. 174) and, consequently, in the regulation of the distribution of water between the intravascular and the interstitial tissue compartments.

Edema is the most prominent feature of reduced plasma albumin concentration, the lowered colloid osmotic pressure tending to diminish the passage of water from the interstitial tissue spaces into the capillaries. The frequently associated increase in globulins is incapable of compensating effectively for the decrease in albumin because, their molecular size being much greater, they exert much less effect than albumin on osmotic pressure (p. 174). By direct measurement, the colloid osmotic pressure (C.O P.) of normal serum or plasma has been found to be 280 to 470 mm. H_2O (average 360 mm.), with slightly lower values in normal pregnancy (average 300 mm.). Various formulas have been evolved for calculating the C.O.P. from the albumin and globulin concentrations; e.g., C.O.P. = 45.2 A + 18.8 G, where A is the albumin and G the globulin concentration in grams per 100 ml. However, such prediction equations are of no clinical significance.

There is a linear relation between the plasma specific gravity and total protein content, normal heparinized plasma (7 g. protein/100 ml.) having a specific gravity of approximately 1.027 and an increase or decrease of 1 g./100 ml. causing a rise or fall of 0.0029. Determination of plasma specific gravity has therefore been proposed as a rapid means of estimating the protein content, especially in dehydration and shock. Inasmuch as albumin and globulin have about the same effect on specific gravity, this procedure cannot reveal simultaneous change in these factors in opposite directions.

Frequent attempts have been made to relate the development of edema to a certain level of hypoalbuminemia. It has been suggested, for example, that patients with nephritis develop frank edema (noncardiac) when the albumin falls below 2.5 ± 0.2 per cent ("edema level"). Experimentally, it has been found that edema appears when the albumin falls below 0.8 g./100 ml. and disappears when it rises above 1 g./100 ml. Similar statements have been made in relation to the occurrence of ascites in cirrhosis. It is stated, too, that edema usually appears when the plasma specific gravity falls below 1.023. However, no categorical statement can be made in this connection. Because of the operation of numerous other complicating factors, edema may be present at higher and absent at lower levels of plasma albumin, C.O.P., or specific gravity in different subjects or in the same subject at different times. These factors include abnormal variations in capillary permeability, capillary (i.e., venous) blood pressure, lymph flow, tissue tension, and sodium intake (p. 526).

As indicated elsewhere (p. 190), globulins are frequently, but not invariably increased in conditions in which the albumin concentration is low. This is not necessarily a relationship of cause and effect (p. 190); more probably, the condition which underlies the depletion of albumin simultaneously stimulates the formation of certain globulins, e.g., antibodies (γ-globulins), lipoproteins (α- and β-globulins). As a result of these changes, the albumin:globulin ratio, normally about 1.5:1, is decreased or may be reversed. This factor is without clinical significance, since it may reflect a decrease in albumin, an increase in globulin, or a combination of both. The factors of clinical importance are the absolute levels of albumin and globulins, not their concentration ratio.

The causes of decrease in plasma albumin concentration, i.e., *hypoalbuminemia,* may be outlined as follows:

(1) Loss of Albumin. This occurs most commonly, clinically, in nephritis and the nephrotic syndrome with excessive and prolonged albuminuria. As much as 60 g. of albumin may be eliminated daily in the urine and a daily loss of 4 to 5 g. may eventually result in a gradual decrease in the level of plasma protein. The plasma protein concentration usually remains within normal limits in the early stages of acute glomerulonephritis. Van Slyke found that cases of this condition which maintain plasma albumin values above 2.2 g. per 100 ml. or total protein values above 5.5 g. per 100 ml. have a better prognosis than those which do not. Of his cases in the group with higher albumin values, only 13 per cent progressed into the chronic or terminal stages of nephritis, in contrast to 72 per cent of the group with low albumin values. It would appear, therefore, that the plasma protein concentration is of prognostic significance during the acute stage of glomerulonephritis. It was also found that a low plasma albumin concentration, usually below 2.5 g. per 100 ml., was the rule in the chronic active stage of glomerulonephritis. As the chronic active stage progressed into the terminal stage, there was a marked tendency for the plasma proteins to increase, normal values being obtained in about half the cases during the terminal stage of the disease. This may be due in part to decreasing albuminuria and in part to hemoconcentration.

Extremely low values may occur in the nephrotic syndrome, falling at times to 0.2 g. per 100 ml. Prolonged albuminuria in amyloid disease of the kidneys and congestive heart failure may contribute to the development of hypoproteinemia. Decrease in albumin, with or without a decrease in the total plasma protein concentration, occurs commonly in multiple myeloma; in some cases the globulin fraction is enormously increased, with consequent hyperproteinemia (p. 183).

The blood plasma is the source of the albumin present in the urine. Similar changes in albumin, without increase in globulin, have been produced experimentally by repeated plasmapheresis, with withdrawal of blood, separation of the plasma and reinjection of the corpuscular elements. The diminution in serum albumin which occurs in renal disease is probably not always due entirely to loss of albumin in the urine. There appears, in some cases, to be a nutritional defect which interferes with the ability of the organism to synthesize serum albumin. In the terminal stages of chronic glomerulonephritis, a possible serum albumin depletion may be masked by the development of a state of hemoconcentration, which may result from vomiting, anorexia, and inability to conserve the salt and water stores of the body. A diminution in serum albumin may occur in acute glomerulonephritis, particularly in the transition between this stage and the chronic active stage of the disease. Here it appears to be referable to loss of albumin in the urine, malnutrition, and impairment of the ability to synthesize serum albumin.

In certain cases of massive ascites, since the ascitic fluid contains a large amount of protein, depletion of the blood proteins may occur with changes similar to those observed in the nephrotic syndrome with prolonged, excessive albuminuria. Edema may be present in such patients in the absence of renal or cardiac lesions. In advanced stages of chronic hepatic disease, including portal cirrhosis, there is also a reduction in total serum protein, the diminution occurring in the albumin fraction. In some cases, particularly in acute forms of liver disease, the serum albumin may be only moderately reduced and the serum globulin increased (p. 620). Such findings are obtained much more commonly in primary hepatocellular disorders than in obstructive jaundice. The decrease in serum albumin in hepatic disease has been attributed by some chiefly to coexisting malnutrition, and by others chiefly to interference with synthesis of serum albumin because of impairment of liver function. In cases in which ascites is present, the loss of albumin from the blood plasma into the ascitic fluid probably plays some part in contributing to the plasma protein depletion.

Large amounts of albumin may be lost also in extensive burns, thoracic duct fistula, severe hemorrhage, and polyserositis (Pick's disease). Hypoproteinemia has been observed in shock due to burns, intestinal obstruction, and generalized peritonitis, due presumably to loss of plasma from the circulation.

Instances have been reported of hypoalbuminemia (with hypogammaglobulinemia) associated with, and presumably resulting from, exudation of albumin (and other plasma proteins) into the gastrointestinal tract. In some cases there may also be a condition of increased permeability of peripheral capillaries, resulting in increase in the extravascular compartment of the total albumin pool. These phenomena may antedate the development of frank intestinal lesions. This condition has been studied by measurement of fecal excretion of I^{131}-labeled albumin or polyvinylpyrrolidone following intravenous injection (p. 607). Similar losses may also contribute to the hypoproteinemia of patients with chronic ulcerative colitis, regional enteritis, and hypertrophic gastritis.

(2) Inadequate Supply of Protein. This may occur in states of inanition, as in advanced tuberculosis and malignancy. It is observed in gastrointestinal disorders accompanied by impaired protein digestion or absorption, prolonged vomiting, or diarrhea. Prominent among such conditions are peptic ulcer, gastrointestinal malignancy, pancreatic disease, intestinal fistula, short-circuiting operations on the intestine, chronic ulcerative colitis, regional enteritis, tuberculous and other forms of chronic enteritis, steatorrhea (celiac disease, sprue), and other causes of the malabsorption syndrome (p. 600), the nephrotic syndrome, congestive heart failure with edema of the gastrointestinal mucous membrane, pellagra. This condition may also result from dietary protein inadequacy. This is illustrated by the many instances of war edema, famine edema or nutritional edema which were observed during World War I and in inmates of prison camps in World War II.

(3) Impaired Synthesis. This factor probably plays an important part in maintaining the low level of plasma albumin in chronic hepatic disease, particularly cirrhosis, in a minority of cases of the nephrotic syndrome, various infections, severe anemias (e.g., uncontrolled pernicious anemia), cachectic states, and pregnancy and lactation. However, the hypoalbuminemia of the nephrotic syndrome is not infrequently associated with an increased rate of synthesis of serum albumin, presumably in response to the stimulus of excessive loss. Observations in experimental animals indicate that, under conditions of simultaneous hepatic damage and dietary protein restriction, the functional capacity of the liver, rather than the protein intake, is the limiting factor in determining the plasma protein concentration (p. 620).

(4) Excessive Protein Catabolism. Excessive breakdown of body protein, in conjunction with either inadequate supply or defective utilization, as in uncontrolled diabetes mellitus, thyrotoxicosis, prolonged febrile illnesses, and trauma (including surgical), may be accompanied by hypoalbuminemia.

(5) Pregnancy and Lactation. A steady decrease in plasma protein concentration may occur during normal pregnancy, due to diminution in the albumin fraction. Actually, subnormal values are rarely observed under normal conditions. However, aggravation of this natural tendency by dietary restriction of protein may lead to harmful consequences. Low values for albumin and total protein have been reported in patients with toxemia of pregnancy, particularly eclampsia. The decrease in normal pregnancy is attributed, in part at least, to increased plasma hydration (p. 344). However, not only the concentration but also the total amount of circulating plasma protein is diminished. This diminution is associated with impairment of ability to synthesize serum protein. Some believe that synthesis of body proteins in the fetus constitutes a drain upon the maternal organism which is of primary importance in causing the lowered serum protein concentration in pregnancy. Similar findings have been obtained during lactation.

(6) Plasma Dilution. Hypoproteinemia may result from sudden dilution of the plasma which follows an acute, massive hemorrhage. A similar phenomenon may accompany sudden recovery from severe dehydration in malnourished subjects, as in diabetic coma and protracted vomiting or diarrhea (especially in children). In the dehydrated state, the actual plasma protein deficit may be masked by hemoconcentration (loss of plasma water). When the latter is corrected by administration of large quantities of solutions of sodium salts (chloride, bicarbonate, lactate, etc.), the plasma protein concentration may fall to a level sufficiently low to produce edema.

The frequently diminished plasma protein concentration in congestive heart failure is due in part to increased plasma hydration. As indicated above, it is also contributed to in such cases by prolonged albuminuria, decreased protein intake, impaired digestion, diminished absorption from the edematous intestinal mucosa, and impaired synthesis (hepatic malfunction).

(7) Genetic Defect. A condition of virtual absence of serum albumin (analbuminemia) may occur as a genetically determined defect in albumin synthesis. The total globulin is increased in such cases. Another congenital disorder, bisalbuminemia, is characterized by a double albumin peak in the electrophoretic pattern. Both of these albumins are immunochemically similar to normal serum albumin. In one instance, it has been claimed that an immunological difference could be demonstrated. One of the two bands migrates as normal albumin, and the other is either slower or faster. The abnormal band can have a varying binding capacity for certain medicaments. It has been claimed that this can sometimes lead to complications in regard to therapeutic dosage. Occasionally, a patient is seen with transient bisalbuminemia.

Prealbumins. Electrophoresis, especially in gels, may reveal a component migrating faster than albumin, designated prealbumin. One such component (tryptophan-rich prealbumin) appears in most cases, together with a second (orosomucoid) in starch or acrylamide gels. The tryptophan-rich component can bind thyroxine.

Globulins. The several approaches to the investigation of possible abnormalities in the plasma globulins are indicated elsewhere (pp. 168 and 196). It will be recalled that the classical Howe fractionation procedure is unreliable, yielding erroneously high values for albumin and low values for total globulins (Table 3-1, p. 168). The fact must be kept in mind that the globulin fraction, as determined by chemical methods, is an extremely heterogeneous group of substances. Over thirty different components have been characterized and identified, but many more have been identified and await characterization. Inasmuch as many if not all of these are functionally and structurally distinct, independent variations may occur in single components, or in small groups of related components, in response to various types of stimuli or as a result of disease. Different globulin fractions may vary in opposite directions, with no significant change in the total globulin concentration. Such abnormalities may be revealed only by more refined analytical procedures, such as electrophoresis. Useful information in this connection may be obtained also by application of so-called "globulin reactions" (p. 205), such as the zinc sulfate turbidity, thymol turbidity, and cephalin-cholesterol flocculation tests.

Although decreased concentrations of individual globulin fractions may be encountered in disease, there is seldom an associated decrease in the total globulin concentration below the lower limit of normal. This is due to (a) the relative quantitative insignificance of the decreased component, and (b) to the fact that other globulins may be increased simultaneously. Exceptions to this generalization may be encountered in cases of agammaglobulinemia (p. 203), idiopathic hypoproteinemia, and occasionally in the nephrotic syndrome, celiac disease, severe malnutrition, Cushing's syndrome, or following administration of adrenocortical hormones.

In the great majority of conditions associated with abnormalities of serum protein concentrations, the common pattern is hypoalbuminemia and hyperglobulinemia, the latter due usually to increase in γ-globulin. Certain generalizations may be made concerning quantitative abnormalities in the various globulin fractions.

Alpha-Globulins. Decrease in the α_1-globulin occurs in those conditions in which the serum mucoproteins are diminished (see later) but this is not always demonstrable. An increase in this fraction not infrequently occurs in conditions characterized by increase in α_2-globulin. These are inflammatory or destructive lesions; acute and chronic infections; extensive malignancy, particularly metastatic; myocardial infarction; severe burns and other skin lesions; trauma, including operative; the nephrotic syndrome; diabetes mellitus; hepatic cirrhosis; periarteritis nodosa; rheumatoid arthritis.

The mechanism underlying this increase is not clear. The fact that glycoproteins are contained in this fraction may be of significance in this connection, in view of the rather consistent correlation with extensive tissue damage.

Beta-Globulins. Decrease in this fraction is seldom demonstrable, but may occur as a congenital disorder associated with intolerance to iron. Increases are often referable to increases in β-lipoproteins. This may occur, therefore, in conditions accompanied by hyperlipemia, e.g., the nephrotic syndrome, diseases of the liver and biliary tract (obstructive jaundice, biliary cirrhosis, acute hepatitis, portal cirrhosis), diabetes mellitus, hypothyroidism, idiopathic hyperlipemia, etc. It may occur also in periarteritis nodosa, malignancy, malaria, sarcoidosis, certain cases of multiple myeloma, and malignant hypertension. Beta-globulin values obtained by paper electrophoresis do not correlate so well with serum lipid or lipoprotein concentrations as do those obtained by boundary electrophoresis.

Gamma-Globulins. This is the fraction that is increased most frequently and, usually, most strikingly. In some cases it may comprise as much as 50 per cent or more of the total plasma proteins. In man, most but not all antibodies are γ-globulins and, therefore, this fraction would be expected to increase in infectious and other diseases evoking an immunological response. The most marked elevations occur usually in (*a*) viral and protozoal diseases, (*b*) bacterial infection, (*c*) extensive granulomatous proliferation, and (*d*) certain other conditions, such as multiple myeloma, hepatic cirrhosis, hepatic necrosis, thyroid disease (auto-immunity) and collagen disorders. Increases may occur also in acute leukemia, chronic granulocytic leukemia, severe burns, lymphomatous diseases, and occasionally in starvation or severe protein malnutrition (decreased in some, p. 202).

It has already been pointed out that not all antibodies are γ-globulins. In fact, antibodies have varying electrophoretic mobility and may appear anywhere from the α_1-globulin zone to a post-γ-globulin position. Antibodies are therefore referred to as immunoglobulins (Ig or γ). These have very similar overall molecular structures. They are built up from two types of polypeptide chain, the *heavy chain,* which contains approximately 420 to 440 amino acids, and the *light chain,* which contains about 210 to 230 amino acids. The basic molecular unit of any immunoglobulin is built up from two identical heavy chains and two identical light chains (Fig. 3-4). At approximately their middle point, the two heavy chains are linked together by a varying number of disulfide bridges in such a fashion that their ends (N-terminal and C-terminal) correspond. The half of each chain on the C-terminal side of the link region is known as the Fc fragment and that on the N-terminal side as the Fd fragment. The C-terminal amino acid (cysteine) of each light chain is attached by a disulfide bridge to a cysteine residue in the Fd portion of one or the other of the two heavy chains. This link is a few amino acid residues on the N-terminal side of the heavy-chain link region. There are at least two internal sulfur bridges in each light chain and in each of the Fd and Fc portions of the heavy chains. These are shown in Figure 3-4. The light chain and the Fd portion of the heavy chain are known as the Fab fragment. This contains the site for combination with the antigen. Since there are two Fab fragments in the basic immunoglobulin molecular unit, each such unit must possess two combining sites. It is possible, by gentle reduction to split the

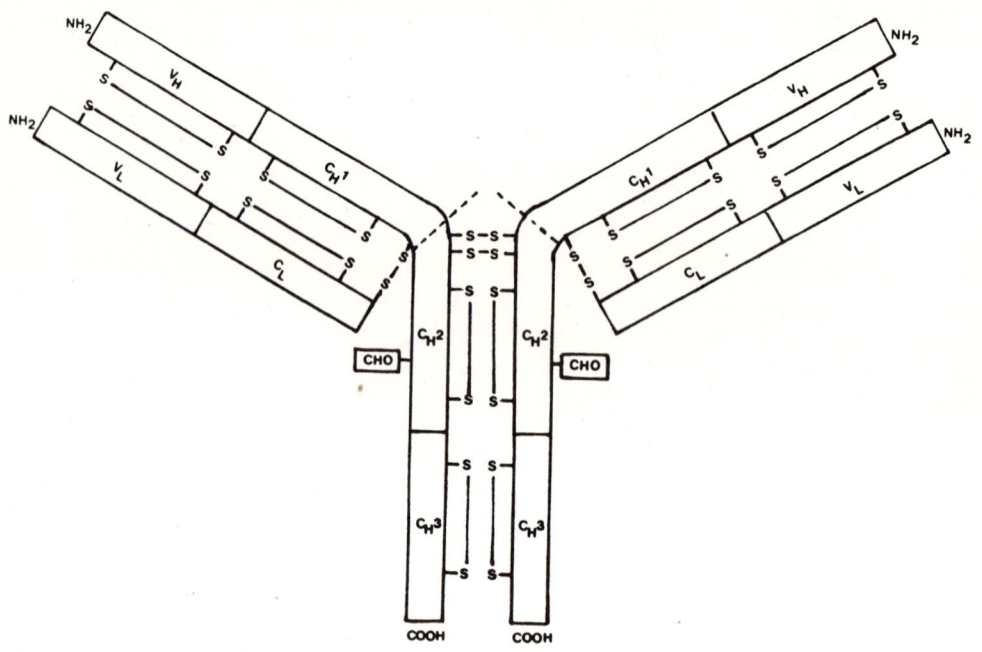

Figure 3-4. Immunoglobulin structure (IgG).

link between the heavy chains and so to obtain two half-units, each containing one combining site. The light chain can also be separated from the heavy chain with the loss of antibody activity. By appropriate enzyme action (papain or pepsin), it is possible to isolate the Fc fragment (as crystals) or the Fab fragments, either separate or combined together. The latter is referred to as (Fab')$_2$ and obviously contains two antigen-antibody combining sites.

By virtue of the fact that Fc fragments could be prepared in a crystalline state from the complex mixture of immunoglobulins present in serum, it was soon realized that the Fc portions must be identical or show very little differences in the immunoglobulin molecules. Rapid advances in this field resulted from the fact that patients, as well as mice, suffering from myelomatosis produced one single immunoglobulin, an M protein, in overwhelming preponderance, and this could therefore be isolated in a highly pure state with relative ease. The same may be said of the Bence Jones proteins in the urine, as soon as it was realized that they corresponded to the light chain of the M protein. Although relatively few myeloma proteins have actually been shown to be antibodies, there is an overwhelming body of evidence that they must be. It was now possible to prepare antisera to light chains of a pure type, as well as to pure immunoglobins and heavy chains, isolated chemically, and the field was open for further advances, including the determination of amino acid sequences in the heavy and light chains. In fact, the complete primary structure of at least one immunoglobulin is now known. All of these substances contain a carbohydrate moiety on each heavy chain.

It is now recognized that, depending on the structure of the heavy chain, all immunoglobulins belong to one of five, possibly six, major groups (IgG or γG, IgA or γA, IgM or γM, IgD or γD, IgE or γE and possibly IgF). The heavy chain is labeled with an appropriate letter of the Greek alphabet, e.g., the heavy chain of IgG is called the γ chain, that of IgA the α chain, that of IgM the μ chain and so on. Depending on the structure of the light chain, there are two major classes K and L and the corresponding light chains are known as κ and λ, respectively. Since each immunoglobulin contains only one type of heavy chain, it is possible to refer to the main classes by simple formulas; e.g., IgG could be $\kappa_2\gamma_2$ (IgGK; γGK) or $\lambda_2\gamma_2$ (IgGL; γGL) and IgA could be $\kappa_2\alpha_2$ (IgAK; γAK) or $\lambda_2\alpha_2$ (IgAL; γAL), and so on. We now know there are subclasses of the main groupings dependent on heavy chain structure. IgG (γG) has four (IgG$_1$, IgG$_2$, IgG$_3$ and IgG$_4$); IgA (γA) and IgM (γM) each have two. As indicated, these subclasses can be represented by a subscript. The situation is even more complicated, since there can be inherited allelic variants of each heavy or light chain, and apparently there are an unknown number of forms (idiotypic) which are unique to individual persons.

The amino acid sequence studies have been most interesting. They indicate that, starting at the C-terminal end, half of the light chain and three-quarters of the heavy chain are relatively constant in structure. The amino acids and their sequence in the remaining portions of these chains are very variable, sufficiently variable, in fact, to account for the formation of antibodies to specific antigens. In other words, the N-terminal half of the light chain together with the corresponding quarter of the heavy chain could account for a specific antigen binding site. The relatively constant portions of the chains are referred to as the C-sequences and the variable portions as the V-sequences. As shown in Figure 3-4, the variant portions of the light and heavy chains are designated V_L and V_H, respectively, and the constant portions C_L and C_H. The latter is subdivided into C_H1, C_H2 and C_H3. Each half molecule of the immunoglobulin, therefore, contains six such designated zones (domains) which are joined together by short lengths of peptide chain. Each zone has an internal disulfide bridge. V_L and V_H have many points of identity in their amino acid sequence. In other words, they are closely homologous to each other. They are, however, very different from C_L, C_H1, C_H2 and C_H3, which are, however, closely homologous to each other. It is the homologous lengths of the polypeptide chain, which are correctly referred to as domains, which in turn are joined by short lengths of fairly straight peptide linkages. The domains themselves, because of their internal disulfide bridges, form loops. This loop formation appears to be highly important in regard to the properties of each domain. V_L and V_H are concerned with antigen-antibody interaction. The other "constant" domains have specific properties; for example, C_H2 contains a carbohydrate moiety, to which complement (p. 175) is probably linked.

From an evolutionary point of view, it seems highly likely that V_L and V_H are developments of the product of what was originally a single gene.

C_L, C_H1, C_H2 and C_H3 appear to be developments of the product of a second, probably unrelated, gene. It seems that different genes are responsible for each of the C-sequences of the light and heavy chains and that some sort of variable gene system for each of the V-sequences. Studies with Bence Jones proteins indicated a recurring sequence of amino acids at the junction of the V_L and C_L portions of the light chains. These are common to the κ and λ types of chain. It is suggested that this may be the recognition point for the joining of the two different gene activities or for firing off the variable portion. Fixed differences in the relatively constant portions could account for the K and L (light chain) types, as well as for the G, A, M, D, E, and possibly F (heavy chain) types. Slight changes, possibly affecting a single amino acid in the C portions, could account for inherited allelic variants and a similar sort of mechanism could also account for the idiotypic forms. The immunoglobulins are synthesized in immunologically competent cells, the immunocytes (plasma cells, small lymphocytes), and one cell is believed to produce only one or, at most, two antibodies. To produce a single antibody in appropriate amount would therefore require a pure cell clone arising from division of a single parent cell capable of producing the antibody in question. The presence of an antigen must stimulate the activity, or size, of such a clone. Alternatively, malignant changes could produce uncontrolled growth and activity and so account for the paraproteins of malignant immunocytomas (myelomatosis, and so forth).

IgG is the major class of immunoglobulin present in normal serum and accounts for about three-quarters of the total immunoglobulin present. It ranges in normal adults from about 540 to 1700 mg./100 ml. serum, with a mean of about 1000 mg./100 ml. The various subclasses of IgG are present in the following rough proportions, IgG_1, 70 per cent, IgG_2 16 per cent, IgG_3 10 per cent and IgG_4 4 per cent. This immunoglobulin class consists of a single immunoglobulin unit and has a molecular weight of approximately 155,000. Its sedimentation constant in the ultracentrifuge is 7S. This type of antibody appears to arise in response to soluble antigens (proteins, bacterial toxins, etc.). At birth, all IgG is derived from the maternal blood, since it can pass the placental barrier. This is destroyed gradually, but more rapidly than it can be replaced in the first few months of life. There is, therefore, a so-called period of physiological hypogammaglobulinemia (Fig. 3-5).

IgA forms approximately 17 per cent of the total serum immunoglobulins. Its normal adult range is approximately 120 to 450 mg./100 ml. serum, with a mean of about 260 mg./100 ml. Its turnover is much more rapid than IgG and so it is actually synthesized at approximately the same rate. As present in serum, it is a 7S immunoglobulin of molecular weight approximately 170,000. It is the major immunoglobulin in saliva, the secretions of the respiratory tract and of the mucous membranes of the alimentary tract. In these situations, however, it has a sedimentation constant of 11S with a molecular weight of approximately 380,000 and corresponds to two molecules of the serum type molecule joined together by

PHYSIOLOGICAL HYPOGAMMAGLOBULINEMIA OF THE NEWBORN

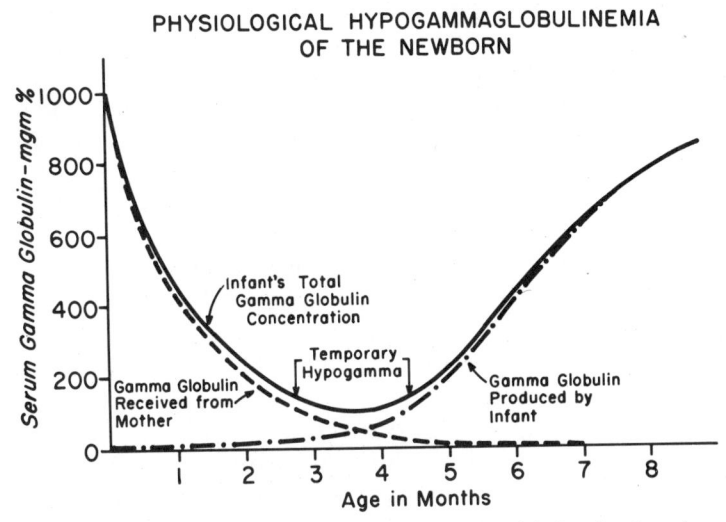

Figure 3-5. Levels of serum gamma globulin in physiological hypogammaglobulinemia of newborn infants. (From Barrett, B., and Volwiler, W.: J.A.M.A., 164:866, 1957.)

so-called secretory piece, containing an appreciable amount of carbohydrate. It appears that at least 60 per cent of the 7S monomer is synthesized in cells in the laminae propriae underlying mucous membranes. It is taken up by the epithelial cells and here converted into the 11S dimer, containing secretory piece, for secretion. Although some 7S IgA is present in the bone marrow and elsewhere, it is possible that this arises in cells, which have migrated from the laminae propriae. The secretory piece protects the molecule from digestion and so 11S IgA can be found in feces. 11S IgA can fix complement, 7S IgA cannot.

IgM is a pentamer of the immunoglobulin unit. It has five such units joined together at their tail ends by sulfur bridges. It has a molecular weight of approximately 950,000 with a sedimentation constant of 19S. Even larger molecules have been reported. An IgM molecule can react with antigen at a cell surface; this affects the molecular conformation of the antibody, a phenomenon which leads to uncovering of the C_H2 sites on the heavy chains capable of binding and activating complement. These activations involve nine different components of serum complement, which interact in sequence (p. 176), during which they give rise to a number of active substances, such as histamine releasers, leukocyte chemotactic agents, and finally produce holes in cell membranes, e.g., those of a red cell or a bacterium. Once the hole is produced in the membrane, lysis occurs (hemolysis, bacteriolysis). To a much lesser extent, some IgG and IgA molecules can also activate complement. IgM is so large a molecule that it is largely, but not entirely, confined to the vascular system, in which it is concerned with defense against foreign cells such as invading bacteria. A deficiency of IgM is therefore manifested by a marked tendency to septicemia. It can be secreted by intestinal epithelial cells and may act on their luminal surface as a second line of defense to IgA. It seems that the major function of IgM is to protect the circulation. About half of the small lymphocytes in the blood have IgM-like sites on their surfaces.

IgD is also a 7S immunoglobulin, with a molecular weight of approximately 160,000. In normal adult serum, its content ranges from 1 to 15 mg./100 ml. It has been found as a myeloma protein, but its function is unknown.

IgE is another example of an antibody made up of one immunoglobulin unit. It has a sedimentation constant of 8S and a molecular weight of approximately 190,000. It is therefore a larger molecule than the 7S varieties and has a longer tail piece (Fc), which enables it to bind, probably to the membranes of mast cells in the capillaries and the tissues (e.g., nasopharynx and bronchi). When it combines with its antigen (allergen), it causes the mast cell to release histamine, serotonin, bradykinin, and so forth, which bring about so-called immediate hypersensitivity or local anaphylaxis. In other words, IgE molecules are reaginic. Normal adult serum contains approximately 0.03 mg./100 ml.

The only immunoglobulin that can cross the placental barrier is IgG. At birth, virtually all of the immunoglobulin in the neonatal circulation is IgG derived from the mother. The neonate gradually produces its own IgG as shown in Figure 3-5. During the first two weeks of life, neonatal serum contains approximately 3 mg. IgA/100 ml. and 1 to 20 mg. IgM/100 ml. At the end of the first year of life, the child's serum contains approximately 300 to 1100 mg. IgG/100 ml., 30 to 100 mg. IgA/100 ml. and 45 to 200 mg. IgM/100 ml. In other words, IgM is already virtually at the adult level. IgG achieves this when the individual is approximately three years of age and IgA some time after the individual is 15. In response to foreign protein material, the human being first produces IgM and then IgG, which interestingly enough coincides with the order of development.

Apart from the characteristically low values that occur normally during the second to fourth month of life, clinically significant decrease in γ-globulin concentration may occur as a result of genetically determined defective synthesis (primary agammaglobulinemia) or acquired disease (secondary hypogammaglobulinemia). These conditions are discussed elsewhere (p. 203).

Glycoproteins and Mucoproteins. Carbohydrate residues of various types can participate in the primary structure of certain conjugated proteins. Nearly all the plasma proteins contain some carbohydrate, and consequently some authorities believe that the terms glyco- and mucoprotein should be reserved for those proteins in which the carbohydrate moiety is a conspicuous feature. The carbohydrate chain linked to the protein is made up of hexoses (galactose or mannose), amino sugars (glucosamine or galactosamine), methylpentose (fucose) and a sialic acid (e.g., N-acetylneuraminic acid) of which there are several related forms. It has become customary to reserve the term *glycoprotein* for those entities containing less than about 4 per cent of amino sugars (hexosamines), and the term *mucoprotein* is used when this figure is exceeded. Total carbohydrate may be responsible for up to 15 per cent of the molecule of a glycoprotein and for 10 to 75 per cent of that of a mucoprotein.

In terms of migration during electrophoresis, mucoproteins may be α_1-globulins (orosomucoid, α_{1x}-glycoprotein) or α_2-globulins (α_2-neuraminoglycoprotein). The most abundant plasma mucoprotein is orosomucoid, which is an α_1-globulin in paper or boundary electrophoresis, but appears as a prealbumin in starch or acrylamide gel electrophoresis.

These substances, which normally comprise about 25 per cent of the α_1-globulin fraction (p. 167), may exhibit changes in either direction independently of the other alpha components. Decreased serum mucoprotein concentrations (normal 75–135 mg./100 ml.) occur chiefly in hepatocellular damage and certain types of endocrine imbalance: acute hepatitis, portal cirrhosis, hyperestrogenism, adrenal hypofunction, panhypopituitarism, hyperthyroidism, hypothyroidism, and diabetes mellitus. Low values may be obtained also in the nephrotic syndrome and, occasionally, in multiple myeloma.

Increased concentrations occur in a great variety of conditions associated with inflammatory, degenerative, proliferative, and traumatic processes. These include such conditions as acute and chronic infections, extensive metastatic malignancy, rheumatic diseases, cardiovascular disease, acute glomerulonephritis, terminal chronic glomerulonephritis (low values in nephrotic stage), and biliary obstruction. In the presence of factors that tend to produce changes in opposite directions, those which promote an increase usually take precedence over those that produce a decrease. For example, subnormal concentrations in a patient with portal cirrhosis often rise to abnormally high levels in the presence of a complicating acute infection, such as pneumonitis. Determination of the serum mucoprotein concentration may only occasionally be useful in distinguishing hepatocellular (low values) from obstructive (high values) jaundice, and in the diagnosis of the cause of hepatomegaly (low values in hepatitis and cirrhosis; high values in metastatic malignancy). The findings have not shown much diagnostic reliability.

The glycoproteins may be α_1-globulins (α_1-antitrypsin, α_1-easily-precipitable glycoprotein, transcortin, 4.6S postalbumin, tryptophan-poor-α_1-glycoprotein), α_2-globulins (haptoglobins, ceruloplasmin, α_2-macroglobulin, α_{2HS}-glycoprotein, Zn-α_2-glycoprotein), β-globulins (transferrin, hemopexin, β_2-glycoprotein) or γ-globulins (the immunoglobulins). α_1-Antitrypsin deficiency is associated with an appreciable proportion of cases of emphysema. Haptoglobins bind any free hemoglobin liberated intravascularly. Hemopexin binds heme. Transcortin and transferrin transport plasma cortisol and plasma iron, respectively. Prothrombin and various other blood-coagulation factors are glycoproteins.

Ceruloplasmin. This is the term applied to a heterogeneous copper-containing fraction of the serum α_2-globulins, which was once believed to play a role in copper transport (p. 334). Normal values range from 25 to 43 mg./100 ml. serum, but values are usually expressed in oxidase units. The increased amounts of copper in certain tissues (e.g., brain, liver, kidneys) of patients with Wilson's disease (hepatolenticular degeneration) is frequently accompanied by subnormal ceruloplasmin (and copper) concentrations. The abnormal tissue deposits are believed to result from impairment of the capacity for incorporating exogenous copper into this nondiffusible complex, due to a genetically determined defect in synthesis of certain specific α-globulins. Subnormal ceruloplasmin concentrations have been observed in apparently normal relatives of patients with Wilson's disease.

Cryoglobulins. This term is applied to proteins which precipitate from abnormal sera spontaneously, usually only in the cold (7 to 11° C.), and which usually, but not invariably, redissolve at room temperatures. Occasionally these proteins may undergo gelation or precipitation on standing at room temperature.

They may occur in multiple myeloma, kala-azar, chronic lymphatic leukemia, malaria, rheumatoid arthritis, and subacute bacterial endocarditis. They have been observed occasionally in asthma, periarteritis nodosa, and other disorders accompanied by eosinophilia, hepatitis, macroglobulinemia, portal cirrhosis, rheumatic fever, disseminated lupus erythematosus, pneumonia, tuberculosis, coronary artery disease, gout and brucellosis. Their association at times with a syndrome characterized by cold sensitivity, purpura, and Raynaud-like symptoms suggests that these clinical manifestations may be due to stasis and sludging of blood in peripheral vessels as a result of cold exposure. The designation "purpura cryoglobulinemia" has been applied to the association of purpuric manifestations with presence of these globulins.

Pyroglobulins. This term has been applied to plasma proteins that gel on exposure to moderate elevation of temperature (e.g., 55° C.); they may not redissolve on returning to room temperature, contrasting with cryoglobulins in this respect. They have been observed in multiple myeloma, being detected usually in the course of inactivating the serum complement in performing serological tests for syphilis.

C-Reactive Proteins. When serum containing these proteins is mixed with a solution of pneumococcus C polysaccharides, a precipitate develops that is not due to the usual antigen-antibody reaction. These proteins are apparently produced as an acute phase response to various types of tissue injury, such as infection, necrosis, trauma, and neoplastic and granulomatous lesions.

The reaction is consistently positive in bacterial infections, and commonly in virus infections. It reflects activity of the disease in rheumatic fever and, less accurately, in tuberculosis. It may aid in distinguishing acute myocardial infarction from nonocclusive coronary insufficiency. It is often positive in rheumatoid arthritis and extensive malignancy, particularly in the presence of metastases.

Macroglobulins. These are proteins of unusually high molecular weight (about 1,000,000), which migrate electrophoretically as a discrete component with the velocity of α_2-, β- or γ-globulins. The electrophoretic pattern may be identical with that seen in certain cases of multiple myeloma, and the presence of macroglobulins can be established with certainty usually only by ultracentrifugation. They are glycoproteins representing polymers or other aggregations of smaller globulin molecules. They comprise less than 5 per cent of the total serum proteins of normal subjects and are believed to originate in lymphoid reticular cells. Abnormally high concentrations may be accompanied by rather characteristic clinical manifestations (macroglobulinemia).

A condition of primary macroglobulinemia (Waldenström) has been described in which these components usually comprise over 15 per cent of the total serum protein. It is characterized by bleeding from the nasal and buccal mucosa, at times with purpura, engorgement of the retinal vessels resembling that seen in multiple myeloma, producing dimness and blurring of vision, usually hepatosplenomegaly and lymphadenopathy, mild anemia with bone marrow infiltration by lymphocytes and plasma cells, and striking increase in the sedimentation rate. The serum albumin concentration is often simultaneously decreased.

Increase in serum α_2-macroglobulins, usually less than 15 per cent of the total serum protein, may occur also in association with various disease conditions (secondary macroglobulinemia). These include: neoplastic diseases (uterine cancer, bronchogenic cancer, multiple myeloma, lymphoma); chronic hepatitis and hepatic cirrhosis; amyloidosis; the nephrotic syndrome; kala-azar and toxoplasmosis; congenital syphilis; connective tissue disorders (systemic lupus, rheumatoid arthritis, periarteritis nodosa).

Myeloma Proteins. Any one of several abnormal proteins may appear in the serum in multiple myeloma. They migrate electrophoretically at rates ranging from one somewhat slower than that of γ-globulin to that of α_2-globulins (rarely). Although the nature of this abnormal component varies in different patients, it apparently remains constant in each individual. The proliferated plasma cells in this disease are the site of production of these proteins.

Serum proteins with similar abnormal electrophoretic mobilities have been reported also, although rarely, in other conditions, including lymphoma, macroglobulinemia, reticuloendothelial disease, neoplastic conditions.

Bence Jones Protein. This term is applied to a group of proteins with a molecular weight of about 22,000, containing 210 to 230 amino acids, the physical state of which is altered in vitro by temperature changes, usually precipitating from solution at 45 to 55° C. and redissolving, partially or completely, at 95 to 100° C. They are more frequently detectable by layering the urine over concentrated hydrochloride acid (Bradshaw). The most sensitive technique, however, is to concentrate the urine by ultrafiltration and to demonstrate the presence of monoclonal light chains by immunoelectrophoresis, using appropriate antisera. Because their small size permits ready passage through the glomerular capillaries, they are usually demonstrable only in the urine, rarely in the blood. In urine, they usually appear as dimers (M.W. 45,000). Although they have been reported in other conditions, the present consensus is that they probably occur in appreciable amounts only in multiple myeloma, macroglobulinemia, soft tissue plasmacytoma, lymphosarcoma and other examples of the type of malignant growth referred to as a malignant immunocytoma. This, of course, includes myelomatosis and the other conditions already mentioned. They correspond to the light chains of the immunoglobulins and appear in the urine when they are produced in excess in this disease. Light chains are present in small amounts in normal urine and plasma, but, unlike Bence Jones protein, they are not monoclonal. As is the case also with the myeloma proteins, different Bence Jones proteins are produced by different subjects, but usually only one appears and remains constant in each case.

Miscellaneous. Quantitative abnormalities of certain specialized serum proteins may be demonstrated by special techniques and may be of diagnostic value. These include blood coagulation factors (p. 628), serum enzymes (p. 547), ceruloplasmin (copper-binding protein) (p. 339), transferrin or siderophilin (iron-binding protein, p. 438), and thyroxine-binding protein (p. 750). Abnormality of each of these factors usually points to some specific disease process. However, such changes are not of sufficient magnitude in themselves to produce clinically significant abnormalities in concentration of the serum protein components with which these factors are associated.

SERUM PROTEIN ABNORMALITIES IN DISEASE

Plasma, or serum, protein abnormalities are of value clinically in indicating the presence of disease but, with a few notable exceptions, are seldom of definitive value in differential diagnosis. These exceptions include: (a) diseases characterized by specific changes in the specialized factors referred to above: (b) multiple myeloma; (c) the nephrotic syndrome; (d) agammaglobulinemia and the dysgammaglobulinemias; (e) macroglobulinemia. However, in conjunction with other findings, they aid considerably in differential diagnosis, e.g., in hepatic and biliary tract disease. It must be remembered, too, that a normal electrophoretic value for one of the major globulin fractions does not preclude the possibility of significant abnormality of one or more of its constituent factors.

Acute and Chronic Bacterial Infections. The globulin increase in these conditions is mainly in the α- and γ-globulins, but occasionally also in the β-globulins. The plasma albumin is usually decreased simultaneously. The degree of hyperglobulinemia does not appear to be related to the height of the fever nor to the severity of tissue inflammation or destruction. Total protein values as high as 10 g./100 ml. have been reported in subacute bacterial endocarditis, with globulin values as high as 8 g./100 ml. Maintenance of a high globulin concentration may be related to the development of amyloid disease in chronic suppurative diseases.

Viral, Protozoal, and Parasitic Diseases. With the exception of multiple myeloma, the largest globulin increases are usually seen in certain diseases in this category, e.g., infectious hepatitis, kala-azar, typhus, lymphopathia venereum. In the latter condition, total protein values over 11 g./100 ml. have been reported. Hyperglobulinemia, usually of lesser degree, may occur also in syphilis, malaria, trypanosomiasis, filariasis, schistosomiasis, and histoplasmosis. The participation of cells of the reticuloendothelial system in the tissue response in certain of these disorders appears to have some relation to the production of the hyperglobulinemia. The increase is mainly in γ-globulins, but also in the α-globulin fraction.

Hepatic Disease. Changes in plasma proteins in various types of liver and biliary tract disease are discussed in detail elsewhere (p. 620 ff.). Increase in γ-globulin in infectious (virus) hepatitis is referred to above. Hypergamma globulinemia occurs rather characteristically also in other forms of acute and chronic liver damage, e.g., acute and subacute hepatic necrosis, kwashiorkor and cirrhosis. Extremely high values (over 6 g./100 ml.) have been reported in hepatic necrosis and in cirrhosis, at times adequate to produce an increase in total protein despite the co-existing hypoalbuminemia. The increase is usually mainly in γ-globulin. Variable increases may occur also in the α- and β-globulins, the latter usually in those cases in which hyperlipemia is present.

Kidney Disease. The most striking changes occur in the nephrotic syndrome. This is usually associated with hypoalbuminemia, increase in α_2- and/or β-globulins, and decrease in γ-globulin. The rise in β-globulins is related to the coexisting hyperlipemia and elevated concentration of β-lipoproteins and S_f 12–20 molecular classes (p. 132). Similar but generally less marked changes may occur in amyloid nephrosis. In the nephrosis of membranous glomerulonephritis, the γ-globulin may be normal or raised. Slight to moderate increases in one or all of the three globulin fractions may occur, inconsistently, in acute glomerulonephritis and in stages of chronic glomerulonephritis other than the nephrotic.

Sarcoidosis. Hyperglobulinemia occurs rather consistently during active stages of this condition, frequently of marked degree, i.e., above 5 g./100 ml., usually in association with hypoalbuminemia. The major increment is in the γ-globulin fraction, although the α-globulins may be elevated somewhat. The total protein concentration may be increased. These changes may aid in differentiating between sarcoidosis and tuberculosis; in the latter condition, hyperproteinemia and marked hyperglobulinemia are not the rule (except with extensive lymphatic involvement), although increases in α- and γ-globulins may occur. However, the abnormalities in sarcoidosis are not always striking in less active forms of the disease.

Multiple Myeloma. One or more of several types of protein abnormality occur commonly in multiple myeloma: (a) hypoalbuminemia; (b) hyperglobulinemia; (c) myeloma protein in serum, which may be a macroglobulin or cryoglobulin; (d) proteinuria. The most characteristic features are the serum globulin abnormalities and the presence of abnormal proteins in the urine.

An abnormal globulin appears on the serum in about 75 per cent of cases. The increase in globulin is of a degree sufficient to produce hyperproteinemia in about 50 per cent of cases. This globulin is a so-called paraprotein and is often referred to as an M (myeloma) protein. As a result of cellulose acetate electrophoresis, it appears in a variety of different positions as a very discrete band. It is most frequently found in the gamma globulin area, but it can be as far toward the anode as the α_2-globulin band. By immunological procedures it has been demonstrated that the M protein belongs to one of the five (possibly six) main classes of immunoglobulins, viz., IgG, IgA, IgD, IgM and IgE, which respectively account for the M proteins in 53, 22, 1.5, 0.5 and 0.1 per cent of cases of myelomatosis. About 20 per cent of cases do not have a paraprotein in their serum but Bence Jones (monoclonal light chain) proteinuria is present. Approximately 1 per cent of cases do not show a paraprotein and do not have Bence Jones proteinuria, and diagnosis depends on the osteolytic bone lesions and the demonstration of the abnormal plasma cells in a bone marrow biopsy. It is remarkable that cases of myelomatosis eventually, almost always, show a severe immune paresis. In spite of a well marked M protein band, the remainder of the gamma globulins are virtually absent. In the early stages of myelomatosis, which is now believed to be of quite long duration, the discrete band may be manifest as part of what is apparently a normal gamma globulin band. This kind of picture may also be found in benign essential monoclonal gammopathy (Waldenström), which tends to be more common in the elderly. It is not accompanied by Bence Jones proteinuria, is usually of no great consequence and may occur in families. On occasion, transition to myelomatosis has been observed.

The occurrence of Bence Jones proteinuria is not a constant finding in myelomatosis. It occurs in 60 per cent of the IgG type, 80 per cent of the IgA and is usually present in the IgD type. It also occurs in the remaining types, but the numbers of these are too small to get a good idea of its frequency. The paraprotein itself may appear in the urine in all except the IgM types.

The hyperglobulinemia can give rise to hypercalcemia and to the increased plasma viscosity syndrome. Myelomatosis is not infrequently associated with renal damage. This may be due to tubular blockage by casts of paraprotein and/or Bence Jones protein. The latter rarely gives rise to a Fanconi type syndrome by its effect on the tubules. Amyloidosis occurs in myelomatosis. The V portions of the light chains are involved and are present in excess when amyloidosis is produced. This can also result in severe kidney damage. It is therefore not surprising that ordinary proteinuria as well as Bence Jones protein can be present in myelomatosis.

Heavy-Chain Disease. This rare disease is characterized by the appearance in blood and urine of a protein fragment corresponding to the Fc fragment of IgG heavy chain but with an additional 8 or so amino acids. It appears as a dimer. There is weakness, loss of weight, transient palatal edema and erythema, adenopathy and hepatosplenomegaly with survival ranging from a few months to several years.

Another disease has been described which is associated with overproduction of the heavy chain of IgA. This arises from a plasmacytoma involving the small intestine, with possible local spread to bone marrow and to tonsils. There is a malabsorption syndrome and an excess of intestinal alkaline phosphatase in the serum.

Waldenström's Macroglobulinemia. This is a disease of the elderly with lymphadenopathy and hepatosplenomegaly. There is weakness, weight loss and vague ill health and a tendency to bleed from mucous membranes, especially the nose. It is compatible with prolonged survival. A paraprotein, IgM, appears in the blood and results in an electrophoretic picture in cellulose acetate or agar gel closely resembling myelomatosis. Sometimes the M protein is a cold agglutinin and can result in varying degrees of hemolysis. The tendency to bleed results from protein-protein interactions involving certain clotting factors and platelets. Occasionally the paraprotein is a cryoglobulin, the intravascular precipitation of which can lead to Raynaud type phenomena on exposure of the patient to cold. There may be Bence Jones proteinuria.

The paraprotein can be identified by immunological techniques and by its inability to travel into acrylamide gels during electrophoresis because of its very high molecular weight. It can also be demonstrated by ultracentrifugation. A simple but not entirely reliable qualitative test, the Sia test, is the formation of a flaky, birefringent precipitate when a drop of serum is slowly added down the sides of a test tube into a few milliliters of 0.01M phosphate buffer, pH 7.1. The IgM must be present at concentrations above 0.7 gm./100 ml.

Agammaglobulinemia; Hypogammaglobulinemia. Conditions characterized by inability to produce adequate amounts of γ-globulins may be classified as follows: (1) physiological (p. 194); (2) congenital; (3) idiopathic; (4) secondary. Temporary hypogammaglobulinemia occurs normally between the first and sixth months of life, usually with minimum levels at 3 to 4 months (p. 195).

The congenital form is characterized by inability to produce more than minute amounts of circulating γ-globulins, usually less than 0.1 g./100 ml.; there is an associated defect in plasma cell formation. The condition occurs only in males and is transmitted as a recessive character by the unaffected mother. Because of inability to produce antibodies, affected children are unusually susceptible to repeated attacks of bacterial infections, characteristically after about 6 months of age (pneumonia, upper respiratory tract infections, otitis media, meningitis). The response to virus infections is usually normal. There is also absence of anti-A and anti-B isoagglutinins in subjects with blood group O.

Idiopathic hypogammaglobulinemia occurs later in life in both males and females who have previously responded normally to bacterial infections. The serum γ-globulin concentration is usually not as low as in the congenital form, usually between 0.1 and 0.2 g./100 ml. The cause is unknown. In some cases, an autosomal recessive inheritance has been demonstrated. In addition to increased susceptibility to bacterial infection, a sprue-like syndrome often occurs. Hepatosplenomegaly may develop.

Secondary hypogammaglobulinemia refers to inadequate production of γ-globulins resulting from various acquired disease states. This may occur in extensive involvement of the reticuloendothelial system in such conditions as multiple myeloma, leukemia, lymphomatous diseases, and, possibly, nitrogen mustard therapy and total body irradiation. It occurs characteristically in the nephrotic syndrome and, in association with low values for other serum proteins, in idiopathic panhypoproteinemia. Although elevated values are not unusual in many conditions of protein malnutrition (p. 191), hypogammaglobulinemia may occur in cases of severe nutritional hypoproteinemia and the malabsorption syndrome, particularly in children with celiac disease. Decreased concentrations may occur in Cushing's syndrome or following administration of adrenocortical hormones. It has been suggested that excessive loss into the intestine may be responsible for the production of low serum albumin and γ-globulin concentrations in patients with chronic ulcerative colitis, regional enteritis, and hypertrophic gastritis, and perhaps also in certain cases of idiopathic hypoproteinemia.

Hypogammaglobulinemia has been observed in certain children with allergic conditions, associated with inadequate responsiveness to antigenic stimuli.

Careful analysis of secondary hypogammaglobulinemia, taking into account normal levels for each age group, has shown that there are two broad types—one in which IgG is low but IgA and IgM are within normal limits, and the other in which all three types are lower than normal. In the latter, there is a tendency for the suppression first of IgM, then IgA and finally IgG;. depending on the cause, the sequence may take months (toxic) or years (lymphoid hyperplasia).

The low IgG type hypogammaglobulinemia is secondary to prematurity, delayed maturity (in relation to development of immunoglobulin production), marrow disorders and a group of conditions in which protein turnover in the liver is stimulated, which results in breakdown of IgG by that organ. The group includes the nephrotic syndrome, protein-losing enteropathies, severe malnutrition, myotonic dystrophy, thyrotoxicosis, the administration of corticosteroids or diazoxide and an idiopathic type of disorder.

The second type of hypogammaglobulinemia can be due to circulating factors suppressing immunoglobulin synthesis. This situation can occur in relation to prolonged uremia, severe infection, cytotoxic therapy, gluten-sensitive enteropathy and diabetes mellitus. Suppression of the three types of immunoglobulin is found in a high proportion of cases of myelomatosis and in a few cases of macroglobulinemia. It is found in an appreciable proportion of patients suffering from the various types of lymphoid hyperplasia (Hodgkin's disease, chronic lymphatic leukemia, etc.).

Dysgammaglobulinemia. In this group of conditions, antibody deficiency is associated with a gammaglobulin, which is apparently normal by electrophoretic investigation. There are a number of different types of dysgammaglobulinemia. The deficiency may affect IgA and IgM (Type I), IgG and IgA (Type II), IgG alone (Type III), IgA alone (Type IV), IgM alone (Type V), IgM and IgG (Type VII), or there may be no apparent change in serum levels but nevertheless there is a γ-globulin correctable immunological deficiency syndrome (Type VI). It has been suggested that the latter type may be due to an immunoglobulin subclass deficiency.

Globulin Reactions

Certain qualitative reactions that yield abnormal results in the presence of abnormalities in one or more globulin fractions have been found to be very useful clinically. The most widely employed currently are the thymol turbidity, cephalin-cholesterol flocculation, and zinc sulfate turbidity tests. Several others have been described, now mainly of historical interest; these include the Takara-Ara (mercuric chloride), Weltmann (calcium chloride), Bauer (magnesium chloride), formol-gel, and CO_2-saturation reactions, and the colloidal-gold curve.

The results obtained with these turbidity or flocculation reactions depend upon the state of balance between certain stabilizing and precipitating factors. In normal serum, the precipitating factors include γ-globulins, lipids, and perhaps β-globulins, the relative importance of each varying with different test procedures. The stabilizing factors in normal serum are albumin and α-globulin. In practice, the test procedures are so adjusted that with normal serum, turbidity or flocculation are at a minimum and within well-defined limits.

Zinc Sulfate Turbidity. This procedure is based upon the fact that the solubility of serum proteins is decreased upon dilution with solutions of low ionic strength; under certain conditions the more insoluble globulins will precipitate, with the development of measurable turbidity. With appropriate concentrations of such heavy metals as zinc or copper, it is possible to produce a situation in which there is minimal precipitation from normal serum and maximal precipitation from sera with increased concentrations of γ-globulin. Turbidity measurements are made by comparison with a standard barium sulfate solution. The normal range is 2 to 9 units (< 1.3 g. γ-globulin).

At any given level of serum γ-globulin concentration, zinc sulfate turbidity is increased by decrease in albumin or increase in lipid concentrations. However, the effect of these factors is not so great as in the case of the cephalin-cholesterol flocculation (albumin) and thymol turbidity (lipid) tests. It therefore provides a more exact indication of increase in γ-globulin concentration than do the latter procedures.

As is true of other globulin reactions, this procedure is applied most frequently in the study of patients with hepatic or biliary tract disease and is considered in more detail in this connection elsewhere (p. 625). In addition to various forms of hepatic disease, increased zinc sulfate turbidity occurs commonly in a number of extrahepatic disorders accompanied by increase of serum γ-globulin concentration. These include: (a) extensive tissue destruction and suppuration; (b) acute infections; (c) viral and protozoal diseases (malaria, filariasis, kala-azar, lymphopathia venereum, schistosomiasis trypanosomiasis); (d) chronic infections (syphilis, tuberculosis, leprosy); (e) granulomatous diseases (sarcoidosis, Hodgkin's disease); (f) collagen diseases (disseminated lupus, periarteritis nodosa, dermatomyositis, rheumatic fever); (g) multiple myeloma.

The zinc sulfate turbidity test has also been used to study the development of antibodies in patients with scarlet fever and rheumatic fever. Increased turbidity has been found to parallel rises in antistreptolysin and antistreptokinase titers in these diseases.

Cephalin-Cholesterol Flocculation Test. This test is based upon the observation that when diluted serum with an abnormally high globulin content is added to an emulsion of a cephalin-cholesterol mixture, the latter substances (plus globulin) precipitate out (flocculation) and the solution is clarified (opalescence decreases). The degree of flocculation and clarification is recorded as zero to 4 plus. This test gives positive results (flocculation) in a large proportion of cases of active hepatocellular damage and has been widely used in the study of patients with jaundice (p. 627). Positive findings are also obtained in most of the other conditions mentioned above in which there is an increase in γ-globulin, especially if the albumin is decreased. A positive reaction in disease may result from one or more of the following conditions:

(1) Increase of gamma globulin in such quantity that the normal components of the serum albumin fraction are unable to inhibit the reaction.

(2) Decrease in serum albumin concentration below that capable of inhibiting the reaction.

(3) Decrease in flocculation-inhibiting properties of the serum albumin, due perhaps to a qualitative change.

(4) Decrease of γ-globulin, a stabilizing factor.

Abnormal results may be obtained in certain extrahepatic disorders, in which, however, the possibility of liver involvement cannot be excluded, e.g., congestive heart failure, infectious mononucleosis, gastrointestinal disorders, malaria. They occur less frequently in various acute and chronic infections accompanied by increase of serum γ-globulin concentration. In biliary obstruction, the serum contains a stabilizing factor; this is important in interpretation of findings in patients with jaundice (p. 626).

Positive reactions may be obtained with normal serum (a) after standing at icebox temperatures for several months, (b) after heating to 56° C. for thirty minutes or (c) if the reaction mixture is exposed to light during the test period.

Thymol Turbidity Test. This test is based upon the observation that addition of thymol solution to certain sera with abnormally high globulin contents results in marked turbidity of the solution (thymol-globulin-lipid complex), whereas little or no such reaction occurs with normal serum. The degree of turbidity (expressed in units) may be measured by comparison against a turbidity standard, e.g., barium sulfate. The normal range is 1 to 4 units, usually read at 30 minutes. After longer periods, flocculation may occur, as in the cephalin-cholesterol test. This observation is usually made at 18 hours, reactions greater than 1+ being regarded as abnormal. The turbidity and flocculation reactions do not always parallel one another. The thymol turbidity test clinically gives results that generally parallel those obtained with the cephalin-cholesterol flocculation test (above). Certain differences have been observed, however, which indicate points of difference in the fundamental mechanism involved. Cephalin-cholesterol flocculation is obtained with the serum of normal laboratory animals, including dogs and rabbits, which gives a negative thymol turbidity test. On the other hand, the latter test gives an abnormal response with certain lipemic sera, as in diabetes mellitus and the nephrotic syndrome, whereas the former does not. Abnormally high values are obtained also with the serum of normal subjects after a high-fat meal, the degree of abnormality roughly paralleling the increment in plasma lipids. This test should be performed, therefore, on serum collected after an overnight fast (ten to twelve hours).

Otherwise, the statements made with respect to zinc sulfate turbidity and cephalin-cholesteol flocculation generally apply also to thymol turbidity. As is true of the latter procedures, this test has been employed most extensively in the study of patients with hepatic and biliary tract disease, but the frequent occurence of abnormal findings in a variety of extrahepatic disorders, as indicated previously (p. 206), must be appreciated.

ABNORMAL BLOOD NONPROTEIN NITROGEN

Alterations in the total nonprotein nitrogen content of the blood depend upon variations in the concentrations of its constituent elements, which will be considered individually (see also Renal Function, p. 507).

Urea Nitrogen (normal 9 to 20 mg. per 100 ml.). **Increased Urea N** (see Renal Function, pp. 509–514). Increase in blood urea N may be due to (*a*) decreased renal excretion, (*b*) absorption from the intestine of excessive amounts of products of protein digestion, or (*c*) excessive protein catabolism, or to a combination of these factors. Elevation of blood NPN or urea N due primarily to factors other than renal or urinary tract disease has been termed "extrarenal azotemia." Decreased renal excretion of urea may be due to (*a*) organic disease of the kidneys, with destruction of a considerable portion of functioning renal tissue, or (*b*) abnormality of the mechanism of glomerular filtration, resulting in significant decrease in the effective filtration pressure (p. 484) (e.g., marked hypotension, increased capsular pressure, hemoconcentration, hyperproteinemia). The most common cause for increased blood urea N (and NPN) is inadequate excretion, due usually to kidney disease, urinary obstruction, or extreme hypotension and oliguria.

(1) **Glomerulonephritis.** This is a common cause for abnormally high blood urea N concentrations, evidence of renal functional impairment in acute and chronic glomerulonephritis. In the terminal state and during acute exacerbations of chronic nephritis, values as high as 200 mg. or more per 100 ml. may be obtained (p. 509).

(2) **Other Kidney Diseases.** These include (a) conditions in which there is extensive destruction or inflammation of the kidneys, as in renal tuberculosis, pyelonephritis, advanced nephrosclerosis, renal cortical necrosis, malignancy, suppuration, and chronic gout; (b) renal conditions accompanied by marked oliguria or anuria, as bichloride of mercury poisoning, postoperative urinary suppression, acute tubular necrosis, and advanced myocardial failure with passive congestion of the kidneys; (c) congenital renal lesions, as hypogenesis or hypoplasia, and polycystic disease of the kidneys; (d) conditions in which the uriniferous tubules are blocked by deposition of some substance which interferes with the passage of fluid through the tubules. Among these are multiple myeloma, amyloid disease of the kidneys, hemoglobinuria, as after transfusion with incompatible blood, and sulfonamide therapy.

(3) **Urinary Tract Obstruction.** Enlarged prostate, urinary lithiasis, malignant obstruction of the ureters from extension of tumors of the bladder or uterus, bowel, retroperitoneal lymph nodes.

(4) **Hepatic and Bilary Tract Disease.** High NPN and urea N values are encountered at times in patients with hepatic disease. This occurs most commonly in terminal stages of acute or chronic liver disorders, after operation on the bile passages, particularly after decompression of obstructed bile ducts, and in traumatic necrosis of the liver. The mechanism operating in such cases to produce the so-called "hepatorenal syndrome" apparently involves production of acute renal tubular necrosis.

(5) **Shock and Hemoconcentration.** Hemoconcentration resulting from loss of plasma water by prolonged vomiting, diarrhea or sweating may result in elevation of blood urea, especially if accompanied by severe hypotension, as in the shock syndrome. These phenomena prevent adequate glomerular filtration by diminishing the effective glomerular filtration pressure and also result in extreme oliguria by prerenal deviation of water. In this category may be placed severe burns, high intestinal obstruction, pyloric obstruction, cholera, typhoid fever, infantile diarrhea, dysentery, Addison's disease, pancreatic fistula, peritonitis.

(6) **Excessive Protein Catabolism.** Slight or moderate increase in blood urea may occur in severe toxic and febrile conditions in which tissue protein catabolism is accelerated and in which there is also some degree of renal functional impairment. This factor may contribute to the increase in blood urea in many of the conditions mentioned above (acute intestinal obstruction, severe infections and burns), and to the increased blood amino N in shock.

(7) **Miscellaneous.** Nitrogen retention may occur as a consequence of renal failure, usually as a terminal event, in patients with hyperparathyroidism (p. 295). This apparently depends in part upon extensive calcification of the renal tubular epithelium or upon the formation of urinary calculi; in acute hyperparathyroidism, hemoconcentration and impaired blood flow through the kidneys may contribute to the development of renal functional insufficiency.

Nitrogen retention is observed not infrequently in diabetic coma. The mechanism of its production is not clearly understood. However, dehydration, hemoconcentration, excessive protein catabolism, and perhaps diminished glomerular filtration due to intercapillary glomerulosclerosis and other forms of renal damage, may be of importance in this connection.

The blood urea may rise soon after hemorrhage into the gastrointestinal tract (peptic ulcer, carcinoma of the stomach, esophageal varices in cirrhosis of the liver, and so on) due probably to absorption of excessive amounts of products of protein digestion (from globin and albumin) and inadequate renal excretion.

Decreased Urea N. (1) **Acute Hepatic Insufficiency.** As urea is normally formed in the liver and, presumably, through hepatic functional activity, it naturally follows that a decrease in blood urea may be expected in conditions associated with acute hepatic insufficiency. Subnormal values (5 to 10 mg. per 100 ml.) have been observed in acute yellow atrophy of the liver, acute toxic hepatic necrosis due to phosphorus, arsphenamine, chloroform, and carbon tetrachloride poisoning, eclampsia, and in acute hepatic insufficiency following operative procedures upon the biliary tract. Such findings are rare in chronic hepatic disease such as cirrhosis, passive congestion, or malignancy, because of the large functional reserve capacity and enormous regenerative power of the liver.

(2) **Normal Pregnancy.** The blood nonprotein nitrogen decreases during the first six months of pregnancy to levels of 20 to 25 mg. per 100 ml., increasing subsequently to reach an average of 25 to 30 mg. per 100 ml. at term. The urea nitrogen may fall as low as 5 mg. per 100 ml. after six months, these low levels being maintained until the eighth or ninth month, when it begins to rise, reaching values of 7 to 9 mg. per 100 ml. at term.

Uric Acid. Abnormalities in blood uric acid are discussed elsewhere (nucleic acid metabolism, p. 253).

Creatinine (normal, 1 to 1.5 mg. per 100 ml.). An increased concentration of creatinine may occur in any condition in which blood urea is increased, but usually only after the latter has risen to comparatively high levels. This is due in part perhaps to extrarenal causes, particularly the solely endogenous origin of creatinine as contrasted to the largely exogenous origin of urea, and also to the fact that the normal creatinine clearance is considerably higher than the urea clearance (pp. 491, 506).

(1) **Nephritis** (see Renal Function. p. 509.) When renal functional impairment is due to chronic nephritis, creatinine retention is of decidedly serious prognostic import, since the nature of the anatomic lesion renders functional improvement impossible. In chronic nephritis with uremia,

figures as high as 35 mg. per 100 ml. may be obtained. In this condition, values above 5 mg. per 100 ml. usually indicate a hopeless prognosis, with a relatively short expectation of life. In acute nephritis, however, and in acute exacerbations of chronic nephritis, extremely high concentrations of creatinine may be found, which, with subsidence of the acute renal lesion, may return to normal (see Renal Function, p. 509).

(2) Urinary Obstruction. Prostatic and bilateral ureteral obstruction may, if the degree of back pressure rises sufficiently, be associated with creatinine, as well as urea and uric acid retention. Similar findings may be observed in unilateral ureteral calculus with reflex anuria. Under such circumstances, the process being purely mechanical in the absence of renal damage, the degree of elevation of blood creatinine is roughly proportional to that of urea; this is in striking contrast to the findings in chronic nephritis, in which condition high urea values may be observed in association with normal creatinine figures. In urinary obstruction, as in acute nephritis, in the absence of irreparable renal damage, extremely high creatinine values (15 to 30 mg. per 100 ml.) may return to the normal level following relief of the obstruction.

(3) Urinary Suppression (see Urea, p. 207).

(4) Cardiac Decompensation (see Urea, p. 207).

(5) Intestinal Obstruction (see Urea, p. 207).

Amino Acid Nitrogen (normal, 3 to 6 mg. per 100 ml.). Increase in the amino acid content of the blood may occur in the following conditions:

(1) Hepatic Insufficiency. Amino acids normally undergo deamination in the liver with the consequent formation of urea. When the urea-forming or deaminizing function of the liver is impaired the concentration of amino acids in the blood increases and that of urea diminishes. The normal ratio of urea nitrogen to amino acid nitrogen is approximately 2 to 1. This ratio is, therefore, decreased in such conditions as acute necrosis of the liver (phosphorus, arsenic, chloroform and carbon tetrachloride poisoning) and, less frequently, in acute hepatitis. Figures as high as 200 mg. per 100 ml. have been reported in cases of acute hepatic necrosis, but the usual range is 10 to 15 mg. per 100 ml., values above 30 mg. being infrequently observed. The increase in amino acid nitrogen is due partly to extensive autolysis of hepatic tissue. High values are not ordinarily found in chronic hepatic disease such as cirrhosis, chronic obstructive jaundice, hepatic syphilis, or malignancy because of the large functional reserve and marked regenerative power of the liver.

(2) Eclampsia. Figures of 6 to 12 mg. per 100 ml. are at times observed in eclampsia, due presumably to the hepatic lesions present in that condition.

(3) Interference with Excretion. The blood amino acid nitrogen may be increased slightly in some cases of advanced nephritis with marked nitrogen retention and also in urinary obstruction or suppression. High values are, however, rare under these circumstances, unless there is an associated condition of hepatic insufficiency. In most cases of nephritis with high blood nitrogen values the amino acid content of the blood is within normal limits.

(4) Miscellaneous. An increase may occur occasionally in myeloid leukemia and, rarely, in diabetes mellitus, acute infections, congestive heart failure, hyperthyroidism, and severe anemia.

Subnormal levels have been reported in pneumococcal pneumonia and after administration of growth hormone, insulin, and androgens (p. 164).

(5) Increase in Individual Amino Acids. Amino acids or their metabolites may accumulate in the blood plasma, singly or in groups, as a result of specific abnormalities of intermediary metabolism. These are due usually to genetically determined enzyme deficiencies (inborn errors of metabolism). Isolated increases in the following substances fall into this category: phenylalanine and certain of its metabolites, phenylpyruvic, phenyllactic, and phenylacetic acids, associated with mental deficiency (phenylketonuria); the branched amino acids, leucine, isoleucine, and valine, with their corresponding keto acids (maple sugar urine disease); glycine; argininosuccinic acid. These conditions are usually identified by detection of increased amounts of these substances in the urine, representing specific types of overflow aminoaciduria; they are, therefore, considered in general detail elsewhere (p. 223).

Ammonia. Increase in blood ammonium ion is of interest clinically mainly in relation to hepatic insufficiency and hepatic coma (p. 619). Although certain of the factors operating are not known, it is probable that abnormal elevations of NH_4^+ in the systemic circulation are usually due to simultaneously increased production and decreased removal from the portal blood (hepatic functional insufficiency and/or portal-systemic shunts in cirrhosis).

Although NH_4^+ may be contributed to the blood by the muscles and kidneys (deamination of glutamine and various amino acids), the increased production in subjects with liver disease is due mainly to the action of bacterial and digestive enzymes on protein and urea in the lumen of the bowel. It is enhanced by such factors as high protein diet, gastrointestinal bleeding, and increase in blood urea concentration (p. 619). In the presence of liver disease, the increased amount of NH_4^+ in the portal blood may escape adequate removal (chiefly for urea synthesis) in the liver, passing into the systemic circulation. Only relatively small amounts are removed by extrahepatic tissues, e.g., amination of keto acids (brain) and exhalation through the lungs. These are inadequate to preserve normal concentrations in the blood under the conditions indicated. Increase in extracellular pH (relative to intracellular) is accompanied by a change in the distribution of NH_4^+ in the opposite direction, i.e., decrease in extracellular and increase in intracellular NH_4^+. Importance has been attached to this relationship in patients with hepatic coma (p. 673).

Total Nonprotein Nitrogen (normal, 15 to 35 mg. per 100 ml.; see Renal Function, pp. 507–514). An increase in the concentration of total nonprotein nitrogen occurs particularly in those conditions in which the blood urea content is increased. Extremely high values (to 400 mg. per

100 ml.) have been observed in chronic nephritis, acute nephritis, urinary obstruction, and renal destruction (tuberculosis, pyonephrosis, polycystic kidneys, etc.). Slight elevations may be found in gout and in myelogenous leukemia, in which conditions the uric acid concentration is increased independently of urea. Normal values are usually present in eclampsia.

ABNORMAL URINARY NITROGEN

Proteinuria

The permeability of the glomerular membrane is such as to permit passage of only very small amounts of the normal plasma proteins. The size and shape of the molecules are determining factors in this connection. The glomerular filtrate normally contains 10 to 25 mg. protein per 100 ml.; accordingly, about 18 to 45 g. protein enter the tubule in twenty-four hours. This is almost completely reabsorbed by the tubular epithelial cells, 100 mg. being regarded as the upper limit of normal daily protein excretion in the urine. Assuming average normal urine volume, this usually provides a concentration too low to be detected by the sulfosalicylic acid test, although positive results may be obtained with the heat and acetic acid test.

Application of electrophoretic and immunochemical techniques has yielded new information concerning proteins in normal and abnormal urine and has resolved several previously controversial issues with respect to mechanisms of proteinuria. Several of the normal plasma proteins have been demonstrated in normal urine, including albumin, ceruloplasmin, transferrin, and γ-globulin. In addition, several urinary proteins have been identified as arising from the kidneys, ureters, bladder, and urethra, and, in the male, also from the prostate. Similarly, in abnormal proteinuria, except in unusual cases of severe inflammatory or degenerative conditions of the lower urinary tract, the pattern of urinary proteins reflects that in the blood plasma. Therefore, in the absence of abnormal plasma proteins (e.g., multiple myeloma, macroglobulinemia), serum albumin is by far the most abundant of the urinary proteins, because of which fact the term "albuminuria" has come to be used synonymously with proteinuria, sometimes erroneously. On the basis of size and shape of the normal plasma protein molecules, with increasing permeability of the glomerular membrane, albumin passes through most readily and fibrinogen least readily, the various globulins occupying intermediate positions in this connection. Although determination of the urinary albumin: globulin ratio is of little practical significance, certain observations are of interest inasmuch as they illustrate to some extent the degree of increase in permeability of the glomerular membrane in the presence of different renal lesions.

Extremely high urinary albumin: globulin ratios have been found in the nephrotic syndrome, figures above 10 being commonly observed.

In chronic glomerulonephritis the ratio is usually lower, the majority ranging from 3 to 5, with a few between 5 and 10. In acute nephritis the figures are usually low (4 to 6), rising as the acute process subsides. Low values are obtained in amyloid disease of the kidneys (0.5 to 3.5). Since the plasma albumin: globulin ratio is normally about 1.5:1, it appears that in amyloid disease the glomerular permeability is so increased that the relatively large globulin molecule passes freely, resulting in a urinary albumin: globulin ratio which approaches that present at the time in the blood plasma. Likewise, in acute nephritis, glomerular permeability is so increased that globulin passes with relative ease; as the acute lesion subsides permeability diminishes and the ratio increases correspondingly. In the nephrotic syndrome, membrane permeability being relatively low, only small amounts of globulin are lost, but the comparatively small albumin molecule passes through in large amounts.

Whereas normal plasma protein is retained in the blood, certain foreign proteins entering the blood plasma are eliminated by the kidney. Thus hemoglobin (m.w. 64,450) when free in the plasma, and egg albumin and other foreign proteins, if introduced into the blood stream, appear in the urine. In the case of hemoglobin, at least, some of the hemoglobin filtered through the glomeruli is reabsorbed in the tubules, appearing in the urine only after the reabsorptive capacity of the tubules has been exceeded (p. 270).

Whether or not this hemoglobinuria is contributed to by associated or induced glomerular damage has not been established, although the observation that there is usually associated excretion of albumin would seem to indicate that such is the case.

Two major mechanisms may operate to produce abnormal proteinuria. (1) Humoral proteinuria: the blood plasma contains abnormal proteins of such small molecular size that they pass readily through normal glomeruli (hemoglobin, Bence Jones proteins, egg albumin). (2) Renal proteinuria: (a) the glomerular membranes are so injured that they become more permeable to the normal plasma proteins, and/or (b) tubular epithelial cell damage results in inadequate reabsorption of proteins from the glomerular filtrate. As has been indicated, considerable amounts of protein normally undergo tubular reabsorption from the glomerular filtrate, a process termed "athrocytosis." Accordingly, abnormal proteinuria occurs when the quantities of proteins reaching the tubules exceed their reabsorptive capacity. Theoretically, two factors may be involved: (1) passage of abnormally large quantities of proteins into the glomerular filtrate, i.e., increased glomerular permeability or abnormal plasma proteins; (2) tubular damage, morphologic or functional. In the great majority of cases of renal proteinuria both factors are probably involved. Except in cases of primary tubular necrosis, and humoral types of proteinuria, it is unwise to assume the absence of glomerular damage.

The pathogenesis of proteinuria may be summarized briefly as follows: (1) in the great majority of instances glomerular abnormalities, structural or functional, primary or secondary, play a fundamental role; (2)

in many cases there is associated tubular involvement; occasionally, this is the primary factor; (3) a few are due to abnormal proteins in the blood plasma, e.g., hemoglobin, Bence Jones proteins; (4) the urinary proteins may originate in the lower urinary tract.

Since the urine normally contains minute quantities of albumin, the term "albuminuria" implies rather a quantitative than a qualitative deviation from the normal. This term is applied to the presence in the urine of abnormally large quantities of albumin, detectable by the commonly employed qualitative tests. Normal amounts of protein may give a positive reaction, especially with sulfosalicylic acid reagents, in conditions of unusually low urine volume, as in dehydration. False positive reactions may be obtained in subjects who have received certain iodine-containing drugs for cholecystography (Priodax, Telepaque) or tolbutamide for treatment of diabetes. These "pseudoalbuminurias" may persist for two or three days after administration of these agents. Albustix strips do not give these false positive reactions.

Albuminuria may be classified as benign or organic.

Benign Albuminuria. Under this heading are placed cases of albuminuria apparently not dependent upon organic disease. The urine is otherwise normal, renal functional activity is unimpaired, the degree of albuminuria is usually slight (below 500 mg. per 100 ml.), the condition is usually transitory and occurs most commonly in young individuals, whose subsequent medical history is negative.

(1) Following Severe Exercise. Temporary albuminuria of varying degree may occur following strenuous muscular exercise in individuals unaccustomed to such activity. It has been observed in raw recruits after forced marches, in football players, bicycle racers, marathon runners, and other athletes if the degree of exertion exceeds that to which the individual is accustomed. This type of albuminuria may be associated with the appearance of casts and even red blood cells in the urine. The presence of temporary renal damage cannot be eliminated with certainty. However, these individuals are otherwise normal in every respect and the condition is transitory.

(2) Prolonged Exposure to Cold (cold baths).

(3) Alimentary Albuminuria. The ingestion of excessive quantities of native protein, particularly egg albumin, may be followed, in two to three hours, by the appearance of albumin in the urine.

(4) Essential Albuminuria. The term "essential albuminuria" has been proposed to include those cases variously designated as albuminuria of adolescence, cyclic albuminuria, postural or orthostatic albuminuria, and intermittent albuminuria. This type of albuminuria is of clinical importance because of its comparative frequency and because of the fact that it is commonly erroneously considered to be an evidence of renal disease. It is usually first discovered in the course of routine examination of applicants for insurance.

The condition occurs most frequently between the ages of fourteen and eighteen, more commonly in boys than in girls, particularly in those presenting manifestations of autonomic imbalance and vasomotor instability. It may be present in several members of the same family. The blood pressure, particularly the diastolic pressure, is frequently low. Lordosis is a prominent feature in certain cases. The term "orthostatic" is applied to those cases in which albuminuria is present during periods of ordinary activity in the erect posture and disappears upon the assumption of the recumbent posture. It must be noted in this connection that exercise or standing may increase the amount of urinary protein in many types of organic albuminuria. Heredity, lordosis, abnormal renal circulation, movable kidney, and vasomotor instability appear to be important etiologic factors in many of these cases. In certain instances the condition persists into adult life, in others it disappears spontaneously.

As in most of the previously mentioned types of benign albuminuria, the essential etiologic factor is probably some disturbance of renal circulation, either in the large vessels (movable kidney, lordosis) or in the renal arterioles (vasomotor instability). The occurrence of temporary albuminuria following experimentally produced circulatory changes in the renal vessels has been frequently demonstrated. The clinical importance of this type of albuminuria lies in the fact that it is not associated with nor followed by renal disease. So far as is known, the majority of these individuals are not predisposed to the development of nephritis or nephrosis, and the expectation of life is not affected by the presence of essential albuminuria.

(5) Premenstrual Albuminuria.

(6) Albuminuria of Pregnancy. Albuminuria may occur in 30 to 50 per cent of women during uncomplicated pregnancy and labor, disappearing immediately after parturition.

Organic Albuminuria. Albuminuria may result from (*a*) renal changes secondary to abnormalities in organs other than the urinary tract (prerenal albuminuria), (*b*) primary renal disease (renal albuminuria), and (*c*) disease of the lower urinary passages and contiguous structures (postrenal albuminuria).

The organic causes of albuminuria may be conveniently discussed under three headings: (*a*) Prerenal, (*b*) postrenal, and (*c*) renal.

(a) Prerenal Albuminuria. The term "prerenal" cannot be applied in a strict sense, for most of the prerenal factors which cause albuminuria do so by producing changes in the kidneys. However, from the standpoint of the primary condition, the use of this term is justifiable and appears in many aspects advantageous.

(1) Cardiac Disease. In the decompensated stages of cardiac disease, with passive congestion of the kidneys, albuminuria of varying degree develops. The quantity of albumin eliminated is directly dependent upon the degree of renal circulatory embarrassment, and, unless the state of passive congestion is unduly prolonged with consequent organic change in the kidneys, the albuminuria disappears upon re-establishment of circulatory efficiency.

(2) **Ascites** due to local intra-abdominal disease, unassociated with nephritis or cardiac failure, may cause albuminuria, presumably by pressure upon the renal veins causing renal congestion.

(3) **Intra-abdominal Tumors** may produce albuminuria if they are so situated as to cause pressure upon the renal veins.

(4) **Febrile Albuminuria.** Fever, regardless of the cause, may be associated with slight albuminuria. This is due to the production of slight glomerular and tubular changes which are not permanent, being essentially nephrotic rather than nephritic in nature (cloudy swelling) and rapidly subsiding following the restoration of normal body temperature. Febrile albuminuria is commonly observed in pneumonia, typhoid fever, rheumatic fever, malaria, and the acute infectious diseases of childhood. In the latter, particularly in scarlet fever, this condition must be carefully differentiated from a complicating nephritis.

(5) **Convulsive Disorders.** Albumin may appear in the urine during and after convulsions from any cause, including brain tumor, tetanus, epilepsy and meningitis. Albuminuria may also occur in coma due to cerebral vascular accidents, particularly cerebral or meningeal hemorrhage, in which conditions large quantities of albumin may be eliminated (1 to 2 g. per 100 ml.).

(6) **Diseases of Blood.** Albuminuria of slight or moderate degree may occur in association with profound anemia, leukemia, and purpura hemorrhagica. It is perhaps due to the effects of hypoxia on the kidneys.

(7) **Hyperthyroidism.** Albuminuria in hyperthyroidism is probably due to nephrosis or to vasomotor instability, as in the case of certain types of functional albuminuria.

(8) **Intestinal Obstruction.** Albuminuria is present in most cases of acute intestinal obstruction. It may be dependent upon renal changes resulting, in part at least, from the state of alkalosis which is a prominent feature of the condition.

(9) **Hepatic Disease and Jaundice.** Jaundice is not infrequently accompanied by the elimination of small quantities of albumin in the urine. This has been attributed to a toxic or irritant effect of the circulating bile pigments or bile acids upon the renal glomerular membrane or the tubular epithelium.

(10) **Drugs** which cause renal irritation may give rise to albuminuria. Among these are turpentine, cantharides, phosphorus, arsenic, mercury, quinine, copaiba, salicylic acid, phenol, bismuth, lead, ether, chloroform, chromates, oxalates, zinc, opiates, apiol, sandalwood oil, Lysol, naphthols, radium, squill, barbiturates, and sulfonamides.

(b) Postrenal Albuminuria. This group includes inflammatory and degenerative lesions of the renal pelvis, ureter, bladder, prostate, and urethra. The possibility of contamination of the urine by vaginal secretions and discharges must be carefully excluded in considering the cause of albuminuria in women. The commonly employed tests for albumin will, of course, yield positive results in the presence of blood, the source of which frequently resides in the lower urinary tract. Inflammatory exudates are rich in protein and are therefore often responsible for albuminuria of this type. Consequently, urine containing pus will practically always contain albumin. It has been estimated that for each 100,000 leukocytes per 2 ml. of urine, 0.1 per cent of albumin may be expected. This observation is at times of importance in determining the presence or absence of albumin of renal origin in urine containing pus resulting from a lower urinary inflammatory process.

(c) Renal Albuminuria. (1) Destructive Lesions of Kidney. Albuminuria may occur in tuberculosis, carcinoma, pyelonephritis, and polycystic disease of the kidney, and in hypernephroma and other lesions which are characterized by destruction or invasion of kidney tissue. In many cases, however, the urinary findings are normal. Renal infarction is usually associated with albuminuria which is at first marked but rapidly subsides.

(2) Glomerulonephritis and Nephrosclerosis. Albuminuria is an almost constant feature of acute glomerulonephritis. It may be of moderate or severe degree; the amount of albumin does not necessarily parallel the severity of the condition. Albuminuria is one of the most persistent manifestations of acute nephritis, being usually present after other manifestations (except microscopic hematuria) have disappeared. In the latent stages of glomerulonephritis, albuminuria may be the only indication of the renal lesion. In the early stages of chronic glomerulonephritis large quantities of albumin may be eliminated over long periods of time. With the development of impairment of renal function the albuminuria diminishes and in the later stages of the disease may be extremely slight. This change is due to the progressive impairment of the ability of the kidney to eliminate not only the substances normally present in the urine but also plasma albumin and other proteins. There is evidence that one of the factors concerned in the pathogenesis of renal insufficiency in many cases of glomerulonephritis is mechanical obstruction of glomeruli and tubules by accumulations of fibrin and hyaline protein coagula, consisting chiefly of precipitated globulins.

In nephrosclerosis (essential hypertension), albuminuria varies both in incidence and in degree. In many patients with essential hypertension, albuminuria is absent except during periods of myocardial failure with renal congestion. It may be intermittent and slight in amount. In so-called "malignant" hypertension, with rapid progression of the vascular changes in the kidney and with consequent nutritional changes in the glomeruli, relatively large quantities of albumin may be eliminated. With increasing renal functional impairment, however, albuminuria may diminish, as in the case of chronic glomerulonephritis.

(3) Nephrotic Syndrome. This is a clinicobiochemical entity defined by the presence of massive proteinuria with edema and hypoalbuminemia. It is most often due to minimal lesions affecting the epithelial cells of the glomerular basement membrane. The etiology is not always obvious. It may be due to hypersensitivity (immune reaction) to pollen, poison ivy and even the sting of a bee. It can be due to adverse drug reactions (penicillamine, gold, tridione, paradione, etc.). Massive proteinuria can also be

associated with renal amyloidosis, diabetes and a variety of other conditions. It can, of course, result from any glomerulonephritis associated with severe lesions of the glomerular basement membrane, when it is frequently associated with an increase in serum γ-globulin demonstrable by electrophoresis. The commoner types of nephrotic syndrome have a hypogammaglobulinemia. In all forms, there is usually a marked diminution in serum albumin and an increase in serum α_2-globulin. The more severe the lesion of the glomerular basement membrane, the more frequent is the occurrence of higher molecular weight serum proteins in the urine.

(4) *Eclampsia.* Large amounts of albumin may be eliminated in true eclampsia. This is probably dependent upon renal vascular damage.

Quantity of Protein in Urine. As has been indicated, the amount of protein eliminated in the urine is exceedingly variable. Addis believed that the upper limit of normal proteinuria is about 30 mg. for the twelve-hour-day period. Others state that as much as 100 mg. may be eliminated in twenty-four hours by normal individuals under normal conditions. In benign, functional or physiologic albuminuria, the mixed twenty-four-hour specimen seldom contains more than 0.2 per cent of protein, although in some cases as much as 0.5 per cent has been observed. Immediately after severe exercise, however, and in orthostatic albuminuria shortly after assuming the erect posture, specimens of urine may contain 1 per cent or more of protein.

The largest amounts of protein are usually present in the urine of patients with the nephrotic syndrome, in which conditions 2 to 5 per cent or more of protein may be eliminated. More than 110 gm. have been eliminated in twenty-four hours, the usual quantity varying from 5 to 60 gm. with an average daily loss of 15 to 20 gm. Some investigators have observed an increase in urinary protein in these cases during periods of excessive protein intake. This is particularly true in those instances in which the daily output is greater than 20 gm. In most cases, however, large amounts of protein may be ingested without an associated increase in albuminuria; in certain instances, albuminuria may diminish following the administration of a high protein diet with consequent increase in the concentration of albumin in the blood plasma.

In acute glomerulonephritis the urinary protein usually varies from 0.2 to 1.0 per cent, but may exceed 2 per cent in rare instances. In chronic glomerulonephritis and in essential hypertension the figures usually vary from 0.2 to 0.5 per cent; in some cases, particularly in those instances of nephritis with a nephrotic component, the urinary protein may be greatly increased and approach the amounts observed in pure nephrosis. With the supervention of renal insufficiency, however, the quantity of protein in the urine in all types of nephritis is usually diminished. In addition to increased severity of damage to the glomerular capillaries, increase in the amount of protein in the urine may be effected in some instances by increase in protein intake, increase in serum albumin concentration, e.g., following intravenous infusion, and even by actual improvement in renal function (increased filtration rate). Such findings emphasize the fact that the quantity of albumin in the urine does not parallel the severity of renal damage.

In cases in which the urine contains large quantities of protein, this substance may have a significant effect upon the urinary specific gravity. One gram of albumin per 100 ml. increases the specific gravity by 0.003. Under such circumstances, correction must be made for the amount of albumin in the urine when performing urine concentration tests, which depend upon accurate estimation of urinary specific gravity. If such correction is not made, results will be obtained which may be misleading.

Other Proteins in Urine. *Nucleoprotein in Urine.* Nucleoprotein is present in the urine in abnormal amounts in inflammatory conditions of the lower urinary passages, particularly cystitis and pyelitis; it may also be found in nephritis. The term "nucleoprotein" is incorrectly applied to include other protein substances such as mucin and phosphoprotein, which may appear in the urine in increased amounts.

Proteoses and Peptones in Urine. Proteoses and peptones (i.e., polypeptides) may be found in the urine in certain febrile disorders, particularly pneumonia, diphtheria, and pulmonary tuberculosis, as well as in peptic ulcer, carcinoma, and osteomalacia.

Bence Jones Proteinuria. This has already been described.

Hemoglobinuria (p. 270). Hemoglobinuria is the term applied to the presence of free hemoglobin in the urine, in contradistinction to hematuria, which refers to the presence of red blood corpuscles in the urine. The excretion of hemoglobin by the kidneys is preceded by hemoglobinemia, an excessive amount of free hemoglobin in the circulating blood plasma. Hemoglobinuria is therefore dependent upon excessive hemolysis. It has been estimated that it occurs only after the plasma contains more than 60 to 150 mg. of free hemoglobin per kilogram of body weight. Up to this point, the hemoglobin has been complexed to haptoglobins for carriage to the reticulo-endothelial system, and possibly some that is filtered through the glomeruli is completely reabsorbed in the tubules; above these levels the excess appears in the urine. The clinical conditions in which this occurs are presented elsewhere (Hemoglobinemia, p. 270).

Before making a diagnosis of hemoglobinuria it is, of course, essential to exclude the possibility that the condition may in reality be hematuria, the red corpuscles having been partly or completely hemolyzed in the bladder or after elimination of the urine. This is particularly apt to occur in alkaline or ammoniacal urine.

Myoglobinuria. The presence of myoglobin in the urine implies its abnormal liberation from muscle cells, the only situation in which it occurs normally. From the standpoint of pathogenesis, two main varieties of myoglobinuria may be distinguished: (1) that secondary to a known etiological factor; (2) that of unknown etiology, frequently designated primary, paroxysmal, idiopathic, paralytic, spontaneous, etc.

The secondary form may occur as a result of damage to a large area of muscle tissue: (a) crush injuries and other forms of muscle trauma; (b) severe electric shock; (c) extensive burns; (d) occlusion of a large artery; (e) dermatomyositis (possibly); (f) alcohol and barbiturate intoxication, with muscle necrosis; (g) an epidemic form, apparently of infectious or toxic etiology (Haff disease).

The primary form, which may at times have a familial incidence, occurs paroxysmally, without known cause, although it may occasionally follow strenuous exertion. There is usually a rather typical clinical picture of an acute onset, with fever, chills and muscular weakness, even to paralysis, involving usually mainly the lower extremities but also other regions. The acute attack is accompanied by extensive necrosis or lysis of striated muscle and is followed by excretion of dark-red or brown urine, containing myoglobin. Kidney damage occurs commonly, associated with circulatory and renal failure.

In addition to the clinical manifestations, the diagnosis is suggested by positive benzidine reactions in the urine, in the absence of significant numbers of red blood cells, and in the blood plasma, in the absence of abnormal discoloration. The normal color of the plasma is attributed to the low molecular size of myoglobin (m.w. 17,000), which permits passage into the urine of relatively large amounts at plasma concentrations (15–20 mg./100 ml. too low to produce visible discoloration. This is in contrast to hemoglobin, which appears in the urine usually only when the plasma concentration exceeds 150 mg./100 ml. However, the diagnosis can be established with certainty only by application of more specific procedures, spectrophotometric, ultracentrifugal, or electrophoretic.

Urinary Nonprotein Nitrogen

Investigation of the nonprotein nitrogenous constituents of the urine in disease has been largely superseded by the quantitative determination of these substances in the blood. However, particularly in nephritis, the study of both blood and urinary nonprotein nitrogenous elements may yield valuable information. Since the amount of urinary nitrogen depends primarily upon the quantity of protein ingested, it is obvious that no clinical significance can be attached to the nitrogen output unless the intake be determined simultaneously. Even under such circumstances accurate information cannot be obtained in many instances because of the fact that nitrogen equilibrium may be established at various levels of blood nonprotein nitrogen. Before drawing conclusions from alterations in urinary nitrogen, therefore, several factors must be carefully investigated.

Urinary Urea. Urea administered to normal individuals is promptly eliminated because the storage capacity of the body for nitrogen is limited. In certain patients with nephritis the elimination of excessive amounts of ingested urea may be delayed for varying periods of time. In some cases, however, no deviation from the normal can be demonstrated. In patients with chronic nephritis in whom retention of nitrogen occurs following the ingestion of excessive quantities of protein, the retained nitrogen frequently cannot be entirely accounted for by the concentration of nonprotein nitrogen in the blood, which may or may not be increased; it is believed that in such cases the excess nitrogen is retained in the tissues, chiefly in the liver and muscles.

In many cases of nephritis the ability of the kidneys to concentrate urea is impaired although the amount eliminated in twenty-four hours may be normal. The normal individual is able to concentrate urea to the extent of 2 per cent or more in the second hour after the ingestion of 15 g. of urea dissolved in 100 ml. of water. If renal function is impaired the urea concentration in the urine is below this figure (see p. 501).

The most important clinical applications of studies of urea elimination have resulted from studies of urea clearance. It has been shown that with urine volumes below 2 ml. per minute (the augmentation limit) the blood urea clearance increases in direct proportion to the square root of the urine volume, although this is not now generally accepted. With urinary outputs below this figure the volume of blood which is cleared of urea in one minute was found to vary, in normal individuals, from 41 to 65 ml., with a mean of 54 ml. (standard clearance). When the rate of urine excretion exceeds 2 ml. per minute, the urea clearance, having attained its maximum at that point, is unaffected by further increase in urine volume. Under such circumstances, the volume of blood which is cleared of urea in one minute was found to vary from 64 to 99 ml. with a mean of about 75 ml. (maximum clearance) (see p. 501). In investigating patients, it is best to ensure, by adequate water administration, that maximum clearance is obtained whenever possible. Decrease in urea clearance may be demonstrable in patients with renal disease before evidence of renal functional impairment can be obtained by any other method. This subject is considered in greater detail in the discussion of tests of renal function (p. 518).

In hepatic insufficiency, owing to decreased formation of urea from amino acids, the proportion of urinary nitrogen occurring as urea may be decreased. Urea normally constitutes about 60 per cent of the total urinary nitrogen on a low protein diet and 90 per cent on a high protein diet, the average figure being about 80 per cent. Figures of 50 per cent may be obtained in acute hepatic disease even with an adequate protein intake. In many cases, however, the urinary nitrogen partition is normal.

A relative decrease in the urinary urea fraction may also occur in conditions associated with acidemia or acidosis. Under such circumstances the low proportion of urea is due to the great increase in urinary ammonium ion, formed by the kidney, which is an important feature of the mechanism of regulation of acid-base balance (p. 411). The urine urea N decreases (positive N balance) following administration of growth hormone, testosterone, or insulin; there is a simultaneous fall in the concentrations of urea and amino acids in the plasma. This is a reflection of stimulation of protein synthesis by these agents. Administration of 11-oxygenated adrenocortical hormones or excessive doses of thyroxine increases urea excretion (negative N balance), a reflection of increased protein catabolism.

An increase (above nitrogen intake) in the amount of urea eliminated in the urine may occur in disorders associated with excessive tissue catabolism, as in prolonged wasting diseases and febrile states, urea (or urea plus ammonium ion) comprising by far the largest portion of the increased nitrogen output. The outstanding exception is severe hepatic necrosis, in which the deamination mechanism in the liver is impaired, resulting in excretion of relatively large amounts of amino acids and correspondingly smaller amounts of urea (as low as 10 to 50 per cent of the total urine N).

Uric Acid in Urine. The factors which influence excretion of uric acid are considered in the discussion of nucleic acid metabolism (p. 253).

Creatine and Creatinine in Urine. The excretion of creatine in the urine under normal conditions has been discussed elsewhere (p. 180). The investigation of this factor is of significance particularly in the study of metabolic changes in certain myopathies, because of the important relation of creatine metabolism to muscular function. Early observations established the presence of an unstable compound containing phosphoric acid and creatine in fresh, excised muscle, which soon began to disappear and, in a few hours after removal of the muscle, was no longer demonstrable; as it disappeared, free phosphoric acid and creatine appeared in molecular ratio to one another. This unstable compound of phosphoric acid and creatine was diminished by stimulation. Stimulation during arrest of the circulation through the muscle caused the compound to break up completely and, when the blood supply was restored and the muscle allowed to rest, the combination of phosphoric acid with creatine was once more renewed, the quantity of the compound increasing steadily during the period of rest. Observations such as these indicated that creatine has an important bearing on the mechanism of muscular contraction (p. 16).

Excessive creatinuria may occur in starvation, febrile and wasting diseases, diabetes mellitus, in eunuchoids and castrates, in postencephalitic and Parkinsonian rigidity, and rather consistently, but not invariably, in certain myopathies (myotonia congenita, myasthenia gravis, various forms of myositis, congenital muscle dystrophies, congenital muscular hypertrophy, and secondary muscle atrophy). Testosterone diminishes creatinuria in eunuchoids and castrates, apparently by increasing the storage, but not the synthesis, of creatine. On the other hand, administration of methyltestosterone, after a latent period of four to sixteen days, is followed by increased excretion of creatine and creatinine, due probably to increased synthesis of creatine. Increased creatine excretion follows intravenous administration of amino acid digests.

Of particular interest in this connection is the influence of glycine upon the clinical and metabolic course of certain of these myopathies, particularly myasthenia gravis. The administration of large doses of glycine results, in certain of these cases, in a 100 to 1000 per cent increase in creatine excretion. A similar increase in creatinuria also occurs in the majority of normal subjects after the ingestion of glycine. After a period of several weeks the creatinuria begins to decrease in most instances, eventually falling to almost control levels despite the continued administration of glycine. The observation has been made that cases of primary myopathy in which the average creatinuria increased more than 50 per cent above the control level after the administration of glycine apparently showed both subjective and objective improvement, provided this increased creatinuria disappeared within a few weeks. Little or no improvement was observed in cases in which the increase in creatinuria was less than 50 per cent.

Excessive creatinuria has also been observed in hyperthyroidism, following the administration of thyroid extract, during the menstrual period and in pregnancy, and when carbohydrate is excluded from the diet. It has been observed after fractures (to 500 mg. daily). It disappears gradually during the period of healing of the fracture and after the resumption of activity.

The output of creatinine is relatively constant under normal conditions and is essentially independent of diet and muscular activity. The elimination of creatinine is increased in conditions associated with increased tissue catabolism, as in fevers, and is decreased in the presence of marked muscular atrophy. It has been found that in the primary myopathies the diminished creatinuria which occurs as a secondary phenomenon following the administration of glycine is accompanied by an increased excretion of creatinine and an improvement in the patient's ability to retain ingested creatine.

Preformed creatinine, administered to normal individuals, is promptly eliminated.

Creatine Tolerance. When 1.32 or 2.64 g. of creatine hydrate (equivalent to 1 and 2 g., respectively, of creatine) are ingested by a normal adult, about 80 per cent is retained by normal men and about 70 per cent by normal, nonpregnant women; about 20 and 30 per cent, respectively, are excreted in the urine during the subsequent twenty-four hours. Diminished tolerance (i.e., excessive excretion) is present in conditions accompanied by spontaneous creatinuria, mentioned above, and in mild cases of these disorders in the absence of creatinuria under ordinary dietary conditions.

Creatine tolerance is decreased in the majority of patients with hyperthyroidism and tends to be increased (excessive retention of administered creatine) in hypothyroidism. This procedure may be of some value in the diagnosis of hyperthyroidism or hypothyroidism in children. In children, correction must be made for normal creatinuria, which averages 4.2 mg. daily per kilogram of body weight on a meat-free diet, providing adequate calories and 2 g. of protein per kilogram.

Amino Acids in Urine. Excessive aminoaciduria may occur in conditions in which the plasma amino acid concentration is increased (overflow aminoaciduria) or in which there is defective reabsorption of these substances in the renal tubules (renal aminoaciduria). These abnormalities may apply to all amino acids or be confined to certain groups or to single amino acids or their metabolites. In some cases the urine amino acid pattern is so characteristic as to identify a specific syndrome or disease (cystinuria, Hartnup syndrome, maple syrup urine disease, phenylketonuria and so forth).

Renal forms of aminoaciduria are identified on the basis of (a) normal plasma amino acid concentrations and (b) increased renal clearance of the amino acids involved, i.e., decreased tubular reabsorption. They may occur either as an isolated phenomenon or in association with other reabsorptive defects. They may represent a congenital abnormality of tubular function or may occur as a result of or in association with acquired disease. Several clinically distinct entities of this type have been described. They include : (a) cystinuria; (b) glycinuria; (c) Fanconi syndrome; (d) cystinosis; (e) Lowe syndrome; (f) Hartnup syndrome; (g) aminoaciduria occurring in subjects with vitamin D deficiency, vitamin C deficiency, galactosemia, hepatolenticular degeneration (Wilson's disease), multiple myeloma, glycogen storage disease, congenital renal tubular acidosis, and poisoning with certain nephrotoxic agents (heavy metals, Lysol, nitrobenzene, salicylate, old tetracycline), after renal transplantation and in muscular dystrophy. These are considered in greater detail elsewhere (p. 534). There may be a distinctive pattern after two-dimensional paper chromatography.

Aminoacidurias of the overflow type are characterized by (a) increased concentration of one or more amino acids in the blood plasma, and (b) normal renal clearance, i.e., normal tubular reabsorption of the amino acids involved. They include: (a) specific abnormalities of intermediary metabolism of a specific amino acid, resulting in increased concentrations of that amino acid and/or certain metabolites in the plasma; (b) more generalized disturbance of intermediary metabolism of amino acids, usually due to extensive liver damage, resulting in increased concentrations of most or all amino acids in the blood plasma; (c) accelerated production of excessive amounts of amino acids as a result of increased protein catabolism.

Excessive Protein Breakdown. Aminoaciduria of this type may occur as a result of extensive tissue injury (trauma, burns), protracted febrile illnesses, malignancy, thyrotoxicosis, muscular atrophy, and following administration of adrenal glucocorticoids.

Hepatic Disease. Normally a large portion of the circulating amino acids is removed and metabolized by the liver (protein synthesis, gluconeogenesis, transamination, urea formation, etc.); consequently, severe impairment of hepatocellular function in any type of hepatic disease, or diversion of portal blood flow from the liver, as occurs in portal cirrhosis, may result in increased concentrations of amino acids in the blood and generalized aminoaciduria. In acute and subacute hepatic necrosis, e.g., carbon tetrachloride poisoning, this is contributed to also by extensive autolysis of liver cell proteins. In severe cases, certain of the amino acids, e.g., leucine and tyrosine, may occur in the urine in crystalline form because of their limited solubilities.

Bone Disease. There is an increase in urinary hydroxyproline in Marfan's disease, Paget's disease, fibrous dysplasia, hyperparathyroidism, hyperthyroidism, and acromegaly. Phosphoethanolamine is increased in hypophosphatasia.

Specific Metabolic Errors. Specific blocks in the metabolism of an amino acid may result in excessive accumulation and excretion of the amino acid and/or certain of its metabolites. Several such conditions have been identified, and in some cases a specific enzyme deficiency has been demonstrated, representing so-called "inborn errors of metabolism."

Abnormal Phenylalanine and Tyrosine Metabolism. Normally, the primary pathway of utilization of phenylalanine involves transformation to tyrosine (hydroxyphenylalanine). The latter undergoes transamination, forming p-hydroxyphenylpyruvic acid, which is oxidized to dihydroxyphenylacetic (homogentisic) acid.

The latter subsequently forms acetoacetic acid and a Krebs cycle intermediate (fumaric acid). Defects may occur at various points in this metabolic sequence, with consequent urinary excretion of increased amounts of certain intermediary metabolites of tyrosine and phenylalanine.

(1) *Alkaptonuria.* This condition is characterized by the urinary excretion of homogentisic acid (dihydroxyphenylacetic acid), due to congenital lack of homogentisic acid oxidase, the enzyme essential for the next step (oxidation) in the metabolism of this substance. This anomaly is often discovered in infancy, since the homogentisic acid in the urine, on exposure to air, is readily oxidized to dark-colored compounds, especially if the urine, on standing, becomes alkaline due to ammoniacal fermentation. It is probably present at birth and persists throughout life, apparently with no directly harmful consequences, despite the fact that several grams of homogentisic acid may be excreted daily. In later life, accumulation of dark pigment in cartilages and tendons may give rise to the condition known as ochronosis, which may be accompanied by arthritic changes.

(2) *Tyrosinosis.* This is a very rare disorder, characterized by the urinary excretion, on an ordinary diet, of p-hydroxyphenylpyruvic acid, with excretion of tyrosine and a little p-hydroxyphenyllactic acid after ingestion of large amounts of tyrosine. The metabolic block was thought to be at the point of conversion of hydroxyphenylpyruvic acid to homogentisic acid. The disturbance is now believed to be caused by an aminotransferase deficiency.

(3) *Neonatal Tyrosinemia.* A transient hypertyrosinemia is not uncommon in the newborn, especially the premature. Tyrosine and its metabolites p-hydroxyphenylpyruvic acid, p-hydroxyphenyllactic acid and p-hydroxyphenylacetic acid appear in the urine. It is caused by a transient deficiency of p-hydroxphenylpyruvic acid oxidase. The condition may last for a few weeks. Administration of ascorbic acid and reduction of protein intake brings about rapid return to normal.

(4) *Hereditary Tyrosinemia.* This by no means uncommon disorder has biochemical features similar to neonatal tyrosinemia, but the proportion of p-hydroxyphenyllactic acid in the urine is greater. The disorder is not suppressed by administration of ascorbic acid and is due to an inherited deficiency of p-hydroxyphenylpyruvic acid oxidase. Liver failure with death can occur approximately six months after birth and the infant has a characteristic odor. Other patients do not have this acute onset but develop nodular cirrhosis, a Fanconi-like syndrome and a tendency to

hypoglycemia. A number of patients develop hypophosphatemic rickets. Hepatic carcinoma may develop.

(5) *Albinism.* This hypopigmentation disorder is due to a deficiency of a tyrosinase normally present in melanocytes.

(6) *Miscellaneous Causes of Tyrosinuria.* An excess of urinary tyrosine may accompany liver disease, hyperthyroidism and scurvy.

(7) *Fibrocystic Disease.* In this condition, there is malabsorption of tyrosine. Intestinal bacteria convert it into p-hydroxyphenylacetic acid and tyramine, which are absorbed and appear in the urine.

(8) *Phenylketonuria.* This condition, also termed phenylpyruvic oligophrenia, imbecillitas phenylpyruvica, and phenylpyruvia, is an "inborn error of metabolism" occurring in certain mentally defective individuals, characterized by the urinary excretion of phenylalanine and its direct metabolites, phenylpyruvic, phenyllactic and phenylacetic acids. Phenylpyruvic usually predominates, which, since pyruvic is a ketone acid, is the reason for the designation, phenylketonuria, a term not applicable to the other acids. The underlying defect is congenital deficiency of phenylalanine hydroxylase, the enzyme involved in conversion of phenylalanine to tyrosine, the major normal metabolic pathway. As a result, phenylalanine accumulates in excess (tissues, plasma) and is metabolized by what is normally a minor pathway, i.e., deamination, with formation of phenylpyruvic acid and its derivatives, indicated above. These are excreted in the urine, together with increased amounts of phenylalanine. The exact relation of this metabolic abnormality to the mental deficiency has not been established. However, early diagnosis (shortly after birth) and drastic restriction of phenylalanine intake has been found effective in preventing serious developmental abnormalities in children with this disorder.

Not all individuals with this metabolic disorder develop mental defect. On the other hand, maternal phenylketonuria, when not effectively treated, can result in the birth of mentally defective children who do not have the enzyme defect. Patients with phenylketonuria tend to have a deficiency of serotonin. This may be connected with the defect of myelin synthesis, which occurs in these patients. The accumulation of phenylalanine also impairs melanin synthesis; children with the defect tend to have fair skin and fair hair. Excess of phenylalanine in the blood leads to excretion of the amino acid into the intestine. Here, it competes with tryptophan for absorption and the latter then becomes subject to the action of intestinal bacteria. The resultant indole derivatives are absorbed and appear in the urine. Some patients have an unusually high tolerance for phenylalanine and some even have a transient phenylketonuria.

A benign persistent hyperphenylalaninemia has also been described. There is a partial deficiency of hepatic phenylalanine hydroxylase. The condition is particularly common in individuals of Mediterranean origin and is not related to classic phenylketonuria. Transient neonatal hyperphenylalaninemia has also been described.

Maple Syrup Urine Disease. This is a congenial, familial disorder characterized by central nervous system manifestations of convulsions and attacks of flaccidity and apnea and the urinary excretion of substances that produce an odor resembling that of maple syrup. These are ketoacids derived from the branched amino acids, leucine, isoleucine and valine. The metabolic block is in the next step in the metabolism of these keto acids, i.e., oxidative decarboxylation to the corresponding simple-branched acids. As a result of this defect, increased amounts of these amino acids can be demonstrated in the blood plasma and, with the keto acids, in the urine.

Methionine Malabsorption Syndrome. In this familial condition, there is diminished intestinal absorption of methionine and, to a lesser extent, some branched chain and aromatic amino acids. Intestinal bacteria act on the unabsorbed methionine to give α-hydroxybutyric acid, which is absorbed and appears in the urine, giving it an odor resembling that of an oasthouse. It is probable that so-called *oasthouse urine disease* is the same condition. It is believed that the severe associated neurological disorders are caused by the hydroxy- acid. The condition improves rapidly on a low methionine diet.

Blue Diaper Syndrome. This condition has been described in young children of one family. There was specific tryptophan malabsorption in the small intestine. This amino acid was converted by intestinal bacteria to indican, which was absorbed and appeared in excessive amounts in the urine and its oxidation produced indigo blue, which stained the diapers. There was also hypercalcemia.

Glycinuria. Cases have been reported of excessive excretion of glycine in the urine only when the plasma glycine concentration was elevated, in subjects with clinical manifestations referable to many systems. Two different conditions have been reported, one with and one without ketoacidemia.

Cystathioninuria. Cystathionine, an intermediate in the biosynthesis of cysteine, is not normally present in the urine. Its occurrence in the urine has been reported as an isolated abnormality in association with mental deficiency, but this is not always present. There is a deficiency of hepatic cystathioninase and a Marfan-like syndrome accompanied by osteoporosis.

Argininosuccinicaciduria. Argininosuccinic acid is a normal intermediate in the ornithine cycle of urea formation. It has been found in the urine, where it does not occur normally, in subjects with mental retardation and other clinical manifestations (p. 230).

Homocystinuria. This abnormality occurs in association with mental deficiency. There is a deficiency of cystathionine synthetase.

Phosphoethanolaminuria. In subjects with hypophosphatasia, phosphoethanolamine has been found in both plasma and urine, where it does not occur normally.

Histidinuria. The amino acid histidine and imidazole pyruvic acid occur in excessive amount in the urine of patients suffering from histidinemia. This is due to a defect in liver histidase. There is usually some mental retardation and difficulty with the development of speech. No formiminoglutamic acid (FIGLU) is present in the urine after histidine loading but occurs after the administration of urocanic acid.

β-Alaninuria. The amino acid β-alanine occurred in excessive amounts in the urine of an infant with hyper-β-alaninemia. There were convulsions, and death occurred at 22 weeks.

Carnosinuria. The dipeptide β-alanyl-histidine appeared in excess in the urine of two mentally retarded patients with myoclonic epilepsy, when on a meat-free diet. One patient had carnosinemia. A deficiency in blood carnosinase was demonstrated.

Prolinuria. The amino acid proline together with hydroxyproline and glycine occurs in hyperprolinemia, which can be caused either by a defect in proline oxidase or in glutamate-γ-semialdehyde dehydrogenase.

Hydroxyprolinuria. The amino acid hydroxyproline in the free state occurred in excess in a mentally retarded patient with hydroxyprolinemia. There was no excess of proline, glycine or hydroxyproline peptides. There was a deficiency of hydroxyproline oxidase.

Valinuria. Valine appeared in excess in one Japanese infant with mental retardation which was observed later. There was hypervalinemia due to a deficiency of a specific aminotransferase.

Serotonin Metabolites. Serotonin (5-hydroxytryptamine), normally a product of a minor pathway of metabolism of tryptophan, is produced in excessive amounts by carcinoid tumors of the intestine. The increased concentration of this biologically active substance in the circulation produces a characteristic clinical picture (carcinoid syndrome). The circulating serotonin is converted (by monamine oxidase), mainly in the liver and lungs, to 5-hydroxyindoleacetic acid, which is excreted in the urine. Diagnosis of carcinoid tumor may be made readily on the basis of the urinary excretion of increased amounts (up to several hundred milligrams per day) of 5-HIAA (normal excretion less than 10 mg.).

Abnormally large amounts of 5-hydroxyindoleacetic acid are excreted in the urine of normal subjects following ingestion of several bananas. In subjects with carcinoid tumors, diversion of a major portion of dietary tryptophan into the serotonin pathway may result in manifestations of tryptophan deficiency, e.g., protein malnutrition and nicotinamide deficiency. The carcinoid syndrome has been observed occasionally in association with certain malignant bronchial adenomas.

Melanuria. Melanin is a dark, brown-black pigment, normally found in the hair, skin, ciliary body, choroid of the eye, pigment layer of the retina, and various nerve cells. It is derived from tyrosine which, through the action of tyrosinase, forms dihydroxyphenylalanine (dopa), the latter, a colorless substance, giving rise to a series of oxidation products terminating in melanin (polymerized indole-5,6-quinone).

Melanuria occurs exclusively in the presence of melanotic sarcoma (in about 20 per cent of cases), usually when extensive liver metastases are present. It first appears as melanogen, which undergoes oxidation.

A syndrome has been described, with a strong familial tendency, characterized by melanotic pigmentation of the oral mucosa, lips and digits, associated with multiple intestinal polyposis, which frequently causes attacks of abdominal pain, intestinal obstruction, and intussusception. Melanuria has not been reported in patients presenting this syndrome.

Ammonia in Urine. The ammonia (as ammonium ion) present in the urine is formed in the kidneys, from amino acids, and glutamine.

Urinary ammonia is increased in many conditions associated with acidosis or acidemia (diabetes, starvation, dehydration, vomiting, diarrhea, etc.). The acidemia of nephritis, in which the ammonia-forming function of the renal tubular epithelium is impaired, constitutes one exception to this general rule. The ammonia content of the urine is usually decreased in alkalosis or alkalemia (p. 412). An increase may result from bacterial decomposition of urea in subjects with bladder retention and urinary tract infection (particularly cystitis).

Negative Nitrogen Balance in Disease. Excessive protein catabolism, with consequent negative nitrogen balance, occurs in a great variety of abnormal states. Appreciation of the increased protein requirement of patients with these conditions has greatly improved their clinical management. It must be recognized that in the majority of instances the deficiency is not in protein alone. Various vitamins, electrolytes, and fluids are also involved, and must be supplied in proper amounts if therapy is to be maximally effective. These therapeutic considerations are, however, beyond the scope of this volume.

Acute Infections. Most of the serious acute infectious diseases are accompanied by negative nitrogen balance, usually proportional to the severity of the disease, although not always to the height of the fever. This is not necessarily true of prolonged chronic infections, e.g., tuberculosis, in which N equilibrium can be maintained usually on a much lower caloric and protein intake than in acute infections with comparable temperature rises. It is perhaps significant in this connection that the BMR does not increase in these conditions in proportion to the rise in temperature, as it does in acute infections.

Immobilization. A negative N balance occurs during periods of immobilization in bed. In healthy individuals, the urinary N begins to increase on about the fifth day, rising to a maximum in about ten days.

Burns. Negative N balance in patients with extensive burns is due in part, but not entirely, to loss of comparatively large amounts of protein in the exudate from the burned areas.

Operations, Fractures, Other Trauma. Major operative procedures are followed promptly by negative N balance, accompanied by a drop in the concentration of plasma albumin, mainly because of water retention, but there is an increase in α_1, and α_2-globulins. The total amino acid content remains unchanged but there is an increase in some amino acids and a fall in others. This negative N balance is not so marked in malnourished individuals and is much more consistent and more pronounced after intra-abdominal and serious intrathoracic operations than following more superficial procedures, e.g., herniorrhaphy.

Increased urinary N excretion occurs within a few days after traumatic fractures and, to a lesser extent, after other types of trauma, normally roughly paralleling the severity of the injury. This is by no means entirely attributable to immobilization.

There is evidence which suggests that the disturbances in protein metabolism in the conditions outlined above are due in part to some metabolic aberration other than those resulting directly from disuse atrophy, tissue destruction or heightened metabolism incident to fever. Hormonal mechanisms may be involved. Participation of the "alarm reaction" is suggested by the observation of the common occurrence of marked eosinopenia promptly after surgical operations and trauma, accompanied by an increase in urinary adrenal corticoids and, less consistently, 17-ketosteroids. Moreover, negative N balance is not produced readily under these conditions in adrenalectomized animals. It appears, however, that factors other than the adrenal hormones are involved in the production of this phenomenon.

Malnutrition. It must be borne in mind that although hypoalbuminemia is a clinically valuable reflection of a state of protein deficiency, its presence may be masked by coexisting dehydration (p. 183). Moreover, the extent of the total body protein deficit is not reflected quantitatively in the decrease of plasma albumin (p. 183). It has been shown that, in simple malnutrition, about 30 grams of tissue protein are lost for every gram lost from the plasma, indicating the high priority held by the plasma proteins in the general nitrogen pool.

Patients with protein malnutrition exhibit a number of undesirable tendencies, some of which are due to decrease in concentration of certain of the plasma proteins, others to impaired tissue protein synthesis. These include: (a) anemia; (b) edema; (c) increased susceptibility to hemorrhagic shock; (d) increased susceptibility to infection; (e) delayed wound healing; (f) delayed callus formation in fractures; (g) poor takes of skin grafts; (h) tendency to decubitus ulcers; (i) anorexia; (j) gastrointestinal motility disturbances; (k) increased susceptibility to liver damage.

DISORDERS OF THE UREA CYCLE

There are a number of hereditary disorders which are known to arise from a deficiency of an enzyme involved in the urea cycle. From the clinical point of view, these conditions are very similar since they are usually characterized by mental retardation, convulsions and other neurological disorders. There may be failure to thrive and even a fatal termination in the neonatal period.

Argininosuccinic Aciduria. This disorder results from a deficiency of the enzyme argininosuccinate lyase, which converts argininosuccinate to arginine and fumarate. In this way, citrulline is made to yield arginine (p. 160). It has been suggested that there might be more than one form of the condition. Argininosuccinic acid and two of its anhydrides can be detected in the urine by two-dimensional paper chromatography. The acid is present in increased amounts in plasma, cerebrospinal fluid and urine. The plasma levels are very much higher in the severe neonatal cases.

There is an increased content of the ammonium ion in plasma and cerebrospinal fluid and a tendency to an increased amount of glutamine and glutamic acid in these fluids as well as an increase in citrulline. Argininosuccinate lyase has been shown not to be present in a liver biopsy, although the other enzymes of the urea cycle are present. The disease is inherited as an autosomal recessive.

Citrullinemia. This results from a deficiency of the enzyme argininosuccinate synthetase, which acts on aspartate and citrulline to form argininosuccinate. The condition is very rare. The level of citrulline is raised in the plasma but is lower in cerebrospinal fluid because citrulline can be utilized by the brain. Plasma ammonium is increased, especially after protein intake. It is likely that plasma glutamine levels are increased. Urinary levels of glutamine and glutamic acid are high. Pyrimidine metabolites, (orotic acid, uridine, uracil) can be present in the urine in fairly large amounts.

Hyperargininemia. This condition, due to a deficiency of arginase, has been reported in two sisters and possibly in one other case. There was aminoaciduria with intermittent hyperargininuria. The arginine content of both blood and cerebrospinal fluid was high. There was a marked increase of the ammonium ion in blood and cerebrospinal fluid.

Hyperammonemia. This is due to a marked deficiency of ornithine transcarbamylase. There is an infantile and adult (one case) type, as is the case in argininosuccinic aciduria. The adult was relatively little affected but the prognosis is poor in children. There is a very high level of ammonium ion in blood and cerebrospinal fluid. Exclusion of protein from the diet can bring about a fall to normal levels. There is a high urinary excretion of ammonium salts as well as glutamine, and the plasma and cerebrospinal fluid glutamine and glutamic acid are increased. These levels are also related to protein intake. Blood citrulline and arginine levels are lowered, but ornithine is not affected.

Hyperammonemia can also occur in other, nonheritable, conditions such as severe liver disease, certain types of respiratory disease (emphysema), lysine intolerance, familial protein intolerance; in association with hyperornithinemia; and in a special group of mentally retarded children with cerebral atrophy.

Congo Red Test for Amyloidosis

Congo red, injected intravenously, is removed from the blood relatively slowly, about 70 to 90 per cent of the quantity injected remaining in the blood at the end of one hour in normal individuals. In the presence of amyloid disease the dye disappears from the blood more rapidly than normally, less than 40 per cent of the quantity injected remaining in the blood stream at the end of one hour. Under normal conditions practically all of the dye is excreted by the liver in the bile. The increased rapidity of its disappearance in amyloid disease is generally attributed to adsorption of the dye by the amyloid material and to its increased filtration from the blood through the damaged capillary endothelium.

This procedure has been widely employed for the diagnosis of amyloid disease. Persistence of Congo red in the blood stream is dependent to a certain extent upon a normal concentration of plasma protein, particularly albumin, to which the dye is adsorbed. In the nephrotic syndrome, with marked albuminuria and a lowered plasma albumin concentration, the dye disappears from the blood with abnormal rapidity, as in the case of amyloid disease, a large proportion appearing in the urine, however, adsorbed to albumin. In the presence of marked albuminuria, therefore, the interpretation of subnormal retention of Congo red in the blood is difficult, but in amyloid disease much less of the dye appears in the urine, as a rule, than in the nephrotic syndrome. Occasionally, findings suggestive of amyloid disease are obtained in patients with low plasma albumin concentration in the absence of albuminuria. The Congo red test may be interpreted as strong confirmatory evidence of amyloid disease when more than 90 per cent of the dye disappears from the blood within one hour, in the absence of elimination of significant amounts in the urine. In exceptional cases of amyloid disease this test may yield essentially normal findings, and findings suggestive of amyloid disease may be obtained occasionally in cases in which no amyloid substance is present. There is apparently no consistent relationship between the rapidity of removal of the dye from the blood and the extent of amyloid deposition in the tissues. The Congo red test is usually negative in primary amyloidosis, a rather rare condition. In the light of information obtained from biopsy techniques, the test is not now considered to be very useful.

ACID MUCOPOLYSACCHARIDES

Connective tissue contains a variety of mucoproteins, which are synthesized in fibroblasts and pass in excretory sacs into the extracellular space. Although fibroblasts are quite active in this respect, it is now believed that similar substances are synthesized in many tissues. The connective tissue mucoproteins consist of one or more acid mucopolysaccharides attached to a protein by a covalent linkage (glycopeptide) between a carbohydrate residue (usually xylose) and a serine or threonine residue of the protein. The protein moiety is very variable in relation to its molecular size. These substances are of varying molecular weight, made up for the most part of repeating disaccharide units composed of an amino sugar (N-acetyl glucosamine or N-acetylgalactosamine; D-GlcN and D-GalN respectively) and a uronic acid (glucuronic or iduronic; D-GlcUA and L-IdUA respectively). At one end of the molecule there are usually three sugar residues (-Gal-Gal-Xyl) enabling a covalent linkage to be made between a xylose residue and a serine or threonine residue on a protein to form a mucoprotein. A typical connective tissue mucoprotein, chondromucoprotein, is made up of a linear protein with acid mucopolysaccharide residues attached along its length.

The mucopolysaccharides of connective tissue are *hyaluronic acid*

(D-GlcN and D-GlcUA), *chondroitin* (D-GalN and D-GlcUA), *chondroitin-4-SO₄*, *chondroitin-6-SO₄*, *dermatan sulfate* (D-GalN and mostly L-IdUA, but also some D-GlcUA), *heparan sulfate* (D-GlcN and mostly D-GlcUA, but also some L-IdUA), *heparin* (D-GlcNH$_2$ and mostly D-GlcUA but some L-IdUA; is also a sulfate) which occurs inside mast cells, *keratan sulfate I* (D-GlcN and Gal instead of a uronic acid) and *keratan sulfate II* (which occurs in the skeleton and also contains D-GalN). The sulfuryl residue usually forms an ester linkage with the amino sugar residue, but in heparan sulfate and in heparin it can also be attached directly to the N of the amino sugar residue. (In heparin, this is probably the only type of linkage). The glycopeptide linkage in the keratans involves N-acetyl glucosamine or N-acetyl galactosamine instead of xylose.

The linkage in the repetitive disaccharide is usually β, but in heparan sulfate and in heparin it can also be of the α type.

The Mucopolysaccharidoses.
These are a group of heritable disorders in which there is excessive deposition of acid mucopolysaccharides in tissues. There may also be excessive secretion in urine (sometimes the latter occurs alone). It is believed that these disorders are due to a defect in catabolism rather than to excessive synthesis.

Mucopolysaccharidosis I (Hurler's Syndrome). Skeletal deformities, corneal opacity and severe mental retardation are present in this disorder. Dermatan sulfate and heparan sulfate are present in urine and tissues and can be demonstrated in fibroblast cultures (metachromatic granules).

Mucopolysaccharidosis II (Hunter's Syndrome). Severe skeletal deformities and moderate mental deterioration occur in this disorder. Deafness occurs early but there is no corneal opacity. The findings in the tissues, urine and fibroblasts are the same as for Type I. This is the only mucopolysaccharidosis inherited as an X-linked recessive characteristic. The others are autosomal recessive.

Mucopolysaccharidosis III (Sanfilippo's Syndrome). Here, there is marked mental retardation, mild skeletal disturbances and little, if any, corneal opacity. Heparan sulfate is present in tissues and urine and dermatan sulfate is found in fibroblasts.

Mucopolysaccharidosis IV (Morquio's Syndrome). In this condition, there is marked skeletal deformity with osteoporosis and dwarfism resulting from defects in the epiphyses of the vertebral bodies, which also results in a shortening of the neck. There may be corneal opacity, but there is no mental retardation. Keratan sulfate and the chondroitin sulfates are present in the urine.

Mucopolysaccharidosis V (Scheie's Syndrome). In this syndrome, corneal opacity is marked, skeletal changes are by no means severe and there is no mental retardation. The urine contains dermatan sulfate.

Mucopolysaccharidosis VI (Maroteaux-Lamy Syndrome). Skeletal deformities and corneal opacities are marked but there is no mental retardation. The urine contains dermatan sulfate.

Mucopolysaccharidosis VII. In this condition, there are changes in the skeleton, which are similar to those seen in Type IV.

Miscellaneous Mucopolysaccharidoses. These include "I-cell" disease, lipomucopolysaccharidosis (increased mucopolysaccharides and lipids in tissues only), and G_{M_1}-gangliosidosis (p. 130).

Relationship to Lipid Storage Diseases. There is undoubtedly a relationship between at least some of the mucopolysaccharidoses and the lipid storage diseases, especially the gangliosidoses (p. 130). Mucopolysaccharidoses I, II and III show decreased β-galactosidase in the tissues and G_{M_1}-gangliosidosis (pseudo-Hurler's syndrome, p. 130) shows features of a mucopolysaccharidosis.

References

Barrett, B., and Volwiler, W.: Agammaglobulinemia and hypogammaglobulinemia. J.A.M.A., 164:866, 1957.
Bigwood, E. J., et al.: Amino aciduria. Advances in Clinical Chemistry, 2:201, 1959.
de Bodo, R. C., and Altszuler. N.: The metabolic effects of growth hormone and their physiological significance. Vitamins and Hormones, 15:206, 1957.
Carson, N. A. J., and Raine, D. N., eds.: Inherited Disorders of Sulphur Metabolism. Edinburgh, Churchill Livingstone, 1971.
Cohen, S., and Milstein, C.: Structure and biological properties of immunoglobulins. In Dixon, F. J., and Kunkel, H. G., eds.: Advances in Immunology. Vol. VII. New York, Academic Press, Inc., 1967, p. 1.
Dorfman, A., and Matalon, R.: The mucopolysaccharidoses. In Stanbury, J. B., Wyngaarden, J. B., and Fredrickson, D. S., eds.: The Metabolic Basis of Inherited Disease. Ed. 3. New York, McGraw-Hill Book Co., 1972, p. 1218.
Edelman, G. M.: Antibody structure and molecular immunology. Ann. N. Y. Acad. Sci., 190:5, 1971.
Farmer, R. G., Cooper, T., and Pascuzzi, C. A.: Cryoglobulinemia. Arch. Intern. Med. 106:483, 1960.
Hobbs, J. R.: Immunoglobulins in clinical chemistry. In Bodansky, O., and Latner, A. L., eds.: Advances in Clinical Chemistry. Vol. 14. New York, Academic Press, Inc., 1971, p. 220.
Hsia, D. Y. Y.: Inborn Errors of Metabolism. Chicago, Year Book Publishers, Inc., 1959.
Levin, B.: Hereditary metabolic disorders of the urea cycle. In Bodansky, O., and Latner, A. L., eds.: Advances in Clinical Chemistry. Vol. 14. New York, Academic Press, Inc., 1971, p. 66.
Owen, J. A.: Paper electrophoresis of proteins and protein-bound substances in clinical investigations. Advances in Clinical Chemistry, 1:238, 1958.
Pollack, H., and Halpern, S. L.: The relation of protein metabolism to disease. In Anson, M. L, and Edsall, J. T., eds.: Advances in Protein Chemistry. New York, Academic Press, Inc,, 6:383, 1951.
Putnam, F. W., ed.: The Plasma Proteins. Vols. 1 and 2. New York, Academic Press, Inc., 1960.
Ruddy, S., and Austen, K. F.: Inherited abnormalities of the complement system in man. In Stanbury, J. B., Wyngaarden, J. B., Fredrickson, D. S., eds.: The Metabolic Basis of Inherited Disease. Ed. 3. New York, McGraw-Hill Book Co., 1972, p. 1655.
Saifer, A.: Rapid screening methods for the detection of inherited and acquired aminoacidopathies. In Bodansky, O., and Latner, A., eds.: Advances in Clinical Chemistry. Vol. 14. New York, Academic Press, Inc., 1971, p. 146.
Schneider, C. L.: Etiology of fibrinopenia. Ann. N. Y. Acad. Sci., 75:634, 1959.
Schultze, H. E., and Heremans, J. F.: Molecular Biology of Human Proteins. Amsterdam, Elsevier Publishing Co., 1966.
Waldenström, J. C.: Monoclonal and Polyclonal Hypergammaglobulinemia. London, Cambridge University Press, 1968.
White, A., Handler, P., and Smith, E. L.: Principles of Biochemistry. Ed. 4. New York, McGraw-Hill Book Co., 1968.
Wuhrmann, F., and Wunderly, C.: The Human Blood Proteins. New York, Grune & Stratton, Inc., 1960.

Chapter 4

NUCLEIC ACID METABOLISM

The importance of the nucleic acids and their derivatives is of quite a different sort from that ascribed to the other classes of metabolites. It is doubtful whether significant quantities of energy are derived from the catabolism of nucleic acids; neither they nor their constituents are dietary essentials. Rather, the chief role of nucleic acids and related compounds appears to be that of metabolic specialists, involved (1) in the control of the general pattern of metabolism through protein synthesis and (2) in aiding the catalysis of certain particular reactions. The former category includes the genes, bacterial "transforming factors," viruses, embryological "evocators," and the cytoplasmic regulators of protein synthesis; in the latter are the nucleoside triphosphates and the free "nucleotides" which act as coenzymes.

From the standpoint of protein chemistry, the nucleoproteins are a class of conjugated proteins, the prosthetic groups of which are the nucleic acids. These are conjugated with histones and other nuclear proteins. As is the case with many other types of conjugated proteins (lipoproteins, glycoproteins, etc.), the chemical and biological properties of the prosthetic groups are at the present time major objects of attention.

Nucleic acids are substances usually of very high molecular weight, containing phosphoric acid, sugars, and purine and pyrimidine bases. The structures of the two parent bases and the two pentose sugars are shown on page 236.

The pyrimidine bases are cytosine (2-oxy-4-aminopyrimidine), uracil (2,4,-dioxypyrimidine) and thymine, which is methyl uracil (5-methyl-2,4,-dioxypyrimidine). The purine bases are adenine (6-aminopurine) and guanine (2-amino-6-oxypurine). Two major chemical types of nucleoproteins have been found in nature, differing from each other in many respects. However, the difference which underlies the current system of nomenclature concerns the nature of the sugar contained in the nucleic acid moiety of the nucleoprotein. In one type of nucleoprotein the sugar is a pentose, which has been shown to be D-ribose. The nucleic acid containing this sugar is called ribonucleic acid (RNA), and the nucleoprotein is named accordingly.

Pyrimidine

Purine

β-Ribose
(β-D-ribofuranose)

β-Deoxyribose
(β-2-deoxy-D-ribofuranose)

The second type of nucleoprotein contains 2 deoxy-D-ribose (deoxyribose). The corresponding nucleic acid is called deoxyribonucleic acid (DNA).

In addition to the different sugars, the two types are distinguished by the pyrimidine bases which they contain; while both contain the purines, adenine (A) and guanine (G), and the pyrimidine cytosine (C), RNA in addition contains the pyrimidine uracil (U), and DNA the pyrimidine thymine (T). Unusual bases are found in certain types of RNA (p. 245).

RNA	DNA
H_3PO_4	H_3PO_4
Ribose	Deoxyribose
Adenine	Adenine
Guanine	Guanine
Cytosine	Cytosine
Uracil	Thymine (methyl uracil)

The nucleic acids consist of multiple *nucleotide* units (ribotides and deoxyribotides), each nucleotide containing phosphoric acid linked to the sugar which is, in turn, linked to a purine (purine nucleotide) or a pyrimidine (pyrimidine nucleotide) base. Being acid in nature, these are named, according to their base component, as follows: adenylic, guanylic, cytidylic, thymidylic, and uridylic acids, respectively. Removal of phosphoric acid by hydrolysis leaves the corresponding *nucleosides* (base plus sugar), also designated ribosides and deoxyribosides, according to the nature of the sugar component or, specifically, adenosine, guanosine, cytidine, thymidine, and uridine, respectively.

Biosynthesis. A key compound in the biosynthesis of each type of nucleotide is 5′-phosphoribosyl-1′-pyrophosphate (PRPP). In this, the pyrophosphate radicle replaces the OH group of ribose and is therefore below the plane of the five-membered ring. This is called the α-glycosidic linkage. For purine nucleotide synthesis, the pyrophosphate is enzymatically replaced, through the action of PRPP amidotransferase, by an amino group derived from the amide nitrogen of glutamine. With this there is a transformation to a β-glycosidic linkage; that is, the amino group is above the plane of the five-membered ring. The nitrogen atom is destined to become that in position 9 of the purine molecule which is built on to it in a series of enzymatic reactions, such that the C atoms in positions 2 and 8 are derived from formyl groupings, that in position 6 from "active" CO_2, those in positions 4 and 5 together with the N at position 7 from glycine, the N at position 1 is derived from aspartate and the N at position 3 from glutamine. This synthetic pathway, in the first place, gives rise to a nucleotide (inosinic acid), which contains the purine hypoxanthine (6-oxypurine). This nucleotide is readily converted to either adenylic or guanylic acids.

For pyrimidine nucleotide synthesis, PRPP reacts directly with orotic acid (6-carboxyuracil), so that the N atom at position 1 forms a β-glycoside linkage with the 1′C atom of the pentose. This gives rise to the nucleotide orotidylic acid, which can be converted biosynthetically to any of the naturally occurring pyrimidine nucleotides. In the pyrimidine ring, the C atoms at positions 4, 5 and 6, together with the N atom at position 1, are derived from aspartate. The C atom at position 2 is derived from CO_2 and the N atom at position 3 from the amide nitrogen of glutamine. In fact, the first step in biosynthesis of the ring is the same as that for the formation of urea, viz., the formation of carbamyl phosphate from glutamine and CO_2. This then condenses with aspartate to form orotic acid (orotate).

In the final nucleotides, therefore, the purine base is connected to ribose by a β-glycoside linkage between its nitrogen at position 9 and the 1′ carbon atom of the sugar, and the pyrimidine base is connected by the same linkage, which utilizes the N atom at position 1. There is a phosphate group in ester linkage with the sugar at the 5′ position. The ribose, in nucleotide form, can be reduced enzymatically to 2-deoxy-D-ribose. In this case, only one hydroxyl group at the 3′ position is left for further ester linkage. This is important in relation to the structure of DNA.

The first enzyme involved in the synthesis of the purine ring (PRPP amidotransferase) is inhibited by the finally produced purine-5′-ribonucleotides. This is a means of feedback control. There are other feedback mechanisms involved not only in pure nucleotide biosynthesis but also in that of the pyrimidine moieties.

There are other pathways for nucleotide biosynthesis. Purine nucleotides can be formed by the interaction of preformed purine (dietary or metabolic) with PRPP. This requires an enzyme known as a phosphoribosyl transferase; for example, adenine phosphoribosyl transferase brings about the formation of adenylic acid from adenine and PRPP. A deficiency of this enzyme has been described in man. There is another phosphoribosyl transferase, which gives rise to inosinic acid by reacting hypoxanthine with PRPP. A deficiency of this enzyme has also been described in man. It gives rise to excessive biosynthesis of purines and a severe neurological disorder. A further phosphoribosyl transferase can combine uracil with PRPP to give uridylic acid, which can in turn be converted to other pyrimidine nucleotides. Uridylic acid can also be formed by phosphorylation of the nucleoside uridine by an enzyme belonging to the phosphokinase group. There is evidence that in man this may be a very important pathway for pyrimidine nucleotide biosynthesis.

Nucleotides can combine with each other by sugar-phosphate diester linkages. In each nucleotide there is, as has already been pointed out, a phosphate group in ester linkage with the 5′C of the pentose. This leaves two free hydroxyl groups in the phosphate radicle, and one of these can form a second ester linkage with the 3′C of the pentose of another nucleotide to form a dinucleotide. In a similar fashion, several nucleotides can combine to form an oligonucleotide. A larger or very large number of nucleotides can combine to form a polynucleotide, which can be of very high molecular weight, indeed. In other words, we have a series of pyrimidine and purine bases attached to a backbone formed by sugar-phosphate diester linkages at the 3′ and 5′ positions of the pentose. This means there is an alternation of linkages at the 3′ and 5′ positions. In RNA the sugar of every nucleotide is ribose and in DNA it is deoxyribose. The various bases in these nucleic acids have already been indicated (p. 236). DNA in the nucleus is made up of two polynucleotide chains forming a right-handed double helix. The chains are held together by hydrogen bonding between a purine base of one chain and a pyrimidine base of the other. In fact, adenine (A) bonds with thymine (T), and guanine (G) with cytosine (C). The sugar-phosphate backbones of each helix are said to be antiparallel. In other words, if the sugar-phosphate linkages in one chain run in the sequence 3′ 5′ 3′ 5′ . . ., those of the other chain form the sequence 5′ 3′ 5′ 3′. . . . Analysis of DNA from a variety of sources has indicated that the number of molecules of adenine equals that of thymine, and the number of guanine equals that of cytosine. This is, of course, in accordance with the base pairs, which form hydrogen bonds with each other. The composition of DNA is virtually infinitely variable, since the purine and pyrimidine bases are not usually in any fixed sequence and the actual percentage content of the base pairs is variable from source to source.

The DNA of mitochondria is similar to that of bacteria. It is a single strand joined end to end and is described as having a "circular" structure.

RNA is single-stranded. It exists in a variety of forms, viz., messenger RNA (mRNA), ribosomal RNA (rRNA) and transfer RNA (tRNA). mRNA can form a hybrid double helix with a strand of DNA. In fact, it is now apparent that DNA forms the template for the biosynthesis of RNA. This is brought about by the action of an enzyme, DNA-dependent

RNA polymerase. Most of the total cell RNA consists of tRNA and rRNA. It is now believed that another form of RNA exists in the nucleus. This is known as heteronuclear RNA (HnRNA) and is thought by some authorities to play an important part in the control of gene expression.

In the test tube, a single strand of DNA can act as a template for the complementary DNA strand, which forms the double helix. For this purpose, the enzyme DNA polymerase is required. It is not now believed that this is the mechanism for increase of DNA prior to cell division, although DNA polymerase may possibly play some part in the repair of "tears" in single chains. Portions of the chain may be damaged for one reason or another, and there is undoubtedly a mechanism for removing this damaged region and replacing it in correct form.

DIGESTION AND ABSORPTION

The nucleic acids are split off from their associated proteins by the action of hydrochloric acid in the stomach, and pepsin commences the breakdown of protein, which occurs in the way already described (p. 148). The unchanged nucleic acids now enter the upper part of the small intestine, where they are acted upon by the enzymes ribonuclease and deoxyribonuclease present in the pancreatic juice. The former acts on RNA and liberates some pyrimidine mononucleotides and also breaks down the nucleic acids to shorter chains (oligonucleotides) with a terminal nucleotide residue, which is a pyrimidine nucleoside 3'-phosphate. The deoxyribonuclease breaks down the DNA to oligonucleotides with terminal nucleotides, which are nucleoside 5'-phosphates. Intestinal mucosal cells contain enzymes (phosphodiesterases), which are capable of producing breakdown to the stage of mononucleotides. There may also be an initial breakdown to very short chain nucleotides by other enzymes (nucleases). It is very likely that, at least after this latter stage, breakdown occurs within the intestinal mucosal cells. These contain nonspecific phosphatases, which hydrolyze the sugar-phosphate bonds and leave pyrimidine and purine nucleosides, which are then absorbed. Spleen, liver, kidney and bone marrow contain enzymes (nucleosidases), which can split off either the purine or the pyrimidine residue from the pentose.

INTERRELATIONS OF METABOLISM OF NUCLEIC ACIDS WITH OTHER FOODSTUFFS

As illustrated in Figure 4-1, the nucleic acids and their derivatives are related metabolically to a number of other foodstuffs. However, no obvious connections are found between the metabolism of lipids and that of nucleic acids, aside from the rather nonspecific contribution of lipids to metabolic CO_2.

The pathways of protein metabolism provide much of the raw material for the synthesis of purines and pyrimidines, viz., glycine, formate (C_1), and "NH_3" for the former, and oxaloacetate (from aspartate) and "NH_2" for the latter (in certain cases, formate also). Carbohydrate supplies oxaloacetate for the synthesis of pyrimidines and is also the ultimate source of the pentoses.

The "free" nucleotides, synthesized from the same raw materials as the "bound" nucleotides, exert a considerable influence on the metabolism of all classes of foodstuffs. As collecting, storing and transferring agents for high-energy phosphate, and as cofactors for biological oxidation and certain other types of reactions (p. 250), the free nucleotides of the tissues may control the rates of many reactions through variations in concentration.

The nucleic acids and nucleoproteins themselves, however, are probably much more important as agents controlling intermediary metabolism, although their exact roles remain speculative. They probably control the genetic pattern of potential enzymes and metabolic pathways in the cell at the nuclear level, and possibly regulate the final realization of that pattern in the cytoplasm. An abnormal variation of these functions may be seen in the mechanism of virus infections.

Thus, it is seen that a reciprocal type of relation exists between the nucleic acids (and derivatives) and other classes of metabolites; the latter provide the raw materials (and probably the energy) for the synthesis of the nucleic acids and derivatives which, in turn, exert an influence over the general pattern and rate of metabolism of the other foodstuffs.

BIOLOGICAL SIGNIFICANCE OF NUCLEIC ACIDS

Information accumulating in recent years has stimulated intense interest in the biochemistry of nucleic acids. The intracellular distribution of deoxyribonucleic acids (DNA) and ribonucleic acids (RNA) may be outlined as follows: (1) nearly all of the DNA is in the nucleus, but some (circular DNA) is present in mitochondria; (2) approximately 70 per cent of RNA of the liver cell is in the cytoplasm, much of it is in the ribosomes; (3) 30 per cent of the RNA is in the nucleus. The association of nucleic acids with chromosomes and viruses, and evidence of their influence in protein synthesis, indicates quite clearly that these substances are of fundamental biological importance. It seems desirable here to point out certain of the directions of current studies in this field, even though at present they have little direct clinical significance, except in relation to cancer chemotherapy.

Viruses. Several plant viruses have been shown to be crystallizable nucleoproteins, and many animal viruses are largely, if not entirely, nucleoprotein. Some (e.g., adenoviruses) contain only deoxyribonucleoproteins, others (e.g., Influenza virus) only ribonucleoproteins. It has been shown that if the crystalline tobacco mosaic virus material, nonliving by usual standards, is introduced into a healthy tobacco plant, the latter develops the disease in typical form. Moreover, the virus nucleoprotein can be subsequently

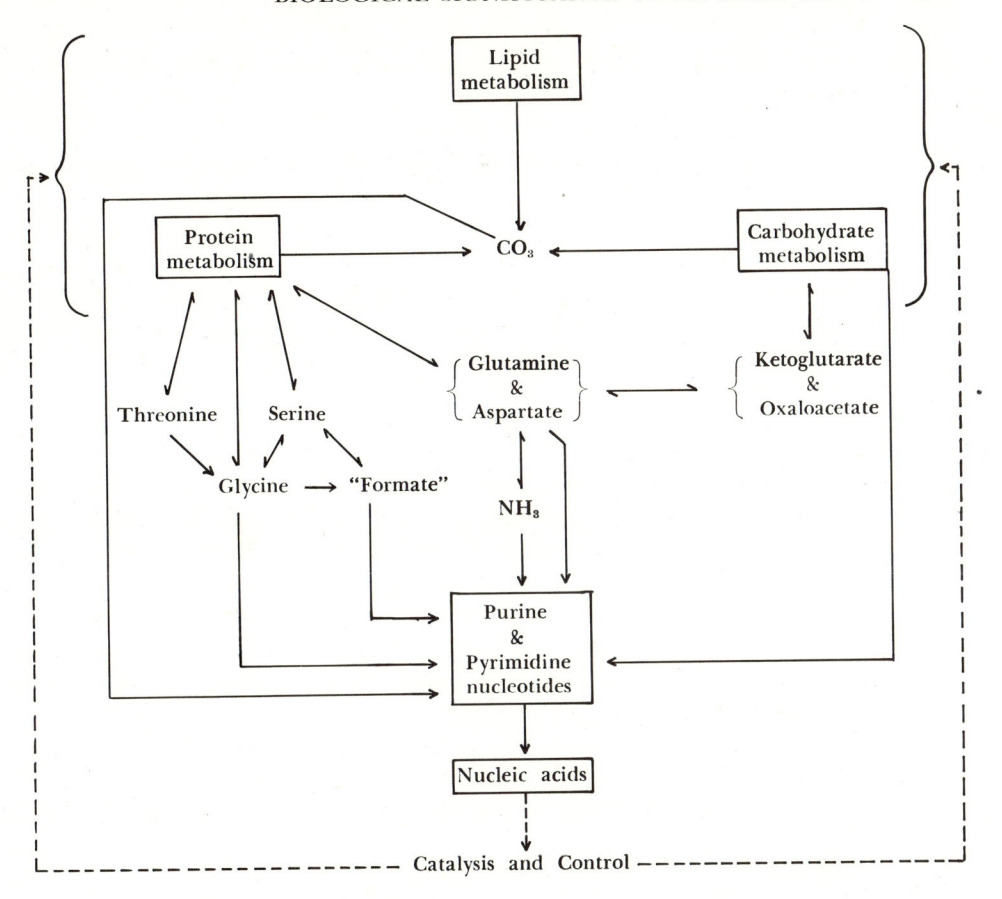

Figure 4-1. Metabolic interrelations of nucleic acids with other foodstuffs.

recovered in amounts greatly exceeding that originally introduced, i.e., it reduplicates (reproduces?) with progression of the disease.

Viruses vary widely in size and structural complexity, certain of the smaller ones exhibiting the physical and chemical behavior of molecules, whereas certain of the larger ones approach microorganisms in size and in structural and functional complexity.

Viruses on entering susceptible cells alter and direct certain metabolic sequences within the cell into new channels. Of particular significance is the fact that metabolic precursors are derived from the host cell for construction of the nucleic acid and protein molecules of multiplying virus. Synthesis of the host cell nucleic acids may actually cease, while the specific virus nucleic acid continues to accumulate, as in the case of bacteria infected by a bacteriophage, a bacterial virus.

Chromosomes—Genes. Present evidence indicates that chromosomes, of both spermatozoa and somatic cells, consist largely of deoxyribonucleoproteins. The obvious implication is that their component genes are made up largely of these substances which, therefore, may be regarded as the "genetic material," concerned in processes of cell reproduction. It is pertinent in this connection that the amount of DNA in spermatozoa containing the haploid number of chromosomes is one-half that of the somatic nuclei of the same species. Moreover, there is a quite striking, although not absolute, uniformity and constancy in the DNA content of the nuclei. The relative metabolic stability of DNA, except during cell division, also is consonant with its role as carrier of genetic information. The chromosomes contain DNA, and, in addition, ribonucleoprotein.

DNA is the bearer of genetic information in the cell. The spatial requirements for bonding between complementary bases in the double helix suggest that the sequence of nucleotides in one strand is determined automatically by that of the other. Thus, if a double helix untwisted, and each single strand attached to itself a new complementary strand of nucleotides, each of the resulting two double helices would be a duplicate of the original DNA, exactly the sort of replication to be expected of genetic material. There is experimental evidence that, in cell division, each DNA molecule is cleaved to two subunits, each incorporated into a different DNA molecule in the offspring; de novo synthesis produces the complementary strand paired to each of these molecules in the progeny. It is no longer thought that there is complete unwinding of the two strands. They are each so long that unwinding would be too time-consuming and complicated. It is now believed by some authorities that breaks occur at intervals along each chain. This allows small lengths to unwind and replicate their complementary component. The severed ends are united by an enzymic repair mechanism. The mechanism involved is not well understood.

Genes resemble viruses in at least two important characteristics: (1) they are autoreproducible, their exact duplication during mitotic division endowing each daughter cell with all of the specific hereditary biochemical potentialities and mechanisms of the original cell; (2) they are capable of undergoing mutation. In determining the biochemical characteristics of the cell, dependent largely on enzymes, the DNA-proteins are responsible for the synthesis of these substances. It was formerly believed that each gene "catalyzes" the production of a single enzyme ("one gene-one enzyme" hypothesis). Genetic evidence indicates that this is the case if the enzyme protein consists of one polypeptide chain, either single or combined with identical entities. Many enzymes are, however, made up of different polypeptide chains in combination with each other. In such cases, more than one gene is required for each enzyme molecule. These genes are not always close together. Where enzymes are related in a particular metabolic sequence, their genes may be connected together to form a so-called polycistron. The order of sequence of genes in this structure does not necessarily correspond to that in which the enzymes are utilized in the metabolic sequence.

It is now known that certain viruses produce an enzyme "reverse transcriptase," which is capable of producing DNA by using the viral RNA as a template. This DNA could now give rise to viral RNA through the action of the DNA-dependent RNA polymerase already present in the host cell. Alternatively, the DNA, or part of it, can become permanently incorporated in the host cell genome. Virus replication may not now occur, but there is a profound effect on host cell genome activity. This can alter certain characteristics of that cell, which is said to be transformed. Such transformation may play an important part in the production of malignant tumors.

The chromosomal DNA carries genetic information in the form of coded sequences of nucleotides, this coded information is passed on, via an intermediary RNA (mRNA) which passes from the nucleus into the cytoplasm.

"Transforming Substance" of Bacteria. Certain strains (e.g., Type 3) of pneumococcus possess capsules containing specific polysaccharides and others (e.g., Type 2) do not. This capacity for formation of a type-specific capsule is a hereditary characteristic. If one adds to an appropriate culture of a nonencapsulated strain an extract of an encapsulated strain, the former type is transformed to the latter, developing its characteristic capsular polysaccharide, as do its progeny subsequently. The active agent (transforming factor) is a deoxyribonucleic acid. It endows the bacterial cell with the capacity for synthesizing an enzyme or enzyme system that it did not previously possess, which, in turn, catalyzes the formation of the type-specific capsular polysaccharide. Inasmuch as this new function is subsequently transmissible, this may be regarded, in effect, as an example of induced mutation. Different specific transforming factors have been found for pneumococci, *Hemophilus influenzae,* meningococci, and *Escherichia coli.* In addition to the factors inducing type-specific encapsulation, others confer upon bacteria the ability to resist certain drugs or antibiotics or to exhibit nutritional characteristics not possessed by the organism before treatment. All appear to be deoxyribonucleates. It is perhaps of fundamental importance that these agents, as is true of genes and viruses, are reduplicated by the bacterial cell and, like viruses, can be recovered in large amounts from descendants of the cell into which they originally entered.

Transduction is a phenomenon somewhat related to bacterial transformation. In this case, bacteriophage, grown initially in bacteria having the ability to carry out a certain chemical reaction (e.g., synthesis of thymine), when infecting organisms lacking that particular chemical reaction, can induce catalysis of the previous absent process. Evidently the phage DNA transfers from the former to the latter organisms the "genetic" information necessary for establishing an otherwise absent metabolic pathway.

Role of Nucleic Acids in Mutation and Carcinogenesis. The fact that genes and viruses, largely if not entirely nucleoprotein in nature, possess in common the highly distinctive properties of autoduplication and susceptibility to mutation, raises the fundamentally important question of the relation of nucleic acids to these phenomena. The former has been referred to previously; genes and viruses are subcellular substances known to be capable of duplicating ("reproducing") themselves. They also can undergo mutation "spontaneously" and as a result of exposure to certain mutagenic agents. The latter include: (1) ultraviolet light, (2) ionizing radiation (e.g., x- and gamma rays and neutrons), and (3) certain chemicals (e.g., colchicine, nitrogen mustards, certain chemical carcinogens, e.g., methylcholanthrene).

This may have relevance to the problem of carcinogenesis which, according to some students of this subject, is one of induced mutation. This opinion is by no means unanimous. It may be significant in this connection, however, that such factors as ultraviolet light, ionizing radiations, and at least some chemicals that are mutagenic are also carcinogenic. Moreover, certain types of malignancy in animals are known to be due to viruses, which not only are largely nucleoprotein in nature, but also impose upon the invaded cell their own specific nucleic acid synthesizing capacities. These observations suggest that the induction of malignancy may be related fundamentally to the development of a changed type of nucleic acid activity in the cell.

Role of Nucleic Acids in Protein Synthesis. Growth and maintenance of the organism require continual synthesis of protein molecules within each cell. The mechanism of protein synthesis involves enzymes, which are themselves protein in nature. The cytoplasmic enzyme pattern is determined largely genetically, i.e., by the DNA (genes) in the nucleus.

RNA is present in greatest abundance in the cytoplasm of actively growing cells (e.g., embryonic, tumor, tip of plant root) and in cells actively engaged in the production and excretion of proteins (e.g., pancreatic acinar cells). Increased protein synthesis, e.g., stimulation of pancreatic secretion by pilocarpine, is accompanied by an increase in RNA. Conversely, when production of protein by liver cells is inhibited by starvation, their RNA content decreases simultaneously.

It is therefore apparent that RNA synthesis is related to protein synthesis. This is dependent on the well-known triplet code for amino acids. Since mRNA is formed on DNA as a template, the code message obviously has its origin along one of the two chains of the DNA double helix. It would appear likely that only one of these two chains is active in the production of RNA, the nonactive chain suffices to reproduce the active partner for cell division, in accordance with the mechanism already described. Since there are four bases in DNA, it is theoretically possible to obtain 64 different combinations if the bases are taken three at a time. As there are only 20 different amino acids to be coded, it is not surprising that each amino acid is coded for by more than one triplet combination. In fact, 61 of these triplet codons correspond to amino acids and three are so-called nonsense codons. The latter are associated with the termination of polypeptide chains on ribosomes, as will be described later. This coding is transferred on to the transcribed mRNA chain, on which, however, uridine (U) takes the place of thymine (T). Thus, A on the DNA chain forms a complementary base pair with U on the mRNA chain, G with C, C with G and T with A. For example, the triplet of bases ATG on DNA would give rise to UAC on the mRNA chain. It will be remembered, however, that the chains of a double helix run in opposite directions. In other words, if one reads each triplet in the same direction, the original ATG of DNA would correspond to CAU (i.e., UAC reversed) of mRNA.

In terms of the mRNA chain, we now know the codons for all the amino acids, as well as the three terminating (or nonsense) codons. The code itself is read eventually (at the ribosomes) in the direction $5' \to 3'$. This is also the direction of growth of the mRNA chain, which starts at its 5' end on the DNA template. The latter also contains start and stop signals, which determine the commencement and termination of the RNA chain itself. The starting signal is recognized by the sigma factor, which is a subunit of DNA-dependent RNA polymerase. Removal of this factor *in vitro* does not interfere with the activity of the enzyme, but synthesis of RNA starts anywhere on the DNA chain and not only at the regular starting signals. The stop signals are recognized by a specific protein known as the rho factor which has a molecular weight of about 60,000. The sigma factor has a molecular weight of 90,000. There is therefore a replication mechanism for the production of mRNA chains, each of which is defined in respect to length and base sequence. Since each carries a coded message for protein synthesis, they belong to the class of RNA known as messenger RNA (mRNA). mRNA molecules have very varying chain lengths. This would be expected from the fact that they code for a large variety of different polypeptide chains. It is also known that some code for more than one peptide and are therefore of great length. These polypeptides probably always have related functions. An mRNA molecule, for example, exists, which codes for ten different enzymes required for the biosynthesis of histidine.

In the cytoplasm, there is another type of RNA, which is of smaller molecular weight, approximately 25,000, and contains some 80 nucleotides linked together in a single chain. This is transfer RNA (tRNA) to which amino acids become attached, so that they can be utilized for polypeptide synthesis. In addition to the bases A, G, C and U, tRNA contains a number of unusual bases, such as pseudo-uridine, ribothymidine, inosine, dihydrouridine, methyl- and dimethylguanosine and methylinosine. These unusual bases occur in varying proportions in different tRNA molecules. There is a specific tRNA molecule for each of the 20 amino acids. In each of these, the more commonly occurring bases (A, C, G and U) form hydrogen bonds among themselves and so distort the chain into a shape roughly resembling a clover leaf. The 5' and 3' ends lie alongside each other in the stalk, but the 3' end somewhat overlaps. This overlapping portion consists of a terminal A, then a C and then another C, and finally a fourth nucleotide. The ACC sequence at the 3' end is present in all the different types of tRNA. Each amino acid becomes attached by its carboxyl group to the terminal A. The unusual bases are present in the loops, which resemble the three portions of a clover leaf. There is no hydrogen bonding in these portions. The middle portion contains a base triplet, known as the anticodon, which binds by base pairing to a codon of mRNA. There is a minor loop between the middle loop and that nearest to the 3' end. The anticodon for each tRNA is such that it will form three base pairs, by hydrogen bonding, with the mRNA codon corresponding to the specific amino acid for the particular tRNA molecule. The amino acid becomes attached by the action of a specific enzyme system for each amino acid. The tRNA not only acts as an adaptor to the appropriate codon, but also activates its attached amino acid so that it is available for peptide bond formation.

mRNA molecules pass out of the nucleus and then become associated with the ribosomes, which are in fact tiny protein factories. Each ribosome consists of two subunits, one of which is about twice as large as the other. In higher organisms (eukaryotes), the whole ribosome has a sedimentation constant of 80 S in the ultracentrifuge and the subunits 60 S and 40 S, respectively. From electron microscope evidence, it appears that the smaller subunit is attached off center to the larger. There is also a cleft in the smaller entity and it is possible that mRNA becomes attached in this cleft and passes through it. In other words, the ribosome passes along the whole mRNA chain and reads its code, thereby giving rise to a polypeptide chain by the mechanism to be described. More than one ribosome may be attached to the same mRNA chain, forming a so-called polyribosome. In higher organisms, each of the ribosome subunits consists of RNA and protein in approximately equal amounts. In *E. coli*, the smaller subunit contains 19 different proteins, and the larger, 35. The components of the smaller subunit have been completely separated *in vitro* and reconstituted to the particle, which is active in protein synthesis. It appears that the ribosomal subunits separate when they come to the end of the mRNA chain. For the initiation of protein synthesis in higher organisms, the latter forms a complex with the small ribosomal unit and "activated" methionine (methionyl-tRNA). The larger subunit now attaches to form a complete ribosome. For each polypeptide to be synthesized, there is only one point of initial attachment on the mRNA chain. This contains the base sequence, which codes for methionine as well as the mechanism for attachment of the smaller ribosome subunit.

The amino acids attached to their adaptor tRNA molecules diffuse to the ribosomes and are available for protein synthesis on the ribosomal surface. This possesses two binding sites for tRNA, the so called peptidyl ("P") and aminoacyl ("A") sites. As already indicated, a methionyl-tRNA is the first aminoacyl-tRNA to attach itself. This is at the "P" site. The ribosome now moves to the next codon of mRNA, and the corresponding aminoacyl-tRNA becomes attached to the "A" site. The methionyl group at the "P" site is transferred from its tRNA and forms a peptide bond with the carboxyl group of the amino acid at the "A" site, which now contains a dipeptidyl-tRNA. The tRNA at the "P" site becomes detached and is replaced by the dipeptidyl-tRNA. This leaves the "A" site vacant. At this stage, the ribosome has moved on to the next codon of the mRNA and a third aminoacyl-tRNA is attached. This, of course, corresponds to the mRNA codon and can give rise to a tripeptidyl-tRNA at the "P" site. By repetition of this process, an appropriate number of amino acids can be linked together to form the appropriate polypeptide coded by the mRNA cistron. At this stage, a terminating (nonsense) codon occupies the "A" site, which triggers the mechanism for releasing the completed polypeptide chain. From the description given, it appears that there is always an N-terminal methionine residue in every protein. In actual fact, the N-terminal methionine residue is usually split off the growing polypeptide chain, while it is still attached to the ribosomal surface. It also appears that the methionine tRNA, which initiates polypeptide synthesis, is not the same tRNA that usually allows incorporation of methionine into the chain.

It is an interesting observation that the polypeptide chain has much of its final three-dimensional shape, while still attached to the ribosomal surface. When released, it immediately achieves its appropriate tertiary structure. It can also associate with other, not necessarily different, polypeptides to form the quaternary structure of protein molecules. One interesting fact that has emerged is that a polypeptide chain may be partially split after release from the ribosome. For example, proinsulin, a continuous peptide chain, is split partially to form insulin, which appears as two peptide chains linked by disulfide bridges. This splitting mechanism is frequently a factor in the production of an active enzyme molecule from an inactive precursor and is also a factor in the formation of fibrin from fibrinogen.

It has already been pointed out that mRNA can carry the code message for more than one polypeptide. It is then described as polycistronic, since each single message occupies a portion of mRNA referred to as a cistron. The various stages of protein synthesis are brought about by enzyme systems. For example, a protein transfer factor is required for the initial attachment at site "A" and another protein transfer factor (translocase) for transferring peptidyl-tRNA to the "P" site. In each case, energy is derived from guanosine triphosphate (GTP). Peptide bond formation is brought about by peptidyl transferase, which is one of the proteins of the larger ribosomal subunit. The mechanism for moving the ribosome along the mRNA is not yet clear.

A number of antibiotics are capable of inhibiting various steps in protein synthesis. Puromycin interrupts chain elongation. It occupies the "A" site and peptidyl transferase can transfer the growing chain on the "P" site to the puromycin forming puromycin terminated chains, which then drop off. Fusidic acid inhibits translocase. Sparsomycin and lincomycin bind to the larger ribosomal subunit and specifically inhibit peptidyl transferase. Streptomycin binds to the smaller ribosomal subunit and inhibits chain initiation. Actinomycin D forms complexes with guanine bases of DNA and so prevents formation of RNA.

All the different types of RNA are formed on a DNA template. In fetal and rapidly dividing tissues, RNA may act as a template for DNA through the action of the enzyme RNA-dependent DNA polymerase (reverse transcriptase).

mRNA can easily be detected in the cytoplasm of higher organisms.

Nevertheless, mRNA is not detectable in the nucleus, where it must be formed. There is, however, a good deal of RNA in the nucleus and some authorities believe that it is involved in control mechanisms affecting gene activity. It never leaves the nucleus and is synthesized and broken down very rapidly. RNA is, therefore, very heterogeneous (heteronuclear RNA or HnRNA). It is believed that mRNA must, however, be hidden among the HnRNA. Evidence for this has recently been forthcoming. It seems that, in higher organisms, mRNA has added to it a repetitive sequence of adenine nucleotides, which may even achieve a length of 200 bases. This is referred to as a poly A stretch. The poly A is apparently added to the 5' end (the beginning of the message) after the mRNA is formed on the DNA template. The poly A sequence is also detectable in HnRNA. It could be that attachment of this sequence to mRNA is necessary for its escape from the nucleus; otherwise, the mRNA is destroyed. It has also been suggested that the poly A sequence may also control the number of times mRNA is utilized by ribosomes before it is eventually destroyed.

Higher organisms also contain much more DNA than is necessary to code for all cellular proteins. Crick has recently suggested that the very small proportion of DNA used as templates is single-stranded and that the remainder is mostly double-stranded and coiled into superhelices. Their state of coiling determines the availability of single strands, possibly obtained by unwinding of double strands. He has suggested that the coiled shape of the superhelices is determined by the nuclear histones. In other words, most of nuclear DNA consists of spring-like control mechanisms for revealing or obscuring single-stranded template DNA. It is obviously the ultimate orderly control of such mechanisms which determines cellular differentiation, and their disturbance could well result in transformation, possibly of a malignant nature.

In bacteria, a *regulatory gene* gives rise to a protein known as a *repressor*. This can bind to a special binding site on DNA, *the operator*, and stop the initiation of transcription of mRNA molecules corresponding to the nearby sequence of DNA nucleotides, known as *the operon*, which may consist of one or more genes. In *E. coli,* there is a protein, which has been isolated and sequenced and which inhibits the formation of the enzyme β-galactosidase. This is a typical repressor. A β-galactoside combines with this repressor and prevents it from binding to the operator. The β-galactoside therefore causes further synthesis of β-galactosidase. It belongs to the class of substances known as *inducers*. On the other hand, repressors are probably inactive when they are formed and must combine with another substance, *a co-repressor,* before they can exert their inhibitory influence at the operator site. The site where RNA polymerase binds is known as *the promoter,* which is on one or other side of the operator. In the case of the β-galactosidase gene, it is between the gene and its operator. The regulatory gene is situated some distance from the operon.

It is possible that the bacterial type of mechanism occurs in eukaryotes, but, as has already been indicated, the gene control mechanisms of these higher organisms are considerably more complicated than those of the bacterial cell.

CANCER CHEMOTHERAPY

Anticancer agents act essentially by interfering with the processes involved in protein synthesis. They may act by inhibiting the synthesis of DNA or mRNA, by interfering with the attachment of amino acids to tRNA or by interfering with the ribosomal mechanism of peptide chain elongation (e.g., puromycin). All these processes interfere with the production of the enzymes, which are in turn responsible for the synthesis of DNA, RNA or their precursors. All known anticancer agents are toxic to normal tissues. It is believed that the more successful of them can be more easily destroyed in normal cells or possibly have more difficulty in entering them. There is, however, virtually always a fine balance between action on cancer cells and action on normal tissues.

Antimetabolites. These are substances which interfere with the formation of nucleotides. Usually they are so similar in structure to normal enzyme substrates that they combine with appropriate enzymes, and further biosynthesis is inhibited.

The earliest stages at which this can happen are synthesis of the purine ring, which requires a derivative of folic acid at two stages, and the methylation of the pyrimidine uracil to form thymine, which also requires a derivative of folic acid. Folic acid antagonists can be made by replacing the hydroxyl group at position 7 (see p. 817) by an amino group (aminopterin) and, in addition to this, methylating the nitrogen at position 10 (methotrexate). The latter compound has been particularly useful in the treatment of acute leukemia and of choriocarcinoma.

6-Mercaptopurine (6MP) has a sulfhydryl group in the same position as the hydroxyl group of hypoxanthine. It is readily converted in the body to the corresponding nucleotide, which closely resembles inosinic acid. It then combines with the enzymes required to convert inosinic acid into adenylic and guanylic acids. In this way, two important nucleotides are made unavailable for DNA synthesis, which is thus prevented. 6MP has also been widely used for the treatment of acute leukemia.

Thioguanine and thioguanosine have shown some activity against chronic myeloid leukemia.

Another antimetabolite, which has proved effective clinically, is the pyrimidine derivative 5-fluorouracil. This has produced regression in a number of solid tumors, including lymphadenoma and carcinomas of the cervix, breast, colon and rectum. The related nucleoside 5-fluoro-2′-deoxyuridine is said to be more effective, especially in carcinomas of the colon and rectum.

A large number of antimetabolites have been prepared and successfully tested in relation to experimentally produced tumors in animals. Many have proved to be too toxic and sometimes ineffective in clinical trials. It is interesting, but perhaps not unexpected, that such studies have led to effective antiviral agents such as 5-iodo-2′-deoxyuridine, which has proved to be so useful in the treatment of herpes simplex.

Alkylating Agents. Research on mustard gas showed that it reduced the white blood cell count. This led to the production of a number of less toxic derivatives for the treatment of leukemia, myelomatosis, the reticuloses and Burkitt's lymphoma. These include nitrogen mustard and a number of related derivatives such as Natulan, Myleran and cyclophosphamide. These act by attaching alkyl groups to the DNA molecule, thus reducing its ability to act as a template either for itself or for RNA. Some alkylating agents must first be activated by intracellular enzymes.

Other Anticancer Substances. It has already been pointed out that actinomycin D prevents the formation of mRNA. It has therefore proved to be very effective in the treatment of choriocarcinoma and, combined with radiotherapy, in the treatment of Wilms' tumor.

The alkaloid vincaleukoblastine prevents the attachment of amino acids to tRNA. It has been used successfully in the treatment of choriocarcinoma and has produced remissions in certain brain tumors occurring in children. It has also been used in the treatment of acute leukemia in childhood.

Some tumor cells are unable to synthesize asparagine, which is an essential requirement for them. It is therefore obtained from the amino acid which is presented in the patient's circulation. Intravenous injection of purified asparaginase will destroy circulating asparagine and so cause death of the tumor. Not all cancers have an essential requirement for asparagine. Asparaginase therapy has not proved as promising as was originally hoped.

Drug Combinations. Much more successful results have been obtained in childhood acute leukemia by using combinations of alkylating agents, antimetabolites, vincaleukoblastine, asparaginase and prednisone. It appears that complete cures are now possible.

FREE NUCLEOTIDES OF BIOLOGICAL IMPORTANCE

In addition to the nucleotides which are components of nucleic acids, a number of substances possessing nucleotide structure exist in the free state in the tissues. They have catalytic functions, and act in conjunction with enzyme systems.

Nucleoside Triphosphates. This designation is applied to compounds such as that containing adenine and ribose (adenosine), plus one, two or three phosphate groups, i.e., adenosine mono- (AMP), di- (ADP), and triphosphate (ATP), respectively. The latter two serve as carriers and transfer agents for high-energy phosphate groups, a function intimately bound up with the energetics of all living cells (p. 6). Similar compounds are obtained by replacing the adenine by guanine, uracil or cytosine. The part played by GTP in protein synthesis has already been indicated; UTP is necessary for the synthesis of glycogen and for the conjugation of bile pigment; CTP is utilized in the synthesis of phospholipids. There are, of course, many functions for each of the nucleoside triphosphates.

Coenzymes. Certain enzymes require for their function the presence of certain organic compounds, rather loosely attached to, and therefore separable from, the enzyme protein. These are called coenzymes, certain of which contain nucleotide units or are structurally related to them. These include nicotinamide-adenine dinucleotide (NAD) and nicotinamide-adenine dinucleotide phosphate (NADP), flavin-adenine dinucleotide (FAD), and coenzyme A, which contain the B vitamins, nicotinamide, riboflavin and pantothenic acid, respectively. These coenzymes are of fundamental importance in biological oxidations, which constitute the core of the energy-producing mechanisms of the body.

Cyclic Nucleoside Monophosphates. Two nucleoside monophosphates in cyclic form (cyclic adenosine monophosphate and cyclic guanosine monophosphate) are of great importance in many biological activities.

The role of cyclic adenosine monophosphate in relation to synthesis and breakdown of glycogen has already been discussed (p. 10). It is concerned with other aspects of carbohydrate metabolism. It has also been shown to be active in relation to the release of hormones (insulin, ACTH, TSH, GH, thyroid hormone and calcitonin) as well as to their activity on target organs (for example, parathormone and renal tubules). It is now recognized as "the second messenger" for a number of hormones. The substance is also of great importance in relation to protein and fat metabolism, including steroidogenesis. It is concerned with the release of parotid and pancreatic amylases, as well as with the secretion of gastric HCl. It is also involved in the synthesis of DNA and mRNA. Its intracellular effects are profoundly modified by the prostaglandins (p. 95). Many other biological functions of this substance have been discovered and there is little doubt that further developments will occur. It is necessary to mention, however, that cyclic AMP plays an important part in cell membrane permeability as well as in the inhibition of cell growth. It also plays an important part in the central nervous system and in the activity of all forms of muscle.

To date, perhaps the most important function of cyclic GMP (cyclic guanosine monophosphate) is in relation to the acceleration of cell growth. It would seem that an increase in the ratio of cyclic GMP to cyclic AMP results in increased cell growth and division, whereas a decrease in the ratio results in decrease of these functions. There seems little doubt that these effects on growth processes are of the greatest importance, and many believe that study of these effects will shed some light on the mechanism of cancer production.

INTERMEDIARY METABOLISM

For convenience, the metabolism of the nucleic acids may be subdivided into (1) the metabolism of components, (*a*) non-nitrogenous and (*b*) nitrogenous, and (2) the metabolism of the macromolecules.

Phosphate is readily obtained from the diet and endogenous sources. It is excreted in the urine as inorganic phosphate at the end of its metabolic career.

Ribose, in the form of ribose-5-phosphate, is derived from glucose-6-

phosphate via the pentose phosphate pathway. PRPP (p. 237) is now formed and utilized for the synthesis of nucleotides. It has already been pointed out that deoxyribose is derived from ribose after the latter has been incorporated in a nucleotide.

The nucleic acids are first broken down by ribonucleases and deoxyribonucleases. The former are more active than the latter since RNA is constantly degraded and resynthesized in one form or another. Deoxyribonucleases, however, may be involved in the repair of DNA, which has been damaged by a variety of causes, including radiation. The enzyme system appears to be active in removing abnormal nucleotides so formed and allows a repair enzyme system to reinsert the correct entities.

The nucleases (mainly RNase) give rise to oligonucleotides and some free 3'-mononucleotides. The former are further broken down by phosphodiesterases to give free 5'-mononucleotides.

The mononucleotides can be utilized for further synthesis of nucleic acids or broken down along two different enzyme pathways. The nucleotide pyrophosphorylases yield PRPP and purines and pyrimidines. The former may be utilized for further nucleotide synthesis or may enter other metabolic pathways and be converted to ribose 5-phosphate. The nucleotidases break down the nucleotides to nucleosides, which are further broken down by nucleoside phosphorylases to yield the purines and pyrimidines and ribose-1-phosphate. The latter is converted by phosphoribomutase to ribose-5-phosphate, which can re-enter the pentose-phosphate pathway and give rise to fructose-6-phosphate and glyceraldehyde-3-phosphate, both of which can be further metabolized by the glycolytic pathway.

Purines and Pyrimidines. Purines liberated by metabolic splitting of nucleic acids and nucleotides are catabolized directly to waste products or may be incorporated into nucleic acids. The purine bases (adenine and guanine) formed in the course of catabolism of nucleic acids and free nucleotides undergo oxidation to uric acid, which is the chief end-product of purine catabolism in man and the higher apes. Other mammals degrade uric acid to allantoin by means of the enzyme, uricase, which is lacking in primates. Adenine (6-aminopurine) is deaminated while still present in adenylic acid. This gives rise to inosinic acid, in which the adenine has been converted to hypoxanthine (6-oxypurine). The deamination is brought about by the enzyme adenylate deaminase, which is quite abundant in skeletal muscle. Adenosine can also be deaminated to give rise to inosine. Both inosinic acid and inosine give rise to free hypoxanthine, which may be re-utilized for nucleic acid synthesis but is most frequently further oxidized to xanthine (2,6-dioxypurine) by the enzyme xanthine oxidase, present in greatest amount in liver and intestinal mucosa. The same enzyme further oxidizes xanthine to uric acid (2,6,8-trioxypurine).

Free guanine (2-amino-6-oxypurine) is deaminated to form xanthine directly. This requires the enzyme guanase, which is very active in most tissues. The liberated xanthine is then converted to uric acid by xanthine oxidase.

It is a matter of some interest that ingested purines, which can be utilized for nucleic acid synthesis, are mostly broken down to uric acid. This is of importance in relation to gout.

The pyrimidine bases give rise to carbamyl derivatives of β-alanine and β-amino-isobutyric acid, which then give rise to malonic and methylmalonic acids or their combinations with CoA, as well as to urea.

Uric Acid. From the clinical standpoint, interest is centered particularly in uric acid. It passes into the blood stream and is normally excreted mainly in the urine, to a lesser extent (about 100 mg. daily) in digestive fluids, and in small amounts in sweat and saliva. Despite some contradictory data, there is no unequivocal evidence that uric acid is destroyed in human tissues apart from leukocytes, which give rise to small amounts of allantoin. However, a variable, and in some circumstances a considerable, portion of the total body uric acid, entering the intestine in the digestive fluids, may be destroyed by bacterial action. This process of intestinal uricolysis gives rise to urea and ammonia, which are absorbed and excreted by the kidneys.

Variations in the concentration of uric acid in the blood plasma are due chiefly to variations in the rate of either its formation or its excretion (kidneys).

Under conditions of normal production and removal, the body contains a "readily miscible uric acid pool" of about 1.2 grams. Fifty to 75 per cent of this (550 to 850 mg.) is turned over daily, representing the rate of formation and disposal of uric acid in normal subjects.

Blood Uric Acid. The normal urate content of plasma or serum is 2.5 to 7.0 mg./100 ml. for males and 1.5 to 6.0 mg./100 ml. for premenopausal females when determined by phostotungstic reagent. With uricase methods, the corresponding figures are 3.8 to 7.1 mg./100 ml. and 2.6 to 5.6 mg./100 ml., respectively, and the range for children is 2.0 to 5.5 mg./100 ml. Saturation of plasma with monosodium urate occurs at a concentration of about 7 mg./100 ml. This means that hyperuricemia indicates supersaturation. Apparently, plasma forms relatively stable supersaturated solutions of this compound. Just over one-third of it is loosely bound to plasma proteins—mostly albumin—but some is bound to an α_1-α_2 globulin. Because the disease gout requires supersaturation, it is much more common in males. Only about 5 per cent of gouty patients are female and most of them are menopausal. This is not surprising, since the plasma urate level at the menopause is not significantly different from that of the male. All determinations should be made in plasma or serum. The plasma urate level is increased by ingestion of purine-rich foods.

Urine Uric Acid. The quantity of uric acid excreted in the urine varies with the purine (nucleoprotein) intake. On a diet very low in nucleoprotein, normal adults excrete less than 450 mg. daily, representing mainly the catabolism of endogenous nucleic acids and nucleotides. The amount varies also with the protein and caloric content of the diet. A high protein and caloric intake results in an increase, and a low protein and caloric intake in a decrease, in urine uric acid. On a high purine diet (meat, liver, kidney, sweetbreads, leguminous vegetables) the output may rise to 1 g. daily; on a mixed diet the average daily excretion is about 0.7 g. Caffeine and other methylpurines (in tea, coffee, cocoa) cause little or no increase. During the first few days of life the urine contains relatively large quantities of uric acid, constituting 7 to 8 per cent of the total nitrogen (1 to 2 per cent in adults). This may lead to the development of "urate infarcts" in the kidneys of infants because of the relative insolubility of uric acid in an acid medium.

In normal subjects, the amount of urate ions excreted in the urine is about 5 to 10 per cent of that present in the glomerular filtrate and a goodly proportion is present as uric acid. As in the case of potassium, the urate of the glomerular filtrate is virtually completely reabsorbed in the tubules under normal conditions, that leaving the body in the urine representing active tubular excretion of urate. Reabsorption is accomplished by an active transport mechanism, a fact of importance in connection with the action of certain uricosuric agents. Such substances as salicylates (large doses), cinchophen, carinamide and probenecid, sulfinpyrazone, and zoxazolamine, which increase urinary elimination of uric acid, produce this effect by inhibiting its reabsorption in the renal tubules, presumably by blocking the enzymatic transport mechanism. The uric acid clearance increases accordingly under the influence of these agents, and the blood level may fall (e.g., in gout). Carinamide and probenecid also block tubular excretion of certain substances (e.g., penicillin, p-amino-salicylic acid, p-aminohippurate, PSP, and Diodrast), and have been employed for the purpose of maintaining therapeutically effective plasma levels of penicillin and p-aminosalicylic acid for prolonged periods. ACTH and the adrenocortical oxysteroids also increase the urinary excretion of uric acid, due apparently to inhibition of renal tubular reabsorption. There is evidence which indicates that there is also an active tubular excretory mechanism. This is inhibited by low doses of salicylate and other uricosuric agents. The clearance of urate by the kidney increases with the plasma urate level. There is a rough linear relationship up to a plasma concentration of 11 mg./100 ml. Above this level, the clearance increases quite rapidly. This is probably because the secretory mechanism becomes more effective.

The urinary excretion is inhibited by certain diuretics (chlorothiazide, ethacrynic acid, triamterene, etc.), lactate, β-hydroxybutyrate and para-aminohippurate. The effect, in this respect, of low dosage of uricosuric agents (salicylate, probenecid, sulfinpyrazone) has already been mentioned. Excretion is also lowered by substances which diminish renal circulation, e.g., catecholamines and angiotensin.

Human urine dissolves two or three times as much uric acid as do aqueous solutions at comparable pH. Urine apparently contains a number of solubilizing substances. These include urea, various proteins and a mucopolysaccharide.

Determination of uric acid excretion in disease is of little practical importance. Changes in gout are indicated elsewhere (p. 256). It may be increased in patients with leukemia (especially chronic myeloid) and polycythemia, presumably because of increased rate of turnover of blood cell nucleic acids. It may also increase during early periods of remission in pernicious anemia, especially if induced by specific therapy.

Hyperuricemia. Theoretically, increase in blood uric acid may be due to one or more of the following factors: (*a*) overproduction; (*b*) decreased destruction; (*c*) decreased renal excretion. Apart from the leukocytes, urate is not destroyed by human tissues; however, considerable amounts, entering the bowel in the digestive fluids, undergo degradation by intestinal bacteria. Decreased destruction does not seem to play any part in the production of hyperuricemia.

(1) **Decreased Excretion.** Increased values are observed in all forms of nephritis with nitrogen retention. Retention of uric acid may precede that of urea. Values as high as 10 mg./100 ml. are frequently observed; figures above 15 mg. are unusual.

Hyperuricemia may occur, in association with retention of other urinary constituents, in conditions associated with urinary obstruction (prostatic, bilateral ureteral calculus, etc.), urinary suppression (acute tubular necrosis, shock, intestinal obstruction, etc.), destructive renal lesions (tuberculosis, pyelonephritis, hydronephrosis, polycystic kidney, etc.), renal hypogenesis and congestive heart failure.

Increase in serum uric acid may occur in eclampsia, in the absence of retention of urea. In this condition there may be decrease in uric acid clearance, with no associated abnormality in other functions of the kidney. On this basis, the simultaneous presence of decreased uric acid clearance and normal urea clearance may aid in differentiating eclampsia from glomerulonephritis complicating pregnancy.

(2) **Hematological Disorders.** The blood uric acid increases commonly in chronic leukemia, particularly myeloid, and also occasionally in polycythemia and multiple myeloma. This is probably a reflection of the greatly increased production of uric acid resulting from the enormous increase in the number of short-lived cells, i.e., increase in nucleic acid turnover and purine catabolism. A similar mechanism probably operates to produce the increase in blood uric acid which occurs early in the course of remissions in pernicious anemia, usually preceding by a day or more the increase in reticulocytes. It may occur also in other blood disorders, e.g., sickle cell anemia, thalassemia, hemolytic anemias, myeloid metaplasia, lymphoblastoma, macroglobulinemia. Renal involvement may contribute to the increase in multiple myeloma.

(3) **Chronic Lead Poisoning.** Hyperuricemia is observed as a late effect in chronic lead poisoning. This is associated with renal damage.

(4) **Hepatic Disease.** Slight elevations of blood uric acid observed occasionally in patients with hepatic disease are probably due to complicating renal functional impairment. Low values have been observed in severe cases of acute hepatic necrosis, suggesting that the liver is a major site of normal production of uric acid in man; it is the main site in birds.

(5) **Gout.** Primary gout is an arthritis characterized by a derangement

of purine metabolism, occurring predominantly in males, the most useful reflection of which, from the standpoint of early diagnosis, is elevation of the serum uric acid concentration. Hyperuricemia not attributable to such causes as renal failure, leukemia, polycythemia, multiple myeloma, etc., should be regarded as highly suggestive of gout, until proven otherwise. The hyperuricemia of primary gout is due both to excessive production of purines and to renal retention of uric acid. The excessive production has been demonstrated by studying uric acid output in the urine on a low purine diet and by implementing tracer studies. Compared with normal individuals, gouty patients tend to show a decreased renal clearance of uric acid with respect to inulin or creatine. About three-quarters of adults with primary gout excrete normal amounts of uric acid; the remainder excrete excessive amounts. Those who excrete normal amounts usually show overproduction of uric acid (excess purine synthesis and breakdown) with underexcretion in relation to inulin. Those who overexcrete uric acid show mainly overproduction.

Excessive purine synthesis has on rare occasions been found to be due to deficiency of hypoxanthine-guanine phosphoribosyl transferase or to defective feedback inhibition of PRPP amidotransferase. Excessive incorporation of ^{15}N from glycine has been demonstrated. This involves the 3 and 9 nitrogens of the purine ring. These are normally derived from glutamine, which gives rise to the suggestion of an abnormal glutamine metabolism in gout. Studies of juvenile gout (see p. 257) indicate the possibility of a defect in the activity of glucose 6-phosphatase.

Diminished excretion of uric acid can be due to an intrinsic defect in the tubular secretory mechanism or the production of a metabolic inhibitor of this mechanism, for example, lactate, as in Type 1 glycogen storage disease. Defective excretion may also result from renal circulatory disturbances. Many gouty patients develop vascular disease at an earlier age than is found in the general population. Hyperuricemia is a factor associated with an increased tendency to myocardial infarction. Thus, the development of a defect in the renal circulation is not surprising, although the characteristic lesion of the nephropathy of gout is deposition of monosodium urate in the renal parenchyma with surrounding interstitial inflammatory reaction.

Increase in serum uric acid, often as high as 10 mg. and occasionally 15 mg./100 ml., occurs practically invariably in acute attacks of gout. Some degree of elevation usually persists between attacks but it may be intermittent and therefore missed. Asymptomatic hyperuricemia has been found in blood relatives of gouty subjects, values in women in this category being somewhat lower than in men.

Despite the rather consistent presence of hyperuricemia in acute gout, this is apparently not the direct cause of the acute symptoms. No such attacks occur in patients with renal failure with high blood uric acid concentrations. Moreover, there is no consistent quantitative correlation between the clinical response to such agents as ACTH, cortisone, cinchophen, salicylates, carinamide, or probenecid, etc., and their uricosuric or blood urate-lowering effects. Remissions induced by colchicine are not accompanied by significant change in uric acid in the blood or urine. Administration of uric acid, orally or intravenously, does not precipitate an attack, even in gouty subjects. However, attacks of acute gout accompanied by hyperuricemia have been precipitated by administration of chlorothiazide and other benzothiodiazene compounds for purposes of inducing diuresis. Attacks have also been produced experimentally by the intra-articular injection of microcrystalline monosodium urate. This can develop into a progressive inflammatory reaction, which can become identical with an attack of acute gouty arthritis and which responds to colchicine. It is believed that leukocytes which have ingested the microcrystals, play an essential role in the production of the acute attack and that colchicine probably acts by inhibiting their metabolism. An easily distinguishable goutlike type of arthritis (pseudogout) is known to be due to intra-articular and periarticular deposition of crystals of calcium pyrophosphate.

The increased concentration of urate in the body fluids is, of course, the direct cause of the local tissue deposits in chronic gout. Because of their comparatively low solubility at the usual pH of the tissues, urates precipitate from the tissue fluids and are deposited particularly in the kidneys, skin, subcutaneous tissues, cartilage, bursae, tendons, ligaments, synovial membranes, and sclerae. Acting as foreign bodies, these deposits incite an inflammatory reaction. Those that are very superficial are recognizable as tophi; if close to bones, they cause erosions that are demonstrable roentgenographically. Deposits in the kidneys may result in inflammatory lesions with impairment of renal function, a common observation in chronic gout and a common cause of death in this condition. Uric acid urinary calculi occur rather frequently.

The "urate pool," i.e., the amount of urate capable of mixing promptly with intravenously injected urate, is increased in subjects with gout. Values 15 to 30 times the normal quantity (1.2 g.) have been observed. As has already been discussed, the precise nature of the metabolic abnormality is not known, but it is apparently genetically determined, at least in part, since some gout is familial. The term "secondary gout" is applied to a condition with similar clinical manifestations occurring in association with various disorders indicated previously as accompanied occasionally by elevation of serum uric acid, e.g., hematological disorders, lymphoblastoma, lead poisoning, multiple myeloma, macroglobulinemia. In contrast to primary gout, the accumulation of uric acid in these conditions is due not to its overproduction directly from simple precursors and/or defective renal excretion but rather to an increased rate of metabolic turnover of nucleic acids.

(6) **Hereditary Disorders.** Excessive purine synthesis with associated hyperuricemia occurs in a number of hereditary disorders.

(a) *Glycogen Storage Disease, Type 1.* The increase in glucose 6-phosphate and consequent possible increase of PRPP may be an explanation of the increased purine synthesis, which accompanies this disorder. There is also renal retention of urate because of the accompanying lactic acidemia.

(b) *Lesch-Nyhan Syndrome.* This presents in childhood as a severe neurological syndrome (choreoathetosis, spasticity, mental retardation and self-mutilation), which is sometimes accompanied by gout. The amount of uric acid in the urine is five to six times normal. The hyperuricemia is due to almost complete deficiency of the phosphoribosyl transferase, which causes hypoxanthine or guanine to form a nucleotide with PRPP (see p. 237). These purines are thus available for uric acid production. The disorder is X-linked in its inheritance.

There is also a syndrome in which there is an incomplete deficiency of phosphoribosyl transferase activity associated with the production of a mutant enzyme (isoenzyme). This enzyme differs from the normal enzyme in relation to heat stability and migration during electrophoresis. Some families produce a more and some a less heat-stable enzyme; within each family, however, the same mutant enzyme is produced. The disease usually presents as severe gout and some cases have a neurological disorder (spinocerebellar syndrome), which is relatively mild. Renal calculi occur in the majority of cases.

(c) *Miscellaneous.* Hyperuricemia has been reported in hereditary nephropathy (renal retention) in Down's syndrome and in nephrogenic pitressin-resistant diabetes insipidus.

(7) **Endocrine Disorders.** Hypothyroidism, hypo- and hyperparathyroidism can be accompanied by hyperuricemia.

(8) **Vascular Disease.** Hypertension may be accompanied by increase in plasma uric acid. Such patients show an increased tendency to myocardial infarction, which may also be a cause of hyperuricemia.

(9) **Miscellaneous.** An increase in blood urate level has also been reported in obesity, starvation, idiopathic hypercalciuria and psoriasis. Chemotherapy and radiotherapy of large malignant tumors can release sufficient purines to give rise to renal tubular blockage by uric acid crystals.

Hyperuricemia may be diminished therapeutically by the administration of substances (salicylate, probenecid, etc.) which increase urinary excretion of uric acid. These are said to be uricosuric. It may be prevented or diminished by the administration of allopurinol, an analogue of hypoxanthine, in which the C atom at position 8 and the N atom at position 7 have been interchanged. Allopurinol combines with xanthine oxidase and renders it unavailable for the production of uric acid. This therefore results in the accumulation of xanthine and hypoxanthine, which are excreted in large amounts in the urine and are deposited as crystals in the muscles. Diminished purine synthesis also results. This is probably due to feedback inhibition resulting from production of excessive amounts of nucleotides by the action of the appropriate phosphoribosyl transferase.

There are a number of synthetic compounds which inhibit xanthine oxidase, but allopurinol has proved to be the most useful as well as the safest of this group.

Hypouricemia. This condition is by no means common. It occurs in the following:

(1) **Hereditary Xanthinuria.** This is a relatively benign and rare disorder, in which there is an inherited deficiency of xanthine oxidase. This leads to a very much diminished level of blood urate (1 mg./100 ml. or less) with less than 50 mg. uric acid in a 24-hour urine collection and the presence of large amounts of xanthine with lesser amounts of hypoxanthine in the urine. Urinary calculi composed of xanthine may be produced.

(2) **Renal Tubular Defects** can result in nonspecific defective reabsorption of urate ions and a lowering of the blood urate level. Such defects occur in Wilson's disease and in the Fanconi syndrome. A specific defect in renal tubular absorption for urate ions has also been described.

(3) **Bronchogenic Carcinoma.** Hypouricemia has been reported in one patient.

Orotic Aciduria. There is a very rare recessively inherited disorder that results in excessive production of orotic acid. Orotic aciduria is caused by deficiency of orotate phosphoribosyl transferase [E.C. 2.4.2.10] and orotidine-5′-phosphate decarboxylase [E.C. 4.1.1.23]. This results in the excretion in the urine of large amounts of the pyrimidine nucleotide precursor. As a result, the urine becomes cloudy on cooling with the deposition of needle-shaped crystals of orotic acid. Affected children have developed a severe megaloblastic anemia and shown physical and mental retardation. Significant improvement has resulted from the administration of uridine.

References

Crick, F.: General model for the chromosomes of higher organisms. *Nature*, 234:25, 1971.
McGilvery, R. W.: Biochemistry: A Functional Approach. Philadelphia. W. B. Saunders Company, 1970.
Robison, G. A., Nahas, G. G., and Triner, L., eds.: Cyclic AMP and Cell Function, *Annals N.Y. Acad. Sci.*, 185, 1971.
Seegmiller, J. E.: Diseases of purine and pyrimidine metabolism. *In* Bondy, P. K., ed.: Duncan's Diseases of Metabolism. Ed. 6. Philadelphia, W. B. Saunders Company, 1969.
Talbott, J. H.: Gout and blood dyscrasias. *Medicine*, 38:173, 1959.
Talbott, J. H., and Terplan, K. L.: The kidney in gout. *Medicine*, 39:405, 1960.
Watson, D. J.: Molecular Biology of the Gene. Ed. 2. New York, W. A. Benjamin, Inc., 1970.
White, A., Handler, P., and Smith, E. L.: Principles of Biochemistry. Ed. 4. New York. Mc-Graw-Hill Book Co., 1968.

Chapter 5

METABOLISM OF HEMOGLOBIN AND PORPHYRINS

Hemoglobin is a conjugate of a protein, *globin,* and four molecules of *heme,* a substance which is responsible for the characteristic color and oxygen-carrying properties of hemoglobin, which are, however, profoundly influenced by the protein moiety. The globin molecule consists of four peptide units arranged in two pairs. The members of each pair have identical chemical composition which differs from that of the other pair. The heme groups, which are planar in shape, lie in pockets on the surface of the molecule, each pocket being formed by folds in one of the peptide chains. Heme is a type of porphyrin (protoporphyrin) to which is attached an atom of iron in ferrous form. Each of the peptides have closely similar three-dimensional shapes. They associate together by a variety of bonding forces to form a structure roughly resembling a tetrahedron. They are packed together so closely that one might almost describe the final shape of the hemoglobin molecule as spheroidal. However, it must be realized that the peptide arrangement would not be correctly envisaged by imagining the molecule being cut into the four separate peptides by any two planes at right angles to each other. The central core of the molecule is hollow and contains water molecules, which play a part in stabilizing the tetrapeptide. The pockets for the heme groups are lined by hydrophobic residues and make an ideal region for insertion of the hydrophobic porphyrin ring, so that the latter forms a very stable complex with the peptide. The iron of the heme molecule is in the ferrous form. It also chelates with a histidyl residue in the pocket. On the opposite side of the pocket there is another histidyl residue, which is highly important in relation to the oxygen-carrying properties of hemoglobin. The heme grouping also plays a definite part in relation to the final shape of the peptide. In respect to shape, oxyhemoglobin differs from the reduced form. The amino acid residues on the outside of the molecule are strongly polar. This gives hemoglobin a high solubility, and hence it can be present in appropriately high concentration in the erythrocytes. Material contributions to better understanding of biochemical aspects of human genetics have resulted from the demonstration of the existence in man of a number of hemoglobins. The differences between these hemoglobin molecules, normal and abnormal, usually reside in differences in a single amino acid, or deletion of one or more, in one or other of the component peptide units (p. 267).

Porphyrins are complex cyclic compounds composed of four pyrrole units linked by methene ($-CH=$) bridges (i.e., cyclic tetrapyrroles) to form a large flat ring structure. The protoporphyrins contain certain side chains attached to carbon atoms in the pyrrole rings, viz., methyl, vinyl and propionic acid groups. Because of the many possible differences in arrangement of the side chains, there are fifteen theoretically possible protoporphyrin isomers, only one of which (protoporphyrin IX) is found in nature.

In addition to protoporphyrin (in heme), two other types of porphyrin are of biological and clinical importance in man, i.e., uroporphyrins (acetate and propionate side chains) and coproporphyrins (methyl and propionate side chains). Of the four theoretically possible isomers of each of these, only two occur in nature, types I and III. The prefixes, "uro" and "copro," indicate the medium from which each of these compounds was originally isolated, but it is now known that both may be excreted in both the urine and the feces.

The parent compound, to which all porphyrins are related, is known as porphin. Its structure is shown in Figure 5-1.

Porphyrins are compounds in which the numbered hydrogens are replaced by various substituent side chains. It has already been pointed

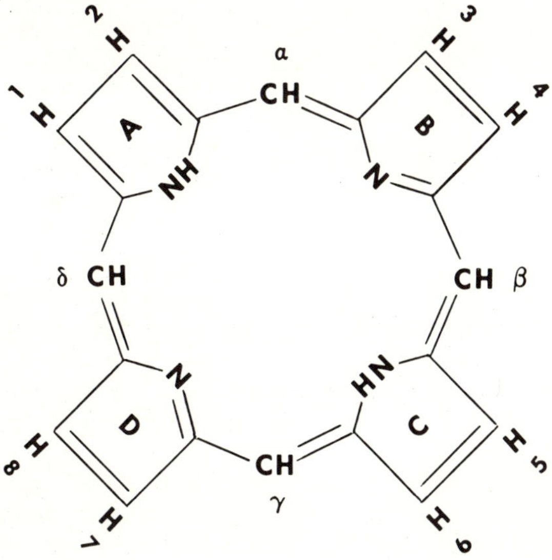

Figure 5-1. The structure of porphin. (From Bondy: Duncan's Diseases of Metabolism, 6th edition. W. B. Saunders Company, 1969.) Note the numbering of the H atoms and the lettering of the methene groups and pyrrole rings.

out, for example, that in uroporphyrin these side chains are acetate ($-CH_2COOH$) and propionate ($-CH_2CH_2COOH$). One of each is attached to each pyrrole ring. If one commences at position 1 and proceeds clockwise around the ring, there are only four possible arrangements for these substituent pairs, as in the following:

Pyrrole ring	A	B	C	D
Type I	AcPr	AcPr	AcPr	AcPr
Type II	AcPr	PrAc	AcPr	PrAc
Type III	AcPr	AcPr	AcPr	PrAc
Type IV	PrAc	AcPr	AcPr	PrAc

Only Type I and Type III uroporphyrins occur in nature. The same may be said of coproporphyrin, in which the acetate side chain is converted to a methyl group by decarboxylation ($-CH_2COOH \rightarrow -CH_3+CO_2$). This actually occurs in the biosynthesis of heme, but the porphyrins are in the form of their precursors, the porphyrinogens. In these compounds, the methene bridges are reduced to methylene bridges ($-CH_2-$) and all four pyrrole nitrogens are combined with hydrogen. In protoporphyrin IX, the two propionate groupings on rings A and B of coproporphyrin III are replaced by vinyl groups ($-CH=CH_2$). Therefore, starting at position 1 and proceeding clockwise, the arrangement of side chains is MeVi MeVi MePr PrMe.

Protoporphyrin is able to form a complex with iron, forming heme, the prosthetic group of a class of hemoproteins which are concerned directly or indirectly in facilitating the reaction which is the basis of practically all bioenergetics, i.e., the union of oxygen and hydrogen to form water. These hemoproteins include hemoglobin, myoglobin (muscle hemoglobin), cytochromes, catalases, and peroxidases.

Hemoglobin is the carrier of atmospheric oxygen from the lungs to the tissue cells. It also participates in the transport of carbon dioxide (p. 441) and in the mechanism of regulation of hydrogen ion concentration (p. 402). Oxygen may be stored temporarily in muscle, attached to myoglobin. The combination of the oxygen from the air with the hydrogen of metabolic substrates to form water is catalyzed by the cytochromes, including cytochrome oxidase. Ancillary roles in biological oxidations are played by the catalases and peroxidases, other heme-containing enzymes. The vital importance of the hemoproteins is indicated by the fact that inhibition of the oxygen transport system (hemoglobin) by carbon monoxide (p. 273), or inhibition of the oxygen utilization system (cytochromes) by cyanide, results in rapid death of the organism. It is significant, too, that chlorophyll, a substance involved in the photosynthetic reactions of plants, is a magnesium-porphyrin which catalyzes the cleavage of water into hydrogen ions and oxygen. Evidently, derivatives of porphyrins are of wide significance in comparative biochemistry, from the initial fixation of solar energy in green plants to its eventual utilization by man.

Biosynthesis of Porphyrins and Hemoglobin. The first,

and rate-controlling, step of porphyrin synthesis is the enzymic condensation of glycine with succinyl coenzyme A. This requires the presence of pyridoxal phosphate. The enzyme involved is 5-aminolevulinate synthase and the product is succinyl glycine [HOOC.CH$_2$.CH$_2$.CO.CH(NH$_2$)COOH]. This loses CO$_2$ and becomes 5-aminolevulinic acid (ALA) (HOOC.CH$_2$.CH$_2$.CO.CH$_2$.NH$_2$). Two molecules of this compound are condensed together by the action of porphobilinogen synthase to form the pyrrole derivative porphobilinogen, which has acetyl, propyl and aminomethyl side chains.

$$\begin{array}{c} \text{COOH} \\ | \\ \text{CH}_2 \\ | \\ \text{CH}_2 \\ | \\ \end{array}$$

Porphobilinogen

The next step probably is the formation of an open tetrapyrrole by the action of the enzyme uroporphyrinogen I synthase. This involves condensation of the carbon of the side chain containing the amino group of one pyrrole ring with the carbon without a side chain in another. In this reaction, three molecules of ammonia are given off, leaving four pyrrole rings combined together by three methylene (—CH$_2$—) bridges. One terminal pyrrole ring still possesses the amino methyl side chain. This compound is capable of spontaneous conversion to uroporphyrinogen I by condensation between the carbon of this side chain with the free carbon of the other terminal pyrrole and loss of ammonia. To some extent, this always happens. The major portion of the open tetrapyrrole is, however, acted upon by a second enzyme, uroporphyrinogen III cosynthase. This also produces a closed porphyrin ring but at the same time brings about the interchange of the acetyl and propionyl side chains of one pyrrole ring, and so gives rise to uroporphyrinogen III. The acetyl side chains of this compound are then converted to methyl groups by uroporphyrinogen decarboxylase, and so gives rise to coproporphyrinogen III. This is then converted to protoporphyrin IX, which then combines with an atom of ferrous iron to form heme. This rapidly enters the pocket of a globin peptide and is thus protected from oxidation to the ferric iron-containing form (hemin). This latter compound, also known as protohemin IX, is soon formed if the production of heme exceeds that of the appropriate protein synthesis. Protohemin IX is a very effective inhibitor of 5-aminolevulinate synthase, which is the enzyme involved in the first step of porphyrin biosynthesis. This

```
succinyl CoA
     +
  glycine
     │         ✳ ←── inhibits
     ↓
5-aminolevulinate
     │
     ↓
porphobilinogen
     │
     ↓
uroporphyrinogen III
     │
     ↓
     ↓
protoporphyrin IX
     +
Fe³⁺  ⇌(NADPH/O₂)  Fe²⁺
     ║
stimulates (?) ║
     ⇓
     ↓
amino acids →→→→   protoheme IX  ──O₂, spontaneous──→ protohemin IX
                        +
                      globin
                        │
                        ↓
                    hemoglobin
```

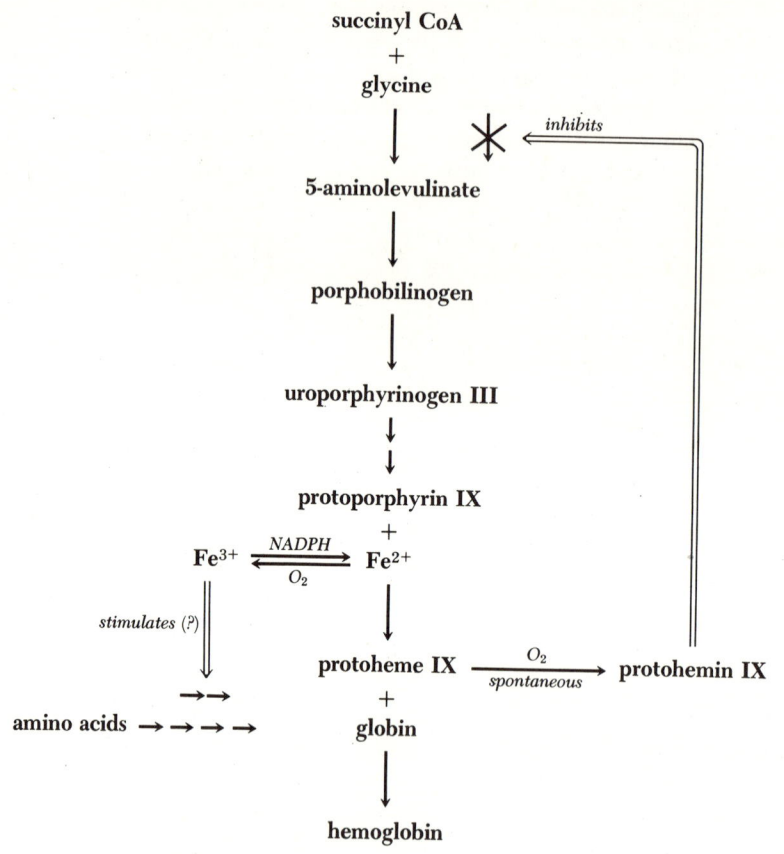

Figure 5-2. Regulation of hemoglobin synthesis according to current concepts. Unless there is globin available to bind newly synthesized protoheme IX, it will be spontaneously oxidized to protohemin IX, which inhibits further formation of porphyrin at the first step peculiar to the process. There is some evidence that the presence of ferric ions stimulates the synthesis of globin. Ferric ions will be present if the supply of iron is adequate to convert all protoporphyrin IX to protoheme IX. If the supply of iron is inadequate, there would be no point in forming more globin. These processes probably regulate the relative balance of the various precursors of hemoglobin, but not the primary adjustments of the total amount of circulating hemoglobin to such factors as environmental oxygen tension. (From McGilvery: *Biochemistry: A Functional Approach.* W. B. Saunders Company, 1970.)

produces a feedback mechanism for the regulation of hemoglobin synthesis (see Fig. 5-2). Heme itself also stimulates globin synthesis. Iron stimulates the production of both heme and globin. Its action on globin synthesis may be a secondary effect of increased heme production. Certain hemoproteins, e.g., cytochrome oxidase, contain porphyrins other than protoporphyrin.

From the clinical standpoint, interest is centered particularly in the synthesis of hemoglobin, which, in the adult, is restricted normally to the red bone marrow. Three components are required: (1) protoporphyrin IX, synthesized as indicated above; (2) globin, produced by the usual mechanisms of protein synthesis; and (3) iron, the sources and metabolism of which are considered elsewhere (p. 319). Synthesis of hemoglobin proceeds concomitantly with maturation of the erythrocyte, rendering a decision difficult as to whether a given cofactor is concerned in one process or the other. In adults (70 kg. man), approximately 8 g. of hemoglobin are synthesized (and degraded) daily, corresponding to about 300 mg. of porphyrin (3.5 per cent of hemoglobin molecule).

The immature erythroid cells contain porphyrins rather than hemoglobin. As the cell matures, the content of porphyrin decreases and that of hemoglobin increases, these changes being reflected in alterations in the staining properties of the cell, i.e., change from basophilia to acidophilia. Hemoglobin production is related to the appearance and amount of ALA synthase and first occurs in the intermediate normoblast. Normal erythrocytes contain small amounts of protoporphyrin and traces of coproporphyrin, mainly Type III. The quantity of protoporphyrin increases in certain conditions in which hemoglobin synthesis is inhibited, e.g., iron deficiency. Increases have been observed also in lead poisoning and hemolytic anemias. Low or normal values are found in pernicious anemia. Increase in coproporphyrin has been observed in conditions accompanied by reticulocytosis, i.e., increased rate of erythropoiesis, e.g., in lead poisoning and excessive hemolysis, and at the onset of remissions in pernicious anemia.

Catabolism of Hemoglobin. Within the mature erythrocyte, hemoglobin undergoes no alteration during the "lifetime" of the cell, about 120 days. Upon disintegration of the cell, the hemoglobin liberated, amounting to about 8 g. daily, undergoes a sequence of changes, as a result of which its iron and globin components are returned to the general community of metabolites, the former being virtually completely reutilized for resynthesis of hemoglobin (and other iron-heme compounds), the latter following the pathways of protein catabolism. The protoporphyrin component is not utilized. It undergoes a series of changes which eventuate in the formation of bilirubin, which enters the blood stream, is removed by the liver, excreted in the bile and converted, in the intestine, to the endproducts of the degradation of heme, i.e., the bilinogens and bilins. These are excreted largely in the feces but also in the urine (after reabsorption from the intestine).

These aspects of hemoglobin catabolism are considered in detail elsewhere (p. 633), as they are of particular importance clinically in relation to the study of liver function.

Haptoglobins. These are present in the circulation and are mucoproteins which migrate as α_2-globulins during electrophoresis in paper. In starch gel, however, they prove to be heterogeneous. Three main types of haptoglobins, which are of genetically determined distribution, have been described. Types 1-1 and 2-2 are homozygous and Type 2-1 is heterozygous. Type 1-1 shows one haptoglobin band; 2-2 does not have this band, but shows a number of bands moving more slowly at pH 8.6. These are polymers of the same protein. Type 2-1 shows the Type 1-1 band and a number of polymers with mobility between those of Type 2-2 and that of Type 1-1. These result from polymers formed from combination of the Type 1-1 protein and that of Type 2-2.

Haptoglobins form complexes with hemoglobin, which are formed wherever there is intravascular hemolysis. They are carried to the reticuloendothelial system where they are broken down and the hemoglobin catabolized, as already described. In hemolytic anemia, little or no haptoglobin can be detected in the circulation, since it has been removed as the complex. Hemoglobinuria will not result until the haptoglobin has virtually disappeared. This happens in so-called black-water fever. A congenital absence of haptoglobins (ahaptoglobinemia) occurs and individuals with this condition frequently have hemoglobinuria.

Varieties of Human Hemoglobin. Normal adult hemoglobin, or hemoglobin A, has a molecular weight of 64,456 and contains two pairs of peptide chains, designated alpha and beta, respectively, each with a characteristic sequence of amino acids, of which the α chain contains 141 and the β 146. Normal adult hemoglobin also contains minor components, at pH 8.6 one migrating electrophoretically more slowly (A_2), another more rapidly, than the major component. Fetal hemoglobin (F) is the more rapid component and is present in very small amounts.

These normal human hemoglobins all possess a common half-molecule, i.e., one identical pair of peptide chains (α chains); the other half of each molecule consists of a pair of different types of peptide chains, one type for each hemoglobin. The structure of each peptide chain is determined genetically, i.e., one gene for each chain type. Hemoglobin A_2 has two delta chains and hemoglobin F, two γ chains; both types of chain contain 146 amino acid residues and are thus of the same length as the β chain. The amino acid sequences in the α, β, γ and δ chains have all been determined. The residues are numbered from 1 to 141 or 1 to 146, starting at the N-terminal ends of the chains. A type of shorthand notation has been developed, which proves very useful when describing abnormal hemoglobins (see below). Hemoglobin A is represented as $\alpha_2^A \beta_2^A$, hemoglobin A_2 as $\alpha_2^A \delta_2^A$ and hemoglobin F as $\alpha_2^A \gamma_2^F$. Adult hemoglobin constitutes approximately 97 per cent, 2.5 per cent and less than 1 per cent, respectively, of these components. In early embryonic life, a fourth hemoglobin $\alpha_2^A \epsilon_2$ exists.

Fetal Hemoglobin. Fetal hemoglobin (F) comprises 50 to 90 per cent of the total hemoglobin in the newborn. It takes up oxygen more readily at low oxygen tensions and releases carbon dioxide more readily than adult hemoglobin (A). It is, therefore, admirably adapted to the conditions of relatively low oxygen tension in the fetus. At birth, efficient gas exchange in the lungs replaces the relatively inefficient placental mechanism, and the erythropoietic activity of the bone marrow increases progressively as that of the liver decreases. Normally, hemoglobin F is gradually replaced by hemoglobin A during the first 6 months of extrauterine life, comprising about 15 per cent of the total hemoglobin at one year, 5 per cent at two years, and less than 1 per cent after four years of age. Hemoglobin F is more resistant to denaturation by alkali and is more susceptible to conversion to methemoglobin by nitrites. The susceptibility of young infants to develop methemoglobinemia, with consequent cyanosis and anoxia, after exposure to nitrites (contaminated water), has been attributed to this cause. High concentrations of hemoglobin F after two years of age may occur in various types of anemia, e.g., sickle cell anemia ($<$ 40 per cent) and thalassemia ($<$ 90 per cent). The presence of high concentrations of hemoglobin F in an obscure anemia is suggestive of a hereditary hemoglobin abnormality; its absence does not exclude this possibility.

Hereditary Persistence of Fetal Hemoglobin. In this condition, homozygotes have an adult hemoglobin which is entirely hemoglobin F and heterozygotes contain 15 to 20 per cent fetal hemoglobin in their adult hemoglobin. The condition is quite benign and compatible with normal health.

Abnormal Hemoglobins. Over one hundred different types of hemoglobin have been described in addition to those occurring normally. The occurrence of these abnormal hemoglobins, some of which are best differentiated by their characteristic electrophoretic mobilities, has given rise to the concept of "molecular disease," according to which a defective gene (mutant) may direct the formation of a molecule similar to a normal molecule but differing from it in shape, composition and electrical charge. This concept differs from that of complete absence of a molecule when an abnormal recessive gene is homozygous, as in inborn errors of metabolism (p. 224). Each of these hemoglobinopathies, which are genetically transmitted, is due apparently to a single mutant gene, resulting usually in the replacement of a single amino acid residue of hemoglobin A by some other amino acid. The substitutions frequently involve replacement of acidic by basic or neutral amino acids; since this change in structure occurs in both of the members of a peptide chain pair, a considerable difference in electrical charge results, accounting for the observed differences in electrophoretic mobility among the abnormal hemoglobins.

Much of the progress in this field is due to the application of the "fingerprinting technique," in which the heat-denatured hemoglobin is subjected to the action of trypsin, which splits the protein at certain specific sites. The resulting peptides are readily separated by a combination of paper electrophoresis and paper chromatography. In this way, peptides containing "abnormal" constituents may be detected and isolated and their constituent amino acids and their amino acid sequences determined. With the exception of hemoglobin S (sickle cell hemoglobin) and hemoglobin M (producing an abnormal methemoglobin), these abnormal hemoglobins have usually been given letter designations in the alphabetical order of their discovery; others are labeled by the name of the place where they were first discovered.

Another very useful technique for identification of hemoglobins is hybridization. This depends on the fact that at pH levels of about 4.7, or in 8M urea solutions, hemoglobins dissociate first into two dimers, each containing one of each different peptide; these further dissociate and then form dimers, each made up of pairs of similar peptides. For example, a hemoglobin with an abnormality in the α chain ($\alpha_2^X \beta_2$) will first dissociate into $\alpha^X \beta$ dimers, and then into α_2^X and β_2 dimers. If neutrality is restored, or urea removed, the dimers will reassociate to form $\alpha_2^X \beta_2$, the original hemoglobin. If such a hemoglobin is mixed with another which has an abnormality in the β chain $(\alpha_2\beta_2{}^X)$, dissociation will finally produce the dimers α_2, $\alpha_2{}^X$, β_2 and $\beta_2{}^X$. If reassociation is then brought about this is random and results in the formation of $\alpha_2\beta_2$, $\alpha_2{}^X\beta_2$, $\alpha_2\beta_2{}^X$ and $\alpha_2{}^X\beta_2{}^X$. The new tetrads have different electrophoretic mobilities which can be used for identification purposes. It is important to note that only like peptides combine together to form the final dimers of dissociation. One never obtains a dimer such as $\alpha\alpha^X$ or $\beta\beta^X$. The same applies to all abnormal hemoglobins; the same abnormality occurs in each member of the similar peptide pair and not in a single peptide. Human hemoglobin can also be made to hybridize with canine hemoglobin and this is also useful for identification.

Individuals who are heterozygous (bearing a single gene for hemoglobin A and another for an abnormal hemoglobin) are said to have the particular "trait" related to the abnormal hemoglobin, e.g., sickle cell trait in the case of a single gene for hemoglobin S. The red cells of such individuals contain both hemoglobin A and the abnormal hemoglobin, the former usually comprising over 50 per cent of the total, since it usually has greater penetrance or expressivity than the abnormal hemoglobin. Individuals who are homozygous (bearing a double complement of genes) for an abnormal hemoglobin are said to have the "anemia" or "disease" related to the abnormal hemoglobin. In these cases, the synthesis of hemoglobin A may be completely suppressed, although hemoglobin F frequently appears together with the abnormal hemoglobin. Individuals have been described who are heterozygous for two abnormal hemoglobins, in which case both abnormal hemoglobins are present in the red cell, hemoglobin A is absent, and hemoglobin F may be present. The geographical distribution of the abnormal hemoglobins is of considerable interest to anthropologists, since certain of these pigments appear to predominate in specific ethnic groups.

A type of hemoglobin not present in normal subjects (hemoglobin S) predominates (about 80 per cent of total hemoglobin) in patients with sickle cell anemia, a hereditary disorder. It comprises 20 to 45 per cent of the total hemoglobin in subjects exhibiting only the sickle cell trait (sickling only in vitro, at lowered oxygen tensions). The reduced form of hemoglobin S is much less soluble than normal reduced hemoglobin. Red cells containing a predominance of their hemoglobin in this form undergo sickling at normal venous oxygen tensions as a result of crystallization of this substance ("tactoid" formation). The anemia, hemolysis and thrombosis which characterize this disease are attributable to the increased fragility and distortion of these cells. If smaller amounts of the hemoglobin are in this abnormal form, sickling does not occur at oxygen tensions present in the blood but can be induced in vitro by exposing the cells to unphysiologically low oxygen tensions (i.e., sickle cell trait). Although subjects exhibiting this trait are usually asymptomatic, there may be inability to concentrate urine adequately and vascular occlusions may occur, e.g., hematuria, particularly at high altitudes. The abnormal gene is present in about 10 per cent of American Negroes and probably in up to 45 per cent of African natives. In some instances, at least, the difference between these two states is explicable on the basis of Mendelian inheritance, sickle cell anemia occurring in homozygous SS individuals and the sicke cell trait in heterozygous SA individuals (hemoglobin less than half type S and more than half type A).

The abnormality in hemoglobin S occurs in the β chain ($\alpha_2^A\beta_2^S$); it involves the replacement of the glutamic acid residue at position 6 by a valine residue ($\alpha_2^A\beta_2^{6Glu \to Val}$ or $\alpha_2^A\beta_2^{6Val}$). It has already been mentioned that heterozygotes of HbA and HbS manifest the so-called sickle trait, which is relatively benign. However, in West Africa, it is relatively common to find individuals who are doubly heterozygous for the two abnormal hemoglobins HbS and HbC ($\alpha_2\beta_2^{6Lys}$). They suffer from so-called sickle cell hemoglobin C disease, which is manifested by a relatively mild anemia with a tendency to infarction. Another abnormal hemoglobin of special interest is HbC$_{Harlem}$. It was first designated as an HbC because of its electrophoretic mobility. It does not, however, have the same amino acid residue abnormality. HbC$_{Harlem}$ is actually a variant of HbS, since it has a value residue at position 6 of the β chain and also an asparagine residue in place of the aspartic residue at position 73. It is unique, insofar as it possesses double variation on the β chain.

Abnormal α chains also occur; there are numerous examples. HbI, for example, is $\alpha_2^{16Lys \to Glu}$.

Apart from the sickling phenomenon already mentioned, these amino acid substitutions may be entirely benign. They may produce an anemia with target cells, or an unstable thermolabile hemoglobin which results in the formation of Heinz bodies, as in patients with congenital nonspherocytic hemolytic anemia; or, a hemoglobin variant may be produced, which affects the carriage of oxygen by the heme iron, resulting in an abnormal methemoglobin (HbM).

Two hemoglobin variants have been described in which, instead of amino acid residue replacement, there has been deletion in the β chain of one or several amino acid residues. The latter makes the chain unable to bind heme.

Hemoglobin M. It has already been mentioned that oxygen-carrying properties of adult hemoglobin are largely conditioned by two histidine residues in the heme pocket of each chain type. The heme iron is attached to one of them and the second is on the other side of the pocket. These are referred to as the proximal and distal histidine residues, respectively, because the proximal appears closer to the iron atom when the molecule is examined by x-ray diffraction. These terms do not refer to proximity to the N-terminal end of the peptide chains. In fact, the proximal histidine residue is at position 87 in the α chain and 92 in the β chain, whereas the distal histidine residues are at positions 58 and 63, respectively. Substitutions in either chain at or near these residues, particularly the distal residue, tend to cause oxidation of the iron to the ferric form, as in ordinary methemoglobin. Individuals with this type of abnormal hemoglobin variant present with chronic cyanosis, but are otherwise asymptomatic. The abnormal methemoglobin is not reduced to hemoglobin by reducing agents such as ascorbic acid, and this helps to distinguish the condition from endogenous and other types of methemoglobinemia. Moreover, the absorption spectrum, which may be distinctive for a specific HbM, differs from that of methemoglobin. Hemoglobin M_{Boston} is $\alpha_2^{58His \to Tyr}\beta_2$ and hemoglobin $M_{Saskatoon}$ is $\alpha_2\beta_2^{63His \to Tyr}$. There are several other types.

Thalassemia. In this condition, there is an inherited failure (which can be complete) to produce sufficient amounts of either the α or the β peptide chain (α or β thalassemia, respectively). Individuals who are homozygous suffer from thalassemia major, whereas heterozygotes have thalassemia minor.

The homozygous state for the α defect is not compatible with life. For the β defect, it is a serious condition, which usually results in death in childhood. There is marked hepatosplenomegaly and severe anemia with target cells and a moderate reticulocytosis. The child often has a mongoloid appearance. The blood shows hyperbilirubinemia and the hemoglobin is virtually all HbF. There may be an increase in HbA_2.

Thalassemia minor may be asymptomatic or it may be present, particularly during pregnancy, as a mild anemia with some splenomegaly, target cells and hyperbilirubinemia. β thalassemia minor is the most frequently seen form of thalassemia and often shows an increase of HbF to 2 to 5 per cent of the total hemoglobin; α thalassemia minor is virtually symptomless. Hb Barts (γ_4) may be present in infancy but disappears quite rapidly. There may be an elevation of HbA_2.

Hemoglobin H ($Hb\beta_4$) disease is a related condition, in which less than half of the total hemoglobin is $Hb\beta_4$.

Detection of Hemoglobinopathy. The various types of hemoglobin may be identified by one or more of the following procedures: electrophoresis, hybridization, alkali denaturation, sickle cell tests (tactoid and gel formation; solubility studies), spectroscopy, immunologic procedures and x-ray diffraction patterns. Electrophoresis cannot accomplish separation of hemoglobin A and F, nor of S and D, nor of a number of other hemoglobins. However, hemoglobin F can be identified readily on the basis of its unique resistance to alkali denaturation.

HEMOGLOBINEMIA; HEMOGLOBINURIA

Hemoglobinemia (hemoglobin free in the plasma) may occur when excessive numbers of erythrocytes are destroyed very rapidly, particularly when the hemolytic process occurs intravascularly. The condition is accompanied by anemia, reticulocytosis, an increase in serum bilirubin and, at times, methemoglobinemia. If the concentration of hemoglobin in the plasma exceeds 90 to 150 mg. per 100 ml., hemoglobinuria may occur (p. 219), and may persist until the plasma Hb has fallen to 50 mg. per 100 ml. Excessive amounts of urobilinogen are excreted in the urine and feces (pp. 651 and 652).

Hemoglobinemia (and hemoglobinuria) may, therefore, occur as a manifestation of severe hemolytic anemias, especially the acute variety, but also as exacerbations of chronic forms. The following factors may cause this condition:

(*a*) *Bacteria.* Hemolytic streptococcemia, severe typhoid and scarlet fevers, clostridia infections.

(*b*) *Protozoa.* Malaria and black-water fever and Bartonella infection (Oroya fever).

(*c*) *Chemical Agents.* Poisoning with arseniuretted hydrogen, arsphenamines, pyrogallic acid, phenylhydrazine, saponin, sulfonamides and, rarely, lead. A number of substances may cause excessive hemolysis in certain subjects rendered hypersusceptible by genetically determined erythrocyte defects, perhaps the most important being decreased glucose-6-phosphate dehydrogenase activity. These substances include fava beans (favism), sulfonamides, acetophenetidin, acetanilid, naphthalene, and primaquine. Some unchanged hemoglobin may appear in the plasma at times in poisoning with substances that cause methemoglobinemia.

(*d*) *Vegetable Poisons.* Poisoning with ricin, crotin, and poisonous toadstools; favism, due to hypersensitivity to the fava bean (see above).

(*e*) *Animal and Endogenous Poisons.* Poisoning with certain snake venoms and occasionally after severe, extensive burns, administration of serum for tetanus, diphtheria, and meningococcus meningitis.

(*f*) *Miscellaneous.* Transfusion with incompatible blood, paroxysmal hemoglobinuria (cold hemolysins and syphilis), paroxysmal nocturnal hemoglobinuria (Marchiafava-Micheli), acute hemolytic anemia of unknown etiology (Lederer's anemia), acute crises in congenital hemolytic jaundice (spherocytic anemia), march hemoglobinuria (strenuous exertion, lordosis), congenital absence of haptoglobin, and occasionally after severe intra-abdominal hemorrhage.

Hemoglobinuria (without hemoglobinemia) may occur in cases of renal infarction and as a result of hemolysis of red blood cells in ammoniacal or extremely hypotonic urine. The former may occur in cases of hematuria with bladder-neck obstruction (e.g., bladder neoplasms), the latter in cases of hematuria in renal tubular necrosis (e.g., following sulfonamide therapy).

MYOGLOBINURIA

The molecular weight of myoglobin (muscle hemoglobin) is about one-fourth that of hemoglobin; it therefore passes through the glomerular membrane more readily than does the latter; it contains 146 amino acid residues and its molecular shape is very similar to the peptides of hemoglobin A. The sequence, however, differs in a large part of the molecule. The heme iron of myoglobin is therefore not so efficient a carrier of oxygen as that of hemoglobin. In contrast to the situation in hemoglobinemia (p. 270), the lower molecular weight permits obvious discoloration of the urine, due to myoglobin, at plasma concentrations (20 mg./100 ml.) too low to produce gross discoloration of the plasma. The diagnosis of myoglobinuria should, therefore, be suspected in cases in which the serum is clear and the urine, containing no red blood cells, gives a positive benzidine reaction. The presence of myoglobin may be established definitively by electrophoresis or spectrophotometry.

Myoglobinuria may occur as a result of extensive muscle damage due to crush injuries, electric shock, burns, beating, arterial thrombosis and, possibly, dermatomyositis. An epidemic form (Haff disease) has been described, apparently of infectious or toxic origin. It may occur in association with hemoglobinuria as a result of excessive physical exertion (march hemoglobinuria). Rarely, it occurs as a primary disorder of unknown origin (idiopathic paroxysmal myoglobinuria). One of the serious complications in all forms of this condition is acute renal failure.

ABNORMAL HEMOGLOBIN DERIVATIVES

Certain derivatives of hemoglobin are of clinical importance because of their production and presence in the blood in abnormal amounts in various disease states. In the majority of instances they result from abnormal exposure to toxic chemical agents.

Methemoglobin. This is an oxidized hemoglobin (oxidized heme), differing from oxyhemoglobin (oxygenated hemoglobin) in that it contains ferric iron in contrast to ferrous iron in the latter and that its oxygen is in firm combination and cannot be removed by exposing the blood to a vacuum. Methemoglobin is consequently incapable of functioning as an oxygen-carrier (p. 443). Cyanosis usually develops when the concentration reaches 3 g. per 100 ml. of blood.

Methemoglobin is present in normal erythrocytes, comprising up to about 1 per cent (average 0.4 per cent) of the total hemoglobin. It apparently exists in reversible equilibrium with hemoglobin, the relative proportions of the two substances being regulated by enzymes (diaphorase; methemoglobin reductase) in the red cells. Under normal conditions, of course, this equilibrium is shifted far in the direction of hemoglobin. When the methemoglobin content increases, the cells at first usually remain intact. With very high concentrations, however, the cells may be injured and destroyed and methemoglobin, often with hemoglobin, may be liberated into the plasma and excreted in the urine. The causes of methemoglobinemia may be classified as follows:

(a) Drugs and Other Toxic Agents. These include nitrites, chlorates, permanganates, aniline, acetanilid, acetophenetidin, antipyrine, nitrobenzene, pyrogallol, sulfonal, primaquine, methylene blue, and certain sulfonamides. The particularly high incidence of toxic methemoglobinemia in infants exposed to aniline dye-stamped diapers and nitrate-contaminated well water is perhaps attributable to the low concentration of methemoglobin-reducing enzyme systems in the erythrocytes of the newborn.

(b) Other Hemolytic Agents. Among conditions falling under this heading are congenital hemolytic icterus (acute crises), paroxysmal hemoglobinuria, infection with anaerobic bacteria (clostridia), black-water fever, and excessive doses of phenylhydrazine.

(c) Enterogenous Cyanosis. In some cases so classified the production of methemoglobinemia is believed to be due to excessive production and absorption of nitrites from the intestine. These nitrites may be elaborated by nitrite-producing bacteria. Methemoglobinemia has been reported in young infants (under 6 months of age) as a result of ingestion of well water containing excessive amounts of nitrites (from vegetation). This has been attributed to the unusual susceptibility of the fetal type of hemoglobin to methemoglobin formation by nitrites.

(d) Congenital. There are three varieties of genetically determined methemoglobinemia, this pigment comprising 15 to 50 per cent of the total circulating hemoglobin. One form is characterized by the abnormality in the molecular structure of the globin component, designated "hemoglobin M" disease. In another type, there is deficiency in the diaphorase enzyme system, normally catalyzing reduction of methemoglobin to hemoglobin. A third form is apparently dependent upon an abnormality in the globin molecule which, however, is indistinguishable spectroscopically from normal hemoglobin, differing in this respect from methemoglobin M.

Because of the cyanosis present in such cases, they may be confused with congenital heart disease. The cyanosis responds to treatment with reducing agents in those forms due to enzyme defects, but not in "hemoglobin M" disease.

Sulfhemoglobin. Reduced hemoglobin combines with hydrogen sulfide to form sulfhemoglobin. Sulfhemoglobinemia is an unusual condition which occurs chiefly as a result of the action of nitrites and coal tar preparations (aniline, acetophenetidin, acetanilid) in the presence of excessive amounts of sulfur. It has also been observed in subjects with marked constipation, in the presence of nitrite-producing bacteria in the intestine. Some believe that aniline may "sensitize" hemoglobin so that it unites with sulfides. This condition has been produced in splenectomized animals by feeding large amounts of sulfur without additional medication. Sulfhemoglobinemia has been reported in patients receiving sulfonamides. As is the case in methemoglobinemia, sulfhemoglobin is usually contained in the erythrocytes, but it may occasionally be liberated into the plasma. Cyanosis usually occurs when the concentration of sulfhemoglobin reaches 3 to 5 g. per 100 ml. of blood.

Carboxyhemoglobin. Carbon monoxide combines with the heme iron in the hemoglobin molecule as does oxygen, the combining capacity of hemoglobin for both being identical. However, the affinity for carbon monoxide is more than 200 times as great as for oxygen, and in the presence of relatively small concentrations of CO in the air a considerable quantity is taken up by the blood, with consequent reduction in the amount of hemoglobin available for transport of oxygen to the tissues. The dissociation of oxyhemoglobin is also diminished. Carboxyhemoglobin has a bright, cherry-red color. It is formed usually as a result of excessive exposure to artificial (but not natural) illuminating gas and to automobile exhaust gases, in closed or poorly ventilated rooms.

Methemalbuminemia. Methemalbumin is a ferric complex of protoporphyrin with albumin. It appears in the blood plasma under certain abnormal circumstances, particularly in the simultaneous presence of excessive hemolysis. This condition has been observed in gas bacillus and other septicemias, severe malaria (black-water fever), severe pernicious anemia, hemolytic anemias, nocturnal hemoglobulinuria, acute hepatic necrosis and, occasionally, in lead poisoning, and after excessive amounts of phenylhydrazine, acetanilid, and aromatic nitro compounds.

Hematinemia unaccompanied by hyperbilirubinemia may occur, occasionally, in conditions in which relatively large amounts of blood are extravasated into epithelial-lined spaces in close communication with the blood stream, e.g., in the kidney. In contact with epithelial cells, hemoglobin is converted to hematin, not to bilirubin.

PORPHYRIA AND PORPHYRINURIA

Normal urine contains coproporphyrin (ranges given vary from 6 to 66 μg./24 hr. to 60 to 180 μg./24 hr.) and uroporphyrin (0 to 35 μg./24 hr.). In each case, the content of Type I is greater than that of Type III. The porphyrins are probably excreted in the first place as their porphyrinogen precursors. The porphobilinogen content is normally 0.38 to 2.42 mg./24 hr. Urines containing excessive amounts of porphyrin may be colored from pink to wine red or may show a reddish-purple fluoresence in ultraviolet light.

The coproporphyrin content of normal feces is less than 20 μg./g. dry weight (often considerably less). That of protoporphyrin is less than 30 μg./g. dry weight. The normal fecal excretion of uroporphyrin amounts to some 10 to 40 μg./24 hr.

Normal erythrocytes contain protoporphyrin and coproporphyrin. Expressed in amounts per 100 ml. packed cells, there is about 36 μg. of the former and 0.5 μg. of the latter.

Conditions in which excessive amounts of porphyrins are excreted in the urine (porphyrinuria) or feces, or both, may be placed in two main categories, (1) congenital (porphyrias) and (2) acquired. The characteristic biochemical features of these conditions are indicated in Table 5-1. Certain generalizations may be made concerning the basic mechanisms which underlie these abnormalities:

1. Formation and excretion of excessive amounts of type I porphyrins usually result from a greatly accelerated rate of synthesis of heme, e.g., in "erythropoietic" porphyrias and hemolytic anemias.

2. Uroporphyrins are excreted only in the porphyrias (i.e., congenital, not acquired porphyrinurias, with the exception of those due to hexachlorobenzene and certain other toxic substances).

3. Accumulation and excretion of excessive amounts of type III porphyrins result from inability to convert them adequately to heme, i.e., interference with heme synthesis.

4. In the absence of an "inborn error" in porphyrin metabolism, increased amounts of coproporphyrin I may appear in the urine in the presence of impaired liver function or biliary obstruction, as a result of

TABLE 5-1 PORPHYRINURIAS

GENERAL TYPE	SPECIFIC CONDITIONS	CHIEF URINARY PORPHYRINS	OTHER CHARACTERISTICS
Hereditary	Erythropoietic porphyria Congenital	Uroporphyrin I Coproporphyrin I	Erythrocyte proto- and coproporphyrin usually raised
	Erythropoietic protoporphyria	Normal	Erythrocyte protoporphyrin very high and coproporphyrin sometimes slightly raised
	Erythropoietic coproporphyria	Normal	Erythrocyte coproporphyrin very high. Proto- and uroporphyrin somewhat increased
	Hepatic porphyria Acute intermittent porphyria	Uroproporphyrin I Coproporphyrin III	Urinary porphobilinogen and ALA raised (very high in attack)
	Porphyria variegata	As above	Urinary porphobilinogen and ALA raised in attack
	Hereditary coproporphyria	Coproporphyrin III Uroporphyrin I sometimes	As above
	Porphyria cutanea tarda	Uroporphyrin I	Urinary porphobilinogen and ALA normal or slightly raised Acquired variety (hexachlorobenzene)

Acquired	Toxic agents:
	  Chemicals
	  Heavy metals
	Acute alcoholism
	Cirrhosis in alcoholics
		Coproporphyrin III	Erythrocyte protoporphyrin raised with Pb

	Certain liver diseases:
	  Infectious hepatitis
	  Mononucleosis
	  Cirrhosis in non-alcoholics
	  Hemochromatosis
		Coproporphyrin I	

	Certain blood dyscrasias:
	  Hemolytic anemias
	  Pernicious anemia
	  Leukemia
		Coproporphyrin I	

	Obstructive jaundice	Coproporphyrin I	

	Miscellaneous conditions:
	  Poliomyelitis
	  Aplastic anemias
	  Hodgkin's disease
		Coproporphyrin III	

diversion of its excretion from the bile to the urine. Under these circumstances, the amount in the feces is correspondingly diminished.

Porphyria. This designation is applied to conditions characterized by excessive production and excretion of porphyrins, with or without their precursors (ALA and porphobilinogen). They are usually congenital in nature, often exhibit a marked familial incidence, and are included among the so-called "inborn errors of metabolism." The nature and time of appearance of the striking clinical manifestations differ in different forms. They include, characteristically: cutaneous photosensitivity, with blistering of exposed surfaces, and subsequent ulceration, scarring and mutilation; a peculiar purplish complexion; gastrointestinal and other abdominal complaints; nervous and mental symptoms. The urine has a pink or red color, which intensifies on standing, exposed to light. It may not be apparent in freshly voided urine. The condititon may remain latent for years and may become manifest as the result of some inciting factor, e.g., alcoholism, barbiturates, or hepatic or some other intercurrent disease.

The porphyrias are subdivided into two main classes according to whether excessive production of porphyrins occurs in the bone marrow (erythropoietic porphyrias) or more commonly in the liver (hepatic porphyrias). The latter group frequently show increased production and excretion of ALA and porphobilinogen.

Erythropoietic Porphyrias. These comprise:

(a) *Congenital Erythropoietic Porphyria.* Photosensitivity and severe skin lesions on exposed areas are present from infancy. These may lead to severe deformities of the fingers, nose, ears and eyes. There may be excess hair on the face and limbs. The teeth and bones may be brownish or pink because of porphyrin deposition. The bones show pink fluorescence in ultraviolet light. There is a tendency to hemolysis and defective erythropoiesis, which frequently results in a normochromic anemia with reticulocytosis. There is a marked increased excretion of uroporphyrin I and, to a lesser extent, coproporphyrin I in both urine and feces. Both these porphyrins are present in the plasma and their concentrations are increased in circulating erythrocytes. The disease probably results from an imbalance between uroporphyrinogen I synthase and uroporphyrinogen III cosynthase. The type of inheritance is autosomal recessive and the disease is rare.

(b) *Erythropoietic Protoporphyria.* Photosensitivity is the major manifestation. Although chronic skin lesions may develop, they are not nearly so severe as in the previously mentioned type of porphyria. However, the condition is much more commonly seen. The teeth do not fluoresce in ultraviolet light. There may be increased protoporphyrin and uroporphyrin in the circulating erythrocytes, the plasma and the feces, or in erythrocytes or feces alone. Inheritance is dominant.

(c) *Erythropoietic Coproporphyria.* This condition has been described in only two members of one family. There was some photosensitivity (swelling and itching of skin) and large amounts of coproporphyrin III in the erythrocytes, which also contained increased amounts of proto- and uroporphyrins.

Hepatic Porphyrias. These comprise:

(a) *Acute Intermittent Porphyria.* This disease is extremely rare before the age of 15 years and after the age of 60. It is slightly more common in females. Periodic attacks of abdominal pain are frequent and can be associated with fever and leukocytosis. There may be obstinate constipation and severe neurological and psychiatric disturbances. There is a high mortality rate. Attacks may be precipitated by certain drugs, particularly barbiturates, menstruation, pregnancy, infections or alcohol. Freshly passed urine is often normal in color but on standing in sunlight comes to look like red wine, or it can even appear black. There is an excess of ALA and porphobilinogen in the urine. It is the latter that spontaneously converts to the wine-colored porphyrin. There may be a diabetic type of glucose tolerance curve, an increase in serum protein-bound iodine (PBI) and some degree of hypercholesterolemia. The condition is due to a marked increase of hepatic ALA synthase. The inheritance is dominant.

(b) Porphyria Variegata. The symptoms may be photosensitivity with cutaneous lesions, or acute attacks resembling acute intermittent porphyria, or both. There is an increased excretion of porphobilinogen in the urine only when acute neurological manifestations are present, but it tends to return to normal in remission. There is an increase in urinary copro- and uroporphyrins and a marked increase in fecal protoporphyrin. Uremia and electrolyte disturbances are frequent. Inheritance is dominant and there is an increase of hepatic ALA synthase. The disease may be latent.

(c) Hereditary Coproporphyria. About half of the 30 patients with this disorder had no symptoms of the disease. The remainder had symptoms similar to those of acute intermittent porphyria, which could be precipitated by barbiturates and other drugs. Photosensitivity occurred in one case. During acute attacks, there is an increased urinary output of porphobilinogen and ALA. There is well marked increased excretion of coproporphyrin III, mainly in the feces. In the urine, this does not always occur. There is no increase in erythrocyte porphyrins. The inheritance is dominant.

(d) Porphyria Cutanea Tarda. Patients suffering from this disease have skin lesions (erythema with vesicles or bullae) in response to minor trauma. There may be hirsutism and areas of pigmentation. Acute photosensitivity reactions usually do not occur but can occur. There may be severe disfigurement similar to that of congenital erythropoietic porphyria. The teeth, however, do not show pink fluorescence in ultraviolet light. Coexistent liver disease is quite common. There may be attacks of abdominal pain with fever. The serum iron is frequently raised. Estrogens can precipitate the skin manifestations and chloroquine may produce the abdominal symptoms. The condition is also precipitated by alcohol. Many of the patients are alcoholics. Barbiturates have no adverse effect. There is an increase in urinary uro- and coproporphyrins. This also occurs at times in the feces, which may also contain protoporphyrin. There is an increase in hepatic ALA synthase.

A hereditary basis for the disease has not been firmly established, although some cases appear to be familial. It is possible that there is a hereditary tendency. The disease is much more commonly acquired. It may be secondary to alcoholism or excessive iron intake. Thousands of cases occurred in Turkey as a result of poisoning with hexachlorobenzene, which had been used as a fungicide for wheat.

Acquired Porphyrinurias. The various types of conditions in which excessive amounts of porphyrins are excreted in the urine as a result of purely acquired disease are indicated in Table 5-1. Porphyrinuria as a manifestation of hepatic and biliary tract disorders is considered in detail elsewhere (p. 655). It will suffice here merely to outline certain characteristic features of the conditions in this category.

1. Impairment of hepatic excretory capacity, whether due to primary liver damage (e.g., hepatitis, cirrhosis) or biliary obstruction (e.g., stone, tumor, stricture), is accompanied by increased amounts of coproporphyrin I in the urine and decreased amounts in the feces. This results, at least in part, from diversion of excretion of this substance from the bile to the urine.

2. Excessive exposure to certain toxic agents, including barbiturates, lead, and alcohol, may result in increased urinary excretion of coproporphyrin III. This occurs also in cirrhosis in alcoholic subjects, in contrast to the increase in Type I in "nonalcoholic" cirrhosis.

3. Increased amounts of coproporphyrin I are excreted in the urine in hemolytic disorders and other conditions accompanied by stimulation of erythropoiesis. Excretion of this substance, formed as a by-product of heme synthesis (p. 263), approximately parallels that of urobilinogen.

4. The basis for the increase in urinary coproporphyrin III in poliomyelitis is not known. Nor is the cause of this phenomenon clear in such conditions as aplastic anemia and Hodgkin's disease. It may conceivably be due in some instances to impaired liver function, and in certain cases of aplastic anemia to unsuspected exposure to toxic chemical agents.

References

Bodansky, O.: Methemoglobin and methemoglobin-producing compounds. Pharmacol. Rev., 3:144, 1951.

Finch, C. A.: Methemoglobinemia and sulfhemoglobinemia. New England J. Med., 239: 470, 1948.

Granick, S.: Iron and porphyrin metabolism in relation to the red blood cell. Ann. New York Acad. Sci., 48:657, 1947.

Hsia, D. Y. Y.: Inborn Errors of Metabolism. Chicago, Year Book Publishers, Inc., 1959.

Ingram, V. M.: Hemoglobin and Its Abnormalities. Springfield, Ill., Charles C Thomas, 1961.

Lehmann, H., and Huntsman, R. G.: Man's Haemoglobins. Amsterdam, North-Holland Publishing Company, 1966.

Tschudy, D. P.: Porphyrin metabolism and the porphyrias. In Bondy, P. K., ed.: Duncan's Diseases of Metabolism. Ed. 6. Philadelphia, W. B. Saunders Company, 1969, p. 600.

Watson, C. J.: The probem of porphyria—some facts and questions. New England J. Med., 263:1205, 1960.

Weatherall, D. J.: The Thalassaemia Syndromes. Oxford, Blackwell Scientific Publications, 1965.

Wolstenholme, G. E. W., and O'Connor, C. M., eds.: Biochemistry of Human Genetics. Boston, Little, Brown and Co., 1959.

Chapter 6

CALCIUM AND INORGANIC PHOSPHATE METABOLISM

Although over 99 per cent of the body calcium (20-25 g./kg. fat-free body tissue) and 80 to 85 per cent of its phosphorus is in the bones, important functions of these elements are exerted in directions other than preservation of skeletal structure. Ca ions (*a*) decrease neuromuscular excitability, (*b*) decrease capillary and cell membrane permeability, and are necessary for (*c*) normal muscle contraction (muscles contain 0.3% total body calcium), (*d*) normal transmission of nerve impulses, and (*e*) blood coagulation. Ca ions also activate certain enzymes, an action which is undoubtedly involved in several of the functions mentioned above; they also play a part in the release of certain peptide hormones and other regulatory substances.

Phosphorus is involved in various important phases of organic metabolism. It is necessary merely to recall here the fundamental role of high-energy phosphate bonds in the storage, liberation and transfer of energy (e.g., ATP, phosphocreatine) (p. 16), the importance of hexose- and triose-phosphates in the intermediary metabolism of carbohydrate (p. 7), and the metabolic significance of such P-containing substances as phospholipids (p. 94), nucleic acids (p. 240), and nucleotides (NAD, NADP, etc.) (p. 250). We shall be concerned here with the role of phosphate in inorganic metabolism.

The metabolism of calcium is intimately related to that of phosphate in many important respects, and it is therefore convenient to discuss the two substances together in order to avoid repetition.

Clinical disturbances of calcium metabolism are reflected mainly in abnormalities of serum calcium, skeletal mineralization, and certain neuromuscular disorders. The factors of greatest importance to calcium homeostasis are those concerned with (*a*) its absorption, (*b*) its deposition in and mobilization from the bones and (*c*) its excretion by the kidneys.

ABSORPTION

Calcium and phosphate are actively absorbed in the small intestine, much more in the upper than in the lower portions. In the course of digestion of nucleoproteins and phosphoproteins, phosphate may be split off and absorbed. If ester forms are present, they may undergo hydrolysis (enzymes of pancreatic and intestinal juice) prior to absorption. It has been suggested that because of this fact a considerable fraction of the dietary phosphate is absorbed later than the major portion of the calcium. This may permit better absorption of both elements, especially in regions of relatively low acidity.

The usual dietary intake of calcium is very variable, commonly ranging from 25 to 50 mEq. (0.5 g. to 1.0 g.) per day. Not all ingested calcium is absorbed; between one-quarter and three-quarters are excreted in the feces. In a diet containing virtually no calcium, the ion still appears in the feces to the extent of approximately 3.0 to 7.5 mEq. (60 mg. to 150 mg.) in 24 hours and does not as a rule vary very much from this range. This is known as endogenous fecal calcium and is derived from intestinal secretions and denuded intestinal mucosal cells.

The best dietary source of calcium is milk—or milk products other than butter. Another main source is green vegetables. In some countries, bread and other foods have been reinforced with added calcium.

Several factors influence the degree of absorption of calcium:

(1) Other factors being equal, percentage absorption of calcium decreases as its intake increases. On the other hand, there is an adaptive process, whereby percentage absorption is greatly increased on a low calcium intake.

(2) *Intestinal pH.* Calcium salts, particularly phosphates and carbonates, are quite soluble in acid solutions and are relatively insoluble in alkaline solutions. Consequently, factors which increase intestinal acidity should favor absorption of Ca, and vice versa. Although this factor may play a part, there is now some doubt as to its significance.

(3) *Other Substances in the Food Mixture.* An excess of Mg apparently diminishes absorption of Ca, especially if the latter is not present in adequate amounts. Considerable interest has centered in the Ca:P ratio, the optimum ratio for absorption of both elements being about 1:1(1:2 to 2:1). An excessively high ratio is accompanied by decreased absorption of phosphate (rachitogenic diet, p. 310). This effect is apparently related to the influence on solubility of the products (calcium phosphates) formed under these conditions. In the rat, such substances as iron, beryllium, lead, manganese, aluminum, and strontium, which form insoluble phosphates, likewise interfere with absorption of phosphate and can induce rickets (p. 310).

Phytic acid (inositol hexaphosphate), which occurs in cereal grains, forms insoluble salts (phytin) with Ca and Mg (insoluble above pH 3 to 4), with consequent impaired absorption of these elements. This is regarded as the basis of the rachitogenic effect of such cereals as oatmeal. Poorly absorbed compounds are formed also with fatty acids (Ca soaps).

Hydroxy-acids, such as lactic, citric and tartaric, shift the precipitation point of Ca phytate toward a higher pH (see also action of vitamin D, below, and p. 290), and favor absorption, which is facilitated, too, by the amino-acids, lysine and arginine, and by a high protein intake (with normal digestion and absorption), as well as by lactose.

(4) *Vitamin D*. Vitamin D promotes absorption of Ca. Intestinal absorption of phosphate is apparently increased somewhat. This may be a result of increased Ca absorption, maintaining an optimum local Ca:P ratio.

Vitamin D apparently counteracts the effect of phytic acid in binding the Ca^{++} ions and thus diminishing Ca absorption. In adequate amounts and with a high Ca intake, it suppresses the rachitogenic and anticalcifying effect of phytic acid.

There are a number of related sterol derivatives, which have vitamin D activity. The most important in man are ergosterol and 7-dehydrocholesterol. The former occurs in plants and is converted by ultraviolet irradiation into vitamin D_2 (calciferol). The latter is of animal origin and is converted by ultraviolet irradiation in the skin to vitamin D_3 (cholecalciferol). This is converted in the liver to another derivative (25-hydroxycholecalciferol), which is converted by the kidney to 1,25-dihydroxycholecalciferol. This is absorbed into the intestinal mucosal cells and stimulates the production of specific mRNA, which is concerned with the production of a calcium-binding protein. This protein apparently plays an important part in calcium transport. A similar mechanism exists in bone. It has also been suggested that the vitamin stimulates biological pump mechanisms for calcium absorption, but these may well involve the specific binding protein. The vitamin is concerned with synthesis of a Ca-dependent ADPase.

(5) *Parathormone* increases the intestinal absorption of calcium. This effect is slow and its physiological significance has not been fully established.

(6) *Endocrine Factors*. In animals, hypophysectomy diminishes intestinal calcium transport, which can be restored by growth hormone. Adrenal glucocorticosteroids diminish intestinal transport.

(7) *Age*. After the age of 55 to 60 years, there is a gradually diminishing intestinal transport of calcium. This may be related to a deficiency of vitamin D.

The active mechanism for phosphate absorption requires the presence of sodium ions. Parathyroid hormone stimulates absorption in the jejunum, where absorption is maximal (calcium is maximally absorbed in the duodenum). As has already been mentioned, vitamin D plays an indirect role in stimulating absorption. The effect of intake is very much like that of calcium. The average diet contains 800 to 900 mg. of phosphorus (as phosphate), and its best sources are meat, cereals and dairy products.

DEPOSITION AND MOBILIZATION OF BONE MINERALS

Bone formation involves two distinct fundamental processes: (1) construction of an organic matrix and (2) deposition of bone salt in this matrix. In areas of developing bone, cells derived from primitive mesenchyme turn into osteoblasts, which appear to be responsible for laying down the intercellular organic matrix (osteoid). This has two main components: (a) the protein, collagen, arranged in bundles of long parallel fibers so that the bundles themselves run in many different directions; (b) so-called ground substance, consisting mainly of mucoprotein and mucopolysaccharide, resembling chondroitin sulfate, but not yet definitely characterized. The collagen fibers are embedded in this mucopolysaccharide ground substance, which may be regarded as a sort of cementing material. Moreover, the mucopolysaccharide component is necessary for the preservation of the normal structure and orientation of the collagen fibers.

The cell types are apparently interconvertible, osteoblasts turning into osteocytes, the mature bone cells. When bone undergoes degradation, the osteocytes and osteoblasts form osteoclasts, which are actively involved in the breakdown of bone; these can, in turn, probably be transformed to osteoblasts.

Collagen is a protein, which has a unique amino acid content. Glycine residues make up approximately one-third of the molecule, proline and hydroxyproline residues each compose approximately 10 per cent, and hydroxylysine residues about 1 per cent of the molecule. Collagen is made up of units of molecular weight 300,000, which are known as tropocollagen. These are rodlike and have a length of 300 nanometers and a width of 1.4 nanometers. Each of these units consists of three peptide chains wound around each other in a right-handed helix, which is stabilized by bonding between the chains. To form collagen, these rods are laid down in the manner shown in Figure 6-1.

At first, this gives rise to soluble collagen, which is maintained as a

GLY—X—Y—GLY—PRO—X—GLY—Y—HYP—

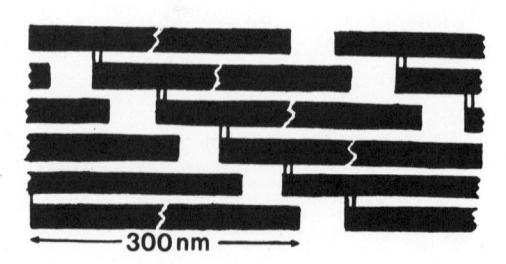

1.5 nm
10.0 nm

300 nm

Figure 6-1. Structure of collagen, showing a typical sequence, one complete turn of the right-handed triple helix, and the assembly of tropocollagen units (shown as broken rectangles) in a collagen fiber. The diagram is not to scale. Note the intramolecular bonding in the triple helix and the intermolecular bonding in the fiber.

discrete pool, but as the overall size increases, eventually the insoluble type of collagen is produced, which forms cartilage and bone osteoid when combined with the appropriate mucoproteins and mucopolysaccharides. The hydroxyproline and hydroxylysine residues are formed by the action of a hydroxylase acting on proline and lysine residues when the collagen peptides have reached an appropriate length. In other words, there is a polypeptide precursor of collagen, which does not contain hydroxy- amino acids. This is known as protocollagen. Ascorbic acid is a co-factor for the hydroxylase. When collagen is broken down, the hydroxy- amino acids are not utilized for further collagen synthesis. Hydroxyproline production and excretion in the urine (as peptides) can therefore be used as a measure of the collagen breakdown which occurs in bone resorption.

The inorganic mineral phase of bone is very similar to hydroxyapatite, $3Ca_3(PO_4)_2.Ca(OH)_2$. Sodium, magnesium, carbonate, citrate and chloride, as well as other ions, are present. It is now virtually the rule for small amounts of radioactive strontium to occur. The mineral crystals of bone tend to be deposited at the regions around the ends of the trophocollagen units. Extracellular fluid is supersaturated with respect to hydroxyapatite, which is maintained in solution by the presence of stabilizing substances, such as pyrophosphate and polyphosphates. If collagen is placed in a supersaturated (metastable) solution of calcium and phosphate ions, it induces the formation of apatite crystals. It is possible that in actual bone formation, phosphate is first covalently bound to the hydroxyl groups of serine residues, and that this is the initiating step for calcification. An appropriate kinase is present in appreciable amounts in collagen-rich tissues. If a collagen fiber is stretched, the zones at the ends of the trophocollagen units become charged. This piezo-electric effect may, at least in part, be responsible for calcification and bone modelling secondary to muscular pull.

The osteocytes are apparently derived from osteoblasts, which have become enclosed within the lacunae of the Haversian systems of compact bone. There are an enormous number of fine radial canalicules, which connect the lacunae with each other. This system creates a very large potential surface area. It has been estimated that the total possible surface area of all the lacunae and canalicules in the skeletal system of a human adult is 1200 sq. meters. It has thus been recognized that the osteocytes play an important part in the relatively rapid release of calcium (osteocytic osteolysis) mediated through the action of parathormone. It has been shown that these cells release alkaline phosphatase (pyrophosphatase), proteases and acid hydrolases.

The osteoclasts are also stimulated by parathormone. They also produce bone resorption and contain the enzyme acid phosphatase. It is currently believed that these cells are concerned with bone remodelling.

The osteoblasts are concerned with the laying down of new bone. They contain glycogen, phosphorylase and the enzyme alkaline phosphatase, which is also a pyrophosphatase. Destruction of pyrophosphate encourages the laying down of bone mineral (hydroxyapatite) from the extracellular fluid, which, as has been pointed out, is already supersaturated with respect to calcium and phosphate concentration. The extracellular fluid, in this respect, is stabilized as a solution by the action of pyrophosphate and other similar molecules, as well as by certain peptides. It is paradoxical that pyrophosphate also stabilizes bone crystals and prevents their solubilization. The alkaline phosphatase of osteocytes could thus bring about solution and since these cells produce enzymes which destroy collagen, the number of possible crystallizing foci is diminished. This prevents the laying down of hydroxyapatite, which should result from destruction of pyrophosphate.

Vitamin D does not really seem to have any direct effect in relation to the laying down of bone. It is apparently necessary in order that parathormone can stimulate absorption and calcium release. Large amounts of the vitamin alone produce bone dissolution.

Calcitonin. This hormone is produced in the parafollicular (derived from the ultimobranchial body) cells of the thyroid, the parathyroid, the thymus and other lymphoid tissue in the neck. Calcitonin inhibits bone resorption and may possibly stimulate the laying down of bone calcium. In this respect, its action is opposite to that of parathormone. It acts, however, at different bone sites from those of parathormone. Moreover, the effect of calcitonin is not mediated through stimulation of adenylcyclase and production of $3',5'$-cyclic AMP, as seems to be the case with parathormone. It is interesting to note that cyclic AMP inhibits pyrophosphatase and this could account, at least in part, for the action of parathormone. Calcitonin stimulates pyrophosphatase activity.

Urinary excretion of hydroxyproline, which reflects breakdown of bone matrix, is diminished by calcitonin. This hormone, like parathormone, is a peptide. Unlike parathormone, its amino acid sequence in the human is already known, and small amounts of the active peptide have been synthesized. There is little doubt that similar success will soon be attained with human parathormone.

The formation of the osteoid matrix is influenced by certain vitamins and hormones. Ascorbic acid is essential for normal function of fibroblasts and osteoblasts and, therefore, for the normal formation of collagen fibers and mucopolysaccharides of the ground substance (p. 821). Adequate amounts of vitamin A are also necessary for normal bone development (p. 805). Growth hormone (somatotropin) has a stimulating influence on the growth of bone (p. 708), and thyroid hormone hastens maturation of the skeleton, i.e., appearance and differentiation of epiphyseal ossification centers and their fusion to the diaphysis. There is evidence that the adrenal glucocorticoids interfere with synthesis of chondroitin sulfate and, consequently, with formation of the bone matrix. Over and above these presumably specific effects, these hormones influence the formation of osteoid tissue by virtue of their effects on protein metabolism in general. Protein anabolism, and therefore osteoid formation, is favored by somatotropin, androgen, and thyroid hormone in physiological amounts; whereas protein catabolism, and therefore inhibition of osteoid formation, is favored by adrenal glucocorticoids and excessive amounts of thyroid hormone. The important action of the parathyroid hormone in stimulating resorption of bone is considered elsewhere (p. 289). The influence of estrogens on the development and metabolism of bone varies widely in different species (p. 291). Their role in this connection in man is suggested by the common occurrence of a form of osteoporosis (defective matrix formation) following the menopause (p. 314).

In common with other tissues, bone undergoes continual metabolic turnover, its various components undergoing degradation, mobilization, and replacement. This applies to both organic and mineral constituents. This turnover is more active in newly formed, incompletely mineralized osteones, the older compact bone being metabolically more stable.

Interrelations of Citrate and Calcium Metabolism. Skeletal citrate comprises a large portion of the total body citrate (about 70 per cent in the mouse). Its intimate relation to the metabolism of calcium may be indicated as follows:

(1) Citrate forms a soluble, poorly ionized complex with calcium and, therefore, can effectively remove calcium ions from solution (basis of anticoagulant and tetanic effects).

(2) Local increases in citrate concentration resulting from cellular activity might conceivably be a factor in promoting dissolution of bone salt, even at comparatively high pH levels. It is believed, however, that the increase in citrate may be a consequence, and not a cause, of the dissolution.

(3) In general, changes in plasma citrate are paralleled by changes in calcium, e.g., after nephrectomy, parathyroid hormone, or vitamin D administration, and after injection of calcium chloride or neutral citrate.

MECHANISM OF REGULATION OF SERUM CALCIUM CONCENTRATION

The concentration of calcium in the blood plasma and other extracellular fluids is maintained within quite narrow limits, even in the face of extensive loss of this element, so long as the regulatory mechanisms and the skeletal stores remain adequate. This must result from close coordination between the amount absorbed from the intestine and that interchanged with bone and excretion by the kidney. Fecal excretion is not important in this respect, since endogenous fecal excretion remains virtually constant over a wide range of serum calcium concentration levels.

The factors affecting intestinal absorption have already been discussed.

Urinary calcium excretion is related exponentially to dietary intake, which in normal subjects means that relatively large variations in intake are accompanied by only small changes in the amount of calcium excreted in the urine. In normal subjects, this may amount to as much as 20 mEq. (400 mg.) in 24 hours. This means that approximately 99 per cent of calcium filtered through the glomeruli is reabsorbed in the tubules. Calcium disappears from the urine when its level in serum falls below 3.75 mEq./l. This statement presupposes that the serum protein levels are normal. Excretion is greater during the day than during the night. It is reduced by parathormone and increased by growth hormone. Increase in plasma calcium is followed by increased urinary excretion. A similar effect results from a fall in plasma pH. The excretion is also inversely related to the dietary intake of inorganic phosphate.

With regard to skeletal interchange, a dual mechanism has been postulated: (1) One which acts rapidly and independently of the parathyroid hormone, concerned with transfer or exchange of ions between the extracellular fluid and the more labile fractions of the skeletal Ca, perhaps representing merely a physicochemical equilibrium between these two media. (2) Another which, mediated by the parathyroid hormone, regulates the rate of degradation and mobilization of the organic and stable mineral components of bone, liberating Ca and other ions from the hydroxyapatite crystal lattice and also the more labile deposits on the crystal surface and in the hydration shell.

Kinetic tracer studies with ^{45}Ca have been used in attempts to determine the avidity with which bone takes up calcium in a variety of different conditions. These indicate the presence of a rapidly miscible pool of Ca (plasma, extracellular fluid and the superficial calcium of bone), which amounts to about 1 per cent of the total calcium present in bone. Other "pools" have been demonstrated. These show slower rates of uptake and in part consist of calcium being incorporated more deeply in bone and in part result from the release of labeled Ca after it has been absorbed. Attempts have been made to assess skeletal calcium accretion, based on measurements at appropriate time intervals of plasma, urinary and fecal calcium and their specific activities, after intravenous injection of ^{45}Ca. In normal young male adults, this was found to vary between 12 and 55 mEq. in 24 hours. The rate of bone resorption corresponded closely to this range, since the subjects did not have any gross disturbances of calcium balance. The figures calculated are almost certainly overestimates because the mathematical models used were oversimplifications of the biological situation. Greater success has been claimed by mathematical analysis using power functions, but it is doubtful whether this approach really is any better, in view of our relatively poor knowledge of what is actually happening to the injected calcium. Variations in bone mineral have also been studied by measurements of bone density by x-ray and by photon beam densitometry. *In vitro* studies on ground sections of bone biopsies have been carried out with low energy x-rays (microradiography). Tetracycline is taken up by osteoid, and after its administration, it can be detected in bone biopsies by virtue of its fluorescence in ultraviolet light. This can be used in relation to bone studies and is especially useful for demonstration of the width of osteoid seams.

As has already been pointed out, measurement of urinary total hydroxyproline can be used to assess the turnover of bone matrix.

Parathyroid secretory activity is regulated by the concentration of Ca ions in the blood plasma, being stimulated by a decrease in the latter and depressed by an increase. Any tendency for the serum calcium concentration to fall is therefore counteracted automatically by mobilization of calcium (a) from the labile deposits in the bones, perhaps by a simple transfer of ions, and (b) from the hydroxyapatite crystal lattice through the medium of increased secretion of parathyroid hormone, other inorganic and organic constituents being mobilized simultaneously as a result of increased degradation of the bone.

BLOOD CALCIUM AND PHOSPHATE

The inorganic phosphate of blood plasma ranges from 3 to 4.5 mg./100 ml. (1.7–2.5 mEq./l.) in adults and from 4.5 to 6.5 mg. (2.5–3.6 mEq./l.) in children. It is somewhat higher in summer than in winter but vitamin D decreases tubular absorption. It decreases during periods of increased carbohydrate metabolism (administration of glucose, insulin, epinephrine or glucagon), due presumably to increased utilization for phosphorylation of carbohydrate metabolites (hexose- and triose-phosphates). A slight increase follows ingestion of large amounts of calcium and a considerable drop follows parenteral administration of magnesium. The kidney is the major means of control of plasma phosphate regulation.

There is virtually no calcium in erythrocytes. The calcium content of plasma (usually determined in serum) is 8.5 to 10.5 mg./100 ml. (4.3 to 5.3 mEq./l.). During infancy and early childhood (to twelve years) the average values approach the upper limits of this range, falling with advancing years.

Calcium exists in the plasma in two fractions, designated (a) diffusible or ionized and complexed (citrate and phosphate) and (b) nondiffusible. The diffusible fraction (50 to 60 per cent of total) is capable of passing through an artificial semipermeable membrane (cellophane, collodion) and, presumably, through the living capillary wall; the nondiffusible fraction (40 to 50 per cent of total) cannot, because of its combination with plasma proteins. Although the two fractions may vary independently of each other under certain abnormal circumstances (e.g., hypoproteinemia, in which the nondiffusible Ca alone may be decreased), they seem to be in a state of rather unstable equilibrium, which is probably controlled to a certain extent by the parathyroid hormone. This agent exerts its influence primarily on the diffusible (ionized) fraction, which probably is the only physiologically active portion of the serum calcium.

The ionized fraction of the serum calcium, normally about 2.25 to 2.50 mEq./l. (4.5 to 5.0 mg./100 ml.), can be measured directly by a biological procedure (i.e., isolated frog heart), which is not suitable for routine clinical purposes. Inasmuch as the nondiffusible (nonionized) fraction is related quantitatively to the serum protein concentration, the calcium ion concentration can be estimated approximately from values for total calcium and total protein (Fig. 6-2). It is, however, best measured by the use of a calcium electrode. The original difficulties presented by this technique have now been overcome.

Total plasma calcium obviously cannot be regarded as a reliable index

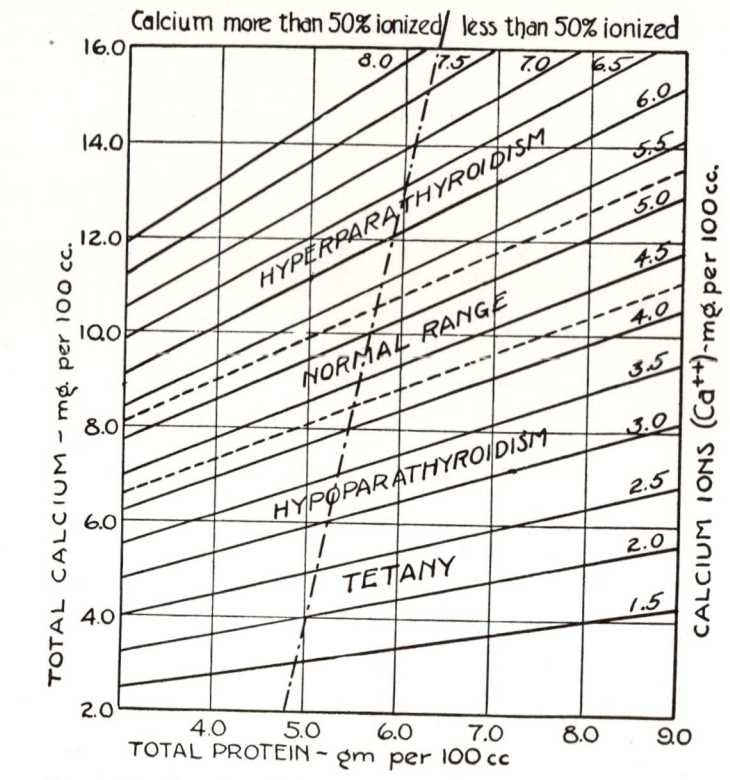

Figure 6-2. Chart for calculation of [Ca^{++}] from total protein and total calcium of serum or plasma. (From McLean and Hastings: Am. J. Med. Sci., 189: 601, 1935.)

of the amount of ionized calcium present. This, of course, is due to the fact that the protein-bound calcium varies independently. Its proportion changes, because of variation in protein concentration, with change in posture, with prolonged venous occlusion or as a result of serum protein changes that can occur in disease. If one could correct for this protein variation, the total serum calcium would be a much more reliable index of the concentration of ionized calcium. Since calcium binds predominantly with albumin, a factor based on the albumin concentration could be used to correct the calcium concentration to what it would be if the albumin concentration were 4.6 g./100 ml. of serum. This figure is a recognized average mean for the serum concentration of this protein. The correction is made by subtracting, from the total calcium value determined, 0.09 mg./100 ml. for every increase of 0.1 g. albumin/100 ml. above the recognized mean value. Conversely, when the plasma albumin concentration is below the mean value, 0.09 mg./ml. is added to the calcium level for every 0.1 g./100 ml. by which the albumin level is below the mean.

Another suggested mode of correction is to calculate the adjusted calcium level in mg./100 ml. from the formula

$$\text{Adjusted Ca (mg./100 ml.)} = \text{Total serum Ca (mg./100 ml.)} - \text{albumin (g./100 ml.)} + 40$$

In relation to SI units (p. 854) the equation to be used is

$$\text{Adjusted Ca (mmol/l)} = \text{Total serum Ca (mmol/l)} - [0.25 \times \text{albumin (g/l)}] + 1.0$$

Maintenance of the plasma Ca concentration within a narrow range is of vital concern to the organism inasmuch as this ion has a profound influence on certain fundamental processes affecting cell function (e.g., membrane permeability, neuromuscular excitability). Certain of the factors concerned in the preservation of the normal level of Ca in the plasma deserve special consideration.

Parathyroid Hormone. In the fasting state (i.e., no absorption from the intestine), the normal plasma Ca concentration is maintained primarily by its rate of excretion and its mobilization from the bones through the action of the parathyroid hormone. Its physiological actions, which are exerted in the metabolism of both calcium and phosphorus, may be illustrated by outlining the consequences of (1) removal of the parathyroid glands, and (2) injection of extracts of these glands (parathyroid hormone).

Effects of Parathyroidectomy. The following phenomena follow removal of all parathyroid tissue: (1) decreased output of phosphate in urine; (2) rise in serum phosphate concentration; (3) decreased urinary calcium output; (4) decrease in serum calcium, largely in the diffusible (ionized) fraction.

Effects of Parathyroid Hormone Injection. The chief consequences of administration of this hormone are: (1) increased urinary phosphate excretion; (2) decrease in serum phosphate; (3) increase in serum calcium concentration; (4) increased urinary calcium but there is decreased renal clearance, because of increased calcium absorption by the renal tubules; (5) increased serum alkaline phosphatase activity. There is an increase, too, in the concentration in the serum of other constituents of bone, e.g., citrate and hexosamine.

The skeleton is the major source of the increased urinary calcium and phosphate. Continued, repeated injections of parathyroid hormone result in the characteristic skeletal picture of diffuse osteitis fibrosa cystica with resorption of trabecular and cortical bone, fractures and deformities, appearance of numerous giant cells and osteoclasts, and necrosis, hemorrhage, and replacement fibrosis of the marrow. Healing processes, which progress simultaneously, are accompanied by the appearance of numerous osteoblasts. The action of parathyroid hormone is exerted not only on the mineral but also on the organic components of bone, as evidenced by an increase of hydroxyproline in the urine. One reflection of this action on the bone matrix is an increase in glucosamine in the blood plasma.

The increased concentration of calcium in the body fluids is often followed by deposition of hydroxyapatite in various tissues, including the skin, kidneys, and cornea. Calculi may develop in the kidneys or lower urinary tract as a result of the high concentrations of calcium and phosphate in the urine. There is sometimes a tendency to deposition of calcium pyrophosphate in joints (pseudogout).

Mechanism of Action. Present evidence suggests that the parathyroid hormone exerts a multiple effect. (1) It decreases renal tubular reabsorption of inorganic phosphate from the glomerular filtrate by stimulating increased renal tubular production and excretion of cyclic AMP; excessive amounts of hormone therefore cause excessive loss of phosphate in the urine and a fall in its concentration in the plasma. There is also increased excretion of Na and K ions and diminished excretion of magnesium ions. (2) It accelerates mobilization of skeletal materials, including calcium, with consequent increase in calcium excretion in the urine, increased calcium concentration in the plasma, and skeletal demineralization (p. 294). This is an early, relatively rapid effect (osteocytic osteolysis). Prolonged action of parathormone results in extensive bone remodelling resulting from the stimulation of osteoclastic activity as well as osteoclastic cell division. In most patients suffering from hyperparathyroidism, however, this is not very marked. There must be some compensatory mechanism (possibly calcitonin). The increase in serum alkaline phosphatase activity in hyperparathyroidism is largely a reflection of increased osteoblastic activity. There is, in most patients, stimulation of osteoid formation.

Control of Parathyroid Secretory Activity. There is no evidence to indicate nervous or hormonal control of parathyroid function. Secretory activity is apparently regulated by the plasma calcium ion concentration, decrease in which stimulates and increase depresses parathyroid hormone secretion. Inasmuch as the effect of this hormone is to raise the level of calcium in the blood, this is an example of a very efficient autoregulatory feed-back mechanism, in which the concentration of a constituent of the blood regulates itself.

Vitamin D. As indicated elsewhere (p. 281), the action of vitamin D in maintaining the normal plasma Ca concentration depends chiefly on its effect in enhancing absorption of Ca from the intestine. In the presence of vitamin D deficiency, the plasma PO_4 usually falls sooner than the Ca concentration. This is due in part to stimulation of parathyroid activity, which tends to maintain the normal Ca level and, simultaneously, to lower the plasma phosphate. In addition, intestinal absorption of phosphate diminishes as calcium accumulates within the lumen.

Calcitonin. An increase in plasma ionized calcium levels is the stimulus for the production of calcitonin, which then causes a deposition of calcium in bone. Under experimental conditions, it has been possible, in animals, to lower plasma calcium levels by injection of this peptide hormone. Nevertheless, in medullary carcinoma of the human thyroid, the plasma ionized calcium levels may remain within normal limits, even though there is a high production of calcitonin. This is not believed to be due to a compensatory parathyroid hyperplasia. Some authorities believe that, in man, the action of calcitonin is to protect bone from the demineralizing effect of parathormone.

Plasma Proteins. Calcium exists in the plasma in higher concentration than would be possible in the absence of protein. As indicated before (p. 287), about half of the plasma (nondiffusible fraction) is bound to plasma proteins, principally albumin, decrease in which may be accompanied by a decrease in the serum Ca concentration. The calcium-binding potentialities of the plasma proteins may not be "saturated" under normal conditions, so that frequently no significant fall in serum Ca occurs at slight or moderate levels of hypoproteinemia. Such variations, since they are confined to the nondiffusible Ca fraction, are not accompanied by manifestations of altered Ca metabolism or neuromuscular excitability (e.g., tetany), and hypocalcemia of this type (i.e., due to hypoproteinemia) does not cause stimulation of parathyroid function.

Plasma Phosphate. There is a roughly reciprocal relationship between the concentrations of Ca and PO_4 ions in solutions in vitro. In the body fluids, this relationship is exhibited to a limited extent and practically only in one direction; i.e., rather marked increase in serum PO_4 may cause a fall in serum Ca concentration, the mechanism being unknown. This occurs clinically in advanced renal insufficiency with high serum PO_4 concentrations (p. 309).

Acid-base Equilibrium. The well-known occurrence of tetany in alkalosis, in the presence of normal total calcium concentrations, suggests that the plasma pH may influence the degree of ionization of the serum calcium. There has been no direct evidence of such an effect. However, the belief has been prevalent for many years that there is a quantitative relationship between the concentrations of hydrogen, bicarbonate and phosphate ions, as expressed in the following formula:

$$\frac{[Ca^{++}][HCO_3^-][HPO_4^=]}{[H^+]} = k \text{ (constant)}$$

According to this concept, the concentration of calcium ions decreases as the bicarbonate and phosphate ion concentrations increase, and as the hydrogen ion concentration decreases.

Miscellaneous. Ingestion of adequate amounts of soluble calcium salts in the postabsorptive state results in elevation of serum calcium, which reaches a maximum in two to three hours and returns to the previous level in about four hours. After intravenous injection, the peak is reached in a few minutes, with a subsequent fall to normal, usually within one to two hours, depending upon the quantity given. Following intramuscular injection (calcium glucogalactogluconate) a maximum level is reached in about one hour, with a subsequent fall over a period of three to four hours. Intravenous or intramuscular injection of magnesium salts may cause a fall in serum calcium, at times to tetanic levels.

MISCELLANEOUS FACTORS

Estrogens and androgens exert an influence in calcium metabolism which varies widely in different species. Marked hypercalcemia occurs in birds during and just before ovulation, and may also be induced by administration of estrogen. This increase is largely in the nondiffusible fraction. Although this phenomenon does not occur in mammals, in rats estrogens produce hyperossification of the proximal epiphyseal zone of the tibia and thickening of the trabeculae, filling the upper fifth of the diaphyseal marrow cavities. This may be prevented by administration of androgens. It is questionable whether these observations are applicable in man. However, after the menopause, a type of osteoporosis often develops, with negative Ca balance, which may be benefited by administration of estrogen and/or androgen.

Osteoporosis and negative Ca balance may occur in subjects with hyperthyroidism or hyperfunction of the adrenal cortex. This may be due to the protein catabolic action of excessive amounts of thyroid hormone and the adrenal 11-oxysteroids, affecting primarily the osteoid matrix and secondarily, therefore, skeletal mineralization.

Decreased alkalinity of the body fluids (acidosis), due to any cause, may result in increased urinary excretion of calcium. The excess is presumably removed from the bones because of the increased solubility of the labile calcium deposits in a more acid medium.

EXCRETION

Normal adults on an adequate calcium intake are in a state of calcium equilibrium. Ca is excreted in the urine, bile, and digestive secretions. Much of that in the feces is food Ca which has escaped absorption. This represents a variable but considerable fraction of the intake, since even under optimal conditions dietary Ca (e.g., in milk) is not more than 75 per cent absorbed. The remainder of the fecal Ca is unabsorbed Ca of the digestive fluids and denuded mucosal cells.

Conditions which interfere with absorption of Ca from the intestine increase the amount excreted in the feces.

The "renal threshold" for Ca excretion, at normal serum protein concentrations, lies between 3.25 and 4.00 mEq./l. (6.5 and 8.0 mg. per 100 ml.) plasma, little being eliminated at lower serum Ca concentrations. Under normal conditions, and despite wide variations in calcium intake, the amount excreted in the urine remains rather constant at about 6.75 to 20.0 mEq. (135–400 mg.)/day. Since approximately 9 g. of calcium leave the blood plasma daily in the glomerular filtrate, about 99 per cent is normally reabsorbed in the tubules. Even in states of marked hypercalcemia, only a relatively small fraction of the increment in the glomerular filtrate escapes in the urine. Moreover, following intravenous injection of as much as 1 g. of calcium within a few hours, the serum calcium concentration returns rapidly to normal, although only 30 to 50 per cent of the injected calcium appears in the urine within the next 24 hours. The kidneys, therefore, possibly do play a role in the mechanism of regulation of the serum calcium concentration. The importance of the kidneys for Ca homeostasis under normal conditions is still the subject of debate. It has been postulated that the renal mechanism is highly important and that in this respect the effect of parathormone on increasing renal tubular absorption of calcium ions is an important factor. Calcitonin, which induces hypercalciuria, could also play a part. This is in direct contrast to the point of view that the bone mechanisms are responsible for calcium homeostasis. Prevailing opinion seems to favor the latter situation, but, as in many such arguments, the truth probably lies somewhere between the two schools of thought. There is no doubt, however, that in disease of the kidney the level of plasma Ca may be considerably altered and that this depends largely on renal tubular dysfunction and altered glomerular filtration rate.

Calcium is present in normal urine in the ionic state and in the form of complexes with citrate and other organic anions.

Inorganic phosphate is excreted in the urine and feces, the relative proportions varying under different conditions, e.g., those which influence absorption from the intestine (q.v.). The source of urinary inorganic phosphate is chiefly that of the blood plasma, although it may be contributed to by hydrolysis of phosphoric acid esters by phosphatase activity in the kidneys. On a balanced diet, urine phosphate constitutes about 60 per cent of the total excretion. As the Ca intake is decreased, the proportion eliminated in the urine increases (increased intestinal absorption), being about 75 per cent on a low Ca, moderately high PO_4 intake. The "renal threshold" for phosphate excretion is about 2 mg./100 ml. of plasma, excretion falling to very low levels at lower concentrations. Reabsorption of phosphate from the glomerular filtrate by the renal tubular epithelium is inhibited by the parathyroid hormone. Administration of this agent (or clinical hyperparathyroidism) causes increased urinary excretion of phosphate. Although vitamin D may possibly have a similar effect under certain experimental conditions, it has an opposite effect (i.e., increases tubular reabsorption of phosphate) in the normal subject. This effect is probably more important than parathormone with respect to phosphate homeostasis, in which the kidneys appear to play the major part.

There is enough calcium and phosphate in normal urine to result in precipitation, since the normal solubility product is often exceeded. Urine, however, contains pyrophosphate and low molecular weight peptides, which prevent precipitation under normal conditions.

CALCIUM AND PHOSPHORUS REQUIREMENT

The adult daily requirement for phosphorus is about 700 mg. Diets adequate in other respects provide an adequate amount of phosphorus because of its wide distribution in common foods (phosphoproteins and phospholipids, inorganic phosphate), especially milk and milk products, wheat, meats and fish.

The minimal daily adult requirement for calcium is 22.5 mEq. (0.45 g.), but the recommended allowance is 50 mEq., i.e., 1.0 g. (10–15 mg./kg. body weight), increasing to 75 mEq. (1.5 g.) or more during pregnancy and lactation. Growing children require about 45 mg./kg. body weight. Even under optimal conditions, not more than 35 to 50 per cent of dietary calcium is absorbed. Milk (1.0 g. per quart) and dairy products constitute the most important dietary source of this element, cereals, meats and most vegetables being low in calcium. Even those which contain relatively large amounts (viz., certain vegetable greens and shellfish) are usually not eaten in sufficient quantities to be of practical value in this regard. The calcium requirement may be supplied satisfactorily in the form of inorganic salts (e.g., phosphate, lactate, chloride, gluconate).

CALCIUM CONTENT OF OTHER BODY FLUIDS

It is probable that the cerebrospinal fluid and normal tissue fluids contain the diffusible fraction of serum calcium, namely, 4.5 to 5 mg. per 100 ml. In the presence of protein in tissue fluids their calcium content is increased by an amount proportional to the quantity of protein present, the added calcium being nondiffusible in nature and the diffusible fraction remaining within the limits mentioned above.

ABNORMAL SERUM CALCIUM

Deviation from the normal state of calcium in the blood serum may be manifested in one of two ways, namely, by an alteration in the total serum calcium concentration or by an alteration in the distribution or partition of calcium in the blood and tissue fluids.

Hypercalcemia

Primary Hyperparathyroidism. One of the most striking results of the injection of parathyroid hormones is an increase in serum calcium, the degree of rise being dependent upon the dosage and the frequency of its administration. This hypercalcemia occurs even though calcium is withheld from the diet, the increased quantity of circulating calcium being derived from the bones. The sequence of metabolic events has been indicated elsewhere (p. 289). The clinical picture of hyperparathyroidism induced in experimental animals by prolonged administration of excessive amounts of parathyroid hormone is virtually identical with one which can occur in man as a result of hypersecretion by neoplastic or occasionally hyperplastic parathyroid glands (generalized osteitis fibrosa cystica; von Recklinghausen's disease of bone). This is by no means the most common clinical presentation. Symptomatic bone disease occurs in 10 to 25 per cent of patients. Kidney disease (nephrolithiasis) is much more common (60 to 75 per cent). Duodenal ulcers occur in about 25 per cent of these patients.

Pancreatitis may also occur. It is possible that hyperparathyroidism may result in pseudogout.

The characteristic, indeed almost pathognomonic metabolic manifestations have been regarded as (*a*) hypercalcemia, (*b*) hypophosphatemia, (*c*) increased urinary excretion of calcium, and (*d*) increased serum alkaline phosphatase activity. In actual practice, hypercalcemia may be the only demonstrable biochemical abnormality. This is by no means constantly present. The diagnosis, in suspected cases, should not be excluded until at least five different serum samples, collected on different days, have shown calcium levels within the "normal range" for the particular laboratory concerned. Determination of the ionic calcium with a calcium electrode is a more sensitive indicator. Unless hypercalcemia has been demonstrated, the diagnosis of hyperparathyroidism should not be made. In contrast to other causes (sarcoidosis, myelomatosis, vitamin D intoxication), the hypercalcemia is resistant to the administration of corticosteroids. Hypercalciuria on restricted calcium intake is relatively common.

Serum calcium values above 10 mEq./l. (20 mg./100 ml.) are rare. In uncomplicated cases, the increase in serum calcium is usually but by no means invariably accompanied by a decrease in serum phosphate to levels of 1.0–2.5 mg./100 ml., the latter phenomenon being of considerable diagnostic significance (p. 311). Occasionally, in cases of primary hyperparathyroidism complicated by renal insufficiency (nephrocalcinosis, nephrolithiasis, pyelonephritis), the serum calcium concentration may fall to normal levels coincidentally with increase in serum phosphate concentration (pp. 309 and 529). Under such circumstances the condition may be confused with "renal rickets" (p. 303) but, as a rule, diagnosis is not difficult.

The serum alkaline phosphatase activity is generally increased when there is significant bone involvement. This is a reflection of osteoblastic activity, i.e., stimulation of bone repair.

As indicated above, renal failure may supervene as a result of complicating nephrolithiasis or renal calcification, often with superimposed infection (pyelonephritis). The clinical picture then includes metabolic manifestations of renal functional impairment, e.g., nitrogen retention, acidosis, hyperphosphatemia. It is interesting that hypertension may occur without renal involvement.

Rarely, the condition may be fulminating (acute hyperparathyroidism), resembling acute experimental parathyroid intoxication, with vomiting, diarrhea, shock, and renal failure.

Recently, radioimmunoassay has been utilized to measure the level of parathormone in the circulation. This is usually increased in patients with hyperparathyroidism. However, a number of patients with proven parathyroid adenomas have had no demonstrable increase. Parathormone determination of multisite samples of blood obtained by catheterization of major veins (neck, thorax, abdomen) is proving to be useful in the location of parathyroid adenomas.

Secondary Hyperparathyroidism. This condition is not as a rule characterized by hypercalcemia but does present many other features of primary hyperparathyroidism. Inasmuch as parathyroid activity is stimulated by a decrease in the concentration of calcium ions in the blood, parathyroid hyperplasia and secretory hyperactivity occur in many conditions associated with a tendency toward diminution in the serum calcium concentration. These include:

1. Chronic glomerulonephritis, renal hypogenesis, and other forms of chronic renal disease associated with "renal rickets" (p. 303).
2. True rickets and osteomalacia (vitamin D deficiency).
3. Various types of chronic steatorrhea (sprue, celiac disease).
4. Pregnancy and lactation.

The parathyroid enlargement and hypersecretion occur in response to a decrease in the plasma calcium ion concentration, the normal stimulus to parathyroid activity (p. 290). In some cases, this results from decreased absorption from the intestine (vitamin D deficiency, sprue, celiac disease), in others from hyperphosphatemia (renal failure). In pregnancy, it is a consequence of the increased requirement for calcium during the period of growth of the fetal skeleton. During lactation, the serum calcium tends to decrease as a result of the excretion of calcium in the milk.

There should be no confusion between primary hyperparathyroidism and these conditions of secondary hyperparathyroidism. Hypercalcemia does not occur characteristically in these disorders. The enlargement and hyperactivity of the parathyroid glands constitute a mechanism designed to maintain a normal plasma calcium level in the face of a tendency toward hypocalcemia. If the blood calcium concentration should rise above normal, parathyroid activity would be depressed automatically, with prompt decrease in the serum calcium. This fact, together with differences in serum phosphate and phosphatase activity, usually aids materially in distinguishing between true, i.e., primary hyperparathyroidism and secondary hyperparathyroidism.

On occasion, however, an adenoma may apparently result from the hyperplasia of secondary hyperparathyroidism. The condition is then referred to as *tertiary hyperparathyroidism*. This gives rise to hypercalcemia, but the other biochemical findings are to a large extent dependent on the condition which gave rise to the secondary hyperparathyroidism.

Familial Hyperparathyroidism. This may occur without any other endocrinopathy. It may be associated (Werner's syndrome) with pituitary and pancreatic adenomas, often with peptic ulceration from associated gastric hypersecretion. This is an example of multiple endocrine adenomatosis (Type I). Another example (Type II), recently described, is the association of hyperparathyroidism with pheochromocytoma and medullary carcinoma of the thyroid.

Diagnostic Aids in Hyperparathyroidism. No difficulty is encountered in typical severe cases with hypercalcemia, hypophosphatemia, increased serum alkaline phosphatase activity, and characteristic renal or skeletal abnormalities. However, mild or atypical cases may present diagnostic difficulties, as may other conditions accompanied by hypercalcemia and/or abnormalities of skeletal mineralization. Additional test procedures have been proposed: (1) renal tubular reabsorption of phosphate (decreased in hyperparathyroidism); (2) response to phosphate deprivation (less than normal increase in tubular reabsorption of phosphate in hyperparathyroidism); (3) calcium tolerance test; (4) parathormone estimation. It is difficult to interpret the results of such tests, however, in the presence of the complicating renal damage, frequently seen.

Hypervitaminosis (Vitamin D). Hypercalcemia may result from the administration of excessive quantities of vitamin D. There appears to be a wide range of individual susceptibility to the introduction of this factor, and the coincident administration of large quantities of calcium salts increases the tendency toward the production of hypercalcemia. Cases have been reported in which skeletal demineralization, with hypercalcemia and metastatic calcification in various situations, especially the kidneys and the blood vessels, followed administration of as little as 50,000 units daily for a few weeks. In others, toxic manifestations appeared only after administration of 500,000 units daily for six to twelve months. Pre-existing renal disease predisposes to the development of this condition. The concentrations of calcium and phosphate in the blood and their excretion in the urine are increased, the calcium and phosphate balances becoming negative unless large amounts of these elements are administered. In children, there is increased deposition of calcium in the zone of provisional calcification. If the intake is inadequate, demineralization occurs in the shafts of the bones, with continued deposition at the epiphyses. Thus, in infants, osteoporosis may progress simultaneously with the healing of rachitic lesions. In experimental animals, bone changes have been produced which superficially resemble those of hyperparathyroidism but are not histologically identical with the latter. Degenerative changes occur in the renal tubular epithelium, blood vessels, heart, stomach, intestines, liver, and bronchi, with metastatic or dystrophic calcification in many of these tissues. Degeneration and abnormal calcification are produced most readily when the diet is high in phosphate and adequate in calcium.

The administration of therapeutic doses of dihydrotachysterol (A.T. 10), a substance chemically related to vitamin D, results in an increase in serum calcium. The physiological effect of this substance appears to be more closely analogous to that of the parathyroid hormone than does that of vitamin D. It is used commonly in the treatment of hypoparathyroidism.

Multiple Myeloma. Serum calcium values ranging from normal to 10 mEq./l. (20 mg./100 ml.) have been observed in patients with multiple myeloma. Hypercalcemia has been reported in about 50 per cent of the cases in which mention was made of the serum calcium concentration, the values being above 6 mEq. in 20 to 45 per cent. The serum phosphate concentration is usually normal or elevated (renal functional impairment) and the serum phosphatase activity is usually normal. The increase in serum calcium may be dependent in some cases upon the frequently increased serum protein concentration. However, the serum calcium is normal in many cases with marked hyperproteinemia and may be elevated in the presence of normal total protein concentrations. It seems that actual destruction of bone by the progressing skeletal lesion, with liberation of bone minerals into the blood stream, is the main underlying cause of hypercalcemia in multiple myeloma. This is probably contributed to by the presence of renal functional impairment, a common feature of this disease.

Neoplastic Disease. The serum calcium concentration is usually within normal limits in the great majority of cases of primary and metastatic neoplasms of bone. However, values as high as 11 mEq./l. have been reported, particularly in cases of extensive osteolytic metastatic involvement of the skeleton. The primary tumors are usually in the breast, bronchus, kidney, or thyroid. In such cases, values for serum phosphorus and serum phosphatase activity are usually within normal limits, except in the case of osteogenic sarcoma, in which the serum phosphatase activity may be increased.

It is now well recognized that certain carcinomas can give rise to hypercalcemia in the absence of metastases. This is because they produce a peptide, probably identical with parathormone, which mobilizes skeletal calcium. The process is referred to as ectopic production of a hormone. Many cancers, not arising from endocrine tissue, can give rise to hormones. This is probably part of the process of de-differentiation associated with malignant neoplasms.

Some breast cancers give rise to hypercalcemia by the production of an osteolytic sterol. Large doses of androgen and estrogen have been employed in the treatment of inoperable mammary carcinoma with skeletal metastases. Severe hypercalcemia may develop in such cases, usually with rapid progression of the osteolytic lesions, presumably the source of the added calcium. The serum calcium may rise rapidly to extremely high levels, i.e., over 20 mg./100 ml. The hypercalcemia is accompanied by nausea, vomiting, and dehydration and may, in severe cases, result in precipitation of calcium in the renal tubules with consequent progressive renal damage and renal functional insufficiency. The serum inorganic phosphate is usually elevated simultaneously.

Immobilization. Hypercalcemia may occur, particularly in children, during prolonged periods of immobilization, e.g., due to fracture or paralysis, especially in the presence of simultaneously impaired renal function. Restriction of activity removes the normal stimulus to anabolic processes in the skeleton, with consequent decreased resynthesis of osteoid tissue and, therefore, defective mineralization. Whereas catabolic processes continue unimpaired in the skeleton, its ability to assimilate calcium is diminished and, with continuing absorption from the bowel, the serum calcium concentration rises. In contrast to hyperparathyroidism, the serum phosphate is usually either normal or elevated. Serum alkaline phosphatase activity is normal or low.

Sarcoidosis. Hypercalcemia occurs occasionally in patients with sarcoidosis, values above 7 mEq./l. having been reported. There may also be hypercalciuria, renal calculi, nephrocalcinosis (with renal insufficiency), generalized calcinosis, band keratitis, and other manifestations of hypercalcemia. These findings may so dominate the clinical picture as to suggest hyperparathyroidism; the skeletal lesions of sarcoidosis may cause additional confusion, and the serum alkaline phosphatase activity may be increased.

There is evidence that hypercalcemia is due to increased sensitivity to vitamin D. In many cases, administration of cortisone (150 mg. cortisone acetate, or its equivalent, daily in divided doses) is followed by decrease in serum and urine calcium. This response does not occur in hyperparathyroidism but has been observed in hypervitaminosis D, infantile idiopathic hypercalcemia and, occasionally, in multiple myeloma and other types of malignancy. The diagnosis of sarcoidosis is supported, too, by the presence of hyperglobulinemia (p. 201) and other clinical manifestations, including hepatosplenomegaly.

Idiopathic Infantile Hypercalcemia. This condition has been detected usually between the ages of 4 months and 2 years. The cause is unknown, but there is evidence that increased sensitivity to vitamin D is a factor in some cases. This is supported by reports of favorable response to administration of cortisone. Two main forms have been described, which may represent separate clinical entities or different degrees of severity of the same disease. (1) A mild form with recovery, apparently complete, in a few months or years. The increased serum calcium concentration is accompanied by the usual clinical manifestations of hypercalcemia, i.e., anorexia, vomiting, constipation, polyuria, and muscular hypotonia. There are no skeletal changes and no significant retardation of physical or mental development. (2) A severe form is characterized by (a) hypercalcemia, with all of its consequences, including renal damage and progressively increasing renal function impairment, (b) osteosclerosis, and (c) retardation of physical and mental development. There is a rather characteristic "elfin" facial appearance, with low-set ears, prominent epicanthic folds, wide mouth, overhanging upper and large lower lip, underdevelopment of the bridge of the nose and, at times, strabismus. Nephrocalcinosis, which occurs frequently, may be complicated by kidney infection, and there may be calcific deposits in the lungs and muscles. There is reduction in size of the brain and skull and increased radiologic density of the base of the skull and the epiphyses, at times of sufficient degree to be mistaken for osteopetrosis. The severe form progresses to death from renal failure or to permanent mental retardation.

Milk and Alkali Intoxication. A rather characteristic syndrome may occur in subjects with peptic ulcer who ingest excessive amounts of milk and absorbable alkalis. This is characterized by hypercalcemia, alkalosis, renal functional impairment and abnormal calcium deposits, without hypophosphatemia or increase in urine calcium. Practically all patients have band keratitis and calcium deposits in the conjunctivae.

It is believed that the hypercalcemia results from the excessively high intake of calcium in the presence of renal impairment induced or aggravated by alkalosis due to the alkali intake. This condition superficially resembles hyperparathyroidism complicated by renal failure, but the differential diagnosis is usually easy. Unless irreversible kidney damage has occurred, the clinical and laboratory manifestations improve dramatically in many instances when milk and alkalis are withdrawn and the alkalosis, and the hypochloremia and hypopotassemia (due to vomiting), are treated by parenteral administration of proper solutions of electrolytes.

It must be remembered, however, that duodenal ulcer occurs in unusually high frequency in subjects with hyperparathyroidism. Administration of milk and absorbable alkalis to such patients may precipitate acute hyperparathyroidism, with nausea, vomiting, lethargy, prostration, and azotemia, which may terminate fatally if not recognized promptly.

Miscellaneous. Ingestion of adequate amounts of soluble calcium salts in the postabsorptive state results in elevation of serum calcium, which reaches a maximum in two to three hours and returns to the previous level in about four hours. After intravenous injection, the peak is reached in a few minutes, with a subsequent fall to normal, usually within two to three hours, depending upon the quantity given. After intramuscular ingestion (calcium glucogalactogluconate), a maximum level is reached in about one hour, with a subsequent fall over a period of three to four hours.

The serum calcium concentration may be elevated in rare instances of advanced nephritis with uremia, although any deviation from the normal in this condition is usually in the opposite direction (p. 303). The hypercalcemia is difficult to explain; it may be due in part to inadequate excretion of calcium and in part to inadequate skeletal assimilation of calcium, and is aggravated by a high calcium intake.

Hypercalcemia has been observed occasionally in leukemia and in Hodgkin's disease, the mechanism of its production being unknown. It can also occur in hyperthyroidism and in Addison's disease. It can occur in Paget's disease of bone, especially when there is prolonged immobilization.

Hypocalcemia

Hypoparathyroidism. Hypocalcemia is one of the most constant features of diminished parathyroid function. Clinically, this condition occurs in two forms: (*a*) postoperative, following thyroidectomy or removal of a parathyroid adenoma, and (*b*) idiopathic, a condition analogous to spontaneous hypothyroidism (autoimmunity). A third rare variety, which is temporary, may occur as a result of irradiation damage after ^{131}I therapy for hyperthyroidism. Temporary hypoparathyroidism can occur in infants with hyperparathyroid mothers. Occurring after thyroidectomy, it is usually due to temporary suppression of parathyroid function as a result of trauma, edema, or hemorrhage, or interference with the blood supply. It is occasionally due to the removal of the parathyroids.

The characteristic metabolic features of hypoparathyroidism are: (*a*) hypocalcemia, (b) increased serum P concentration, (c) decreased urinary excretion of Ca and P, with positive Ca and P balances, (d) normal or occasionally raised serum phosphatase activity, (e) normal acid-base equilibrium, and (f) perhaps increased bone density.

One of the most important manifestations is the diminished serum Ca concentration. This is usually between 3.5 and 4.0 mEq./l. in latent, and between 2.0 and 3.0 mEq./l. in manifest, tetany.

In the absence of parathyroid hormone the organism is unable (a) to excrete normal amounts of inorganic phosphate in the urine and (b) to mobilize calcium from the bones as actively as normally. In subjects with no functioning parathyroid tissue, the serum calcium may fall to 2.5 mEq./l., or even lower, and the serum phosphate may rise to 12 mg. per 100 ml.; calcium may disappear from the urine when the serum calcium falls below 3.5 to 4.0 mEq./l. The hypocalcemia is accompanied by a decrease in the ionized fraction. The nondiffusible fraction may or may not be decreased.

The relation of these chemical changes to the state of neuromuscular hyperirritability which is a dominant characteristic of the parathyroprivic organism is not entirely clear in certain details. The prevailing view is that the increased irritability is due predominantly to a decreased concentration of ionized calcium in the circulating fluids. An increase in inorganic phosphorus accentuates both the hypocalcemia and the tetany, although, as in rickets, hypocalcemic tetany can occur without a significant increase in serum phosphate. As a rule, however, an increase in serum phosphate concentration aggravates the tetanic manifestations at any given level of hypocalcemia, probably by decreasing the calcium ion concentration (p. 291).

Pseudohypoparathyroidism. Cases have been observed presenting the metabolic and clinical picture of hypoparathyroidism, but in which the fundamental disturbance appeared to be diminished end-organ responsiveness rather than merely lack of hormone. There is usually a demonstrable increase in the level of parathyroid hormone in the plasma. In contrast to true hypoparathyroidism, large doses of parathyroid hormone exert little or no effect on the serum calcium and urinary phosphorus excretion in these cases (Ellsworth-Howard test). They are also somewhat resistant to dihydrotachysterol and vitamin D, but not completely so. Large doses of vitamin D_2 raise the serum calcium level and bring about an increased urinary phosphorus excretion after the administration of parathormone.

Children with this condition present a rather characteristic appearance, with short stature, shortness of the extremities in relation to the trunk, round face with thick neck, fat stubby hands, short fingers, and short metacarpal and metatarsal bones. Abnormal glucose tolerance is often present. There is no increase of urinary cyclic AMP after injection of parathormone, which contrasts with the increase occurring in normal subjects.

Pseudo-pseudohypoparathyroidism. This designation has

been applied to a syndrome of multiple congenital defects characterized clinically by the typical morphologic stigmata of pseudohypoparathyroidism, but with none of the characteristic metabolic abnormalities.

Vitamin D Deficiency (Rickets and Osteomalacia). Deficiency in vitamin D (p. 827) results characteristically in rickets in young children and osteomalacia in older children and adults. In the majority of early cases the serum calcium is within normal limits, the serum phosphate being characteristically decreased and the serum alkaline phosphatase activity increased. In some instances, however, especially when the calcium intake is low, hypocalcemia occurs (4 to 7.5 mg. per 100 ml.), with manifestations of tetany (infantile tetany; spasmophilia; osteomalacic tetany).

Although the primary effect of vitamin D deficiency is diminished absorption of calcium from the intestine, hypocalcemia is usually not an early manifestation. This is due presumably to the fact that the tendency toward lowering of the serum calcium concentration stimulates the parathyroids (enlargement and hypersecretion), with consequent increased mobilization of skeletal calcium and lowering of the serum phosphate concentration. Hypocalcemia appears when this compensatory mechanism becomes inadequate, e.g., when the readily mobilizable calcium of the bones has been exhausted. There is also probably diminished intestinal absorption of phosphate (secondary to diminished calcium absorption).

True dietary deficiency does occur in underdeveloped countries and among food faddists. In western communities, the deficiency largely results from a decrease in endogenous vitamin D_3 synthesis due to lack of exposure to sunlight or to the melanin content of the skin, as in the black-skinned members of the community.

Steatorrhea. This condition is encountered in celiac disease, sprue, prolonged obstructive jaundice, and pancreatic disease or pancreatic duct obstruction. It can also occur after partial gastrectomy, after intestinal resections, in the blind-loop syndrome and in regional ileitis, as well as in certain skin diseases. The outstanding feature is defective absorption of fat and fatty acids from the intestine. Defective absorption of calcium results from the formation of large quantities of insoluble calcium soaps and also from the associated defective absorption of the fat-soluble vitamin D, both aggravated by the associated diarrhea. These abnormalities lead to dwarfism and rickets in infants and older children, osteomalacia in adults, and hypocalcemia, frequently hypophosphatemia, and excessive loss of calcium and phosphate in the feces.

Since at least 50 to 80 per cent of dietary fat is actually absorbed, it is no longer believed likely that the steatorrhea itself is mainly responsible. Conditions producing an increase of fecal fat are also frequently associated with defective secretion of bile and reduction in intestinal absorptive surface due to flattening of the intestinal villi. Both these factors play an important part in a general absorption defect, which, among other substances, affects calcium, phosphorus and vitamin D.

Starvation. A diet deficient in total calories, calcium, phosphorus and vitamin D may produce a slowly progressive osteomalacia.

Nephrotic Syndrome.

Serum calcium values as low as 5.7 mg. per 100 ml. have been observed in patients with the nephrotic syndrome in the absence of renal failure. This hypocalcemia is due to a decrease in the nondiffusible fraction which is associated with marked diminution in the concentration of serum protein, which is a characteristic feature of this condition. In no case is there significant alteration in the amount of diffusible calcium, and increased neuromuscular excitability is not observed. In nephrotic children, defective absorption of protein from the intestine is accompanied by poor absorption of calcium; during periods of remission, decrease in fecal loss of nitrogen is accompanied by increased absorption of calcium.

Hypocalcemia in this condition is accompanied by pronounced decrease in urinary excretion of calcium and increase in fecal calcium, with generalized rarefaction of diaphyseal portions of the bones, especially in children.

Mild grades of hypocalcemia have been reported in other conditions accompanied by hypoproteinemia, e.g., kala-azar, malignancy, prolonged obstructive jaundice, and severe protein malnutrition.

Chronic Renal Failure.

Hypocalcemia may occur in chronic glomerulonephritis with nitrogen retention, usually in the later stages, associated with and perhaps dependent upon the increase in serum phosphate which occurs in these cases. In advanced nephritis with uremia, inorganic phosphorus figures of 12 to 20 mg. per 100 ml. may be associated with serum calcium values of 2.0 to 3.0 mEq./l. Tetany may occur as a result of such marked lowering of the serum calcium concentration. As indicated previously, the concentration of calcium in the serum varies roughly inversely with that of phosphorus. A similar situation may develop in chronic renal failure due to any cause, e.g., pyelonephritis, urinary obstruction, renal hypogenesis, polycystic disease, etc. The hypocalcemia, particularly in chronic glomerulonephritis, may be contributed to by a coexisting hypoproteinemia, as indicated above (i.e., nephrotic syndrome).

A state of secondary hyperparathyroidism may develop in such cases, with parathyroid hyperplasia and hyperfunction, initiated by the decrease in serum Ca concentration (p. 295). The associated skeletal abnormalities are due in part to the hyperparathyroidism, but also to the coexisting (*a*) acidemia, (*b*) protein malnutrition, and (*c*) inadequate absorption of Ca resulting possibly in part from the excretion into the intestine of large amounts of phosphate, which would have been eliminated in the urine under conditions of normal renal function but mainly resulting from altered vitamin D metabolism (see p. 304). These abnormalities may, therefore, exhibit considerable variation and may present features characteristic of osteomalacia (p. 314) and/or osteoporosis (p. 313), as well as skeletal lesions indistinguishable from those of primary hyperparathyroidism. There may also be metastatic calcification, particularly in the media of arteries and in the neighborhood of joints.

When it occurs during the period of skeletal growth, the latter may be inhibited, a condition termed "renal rickets" or "renal dwarfism." In both children and adults, the typical clinical picture is one of prolonged renal failure, usually marked acidemia, hyperphosphatemia, initially reduced serum Ca concentration, and increased serum alkaline phosphatase activity. However, deviations from this typical picture are encountered frequently, due probably to the modifying influences of variable factors superimposed upon the condition of secondary hyperparathyroidism, as well as to the variable ability of the parathyroids (and bones) to respond adequately to the stimulus of hypocalcemia. In a completely satisfactory response, increased mobilization from the skeleton may serve to maintain the serum Ca concentration within essentially normal limits; if the renal tubular epithelium is capable of responding to the increased amount of parathyroid hormone, the serum phosphate concentration may fall somewhat from its previously elevated level. The initial hypocalcemia is also counteracted, and may indeed be effectively counterbalanced, by the effect of acidemia in stimulating mobilization of Ca from the bones. Occasionally, the serum Ca concentration may actually rise above the upper limit of normal as a result of the operation of these complicating factors of secondary hyperparathyroidism, acidemia, osteoporosis (with decreased capacity of the skeleton for deposition of Ca), and impaired kidney function. Osteoporosis (p. 313), due to the state of protein malnutrition often present in advanced renal failure, may result in lower values for serum alkaline phosphatase activity than would be anticipated in primary hyperparathyroidism of comparable severity.

It is now believed that the hypocalcemia of chronic renal disease is really secondary to an acquired insensitivity to vitamin D, with a consequent diminished responsiveness to parathyroid hormone. In chronic renal disease, there is undoubtedly a decrease in the intestinal absorption of calcium. When hypocalcemia is present, there is also a diminished, and occasionally absent, response to parathormone in relation to its effect in raising the serum calcium level as well as increasing urinary phosphate. The insensitivity to parathormone is largely due to the acquired insensitivity to vitamin D, since the vitamin is necessary for the activity of the hormone. It is known that the latter exerts its skeletal and renal effects through the medium of cyclic AMP, which in turn requires appropriate amounts of ionic calcium for its action; the ion concentration is, of course, markedly diminished in the presence of an insensitivity to the vitamin.

There is evidence that in chronic renal disease there is a defect both in the metabolism and in excretion of vitamin D. This results in a deficiency of the calcium transport protein of the intestinal mucosa, which requires 25-hydroxycholecalciferol to instigate its synthesis. It has recently been suggested that when calcium intake is low the kidney converts 25-hydroxycholecalciferol to 1,25-dihydroxycholecalciferol and that the latter is the highly active hormone, which regulates calcium transfer in both gut and bone. This would go a long way toward explaining insensitivity to vitamin D in chronic renal disease. It is a matter of some interest that there is evidence suggesting that parathormone suppresses the formation of the 1,25-dihydroxy compound and encourages the formation of the 24,25-dihydroxy derivative, which is much less active on both bone and gut. This compound and not the 1,25 derivative is also formed by the kidney when calcium intake is high.

Maternal Tetany. The requirement for Ca, P and vitamin D is increased in pregnancy, and the occurrence of parathyroid hyperplasia indicates the increased functional demands upon these structures. Under normal conditions the serum calcium tends to fall, but not to subnormal levels, during the late months of pregnancy and during lactation, but it may become subnormal occasionally, with mild manifestations of tetany. The etiology of this condition is not clear, but it appears to be dependent in some cases upon deficiency in calcium and vitamin D and in others upon parathyroid deficiency operating alone or in conjunction with other factors.

Neonatal Hypocalcemia. The most common cause of convulsions in the first month of life is hypocalcemia. There are two peaks of incidence of convulsions, one of which occurs during the first two days and the other between the fourth and tenth days. The former is less common and is associated with hypoglycemia in infants of low birth weight.

The more common form of neonatal hypocalcemia usually occurs in formula-fed babies because of the higher proportion of phosphate to calcium in cow's milk. There may be maternal hyperparathyroidism, which tends to lead to depression of the parathyroid in utero because of high calcium levels in the maternal, and therefore in the fetal, blood plasma. A degree of asphyxia sometimes contributes to the onset of convulsions.

Acute Pancreatitis. Hypocalcemia has been reported to occur in the majority of cases of acute hemorrhagic pancreatitis (up to 70 per cent), the values usually ranging between 3.5 and 4.0 mEq./l., but occasionally falling below 3.5 mEq./l. This has been attributed to the sudden removal of large amounts of calcium from the extracellular fluids, including the blood plasma, as a result of its fixation as insoluble calcium soaps by fatty acids in areas of fat necrosis. The validity of this explanation has not been established. The condition is characterized by marked shock and it has also been suggested that the lowering of serum calcium may be associated with overproduction of cortisol.

Anticonvulsants. Long-term therapy with pheneturide, primidone, phenytoin or phenobarbitone can produce biochemical and other disturbances closely resembling osteomalacia. It appears that these anticonvulsants induce the production of enzymes in the liver, which are concerned with the destruction of vitamin D.

Magnesium, Oxalate and Citrate Tetany. In experimental animals, parenteral administration of magnesium salts may cause a fall in serum calcium, at times to tetanic levels. It is of interest in this connection that there is in man a low magnesium type of tetany, which can be accompanied by a normal serum calcium concentration (p. 318).

Parenteral administration in experimental animals of soluble oxalates and of citrates produces tetany, the former by the production of hypocalcemia through formation of insoluble calcium oxalate, the latter by formation of a poorly ionized calcium citrate compound, without hypocalcemia. The latter may be of practical importance in cases of transfusion of large amounts of citrated blood at a rate exceeding the capacity of the body to metabolize or excrete the citrate. This may be impaired in patients with liver disease, anuria, shock, or induced hypothermia.

TABLE 6-1 CONDITIONS INVOLVED IN DIFFERENTIAL DIAGNOSIS OF ALTERATIONS IN SERUM CALCIUM AND/OR PHOSPHATE CONCENTRATION

CONDITION	TOTAL Ca, MG. PER 100 ML.	DIFFUSIBLE Ca, MG. PER 100 ML.	PHOSPHATE, MG. PER 100 ML.	ALKALINE PHOSPHATASE UNITS	PROTEIN IN GM. PER 100 ML.	pH
Tetany						
Hypoparathyroidism	Decrease	Decrease	Increase	Normal	Normal	Normal
Maternal	Decrease	Decrease	Normal or increase	Normal or increase	Normal or decrease	Normal
Infantile rickets	Decrease	Decrease	Decrease or normal	Increase	Normal	Normal
Osteomalacic	Decrease	Decrease	Decrease	Increase	Normal	Normal
Severe diarrhea	Decrease	Decrease	Decrease	Normal	Normal	Normal or decrease
Gastric	Normal	Normal	Normal	Normal	Normal	Increase
Hyperventilation	Normal	Normal	Normal or decrease	Normal	Normal	Increase
Bicarbonate	Normal	Normal	Normal	Normal	Normal	Increase
Sprue	Decrease	Decrease	Decrease or normal	Normal or increase	Normal or decrease	Normal
Celiac disease	Decrease	Decrease	Decrease or normal	Normal or increase	Normal or decrease	Normal
Hyperparathyroidism	Increase or normal	Increase or normal	Decrease or normal	Increase or normal	Normal	Normal
Nephrotic syndrome	Decrease	Normal	Normal	Normal	Decrease	Normal
Essential hypercalciuria	Decrease	Decrease	Decrease	Normal	Normal	Normal
Renal tubular acidosis	Decrease	Decrease	Decrease	Normal or raised	Normal	Decrease
Renal failure	Decrease	Decrease	Increase	Normal	Decrease	Decrease

Miscellaneous. Persistent hypocalcemia has been reported in certain cases of extensive osteoblastic skeletal metastases. This has been attributed to active deposition of large amounts of calcium in the skeletal lesions. Hypocalcemia can also occur in medullary carcinoma of the thyroid because of excessive calcitonin production.

Hypocalcemia, due to inadequate absorption, may occur in subjects receiving exchange resins over prolonged periods of time.

Alkalemia. It is probable that in alkalemia there is a diminution in the proportion of ionized calcium which is responsible for the tetany, so characteristic of alkalemic states. The serum calcium in these conditions is usually within normal limits. This condition occurs clinically following the administration of excessive quantities of alkali (bicarbonate tetany), in states of hyperventilation (hyperpneic tetany), and in association with pyloric and acute upper intestinal obstruction or excessive vomiting from any cause (gastric tetany).

In bicarbonate and gastric tetany, in addition to the history and clinical findings, significant observations include: increase in plasma CO_2 (p. 424), decrease in plasma chloride (gastric tetany), increase in pH of the blood, and normal serum calcium concentration.

In hyperventilation tetany, the significant chemical changes are: fall in alveolar and plasma CO_2 (p. 422), increase in plasma pH, increase in alkalinity and decrease in ammonium ion in the urine, and normal serum calcium concentration.

ABNORMAL URINE CALCIUM

Increased Urinary Calcium

Hyperparathyroidism. The injection of parathyroid extract is followed by a gradual increase in urinary excretion of calcium. Although in most cases the increased elimination is dependent upon a rise in the level of serum calcium, it may occur in the absence of any such elevation. Similar observations have been made in clinical states of hyperparathyroidism (osteitis fibrosa diffusa). Whereas under normal conditions 70 to 90 per cent of the excreted calcium is eliminated in the feces and 10 to 30 per cent in the urine, in hyperparathyroidism these proportions may be reversed. Patients with this condition cannot be maintained in calcium equilibrium on a daily intake sufficient for normal individuals (22.5 mEq. calcium daily). This negative balance is best demonstrated when the intake of calcium and phosphorus is maintained at a low level for test periods of at least three days (calcium 5.5 mEq. and phosphorus 13.0 mmol. daily). This test procedure is of diagnostic value in early or mild cases of hyperparathyroidism, when the abnormalities in serum calcium and phosphorus are not striking or characteristic (p. 295).

Hyperthyroidism. Increased thyroid secretion is associated with a marked accentuation of calcium excretion, the urinary calcium being increased somewhat out of proportion to the fecal calcium. This appears to be due to a direct stimulating catabolic effect on the bones similar to the general action of thyroid extract upon other body tissues.

Acidemia. Acidemia, whether due to the ingestion of mineral acids or acid-forming substances or associated with disease states, is accompanied by an increase in calcium excretion, the excess calcium, as in the case of hyperthyroidism and hyperparathyroidism, being abstracted from the bones. This increased excretion of calcium is due almost entirely to an increase in urinary calcium.

Hypervitaminosis D. The administration of large quantities of vitamin D or of smaller doses of dihydrotachysterol (A.T. 10) may result in increased excretion of calcium in the urine. The mechanism of action of the vitamin in this connection is not clear. However, its influence is indicated by the fact that, in vitamin D deficiency, urinary excretion of calcium is very low even though the serum calcium may be within normal limits; administration of vitamin D in such cases is followed by an increase in urine calcium with little or no significant change in the serum calcium concentration. Similarly, in cases of hypoparathyroidism, administration of large doses of vitamin D is followed by an increase in urine calcium before the serum calcium concentration begins to rise.

Idiopathic Hypercalciuria. In the majority of cases of nephrolithiasis (calcium oxalate or phosphate calculi) no metabolic abnormality can be demonstrated other than excretion of excessive amounts of calcium. The mechanism underlying this type of hypercalciuria is not known. Excessive urinary excretion of calcium continues even though the calcium intake is reduced. The serum phosphate may be low in some cases. Idiopathic hypercalciuria may be due to some inherent defect in the renal tubular transport mechanism involved in reabsorption of calcium from the glomerular filtrate. Increased absorption of calcium by the gut is quite common. There is no doubt that a proportion of patients said to be suffering from idiopathic hypercalciuria are actually cases of hyperparathyroidism with normal serum calcium levels.

Miscellaneous. Increased urinary excretion of calcium may occur as an incidental phenomenon in certain types of renal tubular reabsorption defects. This occurs rather consistently in renal tubular acidosis (p. 538) and occasionally in hepatolenticular degeneration (p. 537). In the latter condition, it may be associated with urinary calculi, renal calcification, and disturbances in skeletal mineralization. Intravenous infusion of inorganic phosphate for the treatment of severe hypercalcemia, as in sarcoidosis or hyperparathyroidism, can lead to well-marked hypercalciuria.

Decreased Urinary Calcium

Hypoparathyroidism. In hypoparathyroidism, hypocalcemia is associated with a diminution in the urinary excretion of calcium. There is apparently a threshold for urinary calcium excretion since the elimination of calcium in the urine decreases abruptly as soon as the serum calcium falls below 3.25 to 4.25 mEq./l.

Vitamin D Deficiency. Urine calcium is decreased in vitamin D deficiency, as occurs in rickets and osteomalacia and also in certain cases of steatorrhea (sprue, celiac disease). This is due largely to inadequate absorption of calcium from the intestine.

Hypothyroidism. In myxedema, the excretion of calcium has been found to be approximately 40 per cent below the normal average, the diminution in urinary calcium being somewhat greater proportionately than that in fecal calcium.

Miscellaneous. An unusually low excretion of calcium, as compared with normal subjects, has been reported in patients with calcinosis universalis maintained on a low calcium and phosphorus intake. A marked diminution in the output of calcium in the urine is a rather constant finding in chronic nephritis. This is due in part to impaired renal excretion of calcium and in part to inadequate absorption of calcium by the intestine. This, together with the accompanying acidemia, may contribute to development of the condition known as "renal dwarfism" or "renal rickets," an occasional accompaniment of chronic renal failure in children. Secondary hyperparathyroidism plays a part in the pathogenesis of the skeletal changes in this condition (p. 295). Urinary calcium is decreased, and may be virtually absent, in children with the Fanconi syndrome. This is probably due mainly to inadequate absorption of calcium from the intestine, paralleling the defective absorption of amino acids.

ABNORMAL FECAL CALCIUM

An increase in the quantity of calcium eliminated in the feces may be observed in rickets and osteomalacia (vitamin D deficiency), in sprue and in celiac disease. In the latter two conditions this increase is due to defective calcium absorption resulting from steatorrhea and inadequate absorption of vitamin D. Fecal excretion of calcium is excessive in children with the Fanconi syndrome, paralleling the increase in fecal nitrogen.

HYPERPHOSPHATEMIA

Hypervitaminosis (Vitamin D).
The serum phosphate concentration may be increased by large doses of vitamin D in the form of cod liver oil or viosterol. Similar results may be obtained by ultraviolet irradiation.

Hypoparathyroidism.
Diminished parathyroid function is associated with a slight rise in the serum phosphate concentration which is more or less proportional to the diminution in serum calcium. This presumably results from excessive renal tubular reabsorption of phosphate in the absence of adequate amounts of parathyroid hormone. A similar situation exists in pseudohypoparathyroidism, due in this case to inherent inadequacy of responsiveness of the renal tubular epithelium to the hormone (target organ defect).

Renal Failure.
Increase in the concentration of phosphate in the serum of patients with chronic glomerulonephritis is a reflection of renal functional insufficiency. Phosphate retention in nephritis appears to contribute to the acidemia which occurs in the later stages of that condition, high values for serum phosphate being almost invariably accompanied by a decrease in the alkali reserve. However, acidemia may exist in the absence of demonstrable phosphate retention. Hyperphosphatemia in nephritis has approximately the same clinical significance as creatinine retention, values 2.6 mmol./l. (8 mg./100 ml.) in adults being of serious prognostic import. Values as high as 13 mmol./l. (40 mg./100 ml.) have been reported. Similar alterations in the serum phosphate concentration may occur in renal insufficiency associated with nephrosclerosis, multiple myeloma, and destructive kidney lesions such as congenital polycystic kid-

ney, tuberculosis, malignancy, pyonephrosis, pyelonephritis, and hydronephrosis. An increase in serum inorganic phosphorus concentration has also been observed in acute high intestinal obstruction. This may possibly be dependent upon or associated with the state of renal functional insufficiency which occurs in that condition.

Healing Fractures. During the period of healing of fractures in adults the serum phosphate concentration is often slightly increased, values ranging from 5 to 7 mg. per 100 ml. being frequently observed.

Miscellaneous. Hyperphosphatemia can be associated with a high milk intake, phosphate therapy, hyperthyroidism, certain bone tumors, hemolytic disorders, leukemia, hematomas and heparin administration.

HYPOPHOSPHATEMIA

Rickets. In rachitic children, serum phosphate values of 0.3 to 0.6 mmol./l. (1 to 2 mg./100 ml.) are commonly observed. This condition is usually due fundamentally to vitamin D deficiency, the mode of operation of which is discussed elsewhere (p. 302). In some cases of rickets the serum phosphate concentration may be within normal limits, the serum calcium being diminished (low calcium rickets). Under such circumstances the rachitic condition is commonly complicated by tetany. It was pointed out by Howland and Kramer that if the concentration of calcium be multiplied by that of phosphate, each expressed in milligrams per 100 ml. of blood serum, a product is obtained which, in the normal child, ranges from 50 to 60. When this product is below 30, rickets is invariably present and when it is above 40 either healing is occurring or rickets has not been present. The serum alkaline phosphatase activity is increased, roughly in proportion to the severity of the condition (p. 563).

There can be little question that vitamin D deficiency alone is responsible for the production of rickets in children. However, there is no precise information as to whether other dietary features that influence the development of this condition in certain experimental animals are also operative in man. For example, in the rat, the rachitogenic effect of D hypovitaminosis is influenced considerably by the quantities and proportions of calcium and phosphorus in the diet. At any given level of vitamin D deficiency, the development of rickets is facilitated by lowering the intake of these elements. A similar rachitogenic effect may be produced in the rat by administration of excessive amounts of cations which form insoluble, and therefore poorly absorbable, phosphates, e.g., beryllium, lead, iron, aluminum, strontium, magnesium. Clinical experience suggests that factors which promote absorption of calcium or phosphorus from the intestine exert a beneficial influence in rickets, e.g., increased acidity of the intestinal contents, and absence of excessive amounts of phytic acid (p. 280).

Osteomalacia. The metabolic disturbance in osteomalacia is similar to that in rickets, both being usually dependent upon vitamin D deficiency. Diminution in the concentration of serum phosphate is one of the most constant features of this condition.

Idiopathic Steatorrhea. Under this designation are included conditions commonly termed celiac disease, sprue and nontropical sprue. The characteristic fatty diarrhea in these conditions is frequently accompanied by skeletal demineralization, dwarfism, low serum calcium and phosphorus, and characteristic manifestations of rickets and tetany. These abnormalities of calcium and phosphorus metabolism are due to defective absorption of calcium, phosphorus and vitamin D. This results from diminution in the absorbing surface, frequently associated with diminished production of bile.

Hyperparathyroidism. Injection of parathyroid hormone is followed by diminution in the concentration of serum phosphate which results from increased elimination of phosphate in the urine (p. 289). This occurs also in patients with hyperparathyroidism, in the absence of renal complications (p. 295). Following repeated administration of large doses of parathyroid hormone, with a maintained serum calcium concentration between 7.5 and 10.0 mEq./l., the elimination of phosphate in the urine diminishes and the serum phosphate begins to rise. This effect upon serum phosphate is due not directly to the action of the parathyroid hormone but to the development of renal functional insufficiency, which is also manifested by increase in the concentration of blood urea N.

Fanconi Syndrome (p. 537). This designation is applied to a group of conditions, exhibiting a familial tendency, characterized by inadequate renal tubular reabsorption of certain constituents of the glomerular filtrate and due, presumably, to inherent defects in certain enzyme transport mechanisms in the tubular epithelial cells. The following substances, singly or in various combinations, are excreted in the urine in abnormally large amounts: glucose (p. 84); various amino acids (p. 534); lactic acid; β-hydroxybutyric acid; ammonia; phosphate. There are frequently acidemia, hypophosphatemia, and rickets or osteomalacia. The serum calcium concentration is usually within normal limits and the serum alkaline phosphatase activity increased. We are concerned here only with the abnormality in phosphate. The other aspects are discussed elsewhere.

The increased urinary excretion of phosphate, due to its inadequate tubular reabsorption from the glomerular filtrate, is believed to be the direct cause of the lowered serum phosphate concentration. If this falls to a sufficiently low level, the $Ca:PO_4$ ratio becomes unfavorable for deposition of bone mineral, with the consequent development of rickets (infants and growing children) or osteomalacia (adults).

Renal Tubular Acidosis (p. 538). This condition, due to inability of the kidney to produce a urine of normal acidity, is characterized by systemic acidosis (metabolic) with a urine that is alkaline or only weakly acid in reaction. The renal tubular defects include defective reabsorption of phosphate and calcium, as well as other cations and anions, with consequent hyperphosphaturia, hypercalciuria, hypophosphatemia, and rickets or osteomalacia.

Vitamin D-resistant Rickets. This condition, also termed renal tubular rickets, phosphate-losing rickets, and phosphate diabetes, is characterized by increased urinary excretion of phosphate and decreased serum phosphate concentration, due to an inherent defect in renal tubular reabsorption of phosphate. It has a familial incidence and, although the skeletal and blood chemical abnormalities resemble those of the usual vitamin D-deficient rickets, there may be no obvious bone deformity other than shortness of stature. There may be bony abnormalities other than rickets. The serum calcium concentration rarely falls to tetanic levels and the urine calcium may be normal. The intestinal absorption of calcium is diminished.

Evidence is available which indicates that patients with rickets are unable to convert ingested vitamin D to its active metabolite. There is often secondary hyperparathyroidism, which could account for at least some of the urinary loss of inorganic phosphate.

This familial type may be easily confused with other conditions, such as the Fanconi syndrome, renal tubular acidosis or vitamin D-resistant rickets associated with normal or raised serum inorganic phosphate.

There is a form of osteomalacia with hypophosphatemia, which is associated with a hemangioma or giant-cell granuloma. Removal of the tumor results in cure of the osteomalacia.

Miscellaneous. In a hospital population, the most common cause of hypophosphatemia probably is the intravenous administration of glucose, and the next most common cause probably is vomiting. It is also associated with gastric suction, administration of insulin, administration of antacids and with various forms of liver disease.

ABNORMAL URINARY PHOSPHATE

One of the means by which the kidney aids in maintaining the acid-base balance resides in its ability to transform the alkaline phosphate (B_2HPO_4) of the blood to the acid phosphate (BH_2PO_4) which is eliminated in the urine. In the presence of acidemia or factors which tend to produce acidemia (i.e., acidosis), the elimination of acid phosphate in the urine is increased, constituting one of the compensatory measures which aid in the maintenance of the acid-base balance (p. 408). This does not occur in the acidemia of advanced nephritis, which is commonly associated with retention of phosphate as well as other constituents of the blood which are normally eliminated in the urine.

One of the early effects of the administration of parathyroid hormone is an increase in the quantity of phosphate eliminated in the urine (p. 289). This is believed to be the cause of the subsequently occurring hypophosphatemia. Likewise, an increase in urinary phosphate occurs usually in clinical hyperparathyroidism (osteitis fibrosa diffusa). In hypoparathyroidism (parathyroid tetany), on the other hand, the urinary phosphate is diminished in quantity. Urinary phosphate is diminished in ordinary rickets, the greater part of the ingested phosphate being excreted in the feces, owing perhaps in part to inadequate absorption. Administration of vitamin D or irradiation with ultraviolet light increases the urinary elimination of phosphate in rickets and osteomalacia.

Increased urinary excretion of phosphate has been observed in certain cases of poisoning with heavy metals, such as lead, cadmium, and uranium. This has been attributed to depression of renal tubular reabsorption by these agents; there may also be glycosuria and aminoaciduria. Similar findings have been reported in cases of renal tubular damage produced by other agents, e.g., nitrobenzene and phenol.

Changes in urinary phosphate excretion due to other causes have been discussed previously in connection with abnormalities of serum phosphate concentration.

The occurrence of hyperphosphaturia as a cardinal feature of the Fanconi syndrome is referred to in the discussion of abnormalities in serum phosphate (p. 311).

DISTURBANCES OF BONE FORMATION AND MINERALIZATION

Abnormalities may occur in both phases of bone construction, i.e., (a) in the formation of the osteoid matrix, and (b) in the subsequent processes of mineralization and demineralization. Accurate diagnosis and proper therapy require that these two classes of skeletal disorders be clearly differentiated.

Osteoporosis is characterized by loss of bone mass. There is not necessarily any great change in the anatomical volume of the bones, but the amount of actual calcified bone tissue in this volume is diminished. The bones are therefore much more porous (less dense) and contain more marrow space. Trabecular bone is more affected than cortical. Thus, the weight-bearing bones are considerably affected and there is a marked liability to fracture. One or more of the spinal vertebrae may collapse. It is only a matter of degree that distinguishes the condition from the normal loss of bone mass, which commences around the age of 35 years. Such bone tissue as is present seems to differ in no way from that which characterizes normal bone.

It used to be believed that osteoporosis was a primary defect of osteoid formation. It does tend to be more severe where the initial bone mass is less. It therefore becomes apparent earlier in women than in men (fifth and sixth decades, respectively). Kinetic and microradiographic studies have shown that the condition results from excessive bone absorption. Bone formation does not seem too much, if at all, affected. There is good evidence that calcium deficiency can cause osteoporosis. The condition is relatively common in a number of situations in which intake or absorption of calcium is lowered, as in malnutrition, alcoholism, cirrhosis, and in some patients with intestinal lactase deficiency, who voluntarily restrict their intake of milk and its products. In spite of this, no constant biochemical defect has been found and the serum levels of calcium, inorganic phosphate and alkaline phosphatase are usually within normal limits. Osteoporosis may be primary or associated with other disorders (Table 6-2). It has been suggested that it results from increased sensitivity of

TABLE 6-2 OSTEOPOROSIS

1. **PRIMARY**

    Idiopathic (adult and juvenile)
    Senile
    Postmenopausal

2. **SECONDARY**

    Immobilization or disuse
    Calcium deficiency
    Gonadal insufficiency
    Glucocorticoid excess (Cushing's or iatrogenic)
    Hyperthyroidism
    Acromegaly
    Malnutrition
    Scurvy
    Liver disease
    Alcohol
    Rheumatoid arthritis
    Nephrotic syndrome
    Malignancy (e.g., myelomatosis)

3. **OSTEOGENESIS IMPERFECTA**

4. **LOCAL**

    Disappearing bone disease
    Sudeck's atrophy

bone to parathyroid hormone, possibly due to the loss of protection by estrogenic hormones. In actual fact, the mechanism of production remains obscure.

Osteomalacia is characterized by a failure to calcify the osteoid matrix. This results in the presence of an excessive amount of osteoid tissue in relation to the amount of bone mineral. If the skeleton is still growing, the epiphysial cartilage is disorganized, causing the condition known as rickets. Osteomalacia results from deficiency of or resistance to vitamin D (Table 6-3). It can also result from any prolonged systemic acidemia, e.g., ureterocolic anastomosis or renal tubular acidosis, but the mechanism of production is not understood. Osteomalcia may also be found in neurofibromatosis.

Serum calcium levels can be very low, but as a rule they are either slightly lowered or normal. The level of plasma inorganic phosphate in osteomalacia is usually lowered and alkaline phosphatase increased. There is usually a diminution of urinary calcium. Aminoaciduria is occasionally present. There may be some degree of renal tubular acidosis with diminution of plasma bicarbonate.

Osteomalacia and osteoporosis may occur simultaneously.

Hypophosphatasia, the skeletal lesions of which resemble those of vitamin D deficiency, cannot be placed in either category on the basis of the biochemical manifestations (p. 563). The characteristic feature is deficiency or absence of alkaline phosphatase in various tissues, including the bones and blood, frequently with hypercalcemia in severe cases, and normal or occasionally increased serum phosphate concentration.

TABLE 6-3 OSTEOMALACIA

1. **NUTRITIONAL (DISTURBANCE IN ABSORPTION OF Ca, P OR VITAMIN D)**
   Vitamin D deficiency (rickets)
   Calcium and/or phosphorus deficiency
   Steatorrhea
   Chronic diarrhea (ulcerative colitis)
   Gastrocolic fistula, partial gastrectomy, lower intestinal resections, etc.

2. **RESISTANCE TO VITAMIN D**
   Chronic renal failure
   Renal tubular malfunction
     Fanconi syndrome
     Renal tubular acidosis
     Hypophosphatemia

3. **PROLONGED SYSTEMIC ACIDEMIA**
   Ureterocolic anastomosis, etc.

4. **MISCELLANEOUS**
   Prolonged anticonvulsant therapy
   Neurofibromatosis

Osteosclerosis is a rare form of metabolic bone disease in which there is excessive mineralization of bone. It can be general or local. It sometimes occurs as a result of excessive intake of fluoride ("fluorosis").

Osteitis fibrosa is usually due to hyperparathyroidism. There is excessive resorption of previously mineralized bone because of increased osteoclastic activity. The associated biochemical changes have already been described (p. 295).

References

Albright, F., and Reifenstein, E. C., Jr.: Parathyroid Glands and Metabolic Bone Disease. Baltimore, The Williams & Wilkins Company, 1948.
Bauer, G. C. H., Carlson, A., and Lindquist, B.: Metabolism and homeostatic function of bone. *In* Mineral Metabolism, edited by C. L. Comar, and F. Bronner. New York, Academic Press, Inc., Vol. I, Part B, 1961, p. 609.
Berry, E. M., Gupta, M. M., Turner, S. J., and Burns, R. R.: Variation in plasma calcium with induced changes in plasma specific gravity, total protein and albumin. Brit. Med. J., *4*:640, 1973.
Forfar, J. O., and Tompsett, S. L.: Idiopathic hypercalcemia of infancy. Advances in Clinical Chemistry, *2*:168, 1959.
Fourman, P., and Royer, P.: Calcium Metabolism and the Bone. Ed. 2. Oxford, Blackwell Scientific Publications, 1968.
Irving, J. T.: Calcium Metabolism. New York, John Wiley & Sons, Inc., 1957.
Neuman, W. F., and Neuman, M. W.: The Chemical Dynamics of Bone. Chicago, University of Chicago Press, 1958.
Nicolaysen, R.: Physiology of calcium metabolism. Physiol. Rev., *33*:424, 1953.
Nordin, B. E. C.: Metabolic Bone and Stone Disease. Edinburgh, Churchill Livingstone, 1973.
Payne, R. B., Little, A. J., Williams, P. B., and Milner, J. T.: Interpretation of serum calcium in patients with abnormal serum proteins. Brit. Med. J., *4*:643, 1973.
Urist, M. R., ed.: Symposium on clinical physiology and pathology of bone. Clinical Orthopaedics, No. 17, 1960.
Wills, M. R.: Disorders of calcium homeostasis. Brit. J. Hosp. Med., *6*:65, 1971.

Chapter 7

MAGNESIUM METABOLISM

ABSORPTION AND EXCRETION

On the average, 24,000 mEq. (29 g.) of magnesium is present in the body of a man weighing 70 kg. Of this total, approximately 60 per cent is present in the bones. The average daily diet in the U.S.A. contains 20 to 40 mEq. of magnesium. The element is of great importance in relation to the activity of many enzymes, e.g., the phosphatases, various transferases, and oxidative phosphorylation.

The absorption of magnesium from the bowel resembles that of calcium in many respects. An excessively high intake of fat, phosphate, calcium and alkalis appears to diminish the absorption of magnesium, which occurs mainly in the upper intestine. It is of interest also to note that a high magnesium intake appears to increase calcium elimination in the urine. The influence of the above-mentioned factors in this connection is probably dependent upon their influence on the solubility of magnesium salts.

Vitamin D is involved to some extent but its involvement is less important than it is with calcium. The major factor in the control of the concentration of magnesium in body fluids is the output in the urine. Parathyroid hormone increases the absorption from the bowel but diminishes the output in the urine. There is evidence that increased level of plasma magnesium inhibits and a decreased level stimulates the production of parathyroid hormone.

The urinary output of magnesium exhibits diurnal variation. It is lowest during the normal sleeping hours and is maximal in the late morning.

Like calcium, magnesium is excreted in the feces and urine. Under normal conditions, about 50 to 80 per cent is excreted by the bowel (bile and intestinal secretion) and the remainder by the kidneys. After parenteral administration of magnesium salts, 70 to 90 per cent is excreted in the urine in the presence of normal renal function. The administration of acidifying substances, such as ammonium chloride, is followed by increased urinary elimination of magnesium, as of calcium. There is evidence that the urinary excretion of magnesium is interfered with in the presence of renal functional impairment.

BLOOD MAGNESIUM

Magnesium is present in both red corpuscles and plasma. In contrast to calcium, the concentration in the red cells is higher than in the plasma. Perhaps because of variations in the methods employed and the limited number of cases studied, there has been disagreement regarding the normal range of variation in blood magnesium concentration. The normal adult range usually quoted is 1.4–2.5 (av. 2.0) mEq./l. (1.7–3.0 mg./100 ml., av. 17). According to some, the normal range is narrower, 1.5 to 1.8 mEq. (av. 1.7) per liter; this is the range obtained by atomic absorption spectrophotometry. Ultrafiltration and diffusion experiments indicate that 70 to 85 per cent (average 80 per cent) of the serum magnesium is in a diffusible state, the nondiffusible fraction being probably combined with the serum protein as in the case of the nondiffusible fraction of serum calcium. The magnesium content of cerebrospinal fluid is higher than that of blood serum, 2.5–3.0 mEq./l. (3.0–3.6 mg./100 ml.).

Little is known regarding the factors involved in the regulation of the magnesium content of the blood. It is relatively unaffected by phosphate, protein, vitamin D, or parathyroid hormone. There is in some respects a reciprocal relationship between magnesium and calcium in the serum; e.g., in oxalate poisoning the decrease in serum calcium is accompanied by an increase in magnesium, while the hypermagnesemia induced by parenteral administration of magnesium salts is accompanied by a fall in serum calcium, at times to tetanic levels. The renal output is considered the most important factor. The renal mechanisms involved are to some extent similar to those for calcium. This results in competition in regard to reabsorption in the renal tubules, a factor of importance in the hypercalciuria of hyperparathyroidism.

Abnormal Serum Magnesium.
Hypomagnesemia, with negative magnesium balance, has been reported in clinical hyperparathyroidism. Low serum values occur more frequently following parathyroidectomy in hyperparathyroidism; in this case it may be due to increased retention of Mg incident to acceleration of bone mineralization.

Low values for serum Mg have been reported in uremia, normal and abnormal pregnancy, rickets, hyperthyroidism, growth hormone treatment, renal tubular acidosis, hypercalcemia, chronic alcoholism, portal cirrhosis, hyperaldosteronism, the recovery phase of diabetic coma, experimental Mg deprivation, and epilepsy. Many of the values regarded by individual authors as abnormal fall within the range regarded as normal by others, raising question as to the validity of these observations. However, evidence has accumulated which indicates that Mg deficiency, with hypomagnesemia, may be responsible for the production of a rather distinctive clinical syndrome, which can be fatal. This is characterized by manifestations of hyperirritability of the neuromuscular and central nervous systems and, frequently, cardiovascular abnormalities. These include tremors, twitching, athetoid movements, clonic convulsions, hyperactive reflexes, mental aberrations (disorientation, confusion, hallucinations), vertigo, ataxia, tachycardia, and electrocardiographic changes. True tetany probably does not occur in the absence of associated hypocalcemia, although a positive Chvostek sign may occur. There may be intestinal ileus.

This magnesium deficiency syndrome may result from excessive loss of Mg in a variety of gastrointestinal disorders, e.g., the malabsorption syndrome, intestinal fistulas, and protracted vomiting or diarrhea. It may be precipitated by parenteral administration of magnesium-free fluids to subjects with severe dehydration. It has been observed also in patients with chronic pyelonephritis (without uremia), in the Fanconi syndrome, and after diuresis induced by mercurial diuretics and chlorothiazide, but not acetazolamide. Slightly increased values have been reported in various chronic infections, atherosclerosis, diabetes mellitus, hypertrophic arthritis, oxalic acid poisoning, essential hypertension, and glomerulonephritis. Values as high as 13 mg. per 100 ml. have been reported following the ingestion of large quantities of magnesium sulfate by patients with chronic renal failure. Despite contradictory observations, elevated values for serum magnesium probably occur more consistently in chronic renal failure than in any other clinical disorder in which this factor has been studied.

There is some poorly understood relationship between magnesium and potassium. In magnesium deficiency, an increased loss of potassium into the urine has been reported.

In hypercalciuria, the presence of adequate amounts of magnesium in the urine tends to protect against the formation of calculi. These are therefore more common when hypercalciuria is associated with a relatively low magnesium concentration.

References

Hanna, S., et al.: The syndrome of magnesium deficiency in man. Lancet, 2:172, 1960.
Randall, R. E., Jr., et al.: Magnesium depletion in man. Ann. Int. Med., 50:257, 1960.
Stewart, C. P., and Frazer, S. C.: Magnesium. In Sobotka, H., and Stewart, C. P., eds.: Advances in Clinical Chemistry. New York, Academic Press, 6:29, 1963.
Vallee, B. L., et al.: The magnesium deficiency tetany syndrome in man. New England J. Med., 262:155, 1960.
Wills, M. R.: Biochemical Consequences of Chronic Renal Failure. Aylesbury, Harvey Miller and Medcalf, Ltd., 1971.

Chapter 8

IRON METABOLISM

Although the amount of iron in the body is small (about 60 to 70 mg./kg. body weight in the adult, and about 94 mg./kg. in the neonate), the fact that it is an essential constituent of hemoglobin and of cytochrome and other components of respiratory enzyme systems (cytochrome oxidase, catalase, peroxidase) makes it an element of great fundamental importance. Its chief functions lie in the transport of oxygen to the tissues (hemoglobin) and in cellular oxidation mechanisms (cytochrome system).

About 60 to 70 per cent of the total iron is present in hemoglobin; about 15 per cent is in storage as ferritin, 3 per cent as myoglobin and only about 0.1 per cent is carried in the plasma, in combination with the β-globulin transport protein transferrin. The hemoprotein and flavoprotein enzymes together make up less than 1.0 per cent of the total iron. In certain diseases, large amounts are present as hemosiderin.

The major dietary sources are liver, kidney, heart, shellfish, egg yolk and dried legumes. Intermediate sources are meat from muscles, poultry, fish, green vegetables and nuts. Boiling in water can reduce the content of green vegetables by 20 per cent. Poor sources of iron are milk and its products, white flour, polished rice, potatoes and most fresh fruit. Average diets in the U.S.A. provide some 20 mg. of iron daily. This contrasts greatly with the daily 9 mg. present in the diet of the average poor Indian. A diet which is adequate in other respects does not necessarily contain an adequate amount of iron. The source of the food may affect its iron content, as revealed in a study of anemic children in nonurban areas of Florida.

Iron occurs in food in a variety of forms. It may be inorganic or in organic compounds such as hemoglobin or myoglobin or in combination with a number of different proteins.

Human milk contains 0.3 to 0.6 μg. iron/ml. The iron is combined with ferrilactin, which has a similar molecular weight but differs from transferrin chemically and immunologically. Colostrum has an iron content which is three to five times that of milk.

ABSORPTION AND EXCRETION

Iron differs from practically all other electrolytes in that the quantity in the body is controlled by regulation not of its excretion, but rather of its absorption into the organism. The body stores of iron are conserved very efficiently, only minute amounts being excreted in the urine and feces, usually less than 1 mg. daily. Relatively large amounts are, of course, lost in the menstrual flow. That present in the body in various substances, e.g., hemoglobin, is almost completely reutilized following its liberation in the course of metabolic degradation of these substances. The bulk of the iron of the feces (6 to 16 mg./24 hr.) is unabsorbed food iron; a very small amount enters the intestine in the bile and escapes in the feces. Some is derived by desquamation of intestinal mucosal cells, making a total of fecal excreted iron of 0.3 to 0.5 mg./24 hr.

Urinary excretion of iron averages 0.2 to 0.3 mg./24 hr. Iron is also lost from the skin by means of sweat, hair loss, cell desquamation and nail clippings. This averages about 0.5 mg./24 hr. in the adult. In the tropics, iron loss is often much greater. The total amount of iron lost from the body in the feces, urine and sweat averages 0.6 to 1.0 mg./24 hr. Menstrual loss accounts for a higher average of 0.5 to 0.8 mg./24 hr. Breast-feeding accounts for approximately 0.5 mg./24 hr. During pregnancy, iron is, of course, lost to the fetus.

Several factors make absorption of iron difficult: (a) the relatively high pH in the jejunum favors the formation of insoluble basic iron compounds; (b) iron salts of bile acids are relatively insoluble; (c) the presence of relatively large amounts of phosphate favors the formation of insoluble iron phosphates; (d) absorption of iron is interfered with, therefore, in the absence of free HCl in the stomach (achlorhydria) and by administration of alkalis.

Absorption occurs chiefly in the upper duodenum but also, to a lesser extent, throughout the small intestine.

Factors influencing absorption are age, the amount of iron in the body and the form in which the iron is ingested. Other dietary components can have a marked effect, as well as the state of health of the individual or his intestinal mucosa. Absorption apparently occurs in two phases, a fast and a slow phase.

Absorption of inorganic salts of iron compares favorably with that of the iron of foodstuffs. In general, ferrous iron is better absorbed than ferric; this is not invariable.

Normal gastric acidity facilitates the ionization and solution of ingested iron and delays the formation of insoluble, undissociable compounds which may form above pH 5. At the slightly acid reaction usually encountered in the duodenum, ionizable iron compounds are converted chiefly to ferrous or ferric forms, the former being favored at higher acidities and by the presence of reducing agents in the digestive mixture (glutathione, ascorbic acid, SH-groups of proteins and digestion products).

The normal organism absorbs iron only in proportion to its needs, the quantity absorbed being determined by the magnitude of the body reserves of this element. In man, only 5 to 15 per cent of orally ingested iron normally is absorbed. Apparently the intestinal mucosa is the tissue immediately responsible for its acceptance or rejection. The mucosal epithelial cells contain ferritin, an iron-protein complex (20 per cent Fe), containing aggregates of ferric hydroxide and a protein, apoferritin (m.w. 460,000). Ferritin occurs also in other parenchymal cells, e.g., the liver and spleen, and in the bone-marrow and reticuloendothelial cells generally. It is believed to be the chief storage form of iron.

It was previously thought that iron taken into mucosal cells was converted into ferritin. When this was saturated with iron, no more could be taken up until the element was released to the tissue fluids and thence to plasma. In this way a "mucosal block" mechanism was postulated. Although the mechanism of absorption is not understood, the ferritin mechanism is no longer accepted. It would seem that, in the rat, absorption is in some way related to the iron content of the intestinal mucosal cells. In iron deficiency, the element can be absorbed directly across the mucosa and thence into the plasma. When adequate iron is present, the element can be taken up as ferritin and subsequently lost in the feces when the cells are shed from the tips of the intestinal villi. It is interesting that the amount of iron intake is related to the synthesis of ferritin in the mucosal cell. In man, evidence has indicated that there may be no relationship between mucosal cell iron content and its absorption. An iron-binding protein, gastroferrin, has been demonstrated in normal human gastric juice. When iron is bound to this protein it may not be absorbed. It has been suggested that this could be a regulatory mechanism for iron absorption. It is claimed that gastroferrin in gastric juice is reduced in iron deficiency anemia and is completely absent in hemochromatosis. It has also been claimed that a pancreatic extract can inhibit absorption of iron and that this extracted material can be greatly reduced in hemochromatosis. The iron transport protein transferrin possibly may be concerned with intestinal absorption. It has also been suggested that at least some effect is exerted by erythropoietin.

Inorganic iron can form complexes with substances present in normal gastric juice, which remain soluble at the higher pH of the intestine and so render the iron in a state suitable for absorption. Histidine and lysine increase iron absorption. It is suggested that these amino acids form soluble iron chelates.

Iron in hemoglobin, myoglobin and other similar compounds is absorbed in combination with heme. It is then passed into the plasma as the ferric ion, which has been separated in the mucosal cell. Hemoglobin iron is as well absorbed as ferrous iron. Unlike the latter, its absorption is not reduced by sodium phytate or increased by ascorbic acid.

As has already been indicated, a variety of factors can diminish iron absorption. These include achlorhydria, administration of alkalis, relatively high jejunal pH, and phosphate in the diet. Other factors are dietary phytate, copper deficiency and the consequences of partial gastrectomy.

Deficient absorption, of course, leads to iron deficiency. This is often accompanied by an increase of erythrocyte protoporphyrin, not only in iron deficiency anemia but also in diminished iron storage. The normal level of the protoporphyrin has a mean value of 15.5 µg./100 ml. erythrocytes. Values above 35 µg. are considered abnormal. This estimation is of value in confirming iron deficiency as a cause of anemia even after iron has been administered therapeutically. The level, however, is raised in sideroachrestic anemia, in which iron absorption is increased.

Factors which increase iron absorption are alcohol ingestion, diminished iron stores increased erythropoiesis, inosine, succinate, cirrhosis and possibly cystic fibrosis of the pancreas. Absorption is, of course, greatly increased in idiopathic hemochromatosis.

TRANSPORT

In accordance with body needs, the reactions in the mucosal cell bring about iron absorption, the mechanism being unknown.

The iron thus mobilized passes into the blood plasma, undergoes reoxidation to the ferric state and combines with one of the plasma beta-globulins. This compound, termed "transferrin," is present in a concentration of about 0.25 g./100 ml. plasma.

The protein has a molecular weight of approximately 86,000 and each molecule has two iron-binding sites. There are a number of different transferrins, the presence of which is genetically determined. The content in normal human plasma enables each 100 ml. of plasma to bind 250 to 420 µg. of ferric iron. This is known as the total iron-binding capacity (TIBC). The iron content of normal plasma accounts for about one-third of this; that is, the mean normal saturation is 33 per cent of the TIBC. This percentage is considerably lowered in anemia due to iron deficiency but the TIBC is raised above normal. A similar state of affairs can occur in late pregnancy or with oral contraceptives. In anemia due to infection, both the percentage saturation and the TIBC are below normal. In hemochromatosis, there is virtually total saturation but the TIBC may be slightly low.

The turnover rate of normal plasma iron is very rapid. Its half-life is 90 to 100 minutes, which means that 25 to 40 mg. of iron is transported in the plasma every 24 hours. In the process of incorporation of iron into hemoglobin, it is likely that transferrin passes into the developing cell and that its iron is released within the reticulocyte.

It has recently been apparent that the reticuloendothelial system is involved not only in the storage of iron (Fig. 8-2) but also in its release for appropriate utilization. In other words, the system plays some part in maintaining the plasma iron level. It would appear that in the process of transfer from reticuloendothelial ferritin to the extracellular fluid, the ferric ion must be reduced to the ferrous form and subsequently oxidized to the ferric state for combination with transferrin. It is becoming apparent that ceruloplasmin brings about, or at least plays a part in, this oxidation.

The iron content of plasma is 50 to 180 μg./100 ml.; this is the transport iron (ferric). There is a diurnal variation; the level at 8 A.M. is considerably higher than at 6 P.M.

The iron content of whole blood is normally 40 to 60 mg./100 ml., averaging 45 in women and 52 in men. Practically all of this is in organic form, as hemoglobin, which contains about 0.335 per cent iron (ferrous), all of which is in the red blood cells. Free ferritin is always present in very small amounts in plasma. It can be estimated by radioimmunoassay.

After ingestion of inorganic iron, the plasma iron concentration increases three- to fourfold in about two to four hours, falling subsequently over a period of six to twelve hours, depending upon the amount ingested. The rise is much greater in iron deficiency states. The amount given orally has varied from 100 to 250 mg. of ferrous iron. The changes observed are not of much clinical significance. The plasma is the medium of transportation of iron from the intestine to the tissues in which it is utilized and stored. Absorbed iron is transferred to erythrocytes with remarkable rapidity, radioactive iron having been demonstrated in red blood cells in anemic dogs within a few hours after its ingestion. The plasma iron is influenced by and may be regarded as an index of (a) the quantity of iron absorbed from the intestine, (b) the adequacy of the tissue iron reserves; a more accurate indication is the concentration of ferritin in the plasma. (c) the capacity of the bone marrow to utilize iron for hemoglobin synthesis, and (d) the activity of hemolytic processes.

UTILIZATION; STORAGE

In the tissues, as needed, the plasma iron is apparently released from transferrin, passes out of the capillaries and into the cells, where it may be utilized or stored.

Iron is stored as ferritin or hemosiderin; normally the former is about 60 per cent of the storage iron. Total storage iron normally amounts to

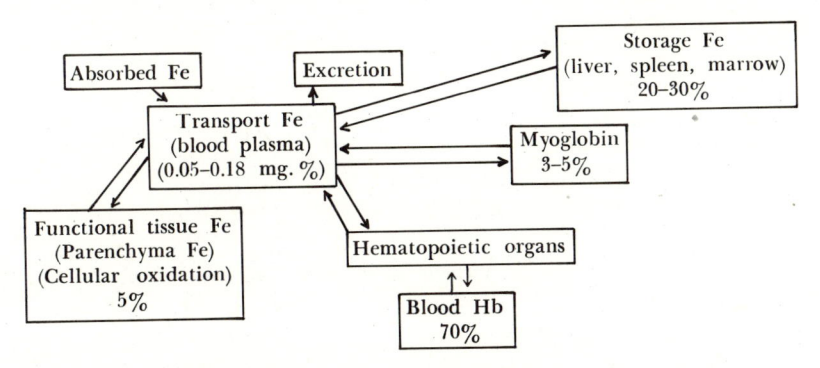

Figure 8-1. Approximate distribution of iron in the body.

1.0 to 1.5 g. In conditions involving excessive iron deposition (p. 326), there is a preponderance of hemosiderin. Ferritin can be obtained crystalline. It is a brown compound which is soluble in tissue fluids. Apoferritin is colorless. Ferritin in tissues does not stain with the usual Prussian blue histochemical stains for iron. On the other hand, hemosiderin does stain. It is an insoluble, relatively amorphous compound which can contain up to 35 per cent of iron, mainly as ferric hydroxide surrounded by protein material. It is present in tissues as brown granules.

The transfer of iron from transferrin to the storage sites apparently requires ascorbic acid, ATP and possibly citrate, as well as other anions.

The liver, spleen and intestinal mucosa are the chief storage sites, but other organs (e.g., pancreas, adrenals) and all reticuloendothelial cells are also involved. When iron is deposited in abnormally large amounts, hemosiderin may be formed in excess.

Iron is utilized chiefly in the synthesis of hemoglobin, myoglobin, and certain respiratory enzymes (cytochromes, peroxidase, catalase). The latter are probably formed in all cells, myoglobin in muscle cells, and hemoglobin in the developing red-blood cells in the erythropoietic tissues, principally the bone marrow in man. The approximate distribution of iron in the body is as follows (Fig. 8-1): about 60 to 70 per cent of the body iron is in the form of hemoglobin; 20 to 30 per cent is in storage form, available for utilization; 5 per cent is functional tissue iron, not readily available for other purposes; and 3 to 5 per cent is in myoglobin.

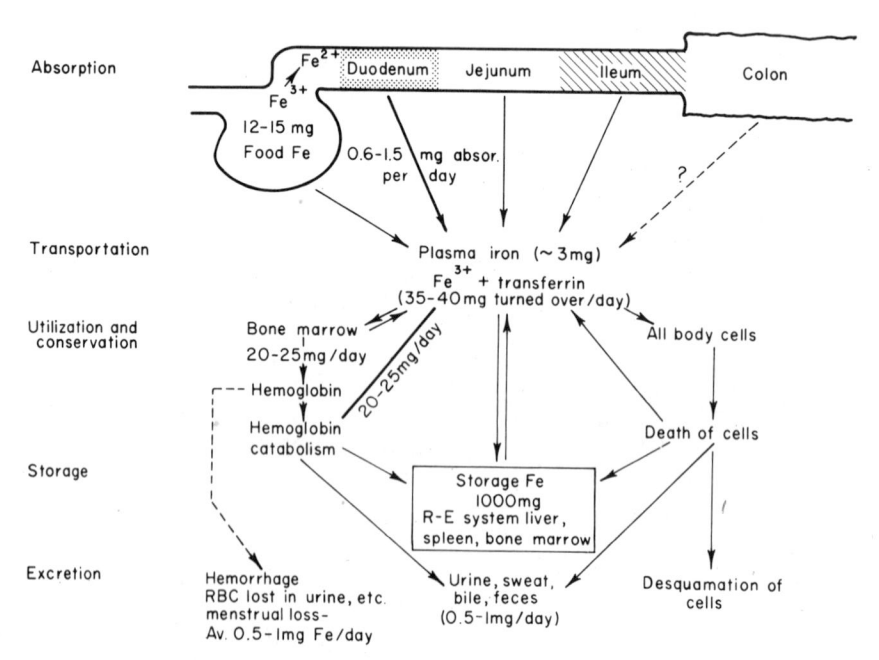

Figure 8-2. Schematic representation of iron metabolism in man. RBC = red blood cells; R-E system = reticuloendothelial system. (From Moore: Harvey Lect. 55:67, 1951.)

The iron of erythrocyte hemoglobin is readily available, but not that of myoglobin. Ferritin iron is more readily available than that in hemosiderin.

The quantity of iron mobilized in the organism daily far exceeds the exogenous supply. About 0.8 per cent of the circulating erythrocytes undergo disintegration daily (life 125 days), liberating about 7 to 8 g. of Hb (0.34 per cent Fe). In the course of its degradation, this gives rise to about 20 to 25 mg. of iron daily, the bulk of which is immediately utilized in the resynthesis of Hb, but some may enter the blood plasma for transport to other tissues to be incorporated in storage iron, myoglobin or iron-containing enzymes.

Mobilization of iron from ferritin in storage sites, e.g., liver, is presumably accomplished by reversal of the reactions involved in its storage; involvement of xanthine oxidase has been suggested in the release mechanism.

REQUIREMENT

Since only minute amounts of iron are excreted, the exogenous requirement should be correspondingly low. However, since only about 10 per cent is absorbed, it is believed that adult men and menopausal women require 5 to 10 mg. daily to meet ordinary and unanticipated demands; 7 to 20 mg. is more satisfactory for women (menstruation); 20 to 48 mg. for pregnant women (diet requires iron supplementation). Young children (4 to 8 years) require about 0.6 mg./kg. body weight and infants (to 1 year) 1 to 2 mg./kg.

ABNORMAL IRON METABOLISM

Disturbances in iron metabolism are evidenced mainly by (a) decreased formation of hemoglobin, (b) decrease in circulating hemoglobin, (c) abnormalities in the serum iron concentration or (d) abnormal deposition of iron-containing pigment (hemosiderin) in the tissues. A consideration of the pathogenesis of various types of anemia is beyond the scope of the present discussion, but it may be stated that certain forms of hypochromic and microcytic anemia are dependent primarily upon inadequate supply or absorption of iron. Absorption is decreased also in animals with severe subcutaneous infection (abscess). Iron deficiency may also result from hemorrhage, with exhaustion of the available tissue reserves of this element. Low values for plasma or serum iron have been reported in hemorrhagic and hypochromic types of anemia. High values occur in forms of anemia characterized by diminished hemoglobin formation not due to iron deficiency, as in pernicious anemia. Increased serum ion concentrations in acute hepatitis may be useful in differentiating this from other causes of jaundice (p. 671).

Inasmuch as there is no efficient excretory mechanism for iron, when excessive amounts are released in or introduced into the body, exceeding the capacity for its utilization, the excess is deposited in various tissues, mainly in the liver (siderosis). This may occur under the following conditions: (a) Excessive parenteral iron therapy, or repeated blood transfusions. A condition of nutritional siderosis occurs in the South African Bantu, owing to high intake of iron (iron-rich foods and iron cooking pots). (b) Excessive breakdown of erythrocytes in hemolytic types of anemia. (c) Inadequate synthesis of hemoglobin due to factors other than iron deficiency (pernicious anemia). (d) In hemochromatosis, the cause of which is unknown. In the last condition, enormous amounts of iron may be deposited in the tissues, especially the liver, pancreas and retroperitoneal lymph nodes, over 50 gm. having been found in the body (exclusive of the blood) in some cases. Studies of iron balance have revealed no significant abnormality in this condition, but it is possible that a slight degree of abnormal retention may exist over a period of many years. The capacity of the plasma to bind iron is saturated to the extent of about 90 per cent in this disease, as contrasted to the normal 30 to 35 per cent saturation. Another reflection of this abnormal situation is the fact that administration of iron is followed by a much smaller rise in serum iron in patients with hemochromatosis than in normal subjects. (e) Idiopathic pulmonary hemosiderosis with excessive deposition in lung tissue.

Iron stores can be evaluated by a variety of techniques: (a) vigorous phlebotomy, (b) isotope dilution (miscible iron store), (c) chemical estimation of nonheme iron in liver biopsies and bone marrow aspirates, (d) histochemical determination of hemosiderin, (e) the desferrioxamine test, in which the urinary excretion of iron is measured after an intramuscular (gluteal) injection of desferrioxamine (10 mg./kg.), (f) determination of serum ferritin by radioimmunoassay.

It has been found that about 20 per cent of European women have depletion of iron stores (sideropenia) without anemia. In iron deficiency, erythrocyte protoporphyrin increases even while the serum iron levels are within accepted normal limits.

References

Castle, W. B.: Disorders of the blood. *In* W. A. Sodeman. ed.: Pathologic Physiology. Ed. 3. Philadelphia, W. B. Saunders Company, 1961.

Finch, S. C., and Finch, C. A.: Idiopathic hemochromatosis and iron storage disease. Medicine, *34*:381, 1955.

Hallberg, L., Harwerth, H.-A., and Vannotti, A. (eds.): Iron Deficiency (Colloquia Geigy). New York, Academic Press, 1970.

Josephs, H. W.: Absorption of iron as problem in human physiology. Blood, *13*:1, 1958.

Ramsay, W. N. M.: Plasma iron. Advances in Clinical Chemistry, *1*:2, 1958.

Sheldon, J. H.: Hemochromatosis. London, Oxford University Press, 1935.

Shorr, E.: Intermediary metabolism and biological functions of ferritin. Harvey Lectures, *50*:112, 1956.

Underwood, E. J.: Trace Elements in Human and Animal Nutrition. Ed. 3. New York, Academic Press, 1971.

Chapter 9

SULFUR METABOLISM

Sulfur is a nutritionally essential element, occurring in substances of physiologic importance, including thiamine, insulin, anterior pituitary hormones, glutathione, cysteine, methionine, and taurocholic acid. However, indications for and methods of investigation of disturbances of sulfur metabolism are so limited that this subject requires only brief discussion here.

ABSORPTION

Sulfur is ingested in (1) inorganic (Na, K and Mg sulfates) and (2) organic forms. The latter may be subdivided into (*a*) nonprotein sulfur (sulfolipids and sulfatides) and (*b*) protein sulfur (the sulfur-containing amino acids, cystine and methionine; glycoproteins, such as mucoitin-sulfuric acid in mucin and ovomucoid and chondroitin-sulfuric acid in cartilage and tendons). The most important source of sulfur is the cystine and methionine of ingested proteins. Inorganic sulfate is absorbed as such from the intestine into the portal circulation, as are the amino acids, cystine and methionine, liberated by digestion of protein. A small amount of sulfide may be formed in the bowel by the action of bacteria but, if absorbed into the blood stream, this is rapidly oxidized to sulfate.

INTERMEDIARY METABOLISM

Sulfur reaching the liver in the portal blood undergoes the following changes:

(1) A portion of the organic S escapes oxidation. Some of this fraction is utilized for the formation of S-containing substances, such as insulin, anterior pituitary hormones, taurocholic acid, melanin, glutathione, and sulfate esters of carbohydrates (mucopolysaccharides). The remainder is excreted in the urine as neutral S.

(2) The bulk of the organic S is oxidized in the liver to inorganic sulfate. A portion of the latter, together with a portion of the inorganic sulfate absorbed from the intestine, enters the systemic circulation and is excreted in the urine.

(3) A portion of the inorganic sulfate is combined in the liver with bilirubin and various phenol derivatives, such as indole, formed in the bowel largely by bacterial decomposition of protein, to form ethereal sulfates, which are excreted in the urine. In addition to the participation of inorganic sulfate in this detoxication process, the amino acid cysteine (formed from cystine) may form conjugation products (mercapturic acids) with certain toxic compounds, such as brombenzene.

SULFUR IN BLOOD

The normal concentration of sulfur in the blood serum is as follows: inorganic S, 0.5 to 1.1 mg. per 100 ml.; ethereal sulfate S, 0.1 to 1.0 mg. per 100 ml.; neutral S, 1.7 to 3.5 mg. per 100 ml. Whole blood contains 2.2 to 4.5 mg. of neutral S per 100 ml., the higher value being due in large measure apparently to the presence of ergothioneine and glutathione in the blood cells. The serum sulfate concentration may be increased in the presence of renal functional impairment, pyloric and intestinal obstruction, and leukemia. Marked sulfate retention in advanced glomerulonephritis may contribute to the development of acidosis. An increase in the blood indican concentration (indoxyl potassium sulfate) may occur in uremia (normal, 0.026 to 0.085 mg. per 100 ml.).

EXCRETION

Sulfur is excreted in the urine in the three forms in which it exists in the blood. The total S output, since it is derived chiefly from protein under ordinary circumstances, varies with the protein intake and the extent of tissue protein catabolism, averaging about 2.5 g. (as SO_4) daily under normal conditions. The urinary N:S ratio normally ranges from 13 to 16 in the fasting state or when meat constitutes the bulk of the protein intake.

The normal total sulfate content (inorganic plus ethereal sulfates) of the urine on an average mixed diet ranges from 1.5 to 3 g. daily (as SO_4), comprising 85 to 95 per cent of the total S output. The proportion as well as the absolute amount excreted varies directly with the protein intake. The total sulfate excretion may be diminished in the presence of renal functional impairment and is increased in conditions accompanied by excessive tissue protein breakdown, such as high fever and increased metabolism.

Inorganic sulfate (Na, K, Ca, Mg and NH_4 sulfates) normally comprises 85 to 95 per cent of the total sulfate of the urine, the remainder (5 to 15 per cent) being ethereal sulfate, which normally ranges from 0.1 to

0.25 g. daily (as SO_4). This fraction consists of sodium and potassium salts of phenolic sulfuric acid compounds (e.g., indoxyl potassium sulfate or indican) and therefore varies with the quantity of phenolic substances produced in the intestine or otherwise entering the body. Increase in urine ethereal sulfate (indican) (absolute and relative) may occur in intestinal obstruction (intestinal stasis and increased putrefaction), paralytic ileus, generalized peritonitis, typhoid fever, and in association with bacterial decomposition of tissue protein and purulent exudates, as in gangrene, empyema, pulmonary suppuration, tuberculosis with cavitation and secondary infection. The greatest clinical importance of urine sulfate partition studies is perhaps the demonstration by this means of abnormal absorption of benzene and the control of industrial exposure to this agent. Benzene is excreted in the urine in the form of an ethereal sulfate (conjugated in the liver), the absorption of abnormally large amounts being reflected in a decrease in the proportion of total sulfate excreted as inorganic sulfate (below 70 to 80 per cent of the total urine sulfate). In acute cases these values may fluctuate rather rapidly, e.g., from 8 to 10 per cent immediately after exposure to 75 to 80 per cent two days later.

The neutral S of the urine, normally comprising about 5 per cent of the total sulfur, consists of such substances as cystine, urochrome, taurine and its derivatives, thiocyanate, and thiosulfate. This fraction is increased in cystinuria (p. 535), melanuria (p. 228), and some cases of obstructive and hepatocellular jaundice (excretion of taurocholic acid in the urine). The inherited disturbances of the sulfur-containing amino acids are discussed elsewhere (pp. 227, 535, and 536).

The sulfur-containing mucopolysaccharides are widely distributed and seem to be of particular importance in connective tissue and in mucous membranes. The mucopolysaccharidoses (p. 233) are a group of disorders in which enzyme defects lead to disturbances in the metabolism of these substances. In the lipid storage disorders (p. 130) there can be abnormal deposits of sphingolipids in the tissue due to genetically determined enzyme deficiencies. One of these conditions, metachromatic leukodystrophy, is concerned with the abnormal deposition of the sulfur-containing lipid sulfonylgalactosyl-ceramide.

Reference

Carson, N. A. J., and Raine, D. N. (eds.): Inherited Disorders of Sulphur Metabolism. Edinburgh, Churchill Livingstone, 1971.

Chapter 10

IODINE METABOLISM

The metabolism of iodine is of particular significance in connection with the formation of thyroid hormones (p. 746). Normal adults require at least 44 µg. daily but the recommended optimal daily intake for adults is 100 to 200 µg.; for children and pregnant women, at least 200 µg. should be given. The requirement is probably increased during pregnancy. The iodine of most common foodstuffs is apparently readily available to the organism but is not superior to inorganic iodides. Because of the high iodine content of sea water, marine vegetation, sea foods, and vegetables and fruits grown on the seaboard are particularly rich in this element. Plants (and animal tissues) grown far inland, especially at high altitudes, may be deficient in iodine because of its low concentration in the water. In such regions, iodide is commonly added to the drinking water or table salt in concentrations of 1 : 5000 to 1 : 200,000 to ensure an adequate intake and to prevent the development of simple goiter (iodine-deficiency goiter).

ABSORPTION AND EXCRETION

Iodine and iodides can be absorbed from any portion of the alimentary tract, most readily perhaps from the small intestine, free iodine and iodates being first converted to iodide. Organic iodine compounds, e.g., diiodotyrosine and thyroxine, are in part absorbed as such and in part broken down in the stomach and intestines with the formation of iodides. Absorption can occur from other mucous membranes and the skin.

Iodine (mostly inorganic) is excreted by the kidneys, liver, skin, lungs, and intestine, and in milk. It is also present in saliva. It is almost entirely in the form of inorganic iodide and may be either endogenous (including liberation from the degradation of thyroid hormones) or exogenous (food, water). Only about 10 per cent of circulating organic iodine is excreted in feces.

The quantity excreted normally varies with the intake. About 40 to 80 per cent is usually excreted in the urine (20 to 70 µg. daily in adults; 20 to 35 µg. in children). The urinary fraction of the total elimination is largest when the intake is lowest. Urine iodine is increased by exercise and other factors that increase metabolism (increased rate of turnover of thyroxine), except in the event of profuse sweating, when relatively large amounts are lost by this route. Urine iodine increases during pregnancy.

Iodine in the feces (2 to 11 μg. daily; 3 to 27 per cent of intake) is almost entirely exogenous (unabsorbed food iodine), the remainder being derived from the bile and intestinal secretions. Bile contains 4–14 μg./100 ml. in the fasting state and about 50 μg./100 ml. after eating. A portion of this is reabsorbed in the intestine (enterohepatic circulation). Biliary iodine is chiefly of alimentary origin (exogenous) but is also in part endogenous (degradation of thyroxine in liver).

Under ordinary atmospheric conditions, negligible amounts of I are excreted by the skin. With profuse perspiration (heat, humidity, exercise), as much as 30 to 60 per cent of the total may be lost by this route. The amount in expired air also varies enormously but may be as much as 10 to 30 μg. daily. The quantity present in milk varies with the iodine intake and may be increased by administration of iodine. Saliva contains 0–350 μg./100 ml., also depending on the intake.

BLOOD IODINE

Practically all of the iodine in the blood is in the plasma, the concentration in the resting state being 4–10 μg./100 ml. plasma (or serum). If no iodine has been administered for several days, 0.08–0.60 μg./100 ml. is in inorganic form (iodide); practically all is organic (4–8 μg./100 ml.), being bound to protein and precipitated with the latter by protein-precipitating agents. The organic form consists mainly of thyroxine (90%). The remainder is tri- and diiodothyronine and very small amounts of tissue breakdown products of thyroxine. About 0.05 per cent of the thyroxine is in the free state unbound to protein. No organic iodine is present in erythrocytes.

Measurement of the protein-bound iodine (PBI), reflecting, like the butanol extractable iodine (BEI), the concentration of circulating thyroid hormone, is of value in the clinical evaluation of the state of thyroid function (p. 756). The PBI rises during pregnancy. There is, however, no increase in the amount of free thyroxine because there is also an increase in the thyroxine-binding globulin (TBG). A similar state occurs in women who are taking oral contraceptives.

In the fasting state the plasma inorganic iodide is quite constant even in the face of wide variations in thyroid activity (which affect only the plasma PBI) but increases strikingly after iodine is ingested, applied to the skin, or administered for purposes of cholecystography, bronchography, or urography.

DISTRIBUTION AND INTERMEDIARY METABOLISM

The body normally contains about 10 to 20 mg. of iodine. Seventy to 80 per cent of this is situated in the thyroid gland. The next largest amount is in the muscles because of their relative bulk, although the concentration is not very high. The concentration in the salivary glands, ovaries, pituitary gland, hair and bile is at least several times greater than that in muscle. It is interesting that relatively high concentrations are present in the orbicular muscle and the orbital fat. Virtually all the iodine in saliva is inorganic, but most tissues contain only a small proportion in this form, most being present as organic iodine. The salivary iodide is proportional to that in plasma. It is known that muscle iodine content increases in hyperthyroidism and decreases in hypothyroidism. The normal iodine value for muscle is about 6 μg./100 g., of which one-sixth to one-third is inorganic. The concentration in the thyroid (200–500 mg./100 g. dry weight) is higher than in other tissues. The amount in all tissues diminishes when the intake is lowered, but the normal thyroid retains its capacity for trapping and storing iodine even under such circumstances. One of the important initial phases of the formation of thyroid hormone involves this remarkable capacity of thyroid cells for trapping and concentrating iodine, which is then rapidly utilized for synthesis of thyroxine. Salivary tissue has a similar concentrating mechanism.

The normal adult thyroid contains 8 to 12 mg. of iodine (0.2 to 0.5% dry weight), varying with the iodine intake and state of thyroid function. By far the largest part (90%) is in organic combination, stored in the follicular colloid as "thyroglobulin," a glycoprotein of mol. wt. 650,000, containing thyroxine, diiodotyrosine, and smaller amounts of triiodothyronine. On demand, these substances are mobilized from this compound and thyroxine and triiodothyronine passed into the systemic circulation. They undergo metabolic degradation apparently chiefly in the liver, the iodine component being excreted in the bile largely as inorganic iodide but also in some organic combination. In the tissues, thyroxine is, at least in part, converted to triiodothyronine, and to acetic and pyruvic acid analogs.

Iodide ions are distributed widely. They enter practically all tissues, including the erythrocytes. There are regions, such as the thyroid and salivary glands, where they are selectively concentrated. After administration of iodide, equilibrium is rapidly established within this iodide pool. The thyroid does not contribute very much to this pool, since its inorganic iodide is rapidly converted to the organic form. The pool is continuously being replenished (to overcome excretory losses) by iodine in the diet and saliva, as well as by the breakdown of the thyroid hormones.

The normal human thyroid daily secretes approximately 52 μg. of iodine as thyroxine and 12 μg. of iodine as triiodothyronine. Iodotyrosines liberated from the thyroglobulin are de-iodinated within the gland by means of an enzyme deiodinase, and the iodine released enters the thyroxine synthetic pathway. In order to maintain normal secretion of thyroid hormones, the gland must trap about 60 μg. of iodide daily. In addition to being bound by TBG, thyroxine is also to some extent bound in plasma by albumin and by a prealbumin.

The thyroid hormones are involved in a variety of biological functions. These include cellular oxidation, growth and differentiation, reproduction and the activity of the central and autonomic nervous systems. The mode of action of the hormones is not yet understood. Much current work is concerned with this. Triiodothyronine is in many respects more active than thyroxine.

The liver usually contains 0.5 to 2.3 mg. of iodine, exhibiting wide variations during periods of absorption and excretion. Iodine entering the portal vein from the intestine in part passes through the liver into the systemic circulation and is in part removed by the hepatic cells and excreted in the bile. Organic compounds containing iodine, exogenous or endogenous, are partly or completely degraded in the liver, most of the iodine being liberated as iodide.

ABNORMAL IODINE METABOLISM

Abnormalities of iodine metabolism are of clinical importance virtually only in connection with disturbances of thyroid function. They are therefore considered in detail in the discussion of thyroid hormones (p. 754).

References

Astwood, E. B., ed.: Symposium on the thyroid. Metabolism, 5:623, 1956; 6:1, 1957.
Keating, F. R., and Albert, A.: The metabolism of iodine in man as disclosed with the use of radioiodine. Recent Progress in Hormone Research. New York, Academic Press, Inc., 4:429, 1949.
Means, J. H., et al.: Thyroid function as disclosed by newer methods of study. Ann. New York Acad. Sci., 50:279, 1949.
Rawson, R. W., Money, W. L., and Greif, R. L.: Diseases of the thyroid. In Bondy, P. K. (ed.): Duncan's Diseases of Metabolism. Ed. 6. Philadelphia, W. B. Saunders Company, 1969, p. 753.
Underwood, E. J.: Trace Elements in Human and Animal Nutrition. Ed. 3. New York, Academic Press, 1971.

Chapter 11

TRACE ELEMENTS

This term is applied to elements which occur in body tissues and fluids in minute amounts. Essential physiological functions have been established for certain of these, i.e., copper, cobalt, zinc, iodine, magnesium, and molybdenum. Others may be accidental contaminants, although there is suggestive evidence of the possible biological importance of fluorine, barium, strontium, and arsenic. In view of the fact that their essential activities are exerted by mere traces of these elements, it would appear that their functions are catalytic in nature, i.e., that they act as direct activators of certain enzymes, or indirectly as essential components of vitamins or hormones. Discussion of these substances will be restricted to aspects of their metabolism that may be of clinical significance.

COPPER

In health, the adult human body contains about 80 mg. of copper. Weight for weight, the concentration in the newborn is about three times that in the adult body. The greatest amount of copper, about 8 mg., is found in the liver. The brain has approximately the same concentration of copper but less in total amount; its concentration increases from birth to maturity. The level tends to fall in other organs. Copper is distributed widely throughout the tissues. In addition to the two organs already mentioned, its concentration is relatively high in the kidneys, heart and hair and exceptionally high in the pigmented parts of the eye. Although the metal tends to be associated with melanin in the eye and bound ionically to protein, the copper content of human hair is not related to its color. There is also a high copper content in dental enamel.

Some of the major sources of dietary copper are liver, kidney, other meats, shellfish, nuts, dried legumes and stone fruits. Milk and its products are very poor sources of copper.

Absorption; Excretion; Metabolism. About 30 per cent of orally administered copper is said to be absorbed in man, but the figure is much lower in animals. Absorption is apparently affected, at least in animals, by dietary zinc, molybdenum and inorganic sulfate. Absorption of copper occurs in the human duodenum. Very little is known about the mechanism of its absorption.

The concentration of copper in the plasma is somewhat higher than that in the erythrocyte. The former averages 106 µg./100 ml. in adult males and 114 µg. in females. The difference is significant. The erythrocytes contain a copper protein, erythrocuprein, of mol. wt. 31,000, the content of which is remarkably constant. It accounts for 60 per cent of erythrocyte copper. Some of the red cell copper is loosely bound to proteins and is quite labile. In plasma, about 80 per cent of copper is firmly bound to a protein, ceruloplasmin, of mol. wt. 151,000. Its concentration in plasma is 20 to 50 mg./100 ml. in adults. It has been shown that the amount of ceruloplasmin copper exchanged daily is very much less than the amount absorbed in the duodenum. This must mean that ceruloplasmin is not the copper transport protein. It is in fact considered to be a true oxidase. The remainder of the plasma copper is loosely bound to a protein, probably albumin. This is probably the true transport form.

The level of copper in the plasma increases during pregnancy from the third month to a level at least twice that of nonpregnant women. Because of their estrogen content, oral contraceptives have a similar effect on the body. In each case, the red cell copper content stays fairly constant. A few weeks after delivery, the level of red cell copper is back to the level it was before pregnancy.

About 2.0 to 5.0 mg. of copper is present in the normal daily diet. Of this, 0.6 to 1.6 mg. is absorbed in the duodenum. The daily biliary excretion of copper is 0.5 to 1.3 mg., and 0.1 to 0.3 mg. is excreted across the intestinal mucosa into the bowel lumen. Only 10 to 60 µg. of copper is present in a 24-hour specimen of normal urine.

Function; Deficiency Manifestations.

Copper is an essential component of the respiratory pigments of certain marine species, phenol oxidases in plants, and tyrosinase in man. All available evidence points to a role in hematopoiesis as the main function of copper, the exact nature of which is unknown. Copper deficiency results in a microcytic anemia, similar to that resulting from iron deficiency, which, however, cannot be corrected by administration of iron. There is impairment of erythropoiesis and decrease in erythrocyte survival time. Associated reductions in tissue catalase and cytochrome C and in cytochrome oxidase activity suggest that copper is involved in the metabolism of iron, acting perhaps as a catalyst for its incorporation into these substances and also into hemoglobin. However, although these abnormalities can be produced in experimental animals, there is no evidence that copper deficiency occurs in human beings, apart from hypocupremia in certain anemias of infancy, in which the primary deficiency probably is of iron. Some cases appear to be due to a true copper deficiency. These were malnourished infants who were being treated with a high-calorie, but low-copper, milk diet.

Elevated values for serum ceruloplasmin, and therefore copper, have been reported in infectious diseases, hepatic and biliary tract disease, leukemia and other forms of malignancy, iron deficiency anemia, hyperthyroidism, myocardial infarction, and certain neurological diseases. The significance of these observations is not known.

Subnormal concentrations have been observed in a variety of conditions associated with hypoproteinemia, e.g., the nephrotic syndrome, protein malnutrition, and the malabsorption syndrome. Clinical interest in abnormal copper metabolism is centered particularly in its relation to the pathogenesis of hepatolenticular degeneration (Wilson's disease).

This condition is characterized clinically by hepatic degeneration and fibrosis (cirrhosis); progressive cerebral damage, involving particularly the basal ganglia; defects of renal tubular reabsorption (p. 537) and glomerular filtration; characteristic corneal rings (Kayser-Fleischer); and occasionally, pigmentation of the nails and skin. These abnormalities are consequences of deposition of unusually large amounts of copper in the affected tissues. The significant biochemical abnormalities are as follows: (1) low serum ceruloplasmin concentration (not invariably); (2) low serum copper concentration, but increase in the amount that is free or loosely bound to albumin; (3) increased urinary excretion of copper; (4) increased amount of copper in certain tissues, e.g., liver, kidney, brain. This is a congenital disorder, with a familial incidence, the low serum ceruloplasmin concentration, when it occurs, being attributable to a genetically determined defect in synthesis of this protein. Apparently normal serum ceruloplasmin concentrations have been explained on the basis of a possible abnormality in the molecular structure of the protein, similar to that known to occur in the case of hemoglobin (p. 267). The prevailing view is that, as a result of this defect in ceruloplasmin synthesis, an unduly large proportion of absorbed copper is carried in the blood plasma in the free state or loosely bound to albumin, and therefore diffuses more readily into the tissues and urine. Inasmuch as the total plasma copper concentration is low, although that of free copper is high, abnormally large amounts continue to be absorbed from the intestine. Objections have been raised to this hypothesis, mainly on the basis that (a) apparently normal plasma ceruloplasmin concentrations are found occasionally in subjects with Wilson's disease, (b) relatives of such patients may be asymptomatic despite low plasma ceruloplasmin levels, and (c) manifestations of Wilson's disease do not occur in subjects with low serum ceruloplasmin concentrations due to protein malnutrition, the nephrotic syndrome, or the malabsorption syndrome. This issue is still controversial. A definitive diagnosis of hepatolenticular degeneration can be made on the basis of a low serum ceruloplasmin concentration, increased urinary excretion of copper, and increase in liver copper (biopsy). Any one of these, if present consistently, usually suffices for diagnostic purposes.

Hepatolenticular degeneration is in effect a form of chronic copper poisoning. Improvement can be obtained by removing the excess of tissue copper by administering the copper chelating agent penicillamine. This brings about marked increase in the urinary secretion of the metal.

COBALT AND ZINC

Cobalt is readily absorbed from the intestine (70 to 80 per cent), about 65 per cent of the amount ingested being excreted in the urine, the remainder in the feces. Only minute amounts are present in the tissues, the largest in the liver (storage site). Its nutritional importance in man apparently arises mainly out of the fact that it is an essential component of vitamin B_{12}, essential for normal hematopoiesis (p. 818).

If cobalt is required in man only as vitamin B_{12}, its requirement is much lower than for any other trace mineral. As little as 1 to 2 μg. of B_{12} daily, containing 0.045 to 0.09 μg. cobalt, suffices to maintain normal bone marrow function in pernicious anemia. An average American diet supplies many times this amount. Consequently, cobalt deficiency in man manifests itself, so far as is known, only as vitamin B_{12} deficiency. This may result from inadequate absorption of the vitamin, due to the malabsorption syndrome (p. 607), absence of intrinsic factor (pernicious anemia, p. 820), or inadequate formation by intestinal bacteria due to oral antibiotic therapy.

Zinc is either a constituent of, or necessary for the action of, many enzymes. These include carbonic anhydrase, alkaline phosphatase, pancreatic carboxypeptidase and a variety of dehydrogenases. Its importance in human nutrition is therefore obvious. It would seem to be concerned with the healing of wounds. In areas of the Middle East, a deficiency of zinc in boys has been related to a form of dwarfism associated with hypogonadism. There are also hepatosplenomegaly, delayed closure of the epiphyses of the long bones and anemia. The level of zinc is low in the plasma and tissues. The deficiency appears to be secondary to a high intake of phytate as well as to the incidence of clay eating. Excessive sweating in the hot climate also results in excessive loss of the metal.

In the whole body of a normal man weighing 70 kg., 1.4 to 2.3 g. of zinc is present. Twenty per cent of this total is present in skin. The concentration is very high in the prostate (102 μg./g. fresh tissue), which is about twice as high as the concentration in liver, kidney and muscle. The element has a wide distribution through the tissues of the body. Probably the highest zinc concentration occurs in the choroid of the eye. Semen contains up to 200 μg./ml. zinc.

The normal plasma concentration of zinc is approximately 100 μg./100 ml. The content is higher in serum. The plasma concentration of this metal ion falls to approximately one-half to two-thirds the quoted normal level among women in the later part of pregnancy and among those taking oral contraceptives. The concentration also falls as a consequence of corticosteroid therapy. Normal plasma contains only about 20 per cent of the zinc present in whole blood, most of which is present in the erythrocytes. About 3 per cent of this metal ion is contained in leukocytes. The individual leukocyte, however, contains considerably more than the individual erythrocyte.

In certain types of chronic leukemia, there is a marked fall in the zinc content of peripheral leukocytes.

The normal daily intake of zinc is 10 to 15 mg., and it is absorbed mainly in the duodenum. It is excreted in pancreatic juice, in bile and across the mucosa of the large intestine. This endogenous fraction forms only a small part of the total zinc excreted in feces, since most is derived from unabsorbed intake. Up to about 0.5 mg. of the ion is excreted daily in the urine. This is a very small amount. Increased urinary excretion of zinc occurs in the nephrotic syndrome, alcoholic cirrhosis, hepatic porphyria and total starvation, and in individuals with hypertension. Administration of appropriate chelating agents, such as EDTA, can also produce a large increase in urinary excretion of zinc.

FLUORINE

Fluorine occurs in many tissues, notably the bones, teeth, and kidneys, but there is no evidence that it is nutritionally essential for general health. Interest in this element rises chiefly out of its efficacy in preventing dental caries, and possibly its use, combined with vitamin D, in the treatment of osteoporosis.

It is absorbed readily from the intestine, being excreted in the urine, in the sweat and by the intestinal mucosa. Intake of abnormally large amounts (in drinking water) during childhood results in "dental fluorosis" ("mottled enamel"), characterized by a patchy, chalky or brownish mottling of the enamel, frequently with pitting of the surface and fracture and chipping of the enamel, which is abnormally fragile. This is a reflection of imperfect formation of this material.

Of special practical importance is the observation that ingestion of amounts of fluoride (1.0 to 1.5 parts per million in drinking water) too small to produce mottling renders the teeth more resistant to caries. This has led to widespread fluoridation of water supplies, with good results. Topical application is also apparently somewhat effective. It is believed that fluorine may undergo surface adsorption by the hydroxyapatite crystals of the enamel, forming a protective layer of acid-resistant fluoroapatite. The fluoride ion may also act by inhibiting the metabolism of oral bacterial enzymes, diminishing local production of acids (from carbohydrates) that are regarded as important in the production of dental caries.

OTHER ESSENTIAL TRACE ELEMENTS

Manganese, molybdenum, chromium, and possibly nickel are now recognized as essential trace elements. Evidence in support of this recognition has been obtained mostly from animals but also from man.

References

Scheinberg, I. H., and Sternlieb, I.: Copper metabolism. Pharmacol. Rev., *12*:355, 1960.
Underwood, E. J.: Trace Elements in Human and Animal Nutrition. Ed. 3. New York, Academic Press, 1971.

Chapter 12

WATER, SODIUM, CHLORIDE, POTASSIUM: PHYSIOLOGICAL CONSIDERATIONS

These substances are so intimately interrelated with respect to amount and distribution in the body, as well as regulatory mechanisms, that it is impossible to discuss any one satisfactorily to the exclusion of the others. Moreover, the important role of the kidneys in the regulation of the volume and composition of body fluids, and the influence of the latter on the state of acid-base balance, add to the complexity of presentation of this subject. This is particularly true in discussions of abnormalities. However, in the interest of clarity, certain of these interrelated topics will be discussed individually insofar as is feasible.

WATER INTAKE

Water is supplied to the body from the following sources: (*a*) dietary liquids; (*b*) solid food; (*c*) oxidation of organic foodstuffs.

Water comprises 70 to 90 per cent of the weight of the average diet of adults, even apparently very solid foods consisting largely of water. Moreover, water is one of the chief products of combustion of protein, fat and carbohydrate in the body, the quantities produced by oxidation of 1 g. of each of these substances being as follows: protein, 0.34 ml.; fat, 1.07 ml.; carbohydrate, 0.56 ml. It may be calculated that 10 to 15 ml. of water are formed per 100 calories of energy produced. An ordinary 3000 calorie diet therefore contains about 450 ml. of water in the solid food and may provide an additional 300 to 450 ml. of water of oxidation. The remainder of the water intake is supplied by dietary liquids (Fig. 12-1).

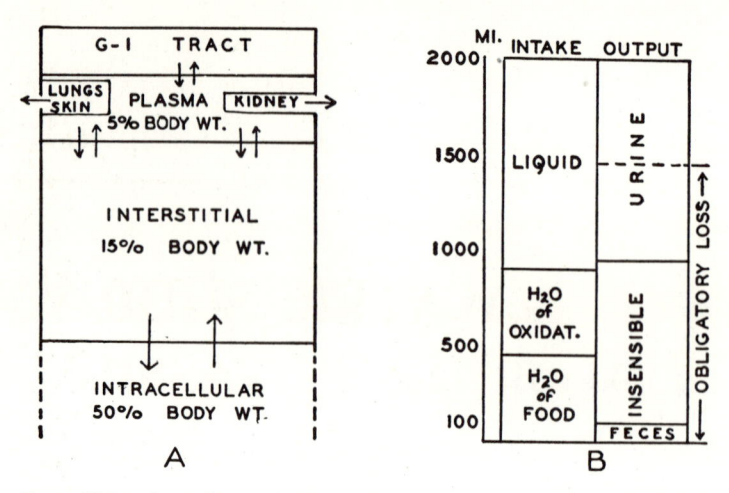

Figure 12-1. A, Body water compartments. B, Sources of water to the organism, and water excretion in a 70-kg. man on an average diet. (After Gamble.)

WATER OUTPUT

Water leaves the body in the (a) urine, (b) feces, (c) perspiration, and (d) so-called insensible perspiration (evaporation from skin and lungs) (Fig. 12-1).

Feces. Under normal conditions, the 3000 to 8300 ml. of digestive fluids (Table 12-1) entering the alimentary tract are almost completely reabsorbed in the intestine. On an ordinary mixed diet, about 80 to 150 ml. of water are excreted daily in the feces of normal adults. This may be increased considerably on a high vegetable diet and particularly in the presence of diarrhea.

TABLE 12-1 NORMAL QUANTITATIVE WATER TURNOVER IN 70 KG. SUBJECT*

FLUIDS EXCRETED AND REABSORBED	VOLUME ML.
Saliva	500– 1,500
Gastric secretion	1,000– 2,400
Intestinal secretion	700– 3,000
Pancreatic secretion	700– 1,000
Liver bile	100– 400
	3,000– 8,300

FLUIDS LOST FROM THE BODY	
Urine	600– 2,000
Feces	50– 200
Insensible perspiration	350– 700
Perspiration	50– 4,000
Milk	0– 900
	1,050– 7,800
Total fluid turnover	4,050–16,100

* (After Adolph, E. F.: Physiol. Rev., *13*:336, 1933.)

Insensible Perspiration. In the absence of active perspiration, the body is continually losing water vapor from the skin surface and lungs in inverse proportion to the relative humidity of the atmosphere. Inasmuch as 0.58 calorie is absorbed in the vaporization of 1 ml. of water, the heat lost by this process, termed "insensible perspiration," at ordinary room temperatures amounts to about 25 per cent of the total heat loss. The latter is proportional to the quantity of heat produced, which is a function of metabolism. The quantity of water lost by insensible perspiration is therefore an obligatory loss, determined by metabolic activity, and, in a normal adult, may be calculated as 500 ml. per sq. meter of body surface area per day. This amounts to about 850 ml. for a 70 kg. man (1.73 sq. meters).

Perspiration. When either heat production or the environmental temperature rises to the point where heat loss by the ordinary means is inadequate, the sweat glands become active. Obviously, the quantity of water lost by this route varies enormously. Moreover, in contrast to insensible perspiration, which consists solely of water, sweat contains 30 to 90 mEq./l. of Na and of Cl (Fig. 12-4, p. 364). It is a hypotonic solution, plasma containing about 140 mEq./l. of Na. Sweating, therefore, removes from the body relatively more water than electrolytes.

Urine. The urine is the important medium of elimination of water provided to the body in excess of its fixed requirements. The kidneys have a remarkable capacity for regulating urinary excretion of water regardless, within wide limits, of the simultaneous requirement for excretion of solids (p. 495). However, their ability to concentrate is not unlimited, a certain minimal quantity of water being required for excretion of a given amount of solute. On an average adequate diet (adults), providing about 50 g. of solids for daily urinary excretion, a minimum of approximately 500 ml. of water (300 ml./sq. meter body surface) is required for their solution. Failure to meet this obligatory urine volume requirement will result in retention of certain urinary constituents, mainly urea, in the body fluids.

The quantitative turnover of water by various organs and the amount excreted in normal circumstances are presented in Table 12-1 and Figure 12-1. When one considers that all of the secretions and excretions mentioned are derived directly from the blood plasma, it appears that a total daily fluid turnover of 4000 to 16,000 ml. is accomplished from a circulating medium of about 3500 ml. (plasma volume). This phenomenon is rendered possible by (1) the efficient and rapid reabsorption of the large volumes of fluid secreted into the gastrointestinal tract (3000 to 8300 ml.); (2) the presence of a large and readily available reservoir represented by the interstitial fluid compartment (about 10,500 ml.); (3) the vast area of the capillary bed of the body (about 68,000 sq. ft.), which constitutes an enormous filtration surface for the rapid interchange of fluid between the vascular and interstitial fluid compartments; (4) so-called pump mechanisms for pushing ions across membranes, especially the sodium pump.

Equilibrium Requirements. As indicated above, in addition to the small amount excreted in the feces (80 to 150 ml.), the minimal daily water requirement is fixed by certain metabolic requirements: (a) loss of heat by insensible perspiration (normally about 850 ml.); (b) renal excretion of excess solid material, mainly urea, and therefore determined largely by the protein intake (normally about 500 ml. of water). Consequently on an average normal diet, approximately 1500 ml. of water must be available for elimination by these routes if the body temperature and blood urea N are to be maintained within normal limits. The "solid" portion of such a diet contributes about 800 ml. of water, approximately one-half of which is preformed, the remainder being produced in the course of oxidation of the organic foodstuffs in the body. The balance of the water requirement must be supplied by intake of liquids if the body fluids are to be maintained at a normal level.

VOLUME AND COMPOSITION OF BODY FLUID COMPARTMENTS

Water comprises 60 to 70 per cent of the adult body weight, values expressed in this manner being somewhat lower in women than in men, and decreasing with advancing age. These age and sex differences after puberty are probably due to differences in the amount of adipose tissue, which has a low water content. Classically, the body water has been regarded as existing in two main compartments, (1) intracellular (approximately 50 per cent body weight), and (2) extracellular (approximately 20 per cent), the latter being further subdivided into (a) blood plasma water (about 5 per cent body weight) and (b) interstitial fluid (about 15 per cent). Although segregation into these anatomical compartments is useful from many standpoints, it cannot be applied strictly to physiological considerations. This is particularly important in relation to the distribution and exchanges of electrolytes between the various body fluids (p. 352); from the standpoint of volume alone it seems desirable to regard the extracellular water as existing in two main subdivisions: (1) blood plasma (about 4.5 per cent body weight); (2) interstitial fluid and lymph (about 12 per cent). These are not in direct contact with the so-called transcellular fluid (C.S.F., joint fluid, fluid in serous cavities, ocular fluid), which can be so dramatically affected in disease states.

The anatomical fluid compartments are illustrated in Figure 12-1. Water entering the organism, i.e., from the gastrointestinal tract, passes into the blood stream, an equivalent amount leaving the latter by way of the urine, lungs, skin and feces. The water of the plasma is in equilibrium with that of the interstitial fluid, and the latter with the intracellular water, across the boundaries between these compartments, i.e., the capillary walls and the cell membranes, respectively. The interstitial fluid can expand or contract considerably, in conditions of excessive retention (edema or loss) (dehydration) of fluid, even in the absence of comparable or even significant alterations in the volume of plasma or intracellular fluid.

The interstitial fluid is no longer regarded as a completely passive member of the trio (i.e., intracellular, plasma, interstitial) which responds one way or another only because of the activities of the other two fluid compartments. It has its own part to play, largely because of its content of hyaluronic acid and protein. Loss of some of its water results in significant changes in its osmotic pressure. This is sufficient to prevent further loss without the expenditure of cellular energy. Moreover, increased concentration of the hyaluronic acid produces a viscosity increase sufficient to affect the passage of ions and small molecules, and this in turn reflects on their passage to and from the other two compartments.

Volume of Body Fluid Compartments

The volume of (*a*) the circulating blood, (*b*) the plasma, (*c*) total extracellular fluid, and (*d*) total body water could be measured by introduction of a substance into the body if that substance would fulfill the following requirements: (1) be retained exclusively in the fluid compartment in question; (2) not leave that compartment during the test period; (3) distribute itself uniformly throughout that compartment; (4) be capable of precise quantitative determination in the blood or plasma. A number of methods have been proposed, none of which is entirely satisfactory; certain of them, however, have proved useful. Their usefulness applies mostly to changes occurring during health. In the presence of pathology, it must be emphasized that changes occur which interfere even with the concept of separate compartments, as well as the determination of them.

Blood and Plasma Volume. *(a) Carbon Monoxide Method.*
Carbon monoxide displaces oxygen from hemoglobin, volume for volume, and can be measured accurately in blood, either as CO or as HbCO. The subject breathes a known amount of CO and the concentration of HbCO in the blood is determined after allowing sufficient time for mixing. If the total hemoglobin concentration is known, one can calculate how much CO would be required to saturate the Hb to its full capacity and can arrive at the total blood volume. By determining the hematocrit value (relative volumes of packed cells and of plasma), the plasma volume can be derived.

(b) Radioactive Iron Method. ^{59}Fe is administered to a normal or anemic subject; it becomes incorporated in the Hb of the circulating erythrocytes. A predetermined amount of the labeled red cells is injected intravenously in the test subject (compatible blood). Measurement of radioactivity in blood samples withdrawn after allowing time for mixing indicates the extent of dilution of the injected cells and permits calculation of the volumes of blood and plasma (from hematocrit value).

(c) Dye Methods. The dye employed most commonly is Evans blue (T-1824), which is adsorbed by plasma proteins. A known amount is injected intravenously and, after allowing time for distribution throughout the blood plasma, a sample of oxalated blood is withdrawn, centrifuged, and the concentration of dye in the plasma determined colorimetrically.

The extent of dilution of the quantity injected can be calculated (plasma volume) and, knowing the hematocrit value, the total blood volume can be determined.

(d) ^{131}I-Labeled Proteins. Serum proteins, labeled with ^{131}I, are injected intravenously and the extent of dilution of the injected protein is determined by measurement of the radioactivity of withdrawn blood samples.

The CO method requires cooperation of the subject, which cannot always be secured under clinical conditions. Moreover, other technical difficulties introduce the possibility of serious errors. Furthermore, the accuracy of this procedure, as well as the ^{59}Fe method, depends upon uniformity of distribution of all of the "labeled" erythrocytes in the circulation. This cannot always be assured, especially in disease states (e.g., shock), in which variable and undeterminable numbers of cells may be "segregated" in such organs as the spleen or in the splanchnic bed. In addition, calculations of total blood volume on the basis of factors measured in either cells or plasma are subject to inaccuracies arising out of variations in the ratio of cells to plasma in different portions of the circulation.

One source of inaccuracy with the use of dyes and isotope-tagged serum proteins is the fact that these proteins normally cross the capillary wall in amounts which vary in different tissues; they are removed from the interstitial fluid mainly by the lymphatic circulation. Approximately 50 per cent of the plasma protein pool is in the interstitial fluid-lymph compartment and, following intravenous injection of tagged protein, the latter is evenly distributed throughout the entire extracellular space in about twenty-four hours. Although under normal conditions loss of protein from the intravascular compartment under the test conditions is not so rapid as to vitiate the usefulness of these procedures, it may become so in many disease states in which capillary permeability is increased.

Generally acceptable average values for blood and plasma volume in normal subjects are as follows: (1) whole blood, 2500–3200 ml./sq. meter body surface or 63–80 ml./kg.; (2) plasma 1300–1800 ml./sq. meter, or 35–45 ml./kg. Values for whole blood are about 7 per cent higher for men than for women, but the plasma values are approximately the same.

A slight increase in blood and plasma volume occurs in normal pregnancy, the increase in plasma volume being relatively greater than that in total blood volume. This disproportionate increase in plasma volume is responsible in part for the physiological decrease in red blood cell count and plasma albumin concentration during this period (p. 189).

Total Extracellular Fluid Volume. Application of the dilution technique to the measurement of the total volume of extracellular fluids theoretically requires the use of a substance which, when injected intravenously, crosses the capillary walls readily and distributes itself rapidly and uniformly throughout all extracellular fluids, does not enter the cells, is not destroyed in the body and is not excreted rapidly. Substances that have been used for this purpose include certain nonmetabolizable carbohydrates (sucrose, mannitol, inulin) and electrolytes (thiocyanate, radioactive sodium, radioactive chloride, sulfate, bromide). There are two major sources of inaccuracy of results obtained with these procedures: (1) the heterogeneous character of the extracellular fluids with respect to composition (p. 350); (2) certain of the substances employed, particularly electrolytes, are by no means excluded from the intracellular compartment, but penetrate different cells to different degrees under normal conditions and particularly in pathological states.

There are marked differences in the rates of equilibration of extracellular fluids in muscle, skin, loose and dense connective tissue, cartilage and bone. Certain of these equilibrate rapidly; others, e.g., dense connective tissue and cartilage and certain regions in bones, very slowly. Furthermore, some of the tracer substances employed (e.g., inulin, mannitol) do not enter the transcellular fluids, the volume of which may vary enormously in certain disease states.

The major electrolytes of extracellular fluid, Na and Cl, are present in only low concentrations in intracellular fluid generally under physiological conditions. However, both Na and Cl enter different cells to a variable degree, and not always to the same extent, the Na : Cl ratio in certain tissues (bone, cartilage, muscle) being higher and in others (erythrocytes, connective tissue, gastrointestinal mucosa) lower than in an ultrafiltrate of blood plasma. Moreover, there are marked differences in the rates at which equilibrium is established between different phases of the extracellular pool of Na. For example, in bone, which contains 35 to 40 per cent of the total body Na, only about 30 per cent is rapidly exchangeable; 10 per cent, slowly exchangeable (adsorbed to the surface of bone crystals); and 60 per cent, nonexchangeable (incorporated in the bone crystal, p. 283).

Values obtained by the use of electrolytes tend to be too high and those with saccharides, too low. For the reasons indicated, it is preferable to use the designations "sodium space," "chloride space," "thiocyanate space," "inulin space," "sucrose space" and "sulfate space," rather than "extracellular fluid volume." The sucrose and sulfate spaces seem to correspond most closely to the extracellular fluid volume when conditions are physiological or approximately so. Employing these procedures, values have been obtained for total extracellular fluid volume (adults) approximating about 20 per cent of the body weight. Of this, about one-fourth is blood plasma and three-fourths is interstitial fluid and lymph (5 and 15 per cent, respectively, of body weight).

Total Body Water. There is no accurate method for determining total body water, inasmuch as no measurable substance has yet been found which, on injection, is distributed uniformly throughout the entire body water and undergoes no metabolic change during the test period. Heavy water (D_2O) and antipyrine are most satisfactory for clinical purposes. For such purposes, however, it is usually sufficient to detect fairly rapid changes in body water by means of changes in total body weight, as well as by appropriate intake and output records.

The water content of several tissues has been determined by direct measurement (animals and human biopsy or autopsy material). Water comprises about 60 to 70 per cent of the body weight of adults (75 per cent in children), varying considerably in different tissues. Intracellular water comprises about 50 per cent of the body weight.

Composition of Body Fluid Compartments

Inasmuch as water serves chiefly as a relatively inert medium in which are conducted the chemical reactions constituting living processes, its significance must be considered in relation to the other components of the body fluids which, in fact, largely determine its volume and distribution in the organism. The marked differences in composition of the intracellular and interstitial fluids and the less striking differences between the latter and the blood plasma are due to differences in permeability of the membranes separating these compartments, to the solutes which these fluids contain, and to intracellular pump mechanisms such as the sodium pump, which is responsible for the active extrusion of sodium ions from cells into the interstitial fluid. This requires energy obtained from ATP, by means of ATPase, which, according to some models, is part of the pump mechanism.

It must also be remembered that lymph drainage plays a part in relation to interstitial fluid.

Milliequivalents (mEq.). Elements which combine chemically with one another do so in fixed weight ratios, which are functions of their atomic weights and valences. Thus, 1.008 g. of hydrogen (at. wt. 1.008) combine with 8 g. of oxygen (at. wt. 16) to form water, and with 35.457 g. of Cl (at. wt. 35.457) to form HCl. Similarly, 35.457 g. of Cl combine with 22.997 g. of Na. 39.096 g. of K and 20.04 g. of Ca (at. wt. 40.08). These numerical values are designated "equivalent weights," because they are equivalent in their combining power. An "equivalent weight" may be defined as the weight of a molecule, or of an atom or radical (group of atoms reacting chemically as a unit), divided by its valence. A "milliequivalent" is 1/1000 of the gram equivalent weight, or the latter expressed in milligrams.

The concentrations of the solid constituents of the body fluids may be expressed in terms of weight per unit volume (e.g., milligrams per 100 ml.) or in terms of apparent chemical combining power (e.g., milliequivalents per liter) (mEq./l.). The former may be converted to the latter according to the following formula:

$$mEq./l. = \frac{mg./l. \ (= mg./100 \ ml. \times 10) \times valence}{atomic \ or \ radicular \ weight}$$

For a variety of reasons, the concept of equivalent weight is no longer considered valid. This is partly because of the fact that many elements are really mixtures of isotopes, some of which can vary, and partly because modern concepts of the way in which atoms are held together to form molecules or radicals do not harmonize with the older concept of valency, which itself allowed that one element, e.g., iron, could have more than one valency. This has led to the introduction of the basic unit called the mole, since "equivalent" is ambiguous.

Molar Concentration. This is the number of moles present in each liter. In order to understand the concept of the basic unit, "the mole," it is important to be aware of the fact that the gram molecular weight of any substance always contains $(6.02252 \pm 0.00028) \times 10^{23}$ actual molecules. This is Avogadro's number (N). Similarly, every gram atomic weight of a pure nuclide (isotope) contains N atoms. Thus, 0.012 kg. (12 g.) of pure ^{12}C would contain N atoms of ^{12}C. (This is the international reference value.)

It is also necessary to become aware of the concept "formula unit." These elementary units can be symbolized by Na^+; Ca^{2+}; $(Ca^{2+})_{0.5}$; Cl^-; SO_4; $(SO_4^{2-})_{0.5}$; PO_4; $(PO_4^{3-})_{0.33}$; C_6H_6; C_2H_5OH; and so on. One mole is the amount which would contain N of these formula units. A formula unit such as PO_4 takes no account of the valency concept. Even with the old-fashioned concepts, it was difficult to express the concentration of phosphate in biological fluids as mEq., since it exists both as HPO_4^{2-} and $H_2PO_4^-$. To express it as moles is quite simple. One mole of PO_4 is one gram atom of phosphorus (30.97 g.) plus four gram atoms of oxygen ($4 \times 16.00 = 64$), i.e., $30.97 + 64 = 94.97$ g. A formula unit may be an atom, a molecule, a radicle, an ion, an electron, etc.

The official definition of one *mole* is an amount of a substance or a system which contains as many formula units as there are carbon atoms in 0.012 kg. of the pure nuclide ^{12}C.

The mole is too big a unit for biological purposes, and so concentrations must be expressed as millimoles (mmol.) per liter (mmol./l.).

It is now suggested that instead of the term "molar concentration," the term "substance concentration" should be used for the purposes of clinical chemistry. The latter is also expressed in terms of mmol./l.

Osmolality. Although absolute osmotic equilibrium is probably never attained in living tissues, all internal body fluids, extracellular and intracellular, tend to have an equal osmotic pressure. Osmotic pressure is a function of the concentration of active chemical components in a solution, i.e., the number of ions and molecules (undissociated compounds). The designation "osmole" is applied to one chemically active undissociated molecule or ion.

A solution containing 1 osmole (expressed in grams) of solute per kilogram of water depresses the freezing point of the water by 1.86° C. The freezing point of blood plasma is about −0.56°. It must, therefore, contain osmotically active solutes in a total concentration of 0.56/1.86 or 0.301 osmole per kilogram plasma water, of which nonelectrolytes contribute about 10 milliosmoles. The usual range quoted is 270 to 310 mosmol./kg. plasma water; i.e., the osmolality is 270 to 310 milliosmoles.

In this connection, the components of the body fluids may be classified conveniently as follows:

(1) Compounds of very large molecular size, e.g., proteins, lipids, gly-

cogen, to which all but a few capillary and cell membranes are almost completely impervious. Inasmuch as osmotic pressure is proportional to the number of active chemical components per unit water, these compounds, because of their large molecular size, exert relatively little effect on osmotic pressure in proportion to their concentration on a weight basis. For example, the concentration of plasma proteins, about 7 g. per 100 ml. of plasma, with molecular weights of about 70,000 to over 1,000,000, represents only about 2 milliosmoles per kilogram plasma water.

(2) Electrolytes, which cross capillary walls more or less freely, but to certain of which cell membranes exhibit a variable degree of impermeability under different conditions, because of changing efficiency of biological pump mechanisms as well as altered "porosity" of the membranes themselves.

Electrolytes dissociate into ions and, therefore, exert an osmotic pressure which is the sum of that due to each of their ions, because the ions behave as units. The osmolality of NaCl, which dissociates almost completely into Na^+ and Cl^-, is practically twice what it would be if it were an undissociated molecule. Similarly, that of $CaCl_2$ is three times what it would be since it dissociates into $Ca^{2+} + 2Cl^-$. In the presence of large amounts of protein, certain elements, e.g., Ca and K, may exist as relatively poorly dissociated salts of protein, being osmotically active only to the extent of the degree of dissociation.

(3) Organic compounds of a molecular size which permits their free diffusion across capillary walls, e.g., urea, glucose, amino acids. Some cell membranes may be relatively impermeable to certain of these, as to certain electrolytes.

EXTRACELLULAR FLUID

All body cells exist in an environment of fluid collectively designated extracellular fluid. This includes the blood plasma, interstitial fluid and lymph. The heterogeneous character of the interstitial fluid is referred to

TABLE 12-2 RADICULAR AND EQUIVALENT WEIGHTS OF IMPORTANT BODY FLUID ELECTROLYTES

ELEMENT OR ION	VALENCE	ATOMIC OR RADICULAR WEIGHT	EQUIVALENT WEIGHT
H	1	1.008	1.008
Na	1	22.997	22.997
K	1	39.096	39.096
Cl	1	35.457	35.457
Ca	2	40.08	20.04
Mg	2	24.32	12.16
HCO_3	1	61.018	61.018
HPO_4	2	95.988	47.994
H_2PO_4	1	96.996	96.996
SO_4	2	96.066	48.033

elsewhere (p. 345); from a physiological standpoint, there are distinct differences in this respect between muscle, skin, dense connective tissue and cartilage, and bone. With these exceptions and reservations, most of the extracellular fluid may be regarded as having an essentially uniform qualitative composition; what quantitative differences exist are due chiefly to differences in the concentrations of proteins, which range from about 7 per cent in plasma and but slightly less in hepatic lymph, to about 0.1 per cent in the subcutaneous interstitial fluid. They are solutions chiefly of NaCl and NaHCO$_3$, with small amounts of Ca, Mg, K, H, phosphate, sulfate and organic acid ions, variable amounts of protein, and some nonelectrolytes (glucose, urea, lipids, etc.), and with pH values ranging from 7.35 to 7.45 under normal conditions.

The total concentration of the ionic constituents is about 310 mmol. per liter of plasma (about 335 mmol. per liter of plasma water). In accordance with the laws of electrical neutrality, cations and anions each comprise half of the total concentration, expressed in terms of chemical equivalence. Proteins, being amphoteric, act as anions in these slightly alkaline fluids, the chemical structures of which are indicated in Figure 12-2. In the light of modern knowledge, much of the intracellular magnesium is not in the ionic form but is bound covalently to protein and other smaller organic molecules. The metal, therefore, does not contribute nearly so much to intracellular cations as is shown in the figure.

For clinical purposes, it has proved to be quite useful in the past to express anion and cation concentrations in terms of milliequivalents per liter. For reasons already given, these expressions of concentration are now being replaced by millimoles per liter. Using milliequivalents, it was possible to draw diagrams representing cationic and anionic concentrations, and by the use of a number of probably unjustified assumptions, to balance the books (Fig. 12-3). The illustration is included because it has historic interest and can be of some help, provided that one is aware of its shortcomings.

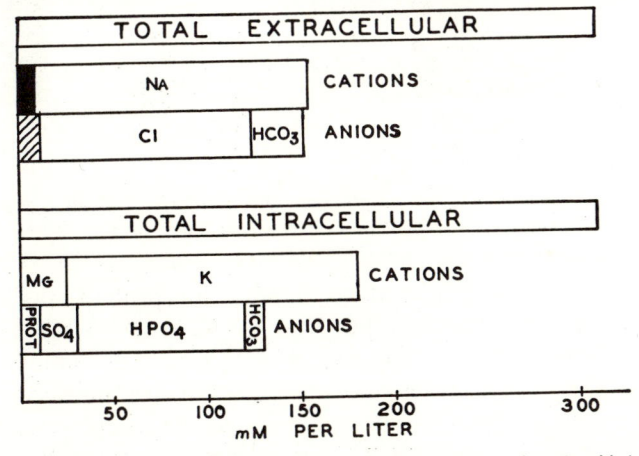

Figure 12-2. Osmolar equality of the extracellular and intracellular fluids. (After Gamble.)

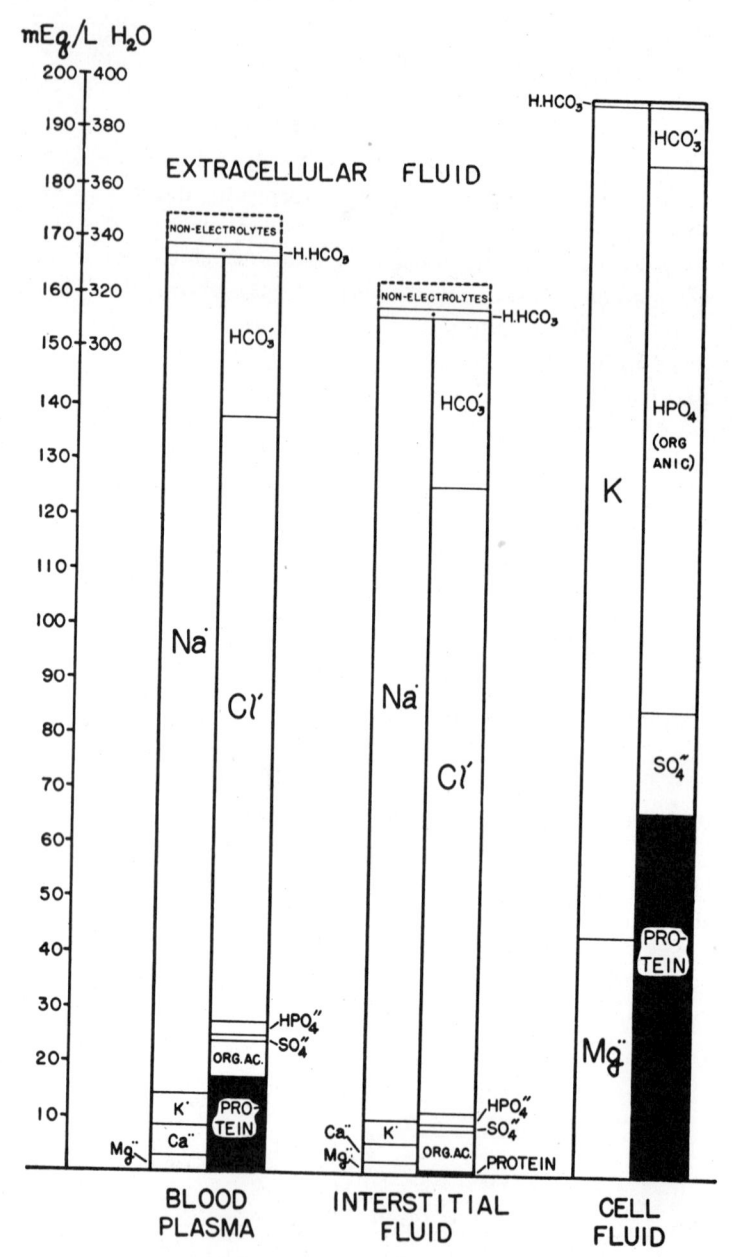

Figure 12-3. Composition of intracellular and extracellular fluids. (From Gamble: *Chemical Anatomy, Physiology, and Pathology of Extracellular Fluid.* Harvard University Press, 1950.)

The milliequivalent value for protein was obtained by multiplying grams of protein per liter by 0.243 (Van Slyke factor). Each millimole (mmol.) of protein was said to represent about 8 milliequivalents, which accounted chiefly for the discrepancy between the ionic osmolar and equivalent values for plasma. At the pH of extracellular fluid, 80 per cent of the phosphate radical carries two equivalents (B_2HPO_4) and 20 per cent, one equivalent of cation (BH_2PO_4). The valence of phosphate was therefore calculated as 1.8. The commonly used designations of carbonic acid and bicarbonate values in terms of "volumes per cent CO_2" was converted to milliequivalents per liter by dividing by 2.22.

The relative importance of the several components of the extracellular fluids in preserving osmotic, anion-cation balance and hydrogen ion regulation was indicated by their osmolar and equivalent concentrations, respectively (Fig. 12-3; Table 12-3). However, those components, particularly cations, which are present in comparatively very low concentrations, viz., K^+, Ca^{++}, Mg^{++}, and H^+, exert profound influences upon a variety of physiological processes.

As illustrated in Figure 12-3, the chief general structural difference between the two main compartments of extracellular fluid, i.e., blood plasma and interstitial fluid, is the relatively large amount of protein in the former. The presence of this nondiffusible component results in certain readjustments of the diffusible ions in order to maintain anion-cation equivalence (Donnan equilibrium, p. 353). Consequently, the interstitial fluid contains a somewhat higher total concentration of diffusible anion and a lower concentration of cation than does the plasma.

Figure 12-3 also shows too large a contribution of magnesium to the intracellular cation content. The indicated protein contribution to anions is more or less an inspired guess. The figure is also by no means accurate in relation to intracellular phosphate.

It is still possible to construct this type of diagram if one expresses concentrations in terms of mmol. of Na^+, K^+, $(Ca^{2+})_{0.5}$, $(Mg^{2+})_{0.5}$, Cl^-, HCO_3^-, SO_4^{2-}, HPO_4^{2-}, and so on. Such diagrams are, however, not really necessary and have in the past been given too much importance. They could also lead to possible confusion, since many workers tend to express calcium and magnesium concentrations in terms of mmol. of Ca^{2+} and Mg^{2+} and not in terms of $(Ca^{2+})_{0.5}$ and $(Mg^{2+})_{0.5}$.

INTRACELLULAR FLUID

In contrast to the rather complete and precise information available regarding the composition of extracellular fluids, knowledge of the chemical structure of intracellular fluids is fragmentary and incomplete. It seems obvious that differences in structure and function of cells of various tissues might be reflected in differences in their chemical constitution, in contradistinction to the comparatively uniform composition of most extracellular fluids. Moreover, the previous concept of an intracellular compartment separated from extracellular fluids by a cell membrane must be revised in the light of present knowledge that various cell constituents may be distributed unequally between the soluble phase of the cytoplasm and the particulate components (nucleus, mitochondria, endoplasmic reticulum, microsomes). The distribution is determined by biological pump mechanisms, hormones, neuromuscular activity and a host of other biological phenomena, including other properties such as "porosity" changes associated with the intracellular membranes.

Nevertheless, it is useful to describe the major differences between typical extracellular fluids and the most abundant of the intracellular fluids, i.e., that of skeletal muscle; with certain exceptions, the latter may be regarded as typical of most cells.

Whereas Na is the major cation in the extracellular fluid (142 mmol./liter), much smaller amounts are present in the intracellular fluids (5–10 mmol./liter), which contain little, but extremely biologically important Ca. The chief cations of the latter fluids are K, about 160 mmol./liter in muscle, and Mg, about 26 mmol./liter in muscle (5 mmol. K^+ and 1.7 mmol. $(Mg^{2+})_{0.5}$ in plasma). As regards anions, the intracellular fluids contain much more phosphate and sulfate ions and protein than the extracellular fluids. Cl, the major anion of the latter, is practically absent from the former, except in the case of erythrocytes (Fig. 12-3), and cells of the kidney tubules, stomach and intestines. Cells of the latter organs contain Na and Cl, since they are engaged in reabsorption or secretion of these elements. However, Na and Cl are not in diffusion equilibrium with the extracellular fluid. Because of the large volume of intracellular fluid, despite their low concentrations, intracellular Na and Cl comprise about one-third of the total exchangeable Na and Cl of the body.

The marked differences in concentration of Na and K in the extra- and intracellular fluids indicate a certain degree of impermeability of the cell membranes to these ions. This impermeability is, however, not absolute nor fixed, both ions being able to cross the membrane more freely under certain physiological and pathological conditions (p. 356), K^+ more freely than Na^+. It would appear that as K leaves the cells in increased amounts, Na and H enter, the total cationic concentration being thereby maintained approximately, but not perfectly.

The bulk of the phosphate is in organic combination and the extent of dissociation of these compounds is not known. Similarly, much of the Mg is undoubtedly present as undissociated compounds of protein and organic phosphate and, therefore, is not in ionic form. Furthermore, the concentration of protein ions has not been established accurately and the estimated HCO_3^- concentration is subject to correction for carbamino CO_2 (p. 441). These factual deficiencies contribute to the present uncertainty regarding the constitution of the intracellular fluids on an osmolar and equivalent basis.

EXCHANGES BETWEEN FLUID COMPARTMENTS

The continual entrance into the body of substances from without (oxygen, water, organic and inorganic foodstuffs, etc.) and production by the cells of a great variety of metabolites, many of which must be distributed to other tissues or be excreted, implies a continual movement of various components of the body fluid compartments across the boundary membranes (capillaries and cell walls). The most important of these exchange systems may be outlined as follows:

(1) Alveolar Air : Blood Plasma (p. 434). This system provides for entrance of oxygen into and loss of CO_2 and water from the body.

(2) Plasma : Erythrocyte (p. 403). This system provides for ready exchange of oxygen, CO_2, water and certain anions (particularly Cl^- and HCO_3^-) in both directions. Cations are exchanged very slowly.

(3) Plasma : Interstitial Fluid. These two media are separated by the capillary walls, which are permeable to water, inorganic ions and small organic molecules (e.g., glucose, amino acids, urea, etc.) but not to large organic molecules, such as proteins.

(4) Interstitial Fluid : Intracellular Fluid. These two compartments are separated by the cell membranes, across which gases in solution, water, and small, unchanged molecules can diffuse. The small molecules, e.g., glucose, may not be subject to simple diffusion, but may be carried across cell membranes by active transport processes. The permeability to electrolytes involves biological pump mechanisms. The sodium pump, for example, actively extrudes sodium ions from cells. If these can be accompanied by an anion such as Cl^-, electroneutrality will be conserved. If, however, this is not the case, then potassium ions or hydrogen ions or both must pass through the membrane to maintain such neutrality. Most membranes seem to be quite permeable to these two ions.

These membranes are also relatively, but by no means entirely, impermeable to large molecules, such as proteins, except in special situations, viz., the liver.

The first two of these systems are discussed in the section on respiration (pp. 434 and 440), since they are intimately concerned with the transport and exchanges of oxygen and CO_2.

Gibbs-Donnan Equilibrium. If two solutions are separated by a membrane which is freely permeable to the solvent and solutes, the concentrations of the latter will be identical when equilibrium is established. Thus, both the chemical composition and the osmotic pressures of the two solutions will be the same.

If, on the other hand, one of the solutions (solution I) contains an electrolyte, e.g., NaCl, to which the membrane is permeable, while the other (solution II) contains a substance, e.g., Na proteinate, in which the anion (protein) is too large to pass through the membrane, the distribution of Na^+ and Cl^- at equilibrium will be unequal on opposite sides of the membrane. Since electrical neutrality must be maintained, in solution II the cation Na^+ must balance two anions, Cl^- and $protein^-$, whereas in solution I it balances only Cl^-. It follows that the concentration of Na will exceed that of Cl in solution II, whereas they will be equal in solution I.

The quantitative ionic relations for the Gibbs-Donnan equilibrium may be stated as follows:

1. The concentration of a diffusible cation (e.g., Na) is higher, and that of a diffusible anion (e.g., Cl) lower, on the side of the membrane (e.g., capillary wall) containing a nondiffusible anion (e.g., protein).

2. At equilibrium, it can be shown that

$$\frac{[Na^+_d]}{[Na^+_n]} = \frac{[Cl^-_n]}{[Cl^-_d]}$$

where the subscripts n and d refer, respectively, to ionic concentrations on the side containing nondiffusible anion and the side containing only diffusible ions.

The Gibbs-Donnan effect finds particular application in exchanges of Cl^- and HCO_3^- between the blood plasma and erythrocyte and in water and ion exchanges between the blood plasma and interstitial fluid (i.e., across the capillary wall).

It must be remembered, however, that the existence of biological pumps, and the like, can easily mitigate the effect in relation to the tissue cells themselves. Therefore, the importance of the Gibbs-Donnan effect is difficult to assess in any given situation. It does, however, play a part in relation to electrical potential differences across cell membranes. If the mobile ion is the hydrogen ion, then the Gibbs-Donnan equilibrium also plays a part in relation to the pH difference across these membranes.

Plasma : Interstitial Fluid Exchange. Exchanges between the blood plasma and interstitial fluid occur across the endothelial lining of the capillaries, which act as semipermeable membranes, allowing free passage of water and crystalloid solutes, inorganic and organic, but not of colloids of large molecular size, viz., proteins. This impermeability to proteins is not absolute nor uniform, the concentration of protein in interstitial fluids varying from 0.05 to 0.5 per cent in subcutaneous tissues and serous cavities to 4 to 6 per cent in liver. The concentrations of diffusible electrolytes on the two sides of the capillary will vary, their distribution depending upon the difference in protein concentration in the two fluids (Gibbs-Donnan effect). The osmotic pressure will be greater in the plasma than in the interstitial fluid, due to the higher protein concentration (colloid osmotic pressure; C.O.P.).

In the absence of opposing forces, water and crystalloid solutes will tend to pass from the interstitial fluid to the plasma, which contains protein in higher concentration, thereby producing an osmotic pressure gradient. This tendency may be counterbalanced by the opposing force of the capillary blood pressure and may also be modified by the concentration gradients of the individual solutes.

The colloid osmotic pressure (C.O.P.) of the proteins of normal plasma is about 22 mm. Hg (30 cm. H_2O). The capillary blood pressure varies considerably in different tissues, but is higher than the C.O.P. at the arterial end of the capillary and lower than the C.O.P. at the venous end. Consequently, filtration from the plasma occurs at the arterial end and reabsorption into it at the venous end of the capillary. Moreover, the increasing concentration of nondiffusible colloids (viz., proteins) which results from loss of plasma water creates a relatively small but actual gradient of C.O.P., which further favors reabsorption as the blood pressure falls in the distal end of the capillary. This mechanism (Starling hypothesis) provides for a continual circulation of fluid between the capillaries and the tissue spaces, a balance being maintained between the quantity of water filtered and that reabsorbed. Exchanges of diffusible solutes between these two fluids depend upon these circumstances and also, independently of them, upon their individual concentration gradients. The factors concerned with the transfer of materials across the capillary wall are therefore those of diffusion, osmosis, and hydrostatic pressure.

Filtration from the capillaries is opposed by the tissue tension, which varies considerably in different situations, being relatively low in loose areolar tissues (e.g., eyelids, external genitalia) and high in dense tissues (e.g., muscle, liver, etc.). Fluid leaves the tissue spaces not only by way of the blood plasma but also by way of the lymph capillaries, ultimately reaching the venous blood by way of the lymphatic circulation.

Interstitial Fluid : Intracellular Fluid Exchange. Exchanges between these two compartments occur across cell membranes, which are generally permeable to gases, water and many small uncharged molecules (urea, etc.) but slightly, and sometimes not at all, to large colloidal molecules, such as proteins. The enormously higher concentration of proteins and other colloids in the intracellular than in the interstitial fluids would cause a much greater osmotic pressure within the cells (Gibbs-Donnan effect) were it not for the fact that the cell membranes are generally not freely permeable to inorganic ions, especially cations as well as to some small molecules. Consequently, the electrolyte composition of cells is quite as distinctive as their organic constitution (Fig. 12-3, p. 350). Because of this limited permeability to inorganic ions, adjustment to deviations of osmotic pressure is usually accomplished largely by transfers of water, not by exchanges of electrolytes, the freely diffusible organic solutes (urea, creatinine) moving with the water. Inasmuch as the concentrations of protein in the body fluids are more stable than the concentrations of inorganic ions, the latter constitute a dominant factor in determining the total osmotic pressure and the exchanges of water between the body fluid compartments.

It is probable that exchanges of water across the outer cell membrane proceed passively, in relation to changes in osmotic pressure, diffusion of various metabolites and active transport of other substances which occur continually in actively metabolizing tissues. However, it is probable, too, that under these circumstances a state of absolute osmotic equilibrium, although constantly approached, is never reached.

The much higher concentrations of Na^+ and Cl^- in interstitial fluid and of K^+ in intracellular fluid are accompanied by a difference in electrical potential, that of resting skeletal muscle cells being about 90 millivolts negative to the interstitial fluid. It is believed that the lipid-protein membrane plays an important role in determining and maintaining these differences in concentration and potential. K^+ ions tend to diffuse out of and Cl^- ions into the cell because of their concentration gradients, but this is almost exactly counterbalanced by a tendency to diffuse in the opposite direction due to the difference in electrical potential, i.e., the relative negativity on the inside of the cell tends to keep Cl⁻ out and K⁺ in. In the case of Na⁺, however, diffusion into the cells is favored by both concentration gradient and electrical potential. Under normal conditions, there must be some mechanism for removing Na⁺ from the cell virtually as rapidly as it enters. Since this is accomplished in opposition to forces of concentration and electrical potential, it involves expenditure of energy, derived from cellular metabolism. This process of active transport of Na⁺ out of the cell is referred to as the "sodium pump," which effectively excludes Na⁺ from the intracellular fluid (resting muscle). It has already been pointed out that the necessary energy is derived from ATP and that ATPase may be part of the pump mechanism. Phosphatides, too, have been proposed as important components of the transport system. The dependence of this function of the cell membrane upon metabolic processes in the cell is illustrated by the fact that if cells are incubated in a glucose-free medium, Na⁺ enters and K⁺ leaves the cell. These changes are reversed upon addition of glucose to the medium.

The relative permeability of cell membranes to inorganic ions is not uniform or constant. HCO_3^- ions and Cl⁻ ions pass freely into and out of erythrocytes (p. 405) and the acid-secreting cells of the gastric mucosa. Although permeability to Na⁺ ions is restricted, Na⁺ may enter cells in increased amounts when they lose K⁺, and the passage of K⁺ across cell membranes appears to be related to metabolic activities of the cells. For example, there is increased entrance of K⁺ into cells during periods of administration of glucose or insulin (p. 47). Changes in the distribution of Na⁺ and K⁺ occur during excitation of nerve and muscle cells, and can be produced by adrenocortical hormones (p. 722), and by dehydration, anoxia, and changes in pH (p. 372) due to alterations in the properties of the cell membrane.

When Na is introduced into the blood plasma, it is distributed rapidly throughout the extracellular fluids, but penetrates the cells only very slowly. Consequently, an increase in Na concentration in the extracellular fluids will cause water to pass out of the cells, equalizing the osmotic pressure in the two fluid compartments. Withdrawal of Na from the extracellular fluids causes passage of water in the opposite direction, i.e., into the cells.

Shifts in water and electrolytes are also caused by changes in pH of the body fluids. Such changes not only affect the integrity of cellular membranes but also lead to altered intracellular metabolism. Small molecules can now escape from the cells and make a significant contribution to osmolality. If the membranes remain more or less intact, this can result in overhydration of the cells themselves. More frequently, perhaps, the membranes are sufficiently affected to allow these small molecules to escape into the extracellular fluid and make their osmotic contribution in this situation—which could result in intracellular dehydration. It is interesting to note that this is the kind of situation in which hyponatremia can occur without any total sodium deficit. It should also be pointed out that not only is there increased membrane permeability but, because of metabolic upset, there is some degree of failure of the sodium pump mechanism. Sodium and chloride ions enter the cells by diffusion and this contributes greatly to the hyponatremia. Under these circumstances, administration of sodium ions can be harmful.

This state constitutes the so-called "sick cell syndrome" and may also be found in anoxia, congestive heart failure, various infections and chronic diseases, and after trauma.

Phosphate is present within cells in much larger amounts than in extracellular fluids. Much of this is in organic combination, e.g., as hexose and triose phosphates, nucleotides, nucleic acid, phosphocreatine (muscles), etc. However, the concentration of inorganic phosphate, too, is higher than in the extracellular fluids and fluctuates continually in relation to the latter at different levels of cell metabolic activity involving formation and splitting of phosphorylated metabolites. In contrast to the case of K^+, the rate of entrance of phosphate into muscle cells is not altered significantly by muscular activity but is increased markedly during the recovery period. It is increased also by administration of glucose and insulin (p. 46). The concentrations of K^+, Na^+, Mg^{2+}, Ca^{2+}, and phosphate in the cell can vary, within limits, in accordance with the metabolic activity of the cell, without influencing exchanges of water because, presumably, considerable amounts of Mg^{2+}, Ca^{2+} and phosphate form undissociated or poorly dissociated combinations with cell proteins, polysaccharides, nucleic acids, etc., and are therefore osmotically inactive.

REGULATORY MECHANISMS

In health, the volume and composition of the various body fluid compartments is maintained within physiological limits even in the face of wide variations in intake of water and solutes. Isotope-labeled water becomes distributed throughout the body fluids in two to three hours after administration; in the absence of increase in volume, this implies continual exchanges of water across compartment boundaries. Fluid exchanges across the cell wall are conducted in a state of osmotic equilibrium (p. 347); assuming a "steady state" of cellular metabolic activity, the osmolality of intracellular fluids is determined mainly by their K concentration, while that of extracellular fluids is determined by their Na concentration. If constancy of osmolality is to be maintained, the volume of intracellular fluid must be proportional to its total K content, and that of extracellular fluid to its total Na content. Consequently, if the volumes of these compartments are to be maintained at constant levels, a mechanism must be provided for adjustments in excretion not only of water, but also of Na and K in response to variations in amounts of each supplied to the organism. It must be remembered, however, that this concept is only a rough approximation, since the idea of a "steady state" is somewhat theoretical in relation to the ever-changing situations which apply under normal conditions to the human being. Moreover, the intracellular content of either potassium or sodium has a definite effect on cell metabolism, which in itself would result in changes of osmolality. It therefore follows that the adjustments in excretion of water, sodium and potassium are also intimately related to adjustments consequent on changes, including those which are cyclical, in cell metabolism.

These adjustments are accomplished mainly by the kidney. Normal renal function provides a steady state with respect to: (1) the amount of water in the body; (2) the amounts of Na and K, which determine the volume of extracellular and intracellular fluids; (3) the osmolar concentrations of these fluids, determined largely by the concentrations of Na and K, respectively. The kidney responds promptly to deviations in osmolality or individual ion concentrations in extracellular fluid, e.g., produced by administration of excessive quantities of water or Na, by augmented excretion of these substances in amounts required for preservation of osmotic equilibrium. However, isosmotic expansion of extracellular fluid volume, e.g., by intravenous infusion of isotonic solutions of salt or serum albumin, is followed promptly by increased urinary excretion of water and salt. Homeostasis of body fluids, therefore, involves a mechanism that responds to fluctuations in volume as well as mechanisms responsive to changes in concentrations of total solute or of individual ions. Urinary excretion of water and salt is conditioned by a number of variations in body fluids; these include such factors as: the volume of plasma and/or interstitial fluid; cellular hydration; intracranial volume or blood flow; cardiac output; pressure changes in various portions of the circulation; renal blood flow; concentrations of total or individual solutes in extracellular fluid. Certain of these variables, which involve significant alterations in renal blood flow or pressure, or in solute concentrations in the blood plasma, may produce their effects by direct action on the kidney. However, others can produce their effects in the absence of significant changes in renal hemodynamics or in solute concentrations in the plasma. These effects on kidney function must, therefore, be mediated by extrarenal factors.

Current concepts of the nature of these regulatory mechanisms include the existence of receptors sensitive to variations in osmolar (osmoreceptors) or individual ion (chemoreceptors) concentrations in extracellular fluids, to local or general variations in intravascular fluids, and to local or general variations in intravascular pressure (baroreceptors) and/or plasma and/or extracellular fluid volume (volume receptors; stretch receptors). The intrarenal mechanisms concerned with excretion of water and solutes may be influenced by stimuli initiated in these receptors either by direct neural connections or through the medium of humoral factors, i.e., alterations in production or release of certain hormones. There is evidence that there are pathways in the nervous system capable of mediating such influences, but more precise information is available concerning the role of certain hormones in this connection. Those involved most fundamentally are the antidiuretic hormone (ADH) and aldosterone, the former regulating the excretion of water, the latter of Na and consequently of K. Adrenal glucocorticoids also exert an influence that is not so well defined. Details are presented elsewhere of the actions of these hormones in promoting renal tubular reabsorption of water and Na, and excretion of K (p. 723). The mechanisms of regulation of production and release of these hormones are also considered elsewhere. It will suffice here to consider the manner in which these two independently controlled systems may be coordinated in the interests of body fluid homeostasis.

The complexity of this regulatory system is indicated (1) by the great variety of conditions to which it makes adjustments (e.g., water and salt intake; extracellular fluid composition and volume; emotional states; blood flow and pressure, general or in certain specific regions) and (2) by the great diversity of its components (e.g., various types of receptors; diencephalic centers; nervous pathways; neurohypophysis; adrenal cortex; renal glomeruli and tubules), many of which have independent controlling mechanisms. Perhaps the simplest concept, applicable to many circumstances but certainly not to all, is that the end-effects of the action of one hormone, e.g., ADH, influence the rate of production or discharge of another, e.g., aldosterone. In addition, since adjustments may be made in either direction, the receptors of the initiating stimuli must be in a state of tonic activity, capable therefore of either excitation or depression.

Adjustment to increasing osmolality of extracellular fluid may occur as follows: (1) stimulation of osmoreceptors in the hypothalamus; (2) this causes increased production (in paraventricular and supraoptic nuclei) and discharge (from neurohypophysis) of ADH; (3) this hormone, reaching the kidney in the systemic circulation, causes increased reabsorption of water in the distal nephron; (4) normal osmolality of the extracellular fluid is restored but its volume is increased; (5) this, directly or indirectly, depresses the activity of stretch receptors (volume receptors; baroreceptors), e.g., in the juxtaglomerular apparatus; (6) because of inhibition of renin secretion, aldosterone production decreases; (7) renal tubular reabsorption of Na diminishes and it is excreted, with water, in the urine in amounts required to restore normal extracellular fluid volume and osmolality.

The converse occurs in response to decreased osmolality. Other relationships are superimposed upon this basic pattern. Secretion of ADH is influenced by changes not only in osmolality but also in volume of extracellular fluid; secretion of aldosterone is influenced by variations not only in volume of extracellular fluid but also in the amounts and concentrations of Na and K (K^+ is required for aldosterone biosynthesis), in some cases at least operating independently of changes in volume. Another hormonal factor, the "third factor," is almost certainly also involved in the renal handling of sodium. The possible existence of a "fourth factor" has been suggested.

The thirst mechanism plays an important role in fluid homeostasis, regulating water intake. It is activated by increasing osmolar concentration of extracellular fluid (osmoreceptors), resulting from either a deficit in water or an excess of solute. It also has a sensory component, mediated by the glossopharyngeal and vagus nerves, stimulated by dryness of the mouth and throat.

The intracellular and extracellular fluids are in osmotic equilibrium. Consequently, the volume of the former is determined largely by the osmolality of the latter. Ultimately, therefore, the kidneys regulate the intracellular fluid volume by adjusting excretion of water in response to changes in osmolality of the extracellular fluid. The volume of circulating blood plasma is of much greater concern to the welfare of the organism than is the volume of other extracellular fluids. It tends to be maintained within normal limits, largely through the operation of cardiovascular mechanisms, even in the presence of marked expansion or contraction of interstitial and transcellular fluids. However, since the plasma proteins do not cross most normal capillary walls readily (p. 354), an increase in plasma volume may follow intravenous infusion of hypertonic albumin solutions. Conversely, reduction in plasma volume may occur in conditions of severe plasma protein depletion.

Renin-Angiotensin System. The juxtaglomerular apparatus of the kidney contains the cells which produce the enzyme renin. This seems to involve a baroreceptor mechanism, which results in a stimulus for renin production when the pressure gradient between the intraluminal arterial and interstitial fluid is diminished. When this gradient is increased, renin production is inhibited. There is also evidence which indicates that the renin response is also controlled by the sodium concentration in the renal tubule affecting the macula densa. An increased concentration causes an increased renin response and a decreased concentration, a decreased response. It is possible that the renal nerve supply is to some extent involved.

Angiotensinogen, an α_2-globulin in blood plasma, is acted upon by renin with the production of angiotensin I, which is a decapeptide. This is acted upon by an enzyme in blood with the production of an octapeptide, angiotensin II. The latter stimulates the production of aldosterone in the zona glomerulosa of the adrenal cortex.

There is evidence that the hormone prolactin is necessary for the action of aldosterone. This could be another regulatory mechanism.

SODIUM, POTASSIUM, CHLORIDE

Sodium, potassium and chloride are concerned in several fundamental physiologic processes: (1) the maintenance of normal water balance and distribution; (2) the maintenance of normal osmotic equilibrium; (3) the maintenance of normal acid-base equilibrium (physiologic neutrality); (4) the maintenance of normal neuromuscular function. The maintenance of normal hydration and osmotic pressure depends primarily upon the total cation concentration of the body fluids (p. 347). Since sodium constitutes the largest fraction of the total cations of the extracellular fluids ($142/155$), it plays a dominant role in this connection (p. 350). At any given H_2CO_3 concentration, the hydrogen ion concentration of plasma and other extracellular fluids is dependent upon the concentration of bicarbonate

$\left(H^+ = k \dfrac{[H_2CO_3]}{[HCO_3^-]}\right)$. Since the bicarbonate content depends upon the amount of total cations present in excess of anions other than HCO_3^-, and since Cl^- constitutes by far the largest fraction of the total anions of the plasma ($103/155$), it follows that the maintenance of the normal pH depends largely upon the presence of normal concentrations of sodium and chloride (p. 401). The importance of a proper balance between Na, K, Mg, and Ca in the maintenance of normal neuromuscular irritability and excitability is well known.

Potassium is the chief cation of the muscles and of most other cells (intracellular fluid), whereas sodium is the chief cation of extracellular fluids of the body. Some movement of K occurs from cells to extracellular fluids (including plasma), especially under abnormal conditions of excessive loss of intracellular fluid, the mobilized K being usually excreted promptly in the urine. Some movement of Na occurs in the opposite direction, i.e., from extra- to intracellular fluid, under these conditions. Any considerable increase or decrease in K concentration in the extracellular fluids is accompanied by serious and at times fatal disturbances in muscle irritability, respiration and myocardial function. Moreover, it would appear that no other cation can replace potassium extensively in the intracellular fluid without impairing cell function. In low concentrations potassium is excitatory and in higher concentrations it is inhibitory, these effects being particularly important in relation to nerve synapses or myoneural junctions. Under normal conditions, its effects resemble those of parasympathetic stimulation and are usually inhibited by calcium (e.g., neuromuscular excitability). The K/Na ratio is also important in this connection. Potassium and calcium modify certain important properties of protoplasm and cells, including the permeability of cell membranes, and thus they play an important role in many "vital" processes.

Absorption and Excretion

An average adequate diet contains about 75 to 100 mmol. (3 to 4 g.) of K and 130 to 250 mmol. (8 to 15 g.) of NaCl daily, the latter being contributed to by the addition of salt to the food. Normally, Na, K, and Cl are practically completely absorbed from the gastrointestinal tract, less than 2 per cent of ingested Na and less than 10 per cent of the K being eliminated in the feces. In subjects with diarrhea, large amounts are lost in the feces, due in part perhaps to transudation into the bowel, but in large measure to relative failure of reabsorption, because of rapid transit, of constituents of digestive fluids, the electrolyte composition of which is indicated in Figure 12-4 (p. 364) and Table 12-3. Na and Cl are eliminated chiefly in the urine and, to a lesser extent, in the perspiration. K is normally eliminated almost entirely in the urine.

Excretion in Urine. Under normal conditions the daily urinary excretion of these electrolytes approximates the intake except under conditions of sudden increase of NaCl intake following a period of restriction. Na and Cl are removed from the circulating plasma by glomerular filtration, over 1000 g. of NaCl entering the renal tubules daily in the glomerular filtrate. Under normal conditions 99 per cent or more undergoes reabsorption into the blood stream by the tubular epithelium. About 80 per cent of this is reabsorbed by an active transport process in the proximal portion of the tubule, the remainder in the distal portion. Adrenocortical hormones, mainly aldosterone, enhance tubular (mostly distal) reabsorption of Na^+ and, secondarily, of Cl^- (p. 461). In the distal portion, a small proportion of the Na^+ reabsorption ocurs in exchange for H^+ ions (also for K^+) (p. 410) as part of the mechanism of acidification of the urine. Under normal conditions the regulation of Na and Cl reabsorption and conservation is so efficient that equilibrium can be maintained on a NaCl intake as low as 0.5 g. daily, Na and Cl virtually disappearing from the urine. The mechanism of excretion of these ions and the factors that influence it are discussed in more detail elsewhere (p. 461).

Na and Cl are substances designated as "threshold bodies" which, being of value to the organism, are, after their passage through the glomeruli, reabsorbed into the circulation in order to maintain their normal concentration in the blood. The normal "renal threshold" value for Na is 110–130 mmol./liter, and of Cl 85–100 mmol./liter, of plasma. In the normal individual, as the plasma concentration descends to these levels elimination of sodium and chloride in the urine decreases and finally virtually ceases. Under normal conditions the quantity of Na and Cl eliminated in the urine approximates their intake; in cases in which there is marked variation in the quantity ingested it may require some time for an equilibrium to be established. For example, if the normal individual is kept upon an adequate sodium chloride intake (8 to 15 g.), the sudden addition of a large quantity of salt is usually followed by complete elimination of the excess within forty-eight hours. However, if the individual has been maintained for prolonged periods of time on a low salt intake, the elimination of salt administered suddenly in large quantity may be delayed for several days, the body retaining the added salt with greater tenacity. In conditions in which the elimination of chloride through other channels such as the skin (excessive perspiration) or gastrointestinal tract (vomiting or diarrhea) is increased, the urine chloride content is correspondingly diminished.

K leaves the plasma in the glomerular filtrate and, like Na and Cl, undergoes extensive active reabsorption in the tubules. However, unlike Na and Cl, it is also excreted into the tubular lumen by the cells of the distal portion of the renal tubules. This is a passive process brought about by an electrochemical gradient. Under conditions of average K intake in normal subjects, the K of the glomerular filtrate is probably virtually completely reabsorbed, tubular excretion accounting for the amount eliminated in the urine. However, increased quantities in the glomerular filtrate, either exogenous or endogenous, may escape complete reabsorption. In the presence of normal kidney function, K is very promptly and efficiently removed from the blood plasma even in the face of a considerably increased supply, preventing undue increase in its concentration in the extracellular fluids.

Salt-active adrenocortical hormones, particularly aldosterone, exert an influence on K excretion the reverse of that exerted on Na excretion, i.e., they produce an increase in urinary K. Inasmuch as the amount of K excreted by the tubules is determined largely by the quantity of Na reabsorbed, it is probable that this influence of aldosterone is the result, not of a direct action on the mechanism of K excretion, but rather of its effect on Na reabsorption.

Although the kidneys conserve the K of the glomerular filtrate more efficiently than the Na, active tubular excretion of K results in persistence of this element in the urine even when none is taken in. This contrasts with the situation with respect to Na and Cl, which virtually disappear from the urine of normal subjects when withdrawn from the diet.

Excretion by the Skin. The Na and Cl concentrations in perspiration are normally lower than in plasma (Table 12-3). Their concentrations in sweat are decreased by aldosterone (p. 378), which is secreted in increased amount during acclimatization to heat. This constitutes an adaptive mechanism for conservation of body Na and Cl, the sweat glands producing a progressively more dilute fluid. In normal subjects, the loss of Na and Cl in the perspiration is compensated by a corresponding diminution in their urinary excretion. However, considerable amounts may be lost from the body under conditions of prolonged strenuous exertion at extremely high temperatures and humidity (p. 360).

Excretion in Digestive Fluids. The approximate electrolyte composition of certain digestive fluids is indicated in Figure 12-4 and in Table 12-3. Under normal conditions these fluids are virtually completely reabsorbed in the bowel, but, in the presence of protracted vomiting or diarrhea, large amounts of water and electrolytes may be lost from the body. This is of importance in the treatment of the dehydration and dislocations of acid-base balance that occur under these circumstances.

The chloride content of the gastric juice of normal individuals may vary considerably but is relatively constant for each individual; this individual variation, however, is in no case so great as would appear to be indicated by the rather wide variation in free acid secretion. Relatively

TABLE 12-3 APPROXIMATE CONCENTRATION OF CERTAIN ELECTROLYTES IN DIGESTIVE FLUIDS AND PERSPIRATION (mmol./LITER)

FLUIDS	Na^+	K^+	Cl^-	HCO_3^-
Gastric juice	20	8	145	0
Pancreatic juice	140	5	40	110
Jejunal juice	138	5	110	30
Bile	140	8	108	38
Perspiration	82	5	85	0
Blood serum	140	4.3	100	25
Interstitial fluid	145	3.3	110	28

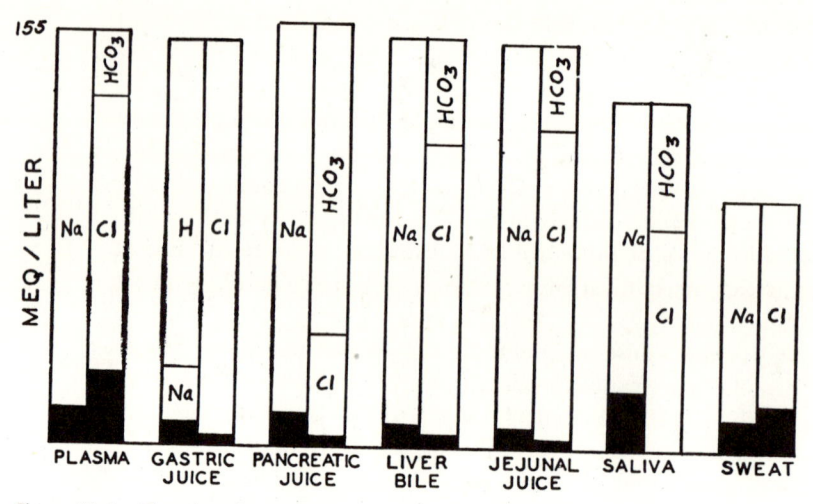

Figure 12-4. The electrolyte composition of certain important body fluids, compared with that of blood plasma (for mEq., read mmol.). (After Gamble.)

normal total chloride values may frequently be observed in cases exhibiting extremely low or absent free HCl in the gastric juice (false achlorhydria), due usually to excessive neutralization by regurgitation of alkaline material from the duodenum.

Normal Blood Na, Cl, and K

The normal amounts of Na, Cl, and K in blood plasma (or serum) are as follows: Na, 132–144 mmol./liter; Cl, 96–105 mmol./liter; K, 3.6–4.8 mmol./liter. In man, erythrocytes contain little or no Na, smaller amounts of Cl than does the plasma, and large amounts of K (76–100 mmol./liter). The distribution of Cl^- between plasma and erythrocytes is intimately related to that of HCO_3^-, in accordance with Donnan's law governing the distribution of diffusible monovalent ions (p. 353). The factors which influence this distribution and the changes that occur under physiological conditions are considered elsewhere (p. 356). Suffice it to recall here that shifts of Cl^- and HCO_3^- between the plasma and the erythrocytes play an important role in the maintenance of a constant pH in the blood.

Because of the unequal distribution of these elements between the blood plasma and erythrocytes, they should always be determined in the plasma (or serum), not in whole blood. In the case of K, care must be exercised to avoid hemolysis, which liberates K into the plasma from the hemolyzed cells. The Cl content of whole blood varies with variations in the number of red cells, being relatively high in anemia (i.e., larger proportion of Cl-rich plasma) and relatively low in polycythemia. Moreover, because of the shift of Cl between the plasma and red cells that occurs with changes in CO_2 tension (p. 405), blood drawn for plasma Cl determinations should preferably be collected under anaerobic conditions, for, if exposed to air, CO_2 escapes from the blood, the plasma Cl concentration being thereby increased.

The concentrations of Na, Cl, and K in the plasma vary relatively slightly under physiological conditions. The plasma Cl may fall somewhat during periods of active gastric secretion of HCl and may rise again in the postdigestive period (reabsorption from intestine). It is affected but slightly by marked variations in Cl intake; the same is true of the plasma Na and K concentrations. The plasma K decreases somewhat during periods of increased carbohydrate utilization, as following administration of glucose or insulin. This is due apparently to its passage from extracellular fluids into the cells in increased amount under these circumstances (p. 47).

The influence of adrenocortical hormones (especially aldosterone) in increasing plasma Na and Cl and decreasing K is considered elsewhere (p. 382). It is the result of their effect on renal tubular reabsorption of Na and Cl and excretion of K and on the distribution of these ions between intra- and extracellular fluids.

Distribution and Intermediary Metabolism

Reference is made elsewhere (p. 346) to differences in the composition of intra- and extracellular fluids. These are illustrated in Figure 12-3 (p. 350). We are concerned here mainly with the striking differences in the concentrations of Na, K, and Cl in these fluid compartments. It should be understood that the concentration of electrolytes varies in the intracellular fluid of individual tissues, but the general pattern is rather uniform. Skeletal muscle is the tissue studied most thoroughly in this connection. The rather strict localization of relatively large amounts of K in the intracellular and of Na and Cl in the extracellular fluid is attributable to the sodium pump. However, this is neither absolute nor fixed. In different functional states and under certain abnormal circumstances the composition of the intracellular fluid may be altered as a result of loss of K from and entrance of Na into the cells. These changes (i.e., abnormal) are often accompanied by alterations in the concentrations of Cl^- and HCO_3^- in the extracellular fluid, which may seriously affect its pH (p. 413).

An equilibrium is maintained normally between the high intracellular (160 mmol./liter) and low extracellular (4 mmol./liter) concentrations of K, and the reverse situation with regard to Na. Factors involved in the preservation of this equilibrium are embodied in the concept of the sodium pump (p. 356), which has a requirement for energy that is supplied by intracellular metabolic processes.

As cell protein undergoes degradation, the associated loss of intracellular fluid (about 3 ml./g. protein) results in release of K to the extracellular fluid (2.7 mmol. K/g. N). Under normal conditions, this catabolic process is balanced by the entrance into the cell of an equivalent amount of K as a part of the process of protein anabolism. Exaggerations of these movements of K occur in conditions of accelerated protein breakdown (starvation, fever, thyrotoxicosis, adrenocortical hyperactivity, etc.) and stimulation of protein synthesis (administration of growth hormone or androgen, convalescence, tissue repair).

Potassium moves into the cells at an accelerated rate during periods of increased glucose utilization, e.g., after administration of glucose or insulin. This may be related to the increase in phosphorylative processes that occurs under these circumstances and may be reflected in a temporary fall in plasma K concentration.

Preservation of electrical neutrality and osmotic equilibrium requires that passage of a K$^+$ ion in one direction across a cell membrane be accompanied simultaneously by passage of a cation in the opposite direction. These exchanges involve movement of Na$^+$ and H$^+$ ions in a direction opposite to that of K$^+$. The relative amounts of Na$^+$ and H$^+$ ions participating in this exchange depend fundamentally on the direction of the movement (i.e., the Na$^+$ concentration gradient) and the H$^+$ ion concentration in the extracellular fluid (i.e., the H$^+$ ion concentration gradient). Accordingly, other things being equal, loss of K$^+$ is accompanied by decrease of H$^+$ ion concentration in the extracellular fluid (alkalemia).

The serum K concentration may vary as a result of primary changes in H$^+$ ion concentration. As a general principle, other things being equal, as the extracellular H$^+$ ion concentration rises, H$^+$ ions move into the cell and K$^+$ ions move out; the reverse occurs with decreased extracellular H$^+$ ion concentration. In other words, the serum K concentration tends to rise in acidosis and to fall in alkalosis. It must be recognized that in many clinical disorders this simple relationship may be obscured because of coexisting situations, e.g., dehydration, Na or K deficits, renal functional impairment. The very fact that alterations in intracellular concentrations of Na$^+$ and/or K$^+$ have occurred results in profound changes in intracellular metabolism; the citric acid cycle, for example, is dependent on appropriate concentrations of these ions. Deficiency of potassium also interferes with aldosterone biosynthesis. Other important examples can be cited: entry into the cell of both glucose and amino acids depends on the maintenance of the gradient between extra- and intracellular Na concentrations; and protein biosynthesis depends on the maintenance of the intracellular K concentration. This means that the very simple explanation that hydrogen ion disturbances result directly from the kind of ionic interchange just described is perhaps a little naive, even though it is frequently used by many authors. Perhaps it is not surprising, therefore, that in the clinical situation a whole variety of permutations of the concentrations of Na$^+$, K$^+$, H$^+$ and Cl$^-$ can occur in the extracellular and probably the intracellular fluids.

In well-marked depletion of body potassium, excessive amounts of hydrogen ion do enter cells, and this results in the liberation of HCO$_3^-$ ions, which enter the extracellular fluid. They are not very well excreted by the kidney, since this organ manifests malfunction when there is a potassium deficit. There is also a Cl$^-$ deficiency, probably due to defect in the renal handling of this ion. This deficiency also plays a part in producing the lowered efficiency in the renal handling of the HCO_3^-.

Accurate determination of the quantities of individual electrolytes in the various body fluid compartments is difficult because of differences in the rates of their transfer between different areas (p. 345). For example, over 40 per cent of the total body Na is in the skeleton, and about 70 per cent of the Na of bone, being incorporated in the bone crystal, is not readily available to the general Na pool. From the functional standpoint, the amounts of more or less readily exchangeable Na, K, and Cl are more important than the total quantities of these ions. These are indicated in Table 12-4.

Under normal conditions, Na and K entering the body are distributed rapidly throughout the extracellular fluids. Very little Na is transferred to the cells; exogenous K reaches equilibrium with the intracellular K after fifteen hours. The addition of Na to the extracellular fluids is followed by a temporary shift of water to that compartment from the cells, osmotic equilibrium being thus preserved. The addition of K, on the other hand, is followed promptly by increased renal excretion of K, preventing its accumulation in excess in the extracellular fluids. Certain of the factors which influence renal excretion of K are discussed elsewhere, viz., adrenocortical hormones (pp. 387 and 463), K concentration in the cells (p. 410), and the hydrogen ion concentration (p. 412).

TABLE 12-4 APPROXIMATE TOTAL AMOUNT OF EXCHANGEABLE Na, K, Cl, AND HCO_3 IN BODY FLUID COMPARTMENTS OF A 70-Kg. MAN (in mmol.)

FLUID	Na	K	Cl	HCO_3	H_2O
Extracellular	2800	70	2000	400	14,000 ml.
Intracellular	100	3500	250	340	35,000 ml.

Chapter 13

WATER, SODIUM, CHLORIDE, POTASSIUM: ABNORMALITIES

The circulating blood volume may increase or decrease independently of changes in the volume of other body fluid compartments as a result of abnormalities of hemodynamics, hematopoiesis, and plasma protein concentration (p. 391). The interstitial fluid volume may also increase, either locally or generally (edema, p. 380), provided that its composition is not unduly altered, i.e., that proportional amounts of water, Na, and Cl are retained in excess. Loss of body fluids (dehydration) results primarily in decreased volume of interstitial fluid; if this is severe and progressive, the plasma volume also decreases. Whether or not the intracellular compartment also undergoes contraction depends upon the extent and nature of the electrolyte loss, and upon the consequent changes in electrolyte concentrations in the extracellular fluids (p. 371).

DEFICITS OF WATER, SODIUM, CHLORIDE, POTASSIUM

Dehydration

Excessive loss of water without electrolytes may occur as a result of: (a) inadequate intake of water, due to its unavailability or to clouding of consciousness, coma, or weakness, incident to severe illness, particularly in children; (b) excessive vaporization from lungs and skin in protracted high fevers. However, in disorders characterized by excessive loss of gastrointestinal fluids, urine or sweat, electrolytes are lost as well as water. Electrolyte depletion may be contributed to also by rigid restriction of intake (e.g., K, in starvation).

Clinical biochemical studies in dehydration are usually concerned mainly with the pattern of electrolyte concentrations in the extracellular fluid, as reflected in the blood plasma, and with the effects of changes in this pattern on hydrogen ion concentration and osmolality. No uniformly applicable rules can be set down for the behavior of electrolytes in various disease states accompanied by dehydration, because of the wide range of individual variability in several important modifying factors, e.g., the nature of the fluid lost, the state of renal function, and the presence of other complications. However, certain general principles may be outlined, which indicate the trend of abnormalities of metabolism of Na, Cl, and K under commonly encountered clinical conditions.

Routes of Loss. Excessive amounts of water and electrolytes may be lost from the body by way of the gastrointestinal tract (vomiting, diarrhea, fistulas, tube drainage), kidneys (diuresis), or skin (excessive sweating). Depletion may occur also as a result of restriction of intake.

Gastrointestinal. Normally, all but about 100 to 200 ml. of the large amounts of water and electrolytes (Table 12-1, p. 340) entering the digestive tract are reabsorbed in the bowel. Under abnormal conditions, however, enormous quantities of these digestive secretions may be lost from the body, viz., vomiting, diarrhea, tube drainage, fistulas. The potential magnitude of these losses is indicated in Table 13-1. These figures may be exceeded in certain diseases because the volumes of gastric and intestinal secretions may be increased due to local irritation, e.g., inflammation. On the other hand, the concentration of electrolytes in certain of these fluids may fall during protracted periods of loss (e.g., ileostomy), perhaps in part due to a process of adaptation, and in part reflecting their diminishing concentrations in the plasma.

Under these circumstances, which are encountered frequently clinically, over 8000 ml of water, 900 mmol. Na, 820 mmol. Cl, and 150 mmol. K may be lost from the body in twenty-four hours. The gravity of such situations becomes apparent when these values are compared with the total body content of these substances (Table 12-4, p. 367). Obviously, losses even remotely approaching these magnitudes cannot be tolerated for longer than very brief periods, emphasizing the importance of prompt and active treatment of clinical states of dehydration.

Skin. Under conditions of ordinary moderate environmental temperature and humidity, there is little or no active sweating. The water loss by "insensible perspiration" (vaporization from skin and lungs) is approximately 500 ml./sq. meter body surface area (p. 341). The volume of "insensible perspiration" is increased in fever, roughly in proportion to the temperature elevation.

The sweating mechanism is invoked by increased atmospheric temperatures and humidity and by physical exertion. Heavy work in hot and humid environments, e.g., steel mills, deep mines, etc., may involve enormous losses of sweat, as much as 2000 ml. per hour. The composition of perspiration varies considerably. Its electrolyte content (40–85 mmol. base/liter) is much lower than that of plasma (155 mmol. base/liter) and, therefore, prolonged, profuse sweating results in loss from the body fluids

TABLE 13-1 VOLUME AND ELECTROLYTE COMPOSITION OF BODY FLUIDS COMMONLY INVOLVED IN PRODUCTION OF DEHYDRATION

(Figures in parentheses are average values.)

	H_2O ml.	mmol. PER LITER Na	Cl	K	POSSIBLE DAILY LOSS (mmol.) Na	Cl	K
Gastric juice (resting)	2500	10–120 (60)	7–160 (85)	0.5–35 (9)	300	400	85
Pancreatic juice	1000	110–150 (140)	20–80 (40)	2–8 (5)	150	80	8
Small bowel	3000	80–150 (135)	40–140 (105)	2–10 (5)	450	420	30
Saliva	1500	(9)	(10)	(25)	15	15	40
Sweat	14,000	40–80	35–70	3–5	110	100	70

of relatively more water than Na and Cl, with consequent hypernatremia and hyperchloremia (p. 378).

The figures in Table 13-1 above represent theoretically possible losses, but these seldom occur under clinical conditions because of the operation of an adaptation mechanism, apparently involving glomerular filtration rate, antidiuretic hormone, and third factor, as well as increased secretion of salt-active adrenocortical hormones. The process of acclimatization to high temperatures and humidity is accompanied by a decrease in the concentrations of Na and Cl in the sweat, e.g., to 2–5 mmol./liter.

Urine. Under normal conditions the urine is the route of elimination of surplus water and electrolytes (and other solids). In the presence of normal kidney function, and in the absence of forces operating directly on renal excretory mechanisms (diuretic drugs, hormones, kidney disease), the quantities excreted by this route therefore vary inversely with the amounts lost through other channels, e.g., by vomiting, diarrhea, or excessive sweating. In this manner the kidneys attempt to preserve the body fluid content and composition in the face of abnormal losses by other avenues of elimination.

Progressive dehydration, with electrolyte depletion, may occur during prolonged periods of diuresis, as in diabetes mellitus, chronic glomerulonephritis, and Addison's disease, and as a result of administration of potent diuretic agents. The nature of the underlying mechanism varies in different conditions and is reviewed in discussions of other aspects of these disorders (q.v.). The most important underlying feature common to most of them is diminished renal tubular reabsorption of Na (also H_2O, and usually Cl) from the glomerular filtrate. As indicated elsewhere (p. 387), excessive amounts of K are also lost frequently by this route.

In diabetes insipidus, enormous quantities of water (up to 30 or more liters daily) may be excreted by the kidneys, without comparable losses of electrolytes (i.e., very low urine specific gravity). If adequate amounts of water are not taken in, this type of dehydration may be accompanied by hypernatremia and hyperchloremia.

General Principles in Electrolyte Abnormalities. *(1) Fluid*

Compartments Affected. Dehydration usually involves loss of excessive amounts of Na and/or Cl, in addition to water. Since these electrolytes must be derived mainly from extracellular fluid, the primary decrease in volume occurs in this compartment. Early, it is restricted to the interstitial fluid; the volume of circulating blood plasma, the other large component of the extracellular fluids, tends to be preserved until the deficits are so great that the regulatory mechanisms can no longer prevent significant plasma losses. If uninterrupted, this leads to progressive fall in blood pressure and shock.

If the loss of Na is disproportionately large, relative to that of water, contraction of the extracellular fluid volume is associated with expansion of the intracellular compartment. The tendency toward hypotonicity of the extracellular fluid (hyponatremia) results in transfer of water into the cells. This may occur in Addison's disease, especially after a water load, and in certain brain and kidney disorders. In all conditions in which the loss of water is relatively greater than that of Na, the primary decrease of extracellular fluid volume is accompanied or followed, shortly, by decrease of intracellular fluid volume. This is the case in the great majority of instances of dehydration encountered clinically. Under such circumstances, for practical purposes, it may be assumed that in dehydration of more than slight degree, due to excessive loss of body fluids (vomiting, diarrhea, diuresis, sweating), the body has lost water, Na, Cl, and K, as well as smaller amounts of other substances (e.g., phosphate, sulfate).

(2) Serum Electrolyte Concentrations. The nature of the changes, if any, in the pattern of concentrations of electrolytes in the extracellular fluids incident to dehydration depends upon a number of variable factors. These include: (*a*) The volume and composition of the fluid lost (e.g., relative concentrations of Na, Cl, and K; pH; hypertonic, hypotonic, or isosmotic). (*b*) The state of renal function. This may be impaired as a result of dehydration or as a part of the underlying disease process. (*c*) Water and electrolyte intake. Inadequate intake of water may be superimposed upon states of dehydration due to other causes, e.g., vomiting and other gastrointestinal disorders, extreme weakness, coma, etc.

The serum Na concentration may be decreased (hyponatremia), increased (hypernatremia), or normal in the presence of even extensive body deficits of this element. The direction and extent of abnormality, if any, in this factor are determined by the relative extent of the associated water deficit. From this standpoint, dehydration may be classified in three categories: (1) loss of sodium in excess of water (hypotonic contraction); (2) loss of water in excess of sodium (hypertonic contraction); (3) isotonic loss of sodium and water (isotonic contraction). ·Each of these types of dehydration may or may not be accompanied also by increase or decrease of serum pH, and of Cl and K concentrations. Regardless of the nature of the primary disturbance, variations in the pattern of concentrations of individual electrolytes in the extracellular fluid occur frequently as a result of coexisting complicating factors; e.g., (*a*) renal failure in patients with diarrhea, vomiting, diabetic acidosis, Addison's disease, etc.; (*b*) vomiting or diarrhea in patients with renal failure; (c) variations in water intake due to variations in responsiveness to stimulation of the thirst mechanism; (d) parenteral administration of electrolyte and glucose solutions.

In the face of losses of water and electrolytes, homeostatic mechanisms, mainly renal, operate to prevent or minimize changes in composition of the extracellular fluids. Thus, unless the disturbance is marked and the homeostatic mechanisms are overwhelmed or renal function is impaired, excess potassium entering the plasma from cells is excreted promptly in the urine, preventing undue rise in plasma potassium concentration. Similarly, excessive loss of sodium tends to be balanced by loss of equivalent amounts of water, thus limiting the decrease in plasma sodium concentration and osmotic pressure that would otherwise occur.

Relatively large amounts of sodium, potassium and water may, therefore, be lost from the body without comparable alteration in the concentrations of sodium or potassium in the plasma. The true extent of the deficits can be estimated only by measuring the total extracellular and intracellular fluid volumes in addition to the plasma electrolyte concentrations. This is not always feasible.

(3) Electrolyte Redistributions. When excessive amounts of K^+ ion move out of the cells into the extracellular fluid (e.g., in excessive loss by vomiting, diarrhea, diuresis, or in adrenocortical hyperfunction), an equivalent number of H^+ and Na^+ ions pass in the opposite direction, replacing K^+ in the intracellular fluid. In the presence of normal renal function, the K^+ increment in the extracellular fluid is excreted promptly in the urine with a comparable amount of anion, chiefly Cl^-, drawn from the extracellular fluid compartment. This extracellular Cl^- deficit is balanced by a simultaneous increase in HCO_3^- ion.

These interrelations may be expressed as follows: (a) the intracellular Na^+ tends to increase as the intracellular K^+ decreases. (b) In the presence of adequate kidney function, movement of K^+ out of the cell tends to produce a decrease in concentration of H^+ ions and Cl^- in the extracellular fluid and increase in serum HCO_3^-, i.e., a hypochloremic type of alkalosis (p. 412). In many cases of dehydration these changes are not observed because of the complicating factor of renal functional impairment. They may manifest themselves, however, as renal function improves during periods of rehydration by NaCl solutions.

Under the conditions outlined, that is, K depletion and adequate kidney function, administration of Na without K (e.g., NaCl solution) results in entrance of additional Na into the cells and passage of additional K from the cells into the extracellular fluid. This aggravates the K depletion and also the associated changes in serum Cl^- and HCO_3^-.

Other things being equal, the concentration of K^+ ions in the extracellular fluid tends to vary directly with the H^+ ion concentration (i.e., inversely as the pH). This is due at least in part to exchanges of K^+ for H^+ ion as the latter moves in directions determined by changes in its concentration gradient. Disturbances of renal mechanisms are also involved.

(4) Osmolality and pH. The osmotic value of extracellular fluid is, of course, determined by the total concentration of its chemical components, usually chiefly that of the electrolytes. Since changes in the concentrations of Cl⁻ and other anions (other than HCO_3^-) are offset by reciprocal changes in the concentration of HCO_3^-, it follows that the total osmotic value of the electrolytes is determined by the total cation concentration. Because Na⁺ constitutes such a large portion of the cation structure of extracellular fluid, the osmotic value of the latter rests almost entirely on this element. Accordingly, hypernatremia causes a tendency toward hypertonicity. Hyponatremia is associated with a tendency toward hypotonicity, unless the decrease in Na concentration is counterbalanced by increased concentrations of other substances, usually nonelectrolytes such as glucose (uncontrolled diabetes) and urea (uremia). It must once again be emphasized that hyponatremia and isotonicity can occur in situations in which there is no real deficit in body sodium ("sick cell syndrome").

The normal value for HCO_3^- depends upon the integrity of the remainder of the electrolyte structure, the most important components of which, from a quantitative standpoint, are Na⁺ and Cl⁻ (p. 350). Consequently, in dehydration, the direction of changes in concentration of HCO_3^- will be determined by the relative amounts of Na and Cl lost. If Cl is lost in excess of Na (gastric juice), the tendency is toward increased HCO_3^- concentration and alkalosis. If Na is lost in excess of Cl (pancreatic, ileal, jejunal secretions; urine), the tendency is toward decreased HCO_3^- concentration and acidosis.

Except in primary respiratory disorders (p. 414), changes in HCO_3^- concentration result from alteration in some other part of the electrolyte structure. For example, if the sum of the other acid anions (chloride, phosphate, sulfate, organic acids) is increased, an equivalent amount of HCO_3^- will be liberated as CO_2 (acidosis). On the other hand, if the sum of other acid anions is decreased, the HCO_3^- concentration increases correspondingly (alkalosis). In other words, the bicarbonate concentration in the plasma is determined by the extent to which the total cation concentration exceeds the sum of the concentrations of acid anions other than HCO_3^-.

(5) Fever and Urea Retention are two important objective manifestations of dehydration. Normal subjects lose about 500 ml. of water per square meter of body surface area per day (850 ml. daily by 70 kg. man) by vaporization from the lungs and skin (p. 341). This is an obligatory loss, determined by metabolic activity (heat production) and is an important part of the mechanism of heat loss. If dehydration progresses to the point where this amount of water cannot be provided, the body temperature rises (dehydration fever). It has been estimated that this occurs when about 60 ml. of water/kg. body weight have been lost.

In normal adults on an average protein intake, about 500 ml. of water are required for adequate renal excretion of solid materials (p. 341). If dehydration progresses to the point where this amount cannot be provided, certain urinary constituents, mainly urea, will be retained in the body fluids (i.e., rise in blood urea nitrogen concentration).

Dehydration per se, and/or decrease in plasma Na and Cl concentrations, which occur commonly in dehydration, may have a deleterious effect on renal function. Severe dehydration, with decrease in plasma volume, may result in decreased renal circulation and consequently decreased glomerular filtration. This may be aggravated by the frequently associated fall in arterial blood pressure (shock syndrome).

(6) Effects of Starvation and Ketosis. Drastic reduction in food intake, prominent in patients with dehydration, may in itself aggravate the loss of water and electrolytes. A fasting normal adult loses about 80 g. protein daily. Since loss of 1 g. protein involves the loss of about 3 ml. intracellular fluid (0.45 mmol. K), there is an associated daily loss of 250 ml. water and 35 to 40 mmol. K.

Inadequate intake of carbohydrate results in ketosis. This may occur in the presence of alkalosis, e.g., in prolonged vomiting (pyloric obstruction), as well as in acidosis (p. 109). Excretion of the increased amounts of organic acid anions in the urine involves excretion simultaneously of corresponding amounts of cations. In the presence of normal renal function, these may consist largely of H^+ and NH_4^+ ions (p. 407). Nevertheless, there is almost invariably an associated loss of increased amounts of Na^+, K^+, C^{++}, and Mg^{++}, as well as of anions other than those of the organic acids (chloride, sulfate, phosphate).

Parenteral administration of glucose performs a dual service in this connection. It abolishes ketosis, thereby conserving Na and water; it reduces the protein loss (protein-sparing action), thereby conserving K and water.

(7) Hyperlipemia. Apparently low concentrations of serum sodium can occur in the presence of an increase of serum lipid. This is due to the fact that an appreciable contribution to the serum volume is made by the hydrophobic lipid molecules. In even a very slightly milky serum, about 3 per cent of the total volume is lipid. Thus, although the total serum concentration appears low, the actual concentration in the serum water, in which all the sodium is dissolved, can be normal.

Biochemical Consequences of Dehydration.

Decrease of plasma water in advanced dehydration is reflected in increased hematocrit and hemoglobin values and plasma protein concentration. However, because of the common occurrence of malnutrition and anemia in conditions associated with dehydration, recognition of the development of hemoconcentration usually requires knowledge of previous values for these factors.

Progressive diminution in extracellular fluid volume leads to renal functional impairment, with increasing concentrations in the blood of urea nitrogen, creatinine, phosphate, and other inorganic and organic acid anions, leading to decreased bicarbonate concentration. Fall in blood pressure (shock) and hyponatremia, by decreasing glomerular filtration, contribute to deterioration of kidney function.

As indicated elsewhere, extensive deficits of Na and water may be accompanied by increased, decreased or normal concentrations of Na in the serum. Regardless of the level of serum Na, reversal of the abnormalities outlined above by administration of Na-containing fluids is good evidence of their dependence upon Na deficiency.

Gastrointestinal Disorders. Some of the most common clinical causes of dehydration and electrolyte deficit fall into this category. They include: (*a*) prolonged vomiting (e.g., peptic ulcer, gastric malignancy, gastritis, intestinal obstruction, peritonitis, gallbladder disease, uremia, pregnancy toxemia, etc.); (*b*) diarrhea (e.g., ulcerative colitis, enteritis of various types); (*c*) external fistulas (e.g., biliary, pancreatic, intestinal; ileostomy, colostomy); (*d*) tube damage (e.g., pre- and postoperative).

In vomiting, the nature of the alteration in the plasma electrolyte pattern depends on the character of the fluid lost, i.e., on the location of the lesion and the acidity of the gastric juice. In stenosing duodenal ulcer, for example, the vomitus (highly acid) contains large amounts of Cl and relatively little Na, resulting in hypochloremic alkalosis (p. 423). With continued vomiting, the Na concentration of the gastric juice often rises considerably, at times to about one-half the Cl concentration. In cases of obstruction lower in the small intestine, e.g., ileum, the vomitus may contain relatively highly alkaline pancreatic juice in addition to the gastric juice, usually of considerably lower acidity than in peptic ulcer. Under such circumstances, the Na loss will tend to be greater and the Cl loss smaller than in the latter condition. The same is true usually in gastric carcinoma, in which the acidity of the gastric juice is generally low. In such cases, the loss of Na tends to parallel that of Cl. Whether alkalosis or acidosis occurs is determined by the pH of the fluid lost, i.e., whether it contains relatively more or less Cl (in relation to Na) than the blood plasma (p. 364). As indicated elsewhere (p. 386), the extent of K loss also exerts an influence in this connection, K depletion tending to lower the Cl^- and raise the HCO_3^- in the extracellular fluids (i.e., to increase the pH). The hypochloremia of vomiting is frequently but not invariably accompanied by hyponatremia, usually of lesser degree.

The same principles apply in diarrhea, external fistulas, prolonged tube drainage, etc. As a rule, however, these conditions cause greater loss of K (p. 386) than occurs in vomiting, and the alteration in extracellular electrolyte pattern is more consistently in the direction of acidosis. This is contributed to by loss of bicarbonate in the feces, starvation ketosis and, in certain cases, hyperchloremia and renal failure. One exception is the rather rare condition of congenital, noninfectious diarrhea in infants, characterized by severe K loss and alkalosis (p. 386).

Greater deficits of electrolytes and water are usually encountered in the diarrheas of infants than of adults, except in such conditions as cholera and dysentery and, occasionally, ulcerative colitis and sprue. In these conditions, water is usually lost in excess of Na and Cl. Consequently, the severe dehydration in diarrhea of infants and in these diseases in adults is frequently characterized by hypernatremia and hyperchloremia. In more chronic disorders, especially in adults, hyponatremia and hypochloremia occur frequently.

Diabetes Mellitus. Na and Cl are lost in large amounts during the progress of uncontrolled diabetes, and particularly in the course of development of diabetic acidosis. This is due in part to diuresis (accompanying the glycosuria and acidosis), and in part to vomiting, which is a common feature of diabetic acidosis. The Cl loss in the urine is contributed to also by the depletion of intracellular K (p. 387). Additional amounts of water are lost by vaporization from the lungs incident to the hyperventilation that characterizes this condition.

In the presence of acidosis, the plasma Na concentration is almost invariably low, whereas the Cl may be normal, subnormal or at times even increased. The supervention of renal insufficiency (due to hyponatremia, dehydration, circulatory collapse, glomerulosclerosis) often modifies the pattern of electrolyte abnormalities, and increases the tendency toward acidosis. In the presence of renal functional impairment, parenteral administration of NaCl solution alone may induce or increase hyperchloremia, delaying correction of the acidosis.

If the patient has a marked lipemia, this may give the impression that the serum sodium is lower than it actually is in the plasma water.

Kidney Disease. *Chronic Renal Failure.* Except in the last stages of the disease, with oliguria, the gradual deterioration of renal function in chronic glomerulonephritis is characterized by progressive loss of the normal flexibility of the kidney; i.e., it becomes increasingly incapable of responding promptly, with conservation of Na, Cl and water, to Na, Cl and water depletion. It is also unable to increase the excretion of these substances in response to excessive loads.

In chronic renal failure several factors contribute to the production of water and electrolyte defects: (a) Loss of Na and Cl by vomiting, diarrhea and previous polyuria. The gastric acidity is often low in uremic subjects, increasing the loss of Na in the vomitus. (b) Impaired capacity for acidification of the urine (i.e., Na^+:H^+ exchange, p. 408) and for ammonia formation (p. 411), with consequent loss of excessive amounts of Na in the urine. (c) As the number of functioning nephrons diminishes, those remaining are essentially in a state of osmotic diuresis; i.e., each tubular unit is exposed to an increased load of filtered solute and water. Under these circumstances excessive amounts of water and Na escape reabsorption. This condition continues until glomerular filtration is reduced to the point where oliguria occurs. Hyponatremia and hypochloremia occur frequently in chronic glomerulonephritis as a result of these causes.

Renal Tubular Acidosis. This condition is due to inadequacy of the renal mechanism for absorption of HCO_3^-. Consequently, excessive amounts of water, Na, Cl, K, Ca, and other electrolytes are lost. An important diagnostic feature is the combination of systemic metabolic acidosis and urine that is alkaline or only weakly acid in reaction. In contrast also to acidosis in kidney disease with glomerular insufficiency, the systemic acidosis is associated usually with a decreased serum phosphate concentration. The increased amounts of Ca in an alkaline urine may lead to precipitation of calcium salts in the renal parenchyma or lower urinary passages (nephrocalcinosis, nephrolithiasis).

Salt-Losing Nephritis. This condition, which is claimed to be more commonly associated with chronic pyelonephritis, is characterized by deficient renal tubular reabsorption of Na and Cl. Abnormally large amounts continue to be lost in the urine in the face of extensive body deficits, dehydration, hyponatremia, and hypochloremia. These may produce a secondary reduction of glomerular filtration, with consequent nitrogen retention. This disorder simulates adrenocortical insufficiency, differing from the latter characteristically in its failure to respond to administration of adrenocortical hormones. Improvement often follows administration of sodium chloride, provided that care is taken not to give too great an amount, which can produce pulmonary edema.

Acute Renal Failure. This occurs usually in acute tubular necrosis, but also occasionally in acute glomerulonephritis. During the initial period of oliguria or anuria, the tendency is toward excessive retention of water and electrolytes rather than excessive loss. Hyponatremia may occur during this phase not only as a result of injudicious parenteral administration of electrolyte-free fluids (glucose solution) but also because excessive metabolism of body fat gives rise to water, which is retained. However, it may occur also as a manifestation of true Na deficit as a result of conditions similar to those operating in chronic renal failure; i.e., extrarenal losses due to vomiting, sweating, diarrhea, intestinal drainage, rigid salt restriction, and ion exchange resins. In spite of hyponatremia, there may be an increase in total exchangeable body Na. Extracellular K concentration is raised but total exchangeable K is diminished. During the subsequent phase of polyuria, there may be serious depletion of water and electrolytes, accompanied by hyponatremia and hypochloremia. Inadequate functional

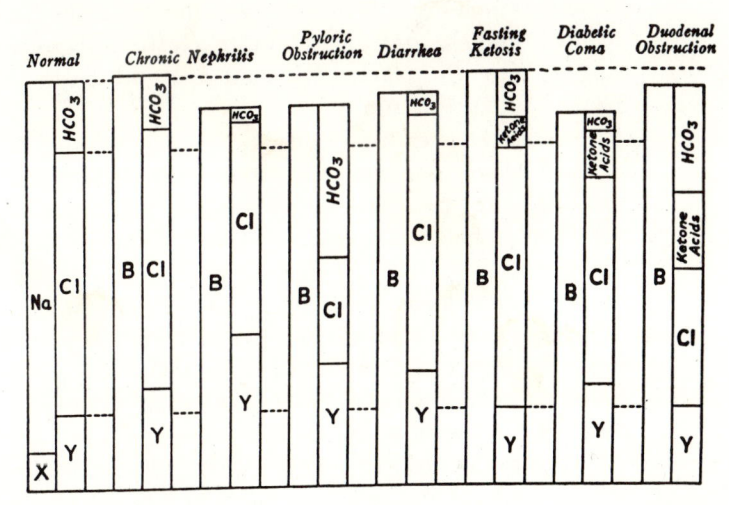

Figure 13-1. Changes in the electrolyte pattern of the plasma in various disease conditions. Dotted lines indicate the normal levels of bicarbonate and chloride. B—total base; Y—unnamed acid radicals. (After Gamble.)

activity of the damaged and also the newly regenerated tubular epithelium results in diminished reabsorption of Na, Cl and water, with escape of excessive quantities in the urine. Moreover, because of the relatively poor tubular as compared to glomerular function, a condition exists of relative osmotic diuresis, characterized by inadequate absorption of filtered water and solutes.

Hypernatremia can occur occasionally during this period, due usually to parenteral administration of excessive amounts of NaCl solution. At times it appears to be associated with rapid mobilization of edema fluid.

Excessive Sweating. Large quantities of sweat (10 to 14 liters daily) may be lost by individuals working in very hot and humid environments (p. 369). Loss of this hypotonic fluid (p. 341) may result in hypernatremia and hyperchloremia (p. 370), particularly since adaptation to these environments is accompanied by decrease in the NaCl concentration of the sweat from the normal 40 to 80 mmol./liter to as low as 2 to 5 mmol./liter. However, if such subjects drink large amounts of water and no provision is made for simultaneous replacement of the lost Na and Cl, the concentration of these elements in the extracellular fluids falls, with the development of manifestations resembling those of adrenal insufficiency, including renal functional impairment and nitrogen retention in the blood. This phenomenon is the basis for the production of so-called "miners'" or "stokers'" cramp, characterized by severe, localized muscle spasms, brought on particularly by exertion and often preceded by weakness of the affected muscles.

Excessive amounts of water alone may be lost by vaporization ("insensible perspiration") in patients with high fever, in the absence of loss of Na or Cl by other routes (e.g., vomiting, diarrhea). The statements made above, concerning excessive sweating, apply here also. Hyperchloremia and hypernatremia are seldom observed in such cases except in the presence of marked oliguria, and then almost exclusively in infants, in whom the body surface area (for vaporization): body fluid ratio is relatively high compared with that of adults.

Excessive sweating occurs during hot weather in children with fibrocystic disease (pancreas and lungs; mucoviscidosis). The sweat in this condition has been found to contain abnormally large amounts of Na (av. 133 mmol./liter) and Cl (av. 106 mmol./liter), the concentrations approximating those in the blood plasma. This abnormality of the sweating mechanism has a genetic basis. The unusual susceptibility to the development of Na deficit and hyponatremia is furthered by an inadequate responsiveness to adrenocortical hormones with respect to their action in lowering the Na and Cl concentrations of the sweat. This interferes with operation of the normal adaptive mechanism.

Congestive Heart Failure. Severe prolonged congestive heart failure is characteristically accompanied by excess of body Na and water, often, paradoxically, with hyponatremia ("sick cell syndrome"). However, occasionally the latter may reflect a true Na deficit, due to excessive vomiting, diarrhea, sweating, or vigorous therapeutic measures, e.g., exchange resins or diuretics.

Cerebral Disorders. Although no neural connections have been demonstrated between the brain and kidneys, there is evidence that cerebral injury may possibly produce alterations in renal blood flow, glomerular filtration and tubular reabsorption. Occasional cases of diffuse cerebral disease, including encephalitis, bulbar poliomyelitis, vascular accidents, metastatic malignancy (bronchogenic carcinoma), and tuberculous meningitis may exhibit Na deficit and hyponatremia. These are frequently instances of sustained increased secretion of antidiuretic hormone, in spite of decreasing osmolality of the blood plasma ("inappropriate ADH secretion"). The malignant tumors can produce their own ADH. Under these circumstances, urine hypertonic to the blood plasma continues to be formed even when the serum osmolality falls below the normal range. In such cases, the hyponatremia may not be accompanied by reduction of extracellular fluid volume.

Pulmonary Neoplasms. Hyponatremia and excessive urinary Na excretion may occur in patients with bronchogenic carcinoma in the absence of evidence of cerebral metastasis. In certain of these cases the negative Na balance was not improved by administration of salt or adrenocortical hormone, but improved with reduction in size of the neoplasm following x-ray and chemotherapy. The mechanism operating under these circumstances is once again ADH secretion by the tumor itself.

Intrathoracic neoplasms (e.g., bronchogenic, thymic) may be accompanied also by syndromes characterized by marked adrenocortical hyperplasia. In some instances, there are no associated clinical or biochemical manifestations of adrenocortical hyperfunction, whereas others present a typical picture of Cushing's syndrome, with severe hypokalemic alkalosis. In those instances without evidence of adrenal overactivity, it has been postulated that the enlargement of the adrenal glands is due to ACTH secretion by the neoplasm.

Adrenocortical Insufficiency (Addison's Disease). The characteristic metabolic manifestations of adrenocortical insufficiency are discussed in detail elsewhere (p. 722). Except in crisis, the serum Na and Cl concentrations (also K) are usually within normal limits. The early abnormalities consist in excessive excretion of Na and Cl (and H_2O) in the urine, usually in such proportions as to preserve the normal electrolyte pattern in the extracellular fluids in the face of progressive dehydration and NaCl depletion. In adrenal crisis, however, there is characteristically a sharp decrease in serum Na and Cl concentrations (and increase in K, p. 387), probably reflecting a break in fundamental regulatory processes, including renal mechanisms.

Excessive Therapy. Deficits of Na and Cl, with associated hyponatremia and/or hypochloremia, may be induced by injudicious or excessive therapeutic measures. This may occur, e.g., after surgery and in the course of treatment of patients with congestive heart failure or portal cirrhosis with ascites, by rigid restriction of Na intake (dietary or with the use of ion exchange resins), together with the use of diuretic agents.

Primary Water Deficit. Dehydration in its purest form occurs as a result of reduction of water intake below the amount of obligatory loss by vaporization from the lungs and skin (800 ml. or more daily).

Water deprivation of this type occurs characteristically in individuals lost in the desert or at sea, in mentally deranged or comatose persons, and in neglected, physically incapacitated individuals. It is also frequently superimposed upon conditions of excessive water loss, as in patients with gastrointestinal disease accompanied by vomiting. Dehydration of this type, i.e., loss of water without comparable loss of Na, results in hypernatremia, hypertonic contraction of the extracellular fluid compartment, and withdrawal of water from the intracellular fluid compartment (cellular dehydration).

The designation "nephrogenic diabetes insipidus" is applied to a condition in which secretion of ADH is apparently normal but the renal tubular epithelium is unable to respond, as normally, by increased reabsorption of water. This condition may be due to a genetically determined renal tubular defect (p. 539) or to acquired disease, e.g., multiple myeloma, obstructive uropathy, or K depletion.

EXCESSES OF WATER, SODIUM, CHLORIDE

As in the case of deficits, excessive retention of water, sodium, and chloride may be accompanied by normal, increased or decreased concentrations of serum Na and Cl, depending upon the relative amounts of water and electrolytes retained. Moreover, as in dehydration, expansion of extracellular fluid volume in overhydration is not always accompanied by similar changes in intracellular fluid volume. Retention of Na in excess of water results in hypertonic expansion of the extracellular fluid (hypernatremia), with consequent movement of water from the intracellular compartment. Conversely, retention of water in excess of Na produces hypotonic expansion of the extracellular fluid volume (hyponatremia), with consequent movement of water also into the intracellular compartment. Edema may be associated with increased extracellular fluid volume of any type.

Clinically, excesses of body Na, Cl, and water occur most frequently in the presence of defects in some portion of the mechanisms concerned with regulation of their excretion. These include: certain types of renal disease, congestive heart failure, hepatic cirrhosis, hyperaldosteronism, and toxemia of pregnancy.

Renal Disease. Na, Cl and water are, of course, retained in excess in types of renal disease accompanied by edema, as in acute glomerulonephritis, the nephrotic syndrome, and complicating congestive heart failure. When these conditions are untreated, water is usually retained in proportion to or in excess of Na, and therefore the serum Na concentration is usually within normal limits or perhaps somewhat diminished. In any form of kidney disease, the situation with respect to water and electrolytes varies considerably due to variations in complicating factors, several of which tend to produce excessive losses (e.g., vomiting, diarrhea, diuresis, p. 527).

Patients with extreme oliguria or anuria are very susceptible to overhydration and Na retention. A well-recognized cause of Na and water excess encountered clinically is injudicious parenteral administration of solutions of NaCl and/or glucose in a misguided attempt to stimulate urine formation in acute glomerulonephritis, acute tubular necrosis, and the terminal stage of chronic glomerulonephritis. The serum Na concentration is usually decreased, a reflection of overhydration. Frequently, administration of glucose solution to patients with Na and water deficits in kidney disease (p. 527) results in a condition of water excess superimposed on Na depletion, aggravating a perhaps pre-existing hyponatremia.

Congestive Heart Failure. Although occasionally Na depletion may occur as a result of complicating factors (p. 378), patients with congestive heart failure usually exhibit excesses of water, Na and Cl, often with edema. As in other conditions (p. 371), this Na excess may, paradoxically, be accompanied by hyponatremia. The mechanism of production of these abnormalities in congestive heart failure is complex. Increased renal tubular reabsorption of Na (plus water and Cl) is an essential feature. In some cases this may result, in part at least, from decreased renal blood flow and glomerular filtration secondary to decrease in cardiac output. However, increased venous pressure is apparently more important in this connection, operating via increase in the concentration of circulating aldosterone. This presumably results from one or both of two causes: (*a*) decreased rate of removal of the hormone from the circulation because of impaired liver function due to chronic passive congestion; (*b*) it was formerly believed that there was also stimulation of aldosterone secretion through the renin-angiotensin mechanism. Studies on patients in congestive heart failure generally have not indicated increased secretion. There is, however, an increased production of antidiuretic hormone.

Liver Disease. Excesses of total body Na and water occur in both acute (e.g., infectious hepatitis) and chronic (e..g., portal cirrhosis) diseases of the liver. Impaired renal excretion of Na and water in these conditions has been attributed to inadequate removal of antidiuretic hormone and aldosterone from the circulation by the damaged liver. Increased amounts have been demonstrated in the urine and also an abnormally high percentage of aldosterone in unconjugated form. The serum Na concentration is usually normal or decreased. Tendency to hyponatremia may be accentuated by complicating renal failure with increase in blood urea nitrogen. Occasionally, a state of Na depletion may occur in patients with cirrhosis as a result of rigid restriction of Na intake, repeated paracentesis, diuresis, and the use of exchange resins.

Toxemia of Pregnancy. Abnormal retention of Na and water, with expansion of the extracellular fluid volume, may occur in toxemia of pregnancy. This has been attributed to progressive decrease in the glomerular filtration rate. Aldosterone secretion and plasma renin levels are decreased.

Adrenocortical Hyperfunction. Administration of aldosterone to normal subjects results in increased retention of Na and Cl (increased renal tubular as well as large intestinal reabsorption) and increased renal excretion of potassium. Other mineralocorticoids produce a similar but much less marked effect. Excessive retention of sodium and water may occur in patients treated with adrenocortical hormones, particularly deoxycorticosterone, and especially if the NaCl intake is not restricted. Glucocorticoids are necessary in order to attain a maximum dilution of the urine. This is possibly due to an effect of these steroids on the permeability to water of the diluting segments of the nephron.

Primary aldosteronism is accompanied by polyuria. This is due to the fact that potassium loss in the urine leads to a fall in the body potassium content, and this in turn results in a diminished effect of antidiuretic hormone.

Increased urinary excretion of aldosterone, presumably due to a condition of secondary hyperaldosteronism, occurs rather consistently in patients with (a) congestive heart failure, (b) acute and chronic liver disease (infectious hepatitis, portal cirrhosis and ascites), and (c) certain types of kidney disease (nephrotic syndrome). In all of these conditions, in contrast to primary hyperaldosteronism, there is expansion of extracellular fluid volume and retention of Na. The serum K concentration is usually within normal limits, except in certain types of renal disease. It is somewhat difficult to understand the reason for continued hypersecretion of aldosterone in the face of the expanding extracellular fluid volume. In congestive heart failure and liver disease, increased concentration of circulating hormone might conceivably be due to decreased removal from the blood stream as a result of impaired liver function.

Postoperative Response. Retention of Na, Cl and water occurs in the immediate postoperative period. This is due to increased production of ADH. The water retention is sufficient to result in hyponatremia. Increased aldosterone production also occurs postoperatively, but the increase does not appear to be concurrent with the electrolyte changes already described. It may, however, be associated with the low level of serum K, which follows an initial high level due to liberation of K from damaged muscle.

Miscellaneous Sodium and Chloride Abnormalities. Hyponatremia and hypochloremia may occur in the rather rare condition of acute hyperparathyroidism, with hemoconcentration, oliguria, and renal failure. These changes are believed to result in part at least from the active diuresis which occurs in such cases.

Hypochloremia with alkalosis may occur in primary K deficiency (e.g., adrenocortical hyperfunction, p. 734) or as a result of administration of large amounts of absorbable alkali (sodium bicarbonate).

Hypochloremia may occur in the course of certain acute infectious diseases, e.g., pneumococcal pneumonia, rheumatic fever, meningococcal meningitis, and less strikingly in erysipelas, typhoid fever, and tuberculosis. The urine Cl is also decreased, as is the serum Na concentration. At the time of crisis in pneumonia, the urine and serum Na increase sharply to essentially normal levels. It appears that hypochloremia and hyponatremia of this type, at least in pneumonia, are the result of expansion of the extracellular fluid compartment, with retention of water in excess of Na and Cl.

Hypochloremia, with no change in Na, occurs in bromide intoxication, the latter ion replacing Cl in the body fluids.

Hyperchloremia, usually without hypernatremia, may occur in patients in whom the ureters have been transplanted into the bowel (ureterosigmoidostomy), e.g., in bladder carcinoma. If urine elimination is not maintained by frequent emptying of the bowel, large amounts of KCl and NH_4Cl may be reabsorbed, with consequent development of a hyperchloremic type of acidosis (p. 420).

Increase in serum Na or Cl or both result from injudicious parenteral administration of hypertonic NaCl solutions in patients with impaired kidney function or with oliguria due to extrarenal deviation of water.

Hyponatremia and hypochloremia are observed as a result of parenteral administration of large amounts of electrolyte-free glucose solutions to patients with Na and Cl deficits, as in vomiting, diarrhea, diabetic acidosis, adrenocortical insufficiency and certain types of kidney disease. Excess of water is superimposed upon salt depletion, and may, if excessive, result in overexpansion of both body fluid compartments.

Hyperchloremia with acidosis may occur as a result of administration of large amounts of NH_4Cl or the use of H^+ or NH_4^+ exchange resins, particularly in the presence of inadequate renal function.

POTASSIUM ABNORMALITIES

General Principles. (1) As in the case of Na, the serum K concentration may remain within normal limits in the face of extensive losses of K from the body. It may, in fact, be increased if renal function is impaired in such cases (p. 389). Under these circumstances, which are encountered frequently clinically, the existence of a state of K depletion may be revealed only by the development of hypokalemia after administration of K-free fluids, e.g., NaCl solution, with consequent expansion of the extracellular fluid compartment. However, if one can exclude overhydration and transfer of K into cells (familial periodic paralysis, alkalosis), hypokalemia almost certainly indicates K deficit.

(2) The small size of the extracellular pool of K (70 mmol.), regulation of which depends upon proper exchanges with the large intracellular pool (3500 mmol.) and excretion in the urine, makes it extremely susceptible to wide fluctuations when these latter mechanisms are disturbed. Even in the absence of exogenous supply, the extracellular fluid may receive relatively large increments from the intracellular compartment in a variety

of disease states associated with regressive changes in cells, hypoxia, dehydration, depressed carbohydrate utilization, increased protein catabolism, etc. Under such circumstances, impaired renal excretion of K results in an abrupt increase of the serum K concentration.

(3) Perhaps because of its active excretion by the kidney, K continues to be lost in the urine even when none is taken in and there is a serious K deficit. Consequently, again because of the small size of the extracellular pool, the serum K concentration may fall rather readily as a result of a combination of excessive extrarenal loss (vomiting, diarrhea, etc.) and continuing excretion in the urine.

(4) There are important interrelationships between the concentration of K and the pH of the serum. As indicated elsewhere (p. 423), primary K deficit, per se, results in hypochloremic alkalosis. Potassium depletion should, therefore, be suspected whenever there is an unexplained low serum Cl^- and high HCO_3^- concentration in the presence of normal kidney function.

Changes in the concentration of H^+ ions in the extracellular fluid, by altering the rate of their passage into the cells, are accompanied by changes in the same direction in the concentration of K^+ ions in the extracellular fluid. This is due to the fact that passage of a H^+ ion in one direction, e.g., from the extracellular to the intracellular space, is accompanied by passage of a K^+ ion in the opposite direction, i.e., from the intracellular to the extracellular fluid. Accordingly, other things being equal, purely as a result of changes in pH, the serum K concentration is increased in acidosis and is decreased in alkalosis. This is of practical importance in that a serum K concentration within normal limits in severe acidosis can signify an actual K deficit; and hypokalemia, extensive depletion. Clinical manifestations of K depletion may occur at higher levels of serum K concentration in patients with acidosis than in those with alkalosis or normal serum pH.

(5) In patients with dehydration and Na and K deficit, hypokalemia may be induced rather abruptly by expansion of the extracellular fluid volume by means of parenteral administration of K-free fluids. This is due to: (*a*) dilution of the extracellular fluid; (*b*) increased renal excretion of K as renal function improves with rehydration; (*c*) increased rate of transfer of K into cells with improvement in metabolic activity, particularly if the therapeutic fluids contain glucose.

(6) Calcium and digitalis are K antagonists with respect to certain important myocardial functions. This is of practical importance in two connections: (*a*) patients receiving digitalis may exhibit manifestations of digitalis intoxication as the serum K concentration falls; (*b*) patients with simultaneous hypokalemia and hypocalcemia (e.g., in sprue or celiac disease) may exhibit no clinical evidence of neuromuscular hyperexcitability but may develop frank tetany when the serum K concentration rises as a result of adequate treatment.

(7) Potassium depletion results in rather characteristic structural and functional changes in the kidney. Renal blood flow, glomerular filtration

rate, and urine-concentrating capacity are decreased. In addition, in a primary K deficiency, as in hyperaldosteronism, there is an alteration in the mechanism of acidification of the urine, resulting in a tendency toward hypochloremic alkalosis (p. 423). The morphologic consequences of K deficiency consist essentially in vacuolar and degenerative changes in the convoluted tubules.

Starvation; Malnutrition. Deposition of 1 g. protein involves retention of approximately 3 ml. intracellular water, containing about 0.45 mmol. K per gram protein N. In protein undernutrition, therefore, one would expect the negative N balance to be accompanied by increased urinary excretion of K in approximately this ratio (mmol. excess K : g. excess N = 2.75 : 1). However, considerably higher ratios are observed in clinical conditions of malnutrition, even in the absence of such obvious causes of associated dehydration as diabetic acidosis, in which the ratio may be as high as 10 mmol./g. N.

This excess K loss is due, in part at least, to the fact that the kidneys do not conserve K as efficiently as Na. When subjects on a previously normal intake are suddenly deprived of K, they continue to excrete it in the urine in excess of the calculated K : N ratio. This extra loss may amount to 20 to 50 mmol. or more on the first day, decreasing gradually subsequently, representing loss from intracellular fluids over and above that bound to protein. As indicated elsewhere, this phenomenon is of particular importance in relation to the postoperative management of dehydrated, undernourished patients (p. 388), in whom hypokalemia may be induced by dietary restriction, in addition to injudicious parenteral administration of NaCl and glucose solutions without adequate amounts of K, as well as the usual postoperative response.

Administration of NaCl and Glucose Solutions. When solutions of NaCl and glucose are given parenterally to K-depleted (i.e., dehydrated) subjects, the serum K often falls from the previously normal levels that had been maintained as a result of loss of corresponding amounts of water (p. 384). Expansion of the extracellular fluid volume by isotonic NaCl solution unmasks the state of K depletion, with consequent hypokalemia, often of serious degree and sudden development.

In addition to this dilution effect, infusion of NaCl solution in such cases results in increased urinary excretion of K. This is due to acceleration of transfer of cell K to the extracellular fluids and its replacement by Na. The excess mobilized K is excreted in the urine with extra Cl, resulting in a tendency toward a hypochloremic type of alkalosis (p. 412) if renal function is adequate. This process can be reversed only by administration of K (p. 372).

Parenteral administration of glucose solution to K-depleted subjects results in increased urinary excretion of K, although not to the same extent as that produced by NaCl. Moreover, the stimulation of cell oxidation processes by the added glucose is accompanied by accelerated passage of K from the extracellular fluids into the cells (p. 47), with consequent fall in the serum K concentration. This is accentuated by simultaneous insulin therapy and is particularly important in connection with the precipitation of serious hypokalemia during treatment of patients in diabetic coma (p. 387).

Gastrointestinal Disorders. Under normal conditions, of the 50 to 150 mmol. of K ingested daily, 10 mmol. or less are excreted in the feces, the remainder in the urine. Under abnormal conditions, 100 to 200 mmol. or more may be lost via the gastrointestinal tract (e.g., vomiting, diarrhea, fistulas, exchange resins, etc.). The gravity of this situation is emphasized by the fact that, in such circumstances, renal excretion of K continues, although often at a low level, and that the food intake (and therefore K) is often reduced to practically nothing.

Vomiting. This is a frequent cause of K deficiency in surgical cases, together with tube drainage of the gastrointestinal tract, which produces similar effects. Large amounts of water, Na, Cl, and K (p. 369) may be lost by these routes. The extent of depletion of K is increased by the usually simultaneously lowered food intake (p. 385). As in other conditions, the serum K often remains within normal limits until active rehydration with K-free solutions is instituted.

If alkalosis develops as a result of vomiting (p. 423), another mechanism comes into play that tends further to lower the serum K concentration. As the urinary acidity decreases in response to the alkalosis, i.e., decreased tubular excretion of H^+ ions, tubular excretion of K^+ in exchange for Na^+ increases correspondingly (p. 410). As the condition progresses, depletion of K^+ in the renal tubular cells leads to depression of K^+ excretion and corresponding enhancement of H^+ excretion and HCO_3^- reabsorption. This tends to maintain the condition of metabolic alkalosis initiated by a different mechanism, i.e., loss of HCl by vomiting. In the presence of K deficiency, administration of K is required for correction of alkalosis that might otherwise be treated adequately with solutions of NaCl. In the presence of K deficiency, such treatment may actually aggravate the alkalosis.

Diarrhea. The factors reviewed in discussing the effects of starvation, fluid administration and vomiting operate also in dehydration due to diarrhea, bile, pancreatic, and intestinal fistulas and use of cation exchange resins. In these conditions, alteration in acid-base balance is usually in the direction of acidosis (p. 421). K depletion may be severe and may manifest itself in hypokalemia only after rehydration by K-free solutions (p. 385). In infantile diarrhea with severe dehydration, the serum K concentration may be increased, as well as the Na and Cl; the acidosis is contributed to by loss of HCO_3^- in the feces, ketosis due to starvation, renal failure, and hyperchloremia.

An apparently rare condition of congenital alkalosis with diarrhea has been described in infants in which there is a fundamental defect in intestinal absorption of electrolytes. The voluminous, watery feces contain, in addition to excessive quantities of K, a relative excess of Cl over Na. This tends to produce a state of alkalosis, which increases renal excretion of K, as indicated above. Inasmuch as the excess K is lost in the urine with extra

Cl drawn from the extracellular fluids, the plasma Cl decreases and the HCO_3 increases, aggravating the condition of hypochloremic alkalosis.

It is perhaps pertinent to mention here that subjects with alkalosis (e.g., due to vomiting) or hypocalcemia (e.g., in sprue, celiac disease) may not exhibit clinical manifestations of tetany, i.e., neuromuscular hyperirritability, in the presence of hypokalemia. This is due to the opposing influences of Ca^{++} and K^+ ions on this function.

Diabetic Acidosis. In addition to extensive losses of water, Na, and Cl (p. 376), subjects with inadequately controlled diabetes mellitus and increasing acidosis lose large amounts of K in the urine (diuresis). As acidosis progresses, vomiting occurs commonly, aggravating the K deficiency, which is contributed to also by increased protein catabolism (pp. 34 and 385) and reduction in food (and K) intake. Despite this almost invariable state of advanced K depletion, severely dehydrated acidotic patients in diabetic coma usually present normal or, at times, somewhat elevated serum K concentrations. This is due in part to the associated loss of corresponding amounts of water, and in part to impairment of kidney function, which frequently supervenes.

If such patients are treated by parenteral administration of large volumes of NaCl solutions and large doses of insulin, with or without glucose, the serum K concentration usually falls in 8 to 16 hours, often to dangerously low levels. This is due to: (1) expansion of the extracellular fluid volume, unmasking the existing K deficiency; (2) persistence of K excretion in the urine; (3) passage of increased amounts of K into cells from extracellular fluids as a result of stimulation of glucose oxidation; (4) retention of increasing amounts of K in the cells as a result of decreased protein catabolism. The development of hypokalemia can be prevented by administration of adequate amounts of K as soon as this is regarded as safe.

Adrenocortical Hormones. Administration of excessive amounts of "salt-active" hormones of the adrenal cortex results in increased urinary excretion of K and the development of a hypochloremic type of alkalosis. This occurs also in Cushing's syndrome and hyperaldosteronism (p. 734). These hormones favor transfer of K from intra- to extracellular fluids and increase its excretion in the urine, with extra Cl derived from the extracellular fluids. The plasma Cl^- concentration falls and the HCO_3^- concentration increases correspondingly (i.e., hypochloremic alkalosis). A fall in serum K also results from prolonged administration of hydrocortisone or synthetic glucocorticoids, such as prednisone.

Adrenocortical insufficiency (Addison's disease) results in decreased renal excretion of K (p. 722) and increased excretion of Na, Cl and water. Despite the associated dehydration, however, frank and serious hyperkalemia is seldom encountered in this condition except during periods of crisis (p. 739).

Diuretic Drugs. Hypokalemia may occur in patients with congestive heart failure, portal cirrhosis, and the nephrotic syndrome in whom diuresis has been induced. It is particularly liable to occur in cases in which hypochloremic alkalosis is also induced by such therapy. The increased urinary excretion of K produced by these agents is superimposed frequently upon a background of secondary aldosteronism (p. 734) and K deficit due to extrarenal losses, e.g., vomiting, diarrhea, starvation. The serum K concentration is generally within normal limits in the great majority of patients with these diseases uncomplicated by renal insufficiency or gastrointestinal losses and not receiving active diuretic therapy.

Patients with congestive heart failure receiving well-tolerated maintenance doses of digitalis may develop evidences of digitalis intoxication if the serum K concentration falls. This is due to the antagonistic effects of K^+ and digitalis on myocardial function.

Familial Hyperkalemic Paralysis. This condition is characterized by periodic attacks of muscular weakness associated with elevations of serum K concentration and aggravated by administration of K. This is in contrast to the situation in familial periodic paralysis (p. 390), which is promptly benefited by such therapy. Inasmuch as there is no associated decrease in urinary excretion of K, the hyperkalemia is attributed to an unexplained increased passage of K^+ from the intracellular to the extracellular space.

Potassium Therapy. If care is not exercised, a condition of K depletion and hypokalemia may be converted abruptly to one of hyperkalemia by too rapid parenteral administration of solutions containing unduly high concentrations of K to dehydrated subjects with low urine volumes and/or renal functional impairment due to kidney disease or to the dehydration. This dangerous condition (hyperkalemia) must be carefully guarded against by cautious K therapy under such circumstances, with frequent determinations of serum K concentration and urine output, repeated electrocardiograms and close clinical observation. Except for the correction of serious hypokalemia, it is advisable to delay such therapy until rehydration has resulted in beginning improvement in kidney function and increase in urine volume. It should be noted in this connection that bank blood may contain as much as 40 mmol. K/liter plasma (diffused from the red cells), the concentration increasing with the duration of storage.

Postoperative States. Hypokalemia may occur following operations, most often on the gastrointestinal tract, due usually to a combination of circumstances which result in inadequate intake and excessive loss of K. It is favored by a prolonged preoperative period of undernutrition, e.g., due to pain, anorexia, or nausea, and by inadequately corrected preoperative dehydration, e.g., due to vomiting, diarrhea, fistula, or diuresis. Such individuals are in a state of water, Na, Cl, and K depletion, but the serum K concentration is usually within normal limits (due to corresponding water loss and, often, to impaired kidney function).

During the first twenty-four to forty-eight hours following major operations, there is a rather consistent decrease in urine volume, Na, and Cl, and an increase in urinary K in excess of the amount anticipated on the basis of the negative N balance. These phenomena, which occur also after various types of trauma and other forms of stress, are at least in part manifestations of stimulation of adrenal cortical secretion and of antidiuretic hormone. Eosinopenia and increased urinary excretion of adrenocortical hormones (corticoids) occur simultaneously during this period. Aldosterone may possibly be involved, but increased secretion is not quite coincidental with the electrolyte changes. This physiological mechanism would tend to aggravate a pre-existing state of K depletion (i.e., vomiting, diarrhea, starvation, etc.). If, on this background, other factors are superimposed which cause loss of K, hypokalemia may develop. These include: postoperative food restriction (protein and K) (p. 385); tube drainage of the stomach and/or intestine (p. 386); vomiting (p. 386); and parenteral administration of large quantities of NaCl and glucose solutions (p. 385).

Similar electrolyte changes occur after bilateral adrenalectomy. It is likely, therefore, that some other mechanism is involved. This mechanism could very likely be renal and could involve such phenomena as renal circulation, glomerular filtration rate and tubule permeability. In the normal patient with intact adrenals, it is likely that the increase postoperatively of adrenocortical activity is to a large extent "permissive" as far as the electrolyte changes are concerned.

Persistence and aggravation of K deficiency after operations may lead to troublesome and at times serious consequences. This condition should be suspected under the following circumstances: (1) in the presence of clinical manifestations of hypokalemia (symptoms; ECG changes); (2) when hypochloremic alkalosis appears postoperatively; (3) when a pre-existing hypochloremic alkalosis does not respond satisfactorily to administration of NaCl solution.

Renal Disease and Malfunction.
Several variable factors influence the serum K concentration in diseases of the kidney. In the absence of losses of gastrointestinal fluids, the kidney must be responsible for removing any excess K that enters the small extracellular pool (70 mmol.) from the large intracellular compartment (3500 mmol.). Movement of K in this direction may occur in dehydration, protein and carbohydrate undernutrition (inadequate intake), and increased adrenocortical function in response to stress, all of which may be present in patients with renal disease. Under such circumstances, impaired excretion of K may cause a rapid increase in the serum K concentration. For example, it has been calculated that a normal adult man, deprived of all food, loses about 80 g. protein daily. This involves the loss simultaneously of approximately 240 ml. intracellular fluid, containing 36 mmol. K, which, if retained in the extracellular compartment, represents an increment of about 50 per cent in the total extracellular K.

In general, hyperkalemia tends to occur in conditions of marked oliguria or anuria, regardless of the cause. It is seen in the oliguric stage of acute glomerulonephritis and acute tubular necrosis, in the terminal stage of chronic glomerulonephritis, in advanced renal failure due to destructive kidney lesions or to urinary obstruction, and in renal failure secondary to dehydration or shock. In chronic renal failure, the serum K concentration may increase due to acidosis (p. 385), even in the presence of adequate urine volume. As a rule, however, it tends to remain within normal limits.

Patients with renal failure and oliguria may present normal serum K concentrations or, occasionally, hypokalemia. This may result from continuing large extrarenal losses incident to vomiting, diarrhea, tube drainage of the gastrointestinal tract, or the use of cation exchange resins or the artificial kidney. This tendency toward lowering of the serum K concentration in acute renal failure is seen most frequently perhaps in association with severe alkalosis due to vomiting or administration of excessive quantities of alkali (p. 423). It may be contributed to, also, by expansion of the previously contracted extracellular fluid volume by administration of K-free NaCl solutions. In types of kidney disease with adequate urine output, renal excretion of K may suffice to prevent its accumulation in the extracellular fluid, with the development of progressive depletion of K associated with a normal serum K concentration. Hypokalemia may occur under certain circumstances. It is seen during the polyuric phase of acute tubular necrosis and also in chronic renal failure, due to one or more of the following causes: (*a*) excessive loss in the urine, either spontaneous or induced by diuretic agents; (*b*) large extrarenal losses, as by vomiting, diarrhea, or the use of cation exchange resins; (*c*) unmasking of K deficit by expansion of the extracellular fluid volume with K-free solutions, e.g., NaCl solution.

Hypokalemia has been reported in certain cases of renal tubular acidosis, associated with hyperchloremia and alkaline urine (p. 538). In such cases it is attributable to excessive renal tubular excretion of K^+ owing to a decreased capacity for absorbing HCO_3^- ions. Similar findings may be obtained in patients with the Fanconi syndrome (p. 537).

Certain cases of chronic renal disease referred to as "potassium-losing nephritis" have been demonstrated to be instances of hyperaldosteronism. It is recognized that severe K deficiency may result in rather characteristic morphologic changes in the renal tubular epithelium, as well as in reduction of renal plasma flow and glomerular filtration rate, with hyposthenuria; it is probable that in such cases the renal lesion and functional impairment are the result rather than the cause of the hypokalemia.

Familial Periodic Paralysis. In this disorder, attacks of paralysis are associated with sudden drops in serum K concentration, at times to very low levels (e.g., below 2 mmol./liter). Urinary excretion of K decreases simultaneously, which, in the absence of evidence of expansion of the extracellular fluid volume, indicates that large amounts of K must have passed suddenly from the extracellular fluids into the cells. The paralytic attacks may be induced in such cases by administration of glucose, insulin or epinephrine, which promote the passage of K in this direction.

Miscellaneous. A fall in circulating potassium occurs in hypothermia. On the other hand, an apparent rise in blood potassium, in conditions with increased platelet counts, may not reflect the state of affairs in the intact patient, since the extra potassium could result from disruption of platelets after the blood is collected.

ABNORMAL BLOOD VOLUME

Pathologic variations in total blood and plasma volumes may be dependent upon primary changes in the number and volume of the corpuscles. Thus, in polycythemia vera the enormous increase in red blood cells may result in a marked increase in total blood volume with a normal, increased or decreased plasma volume. On the other hand, a marked diminution in the number of red blood cells may be accompanied by a decrease or normal total blood volume and a normal, increased or decreased plasma volume. Moreover, in the presence of increased acidity of the blood, the corpuscles tend to swell, due to the imbibition of water, at the expense of the plasma volume; decreased acidity of the blood tends to produce the opposite effect.

Increased Blood Volume. An actual increase in total blood volume in relation to body weight or surface area is seldom encountered clinically. It occurs in polycythemia vera, at times in chronic leukemias before the development of anemia, occasionally during periods of water intoxication in diabetes insipidus and, similarly, in experimental water intoxication produced by excessive water administration plus the administration of vasopressin. In the last condition, the increase is due to a primary increase in plasma volume. A syndrome highly suggestive of water intoxication has been reported in very heavy beer drinkers. Beer provided a large fraction of the daily caloric requirement but it contains very little Na, hence the dietary intake of Na was markedly deficient. There was a well-marked fall in the serum Na level. A temporary increase in total blood volume, due to an increase in plasma volume, may occur immediately following the administration of hypertonic solutions (e.g., salt, sugars) to patients with edema. This is due to a temporary transfer of excessively large quantities of fluid from the tissues to the blood stream; as a result of the usually prompt diuretic response, however, the plasma volume returns to normal or may actually decrease for a time if excessively large amounts of fluid are eliminated. A slight increase in blood volume has been reported occasionally in patients with essential hypertension, Raynaud's disease thromboangiitis obliterans, hyperthyroidism, the nephrotic syndrome, and congestive heart failure. However, these conditions are rarely accompanied by any significant deviation from the normal in the absence of complicating factors, the majority of which tend to lower rather than to raise the plasma volume. Elevated blood and plasma volumes have been reported in patients with splenic anemia and portal cirrhosis.

Decreased Blood Volume. Diminution in blood volume may be due to a decrease in the volume of corpuscles or of plasma or both Immediately after a severe hemorrhage the blood volume is decreased due to loss of both cells and plasma. The plasma volume is quickly restored, particularly if fluids are administered in adequate amount and, after the first twenty-four hours, the plasma volume may be actually greater than normal. However, the total blood volume is not restored for some time, depending upon the severity of the blood loss. This restored plasma is usually deficient in protein, the plasma protein concentration usually being restored to normal before the red blood cells and hemoglobin concentration. In the majority of chronic anemias the total blood volume is usually normal or slightly subnormal, the plasma volume being normal or increased. In pernicious anemia the total blood volume is low, the plasma volume being usually normal or subnormal. A decrease in total blood volume has also been observed in hemolytic jaundice, the plasma volume usually being somewhat increased.

A marked decrease in circulating blood volume is a prominent feature of the shock syndrome, which may be due to a variety of causes. Except in the presence of hemorrhage, this decrease is usually due almost entirely to a decrease in plasma volume, contributed to by the transudation of fluid, and usually also some protein, through the abnormally permeable capillaries, and by sequestration of blood in large capillary beds (e.g., splanchnic). This results at times in a state of hemoconcentration, as evidenced by high hematocrit values and red blood cell counts. Subnormal total blood and plasma volumes, with hemoconcentration, may also be observed under the following conditions: (*a*) prolonged water restriction; (*b*) excessive water loss (diuresis, diarrhea, vomiting, intestinal fistula, pancreatic and biliary fistula, sweating); (*c*) acidosis, in which the loss of plasma water may be due in part to diuresis and in part to transfer of water from the plasma to the cells; (*d*) advanced diabetes mellitus, in which the low plasma volume may be due in part to diuresis and in part to acidosis; (*e*) excessively high external temperatures, in which large quantities of water may be lost through the skin and lungs; (*f*) occasionally in uremia, due to such complicating factors as acidosis, vomiting, diuresis, diarrhea, etc.; (*g*) ether anesthesia; (*h*) adrenal cortical insufficiency, in which the low plasma volume results mainly from the excessive loss of sodium and chloride ions, with water, in the urine, or their passage from the plasma into the tissues (p. 723). It should be emphasized that, with the possible exception of the shock syndrome and adrenal cortical insufficiency, marked decrease in plasma volume occurs only when the factors which operate to produce it act over long periods of time. As was suggested above, any tendency toward the loss of fluid from the blood plasma is usually promptly counteracted by the passage of fluid from the interstitial fluid compartment into the vascular compartment. Since the reserve supply of fluid in the interstitial tissues is about three times as great as the normal plasma volume, it is obvious that significant diminution in the latter usually occurs only with extensive depletion of the body fluids. Adrenal cortical insufficiency and clinical Addison's disease constitute exceptions to this rule because of the fact that in the absence of adequate amounts of cortical hormones the organism is unable to maintain the normal concentration of sodium ion in the plasma, its loss involving necessarily the simultaneous loss of considerable amounts of water (p. 379).

TRANSUDATES AND EXUDATES

The chemical examination of abnormal accumulations of fluid in the subcutaneous tissues and serous cavities frequently yields information of diagnostic importance. The chemical constituents of edema fluids are derived chiefly from the blood plasma, the composition of such fluids being dependent to some extent upon the permeability of the capillaries in the region involved, a factor which varies considerably in normal individuals in different vascular areas, and pathologically in response to abnormal stimuli. Alterations in the chemical composition of cerebrospinal fluid are considered elsewhere (p. 835). No attempt will be made to discuss in detail the chemical composition of edema fluid, attention being directed particularly to alterations which are of significance from the standpoint of diagnosis.

Specific Gravity. Crystalloids, being relatively readily diffusible, exist in practically the same concentration in edema fluids in different situations formed under similar conditions; such fluids, however, vary in specific gravity due to differences in their protein content, which are dependent upon variations in the permeability of the capillaries in different portions of the body. Normally, the lymph coming from the lower extremities contains 2 to 3 per cent of protein, that coming from the intestines 4 to 6 per cent, while that from the liver contains 6 to 8 per cent. Similarly, it has been found that, in general, the quantity of protein in edema fluid in different localities varies in decreasing order as follows: (1) pleural, (2) peritoneal, (3) cerebrospinal, (4) subcutaneous. Some observers would vary this order slightly, reversing the positions of peritoneal and pleural effusions and of subcutaneous and cerebrospinal fluid. The statement is frequently made that the specific gravity of transudates is usually below 1.015 and that of exudates above 1.018; however, in view of the fact that the crystalloid content, with the exception of minor differences due to the existence of a Donnan equilibrium (p. 353), varies but slightly, the specific gravity varying more or less in direct proportion to the concentration of protein, there must be many exceptions to this rule since under identical conditions the protein concentration of edema fluids in various situations differs, as stated above. However, with these reservations in mind, statements made with regard to the difference in specific gravity of transudates and exudates may be accepted as a fairly satisfactory working basis for distinguishing between these two types of edema fluid. Values as high as 1.035 may be observed in inflammatory pleural and peritoneal exudates and values as low as 1.005 in subcutaneous transudate fluid.

Protein. Under normal conditions the fluid in most of the tissue spaces and in the serous cavities may be regarded as almost protein-free filtrates of the blood plasma. The cerebrospinal fluid, which is the only one normally present in amounts large enough to enable quantitative analysis to be made with any degree of accuracy, normally contains 15 to 45 mg. of protein per 100 ml. If these figures may be accepted as representative of tissue fluids throughout the body, it is evident that their normal protein content is practically negligible as compared with that of blood plasma (6 to 8 g. per 100 ml.).

The extremely low protein content of these fluids is due to the relatively poor diffusibility of protein through normal capillaries. Since inflammatory processes are associated with a marked increase in the permeability of the capillaries in the involved area, the degree of such increase being dependent upon the intensity of the process, the protein content of inflammatory exudates is relatively high. In purulent exudates resulting from a severe inflammatory process, as illustrated by empyema, the protein content of the serous portion of the fluid, obtained by centrifugation, may be approximately the same as that of the blood plasma. In the case of exudates resulting from inflammatory processes of lesser intensity, such as tuberculous pleurisy, tuberculous peritonitis, pneumonic pleurisy, meningitis, etc., the total protein concentration usually ranges from 0.1 to 5 g. per 100 ml., being usually lower in the cerebrospinal fluid than in the peritoneal and pleural exudates. In contradistinction to the relatively high values observed in inflammatory fluids, the protein content of noninflammatory edema fluids or transudates is relatively extremely low since their pathogenesis is usually independent of alterations in capillary permeability. Thus the protein content of subcutaneous edema fluid is frequently below 0.1 g. per 100 ml., that of pleural and peritoneal transudates occurring as a result of myocardial failure, the nephrotic syndrome, uncomplicated cirrhosis of the liver, or the like, being correspondingly low (0.1 to 1.0 g. per 100 ml.). It must be realized, however, that, particularly in the case of pleural and peritoneal transudates, if the effusion has existed for some time, water may be reabsorbed more rapidly than solids, resulting in a slowly increasing concentration of protein which may eventually approach that of true exudates. It has been estimated that in the presence of a normal serum protein concentration, the presence of increased capillary permeability can be assumed with some degree of certainty only if the protein content of the edema fluid exceeds 4.1 g. per 100 ml.

A relatively high protein content has frequently been found in the subcutaneous edema fluid of acute nephritis, suggesting its dependence upon generalized capillary injury rather than upon other forces which are believed to be responsible for the production of edema in chronic nephritis and myocardial failure. The protein concentration of blister fluid is also very high. Of particular interest is the observation that the fluid of angioneurotic edema contains large amounts of protein, the inference being that this condition is associated with a marked increase in capillary permeability.

Albumin usually constitutes by far the largest part of the protein present in edema fluid, globulin occurring in small amounts and fibrinogen being seldom observed except in acute inflammatory exudates. The ability of these proteins to pass through the capillary wall is dependent upon the size of their molecules, those of greater size passing with greater difficulty. Therefore, in transudates, albumin may be present alone or may be accompanied by a small amount of globulin; fibrinogen is usually present only in inflammatory exudates, the last named being indicative of an intense inflammatory reaction. Pneumococcus exudates appear to be particularly rich in fibrinogen, which is at times present in such high concentrations that the fluid may coagulate spontaneously.

In some cases a protein substance closely resembling mucin has been found in inflammatory exudates. Whereas the proteins mentioned above are believed to be derived from the blood plasma, this substance (mucin) appears to be a product of the inflamed cells. It has been observed frequently in joint fluids which, whether transudates or exudates, usually contain larger quantities of protein than similar fluids in other situations, the concentration of mucin being quite constant and apparently independent of the nature of the pathologic process.

Glucose. With the exception of cerebrospinal fluid, the fluid of the tissues and that of the pleura, peritoneum, and other serous cavities contains sugar in practically the same concentration as that of the blood. The sugar present in these fluids appears to be chiefly glucose. Alterations in the glucose concentration of the blood are reflected in parallel changes in the glucose content of the various tissue fluids. The glucose content of pleural and peritoneal transudates is usually practically the same as that of the blood. The glucose content of inflammatory exudates is relatively low due to the destruction of glucose by the action of bacteria and cells present in the fluid, the degree of this reduction being dependent somewhat upon the intensity of the inflammatory process. The significance of this observation in the differential diagnosis of meningeal and cerebrospinal lesions is discussed in the section on cerebrospinal fluid (p. 838).

Chloride. The factors which determine the relationship between the chloride content of blood plasma and tissue fluid are discussed elsewhere (p. 353). The chloride content of noninflammatory edema fluid, or transudates, is higher than that of the blood plasma, ranging from 123 to 128 mmol. per liter, the difference being due to the existence of a Donnan equilibrium dependent upon the higher concentration of protein in the plasma as compared with normal tissue fluid and transudates. The chloride content of inflammatory exudates, which are relatively rich in protein, is lowered, approaching that of blood plasma, the degree of diminution varying roughly directly with the increase in the concentration of protein, in accordance with the laws governing the concentrations of readily diffusible substances on two sides of a semipermeable membrane under such circumstances (see p. 354). The chloride content of pleural effusions in pneumococcus pneumonia is particularly low because of the low chloride concentration of the blood plasma.

Lipid. Neutral fat and fatty acids are not usually present in transudates or inflammatory exudates. A small amount of lecithin, varying from

20 to 100 mg. per 100 ml., may practically always be demonstrated, existing partly in the free state and partly in combination with protein. Cholesterol does not appear to be a constant constituent of transudate fluid but is practically always present in inflammatory exudates, particularly those of long standing, being probably derived in part from degenerative changes either in the cells present in the fluid or in those lining the serous sac or the abscess cavity. The cholesterol content of such long-standing effusions may decrease markedly following repeated tapping, values ranging from 1 to 4.5 g. per 100 ml. having been observed to fall to 20 to 50 mg. following repeated aspiration. In some cases exhibiting high cholesterol values fat is also present, having been observed particularly in tuberculous pleural and peritoneal effusions. Capillaries appear to be permeable to cholesterol to about the same extent as to protein, this fact accounting for the very low cholesterol content of transudate fluids. This is true usually even in cases of the nephrotic syndrome, in which the concentration of cholesterol in the blood plasma may be enormously increased. In inflammatory exudates associated with increased capillary permeability, the lipid content of the edema fluid roughly parallels its protein content.

Effusions which contain lipids in sufficient quantity to cause a milky appearance are designated chylous, chyliform and pseudochylous, depending upon their fat content and their pathogenesis. True chylous effusions are due to the escape of chyle from a ruptured or obstructed thoracic duct into the pleural or peritoneal spaces. The fat content of chylous fluid naturally varies with the quantity of fat ingested, being modified to a certain extent by processes of effusion or resorption occurring in the pleura or peritoneum. Values from 0.05 to 3.85 g. per 100 ml. have been reported. The concentration of cholesterol and lecithin, although increased, is usually low in proportion to that of neutral fat and fatty acids. Such fluids usually contain relatively high concentrations of protein, this factor also being largely dependent upon the composition of the diet, and spontaneous coagulation occurs not infrequently.

The term "chyliform effusion" is applied to fluids which may be identical in appearance with those described above, the fat content of which, however, is due not to chyle but to fatty degeneration of the cells present in the effusion or of those lining the walls of the cavity involved. Differentiation between chylous and chyliform effusions is frequently difficult since in both instances the turbidity is probably due to emulsified fat. However, in some cases the cholesterol and lecithin content of chyliform effusions is relatively higher and the fat content relatively lower than those of true chylous fluids. The term "pseudochylous effusion" is applied to fluids which may be turbid or milky in appearance but contain little or no fat, the turbidity being due chiefly to lecithin and cholesterol. It has also been demonstrated that albumin in a highly dispersed state may impart a milky appearance to such fluids. Subcutaneous pseudochylous fluid in the lower extremities and the scrotum may result from obstruction of lymphatic vessels by filariae. Similar fluids in the pleural and peritoneal cavities have been observed in lipoid nephrosis and in chronic glomerulonephritis with superimposed nephrotic lesions. They have also been observed in association with carcinoma of the peritoneum and in tuberculous pleurisy and peritonitis. Pseudochylous fluids may appear to be relatively clear when first removed, the turbidity and milky appearance increasing upon cooling. Relatively large amounts of protein may be present, varying from 0.1 to 4.2 g. per 100 ml., and spontaneous coagulation, though not so commonly observed as in true chylous effusions, may nevertheless occur.

Other Constituents. Creatinine, uric acid, and particularly urea are present in exudates and transudates in practically the same concentrations as in the blood. Some observers have found that in nephritis edema fluid may at times contain more nonprotein nitrogen than the blood, an observation which may be of significance but the practical importance of which has not been demonstrated. The calcium content of transudates, ranging from 1.12 to 1.37 mmol. per liter in the case of fluid with a low protein count, apparently represents the normal diffusible fraction of serum calcium. With increasing values for protein in both transudates and exudates the calcium concentration increases, the increase representing a nondiffusible fraction, which is probably in combination with protein. Transudates contain approximately the same concentration of inorganic phosphorus as the blood serum, and the concentration of bicarbonate is somewhat higher and that of sodium somewhat lower than in serum. The concentration of magnesium in transudate fluids averages about 65 per cent and that of potassium about 80 per cent of that in the serum. In the presence of increased protein concentration in the fluid, the concentration of magnesium increases.

Small amounts of bilirubin may be present in transudate fluid in the presence of hyperbilirubinemia. Larger amounts may be found in exudate fluids under such circumstances in concentrations varying roughly in proportion to the concentration of protein in the fluid. Bilirubin has been demonstrated in pleural and peritoneal effusions in patients with congestive heart failure and cirrhosis of the liver without hyperbilirubinemia (degradation of extravasated hemoglobin). The diagnostic significance of enzyme studies in these fluids is discussed elsewhere (p. 573).

References

Berliner, R. W.: Relationship between acidification of the urine and potassium metabolism. Am. J. Med., *11*:274, 1951.
Black, D. A. K.: Current concepts of potassium metabolism. J. Pediat., *56*:814, 1960.
Bland, J. H.: Clinical Recognition and Management of Disturbances of Body Fluids. Philadelphia, W. B. Saunders Company, 1956.
Christensen, H. N.: Diagnostic Biochemistry. New York, Oxford University Press, 1959.
Darrow, D. C., and Hellerstein, S.: Interpretation of certain changes in body water and electrolytes. Physiol. Rev., *38*:114, 1958.
Davis, J. O.: Mechanisms of salt and water retention in congestive heart failure. Am. J. Med., *29*:486, 1960.
Dingman, J. F.: Hypothalamus and the endocrine control of sodium and water metabolism in man. Am. J. Med. *235*:79, 1958.
Edelman, I. S., and Leibman, J.: Anatomy of body water and electrolytes. Am. J. Med., *27*:256, 1959.

Farrell, G.: Regulation of aldosterone secretion. Physiol. Rev., *38:*709, 1958.

Gamble, J. L.: Companionship of water and electrolytes in the organization of body fluids. Lane Medical Lectures, California, Stanford University Press, 1951.

Josephson, B.: Chemistry and therapy of electrolyte disorders. Springfield, Ill., Charles C Thomas, 1961.

Kleeman, C. R., and Vorherr, H.: Water metabolism and the neurohypophysial hormones. *In* Bondy, P. K. (ed.): Duncan's Diseases of Metabolism. Ed. 6. Philadelphia, W. B. Saunders Company, 1969, p. 1103.

Leaf, A., and Santos, R. F.: Physiologic mechanisms in potassium deficiency. New Eng. J. Med., *264:*335, 1961.

Merrill, J. P.: Electrolyte changes in renal failure. Metabolism, *5:*419, 1956.

Moore, F. D., ed.: Symposium on water and electrolytes. Metabolism, *5:*367, 1956.

Moyer, J. H., and Fuchs, M., eds.: Edema: Mechanisms and Management. Philadelphia, W. B. Saunders Company, 1960.

Scribner, B. H., and Burnell, J. M.: Interpretation of the serum potassium concentration. Metabolism, *5:*468, 1956.

Snavely, J. R.: Hepatic factors in salt and water metabolism. Am. J. Med. Sci., *223:*100, 1952.

Strauss, M. B.: Body Water in Man. Boston, Little, Brown and Company, 1957.

Tietz, N. W.: Electrolytes. *In* Tietz, N. W. (ed.): Fundamentals of Clinical Chemistry. Philadelphia, W. B. Saunders Company, 1970, p. 612.

Wesson, J.: Glomerular and tubular factors in the renal excretion of sodium. Medicine, *36:*281, 1957.

Weston, R. E., et al.: Homeostatic regulation of body fluid volume in nonedematous subjects. Metabolism, *9:*157, 1960.

White, A. G.: Clinical Disturbances of Renal Function. Philadelphia, W. B. Saunders Company, 1961.

Wills, M. R.: Biochemical Consequences of Chronic Renal Failure. Aylesbury, Harvey Miller and Medcalf, Limited, 1971.

Chapter 14

HYDROGEN ION CONCENTRATION IN BODY FLUIDS

The topics to be discussed in this chapter have been described in the past under the general titles of "Neutrality Regulation," "Acid-base Balance" or "Regulation of Hydrogen Ion Concentration." None of these is really satisfactory. The first two seem to imply that body fluids are at neutral pH, which is certainly not the case. The last title is perhaps somewhat more apt, but unfortunately it is possible that the hydrogen ion concentration of body fluids is determined not directly but secondarily to the regulation of the concentration of other ions, of which HCO_3^- is a good example.

Under normal conditions the pH of extracellular fluids usually does not vary for clinical purposes beyond the range 7.35 to 7.45 and is maintained approximately at 7.4 (40 nmol. H⁺ ion/liter). If the pH of blood plasma falls below 7.35, there is a state of *acidemia*; above 7.45, there is *alkalemia*. The body has compensatory mechanisms against these changes, which in the presence of a tendency toward acidemia or alkalemia maintain the pH within the normal range. The persistence of such a compensated tendency is correctly referred to as *acidosis* or *alkalosis*. Nevertheless, those two terms are used very loosely in the clinical situation and frequently should really be replaced by "acidemia" or "alkalemia," respectively. For the purpose of this chapter, *acidosis* and *alkalosis* are used in the loose clinical sense.

Inasmuch as large amounts of H⁺ are continually contributed to these fluids from intracellular metabolic reactions, preservation of this degree of constancy requires that they be removed from the fluids effectively and promptly. The mechanisms concerned maintain a state of equilibrium between production (and introduction) and removal of H⁺ ions.

Virtually all precise information concerning mechanisms affecting the

H⁺ ion concentration in body fluids is derived from studies of the blood. In view of the ready diffusibility across the capillary wall of most of the important factors concerned, relevant aspects of the structure of the blood plasma may be regarded as reflecting quite accurately the status in this connection of the other major extracellular fluids. However, the pH of interstitial fluid is probably somewhat lower than that of blood plasma, since it occupies a position intermediate between the latter and the site of production of acids within the cells. There is increasing information concerning the intracellular pH. This undoubtedly differs considerably in different tissues, and in different parts of the same cell; it is generally lower in most cells, 6.8 or less.

Intracellular pH is measured by microinjection of indicators or by the insertion of microelectrodes. There have been innumerable difficulties with both these techniques. The thickness of the microelectrode, for example, can affect the pH measurement obtained. It appears likely, however, that the pH of the mitochondrial matrix is probably very similar to that of cytoplasm. There may be loci of heterogeneity near the cristae. The pH in the cell nucleus appears to be appreciably higher. One measurement, admittedly of a malignant cell, gave a value of 7.6.

It has been argued that the whole concept of pH is not really applicable to the microenvironments within the cell. At pH 7.00, there would be only four H⁺ ions free in the matrix fluid of a single mitochondrion. In other words, a pH measurement in this situation would really be a statistical average of the number of free protons (H⁺). On the other hand, there are about 5×10^5 HCO_3^- ions and about one-tenth of this number of H_2CO_3 molecules. Measurements of the ratio of these ions and these molecules have indicated calculated pH values closely similar to those found by direct pH measurement.

Many authorities consider that the hydrogen ion concentration of tissue fluids should be expressed as nanomoles (nmol.) per liter, and not as pH. The average extracellular pH of 7.4 would correspond to 40 nmol. H⁺/l. This assumes an activity coefficient of 1.0, which is an approximation.

Acids and Bases. An acid is defined as a substance (ion, molecule) that yields H⁺ ions (protons) in solution, and a base is one that accepts H⁺ ions. Accordingly, whereas H_2CO_3 is an acid, dissociating into H⁺ and HCO_3^- ions, its anionic component, HCO_3^-, is a base. Other examples are:

$$\begin{array}{cc} \text{Acid} & \text{Base} \\ HCl \rightleftarrows H^+ & +\ Cl^- \\ CH_3COOH \rightleftarrows H^+ & +\ CH_3COO^- \\ H_2PO_4^- \rightleftarrows H^+ & +\ HPO_4^{2-} \end{array}$$

HCl is a strong acid by virtue of its extensive dissociation into H⁺ and Cl⁻ ions; accordingly, Cl⁻ is an extremely weak base, because it has very little capacity for combining firmly with H⁺ ions. On the other hand, such anions as HCO_3^-, HPO_4^{2-}, $H_2PO_4^-$, and protein are comparatively strong bases because they have a relatively strong affinity for H⁺ ions, thereby forming weak acids (i.e., relatively slight dissociation). The free H⁺ ions donated by strong acids would, to a large extent, become fixed in the presence of the strong bases. In other words, HCO_3^-, HPO_4^{2-}, $H_2PO_4^-$ and protein act as buffers against increased concentration of free H^+ ions.

The concentration of buffer anions, the effective bases of the extracellular fluids, is represented by the difference between the concentrations of total anions and of anions which are weak bases, e.g., chloride.

The introduction of acids stronger than carbonic to the extracellular fluids results in decrease in bicarbonate, the major extracellular buffer anion for these acids, in an amount roughly proportional to the amount of acid added. However, bicarbonate is not the only available defense (as a buffer) against the introduction of additional acid. Hemoglobin is the most important buffer for carbonic acid, and the defenses of the organism against acidosis include also phosphate and protein anions, as well as bicarbonate. Nevertheless, since in the case of accumulation of acids other than carbonic the bicarbonate buffer system is in equilibrium with the other buffer systems, changes in the plasma bicarbonate concentration may be regarded as reflecting changes in the capacity of the organism to resist further additions of these acids. This is not true of conditions of primary increase or decrease in carbonic acid (respiratory acidosis and alkalosis, p. 415). "The buffer base" of the plasma may be calculated as approximately equal to the sum of the concentrations of the cations (Na^+, K^+, Mg^{2+}, Ca^{2+}) minus the concentration of Cl^-. This is so weak a base that it can be disregarded for practical purposes. For the calculation, all concentrations are expressed as mEq./l.

H_2CO_3 is the chief acid formed in the course of cellular oxidations, about 10 to 20 or more moles being produced daily (average 13). Sulfuric (oxidation of S of protein) and other acids, e.g., lactic and β-hydroxybutyric, are produced in a total amount of about 80 to 120 millimoles daily under ordinary conditions. The metabolic products of foodstuffs provide an excess of acids over bases. Both the H^+ ions and the anions produced by these acids must be disposed of, i.e., ultimately excreted from the body, in such manner that their temporary sojourn in the extracellular fluids does not unduly affect the pH under normal circumstances. The means whereby these ends are accomplished determine the H^+ ion concentration in tissue fluids. These may be outlined as follows:

1. Dilution. The acids introduced into and formed in the body are distributed throughout the extracellular fluid volume. Although this may not properly be regarded as a regulatory mechanism, entrance of a given amount of acid into a smaller volume of fluid, as in conditions of severe dehydration, results in relatively greater rise in H^+ concentration and decrease in effective buffer base.

2. Buffer Systems. Restriction of pH change in body fluids.

3. Respiration. Regulation of excretion of CO_2 and, therefore, regulation of the H_2CO_3 concentration in extracellular fluids.

4. Renal Mechanism. Ultimate excretion of excess "acid" or "base"; ultimate regulation, therefore, of the concentration of H^+ and HCO_3^- ions in the extracellular fluids.

PHYSIOLOGICAL BUFFER SYSTEMS

The capacity of the extracellular fluids for transporting hydrogen ions from the site of their formation (cells) to the site of their disposal (e.g., lungs, kidneys) without undue change in pH is dependent chiefly upon the presence of efficient buffer systems in these fluids and in the erythrocytes. Each buffer system consists of a mixture of a weak acid, HA, and its salt, BA (p. 405). The most important of these are as follows:

$$\textit{Plasma:} \quad \frac{H_2CO_3}{BHCO_3}, \frac{H \cdot \text{protein}}{B \cdot \text{protein}}, \frac{BH_2PO_4}{B_2HPO_4}, \frac{H \cdot \text{organic acid}}{B \cdot \text{organic acid}}$$

$$\textit{Erythrocytes:} \quad \frac{H_2CO_3}{BHCO_3}, \frac{HHb}{BHb}, \frac{HHbO_2}{BHbO_2}, \frac{BH_2PO_4}{B_2HPO_4}, \frac{H \cdot \text{organic acid}}{B \cdot \text{organic acid}}$$

The buffer systems in the interstitial fluids and lymph are much the same as in the blood plasma, except that proteins are generally present in much smaller quantities. The main buffer systems in intracellular fluids are also qualitatively much the same as in the plasma, but the cell fluids contain much higher concentrations of protein and phosphate.

Although all of these buffer systems are operative to a certain extent within the physiological range of pH values, only a few of them exist in sufficiently high concentrations to be of distinct quantitative significance in the determination of H^+ ion concentration. In the blood plasma, the bicarbonate and plasma protein systems, and in the erythrocytes, the bicarbonate and hemoglobin systems play the most important roles in this connection.

Buffer Action of Hemoglobin.

The buffering capacity of Hb, as of any protein, depends upon the number of ionized buffering groups, acidic (carboxyl) or basic (amino, guanidino and imidazole), which varies with the pH of the medium. Within the pH range 7.0 to 7.8, most of the buffering action of Hb is due to the imidazole groups of histidine.

The degree of dissociation of a buffering group in a protein, which determines its buffering capacity, is influenced by adjacent groups in the molecule. In the case of the imidazole group of histidine, which is intimately associated with the iron of hemoglobin, its strength as a buffer is affected by changes in the degree of oxygenation of hemoglobin. When oxygen is removed, the imidazole groups are rendered more basic, consequently removing hydrogen ions from solution and becoming electrically positive (imidazolium). This effect is reversed with increased oxygenation of the hemoglobin molecule. This reaction indicates not only that the buffering capacity of Hb is related to the degree of oxygenation, but also that its ability to accept or liberate oxygen is influenced by the acidity of the medium. A decrease in acidity facilitates oxygenation of Hb and an increase in acidity facilitates liberation of oxygen, producing characteristic changes in the oxygen dissociation curve of Hb (p. 438).

The implications arising from these considerations may be summarized as follows:

1. Oxygenation of Hb increases its acidity, causing it to give up hydrogen ions to the medium, i.e., oxyhemoglobin is a stronger acid than reduced hemoglobin.

2. Reduction of oxyhemoglobin (i.e., deoxygenated Hb) decreases its acidity, removing hydrogen ions from the medium.

3. Introduction of an acid, e.g., H_2CO_3, into a medium containing oxyhemoglobin facilitates loss of oxygen, i.e., formation of reduced hemoglobin. The latter, being a weaker acid than oxyhemoglobin, is better able than oxyhemoglobin to counteract the acidifying effect of the added acid (p. 404).

4. Addition to the blood of CO_2 (in the tissues) or O_2 (in the lungs), in the course of normal metabolism and respiration, results in release of significant amounts of H^+ ion; in the case of CO_2 from H_2CO_3, in the case of O_2 from hemoglobin. Under physiological conditions, hemoglobin serves as a H^+ acceptor when H_2CO_3 acts as a H^+ donor. In the lungs, the additional hydrogen ions added to the blood by oxyhemoglobin would be exactly balanced by the loss of carbon dioxide in expired air if the respiratory quotient were approximately 0.7. However, it is usually higher than this, and therefore hydrogen ions derived from other sources are also neutralized.

The buffer mechanisms involved in the case of H_2CO_3, the anhydride of which (CO_2) is volatile, differ from those for the stronger, nonvolatile, fixed acids (sulfuric, phosphoric, lactic, etc.).

Buffer Systems for H_2CO_3 (CO_2). H_2CO_3 is buffered chiefly by the imidazole group of Hb in the erythrocytes and by proteins in the plasma. Hb exerts by far the greater effect in this connection because its capacity for combination with H^+ ions greatly exceeds that of the plasma proteins. The over-all reactions may be indicated as follows:

(Plasma)
$$CO_2 + H_2O \rightleftharpoons H_2CO_3$$
$$H_2CO_3 + \text{protein} \rightleftharpoons HCO_3^- + H^+ \cdot \text{protein}$$

(Erythrocytes)
$$CO_2 + H_2O \underset{\text{anhydrase}}{\overset{\text{carbonic}}{\rightleftharpoons}} H_2CO_3$$
$$H_2CO_3 \rightleftharpoons HCO_3^- + H^+$$
$$H^+ + Hb^- \rightleftharpoons HHb$$
$$O_2Hb^- + CO_2 + H_2O \rightleftharpoons HHb + O_2 + HCO_3^-$$

The effect of addition of CO_2 is to increase both H_2CO_3 and $BHCO_3$, the latter owing to removal of additional H^+ ions by Hb and, to a minor extent, by the plasma proteins. The liberation of oxygen by Hb in the tissues, which occurs simultaneously with the addition of CO_2, increases the buffering capacity of the Hb, as indicated above. The reciprocal exchanges of Cl^- and HCO_3^- ions between the plasma and erythrocytes (chloride shift) constitute an additional important phase of the extremely efficient mechanism for buffering and transporting H_2CO_3 (p. 405) (Fig. 14-1).

CO_2 from Tissues Enters Plasma (p. 440). As indicated below, the largest portion of the added CO_2 diffuses into the erythrocytes. The remainder stays in solution in the plasma, reacting with water as follows:

$$CO_2 + H_2O \leftrightarrows H_2CO_3$$

This reaction occurs slowly; the equilibrium is far to the left, the concentration of dissolved CO_2 being about 1000 times that of H_2CO_3. The small increase in the latter which occurs when CO_2 is added to the plasma ionizes ($H_2CO_3 \rightleftarrows H^+ + HCO_3^-$), the added H^+ ions being buffered by the relatively weak plasma buffer systems (viz., proteins). The pH falls slightly.

CO_2 Enters Erythrocytes (p. 441). The bulk of the added CO_2 diffuses into the erythrocytes chiefly because the reaction, $CO_2 + H_2O \rightleftarrows H_2CO_3$, progresses very rapidly to the right by virtue of the catalytic action of the enzyme, carbonic anhydrase. A small amount of CO_2 remains as such in solution. Considerably more forms carbamino compounds with Hb ($R - NH_2 + CO_2 \rightleftarrows R - NHCOO^- + H^+$). The reaction is facilitated by the simultaneous liberation of oxygen, with an increment in reduced Hb, which is a weaker acid than oxyhemoglobin.

Most of the H_2CO_3 newly formed in the erythrocytes undergoes ionization ($H_2CO_3 \rightleftarrows H^+ + HCO_3^-$). This reaction proceeds to the right chiefly because both ions are removed rapidly, H^+ being buffered by Hb, and HCO_3^- diffusing into the plasma.

Buffering of CO_2 by Hb. Reduced Hb, being a considerably weaker acid than oxyhemoglobin, has a higher H^+-binding capacity than the latter. At an erythrocyte pH of 7.25, 1 mmol. of HbO_2 yields 1.88 mmol. of H^+ ion, whereas 1 mmol. of Hb yields only 1.28 mmol. Consequently, in the tissue capillaries, the liberation of each millimole of O_2 (22.4 ml.) permits binding of 0.6 mmol. more of H^+ ion, permitting the formation of 0.6 mmol. of $BHCO_3$ from 0.6 mmol. of CO_2 (13.4 ml.) with no change in pH. At a

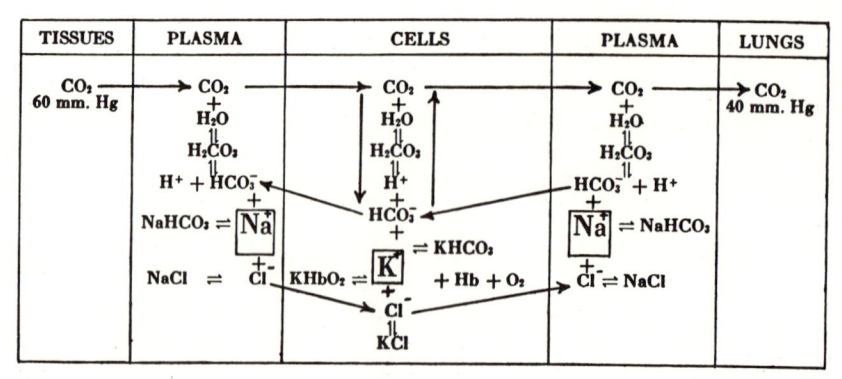

Figure 14-1. Exchange of HCO_3^- and Cl^- between the red blood cells and the plasma in the tissues and lungs (chloride shift). The heavy arrows indicate the direction of the changes in these situations.

respiratory quotient of 0.6, i.e., $\frac{0.6 \text{ mmol. } CO_2 \text{ produced (p. 445)}}{1.0 \text{ mmol. } O_2 \text{ consumed}}$, 0.6 mmol. of H⁺ ions can become available (from 0.6 mmol. H_2CO_3), which will be completely buffered as a result of reduction of Hb. At the usual respiratory quotient of 0.82, therefore, only 0.82 minus 0.6, or 0.22 mmol. of H⁺ ions per mmol. of O_2 consumed must be buffered by other means. Under average conditions of hemoglobin content and saturation and CO_2 transport, over 73 per cent of the total amount of CO_2 entering the blood in the tissues is handled in this manner.

Chloride-Bicarbonate Shift. Because of the great rapidity with which hydration of CO_2 (transformation to H_2CO_3) occurs in the erythrocytes as compared with the plasma, the concentration of HCO_3^- rises in the former faster than in the latter. Inasmuch as the erythrocyte wall is permeable to HCO_3^-, this ion diffuses from the erythrocytes into the plasma. In order to preserve electrical neutrality of the cell and plasma fluids, Cl⁻ passes in the opposite direction; the numbers of negative ions (i.e., Cl⁻ and HCO_3^-) exchanged must be identical. The Cl⁻ ions passing into the erythrocytes are balanced electrically by the K⁺ ions previously balanced by the HCO_3^-, while the latter ions entering the plasma are balanced by the Na⁺ ions previously balanced by the diffused Cl⁻. As a result of these changes, most of the HCO_3^-, although formed originally in the erythrocytes, is actually transported in the plasma (Table 15-4, p. 441).

All of the above reactions are reversed in the lungs, the quantity of CO_2 entering the blood in the tissues passing into the alveoli and the quantity of oxygen lost in the tissues being restored from the alveolar air.

Buffer Systems for Nonvolatile Acids. Neutralization of nonvolatile acids entering the extracellular fluids is accomplished chiefly by the $H_2CO_3/BHCO_3$ buffer system. Such acids (e.g., HCl. H_2SO_4 lactic, etc.) react with $BHCO_3$ as follows:

$$HCO_3^- \text{ (from } NaHCO_3) + H^+ \text{ (from HCl)} \rightleftharpoons H_2CO_3$$

Protein and phosphate buffer systems may also be involved, but to a relatively minor extent.

$$\text{Protein} + H^+ \rightleftharpoons H^+ \cdot \text{Protein}$$
$$HPO_4^{2-} + H^+ \rightleftharpoons H_2PO_4^-$$

These reactions result in diminution of buffer base, which requires restoration by other mechanisms. Bicarbonate is particularly efficient in this connection because (1) it is present in higher concentration than the other buffer salts, and (2) the acid product of the reaction, H_2CO_3, is effectively buffered (p. 403) and is readily disposed of by the lungs (p. 435) by virtue of the volatility of its anhydride, CO_2, rendering the reaction irreversible.

$$H_2CO_3 \rightleftharpoons H_2O + CO_2 \uparrow \text{ (expired air)}$$

When a base (e.g., NH_3) enters the extracellular fluid, it reacts with the acid components of these buffer systems, chiefly H_2CO_3 and $H_2PO_4^-$:

$$H_2CO_3 + NH_3 \rightleftharpoons HCO_3^- + NH_4^+$$
$$H_2PO_4^- + NH_3 \rightleftharpoons HPO_4^{2-} + NH_4^+$$

Net Effect of Buffer Mechanisms. Carbonic acid is by far the most abundantly produced organic acid. The anions of nonvolatile acids, e.g., SO_4^{2-}, enter the extracellular fluid. Because of the relative abundance of bicarbonate in extracellular fluid, more carbonic acid is formed by interaction with extra hydrogen ions resulting from the metabolic oxidation of cysteine and methionine, which produces the SO_4^{2-}. This means that there is a decrease in HCO_3^- concentration, which leads to an increase in hydrogen ion concentration of extracellular fluid, because the ratio of HCO_3^- to H_2CO_3 has decreased (p. 414). The metabolic formation of lactate and other anions of nonvolatile acids results in interchange with extracellular bicarbonate, which also decreases the ratio of extracellular HCO_3^- to H_2CO_3.

Under normal conditions this increase in hydrogen ion concentration is but slight because (1) H_2CO_3, a weak acid (i.e., weakly dissociated), forms relatively few H^+ ions, (2) it is effectively buffered (p. 403), and (3) it is readily eliminated by the lungs as CO_2, the increase in concentration of both H_2CO_3 and H^+ being thereby minimized.

$$CO_2 + H_2O \leftarrow H_2CO_3 \rightleftharpoons H^+ + HCO_3^-$$

The carbonic acid produced in the course of oxidative metabolic processes is thus effectively disposed of, chiefly by the hemoglobin mechanism, with very little change in pH (p. 404). The nonvolatile acids can be buffered efficiently by bicarbonate as long as an adequate amount of the latter is present in the extracellular fluid.

Bone. Some of the Na^+ ions of bone have been shown to interchange with H^+ ions of interstitial fluid. It is very likely, therefore, that the skeletal system can play a part in relation to the concentration of H^+ ions in extracellular fluid. It has also been suggested that the phosphate ions of bone possibly play a part in supplying an appropriate buffer mechanism. In this way, it is possible that parathormone and calcitonin play a part in the regulation of extracellular fluid H^+ ion concentration. The latter hormone has been shown to produce not only a calciuria but also an increased excretion of sodium, phosphate and magnesium in the urine. The Na^+ excretion could also be involved in relation to H^+ ion concentration.

Respiratory Regulation. Participation of the respiratory mechanism in regard to H^+ ion concentration is dependent upon (1) the sensitivity of the respiratory center (medulla) to very slight changes in pH and in pCO_2, and (2) the ready diffusibility of CO_2 from the blood, across the pulmonary alveolar membrane, into the alveolar air (p. 435).

An increase in blood pCO_2 of only 1.5 mm. Hg (0.2 per cent increase in CO_2) results in a 100 per cent increase in pulmonary ventilation (stimulation of respiratory center), which increases also with slight increases in H⁺ ion concentration of the blood. The excess CO_2 is thereby promptly removed from the extracellular fluids in the expired air. Decrease in blood pCO_2 or H⁺ ion concentration causes depression of activity of the respiratory center, with consequent slow, shallow respiration, hypoventilation and retention of CO_2 in the blood until the normal pCO_2 and pH are restored. This respiratory mechanism therefore tends to maintain the normal $H_2CO_3/BHCO_3$ ratio in the extracellular fluids in the face of the continual addition of H_2CO_3 as a result of the metabolic production of both CO_2 and nonvolatile acids. According to the Henderson-Hasselbalch equation (p. 414), at pH 7.4, the $BHCO_3/H_2CO_3$ ratio is 20:1, which conforms to the observed concentration ratios of these substances in normal plasma, viz.. 27 mmol. $BHCO_3$/1.35 mmol. H_2CO_3 per liter.

However, although the pH change incident to the entrance of nonvolatile acids has been minimized by the respiratory mechanism, this is only a temporary expedient. It has been accomplished at the expense of an equivalent amount of HCO_3^- and, the concentration of this important buffer being decreased, the capacity of the extracellular fluids to resist additional increments of such acids is reduced.

Renal Regulation. Under ordinary dietary conditions, the amount of nonvolatile acid produced daily exceeds the intake of available base by about 50 to 100 mmol. As indicated above, these acids are effectively buffered, at the expense, however, of a decrease in HCO_3^-, the chief component of the buffer base available for this purpose (total HCO_3^- is about 1000 mmol.). This would eventually be exhausted if the increment in anions, temporarily replacing HCO_3^- ions, were not removed from the body and the HCO_3^- restored. This is accomplished by the kidneys, which are, consequently, the ultimate regulators of the H⁺ ion concentration, providing the most important final defense of the pH of the body fluids.

In the course of urine formation, a protein-free filtrate of blood plasma passes through the glomerular capillaries, the final composition of the excreted urine resulting from subsequent changes in this filtrate in the tubule (p. 459 ff.). In normal subjects, under ordinary dietary conditions, the urinary pH is about 6.0. The difference in titratable acidity between the pH of the urine and that of plasma (7.4) represents the amount of acid removed by renal action from the plasma. Under conditions of extreme requirement for removal of excess acid or base, respectively, the H⁺ ion concentration of the urine can vary from pH 4.5 to 8.2 if the kidneys are functioning normally. This ability to excrete urine of variable acidity (or alkalinity) removes from the blood the quantity of excess acid or base required to preserve the normal H⁺ ion concentration of the body fluids.

Renal Excretion of Acid. Even at the maximal attainable urinary acidity, pH 4.5, all SO_4^{2-} and Cl⁻ anions must be electrically balanced by B⁺ cations, chiefly Na⁺ and K⁺, i.e., they cannot exist as free acids (sulfuric, hydrochloric). The same is true of over 90 per cent of the lactate anions. Consequently, at any urinary pH, these anions remove from the body practically the same amount of B^+ as is required to balance them electrically during their transport in the blood plasma; no significant deviation from the pH of the plasma (7.4) could be effected inasmuch as no H^+ ions could be carried out by these anions (SO_4^{2-} and Cl^-) derived from strong acids.

Increase in urinary acidity is accomplished by excretion of increased amounts of the weakly acidic BH_2PO_4 and certain weak organic acids, e.g., β-hydroxybutyric, acetoacetic, citric. In the plasma, at pH 7.4, phosphate exists as a mixture of about 80 per cent B_2HPO_4 and 20 per cent BH_2PO_4 ($Na_2HPO_4/NaH_2PO_4 = 4/1$). At pH 4.8, about 99 per cent is in the form of BH_2PO_4 ($Na_2HPO_4/NaH_2PO_4 = 1/99$). According to these ratios, every five phosphate anions circulating in the plasma and entering the glomerular filtrate carry nine negative charges, which must be balanced by nine positive charges, i.e., nine Na^+ ions. In the urine at pH 4.8, these five phosphate anions, being virtually all in the form of $B^+H_2PO_4^-$, carry only slightly more than five negative charges and require only about five Na^+ ions for their neutralization. This degree of acidification of the urine, accomplished, as indicated below, by the addition of H^+ ions from the tubular cells, effects a saving of about four Na^+ ions for every five phosphate anions excreted in the urine, the Na^+ being returned to the blood plasma.

A similar situation obtains with respect to weak organic acids, such as β-hydroxybutyric, acetoacetic, and citric. Anions of these acids must be transported in the plasma, at pH 7.4, completely covered by Na^+ cations, but a considerable portion can exist as the free acids in urine at pH 4.8. For example, at pH 4.8, about 45 per cent of the β-hydroxybutyrate anion in the urine exists as free acid and, consequently, excretion of each mole of this anion at this urinary pH releases and restores to the blood 0.45 mole of the Na^+ with which it had been carried in the plasma.

The process of acidification of the urine is explicable most plausibly by the ionic exchange theory. CO_2, formed in the cells of the tubules (where acidification of the urine occurs), under the influence of carbonic anhydrase, is rapidly transformed to H_2CO_3, which undergoes ionization:

$$CO_2 \text{ (metabolic)} + H_2O \underset{\text{anhydrase}}{\overset{\text{carbonic}}{\rightleftharpoons}} H_2CO_3 \rightleftharpoons H^+ + HCO_3^-$$

The H^+ ions thus formed are passed into the lumen of the tubule, an equal number of Na^+ ions passing in the opposite direction into the cells (Fig. 14-2). Thus, a number of Na^+ ions, previously balancing HPO_4^{2-}, citrate$^-$, β-hydroxybutyrate$^-$, etc., ions in the glomerular filtrate, are exchanged for H^+ ions.

$$(Na^+)_2HPO_4^{2-} + H^+ \rightarrow Na^+H_2PO_4^- + Na^+$$
$$Na^+\text{β-hydroxybutyrate}^- + H^+ \rightarrow \text{β-hydroxybutyric acid} + Na^+.$$

PHYSIOLOGICAL BUFFER SYSTEMS

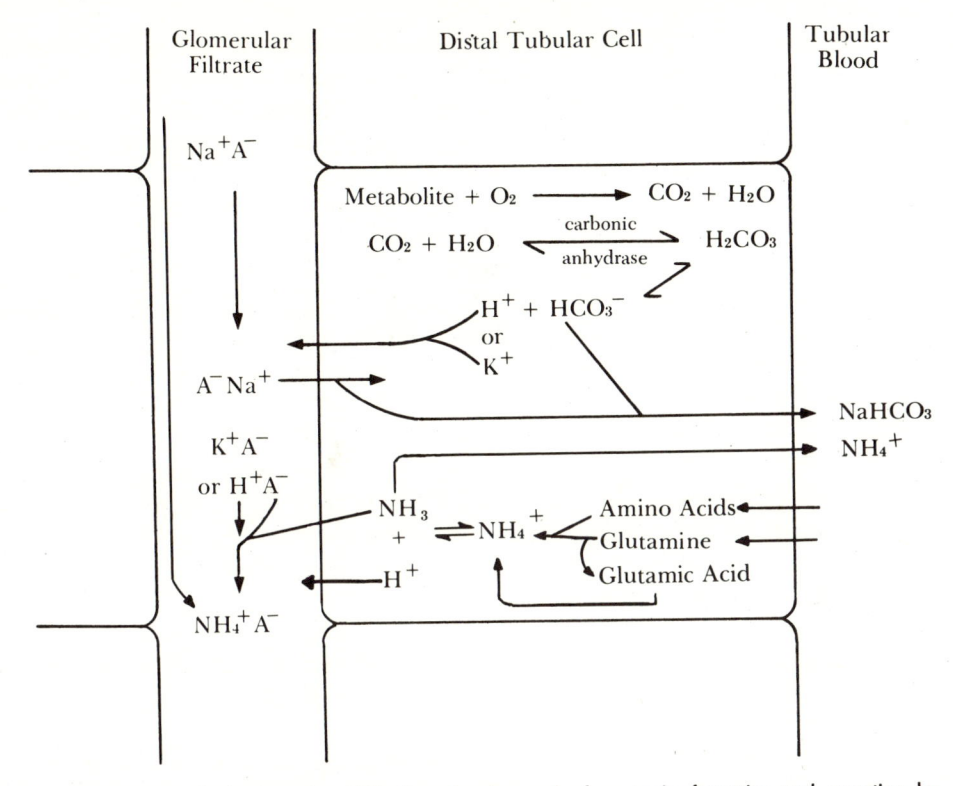

Figure 14-2. Mechanism of acidification of urine and of ammonia formation and excretion by the kidney, indicating exchange of H⁺ and/or K⁺ for Na⁺.

The Na⁺ ions, entering the cell, are absorbed into the blood stream with HCO_3^- ions formed from the H_2CO_3, thereby supporting the bicarbonate structure of the extracellular fluids (Fig. 14-2). This interchange takes place for the most part in the distal tubule, but it can occur elsewhere in the nephron.

It is now known that a major part of renal hydrogen ion excretion takes place in the proximal tubule, the cells of which also contain carbonic anhydrase. In this region, the excreted H⁺ ion reacts with HCO_3^- in the lumen to form CO_2 and water. The sodium ion absorbed into the cell to maintain electroneutrality is reabsorbed into the plasma along with the HCO_3^- ion left in the proximal tubule cell, when the hydrogen ion is excreted into the lumen. The net result is reabsorption of $NaHCO_3$ into the plasma. The amount of HCO_3^- ion is the same as that originally removed from the lumen by reaction with the H⁺ ions excreted. It must be emphasized, however, that this reabsorbed HCO_3^- has been formed in the proximal renal tubule cell and is not that which was originally present in the lumen.

The amount of HCO_3^- absorbed into the plasma by the renal tubular system is proportional to that of the pCO_2 in the plasma. This indicates that bicarbonate absorption is dependent on the formation of carbonic acid.

In nonrespiratory alkalosis, there is an excess of HCO_3^- in the plasma. Under these conditions, since there is an upper limit to the rate at which HCO_3^- can be reabsorbed from the tubular lumen of the nephron, HCO_3^- is excreted in the urine, rendering it alkaline. Since the cells of the collecting ducts are also capable of excreting H^+ ion into the lumen, some of this reacts with the bicarbonate to form carbonic acid. This explains why the pCO_2 of very alkaline urine is higher than that of the plasma.

The number of HCO_3^- ions originating in the renal tubular epithelial cells and restored to the blood stream, under conditions of physiological equilibrium, equals the number of HCO_3^- ions originally displaced (and excreted by the lungs as CO_2) by the entrance of nonvolatile acids. It is important to appreciate the fact that the return of each HCO_3^- ion to the blood is accompanied by the passage of a H^+ ion into the lumen of the tubule. Consequently, the kidney excretes one H^+, originating in the tubular epithelial cells, for every H^+ ion originally contributed to the blood stream by the nonvolatile acids entering in the tissues. This is the final step in the maintenance of H^+ production (and introduction) and excretion.

A portion of the urinary potassium enters the uriniferous tubule by active excretion by the distal tubular epithelium, in addition to the larger amount which filters through the glomeruli (p. 463). In this region (mainly distal tubule), K^+ and H^+ ions compete in a common excretory mechanism, either ion being exchanged for Na^+ (Fig. 14-2). With increased concentration of K^+ in the renal tubular cells, more K^+ than H^+ ions are exchanged for Na^+, the acidity of the urine falls and that of the body fluids increases. Conversely, when the tissue cells (including kidney) are depleted of K^+, more H^+ and fewer K^+ ions are exchanged for Na^+, and the urine becomes more acid even though a state of alkalosis may exist. Inhibition of renal carbonic anhydrase results not only in lowered ability to acidify the urine but also in striking increase in excretion of potassium, another indication of the competition between these two ions for tubular excretion (p. 463).

With increasing acidification of the glomerular filtrate, progressively more of the filtered HCO_3^- is converted to H_2CO_3 and, consequently, to H_2O and CO_2, the latter diffusing across the tubule into the blood, the pCO_2 of which is in equilibrium with that of the fluid in the uriniferous tubule. The bicarbonate content of the urine may fall to zero as the acidity increases.

The phenomena described above occur in response to a stimulus provided by the kidney by an increase in H^+ ion concentration of the plasma and glomerular filtrate (increased requirement for excretion of acid and reabsorption of bicarbonate). If the plasma pH rises (alkalosis), the renal mechanism is stimulated in the opposite direction, i.e., decreased excretion of acid, decreased reabsorption of bicarbonate. The urinary acidity falls as a result of decreased exchange of H^+ and Na^+ ions, increasing amounts of phosphate being excreted in the form of Na_2HPO_4 and of organic acids as sodium salts. With increasing pH, up to 8.2, there is increased urinary excretion of HCO_3^- and of Na^+, the cation with which it is chiefly associated. Changes in the reverse direction occur in the concentrations of these substances in the blood, tending to restore the normal pH of the plasma.

Excretion of Ammonium Ion. A normal subject, under ordinary dietary conditions, excretes 30 to 50 mmol. ammonium daily. If the plasma H$^+$ ion concentration rises, the urinary ammonium increases, even tenfold in severe diabetic acidosis. A decrease occurs in alkalosis. NH$_4^+$ is formed along the entire length of the uriniferous tubule. About 40 per cent is derived from the amide group of glutamine (by the enzyme glutaminase) and about 30 per cent from the α-amino group of certain amino acids (glutamic, alanine and glycine). The remaining 30 per cent is derived from plasma ammonium by glomerular filtration.

It is likely that the amino groups of alanine and glycine are first transferred to α-ketoglutarate to give glutamate, which gives rise to NH$_4^+$ by the action of glutamate dehydrogenase. It is important to emphasize that it is virtually all NH$_4^+$ and very little NH$_3$ which is formed at intracellular pH. There is a dissociation equilibrium between NH$_4^+$ and NH$_3$ + H$^+$, greatly in favor of the formation of undissociated NH$_4^+$. It so happens that the cell membrane is permeable only to NH$_3$ and not to the positively charged NH$_4^+$ ion. NH$_3$ therefore diffuses into the tubular lumen and the remaining NH$_4^+$ dissociates further, and so on. It is important to realize that, for every NH$_3$ molecule formed, an H$^+$ ion is liberated. This could pass into the plasma or into the lumen. It is wrong, therefore, to regard ammonium ion in the urine as a direct regulating mechanism for H$^+$ ion concentration in extracellular fluid. However, each NH$_3$ molecule in the tubular lumen combines with a hydrogen ion to form NH$_4^+$. This could replace a Na$^+$ ion in the lumen and allow it to interchange with an additional hydrogen ion from the tubular cell. This would, of course, play a part in the control of hydrogen ion concentration, since the interchange resorption of Na$^+$ with H$^+$ would lead to the formation of more HCO$_3^-$ in the renal tubular cell.

It is likely that the ammonium forms continuously and that some ammonia molecules diffuse into the plasma to become plasma ammonium, which is removed by glomerular filtration or converted to glutamine or urea in the liver. When the urine is very acid, the diffusion gradient tends toward the lumen more than toward the plasma, since the highly acid urine rapidly converts NH$_3$ to NH$_4^+$ and there is virtually no NH$_3$ in the lumen.

It is a very common error to regard the increase of urinary ammonium in acidosis as evidence that the formation of ammonium salts in the urine directly results in the removal of H$^+$ ions from the body. This is because it is mistakenly believed that NH$_3$ and not NH$_4^+$ is formed intracellularly.

Why, then, is urinary ammonium markedly increased in acidosis? The answer is really very simple. When the liver converts ammonium or amino acids into urea by the arginine-citrulline cycle, two additional H$^+$ ions are formed for every urea molecule synthesized. In acidosis, however, the urea mechanism can be diminished, with the production of fewer H⁺ ions and the excretion of nitrogen occurs as urinary ammonium instead of urea. It is not known if this mechanism involves a diminution in the liver enzymes concerned with the urea cycle, or even if the kidney is stimulated to produce more ammonium. It is of interest to note that patients suffering from hereditary disorders of the urea cycle with consequent hyperammonemia have blood pH levels which are frequently near the upper limit of normal, and sometimes even above it. This would fit in with the concept that the urea cycle is an important mechanism for supplying hydrogen ions to the body.

In acidosis, the urinary NH_4^+ ion begins to increase shortly after the H⁺ ion concentration of the blood rises, but the increase continues for some time after the blood and urine pH and bicarbonate content have reached their lowest level.

INTRACELLULAR-EXTRACELLULAR pH INTERRELATIONS

Under physiological conditions, acids produced within the cells in the course of their metabolic activity, following initial buffering by intracellular systems, enter the extracellular fluids as H⁺ ions and anions (e.g., bicarbonate, lactate β-hydroxybutyrate) or in an unchanged state (e.g., H_2O and CO_2, H_2CO_3). Their subsequent disposition has been discussed previously.

By virtue of exchanges of Na⁺ for H⁺ and/or K⁺ ions across the cell membrane, under certain circumstances the intracellular and extracellular fluids act in conjunction in regulation of H⁺ ion concentration. In alkalosis resulting from primary increase in plasma HCO_3^- (p. 424), H⁺ passes into the extracellular fluids from the cells and K⁺ passes into the cells from the extracellular fluid. An amount of HCO_3^- equivalent to that of the entering H⁺ is converted to CO_2 and removed from the body by the lungs, with consequent reduction in the concentration of HCO_3^- in the extracellular fluid. Conversely, in acidosis associated with primary decrease in plasma HCO_3^-, H⁺ passes into the cells from the extracellular fluid and K⁺ passes in the opposite direction. The HCO_3^- concentration of the extracellular fluids is thereby increased in an amount equivalent to that of the H⁺ entering the cells, now replaced in the extracellular fluid by K⁺ which replaces the H⁺ ion of carbonic acid.

However, under certain other conditions, the pH changes in extracellular and intracellular fluids may occur in opposite directions. In conditions of primary K depletion, e.g., induced by aldosterone or dietary K restriction, passage of excessive amounts of K⁺ from the cells into the extracellular fluid is accompanied by passage of Na⁺ and H⁺ in the opposite direction. Accordingly, the H⁺ ion concentration increases within the cells and decreases in the extracellular fluid. Moreover, the lowered K content of the distal renal tubular cells results in an increased $Na^+ - H^+$ exchange (p. 410) and return of a correspondingly increased number of HCO_3^- ions to the peritubular blood. Renal functional adequacy is necessary for the development of this rather anomalous situation in states of K depletion, in which extracellular alkalosis with an increased HCO_3^- concentration is associated with intracellular acidosis and increased urinary acidity.

ABNORMALITIES OF HYDROGEN ION CONCENTRATION

It is obvious from the preceding discussion that a number of interrelated factors and mechanisms are involved in the maintenance of the normal electrolyte pattern and pH of the body fluids. Several of these do not lend themselves to direct quantitative study. However, factors operating in the blood can be investigated more or less readily because of accessibility of this medium. The most important of these are:

1. *Total "base"* (B^+): the sum of the concentrations (in mEq./liter) of all anions.

2. *H_2CO_3 content:* the concentration of H_2CO_3 (in mmol./liter), which is in equilibrium with the pCO_2 (CO_2 tension) [pCO_2 (in mm. Hg) $\times 0.03 = H_2CO_3$ (in mmol./liter)].

3. *CO_2 content:* the concentration (in mmol./liter) of total CO_2 in solution (HCO_3^- plus H_2CO_3 plus dissolved CO_2).

4. *Plasma HCO_3^-:* the *CO_2 content* minus dissolved CO_2.

5. *Standard HCO_3^-:* the *plasma HCO_3^-* of whole blood that has been equilibrated at a pCO_2 of 40 mm. Hg at 38° C., with the hemoglobin fully saturated with oxygen.

6. *CO_2 combining power:* the *plasma HCO_3^-* of plasma, which has been equilibrated at body temperature with a gas mixture having a pCO_2 of 40 mm. Hg.

7. *pH:* the logarithm of the reciprocal of the H^+ ion concentration.

8. *Buffer base:* the "base" equivalent to the sum of the concentrations (in mEq./liter) of buffer anions (whole blood = HCO_3^- and HPO_4^{2-} + Hb and plasma proteins; plasma = HCO_3^- and HPO_4^{2-} plus plasma proteins).

9. *Base excess:* the base concentration of whole blood as measured by titration with strong acid to pH 7.40 at a pCO_2 of 40 mm. Hg at 38° C. Astrup and his co-workers regard this as representing the nonrespiratory component (p. 414) of a disorder of hydrogen ion regulation. It is expressed as mEq./liter whole blood. It does not indicate the actual amount of acid or base required for therapeutic purposes. It does give a qualitative indication of the degree of severity of the nonrespiratory disturbance.

The important role of Hb as a buffer has been pointed out elsewhere (p. 402), as has the difference in this connection between oxyhemoglobin and reduced hemoglobin. For most precise analysis of the state of H^+ ion concentration, the following factors must be determined: (1) hematocrit (relative proportions of erythrocytes and plasma); (2) pH; (3) CO_2 content; (4) buffer base; (5) H_2CO_3 content (calculated from pCO_2 of blood or alveolar air); (6) Plasma HCO_3^-; (7) Base excess. However, for practical clinical purposes, certain of these may usually be dispensed with (viz., hematocrit, buffer base) inasmuch as significant abnormalities due to changes in hemoglobin are reflected in the pH and HCO_3^- content.

As indicated below, clinical aberrations of H^+ ion concentration are often primary disturbances in either HCO_3^- (bicarbonate) or H_2CO_3, the relation of which to the pH is expressed by the Henderson-Hasselbalch equation: $pH = pK + \log \dfrac{HCO_3^-}{H_2CO_3}$. At pH 7.4, the $HCO_3^- : H_2CO_3$ ratio is 20 : 1 (27 mmol. : 1.35 mmol./liter of plasma water). Knowledge of any two of these three factors (pH, HCO_3^-, H_2CO_3) therefore permits derivation of the third. Determination of the total CO_2 content (HCO_3^- plus H_2CO_3) and any one of the three factors mentioned permits derivation of the other two. Consequently, the most widely employed approach to the investigation of clinical disturbances of H^+ ion concentration centers on these factors.

Primary change in acid or base causing change in HCO_3^- (nonrespiratory acidosis and alkalosis, p. 415) is by far the most common cause of acidosis and alkalosis of clinically significant degree. For this reason, reliance is frequently placed on determination of this factor alone for the diagnosis of acidosis (decreased HCO_3^-) and alkalosis (increase). Determination of the "CO_2 combining power (CO_2 capacity)" used to be employed as a substitute for the CO_2 content. This yielded results of a varying degree of inaccuracy, particularly in respiratory acidosis and alkalosis, because it involved artificial equilibration of the blood at a physiological level of CO_2 tension, regardless of the status in this respect of the subject from whom the blood was obtained. It must be noted that disturbances due to primary changes in H_2CO_3 (respiratory acidosis and alkalosis) are accompanied by alterations (secondary, compensatory) in HCO_3^- and CO_2 content in directions opposite to those which occur in "non-respiratory" acidosis and alkalosis (p. 415 ff.; Table 14-1). Unless the true nature of the disturbance is recognized, which is often possible (e.g., diabetes mellitus, nephritis, vomiting, diarrhea, hyperventilation, etc.), an incorrect diagnosis may be made on the basis of determination of this isolated factor.

Abnormalities of H^+ ion concentration are secondary to two broad groups of disturbances, which can be classified as *respiratory* and *non-respiratory*. The latter group can be subdivided into *metabolic, alimentary* (ingestion, diarrhea, vomiting) *renal*, and *volemic* (i.e., dilution or concentration of extracellular fluid).

Fundamentally, acidosis results from a primary excess of acid or a primary deficit of base, whereas alkalosis results from a primary deficit of acid or a primary excess of base. These concepts are taken into account in the classification of disturbances of acid-base status shown on the opposite page (Table 14-1).

TABLE 14-1 DISTURBANCES OF HYDROGEN ION CONCENTRATION

1. **ACIDOSIS**
   A. **Primary Excess of Acid**
      (1) RESPIRATORY ACIDOSIS
         (a) *Rebreathing*
         (b) *Hypercapnia:* Mechanical asphyxia; morphine or other respiratory depressants; pneumonia; emphysema; pulmonary fibrosis; cardiac decompensation; central nervous system lesions; disorders of respiratory muscles; extreme obesity.
      (2) NONRESPIRATORY ACIDOSIS
         (a) *Metabolic causes:* Diabetes mellitus; starvation; liver disease; lactic acidosis.
         (b) *Alimentary causes:* Ammonium chloride; methyl alcohol.
         (c) *Renal causes:* Renal failure; ureterosigmoidostomy.
   B. **Primary Reduction of Base**
         (a) *Gastrointestinal disturbances*
         (b) *Renal tubular acidosis*
         (c) *Dilutional acidosis*

2. **ALKALOSIS**
   A. **Primary Acid Deficit**
      (1) RESPIRATORY ALKALOSIS: Hysteria; fever; high external temperature; hypoxic anoxemia; encephalitis; salicylates; pregnancy.
      (2) NONRESPIRATORY ALKALOSIS
         (a) *Metabolic cause:* Potassium depletion.
         (b) *Alimentary causes:* Excessive loss of gastric HCl; defective intestinal absorption of potassium.
         (c) *Iatrogenic:* Rehydration with fluids containing insufficient or no potassium.
   B. **Primary Excess of Base**
         (a) *Alimentary causes:* Alkali administration; milk-alkali syndrome.
         (b) *Bone resorption*
         (c) *Excessive diuresis*

3. **MIXED DISTURBANCES**
         (a) *Multiple pathology*
         (b) *Salicylate poisoning*
         (c) *Pregnancy toxemia*
         (d) *Iatrogenic*

ACIDOSIS

The term "acidosis" is applied to the condition resulting from the formation or absorption of acids at a rate exceeding that of their neutralization or elimination. It may also, although much less frequently, be due to the loss of excessive quantities of base from the body. On the basis of the equation:

$$cH = k \frac{H_2CO_3}{BHCO_3},$$

it is evident that an increase in the hydrogen ion concentration of the blood (acidosis) may be caused by either a primary increase in the concentration of H_2CO_3 or a primary decrease in the concentration of $BHCO_3$. If these changes are of such magnitude that the hydrogen ion concentration rises above the upper limit of normal (pH below 7.35) the condition is one of uncompensated acidosis. However, as has been indicated above, because of the existence of a remarkably efficient compensatory mechanism, if the concentration of carbonic acid rises, the concentration of blood bicarbonate also tends to increase in order to maintain the normal equilibrium; likewise, as the concentration of blood bicarbonate diminishes, increased quantities of carbon dioxide are removed through the lungs with a consequent compensatory decrease in the concentration of carbonic acid in the blood. Because of these compensating processes, primary alterations in either of these two factors are, for a time at least, balanced to a certain degree by secondary changes in the other factor, as a result of which there may be little or no perceptible alteration in the hydrogen ion concentration. If under such circumstances the hydrogen ion concentration of the blood is maintained below the upper limit of normal (pH above 7.35) the organism is in a state of compensated acidosis. The clinical conditions in which a state of acidosis is commonly observed will be considered under two headings: (1) those associated with a primary excess of acid; (2) those associated with a primary decrease in base.

Primary Excess of Acid

This may be due to respiratory or nonrespiratory causes.

Respiratory Acidosis. Increase in the carbonic acid content (CO_2 tension) of the blood may occur in one of two general ways:

(1) Rebreathing, or breathing air conditioning abnormally high percentages of CO_2.

(2) Conditions in which the elimination of CO_2 through the lungs is retarded (hypercapnia). In this group may be placed conditions causing mechanical asphyxia, morphine and other drugs causing diminished minute ventilation, pneumonia (particularly bronchopneumonia), pulmonary emphysema and pulmonary fibrosis (in which the diffusion of CO_2 is impaired), cardiac decompensation (in which pulmonary congestion and slowed circulation combine to cause increased CO_2 tension in the blood), disorders of the central nervous system, or the respiratory muscles, and extreme obesity.

Compensatory Mechanisms. As the carbonic acid content of the blood increases, certain compensatory mechanisms come into play by means of which the organism attempts to maintain the hydrogen ion concentration of the blood within normal limits.

(1) *Increased Ventilation.* The increased stimulation to the respiratory center caused by the increased CO_2 tension of the blood results in increased depth and rate of respiration with consequent increased ventilation. As stated above, the sensitivity of the respiratory center to a relatively slight increase in hydrogen ion concentration or CO_2 tension of the blood renders this mechanism extremely efficient in counteracting the effects of primary CO_2 excess. This mechanism cannot of course be invoked in conditions in which the fundamental defect lies in depressed activity of the respiratory center, e.g., in morphine narcosis.

(2) Increase in Bicarbonate. In conditions associated with primary CO_2 excess the blood bicarbonate has been found to be increased. This simultaneous change in both elements of the bicarbonate system tends to limit the increase in hydrogen ion concentration of the blood which would otherwise occur. The increase in blood bicarbonate is accomplished by stimulation (by the increasing acidity) of the renal mechanism (p. 407), with consequent increased reabsorption of Na and HCO_3 and increased urinary acidity and urinary NH_4^+ ion concentration.

(3) Increased Ammonium Formation. An increase in the rate of NH_4^+ ion formation and excretion of NH_3 by the kidney tends to diminish the loss of cations (mainly Na^+) from the body (p. 411). This enables more bicarbonate ions to be formed.

(4) Increased Urinary Acidity. The increasing acidity of the blood plasma stimulates the Na-H exchange in the renal tubules (p. 408), with consequent increase in urinary acidity.

Because of the efficiency of these compensatory mechanisms, primary CO_2 excess does not necessarily result in a state of severe uncompensated acidosis, as evidenced by the hydrogen ion concentration of the blood. From this standpoint, therefore, conditions in this group can be of relatively minor importance clinically, symptoms of acidosis, apart from dyspnea, not always being encountered. These include mental confusion or even coma, due to the increased pCO_2.

Biochemical Characteristics. These are shown in Table 14-2. There is also an increased loss of chloride and potassium ions in the urine. The loss of chloride (together with the chloride shift) results in a fall of the plasma chloride ion level; the loss of potassium causes some body deficit of K^+ ion, but the plasma level does not fall significantly, since in the presence of an acidosis, K^+ ions leave the cells and enter the extracellular fluid.

Nonrespiratory Acidosis. There are three main causal groups of disturbances.

(1) Metabolic Causes: Diabetes Mellitus (see p. 33). Acidosis is a constant feature of advanced diabetes mellitus. It is due largely to the existing ketosis and in part to the excessive loss of water with the associated elimination of excessively large quantities of base. The polyuria which attends the development of glycosuria is accompanied by a pronounced increase in the excretion of electrolytes, particularly sodium, potassium and chloride. This initial electrolyte loss is supplemented by a secondary increase in the loss of water and electrolytes simultaneously with the development of ketosis and is aggravated by vomiting. These changes eventually lead to dehydration and depletion of body electrolytes and base. The diminution of renal circulation results in uremia.

The mechanism of production of ketosis in diabetes mellitus is discussed in detail elsewhere (p. 109). This results from excessive production of ketones because of inadequate carbohydrate utilization and excessive catabolism of fat and certain amino acids. The "ketone bodies," including acetoacetic acid, beta-hydroxybutyric acid, and acetone, accumulate in the blood and tissues. The first two are quite strong acids.

TABLE 14-2 BIOCHEMICAL CHARACTERISTICS OF UNCOMPENSATED AND COMPENSATED ACIDOSIS AND ALKALOSIS*

PLASMA	NORMAL	ACIDOSIS NONRESPIRATORY Uncomp.	ACIDOSIS NONRESPIRATORY Comp.	ACIDOSIS RESPIRATORY Uncomp.	ACIDOSIS RESPIRATORY Comp.	ALKALOSIS NONRESPIRATORY Uncomp.	ALKALOSIS NONRESPIRATORY Comp.	ALKALOSIS RESPIRATORY Uncomp.	ALKALOSIS RESPIRATORY Comp.
HCO_3^-	mmol./liter 21.3–28.5 (24.9)	– –	–	+	++	++	+	N(–)	–
CO_2 content (AutoAnalyzer)	mmol./liter 25–32 (28.5)	– –	–	+	++	++	+	–	–
Standard HCO_3^-	mEq./liter 22.1–25.8 (males) 21.3–25.0 (females)	– –	–	N	N	++	+	N	N
pCO_2	mm. Hg 35.8–46.6 (males) 32.5–43.7 (females)	N(–)	–	++	+	N(+)	+	– –	–
Base excess	mEq./liter –2.4 to +2.3 (males) –3.3 to +1.2 (females)	– –	N	N	N	++	N	N	N
pH	7.36–7.42	–	N	–	N	+	N	+	N
Urinary titratable acidity and NH_4^+	mEq./liter 20–40 (acid) 20–70 (NH_4^+)	+	+	N(+)	+	–	–	N(–)	–

* + = increase; – = decrease; N = normal

Mean values are shown in parentheses. Because a pure uncompensated state rarely exists, in the columns headed "Uncomp.", signs in parentheses have been included to indicate the usual findings because partial compensation occurs rapidly. Apart from CO_2 content and the urinary values, normal values quoted are those obtained by the Astrup microtechnique on capillary blood.

Starvation. The carbohydrate reserves of the body are relatively rapidly exhausted in starvation. This means that energy is soon derived almost entirely from the breakdown of fat and protein. This leads to the production of an excess of "ketone bodies" (ketosis), which give rise to acidosis. Conversely, ketosis can be produced as a result of diets containing an excessive amount of fat.

Liver Disease. Severe disease of the liver can give rise to severe ketosis and acidosis.

Lactic Acidosis. The production of excessive amounts of lactic acid has been discussed previously (p. 77).

(2) Alimentary Causes: Ammonium Chloride. Ingestion of the NH_4^+ ion in excessive amounts leads to acidosis. The ion is itself an acid and, moreover, when it is converted into urea by the liver, two H^+ ions are liberated for every urea molecule formed. The response to ammonium chloride intake is virtually equivalent to that of equimolar hydrochloric acid. If the amount of NH_4Cl taken gives rise to an amount of H^+ ion in excess of that which can be excreted by the kidney (200 to 250 mmol./24 hr.), acidosis will result. There is diminution in the plasma bicarbonate and pH and a fall in pCO_2 because of hyperventilation. The chloride level increases, as does that of the K^+ ion, since acidosis of this severity results in a shift from the intracellular to the extracellular fluid.

Methyl Alcohol. The body converts methyl alcohol to formic acid. Intoxication with this alcohol will therefore lead to a severe acidosis and results very similar to those produced by ammonium chloride. The condition may, however, be complicated by hypoglycemia.

(3) Renal Causes: Renal Failure. In chronic renal disease with glomerular insufficiency, there is a diminution of function renal mass, in spite of the fact that the surviving nephrons can hypertrophy. This diminution results in a decrease in the H^+ ion secretion, and so, to some extent at least, the acidosis of chronic renal disease of this type is a primary excess of acid. It is interesting that acid urine can be produced, which is of comparable acidity to normal urine. This maximal acidity, however, occurs at a very much lowered plasma HCO_3^- concentration. Titrable acid is, however, diminished. There is also retention of sulfate, phosphate and organic radicles, which also leads to a primary excess of acid as well as a fall in the Cl^- ion concentration of extracellular fluid, in spite of low HCO_3^- ion concentration.

The position is much more complicated, since it appears that at least a large part of the acidosis of all chronic renal disease is associated with defective excretion of H^+ ions, associated with defective reabsorption of HCO_3^- ions, in the renal tubular system. This leads to HCO_3^- wasting in the urine, which causes a primary deficiency of base.

The acidosis is remarkably stable for long periods. It is also known that significantly less acid appears in the urine than is produced metabolically. This means there must be another compensatory mechanism, and there is now much evidence that it involves bone, either by interchange of H^+ ions and exchangeable Na^+ ions or by liberation of base in the form of phosphate. Both may occur. It does not seem very likely, however, that this compensatory mechanism can account for azotemic osteodystrophy. In special cases of renal disease, it may possibly play some part. A major factor in the production of renal osteodystrophy is probably secondary parathyroid hyperplasia resulting from the low level of serum calcium due to insensitivity to vitamin D (p. 304).

Ureterosigmoidostomy. Transplantation of the ureters into the sigmoid colon (e.g., for bladder carcinoma) is often followed by the development of acidosis, since the intestinal mucosa reabsorbs chloride in excess of sodium. To maintain electroneutrality, the excess Cl^- is absorbed with NH_4^+ or K^+ or both. The excess of Cl^- leads to lowered absorption of bicarbonate and lowered excretion of H^+ ion by the renal tubules. This, of course, results in acidosis. Although included as a renal type of acidosis, there is obviously also an alimentary aspect. As a matter of fact, the chloride absorption depends on the time of contact with the intestinal mucosa. The acidosis is kept at minimal levels by frequent bowel evacuation or by enemas.

Compensatory Mechanisms in Nonrespiratory Acidosis. As the buffer base of the body diminishes, certain compensatory mechanisms are set in operation in an attempt to maintain the hydrogen ion concentration within normal limits.

(1) *Increased Pulmonary Ventilation.* The respiratory center responds to an increase in hydrogen ion concentration just as it does to increased CO_2 tension, namely, by causing increased rate and depth of respiration. As a consequence of this increased pulmonary ventilation, CO_2 is washed out of the blood and the ratio of blood carbonic acid to blood bicarbonate tends to be restored to normal, the hydrogen ion concentration being consequently but little affected. During the early stages, therefore, the organism is in a state of compensated acidosis, but as the condition progresses and the buffer base (HCO_3^-) deficit becomes more pronounced the compensatory mechanism fails and the condition becomes one of uncompensated acidosis with an increase in the hydrogen ion concentration of the blood.

(2) *Increased Ammonium Formation.* As the buffer base diminishes, more ammonium ion appears in the urine. Insofar as it is possible that this can displace some Na^+, which can then be exchanged for H^+ ion in the renal tubule, it is conceivable that this is a compensatory mechanism. The concept that NH_3 and not NH_4^+ is formed in the renal tubule is no longer considered to be correct. It has already been pointed out that the combination of NH_3 with H^+ ions in the lumen cannot in itself directly affect H^+ ion concentration in extracellular fluid. In nephritis, however, this mechanism fails to a certain degree, due to the fact that the ammonium-forming function of the kidney is impaired.

(3) *Increased Acid Excretion.* In the majority of cases of acidosis due to primary increase of acid, increased quantities of acid are eliminated in the urine (increased $H^+ \rightarrow HCO_3^-$ and increased Na^+-H^+ exchange) (p. 408). This does not occur in the acidosis of advanced nephritis, which is in itself partly due to failure of elimination of acids by the kidneys. The increased

elimination of electrolytes in the urine in this condition and the diminution in the concentration of base and bicarbonate in the blood and tissue fluids are associated with the elimination of increased quantities of water in an attempt to maintain the normal electrolyte concentration of the blood and tissue fluids. This diuresis results in dehydration, which is one of the more constant features of this type of acidosis.

Biochemical Characteristics. These are shown in Table 14-2.

Primary Reduction of Base

Nonrespiratory Acidosis is the only possibility.

(1) Alimentary Causes: Gastrointestinal Disturbances. Loss of base in gastrointestinal secretions accompanies diarrhea, ileostomy or fistulae (especially biliary). Amounts of bicarbonate may be lost that are sufficiently significant to produce an acidosis. The amount of bicarbonate lost must be more than that which can be regenerated in the renal tubules. The acidosis is accompanied by a deficit of body Na^+ and K^+, which are also lost with the fluids.

(2) Renal Cause: Renal Tubular Acidosis. This is the term applied to a syndrome in which glomerular function is normal, but in which there is a defect in the ability to acidify the urine, accompanied by excessive excretion of K, PO_4 (p. 311), and Ca (p. 308). This disorder is apparently dependent upon a specific defect in the tubular mechanism of secretion of H ions and the production of HCO_3^- for absorption, rather than upon nonspecific tubular damage, and resembles the condition produced experimentally by inhibition of carbonic anhydrase activity in the renal tubular epithelium. Depression of carbonic acid (and therefore, H^+) formation in the tubular cells results in excessive excretion of K into the lumen. The excess of bicarbonate in the urine also leads to a diminution in the production of urinary titratable acid as well as of ammonium salts. The renal loss of sodium ion can lead to secondary hyperaldosteronism.

(3) Volemic Cause: Dilutional Acidosis. If large amounts of fluids are ingested, retained or infused, there is a fall in HCO_3^- concentration by simple dilution. The H_2CO_3 concentration in the extracellular fluid is, however, rapidly restored by the addition of metabolic CO_2. The resultant decrease in the HCO_3^-/H_2CO_3 ratio results in acidosis. The situation is further aggravated if the infusion fluid contains Cl^-, since this ion depresses the renal tubular production of bicarbonate by the H^+ ion secretion mechanism. It will be remembered that such depression leads to loss of HCO_3^- in the urine.

ALKALOSIS

Alkalosis is a state in which either excessive amounts of acid are lost from the body without a comparable loss of base, or base is formed in or supplied to the body at a rate exceeding that of its neutralization or elimination. In terms of the bicarbonate system, alkalosis may result from either a primary decrease in carbonic acid or a primary increase in bicarbonate. As in the case of acidosis, a primary change in one of these factors is almost invariably associated with or followed by a secondary change in the same direction in the other factor, the hydrogen ion concentration of the blood under these circumstances being but little affected, the condition being one of compensated alkalosis. As the metabolic error progresses, however, the compensatory secondary change becomes insufficient to maintain the normal balance and the hydrogen ion concentration of the blood diminishes below the lower limit of normal (pH above 7.45), constituting a state of uncompensated alkalosis.

Primary Acid Deficit

This may be due to respiratory or nonrespiratory causes.

Respiratory Alkalosis. Excessive quantities of CO_2 may be washed out of the blood by hyperventilation of the pulmonary alveoli (hypocapnia). This condition may be induced voluntarily by excessively rapid and deep respiration. Clinically, respiratory alkalosis is observed in the following conditions:

Hysteria. Alkalosis due to primary H_2CO_3 deficit is occasionally observed as a result of the hyperventilation which occurs at times during hysterical attacks.

Fever. Hyperventilation may occur as a result of the increased respiratory rate associated with an increase in body temperature.

High External Temperatures. Hyperventilation may be induced by exposure to high external temperatures such as hot baths. If prolonged, this exposure may result in alkalosis due to primary H_2CO_3 deficit.

Hypoxic Anoxemia. Hyperpnea occurring in untrained individuals ascending to high altitudes where the atmospheric oxygen tension is low (hypoxic anoxemia) commonly results in primary H_2CO_3 deficit and alkalosis.

Encephalitis. Alkalosis due to hyperventilation has been observed in some cases of encephalitis manifesting hyperpnea over prolonged periods of time.

Drugs. Large amounts of sodium salicylate, such as are sometimes given in the treatment of acute rheumatic fever, produce stimulation of the respiratory center, with consequent hyperventilation and a tendency toward alkalosis of this type. Subsequent conversion to salicylic acid may result in a metabolic type of acidosis (p. 425). Because the formation of H^+ ion is decreased in the tubules (low pCO_2), the mechanism for reabsorption of HCO_3^-, mainly in the proximal tubules, is affected and consequently some HCO_3^- can appear in the urine in spite of the low plasma level. The plasma chloride is increased.

Renal compensation, when achieved, begins within a few hours and increases to a steady state in about 20 hours, at which the plasma HCO_3^- is low but the HCO_3^-/H_2CO_3 ratio is virtually normal. H^+ ion is now excreted at a rate which balances its production by metabolic processes. The lowered HCO_3^-, however, increases susceptibility to any administered acid. Conversely, there is also an increased susceptibility to administered bicarbonate.

Pregnancy. A lowered pCO_2 with a tendency to a mild alkalosis occurs toward the end of normal pregnancy.

Compensatory Mechanisms in Respiratory Alkalosis. In view of the pathogenesis of this condition, the task of compensating for this defect falls on the kidneys. These organs, unless damaged, respond to the increasing alkalinity of the blood plasma by decreasing H^+ ion secretion and returning less Na and HCO_3 to the blood plasma. There is increased retention of chloride ions.

As the condition of alkalosis progresses, ketone bodies (acteoacetic and beta-hydroxybutyric acid and acetone) may accumulate in the blood due presumably to a disturbance of carbohydrate utilization in the liver. Ketonuria may occur under such circumstances. This ketosis, however, has but little effect upon the acid-base equilibrium.

Biochemical Characteristics. These are shown in Table 14-2.

Nonrespiratory Alkalosis. This occurs in the following conditions, which are among the more frequent causes of clinically observed alkalosis:

(1) Alimentary Causes: Excessive Loss of HCl from the Stomach. The loss of excessive quantities of hydrochloric acid from the stomach is encountered most frequently in individuals with pyloric or high intestinal obstruction and following protracted gastric lavage without proper provision for acid replacement. It is also at times observed in infants with pylorospasm and in patients with generalized peritonitis. Because of the loss of hydrochloric acid, an equivalent amount of bicarbonate secreted into the intestinal lumen is not neutralized and is reabsorbed into the extracellular fluid, thereby increasing above normal the HCO_3^- load on the kidney. Renal mechanisms for handling this ion are depressed, partly because of loss of chloride ion in the stomach contents and partly because loss of fluid results in impaired renal circulation, with reduction of glomerular filtration rate. The position is even further aggravated by the loss of K^+ ions in the stomach contents (see p. 424).

(2) Metabolic Causes: Potassium Depletion. Primary depletion of cell potassium results in a tendency toward alkalosis, with decrease in plasma Cl and increase in plasma HCO_3. Under these circumstances, excessive amounts of K move out of the tissue cells (including renal tubular epithelium) into the extracellular fluids and therefore enter the glomerular filtrate in excess, escaping in the urine. In the presence of a decreased concentration of K in the distal tubular cells, excessive amounts of H ions (from carbonic acid formed in the cells) are excreted in exchange for Na, which, with the residual HCO_3 ions, passes into the blood stream in excessive amounts. The urinary excretion of the increased amount of K (originating in the intracellular fluid) must be balanced by corresponding amounts of anions, mainly Cl, drawn from the extracellular fluid compartment. This results in decrease in plasma Cl and an increase in plasma HCO_3, i.e., hypochloremic alkalosis. Potassium depletion causes a defect in the reabsorption of chloride in the renal tubules. Because K^+ ion has moved out of body cells, extracellular H^+ ion migrates into the cells and gives rise to HCO_3^-, which migrates back to the extracellular fluid. These phenomena add to the alkalosis. The situation is aggravated by the K^+ and Cl^- deficiency, each of which can apparently inhibit renal mechanisms for the excretion of HCO_3^-.

The intracellular migration of H^+ ion could be regarded as the primary trigger mechanism. It is for this reason that potassium deficiency has been included under the heading of primary acid deficit.

Alkalosis of this type also occurs in an apparently rare condition of noninfectious diarrhea in infants, in which there is probably a congenital defect in the mechanism of intestinal reabsorption of potassium. Large amounts of KCl are lost in the feces, with consequent depletion of tissue cell potassium and the sequence of metabolic aberrations outlined above. A similar situation develops in subjects with alkalosis, dehydration, and K depletion due to excessive vomiting, who have been rehydrated with fluids containing no potassium (i.e., NaCl, NH_4Cl and glucose solutions). This results in additional loss of cellular K, which is replaced by Na (p. 372). As rehydration progresses, the concentration of K within the cells therefore falls, and the same sequence of events ensues. This prevents complete correction of the alkalotic state in such patients when only K-free solutions are administered.

Primary Excess of Base

Nonrespiratory Alkalosis is the only possibility.

(1) Alimentary Causes: Alkali Administration. Alkalosis may follow the administration of excessive amounts of alkaline substances, particularly sodium bicarbonate, which is frequently taken in large doses. The administration of sodium bicarbonate in the treatment of acidosis in chronic nephritis is particularly liable to result in alkalosis due to the difficulty of elimination of the excess base by the diseased kidneys.

Milk-Alkali Syndrome. Alkalosis, along with hypercalcemia and renal tubular calcification, is particularly liable to occur when large amounts of bicarbonate are taken along with large amounts of milk.

(2) Metabolic Cause: Bone Resorption. In the presence of excessive bone resorption, excessive amounts of phosphate and bicarbonate can be liberated. These can give rise to alkalosis. This occurs in hyperparathyroidism, vitamin D intoxication and in the presence of certain metastases in bone.

(3) Volemic Cause: Excessive Diuresis. This can give rise to alkalosis associated with reduction in volume of the extracellular fluid, under conditions in which its HCO_3^- concentration is increased. The respiratory mechanisms maintain the pCO_2 at normal levels, and consequently the HCO_3^-/H_2CO_3 ratio is increased to give the alkalosis. This is the converse to dilutional acidosis.

Compensatory Mechanisms in Nonrespiratory Alkalosis. If functioning normally, the respiratory apparatus and the kidneys attempt to maintain the blood pH within normal limits. The increasing alkalinity of the blood plasma depresses the respiratory center (p. 436), with consequent slowing of respiration and hypoventilation and decreased excretion of CO_2. The blood H_2CO_3 content increases accordingly.

The kidneys (unless damaged or subject to K^+ deficiency) attempt to rid the body of the excess HCO_3 by decreasing the extent of Na-H exchange (p. 410), thus returning less HCO_3 (more H_2CO_3) to the blood stream. Urinary acidity and ammonium ion are decreased; in severe cases the urine may become alkaline. It may be acid in K^+ deficiency. Ketosis and ketonuria occur.

Biochemical Characteristics. These are shown in Table 14-2.

Alkalosis of this as well as the respiratory type can be accompanied by tetany, due perhaps to decreased ionization of calcium salts in the extracellular fluids (p. 306). Kidney damage, mainly degenerative changes in the tubules, occurs fairly frequently, with oliguria and nitrogen retention. Another important feature, potassium depletion, at times manifested by decrease in serum K concentration, is discussed in detail elsewhere (p. 372).

MIXED DISTURBANCES OF ACID-BASE STATUS

Multiple Pathology. Although the previous discussion has dealt with primary excess or deficit of acid or base, under clinical conditions mixed disturbances are possible. These may be additive or may cancel each other out. A diabetic may suffer from chronic respiratory or renal disease or both.

Salicylate Poisoning. In the early stages, when relatively small amounts have been absorbed, salicylate stimulates the respiratory center, with consequent hypocapnia and respiratory alkalosis with lowered pCO_2. This is then virtually compensated for by the excretion of bicarbonate through renal mechanisms, and the urine becomes alkaline. The situation may not alter any further with moderate dosage and after a day or two may gradually resolve. If, however, large doses of salicylate have been swallowed, there is interference with the citric acid cycle and the consequent liberation of a large amount of H^+ ions with ketosis. This results in profound acidosis, since the stage of renal compensation has depleted the body stores of bicarbonate.

Pregnancy Toxemia. In the toxemia of late pregnancy an acidosis develops. This is in part due to fluid retention (dilutional acidosis), in part to metabolic disturbance, including lactic acidosis, and in part to renal failure and diminished glomerular filtration rate.

In toxemia of early pregnancy, the persistent vomiting tends to be associated with an acidosis rather than an alkalosis.

METHODS OF STUDYING ACID-BASE STATUS

The difficulty of investigating accurately the condition of a system which may be disturbed in one or more of so many ways must be apparent. Perhaps the most satisfactory method of approach, from a clinical standpoint, is to consider all disturbances of the acid-base status in terms of the bicarbonate system, as has been done above, and to consider the hydrogen ion concentration of the blood as dependent upon the ratio between the concentrations of carbonic acid and bicarbonate in the blood. In other words,

$$[H^+] = K \frac{[H_2CO_3]}{[HCO_3^-]}$$

It must be remembered that a primary change in the concentration of either H_2CO_3 or HCO_3^- is associated with or followed by a secondary compensatory change in the same direction in the other factor, in an attempt to maintain the hydrogen ion concentration within normal limits. For example, if the HCO_3^- concentration is diminished, as in diabetes mellitus and nephritis, the development of uncompensated acidosis with an increase in the hydrogen ion concentration of the blood is for a time prevented by a compensatory decrease in the H_2CO_3 concentration of the blood (compensated acidosis). The subnormal carbonic acid content of the blood is, in this instance, a manifestation of acidosis. On the other hand, a primary decrease in the H_2CO_3 concentration is observed in conditions associated with hyperventilation, such as occurs at high altitudes, encephalitis, hysteria, etc., resulting in a tendency toward a decrease in the hydrogen ion concentration which is for a time balanced by a compensatory secondary diminution in blood bicarbonate. In this case the decreased bicarbonate and carbonic acid concentrations of the blood are indicative of a state of alkalosis (compensated). It is evident that a clear distinction must be made between primary and secondary changes in these factors, primary increase in carbonic acid and primary decrease in bicarbonate being indicative of a state of acidosis, whereas secondary increase in carbonic acid and decrease in bicarbonate are indicative of a state of alkalosis. On the other hand, primary decrease in carbonic acid and primary increase in bicarbonate are indicative of a state of alkalosis, whereas a secondary decrease in carbonic acid or increase in bicarbonate is indicative of a state of acidosis.

In considering the problem from this standpoint, three variable factors must be considered: (1) The hydrogen ion concentration (pH) of the blood, (2) the H_2CO_3 concentration (pCO_2), and (3) the HCO_3^- concentration. Obviously, a distinct disturbance of the acid-base status may exist, as evidenced by a primary change in the concentration of either

H_2CO_3 or HCO_3^-, which, by virtue of a compensatory change in the other factor, is associated with no significant alteration in the hydrogen ion concentration. In other words, the primary disturbance is compensated. Under such conditions the true acid-base state can be determined accurately only by the determination of at least two of the three components of the equation cited above (hydrogen ion concentration, blood carbonic acid, and blood bicarbonate). However, in the great majority of clinical conditions associated with significant disturbances of the acid-base status the fault results primarily in a decrease or an increase in plasma bicarbonate, so that for most practical purposes investigation of this factor furnishes satisfactory although not exact information as to the state of affairs in relation to the acid-base status. Inasmuch as primary changes in either direction in either $[H_2CO_3]$ or $[HCO_3^-]$ are followed by secondary changes (compensatory) in the same direction in the other factor, determination of the CO_2 content of the plasma (CO_2 present in both H_2CO_3 and HCO_3^-) may be used as an index of the nature and magnitude of such changes. This determination is preferable to that of the CO_2 combining power. It must be realized that serious error may result from a failure to differentiate clearly between primary and secondary changes and that in doubtful cases too much reliance should not be placed upon the determination of any single factor.

CO_2 Content of Plasma. This is a measure of the amount of H_2CO_3 plus HCO_3^- and dissolved CO_2 present in the plasma of the subject, expressed as millimoles per liter (25 to 32, av. 28.5). The determination is made on venous blood obtained under anaerobic conditions in order to avoid changes in pCO_2 that occur upon exposure of blood to air. With this procedure, both the H_2CO_3 and the HCO_3^- are maintained in the same condition in which they existed in the body. The actual determination is carried out with an AutoAnalyzer or with a Natelson Microgasometer. With the former method, experience has shown that once the plasma specimen has been separated, provided it is not treated roughly, strict anaerobic precautions are not really necessary for clinical purposes. It also used to be the custom to cover the specimen of blood with a layer of paraffin. Unfortunately, CO_2 is soluble in paraffin, and it is advisable not to use it. In actual fact, the venous specimen collected should be put into a heparinized tube so that the tube is full, stoppered and centrifuged with the stopper in place. The plasma is then gently separated and put into the AutoAnalyzer cup, which is capped prior to its use for analysis.

As indicated in Table 14-2 (p. 418), the plasma CO_2 content is decreased in nonrespiratory acidosis (p. 417) and respiratory alkalosis (p. 422) and is increased in nonrespiratory alkalosis and respiratory acidosis. The complete quantitative picture can be obtained by determining, in addition, the pH of the blood plasma sample, which is a function of the ratio of HCO_3^- to H_2CO_3, as expressed in the Henderson-Hasselbalch equation. The ratio at any pH level may be calculated or be obtained from available tables or charts. Knowledge of the ratio and of the total amount of CO_2

present (H_2CO_3 plus HCO_3^-) permits calculation of both H_2CO_3 and HCO_3^- with precision.

This procedure, i.e., determination of (a) CO_2 content and (b) pH, was formerly employed in all cases in which a higher degree of accuracy was desired or in which the disease diagnosis was uncertain. In the latter case, there might be some question as to whether a low CO_2 content was a reflection of a nonrespiratory type of acidosis or a respiratory type of alkalosis. Similar difficulty might be experienced in interpreting an increase in CO_2 content without the aid of the pH determination. However, in the great majority of cases, the nature of the underlying disease was known, e.g., diabetes mellitus, glomerulonephritis, pyloric obstruction, diarrhea, etc., and the direction of an associated disturbance in acid-base balance, i.e., acidosis or alkalosis, could be anticipated. Under these circumstances, clinical interest is centered in whether and to what degree the anticipated change is present. Determination of the CO_2 content alone may suffice for this purpose.

More modern methods determine exactly the pCO_2 and pH of arterial capillary blood at body temperature (Astrup technique) and the plasma bicarbonate can be calculated exactly. It is also possible to determine the actual plasma CO_2 by means of a CO_2 electrode and to measure the pH exactly. Once again the bicarbonate can be calculated. These modern methods have virtually superseded the older methods of investigation. For most clinical purposes, where the diagnosis is known, it is sufficient to determine the plasma CO_2 content by an AutoAnalyzer.

Carbon Dioxide Combining Power of the Plasma. The carbon dioxide combining power of the blood plasma is expressed as the number of millimoles of CO_2 which can be bound as bicarbonate by 1000 ml. of blood plasma at body temperature at the normal pCO_2 of 40 mm. Hg. The normal values for adults range from 23 to 28, av. 26 mmol./liter. The normal values for infants are about 10 volumes per cent lower than those for adults. Decrease or increase in the CO_2 combining power is indicative of a corresponding change in the bicarbonate level. For clinical purposes, this estimation has nowadays been superseded by the far more satisfactory modern methods.

Although the values obtained by this method may be interpreted in a manner satisfactory for most clinical purposes, they do not always indicate accurately the true quantitative variation in blood bicarbonate. This is due to the fact that the CO_2 capacity is determined after artificially saturating the blood plasma with CO_2 at its normal tension in arterial blood, i.e., 40 mm. Hg. If the condition is one of severe metabolic acidosis, the actual in vivo pCO_2 is much lower, and in metabolic alkalosis, higher. If there is significant deviation from the normal, therefore, plasma CO_2 combining power values do not reflect the true in vivo state of the $H_2CO_3 : HCO_3^-$ balance with the same degree of quantitative accuracy as do values for CO_2 content, which is measured under conditions which preserve the in vivo relationships.

Blood CO₂ Tension. Procedures are now available that permit estimation of the blood pCO_2 with relative ease and accuracy (pCO_2 electrode, etc.). The concentration of H_2CO_3 may be calculated from this factor (H_2CO_3 (millimoles/liter = pCO_2 (in mm. Hg) × 0.03). The blood pH is measured simultaneously. These values, taken in conjunction with the CO_2 content, provide a complete picture of the state of the bicarbonate buffer system of the blood and, therefore, of the state of acid-base balance. Normal values for plasma pCO_2 range from 32.5 to 46.6 mm. Hg.

Astrup Microtechnique. This is a procedure for determining accurately the pH of arterialized capillary blood obtained by incision of an ear lobe or finger which has been previously warmed. The blood is collected in heparinized capillary tubes and there is a special arrangement for mixing before measurement. The pH at 38° C. is measured by means of a microelectrode. This is done on the sample itself and on two other samples. One of these has been equilibrated with an oxygen and carbon dioxide mixture containing CO_2 at a relatively low tension; the other, with a mixture containing CO_2 at a relatively high tension. The method depends on the fact that pH plotted against log pCO_2 of whole blood gives an approximately straight line. One can reproduce this line for the blood sample being investigated by joining each of the points obtained after equilibration with the known gas mixtures. This line can now be used to determine the pCO_2 of the original blood sample, since its pH at 38° C. has been accurately measured. A nomogram is available, whereby the straight line plot can also be used to determine *buffer base, base excess,* and *standard bicarbonate; CO_2 combining power* and the actual bicarbonate of plasma can also be determined if the pH/log pCO_2 line is also determined for separated plasma. Using these data, a number of nomograms are available for indicating the type of acidosis or alkalosis from which a patient is suffering. They are not entirely reliable nor really necessary. The Astrup microtechnique and related methodologies have proved particularly useful in cardiac surgery, as well as in intensive care units.

Whitehead plots the log of H^+ ion concentration (in nmol./liter) against log pCO_2 from the pH and pCO_2 values using the Astrup microtechnique. In this way, he obtains the respiratory and nonrespiratory components of the patient's acid-base status.

Determination of pH of Blood, Plasma, or Serum. The pH may be determined by electrometric methods. The normal ranges from pH 7.35 to 7.45, the average value being pH 7.4. The hydrogen ion concentration of the plasma of venous blood is very slightly greater than that of arterial blood, the pH of the former being about 0.03 lower than that of the latter. In interpreting hydrogen ion concentration, it must be remembered that abnormal values are obtained only during uncompensated stages of acidosis and alkalosis and, therefore, no information of positive value is afforded during the earlier, compensated stages, although during these periods evidence of a disturbance of the acid-base balance may be obtained by means of the estimation of the CO_2 content of plasma, etc. The chief value of the determination of the hydrogen ion concentration of the blood in clinical conditions lies in the fact that when done in conjunction with one of the methods discussed above it indicates the degree of compensation or decompensation of the existing disturbance and makes possible differentiation between nonrespiratory acidosis and respiratory alkalosis as the cause of decrease in plasma CO_2, and between nonrespiratory alkalosis and respiratory acidosis as the cause of increase in this factor.

Values below pH 7.3 indicate a state of uncompensated acidosis, figures as low as 6.95 having been reported in advanced diabetes. Values below pH 7.0 are not very common. Values above pH 7.5 indicate a state of uncompensated alkalosis, the highest values reported in clinical conditions being in the neighborhood of pH 7.6 although figures as high as pH 7.8 to pH 7.9 have been produced by experimental hyperventilation.

Other Methods of Investigation. **Determination of Urinary Ammonium.** A normal individual with an average diet eliminates approximately 0.7 g. of ammonium ion in the urine daily, constituting 2.5 to 4.5 per cent of the total urinary nitrogen. As stated previously, the formation of ammonia in the kidney and its elimination in the urine in combination with acid radicals could constitute one of the means whereby the body conserves its supply of available base, because NH_4^+ could make available more urinary Na^+ ion for interchange with renal tubular cell H^+ ion, and hence the production of more HCO_3^- (p. 411). The necessity for elimination of increased quantities of H^+ ion, as in acidosis, could, in this way, partly be met by an increase in the quantity of ammonium formed by the kidney, the urinary NH_4^+ ion being correspondingly increased. An increase in urinary NH_4^+ ion, therefore, may be indicative of a state of acidosis, a decrease occurring in alkalosis. Values as high as 7 g. of ammonium ion, comprising 50 per cent of the total urinary nitrogen, have been reported in diabetic acidosis.

Advanced grades of acidosis may be present in nephritis (uremia) with no comparable increase in the ammonium content of the urine. This is probably due to the fact that the ammonium-forming function of the kidney has been impaired. Variations in urinary ammonium ion may also be produced by dietary factors, the ingestion of acid-forming foods causing an increase and base-forming foods a decrease in the daily output. In advanced hepatic disease there may be an increase of urinary ammonium ion due to impairment of urea formation by the liver (see p. 618). Under such circumstances the urinary urea is correspondingly decreased.

Titratable Acids in Urine. The ability of the kidneys to excrete increased quantities of acid constitutes one of the mechanisms whereby the body compensates for any tendency toward an increase in the hydrogen ion concentration of the blood plasma. The determination of urinary ammonium plus the titratable acid of the urine is employed to give some information as to the state of the acid-base balance. The following figures, expressed in terms of twenty-four-hour excretion of N/10 acid plus NH_4^+ have been given by Van Slyke: normal resting adult, 0 to 27 ml. per kilo body weight; mild acidosis (compensated), 27 to 65 ml. per kilo body weight; moderate to severe acidosis, 65 to 100 ml. per kilo body weight; severe acidosis, over 100 ml. per kilo body weight. This determination is fairly reliable in indicating the presence or absence of acidosis in diabetes but cannot be utilized as a measure of the degree of acidosis in advanced cases. Moreover, the results may be misleading in conditions associated with primary disorders of the mechanism of urine acidification. In renal tubular acidosis, for example, the urine exhibits decreased acidity or even alkalinity in the presence of the characteristic features of metabolic acidosis in the blood plasma. Conversely, in states of primary potassium depletion (p. 410) the characteristic plasma features of metabolic alkalosis may be accompanied by increased urinary acidity.

Determination of Ketone Bodies. Ketosis, or the accumulation of ketone bodies (acetoacetic acid, beta-hydroxybutyric acid, and acetone) in the body, is frequently, although not invariably, associated with acidosis. The concentration of ketone bodies in the blood of normal individuals ranges from 0.2 to 0.8 mg. per 100 ml., expressed as acetone. On a mixed diet small quantities of ketone bodies are eliminated in the urine of normal individuals, less than 125 mg. (expressed as betahydroxybutyric acid) being eliminated in twenty-four hours if sufficient quantities of carbohydrate are present in the diet. Factors which influence the rate of production, accumulation, and excretion of ketone bodies are discussed elsewhere (p. 109).

Ketosis and excessive ketonuria are commonly observed in starvation, during periods of carbohydrate privation, in normal pregnancy and in the toxemias of pregnancy, following ether anesthesia, in diabetes mellitus and, at times, in certain conditions associated with alkalosis, such as hyperventilation, intestinal obstruction and excessive alkali administration. In all of these conditions the fundamental cause of ketosis is probably the same (p. 109). Whether or not acidosis results from ketosis in any given case depends upon the quantity of ketone bodies produced, upon the condition of the buffer base, and upon the other compensatory mechanisms whereby the body attempts to maintain the normal hydrogen ion concentration of the blood and tissue fluids, namely, hyperventilation, acid elimination in the urine, and possibly ammonium formation by the kidney. Acidosis rarely attains a maximum degree of severity as a result of ketosis of normal pregnancy, starvation, or carbohydrate privation. It is in diabetes mellitus that the presence of ketosis is most significant, being practically always associated with true acidosis. The concentration of ketone bodies in the blood of patients with diabetic acidosis may reach values of 350 mg. or more per 100 ml. Large quantities of these substances may be eliminated in the urine, a twenty-four-hour output of 50 g. or more being not infrequently observed. As stated above, ketonuria is frequently observed in alkalosis due to various causes. In that form dependent upon pyloric or upper intestinal obstruction with continued vomiting, starvation and carbohydrate privation play an important part in its pathogenesis. Moreover, alkalosis itself may increase ketogenesis by diminishing glucose utilization in the liver.

The acidosis of nephritis is not consistently associated with ketosis for obvious reasons. Clinically, the presence or absence of this condition is determined usually by the application of qualitative tests for the presence of acetone or actoacetic acid in the urine (ketonuria). These qualitative tests serve as roughly quantitative procedures and are satisfactory for clinical purposes. Their chief value lies in the fact that ketonuria is one of the first clinical manifestations of beginning acidosis in diabetes and serves as a valuable therapeutic guide.

References

Astrup, P., and Siggaard-Andersen, O.: Micromethods for measuring acid-base values of blood. *In* Sobotka, H., and Stewart, C. P. (eds.): Advances in Clinical Chemistry, Vol. 6. New York, Academic Press, 1963, p. 1.

Darrow, D. C., and Hellerstein, S.: Interpretation of certain changes in body water and electrolytes. Physiol. Rev., *38:*114, 1958.

Davenport, H. W.: The ABC of Acid-Base Chemistry. Ed. 5 (Revised). Chicago, University of Chicago Press, 1969.

Gamble, J. L.: Companionship of Water and Electrolytes in the Organization of Body Fluids. Lane Medical Lectures, California, Stanford University Press, 1951.

Levitin, H.: Acid-base balance. *In* Bondy, P. K. (ed.): Duncan's Diseases of Metabolism. Ed. 6. Philadelphia, W. B. Saunders Company, 1969, p. 1150.

McGilvery, R. W.: Biochemistry: A Functional Approach. Philadelphia, W. B. Saunders Company, 1970.

Robinson, J. R.: Fundamentals of Acid-Base Regulation. Ed. 2. Oxford, Blackwell Scientific Publications, 1965.

Robson, J. S., Bone, J. M., and Lambie, A. J.: Intracellular pH. *In* Bodansky, O., and Stewart. C. P. (eds.): Advances in Clinical Chemistry. Vol. 11. New York, Academic Press, 1968, p. 213.

White, A. G.: Clinical Disturbances of Renal Function. Philadelphia, W. B. Saunders Company, 1961.

Whitehead, T. P.: Blood hydrogen ion: terminology, physiology, and clinical applications. *In* Sobotka, H., and Stewart, C. P. (eds.): Advances in Clinical Chemistry. Vol. 9. New York, Academic Press, 1967, p. 195.

Chapter 15

RESPIRATORY EXCHANGE AND BASAL METABOLISM

The maintenance of cell functions and of life is dependent upon the continuous supply of adequate amounts of oxygen to the tissues. In the course of metabolic activities of the cells, large quantities of CO_2 are produced, the bulk of which must be removed from the body. Both substances being gases, these exchanges of O_2 and CO_2 between the organism and the environment are accomplished by way of the lungs, comprising the beginning and end, respectively, of the process of respiration. Proper understanding of some of the fundamental aspects of these exchanges requires an understanding of certain of the so-called "gas laws."

1. At the same pressure and temperature, equal volumes of all gases contain the same number of molecules.

2. In the absence of chemical reaction between gas and solvent, the amount of a gas which dissolves in a liquid is directly proportional to the pressure of the gas.

3. The pressure exerted by a gas at a given temperature depends upon the number of molecules of gas in a given volume. This is expressed as the "partial pressure" or "tension" (p) of the gas, e.g., pO_2, pCO_2, pN_2.

4. The total pressure of a mixture of gases is equal to the sum of their partial pressures. Thus, the atmospheric (barometric) pressure (BP) is represented by $pO_2 + pCO_2 + pN_2$. This applies only to dry atmospheric air. The partial pressure of water vapor is a function of temperature and is independent of the presence of other gases. Consequently, in calculating pO_2 and pCO_2, the following equation is applicable:

$$BP - pH_2O \text{ vapor} = pO_2 + pCO_2 + pN_2.$$

Inasmuch as the pressure exerted by a gas is determined by the number of molecules of gas, and since the same number of molecules of gases, under the same pressure, occupy equal volumes, it follows that the partial pressure of each component of a mixture of gases will be determined by the fraction which it occupies of the total volume of the mixture. In air, therefore,

$$pO_2 = \frac{(BP - pH_2O \text{ vapor}) \times \% O_2}{100}$$

$$pCO_2 = \frac{(BP - pH_2O \text{ vapor}) \times \% CO_2}{100}$$

$$pN_2 = \frac{(BP - pH_2O \text{ vapor}) \times \% N_2}{100}$$

In dry atmospheric air at a barometric pressure of 760 mm. Hg, the partial pressures of the constituent gases would be: $pO_2 = 20.9$ per cent of 760, or 158 mm.; $pCO_2 = 0.04$ per cent of 760, or 0.3 mm.; $pN_2 = 79$ per cent of 760, or 600 mm. (Table 15-1). Although throughout this book, pressure is expressed in mm. Hg, the International Unit is the *bar* or the *pascal*. The *pascal* is the unit of pressure which corresponds to a force of one newton per square meter. The newton is that force which, when applied to a body with a mass of one kilogram, gives it an acceleration (*in vacuo*) of one meter per second per second. The *bar* is 10^5 *pascals*. 760 mm. Hg is equivalent to 1.01325 bar or 1.01325×10^5 pascals. 1 mm. Hg is equivalent to 1.33322 millibar (mbar.) or 133.322 pascals.

The air in the pulmonary alveoli similarly contains O_2, CO_2 and N_2 but is also saturated with water vapor, which evaporates from the surface of the lining membranes at body temperature, exerting a partial pressure of about 48 mm. Hg. Inasmuch as the total gas pressure in alveolar air is the same as in the inspired (atmospheric) air, the partial pressure of each of the dry gases is exerted in a total of 760 minus 48 mm., or 712 mm. Hg. The same is true of the expired air. The percentage composition of, and pO_2 and pCO_2 in, alveolar and expired air in a normal subject at rest are indicated in Table 15-1.

In the lungs the blood gases come into approximate equilibrium with those of alveolar air. The separating membrane is very thin (1 to 2 μ), permitting ready diffusion of gases, and its surface area is very large (50 to 100 sq. m.). Blood passes through the lungs in about 0.75 second in a resting subject and in about 0.3 second during severe exercise, gas equilibrium being approximated more closely in the former state, which provides longer exposure of the blood to alveolar air. Under normal resting conditions,

TABLE 15-1 PERCENTAGE COMPOSITION OF, AND pO_2 and pCO_2 IN, INSPIRED, EXPIRED AND ALVEOLAR AIR IN NORMAL SUBJECT AT REST

	BAROMETRIC PRESSURE (mm. Hg)	H_2O VAPOR (mm. Hg)	OXYGEN CONTENT (vol. %)	pO_2 (mm. Hg)	CARBON DIOXIDE CONTENT (vol. %)	pCO_2 (mm. Hg)
Inspired air (dry)	760	0	20.9	158	0.04	0.3
Expired air	760	48	16.1	115	4.4	31
Alveolar air	760	48	14.2	101	5.6	40

CHEMICAL CONTROL OF RESPIRATION   435

therefore, the pO_2 and pCO_2 of blood leaving the lungs (arterial blood) are about 100 mm. Hg and 40 mm. Hg, respectively.

In the tissue cells, where oxygen is being utilized and carbon dioxide produced, the pO_2 is relatively low (< 30 mm. Hg) and the pCO_2 high (50 to 70 mm. Hg).

Inasmuch as a gas flows from a higher to a lower pressure (diffusion gradient), O_2 passes from the arterial blood to the tissue cells and CO_2 passes in the opposite direction. Under ordinary conditions of blood flow in a normal resting subject, the pO_2 of the blood leaving the tissues (venous blood) is thereby lowered to about 40 mm. Hg and the pCO_2 increased to about 46 mm. Hg. The venous blood is then returned to the lungs, where it is arterialized, drawing O_2 from and losing CO_2 to the alveolar air, with which it comes into approximate equilibrium. These pressure relationships are illustrated in Figure 15-1.

At rest, a normal man absorbs about 250 ml. O_2 and eliminates about 200 ml. CO_2 per minute. During severe exercise, the volume exchange of these gases may increase more than tenfold. The mechanisms of control of pulmonary ventilation are discussed in detail in texts on physiology, and the chemical aspects are summarized here only briefly.

CHEMICAL CONTROL OF RESPIRATION

The respiratory movements and pulmonary ventilation are controlled by nerve impulses arising in the respiratory center in the medulla. The activity of the latter is influenced, directly or indirectly, by changes in: (1) pCO_2, (2) pH, (3) pO_2, (4) blood flow, and (5) temperature. The indirect influences are mediated through chemoreceptors in the carotid and aortic bodies, which are stimulated chiefly by a decrease in pO_2 but also by an increase in pCO_2 and in H^+ ion concentration. Activity of the respiratory center is stimulated directly by an increase in pCO_2, and in H^+ ion concentration, the pCO_2 being the most important chemical factor in regu-

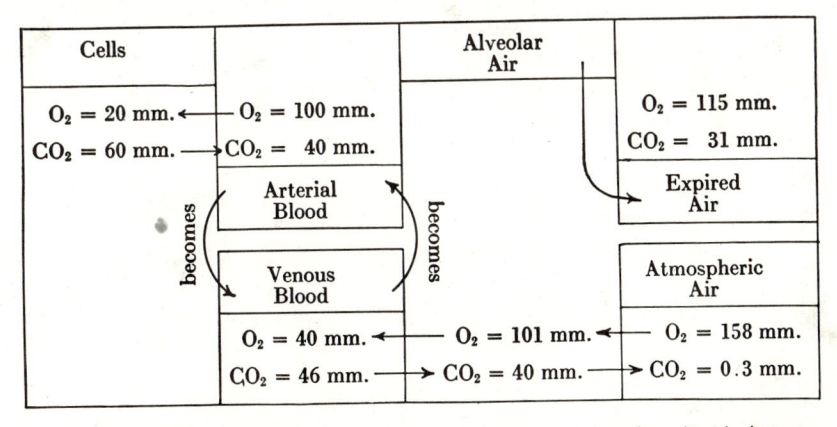

Figure 15-1. Pressure relationships and direction of flow of oxygen and carbon dioxide between tissue cells, blood, lungs and atmosphere.

lation of respiration. The blood brain barrier is freely permeable to CO_2 in solution, but by no means so freely permeable to HCO_3^- ions. This means that in acidosis due to a primary excess of acid, the pCO_2 of cerebrospinal fluid rises and the pH falls rapidly. Moreover, there is not very much buffering capacity in the form of protein. In acidosis due to a primary deficiency of base, the cerebrospinal fluid may remain more alkaline than the extracellular fluid for some time, and the pCO_2 falls. The respiratory center, therefore, may not respond effectively. In alkalosis due to an excess of base, the reaction of the cerebrospinal fluid can remain for some time near normal levels, and hence once again response of the respiratory center may be slow.

Influence of CO_2 Tension and pH. Increase in the pCO_2 or acidity, either in the blood or locally in the respiratory center or aortic and carotid body chemoreceptors, results in stimulation of respiration and increased pulmonary ventilation. The reverse occurs with decrease in pCO_2 and H^+ ion concentration. The pCO_2 apparently exerts an influence in this connection beyond that which can be accounted for by the associated change in pH alone.

Under normal conditions, the extreme sensitivity of the respiratory center to minute changes in pCO_2 in the blood serves to maintain the latter within very narrow limits. The increased ventilation which accompanies any tendency toward a rise in pCO_2 (or H^+ ion concentration) results in prompt elimination of excess CO_2. Similarly, the slowed and shallower breathing, with diminished ventilation, which accompanies any tendency toward a drop in pCO_2 (or H^+ ion concentration) results in decreased excretion of CO_2. This constitutes an important part of the mechanism of acid base regulation (p. 406).

Influence of O_2 Tension. Reduction in the pO_2 of arterial blood causes depression of the respiratory center but stimulation of the carotid and aortic chemoreceptors. The latter effect exceeds the former, the predominant consequence being stimulation of respiration and increased ventilation. This is of importance in physiological adjustment to diminished atmospheric O_2 tensions during ascent to high altitudes. The stimulating effects of decrease in pO_2 are usually not manifest in the resting subject until the pO_2 has fallen to rather low levels but are more pronounced at higher levels during exercise.

TRANSPORT OF OXYGEN

The tissue cells of a normal man utilize about 250 ml. of oxygen per minute in the resting state and may consume over ten times that amount during strenuous exercise. At the pO_2 of alveolar air, a liter of blood is able to take up in physical solution only about 2 ml. O_2. Inasmuch as the maximum attainable blood flow is about 25 liters per minute, the maximum amount of O_2 that can be supplied to the tissue cells in this form is about 50 ml./minute, less than 2 per cent of the maximum requirement.

Obviously, the oxygen must be transported in the blood chiefly in some form other than physical solution. This is accomplished by its combination with hemoglobin.

Under conditions of complete saturation, the blood of a normal man (16 g. Hb/100 ml.) holds a little more than 21 ml. of O_2/100 ml., only about 0.3 ml. of which is in physical solution. Although of small magnitude, the latter factor is of considerable significance because, being a reflection of the pO_2, it comes into equilibrium in the lungs with the alveolar air O_2, and in the tissues with the interstitial fluid O_2. It is, therefore, one of the determinants of the amount of O_2 taken up and liberated by Hb, according to the equation

$$O_2 + Hb \rightleftarrows HbO_2.$$

One mole of oxygen combines with 16,700 grams of Hb. Inasmuch as a mole of O_2 (at 0° C. and 760 mm. Hg) occupies 22,400 ml., 1 gram of Hb can combine with 22,400/16,700 or 1.34 ml. of O_2 (Hüfner factor). Consequently, the 16 grams of Hb in 100 ml. of blood, when fully saturated, can carry 21.4 ml. of O_2. However, under the conditions existing in the circulating blood, the Hb is not completely saturated with O_2, the amounts present in normal arterial and venous blood being indicated in Table 15-2. The degree of saturation depends upon the pO_2, this relationship being expressed in the dissociation curve of oxyhemoglobin (Fig. 15-2).

Dissociation of Oxyhemoglobin. The equation $O_2 + Hb \rightleftarrows HbO_2$ is shifted to the right or left as the pO_2 increases or decreases, with corresponding increasing saturation of Hb with oxygen (HbO_2) and increasing dissociation of HbO_2, respectively. However, the degree of saturation (or dissociation) is not directly proportional to the pO_2. If the blood is equilibrated with air containing O_2 at different partial pressures, and the quantity of O_2 (or HbO_2) in the blood is plotted against the pO_2, an S-shaped curve is obtained (Fig. 15-2). This is referred to as the dissociation curve of oxyhemoglobin and indicates the relative amounts of oxyhemoglobin (HbO_2) and reduced hemoglobin (Hb) present at different levels of oxygen tension.

Certain features of this curve are of great physiological significance.

(a) Influence of pO_2. At a pO_2 of 100 mm. Hg, as in arterial blood

TABLE 15-2 OXYGEN AND CARBON DIOXIDE CONTENTS AND PRESSURES IN BLOOD

	OXYGEN		CARBON DIOXIDE	
	CONTENT (ml./100 ml.)	pO_2 (mm. Hg)	CONTENT (ml./100 ml.)	pCO_2 (mm. Hg)
Arterial blood	17–22 (20)*	(100)	44–50 (48)	(40)
Venous blood	11–16 (13)	(40)	51–58 (55)	(46)
A-V difference	4–8 (6.5)		4–8 (6.5)	

* Mean values indicated in parentheses.

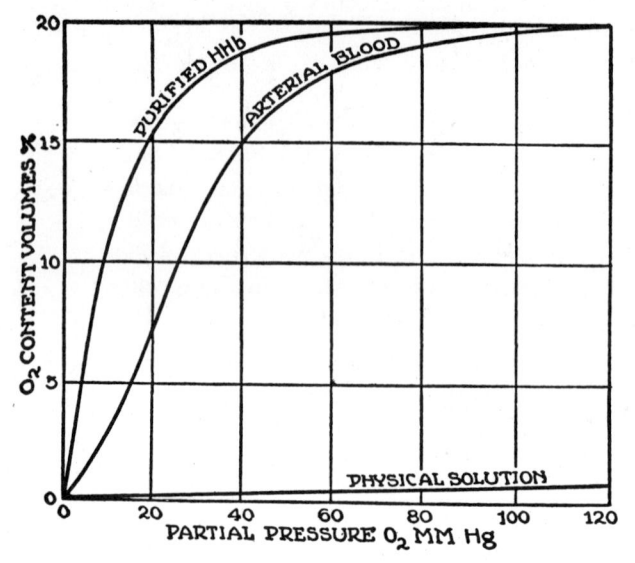

Figure 15-2. Dissociation curve of oxyhemoglobin. The O_2 content of arterial blood and plasma (physical solution) and a solution of purified HHb at various partial pressures of O_2. (After Barcroft. From Fulton: Howell's Textbook of Physiology, 16th edition. W. B. Saunders Company, 1950.)

(at pCO_2 40 mm. Hg and pH 7.4), Hb is 95 to 98 per cent saturated with O_2, i.e., 95 to 98 per cent is in the form of HbO_2 and 2 to 5 per cent in the reduced form. At a pO_2 of 70 mm. Hg, Hb is still 90 per cent saturated. The ability of the blood to carry oxygen, therefore, varies relatively slightly with variations in pO_2 above 70 mm. Hg. In fact, there is comparatively little reduction in Hb saturation, i.e., comparatively little dissociation of HbO_2, until the pO_2 falls below 50 mm. Hg, dissociation of HbO_2 increasing greatly, i.e., Hb saturation decreasing greatly, as the pO_2 falls below this level (Fig. 15-3; Table 15-3).

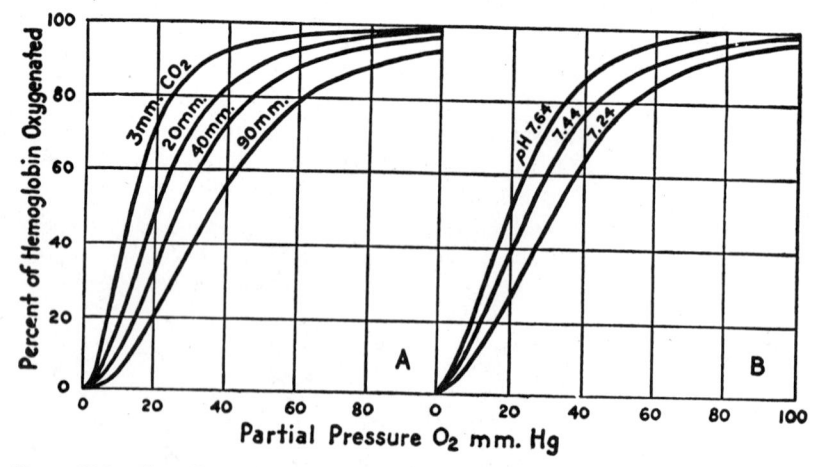

Figure 15-3. The effects of increased pCO_2 and decreased pH upon the dissociation of oxyhemoglobin. Increased temperature also shifts the curve in the same direction as increased pCO_2. (After Peters and Van Slyke. From Fulton: Howell's Textbook of Physiology, 16th edition. W. B. Saunders Company, 1950.)

TABLE 15-3 EFFECT OF OXYGEN AND CARBON DIOXIDE TENSIONS ON OXYGENATION OF HEMOGLOBIN*

O₂ TENSION mm.	PROPORTION OF HEMOGLOBIN COMBINING WITH OXYGEN AT FOLLOWING CO₂ TENSIONS			
	CO₂ = 3 mm. (%)	CO₂ = 20 mm. (%)	CO₂ = 40 mm. (%)	CO₂ = 80 mm. (%)
0	0	0	0	0
5	13.5	6.8	5.5	3.0
10	38.0	19.5	15.0	8.0
20	77.6	50.0	39.0	26.0
30	92.0	72.2	60.6	49.8
40	96.7	87.0	76.0	63.5
50	98.5	93.3	85.5	76.9
60	100	96.3	90.5	85.0
70	100	98.0	94.0	90.3
80	100	99.0	96.0	93.7
90	100	100	97.5	95.7
100	100	100	98.6	97.1

* (Henderson, Bock, Field, and Stoddard: J. Biol. Chem. 59:379.)

Because of this peculiar behavior of Hb in relation to the pO_2, adequate uptake of oxygen by the blood in the lungs is assured as long as the alveolar air pO_2 is above 80 mm. Hg. Adequate liberation of oxygen from the blood is assured in the tissues, where the pO_2 is usually below 30 mm. Hg.

(b) Influence of pCO_2 and pH. With increasing CO_2 tension in the blood, the affinity of Hb for O_2 decreases (i.e., increasing dissociation of HbO_2, or decreasing saturation of Hb), the equation, $O_2 + Hb \rightleftarrows HbO_2$, being shifted to the left. This effect is more pronounced at relatively low than at high levels of pO_2 because of the shape of the dissociation curve (Fig. 15-3, Table 15-3). Thus, at 90 to 100 mm. Hg pO_2 (alveolar air), variations in pCO_2 from 20 to 80 mm. Hg have comparatively little effect on HbO_2 dissociation (95 to 100 per cent saturation of Hb). On the other hand, at a pO_2 of 20 mm. Hg (tissues), increasing the pCO_2 from 20 mm. to 80 mm. Hg causes the percentage saturation of Hb to fall from about 40 to less than 20.

Acids other than H_2CO_3, e.g., lactic acid, exert a similar effect, i.e., increase in the H^+ ion concentration of the blood increases dissociation of HbO_2. As in the case of the pCO_2, this influence is more pronounced at relatively low than at high pO_2 levels (Fig. 15-3).

Rise in temperature also increases dissociation of HbO_2, shifting the curve in the same direction as do increased pCO_2 and acidity.

As a result of these phenomena, adequate uptake of oxygen from the alveolar air is assured, despite the presence of CO_2, and its delivery to the tissue cells, facilitated primarily by the relatively low pO_2, is enhanced by the relatively high pCO_2, acidity and temperature in the actively metabolizing tissues.

TRANSPORT OF CARBON DIOXIDE

In the resting state, about 200 ml. of CO_2 are produced per minute in the course of oxidative processes in the tissues and are carried in the blood to the lungs where they are excreted. During strenuous exercise, the quantity produced and transported increases enormously. The manner in which this relatively large amount of acid ($CO_2 + H_2O \rightleftarrows H_2CO_3$) is carried in the blood with only slight change in pH is discussed elsewhere (p. 403). Attention will be directed here chiefly to the various forms in which CO_2 exists in the blood and their quantitative distribution.

The direction of flow of CO_2 is determined by the pCO_2, which is 50 to 70 mm. Hg in the tissues, is about 40 mm. in the arterial blood, rises to 46 mm. during passage of blood through the tissues (venous), and is 40 mm. in the alveolar air. The diffusion constant for CO_2 is much higher than for O_2 because of its greater solubility in body fluids; consequently, it diffuses readily from the venous blood into the pulmonary alveoli despite the rather low pressure gradient (Fig. 15-1, p. 435).

CO_2 in Arterial Blood (Table 15-4). At rest, 100 ml. of arterial blood contain 45 to 55 ml. of CO_2 (volumes per cent), about 75 per cent of which is in the plasma and 24 per cent in the erythrocytes.

The 35.6 ml. (average) in the plasma are present in three forms: (1) physical solution (1.6 ml.); (2) HCO_3^- ions (34 ml.); (3) carbamino compounds of plasma proteins (p. 441), in very small amounts (less than 0.7 ml.) because the plasma proteins contain relatively few NH_2 groups capable of combining with CO_2 under conditions existing in the blood. The 12.6 ml. (average) in the erythrocytes are present in the same forms: (1) physical solution (0.8 ml.); (2) HCO_3^- ions (9.6 ml.); carbamino compounds of Hb (2.2 ml.). About 90 per cent of the CO_2 in arterial blood is in the form of HCO_3^- ions (bicarbonate).

Entrance of CO_2 in Tissues (Table 15-4). At rest, 3.5 to 4.5 ml. (average 3.7) of CO_2 enter each 100 ml. of blood in the tissues, diffusing across the capillary walls in consequence of the tissue : blood CO_2 diffusion gradient. It enters the blood plasma in physical solution and diffuses readily from the plasma into the erythrocytes.

Carbon dioxide reacts with water as follows:

$$CO_2 + H_2O \rightleftarrows H_2CO_3 \rightleftarrows H^+ + HCO_3^-$$

In the plasma, this reaction proceeds very slowly, the equilibrium being far to the left. Consequently, the increment in CO_2 occurring in passing through the tissues drives the reaction to the right only slightly.

Within the erythrocytes, the reaction

$$CO_2 + H_2O \rightarrow H_2CO_3 \rightarrow H^+ + HCO_3^-$$

proceeds very rapidly, owing to the presence of carbonic anhydrase, an enzyme which catalyzes this reaction. Because of prompt removal of the ion end-products (H^+ and HCO_3^-), as described on the following page, the fraction of the added CO_2 which diffuses into the erythrocytes is much larger than that which remains in the plasma (Table 15-4).

The relatively small capacity of plasma proteins for forming carbamino compounds is not influenced significantly by changes which occur in the blood during its passage through the tissues. Consequently, virtually none of the added CO_2 is transported in this manner (i.e., plasma carbamino CO_2 is unaltered). Within the erythrocytes, however, a considerable fraction of the added CO_2 is held in this form because of the fact that the capacity of reduced hemoglobin for forming carbamino compounds is more than three times that of oxyhemoglobin. Moreover, this capacity increases with increasing levels of pCO_2. In consequence, therefore, of the liberation of O_2 and addition of CO_2 in the tissues, about 20 to 25 per cent of the increment in blood CO_2 is transported by hemoglobin as carbamino-CO_2.

The largest fraction, by far, of the added CO_2 which enters the erythrocytes undergoes the reaction

$$CO_2 + H_2O \rightarrow H_2CO_3 \rightarrow H^+ + HCO_3^-$$

which proceeds rapidly because the end-products, H^+ and HCO_3^- ions, are removed promptly. The added H^+ ions are very effectively "neutralized" by the enhanced buffering capacity of hemoglobin incident to the transformation of HbO_2 to Hb, reduced hemoglobin being a weaker acid than oxyhemoglobin (p. 403). The added HCO_3^- ions, being produced within the erythrocytes more rapidly than in the plasma, diffuse from the erythrocytes into the plasma, an equal number of Cl^- ions passing in the opposite direction (p. 405). Within the erythrocytes, the increase in negatively charged ions (HCO_3^- and Cl^-) is balanced by the K^+ and some Na^+ ions "released" by the transformation of HbO_2 to Hb, in place of the H^+ ions, which the latter is buffering. In the plasma, the HCO_3^- ions entering from the erythrocytes are balanced by the Na^+ ion freed by the diffusion of Cl^- ions into the erythrocytes (Fig. 14-1, p. 404).

About 70 per cent of the added CO_2 is carried in the blood as bicarbonate (HCO_3^-), about 60 per cent in the plasma, the remainder in the erythrocytes. However, about 90 per cent of the increment in plasma HCO_3^- originates within the erythrocytes, as indicated above, largely as a direct

TABLE 15-4 AVERAGE DISTRIBUTION OF CARBON DIOXIDE IN 100 ML. NORMAL BLOOD

CARBON DIOXIDE	ARTERIAL ml.	% of total	VENOUS ml.	% of total	DIFFERENCE ml.	% of total
Total	48.2		51.9		3.7	
Total in plasma (60 ml.)	35.6	74	38.0	73	2.4	65
as dissolved CO_2	1.6	3	1.8	3	0.2	5
as HCO_3^-	34.0	71	36.2	70	2.2	60
Total in erythrocytes (40 ml.)	12.6	26	13.9	27	1.3	35
as dissolved CO_2	0.8	1.5	0.9	1	0.1	3
as carbamino-CO_2	2.2	4.5	3.1	6	0.9	24
as HCO_3^-	9.6	20	9.9	20	0.3	8

consequence of the enhanced buffering capacity of the hemoglobin. Directly and indirectly, Hb is responsible for the transport of over 80 per cent of the CO_2 added to the blood in the tissues.

HYPOXIA

Hypoxia, meaning oxygen deficiency, should properly be applied to any condition of insufficiency of tissue oxidation processes. According to the factors involved, four types may be differentiated. These have been termed (1) hypoxic hypoxia, (2) anemic hypoxia, (3) stagnant hypoxia, and (4) histotoxic hypoxia.

Hypoxic Hypoxia. This group includes conditions characterized by normal oxygen capacity but diminished oxygen tension in the arterial blood with a consequent varying degree of hemoglobin unsaturation.

(a) High Altitudes. The condition commonly designated mountain sickness is in reality a state of hypoxic hypoxia caused by diminished oxygen tension in the atmospheric air and consequently in the alveolar air and blood stream. The resultant stimulation of respiration leads to a respiratory alkalosis. This increases the ability of hemoglobin to take up oxygen at the lowered partial pressure (Bohr effect) and results in increased arterial saturation with oxygen.

(b) Rapid, Shallow Respiration. Shallow breathing is conducive to inefficient oxygenation of the blood since, if the volume of tidal air is greatly decreased, comparatively little fresh air passes the physiologic dead space (150 ml.) to enter the alveoli. It is questionable whether this factor, in itself, is capable of producing hypoxia in disease states but it is unquestionably a contributory factor in such conditions as pneumonia, in which hypoxia is dependent largely upon other factors.

(c) Mechanical Interference with Oxygen Absorption. This condition exists in pneumonia, pulmonary edema, pulmonary congestion, emphysema, bronchial asthma, bronchitis, other diseases of the respiratory tract and in weakness of the respiratory muscles. This type of hypoxic hypoxia is the one most commonly observed clinically. Obviously, any condition which interferes with the passage of atmospheric air into the pulmonary alveoli or with the diffusion of oxygen from the alveoli into the blood will result in diminution in the oxygen content of the blood. It must be recognized, however, that a considerable portion of the ventilating surface of the lung may be functionally incapacitated with no alteration in oxygenation of the blood. Whether or not hypoxia occurs in these conditions is determined largely by (1) the circulation of blood in the affected area and (2) the rate of blood flow through the unaffected portions of the lung.

(d) Congenital Heart Disease. In certain cases of congenital cardiac septal defects a portion of the blood may flow directly from the right to the left side of the heart without having passed through the lungs, the mixture of aerated and nonaerated blood in the systemic circulation resulting in a state of hypoxic hypoxia. The condition is apparently further aggravated by incomplete oxygenation of the blood that does flow through the lungs. Hypoxic hypoxia is relatively infrequently observed in congenital heart lesions associated with septal defects because the pathologic intracardiac deviation of blood flow is usually from the left to the right rather than from right to left.

The characteristic chemical feature of this type of hypoxia is an abnormally low oxygen saturation of the hemoglobin of arterial blood. Because of the compensatory polycythemia which usually occurs, particularly if the underlying condition is of a chronic nature, the oxygen content of arterial blood may be within normal limits or even actually increased. The oxygen content of venous blood may be normal or diminished, depending upon the arterial blood values and upon the degree of oxygen utilization in the tissues.

Anemic Hypoxia. This type of hypoxia is characterized by a diminution in the oxygen capacity of arterial blood due to a decrease in the amount of functioning hemoglobin.

(a) Anemia. The occurrence of hypoxia in anemia is readily understandable, the degree of oxygen saturation of arterial blood being normal but its oxygen content being diminished in proportion to the decrease in hemoglobin.

(b) Carbon Monoxide Poisoning (p. 273). Carbon monoxide combines with the same site in the hemoglobin molecule as does oxygen, the combining capacity of hemoglobin for both gases being identical. However, the affinity of hemoglobin for carbon monoxide is more than 200 times as great as its affinity for oxygen and, therefore, in the presence of relatively small concentrations of carbon monoxide in the air, a considerable quantity is taken up by the blood with a consequent reduction in the amount of hemoglobin available for the transportation of oxygen to the tissues. Furthermore, the presence of carbon monoxide in the blood apparently diminishes the facility of dissociation of oxyhemoglobin and, therefore, increases the existing hypoxic tendency by interfering with the liberation of oxygen from the blood in the tissues.

(c) Methemoglobinemia (p. 272). Methemoglobin is a substance formed from reduced hemoglobin through the action of an oxidizing agent, which has the property of not being capable of combining with oxygen. Hypoxia associated with methemoglobinemia is of the anemic type inasmuch as is it due to a decrease in the quantity of functioning hemoglobin in the circulating blood.

(d) Sulfhemoglobinemia (p. 273).

Stagnant Hypoxia. Stagnant hypoxia is due to circulatory inefficiency, the rate of blood flow through the tissues being retarded, with resulting increase in the percentage volume of oxygen removed from the blood in its passage through the capillaries. It is observed most commonly in circulatory failure associated with decompensated heart disease, in shock, in systemic arteriovenous shunts and in conditions associated with vasospastic phenomena such as Raynaud's disease. In this type of hypoxia the arterial oxygen tension, capacity and content may be normal, but because of the excessive oxygen loss in the tissues the venous oxygen tension and content are subnormal. In myocardial failure, hypoxic anoxia is superimposed upon stagnant hypoxia because of the impaired diffusion of oxygen from the alveolar air into the blood through the congested pulmonary alveoli.

Histotoxic Hypoxia. Histotoxic hypoxia is a term suggested by Peters and Van Slyke to indicate a condition in which the oxygen supply is normal in every respect but the degree of oxygen utilization by the tissues is diminished because the tissue cells are poisoned in such a manner that they cannot use oxygen properly. Histotoxic hypoxia occurs in poisoning by alcohol, cyanide, and perhaps formaldehyde and acetone. In the absence of complicating factors such as shock, the arterial and venous oxygen tension, capacity, and content are within normal limits.

Relative Hypoxia. In certain situations, such as thyrotoxicosis and severe muscular exercise, the increased requirement of the tissues for oxygen can exceed the rate of supply. Here again, the arterial oxygen tension and capacity are within normal limits.

ENERGY METABOLISM

Throughout the discussions of the metabolism of organic foodstuffs, attention has been focused particularly on the mechanisms and the nature of their chemical transformation. Various components of the tissues are undergoing degradation (catabolism) and resynthesis (anabolism) continually. Certain of the chemical reactions involved in these metabolic processes are exergonic, i.e., they are accompanied by liberation of energy, whereas others are endergonic, i.e., they require the introduction of energy.

Obviously, even under resting conditions, the maintenance of life and normal function requires the constant performance of work by a variety of tissues. This implies the utilization of energy, which must be supplied from external sources if the body stores of energy are to be preserved at a constant level. We shall concern ourselves here with the magnitude of the energy requirement under normal conditions and the factors that may influence it.

Caloric Value of Foods. Following the demonstration by early investigators in this field that production of heat in the body was dependent primarily on oxidative processes, attempts were made to relate the quantity of heat produced to the amounts of foodstuffs metabolized. This was feasible because the heat produced by burning (oxidizing) these substances outside the body can be determined readily in a combustion calorimeter, the average values (in Calories) obtained per gram being: glucose, 3.96; starch, 4.2; animal fats, 9.5; animal proteins, 5.6 (vegetable proteins lower). A "Calorie" (kilocalorie) is the amount of heat required to raise the temperature of 1 kg. of water 1° C. (e.g., from 15° to 16° C.). In the body, protein does not undergo complete oxidation, a portion of its amino groups being converted to and excreted as ammonia and urea; this involves a loss of about 1.3 Cal. per gram, leaving 4.3 Cal. produced per gram of protein metabolized. Taking into consideration the variations in caloric value of individual carbohydrates, fats and proteins, their average energy value when metabolized may be represented as follows (Calories per gram): fat, 9.3; carbohydrate, 4.1; protein, 4.1. When these corrected figures are applied, it is found that the amount of energy (Calories) produced by a given quantity of these foodstuffs in the body is the same as that (heat) produced by their combustion outside the body, within the limits of experimental error. Inasmuch as ingested foods are not completely assimilated, the caloric values of the fats, carbohydrates, and proteins of the diet are usually calculated as 9, 4 and 4 Cal./gram, respectively.

Heat Production. In living tissues, the bulk of the energy derived from the metabolism of organic compounds is liberated not as a single burst of heat, as occurs in their combustion outside the body, but rather in a stepwise fashion through a series of integrated enzymatic reactions. Therefore, although in the complete oxidation of these compounds (e.g., glucose to CO_2 and H_2O) their full potential energy value is ultimately realized, relatively small fractions of the total are made available at each stage of their degradation. Presented in small parcels, energy is utilized much more efficiently than if it had been liberated explosively.

The energy thus produced is disposed of as follows: (1) A portion is primarily converted to heat for the maintenance of body temperature. (2) A portion is utilized for the performance of work, e.g., mechanical (muscle contraction), electrical (nerve impulse), secretory (glandular, intestinal, renal tubular cells, etc.), chemical (endergonic metabolic reactions). Such work is ultimately reflected largely in heat production. (3) A portion may be stored very temporarily in "energy-rich" phosphate bonds, or for longer periods in the form of substances (e.g., fat, glycogen) that may be called upon to provide energy at some future date. The over-all body composition and weight of normal adults does not vary appreciably from day to day; i.e., there is little or no accretion or loss of tissue, new components being formed only in amounts required for replacement (metabolic turnover, "dynamic state"). On an adequate diet, the organism is therefore in a state of equilibrium with its environment; this applies to its intake of energy as well as of H_2O, N, and inorganic elements. Consequently, an amount of heat must be lost, either actual (radiation, conduction, vaporization of water) or potential (constituents of urine and feces), that is equal to the caloric value of ingested proteins, fats and carbohydrates (i.e., caloric equilibrium). This has been demonstrated to be the case by direct measurement of heat loss (direct calorimetry).

Respiratory Quotient. In the process of oxidation, carbohydrates, fats and proteins react with definite amounts of oxygen, forming definite amounts of carbon dioxide and water. The designation "respiratory quotient" or "R.Q." is applied to the ratio of the volume of CO_2 produced to that of O_2 utilized in this process (R.Q. = Vol. CO_2/Vol. O_2).

TABLE 15-5 CALORIC, O_2, AND CO_2 EQUIVALENTS OF CARBOHYDRATE, FAT AND PROTEIN

	CARBOHYDRATE	FAT	PROTEIN
Calories per gram	3.7–4.3	9.5	4.3
Liters CO_2 per gram	0.75–0.83	1.43	0.78
Liters O_2 per gram	0.75–0.83	2.03	0.97
Respiratory quotient	1.0	0.707	0.801
Caloric value per liter O_2	5.0 Cal.	4.7 Cal.	4.5 Cal.

Carbohydrate R.Q. Complete oxidation of glucose may be represented as follows:

$$C_6H_{12}O_6 + 6\ O_2 \rightarrow 6\ CO_2 + 6\ H_2O$$
(1 mole) + (6 moles) → (6 moles) + (6 moles)
(108 grams) + (192 grams) → (264 grams) + (108 grams)

The glucose R.Q. is therefore 6/6, or 1.0. Inasmuch as 1 molecule of gas occupies a volume of 22.4 liters at normal temperature and pressure, the volume of O_2 and CO_2 involved in this reaction is 22.4 × 6, or 134.4 liters. Consequently, 1 gram of glucose reacts with 134.4/180, or 0.75 liter of oxygen, forming 0.75 liter of CO_2. Since oxidation of 1 gram of glucose liberates 3.74 Cal., 1 liter of O_2 is the equivalent of 3.75/0.75, or 5.0 Cal. when glucose is oxidized. This (5.0 Cal.) is the caloric value of 1 liter of oxygen for glucose oxidation in the body.

Fat R.Q. Complete oxidation of tripalmitin may be represented as follows:

$$2\ (C_{51}H_{98}O_6) + 145\ O_2 \rightarrow 102\ CO_2 + 98\ H_2O$$
(2 moles) + (145 moles) → (102 moles) + (98 moles)
(1612 grams) + (4640 grams) → (4488 grams) + (1764 grams)

The R.Q. for tripalmitin is therefore 102/145, or 0.704. That for triolein is 0.712. The R.Q. for mixed fats in the body is regarded as 0.707. The volumes of O_2 and CO_2 involved in this reaction are 3248 liters (145 × 22.4) and 2284.8 liters (102 × 22.4), respectively. Consequently, 1 gram of tripalmitin reacts with 3248/1612, or 2.01 liters of O_2. Since oxidation of 1 gram of this fat liberates 9.5 Cal., 1 liter of O_2 is the equivalent of 9.5/2.01, or 4.7 Cal. when tripalmitin is oxidized. Essentially the same value applies to other animal fats and, therefore, 4.7 is the caloric value of 1 liter of oxygen for fat oxidation in the body.

Protein R.Q. Amino acids do not undergo complete oxidation in the body and, therefore, the type of equation written for carbohydrates and fats is not applicable to proteins. However, the approximate quantitative relationships between O_2 and CO_2 in the metabolic oxidation of amino acids may be calculated from data on the average composition and metabolism of meat protein. These are indicated in Table 15-6.

In its oxidation to CO_2, 12 grams of C combine with 32 grams of O_2; in forming H_2O, 2 grams of H combine with 16 grams of O_2. Consequently, 41.5 grams of C require 110.66 grams of O_2 and will produce 152.17 grams (3.46 moles) of CO_2, while 3.4 grams of H require 27.52 grams of O_2 [total O_2 utilized is 138.18 grams (4.32 moles)]. The R.Q. for protein oxidation is, therefore, 3.46/4.32, or 0.801. The volumes of O_2

TABLE 15-6 AVERAGE ELEMENTAL COMPOSITION AND METABOLISM OF 100 GRAMS OF MEAT PROTEIN (LOEWY)

	100 GRAMS MEAT PROTEIN	EXCRETED IN FECES AND URINE	INTRAMOLECULAR WATER	REMAINING FOR OXIDATION
	(grams)	(grams)	(grams)	(grams)
C	52.38	10.877		41.5
H	7.27	2.87	0.96	3.4
O	22.68	14.99	7.69	
N	16.65	16.65 (16.28 urine)		
S	1.02	1.02		

and CO_2 involved were 96.7 liters (4.32 × 22.4) and 77.46 liters (3.46 × 22.4), respectively (per 100 grams protein). Since oxidation of 1 gram of protein liberates 4.3 Cal., 1 liter of O_2 is the equivalent of 4.3/0.967, or 4.5 Cal., which represents, therefore, the caloric value of 1 liter of O_2 for protein oxidation in the body.

As indicated in Table 15-6, the metabolism of 100 grams of protein gave rise to urinary excretion of 16.28 grams of nitrogen. Each gram of urinary N therefore represented the metabolism of: 100/16.28, or 6.15 grams of meat protein; 96.7/16.28, or 5.94 liters of O_2; 77.46/16.28, or 4.76 liters of CO_2. On the basis of analytical data for the average protein, it is generally estimated that 1 gram of urinary N represents: (1) the metabolism of 6.25 grams of protein; (2) utilization of 5.91 liters of O_2; (3) production of 4.76 liters of CO_2; (4) liberation of 26.51 Cal.

Example. The following data were obtained under basal conditions (overnight fast): (1) urinary N, 0.18 gram/hour; (2) O_2 consumption, 12.2 liters/hour; (3) CO_2 production 9.2 liters/hour. Information desired: (1) heat (Calories) production; (2) quantities of protein, carbohydrate and fat metabolized; (3) percentages of total heat produced by oxidation of protein, carbohydrate and fat, respectively.

0.18 gram of urinary N represents:
    0.18 × 6.25 = 1.125 grams of protein metabolized
    0.18 × 5.91 = 1.06 liters of oxygen
    0.18 × 4.76 = 0.85 liter of carbon dioxide
    0.18 × 26.51 = 4.77 Calories
Total O_2 (12.2 liters) minus protein O_2 (1.06 liters) = nonprotein O_2 (11.14 liters)
Total CO_2 (9.2 liters) minus protein CO_2 (0.85 liter) = nonprotein CO_2 (8.35 liters)
Nonprotein R.Q. = 8.35/11.14 = 0.75

As indicated in Table 15-7, a nonprotein R.Q. of 0.75 represents the liberation of 4.739 Cal. per liter O_2, 15.6 per cent of which comes from carbohydrate and 84.4 per cent from fat oxidation.

    11.14 liters O_2 × 4.739 = 52.79 Cal. (nonprotein)
    15.6 per cent of 52.79 = 8.24 Cal. from carbohydrate (2.01 grams)
    84.4 per cent of 52.79 = 44.55 Cal. from fat (4.79 grams)
    52.79 + 4.77 (protein Cal.) = 57.56 total Cal./hour, of which
        1.125 grams protein provided 4.77/57.56, or 8.3 per cent,
        2.01 grams carbohydrate provided 8.24/57.56, or 14.3 per cent,
        4.79 grams fat provided 44.55/57.56, or 77.4 per cent.

Significance of R.Q. The chief practical value of the calculations discussed above lies in the fact that they permit estimation of the amounts of protein, fat and carbohydrate metabolized during a given period, from knowledge of (1) the urinary N excretion, (2) the O_2 consumption, and (3) the CO_2 production.

If the organism (or isolated tissue, in vitro) were utilizing carbohydrate exclusively, the R.Q. would be 1.0; if utilizing fats exclusively, it would be 0.7; and if utilizing proteins exclusively, it would be 0.8. Conversion of carbohydrate to fat results in R.Q. values greater than 1.0 because some of the oxygen of the carbohydrate, which contains relatively more of this element in proportion to carbon than do fats, becomes available for oxidative processes, decreasing the quantity of O_2 required from the inspired air. On the same basis, theoretically, conversion of fat to carbohydrate would produce R.Q. values below 0.7. Such values have been observed (e.g., in diabetes mellitus), but fatty acids have not been shown to produce a net increase in carbohydrate. The R.Q. for conversion of protein to glucose has been estimated as 0.632 to 0.706.

It has been found that in the average normal adult, studied under "basal" conditions, i.e., at complete mental and physical rest, fourteen to eighteen hours after taking food, the R.Q. is approximately 0.82. At this level, the caloric value of oxygen is 4.825 Cal./liter (Table 15-7). Progressive decrease in the proportion of carbohydrate being oxidized in the "metabolic mixture" (i.e., carbohydrate deprivation, diabetes mellitus) is reflected in progressive lowering of the R.Q., approaching 0.7 and occasionally falling below this value (excessive gluconeogenesis from protein).

CALORIMETRY

On the basis of what has been stated concerning heat production, it is evident that measurement of the amount of heat lost over any given period of time affords an approach to the estimation of the energy production (oxidative metabolism) of the body during that period. This may be accom-

TABLE 15-7 RESPIRATORY QUOTIENT AND CALORIC EQUIVALENT OF OXYGEN FOR DIFFERENT MIXTURES OF FAT AND CARBOHYDRATE

R.Q.	PERCENTAGE OF TOTAL O_2 CONSUMED BY Carbohydrate	Fat	PERCENTAGE OF HEAT PRODUCED BY Carbohydrate	Fat	CALORIES PER LITER O_2
0.707	0	100	0	100	4.686
0.75	14.7	85.3	15.6	84.4	4.739
0.80	31.7	68.3	33.4	66.6	4.801
0.82	38.6	61.4	40.3	59.7	4.825
0.85	48.8	51.2	50.7	49.3	4.862
0.90	65.9	34.2	67.5	32.5	4.924
0.95	82.9	17.1	84.0	16.0	4.985
1.00	100	0	100	0	5.047

plished either directly (direct calorimetry) or indirectly (indirect calorimetry).

Direct Calorimetry. The subject is placed in an insulated chamber (calorimeter) so constructed as to permit direct and precise measurement of heat loss from the body. Water is circulated through tubes within the chamber; the rate of flow is recorded automatically, as is the temperature of the water entering and leaving the chamber. Air is circulated through the calorimeter and measurements are made of the rate of flow, and of changes in temperature and water vapor. On the basis of these data, one may calculate the actual heat loss very accurately. Determinations may be made simultaneously of the consumption of oxygen and excretion of carbon dioxide by analyses of the air entering and leaving the chamber, permitting calculation of the respiratory quotient.

This procedure requires elaborate and expensive apparatus and is now chiefly of historical interest. It was found that calculation of heat production from data on O_2 consumption and CO_2 production gave results which approximated very closely those obtained by direct calorimetry. This observation formed the basis for the application of the widely employed procedure of indirect calorimetry.

Indirect Calorimetry. As illustrated by the example cited on page 447, the amount of heat loss during an experimental period can be calculated accurately from the following data: (1) urinary N excretion during that period; (2) O_2 consumption; (3) CO_2 excretion. The validity of this calculation rests upon the following facts: (1) 1 gram of urinary N represents the metabolism of 6.25 grams of protein, the consumption of 5.91 liters of O_2, the production of 4.76 liters of CO_2, and the liberation of 26.51 Cal.; (2) oxygen has a known fixed caloric value per liter at any given R.Q. level.

For usual clinical purposes, determination of urinary N may be dispensed with, inasmuch as 90 to 95 per cent of the heat produced is derived from oxidation of carbohydrate and fat. Two general types of apparatus may be employed: (1) the open-circuit system; (2) the closed-circuit system. The former is the more accurate; the latter is much more convenient, is sufficiently accurate for clinical purposes, and is used most widely.

Open-Circuit System. The subject breathes atmospheric air of determined composition, the expired air being collected in a rubber bag or a spirometer. Determination of the volume and O_2 and CO_2 contents of the expired air permits calculation of the volumes of O_2 absorbed and CO_2 produced and also, therefore, the R.Q. From the R.Q., one can ascertain the caloric value of 1 liter of oxygen, and, therefore, the heat production during the test period.

Closed-Circuit System. Several different types of apparatus are available, all of which are based on essentially the same principle. The subject breathes from and into a system filled with pure oxygen, the expired CO_2 and H_2O being trapped by soda lime. Decrease in the total volume of gas in the closed system is therefore due to, and is a direct measure of, oxygen consumption. The test period usually occupies six or eight minutes.

Inasmuch as CO_2 production is not measured, the R.Q. cannot be determined. However, the assumption that the R.Q. is approximately 0.82 (p. 448) under basal conditions is quite satisfactory for most clinical purposes. The caloric value per liter of O_2 is known to be 4.825 at this R.Q. Consequently, determination of the amount of O_2 consumed permits calculation of the approximate heat production during the test period.

BASAL METABOLISM

The terms "basal metabolism" or "basal metabolic rate" (BMR) are applied to the heat produced per unit time under "basal" conditions, i.e., at complete physical and mental rest, and in the postabsorptive state fourteen to eighteen hours after taking food. It was indicated previously that energy is derived by the organism from the metabolism (mainly oxidation) of exogenous (dietary) and endogenous proteins, carbohydrates, and lipids. This energy is dissipated in the form of heat and work or is stored in the body. Under the specified "basal" conditions, the exogenous source of energy is removed and loss of energy in the form of work by voluntary muscles is virtually abolished. Under these conditions energy is expended in the maintenance of respiration, circulation, muscle tonus, gastrointestinal contractions, and body temperature, and the functional activities of various organs (e.g., kidneys, liver, endocrine glands, etc.). Although the respiratory, heart, and other involuntary muscles continue to function, their activity does not result in permanent increase in the potential or kinetic energy of the materials on which they act, viz., the blood, respiratory air, lungs, etc. Consequently, the organism, under these conditions, is deriving its energy solely from stored sources and is dissipating it almost exclusively in the form of heat. Inasmuch as the amount of energy lost must be equal to that produced, measurement of the heat loss over a given period of time is an index of the rate of energy production, i.e., metabolism. As indicated previously, this is usually done clinically by the indirect procedure, using a closed system.

Physiological Variations in BMR. Basal metabolism varies with body size, age, and sex (Table 15-8). As would be anticipated, the magnitude of energy exchange increases with body size, but more directly in relation to surface than to either height or weight. Tables are available for derivation of surface area from the latter two factors. The BMR is therefore commonly expressed in one of three ways: (1) kilocalories/sq. meter body surface/hour; (2) liters of oxygen consumed/sq. meter/hour; (3) percentage above or below the mean normal for the subject in question. The last mode of expression was the one used most widely clinically, but it could occasionally obscure certain important facts, e.g., in obesity.

The average normal BMR for young adult men (20–30 years old) is about 40 Cal./sq. meter/hour (8.3 liters O_2/sq. meter/hour), increasing progressively below this age level (to 5 years) and diminishing progressively with increasing age. Normal values for females are about 6 to 10

TABLE 15-8 CALORIES PER SQUARE METER PER HOUR (DUBOIS STANDARDS) (MODIFIED BY BOOTHBY AND SANDIFORD)

AGE	MALES	FEMALES	AGE	MALES	FEMALES
5	53.0	51.6	20–24	41.0	36.9
6	52.7	50.7	25–29	40.3	36.6
7	52.0	49.3	30–34	39.8	36.2
8	51.2	48.1	35–39	39.2	35.8
9	50.4	46.9	40–44	38.3	35.3
10	49.5	45.8	45–49	37.8	35.0
11	48.6	44.6	50–54	37.2	34.5
12	47.8	43.4	55–59	36.6	34.1
13	47.1	42.0			
14	46.2	41.0	60–64	36.0	33.8
15	45.3	39.6	65–69	35.3	33.4
16	44.7	38.5			
17	43.7	37.4	70–74	34.8	32.8
18	42.9	37.3			
19	42.1	37.2	75–79	34.2	32.3

per cent lower than for males (Table 15-8). For average individuals (e.g., 70 kg. man, 1.73 sq. meters), this would amount to 1300 to 1600 Cal. daily.

Specific Dynamic Action (SDA) of Foods

Ingestion of food by subjects in an otherwise "basal" state results in an increase in heat production. This phenomenon is referred to as the specific dynamic action (SDA) or calorigenic action of foods. The extent of this increase above the basal level depends on the nature and quantity of the food. It is greatest for protein, less for carbohydrate and least for fat. It is usually stated that ingestion of protein (25 g.) equivalent to 100 Cal. gives rise to 130 Cal., and that ingestion of equicaloric amounts of carbohydrate and of fat gives rise to 106 and 104 Cal., respectively. The increment must be derived from body tissue sources and is "waste heat," not available for work.

The SDA of foodstuffs, therefore, is an expression of the "cost" of their metabolism, and adequate provision must be made to meet this expenditure if energy equilibrium is to be maintained. The values indicated above are of academic rather than practical importance, inasmuch as these foodstuffs administered together do not exert their SDA in an additive manner, the observed values being invariably lower than the calculated values. In calculating the dietary caloric requirement (mixed diet), it is recommended that the SDA be provided for as follows: 5 to 6 per cent of the total food calories should be added for a maintenance diet and 6 to 8 per cent for a liberal diet. More should be added (about 15 per cent) if the diet is very high in protein. On this basis, a subject with a BMR of 1600 Cal./day must be given an average mixed diet providing at least 1680 to 1696 Cal. in order to preserve caloric equilibrium.

The nature of the specific dynamic action of foods is not clear. It is not a reflection of the energy expended in digestive and absorptive processes, since it manifests itself after intravenous injection of amino acids. Amino acids vary in the magnitude of their SDA, which is apparently related to the metabolism of the non-nitrogenous moiety. For example, oxidative deamination, with transformation to urea, is accompanied by a larger heat increment than transamination. Phenylalanine, glycine, and alanine exert a particularly marked effect in this connection. There is evidence also that the increment is greater during the process of lipogenesis from glucose than during its oxidation to CO_2 and H_2O.

Total Metabolism (Caloric Requirement)

If a normal subject is to remain in energy (caloric) equilibrium, the daily diet must provide at least as much energy as is expended each day. The daily expenditure may be classified in the following categories: (1) basal metabolism (BMR), which is quite constant for any given individual, and is influenced by age (growth), sex, body size, environmental temperature (and barometric pressure), and such physiological states as pregnancy and lactation; (2) specific dynamic action of food; (3) muscular activity, the most important variable under normal conditions.

Extremely strenuous exercise may increase the energy expenditure more than tenfold above the "basal" level (Table 15-10). In the case of a painter (1.7 sq. meters body surface) the daily caloric requirement might be calculated as follows:

```
8 hours sleep, at 65 Cal./hour            =  520 Cal.
8 hours work, at 210 Cal./hour            = 1680 Cal.
2 hours light exercise, at 170 Cal./hour  =  340 Cal.
6 hours sitting quietly, at 100 Cal./hour =  600 Cal.
                                            3140 Cal.
Specific dynamic action (6 per cent)      =  188 Cal.
Daily caloric requirement                 = 3328 Cal.
```

TABLE 15-9 RECOMMENDED DAILY CALORIC REQUIREMENTS

	CALORIES PER DAY	
	MALE	FEMALE
Adults*		
Sedentary	2400	2000
Moderate activity	3000	2400
Strenuous activity	4500	3000
Pregnancy (latter half)		2400
Lactation		3000
Children and young adults		
16–20 years	3800	2400
13–15 years	3200	2600
10–12 years	2500	2500
7–9 years	2000	2000
4–6 years	1600	1600
1–3 years	1200	1200
Under 1 year	110/kg.	110/kg.

* Men 70 kg., women 56 kg.

TABLE 15-10 APPROXIMATE INCREMENTS IN HOURLY CALORIC REQUIREMENTS (ABOVE BASAL) FOR DIFFERENT OCCUPATIONS

OCCUPATION OR ACTIVITY	INCREASED REQUIREMENT CAL./HOUR
Sitting quietly	35
Reading aloud	40
Standing quietly	40
Tailor	70
Typing	75
Housework	110
Painter	145
Carpenter	150
Walking (moderate)	235
Sawing wood	380
Walking (fast)	550
Walking up stairs	1000

In Table 15-10 are indicated the recommendations of the Food and Nutrition Board of the National Research Council with regard to the average caloric requirements at various levels of activity.

Clinical Significance of BMR

The value of the determination of basal metabolism in clinical practice was largely in the exclusion of the thyroid gland as the cause of the patient's symptoms. This no longer applies.

References

DuBois, E. F.: Basal Metabolism in Health and Disease. Ed. 3. Philadelphia, Lea & Febiger, 1936.
Keys, A.: Energy requirements of adults. *In* Handbook of Nutrition. American Medical Association, Philadelphia, Blakiston Company, 1951, p. 259.
Lundsgaard, C., and Van Slyke, D. D.: Cyanosis. Medicine, 2:1, 1933.
McGilvery, R. W.: Biochemistry: A Functional Approach. Philadelphia, W. B. Saunders Company, 1970.
McIntyre, J. P.: Hypoxia. Brit. J. Hosp. Med., 2:1113, 1969.

Chapter 16

RENAL FUNCTION

As stated by Homer Smith, the composition of the blood plasma is determined by what the kidneys keep rather than by what the mouth ingests. These organs are the chief ultimate regulators of the internal environment of the body, and the urine is a by-product of their regulatory activities.

In accomplishing their purpose, i.e., maintenance of a reasonable constancy of composition of the extracellular and, to a certain extent, the intracellular fluids, the kidneys are involved in:

1. Elimination of water formed in or introduced into the body in excess of the amount required.

2. Elimination of nonvolatile end-products of metabolism.

3. Elimination of inorganic elements in accordance with the needs of the organism.

4. Elimination of certain foreign substances that gain access to the body.

5. Retention in the body of substances necessary for the maintenance of normal functions, e.g., plasma proteins, glucose, amino acids, hormones, vitamins, etc.

6. Formation and excretion of certain substances, e.g., ammonium and hydrogen ions.

As a result of these activities the kidneys play an important part in:

1. The regulation of water balance (p. 358), electrolyte balance (p. 361) and osmotic pressure of the body fluids (p. 359).

2. The regulation of acid-base status (p. 407).

3. The removal of metabolic waste products and certain toxic substances.

4. It is now recognized that the kidney plays a major part in the metabolism of vitamin D. On a low-calcium diet, the kidney converts the 25-hydroxycholecalciferol (formed from the vitamin, in the liver) into 1,25-dihydroxycholecalciferol, which is highly active and is probably the

major functional form of the vitamin (p. 827). On a high-calcium diet or under the influence of parathormone, this reaction is suppressed and the kidney now produces 24,25-dihydroxycholecalciferol, which is much less active than the compound from which it was formed. In this way, the kidney plays a major part in calcium regulation.

Morphological Features of Functional Importance. The nephron (uriniferous tubule) is the functional unit, each kidney containing about 1,200,000 of these structures. The nephron consists of two functionally distinct units, (*a*) the glomerulus, primarily a vascular channel, and (*b*) the tubule, lined by epithelial cells of different types in different regions.

After entering the kidney, the renal artery divides, successively, into interlobar, arcuate, and interlobular branches, the latter eventuating in the afferent arterioles of the glomerulus. The afferent arteriole breaks up into a set of capillary loops, each of which is enveloped in and intimately bound to the internal (basement) layer of Bowman's capsule, the invaginated glomerular end of the tubule, lined by flattened epithelial cells. The

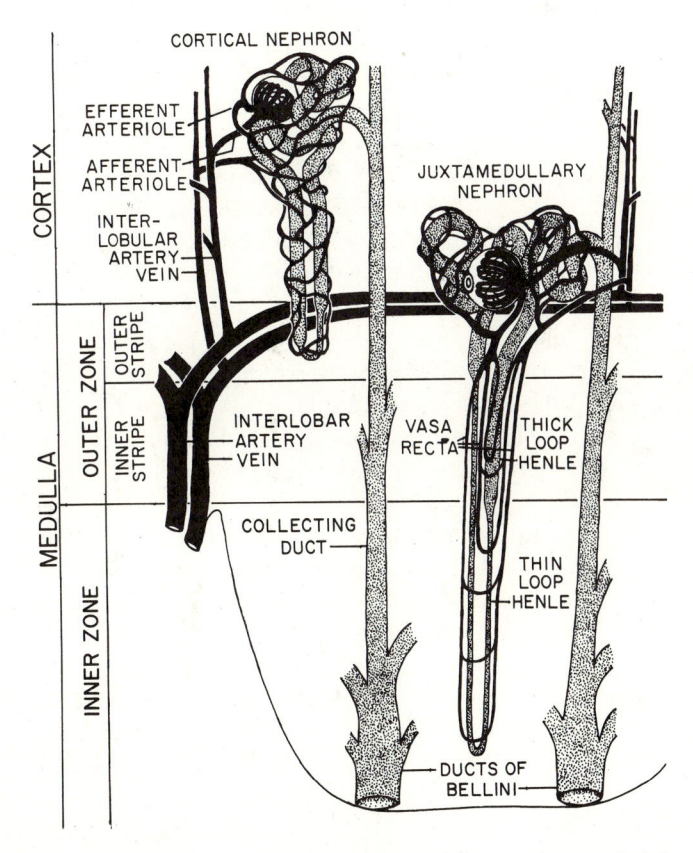

Figure 16-1. Diagrammatic representation of the cortical and juxtamedullary nephrons and their vascular supply. (From Pitts, R. F.: *Physiology of the Kidney and Body Fluids.* 3rd ed. Copyright 1974 by Year Book Medical Publishers, Inc., Chicago. Used by permission.)

attached walls of the glomerular capillaries and of Bowman's capsule together form a semipermeable membrane, about 1 μ thick, across which substances pass from the blood plasma into the lumen of the tubule by a process of simple filtration.

The glomerular capillaries reunite to form the efferent arteriole of the glomerulus, the diameter of which is approximately one-half that of the afferent arteriole. This difference in caliber serves to maintain an effective "head" of filtration pressure in the capillaries under varying conditions of arteriolar tonus. The efferent arterioles pass to the several portions of the tubule, breaking up into an elaborate network of capillaries which lie in close approximation to the outer tubular lining, i.e., basement membrane of the tubular epithelial cells. They subsequently unite to form a venous plexus which leads, successively, into the interlobular, arcuate, and interlobar veins and, finally, the renal vein.

These vascular features have an important bearing on certain important functional considerations, normal and abnormal. By far the bulk of the blood supplying the tubules passes first through the glomeruli, as indicated above. Consequently, interference with blood flow through glomeruli, e.g., in glomerulonephritis, is followed by degenerative changes (ischemia, anoxia) in the tubular epithelial cells supplied by the corresponding efferent arterioles. This accounts for the almost invariable secondary occurrence of renal tubular lesions as a complication of primary glomerular damage, unless of brief duration. Moreover, the blood plasma reaching the tubules has an increased colloid osmotic pressure owing to the fact that, in passing through the glomeruli, it has lost about 18 to 20 per cent of its water and diffusible components, but virtually no protein. The resulting increased viscosity of the blood (hemoconcentration) helps to maintain the relatively high pressure in the glomerular capillaries, and the increased plasma colloid osmotic pressure aids in the extensive reabsorption processes that occur in the tubule.

In passing through the glomeruli, the blood loses an essentially protein-free plasma filtrate. In the course of its subsequent passage through the tubules, this filtrate is modified extensively by reabsorptive, excretory and other activities of the tubular epithelial cells, the final product being urine. Largely as a result of the development of the clearance concept, various phases of renal function can be studied by rather precise quantitative methods. These are of value, not only in furthering understanding of the pathological physiology of certain disease states, but also in differential diagnosis, prognosis and treatment of diseases of the kidneys or other conditions affecting renal function. It must be borne in mind, however, especially in relation to the study of the pathological physiology of disease states, that there is a tendency to assume that the diseased kidney differs quantitatively, but not qualitatively, from the normal organ. This can, for obvious reasons, be regarded only as an approximation of the true state of affairs.

GLOMERULAR FILTRATION

Studies of glomerular fluid and serum filtrate (protein-free) have shown that the pH, vapor pressure, conductivity and concentrations of urea, glucose, chloride, inorganic phosphorus, uric acid, creatinine, bicarbonate, etc., in these fluids are practically identical. These observations indicate that the glomeruli act merely as ultrafilters and that glomerular urine is formed by a process of filtration alone, i.e., a purely mechanical process. The effective filtration pressure is the resultant of the blood pressure in the glomerular capillaries and the opposing forces of the colloid osmotic pressure of the blood plasma and the tension within Bowman's capsule (capsular pressure). The mean glomerular pressure may be regarded as about 70 per cent of the mean systemic arterial pressure (90 mm. Hg), thus averaging about 65 mm. Hg. This is subject to regulation by variation in the relative degree of constriction of the afferent and efferent arterioles of the glomerulus. Under stimulation, the latter may be constricted more than the former, with consequent increase in the pressure within the glomerular capillaries. If it is assumed that the colloid osmotic pressure is 25 mm. Hg in the blood entering the glomeruli, and that the capsular pressure is about 15 mm. Hg, it is obvious that the effective filtration pressure is approximately 65 minus (25 + 15) or 25 mm. Hg. As fluid (and diffusible solids, but little or no protein) leaves the plasma within the capillaries as a result of this force, the local plasma protein concentration increases and may reach a point where the effective filtration pressure is so reduced as to impair glomerular filtration. Further constriction of the efferent arteriole may then still increase the glomerular blood pressure, allowing filtration to continue. In addition to the influence of these pressure factors, the volume of glomerular filtrate is influenced significantly by (a) the surface area of the filter (glomerular capillary surface, normally 1.56 sq. meters) and (b) the minute volume flow of blood plasma over this surface (normally about 700 ml. plasma or 1200 ml. blood per 1.73 sq. meters body surface). Under normal conditions, the volume of glomerular filtrate averages about 125 ml. per minute per 1.73 sq. meters. This is discussed in detail in connection with the discussion of inulin clearance (p. 482).

It is apparent that glomerular filtration may be diminished with a tendency toward retention of waste products in the blood, in the absence of primary morphologic changes in the kidneys, by (1) extrarenal factors which diminish the renal blood flow and (2) factors which lower the effective filtration pressure by (a) decreasing the glomerular blood pressure, (b) increasing the plasma oncotic pressure (p. 485) or (c) increasing the capsular pressure. Among the most important of these are clinical conditions characterized by marked lowering of the systemic blood pressure, dehydration or hemoconcentration and cardiac failure (coronary artery occlusion), duodenal and pyloric obstruction, peritonitis, severe hemorrhage, Addison's disease, pneumonia, severe diarrhea, acute pancreatitis, perforated

458   RENAL FUNCTION

peptic ulcer or gallbladder, diabetic coma, extensive burns, left ventricular failure, urinary tract obstruction and various clinical states of shock. In such conditions, as in any condition of renal functional impairment, the tendency toward nitrogen retention may be aggravated by a simultaneous increase in protein catabolism or absorption of excessive amounts of protein end-products into the organism (e.g., in massive gastrointestinal hemorrhage).

In shock, both glomerular filtration and renal blood flow are decreased,

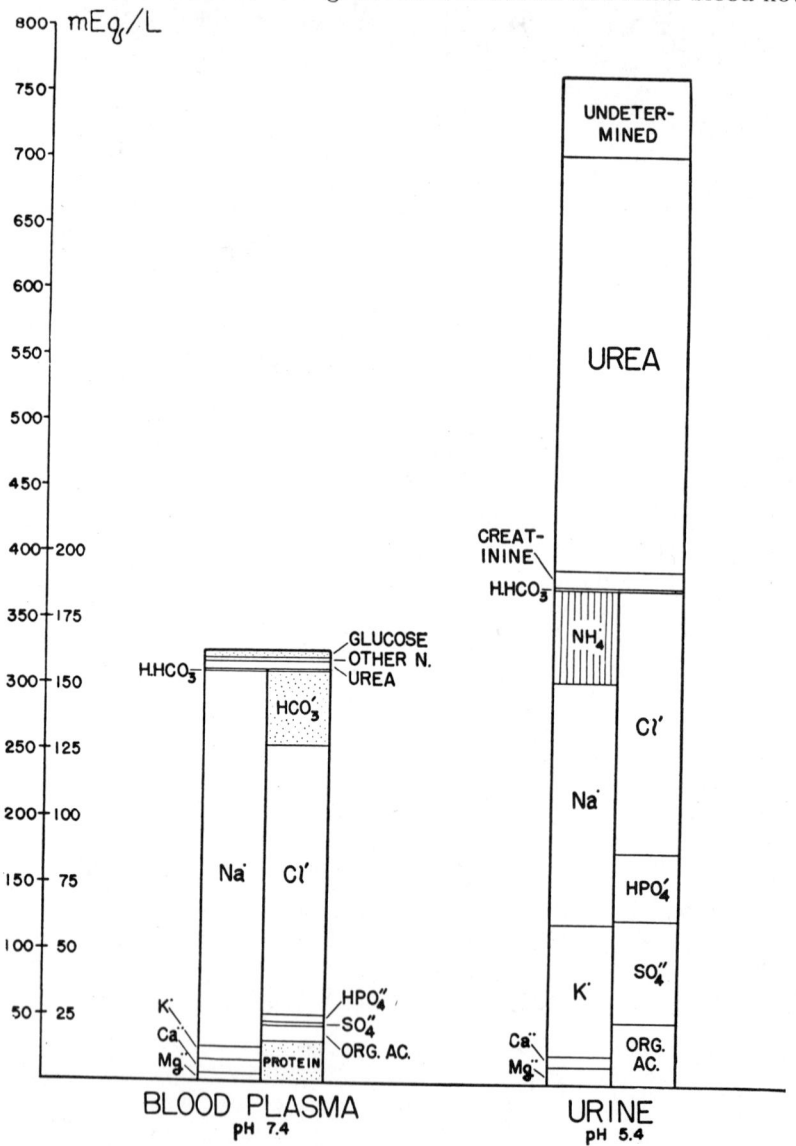

Figure 16-2. Illustrating the manner in which the kidney defends the chemical pattern of the blood plasma, producing a substance (urine) differing widely and variably from blood plasma in the relative quantities of substances, osmotic value, and reaction. The figures to the left (ordinate) indicate the sum of the milliequivalents of acidic and basic ions. (After Gamble.)

approximately in proportion to the degree of shock. This decrease, which is greater than can be explained solely on the basis of the fall in blood pressure, results from active renal arteriolar constriction. Renal vasoconstriction plays a homeostatic role in shock, tending to maintain circulatory efficiency, for normally about 25 per cent of the resting cardiac output perfuses the kidneys. This, however, is effected at the expense of renal function, for oliguria and diminution in concentrating power result from the decreased renal circulation.

Decreased filtration due to increase in capsular pressure occurs commonly as a result of intrarenal or extrarenal obstructive lesions (e.g., urinary calculi, prostatic enlargement). Significant elevation of plasma colloid osmotic pressure is rarely encountered as a primary event, since increase in the plasma protein concentration, when it occurs, is virtually invariably in the globulin fraction, which exerts relatively little influence on osmotic pressure.

TUBULAR FUNCTION

Reabsorption. If, as has been estimated, the rate of formation of glomerular filtrate under normal conditions is approximately 125 ml. per minute, while the rate at which urine passes into the bladder under the same conditions is approximately 1 to 2 ml. per minute, it is obvious that in its passage through the uriniferous tubules about 99 per cent of the water of the glomerular filtrate must have been reabsorbed. Furthermore, since the glomerular filtrate contains glucose in practically the same concentration as the blood plasma, whereas the bladder urine contains none, or very little, this substance, too, must have undergone practically complete reabsorption in the tubules. Quantitative studies of the excretion of other solids, such as chloride, phosphate, urea, uric acid, sodium, potassium, etc., in glomerular and bladder urine indicate a variable degree of reabsorption during their passage through the uriniferous tubules. Observations made in experimental animals have thrown considerable light upon the site of reabsorption of several of these solids in the renal tubules. It was found (in amphibians) that reabsorption of glucose occurs entirely in the proximal segment, reabsorption of chloride and sodium throughout the tubule, and acidification of the urine in the distal segment. The proximal tubule appeared to be capable of actively reabsorbing phosphate.

Insofar as excretory functions are concerned, therefore, the function of the renal glomeruli may be regarded as that of ultrafiltration (essentially protein-free filtrate of the blood plasma) and that of the tubular epithelium, in part at least, as "selective" reabsorption (water, glucose, chloride, etc.). Inasmuch as the average rate of glomerular filtration is 120 to 130 ml. per minute (about 70 ml. per square meter of body surface) (p. 482), it follows that more than 170 liters are filtered through the glomeruli from the plasma daily, the tubules subsequently reabsorbing about 168.5 liters of water, 1000 g. NaCl, 360 g. $NaHCO_3$, 170 g.

glucose and smaller amounts of phosphate, sulfate, amino acids, urea, urate, etc., in order to excrete about 60 g. of NaCl, urea and other waste products in about 1500 ml. of urine.

Several of the factors involved in the regulation of tubular reabsorption of various components of the glomerular filtrate are not entirely understood. However, information is available concerning certain constituents which aids materially in understanding abnormalities in disease. The extent to which any component of the glomerular filtrate is reabsorbed may be ascertained by determining its "clearance" simultaneously with the clearance of inulin or mannitol (p. 482), since the latter affords an index of the amount of material filtered through the glomerulus per unit of time.

Glucose. In normal men, under test conditions, the tubules can reabsorb 375 ± 80 mg. of glucose per minute (women 305 ± 55 mg.). At a glomerular filtration rate of 125 ml./min., and a plasma glucose concentration of 280 mg./100 ml., 350 mg. of glucose would be entering the tubules each minute. The same quantity would be entering if the volume of glomerular filtrate were reduced to 75 ml./min. and the plasma glucose concentration were raised to 465 mg./100 ml. Under these circumstances, if tubular reabsorption was not impaired simultaneously, glucose would not necessarily escape in the urine despite the considerable elevation of blood sugar concentration (p. 85). The reabsorptive burden imposed upon the tubular cells is determined by the total quantity of substance presented to them per unit of time, i.e., concentration × volume, and not by its concentration alone. Reabsorption of glucose is largely an active process, apparently involving phosphorylation in the tubular epithelial cells. This is blocked by phloridzin (p. 83), with consequent glycosuria. Similar types of renal glycosuria occur clinically (p. 83 ff.)

Water. Over 99 per cent of the enormous volume of water (about 180,000 ml. daily) that leaves the blood plasma in the glomeruli is normally returned to the blood in the tubules. Water is reabsorbed in different segments by different mechanisms, certain details of which have not yet been established but which may be represented as follows:

(1) **Proximal Convolution.** This segment is permeable to Na, Cl and water. Na and Cl are reabsorbed by an active transport mechanism, with passive isosmotic reabsorption of water. Under normal conditions, about 85 per cent of the filtered solutes and water (125 ml./minute/1.73 sq. M. body surface) are reabsorbed in this region. The volume of fluid leaving the proximal convolution is, therefore, normally about 16 ml./minute, the actual amount in any case being conditioned primarily by the volume of glomerular filtrate and the rate of reabsorption of solutes. This phase of passive water reabsorption, representing over 85 per cent of the total, has been designated "obligatory" reabsorption. Although the fluid at this point in the tubule is approximately isotonic with the original glomerular filtrate (i.e., same total osmolar concentration of solutes), it differs considerably from the latter in respect to the concentrations of individual components, many of which have been reabsorbed in varying degrees.

(2) **Loop of Henle.** The descending limb is freely permeable to water,

sodium ions and urea. The ascending limb is relatively impermeable to water. Active reabsorption of Na here results in dilution of the tubular contents. The fluid leaving the loop of Henle is, therefore, hypotonic with respect to the original glomerular filtrate; the interstitial fluid in this region (renal medulla) is correspondingly hypertonic.

(3) **Distal Convolution.** Active reabsorption of the Na (and other solutes) continues in this segment, which, however, is relatively impermeable to water in the absence of the antidiuretic hormone. Under such circumstances (diabetes insipidus), the tubular fluid would here become progressively more dilute. In the presence of ADH, water is absorbed isosmotically with Na and also by flowing along the osmotic gradient into the relatively hypertonic cortical interstitial fluid. Accordingly, depending upon the amount of ADH present, the fluid leaving the distal convolution may vary from hypotonicity to isotonicity with the original glomerular filtrate.

(4) **Collecting Ducts.** This segment is permeable to Na by an active process and, in the absence of ADH, relatively impermeable to water. In the presence of adequate amounts of ADH, water flows along the osmotic gradient into the surrounding hypertonic medullary interstitial fluid. Under normal conditions, the consequent concentration of the tubular fluid in this segment (urine) may reach a maximum of 1100 milliosmoles per liter, and 1400 milliosmoles if no water is taken for 72 hours. The term "facultative reabsorption" has been applied to the phase of water reabsorption controlled by the antidiuretic hormone and occurring independently of the amount of solute being reabsorbed simultaneously. The mechanism of regulation of secretion of ADH is discussed elsewhere (p. 711). In its absence (diabetes insipidus), the specific gravity of the urine is very low (1.001 to 1.003) and the volume may be extremely large (as much as 20 liters daily).

Sodium; Chloride. Under average normal conditions, more than 99 per cent of the Na that leaves the plasma in the glomerular filtrate is reabsorbed in the tubules, 85 to 90 per cent in the proximal segment, the remainder in the loop of Henle, the distal segment and the collecting duct. Several details of the mechanisms involved are not clearly understood, and there are differences of opinion on certain important points. Reabsorption of Na by the tubular epithelial cells (mainly, but not entirely, distal) is generally regarded as an active process, stimulated by "salt-active" hormones of the adrenal cortex, the most potent of which is aldosterone (p. 723); this plays an important role, therefore, in the regulation of Na balance (p. 357). The mechanism of regulation of aldosterone secretion in accordance with requirements for retention or excretion of Na is considered elsewhere (p. 729). The "third factor," as yet unidentified, inhibits absorption in the proximal tubule.

As Na^+ ions pass from the tubular fluid into the tubular epithelial cells, preservation of electrical neutrality on both sides of the cell membrane requires either (a) passage of an anion in the same direction or (b) passage of a cation in the opposite direction. Both of these occur, the former throughout the area of Na⁺ reabsorption, the latter apparently in specialized segments. About 90 per cent of the Na⁺ ions reabsorbed from the tubular fluid are accompanied by anions. In some instances, this may be a passive process; in others, an active transport mechanism is involved.

Inasmuch as the quantity of Na excreted is the resultant of the amount filtered through the glomeruli and that reabsorbed, it is subject to influence by a number of variable factors, renal and extrarenal, which affect these functions. These include: (1) the glomerular filtration rate; (2) the concentration of Na in the plasma; (3) the Na intake; (4) the renal plasma flow; (5) the concentrations of other unabsorbed solutes in the tubular fluid; (6) endocrine influences, particularly adrenocortical and antidiuretic hormones; (7) the relative concentrations of associated anions, particularly Cl^- and HCO_3^-; (8) the rate of flow through the tubules; (9) acid-base status. Several of these factors are obviously interrelated and, under normal conditions, operate as components of a homeostatic mechanism which effectively regulates Na excretion in accordance with existing requirements.

Other things being equal, the rate of excretion of Na varies in the same direction as the quantity entering the tubules. It is, therefore, influenced by the plasma Na concentration and by the glomerular filtration rate, the latter being itself influenced by the renal plasma flow, the glomerular capillary surface area, the intratubular (capsular) pressure, and the colloid osmotic pressure of the blood plasma (p. 457).

Enhancement of tubular reabsorption of Na by aldosterone (p. 723) and by stimulation of the hydrogen exchange mechanism, as occurs in acidosis (p. 410), is considered elsewhere. The influence of Cl^- and HCO_3^- ions is discussed below. Reabsorption of Na is depressed (excretion increased) in the presence of increased concentrations, in the tubular fluid, of relatively or absolutely unabsorbable solutes. In the case of anions in this category, such as sulfate, increased amounts of Na are retained in the tubular fluid to preserve a state of electrical neutrality. In the case of nonelectrolytes, such as mannitol, citrate and urea, the consequent osmotic retention of water in the proximal segment, by diluting the Na in the tubular fluid to a concentration below that of the plasma, may establish a progressively increasing concentration gradient which reduces the rate of Na reabsorption. The consensus is that regulation of Na excretion in man is accomplished mainly by influences exerted primarily on tubular reabsorption, although variations in glomerular filtration, when they occur, undoubtedly produce a significant effect.

Excretion of Cl usually parallels that of Na, and it is generally believed that tubular reabsorption of Cl occurs passively in association with reabsorption of Na (cation-anion pairing). In man, Cl is reabsorbed at the rate of approximately 105 mmol./liter of filtrate under ordinary circumstances.

Acidification of Urine; Ammonium Excretion. The mechanism of acidification of urine is discussed in detail elsewhere (p. 407). This is accomplished by the passage of H⁺ ions from the tubular epithelial cell (derived from H_2CO_3) into the lumen of the tubule in exchange for Na⁺ ions, which pass from the tubular fluid into the epithelial cells, then into the blood stream paired with the HCO_3^- ions "released" by the excretion of H^+ ions (Fig. 14-2, p. 409).

The greater part of the $Na^+ - H^+$ exchange mechanism operates in the proximal tubule; the H^+ ions react with HCO_3^- from the glomerular filtrate. A portion of the H^+ ions entering the tubular fluid is, therefore, disposed of by the formation of H_2CO_3, which is converted to H_2O and CO_2, the latter diffusing across the epithelial cell into the blood stream. An additional fraction of the H^+ ions is removed by exchange with Na^+ ions released by the formation of NH_4^+ ions from NH_3 secreted into the tubular fluid in the distal segment. The NH_3 is formed from NH_4^+ synthesized in the tubular epithelial cells by deamination of glutamine and, to a lesser extent, α-amino acids (Fig. 14-2, p. 409). NH_4^+ ions are also present in the glomerular filtrate.

The remaining H^+ ions are largely buffered by the HPO_4^{2-} ions in the tubular fluid (p. 408). The maximal attainable acidity of the urine is approximately pH 4.5.

Bicarbonate. Under conditions of normal urinary acidity the urine contains no HCO_3^- ions. The major portion of the HCO_3^- of the glomerular filtrate (27 mmol./liter) is removed in the proximal segment. The process of $Na^+ - H^+$ exchange occurs in this segment; the H^+ entering the tubular fluid reacts with the HCO_3^- to form H_2CO_3 and, in turn, H_2O and CO_2, the latter diffusing across the lining epithelium to the peritubular blood.

The HCO_3^- ions in the tubular fluid entering the distal segment are also neutralized as a result of the entrance of H^+ ions.

In man, quantities of HCO_3^- in excess of 28 mmol./liter of glomerular filtrate usually escape, the urine becoming correspondingly alkaline.

Potassium. The mechanism of excretion of K differs from that of Na in that it is secreted into as well as reabsorbed from the tubular fluid. It is probable that under physiological conditions, and by an active process, the K^+ entering the tubule in the glomerular filtrate undergoes complete reabsorption, paired with an anion (for example, Cl^-), before reaching the distal segment. Excessive amounts entering in the glomerular filtrate, e.g., in dehydration, may escape reabsorption. The cells of the distal tubules remove K^+ from the peritubular blood and it passes into the tubular fluid down an electrochemical gradient created by reabsorption of Na^+ ions. Thus, K^+ enters the lumen by a passive process rendering it competitive with H^+.

Accordingly, the amount of K excreted by this mechanism depends upon the quantity of Na reabsorbed, stimulation of the latter (e.g., by aldosterone) resulting in increased urinary excretion of K (and decrease in urinary Na). It is probable that alterations in K excretion induced by adrenocortical hormones occur as a result of their effects upon Na reabsorption and not, as believed formerly, as a result of a direct action of the hormones on the mechanism of K excretion.

Although the kidneys conserve the K of the glomerular filtrate more efficiently than the Na (more complete reabsorption), because of secretion of K into the tubular fluid the urine almost invariably contains some K, even when none is taken in (25 to 50 mmol./liter urine). Na and Cl, on the other hand, virtually disappear from the urine when withdrawn from the diet of subjects with normal renal function.

Urea. About 40 per cent of the urea which leaves the blood in the glomerular filtrate is reabsorbed in the tubules under average normal conditions. This is a passive process, i.e., diffusion, the extent of urea reabsorption depending, therefore, on the amount of water being reabsorbed and on the concentration gradient between the intratubular fluid (i.e., modified glomerular filtrate) and the fluid in the tubular cells and kidney interstitial spaces.

It is to be expected, therefore, that the amount of urea reabsorbed (and excreted) would vary, within certain limits, with changes in urine volume. At low urine volumes, i.e., more water reabsorption, more urea is reabsorbed and less is excreted; at high urine volumes, i.e., less water reabsorption, more urea is excreted. Maximum urea reabsorption (about 70 per cent) occurs at urine volumes of 0.35 ml. or less per minute. At a urine volume of 1 ml. per minute, about 60 per cent is reabsorbed, and at 2 ml. or more per minute 40 per cent is reabsorbed. If marked diuresis is induced by intravenous injection of an osmotically active substance, e.g., glucose or sucrose, the extent of urea reabsorption may fall considerably below 40 per cent, and values for urea clearance will approach those for inulin clearance.

Miscellaneous. Practically all of the normal components of the glomerular filtrate are reabsorbed in the tubules to extents varying from practically 100 per cent (e.g., glucose) or over 99 per cent (e.g., Na, water) to negligible amounts (e.g., creatinine). Many if not most of these are reabsorbed by active "tubular transport mechanisms," i.e., enzyme systems which effect their passage from the tubular lumen into the cells, and from the latter into the adjacent interstitial fluid. Reference will be made to a few of these which are of some clinical interest.

Phosphate is absorbed probably mostly in the proximal segment. This process is influenced by vitamin D and the parathyroid hormone, which increase phosphate excretion (p. 289) by decreasing its reabsorption. It is likely that vitamin D is the major factor. The mechanism for absorption of phosphate may possibly be linked with that for glucose, but little is known about the mechanism itself.

It is generally believed that 90 to 95 per cent of the filtered uric acid is reabsorbed in the tubules. This is an active process, interest in which has been stimulated by the demonstration of the uricosuric effect of agents which block uric acid reabsorption, e.g., probenecid, carinamide. Such agents are useful in the treatment of acute gout, lowering the plasma uric acid concentration as a result of their uricosuric action. However, there is evidence that the urate of the glomerular filtrate is practically completely reabsorbed by the tubules, that excreted in the urine being actively secreted into the tubular fluid.

Under normal conditions, only small amounts of amino acids are excreted in the urine, the major fraction of what enters the glomerular filtrate being reabsorbed by active tubular transport mechanisms. Clinical conditions of excessive aminoaciduria, e.g., Fanconi syndrome (p. 537), cystinuria (p. 535), occur in the presence of defects in these reabsorptive mechanisms.

Tubular Excretion and Synthesis. The term excretion is applied to the active transport of substances by the tubule cells from the blood plasma to the lumen of the tubule. Synthesis refers to formation, in the tubule cells, of substances which can then be passed into the lumen. Under physiological conditions probably very few normal components of the plasma undergo significant excretion by the renal tubular cells. Potassium is an important exception, and urate and a portion of the urinary creatinine may be excreted by this route. A number of foreign substances that may be introduced into the body for therapeutic or diagnostic purposes are removed from the blood plasma extensively, and in certain instances even predominantly, by the tubular epithelium. Among these are penicillin, p-aminosalicylic acid, phenolsulfonphthalein (PSP), p-aminohippurate, and Diodrast. Recognition of this fact in the case of penicillin led to the use of transport-blocking agents for the purpose of maintaining effective plasma penicillin levels for longer periods by decreasing its excretion in the urine. The use of the other substances named in the study of renal function is referred to elsewhere (p. 487 ff.).

There is evidence that, in certain animals, compounds present in the urine are synthesized in the kidneys from precursors removed from the blood plasma, e.g., hippuric acid from glycine and benzoic acid. In man, such processes occur in the kidney to a very limited extent, if at all. However, the distal tubule cells do form two substances, hydrogen ions and ammonia, which, excreted into the lumen, play an extremely important part in the regulation of acid-base balance. This subject is discussed elsewhere in detail (p. 407).

Competition for Transport Mechanisms. There is evidence that several substances that are excreted by renal tubular cells are handled by a single mechanism, at least in one phase of their transport across the cell. The same is true of reabsorption of certain substances by these cells. The same mechanism, or one of its components, may be involved in reabsorption of certain compounds and excretion of others. Such substances may, under certain conditions, compete for these mechanisms, so that excretion or reabsorption of one or more is depressed by the presence of large amounts of another which utilizes the same mechanism.

Diodrast, p-aminohippurate, phenolsulfonphthalein, penicillin, and p-aminosalicylic acid, which are all actively excreted by tubule cells, exert mutually depressing effects on their urinary excretion when administered in sufficiently high dosage. Similar competition, in this instance for reabsorption, is exhibited between creatine and certain α-amino acids, between glucose and xylose, and between ascorbic acid and sodium chloride.

Reabsorption of ascorbic acid is depressed also by p-aminohippurate, which is excreted by the tubules. Carinamide and probenecid, although they are not actively excreted by the tubule, depress tubular excretion of Diodrast, phenolsulfonphthalein, p-aminohippurate, penicillin, and p-aminosalicylic acid. They also decrease reabsorption of uric acid. Observations such as these have led to improvements in therapy (infections; gout).

WATER HOMEOSTASIS

The kidney plays an extremely important part in relation to the conservation of water. It has already been pointed out that about 180 liters pass the glomerular ultrafilter daily and that more than 99 per cent of this is reabsorbed by the kidney. About 80 per cent is reabsorbed iso-osmotically in the proximal tubule because of the active absorption of sodium and potassium ions and the consequent passive absorption of chloride ions. Water rapidly diffuses along the osmotic gradient, then passes into the interstitial portion of the kidney and is removed, as rapidly as it enters, by the capillaries round the tubule (Fig. 16-1, p. 455), and so is returned to the blood stream to be made more available to the body.

Of the total daily amount filtered by the glomeruli, about 30 liters remain to be dealt with by those portions of the nephrons distal to the proximal tubule so that the daily urinary output is about one and one-half liters. This is brought about by the active absorption of sodium ions into the renal interstitium. These are accompanied by anions, mainly chloride, to maintain electroneutrality. Because of the absorption of sodium salts, there is an increase of osmotic pressure, and so water passively diffuses into the renal interstitial tissue down an osmotic gradient. This water, along with its contained salts, is reabsorbed into the circulation by way of the vasa recti.

If the mechanisms were as straightforward as described, very large amounts of energy would be required to maintain a large osmotic gradient. This problem has been overcome by the evolution of a "countercurrent" system in the mammalian kidney. This was first described by Wirz and depends largely on the fact that the descending limb and the tip of the loop of Henle are freely permeable to water, sodium salts and urea, and consequently their contents are isotonic with the fluid in that part of the renal interstitium, in which they are situated. Because of the mechanism to be described, it so happens that osmolality of the interstitial fluids increases as we pass from the cortex through the outer medulla to the inner medulla. In the cortex, it is 300 milliosmoles per kilogram water; and in the inner medulla it varies from 700 to about 1200 between the extremes of maximal diuresis and maximal antidiuresis, respectively (Fig. 16-3). The ascending limb of the loop of Henle is relatively impermeable to water but it has an active transport mechanism for sodium ions, as is the case for the distal convoluted tubule and the collecting duct. The water permeability of these two latter portions of the nephron, however, can be increased by the action of antidiuretic hormone. The countercurrent system in the loop of Henle was so clearly described by Wirz* that the following account is virtually his, apart from minor modifications.

There is only one explanation which covers all the known facts and moreover gives a satisfactory reason for the existence of the loops of Henle in the mammalian kidney. This theory is based on the combination of the flow along a tube that doubles back upon itself and an active process of sodium transport setting up a small osmotic pressure difference between the two limbs of this tube, the content of the descending limb (and interstitial tissue) being of higher tonicity. The concentrating effect of this system starts being multiplied as soon as the hypertonic content of the descending limb is carried by bulk flow around the bend into the ascending limb. The lower ends of both limbs (and intervening interstitium) now being equally hypertonic, the osmotic pressure difference is momentarily abolished. If active sodium transport from the ascending limb continues, a new small osmotic pressure difference is set up, the content of the lower part of the descending limb (and interstitium) now being twice as hypertonic as in the first step. This fluid is again transported into the ascending limb, again abolishing the osmotic pressure differences. The process continues, until finally a steady state arises, the fluid entering the system at the upper end of the descending limb being more and more concentrated toward the hairpin bend and rediluted on its way back up the ascending limb. Osmotic pressures considerably higher than initially may be produced in the lower portion of the loop.

The pressures in the descending limb are mirrored by the interstitium, since this portion of the loop is freely permeable to water and salts. Because of their anatomical situation and loop formation, the vasa recti also act as countercurrent exchangers, with diffusion of salt and urea from the ascending to the descending portion of the loop and water going in the opposite direction. The latter is probably to some extent helped by the fact that there is a slightly raised protein content in the vascular loop arising from ultrafiltration in the glomerulus. This countercurrent effect produces a loss of sodium (and urea) from the medulla, which is not so great as to disturb the osmotic gradient shown in Figure 16-3. In a sense, urea is trapped in the medulla. In actual fact, there is to some extent a dynamic equilibrium between urea passing into the interstitial tissue from the collecting duct and into the loop of Henle and that which is reabsorbed in the distal tubule. It appears probable that the lymphatics of the kidney also manifest a countercurrent effect.

To summarize the whole countercurrent process, the osmolality of the medullary fluid and the gradient difference between cortex and inner medulla are proportional to the number of loops of Henle and dependent upon their length (some do not enter the medulla very deeply). They also vary directly with sodium transport across the ascending limbs of the loops and the urea load in the collecting ducts. There is an inverse relationship to the rate of flow of fluid in the loops of Henle, the medullary blood flow and the rate of water transport from the collecting ducts and the descending limbs (Fig. 16-3).

---

* Wirz, H.: The location of antidiuretic action in the mammalian kidney. *In* Heller, H., ed.: The Neurohypophysis, Proceedings of the 8th Symposium of the Colston Research Society. New York, Academic Press, Inc., 1957, p. 157.

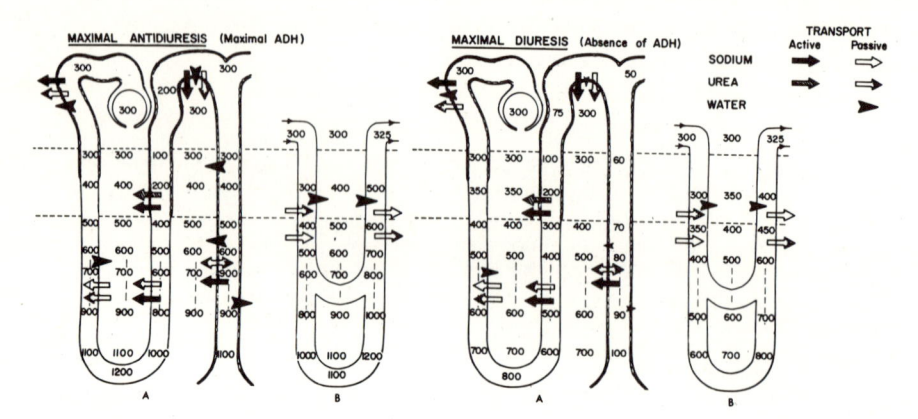

Figure 16-3. Countercurrent mechanism as it is believed to operate in the juxtamedullary nephron (A) and vasa recta (B). The numbers represent illustrative osmolality values during maximal antidiuresis (left) and maximal diuresis (right). No quantitative significance is to be attached to the number of arrows; only net movements are indicated. As with the vascular loops, not all loops of Henle reach the tip of the papilla and, hence, the fluid in them does not become as concentrated as that of the final urine, but rather only as concentrated as the medullary interstitial fluid at the same level. (Modified from Gottschalk and Mylle: Amer. J. Physiol., 196:927, 1959.)

Under normal circumstances, the fluid at the upper end of the ascending limb of the loop of Henle is hypo-osmotic with respect to plasma, but becomes iso-osmotic in the distal tubule. In the absence of antidiuretic hormone (ADH), it remains hypo-osmotic both here and in the collecting duct. ADH renders the distal tubule and collecting duct more permeable to water. Under maximal antidiuresis, therefore, the fluid in both these portions of the renal tubular system is in equilibrium with the interstitial fluid of the parts of the kidney in which they are situated. There is, therefore, an increase in osmolality of the collecting duct fluid as it passes from cortex to medulla.

The mechanism of action of ADH is not understood. It has been suggested that it involves reduction of disulfide bonds, secretion of hyaluronidase or stimulation of synthesis of cyclic AMP or, possibly, all of these.

DIURETICS

Diuretics are drugs which increase urine flow and concomitantly increase excretion of sodium and chloride and sometimes of potassium ions. Frequent use of diuretics in modern medicine necessitates an understanding of their mode of action. Such understanding can also give an insight into the electrolyte disturbances which can result from the very powerful diuretics now available.

One obvious way of increasing urine flow is to reduce the production of ADH. This can be done by excessive water intake or by the action of substances like alcohol, which inhibit the activity of the neurohypophyseal system responsible for production of the hormone. There is no excessive loss of electrolytes. Any substance which increases the solute load of the glomerular filtrate osmotically interferes with absorption of water in the renal tubules; urea and mannitol are typical examples. Hyperglycemia also produces an increased osmotic effect through the medium of filtered glucose. Ammonium chloride or calcium chloride administration produces hyperchloremic acidosis with an excess of chloride ion in the renal tubules. To maintain electroneutrality, this must be excreted along with sodium or potassium ions. There is thus an increased osmotic load and loss of Na^+ and K^+. The effect, however, is only temporary because of renal compensatory mechanisms. Theophylline, caffeine and theobromine increase glomerular circulation and hence glomerular filtration rate. They also inhibit renal tubular reabsorption of Na^+ and Cl^-, possibly by inhibiting breakdown of cyclic AMP by the enzyme phosphodiesterase. Organic mercurials (Salyrgan, Mercuhydrin, etc.) are believed to inhibit the action of transport ATPase, and hence the renal pump mechanisms for sodium. This results in increased loss of Na^+ and Cl^- but not so much of K^+, since there is also inhibition of K^+ excretion by the distal tubule mechanism.

The modern era was ushered in by the use of carbonic anhydrase inhibitors (acetazolamide, methazolamide, etc.). By slowing the formation of H_2CO_3 from H_2O and CO_2, these substances result in diminished H^+ ions within the renal tubule cells. For reasons already discussed, this results in increased urinary secretion of Na^+, K^+ and HCO_3^-, and hence also of water.

Aldosterone antagonists (spironolactone) result in less tubular absorption of Na^+ (mainly distal tubules), and hence of Cl^- and H_2O. There is possibly some K^+ retention.

The thiazide diuretics (chlorothiazide, hydrochlorothiazide, etc.) are weak carbonic anhydrase inhibitors and so can result in increased urinary HCO_3^-. By a mechanism which has not been elucidated, they decrease tubular absorption of Na^+ and Cl^-, and to some extent K^+. The thiazides compete for uric acid secretion sites in the renal tubules and consequently produce a mild hyperuricemia. They also have a diabetogenic action.

Ethacrynic acid is a powerful inhibitor of Na^+ reabsorption in the renal tubules and hence of Cl^- and water. Its action seems to be an interference with energy mechanisms involving ATP. It can also cause some loss of K^+.

Furosemide has a double action. It exerts its effect as a thiazide as well as through an additional mechanism, which is quite unknown. There is decreased renal tubular reabsorption of Na^+ and even more of Cl^-, and hence of H_2O. There is some urinary loss of K^+. It can produce hyperuricemia and diabetes.

The pteridines (triamterene) increase the output of Na^+ and HCO_3^- in the urine. They also cause K^+ retention and can produce hyperkalemia. Their mode of action is unknown.

Many of these diuretics, when used as mixtures, show additive effects. There are other advantages; e.g., mixing a thiazide with spironolactone or triamterene results in less loss of K^+.

CHARACTERISTICS OF NORMAL URINE

As indicated in the preceding discussion, the ultrafiltrate of the blood plasma, which leaves the circulation in the glomeruli, passes through the tubules and is subjected to various modifying influences, the nature and magnitude of which are determined, under normal conditions, by the requirements of the organism. The fluid which emerges from the collecting tubules is urine. It seems desirable to review certain important features of the composition of normal urine, an understanding of which may aid in the interpretation of changes that occur in disease.

Volume. The quantity of urine excreted daily by normal subjects varies widely and is determined chiefly by the fluid intake. Under ordinary dietary conditions it ranges from 1000 to 2000 ml. It is influenced somewhat by the protein and NaCl intake, the excreted urea and salt acting as diuretic agents. Excessive perspiration and strenuous exercise decrease the urine volume by prerenal deviation of water (skin, lungs); and, under abnormal conditions, the same is accomplished by vomiting, diarrhea and edema. The amount excreted during the day (8 A.M. to 8 P.M.) is two to four times that excreted during the night (8 P.M. to 8 A.M.). The minimum volume required for excretion of a given quantity of solids is determined by the concentration capacity of the kidneys (p. 495). Under conditions of average adequate food intake, it is about 300 ml./ sq.m. body surface (i.e., about 500 ml. for a 70 kg. man) (p. 341).

Specific Gravity. The specific gravity of the urine is directly proportional to the concentration of solute and, therefore, with any given total solid excretion, varies inversely with the volume under usual normal conditions. The grams of solute per liter may be calculated roughly by multiplying the last two figures of the specific gravity (to the third decimal) by Long's coefficient, 2.6. This is only an approximation, particularly since all substances do not exert the same effect on specific gravity. For example, the specific gravity of a liter of urine is raised 0.001 by 3.6 grams of urea, 1.47 grams of NaCl, and 3.8 grams of NaH_2PO_4 (also 2.7 grams of glucose and 3.9 grams of albumin).

Under ordinary dietary conditions, the specific gravity of the mixed twenty-four-hour urine usually varies between 1.014 and 1.026. However, it may be as low as 1.001 on a very high water intake or as high as 1.040 if fluid is restricted or excreted in excess by other routes (skin, bowel, etc.). In normal subjects it rarely exceeds 1.040, the maximum attainable osmotic concentration in man being about 1.4 osmoles per liter.

Normal kidneys eliminate the required amount of solids regardless, within wide limits, of the amount of water available for their solution. When their function is impaired, the kidneys lose this ability to dilute and concentrate the urine, and, with increasing damage, the osmolar concentration and specific gravity of the urine approach values for protein-free blood plasma (glomerular filtrate). The maximum attainable specific gravity falls and the minimum rises, the urine specific gravity being relatively fixed within narrowing limits below and above 1.007. The importance of the urinary specific gravity arises out of these facts, determination of the concentrating ability of the kidneys constituting an important test of renal function.

Acidity. Urinary acidity may be expressed in two ways: (1) its H^+ ion concentration (true acidity); (2) its titratable acidity. These two factors (and also urinary ammonia) are responsive to the same influences but are not necessarily related quantitatively. The pH may be regarded as an "intensity" factor and the titratable acidity as a "capacity" factor. Secretion of ammonia into the tubular urine decreased its H^+ concentration.

pH of Normal Urine. The extreme limits of urinary pH in health are 4.8 and 8.0, usually, under ordinary conditions, ranging between 5.0 and 7.0, averaging 6.0 in the mixed twenty-four-hour sample. The mechanism of acidification of the glomerular filtrate (pH 7.4) is discussed elsewhere (p. 407).

Titratable Acidity. The titratable acidity of the urine can be expressed as milliliters of N/10 alkali required to bring to pH 7.4 the twenty-four-hour output of urine. It can also be expressed as millimoles (milliequivalents). This usually varies normally from 20 to 40 mmol. Because of the intimate relationship between urinary acidity and urinary ammonium ion (p. 430), the latter (usually 20–70 mmol.) is included with the former in studies of acid-base status (i.e., titratable acidity plus ammonium ion).

The urinary acidity and pH are reflections chiefly of the ratio $NaH_2PO_4 : Na_2HPO_4$, which is 1 : 4 at pH 7.4 (blood plasma; glomerular filtrate), about 50 : 1 at pH 4.8, and 9 : 1 at pH 6.0. This transformation of the dibasic to the monobasic salt effects considerable conservation of Na, which is returned to the plasma (p. 407). Acid urine contains virtually no bicarbonate (HCO_3^-). In alkaline urine, however, this ion is present and the pH may be determined largely by the ratio $H_2CO_3 : NaHCO_3$. The urinary acidity is normally also contributed to, but to a minor extent, by organic acids (lactic, uric, hippuric, β-hydroxybutyric).

The acidity of the urine is normally influenced chiefly by the nature of the diet. Ingestion of large amounts of protein (meats, breads, cereals), which yields acids in the course of metabolism (sulfuric, phosphoric), increases urinary acidity and ammonium ion. Most vegetables and fruits (orange, lemon, grape, apple, peach, pear) contain organic acids (e.g., citrate, oxalate) which form bicarbonate in the body and hence decrease urinary acidity. Certain fruits (plums, cranberries) contain benzoic acid and quinic acids, which are metabolized to and excreted as hippuric acid, increasing the urinary acidity and ammonium ion content. These are increased also by starvation or dietary carbohydrate restriction, owing to increased catabolism of body protein and fat, with increased urinary excretion of sulfate, phosphate, and ketone acid (β-hydroxybutyric and acetoacetic).

Shortly after eating (within an hour) there is a decrease in urinary acidity and ammonia and H^+ concentration. This "postprandial alkaline tide" has been attributed generally to the increased secretion of HCl by the stomach, temporarily increasing the plasma bicarbonate. It is said not to occur in subjects with achlorhydria. This explanation is not accepted by some. On standing, normal urine becomes alkaline and ammoniacal owing to the action of bacteria on urea (ammoniacal fermentation).

Nonprotein Nitrogenous Constituents. The term "nonprotein nitrogen" (NPN) is applied to nitrogen present in compounds not precipitable by the usual protein-precipitating agents. These substances, occurring in the blood and other body fluids and in the urine, consist chiefly of intermediary- or end-products of protein metabolism. They include urea, ammonium ion, amino acids, creatinine, creatine, uric acid, and a number of other compounds present in small amounts, not readily determined quantitatively, and referred to as "undetermined nitrogen" (peptides, hippuric acid, etc.).

The several components of this heterogeneous mixture have different origins and significance and are influenced by different factors; the amount of each excreted in the urine varies under different conditions of diet and metabolism. The most important features of the biogenesis and metabolic significance of these substances are considered in the section on intermediary metabolism of proteins (p. 158 ff.) and nucleic acids (p. 257).

Total Nonprotein Nitrogen. Normal adults on an adequate protein diet are in a state of nitrogen equilibrium, the quantity excreted daily being equivalent to that ingested. This is not the case during periods of tissue growth (children, pregnancy, convalescence) (positive nitrogen balance). Inasmuch as the amount of nitrogen excreted in the feces is relatively constant (p. 149) and that lost by other extrarenal channels is usually small and constant, the quantity in the urine may be regarded as reflecting the amount of protein ingested and the state of protein catabolism. It may vary from as little as 3 grams daily on a low protein intake (20 g.) to as much as 24 grams or more on a very high protein diet (160 g.); it ranges between 11 and 15 grams on usual protein intakes (70 to 80 g.).

Urea. Urea is the chief end-product of the catabolism of amino acids (p. 160) and is the substance in which is incorporated, for purposes of excretion, the bulk of the nitrogen provided to the organism in excess of its needs. Under usual dietary conditions it comprises about half of the total urine solids and is, therefore, the most abundant urinary constituent (except water). Because of these facts, and because it is a metabolic end-product, virtually not utilizable, the quantity of urea excreted in the urine is the most significant index of the extent of protein catabolism and, consequently, is influenced more directly by the protein intake than are the other nitrogenous compounds.

The mechanism of excretion of urea is discussed elsewhere (p. 464). On a high protein diet, urea nitrogen (16 to 25 g.) comprises 80 to 90 per cent of the urinary NPN. On very low protein intakes it may fall to 60 per cent or less of the total NPN, since variations in dietary protein are reflected almost exclusively in urinary urea and ammonium ion content, the amounts of the other nitrogenous constituents remaining relatively constant (Table 16-1).

Ammonium Ion. Preformed ammonium ion, which escapes the liver,

TABLE 16-1 URINARY NITROGEN PARTITION IN SAME INDIVIDUAL ON HIGH AND LOW PROTEIN DIETS*

	HIGH PROTEIN (grams/24 hours)	(% of total N)	LOW PROTEIN (grams/24 hours)	(% of total N)
Urine volume	1170 ml.		385 ml.	
Total N	16.80		3.60	
Urea N	14.70	85.5	2.20	61.7
Ammonia N	0.49	3.0	0.42	11.3
Uric acid N	0.18	1.1	0.09	2.5
Creatinine N	0.58	3.6	0.60	17.2
Undetermined	0.85	4.8	0.29	7.3

* (After Folin, O.: J.A.M.A., *69*:1209)

is removed by glomerular filtration from the blood stream, in which ammonium ion is present in only minute amounts. It is formed in the renal tubular epithelium, chiefly from glutamine, the remainder from other amino acids (p. 411). It is passed as NH_3 largely into the tubular urine where, acting as a base, it combines with H^+ ions, forming NH_4^+, thus serving to lower the H^+ concentration and permitting further passage of H^+ ions into the lumen in exchange for Na^+ (p. 411). This mechanism is probably a factor in the renal regulation of "acid-base" status.

Under usual dietary conditions, normal adults excrete about 0.3 to 1.2 grams (average 0.7 g.) of ammonium N daily (20 to 70 mmol.), comprising 2.5 to 4.5 per cent of the total NPN. Inasmuch as the urea and NH_4^+ of the urine are both derived ultimately from amino acids, an increase in ammonium ion is accompanied by a corresponding decrease in urea, at any given level of NPN excretion. The quantity of NH_4^+ in the urine of normal subjects is determined chiefly by the state of acidification of the urine.

The output of ammonium ion increases with increasing levels of dietary protein, owing probably to the acid-producing properties of protein (oxidation of S and P). It is increased also by ingestion of acids or acid-forming substances (NH_4Cl, NH_4NO_3) and in association with acidosis (except renal) and is decreased by ingestion of alkalis or base-forming foods and in alkalotic subjects. As much as 10 grams of NH_4^+ may be excreted daily by individuals with severe diabetic acidosis. Administration of ammonium salts of mineral acids (viz., NH_4Cl, NH_4NO_3) is at least equivalent to administration of corresponding amounts of the free acids. These ammonium salts therefore increase urinary acidity and can induce acidosis. If urine is allowed to stand without a preservative, its ammonia content increases markedly as a result of hydrolysis of urea by bacteria ("ammoniacal fermentation") and loss of CO_2, causing a rise of pH.

Creatine and Creatinine. The metabolic origin and significance of these substances are considered elsewhere (p. 180).

Creatine is not excreted in significant amounts in the urine of normal adult males. This is apparently due to the fact that it is completely reabsorbed from the glomerular filtrate in the renal tubules at normal plasma creatine levels (< 0.6 mg./100 ml.). It is present in the urine (increased plasma level) of prepubertal children (4.2 mg./kg. daily) and certain normal women, periodically or constantly, and occasionally during pregnancy and the early puerperium.

Under normal conditions, creatine formed in or otherwise provided (food) to the organism over and above the requirements or capacity for storage in the muscles (as phosphocreatine, p. 16) is either transformed to creatinine or excreted as such. The creatine tolerance test is a measure of the capacity for utilization of creatine. Normal adult men excrete in the urine (in twenty-four hours) not more than 20 per cent and women not more than 30 per cent of a test dose (1.32 or 2.64 g.) of creatine hydrate. Urinary creatine is increased in conditions accompanied by muscle wasting, muscle dystrophies, and thyrotoxicosis.

Creatinine (creatine anhydride) is formed from creatine in quite constant amounts daily (men, 1.0 to 1.8 g.; women, 0.7 to 1.5 g.), corresponding to about 2 per cent of the body creatine, practically all of which is eliminated in the urine. The excretion of creatinine is, therefore, constant from day to day under quite varied conditions, being relatively uninfluenced by diet or urine volume, and being determined chiefly by the muscle mass (creatine content of body). Because of this fact, the daily excretion of creatinine is determined commonly in balance studies as a check on the accuracy of twenty-four-hour urine collections but is not entirely reliable. Endogenous creatinine is excreted in men by glomerular filtration and active tubular excretion, the value for creatinine clearance (about 145) exceeding that for inulin (p. 482).

The term "creatinine coefficient" is applied to the number of milligrams of creatinine (plus creatine, if any) nitrogen excreted daily per kilogram of body weight. Normal values are 20 to 26 for men and 14 to 22 for women. The corresponding values in terms of nitrogen are 7.5 to 10 for men and 5 to 8 for women. The creatinine coefficient may be regarded as an index of the amount of muscle tissue and is, therefore, more directly proportional to the ideal than to the actual weight.

Amino Acids. Amino acids are excreted in the urine in both free and combined form. The latter are present in acyl linkage, as in compounds such as hippuric acid (glycine). The free amino nitrogen ranges normally from 0.1 to 0.15 gram daily, comprising 0.5 to 1.0 per cent of the total nitrogen. The total amino nitrogen output is 0.5 to 1.0 gram daily (2 to 6 per cent of the total nitrogen). The organism conserves amino acids efficiently, the amount excreted by normal subjects varying but slightly with wide fluctuation in protein intake and even when amino acids are administered parenterally. Severe liver damage results in an increase in urinary amino acids and a corresponding decrease in urea (impaired deamination and urea formation).

Uric Acid. Uric acid is the chief end-product of the metabolism of purines. Its metabolic origin is considered elsewhere (p. 253). Normal values for urate clearance range from 11 to 15 ml. per minute.

In urine of average pH (6.0 or higher) uric acid is present largely as the soluble sodium and potassium urates, whereas in highly acid urine the relatively insoluble free acid may predominate and may precipitate from solution.

The amount of uric acid in the urine of normal subjects depends on the quantity of nucleoprotein ingested (exogenous) and that formed from tissue nucleic acids and nucleotides (endogenous). On purine-free diets, the urinary uric acid is of solely endogenous origin and is rather constant for the individual, usually 0.1 to 0.5 gram daily. This may be increased somewhat by strenuous exercise and by a high protein or high caloric intake (stimulating endogenous metabolism).

On a high purine diet (meat, liver, kidney, sweetbreads, leguminous vegetables, etc.) the uric acid output may be as much as 2 grams daily. In adults on the usual mixed diet the average daily excretion is about 0.7 gram, comprising 1 to 2 per cent of the total nitrogen. The urine of newborn infants contains relatively large amounts of uric acid (0.2 g./100 ml.), comprising about 7 to 8 per cent of the total nitrogen. Crystals of free uric acid and ammonium urate may be present and urate "infarcts" of the kidneys are not uncommon. A sharp drop to normal adult levels occurs after seven to ten days. A considerable increase follows administration of cinchophen and salicylates. Caffeine and other methylpurines (in tea, coffee, cocoa) cause little or no increase, either being transformed largely to urea or undergoing more complete destruction in the body.

Urinary excretion of uric acid, reflecting nucleoprotein catabolism, is increased by 11-oxygenated adrenocortical hormones and by ACTH.

Allantoin. Allantoin is formed from uric acid in the liver by the action of the enzyme uricase (p. 252). In man and anthropoid apes, the livers of which exhibit little or no uricase activity, allantoin is excreted in very small amounts (5 to 25 mg. daily). In other mammals it is the chief end-product of purine metabolism, accounting for 90 per cent or more of the excretion of purine metabolites. It was formerly believed that the metabolism of purines in the Dalmatian dog resembled that in man and apes (differing from that of other mammals) in having uric acid as its chief metabolic end-product. Present evidence indicates that the urinary excretion of relatively large amounts of uric acid is due to its incomplete reabsorption by the renal tubular epithelium and not to a difference from other breeds of dogs in capacity for forming allantoin from uric acid.

Oxalic Acid. Small amounts of oxalate (10 to 25 mg.) are excreted daily in the urine of normal subjects. Its origin is not well understood. Calcium oxalate, which precipitates from neutral or alkaline urine in the presence of abnormally high oxalate excretion, is an important constituent of urinary calculi. Under normal conditions, the main source of urinary oxalate is probably dietary, it or certain metabolic precursors being present in many common foodstuffs (tomatoes, rhubarb, cabbage, spinach, asparagus, apples, grapes, etc.). Oxalate is one of the end-products of the metabolism of ascorbic acid and may also be formed, perhaps, in the intermediary metabolism of carbohydrate, presumably not significantly under normal conditions. Urinary oxalate is greatly increased in the hereditary disorder *primary hyperoxaluria* (150 to 650 mg. daily).

Glucuronic Acid.

Glucuronic acid occurs in normal urine conjugated with a wide variety of compounds in two types of linkage, (1) ester and (2) glycosidic. The carboxyl groups of such compounds as phenylacetic acid and benzoic acid form ester linkages, whereas the hydroxyl groups of such substances as aliphatic or aromatic alcohols (chloral, camphor, phenol, menthol, indole, skatole, morphine, pregnanediol, estriol, etc.) form glycosidic linkages. This conjugation takes place in the liver, which is also the site of formation of glucuronic acid.

The normal daily excretion of glucuronic acid is about 0.3 to 1.0 gram. This may be increased by administration of the substances indicated above, as well as acetylsalicylic acid, sulfonamides (except sulfanilamide), turpentine, antipyrine, and phenolphthalein. When present in relatively high concentration, the ester types reduce the usual copper reagents (e.g., Benedict's), probably owing to concomitant hydrolysis, and may therefore interfere with the interpretation of urinary tests for glucose. The glycosidic types do not reduce these reagents unless previously hydrolyzed.

Hippuric Acid (Benzoylglycine).

Hippuric acid is formed by the conjugation, in peptide linkage, of benzoic acid and glycine. In man this occurs largely, if not solely, in the liver; in certain species (e.g., dog) it occurs also in the kidney.

Benzoic acid (or benzoate) is present, as such, in many vegetables and fruits (e.g., cranberries, plums) and addition of sodium benzoate as a preservative to certain prepared foodstuffs is legally permissible. Quinic acid, another constituent of vegetables and fruits, is converted to benzoic acid in the body; the latter may arise also from intestinal bacterial action on aromatic compounds, viz., tyrosine and phenylalanine.

Glycine is normally invariably present in amounts adequate to convert all of the benzoic acid formed to hippuric acid. The latter, formed in the liver cells, enters the systemic circulation and is excreted in the urine in amounts ranging from 0.1 to 1.0 g. (usually 0.5 to 0.7 g.) daily, depending largely on the dietary intake.

Citric Acid.

Citrate is excreted in the urine in amounts of 0.2 to 1.2 g. daily. It is increased by administration of estrogen and decreased by androgen. Urinary citrate varies during the menstrual cycle, increasing during the first half (increasing elaboration of estrogen), persisting at a high level during the luteal phase (synergistic action of estrogen and progesterone), and falling in the premenstrual phase (drop in estrogen). Administration of alkalis or alkaline ash diets increases, and acidification decreases urinary citrate.

Citric acid is formed in large amounts in the course of oxidative metabolism of carbohydrate. Whether or not this is the main source of urine citrate is not known. Citrate is present in the bones, apparently as a superficial deposit on the crystal lattice of apatite, which is the chief mineral complex in bone (p. 283). Citrate forms soluble and undissociable complexes with calcium, and these two ions exhibit certain interrelationships in their urinary excretion. Administration of parathyroid hormone or of very large doses of vitamin D, producing hypercalcemia, increases urine calcium and citrate (mobilization from skeleton?). Injection of citrate increases urinary excretion of calcium, perhaps by increasing its solution from bone.

Other Organic Acids. Lactic acid, the end-product of glycolysis in muscle, passes into the blood stream, undergoing ultimate reconversion to glycogen or glucose in the liver (p. 13). About 50 to 200 mg. are excreted daily in the urine under ordinary conditions of activity. Much more may appear during periods of prolonged, strenuous muscular exercise or anoxia.

Minute amounts (up to 50 mg. daily) of fatty acid anions and possibly free acids may be excreted in the urine. These are mainly short-chain acids (formic, acetic, butyric), which arise chiefly during intermediary metabolism and as a result of intestinal bacterial action.

Certain aromatic hydroxyacids may appear in the urine of normal subjects in small amounts. The most important are p-hydroxyphenylacetic and p-hydroxyphenylpropionic acids, which are formed in the intermediary metabolism and intestinal putrefaction of tyrosine. Other aromatic hydroxyacids (p-hydroxyphenylpyruvic, p-hydroxyphenyllactic, dihydroxyphenylacetic acids) may be excreted in abnormal states (tyrosinosis, vitamin C-deficient infants, certain cases of mental deficiency, alkaptonuria).

Ketone Bodies. The substances designated "ketone bodies" include acetone, acetoacetic acid, and β-hydroxybutyric acid. These are formed in the liver in the course of metabolism of fatty acids (p. 104) and are conveyed in the blood stream to the tissues for further oxidation. Under normal conditions, on the usual mixed diet, less than 125 mg. are excreted daily in the urine. Larger amounts may be eliminated on high-fat diets, during starvation or severe carbohydrate restriction, and during normal pregnancy. Marked increases are encountered in a variety of abnormal states associated with excessive ketogenesis (p. 109), e.g., diabetes mellitus (p. 109).

Sulfur-containing Compounds. Sulfur appears in the urine as (1) so-called neutral sulfur and (2) sulfate, the latter in (a) inorganic and (b) organic (ester; ethereal sulfate) combination. The bulk of the urinary sulfur is derived from the metabolism of the amino acids, methionine, cystine, and cysteine, the amount excreted by normal subjects therefore varying with the protein intake.

The total sulfur excretion usually ranges from 0.7 to 1.5 grams daily, averaging about 1.0 gram. The urinary N : S ratio is about 15:1. Neutral sulfur comprises about 5 to 15 per cent of the total (0.04 to 0.15 g.), and sulfate-S 85 to 95 per cent. About 5 to 15 per cent of the latter is organic (ethereal) sulfate (0.06 to 0.12 g.), the remainder inorganic (0.6 to 1.0 g.).

Neutral sulfur includes such substances as cystine, methionine, methylmercaptan, taurine and its derivatives (taurocholic acid), thiocyanate, thiosulfate, and urochrome.

Ethereal (organic) sulfate consists largely of phenolic sulfates, formed in the liver from aromatic compounds arising from intestinal bacterial action on tyrosine and tryptophan (indole, skatole, cresol, phenol) or from benzene or other phenolic compounds entering the organism (as under conditions of industrial or accidental exposure). Certain hormones, including estrone and androsterone, are excreted partly as sulfates and fall into this category. Indole, which arises almost entirely from intestinal putrefaction, is converted to indoxyl in the liver, is conjugated with sulfate, and appears in the urine as potassium (and sodium) indoxyl sulfate (indican). This is a rough index of the extent of intestinal putrefaction (< 20 mg. daily normally).

Further details are considered in the section on sulfur metabolism (p. 328).

Phosphate. The metabolism of phosphorus is considered elsewhere. Phosphorus is excreted (urine and feces) largely as inorganic phosphate, and to a small extent in organic form (< 4 per cent of total). Inasmuch as the phosphate is derived chiefly from the metabolism (oxidation) of phosphorus-containing organic foodstuffs (and tissue components), such as phosphoproteins, nucleoproteins, nucleotides, and phospholipids, the quantity excreted depends in large measure on the nature of the diet. It is, therefore, extremely variable.

The proportion of the total phosphorus excretion eliminated in the urine depends, under normal conditions, on dietary and intra-intestinal factors. On the usual mixed diet the daily urinary phosphorus excretion is 0.7 to 1.5 grams, averaging about 1.1 grams, almost entirely in the form of inorganic phosphate, comprising about 60 per cent of the total excretion (feces 40 per cent). If intestinal absorption of phosphate is diminished, as by high intake of calcium or magnesium (also iron), which tend to form relatively insoluble phosphates, unusually high intestinal alkalinity, and low vitamin D intake, the amount and proportion of phosphorus eliminated in the urine decreases (increases in feces). Conversely, on a low calcium, moderately high phosphorus intake, about 75 per cent of the total excretion is urinary. Vitamin D, as well as the parathyroid hormone, increases renal excretion of phosphate by diminishing its reabsorption in the tubules (p. 312). The "renal threshold" for phosphate excretion is about 2 mg. P/100 ml. plasma, excretion falling to a minimum at lower concentrations.

The phosphate ion appears in the urine in two forms, viz., acid phosphate, $H_2PO_4^-$, and basic phosphate, HPO_4^{2-}. The ratio $BH_2PO_4 : B_2HPO_4$ varies with and largely determines the pH of the urine, since acid and basic phosphates constitute the major buffer system in that fluid. Relatively insoluble phosphates tend to form in alkaline urine and are often important components of urinary calculi. The turbidity which develops in urine on standing (ammoniacal fermentation) is contributed to largely by precipitation of phosphates, including $MgNH_4PO_4$ ("triple phosphate"), insoluble at alkaline reactions.

Chloride. In normal subjects on an average diet (8 to 15 g. NaCl intake), the quantity of chloride in the urine is second only to urea, among the solid constituents, approximating the intake (5 to 9 g. chloride ion; 110 to 225 mmol.). The normal "renal threshold" for chloride is about 340 mg./100 ml. plasma; in normal subjects, as the plasma chloride concentration descends to this level, urinary excretion of chloride decreases (renal tubular reabsorption). This is influenced by the level of adrenal cortical function.

If the chloride intake is changed suddenly, some time may be required for the re-establishment of equilibrium between intake and output. For example, the sudden addition of a large amount of salt is usually followed by elimination of the excess within forty-eight hours. However, if the previous intake has been unusually low, elimination of the increment may be delayed several days.

In conditions in which elimination of chloride through other channels is increased, e.g., excessive perspiration, diarrhea, vomiting, the urine chloride decreases correspondingly.

Sodium; Potassium. The kidney is the chief regulator of excretion, consequently of the body equilibrium of sodium and potassium, the chief cations of extracellular and intracellular fluids, respectively. Operation of these excretory mechanisms, therefore, plays an important role in the maintenance of water and "acid-base" status (pp. 357 and 407). The daily urinary output of these elements by normal adults approximates their intake. Under usual dietary conditions, normal adults excrete in the urine about 3 to 5 g. (130 to 215 mmol.) sodium and 2 to 4 g. (50 to 100 mmol.) potassium daily. During periods of fasting or inadequate protein intake, excessive tissue protein catabolism, with liberation of intracellular fluid components, results in an increase in urinary potassium and a change in the Na : K ratio.

Adrenal cortical hormones (aldosterone), under ordinary circumstances, stimulate reabsorption of sodium and may depress that of potassium by the renal tubular epithelial cells (p. 382). Potassium enters the urine from the plasma by glomerular filtration and also by active tubular secretion, the latter mechanism apparently competing with that for urine acidification. Details of this process are considered elsewhere (p. 463). The amount of potassium in the urine is, therefore, influenced by the requirement for urine acidification, increasing when this requirement is low (e.g., alkaline ash diet; alkalosis) and decreasing when it is high (highly acid ash diet; acidosis).

Abnormal loss, through extrarenal channels, of sodium (excessive sweating, diarrhea, etc.) or of potassium (diarrhea, vomiting) results in decrease in the urinary excretion of these elements.

Calcium; Magnesium. Under usual dietary conditions in normal subjects, urinary excretion accounts for about 15 to 40 per cent of the total calcium and about 20 to 50 per cent of the total magnesium elimination, the remainder being excreted in the feces. At low or moderate levels of calcium intake (0.1 to 0.5 g. daily), about 30 to 50 per cent is eliminated in the urine, whereas at high levels of intake (1.0 g.) about 10 to 25 per cent is so eliminated. The dependence of the urinary excretion of these elements on their absorption from the intestine, and the factors involved (vitamin D, phosphate, etc.), are considered elsewhere (pp. 280 and 316).

The "renal threshold" for excretion of calcium lies between 3.25 and

4.00 mEq. (1.63 and 2.00 mmol.) per liter plasma, little being eliminated at lower plasma calcium concentrations. The influence of the parathyroid hormone in increasing renal tubular absorption has already been mentioned (p. 289); calcitonin appears to have an opposite effect. Urinary calcium may be but is not always increased in hyperparathyroidism because of the increased filtered load on the tubules.

It has been recently established that the kidney possesses a steep calcium gradient, just as for sodium, from cortex to medulla. Necropsy material in man has indicated a calcium concentration of 8.9 ± 3.7 mmol. per kg. wet tissue at the cortex, increasing to 38.9 ± 48.4 at the papilla tip. The calcium concentration at the papilla seems to be far higher than in urine. On the countercurrent multiplication principle, it must be assumed that the kidney possesses a calcium pump mechanism, which may or may not be the same as that for sodium. There does seem to be some direct relationship between the renal clearance of sodium and calcium. They tend to rise and fall in parallel. It is of great interest that no renal calcium concentration gradient could be demonstrated in the kidneys of children under one year of age.

Carbohydrates (p. 82). Normal urine may contain reducing substances (< 1.5 grams/24 hours), up to 40 per cent of which are fermentable; concentrations over 0.25 per cent (Benedict's test) should be regarded as probably abnormal. A portion of this is glucose, the remainder being derived chiefly from the diet (galactose from lactose of milk; pentose from fruits; caramelized sugar and dextrins; proteins; products of intestinal bacterial action).

Glucose may appear in the urine in small amounts during normal pregnancy, particularly in the later months. In certain normal individuals, glucose may be excreted after ingestion of large amounts of sucrose, glucose or, at times, starch (alimentary glucosuria). Lactose is excreted frequently during lactation. Pentose (L-arabinose) may appear in the urine of normal subjects after they have eaten large quantities of foods rich in pentose, e.g., cherries, grapes, plums, prunes (alimentary pentosuria). There is a chronic, familial type of pentosuria, in which L-ketoxylose is excreted independently of the nature of the diet. Galactose and fructose rarely appear in the urine of entirely normal subjects except after administration of supertolerance doses of these sugars.

Miscellaneous. The urine of normal subjects may contain protein in amounts too small to be detected except by very sensitive procedures. The normal urinary protein does not exceed 90 mg. daily, and consists of serum albumin, which passes through the glomerular filter, enzymes (e.g., pepsin, trypsin, lipase, amylase, lactate dehydrogenase and alkaline phosphatase), and nucleoprotein and "mucin" derived from the lower urinary tract.

The twenty-four-hour urine may contain minute quantities of ionic iron (< 0.3 mg.), arsenic (< 0.5 mg.), copper (< 0.1 mg.), zinc (< 0.6 mg.), iodine (< 0.07 mg.), and trace amounts of ionic cobalt, nickel, fluorine, silicate and lead.

A number of the water-soluble vitamins and their metabolites are excreted in the urine, usually in proportion to their intake. The quantities eliminated in the urine under controlled conditions have been employed as a basis for clinical evaluation of the state of vitamin nutrition and are considered in the discussion of the individual vitamins (q.v.).

Certain hormones or their metabolites are excreted in the urine, e.g., gonadotrophins (p. 691 ff.), estrogens (p. 775), androgens (p. 794), pregnanediol (p. 779), adrenocortical hormones (p. 714). Determination of the quantities eliminated is employed in the clinical evaluation of the state of functional activity of the respective endocrine glands (q.v.).

Normal human urine contains only minute amounts of lipids, in contrast to that of the tiger. Fat may appear in the urine as a result of lipoproteinuria or of chyluria, which is due to lymphatic rupture into the urinary tract.

CLINICAL STUDY OF RENAL FUNCTION

Present understanding of renal physiology permits rather precise evaluation of certain individual phases of kidney function. Measurements can be made of the rate of glomerular filtration, the renal blood flow, and the maximum capacity of the tubules for excretion and for reabsorption. Certain of the test procedures required for precise quantitative determination of these factors do not lend themselves as yet to routine general clinical use, although they are so used in many clinics devoted to the study of patients with renal and hypertensive diseases. However, procedures are available for general use which provide satisfactorily quantitative information on essential aspects of renal functional activity. The methods of study employed most widely at the present time fall into three categories: (a) tests based on the "clearance" concept; (b) studies of excretion of certain dyes preferentially excreted by the kidneys; (c) studies of the concentrating capacity of the kidneys.

CLEARANCE TESTS

In the performance of its excretory functions, the kidney may be said to "clear" the blood of certain of its constituents. Urea clearance is defined as the volume of blood plasma which one minute's excretion of urine suffices to clear of urea. Inasmuch as all of the blood flowing through the kidneys is only partially cleared of urea (and other substances), a more exact definition would be "the number of milliliters of plasma which contain the amount of urea removed per minute by renal excretion." The clearance also represents the minimum volume of plasma required to furnish the amount of substance excreted in the urine in one minute. This concept of renal clearance has contributed largely to our present understanding of renal function in health and disease.

If U indicates the concentration (mg. per 100 ml.) of substance (e.g., urea) in urine, and V the volume (in ml.) of urine formed per minute, U × V equals the quantity of substance excreted per minute. If P indicates the concentration (mg. per 100 ml.) of substance in the plasma, UV/P indicates the virtual volume of plasma "cleared" of substance per minute, i.e., the "clearance." It is essential that separated plasma be used instead of whole blood for clearance determinations if the substance investigated is not distributed uniformly between plasma and corpuscles (e.g., inulin, Diodrast, phenol red).

Theoretically, a substance may be excreted by (a) glomerular filtration alone, (b) filtration plus tubular excretion or (c) filtration plus tubular reabsorption. If a substance is completely filtered at the glomerulus and is subsequently completely reabsorbed by the tubules, its clearance will be zero (e.g., glucose). As the degree of tubular reabsorption diminishes, the substance appears in the urine and its clearance increases (e.g., urea), until, if there is no reabsorption (e.g., inulin), the clearance will be equivalent to the rate of glomerular filtration. If a substance, in addition to being filtered through the glomeruli, is also excreted by the tubular epithelium (e.g., phenol red, Diodrast), its clearance will exceed the rate of glomerular filtration by an amount equal to the extent of tubular excretion. Inasmuch as the kidneys cannot excrete more of a substance per unit of time than is brought to them in the blood, the upper limit of renal clearance is determined by the renal blood flow. For example, if a substance undergoes glomerular filtration and tubular excretion, and if all that is contained in the blood passing through the kidneys is removed and is concurrently transferred to the urine, its clearance will be complete (e.g., Diodrast clearance); i.e., it will be equivalent to the volume of plasma flowing through the kidneys per minute. These facts constitute the basis for quantitative determination of various aspects of renal function.

Glomerular Filtration. Inulin is a polysaccharide which is not metabolized in the body and, following its intravenous injection, is excreted quantitatively by the kidneys within a short time. It is excreted entirely by glomerular filtration and undergoes no reabsorption in the tubules. Inasmuch as inulin, being freely filtrable at the glomerulus, exists in the blood plasma and glomerular filtrate in identical concentration, and since the quantity of inulin excreted per minute in the bladder urine is equal to the amount entering the glomerular filtrate per minute, it follows that the inulin clearance (UV/P) represents the volume of glomerular filtrate formed per minute. Employing this procedure, it has been found that the average rate of glomerular filtration is 125 ml. per minute (about 70 ml. per sq. meter of body surface). Similar figures are obtained with certain polyhydric alcohols, viz., mannitol, sorbitol, and dulcitol. Lower values (by 25 to 45 per cent) may be obtained in elderly (70 to 85 years), apparently normal subjects.

Determination of the rate of glomerular filtration (inulin clearance) has several important physiologic applications: (a) the glomerular filtration rate minus the rate of urine flow (bladder) equals the quantity of water reabsorbed in the tubules per minute; (b) the inulin clearance minus the clearance of another substance (X) divided by the inulin clearance equals the proportion of substance X reabsorbed in the tubules. For example: Inulin Clearance (125) minus Urea Clearance (75), divided by Inulin Clearance (125), equals 0.4, indicating that 40 per cent of the urea of the glomerular filtrate is reabsorbed during its passage through the tubules; (c) when a substance has a clearance value higher than that of inulin, it is excreted partly or entirely by the tubules. Such data have been found to be useful in studying the mode of action of diuretic agents and abnormalities in excretion of various substances.

Inulin clearance may also be measured by labeling it radioactively and then determining the change in blood radioactivity after a steady state has been achieved. This obviates the necessity for collecting urine specimens but does make certain assumptions in regard to the appropriate final calculation. A variety of labeled inulin compounds have been used for this purpose. Although the classic inulin clearance must remain the reference method for glomerular filtration procedures, glomerular filtration rate has also been determined by the use of ^{58}Co-labeled vitamin B_{12} as well as by radioactive chromium-labeled EDTA. The B_{12} method has proved to be somewhat disappointing because of difficulties associated with protein binding.

Glomerular filtration is influenced by: (a) renal blood flow; (b) glomerular blood pressure and plasma colloid osmotic pressure (p. 457); (c) capsular pressure (p. 457); (d) the filtration surface area, i.e., the functioning glomerular surface area. Abnormalities in the rate of glomerular filtration (i.e., inulin or mannitol clearance) may occur, therefore, in a variety of clinical disorders in which these factors are altered.

1. Diminished Renal Blood Flow. Under normal conditions an enormous volume of blood (p. 487) passes through the glomeruli each minute, about 18 to 20 per cent of the plasma (except protein) filtering into Bowman's capsule. If other factors remain unchanged, a significant decrease in blood flow through the kidneys may result in a decrease in the volume of glomerular filtrate. The important causes of diminished renal blood flow are outlined elsewhere (p. 487). It is necessary to indicate here, however, that this does not invariably result in decreased filtration because of the simultaneous occurrence, in certain conditions, of other changes which tend to increase the filtration pressure.

Serious fall in systemic arterial blood pressure due to left ventricular (e.g., coronary occlusion) or peripheral circulatory (e.g., shock, hemorrhage) failure is accompanied by reduction in renal blood flow and glomerular filtration. However, the decrease in filtration is not proportional to that in blood flow, and in mild arterial hypotension it may not be altered significantly, due to simultaneous constriction of the efferent glomerular arterioles, the filtration pressure being thereby maintained or increased. A similar phenomenon occurs in congestive heart failure, in which the reduction in renal blood flow exceeds that in filtration. This is probably due to renal arteriolar constriction, affecting the efferent arterioles pre-

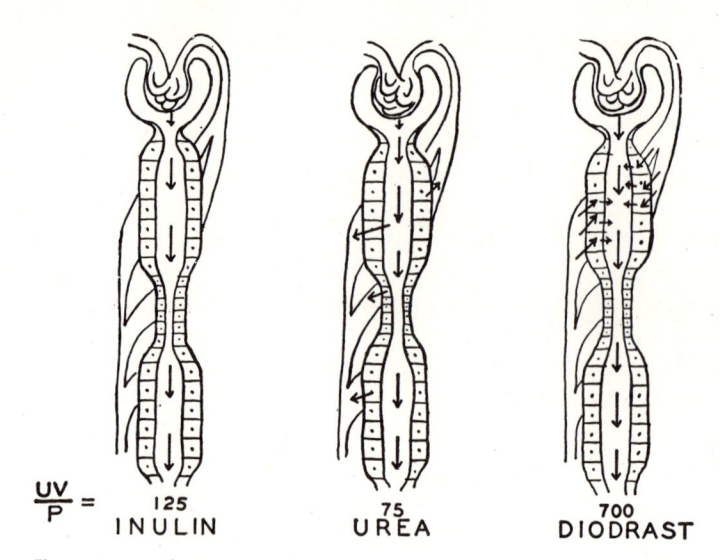

Figure 16-4. Characteristic clearance values. Scheme to illustrate mechanism of excretion: Inulin is excreted by glomerular filtration alone, with no tubular reabsorption. Urea is excreted by glomerular filtration and is partially reabsorbed by the tubule. Diodrast is excreted by both glomerular filtration and tubular excretion.

dominantly, in response to the increased venous pressure, with consequent increase in the filtration fraction (p. 489).

In essential hypertension, constriction or sclerosis of the glomerular arterioles, predominantly the afferent, results in diminution in renal blood flow and glomerular filtration pressure and, consequently, in the volume of glomerular filtrate.

2. Glomerular Damage. Significant decrease in the surface area of the glomerular filter should result in reduction in the volume of filtrate, unless the renal blood flow and/or filtration pressure were increased correspondingly, a situation which is virtually never encountered clinically. In fact, the renal blood flow is diminished in practically all types of diffuse glomerular damage of significant degree because of the obstacle imposed to the flow of blood, almost all of which must pass through the glomeruli (p. 455).

Glomerular filtration may be decreased, therefore, in diseases of the kidneys characterized by diffuse glomerular damage. These include glomerulonephritis, nephrosclerosis, intercapillary glomerulosclerosis, renal amyloidosis, advanced bilateral chronic pyelonephritis, etc.

3. Increased Capsular Pressure. Resistance to filtration is increased in the presence of obstruction to the passage of fluid through the uriniferous tubule or lower urinary passages. This raises the pressure in Bowman's capsule and decreases the effective filtration pressure (p. 457). Conditions in which this factor may operate include: blockage of renal tubules by desquamated cells, casts, and other precipitated material, e.g., hemoglobin, Bence-Jones protein, uric acid, calcium salts; obstruction in any portion of the lower urinary tracts by calculi, inflammation, stricture, neoplasm, etc.

4. Altered Plasma Colloid Osmotic Pressure. Significant elevation of plasma C.O.P. is rarely encountered, since increase in the plasma protein concentration, when it occurs, is virtually invariably in the globulin fraction (p. 183), which exerts relatively little influence on osmotic pressure. On the other hand, decrease in plasma C.O.P. occurs commonly, due to the frequent occurrence of hypoalbuminemia (p. 185). This should result in an increase in effective glomerular filtration pressure (p. 457) and increased filtration, if other factors remain unchanged. Whether this actually occurs is uncertain, but an increased glomerular filtration rate has been observed in children with chronic nephrotic syndrome (severe hypoproteinemia) which may be attributable to this cause, although increased renal blood flow has also been incriminated.

Determination of glomerular filtration (inulin or mannitol clearance) is of obvious value in investigating the pathological physiology of renal function. Inasmuch as all of the normal constituents of the urine probably leave the blood stream exclusively via the glomerular filtrate, with the exception of potassium (p. 463) and a relatively small fraction of the creatinine, reduction in glomerular filtration is usually the most important primary cause of abnormal retention of urinary waste products in the body fluids.

In interpreting deviations from normal values for inulin or mannitol clearance, i.e., glomerular filtration rate (Table 16-2), certain points should be kept in mind. Determinations should be made under carefully controlled and standardized conditions, including performance in the morning after an overnight fast (diurnal variation), adequate antecedent diet (dietary influences), avoidance of emotional disturbance (hormonal influences), and bed rest (exercise influence). Moreover there may be some question of the validity of measurements of glomerular filtration in certain kidney diseases, particularly in the presence of severe acute tubular damage, e.g., mercury or carbon tetrachloride poisoning, severe pyelonephritis. In such conditions, a portion of the filtered inulin or mannitol may diffuse out of the lumen across the damaged, nonfunctioning tubular cells, with the consequent production of spuriously low filtration values. However, this occurs rarely, even in advanced stages of the types of renal disease commonly encountered.

The usefulness of inulin and mannitol clearance tests as routine clinical diagnostic procedures is limited by the necessity of injecting the test substances intravenously and by certain other technical difficulties. Inasmuch as endogenous creatinine is not reabsorbed in the tubules, determination of endogenous creatinine clearance is often substituted clinically for these procedures. However, this is open to the criticism that the small amount of creatinine that enters the urine under normal conditions by tubular secretion may increase to a considerable and variable degree in conditions of impaired glomerular filtration. In normal subjects, and often in disease states, under average conditions of urine flow, values for urea clearance usually parallel those for inulin clearance, and the latter may be calculated from the former according to the formula: Urea Clear-

TABLE 16-2 NORMAL VALUES FOR VARIOUS PHASES OF RENAL FUNCTION DERIVED FROM CLEARANCE DATA AND CHARACTERISTIC VALUES IN CERTAIN TYPES OF KIDNEY DISEASE*

FACTORS	MEN	WOMEN	GLOMERULONEPHRITIS ACUTE	NEPHROTIC	CHRONIC	TERMINAL	CHRONIC PYELO-NEPHRITIS	BENIGN NEPHRO-SCLEROSIS
Glomerular filtration rate (GFR) Inulin clearance } (ml./min.) Mannitol clearance	130 ± 20	115 ± 15	15–100 (50)	40–80 (55)		0–40 (10)	50–90 (75)	50–150 (100)
Renal plasma flow (RPF) PAH clearance } (ml./min.) Diodrast clearance	700 ± 135	600 ± 100	60–800 (400)	400–700 (500)		20–500 (100)	300–700 (425)	450–700 (500)
Renal blood flow (ml./min.)	1275 ± 245	1090 ± 180						
Filtration fraction (%)	18–20	18–20	10–25 (15)	10–20 (15)		10–30 (25)	15–20 (17)	10–30 (22)
Maximum tubular excretion (Tm) PAH Tm (mg./min.)	75 ± 13	70 ± 10	10–80 (50)	30–40 (35)		10–60 (30)	10–50 (35)	50–100 (75)
Diodrast Tm (mg. I/min.)	55 ± 15	50 ± 15						
Maximum tubular reabsorption (Tm) Glucose Tm (mg./min.)	375 ± 80	300 ± 55						
Plasma flow per unit excretory mass RPF/Diodrast Tm (ml./mg.) RPF/PAH Tm (ml./mg.)	9 ± 1.5 9 ± 1.5	8 ± 1.5 8 ± 1.5	6–18 (8)	10–20 (12)		6–18 (10)		7–10 (8)
Plasma flow per unit reabsorption mass RPF/Glucose Tm (ml./mg.)	2 ± 0.4	2 ± 0.5						
Filtration per unit functioning nephrons Inulin clearance/PAH Tm (ml./mg.)	1.7 ± 0.4	1.5 ± 0.3	0.8–2.2 (1.2)	0.9–1.8 (1.2)		1.0–4.5 (1.5)		1.0–2.2 (1.5)

* Expressed per 1.73 square meters body surface area. Figures in parentheses are mean values.

ance/0.6 = Inulin Clearance. This proportionality arises out of the fact that all of the urinary urea is removed from the blood by glomerular filtration and, at urine volumes of 2 ml. or more per minute, about 60 per cent is excreted in the urine (40 per cent reabsorbed) (p. 464). This substitution of urea clearance for inulin clearance is not valid at low urine volumes. Nevertheless, except under unusual circumstances, determination of urea clearance is satisfactory in the clinical study of renal function and for purposes of diagnosis and prognosis (p. 503). There is, however, a growing tendency to replace urea clearance by creatinine clearance for these purposes.

Renal Blood Flow. If a substance is completely removed from the blood plasma during its passage through the kidney, the "clearance" of that substance (i.e., volume of plasma "cleared" per minute; $\frac{UV}{P}$ (p. 482)) would equal the renal blood plasma flow. Knowledge of the hematocrit value would permit calculation of the renal blood flow. This has been found to be true of Diodrast and p-aminohippurate, over 90 per cent of the plasma content of these substances being removed during a single passage through the kidneys ("extraction ratio") at relatively low plasma levels. Consequently, at least in normal subjects under standardized conditions, values for Diodrast or p-aminohippurate are approximately equivalent to the "effective" renal plasma flow, i.e., the flow through active renal excretory tissue. These have been found to be about 700 ± 135 ml. for men, and 600 ± 100 ml. for women, per 1.73 square meters of body surface. Corresponding values for renal blood flow are about 1275 ± 245 ml. and 1090 ± 180 ml., respectively. Lower values are obtained in elderly individuals.

Renal blood flow exhibits a diurnal variation, decreases during exercise, and is influenced by emotional states. These determinations must therefore be made under strictly standardized conditions. p-Aminohippurate (PAH) has largely replaced Diodrast for estimation of renal blood flow, because it possesses the following advantages over the latter substance: (a) it does not penetrate the erythrocytes; (b) it is less extensively bound to plasma proteins; (c) its quantitative determination is a simpler procedure. Values for PAH clearance in experimental animals correspond closely to those obtained by direct measurement of renal blood flow, amounting to about 30 per cent of the cardiac output and about 25 per cent of the total blood volume. This reflects the vital importance of these organs in maintaining a constant extracellular fluid composition.

Abnormal decrease in "effective" renal blood flow (PAH clearance) may result from:

1. Decrease in cardiac output or arterial blood pressure, e.g., congestive heart failure, coronary artery occlusion, shock syndrome.

2. Organic disease of the renal vascular system, e.g., renal arteriosclerosis, arteriolosclerosis, (nephrosclerosis), intercapillary glomerulosclerosis, glomerulonephritis, periarteritis nodosa.

3. Increased local resistance to the flow of blood, resulting from vaso-

constriction of the afferent and/or efferent glomerular arterioles, e.g., early essential hypertension, systemic arterial hypotension (p. 457).

4. Decrease in the mass of functioning kidney tissue, e.g., renal hypoplasia, polycystic disease, pyelonephritis, tuberculosis, malignancy.

An absolute increase in renal blood flow may occur after administration of pyrogenic agents, e.g., typhoid vaccine, due to renal arteriolar dilatation. It may be increased temporarily by a high meat diet. Following unilateral nephrectomy, the blood flow through the remaining kidney may increase more than 70 per cent in a short time.

The significance of diminished renal blood flow in contributing to reduction in glomerular filtration is referred to elsewhere (p. 483). The fact must be borne in mind that measurement of renal blood flow by the clearance procedure is valid only when renal extraction of PAH or Diodrast is almost complete, i.e., when over 90 per cent is removed during a single passage through the kidneys. Inasmuch as over 80 per cent of the normal excretion of these substances is accomplished by direct tubular cell action and less than 20 per cent by glomerular filtration, severe impairment of tubular excretory function may result in reduction in PAH or Diodrast clearance in the absence of corresponding diminution in renal blood flow. This will be reflected in decrease in the PAH extraction ratio, which can be determined by simultaneous measurements of PAH in the systemic arterial blood and in the renal vein blood obtained by catheterization. By this means, low extraction ratios have been observed in subjects with severe kidney damage due to glomerulonephritis or nephrosclerosis. In such cases, the renal blood flow may be calculated by substituting the arterial-venous plasma difference in PAH (or Diodrast) concentration for "P" in the clearance formula, i.e. $\dfrac{U_{PAH} \cdot V}{P(\text{art-ven})_{PAH}}$. However, PAH extraction is usually normal in early stages of these conditions, and also in congestive heart failure.

Maximum Tubular Excretory Capacity. The process of renal tubular excretion is limited by the mass of functioning tubular tissue available for the transfer of a substance from the blood to the urine. Consequently, the measurement of the maximal rate of excretion of a substance excreted by this mechanism reflects the "tubular excretory mass" of the kidneys for that substance. This value, designated Tm, is calculated from simultaneous determinations of inulin and Diodrast or p-aminohippurate clearances at high plasma Diodrast or PAH levels. It is expressed as milligrams of Diodrast iodine or PAH eliminated per minute by tubular excretion, i.e., total excretion minus the amount excreted by glomerular filtration. Normal values for Diodrast and PAH Tm are indicated in Table 16-2 (p. 486).

It has been suggested that the Tm may be calculated indirectly as follows:

$$\text{Tm} = \frac{\text{Sp. Gr.} - 3.4}{4.8} \sqrt{UC}$$

where Sp. Gr. represents the second and third decimal place figures of the maximum urinary nonprotein specific gravity obtained by the concentration test (p. 497) (viz., 1.028 = 28.0) and UC is the urea clearance in terms of per cent of average normal.

The value of this determination lies in the fact that, at effective plasma PAH or Diodrast concentrations and with adequate but not necessarily normal renal blood flow, the Tm is independent of glomerular activity and reflects the amount of functioning renal tubular tissue. For example, if one kidney were removed, Tm would be diminished by 50 per cent; if a portion of the excretory tissue were destroyed (either tubular destruction or obliteration of circulation), Tm would be reduced in proportion to the extent of destruction. If the glomeruli were entirely obliterated without impairing the circulation to the tubules, Tm would be unaltered.

The ratio of renal plasma flow to maximum tubular excretory capacity, i.e., of PAH or Diodrast clearance to PAH or Diodrast Tm, is an expression of the plasma flow per unit of functioning excretory renal tissue. Normal values are indicated in Table 16-2 (p. 486). Increased values may be obtained in the presence of fever (active hyperemia), and decreased values (relative ischemia) in the presence of intrarenal vasoconstriction, e.g., in shock, essential hypertension, peripheral circulatory failure, and in severe anemia.

Percentage Tubular Reabsorption. If another substance as freely filtrable through the glomeruli as inulin has a clearance value lower than the latter, it has undergone reabsorption in the tubules. The extent of this reabsorption may be calculated as follows:

$$\text{Reabsorption} = \frac{\text{Inulin Clearance (e.g., 125) minus Clearance of X}}{\text{Inulin Clearance (125)}}.$$

By means of this calculation it is evident that about 40 to 50 per cent of the urea present in the glomerular filtrate is normally reabsorbed in the tubules under average conditions of urine flow (p. 464). $\left(\frac{125 - 75}{125} = 0.4 \text{ or } 40\%.\right)$

Filtration Fraction. The ratio of plasma inulin clearance (i.e., volume of glomerular filtrate) to plasma PAH or Diodrast clearance (i.e., renal plasma flow) represents the fraction of plasma filtered through the glomeruli. Under normal conditions this is about 0.18 (i.e., $125/700$), indicating that approximately 18 per cent of the water of the plasma flowing through the kidneys is filtered through the glomeruli into the lumen of Bowman's capsule.

Experimental studies have shown that increase or decrease in the filtration fraction is usually due to increase or decrease, respectively, in the tone of the efferent as compared to that of the afferent glomerular arteriole, with consequent increase or decrease, respectively, in intraglomerular blood pressure. Renal hyperemia produced by a pyrogen is probably due to predominantly efferent arteriolar dilatation, since the renal blood flow increases and the filtration fraction decreases. The tone of the efferent arteriole is increased by administration of epinephrine and in orthostatic and psychogenic vasoconstriction, with consequent decrease in renal blood flow and increase in the filtration fraction. In early essential hypertension, as well as in hypertension induced by administration of renin or angiotonin (experimental), the renal blood flow (Diodrast clearance) is usually decreased, glomerular filtration (inulin clearance) is often normal, and the filtration fraction is frequently increased; this combination of circumstances can be produced practically only by predominantly efferent arteriolar constriction. Because of the increased filtration fraction, glomerular filtration and urea clearance may be maintained within normal limits until late in the course of essential hypertension (p. 523); under such circumstances, the normal urea clearance is not an expression of absolute integrity of renal function but is usually an indication of renal vasoconstriction. An increase of filtration fraction from 0.2 to 0.3 is adequate to maintain urea clearance at 100 per cent of normal at a time when renal blood flow has fallen from a normal level of 1000 ml. per minute to an ischemic level of 600 ml. per minute.

Maximum Tubular Reabsorption Capacity. This may be studied in a manner similar to that employed for determination of excretory Tm, except that in this case a substance must be employed that is reabsorbed extensively rather than excreted by the tubular epithelial cells. Glucose is used for this purpose. When the plasma glucose concentration is maintained at a sufficiently high level, the amount of glucose entering the tubule will exceed the reabsorptive capacity and some will escape in the urine. Simultaneous determination of inulin clearance and mean blood glucose concentration permits calculation of the amount of glucose entering the tubule per minute, since all of it enters in the glomerular filtrate. Subtraction of the quantity escaping in the urine per minute gives the amount that has been reabsorbed.

Normal values for glucose Tm are indicated in Table 16-2 (p. 486). The significance of abnormalities in this factor is much the same as in the case of the excretory Tm (PAH or Diodrast), applying in this instance to tubular reabsorptive mass. The plasma flow per unit of functioning tubular reabsorptive mass may be calculated by determining the ratio of renal flow (PAH or Diodrast clearance) to the glucose Tm (Table 16-2).

Glomerular function in terms of active nephrons may be expressed by the ratio of glomerular filtration rate (inulin clearance) to maximum tubular excretory capacity (PAH Tm). Normal values are indicated in Table 16-2. Increased values reflect hyperfiltration and decreased values hypofiltration in residual nephrons.

Other Clearance Procedures. Determination of urea clearance has been employed extensively clinically as a measure of renal functional efficiency. At urine volumes of 2 ml. or more per minute, normal urea clearance is 75 ± 10 ml. per minute (per 1.73 sq. meters body surface), indicating tubular reabsorption of about 40 per cent of the urea of the glomerular filtrate. With decreasing urine volumes below 2 ml. per minute (augmentation limit), the proportion of the filtered urea undergoing reabsorption increases progressively (p. 502). Under these circumstances, the volume of blood cleared of urea per minute is not directly proportional to the urine volume, and the usual clearance formula (UV/P) is not applicable. It has been proposed (Van Slyke) that with urine volumes below 2 ml. per minute an approximately accurate value for urea clearance may be obtained by applying the formula $U\sqrt{V}/P$ designated "Standard Clearance." This is normally 55 ± 10 ml. per minute. This is no longer considered reliable for clinical purposes. By administering orally 570 ml. of water, one-half hour before the test is carried out (p. 503), and again during the test, more than 2 ml. urine per minute usually results.

At physiological levels of plasma creatinine (i.e., endogenous), values for creatinine clearance correspond rather closely to those for inulin clearance in normal subjects. However, when the concentration in the plasma is raised by intravenous injection of creatinine, or in renal disease (glomerular damage), the clearance values exceed those for inulin (rising to about 175 ml.), indicating partial tubular excretion of this substance under these conditions. Consequently, the creatinine clearance is not an invariably reliable index of the glomerular filtration rate.

Phenolsulfonphthalein has an average clearance value of 400 ml., indicating that it is eliminated mainly (about 70 per cent) by tubular excretion. This fact serves to explain occasional discrepancies between values for PSP excretion and urea clearance in disease states (p. 517), urea being excreted entirely by glomerular filtration, a portion being reabsorbed subsequently.

Renal clearances of other substances are determined occasionally in studying the pathological physiology of certain conditions, e.g., uric acid in gout and eclampsia, sodium in congestive heart failure. However, they have little place in the clinical evaluation of renal functional efficiency. It is now possible to determine renal clearance for individual plasma proteins in patients suffering from proteinuria. This gives an indication of the degree of permeability of the damaged glomerular membrane.

ELIMINATION OF WATER

Comparison of Fluid Intake and Fluid Output (Fig. 12-1, p. 340). The quantity of water eliminated by the kidneys in twenty-four hours depends upon two factors, namely, the amount of water supplied to the body and the amount lost through extrarenal channels or otherwise rendered unavailable for excretion by the kidneys.

The organism receives water from three main sources. These are considered elsewhere in detail and need be outlined here only briefly.

(a) Ingested liquids.

(b) Ingested solids. This source of supply is too frequently overlooked. The water content of many solid foods is extremely high and, in some instances, is higher than that of certain foods commonly considered to be liquid. For example, tomatoes contain a greater percentage of water than does milk. In any accurate study of water balance the water content of so-called "solid foods" must be taken into consideration.

(c) Water of combustion, or metabolic water. One hundred g. of dry starch metabolized in the body form 55.5 g. of water, and a similar quantity of glucose, 60 g. of water. One hundred g. of fat metabolized in the body form approximately 110 g. of water. This source of supply of water to the organism usually represents but a small proportion of the total supply, however, and does not introduce a serious error in the calculation of the water intake.

The kidneys ordinarily excrete more than one-half of the water eliminated from the body. The basal requirement of water for urine formation is dependent chiefly upon the intake (or catabolism) of protein and NaCl, inasmuch as urea and NaCl comprise the bulk of the solid constituents of the urine. Under normal conditions the remainder is lost largely through the vaporization of water in the respiratory passages and by the skin. The average quantity lost daily in this way is 850 ml. In the presence of active perspiration or fever this amount is increased and that eliminated by the kidneys is correspondingly diminished. The amount of water excreted by normal kidneys may be greatly decreased in the presence of conditions which cause depletion of the water reserves or abnormal retention of water in the body, as indicated below.

These extrarenal factors must be taken into consideration when studying renal function on the basis of the quantitative relationship between water supply and urine volume. Ordinarily the latter is simply compared with the fluid intake. The total water supply, in individuals on an average diet, exceeds the fluid intake by about 700 ml.; in the absence of the pathological factors mentioned above, the water loss exceeds the urine volume by approximately the same amount. Therefore, under such circumstances, the consideration of fluid intake as total water supply and urine volume as total water elimination involves an error of approximately only 5 per cent. Interpretations based upon such observations should be made, however, with full realization of the inherent possibility of error, particularly in the presence of extrarenal factors which may influence the water balance.

Oliguria; Anuria. Decrease in urine volume (oliguria) occurs in a variety of clinical disorders and, if marked, may of itself produce renal insufficiency in the absence of intrinsic kidney damage, or may aggravate renal functional impairment due to kidney disease. Oliguria (or anuria) may be due to prerenal, renal, or postrenal (lower urinary tract) causes.

Dehydration (p. 368). Excessive loss of water and electrolytes by extrarenal routes (prerenal deviation of water) results in a corresponding decrease in urine volume unless compensated by an increased intake. Oliguria of this type occurs in a number of conditions accompanied by prolonged vomiting or diarrhea, excessive sweating, profuse expectoration (lung abscess, bronchiectasis), profuse inflammatory effusions, external intestinal or bile fistulas, etc. In the absence of significant kidney damage, the specific gravity of the small volume of urine is high. If the quantity of water available for solution of the required amount of urinary solids is inadequate for this purpose, some will be retained in the body, as reflected in an increase in blood urea N. A similar situation may develop as a consequence of prolonged diuresis, e.g., in uncontrolled diabetes mellitus, with depletion of available body water reserves (p. 370).

Arterial Hypotension. Maintenance of glomerular filtration requires an adequate effective filtration pressure; this is dependent primarily on the glomerular blood pressure, which, in turn, depends on the general arterial pressure (renal artery). Filtration cannot be maintained at mean arterial pressures much below 70 mm. Hg, and anuria develops. This may occur in peripheral circulatory collapse (shock), as seen frequently after severe trauma and extensive burns, postoperatively, following acute hemorrhage, in acute coronary occlusion, overwhelming infections, and as a consequence of severe dehydration.

Conditions in this category are among the most common causes of oliguria and renal functional insufficiency, often with blood urea N retention. The kidneys may be entirely normal initially. The fall in blood pressure, whether due to blood loss, decreased plasma volume, peripheral vasodilatation, or decreased cardiac output, is followed by intrarenal arteriolar constriction, the efferent glomerular arterioles being affected disproportionately, with consequent rise in glomerular blood pressure and increase in the filtration fraction (p. 489). This mechanism tends to maintain glomerular filtration at lower levels of arterial pressure than would be possible otherwise. However, this is accomplished at the expense of renal blood flow, which, diminished originally, as a result of the hypotension or of the underlying phenomena, is further reduced by the increasing intrarenal vasoconstriction. Under these circumstances, glomerular filtration ultimately becomes inadequate and an unduly large fraction of the filtrate that is formed is reabsorbed in the tubules.

Moreover, tubular damage, even to the extent of severe necrosis, may develop as a consequence of the renal ischemia (anoxia) resulting from the reduced renal blood flow. Under these conditions, e.g., "acute tubular necrosis," oliguria and anuria are contributed to by the unselective reabsorption of the glomerular filtrate across the now nonfunctioning tubular epithelial lining. Superimposition of this type of kidney damage upon the original circulatory disturbance is of serious prognostic significance, the gravity of this condition increasing with increasing duration of the hypotensive state (i.e., shock).

Edema. In edema due to any cause, prerenal deviation of water tends to result in decrease in urine volume. The several factors involved in abnormal retention of water in the body in various clinical disorders are reviewed elsewhere (p. 380). Intrinsic kidney disease may or may not be present. However, disturbance of tubular function, i.e., increased reabsorption of Na and water, may occur in the absence of significant morphological kidney damage, e.g., in congestive heart failure and hepatic cirrhosis (p. 381).

Kidney Disease. Oliguria may result from diffuse glomerular damage (e.g., glomerulonephritis) of sufficient degree to reduce glomerular filtration to very low levels. It occurs also even in the presence of adequate glomerular filtration, if there is extensive tubular degeneration (toxic necrosis of tubules), as in poisoning by mercury, arsenicals, carbon tetrachloride, and sulfonamides, and in the shock syndrome and tubular necrosis (severe trauma, burns, etc.). In such conditions, the tubular functions of selective reabsorption are lost in the damaged areas, and back-diffusion of the glomerular filtrate occurs as a result of loss of the physiological barrier (tubule cells) between the fluid in the lumen of the tubules and the concentrated blood in the peritubular capillaries. In severe cases the continuity of the tubule may actually be disrupted at some points, permitting direct passage of the glomerular filtrate into the peritubular interstitial spaces.

Urinary Obstruction. This may be intrarenal or extrarenal. Flow of the fluid through the tubules may be impeded as a result of occlusion of the lumen by casts or other precipitated material (crystals, hemoglobin derivatives, Bence-Jones protein), desquamated epithelial debris, or compression by edematous renal interstitial tissue. Back-pressure may be increased also as a result of bilateral lower urinary tract obstruction (e.g., urinary calculi; ureteral stricture, tumor, or compression; prostatic enlargement; bladder tumors; urethral stricture, etc.). Under these circumstances the capsular pressure (p. 457) is increased, filtration is reduced, and an unduly large fraction of the smaller filtrate undergoes reabsorption as a result of its slower passage through the tubule.

Water Function Test (Dilution Test). This test possesses the advantage of putting a strain upon the water excretory function of the kidneys. It may be performed as follows:

(*a*) The fasting individual, after emptying the bladder, ingests 1200 ml. of water in one-half hour.

(*b*) The bladder is emptied at hourly intervals for four hours.

In the normal subject approximately 1200 ml. are eliminated within four hours, the larger part being excreted in the first two hours. The specific gravity of one of the hourly specimens should fall to at least 1.003. If renal function is impaired the quantity eliminated in four hours may be quite small and the specific gravity often 1.010 or higher, although lower figures are obtained not infrequently. Identical results may be obtained if any of the extrarenal factors mentioned above are operative. The differentiation between renal and extrarenal causes of defective water excretion must depend upon the results of other tests of renal function. The value of the dilution test lies in its clinical availability; it rarely adds any significant information to that obtained by other methods.

The ability of the kidney to eliminate water may be impaired in all types of renal disease, acute or chronic. This impairment may appear to be most pronounced in those cases associated with oliguria and edema. However, inability on the part of the kidneys to excrete water is not the fundamental factor in the pathogenesis of edema.

The edema of acute or chronic nephritis and the nephrotic syndrome is probably due largely to prerenal deviation of fluid into the tissues, caused by increased capillary permeability, decrease in the colloid osmotic pressure of the blood plasma (decreased plasma albumin concentration), or increase in capillary blood pressure (congestive heart failure) (p. 354). The marked degree of oliguria which is an outstanding feature of such cases is largely secondary to the extrarenal factors mentioned above.

The existence of polyuria in chronic renal failure may mask a definitely impaired ability to eliminate water. Polyuria in such cases must be regarded as a compensatory mechanism whereby the kidney, its ability to excrete solids in their normal concentration being impaired, must dilute these substances in order to secure their adequate elimination. As in the case of other organs, such as the heart, this compensation is effected at the expense of the functional reserve capacity. Therefore, when such a test as the dilution or water function test is performed and 1200 ml. of water are ingested within one-half hour, only a small fraction may be eliminated in the succeeding four hours. If such a patient can excrete only 400 ml. in the four hours, a markedly deficient water excretion, he might still maintain a polyuria of 2400 ml. in twenty-four hours. The polyuria is due to the increased osmotic load on each surviving nephron, which results from the increased urea content of the tubular fluid and the increased glomerular filtration per nephron. The effect of ADH is diminished, but aldosterone seems to play little or no part. The "third factor" inhibits sodium absorption in the surviving proximal tubules, and the consequent increase in salt loss also adds to the osmotic load. In chronic renal failure, there is loss of the normal diurnal variation in urine output. Consequently, the polyuria gives rise to nocturia.

ELIMINATION OF SOLIDS—URINE SPECIFIC GRAVITY

Considered in a broad sense, the chief excretory function of the kidney is to eliminate solid substances in solution in water. Many of these substances exist in the urine in much higher concentration than in the blood, the ratio of the average concentration in the urine to its concentration in the blood during the same period (concentration ratio) varying greatly for each of the urinary constituents. Thus, in the normal performance of its excretory functions the kidney must concentrate the eliminated substances, the necessary degree of concentration at any moment depending upon the relative quantities of solids and water present at that moment in the blood passing through the glomerular capillaries. One of the most important characteristics of the healthy kidney is its ability to eliminate the required quantity of solids regardless, within wide limits, of the amount of water available for their solution. In other words, the normal kidney exhibits a remarkable flexibility in its concentrating ability. Consequently, the concentration of solid constituents of normal urine, as evidenced by the specific gravity, varies considerably during the day in accordance with the ingestion of fluids and solid food and with the metabolic activity of the tissues. If large amounts of fluids are ingested the urine is of large volume and low specific gravity; if little water is ingested, or large amounts are lost through other channels such as the skin and gastrointestinal tract, the urine is of small volume and high specific gravity. Investigation of the concentrating ability of the kidney constitutes a most valuable measure for the determination of renal functional integrity.

Concentration Test of Renal Function

One of the earliest reflections of renal functional impairment is diminution in the ability of the kidney to eliminate the solid constituents of the urine in maximal concentration. Several tests have been proposed for the determination of the concentrating ability of the kidneys. Perhaps the most satisfactory procedures are those which involve the determination of urinary specific gravity under different conditions. Normally, urine of low volume has a high specific gravity, that of large volume a low specific gravity. When renal function is impaired, the ability of the kidneys to elaborate a concentrated urine is diminished. This functional aberration is termed hyposthenuria. With increasing renal damage the kidney loses its ability, not only to concentrate, but also to dilute urine. As renal functional impairment progresses, the molecular concentration of the urine approaches that of the protein-free blood plasma (isosthenuria). In other words, the maximum possible specific gravity diminishes and the minimum increases, the specific gravity of the urine being relatively fixed within narrowing limits above and below 1.010, the approximate specific gravity of protein-free blood plasma. Diminished ability to concentrate the urine occurs only as a result of renal functional impairment and is not affected by extrarenal factors. Investigation of the concentrating ability of the kidneys constitutes one of the few simple measures by which renal functional impairment can be demonstrated during compensated stages. In estimating the urinary specific gravity in specimens containing albumin or glucose, a correction of 0.003 must be subtracted from the observed specific gravity for each gram of albumin and 0.004 for each gram of glucose per 100 ml. of urine. The values stated subsequently in this discussion refer to nonprotein urinary specific gravity.

Two procedures that have been employed extensively for the investigation of the concentrating power of the kidney are (1) the two-hour specific gravity test and (2) the concentration test.

The Two-hour Specific Gravity Test.

This test, as modified by Mosenthal, consists essentially of the investigation of the urinary response to a diet containing a reasonable amount of fluid, salt and protein. The dietary is divided into three meals taken at 8 A.M., 12 noon and 5 P.M. No food or fluid of any kind is taken between meals or during the night. The bladder is emptied at 8 A.M. immediately before breakfast, the urine specimen being discarded. The urine is collected at two-hour intervals during the twelve-hour day period from 8 A.M. until 8 P.M. The night urine is collected as a single specimen for the twelve-hour period from 8 P.M. until 8 A.M. of the succeeding day.

In normal subjects, the specific gravity of individual specimens varies 10 points or more from the highest to the lowest; the specific gravity of the night urine is 1.018 or more; the total quantity of urine passed during the twelve-hour day period is three to four times as great as that eliminated during the twelve-hour night period. As the renal lesion progresses and the ability of the kidneys to concentrate solid substances in the urine diminishes, the specific gravity tends to become fixed at relatively low levels, regardless, within certain limits, of the urine volume.

This procedure, although quite satisfactory for the estimation of renal functional efficiency, possesses no advantage over the simpler urine concentration test outlined below insofar as the practical demonstration of renal functional impairment is concerned.

The Urine Concentration Test.
The test suggested by Addis and modified by Fishberg is as follows:

At 6 o'clock on the evening before the test the patient ingests a meal which should not contain more than 200 ml. of fluid but which has a high protein content. After this no fluid or food is taken until the test period is over. The bladder is emptied before retiring and the urine is discarded as is all urine passed during the night. The bladder is emptied at 8 A.M., 9 A.M. and 10 A.M., each specimen being kept in a separate bottle. The specific gravity of each of the specimens is taken. In normal individuals the specific gravity of at least one of the specimens will exceed 1.025, figures as high as 1.032 being frequently obtained. With increasing renal functional impairment the maximum specific gravity diminishes, approaching 1.007, the specific gravity of protein-free blood plasma. Fishberg states that he has been unable to find any patients with renal disease whose ability to concentrate the urine was so severely diminished that they could not, when placed under appropriate conditions, reach a maximum specific gravity of at least 1.010.

The following procedure has been recommended by Lashmet and Newburgh:

(1) At 10 P.M. stop all fluids and food except special diet until the end of the test period (38 hours).

(2) At 8 A.M. the following morning, empty bladder and discard urine. A special diet, containing 40 g. protein, 104 g. fat, 204 g. carbohydrate, 1900 calories, and 1 g. added sodium chloride, divided into three meals, is taken on this day. All urine is collected from 8 A.M. until 8 A.M. the following day.

(3) No food or water is taken on this day until the completion of the test. Urine is collected at 10 A.M. and again at 12 noon.

Under these conditions, normal subjects are able to concentrate urine to a specific gravity of 1.029 to 1.032. Impairment of renal function is indicated by decreasing urinary specific gravities below 1.028. Correction should be made in this, as in other concentration tests, for albumin in the urine. This procedure possesses little if any advantage over the simpler concentration test outlined above.

As in the case of the two-hour specific gravity test, the urine concentration test aids in differentiating prerenal from extrarenal factors and is capable of indicating renal functional impairment early in the course of renal disease when the functional defect is well compensated. In patients with edema, if the edema fluid is being evacuated, the test period urine may be of low specific gravity, simulating diminished concentrating ability. Obviously, under such conditions a low maximum specific gravity is not necessarily indicative of renal disease.

This simple test is one of the most valuable tests of renal function. A great deal of information may be obtained by the use of the concentration test in conjunction with the estimation of the nonprotein nitrogenous constituents of the blood. The former indicates the presence or absence of renal functional impairment; the latter indicates whether this is in the compensated or uncompensated stage.

The relative values of the concentration tests and the determination of blood urea clearance are discussed elsewhere (p. 418 ff.). The latter gives more definite information regarding the actual degree of renal functional impairment, but the former is usually quite as sensitive in detecting the presence of such impairment. From a practical standpoint, if normal results are obtained by the concentration test, renal function may generally be regarded as normal, and determination of the blood urea clearance is usually superfluous. Exceptions to this generalization are indicated elsewhere.

Significance of Hyposthenuria. Observations in experimental animals indicate that the fluid entering the distal segment of the uriniferous tubule is approximately isotonic with protein-free plasma (sp. gr. about 1.010), although the concentrations of certain components have been changed. Inasmuch as the capacity for concentrating this fluid must therefore reside mainly in functions of the distal tubular cells, diminished concentrating ability (hyposthenuria) is a reflection chiefly of impaired function of the distal portion of the tubule. However, this is not necessarily dependent upon demonstrable morphological changes in these cells. Although the mechanisms underlying hyposthenuria in certain conditions are not entirely clear, it may usually be attributed to one or more of the following factors:

1. Tubular Epithelial Damage. Tubular degeneration may result from: (a) nephrotoxic agents, such as mercury, carbon tetrachloride, uranium, and sulfonamides; (b) severe alkalosis; (c) the shock syndrome, due to a variety of causes (acute tubular necrosis); (d) secondary to glomerular damage, due to impairment of tubular blood supply (p. 456), as in glomerulonephritis and advanced nephrosclerosis.

If the damage is very severe, e.g., in acute tubular necrosis, anuria may occur initially (back-diffusion). In less severe cases there may be a low output of urine of low specific gravity. During recovery from severe damage, polyuria may occur (low urine specific gravity) and, subsequently, hyposthenuria may persist for several weeks or months. This is apparently a reflection of functional inadequacy of the newly regenerated tubular epithelial cells. In the anuric stage, there is retention of potassium ion and the serum K^+ level is raised to dangerous levels. With the onset of diuresis, there is marked urinary loss of K^+, which easily leads to a highly dangerous hypopotassemia. The events after exposure to a cause of acute tubular necrosis have been subdivided into five stages. These are: stage of onset, the oliguric-anuric stage (urine output less than 400 ml./day), the diuresis stage (urine output greater than 400 ml.), the recovery stage (blood urea begins to fall) and finally the convalescent stage.

In so-called "lipoid" nephrosis, the urine volume is characteristically low (edema) and the specific gravity high. This indication of satisfactory tubular function despite visible morphological damage is supported by observations of normal ammonium ion formation and urine acidification in this condition.

2. Decreased Renal Blood Flow. A considerable amount of osmotic work must be performed (i.e., energy produced) by the tubule cells in forming urine of a specific gravity higher than that of protein-free blood plasma. This metabolic activity is reflected in the relatively high oxygen consumption of kidney tissue. It would be expected, therefore, that these cells should be particularly vulnerable to anoxia and that their functional activity should be impaired by reduction in renal blood flow. Hyposthenuria occurs under these circumstances and also in severe chronic anemias. The conditions in which renal blood flow is diminished are outlined elsewhere (p. 487). In many of these, initially, there may be no visible morphological damage, but regressive changes develop if the circulatory impairment (or anemia) is marked and prolonged.

3. Decreased Number of Functioning Nephrons. Removal of a large portion of kidney tissue in experimental animals results in polyuria and hyposthenuria. Although the mechanism is not known exactly, there is evidence that the number of glomeruli is reduced proportionately more than the renal blood flow. Consequently, each glomerulus is perfused with an abnormally large volume of blood, and the minute volume of glomerular filtrate may be increased to such an extent that the reabsorption capacity of the tubules is exceeded (relative tubular insufficiency). This results in hyposthenuria. A similar situation may occur at times in certain cases of chronic pyelonephritis and other destructive kidney lesions (e.g., tuberculosis, malignancy), and occasionally in glomerulonephritis and nephrosclerosis.

Decreased concentrating capacity, with polyuria, occurs in association with hypercalcemia and hyperkalemia, apparently a consequence of these electrolyte abnormalities. Before the development of severe morphologic changes in the kidney, this functional defect is readily reversed by restoration of the normal blood calcium or potassium concentrations. In the case of hypokalemia, there is a reduction in glomerular filtration and in p-aminohippurate transport; under these circumstances, the latter cannot be regarded as an index of renal plasma flow. There is also impairment of the mechanism of acidification of the urine. Later, rather characteristic degenerative changes occur in the tubular epithelium. The mechanism is not clear in the case of hypercalcemia. Polyuria and hyposthenuria in this condition occur before the development of nephrocalcinosis and/or nephrolithiasis, and can be induced in normal subjects by intravenous injection of calcium salts. It would appear that the increased concentration of calcium exerts a depressing influence on the renal concentrating mechanism in the tubules.

Hyposthenuria occurs consistently in sickle-cell disease and, occasionally, in subjects with sickle-cell trait. This inability to concentrate the urine is reversible in early stages of the disease, normal function in this regard being restored in young patients by transfusion of normal blood. The mechanism of production of this functional defect is not clear, but there is evidence that it may be dependent upon renal hypoxia and small intrarenal vascular occlusions incident to the sickling phenomenon. It may result from increased lymphatic drainage disturbing the osmotic gradient in the renal medulla. Acute renal failure may occur as a complication of sickle-cell disease.

As indicated above, hyposthenuria occurs in almost all diseases of the kidneys and in certain types of renal functional impairment due to extrarenal causes. Investigation of the maximal concentrating capacity of the kidneys (concentration test) is therefore a valuable approach to the study of kidney function. However, it has certain limitations.

Misleading low specific gravity findings may be obtained during periods of elimination of edema fluid, e.g., improvement in congestive heart failure, chronic nephrosis, hepatic cirrhosis, etc. This is, of course, not a reflection of impaired concentrating ability but a consequence of the increased water excretion during the test period.

Variable results are often obtained in essential hypertension (benign nephrosclerosis). In many cases, hyposthenuria cannot be demonstrated in the early stages of the disease and, indeed, in some instances, for prolonged periods during the evolution of the renal lesion (nephrosclerosis). Clearance studies during this period reveal reduction in effective renal plasma flow, glomerular filtration, and maximal tubular excretory capacity, reflecting significant renal functional abnormalities that may not be revealed by urine concentration tests in this condition. Hyposthenuria occurs almost invariably in the later stages of nephrosclerosis.

An essentially normal concentrating capacity may be exhibited in conditions in which glomerular damage is not accompanied by significant tubular damage or functional impairment (uncomplicated glomerular insufficiency). This occurs at times in early stages of acute glomerulonephritis and also in prerenal deviation of water (p. 507). In such cases, the reduction in glomerular filtration may result in marked oliguria and elevation of blood urea N, despite the fact that the specific gravity of the inadequate volume of urine may be quite high. As a rule, hyposthenuria occurs later in the course of these conditions, if they persist or increase in severity, because of the superimposition of tubular damage as a result of prolonged curtailment of blood supply (p. 456).

Whereas decrease in maximal concentrating capacity, with certain exceptions indicated above, denotes impairment of kidney function, the extent of such decrease (i.e., of hyposthenuria) does not always reflect accurately the degree of renal functional inadequacy. In chronic glomerulonephritis, for example, hyposthenuria may progress rather rapidly to isosthenuria, i.e., fixation of specific gravity at about 1.010, long before the urea clearance has fallen to its ultimate terminal levels. During recovery from acute glomerulonephritis, hyposthenuria often persists for some time (weeks or months) after glomerular filtration and renal blood flow have returned to normal. This may reflect functional inadequacy of the newly regenerated tubular epithelial cells (p. 514).

ELIMINATION OF NONPROTEIN NITROGENOUS SUBSTANCES

The methods of study of the ability of the kidneys to eliminate the nonprotein nitrogenous end-products of protein metabolism may be classified as follows:
(1) Urinary studies.
(2) Simultaneous urinary and blood studies.
(3) Blood studies.

Urinary Studies

Studies of urinary nonprotein nitrogenous elements from the standpoint of the investigation of renal function have been largely directed toward urea and creatinine. These studies have been essentially of two varieties: those designed to determine the ability of the kidneys to eliminate a known quantity of preformed urea or creatinine (balance studies), and those designed to determine the ability of the kidneys to concentrate urea (urea concentration tests). This type of study is not employed extensively at the present time, having been largely replaced by the more informative clearance studies.

Simultaneous Study of Blood and Urine

Blood Urea Clearance. The most exact information regarding the urea excreting ability of the kidney requires comparisons of the blood urea concentration and the urea excretion in the urine. With fairly large urine volumes the rate of elimination of urea is directly proportional to the blood urea content. However, this direct relationship holds only when the urine volume exceeds a certain limit, 2 ml. per minute (adults), the "augmentation limit." With urine volumes below this figure, the rate of urea elimination falls, and is, on the average, proportional to the square root of the urine volume. The most satisfactory medium of expression of the relationship between the blood urea concentration and the urinary urea excretion is by means of the "blood urea clearance," which is expressed as the number of milliliters of blood completely cleared of urea by renal excretion. The blood, of course, is not completely cleared of urea in its passage through the glomerulus. About 10 per cent of the blood urea is so removed. Therefore, if, under conditions of maximum blood urea clearance, about 750 ml. of blood plasma pass through the kidney per minute, the amount of urea removed would be equivalent to that contained in 75 ml. of plasma. A more exact definition of blood urea clearance would be "the number of milliliters of plasma which contain the amount of urea removed per minute by renal excretion."

In normal adults, when the excretion of urine proceeds at a rate of 2 ml. or more per minute, a certain volume of blood plasma will be freed of urea each minute. This volume of plasma, termed the "maximum clearance" (Cm), normally ranges from 64 to 99 ml., the mean value for an adult of average size (surface area of 1.73 square meters) being 75 ml. of blood per minute. Expressed as a formula,

$$Cm = \frac{UV}{P},$$

where U designates the milligrams of urea in 100 ml. of urine, P the milligrams of urea in 100 ml. of blood plasma and V the urine volume in milliliters per minute.

Similarly, when the urine output is below the augmentation limit (less than 2 ml. per minute), a certain volume of blood plasma will be freed of urea each minute. This volume of plasma is termed the "standard clearance" (Cs) and varies normally from 41 to 65 ml., the mean value for an adult (surface area 1.73 square meters) being 54 ml. per minute. Expressed as a formula,

$$Cs = \frac{U}{P}\sqrt{V}.$$

Because the blood urea clearance, urine volume and augmentation limit vary directly with variations in the body surface area, a correction factor should be introduced into the above formula in the case of individuals distinctly above or below the average adult size. No correction need be made for persons between 62 and 71 inches in height since the error involved does not exceed 5 per cent. The corrected formulas are expressed as follows:

$$Cm = \frac{U}{P} \times V \times \frac{1.73}{A},$$

$$Cs = \frac{U}{P}\sqrt{V \times \frac{1.73}{A}},$$

where A is the body surface in square meters, calculated according to the height and ideal weight, available in standard tables provided for such determinations.

While extensively used in the past, the "standard clearance" is no longer regarded as reliable for clinical purposes. In order to attempt to obtain a "maximum clearance," the patient is made to drink an appropriate amount of water before the test begins and during it.

Procedure. The patient is not subjected to any preparatory routine, except that vigorous exercise is avoided. The test is best carried out during the morning with the patient fasting. 570 ml. of water is given one-half hour before the test is to begin. At the beginning of the test the bladder is emptied completely, the exact time being noted. This urine is discarded. Exactly one hour later the bladder is again completely emptied, the urine being saved for examination. At this time a specimen of blood is withdrawn for plasma urea determination and another pint of water given. Exactly one hour later the bladder is again completely emptied, the urine being saved for examination. If the urine specimens are not obtained exactly one hour apart, the exact time at which they are obtained should be noted. The volume of each specimen is measured carefully and determinations are made of the concentration of urea in the blood and urine specimens. Inasmuch as the urea is, of course, removed only from the plasma in the process of urine formation, plasma and not whole blood should be used for clearance determinations. The patient remains quiet during the period of the test. The chief source of error lies in the possibility of incomplete emptying of the bladder. In cases in which conditions are present which interfere with complete emptying of the bladder, such as prostatic hypertrophy, tumors of the uterus, advanced pregnancy, etc., the bladder should be emptied by catheter at the beginning and end of the two-hour test period. If desired, the test period may be shortened to one hour, the blood sample being obtained at the mid-point.

If the urine volume (corrected) exceeds 2 ml. per minute the maximum clearance is calculated. If the urine volume (corrected) is less than 2 ml. per minute the test is unreliable.

Results are expressed in terms of milliliters of blood cleared of urea per minute or in terms of percentage of average normal clearance.

Example of Calculation of Maximum Clearance

Plasma urea N = 14 mg. per 100 ml. = P
Urine area N = 210 mg. per 100 ml. = U
Urine volume = 180 ml. per hour
= 3 ml. per minute = V

$$C_m = \frac{U}{P} \times V = \frac{210}{14} \times 3 = 45 \text{ ml. of blood cleared of urea per minute.}$$

Percentage of normal function = $\frac{45}{75} \times 100 = 60$ per cent.

Interpretation. Under average normal conditions, with urine volumes of 2 ml. or somewhat more per minute, about 40 per cent of the urea which leaves the blood in the glomerular filtrate is reabsorbed in the tubules. This is a passive process, being dependent, therefore, on the amount of water undergoing reabsorption. Under conditions of marked osmotic diuresis, i.e., marked decrease in water reabsorption, the extent of urea reabsorption diminishes considerably, and values for urea clearance approach those for inulin clearance. With low urine volumes, i.e., increased water reabsorption, there is also more urea reabsorption; e.g., about 60 per cent reabsorption at a urine volume of 1 ml. per minute, and about 70 per cent reabsorption at urine volumes of 0.35 ml. or less per minute. It should be noted that the true clearance formula (UV/P) is applicable only at urine volumes of 2 ml. or more per minute. At these rates of urine flow, conditions are optimal for the use of urea clearance as a measure of renal function. As the urine volumes decrease, below 2 ml. per minute, the situation in which the "Standard Clearance" formula used to be applied, the accuracy of this procedure as a measure of renal functional capacity decreases progressively. This is true particularly at urine volumes below 1 ml. per minute, and little significance can be reliably attached to low clearance values in subjects with urine volumes below 0.5 ml. per minute. This diminishes considerably the clinical usefulness of this procedure because of the frequency of oliguria in conditions in which studies of renal function are indicated.

Under these conditions of severe oliguria, as in shock, acute glomerulonephritis, dehydration, congestive heart failure, acute tubular necrosis, etc., determination of the ratio of the concentration of urea N in urine and serum may provide useful information. This factor is a reflection of the ability of the kidney to excrete urea and may aid in differentiating between renal and prerenal causes of oliguria and azotemia. Urine : serum urea N ratio values above 15 are usually obtained in transient oliguria and azotemia due to dehydration or inadequate renal circulation. Ratios below 10 occur in subjects with acute tubular necrosis and in chronic renal insufficiency.

Urea clearance is increased by the ingestion of caffeine, milk, a mixed meal and by small doses of epinephrine. It is decreased by the administration of vasopressin (ADH) or large doses of epinephrine. In a study of the diurnal variation of the standard blood clearance, it was found to be depressed during the first hour after arising; following this, commencing before breakfast, there was a regular increase, the higher level continuing through the morning. A drop occurred after lunch with a subsequent rise during the late afternoon and evening. The least variation occurred during the period between breakfast and lunch (9 to 12 A.M.). Strenuous exercise causes a decrease in blood urea clearance, particularly in patients with impaired renal function.

The urea clearance may be abnormally low during periods of subsistence on a low protein diet. Low values have been reported during relapse in pernicious anemia and in other severe anemias. The urea clearance in pregnancy appears to remain within normal limits in subjects without severe anemia and with an adequate protein intake. Reports of low values are probably due to low protein intake. No alteration has been observed in glomerular filtration rate, renal blood flow or tubular excretory mass in normal pregnancy. Normal values are the rule in uncomplicated eclampsia and preeclampsia.

Particularly in patients under forty years of age, elevation of blood urea clearance (over 25 per cent above the average normal) may occur during the course of acute infections such as pneumonia and rheumatic fever. This may be due to increase in renal blood flow and glomerular filtration resulting from renal arteriolar dilatation (p. 488). The absence of this phenomenon in older patients is attributed to a decrease of renal resiliency with increasing age. An increase has been observed also after injection of pyrogens, but not in hyperthermia induced by diathermy. Erroneously low values may be obtained when the urine volume is less than 0.35 ml. per minute and erroneously high values during periods of marked diuresis. Unilateral nephrectomy is followed by an average increase of 43 per cent in urea clearance by the remaining kidney (increased blood flow) (p. 488). In children with the nephrotic syndrome, high values (up to 200 per cent of average normal) may be obtained. This is probably due chiefly to an increase in renal blood flow but has also been attributed, in part at least, to increase in glomerular filtration resulting from the decreased plasma colloid osmotic pressure (p. 485).

Inasmuch as all of the urea in the urine leaves the blood plasma in the glomerular filtrate, all causes of decreased glomerular filtration are potential causes of decreased urea clearance. The underlying mechanisms, which have been discussed elsewhere (p. 482), include: (*a*) diminished renal blood flow; (*b*) glomerular damage; (*c*) increased capsular pressure; and, (*d*) theoretically at least, increased plasma colloid osmotic pressure. One or more of these mechanisms is usually operative in virtually all clinically important forms of diffuse kidney disease, e.g., acute and chronic glomerulonephritis, advanced chronic pyelonephritis (bilateral), nephrosclerosis, intercapillary glomerulosclerosis, renal hypogenesis, and advanced bilateral destructive lesions (polycystic disease, tuberculosis, malignancy). They are operative also in congestive heart failure, bilateral urinary tract obstruction due to any cause, and extrarenal conditions in which glomerular filtration and renal blood flow are seriously reduced, as a result of marked hypotension (shock, Addison's disease, etc.) (pp. 483, 487).

Theoretically, decreased urea clearance could conceivably be due to excessive tubular reabsorption. This does occur in acute tubular necroses, in which conditions all components of the glomerular filtrate are returned to the blood stream in excess, unselectively, across nonfunctioning or disrupted portions of the tubule walls (p. 522). However, this mechanism apparently does not operate in acute or chronic glomerulonephritis nor in the common forms of chronic diffuse kidney disease mentioned above. In these conditions, tubular reabsorption of urea is usually decreased.

In routine clinical practice, the urea clearance is generally a reliable quantitative index of the state of renal function at urine volumes of 2 ml. or more per minute. Its usefulness stems from several facts. The test procedure requires no intravenous injection, as do other clearance tests (inulin, PAH); the test period may be as short as one hour, and quantitative urea determinations can be readily and accurately performed in a routine clinical laboratory. Except under conditions of unusual glomerulotubular functional imbalance, or of very small or very large urine volumes, the urea clearance is roughly proportional to the rate of glomerular filtration (p. 503). Since this is decreased in the great majority of conditions in which renal function is impaired (p. 483), the urea clearance is usually diminished correspondingly. Moreover, not only is urea normally the most abundant urinary solid, but also it is excreted by a mechanism similar to that which is involved in the excretion of the bulk of the other important components of the urine; i.e., glomerular filtration and rather extensive tubular reabsorption, although the latter is passive in the case of urea.

In interpreting urea clearance values in disease, it must be remembered that this factor is not a precise measure of individual phases of renal function, i.e., glomerular filtration rate, renal blood flow, and tubular reabsorption or excretion. These can be measured accurately only by determination of inulin (or mannitol) and p-aminohippurate (or Diodrast) clearance, and of tubular maxima (Tm). Nevertheless, although these more refined procedures may reveal discrete abnormalities of renal physiology in the presence of essentially normal urea clearance, the latter affords a quantitative index of the efficiency of renal excretion of an important and representative constituent of the urine. It is therefore of great value in the clinical assessment of over-all renal function. Characteristic findings in disease are discussed elsewhere (pp. 518–526).

The urea clearance is of greatest value in assessing renal function in patients with normal or slight increase in plasma urea levels. The investigation has a number of disadvantages, among them the difficulty of obtaining an accurate urine collection over so short a period, necessitated by the fact that the level of plasma urea varies over relatively short intervals. As has been pointed out, there can also be significant variations due to diet. Blood urea clearance is now being largely superseded by the creatinine clearance because of these points.

Endogenous Creatinine Clearance. This is frequently employed clinically as a measure of glomerular filtration. As a routine procedure, it has an advantage over the inulin clearance in that it requires no intravenous injection and the analytical techniques are relatively simple. It has an advantage over the urea clearance in that it is less influenced by decrease in the rate of urine flow; unlike the blood urea N concentration, it is independent of the protein intake. Under normal conditions, the endogenous creatinine clearance ranges from 75 to 125 per cent of the inulin clearance, averaging about 100 ml.

This procedure has certain disadvantages. Unless specific enzymatic methods are employed, which is not always feasible, determination of creatinine, dependent upon a nonspecific color reaction, is not precise and is more likely to be in error than is the urea determination. Moreover, although creatinine enters the urine normally largely in the glomerular filtrate, a portion is actively excreted by the tubular epithelium. This latter fraction may increase as the blood creatinine concentration rises as a result of impaired glomerular filtration. Under these circumstances, the degree of impairment of glomerular function would not be reflected as accurately by the creatinine clearance as by the inulin clearance. If these reservations are kept in mind, this is a very useful procedure.

The preparation for the test is similar to that for urea clearance. The patient drinks 600 ml. of water. The bladder is emptied and a five-hour urine specimen collected. At the same time, a sample of blood is taken for plasma creatinine determination. The value of the test is enhanced not only by correction for body surface area, but also for lean body mass. The normal range is then the same for males and females. Without lean body mass correction, the range for men is 140 ± 27.2 ml. per minute and that for women, 112 ± 20.3. Some authorities do not consider that the correction for body surface area is necessary.

Blood Nitrogen Studies

The several conditions which may be associated with elevation of the level of nonprotein nitrogen in the blood are reviewed elsewhere (p. 509 ff.). The following discussion is concerned only with blood nitrogen retention in its relation to renal functional impairment.

In the presence of renal disease, blood nitrogen retention depends essentially upon three factors:

(a) The degree of renal functional impairment.
(b) Prerenal deviation of water.
(c) Excessive protein catabolism.

Since the diagnostic and prognostic significance of nitrogen retention depends upon the part played by each of these factors in any given case, they will be considered individually in greater detail.

Renal Functional Impairment. The concentration of nonprotein nitrogenous substances in the urine greatly exceeds that in the blood. Under ordinary conditions the concentration of urea in the urine is 60 times as high as in the blood, uric acid 20 and creatinine 50 times as high. The ability to concentrate these and other solid substances constitutes one of the most important functions of the kidney. The normal kidney can satisfactorily eliminate the required quantity of solids even though only relatively small quantities of water are available for their solution. Inability of the kidney to eliminate nitrogenous substances adequately may be regarded as dependent essentially upon diminution in its concentrating ability. The only method by which the kidney can compensate for this defect is by the elimination of increased quantities of water, the urine during this period of compensation being of large volume and relatively low specific gravity.

During this compensatory stage, in spite of the presence of renal functional impairment, the blood nitrogen values may remain within essentially normal limits. They rise in the absence of extrarenal factors only when the renal lesion has progressed to the point at which the kidney fails to eliminate sufficient water to compensate for the diminution in its power of concentration. Under such circumstances the urine is of consistently low specific gravity regardless of the volume of water eliminated. The variable significance of nitrogen retention dependent entirely upon renal functional insufficiency will be considered in the discussion of various types of renal disease.

Prerenal Deviation of Water. Since, particularly in the presence of renal disease, the ability of the kidneys to eliminate nonprotein nitrogenous elements depends largely upon the amount of water available for excretion by the kidneys, blood nitrogen retention may be induced by factors which cause prerenal deviation of water. The most important of these may be classified under two headings: (1) those producing edema, and (2) those resulting in dehydration.

(1) Edema. Water retention in the tissues, due to any cause, imposes a burden upon the nitrogen-excretory function of the kidneys, particularly if their concentrating ability is already impaired. This factor plays an important part in the production of nitrogen retention in acute nephritis. The occurrence of myocardial insufficiency with edema, so frequently a complication of chronic hypertensive nephritis, may precipitate renal insufficiency in an individual with a previously well compensated renal functional defect. The same phenomenon may be observed in association with edema due to a nephrotic lesion or to malnutrition with hypoproteinemia complicating chronic nephritis.

(2) Dehydration. Excessive loss of water or water privation exerts a similar influence. Protracted vomiting, so common a feature of advanced nephritis, and, less commonly, profuse diarrhea, may result in nitrogen retention in patients with previously compensated renal functional impairment. Excessive perspiration or vaporization of water from the body surface in complicating febrile disorders may act in a similar manner.

Excessive Protein Catabolism. With a constant state of renal functional activity, an excessive rate of protein catabolism will cause a tendency toward the accumulation of nitrogenous elements in the blood. This factor may play a significant role in the development of nitrogen retention in nephritis. Among the more common conditions associated with increased protein destruction and a consequent increase in the quantity of nonprotein nitrogenous substances that must be eliminated by the kidneys are the following:

(1) High Protein Intake. Under conditions of normal renal function, normal water balance and adequate fluid intake, the ingestion of large quantities of protein is followed by prompt elimination of the excess nitrogen with relatively little consequent significant elevation of the level of nitrogen in the blood. However, if renal function is impaired, and especially if, in addition, the urine volume is diminished, the nonprotein nitrogenous elements in the blood may increase quite significantly during periods of high protein intake and return to normal following the administration of a diet low in protein.

(2) Infections. Infectious processes, occurring in patients with renal functional impairment, may result in an increase in blood urea nitrogen. Infections such as tuberculosis, even in the absence of fever, cause toxic destruction of protein. In febrile disorders, increased protein catabolism results from both toxic destruction and increased energy requirement incident to the elevation of body temperature. The tendency toward urea nitrogen retention in such conditions is aggravated by the frequent coexistence of other factors exerting a similar influence, such as deficient fluid intake, excessive evaporation from the skin and respiratory tract, diarrhea, vomiting, edema, and so on. Excessive protein catabolism may perhaps contribute to the nitrogen retention frequently observed in acute nephritis and bichloride tubular necrosis.

(3) Dehydration. Dehydration due to water privation, vomiting, diarrhea, or the like, so alters tissue metabolism as to result in an increase in the rate of protein catabolism. The increase in the level of blood nitrogen which occurs under such circumstances is due to the combined effects of this factor, diminution in the quantity of water available for elimination by the kidneys, and inadequate renal blood flow.

(4) Increased Catabolism of Protein. This occurs in uncontrolled diabetes mellitus, thyrotoxicosis, adrenal cortical hyperfunction, infectious diseases, malignancy, etc.

Obviously, the interpretation of the significance of blood nitrogen retention in individuals with renal disease depends upon the proper evaluation of the relative importance of renal and extrarenal factors in its production.

Blood Nonprotein Nitrogen in Renal Disease. Nitrogen retention in renal disease depends upon the nature and extent of the renal lesion and upon extrarenal factors, chiefly prerenal deviation of water and excessive protein catabolism. Since the relative importance of each of these factors may vary in different types of renal disease, the significance of blood nitrogen retention must vary accordingly.

Acute Glomerulonephritis. Increased blood nitrogen values are frequently observed in acute nephritis, the degree of retention varying from slight elevations (e.g., NPN 45 mg., urea N 25 mg., creatinine 2 mg. per 100 ml.) to extremely high figures (e.g. NPN 200 mg., urea N 160 mg., creatinine 25 mg. per 100 ml.). Nitrogen retention in this condition is the resultant of several factors, including (*a*) impaired renal function, (*b*) prerenal deviation of water, and (*c*) excessive protein catabolism and/or diminished anabolism.

There is considerable variation in the degree of renal functional impairment in acute glomerulonephritis. In some cases renal function is essentially intact. In others, it is markedly diminished (p. 518). The majority exhibit some decrease in glomerular filtration and, to a lesser extent, and less consistently, in renal blood flow. This, in itself, is not always sufficient to cause the marked elevations of blood nitrogen which are frequently observed and which are dependent upon the operation of extrarenal factors superimposed upon a relatively mild renal functional defect. In some cases of acute, and also chronic, glomerulonephritis, one of the factors contributing to the development of renal functional impairment is obstruction of glomeruli and tubules by accumulations of fibrin and other protein coagula.

Oliguria is a common feature of acute nephritis. Whereas in a proportion of cases this is a manifestation of renal functional impairment (glomerular occlusion), it is frequently dependent upon prerenal deviation of water, due to edema, vomiting, diarrhea, or fever. The edema of acute nephritis is probably not dependent primarily upon the renal lesion per se but rather upon generalized capillary injury with consequent increased permeability, upon a diminution in the concentration of plasma proteins, or upon myocardial weakness (p. 508). Edema may be latent or frank; as much as 4000 ml. of water may be retained without being detectable by ordinary methods of physical examination.

This prerenal deviation of water is an important factor in determining the presence or degree of blood nitrogen retention in acute nephritis. Occasionally, edema due to myocardial insufficiency may exert an influence in this connection as may, more frequently, limitation of fluid intake and active purgation, which have been common practices in the presence of edema.

It is evident, therefore, that several factors must be considered in determining the prognostic significance of high blood nitrogen values in acute nephritis. In most cases the condition is a temporary one, the nitrogen values returning to normal upon subsidence of the acute infection. Occasionally, with improvement in the clinical manifestations, a transient rise in blood nitrogen may occur during periods of diuresis and diminution in edema. Nitrogen retention associated with urine of a low specific gravity, indicating renal functional impairment, is much more significant than that associated with urine of a high specific gravity. Because of the essential character of the condition, with its inherent capabilities of retrogression and eventual recovery, anatomical and functional, high nitrogen values have less prognostic significance in acute than in chronic nephritis. This is particularly true if they are dependent largely upon extrarenal factors, as is commonly the case. Despite these complicating circumstances, analyses of large groups of cases reveal that the prognosis for complete recovery from acute glomerulonephritis is generally not so good in subjects with severe azotemia as in those with blood urea N values below 30 mg./100 ml. This correlation is by no means consistent, however, exceptions occurring in both directions.

The proper interpretation of the retention of nonprotein nitrogen in the blood in acute, as indeed in chronic, nephritis depends upon the relative part played by the renal lesion per se, whether it subsides or progresses into chronic nephritis, and upon the relative importance of extrarenal factors. Single determinations are less valuable than serial determinations because of marked temporary variations due to extrarenal influences. In the majority of cases nitrogen retention may be regarded as a result of the combined effects of mild or moderate renal functional impairment, prerenal deviation of water and excessive protein catabolism, the last two factors being often more important than the first. Even advanced grades of nitrogen retention in acute nephritis may possess relatively little significance from the standpoint of the estimation of the extent of renal damage. They must be interpreted in the light of information afforded by other tests of renal functional efficiency, particularly the concentration and clearance tests, which are relatively uninfluenced by extrarenal factors.

Chronic Glomerulonephritis. The factors involved in the production of azotemia in chronic nephritis are essentially the same as those discussed in connection with acute nephritis with the significant exception that the pathologic process in the kidneys is chronic and progressive and that, as a result, renal functional impairment plays a more important, and extrarenal factors a less important, part than in acute nephritis. For a variable period of time, depending upon the rapidity of progression of the renal lesion, the blood nonprotein nitrogen may remain within normal limits. This is almost invariably the case during the so-called latent period, although occasionally the progress of the disease may be halted temporarily at a more advanced stage of kidney damage, with mild azotemia. The blood nonprotein nitrogenous components are usually within normal limits also in the so-called "nephrotic," i.e., hypoproteinemic edematous (chronic active) type of chronic glomerulonephritis. In this connection, it must be remembered that although blood urea N (normal 8–23 mg./100 ml.) and NPN (normal 16–40 mg./100 ml.) concentrations within the rather wide normal range must be regarded as normal, values for urea N above 15 mg./100 ml. and for total NPN above 35 mg./100 ml. are often in reality abnormal. During these relatively nonazotemic periods, abnormal retention of urea is minimized owing to the fact that the diminished ability of the kidneys to excrete this substance (and others) in adequate concentration is compensated, more or less successfully, by excretion of larger volumes of urine (polyuria), except in edematous phases.

As the kidney lesion progresses and renal functional efficiency becomes more distinctly impaired, a point is reached where increased water elimination is no longer able to compensate for the increasing inability to excrete solids. Increasing retention of nonprotein nitrogen ensues and, at first mild, progresses steadily but with variable rapidity, until in the terminal stages of uremia extremely high levels may be reached. In patients suffering with this form of the disease, nonprotein nitrogen values of more than 100 mg. per 100 ml., urea nitrogen of 80 mg. and creatinine of 5 mg. per 100 ml. foreshadow a rather speedy fatal termination. Blood urea N values above 900 mg./100 ml. have been reported. Urea usually constitutes the largest proportion of this increase in nonprotein nitrogen (60 to 90 per cent), the blood creatinine commonly remaining within normal limits for a considerable period of time during which the urea nitrogen concentration has exhibited a progressive increase.

As in the case of acute nephritis, marked elevations of blood NPN frequently occur during the course of chronic nephritis due to the superimposition of extrarenal factors. Under such circumstances, being not necessarily indicative of the actual extent of the renal lesion, nitrogen retention loses its serious significance. As a rule, the course of the disease is punctuated by periods of excessively high blood nitrogen values which subside after correction of extrarenal influences. Among the most common of these are intercurrent infections, excessive vomiting and diarrhea, excessive protein ingestion, water privation, acute exacerbations of the renal process, and congestive heart failure. These factors have been considered in detail above; they operate largely in one or both of two ways, i.e., increased protein catabolism and prerenal deviation of water. Thus, renal insufficiency is precipitated by the imposition of an excessive burden upon a diminished renal functional reserve. Upon removal or cessation of operation of these factors the blood nitrogen concentration falls to the level determined by the degree of renal functional impairment and the efficiency of the compensating mechanism (polyuria).

It must be apparent that too much reliance must not be placed upon the concentration of the nonprotein nitrogenous constituents of the blood as an indication of the extent of renal damage in chronic nephritis. In the absence of complicating conditions the pathologic process in the kidneys may be rather far advanced before the blood urea N (or NPN) rises above the upper limit of normal; in the late or terminal stages it may have some prognostic significance, particularly if creatinine retention is marked. However, at any time during the course of the disease, advanced grades of nitrogen retention may occur because of the operation of one or more of the several extrarenal factors mentioned above. In the interpretation of high nonprotein nitrogen values, these complicating influences must be carefully considered and their part in the production of the existing condition properly evaluated.

Nephrosclerosis. The concentration of nonprotein nitrogen in the blood is often within normal limits during the entire course of essential hypertension and nephrosclerosis. Nevertheless, even in early stages, the presence of abnormality of renal function (glomerular dynamics) is evidenced by decrease in glomerular filtration and reduction in renal blood flow (p. 523). Statements made above concerning the difficulty of interpreting blood urea N and NPN values in the upper range of normal in glomerulonephritis apply equally here. In the late stages of this condition, and in so-called "malignant hypertension," renal failure may occur and progress rapidly to a fatal termination, if untreated. Because of the frequent occurrence of vascular accidents or cardiac failure incident to the prolonged state of hypertension, renal insufficiency is observed less frequently than in chronic glomerulonephritis. However, if it does occur, the metabolic features and blood chemical findings are identical with those which characterize the latter condition.

Destructive Renal Lesions. Under this heading may be considered such conditions as renal tuberculosis, malignancy, pyelonephritis, hydronephrosis, polycystic disease, etc. The estimation of renal functional efficiency in these disorders is of particular importance from the standpoint of the advisability of surgical intervention and the determination of the operative procedure to be employed.

Since adequate elimination may be carried on by two-thirds of one kidney, the presence of nitrogen retention in destructive lesions of the kidney indicates extensive renal damage. It may be interpreted as signifying either that both kidneys are involved in the pathologic process or that the uninvolved kidney is overburdened by the necessity of eliminating excessive quantities of nitrogen resulting from increased toxic destruction of protein. In this group of disorders the influence of extrarenal factors must be carefully evaluated. High protein intake, fever, water privation, vomiting, edema, diarrhea, etc., must be taken into consideration.

The consensus is that little information of significance in prognosis can be gained from single nonprotein nitrogen determinations. In destructive renal lesions it is the persistence rather than the degree alone of nitrogen retention to which attention should be directed and by which the extent of renal damage should be estimated. The persistence of elevated blood nonprotein nitrogen values in such cases following the institution of proper therapeutic procedures is of definitely grave prognostic import. Because of the fact that individuals with marked nitrogen retention are poor operative risks, repeated preoperative estimations of the nonprotein nitrogen concentration of the blood are of value in determining the time at which radical surgical measures may be attempted with a minimum of risk to the patient. In surgical disorders of the kidneys, the functional efficiency of each kidney should be studied by estimating its capacity for dye elimination (p. 515), either by ureteral catheterization or by direct vision through the cystoscope.

Tubular Necroses. Acute renal failure, with azotemia, occurs in the presence of severe degenerative changes (necrosis) in the renal tubular epithelium. From the standpoint of etiology, there are two main classes of conditions in this category: (1) those due to nephrotoxic poisons, e.g., mercury, uranium, carbon tetrachloride, sulfonamides, chlorate, etc.; (2) those due fundamentally to severe renal ischemia (p. 522), e.g., shock, acute tubular necrosis (severe trauma, burns, mismatched transfusions, postoperative, etc.).

In these conditions, severe oliguria (or anuria) and azotemia develop because of removal (necrosis, desquamation) of portions of the selective barrier to the free passage of components of the glomerular filtrate back into the blood stream. As a result of disruption of continuity of this functional (and morphological) barrier, excessive amounts of all components of the filtrate diffuse back, unselectively (p. 493). In all conditions in which arterial hypotension is a dominant feature (shock syndrome, acute tubular necrosis), there is also a diminished volume of glomerular filtrate (and decreased renal blood flow), which contributes to the inadequacy of renal function. In the past, the terms "renal crush syndrome," "hemoglobinuric nephrosis," "gestation nephrosis," "hepatorenal syndrome," etc., have been applied to acute tubular necrosis and renal failure occurring under circumstances implied by each of these designations. In some cases, i.e., crush syndrome and hemoglobinuric nephrosis (e.g., mismatched transfusion, black-water fever), it has been suggested that factors other than ischemic necrosis may be involved, viz., obstruction of tubules by precipitated myoglobin or hemoglobin (especially if urine is highly acidic), and toxic effects on the kidney of circulating methemoglobin and myoglobin. At the present time, it is believed that severe hypotension, with intrarenal vasoconstriction and renal ischemia, is the most important etiologic factor in this important group of conditions.

Acute renal failure occurring as a complication of pregnancy is one of the most important examples of this condition from the standpoints of both frequency and gravity. It has many causes: severe intrauterine hemorrhage; afibrinogenemia (p. 184); criminal abortion (shock, nephrotoxic chemicals, infection); unusually difficult labor, which may possibly produce renal vasoconstriction.

In some instances, azotemia occurs in the absence of significant oliguria. The combination of moderate decrease in glomerular filtration, renal blood flow, and tubular concentrating functions, and increased protein catabolism (severe trauma, operations, burns, etc.) may result in inadequate excretion of urea (and other urine solids) despite urine volumes of 800 to 1200 ml. (low specific gravity). Azotemia may also persist during periods of recovery from acute tubular necrosis, even in the presence of polyuria (with hyposthenuria), due to functional inadequacy of the newly regenerated renal epithelium (p. 522).

Urinary Obstruction. In patients with prostatic or urethral obstruction or with bilateral ureteral calculus, and in some cases of unilateral ureteral calculus (with "reflex" anuria), the back pressure may be so great that effective filtration through the glomerular membrane cannot occur. Under such circumstances, remarkable grades of nitrogen retention may be observed. It is believed by some that in these conditions the blood creatinine frequently remains normal in the presence of extremely high urea nitrogen values, in contradistinction to the findings with similar grades of urea nitrogen retention in nephritis. In the experience of the authors, this is not generally the case.

In the absence of permanent renal disease, relief of the obstruction and institution of adequate drainage are usually followed by restoration of the normal blood nitrogen level. This may occur in the presence of extremely high nitrogen values. In a patient with prostatic obstruction, with a total blood nonprotein nitrogen concentration of 420 mg. per 100 ml. and creatinine 32 mg. per 100 ml., following prostatectomy in two stages the nonprotein nitrogen fell to 54.8 mg. and the creatinine to 2.6 mg. per 100 ml. Nitrogen retention in urinary obstruction amenable to correction does not possess the significance attributed to it in primary kidney disease.

Urea in Other Body Fluids

Because of its extreme diffusibility, urea exists in the lymph, spinal fluid, bile and pancreatic juice in practically the same concentration as in the blood. Considerable interest has been attached to the investigation of the nonprotein nitrogen and urea contents of the sweat because of the possibility of stimulating elimination through this channel in patients with renal insufficiency. It has been found that the nonprotein nitrogen content of the sweat of normal individuals averages about 30 per cent greater than that of the blood, about 65 per cent of the total NPN consisting of urea nitrogen. Whereas under ordinary circumstances the amount of nitrogen lost through the skin is quite small, it may be greatly increased if perspiration is profuse. However, it is doubtful that this channel of elimination can be used to practical advantage in patients with nitrogen retention since the coincident elimination of relatively large amounts of water by the skin may tend to increase rather than to decrease the level of blood nitrogen. The significance of the concentration of the various nonprotein nitrogenous substances in the cerebrospinal fluid has been considered elsewhere (p. 840).

The urea content of saliva is approximately 80 per cent of that of the blood. The combined urea and ammonia nitrogen content of the saliva more closely approximates the concentration of urea nitrogen in the blood because of the fact that urea in the saliva is rather readily broken down into ammonium carbonate through bacterial action. Because of the mercury binding capacity of urea, determination of the mercury combining power of saliva has been suggested as an index of the degree of renal functional insufficiency in cases in which determination of the blood urea nitrogen concentration is not practicable. It has been found that 100 ml. of saliva normally contain enough urea to combine with 30 to 50 ml. of a 5 per cent solution of mercuric chloride. This constitutes the mercury combining index or the salivary index. It is stated that the probable blood urea concentration may be roughly calculated from the salivary index as follows:

$$1.43 \times \text{salivary index} - 34 = \text{probable blood urea in mg. per 100 ml. of blood.}$$

Presumably, when the salivary index is below 50 there is no retention of urea nitrogen in the blood in 90 per cent of cases; in 10 per cent there may possibly be a mild degree of retention but the blood urea concentration is never above 60 mg. per 100 ml. under such circumstances. This method is by no means accurate and should be used only as a preliminary diagnostic procedure and never as a substitute for the determination of blood urea nitrogen.

ELIMINATION OF FOREIGN SUBSTANCES

The kidney normally possesses the ability to eliminate certain foreign substances from the blood stream. This fact has served as the basis for many tests of renal function, various substances being used, including potassium iodide, methylene blue, fuchsin, lactose, indigo-carmine, and, most widely employed, phenolsulfonphthalein.

The Phenolsulfonphthalein Test.
Rowntree and Geraghty, in 1912, introduced the use of phenolsulfonphthalein as a test of renal functional efficiency. This dye is promptly and almost completely eliminated in the urine. The PSP test is one of the most widely employed measures for the estimation of renal functional capacity. A variety of drugs, including aspirin and penicillin, compete with phenolsulfonphthalein for excretion by the renal tubules. It is therefore advisable to stop all medicinal treatment for 24 hours before the test is carried out.

The bladder having been emptied, the patient drinks 600 ml. of water. One milliliter of phenolsulfonphthalein solution, containing 6 mg. of the dye, is injected intravenously thirty minutes later with the patient at rest. Urine is collected fifteen, thirty, sixty and one hundred and twenty minutes after the injection. The fifteen-minute specimen alone is sufficient for routine practical purposes. Each specimen of urine is diluted to a suitable volume, sodium hydroxide (10 per cent solution) is added until a maximum red color is developed, and the amount of dye (PSP) determined by comparison with a standard solution containing 6 mg. of PSP in 1000 ml. of water (alkalinized). It is, of course, essential that there be no bladder retention of urine and that the bladder be completely emptied at the end of each period.

Normal excretion values are as follows:

Specimen	%	Average %
15-minute	28–51	35
30-minute	13–24	17
60-minute	9–17	12
120-minute	3–10	6
Total 2-hour	63–84	70

Although flattening of the curve of elimination is of some significance, elimination of less than 25 per cent of the dye during the first fifteen minutes (30 per cent in twenty minutes) is the most significant evidence of renal functional impairment yielded by this test. If the bladder cannot be emptied at fifteen minutes, the thirty-minute sample should normally contain not less than 40 per cent of the administered dye.

In a less satisfactory procedure, the dye (6 mg.) is injected intravenously or intramuscularly (lumbar region) and the bladder is emptied one hour and two hours later, allowing ten additional minutes in the case of intramuscular injection. Normal subjects excrete 40 to 60 per cent of the dye in the first hour and 20 to 25 per cent in the second hour. However, especially with intramuscular injection, values above 50 per cent in the two-hour period must be regarded as within normal limits. Mild functional impairment is not revealed by this procedure as readily as by the fractional test outlined above. Inability to excrete a normal amount of the dye during the first fifteen minutes, due to inability to achieve a high concentration in the urine, may be obscured by prolonging the period of collection of urine to one hour. Abnormal results have been obtained with the fractional procedure in as many as 30 per cent of subjects with mild renal disease in whom normal results were obtained by this method (one- and two-hour collections).

Certain facts must be kept in mind in interpreting results obtained by the phenolsulfonphthalein test. Low values may be obtained in the presence of factors which tend to cause prerenal deviation of water. Whereas in normal individuals the quantity of dye eliminated is independent of the urine volume, in the presence of severe impairment of renal function the output of phenolsulfonphthalein varies more or less directly with the volume of the urine and can be increased by liberal administration of water. As is the case with clearance tests, PSP excretion is lower (10 to 15 per cent) in the erect than in the recumbent position. This is due presumably to the intrarenal (and general intra-abdominal) vasoconstriction which occurs on assuming the erect position. This postural difference in PSP excretion is exaggerated in subjects with glomerulonephritis and nephrosclerosis.

The average normal value for PSP clearance is about 400 ml., which indicates that the largest proportion of the dye is removed from the blood plasma by tubular excretion and a comparatively minor fraction by glomerular filtration. Consequently, excretion of phenolsulfonphthalein is diminished more directly by specific impairment of tubular excretory capacity (tubule cell damage) and reduction in renal blood flow than by primary glomerular damage or reduction in glomerular filtration out of proportion to decrease in effective renal plasma flow. The fundamental differences in the mechanisms of excretion of urea and PSP account for discrepancies observed occasionally between urea clearance and PSP excretion values in certain cases of renal disease or other conditions (e.g., congestive heart failure) affecting renal function. Stated in general terms, urea clearance is primarily a reflection of glomerular filtration, and PSP excretion a more direct reflection of tubular excretory function and effective renal plasma flow. However, because of the physiological interrelationship between these factors and because of the usual involvement of all of them, although not to the same extent, in the majority of clinical disorders affecting kidney function, there is generally good qualitative if not quantitative agreement between values for PSP excretion and urea clearance.

Under normal conditions a certain proportion of the injected phenolsulfonphthalein is removed from the blood stream by the liver. In patients with impaired hepatic function this fraction is diminished, the urinary excretion of the dye being correspondingly increased. Consequently, in cases in which hepatic disease is present together with renal disease, the urinary elimination of phenolsulfonphthalein may be greater than if the liver function were normal. Under such circumstances, borderline normal results may be misleading.

This test may be employed to determine the functional capacity of each kidney. This is one of its most useful applications. Ureteral catheters are introduced and the test performed as outlined above. In normal individuals the dye appears from both sides usually in three to five minutes following intravenous injection. An approximately equal amount is eliminated by each kidney. In the presence of unilateral renal disease the appearance time is delayed on the diseased side and the quantity eliminated is diminished in proportion to the degree of renal functional impairment. It is important to keep in mind the fact that the opposite kidney, if functioning normally, should excrete not only the 50 per cent of the total normally excreted by both kidneys, but also that portion not eliminated by the diseased kidney.

LOCALIZATION OF RENAL FUNCTIONAL DEFECT

As indicated in the discussion of clearance tests (p. 481), it is possible, by currently available methods, to measure certain discrete phases of renal function, e.g., glomerular filtration rate, effective renal blood flow, and maximum tubular capacity for excretion and reabsorption. This can be done with satisfactory precision in normal subjects and in the majority of disease states. It must be remembered, however, that interpretations of inulin clearance values in terms of glomerular filtration rate are invalid under conditions of extensive back-diffusion of components of the filtrate across damaged tubule walls (p. 485). Interpretations of PAH or Diodrast clearance values in terms of effective renal plasma flow are invalid under conditions of subnormal extraction of these substances from the blood plasma (low extraction ratios, p. 488). The latter source of error can be eliminated by determining the extraction ratio (renal vein catheterization). These procedures and concepts have contributed enormously to present understanding of normal and pathological physiology of the kidney, and no other available test procedures can approach them in accuracy of measurement of the functions in question. In no other way can the factors involved in the production of renal functional impairment in a given case be dissected and their relative contribution to the deficiency evaluated. However, because of technical difficulties, the use of these methods is almost necessarily restricted to specialized clinics and laboratories, and they cannot readily be employed as routine diagnostic procedures in general practice nor in the average laboratory. Fortunately, the type of information usually required by the clinician may be obtained satisfactorily by methods that are readily applicable in routine practice. The relations of these clinically available procedures to various aspects of renal function are indicated in Table 16-3. The studies used most extensively in the investigation of renal function are the urea or creatinine clearance, phenolsulfonphthalein excretion, and concentration tests, and the blood urea N (or NPN) concentration. Although the findings with these procedures may vary considably, certain patterns of abnormality are exhibited in common types of kidney disease that reflect the nature of the functional aberration and, to a certain extent, of the morphological lesion.

Acute Glomerulonephritis (Table 16-2, p. 486). This is characterized primarily by inflammatory changes in the glomeruli, which are engorged, and many of which are blocked, with consequent reduction in filtration surface urea and, therefore, in filtration rate. Reduction in the latter is probably contributed to also by the increased interstitial pressure in the kidneys, which decreases the blood flow and increases the capsular pressure (p. 483). The relatively greater decrease in filtration rate than in effective renal plasma flow results in decrease in the filtration fraction (p. 489). The impaired glomerular function is reflected in abnormally low values for inulin clearance and urea clearance in all but very mild cases. This is probably due in part also to the fact that slowing of filtration prolongs the period of contact of the smaller volume of filtrate with the tubular reabsorptive surface, permitting increased reabsorption of components of

TABLE 16-3 FUNCTIONAL SIGNIFICANCE OF PROCEDURES COMMONLY EMPLOYED IN KIDNEY DISEASE

FUNCTIONS	SPECIFIC STUDIES	CLINICAL TESTS
Glomerular filtration	Inulin clearance Mannitol clearance	Urea clearance Creatinine clearance
Renal plasma flow	PAH clearance Diodrast clearance	PSP excretion
Proximal tubule activity	Excretion-PAH Tm Absorption-glucose Tm Concentration Dilution Electrolyte balance	PSP excretion Concentration test Dilution test Serum electrolytes
Distal tubule activity	Acid-base status Ammonium ion formation Urine acidification	Serum CO_2 content Serum pH Urine ammonium ion Urine titratable acidity Urine pH

the filtrate (water, electrolytes, urea, inulin, etc.). In the erect position, values for inulin, urea, and PAH clearances are considerably lower than in recumbency. This phenomenon, indicative of decreased filtration and renal blood flow, is due to intrarenal vasoconstriction, the renal vascular adjustment to orthostasis. The associated decreases in urine volume and in electrolyte and urea excretion predispose to the increase in edema and nitrogen retention in patients with acute nephritis who are not confined to bed.

In the great majority of cases of acute nephritis the blood urea clearance diminishes during the first two months, reaching values of 50 per cent or less of normal. Occasionally, however, essentially normal values may be maintained throughout the acute stage of the disease. Prognosis as to eventual recovery appears to depend not so much upon the absolute values observed during the early stage as upon the duration of the period of impaired renal function. It has been observed that recovery occurs in those cases in which a consistent rise in blood urea clearance values, progressing to within normal limits, begins within four months after the acute onset of the condition. In those cases in which no such tendency is noted within six months, the condition almost invariably progresses to the chronic or terminal stages of nephritis with eventual renal insufficiency. Low values during the first four months of acute glomerulonephritis are not at all inconsistent with complete recovery, both functional and anatomical. During this period, therefore, the urea clearance is of little value in determining the eventual outcome of the condition, since progression into chronic glomerulonephritis may occur in cases with normal clearance values during the acute stage of the condition.

In the initial stages, except in fulminating cases, there may be little indication of tubular malfunction. The urine specific gravity is often high and PSP excretion may be essentially normal, even in the presence of increase in blood urea N. This is a reflection of dissociated impairment of glomerular function. The urine volume is characteristically low, due primarily to decreased filtration and excessive reabsorption, as indicated above, and in part also, in some cases, to prerenal deviation of water (e.g., vomiting). With increasing duration or severity of the disease, degenerative changes invariably develop in the tubules (inadequate blood supply, p. 456). This is reflected in decreases in PAH Tm and PSP excretion, and in the appearance of hyposthenuria (p. 498). At this stage, low values are usually obtained for PAH or Diodrast clearance, suggesting decrease in renal blood flow. However, these are largely attributable to low extraction of these substances, due presumably to the decreased tubular function (p. 488). If these clearance values are corrected for the diminished extraction ratio (p. 488), the findings suggest an essentially normal or even increased blood flow to the functioning tubular tissue. Viewed in conjunction with the data on glomerular filtration and PAH Tm, these observations may be interpreted as signifying glomerular and tubular damage, the former predominating, with active hyperemia of the functioning nephrons.

The occurrence of blood nonprotein nitrogen retention (NPN, urea N) in acute glomerulonephritis is discussed elsewhere (p. 509) and requires only brief mention here. It is the general experience that significant elevation of blood urea N, i.e., above the upper limit of normal (20–23 mg./100 ml.), is often not observed in acute (and chronic) nephritis until the urea clearance has fallen below 50 per cent, and in some instances to 20 per cent of the average normal. However, it must be recognized that any significant decrease in urea clearance, other things being equal, must inevitably result in abnormal retention of urea N in the body. Inability to detect this retention is due to inability to interpret as abnormal any values for blood urea N within the normal range. It should be pointed out, too, that at a plasma urea N of 40 mg./100 ml. and a urea clearance of 35 ml. (50 per cent), the kidneys are excreting as much urea as at a plasma level of 20 mg./100 ml. and a urea clearance of 70 ml. (100 per cent). Variation in blood urea N due to extrarenal factors must also be borne in mind (e.g., diet, protein catabolism, vomiting, diarrhea, edema, etc.).

The inherent reversibility of the renal lesions is indicated by the fact that all functional defects may disappear, and usually do so, in some cases very rapidly. They persist in cases which progress to chronic glomerulonephritis, or those which proceed to a fatal termination in the acute stage.

During the period of recovery, values for urea clearance and blood urea N return to normal usually before those for PSP excretion and maximum concentrating capacity (concentration test). Hyposthenuria may persist for weeks or months after complete subsidence of the acute renal lesion, as indicated by absence of symptoms and urine abnormalities and by normal clearance values. This is attributed to temporary functional inadequacy of the newly regenerated tubular epithelial cells.

Chronic Glomerulonephritis (Table 16-2, p. 486). The clinical and laboratory manifestations of chronic glomerulonephritis vary widely in different patients and at different times in individual cases. This is due mainly to the fact that the progress of the disease is characterized by stages exhibiting different anatomical lesions and functional abnormalities (and clinical manifestations), the duration of which varies markedly in different subjects. These are commonly designated as (1) latent, (2) chronic active, or nephrotic, and (3) terminal.

Latent Stage. As the term implies, this is a period during which the renal lesion is not progressing actively (temporary arrest). This usually occurs immediately following an attack of acute glomerulonephritis, and the manifestations therefore depend upon the condition of the kidneys at the time of transition from the acute to the chronic state. The usual functional findings are: (1) decreased inulin and urea clearances (decreased glomerular filtration); (2) decreased PAH clearance and PSP excretion (decreased renal plasma flow and/or decreased tubular excretion); (3) hyposthenuria. Glomerular filtration (inulin and urea clearances) is characteristically diminished relatively more than PSP excretion. These abnormalities may be of minimal degree; their extent and the presence or degree of azotemia vary considerably and, as indicated above, depend upon the severity of glomerular and tubular damage at the moment. The chronic process, although fundamentally progressive in nature, may become arrested temporarily, with a variable period of latency, at any time, except perhaps in the terminal stage.

Nephrotic Stage (Chronic Active). This is characterized by oliguria, marked albuminuria, hypoproteinemia (p. 187), hypercholesterolemia (hyperlipemia) (p. 138), and edema. Despite the uniformity of the clinical picture and these urinary and blood findings, the results of studies of renal function vary widely (Table 16-2, p. 486). As in the case of the latent stage, this is due mainly to differences in the rate of progress of the kidney lesions in different cases and at different times in individual cases.

All measurable functions are usually depressed, glomerular filtration (urea clearance) generally more than renal blood flow (PAH clearance). The PAH Tm is rather consistently decreased, reflecting the tubular damage. As in acute nephritis, the ratio of PAH clearance to PAH Tm, i.e., the blood flow per unit of functioning nephrons, is greater than normal, implying a state of hyperemia of these nephrons. PSP excretion is decreased and concentrating ability is impaired (hyposthenuria). All of these functions deteriorate as the disease progresses, with increasingly abnormal findings with all renal function tests and, usually, increasing azotemia.

Terminal Stage (Table 16-2, p. 486). Transition into the terminal stage may be identified, for purposes of convenience, by a fall in urea clearance to values below 20 per cent of the average normal. Obviously, in this stage, there is severe impairment of all functions of the kidney, with isosthenuria and increasing azotemia. The clinical syndrome of uremia appears, as well as other evidences of renal failure, e.g., dehydration (p. 526) and acidosis (p. 529). The more precise tests of renal function, including clearance procedures and the PSP tests, are almost never required and are of academic interest almost exclusively. The progress of the disease is reflected satisfactorily for practical purposes during this terminal period in such factors as the blood urea N concentration, serum CO_2 content, anemia, etc.

Tubular Necroses (Acute Renal Failure). The pathological physiology of this group of conditions is discussed elsewhere (p. 513). They are characterized by extensive necrosis of tubular epithelial cells, due either to nephrotoxic agents (e.g., mercury), or to ischemia resulting from acute depression of renal blood flow, associated usually with the shock syndrome (acute tubular necrosis).

In severe cases, in the initial stage of anuria or marked oliguria, virtually all of the modified glomerular filtrate is returned to the blood plasma by back-diffusion, and, obviously, all clearance studies yield extremely low or zero values. What little urine may be excreted has a specific gravity of about 1.010, and the blood urea N increases progressively.

In cases which ultimately recover, the initial stage (first few days) is followed usually by a steady increase in urine output, which, however, may not return to normal for one to two weeks (period of oliguria). The urine specific gravity is subnormal (hyposthenuria) and low values are obtained for urea, inulin, and PAH clearances, and for PSP excretion. Azotemia persists but may diminish somewhat. The persistence of these abnormalities is due apparently to continued back-diffusion of excessive amounts of the modified glomerular filtrate and persisting diminution in renal blood flow despite restoration of normal blood pressure and volume (interstitial edema?). It is important to appreciate the fact that there may be no initial period of severe oliguria in certain cases of ischemic necrosis. Azotemia may occur with urine volumes as large as 1000 ml. daily (low specific gravity), especially in diabetic coma, protracted severe diarrhea, and alkalosis due to injudicious alkali therapy in patients with peptic ulcer.

In the majority of cases, the period of oliguria is followed by one of diuresis, with rather rapid restoration of normal values for urea, inulin, and PAH clearances, and disappearance of azotemia. Hyposthenuria and decreased PSP excretion usually persist for some time after the urea clearance has returned to normal. This is probably attributable to functional immaturity of the newly regenerated tubular epithelial cells, which accounts also perhaps for the diuresis.

Pyelonephritis. In acute pyelonephritis, the renal lesion involves primarily the interstitial kidney tissue, mainly in the medulla, and there is usually no associated demonstrable abnormality of renal function, except in fulminating cases, in which diminished glomerular filtration and renal blood flow may result from severe interstitial inflammatory edema (pressure). If the process becomes chronic, functional renal tissues (glomeruli and tubules) may be destroyed in the areas of involvement and be replaced by fibrous tissue. In this stage, too, all renal functions may be well preserved because of the patchy distribution of the inflammatory lesions. There is some evidence that during the progress of the condition, tubular function (PAH clearance and Tm) undergoes proportionately greater reduction than does glomerular filtration (inulin and urea clearances).

When destruction of or damage to nephrons progresses to the point where the residual functioning elements are unable to maintain normal kidney function, the usual function tests yield essentially the same findings as in chronic glomerulonephritis, except that nephrotic manifestations (see below) are extremely unusual (i.e., severe hypoproteinemia, hypercholesterolemia, edema). PSP excretion and concentrating ability may be affected earlier and to a greater degree than urea clearance, until terminal stages. Other evidences of tubular functional impairment in chronic pyelonephritis are discussed elsewhere, e.g., renal tubular acidosis (p. 529) and osteomalacia (p. 529).

Benign Nephrosclerosis (Essential Hypertension). The characteristic findings with clearance procedures in benign nephrosclerosis are indicated in Table 16-2 (p. 486). In the great majority of cases, until rather late in the course of the disease, glomerular filtration is well maintained (i.e., inulin and urea clearances fairly normal), despite a reduction in renal blood flow (PAH clearance). This results in an increased filtration fraction, implying disproportionate constriction of efferent glomerular arterioles. The renal ischemia also implied by these findings is usually reversible for a rather long period in the course of this disease, as indicated by the response to therapeutic measures designed to relieve arteriolar spasm, e.g., sympathectomy, hypotensive drugs. During this period, i.e., before the development of extensive, irreversible glomerular damage (arteriolosclerosis), all of the usual renal function tests may yield essentially normal results (i.e., urea clearance, PSP excretion, concentration test) in the absence of complications, e.g., congestive heart failure due to hypertension, or acute left ventricular failure due to the same cause or to coronary occlusion.

As the disease progresses, all function tests tend to become abnormal, although advanced renal failure, with azotemia, is unusual in uncomplicated cases. The patient usually succumbs to some consequence of hypertension, e.g., heart failure, coronary occlusion, cerebral vascular disease, before the kidney lesion progresses to this point. This statement does not apply to the rapidly progressive condition referred to as "malignant" nephrosclerosis.

Malignant Nephrosclerosis (Necrotizing Arteriolitis). In this condition, the functional findings are essentially identical with those obtained in rapidly progressing, severe chronic glomerulonephritis. There is steady and rapid deterioration of urea clearance, PSP excretion, and concentrating ability, and increasing blood NPN retention.

The Nephrotic Syndrome. This designation is applied to a clinical state of varied etiology (p. 217) characterized by massive albuminuria, hypoalbuminemia, and hyperlipemia (hypercholesterolemia) and, almost invariably, edema; other indications of renal disease may or may not be present, e.g., azotemia, hematuria, hypertension. It may occur as a manifestation of a number of diseases: chronic glomerulonephritis (p. 521); diabetic glomerulosclerosis; systemic lupus erythematosus; amyloidosis; syphilis; circulatory abnormalities (renal vein thrombosis, constrictive pericarditis); nephrotoxic chemical agents (mercurials, bismuth, gold, certain carcinostatic and antibiotic drugs); allergic reactions (serum sickness, plant sensitivity, insect stings, etc.). It has been suggested, on the basis of electron microscopic studies, that the fundamental abnormality is a change in the glomerular epithelium (foot process fusion), with the occurrence, only in the adult type, of thickening of the basement membrane. There are indications, however, that the observed changes may be consequences rather than causes of increased glomerular permeability.

The characteristic biochemical features of this condition include: oliguria (high specific gravity); marked albuminuria (p. 217) (with hyaline granular and fatty casts, but not hematuria); hypoalbuminemia, with extensive edema (p. 187); hypercholesterolemia (hyperlipemia) (p. 138); frequently, hypocalcemia (non-diffusible fraction) (p. 303); hypogammaglobulinemia (p. 204); hypometabolism (decreased BMR, p. 450). The low BMR is accompanied by a low concentration of serum protein-bound iodine (decreased thyroxine-binding protein, p. 750); nevertheless, thyroid uptake of I^{131} is normal, and there is no clinical evidence of hypothyroidism. There is also a decrease in serum transferrin (iron-binding protein, p. 322), which contributes to anemia in this condition. The mechanism of production of these abnormalities is not clear, nor are their interrelationships; this is particularly true of the hyperlipemia, which is probably due in part at least to the hypoalbuminemia. Some of the other abnormalities, at least in part, may well be secondary to deficiency of certain serum proteins because of loss in the urine. These include albumin, thyroxine-binding globulin and transferrin among others. Several factors contribute to the development of edema in the nephrotic syndrome, perhaps its outstanding clinical manifestation; an outline of its probable pathogenesis is indicated in Figure 16-5.

Decreased intestinal absorption of amino acids and calcium is reflected in increased excretion of nitrogen and calcium in the feces (negative balances). There may be evidence of impaired renal tubular function, e.g., glucosuria, aminoaciduria, phosphaturia and, in some cases, a complete Fanconi syndrome (p. 537).

There is no abnormality of PSP excretion, concentrating ability or, usually, of acid-base status (normal ammonium ion formation and urine acidification) despite the presence of large amounts of albumin and large numbers of casts (including granular casts) in the urine. High urea clearance values (up to 200 per cent of the average normal) may be obtained, particularly during periods of induced or spontaneous diuresis. The blood nonprotein nitrogenous constituents are within normal limits. Differentiation of this condition from chronic glomerulonephritis (nephrotic stage) is based largely on the absence of evidence of significant glomerular functional impairment and damage (hematuria, hypertension) as well as on differing patterns of serum proteins, and, if necessary, renal biopsy, with electron microscopy where appropriate.

Toxemias of Pregnancy. Albuminuria may occur in the pernicious vomiting of pregnancy and preeclamptic and eclamptic toxemias. It is usually most marked in eclampsia but may, in rare instances, be absent in that condition. Some degree of edema, latent or frank, is almost invariably present in true toxemia of pregnancy. This has been shown to be associated with a decrease in the plasma protein concentration. This

LOCALIZATION OF RENAL FUNCTIONAL DEFECT

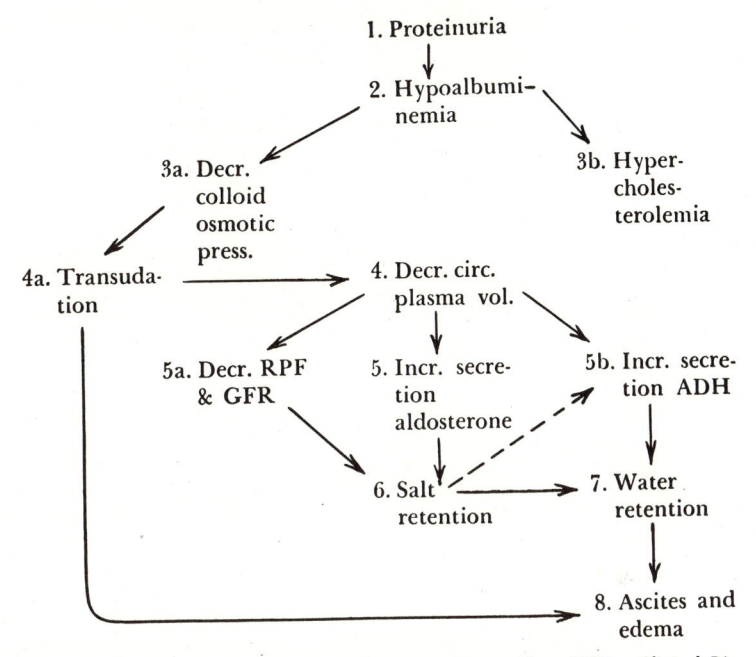

Figure 16-5. Pathogenesis of ascites and edema in nephrotic syndrome. (From White: *Clinical Disturbances of Renal Function.* Philadelphia, W. B. Saunders Company, 1961.)

decrease occurs practically entirely in the albumin fraction and is more marked in eclampsia than in the nonconvulsive forms of pregnancy toxemias. As in the case of normal pregnancy, this decrease in plasma protein concentration is aggravated by deficient protein intake. Although the degree of hypoproteinemia may not in itself be sufficient to produce edema, it favors the development of this phenomenon in the presence of other contributing factors.

In mild and moderately severe cases of pernicious vomiting of pregnancy, the nonprotein nitrogenous constituents of the blood remain within normal limits. In severe cases, however, they may be increased. This increase may be associated with a state of dehydration and electrolyte depletion which is at times dependent upon excessive and prolonged vomiting (p. 375). The blood amino acids may be simultaneously increased. In preeclampsia and eclampsia the total NPN and urea N are usually within normal limits. Most observers report an increase in blood uric acid, and uric acid clearance (p. 255) may be decreased. This is frequently the only demonstrable abnormality in blood nonprotein nitrogenous constituents. In occasional cases with urinary suppression the nonprotein nitrogenous constituents of the blood may be considerably increased. Similar increases are also observed in (*a*) nephritis complicating pregnancy, which may be confused with eclampsia, (*b*) renal cortical necrosis, which at times occurs during or immediately following pregnancy, and (*c*) occasionally in severe hepatic necrosis with an associated "hepatorenal syndrome" (p. 513).

The hypertensive toxemias of pregnancy are commonly regarded as manifestations of a generalized vascular reaction eventuating in chronic vascular disease in about 25 per cent of cases. The special predilection of pregnant women for the development of these manifestations may be dependent, in part at least, upon certain physiologic changes which occur during this period. Among these are physiologic hydronephrosis, which predisposes to pyelonephritis and prevents proper drainage and elimination of infection, and the physiologic hypoproteinemia, which enhances the tendency to edema. A slight temporary increase in renal blood flow has been observed during pregnancy in women with essential hypertension, with no change in glomerular filtration and tubular excretory mass. The blood urea clearance is usually within normal limits in eclampsia. However, re-examination of such cases about three months post partum reveals a significant decrease in blood urea clearance in 25 to 50 per cent of cases, suggesting a permanent renal or renal vascular damage. A decrease in glomerular filtration rate and an increase in renal blood flow have been observed in toxemia of pregnancy. The renal blood flow returned to normal following delivery in about one-half the cases and fell to ischemic levels in the remainder, which developed persistent hypertension. Renal function tests are of particular value in distinguishing between true eclampsia and glomerulonephritis complicating pregnancy.

Congestive Heart Failure. In congestive heart failure, the urine is characteristically of small volume and contains albumin, casts and red blood cells. The urine specific gravity is usually elevated in proportion to the diminution in volume, in the absence of complicating kidney damage. Urea clearance and PSP excretion are usually decreased; diminution in glomerular filtration is indicated by low inulin clearance, and a relatively smaller reduction in renal blood flow by decreased PAH clearance. As indicated elsewhere (p. 505), these changes probably result from decreased cardiac output and increased intrarenal arteriolar constriction (as in shock).

OTHER BIOCHEMICAL MANIFESTATIONS OF CHRONIC RENAL INSUFFICIENCY (CHRONIC RENAL FAILURE)

As might be anticipated in view of the fundamental role of the kidney in the preservation of the normal composition of the body fluids, renal functional insufficiency results in widespread metabolic disturbances. Most of these are indicated elsewhere in discussions of various aberrations of metabolism (protein, lipid, water, electrolyte, acid-base status, etc.) and it will, therefore, suffice here merely to review briefly the more important of these abnormalities.

Na and Cl Depletion; Dehydration. The progression of chronic renal disease is accompanied by increasing loss of Na, Cl, and water, eventuating in severe dehydration in the terminal stage of the disease. This is due basically to urinary excretion of excessive amounts of Na, and approximately corresponding amounts of Cl and water, as a result of progressively diminishing tubular capacity for urine acidification and ammonium ion formation, with consequent decrease in Na reabsorption

OTHER BIOCHEMICAL MANIFESTATIONS OF RENAL INSUFFICIENCY

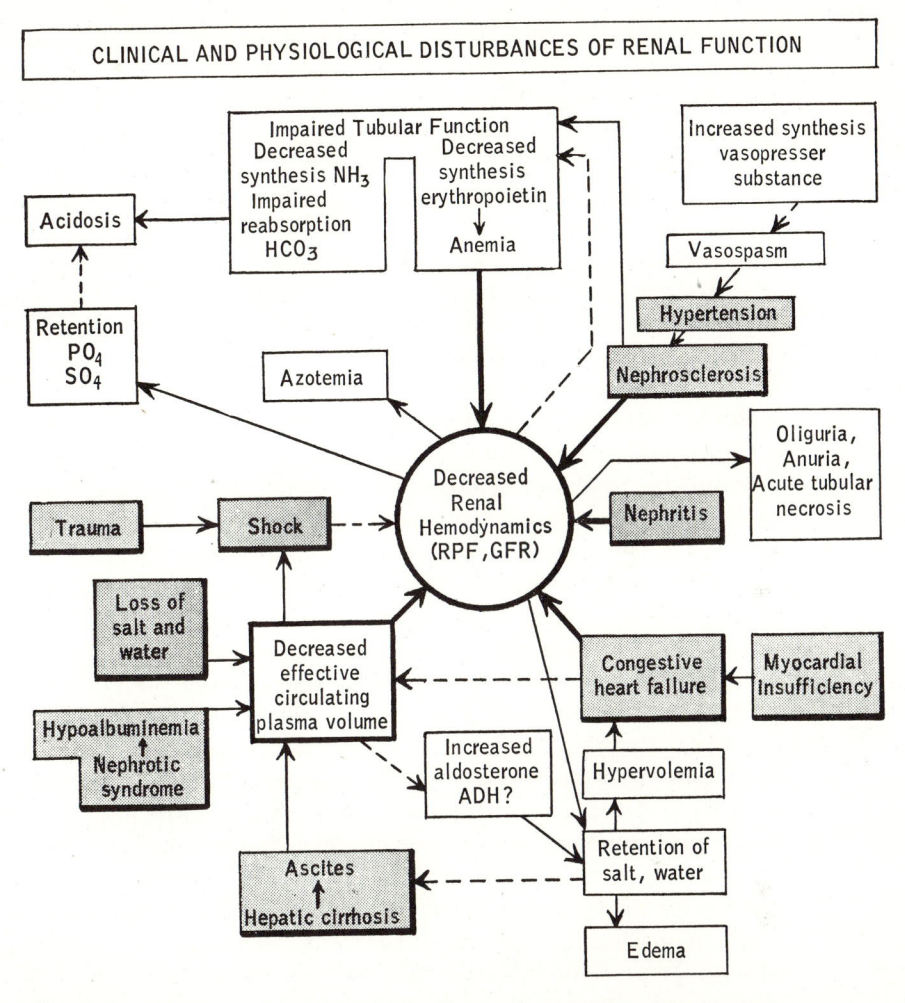

Figure 16-6. Clinical and physiological disturbances of renal function. RPF, renal plasma flow; GFR, glomerular filtration rate; ADH, antidiuretic hormone. (From White: *Clinical Disturbances of Renal Function.* Philadelphia, W. B. Saunders Company, 1961.)

(p. 407). This condition is also frequently contributed to by vomiting and diarrhea, which occur commonly in patients with advanced renal insufficiency (uremia).

In the majority of cases, the state of depletion of Na and Cl may be masked by the simultaneous water loss, and the serum Na and Cl may remain within normal limits (p. 371). Hyponatremia, if present, is usually mild. However, in certain cases, especially in advanced stages, the electrolyte loss is disproportionately great, and hyponatremia and hypochloremia (especially with vomiting) develop. Cases are encountered occasionally that present a syndrome resembling adrenocortical insufficiency (Addison's disease); namely, hyponatremia, hypochloremia, hypotension, and collapse. There is an inability for the surviving nephrons to lower the concentration of sodium and chloride ions in the tubular fluid below a limiting level, which is actually above that found normally. The reasons for this have already been discussed in relation to loss of ability to retain body water (p. 376). They are concerned with osmotic load and the natriuretic effect of "third factor."

Hyperchloremia is observed occasionally. This is unusual in advanced glomerulonephritis except when excessive quantities of NaCl are administered. It may occur also occasionally in such cases during periods of rapid mobilization of large amounts of edema fluid.

Serum K Abnormalities. As in dehydration due to other causes, that in chronic renal insufficiency, e.g., in chronic glomerulonephritis, produces depletion of body potassium (p. 390). Usually, the serum K concentration remains within normal limits. It may fall during administration of K-free fluids (p. 385). On the other hand, it may rise sharply to dangerous levels during injudicious administration, especially if parenteral, of potassium salts, due to inadequate renal excretion. Hyperpotassemia occurs spontaneously at times in the course of severe renal insufficiency, acute or chronic, accompanied by excessive cell destruction or protein catabolism (severe anoxia, dehydration, fever, necrosis, suppuration, etc.), with consequent mobilization of intracellular K faster than it can be eliminated by the damaged kidneys.

In the normal kidney, all the potassium ions in the glomerular filtrate are reabsorbed before the fluid reaches the distal tubule. This is against an osmotic gradient and is an active process. Any potassium ions which appear in the urine are the result of a distal tubular mechanism, which is passive since the ions are passing down an electrochemical gradient. The latter mechanism remains effective in chronic renal failure and consequently hyperkalemia is rare, except in the terminal stages. This is a remarkable phenomenon, since the acidosis, as well as the increased breakdown of intracellular protein, causes intracellular potassium ions to pass into the extracellular fluid. It has been shown that in patients with chronic renal disease, there is increased fecal excretion of potassium salts, apparently by colonic excretion stimulated by aldosterone.

Hypermagnesemia. It is quite common for hypermagnesemia to occur in acute renal failure. It has been reported as a feature of chronic renal failure, but others have found normal or only slightly increased levels of plasma magnesium. In the chronically diseased kidney, with its diminished number of nephrons, there may well be an increased secretion of magnesium ions by individual nephrons. There is much disagreement as to whether a state of hypermagnesemia contributes to the neurological manifestations of chronic renal failure.

Hypocalcemia; Hypercalciuria; Osteodystrophy. Some degree of hypocalcemia occurs commonly in advanced renal failure. The decrease in serum calcium concentration is due to intestinal malabsorption resulting from the decreased sensitivity to vitamin D because of diminished formation of 1,25-dihydroxycholecalciferol by the diseased kidney. This could also result in diminished reabsorption of calcium ions in the renal tubules. The situation is aggravated by the increased osmotic load on each nephron. It is no longer generally believed that the hyperphosphatemia of chronic renal failure contributes significantly to the production of hypocalcemia. It is even possible that the latter, to some extent, contributes to hyperphosphatemia. This is because the kidney tubules become insensitive to the phosphate excretion effect of parathormone, when hypocalcemia is marked. It will be remembered that it is now believed that parathormone exerts its effect through the formation of cyclic AMP, which requires an appropriate concentration of calcium ions to exert its effect.

The occurrence of skeletal lesions (renal osteodystrophy, i.e., osteomalacia, osteitis fibrosa and osteosclerosis) in subjects with progressive renal insufficiency is discussed elsewhere (p. 303). In such cases the serum phosphate is elevated, and the serum calcium concentration tends to be low.

Acidosis. Acidosis of the "nonrespiratory" type (primary reduction in bicarbonate) develops almost inevitably as renal insufficiency progresses. It is an invariable feature of advanced renal failure. This is a reflection of inadequacy of the important function of the kidneys in the maintenance of acid-base status. It is characterized by a decrease in the concentrations of sodium (p. 376) and bicarbonate (p. 419) in the blood plasma. The methods of study and characteristic findings are discussed in the section on hydrogen ion concentration (p. 426), and it will suffice here merely to review the factors involved in the pathogenesis of acidosis in renal insufficiency.

Decreased Ammonium Ion Formation and Urine Acidification. Under normal conditions, cells in the distal segment of the tubule form H ions (from H_2CO_3) (p. 407) and NH_4^+ (from glutamine and amino acids) (p. 411), which are passed into the lumen of the tubule in exchange for Na ions; the latter are reabsorbed, with HCO_3^- ions (produced within the cells from H_2CO_3), into the blood plasma. This mechanism ($Na^+ : H^+$ exchange) plays a role in the maintenance of the normal concentrations of Na and HCO_3 in the extracellular fluids (p. 407).

These functions may be impaired in diseases of the kidney in which the tubular epithelium is damaged or in which a large portion of the functioning renal tissue is destroyed, e.g., glomerulonephritis, "tubular nephritis" ("tubular acidosis") (p. 538), chronic pyelonephritis, malignant nephrosclerosis, polycystic disease, and other destructive lesions (tuberculosis, malignancy). Under these circumstances, the titratable acidity and ammonium ion content of the urine are decreased in contrast to the increase which occurs in these factors in acidosis due to other causes (p. 411).

Retention of Acid Anions. The serum inorganic phosphate is increased in renal insufficiency also exhibiting an increase in blood NPN or urea N (p. 509). It has already been discussed in chronic renal failure; in acute failure, such as tubular necroses, it is, in part at least, due to excessive back-diffusion. In most cases the degree of hyperphosphatemia roughly parallels the increase in blood creatinine. Values as high as 24 mg./100 ml. have been observed in chronic glomerulonephritis, concentrations of 10 mg./100 ml. in this condition (in adults) usually indicating an early fatal termination.

Abnormal retention of phosphate in the extracellular fluids, displacing and thereby lowering bicarbonate, is a factor in the production of acidosis in renal failure. Whereas acidosis may occur in renal disease in the absence of hyperphosphatemia, it is almost invariably present when phosphate retention exists.

What has been said concerning phosphate holds true also for sulfate (p. 328). Inasmuch as the urine is the main channel of elimination of this ion also, its concentration in the blood plasma may increase in renal insufficiency, usually of advanced degree (i.e., urea N retention). Values above 5 mg./100 ml. have been reported but are rare.

The base deficit in renal failure is usually accountable for largely by the simultaneous decrease in total base (mainly HCO_3^-) (p. 419) and increase in phosphate and sulfate. In some cases, however, a considerable fraction of the plasma anion increment consists of organic radicals, the exact nature of which is not known. They may include amino acids, keto acids, etc.

Chloride Retention. As indicated elsewhere (p. 376), the dehydration which occurs in association with renal insufficiency results in depletion of body chloride and, frequently, in hypochloremia. If vomiting is excessive, large amounts of free HCl may be lost by this route, which may, in some measure, offset the tendency toward acidosis. Rarely, it may result in actual alkalosis; a blood pH of 7.6 has been reported under such circumstances. However, this does not occur frequently because the vomitus in patients with uremia usually contains but little HCl.

Parenteral administration of NaCl solution to subjects with impaired kidney function may result in retention of abnormal amounts of Cl^-, and displacement of equivalent amounts of HCO_3^- from the extracellular fluids, aggravating the HCO_3^- deficit ("chloride acidosis"). This is due to the fact that correction of the existing abnormality in acid-base status requires a larger increment in HCO_3^- than in Cl^- in the extracellular fluids. When NaCl solutions are given, the functionally impaired kidneys cannot effectively perform their normal function of excreting the superfluous Cl^- ions.

Hypoproteinemia. Changes in plasma proteins in glomerulonephritis and the nephrotic syndrome are discussed elsewhere in detail (p. 186 ff.). They are characterized by variable decrease in albumin (p. 187) and increase in globulin, the former usually predominating, with consequent hypoproteinemia. These changes, although they may occur at any stage of glomerulonephritis, are most pronounced in the nephrotic (chronic active) stage and also in "lipoid" nephrosis. Hypogammaglobulinemia occurs rather consistently in the idiopathic nephrotic syndrome (p. 204).

In advancing renal insufficiency, especially in the terminal stage of chronic glomerulonephritis, the plasma albumin (and total protein) concentration, if previously low, tends to increase and may return to normal. This is due largely perhaps to the associated dehydration and hemoconcentration. It may be dependent in part upon the diminished elimination of albumin in the urine which not infrequently occurs as the chronic active stage of glomerulonephritis progresses into the terminal stage of uremia. This is due presumably to the complete occlusion of increasingly large numbers of glomeruli, with consequent diminution in the filtration surface area which previously, while blood was able to flow through the inflamed glomerular capillaries, allowed the passage of relatively large quantities of albumin into the glomerular filtrate. This diminution in albuminuria is not observed in all cases.

Plasma Insulin. Insulin, which passes readily into the glomerular filtrate, is rapidly reabsorbed in the proximal tubules. Some of this, as well as some of the filtered insulin, is utilized in relation to the metabolism of the renal cells themselves. This results in destruction of this insulin fraction and consequently less insulin leaves by the renal vein than has entered by the renal artery.

Patients in chronic renal failure frequently have glucose intolerance but the fasting blood glucose level is only rarely raised. Hemodialysis returns the glucose tolerance to normal, and it has been suggested that a circulating low molecular weight insulin antagonist has been removed. Plasma and urine insulin levels are raised in chronic renal failure, both in fasting and as a consequence of a glucose load. Further evidence for the presence of insulin antagonists is the finding that there is a lowered response to injected insulin.

On the other hand, it is now considered likely that there may well be an alteration in peripheral glucose utilization resulting from metabolic changes produced by renal failure. It could well be that the diminution of intracellular potassium may play a part in relation to diminished synthesis of insulin and its release as well as to altered peripheral response to the hormone.

The concentrations of sodium, calcium and magnesium are also important in relation to the release and action of insulin. They are all affected by chronic renal failure and could also play a part in relation to diminished glucose tolerance.

Plasma Lipids. The plasma cholesterol concentration varies considerably in patients with chronic glomerulonephritis and other forms of renal disease.

Although in one series it was reported that patients with chronic renal failure showed an increase in free cholesterol and a normal level of the esterified form, the total serum cholesterol is usually within the normal range in patients with non-nephrotic chronic renal failure. They do, however, as a rule, have an increase in serum total lipids and triglycerides. The hyperlipemia (increased pre-β-lipoprotein) resembles Type IV (p. 135) and is endogenous. It seems that there is probably both increased synthesis and decreased removal of triglyceride. A diminution in the activity of lipoprotein lipase has been demonstrated. The antagonism to endogenous insulin could well play an important part in all these changes.

Phenol and Other Organic Substances. It has been shown that uremia is usually accompanied by increased concentrations of phenols and related substances in the body fluids. The interpretation of this observation is rendered difficult by the nonspecificity of the methods employed for the determination of phenols. Phenol, paracresol, indole and other related substances are formed in the body as a result of processes involving deamination, decarboxylation and oxidation of the aromatic amino acids, tyrosine, phenylalanine and tryptophan. These phenols appear to be formed chiefly, if not entirely, as a result of bacterial action on protein derivatives in the intestines. Under normal conditions, these substances are absorbed from the gastrointestinal tract and are detoxified by sulfate or glucuronate conjugation. The adequacy of this detoxification depends upon the functional integrity of the liver, intestines and other organs that may be involved in these processes. It has been suggested that certain of the manifestations of the uremic syndrome are dependent upon the accumulation in the blood and other body fluids of phenols and other aromatic compounds. Some observers attribute considerable significance to high values for indican in the blood serum as an indication of approaching uremia. Concentrations of 5–10 mg./100 ml. are not unusual in renal failure (normal 0.05 to 0.15 mg.).

Urea itself has been incriminated as at least a partial contributor to the symptoms of uremia. It is known that at high concentrations urea inhibits certain enzymes. There is, however, evidence for and against its being the primary agent in uremia. Other organic substances which accumulate in the blood of uremic patients are creatine, creatinine, uric acid, certain amino acids, indican, hippuric acid, acids of the tricarboxylic acid cycle, guanidine derivatives, acetoin and 2,3-butyleneglycol. Aliphatic amines have also been demonstrated, especially in tissue fluids. Much attention has recently been given to the so-called "middle molecular weight" (500–2000) substances which, according to some authorities, may be important in causing the symptoms of uremia. These could include pharmacologically active polypeptides.

As was the case with phenol, the evaluation of the significance of guanidine studies was complicated by the lack of specificity of the methods employed for the determination of this substance. However, several investigators recently have reported an increase in the methylguanidine content of the blood in patients with uremia and have attributed to this increase some of the toxic manifestations of the uremic state.

A modification of the van den Bergh reaction (p. 636) was used occasionally in differentiating uremic coma from coma due to other causes. The test (diazo test) is performed as follows:

To 1 ml. of blood plasma or serum add 2 ml. of 95 per cent alcohol. To 1 ml. of the supernatant fluid obtained following centrifugation add 0.25 ml. of the van den Bergh reagent and boil gently for thirty seconds. In the serum of normal individuals there is either no color change or the development of a faint pink color (bilirubin). A positive uremic reaction consists in the appearance of an orange-buff color. Add a few drops of 40 per cent NaOH. In the case of normal individuals (nonuremic) a green color appears; a positive uremic reaction consists in the appearance of a definite pink or cherry-red color which lasts from a few seconds to several hours, depending upon the concentration of the responsible agent.

In the great majority of instances the positive reaction is obtained only in the presence of advanced renal insufficiency, as evidenced by marked retention of nonprotein nitrogen and acidosis. On the other hand, cases have been occasionally reported in which a positive reaction is obtained before significant nitrogen retention has occurred. The nature of the substance responsible for this reaction has not been established with certainty. Alcoholic extracts of normal urine give the same reaction, and it is believed by some that the responsible substance is an indoxyl compound, probably indican, or possibly, in part, indoxyl glycuronate. A positive reaction is of value chiefly from the standpoint of prognosis, usually indicating a rapidly fatal outcome. This test cannot be applied in the presence of hyperbilirubinemia, which causes the development of a green or violet color following the addition of NaOH, which will obscure the light cherry-red uremic reaction.

Anemia. This is common in chronic, as well as acute, renal failure. It is thought to be largely due to the action of accumulated toxic metabolites on the bone marrow, since erythrokinetic studies in chronic renal failure have indicated transient beneficial effects resulting from hemodialysis (p. 543). In the early stages of acute renal failure, the anemia is dilutional and is secondary to fluid retention. It is interesting that red cells from patients with this type of renal failure have a diminished life span when put into normal recipients. This is not the case in chronic renal failure; excessive hemolysis does occur in the patient. Erythropoietin is deficient in chronic renal disease, and this factor also plays a part in the causation of the anemia.

RENAL TUBULAR FUNCTIONAL DEFECTS

Development and elaboration of the clearance concept and improvements in analytical techniques have made possible detailed studies of tubular functions. The use of various procedures involving chromatography has served to focus attention on a number of clinical syndromes dependent upon or associated with isolated or multiple aberrations of renal tubular function. These functions may be classified broadly under three headings:

(1) Selective reabsorption of virtually all components of the glomerular filtrate.

(2) Acidification of the urine and support of the plasma bicarbonate (p. 407). This involves exchange of H^+ for Na^+ ions of the tubular fluid and formation of ammonium ions (p. 411).

(3) Secretion, or active excretion, of exogenous creatinine and certain other substances such as potassium, normally, and Diodrast, PSP and other dyes under experimental conditions.

These functions may be disturbed as a result of (*a*) genetically determined defects in transport mechanisms; (*b*) endocrine abnormalities; (*c*) acquired disease and morphologic damage of the renal tubular epithelium.

In the past, clinical attention was focused upon the functions of the kidney as an organ of excretion of waste products, e.g., urea, and upon abnormal retention of such products as a consequence of renal functional impairment. Improved understanding of the activity of the kidney as an organ of primary importance in preserving the volume and composition of body fluids has served to indicate that it is even more important in keeping things in the body than in getting them out. The magnitude of the task performed by the renal tubular reabsorptive mechanisms is indicated in Figure 16-2. These mechanisms make possible precise adjustments of body fluid volume and composition by very slight quantitative changes in reabsorption; e.g., a decrease of 1 per cent in water reabsorption, i.e., from 99 per cent to 98 per cent, results in 100 per cent increase in urine volume. Excretion of Na, Cl, Ca, etc., is controlled with equal precision under normal conditions. It is also clear, however, that abnormalities of these tubular mechanisms can result in profound disturbances in the body fluids. Relatively slight impairment of reabsorption of Na, K, Ca, PO_4 and water can result in serious depletion with respect to these substances in a relatively short time. Because of the almost invariable presence of tubular damage in conditions of primary diffuse glomerular disease, as in glomerulonephritis, disturbances of both glomerular and tubular functions contribute to the over-all picture of renal functional impairment in these conditions. However, tubular damage or dysfunction may occur in the absence of glomerular damage, and certain manifestations of such disturbances have come to be recognized as clinical entities.

Renal Glycosuria. This condition may occur as a manifestation of an isolated tubular reabsorptive defect and also in association with other defects, as in the Fanconi syndrome (p. 84); it is discussed in detail elsewhere (p. 537).

Albuminuria. There is evidence that the glomerular filtrate contains up to 30 mg. albumin per 100 ml. On this basis, up to 54 g. albumin may enter the tubules daily, virtually all of which is reabsorbed under normal conditions, the daily urinary excretion being usually less than 90 mg. Although conceivably albuminuria could be due to defective tubular reabsorption alone, there is no evidence that this occurs clinically. For practical purposes it may be assumed, in the presence of proteinuria, that glomerular damage has permitted the amount of protein entering the tubules to exceed their reabsorptive capacity (p. 213).

Renal Aminoaciduria. Renal conservation of the free amino

TABLE 16-4   RENAL TUBULAR DYSFUNCTION

Inability to reabsorb:	
Water	Renal diabetes insipidus (p. 539)
Phosphate	Vitamin D-resistant rickets (p. 538)
Calcium	Idiopathic hypercalciuria (p. 538)
Glucose	Renal glycosuria (p. 534)
Amino acids	Cystinuria; hemochromatosis; galactosemia; metal poisonings, etc. (p. 534)
Sodium, chloride	"Salt-losing nephritis" (p. 539)
Potassium	"Potassium-losing nephritis" (p. 539)
Inability to acidify urine	Renal tubular acidosis (p. 538)
Excessive reabsorption of phosphate	Pseudohypoparathyroidism (p. 538)
Multiple reabsorption defects	Fanconi syndrome (p. 537)

acids of the plasma is quite efficient; under normal conditions less than 1 to 2 per cent of the amount entering the tubules in the glomerular filtrate escapes reabsorption into the blood stream. When the exogenous supply is increased, only a fraction of the increment is excreted in the urine, the bulk of the excess being metabolized, mainly to urea in the liver. There is no uniform correlation between the amounts of individual amino acids in the urine and their concentrations in the blood plasma; i.e., renal clearances of the amino acids vary widely under normal conditions. Moreover, in certain disease states, e.g., cystinuria, excretion of certain amino acids may be selectively affected. This suggests that there are independent renal tubular transport mechanisms for various amino acids, individually or in groups. In certain instances, the disease state may be identified by the characteristic pattern of urinary amino acids.

Pathological aminoaciduria is conveniently classified as of the (*a*) overflow or (*b*) renal types. The overflow variety is characterized by (*a*) sustained increase in the concentration of one or more amino acids in the blood plasma, and (*b*) normal renal tubular reabsorption, i.e., renal clearance, of amino acids. Aminoaciduria of this type results from some defect in intermediary metabolism and occurs commonly in severe hepatic disease (p. 224). There are also instances of defective metabolism of individual amino acids, e.g., phenylketonuria, in which a genetically determined inability to hydroxylate phenylalanine to tyrosine results in increased concentration of phenylalanine in the plasma and its excretion in the urine in increased amounts, in association with certain of its metabolites (phenylpyruvic and phenyllactic acids, p. 226).

Renal aminoaciduria is characterized by (*a*) normal plasma amino acid concentrations and (*b*) decreased renal tubular reabsorption of certain or all amino acids, i.e., increased renal clearance of the amino acids involved. It may be due to a variety of causes, occurring either as an isolated phenomenon or in association with other abnormalities. The abnormality of tubular reabsorption may represent a genetically determined defect in the transport mechanisms (e.g., cystinuria) or a functional response to the action of some toxic agent or a vitamin deficiency state. It has been observed occasionally in children with the nephrotic syndrome but occurs rarely in the common types of generalized kidney disease, e.g., glomerulonephritis, nephrosclerosis, chronic pyelonephritis.

Cystinuria. This is a genetically determined defect (autosomal recessive) in the renal tubular and gastrointestinal tract transport mechanism for four amino acids, cystine, lysine, arginine, and ornithine, which are consequently excreted in increased amounts in the urine. This defect may exist in varying degrees of severity, as well as in different subgroups, and may occasionally be demonstrated in clinically well relatives of patients with cystinuria of severe degree. The emphasis on the cystine component of this multiple defect is due to the fact that whereas the other three amino acids are quite soluble, cystine is relatively poorly soluble at pH 5 to 7 and, consequently, tends to precipitate and to form calculi in subjects who excrete more than 0.5 g. daily.

It is a matter of some interest that the renal tubular reabsorption of cystine and the dibasic amino acids is really quite complicated. Evidence exists which indicates that, in addition to the common tubular absorption mechanism already discussed, there is also a separate site for cystine absorption (possibly as cysteine) and another separate site for the absorption of the dibasic amino acids.

Hartnup Syndrome. This is a hereditary condition, one of the most constant features of which is increased urinary excretion of a number of amino acids, due to a specific disturbance in renal tubular reabsorption. These are the neutral monocarboxylic amino acids, with the exception of the iminoacids and methionine. The existence of a fundamental disturbance of tryptophan metabolism was suggested by the presence of increased amounts of indican and indolylacetic acids and other indolic acids in the urine. The clinical picture includes a pellagra-like skin eruption, transitory manifestations of cerebral ataxia and mental deficiency. It is now known that there is defective intestinal absorption of tryptophan, and the indolic derivatives result from bacterial metabolism of this amino acid in the colon.

Prolinuria (Familial Iminoglycinuria). This is inherited as an autosomal recessive characteristic. Proline, hydroxyproline and glycine appear in the urine because of defective renal transport. Some cases also have defective intestinal transport of L-proline. It is a benign condition.

Iminoglycinuria is normal during the first three months of life. It can also occur in relation to more serious conditions, some characterized by convulsions (Joseph's syndrome) and others by mental retardation, liver cirrhosis or optic atrophy. The association is probably fortuitous.

Iminoglycinuria also occurs as part of the Fanconi syndrome.

Methionine Malabsorption (Oasthouse Syndrome). In this condition, which is probably genetically recessive, there is mainly deficient transport of methionine, but also aromatic and branch chain amino acids in the renal tubule cells as well as in the mucosa of the small intestine. Some of the methionine is acted upon by intestinal bacteria to form α-hydroxybutyric acid, which is absorbed and secreted in the urine to which it gives a smell resembling an oasthouse. It is possible that this acid may be responsible for the severe neurological disorders which occur. A low-methionine diet gives rise to considerable improvement.

Renal Glycinuria. This is a genetically dominant increased urinary excretion of glycine which has been reported (*a*) in association with hypophosphatemic rickets, (*b*) in association with glucosuria and (*c*) apparently as a defective renal tubular transport of glycine alone.

Glycinuria occurs, of course, as part of an iminoglycinuria, e.g., familial iminoglycinuria, or the Fanconi syndrome. It also results from hyperprolinemia, since proline competes with glycine for renal tubular transport. The hereditary disorder of glycinuria associated with nephrolithiasis is probably a heterozygous phenotype of familial iminoglycinuria.

Galactosemia (p. 89). Galactosuria in subjects with this disorder is accompanied by aminoaciduria. Possibly the renal tubular transport mechanism for amino acids is sensitive to the accumulation of galactose-1-phosphate which results from the metabolic defect in this condition. Removal of lactose from the diet is followed by disappearance of the aminoaciduria.

Hepatolenticular Degeneration (Wilson's Disease) (p. 336). The primary disturbance in this condition is a genetically determined disorder of copper metabolism. Generalized aminoaciduria occurs in many but not in all cases, due presumably to functional damage to the tubular transport mechanism by the accumulation of copper in the tubular epithelial cells. An overflow type of aminoaciduria may possibly occur in advanced cases due to the associated liver cirrhosis. There may occasionally be other evidences of defective renal tubular reabsorption, such as a renal type of glycosuria, and hyperphosphaturia, with hypophosphatemia and osteomalacia as well as hypouricemia resulting from excessive output of uric acid in the urine. The urine tends to be alkaline (excessive HCO_3^-).

Fanconi Syndrome (p. 311). A generalized type of renal aminoaciduria is one of the constant features of this syndrome, which is characterized by multiple tubular reabsorptive defects. In its complete form, the syndrome includes also renal glycosuria, hyperphosphaturia (with hypophosphatemia and vitamin D-resistant rickets), metabolic acidosis with increased urinary excretion of Na, Ca, K (with hypokalemia) and HCO_3^-, polyuria, isosthenuria and an alkaline urine. In some cases there is also cystinosis, presumably an abnormality of cystine metabolism, independent of the renal tubular disorder, in which cystine crystals are deposited in macrophages in the liver, spleen, bone marrow, lymph nodes, kidneys, and subcutaneous tissues, and also in the cornea, where they may be seen by slit-lamp microscopy.

Much confusion has resulted from regarding the Fanconi syndrome as a disease. It can result from a variety of causes. These include the nephropathic type of cystinosis already mentioned, an idiopathic type, Lowe's syndrome (p. 538) and tyrosinemia. The Fanconi syndrome is a major part of all these conditions. The syndrome tends to occur, to a lesser degree, in glycogen storage disease (Type I), hepatolenticular degeneration and hereditary fructose intolerance. It occurs as a result of certain toxic agents, including lead, uranium, cadmium, Lysol and old tetracycline. It has been reported in association with myelomatosis and after kidney transplantation.

Vitamin Deficiencies. Renal aminoaciduria may occur in vitamin C and vitamin D deficiency in infants. Interest has been centered in the latter particularly because aminoaciduria occurs also in other conditions associated with hypophosphatemia, e.g., the Fanconi syndrome and renal tubular acidosis. In these disorders, as in vitamin D deficiency, the aminoaciduria responds to treatment with vitamin D, suggesting that there may be some relationship between the renal transport mechanisms for amino acids and phosphate. However, there is abundant evidence that each can be influenced independently of the other.

Toxic Agents. Defective renal tubular reabsorption of amino acids, glucose, and phosphate may occur in lead poisoning and as a consequence of renal tubular damage induced by other heavy metals (uranium, cadmium) and other nephrotoxic agents, such as Lysol and nitrobenzene.

Cystinosis. In contrast to cystinuria, in which clinical manifestations result only from the formation of cystine calculi in the urinary tract, this condition presents serious systemic manifestations associated with deposition of cystine crystals in macrophages of the reticuloendothelial system in various situations, including the liver, spleen, bone marrow, lymph nodes, subcutaneous tissues, and kidneys. The presence of crystals in the cornea may be revealed by slit-lamp microscopy. In the nephropathic type, which has its onset in infancy, there is a Fanconi syndrome. Death occurs usually as a result of renal failure with severe acidosis. Cystinosis is a lysosome storage disorder probably resulting from defect of the transport process for the passage of cystine across the lysosomal membrane.

Lowe's Syndrome. This is a congenital disorder characterized biochemically by generalized aminoaciduria, phosphaturia with rickets, increased excretion of organic acids, and decreased ability to form ammonia and to acidify the urine, resulting in systemic acidosis. It therefore presents many of the features of the Fanconi syndrome. There are also mental retardation and ocular pathology, such as cataracts. It is inherited as an X-linked recessive.

Idiopathic Hypercalciuria. Many subjects with urinary calculi excrete abnormally large amounts of calcium in the urine without any demonstrable cause. Possibly, renal tubular reabsorption of calcium is impaired, perhaps on a genetic basis. This is influenced to only a slight extent by dietary restriction of calcium.

Pseudohypoparathyroidism. This hereditary condition, which presents the characteristic biochemical features of hypoparathyroidism (hypocalcemia, hyperphosphatemia, decreased urinary phosphate), is due to lack of responsiveness of the target organs (bones, renal tubular epithelium) of the parathyroid hormone (p. 289) associated with failure to produce appropriate amounts of cyclic AMP. It is distinguished from true hypoparathyroidism by its failure to respond to administration of parathyroid hormone; it does respond to administration of dihydrotachysterol and to large amounts of vitamin D and increased dietary calcium.

Vitamin D-resistant Rickets. This condition, which may be one of the components of the Fanconi syndrome, may exist also as an independent entity, i.e., an isolated defect in renal tubular reabsorption of phosphate. The consequent excessive loss of phosphate in the urine results in hypophosphatemia; this leads to a form of rickets which can be controlled only by relatively enormous doses of vitamin D. There is decreased conversion of the vitamin to its active forms.

Renal Tubular Acidosis. One form of this condition is due to inability to produce a urine of normal acidity, i.e., to a defect in the $Na^+ - H^+$ exchange mechanism in the distal portion of the renal tubule (p. 407). Another form is due to diminished absorption of bicarbonate in the renal tubules (mostly proximal) with lysozymuria. This results in the rather unique association of systemic manifestations of metabolic acidosis (low plasma CO_2, decreased pH) with a urine of decreased acidity (at times neutral or alkaline). Inadequacy of the $Na^+ - H^+$ exchange mechanism leads to loss or excessive quantities of Na (p. 526) and also K (p. 528) in the urine. The acidosis is also accompanied by excessive mobilization and urinary excretion of calcium (p. 528). These abnormalities may lead to clinical manifestations of dehydration, potassium depletion and hypokalemia, defective skeletal mineralization, and nephrocalcinosis.

A variety of this condition which occurs during the first year of life is different in certain respects from that occurring in adolescents and adults.

Salt-losing Nephritis. In this condition, defective renal tubular reabsorption of Na and Cl results in their excessive excretion in the urine, which continues even in the presence of severe dehydration, hyponatremia, and hypochloremia. As in sodium depletion due to other causes (p. 375), the glomerular filtration rate may be reduced, with consequent increase in the blood urea nitrogen concentration. These biochemical abnormalities resemble those of Addison's disease, from which this condition may be distinguished by (*a*) its resistance to adrenocortical hormone therapy and (*b*) the presence of normal amounts of these hormones in the urine and blood. The failure to respond to adrenocortical hormone therapy contrasts sharply with the usually prompt response to parenteral administration of NaCl solution. There is evidence that chronic pyelonephritis is commonly present as an underlying condition.

Potassium-losing Nephritis. Excretion of excessive amounts of potassium in the urine may lead to hypokalemia and clinical manifestations of potassium depletion. This may occur in the Fanconi syndrome (p. 537) and in renal tubular acidosis (p. 538). Renal functional consequences of potassium depletion include: reduction of renal plasma flow and glomerular filtration rate; inability to concentrate the urine, associated with polyuria (decreased water reabsorption), and a "diabetes insipidus-like state." Severe potassium deficiency may result in rather characteristic morphologic changes in the renal tubular epithelium. Certain cases previously included in this category of "potassium-losing nephritis" can be demonstrated to be due to hypersecretion of aldosterone, i.e., primary aldosteronism (p. 734).

Nephrogenic Diabetes Insipidus ("Water-losing Nephritis"). This condition is due to defective reabsorption of water in the distal portion of the tubule and may therefore resemble true diabetes insipidus. It is distinguished from the latter by (1) failure to respond to adequate doses of antidiuretic hormone and (2) the presence of this hormone in the urine. There is a congenital form that is sex-linked, occurring only in males and transmitted by females. Administration of vasopressin produces the usual effects on smooth muscle but no change in urine volume or concentration. An acquired form may occur in association with states of potassium depletion (p. 384) but has been observed particularly in conditions accompanied by hypercalcemia and nephrocalcinosis. It has been reported in cases of hyperparathyroidism, sarcoidosis, and multiple myeloma. This abnormality disappears when the normal serum calcium concentration is restored, unless irreversible morphologic renal damage has occurred.

RENAL AND URINARY LITHIASIS

Kidney and bladder stones commonly contain calcium phosphate, hydroxyapatite, magnesium ammonium phosphate or calcium carbonate. They are frequently mixtures of these substances. Calcium phosphate is particularly liable to form staghorn calculi in the renal pelvis. Calcium oxalate may also form calculi, which tend to be smaller than phosphate stones but are also relatively common. Calcium oxalate together with one or more of the other substances already mentioned may just as commonly form the so-called mixed stones. Less commonly, calculi are composed of uric acid (4-10% of patients with urolithiasis). Even less common are cystine stones (less than 1%). Xanthine stones are very rare.

In order for a stone to be formed, it is necessary for the urine to be saturated or supersaturated in regard to the salts or the organic substance of which the stone is composed. This can be brought about by alteration in the pH of the urine. One of the most common causes of phosphate stone formation is a urinary infection with urea-splitting bacteria, which form ammonia and cause the urine to become alkaline. Phosphates are much less soluble in alkaline urine, and so they crystallize. Conversely, uric acid stones form in acid urine. The formation of calcium oxalate crystals is unaffected over the physiological range of urinary pH. Another factor in regard to urinary crystal formation is dehydration. Hence, urinary stones are more common in tropical countries. Their formation is also favored by stasis of urine in the renal pelvis because of lack of movement. They are thus prone to occur in patients who have to spend a long time immobile in bed (e.g., after certain fractures).

In regard to the substances which more commonly form stones (phosphate and oxalate), normal urine contains inhibitors of crystallization such as organic acids, magnesium ions, pyrophosphates and two small peptides. Certain polyvalent cations are capable of blocking these solubilizers. They are, however, present only in small amount in normal urine. It has recently been suggested that normal urine also contains an inhibitor of calcium oxalate crystal growth.

Metabolic factors favoring stone formation are hypercalciuria, hyperuricuria, hyperoxaluria, cystinuria and xanthinuria.

Hypercalciuria. On a diet containing a daily intake of 700 mg. (17.5 mmol.) of calcium ion, the upper limit of daily urinary calcium ion is 250 mg. (6.25 mmol.) for females and 275 mg. (6.88 mmol.) for males. Higher values constitute hypercalciuria. A raised calcium level tends to cause crystallization of calcium salts, because the solubility product would be increased.

Hypercalciuria of Known Etiology. Hypercalciuria and increased stone formation can result from primary hyperparathyroidism, renal tubular acidosis, Paget's disease of bone, immobilization and the medullary sponge kidney (cystic dilatation of collecting tubules).

Idiopathic Hypercalciuria. In this condition, there is hypercalciuria without demonstrable cause. The plasma phosphate is usually low. There is possibly overabsorption of dietary calcium. As a cause of the hypercalciuria, this is undoubtedly not nearly so important as the excessive renal loss.

Idiopathic Renal Lithiasis.
There is no doubt that the majority of kidney and bladder stones cannot be related to the known causes of hypercalciuria. These stones, of course, are constituted mainly of calcium oxalate, calcium phosphate and hydroxyapatite. They are more common in males, and at least 70 per cent of the patients who have recurrent stones containing calcium belong to this idiopathic group. Idiopathic hypercalciuria is present in at least half the patients and there may be hypophosphatemia. The serum calcium level is within the normal range.

Those with hypercalciuria may differ etiologically from the others. Restricting calcium in the diet of the hypercalciurics diminishes the tendency to stone formation, but has no effect on the other group. Significant hypercalciuria is by no means always present in those patients showing most active stone formation.

Urine contains an amount of calcium phosphate, which could not be held in solution in an equal volume of water. Solution in urine is maintained by the substances which prevent crystallization (organic acids, magnesium ions, pyrophosphates and peptides). Although hypercalciuria obviously plays a part in stone formation, it appears that at least an equally important part is played by an imbalance of the solubilizing substances. In a system containing rachitic rat cartilage, normal urine did not induce mineralization. Urine from patients with idiopathic nephrolithiasis did mineralize the cartilage. It has been claimed that two small peptides have been isolated from normal urine, which are highly effective in preventing the mineralization. It could be possible that a urinary deficiency of these could lead to stone formation. Other workers, however, have stated that the active material does not have a peptide structure but this does not affect the hypothesis.

It will be remembered that the kidney possesses a calcium gradient (p. 480). This did not differ from the normal gradient in patients who had renal stones. Moreover, it has also been mentioned that the renal clearance of calcium and sodium tend to run in parallel. Studies on the South African Bantu compared with White South Africans indicate that the former have a much lower incidence of renal and bladder lithiasis but a considerably higher urinary output of sodium. It appears that a high urinary sodium output protects against crystallization of calcium salts. It would appear, in fact, that it is the high urinary osmolality, rather than the sodium, which protects. It is possible that calcium stone formation is related to low urinary osmolality.

Hyperoxaluria. Oxalate taken with the diet is excreted unchanged in urine and feces (mostly in urine). Much less is absorbed on a high-calcium intake (dairy foods) because of the formation of insoluble calcium oxalate in the intestine. Hyperoxaluria, with increased tendency to the formation of calcium oxalate stones, can occur on a high oxalate but low calcium intake. Certain vegetables, such as spinach and rhubarb among others, have a high oxalate content. Consequently, after the first World War, an increased incidence of calcium oxalate stones occurred in Central Europe because dairy foods were scarce and the diet was largely vegetarian.

Oxalate ($COO^-.COO^-$) can also be produced endogenously by two major mechanisms. The one involves oxidation of ascorbic acid and the other of glyoxylate ($CHO.COO^-$). Endogenously produced oxalate is virtually all excreted in the urine. The normal adult output is 10 to 55 mg. daily. The output of children is comparable to that of adults, when corrected for body surface area. Ascorbic acid accounts for 35 to 50 per cent of the total output. Administration of large amounts of ascorbic acid does not significantly alter this fraction. The occurrence of excessive amounts of endogenously produced oxalate must therefore depend on the accumulation of glyoxylate. This can result from excessive production or diminished removal. Excessive production does occur in ethylene glycol poisoning, but this need not be considered here. It may also occur in Type II primary hyperoxaluria (see below).

Glyoxalate is mostly removed by its conversion to glycine (aminotransferases) or by conversion to α-keto-β-hydroxyadipate. The former requires pyridoxal phosphate and hence is diminished in pyridoxine deficiency. The latter requires thiamine and is diminished in deficiency of this vitamin. Hyperoxaluria has been reported in pyridoxine deficiency.

Hyperoxaluria occurs in association with ileal disease. This is due to glyoxalate formation by bacterial action in the large intestine on the glycine from unabsorbed glycine conjugated bile salts.

Primary Hyperoxaluria. This condition usually becomes manifest before the age of five years. It presents with renal colic or hematuria. These are due to the formation of calcium oxalate calculi and deposition of calcium oxalate in the kidney itself. There is, of course, a great increase in urinary oxalate output. The condition has a high mortality rate and the vast majority of patients with this disorder die before the age of 20 years from renal failure. There is frequently hyperuricemia. In addition to nephrocalcinosis, there may be deposits of calcium oxalate in other body tissues (oxalosis). There are two genetic types, each of which is inherited as an autosomal recessive characteristic.

Type I hyperoxaluria is more common and shows excessive amounts of glycolic and glyoxylic acids as well as oxalate in the urine. There is a defect in the biochemical mechanism for irreversibly converting glyoxalate into α-oxo-β-hydroxyadipate. Hence, glyoxalate accumulates and gives rise to excess oxalate as well as to excess of glycolate. The enzyme responsible for conversion to α-oxo-β-hydroxyadipate (α-oxoglutarate: glyoxylate carboligase) has been shown to be deficient in liver, spleen and kidney.

Type II hyperoxaluria does not manifest excessive secretion of glyoxalate, but in addition to hyperoxaluria, L-glycerate is present in the urine. Both D-glycerate and L-glycerate are formed from hydroxypyruvate, which is derived largely from serine. The formation of D-glycerate requires the enzyme D-glycerate dehydrogenase, whereas L-glycerate requires lactate dehydrogenase. The former enzyme has been shown to be deficient in the red cells of patients with Type II primary hyperoxaluria. A general deficiency would result in the formation of excess hydroxypyruvate and hence excess L-glycerate. The latter reaction helps to maintain the coenzyme of lactate dehydrogenase in the oxidized state (NAD) and so presumably facilitates the enzymic oxidation of glyoxalate to oxalate.

Cystinuria. This condition has already been described (p. 535). It results from defective transport of cystine and other amino acids across the renal tubule cells. There is also defective transport across the intestinal mucosa. Consideration of intestinal transport patterns indicates the existence of three genetic groups.

Hyperuricuria. Excess uric acid in the urine can occur in gout, in polycythemia vera and in leukemia. This can induce stone formation. The commonest causes of urinary supersaturation with uric acid are dehydration and an increased acidity or both.

Xanthinuria. This very rare autosomal recessive defect results from a deficiency of xanthine oxidase (p. 252). Xanthine replaces uric acid in the urine and can lead to the rarest form of renal stone.

HEMODIALYSIS

The advent of hemodialysis has profoundly affected the outlook in renal failure of acute or chronic type. The principle involved is the connection of an artery to a vein by means of a long length of dialysis tubing (cellophane, etc.), which has previously been filled with blood and is immersed as a twin coil in a dialysis bath. Alternatively, sheets of cellophane are compressed between plastic plates. The blood flows between the cellophane sheets and the dialyzing fluid between the sheets and the plates. Quite small, but highly efficient, disposable dialysis devices are now available (Hollow fiber type). The dialysis fluid employed varies somewhat from clinic to clinic. One recipe contains in each liter of deionized tap water 1.25 mmol. Ca^+, 0.75 mmol. Mg^+, 2.0 mmol. K^+, 126 mmol. Na^+, 105 mmol. Cl^-, 33 mmol. CH_3COO^- and 11 mmol. glucose. There is a tendency to omit glucose to avoid bacterial contamination.

Disequilibrium Syndrome. During or toward the end of dialysis, the patient sometimes complains of headache. This may be followed by confusion, muscle twitching and even convulsions. Coma may occur. The condition has, on rare occasions, gone on to death. The clinical picture resembles water intoxication. It has been named the dialysis disequilibrium syndrome, since it was thought to be due to a more rapid removal of urea from fluid compartments other than the cerebrospinal fluid and the brain. This would mean that water would move from the rest of the body to the brain. This explanation is no longer acceptable. There is a possibility that some degree of hyponatremia could play a part, but there is no proof as to the actual cause.

Acute Renal Failure. Acute failure is of three types. There may be *prerenal failure*. This implies an impairment only of the circulation (hypovolemia, shock), with virtually normal kidney tissue. The condition may progress to acute tubular necrosis or bilateral renal cortical necrosis. This is now *acute renal failure*, of which other examples are glomerulonephritis (secondary to streptococcal infection, lupus nephritis, polyarteritis, drug hypersensitivity, etc.) and acute interstitial nephritis (caused by sulfonamides, penicillin, Dilantin, mercurohydrin, etc.). The third type of acute renal failure is *postrenal failure* due to obstruction of the urinary tract (calculi, enlarged prostate, crystals, etc.). This of course requires surgical intervention to relieve the obstruction.

Acute renal failure is usually accompanied by oliguria, but this is by no means always the case. If oliguria is present, dialysis is necessary; acute nonoliguric renal failure patients often do well without dialysis.

In order to assess the type and cause of renal failure, appropriate biochemical and other types of investigations are necessary, possibly including renal biopsy. The biochemical investigations involve routine examination of urine, blood urea, creatinine, HCO_3^-, Na^+, K^+ and glucose. The osmolality of both plasma and urine should be determined. In prerenal failure, the ratio of urine osmolality to that of plasma is usually above 1.3, whereas in acute tubular necrosis, it is or approximates 1.0. In prerenal and postrenal failure, the urinary Na^+ is low (less than 20 mmol./liter); in acute tubular necrosis it is usually above 30 mmol./liter.

In acute renal failure, dialysis is definitely indicated if the blood urea nitrogen is greater than 150 mg./100 ml. or the plasma CO_2 content is lower than 12 mmol./liter, or the serum K^+ is above 7.0 mmol./liter. This does not mean that dialysis should not be carried out even earlier, as a prophylactic measure. The rate at which the blood urea N, CO_2 content and K^+ ion concentration increase before dialysis indicates the degree of seriousness of the situation as well as the probable frequency of dialyses required. Obviously, it is important to monitor the blood electrolyte levels during the dialysis itself.

Acute dialysis is also necessary in various types of severe poisoning, e.g., barbiturate. This means that the poison must be identified by examination of blood, urine and gastric content. The nature of the poison indicates whether dialysis could be effective treatment, and in appropriate cases the blood level indicates whether this form of treatment should be undertaken, as well as how often it should be repeated.

Chronic Dialysis. This is a life-saving measure in the uremic syndrome. This syndrome is characterized by appropriately raised blood urea nitrogen accompanied by weakness and manifestations due to involvement of a variety of body systems (gastrointestinal, cardiovascular, hematological, neurological and dermatological). Dialysis is repeated 2 or 3 times a week. The time taken for the procedure depends on the type of "artificial kidney" as well as on the properties of the membrane employed.

Improvement can be demonstrated by diminution in the blood urea level. Urea is a small and easily dialyzable molecule, so the levels of blood creatinine are probably better indicators of the degree of improvement. The serum electrolytes (Na^+, K^+ and Cl^-) tend to be normal prior to dialysis. Na^+ may be lowered because of too great an intake of fluid or occasionally because of loss in the urine. K^+ may be raised, usually slightly, as a result of acidosis, infections, dietary indiscretions or dehydration. (Hence, the K^+ concentration in dialysis fluid is usually kept low). A low serum Cl^- concentration can result from vomiting. There is a tendency to acidosis prior to dialysis (pH 7.19 to 7.30 in one series) and a slight alkalosis just after dialysis. The serum HCO_3^- concentrations mirror these changes. Plasma inorganic phosphate is always raised prior to dialysis and then the level falls somewhat. The behavior of the low serum calcium is variable. It may be restored to normal by the dialysis, or this may result in calcium increase. The behavior in this respect is probably dependent on the properties of the tap water. In chronic renal failure, there is a greater production of parathormone which is associated with renal osteodystrophy. There is no doubt, however, that in some dialysis clinics there is a relationship between the dialysis and the frequency of occurrence of the osteodystrophy (osteomalacia). It is not known why this should occur in some clinics but not in others. About half of all individuals undergoing chronic intermittent dialysis show a raised level of serum alkaline phosphatase. The serum magnesium level is almost always raised before dialysis and then falls somewhat.

Patients with chronic uremia have an increase in fasting plasma triglycerides and a deficiency of a number of plasma amino acids. Dialysis aggravates the triglyceridemia but, associated with a better food intake, most of the amino acid deficiencies tend toward correction. Threonine and cystine deficiency tend to be aggravated by the procedure.

Patients on intermittent dialysis must be given an adequate supply of vitamins. It is possible to remove water-soluble vitamins by dialysis. Frank scurvy has been produced.

RENAL HOMOTRANSPLANTATION

With the advent of successful renal homotransplantation, it is necessary to monitor the recipient by a variety of investigations, some of which are biochemical. Rejection is indicated by clinical symptoms, such as edema, malaise, fever, pain, tenderness and swelling in relation to the grafted kidney, as well as by diminished urinary output. There is a rise in the blood urea and serum creatinine and a decrease in creatinine clearance. There is also decreased urinary sodium content. These changes can only be noted if the appropriate investigations are carried out at regular intervals after the initial transplantation. A fall in PAH clearance has been used to detect early rejection, but the investigation is not too easily undertaken in the clinical context. The serum creatinine and creatinine clearance are for practical purposes the most useful biochemical investigations.

References

Adams, D. A.: The pathophysiology of the nephrotic syndrome. Arch. Int. Med., *106:* 117, 1960.

Black, D. A. K. (ed.): Renal Disease. Ed. 2. Edinburgh, Blackwell Scientific Publications. 1967.

Corcoran, A. C.: The Kidney. *In* W. A. Sodeman, ed.: Pathologic Physiology. Philadelphia, W. B. Saunders Co., 1961, p. 811.

Franklin, S. S., and Merrill, J. P.: Acute renal failure. New England J. Med., *262:*711, 761, 1960.

Holt, L. E., and Snyderman, S. E.: Disturbances of amino acid metabolism. Bull. N. Y. Acad. Sci., *36:*431, 1960.

Josephson, B., and Ek, J.: The assessment of the tubular functions of the kidneys. Advances in Clinical Chemistry, *1:*41, 1958.

Lotspeich, W. D.: Metabolic Aspects of Renal Function. Springfield, Ill., Charles C Thomas, 1959.

Pitts, R. F.: The Physiological Basis of Diuretic Therapy. Springfield, Ill., Charles C Thomas, 1959.

Potts, J. T., Jr., and Deftos, L. J.: Parathyroid hormone, thyrocalcitonin, vitamin D, bone and bone mineral metabolism. *In* Bondy, P. K., ed.: Duncan's Diseases of Metabolism, Ed. 6. Philadelphia, W. B. Saunders Company, 1969, p. 904.

Smith, H. W.: Principles of Renal Physiology. New York, Oxford University Press, 1956.

Sunderman, F. W., and Sunderman, F. W., Jr., eds.: Laboratory Diagnosis of Kidney Diseases. St. Louis, Waren H. Green, Inc., 1970.

Van Slyke, D. D., et al.: Observation on courses of different types of Bright's disease and on resultant changes in renal anatomy. Medicine, *9:*257, 1930.

White, A. G.: Clinical Disturbances of Renal Function. Philadelphia, W. B. Saunders Company, 1961.

Wills, M. R.: Biochemical Consequences of Chronic Renal Failure. Aylesbury, Harvey Miller and Medcalf, Limited, 1971.

Chapter 17

ENZYMES

In this chapter, a variety of so-called normal ranges are quoted in relation to various enzymes. These are values quoted in the literature. It cannot be emphasized too strongly that each laboratory must determine its own ranges of reference values. We do not yet have standard methods for enzyme analyses, but they are being developed on an international basis. Since human beings are what they are, there is little doubt that even with standard methodologies, ranges will still vary from laboratory to laboratory.

Quantitative determinations of enzyme activities in clinically available material are employed as aids in diagnosis, therapy and prognosis. Blood serum is the medium most commonly studied; in special circumstances, determinations are made in cerebrospinal fluid, erythrocytes, leukocytes, biopsy material and cultures.

Abnormalities of cell enzyme activity may yield specific disease diagnostic information, e.g., decreased erythrocyte glucose-6-phosphate dehydrogenase in hemolytic reactions to drugs, and decreased glucose-6-phosphatase activity in the liver in glycogen storage disease. However, this is not true of serum enzymes, with a few notable exceptions. Apart from coagulation factors, complement components and lecithin-cholesterol acyltransferase, enzymes present in the plasma (or serum) are not functioning as such in that medium, but represent rather an "overflow" from their tissues of origin. They are removed from the blood stream partly by excretion and partly, in common with other plasma proteins, by metabolic degradation. In some cases, removal appears to involve immunological mechanisms. "Overflow" can be increased by disease (cell death, increased membrane permeability, enzyme induction). The diagnostic specificity of a change in the level of activity of a serum enzyme is in inverse relation to the extent of its normal tissue distribution. Accordingly, those which reflect highly specialized activities of a single or very few organs have the greatest diagnostic significance; these include "prostatic" acid phosphatase, amylase, isoenzymes of alkaline phosphatase, and lipase. On the other hand, those which perform important functions in intermediary metabolism in general and are, therefore, necessarily present in practically all cells will have somewhat less precise significance with respect to localization of disease in a specific organ; these include enzymes involved in transamination and intermediary metabolism of glucose.

The extent of increase in concentration of an enzyme in the serum is a resultant of (a) the rate of its release from the tissue of origin and (b) the rate of its removal from the circulation. Consequently, the serum enzyme pattern does not invariably reflect the enzyme pattern of the damaged tissue, e.g., myocardium or liver, different enzymes being released and removed at different rates. Moreover, what is being measured is enzyme activity, and the results obtained may be influenced by variations in activating or inhibiting factors, independently of variations in enzyme concentration. Nevertheless, clinical experience has demonstrated the usefulness of these determinations.

International Unit of Enzyme Activity. It has been customary to express the amount of enzyme activity in terms of units defined by the originators of the particular method employed. For example, alkaline phosphatase could be expressed as Bodansky units, King-Armstrong units or even Jenner and Kay units. This has led to confusion in the literature, and so an International Unit has been introduced. It is defined as that amount of enzyme which will catalyze the transformation of one micromole of the substrate per minute under standard conditions. These conditions involve definition of the technique employed as well as the temperature at which the reaction is carried out. It is now suggested that the temperature should be 30° C. This means that there are at present as many "normal ranges," measured in International Units, as there are methods of measurement. Standard reference techniques are now being evolved so that the method employed by a particular laboratory can be compared with the standard method. Thus, the laboratory results can then be corrected by an empirical factor, which will convert them into truly standard International Units.

There are many difficulties still to be overcome. Consequently, units given in this chapter will be those defined by the original authors of a method and will be accompanied, where possible, by conversion of these measurements to International Units for that method. The symbol for the International Unit is U. It is now considered preferable to express the enzyme concentration as the number of International Units per liter (U/l.).

It is now proposed that enzyme activity should be expressed as a catalytic amount. The proposed unit is the *katal*. This is the catalytic amount which will result in the conversion of one mole of substrate per second under standard conditions. The katal is not yet used in the clinical laboratory. One International Unit is equal to 16.67 nanokatals (i.e., U = 16.67 nkat.).

Isoenzymes (Isozymes). It is now known that biologically active proteins can exist in more than one form and still retain their biological activity. Hemoglobin (p. 266) is a good example of this so-called protein polymorphism. Enzymes are proteins, and so it is not surprising that many of them exist in multiple forms, which are then called *isoen-*

zymes. When multiple forms of an enzyme occur in the same species, especially in one tissue of an individual animal, there is a tendency to refer to the different forms as isoenzymes. This, it must be confessed, is a very loose definition. Similar acting but nevertheless very different enzymes can occur within a single cell. Mitochondrial malate dehydrogenase is certainly very different from that which exists in the soluble cytoplasm. The mitochondrial component, however, exists in multiple forms, which are obviously isoenzymes. Nevertheless, hybrids of the mitochondrial and cytoplasmic forms have been prepared *in vitro*. The major multiple forms of human alkaline phosphatase are referred to as isoenzymes, in spite of the fact that each occurs in a different tissue (liver, bone, intestine, placenta, etc.).

Much more stringent definitions of the isoenzyme concept are now proposed, since a number have been isolated and much is known about their over-all molecular structure, especially in lower animals and in plants. However, very little is known about isoenzymes in humans, where a closer definition can at present be made only by analogy with the state of affairs in other species, by hybridization techniques and by the use of genetic studies.

Isoenzymes can be detected by their distinctive behavior during electrophoresis (i.e., surface charge) and their different reactions to heat, inhibitors, activators, ion-exchange chromatography, isoelectric focusing and sometimes antisera. The most popular of these detection methods has undoubtedly been electrophoresis (using cellulose acetate, agar, starch or acrylamide gel), but demonstration of more than one zone of enzyme activity is not necessarily proof of the presence of isoenzymes. This is especially the case in relation to the esterases. Heat resistance or inactivation can be used in the laboratory to distinguish, for clinical purposes, between isoenzymes of the same enzyme system. Sometimes it is possible to detect differences between isoenzymes by using a different substrate. In this way it can be shown, with the five demonstrable isoenzymes of lactate dehydrogenase, that the one which moves fastest in the standard electrophoresis techniques is far more effective as a hydroxybutyrate dehydrogenase than the slowest-moving form. This reaction is used commonly in clinical evaluations.

Occasionally, isoenzymes are distributed differently in different tissues. The pattern of distribution can be recognized by subjecting a tissue extract to electrophoresis. In this way, a specific tissue pattern may be recognized. When there is increased tissue enzyme "overflow" in disease, it is sometimes possible to detect the specific tissue distribution in a sample of serum by means of the various methods previously indicated.

DIGESTIVE ENZYMES

A few digestive enzymes appear in the plasma of normal subjects in measurable amounts. These include amylase, lipase, pepsinogen, and perhaps trypsin. Determination of serum amylase activity is commonly employed clinically, particularly in the diagnosis of pancreatic disease; less frequently, serum lipase activity is determined for the same purpose.

Serum Amylase (Diastase). An enzyme that hydrolyzes starch, glycogen, and dextrins is present normally in the blood plasma and also in urine, lymph, feces, and milk. The amylase activity of the blood plasma or serum is usually expressed in units which represent the milligrams of "apparent glucose" formed by digestion of the starch substrate. The amylolytic (diastatic) activity of normal serum or plasma (Somogyi or related method) is 40–160 units/100 ml., 80 per cent of normal subjects falling between 80 and 150 units. Values above 200 units and below 60 units are abnormal. There is evidence that the liver is the main source of the plasma amylase in normal subjects, although clinically significant increases are usually due to pancreatic (or salivary) amylase. Amylases have also been demonstrated in skeletal muscle, fallopian tubes, and adipose tissue.

Methods for estimation of amylase by estimating "apparent glucose" liberation are known as saccharogenic methods. It is also possible to measure the enzyme activity by determining the amount of starch hydrolyzed. This can be done by determining the decrease in turbidity of the starch suspension by nephelometric methods or estimating the amount of dye liberated from a starch-dye substrate. Methods which determine starch breakdown are known as amyloclastic procedures. For the purposes of description in this text, Somogyi units (saccharogenic) are used.

Values below 60 units/100 ml. may occur in severe hepatocellular damage. Low values in other conditions are perhaps attributable to associated liver damage; these include severe burns, shock, diabetes mellitus, acute alcoholism, thyrotoxicosis, and toxemia of pregnancy.

The diagnostic significance of increased serum amylase activity, its chief clinical application, is considered in greater detail elsewhere (p. 594). Suffice it here to review certain general principles involved in the interpretation of these abnormalities. In common with other proteins, serum amylase undergoes metabolic degradation, but it is removed from the blood stream mainly by excretion in the urine. A range of 1000 to 6500 units may be excreted in 24 hours, with a mean of about 2000 units. In view of the relatively small amounts involved, large increases in serum amylase activity would not be expected to result from decreased excretion alone, although increments of several hundred units per 100 ml. serum have been reported as due to renal insufficiency. However, the state of renal function may influence considerably the extent and duration of increase of serum amylase in conditions of accelerated entrance into the blood stream, e.g., in acute pancreatitis.

Large amounts of pancreatic (and salivary) amylase may gain entrance to the blood stream by one or both of the following mechanisms: (*a*) continuing secretion against obstruction of the duct system (calculus, edema, inflammation, etc.); however, the associated acinar damage in such cases usually does not permit such sustained secretion; (*b*) disruption of acini and ductules; this occurs commonly in acute pancreatic disease, resulting in direct passage of preformed amylase into the blood and lymph and into the peritoneal space, from which it is subsequently absorbed into the circulation. This is the dominant factor in the majority of cases.

The highest values have been obtained in various types of acute pancreatitis and in acute exacerbations of chronic (recurrent) pancreatitis; elevations occur less consistently in other forms of pancreatic disease (p. 594). Moderate increase has been observed occasionally in certain diseases of the salivary glands, including mumps, suppurative parotitis, and calculous obstruction of the salivary duct.

Absorption of large amounts of enzyme from the peritoneum may account for the occasionally high values observed in perforated duodenal ulcer and in ruptured ectopic pregnancy. High values in posterior penetrating ulcers (gastric or duodenal) may be due to pancreatic involvement (inflammation, edema) and/or compression of the pancreatic duct system. Elevations occasionally seen in biliary tract disease (calculous obstruction, cholelithiasis) have been attributed to associated pancreatic disease. Moderate but occasionally high values have been reported in cases of strangulated intestinal obstruction, presumably resulting from direct absorption from the lumen of the devitalized bowel into the blood stream or lymph.

The use of auxiliary measures in conjunction with serum amylase determinations in the diagnosis of pancreatic disease is considered elsewhere (p. 595).

Macroamylasemia. In some individuals, a form of amylase with a high molecular weight occurs in the circulation. It cannot pass the glomerular filter and consequently accumulates in the bloodstream. Macroamylasemia should be suspected when there is an increase in serum amylase and no increase in urinary amylase output. Macroamylase can be formed by combination of ordinary serum amylase with an antibody. It can possibly result from polymerization of the enzyme molecule.

Serum Lipase. The degree of lipolytic activity of the serum may be measured (method of Tietz and Fiereck) as the amount of olive oil in emulsion hydrolyzed by a given amount of serum in three hours at 37°C. The values for lipase may be expressed in terms of the amount of 0.05 M NaOH required for neutralization of the fatty acids produced by 1.0 ml. of serum. The upper limit of normal is about 1.0 ml. of 0.05 M NaOH. This is equivalent to 277 U (International Units) of lipase per liter.

The remarks concerning changes in serum amylase activity in pancreatic disease apply equally, generally, to serum lipase. Although serum amylase determinations are performed much more commonly, serum lipase determinations are preferred by some (p. 595).

Uropepsin. About one per cent of pepsinogen (propepsin) formed in the gastric mucosa is absorbed into the circulation and then excreted by the kidney. The mild acidity of urine converts it, at least in part, to the active enzyme which is now called uropepsin. Study of the amount excreted in the urine has found some application in relation to the investigation of gastric function (p. 587).

SERUM CHOLINESTERASE

The blood plasma contains an enzyme capable of splitting acetylcholine and certain other choline esters (nonspecific cholinesterase). This enzyme, designated Type II cholinesterase, originates largely in the liver, as well as in the pancreas, the heart and the white matter of the brain. Activity is expressed in terms of change in acidity (pH) resulting from liberation of acetic acid from the acetylcholine substrate. Normal serum values range from 0.35 to 1.1 pH units according to some, and 0.7 to 1.4 pH units according to others. This method has been largely replaced by techniques using acetylcholine or acetylthiocholine as substrate and estimating, respectively, the amount of acetylcholine remaining or the amount of thiocholine liberated. Using acetylcholine bromide as substrate, the normal range for plasma is 2.3 to 5.2 U/l at 37° C. (de la Huerga). Benzoylcholine has also been used as a substrate and yields a normal adult range of 1.08 to 2.38 U/l. at 37° C. (Kalow and Lindsay).

In general, reasonably normal values are the rule in uncomplicated extrahepatic obstructive jaundice, and subnormal values are usually obtained in patients with severe hepatocellular jaundice, and also in advanced cirrhosis. However, the serum cholinesterase activity is influenced in this way by conditions other than primary hepatic disease, including fever, inanition, and thiamine deficiency. Determination of this factor is of little value in the differential diagnosis of hepatic and biliary tract disease.

Low values may occur also in patients with acute myocardial infarction, persisting at times for two weeks. Although the mechanism has not been established, it may be a reflection of hepatic functional impairment due to hypoxia. Low levels of serum cholinesterase have also been found in association with pulmonary embolism and muscular dystrophy and during the postoperative period. They also occur as a result of poisoning with organic phosphorus compounds used as insecticides, since these inhibit the activity of cholinesterase.

A condition of congenitally determined deficiency in serum cholinesterase, reflected in low levels in the serum, may be important clinically under certain circumstances. Subjects with this defect exhibit unusual and occasionally fatal sensitivity (respiratory paralysis) to succinyldicholine (suxamethonium), employed by anesthetists to induce muscle relaxation. This hypersensitivity is due to inability to destroy rapidly this highly active choline ester. These subjects possess an abnormal variant of the serum enzyme, which is less effective in hydrolyzing choline esters, and in particular succinyldicholine. It is also more resistant to certain cholinesterase inhibitors than the normal enzyme. Using dibucaine hydrochloride (10 μmol./l.) as an inhibitor, it is possible to evaluate percentage inhibition of the serum activity, which is expressed as the so-called dibucaine number. In general, a value above 71 indicates that the normal enzyme is present, and a value below 30 indicates the atypical variant.

Moderately elevated levels of serum cholinesterase are found in the nephrotic syndrome. Some increase may be found in hyperthyroidism, in anxiety states and in obese diabetics.

AMINOTRANSFERASES (TRANSAMINASES)

Aminotransferases catalyze the transfer of the amino group from an amino acid to a keto acid, one of the important general reactions of protein metabolism; a new amino and keto acid are formed in the process.

The two most important clinically are (a) glutamic oxalacetic transaminase (GOT) (glutamic + oxaloacetic ⇌ ketoglutaric + aspartic), now designated aspartate aminotransferase; (b) glutamic pyruvic transaminase (GPT) (glutamic + pyruvic ⇌ ketoglutaric + alanine), now designated alanine aminotransferase. These enzymes have a wide tissue distribution and are normally present also in the blood serum in concentrations (activity) up to 19 U/l. (Henry modification of Karmen technique at 32° C.).

Increased serum aminotransferase activity occurs in certain diseases involving tissues rich in these enzymes, notably the liver and myocardium. This is due presumably to the liberation of abnormally large amounts from the damaged tissues. However, there is no consistent parallelism between the extent of increase in the serum and the amount of enzyme in the affected tissue, suggesting that other factors may be involved. Higher values are obtained normally during the first few months of life (GOT < 60 U/l.; GPT < 45 U/l.).

Liver and Biliary Tract Disease. Increase in serum aminotransferase activity occurs in inflammatory, degenerative, and neoplastic lesions of the liver. In general, the extent of rise reflects the severity and extent of hepatocellular damage. The highest values have been obtained in acute hepatic necrosis, e.g., carbon tetrachloride poisoning ($< 15,000$ U/l.); striking increases occur also in viral hepatitis (< 2500 U/l.). Lower values (< 250 U/l.) are usually obtained in uncomplicated portal cirrhosis, biliary cirrhosis, and extrahepatic biliary obstruction, in most types of toxic hepatitis (chlorpromazine, salicylates, iproniazid, etc.), and in hepatic malignancy, primary or metastatic. The use of this determination in the study of patients with jaundice is discussed elsewhere in greater detail (p. 575).

Heart Disease. Increase in serum GOT activity may occur, but inconsistently, in acute right- or left-sided heart failure; the values are usually below 100 U/l., but occasionally as high as 1500 U/l. Liver damage due to hypoxia may be a contributory factor. Slight to moderate increases may occur after intracardiac operations. With occasional exceptions, essentially normal values are obtained in rheumatic carditis, pericarditis, and pulmonary infarction.

The form of heart disease in which this determination is of greatest value is acute myocardial infarction. This is characteristically accompanied by significant increase in serum GOT activity; usually there is little or no increase in GPT except in the case of very large infarctions or associated liver damage (congestive heart failure, hypoxia). The rise in serum GOT begins typically six to twelve hours after occlusion of the coronary artery, reaching a maximum at twenty-four to forty-eight hours and returning to normal at about the third to the fifth day, except in the event of extension or fresh infarction. The GPT level may remain relatively unaffected. This fact, together with the usual parallelism between the extent of increase of serum GOT and the size of the infarcted area, suggests that the damaged tissue is the source of the enzyme increment in the blood plasma. The extent and duration of the increase are, therefore, of prognostic significance. The peak of rise is usually below 250 U/l.; values above 150 U/l. suggest a poor prognosis, but lower values, reflecting a smaller lesion, may also be serious, depending upon its location. The smaller increases are found most commonly.

The elevation of GOT activity reflects only the extent of acute myocardial damage and is not consistently related to the site of infarction, the state of myocardial function or ECG changes. This determination is of particular diagnostic significance in the following circumstances:

(a) Differentiation between acute myocardial infarction and coronary insufficiency (myocardial ischemia); no increase usually occurs in the latter condition.

(b) Diagnosis of acute myocardial infarction when the ECG changes are not definitive or are difficult to interpret because of previous infarction, bundle branch block, digitalis effects, Wolf-Parkinson-White syndrome, etc.

(c) Diagnosis of extension of the original infarction or of recurring acute infarction.

Miscellaneous. Increase in serum GOT may occur as a result of muscle trauma (accidental or surgical) and disease (muscular dystrophies, myositis, gangrene). The values obtained (usually below 250 U/l.) in general reflect the extent of involvement. An increase occurs occasionally in patients with acute pancreatitis, leukemia, and toxemia of pregnancy; liver damage may be a contributory factor in such cases. Increases in serum GOT also occur in patients with acute hemolytic anemia, in acute renal necrosis, after severe burns, as a result of cardiac catheterization and in patients with recent brain injury. In normal individuals, an increase can result from prolonged severe exercise. Little or no increase occurs, as a rule, in muscular diseases of nervous origin. A number of drugs can cause increases in serum GOT. Some, such as erythromycin, interfere with the analytical method which makes use of hydrazone formation. Some drugs produce the effect by hepatocellular damage, others by cholestasis.

Isoenzymes. Aspartate aminotransferase (GOT) exists in at least two isoenzyme forms, which can be demonstrated by electrophoresis. One of these is derived from the soluble cytoplasm, and the other from mitochondria. There is evidence that when the latter appears in serum, cell death has occurred. This indicates a severe lesion and affects the prognosis.

ENZYMES OF CARBOHYDRATE METABOLISM

Certain intracellular enzymes of intermediary glucose metabolism appear in the blood plasma in measurable amounts. A few of these exhibit significant increase in disease that may be of diagnostic significance. These include: (a) the glycolytic enzymes, phosphohexose isomerase, aldolase, lactate dehydrogenase; (*b*) the Krebs cycle enzymes, isocitrate dehydrogenase and malate dehydrogenase. Of these, perhaps lactate dehydrogenase has been studied most extensively.

Lactate Dehydrogenase. This enzyme catalyzes the interconversion of lactate and pyruvate. It has a wide tissue distribution and is particularly abundant in kidney, skeletal muscle, liver and myocardium. Lactate dehydrogenase (LD) activity in both serum and tissues is due to a heterogeneous group of components (the isoenzymes LD_1, LD_2, LD_3, LD_4 and LD_5) and not to a single homogeneous substance, as shown in Figure 17–1. Numerical values for normal serum concentrations vary considerably, depending upon the analytical procedure employed. Using the method of Bergmeyer at 25° C., the upper limit for total serum LDH has been found, in the author's laboratory, to be 220 U/l.

Myocardial Infarction. The pattern of change in serum LD in myocardial infarction is essentially the same as that of serum GOT (p. 553), but differs in time relationships. There is a rise, commencing about 12 hours after infarction, to a maximum at 48 to 72 hours, with a steady fall subsequently to normal at 7 to 10 days, sometimes a little longer. There is usually little or no increase in the majority of conditions that must be considered in the differential diagnosis of coronary occlusion, e.g., coronary insufficiency (myocardial ischemia), pleurisy, pericarditis, acute cholecystitis, etc. The statements made concerning the diagnostic value of serum aminotransferase determinations in acute myocardial infarction apply equally to serum LD determinations.

Lactate dehydrogenase isoenzyme studies are employed in relation to the diagnosis of myocardial infarction. The molecule of lactate dehydrogenase is a polypeptide tetramer made up from two different polypeptide monomers, designated H (heart) and M (muscle). There are thus five possible tetramers H_4, H_3M, H_2M_2, HM_3 and M_4. These correspond to LD_1, LD_2, LD_3, LD_4, and LD_5 respectively. Under the defined conditions of electrophoresis at pH 8.6, LD_1 travels fastest, followed by LD_2 and so on to

Figure 17-1. Five isoenzymes of lactate dehydrogenase (embryonic human lung).

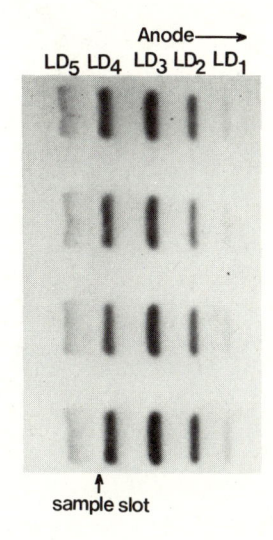

LD$_5$, which is the slowest-moving form. LD$_1$ and LD$_2$ predominate in heart muscle, whereas LD$_5$ is most marked in liver (or muscle), as is shown in Figure 17–2, which shows the patterns obtained when extracts of human heart muscle or human liver are mixed with normal human serum and subjected to starch gel electrophoresis followed by appropriate staining for enzyme activity. After a myocardial infarction has occurred, the heart muscle type of pattern appears in the patient's serum within 6 to 12 hours and persists for up to three weeks (Fig. 17-3). In other words, it is present when the total serum lactate dehydrogenase level has returned to normal levels.

Unlike LD$_5$, LD$_1$ is relatively heat-stable, is active in the presence of 2 moles/liter of urea and has well-marked α-hydroxybutyrate dehydrogenase activity. Increased serum LD$_1$ in myocardial infarction can therefore be

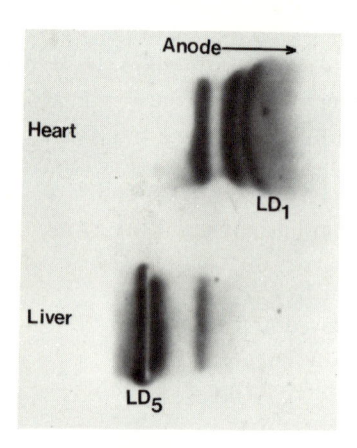

Figure 17-2. Isoenzyme patterns of lactate dehydrogenase from extracts of human heart muscle and of human liver, both of which have been mixed with human serum.

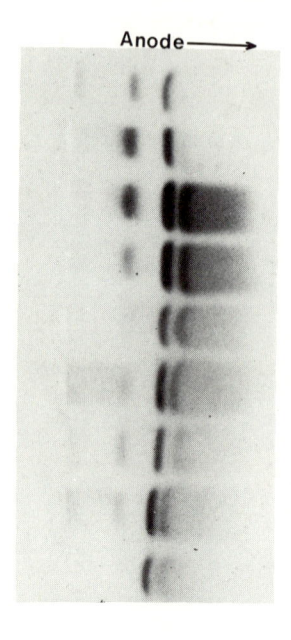

Figure 17-3. Isoenzyme patterns in serum after myocardial infarction. The top pattern is normal serum. The one below is the pattern on admission, which is followed by that obtained 12 hours after infarction. The lowest pattern was obtained three weeks after admission and abnormality could still be detected. Note resemblance of the serum patterns to that shown for heart muscle in Figure 17-2.

demonstrated by measuring heat-stable serum LD or urea-stable serum LD or by estimating the serum hydroxybutyrate dehydrogenase activity.

Hepatic Disease. In contrast to the sharp increase in serum aminotransferase activity, patients with acute hepatitis exhibit usually a smaller relative increase in lactate dehydrogenase activity. Inasmuch as this enzyme is present in liver in amounts comparable to those in myocardium, factors other than tissue content and tissue damage must be involved in determining the pattern of change in activity of certain serum enzymes. Nevertheless, the elevation of lactate dehydrogenase in patients with acute hepatitis persists longer than that of aminotransferase. LD_5 is present in the serum (Fig. 17-4).

Malignancy. An increase in serum LD activity occurs in about two-thirds of patients with leukemia and those with solid tumors, particularly widespread or rapidly growing. In experimental animals, an increase is often demonstrable within the first 24 hours after transplantation of a tumor, and in the absence of demonstrable tissue necrosis; this suggests an increased rate of release from cells not necessarily due to degeneration, inflammation, or trauma. It may be a reflection of the alteration in metabolic activity of neoplastic cells.

Miscellaneous. Increase in serum LD activity may occur in cases of extensive skeletal muscle trauma (including surgical), acute pancreatitis, pulmonary infarction, severe intravascular hemolysis (LD derived from erythrocytes), megaloblastic anemias (LD derived from bone marrow, mostly LD_1), renal infarction, shock and cerebral infarction.

Malate Dehydrogenase. This enzyme catalyzes the interconversion of malate and oxaloacetate (tricarboxylic acid cycle). Changes in this enzyme in the serum generally parallel those in lactate dehydrogenase in liver and some other diseases.

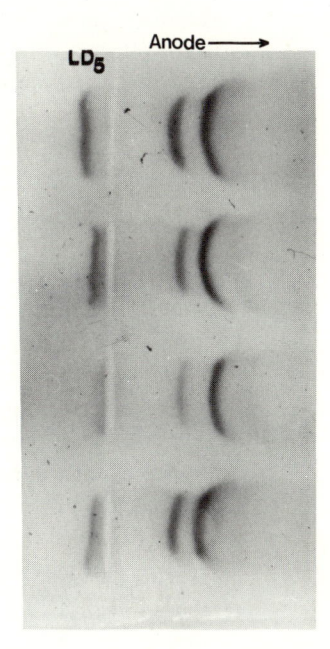

Figure 17-4. Serum lactate dehydrogenase isoenzymes in hepatitis. Note the occurrence of LD_5.

Isocitrate Dehydrogenase (ICD). This enzyme catalyzes the interconversion of isocitrate and oxalosuccinate (tricarboxylic acid cycle).

Hepatic Disease. This enzyme differs from most of the others considered here in that an increase in ICD activity in the serum seldom occurs in the absence of liver disease. (An exception is megaloblastic anemia.) This is perhaps attributable to its very rapid clearance from the blood stream; the production of demonstrable increase in the serum, therefore, requires a rather protracted period of accelerated liberation from damaged tissue. High values are obtained rather consistently in viral hepatitis, occasionally in metastatic liver cancer, but seldom in portal cirrhosis and extrahepatic obstructive jaundice.

The increase in ICD activity and the relatively smaller increase in LD activity in acute hepatitis, despite the fact that liver is rich in both enzymes, add support to the view, expressed above, that factors other than tissue enzyme content and tissue damage may influence the pattern of change in serum enzyme activity.

Myocardial Infarction. Although a transient slight increase may occur during the first twenty-four hours following a major coronary occlusion, normal values are usually obtained in patients with acute myocardial infarction. As indicated above, this is probably due to its rapid removal from the circulation. It must be pointed out, however, that heart ICD is heat-labile at 56° C., whereas liver ICD is stable. It can be shown by electrophoresis that heart ICD and liver ICD are two different forms, which are regarded as isoenzymes.

Aldolase. This enzyme catalyzes the interconversion of fructose-1-6-diphosphate and the triose phosphates (glycolysis). It has a widespread tissue distribution and is abundant also in neoplastic tissue. It exists in multiple forms, of which there are three classes: A aldolases (muscle), B aldolases (liver) and C aldolases (brain). The ratio of activity with fructose 1-phosphate to that with fructose 1,6-diphosphate is high with B aldolase, low with A aldolase and intermediate with the C form. By electrophoresis, it can be shown that each of these forms is made up of isoenzymes. With starch gel electrophoresis, the human brain enzyme gives four bands migrating toward the anode; the heart enzyme gives three bands, the two fastest corresponding in position with the two slowest brain bands. Kidney, liver and muscle each yield two bands corresponding in position with the two slowest bands of the heart enzyme. With kidney and liver, the slowest band shows most of the activity. With muscle, however, the two bands show approximately equal activity. An aldolase exists in two forms. Aα and Aβ. In *hereditary fructose intolerance,* an abnormal B aldolase is produced.

Hepatic Disease. A usually marked increase in serum aldolase activity occurs early in viral hepatitis and hepatic necrosis, and elevated values may be obtained in malignancy of the liver. There is usually no significant change in portal cirrhosis or obstructive jaundice.

Myocardial Infarction. An increase in serum aldolase occurs occasionally in patients with acute myocardial infarction, but the changes are not so striking nor so consistent as those in GOT or LD activity.

Skeletal Muscle Disease. Serum aldolase activity may be significantly increased in progressive muscular dystrophy, even at an early stage, when diagnosis may be difficult. This determination is particularly valuable because normal values are the rule in muscle diseases of nervous origin (progressive spinal atrophy, poliomyelitis, etc.). High values may be obtained also in myositis and severe muscle trauma.

Malignancy. Serum aldolase may be increased in patients with extensive malignancy, particularly in the presence of hepatic or skeletal metastases. In prostatic cancer, changes in this enzyme, as in serum acid phosphatase, may reflect extension or regression of metastatic lesions in the skeleton.

Miscellaneous. Elevations of serum aldolase may occur in acute pancreatitis and in massive pulmonary infarction. These elevations may be a result of hemolysis. Serum levels are high in neonates.

Glucose Phosphate Isomerase. This enzyme catalyzes the interconversion of glucose-6-phosphate and fructose-6-phosphate (glycolysis). It has a wide tissue distribution and occurs abundantly also in neoplastic cells.

Changes in the activity of this enzyme in the serum usually parallel those in serum aldolase. High values may result from active intravascular hemolysis, as a result of liberation from erythrocytes.

Hexokinase. A deficiency of this enzyme in erythrocytes is associated with hemolytic anemia (p. 573). The hexokinase activity of circulating leukocytes is decreased in chronic myelocytic and lymphocytic leukemias.

Glucose-6-phosphatase. This enzyme catalyzes the conversion of D-glucose-6-phosphate to glucose. It occurs mainly in the liver, and little or none is present in normal serum. Glucose-6-phosphatase appears in large amounts in serum of patients with severe hepatocellular disease, especially after ingestion of liver toxins, such as chloroform and carbon tetrachloride. Lower levels of the enzyme are found in patients with cirrhosis. Raised levels can occur in those with kidney disease, presumably because of decreased clearance. Hemolyzed blood must not be used for assay.

PHOSPHATASES

Phosphatases are enzymes capable of hydrolyzing phosphoric esters, both aliphatic and aromatic, including a portion of the phosphoric esters of the circulating red blood cells and those present, in small amounts, in blood plasma. These agents, which hydrolyze phosphoric esters with the liberation of inorganic phosphate, have been identified in practically all tissues of a variety of animals. In the fetus and growing animal the greatest relative quantity of phosphatase is found in the bones and teeth. In the adult animal, the intestinal mucosa contains the greatest amount per unit of weight (wet issue).

Three types of phosphomonoesterases of clinical significance may be differentiated on the basis of their activity in different pH ranges: (1) a type (alkaline phosphatase) with optimum activity at about pH 9.3, present in blood plasma or serum, bone, liver, kidney, intestine, mammary gland, spleen, lung, leukocytes, adrenal cortex and seminiferous tubules; (2) a type (acid phosphatase) with optimal activity at pH 6, occurring in mammalian erythrocytes; (3) a type (acid phosphatase) with optimum activity at pH 5, occurring in prostatic epithelium, spleen, kidney, blood plasma, liver, and pancreas.

From a clinical standpoint, interest is centered particularly upon the phosphatase activity of the blood plasma or serum, i.e., alkaline phosphatase and acid phosphatase. Leukocyte alkaline phosphatase is also of clinical importance.

ALKALINE PHOSPHATASE

According to the method of Bodansky, a unit of phosphatase activity (equivalent to approximately 5.5 U) is defined as "equivalent to the actual or calculated liberation of 1 mg. of phosphorus as the phosphate ion during the first hour of incubation at 37° C. and pH 8.6, with the substrate containing sodium beta-glycerophosphate, hydrolysis not exceeding 10 per cent of the substrate." Employing this procedure, the range of normal values for plasma or serum alkaline phosphatase activity in adults is 1.5 to 4.0 units per 100 ml. (8.0–22 U/l.), that for children being 5 to 14 units (27–86 U/l.). In our experience, values in adults of less than 7 units are of little clinical significance in hospital patients, in the absence of other significant abnormalities. The alkaline phosphatase activity of plasma is within the adult range at birth, and rises rapidly to a maximum during the first month of life (4–14 units), the range remaining fairly constant until the third month. The upper limit now falls very slowly until it is about two-thirds of the previous maximum, at about 10 years of age. With the onset of puberty, there is a great increase in bone growth. Consequently, within the ages of 11 to 16 years, the maximum may again rise to 14 Bodansky units. In other words, there is a peak value and then the range falls to that of the adult at about 16 years of age, depending on the age of onset of puberty. Using the method of Kind and King, an investigation of alkaline phosphatase levels in the sera of school children in London and Hertfordshire has indicated that the "puberty" peak value showed a mean of 9.1 King-Armstrong units at the age of 13 years in boys. The corresponding value for girls was 17.9 King-Armstrong units at the age of 12 years. Normal values obtained by the King-Armstrong method range from 3.5 to 13 units or 25 to 92 U/l. (adults).

There is some evidence that senility, malnutrition and anemia tend to lower the serum phosphatase activity; this should be taken into consideration in interpreting variations in this factor in clinical conditions. Some observers have found that high protein diets cause a decrease and high carbohydrate diets an increase in alkaline phosphatase activity of the serum; an increase occurs during alimentary hyperglycemia.

The alkaline phosphatase of normal plasma (or serum) originates mainly in the liver. Some can also be intestinal in origin. In growing children, where it is formed in large amounts by actively functioning osteoblasts, the main source is bone. Its possible role in bone mineralization is discussed elsewhere (p. 283). Despite the present state of uncertainty regarding the specific function of this enzyme in bone formation, measurement of serum alkaline phosphatase activity is of value clinically mainly in the study of patients with diseases of two systems, viz., (1) the skeleton, and (2) the liver and biliary tract.

Isoenzymes. By starch gel electrophoresis at pH 8.6, it can be easily demonstrated that different tissues contain alkaline phosphatases, which migrate at different rates. In the author's laboratory, they have been found to take up the positions shown diagrammatically in Figure 17–5. The liver isoenzyme travels fastest toward the anode and occupies the same position as the fast α_2-globulin. It is closely followed by the bone isoenzyme β-globulin region, and there is clear separation from the intestinal isoenzyme (approximately slow α_2-globulin region). It must be emphasized however, that in our hands there is clear differentiation between the liver and bone isoenzyme positions and that this technique is in daily use in my laboratory for routine diagnostic purposes. Normal sera always show the main liver band. A high proportion (ABO secretors) also shows the intestinal band.

Nearly all tissues show a subsidiary band near the point of insertion. This approximates in position to the serum β-lipoprotein of starch gel electrophoresis. Liver shows a second subsidiary band (derived from the biliary system) between the β-lipoprotein and the slow α_2-globulin. The two subsidiary bands form a doublet, which is of some diagnostic significance, since it is frequently present in sera of patients with jaundice resulting from extrahepatic biliary obstruction.

Figure 17-5. Diagrammatic representation of isoenzymes of alkaline phosphatase from different sources.

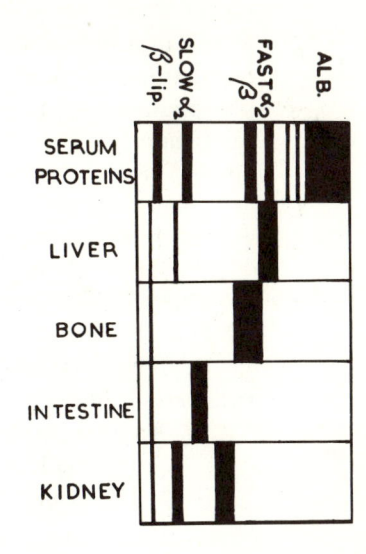

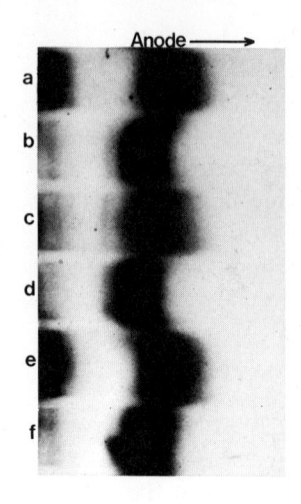

Figure 17-6. Serum alkaline phosphatase patterns in liver and bone disease. *a, c and e:* liver disease (*a* and *e:* biliary obstruction; *c:* hepatocellular disease); *b, d* and *f:* bone disease (parathyroid osteitis, Paget's disease, osteomalacia).

Kidney also shows a somewhat more prominent subsidiary band. Isoenzymes derived from kidney can be found in the urine of patients with certain kidney disorders. The placental pattern is somewhat variable and is genetically determined. It is prominent in normal sera during the last trimester of pregnancy.

The placental isoenzyme, unlike the others, is heat-stable (70° C. for 30 min.). This property can be used for its detection and estimation independently of electrophoresis. Intestinal and placental alkaline phosphatases are inhibited by L-phenylalanine. Urea, at concentrations of 2 mmol./l. or greater, inhibits alkaline phosphatases. The placental and intestinal isoenzymes are more resistant than the others. Placental alkaline phosphatase, unlike the others, resists inactivation by EDTA.

Skeletal Diseases

In this and the subsequent section, activities are expressed in Bodansky units. They can easily be converted to International Units by multiplying by the appropriate factor (approximately 5.5). It must, however, be pointed out that many laboratories now use other techniques for estimating alkaline phosphatase, and the ranges vary with each method. The qualitative changes in disease are the same with all reliable methods and, apart from the figures quoted, the following account applies regardless of methodology.

The clinical significance of serum alkaline phosphatase activity in skeletal disorders may be indicated by the following generalizations:

1. The bone enzyme originates mainly in functioning osteoblasts.
2. An increase (serum) reflects osteoblastic proliferation or increased activity.
3. The usual stimulus to osteoblastic proliferation is bone stress, strain, weakness, or injury.
4. An increase in bone alkaline phosphatase activity usually implies:
    (*a*) Inadequate mineralization, or
    (*b*) injury, with

(c) attempts at active repair, which implies

(d) adequate bone matrix.

5. Therefore, an increase occurs in osteomalacia, not in osteoporosis (p. 313).

Rickets. This determination is one of the most important aids in the differential diagnosis of skeletal disorders, particularly those accompanied by inadequate mineralization.

The serum alkaline phosphatase activity is consistently considerably increased in active rickets. Normally 5 to 14 units per 100 ml. in children (Bodansky method), it may be 20 to 30 units in mild rickets, as high as 60 units in marked rickets and over 60 in very marked rickets. Values as high as 190 units have been obtained in the last group. These figures may be regarded as reliable criteria of the severity of the condition at the time of first observation. Following the institution of antirachitic therapy the serum phosphatase activity decreases, usually after an interval of four to twelve days. A high normal figure may be reached within two months and is frequently maintained for some time during the period of active repair. When bone reconstruction is complete the serum phosphatase is within normal limits. It has been found that the serum phosphatase activity may remain above normal even after healing is apparently complete roentgenographically. There is usually a reciprocal relationship between serum phosphatase activity and inorganic phosphate concentration in rickets. Determination of serum phosphatase activity may be regarded as a reliable means of detecting latent and active rickets and may be accepted as an index of improvement in this condition.

Hyperparathyroidism (Generalized Osteitis Fibrosa Cystica). A moderate increase in serum phosphatase activity can occur in clinical and experimental hyperparathyroidism but is not invariably present (p. 295). Values as high as 65 units have been reported, but the values usually range between 20 and 40 units. A gradual fall to normal levels has been observed following parathyroidectomy in such cases, associated with increasing mineralization of the affected bones.

Osteitis Deformans (Paget's Disease). Increased serum alkaline phosphatase activity occurs rather consistently in patients with osteitis deformans involving several bones. Higher values have been obtained in this condition than in any other clinical disorder. Bodansky and Jaffe found values of 15 to 125 units in ten cases of the polyostotic variety and 4.9 to 23.1 units in thirteen cases with localized involvement of one or two bones. Spontaneous healing by sclerosis was associated with relatively low values for serum phosphatase activity. Age appears to be a factor, the oldest patients showing the lowest values. The finding of moderately elevated serum phosphatase activity, together with normal values for serum calcium and inorganic phosphorus, is of diagnostic significance in this condition. The degree of increase is also of prognostic value, reflecting the activity of the bone lesion, which can otherwise be gauged only by serial x-ray studies.

Hypophosphatasia. This is a genetically determined disorder in which alkaline phosphatase is usually greatly diminished or absent in the serum, leukocytes, and tissues, including bone. The fundamental skeletal defects resemble those of rickets: (1) deficient mineralization; (2) excess of osteoid tissues; (3) irregularity of endochondral ossification. The clinical manifestations include genu valgum and varum, prominent sternum, Harrison's groove, costochondral enlargements (rachitic rosary), large wrists, protuberant abdomen, premature loss of deciduous teeth, and premature synostosis in some parts of the cranium and widening of sutures in others. In addition, the ribs and long bones are unusually short and the cranial vault extremely soft, although the base of the skull may be well calcified.

The severity of the condition varies considerably. Infants are often stillborn, or death may occur shortly after birth or in early childhood. In milder cases, survival into adult life may be accompanied by skeletal deformities, increased fragility of the bones, with frequent fractures, or x-ray evidence of multiple symmetrical "pseudofractures." There may be hypercalcemia (12–14 mg./100 ml.); the serum inorganic phosphate is usually normal, occasionally increased. These findings, together with the decreased alkaline phosphatase activity in the serum, also demonstrable in the granulocytes, readily permit differentiation from rickets. Another characteristic biochemical abnormality is the constant excretion of phosphoethanolamine in the urine; this substance is perhaps one of the natural substrates for alkaline phosphatase, accumulating in excess as a result of deficiency in this enzyme. Although phosphoethanolamine occurs rarely in normal urine, it may be excreted by patients with hypothyroidism, celiac disease, and scurvy. Similar biochemical abnormalities have been reported in clinically well relatives of patients with hypophosphatasia.

Miscellaneous Bone Disorders. Slightly elevated values for serum alkaline phosphatase activity (5 to 15 units) can occur in patients with various forms of osteomalacia (p. 314), metastatic carcinoma involving bone, osteogenic sarcoma, healing fractures, Gaucher's disease with bone resorption, and osteosclerosis fragilis generalisata (marble bones). In the case of bone tumors, either primary or metastatic, increases occur principally in those that are primarily osteoblastic in nature, e.g., certain types of osteogenic sarcoma and prostatic carcinoma. In those in which the lesions are predominantly osteolytic, the level of serum alkaline phosphatase depends upon the extent of osteoblastic proliferation, i.e., attempt at repair, in the adjacent osteoid tissue. A favorable response to therapy may be associated with an initial rise, a reflection of the reparative process that follows destruction of the lesion.

Essentially normal findings are obtained in most cases of chronic arthritis, various forms of osteoporosis (p. 313), osteomyelitis, bone cysts, tumors not involving bone, achondroplasia, cretinism, calcinosis universalis, and multiple myeloma. High values are encountered occasionally in cases of carcinoma with extensive and widespread metastases involving the skeleton. Slight increases have been reported also in occasional cases of renal rickets and multiple myeloma. Decreased serum alkaline phosphatase activity has been reported in idiopathic infantile hypercalcemia, hypothyroid children, and scurvy. Elevated values have been reported during periods of calcification of hemorrhages in scurvy, with subsequent return to normal as the calcified areas are absorbed.

Hepatic and Biliary Tract Disease

Increased serum phosphatase activity occurs in a large proportion of patients with both obstructive and, less frequently, hepatocellular jaundice (p. 575). Values as high as 60 units (Bodansky) have been observed in such cases. Phosphatase activity tends to parallel the degree of hyperbilirubinemia in extrahepatic (biliary system) obstructive jaundice more closely than in hepatocellular jaundice, although this finding is by no means constant. However, although phosphatase activity values above 20 units are obtained about twice as frequently in patients with extrahepatic obstructive jaundice as in those with hepatocellular jaundice, the value of this factor in differential diagnosis is limited by overlapping of values in the two groups of cases. The presence and extent of elevation is probably determined by the relative degree of hepatocellular damage and of obstruction, intrahepatic or extrahepatic. High values, comparable to those seen in common duct obstruction, may occur in certain types of toxic hepatitis (e.g., resulting from chlorpromazine and other drugs causing cholestasis), in which there is minimal hepatocellular damage and extensive intracanalicular obstruction. Conversely, relatively low values may be obtained in cases of common duct obstruction with superimposed severe hepatocellular damage.

In addition to the nonspecific alkaline phosphatases, the serum contains one with a higher degree of substrate specificity, a 5'-nucleotidase, acting upon such substrates as adenosine-5-phosphate. The liver seems to be the source of the serum enzyme and increases in the level in the serum have been observed mainly in diseases of the liver and biliary tract. The normal range for adults and children is 2 to 15 U/l. (method of Campbell). The general pattern of increase in cases of obstructive and hepatocellular jaundice resembles that in nonspecific alkaline phosphatase. This determination may be of diagnostic value in doubtful cases and is considered to be useful in differentiating hepatocellular from extrahepatic obstructive jaundice, in which the levels tend to be higher. It has the added advantage of being unaffected by osteoblastic activity, which influences the level of nonspecific alkaline phosphatase activity in the serum. This subject is discussed elsewhere in more detail (p. 283 ff.).

Patients with portal cirrhosis may present high values for serum phosphatase activity with little or no increase in serum bilirubin concentration; others present normal phosphatase values with relatively high concentrations of serum bilirubin. Increased values may also be obtained in patients with metastatic carcinoma of the liver, the increase occurring in many instances without demonstrable elevation of the serum bilirubin concentration. In the presence of known intra-abdominal malignancy, e.g., of the stomach, if skeletal metastases can be excluded, high serum alkaline phosphatase values suggest metastasis to the liver. An increase occurs frequently also in other more or less localized lesions of the liver producing obstruction to the bile drainage from relatively small areas of the liver. These include such conditions as solitary abscess, echinococcus cyst, obstruction of single hepatic ducts (stone, stricture, lymph nodes), and granulomatous lesions (sarcoidosis, Hodgkin's disease). Increase in alkaline phosphatase activity in chronic leukemias and lymphomatous diseases is probably a reflection of hepatic or skeletal involvement.

Miscellaneous Disorders

Serum levels are usually raised in patients with intestinal malabsorption. They may be increased in those with peptic ulcer as well as those with ulcerative colitis. Increased levels may also be found in those with infarction of the kidney, lung or spleen. The intravenous administration of placental albumin can cause a rise in serum levels. A number of malignant neoplasms (bronchogenic carcinoma, etc.) have been associated with the ectopic production of a heat-stable alkaline phosphatase (Regan isoenzyme), which is identical with placental alkaline phosphatase.

Serum levels may be lowered in the presence of hypothyroidism, cachexia, magnesium deficiency and pernicious anemia.

The determination of leukocyte alkaline phosphatase is of value in certain hematological disorders, such as the acute leukemias. The alkaline phosphatase activity of circulating granulocytes is increased in individuals with suppurative diseases and myeloid metaplasia and is decreased or virtually absent in those with chronic myelocytic leukemia. It is also diminished in those with acute monocytic leukemia, acute myelocytic leukemia, infectious mononucleosis and paroxysmal nocturnal hemoglobinuria. It can also be diminished in the presence of nonhematological conditions, e.g., sarcoidosis and hypophosphatasia.

ACID PHOSPHATASE

Normal plasma contains small amounts of acid phosphatase (pH 5.0). With the Bodansky procedure, the normal range is 0.1 to 0.4 unit/100 ml. With the modified King-Armstrong procedure (Gutman and Gutman), normal values range from 0.6 to 3.1 King-Armstrong units/100 ml. or 1.1 to 5.5 U/l. at 37° C. With the Bessey, Lowry and Brock method, the range is 0.15 to 0.70 unit/ml. or 2.5 to 12.0 U/l. at 37° C. (slightly lower for females). With the method of Shinowara, Jones and Reinhart, the range is 0.0 to 1.5 units/100 ml. or 0.0 to 6.0 U/l. at 37° C. Because of the presence of acid phosphatase in red blood cells, care must be taken to avoid hemolysis in samples in which this determination is to be made. The acid phosphatase of normal plasma probably originates mainly in erythrocytes and platelets. It is, however, present, largely in the lysosomes, in many tissues and in greatest concentration in the prostate gland. These probably make some contribution to plasma. Liberation from platelets makes serum acid phosphatase levels higher than those of plasma. It is certainly not largely of prostatic origin, since it is present in the plasma of women and children almost in the same amounts as in normal men.

The clinical significance of serum acid phosphatase lies mainly in its relation to the diagnosis of metastasizing prostatic carcinoma. Acid phosphatase activity in the adult prostate is at least 100 times that in other tissues. It has been shown that this enzyme is formed by mature prostatic epithelial cells and is passed into the prostatic secretion and ultimately into the seminal fluid. Immature prostatic cells (i.e., preadolescent boy or castrate) do not form this enzyme, but do so under stimulation by androgens.

In a large proportion of cases of prostatic cancer the malignant cells retain the capacity to form large amounts of this enzyme, even after they have metastasized to other tissues, e.g., periprostatic structures, bones. In these extraprostatic locations, the enzyme gains access to the blood stream and the serum acid phosphatase activity is increased. Inasmuch as virtually no other tissue is capable of forming sufficient amounts of this enzyme to raise its concentration in the blood plasma to a high level, such increases may be regarded as presumptive, and at times conclusive, evidence of the presence of prostatic cells outside the prostatic capsule, i.e., carcinomatous metastases. Although, theoretically, similar elevations may occur as a result of ulcerative lesions or surgical procedures involving the prostate, they are practically never observed under such circumstances.

Acid phosphatase of the prostate (also liver and spleen) is inhibited by L-tartrate, whereas that of other tissues is not. Measurements of acid phosphatase activity before and after tartrate inhibition yield values for specific "prostatic" phosphatase in normal subjects ranging from 0.0 to 0.5 King-Armstrong unit/100 ml. serum (0.0–0.9 U/l. at 37° C.). There has been some difference of opinion as to whether this determination is generally more useful than that of total serum acid phosphatase activity. It would appear to be of distinct advantage at least in certain cases. The following statements seem warranted on the basis of the majority of reported observations: (*a*) total serum acid phosphatase values above 5 Bodansky units or 10 King-Armstrong units are obtained in about 85 per cent of cases of metastasizing carcinoma of the prostate and normal values in about 95 per cent of cases of nonprostatic disease; (*b*) values of 3 to 5 Bodansky or 5 to 10 King-Armstrong units may be regarded as indicative of metastatic prostatic cancer in the absence of substantial evidence to the contrary; (*c*) values of 1.5 to 3 Bodansky or 3 to 5 King-Armstrong units may be of diagnostic value under certain circumstances, e.g., they suggest metastasis in cases of known prostatic cancer.

According to some observers, clinically significant increases of total serum acid phosphatase occur in only about 40 per cent of cases of metastatic prostatic cancer, whereas abnormally high values for "prostatic" acid phosphatase are obtained in 90 to 95 per cent of such cases. Observations in cases under treatment indicate that changes in values for total serum acid phosphatase activity are due virtually exclusively to changes in "prostatic" phosphatase, and that increase in the latter may be demonstrable before the former has risen to abnormally high levels. This determination may therefore be useful in doubtful cases. It is claimed that increase in serum "prostatic" phosphatase occurs in up to 80 per cent of cases of prostatic cancer without demonstrable metastasis; however, some believe that abnormal values almost invariably point to metastasis, even though this may not be demonstrable. Slight increase in the specific enzyme has been reported in up to 30 per cent of cases of benign prostatic hypertrophy, and a transient rise (1 to 1.5 units, i.e., 1.8 to 2.7 U/l. at 37° C.) may occur following massage or palpation of a normal prostate. Spuriously low values may be obtained in the presence of fever.

In some cases of highly anaplastic carcinoma (i.e., very immature cells), especially in older men (i.e., little androgen production), the serum acid phosphatase may be normal in the presence of metastases. Under such circumstances, the injection of 25 mg. of testosterone propionate daily for five days may stimulate production of the enzyme by these cells (Sullivan test). Significant increase in serum acid phosphatase after such treatment may be of diagnostic value.

In addition to its usefulness in diagnosis, this determination may be of value in connection with treatment of prostatic cancer. In the majority of cases, castration is followed promptly (24 to 48 hours) by a sharp drop (av. 55 per cent) in a previously elevated serum acid phosphatase level, the decline continuing for several months, falling in favorable cases to essentially normal values. Similar but less dramatic results are obtained with estrogen therapy. This type of response reflects a satisfactory therapeutic effect, and is of prognostic value. Recurrence or progression of the lesion, after a period of remission, is often reflected in a rise in serum acid phosphatase before other clinical manifestations appear.

The significance of an increase in serum acid phosphatase may be summarized as follows:

1. It corroborates the diagnosis of metastatic prostatic cancer, suggested by other methods of examination.

2. It may provide evidence of metastasis before this is demonstrable otherwise.

3. It indicates the prostatic origin of metastatic lesions when the primary tumor may not be detectable clinically.

4. It detects recurrence or progression after a period of quiescence.

5. It aids in the selection of cases suitable for castration or estrogen therapy.

There are other possible causes for an increase in serum acid phosphatase. There is an occasional rise in individuals with Paget's disease of bone, osteopetrosis, hyperparathyroidism, and osteolytic metastases from breast and other carcinomas. There is a marked rise in association with thrombocytosis (chronic granulocytic leukemia, myeloproliferative disorders, etc.), but this does not appear in platelet-poor plasma from these patients. There is also evidence that granulocytes may contribute to the high serum level in myeloproliferative disease. No rise in serum acid phosphatase occurs in cases of lymphocytic leukemia or lymphoma, although an increased level has been reported in one case of lymphoblastic leukemia. There is a rise in cases of hemolytic anemia. Small elevations of serum acid phosphatase occur in association with thromboembolic disorders, e.g., pulmonary embolism, as well as myocardial infarction. Plasma acid phosphatase is raised in patients with thrombocytopenia, in whom there are normal or increased numbers of megakaryocytes in the bone marrow. It is low if there is an associated deficiency of bone marrow megakaryocytes.

With phenyl phosphate as substrate, virtually all patients with Gaucher's disease show elevated serum levels. Occasional elevations occur in those with Niemann-Pick disease.

Serum acid phosphatase is raised in patients undergoing cardiopulmonary bypass for open-heart surgery. This could be due to hemolysis, but may be related to the associated hemorrhagic enteropathy.

Leukocyte Acid Phosphatase. The activity in leukocytes is decreased in cases of acute granulocytic and lymphatic leukemias. In these conditions, the acid phosphatase isoenzyme distribution in leukocytes differs greatly from the normal distribution.

Lysosomal Acid Phosphatase Deficiency. This occurs as a familial metabolic disorder. The deficiency can most easily be demonstrated in cultured fibroblasts from skin biopsies. In the hereditary disorder characterized by multiple lysosomal enzyme deficiency, it is interesting to note that lysosomal acid phosphatase levels are normal.

CREATINE KINASE

This enzyme catalyzes the formation of phosphocreatine from ATP (adenosine triphosphate) and creatine. It is frequently erroneously referred to as creatine phosphokinase (CPK), since all kinases are phosphotransferases. The enzyme is found in greatest amount in skeletal muscle but is present in quite appreciable amounts in cardiac muscle and in brain. Small amounts are present in the lung, thyroid and adrenals. The enzyme is apparently not present in liver and kidney and is certainly not detectable in erythrocytes; consequently, serum levels are not affected by hemolysis. By the method of Rosalki, the normal range at 37° C. is 12 to 99 U/l. for males and 10 to 60 U/l. for females. These correspond to 7 to 54 U and 6 to 40 U, respectively, when converted to activity at 30° C. The serum enzyme is unstable, and estimation should be carried out soon after the sample is collected. There may be a 50 per cent loss after 6 hours at room temperature or 24 hours at 4° C. The enzyme can be stabilized for some days at 4° C. by the addition of cysteine to the serum.

The enzyme exists in isoenzyme forms. On agar-gel electrophoresis at pH 9.0, brain extract gives a single band migrating rapidly towards the anode, skeletal muscle shows a single band with the mobility of a γ-globulin, and cardiac muscle shows this band, as well as one or more intermediate forms. It has been demonstrated that the molecule of creatine kinase is a polypeptide dimer.

Skeletal Muscle Disease. There is a marked increase in serum creatine kinase levels, up to 50 times normal, in patients suffering from the Duchenne type of muscular dystrophy. The rise occurs before the clinical manifestations and serum enzyme levels are used to detect carriers. There is a small elevation in the serum enzyme activity in patients suffering from certain other forms of muscular dystrophy. These include the limb-girdle, facioscapulohumeral and myotonic types. Enzyme levels are normal in muscular atrophy resulting from neurological disorders or associated with hyperthyroidism.

Increased levels of serum creatine kinase also occur in muscle trauma, including surgery, in polymyositis, in McArdle's syndrome, after severe exercise and after intramuscular injections resulting in damage to muscle.

Heart Disease. An increase in serum creatine kinase activity occurs within 12 hours after the onset of an acute myocardial infarction; the rise may occur as soon as 4 hours, and a peak level is reached in 24 hours. The levels return to normal in 3 or 4 days. The changes mirror those of aspartate aminotransferase, but are slightly more rapid. The absence of any effect resulting from hemolysis is a decided advantage. It has been recently suggested that the test is even more precise if the increased serum levels are associated with the appearance of the slow-moving isoenzyme.

Miscellaneous Disorders. Increased serum levels can occur in individuals with hypothyroidism, in patients with diabetic ketoacidosis, during the postpartum period, after pulmonary infarction, in the presence of convulsive disorders, and in malignant hyperthermia. Increased levels have also been reported in patients with schizophrenia and during heart transplant rejection.

GAMMA-GLUTAMYL TRANSPEPTIDASE

This enzyme catalyzes the transfer of the γ-glutamyl group from one peptide to another or to an amino acid. Therefore, it is probably of importance in the regulation of tissue glutathione content and in protein synthesis. Gamma-glutamyl transpeptidase has been suggested as part of the mechanism for the transport of amino acids across cell membranes. The highest tissue activity of this enzyme is found in the kidney, but activity is relatively high in the liver, the lung, the pancreas and the prostate. Some activity is present in the intestinal mucosa, the thyroid and the spleen. High levels of γ-glutamyl transpeptidase are present in bile and in seminal fluid, and activity is present in plasma, urine and cerebrospinal fluid. Normal serum levels at 37° C., with γ-glutamyl p-nitroanilide as substrate and glycylglycine as acceptor, have been reported by Rosalki and his colleagues to give an upper limit of 45 U/l. for normal males and 35 U/l. for normal females.

Elevated levels of serum activity are found in association with hepatobiliary disorders, including cancer. They also occur in patients with pancreatic disease. In some patients, there is a rise about four days after an acute myocardial infarction. A rise of serum activity also occurs in congestive cardiac failure and in acute coronary insufficiency. These raised levels in the presence of heart disease are probably due to secondary effects on the liver. Elevated serum activity occurs in epileptic patients, probably due to enzyme induction by drug therapy.

An important finding is the frequency of increased enzyme levels in alcoholics. This finding is a sensitive test for determining hepatic involvement and is useful in relation to patients who deny that they are heavy drinkers.

SORBITOL DEHYDROGENASE

This enzyme (L-iditol dehydrogenase) catalyzes the conversion in the liver of sorbitol to fructose. It occurs mainly in the liver and in appreciable amounts in kidney and prostate; it is present in small amount in other tissues. Normal serum shows little or no demonstrable sorbitol dehydrogenase activity.

Raised serum levels are virtually specific for acute hepatocellular disease, e.g., acute hepatitis. In patients with chronic hepatitis, in cirrhosis, in extrahepatic obstructive jaundice, as well as in hepatic metastases, serum levels are normal or very slightly raised. Raised levels can be found in damage to hepatocytes resulting from prolonged biliary obstruction.

LEUCINE AMINOPEPTIDASES

These enzymes hydrolyze L-peptides, splitting off N-terminal residues containing a free α-amino group, especially when the residue is leucine or a related amino acid. There is some degree of substrate specificity, and for clinical biochemical purposes the name leucine aminopeptidase is reserved for the activity shown with the substrate L-leucyl-β-naphthylamide. When the substrate is L-cystine-di-β-naphthylamide, the name cystine aminopeptidase or oxytocinase is employed.

Leucine Aminopeptidase.
This enzyme is found in highest concentration in the duodenum, kidney, liver and pancreas. Using the method of Arst and his colleagues, as modified by King, the normal serum range is 15 to 50 U/l. Increased levels are found in patients with hepatobiliary disease, including drug cholestasis. The levels are especially high in those with jaundice from biliary tract obstruction and consequently in patients with carcinoma of the head of the pancreas with obstruction of the common duct. It is claimed that the serum enzyme levels are particularly useful in differentiating biliary atresia in the jaundiced newborn from neonatal hepatitis.

The serum levels of this enzyme increase with pregnancy, but more specific findings result from the estimation of oxytocinase.

Oxytocinase (Cystine Aminopeptidase).
The placenta is the main source of this enzyme, which hydrolyzes oxytocin and so prevents

premature uterine contractions. In normal pregnancy, serum enzyme levels rise until just before labor. An unduly low level or a premature drop in serum enzyme level is regarded as evidence of placental dysfunction. It is therefore necessary to undertake serial assays to detect this.

Isoenzymes. Both leucine aminopeptidase and oxytocinase exist as isoenzymes. They can be separated by electrophoresis or DEAE-cellulose column chromatography. These studies have shown that the liver is probably the source of serum leucine aminopeptidase, and unusual patterns occur in pathological sera. Oxytocinase exists in two forms, CAP_1 and CAP_2. The latter predominates in late pregnancy.

ERYTHROCYTE ENZYME DEFECTS

Interest in metabolic abnormalities of erythrocytes has been stimulated by the demonstration of a genetically determined enzyme deficiency in the red blood cells of subjects unusually susceptible to the induction of hemolysis by certain drugs. Immature cells of the erythroid series, one of whose major functions is synthesis of hemoglobin, must contain a mechanism adequate for this purpose, which includes RNA (for protein synthesis) and enzymes of the Krebs cycle system (for energy production), as well as those of glycolysis. The mature erythrocyte, on the other hand, serves no such biosynthetic function; it contains little or no RNA and the Krebs cycle system of enzymes is not complete. The energy it requires for preservation of its biconcave shape and for maintaining the Fe^{2+} of hemoglobin in the reduced state (continued reduction of methemoglobin Fe^{3+} by methemoglobin reductase) is probably derived primarily from glycolysis, including oxidation via the hexose monophosphate pathway.

As a result of inability of the mature erythrocyte to synthesize protein, the initially active enzymes, as they undergo degradation, are not replaced. The level of activity of certain key enzymes therefore declines as the erythrocyte ages. The decrease in metabolic activity results in decreased production of ATP and also in decreased concentration of certain structurally important components, e.g., lipids and lipoproteins. These changes probably contribute to the lowered resistance of the older cells to the action of physiological factors that accomplish their destruction normally after an average period of 120 days. They are also less resistant than younger erythrocytes to hemolysis by a number of chemical agents.

Genetically determined metabolic defects in the red blood cells have been demonstrated in certain hemolytic disorders and probably play an important role in their pathogenesis. The occurrence of hemolytic anemia in about 10 per cent of American Negroes receiving therapeutic doses of primaquine (antimalarial agent) has been found to be associated with a low erythrocyte content of reduced glutathione and glucose-6-phosphate dehydrogenase. Formation of reduced glutathione from the oxidized form, by the action of glutathione reductase, requires NADPH (reduced form of nicotinamide adenine dinucleotide phosphate), production of which depends upon glucose-6-phosphate dehydrogenase activity (p. 12). The precise relation of these biochemical abnormalities to the increased susceptibility to hemolysis is not known. Similar changes in the erythrocytes have been found in Caucasians of Mediterranean origins, who exhibit unusual susceptibility to the hemolytic effect of a variety of substances, including fava beans (favism), sulfonamides, naphthalene, acetanilid and acetophenetidin.

A number of erythrocyte enzyme deficiencies are related to the occurrence of chronic nonspherocytic hemolytic anemia. Glucose-6-phosphate dehydrogenase and hexokinase have already been discussed. Other red cell deficiencies involve any one of the following enzymes: glucose phosphate isomerase, phosphogluconate dehydrogenase, pyruvate kinase, triose phosphate isomerase, phosphoglycerate kinase, glyceraldehyde phosphate dehydrogenase, diphosphoglyceromutase, glutathione reductase, glutathione peroxidase, glutathione synthetase and ATPase.

The hemolytic anemias have been subdivided into two types (Dacie and Selwyn). In Type I, *in vitro* hemolysis is corrected by adding glucose. In Type II, *in vitro* hemolysis is corrected by adding ATP, and not glucose.

The erythrocytes in hereditary spherocytosis (congenital hemolytic jaundice) exhibit a decreased capacity for forming ATP and a decreased content of diphosphoglyceric acid. This inadequate energy production may contribute to their inability to maintain the normal elliptical shape, the spherical cells being more readily trapped and destroyed in the spleen.

The basic abnormality in paroxysmal nocturnal hemoglobinuria is believed to reside in the erythrocyte. Certain biochemical abnormalities have been demonstrated. These include abnormalities in phospholipid content and in the lipid pattern of the stromal lipoprotein and a decrease in erythrocyte acetylcholinesterase activity. Relatively slight and inconsistent reduction in this erythrocyte enzyme has also been reported in pernicious and aplastic anemia and in anemias associated with myelosclerosis and kidney disease. The relationship of this enzyme defect to the hemolytic episodes of paroxysmal nocturnal hemoglobinuria is not known.

Congenital deficiency in galactose-1-phosphate uridyl transferase, resulting in galactosemia (p. 89), is demonstrable in erythrocytes. Red cells show a deficiency of galactokinase in the other type of congenital galactosemia.

Congenital deficiency of red cell methemoglobin reductase leads to an accumulation of methemoglobin, since this forms spontaneously in the erythrocyte. There is a persistent cyanosis and moderate erythrocytosis. The enzyme activity can be inhibited by a number of drugs.

ENZYMES IN PLEURAL AND PERITONEAL EFFUSIONS

The primary purpose of these studies is to aid in the diagnosis of intrathoracic and intra-abdominal malignancy associated with pleural effusion and ascites. Lactate dehydrogenase is frequently increased in these fluids in such cases, the level of activity often being higher than in the serum; the enzyme apparently originates in malignant cells in the fluid or on the serous surface. Increased values are obtained also in purulent fluids and, occasionally, in massive pulmonary infarction, as well as in bloody effusions. LDH activity in ascitic fluid may be helpful in the diagnosis of carcinoma of the liver in patients with cirrhosis and ascites, inasmuch as it is usually low in cirrhosis in the absence of complicating malignancy. Transudates have a lactate dehydrogenase content of less than 200 U/l.; that of exudates is higher.

Aspartate aminotransferase activity is not consistently increased in the ascitic fluid in patients with malignancy, but it is usually higher in proportion to its level in the serum than is found in nonmalignant conditions, e.g., hepatic cirrhosis.

Both peritoneal and pleural fluids associated with acute pancreatitis contain raised levels of amylase. The enzyme is believed to enter the pleural cavity *via* the lymphatics and is normally present in the thoracic duct. On occasion, because of lymphatic obstruction, pleural effusions associated with bronchogenic carcinomas can contain elevated amylase levels. Peritoneal fluid amylase elevation can result from a ruptured ectopic pregnancy or from perforation of the intestine.

CEREBROSPINAL FLUID ENZYMES

Changes in serum enzyme activities are not reflected in the cerebrospinal fluid; conversely, increase in cerebrospinal fluid enzyme activities is not accompanied by parallel change in the blood. Changes may occur in the cerebrospinal fluid enzymes in certain diseases of the central nervous system but they are seldom if ever of diagnostic value.

Increase in cerebrospinal fluid aspartate aminotransferase and lactate dehydrogenase occur frequently in meningitis, cerebral thrombosis and hemorrhage. In malignant tumors of the brain, the cerebrospinal fluid glucose phosphate isomerase is usually increased, lactate dehydrogenase occasionally, and aspartate aminotransferase seldom. Isomerase activity is increased also frequently in meningitis and cerebral thrombosis.

ENZYME ABNORMALITIES IN DIAGNOSIS

The determinations of serum enzyme levels can be very useful as diagnostic aids. However, care must be taken in evaluating the meaning of a serum enzyme pattern, since one must take into account the turnover rates of individual enzymes in the circulation. Little is known about this aspect, or even the exact mechanism whereby enzymes are removed from the blood, or what really controls their rate of supply. Much must obviously depend on such factors as enzyme lability, as well as the functional state of organs such as the liver or kidney. Experience has shown, however, that clinical enzymology has a part to play in the diagnostic procedure, provided that the results it supplies are, like other physical signs of disease, considered in conjunction with a careful history of the illness and the findings of a full clinical examination. This includes such accessory aids as radiology, electrocardiography, and endoscopy.

Liver Disease. The serum enzyme levels which have proved to be most useful in practice are alkaline phosphatase, aspartate aminotransferase, lactate dehydrogenase, 5'-nucleotidase, γ-glutamyl transpeptidase (D-glutamyl transferase), and more recently sorbitol dehydrogenase.

In the presence of jaundice, high levels of serum alkaline phosphatase and/or 5'-nucleotidase tend to favor a diagnosis of biliary tract obstruction. It must be remembered, however, that a virtually identical picture can be produced by the cholestatic effect of certain drugs. In the presence of jaundice, high levels of serum aspartate aminotransferase, lactate dehydrogenase and easily detectable sorbitol dehydrogenase tend to favor a diagnosis of hepatocellular disease. Gamma-glutamyl transpeptidase tends to come in between these two extremes, since it can be raised in almost any kind of liver disease. It must be kept in mind, however, that prolonged biliary tract obstruction leads to liver cell dysfunction, or even death, which makes interpretation difficult, no matter what enzyme system is employed. It has been suggested that a good index of hepatocyte necrosis is given by the ratio of the sum of the serum levels of the aminotransferases to the level of serum glutamate dehydrogenase, which is a mitochondrial enzyme. Unfortunately, this has turned out to be only a pious hope.

In the absence of jaundice, increase in the levels of any of the enzymes mentioned can indicate the possible presence of hepatobiliary disease, and consequently serum enzyme levels (usually aminotransferases and alkaline phosphatase) are employed in relation to testing drugs for hepatotoxicity. The aminotransferases can reach quite high levels in the absence of jaundice. This occurs in the pre-icteric phase of infective hepatitis. The levels can be very high when a liver toxin is so potent that the patient dies before appreciable jaundice develops.

Metastases in the liver are not infrequently associated with an increase in serum alkaline phosphatase, as well as in serum lactate dehydrogenase.

The levels of serum γ-glutamyl transpeptidase are very useful in detecting the alcoholic liver, but they are obviously not pathognomonic. The same may be said of any serum enzyme level or enzyme pattern in any liver disease situation.

Serum isoenzyme studies may be of use in the clinical situation. Serum LD_5 is readily detectable in hepatocellular disorders. In the author's experience, it is virtually always present in the serum of patients in the pre-icteric phase of infective hepatitis and is then detectable throughout the disease. The situation with isoenzymes of alkaline phosphatase is somewhat different, since the liver isoenzyme is constantly present in the serum of adults. However, an increased amount can be detected in the sera of many patients with hepatobiliary disorders. In the presence of jaundice arising from biliary tract obstruction, starch-gel electrophoresis not infrequently demonstrates an isoenzyme doublet near the origin. Occasionally, jaundiced patients, regardless of their sex, show the presence of the heat-stable placental alkaline phosphatase isoenzyme (Regan isoenzyme) in their circulation. Apart from pregnancy, this indicates the presence of malignant metastases in the liver (bronchogenic carcinoma, and others.).

Serum leucine aminopeptidase levels are stated to be helpful in differentiating neonatal hepatitis from biliary atresia, since they tend to be much higher in the latter condition.

Heart Disease. The changes in the serum levels of creatine kinase, aspartate aminotransferase and lactate dehydrogenase after myocardial infarction have already been discussed. Given an appropriate history of the illness, these changes, together with isoenzyme studies of serum lactate dehydrogenase, are virtually diagnostic and have proved to be even more sensitive than the electrocardiogram in an initial attack. They are also very useful in detecting a second infarction or an extension, as well as detecting myocardial infarction in the presence of previously existing electrocardiographic abnormalities. The enzyme elevation also has prognostic importance. Pulmonary infarction is of importance in the differential diagnosis of myocardial infarction. About one-fourth of patients with the lung disease develop a small increase of serum aspartate aminotransferase, but this usually occurs about four days after the onset of pain. There may be a minor degree of increase in serum lactate dehydrogenase, but this is by no means constant. Relatively frequently, it can be shown that the increase is not due to the release of LD_1. Creatine kinase levels are not affected by pulmonary infarction, liver disease, blood disease or malignant disease. Serum creatine kinase levels have, therefore, a higher degree of specificity for the detection of myocardial infarction.

In about 10 per cent of cases of acute coronary insufficiency, whatever this might mean, there is an increase in one or other of the appropriate serum enzymes. When lactate dehydrogenase is elevated, LD_1 can be demonstrated in the serum. Gamma-glutamyl transpeptidase levels are fairly frequently raised in acute coronary insufficiency.

The enzyme pattern in acute rheumatic or other acute myocarditis can resemble that of myocardial infarction but persists as long as the disease is active.

In paroxysmal arrhythmias lasting for more than one-half hour, about one-fifth of patients with a pulse rate above 140 per minute show elevation of aspartate aminotransferase and occasionally of lactate dehydrogenase (LD_5). The same state of affairs can be found in chronic congestive cardiac failure. In each instance, the serum enzyme changes are secondary to changes in the liver. It is, in fact, not uncommon to be able to detect LD_5 in the serum of patients who have suffered a myocardial infarction about one week previously.

The cardiac type of serum isoenzyme pattern can be found in a number of noncardiac disorders (megaloblastic anemia, intravascular hemolysis and renal infarction), but the differential diagnosis is not difficult, nor is the serum creatine kinase raised.

Placental Insufficiency (Placental Dysfunction). This

condition is an important consideration in the practice of obstetrics. It involves diminished transmission of nutrients to the fetus, which can result in retarded fetal growth, fetal anoxia or fetal death. Attempts have been made to assess placental insufficiency by the assay of the serum levels of three enzymes, viz., placental alkaline phosphatase, oxytocinase and diamine oxidase. Placental alkaline phosphatase levels in normal pregnancy increase progressively during the last 12 weeks. Attempts to relate abnormally high or abnormally low levels to placental insufficiency have proved to be somewhat disappointing. Oxytocinase serum levels increase in normal pregnancy from the beginning of the fourth month. Increasing levels are said to indicate good fetal prognosis, whereas falling levels indicate the imminent onset of labor or intrauterine fetal death. There is a wide range of serum levels of this isoenzyme, and it is really not surprising that estimations detect only about two-thirds of the patients with placental dysfunction. On the whole, however, it is fairly safe to assume that rising serial estimations indicate that the fetus is not in danger. Diamine oxidase levels increase rapidly during the first three months and then more slowly during the rest of the pregnancy. Here again, normal levels have a wide range. A falling or low level indicates poor fetal prognosis but the assay has not proved reliable in regard to fetal prognosis in high risk pregnancies.

Central Nervous System Disease. In cerebral thrombosis and cerebral hemorrhage, there is a large increase in the aspartate aminotransferase level in the cerebrospinal fluid. The level reaches a peak in four or five days and may stay high for months. Some of the enzyme may leak into the bloodstream, but because of the blood-brain barrier the plasma levels are not as high as those in the cerebrospinal fluid. The isoenzymes of lactate dehydrogenase LD_2 and LD_3 are present in the C.S.F. (cerebrospinal fluid) in increased amounts. There is an increase of creatine kinase in the blood. This commences approximately two days after the cerebral catastrophe, and the rise lasts for several days. Serum shows an increase in LD_5.

In bacterial meningitis, including the tuberculous variety, there is an increase in lactate dehydrogenase and aspartate aminotransferase, possibly derived from inflammatory cells. There is an increase of creatine kinase with abscess formation. In viral meningitis, the picture is somewhat more complicated because viral hepatitis and viral myocarditis may also be present, producing appropriate changes in the plasma enzyme levels.

In Korsakoff's psychosis and Wernicke's encephalopathy, there can be a decrease in the transketolase activity of the cerebrospinal fluid. Transketolase requires thiamine for its activity, and deficiency of this vitamin results also in a fall in activity in the blood.

Varying and somewhat contradictory results have been obtained in enzyme studies of the demyelinating disorders.

Genitourinary Disease. Carcinoma of the prostate has already been discussed (p. 567).

In renal tubular necrosis, the levels of serum aspartate aminotransferase and lactate dehydrogenase rise within 24 hours. Serum alkaline phosphatase sometimes shows a rise commencing in the first five days and remaining elevated for two or three weeks. There are often increases in urinary alkaline phosphatase levels, if urine is passed. There is said to be an isoenzyme in the urine, which travels very fast during electrophoresis. Lactate dehydrogenase activity is also increased in the urine. It is suggested that the isoenzyme pattern will differentiate between necrosis of cortical tubules and necrosis of medullary tubules. Urinary lysozyme activity is also increased.

Serum and urinary levels of lactate dehydrogenase are increased with rejection of transplanted kidneys. They are similarly affected in renal artery stenosis.

A variety of kidney diseases, including acute glomerulonephritis, show an increased level of lactate dehydrogenase and alkaline phosphatase in the plasma. The presence of these diseases also affects the levels of a variety of urinary enzymes. There has been very little clinical application of enzyme determinations in renal disorders.

Malignant Disease. It is not generally appreciated that serum lactate dehydrogenase levels are raised in about two-thirds of patients with malignant disease. LD_5 is usually the predominant isoenzyme in the tumors themselves; however, because of the short half-life of this isoenzyme in the circulation, the serum pattern in malignancy is different from the tumor itself, with one or more of the isoenzymes LD_2, LD_3 and LD_4 predominating. In certain solid tumors (seminoma, teratoma testis, ovarian dysgerminoma), there is an increase in serum LD_1 and LD_2. These isoenzymes are also present in the blood of patients with acute leukemia.

Serum alkaline phosphatase activity is quite frequently increased in patients with bone or liver metastases, and it is possible to recognize the presence of the bone or liver isoenzyme, respectively. Malignant tumors themselves produce a variety of alkaline phosphatase isoenzymes. Perhaps the best known is the Regan isoenzyme, which is identical with placental alkaline phosphatase.

The tartrate-labile acid phosphatase has already been discussed in relation to carcinoma of the prostate (p. 567). In the presence of advanced carcinoma of the breast, there may be an increase in a tartrate-stable and formaldehyde-stable serum acid phosphatase.

Serum lysozyme levels are elevated in most patients with acute or chronic granulocytic leukemia and very high in those with monocytic leukemia. They tend to be normal or low in the presence of lymphocytic leukemias.

Serum levels of the various enzymes tend to revert toward, or return to, normal following any form of successful therapy (surgical, radiotherapy, chemotherapy and possibly immunotherapy). If relapse occurs, the enzyme levels become abnormal again. Enzyme and isoenzyme levels in the serum are therefore of importance in monitoring therapy. *In vitro* enzyme studies can be used to anticipate the possibility of response to treatment of leukemia with cystosine arabinoside. Such studies are being extended.

Miscellaneous Disorders. Enzyme studies in hemolytic anemia (p. 572), bone disease (p. 562), pancreatic disease (p. 589) and skeletal muscle disease (pp. 559, 569) are discussed elsewhere.

Inherited Disease. It is now possible to make a prenatal diagnosis of a number of inherited disorders by the collection of amniotic fluid and culture of its cells. These disorders include mucopolysaccharidosis, cystic fibrosis, Lesch-Nyhan syndrome, galactosemia and Pompe's disease, among others.

The confirmation of the diagnosis of an inherited disorder can also be made by studying enzyme deficiencies in cultures of skin fibroblasts, in erythrocytes, as well as in isolated leukocytes. Enzyme studies on tissue biopsies are also employed.

References

Bodansky, O.: Acid phosphatase. Advances Clin. Chem., *15*:44, 1972.
Fraser, D.: Hypophosphatasia. Am. J. Med., *22*:730, 1957.
Greenberg, D. M., and Harper, H. A., eds.: Enzymes in Health and Disease. Springfield, Ill., Charles C Thomas, 1960.
Gutman, A. B.: Serum alkaline phosphatase activity in diseases of the skeletal and hepatobiliary systems. Am. J. Med., *27*:875, 1959.
Janowitz, H. D., and Dreiling, D. A.: The plasma amylase. Am. J. Med., *27*:924, 1959.
King, J.: Practical Clinical Enzymology. London, D. Van Nostrand Co. Ltd., 1965.
Latner, A. L., and Skillen, A. W.: Isoenzymes in Biology and Medicine. New York, Academic Press, 1968.
McCance, R. A., et al.: Genetic, clinical, biochemical, and pathological features of hypophosphatasia. Quart. J. Med., *25*:523, 1956.
McGowan, G. K., and Walters, G., eds.: Symposium on Enzyme Assays in Medicine. J. Clin. Path., *24*:Suppl. (Assoc. Clin. Path.), 1971.
Mullan, D. P.: Studies in Clinical Enzymology. London, Wm. Heinemann, 1969.
Wilkinson, J. H.: An Introduction to Diagnostic Enzymology. London, Arnold, 1962.
Wilkinson, J. H.: Isoenzymes. 2nd ed., London, Chapman and Hall Ltd., 1970.
Wolf, P. L., Williams, D., and Von der Muehll, E.: Practical Clinical Enzymology and Biochemical Profiling. New York, John Wiley and Sons, 1973.

Chapter 18

CHEMICAL INVESTIGATION OF GASTRIC FUNCTION

Chemical examination of the gastric contents may yield valuable information regarding both the secretory and the motor activities of the stomach and may reveal the presence of abnormal substances indicative of pathologic conditions. In order to obtain complete evidence regarding gastric function, the contents of the stomach must be examined both during the interdigestive period (fasting stomach) and during the period of digestion, or stimulation.

The most important specific secretory products of gastric activity are (1) hydrochloric acid, (2) enzymes (pepsin and a weak lipase), (3) mucus and (4) intrinsic factor. Gastric juice contains also amino acids, histamine and other amines, urea, ammonium salts and neutral inorganic salts, such as chlorides. The true gastric glands, i.e., those which secrete HCl and enzymes, occur in the body and fundus. There are three cell types: (*a*) parietal or border cells, which secrete HCl at a rather constant concentration of about 0.155M; (*b*) chief cells of the body of the glands, which secrete pepsinogen, which is converted to pepsin by an autocatalytic process in the presence of HCl; (*c*) cells of the neck of the glands, which secrete mucus ("dissolved mucus").

In the gastric antrum (Fig. 18-1) the pyloric cells secrete an alkaline mucus. The antral mucosa also possesses special cells (G cells) which, under appropriate stimuli, produce the hormone gastrin, heptadecapeptides of known amino acid sequence (p. 771 ff.) which are absorbed into the bloodstream and thence stimulate the production of HCl in the body of the stomach (parietal cell mass). When this acid reaches the antrum it neutralizes the alkaline mucus and eventually lowers the pH to a level which will bring about inhibition of gastrin production, and so prevent overproduction of HCl in the body of the stomach.

Gastric secretion is commonly described as consisting of three phases, (1) a psychic or cephalic phase, (2) a gastric phase and (3) an intestinal phase.

Psychic or Cephalic Phase. This is due to a nervous mecha-

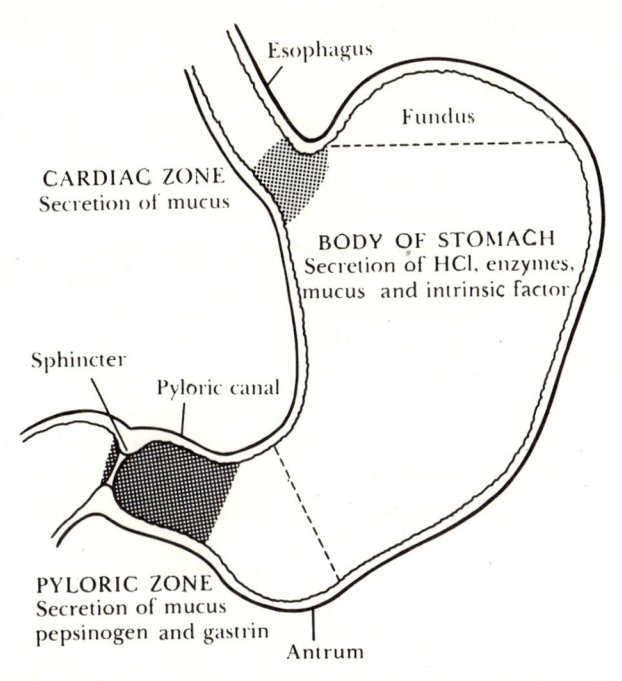

Figure 18-1. Schematic drawing of the stomach, with major zones. (From Tietz: *Fundamentals of Clinical Chemistry*. W. B. Saunders Company, 1970.)

nism mediated through the hypothalamus bringing about stimulation of the vagus. It is evoked by pleasurable sensations accompanying the thought, sight, smell and taste of palatable food and has been elicited by hypnotic suggestion. Vagal stimulation brings about HCl production in two ways. There is direct nervous stimulation of both the parietal cells in the body and the specialized G cells of the pyloric antrum; the G cells produce gastrin, which in turn stimulates HCl production. Inhibition of secretion may result from psychic influences such as worry and anxiety and the sight or smell of disagreeable food.

Gastric Phase. Food brings about distention of the body of the stomach, which elicits reflex activity, resulting in further stimulation of the vagus and further production of HCl. Stretching of the antrum by food stimulates stretch receptors, and at the same time chemical stimulation occurs. Both these factors bring about the production of more gastrin by the G cells, which again stimulates acid production.

Intestinal Phase. Certain products of gastric digestion, when they enter the duodenum, act as chemical excitants to gastric secretion. This is true also of water, meat extracts, peptides, saponin, soaps and magnesium sulfate. This aspect of the intestinal phase is probably due to the stimulation of nerve endings resulting in gastrin production.

The release of the hormone cholecystokinin-pancreozymin (CCK-PZ) from the antrum and the duodenum is stimulated by both acid and food (p. 773). Apart from its action on the gallbladder and pancreas, the hormone is an inhibitor of gastric acid production. The release of the hormone secretin by the duodenum is also stimulated by the presence of acid and food. This also exerts an inhibitory effect on acid production by the stomach. Acid, fat and hypertonic solutions in the duodenum bring about inhibition of gastric secretion. A similar effect is produced by injection of extracts of intestinal mucosa or urine; the factors responsible for this effect have been termed enterogastrone and urogastrone, respectively. The latter factor is not produced by the urinary excretion of enterogastrone. The inhibitory activity of the intestinal phase is probably more important than its excitatory action.

Throughout the various phases, the secretion of pepsinogen usually runs parallel with that of HCl. This is related to the fact that stimulation of the basal production of this enzyme precursor is brought about by both a vagal and a gastrin mechanism.

Other Factors. Insulin (hypoglycemia) stimulates gastric secretion through the vagus mechanism, the volume, acidity and peptic activity of the gastric juice being increased. Mecholyl, pilocarpine, and nicotine increase the volume considerably and the acidity somewhat. Alcohol is a powerful stimulant (acidity and volume), while acids and atropine are secretory depressants.

Different foodstuffs influence the secretion in different ways. For example, meat produces a juice of high acid and moderate pepsin content, bread a juice of low acid and high pepsin content, and milk a juice of moderate acid and low pepsin content. Fat depresses peptic activity relatively more than acidity or volume.

NORMAL GASTRIC JUICE

Gastric juice is secreted continually. The acidity of the parietal cell secretion is rather constant (about 0.155M) and is dependent upon the electrolyte composition of the blood plasma. Variations in acidity of the mixed gastric juice (normal and abnormal) are due, therefore, not to variations in the concentration at which HCl is secreted, but (a) to variations in the relative proportions of the different gastric secretions (e.g., by mucous cells, chief cells, parietal cells), and (b) to such neutralizing substances as saliva, food, mucus, and regurgitated materials (e.g., pancreatic juice, bile, duodenal secretion) from the duodenum.

Gastric Residuum. Valuable information may be obtained by the examination of the stomach contents during the interdigestive period (gastric residuum). The following chemical and gross physical characteristics are of importance from the standpoint of functional diagnosis.

Amount. The quantity of material normally found in the fasting stomach varies from 20 to 100 ml., averaging about 50 ml. An increase in the volume of the gastric residuum may be due to hypersecretion, retention or regurgitation from the duodenum. These conditions may be differentiated on the basis of other observations such as the presence of food particles in retention, the presence of abnormal quantities of bile and duodenal enzymes in regurgitation, and the absence of these factors in true hypersecretion, in which condition the gastric contents, apart from their increased volume and perhaps increased acidity, may be essentially normal. If intubation (p. 584) produces more than 250 ml. of residuum four hours after a meal, pyloric stenosis is likely to be the cause.

Color. Freshly secreted gastric juice is colorless. However, in about 55 per cent of normal individuals the gastric residuum is either yellow or green, due to regurgitation of bile from the duodenum, which occurs in about 25 per cent of normal individuals, or to the presence of molds or the *Cryptococcus salmonicus*. Increased quantities of bile in the stomach, when not due to retching incident to the passage of the tube, result from intestinal obstruction or ileal stasis. A bright or dark red, brown or black color in the residuum is usually indicative of the presence of blood (see below).

Consistency. The normal gastric residuum is rather fluid in consistency, containing no solid food particles and only a small quantity of ropy mucus which may be derived from the nasopharynx. The presence of an increased quantity of sediment is usually indicative of retention, and increased quantities of mucus are found in catarrhal inflammations of the stomach.

Blood. The benzidine test is the one most commonly employed for the detection of blood in the gastric contents. A trace of bright red, aerated blood in specimens extracted through the ordinary metal-tipped tube is most commonly due to accidental trauma to the gastric mucous membrane. The incidence of accidental blood may be diminished by the use of a paraffin- or rubber-coated tip. Pathologically, the gross appearance of specimens containing blood depends upon the extent of the hemorrhage, the length of time the blood has remained in the stomach and the degree of acidity of the gastric contents. In the presence of hydrochloric acid the red cells are hemolyzed, the hemoglobin being converted to acid hematin, which is dark brown in color. Blood which has resided in the stomach for relatively long periods of time is therefore usually dark and well mixed with the gastric contents. If coagulation has occurred, as is frequently the case in carcinoma, the clot may be partially disintegrated, having the so-called characteristic "coffee-grounds" appearance. The presence of blood in the gastric residuum may be due to such lesions as carcinoma of the stomach, portal cirrhosis, chronic passive congestion of the stomach, peptic ulcer, acute gastritis, and hemorrhagic blood dyscrasias, including purpura hemorrhagica, acute leukemia, agranulocytosis, aplastic anemia, and so on. It must be remembered that blood derived from the gums and from nasopharyngeal, laryngeal, tracheobronchial and pulmonary lesions may be swallowed.

Organic Acids. Lactic, butyric and other fatty acids may be found in the gastric contents. Some observers believe that lactic acid is secreted by the gastric mucosa in carcinoma of the stomach; however, the consensus is that organic acids result from stagnation of the gastric contents with consequent bacterial action and fermentation. Since these processes are inhibited by high free acidity, lactic acid is most commonly found in association with gastric retention and hypochlorhydria, the combination of the three being rather significant of carcinoma of the stomach.

Enzymes. The normal gastric residuum contains pepsin and rennin (?). Trypsin is frequently found to be present as a result of regurgitation from the duodenum. The absence of pepsin, occurring in conjunction with achlorhydria, constitutes true achylia, which is encountered typically in pernicious anemia and in subacute combined degeneration of the spinal cord. This enzyme is practically never absent if free hydrochloric acid is present in the gastric secretion.

INVESTIGATION OF GASTRIC SECRETORY ACTIVITY

Many procedures have been employed for the evaluation of gastric secretory activity. They have been designed chiefly to aid in the diagnosis of peptic ulcer (gastric and duodenal), carcinoma of the stomach, and pernicious anemia (achlorhydria). Their usefulness is based upon the observations (a) that hypersecretion, i.e., hyperacidity, occurs more commonly in duodenal ulcer, (b) that hyposecretion, i.e., hypoacidity, is a common feature of atrophic gastritis and, by no means invariably, of gastric cancer, and (c) that pernicious anemia is virtually invariably accompanied by achlorhydria, except in the rare juvenile type.

Until quite recently, it was customary to pass a nasogastric tube blindly into a fasting patient's stomach and suck out gastric contents with a syringe. The stomach was emptied as much as possible, yielding the so-called gastric residuum, which has already been described. A test meal (Ewald, Riegel, and others) was then given or a stimulus to gastric secretion applied (oral alcohol, oral caffeine sodium benzoate, intramuscular histamine or intramuscular Histalog: 3-β-aminoethylpyrazole dihydrochloride). Samples of the gastric contents were then removed at 15-minute intervals, freed of their mucus and solid matter and then titrated against a 0.1 molar solution of NaOH to determine the concentration of HCl in each sample. This enabled the observer to construct a curve in which time was plotted against acidity.

Experience has shown that this type of investigation was not very useful and even, on relatively rare occasions, could be dangerous. It is now realized that passing a nasogastric tube is an expert procedure. A radiopaque tube of large bore is used and its end eventually passed into the most dependent part of the stomach under fluoroscopic guidance. The stomach is emptied by means of a 50-ml. syringe and then pump aspiration is applied to collect the *basal secretion,* which is either a continuous 60-minute sample or four consecutive 15-minute samples. It is necessary at intervals to use manual syringe suction and to inject air down the tube in order to keep the bore unobstructed. Maximum acid output is now determined by an intramuscular injection of 6 μg./kg. of pentagastrin (N-t-butyl-oxycarbonyl-β-Ala. Try. Met. Asp. Phe. NH$_2$); this contains the C-terminal tetrapeptide amide of gastrin, which is responsible for the physiological action of the natural hormone. Immediately after the injection, samples of gastric juice are aspirated over 10-minute or 15-minute collection periods for a total time of one hour. The volume and pH of each sample is determined and its total acid measured by titration against 0.1 molar NaOH to pH 7.0 (or 7.4). The results on the two consecutive samples of highest acid content are added together to give a peak 20 minutes or half-hour excretion, depending on whether samples are collected over 10 or 15 minutes. The figures obtained are multiplied by 3 or 2, respectively, to give the number of millimoles of acid secreted in one hour.

The normal ranges for men are basal 0 to 5 mmol., or slightly more, and maximum acid output (or peak output) 1 mmol., or somewhat less, up to 45 mmol.; the respective figures for women are 0 to 5 mmol., and somewhat less than 1 up to 30 mmol.

Insulin (Hypoglycemia) Stimulation. Hypoglycemia causes stimulation of the vagus centers in the brain, one manifestation of which is increased gastric secretion. This procedure is used chiefly to ascertain the effectiveness of vagotomy in patients with ulcer.

After a basal collection period (fifteen-minute samples for sixty minutes), blood is drawn (for sugar determination) and 20 units of regular insulin are injected intravenously. The stomach is emptied at fifteen-minute intervals for two hours, and blood is drawn at thirty and forty-five minutes. Saliva must be expectorated. Glucose for intravenous administration should be at hand, and sweetened orange juice should be given at the conclusion of the test.

A positive response, i.e., indicating vagal innervation of the stomach, consists in a rise above the basal level. A negative response is not significant if the blood sugar has not fallen to at least 40 mg./100 ml.

Serum Gastrin. The hormone gastrin can now be estimated in serum by radioimmunoassay. The normal upper limit varies from laboratory to laboratory. One reliable value seems to be 120 pg./ml. with a mean value of 46 pg./ml.

Intrinsic Factor. Cells in the body of the stomach also secrete intrinsic factor, which is mucopolypeptide in structure. Intrinsic factor combines with vitamin B_{12} in the stomach, and the complex is carried to acceptor sites in the lower part of the small intestine, where the vitamin is absorbed.

Abnormal Response

From the standpoint of acidity, three main types of response may be obtained under pathologic conditions.

Achlorhydria. The term "achlorhydria" is applied to the absence of free hydrochloric acid after stimulation with pentagastrin. The term is also applied to failure to produce gastric juice below pH 6.5. Achlorhydria is most commonly observed in pernicious anemia, gastric carcinoma (20% of cases), atrophy of the gastric mucosa and severe iron deficiency anemias. Pernicious anemia can be excluded, with the exception of the juvenile type, if any acid is present. In the presence of achlorhydria, the diagnosis can be confirmed by demonstration of the absence or gross diminution of intrinsic factor, which can now be estimated by radioimmunoassay, depending on radioactive-vitamin B_{12} binding, as well as by biological methods using the labeled vitamin. The occurrence of achlorhydria in the presence of an ulcerated gastric lesion should arouse suspicion of carcinoma of the stomach. The latter can, however, occur in the presence of quite large peak outputs of acid.

There is in some individuals a condition of primary achlorhydria, associated with none of the recognized causes for the absence of free HCl, occurring probably on the basis of some constitutional defect, and showing a definite hereditary and familial tendency. Such individuals may be predisposed to the development of pernicious anemia.

Hyposecretion. The term hyposecretion is difficult to define, since the lower level of normal maximum acid output approaches zero. Levels below the normal mean peak output can, for clinical purposes, be regarded as presumptive evidence of hyposecretion. This usually indicates the presence of atrophic gastritis. The condition increases with advancing age and is possibly associated with autoimmunity. It can also occur with iron deficiency. No matter how low the acid output, intrinsic factor is present in appreciable amounts.

In patients with gastric ulcer, normal levels of acid are commonly found, but those with ulcers situated high up in the body of the stomach may show low peak output levels. In patients with carcinoma of the stomach, hyposecretion is fairly common. It can also occur in individuals with chronic constipation, chronic appendicitis, gastric and other neuroses, the irritable bowel syndrome, secondary anemia, chronic debilitating disease, particularly tuberculosis, in hyperthyroidism, and in a proportion of normal individuals. Subnormal secretion of free HCl and pepsin has been observed in normal pregnant women.

Hypersecretion. This can be defined as a maximum acid output above the normal upper limit.

Patients with duodenal ulcer have an increased parietal cell mass and consequently tend to show an increase in the basal and peak acid outputs. About one-third to one-half of these patients have hypersecretion. The remainder almost invariably shows peak outputs above the normal mean level. It should be noted that the basal output is increased in only about one-third of these patients, and can be zero. It is now fairly generally accepted that measurements of acid output in duodenal ulcer have no real bearing on the clinical course. They are not indications for surgery or the type of operation to be performed. Measurements of acid output may indicate the possibility of duodenal ulcer in x-ray negative dyspepsia.

In the syndrome known as hypertrophic, hypersecretory gastropathy, there is frequent occurrence of hypersecretion.

An important but rare condition in which hypersecretion occurs is the *Zollinger-Ellison syndrome* (p. 597), in which there is recurrent gastric ulceration due to a gastrin-producing tumor of the pancreas. Here, the basal acid output is very high. It is often 40 per cent or more of the maximum output. In fact, on occasion, stimulation with pentagastrin will not produce any further increase of acid output. The syndrome is, however, best diagnosed by the assay of serum gastrin levels.

Hypersecretion can also result from hypercalcemia, as in hyperparathyroidism, in which a peptic ulceration syndrome is recognized.

"Tubeless Gastric Analysis." This procedure was introduced for the purpose of demonstrating gastric HCl secretion without intubation of the stomach. It involved oral administration of an ion exchange resin tagged with a dye, Azure A. In the presence of free HCl, the resin releases the dye in exchange for the H^+ ions of the acid. The dye is absorbed from the intestine and is excreted in the urine, where it can be readily determined quantitatively. Usually, either caffeine sodium benzoate or, preferably, histamine was employed as a stimulant to gastric secretion. Urine was usually collected for two hours following ingestion of the resin.

Measurement of the concentration of Azure A in the blood could be substituted for that in the urine in cases in which the urine collection might be unreliable. This procedure may yield misleading results and should no longer be employed for clinical purposes.

Peptic Activity

The normal gastric juice contains pepsin. Trypsin may also be present in the gastric contents as a result of regurgitation from the duodenum. Pepsin is never absent from the gastric juice if free hydrochloric acid is present. However, in the absence of free hydrochloric acid, the determination of pepsin is of importance in distinguishing between achlorhydria and achylia.

Achylia is the term applied to the absence of both free hydrochloric acid and gastric enzymes after pentagastrin injection. True achylia may occur in pernicious anemia, subacute combined degeneration of the cord, severe secondary anemia and following gastroenteritis in children.

A proteolytic enzyme, apparently pepsin, is excreted normally in the urine, and has been termed "uropepsin." It apparently enters the blood stream (as pepsinogen) directly from the gastric cells. It has been suggested that measurement of uropepsin excretion might be employed as a substitute for gastric analysis by intubation. Despite occasional satisfactory results, this procedure does not appear to afford a reliable index of gastric secretion in duodenal ulcer, gastric ulcer and gastric carcinoma.

Serum Gastrin

Very high levels of fasting serum gastrin (10 times normal or higher) are found in patients with the Zollinger-Ellison syndrome, and in patients with pernicious anemia. The latter is secondary to the achlorhydria and is due to loss of the normal inhibition of gastrin production by HCl. The fasting serum level is usually raised in patients with duodenal ulcer but there is overlap with normal values. A rise also occurs in chronic atrophic gastritis, seropositive for parietal cell antibody. The fasting level in this form of gastritis is inversely related to the degree of pathological change in the gastric antrum (antral gastritis).

Hypoglycemia causes a rise and hyperglycemia, produced by intravenous glucose, lowers the serum level.

References

Baron, J. H.: The clinical use of gastric function tests. Scand. J. Gastroent., 5:(Suppl. 6), 9, 1970.
Baron, J. H., and Lennard-Jones, J. E.: Gastric secretion in health and disease. Brit. J. Hosp. Med., 6:303, 1971.
Grossman, M. I.: Gastrointestinal hormones. Physiol. Rev., 30:33, 1950.
Segal, H. L.: Clinical measurement of gastric secretion: significance and limitations. Ann. Int. Med., 53:445, 1960.

Chapter 19

PANCREATIC FUNCTION

The metabolic significance of the pancreas lies in its internal and external secretions. The internal secretion, insulin, is produced in the beta cells of the islands of Langerhans, absorbed directly into the blood and plays an important part in the metabolism of carbohydrates, as is indicated in the consideration of carbohydrate metabolism (p. 18) and diabetes mellitus (p. 47 ff.). A hyperglycemic factor, glucagon (p. 26), is secreted by the alpha cells of the islets. The external secretion, produced by the acinar cells of the pancreas and the duct epithelium, is strongly alkaline in reaction due to the presence of bicarbonate and contains a small amount of coagulable protein and inorganic constituents. The most important constituents, however, from a functional standpoint, are enzymes or their zymogens, namely, trypsinogen, chymotrypsinogen A and B, procarboxypeptidase A and B, which are proteolytic proenzymes; pancreatic amylase, an amylolytic enzyme; lipase, a lipolytic enzyme; ribonuclease; and desoxyribonuclease. The determination of pancreatic functional efficiency will be discussed from the standpoint of its external secretion, since its internal secretory function is considered in detail elsewhere (p. 18 ff.).

Pancreatic exocrine secretion is under hormonal and nervous control, with cephalic gastric and intestinal phases. *Secretin* is a pancreatic secretagogue obtained by extraction of the upper intestinal mucosa. Entering the blood stream, it exerts a rather specific effect, mainly on the intercalated duct cells and possibly, to some extent, the acinar cells themselves. It is the duct cells which produce the HCO_3^- ions and much of the water present in the pancreatic juice. Consequently, stimulation by secretin largely brings about an increased production of HCO_3^- ions and an increased flow of juice. *Cholecystokinin-pancreozymin* stimulates the output of enzymes. With simultaneous stimulation by both these hormones, the effect on secretion, including the enzymes, is greater than with either hormone alone. *Gastrin* stimulates both volume flow and enzyme output. The secretion stimulatory effect of each of these hormones is preceded by an increased blood flow to the pancreas.

Vagus stimulation results in a pancreatic secretion of small volume with a high enzyme content. The vagus effect may be produced by administration of insulin (hypoglycemia), mecholyl chloride, pilocarpine or prostigmine.

As with gastric secretion, pancreatic secretion has cephalic, gastric and intestinal phases.

There seems to be some relation between the islets of Langerhans and the acinar cells of the exocrine pancreas. Insulin promotes the synthesis of trypsinogen and amylase, and glucagon inhibits pancreatic secretion. This has had clinical application in relation to pancreatic fistulae and acute pancreatitis.

Investigation of pancreatic exocrine function involves studies of the following types: (1) examination of the feces for evidence of incomplete digestion (proteolytic and lipolytic deficiencies); (2) determination of pancreatic enzyme activity in the duodenal contents (Lundh test meal); (3) determination of the exocrine secretory capacity of the pancreas by measurements of increments in the volume, bicarbonate content, and enzyme activity of pancreatic juice, obtained by duodenal intubation, following stimulation by secretin and/or pancreozymin; (4) studies of serum amylase or lipase following administration of secretin and pancreozymin (or opiates). Some of these procedures are particularly useful in the diagnosis of certain types of pancreatic disease, e.g., acute pancreatitis and pancreatic duct obstruction. These include studies of amylase or lipase activity in the serum and urine, with or without secretin stimulation, and of pancreatic enzyme activity in the duodenal contents.

EXAMINATION OF PANCREATIC JUICE: SECRETIN TEST

Quantitative analysis of the pancreatic juice theoretically should afford a rational and accurate method of studying the external secretory function of the pancreas. Use of a double-lumen gastroduodenal tube, properly placed under fluoroscopic guidance, has largely overcome the difficulty of obtaining pancreatic juice uncontaminated by variable amounts of gastric juice. A more or less pure pancreatic juice is obtained by stimulation with a purified preparation of secretin.

The long end of the tube is passed to the third portion of the duodenum, in which position the shorter end (ten inches shorter) is in the stomach. Continual gentle suction is then applied, the negative pressure not exceeding 50 mm. Hg, and the gastric juice and duodenal contents are collected simultaneously. Usually after about twenty to twenty-five minutes, the duodenal juice becomes clear and is no longer contaminated with gastric juice. The duodenal contents consist principally of pancreatic juice and usually some bile. When admixture with gastric juice has ceased, a twenty-minute basal sample is obtained, and secretin is injected intravenously in a dosage of 1 unit of a commercial preparation per kilogram of body weight. Collection is continued for one hour, separated, if desired, into ten- or twenty-minute samples. Blood specimens may be collected before and two hours after the injection of secretin (p. 773). Urine may be collected for three hours before, and a second specimen for three hours after the injection. Normally, there is no change in urinary amylase output. An increase indicates the possibility of pancreatic pathology.

Immediately after injection of secretin, the volume of material aspirated from the duodenum increases strikingly, usually with a change in color. Prior to injection this is usually light yellow-brown unless the gallbladder is emptying. Promptly after stimulation of pancreatic secretion dark bile appears briefly and then the color becomes paler or all trace of bile pigment disappears. In most instances the liver bile accumulates in the gallbladder during the experimental period. The occurrence of a deeply bile-stained secretion throughout the test period suggests that the gallbladder is absent, diseased or nonfunctioning. Regurgitation of duodenal contents into the stomach occurs occasionally, especially in cases of achylia gastrica, when the pyloric mechanism is incompetent. This is evidenced by a sudden rise in the volume of a gastric fraction, by bile discoloration and by reduction in acidity of the fraction. By measuring the volume and the concentrations of bicarbonate and bilirubin, and comparing the latter with their concentrations in the corresponding duodenal fraction, the quantity of regurgitated material may be estimated with reasonable accuracy. The following quantitative determinations may be made: (1) volume; (2) bicarbonate; (3) amylase; (4) trypsin; (5) lipase. Values are expressed as quantities excreted in the one-hour test period and as concentrations.

This procedure has important practical limitations. The double-lumen tube is comparatively large and produces a variable degree of discomfort. The tips must be placed properly under fluoroscopic guidance, and even then varying amounts of gastric juice may enter the duodenum and contaminate the samples of pancreatic juice. There is evidence that pancreatic secretion may be stimulated mechanically by the presence of the tube in the duodenum, contributing to the rather wide variation in values obtained in normal subjects. Moreover, there has been difficulty in obtaining uniformly active preparations of secretin suitable for intravenous injection.

Volume. In normal subjects, 135 to 250 ml. of pancreatic juice are obtained in the sixty-minute period, or 2.1 to 4.5 ml. per kilogram of body weight. In the majority of instances maximum secretion occurs during the first ten to twenty minutes, with a subsequent rapid fall, approximating the control rate of secretion in sixty to one hundred and twenty minutes. The maximum rate of secretion varies from about 2 to 8 ml. per minute.

Bicarbonate. The normal total bicarbonate output in the one hour period is 90 to 130 mmol., the maximum concentration occurring usually in forty to sixty minutes and falling more slowly than does the volume. In order to overcome the difficulty of a wide range of response to the secretin injection, results have been expressed in different ways. It has been found that the maximum bicarbonate concentration in any one of the 10-

or 20-minute samples is the least variable from observer to observer. In different clinics, the lower levels of maximum HCO_3^- output ranged from 65 to 94 mmol./l.

Amylase. About 300 to 1200 Somogyi units of amylase are excreted in the one-hour period (5.5 to 11 units per kilogram of body weight). The curve of excretion reaches a peak during the first twenty minutes, falls sharply after twenty minutes and then gradually during the next one to two hours. The concentration of the enzyme is lower than in the resting juice except during the first ten minutes, during which the pancreas is probably discharging its store of preformed amlyase. Subsequently, the stimulated gland forms a rather constant amount of enzyme, independent of the volume of secretion.

Trypsin. About 20 to 40 units are excreted normally in one hour (0.35 to 0.7 unit per kilogram of body weight). What has been said of amylase applies to trypsin.

Lipase. The normal one-hour excretion is 7000 to 14,000 units (135 to 225 units or 1.2×10^4 to 2.1×10^4 U per kilogram of body weight).

Abnormal Findings with the Secretin Test

Abnormal findings have been obtained with this procedure in patients with acute and chronic pancreatitis, pancreatic cysts, hemochromatosis, cystic fibrosis. calcification, carcinoma of the pancreas and pancreatic edema. Abnormal results, indicating clinically unsuspected complicating pancreatic disturbance, have been obtained also in patients with diabetes mellitus, cholelithiasis, cirrhosis of the liver, acute hepatic necrosis and late syphilis. The procedure may be useful in distinguishing steatorrhea of pancreatic origin from the idiopathic form (sprue, celiac disease), in which the secretin test yields normal results.

The general statement has been made that in cases of obstruction of the pancreatic duct or extensive destruction of pancreatic tissue (e.g., acute necrosis, carcinoma or fibrosis) there is usually a decrease in total values for volume, bicarbonate and enzymes, under the conditions of the test, the decrease being proportional to the degree of duct obstruction or extent of tissue destruction. However, whereas this may be true of advanced lesions, such findings are not commonly observed in cases of mild pancreatic dysfunction. The earliest functional abnormality appears to be reflected usually in decreased excretion of amylase and lipase and later by a decrease in trypsin, the volume and particularly the bicarbonate being less easily disturbed. Nondissociated disturbance of function, as evidenced by a decrease in all factors, may perhaps be interpreted as indicating obstruction of the pancreatic duct or a significant diminution in the mass of functioning pancreatic tissue. Dissociated disturbance, first consisting usually in a decrease in amylase or lipase, the other factors being relatively normal, represents a milder form of functional disturbance and is seen characteristically in mild forms of acute and chronic pancreatitis.

Experience has shown that obstruction of the main duct by a carcinoma of the head of the pancreas usually gives a low-volume response with normal HCO_3^- and enzyme concentration. A normal-volume response with normal enzyme concentration but lowered maximal HCO_3^- is more often found in individuals with chronic pancreatitis. The HCO_3^- is very low in cases with calcification of the pancreas.

High-volume output with low HCO_3^- concentration can occur in patients with acute pancreatitis, chronic alcoholism or cirrhosis.

In about half the patients with acute pancreatitis, normal function returns in a few days, with enzyme secretion the last to return to normal. Chronic damage is likely to be present if function has not returned to normal within two months.

It has been suggested that more reliable results can be obtained by attempting to achieve a maximal secretin response; this can be done by administering the hormone by constant-rate intravenous infusion. Ten units of secretin per kg. body weight per hour usually suffices, but occasionally the dosage required may be as high as 25 units/kg./hr.

The chief disadvantages of the secretin test as a routine diagnostic procedure are the practical limitations referred to previously. Findings within the normal range are commonly obtained in patients with mild pancreatitis. Although advanced disease is usually reflected in moderate or severe impairment of exocrine function, as indicated by this procedure, it does not often provide information of value in differential diagnosis apart from the rather loose generalizations already stated. Similar findings can be obtained in patients with pancreatic fibrosis, cysts, calcification, edema, inflammation, carcinoma, duct obstruction and other conditions.

The response to insulin, pancreozymin, and mecholyl chloride has been utilized in the same manner as the secretin test. These agents normally produce a marked increase in the output (concentration) of pancreatic enzymes, the bicarbonate being unaffected. They have also been given in conjunction with secretin, the resulting output of enzymes being much greater than with secretin alone. However, the findings in abnormal states appear to parallel those obtained with secretin alone, and the use of these agents adds little information of diagnostic value. It has been found by some that large doses of cholecystokinin-pancreozymin given by constant-rate intravenous infusion produce a maximal and more reliable amylase response.

LUNDH TEST MEAL

Under fluoroscopic control, a thin polythene tube is introduced so that its tip is in the distal duodenum. The patient then drinks 300 ml. of a liquid test meal containing standard quantities of carbohydrate, protein and vegetable oil. Small samples of duodenal juice are collected every 20 minutes for 2 hours and stored in iced containers. Trypsin, amylase and lipase concentrations are then determined in each sample. The results are expressed as peak enzyme concentration or as the mean concentration. In patients with steatorrhea or obstructive jaundice due to pancreatic disease (including cancer), enzyme levels are very low. This test is not a substitute for the secretin test.

SERUM ENZYMES

Diagnosis of certain types of pancreatic disease may be facilitated by studies of the concentrations in the serum of digestive enzymes of the pancreas. General principles underlying the production of abnormalities in these substances are considered elsewhere (p. 550). We are concerned here with their specific applications in the diagnosis of pancreatic disorders.

Serum Amylase. The chief clinical value of serum amylase determinations lies in the diagnosis of acute pancreatitis, whether edematous, hemorrhagic, suppurative or necrotic. In this condition, the serum amylase practically invariably increases almost simultaneously with the onset of symptoms, usually rising above 500 Somogyi units/100 ml. (normal 40 to 160) and occasionally to over 3000 units, the peak being reached usually in twelve to twenty-four hours, but occasionally as late as forty-eight hours. After reaching the peak, there is usually a precipitous but occasionally a gradual fall to a normal or subnormal level within two to six days after the onset. The absence of an increase in serum amylase within the first six to twenty-four hours after the onset of acute symptoms is strong evidence against the possibility of acute pancreatitis as the cause of the symptoms, except in rare instances of extremely rapid and extensive destruction of acinar cells. Because of the rapidity with which normal values are restored, negative findings after forty-eight to seventy-two hours do not exclude this possibility. With the possible exception of serum lipase determinations, this procedure constitutes the most valuable method available for the diagnosis of acute pancreatitis.

The increase is due to inflow of preformed amylase from the damaged pancreas into the circulation via the portal vein and lymphatics (pancreas, mesentery, peritoneum).

The rise in the blood is the resultant of (a) the rate of its entrance into the circulation, usually proportional to the rapidity and extent of pancreatic damage, and (b) the rate of its removal, by metabolism, and, mainly, renal excretion. Following the initial event, the quantity of enzyme liberated from the pancreas decreases rapidly, whereas it continues to be removed from the blood stream. Consequently, the concentration in the serum falls steadily, usually to a normal level, in a relatively short time.

The serum amylase may increase in patients with peptic ulcer. In some cases, this may be due to acute pancreatic damage by a posterior penetrating duodenal ulcer; in others, it may result from entrance of duodenal contents into the peritoneal space as a consequence of perforation of a duodenal ulcer. The values obtained are usually considerably lower than in acute pancreatitis.

Elevations occasionally seen in biliary tract disease, usually with common duct stone, have been attributed to associated pancreatitis. Elevations have been reported also in cases of strangulated high intestinal obstruction, presumably due to passage of the enzyme from the lumen of the devitalized bowel into the blood stream, lymph or peritoneal space.

Transient slight increases may occur following administration of opiates; this has been attributed to spasm of the sphincter of Oddi and increased absorption by lymphatics as a result of the increased intraductal pressure. Although moderate elevation (200 to 1000 units) has been reported to result from renal functional impairment, due to deficient excretion of amylase, this factor is more important clinically in augmenting and prolonging the increase in subjects with coincident acute pancreatitis.

Moderate increase has been observed occasionally in certain diseases of the salivary glands, including mumps, suppurative parotitis and calculous obstruction of the salivary duct. High values have been reported in occasional cases of ruptured ectopic pregnancy, presumably due to absorption from the peritoneum of amylase derived from the Fallopian tubes.

Normal findings are the rule in chronic pancreatic lesions, e.g., carcinoma, atrophy, cysts, fibrosis, calcification. However, values comparable to those in acute pancreatitis are observed rather consistently during acute exacerbations in chronic recurrent (relapsing) pancreatitis and, occasionally, in carcinoma of the head of the pancreas (duct obstruction). In acute obstruction of the pancreatic duct, e.g., calculous, there may be an immediate transient increase in serum amylase; this may recur with repeated attacks and may be detectable only by repeated studies. Macroamylasemia has already been discussed (p. 551).

Serum Amylase after Stimulation. Procedures have been described for use in diagnosis of nonacute pancreatic disorders that involve measurement of serum amylase (or lipase) activity following stimulation by secretin, or with pancreozymin injected just after the secretin, or 10 to 20 minutes beforehand. The test procedures described vary considerably, but the underlying principles are essentially the same.

If secretin is injected in a dosage that normally does not produce an increase in serum amylase, the presence of an increase suggests obstruction to the flow of pancreatic juice through the duct system and a pancreas capable of responding to a normal stimulus. Injection of secretin and pancreozymin normally produces an increase in serum amylase. Absence of this increase suggests pancreatic secretory inadequacy, such as may occur in chronic pancreatitis, fibrocystic disease, carcinoma, atrophy, and calcification.

Rather limited experience with these procedures does not permit satisfactory evaluation of their clinical significance. Reported results vary considerably, due in part at least to commercial unavailability of purified secretin of standardized activity.

Serum Lipase. The lipolytic activity of the serum may be determined by the amount of olive oil emulsion hydrolyzed by a given quantity of serum in a given time at 37° C. (p. 551). The values for lipase may be expressed as the amount of 0.05 M NaOH required to neutralize the fatty acid produced by 1 ml. of serum. The upper limit of normal is about 1.0 ml. of 0.05 M NaOH (277 U/l.). Measurement of esterase activity (tributyrin substrate) may be substituted for that of lipase. It has certain technical advantages and yields similar results.

Increase in serum lipase is a reflection of pancreatic disease, the enzyme entering the blood stream in the same manner as pancreatic amylase. The findings with regard to lipase differ in certain important respects from serum amylase in pancreatic disease. As in the case of the latter, in acute pancreatitis the serum lipase increases promptly at the time of onset of symptoms, values as high as 2800 U/l. having been reported. The subsequent fall is more gradual than in the case of amylase, elevated values persisting in some cases for ten to fourteen days or longer (less rapid removal from circulation). Because of this fact, accurate diagnosis is often possible by means of lipase determinations after the serum amylase has fallen to an essentially normal level. This procedure is preferred by some for this reason. In general, findings in nonacute pancreatic diseases parallel those for serum amylase, although, according to some observers, lipase abnormalities occur more consistently in these conditions.

Serum Trypsin. Studies of serum trypsin have not been reliable because of the presence in normal blood of substances with inhibiting actions. A procedure has been proposed which employs a synthetic polypeptide substrate that is not affected by such substances. Positive results, i.e., increase in serum trypsin, have been reported in a limited series of cases of pancreatitis and pancreatic cancer. This procedure, however, has not proved to be reliable.

Amylase Isoenzymes. One of the problems regarding serum amylase estimations is confusion between a salivary gland source or a pancreatic source. Each source, however, yields a different isoenzyme, and separation can be successfully effected by electrophoresis. This finding can be useful clinically.

AMYLASE IN URINE

In the absence of significant impairment of renal function, variations in amylase and lipase in the urine tend to parallel those in the blood. Usually, in acute pancreatitis, the urine values rise shortly after those in the blood, the increase persisting for several days. Values as high as 8000 Somogyi units/ml. (normal, 5–50 units/ml.) have been reported in acute pancreatitis, 300 units or more being of diagnostic significance. Elevated values have been obtained in chronic pancreatitis (during acute exacerbations), stone in the common bile duct, perforating duodenal ulcer, obstruction of the pancreatic duct, and occasionally in carcinoma of the pancreas.

The reliability of this procedure is impaired because of the common occurrence of variations in renal function and in the concentration and pH of the urine. It is best, however, to have a timed collection of urine, preferably a 24-hour specimen. The normal output ranges from 43 to 245 units/hour.

MISCELLANEOUS FINDINGS IN PANCREATIC DISEASE

Hyperbilirubinemia, often without visible jaundice, may occur in patients with acute or chronic pancreatitis. This is due in some cases to associated hepatic or biliary tract disease and in others to common duct obstruction incident to the pancreatic lesion. Hyperbilirubinemia occurs at some time during the course of the disease in over 80 per cent of cases of carcinoma of the head of the pancreas, frequently as the initial clinical manifestation. It may be absent in cases of even extensive malignant involvement of the body and tail. The determining factor is the anatomical relation of the pancreatic lesion to the terminal portion of the common bile duct.

Glucose tolerance is decreased in many cases of pancreatic carcinoma and also in chronic (about 30 per cent) and acute pancreatitis. Fasting hyperglycemia has been observed in 15 to 100 per cent of different series of cases of acute pancreatitis, with glycosuria in 5 to 25 per cent. The incidence is lower in chronic pancreatitis (hyperglycemia in 15 to 30 per cent).

Hypocalcemia has been reported in acute pancreatitis (up to 70 per cent), developing usually twenty-four to seventy-two hours after the acute attack (p. 305). The decrease is of slight degree (8–8.5 mg./100 ml.) in the majority of cases, but values below 7 mg./100 ml., accompanied by tetany, are observed occasionally and may indicate a poor prognosis. This hypocalcemia has been attributed, questionably, to sudden withdrawal of large amounts of calcium from the extracellular fluids by fixation as calcium soaps (i.e., by fatty acids) in areas of fat necrosis (p. 305). It may be due to excessive release of cortisol. A normal serum calcium can be an indication of an underlying hyperparathyroidism, which is associated with an increased frequency of acute pancreatitis.

A type of chronic relapsing pancreatitis has been described which is associated with an increase in plasma lipids, mainly neutral fat, but also cholesterol and phospholipids. Other manifestations of hyperlipemia may be present, viz., skin xanthomatosis and lipemia retinalis, and the recurring acute attacks resemble those in idiopathic familial hyperlipemia (p. 134). The pancreatic involvement, as well as other manifestations, has been attributed to fat embolization due to change in the physical state of the plasma lipids. The hyperlipemia often persists during symptom-free periods.

A syndrome (Zollinger-Ellison) has been described of gastric hypersecretion, peptic ulceration, and gastrin-secreting tumors ("ulcerogenic") of islet cells apparently other than the alpha and beta cells; there is consequently no associated disturbance of blood sugar concentration. Some cases exhibit multiple endocrine tumors, i.e., adenomas of the pituitary, parathyroids and adrenals.

Disturbance of the sweating mechanism occurs in children with fibrocystic disease of the pancreas, characterized by unusually profuse perspiration in hot weather. There is a striking increase in the Na and Cl content of the sweat, which does not exhibit the normal decrease following administration of deoxycorticosterone. This abnormality predisposes to the development of heat prostration in such cases (p. 378).

Measurement of the increment in plasma glycine concentration after ingestion of gelatin has been proposed as an index of tryptic activity in the intestine. Good results (subnormal rise) have been reported in pancreatic fibrosis. The same applies to studies of radioactivity in the blood (and feces) following ingestion of ^{131}I-labeled albumin. Similar studies with ^{131}I-labeled triolein are discussed in the section on the malabsorption syndrome (p. 605). An oral starch tolerance test has also been used.

EXAMINATION OF FECES

Examination of the feces is of little practical value in the diagnosis of impaired exocrine function of the pancreas except in far advanced form. This is due to several factors. (1) The pancreas has a large reserve capacity, and significant digestive abnormalities may occur only when secretory activity is virtually abolished. (2) Secretion of pancreatic enzymes is influenced by the intake of protein and intestinal absorption of amino acids. Inadequate absorption due to extrapancreatic disorders may, therefore, significantly influence pancreatic function. (3) The action of other gastrointestinal and bacterial enzymes may mask the deficiency in pancreatic digestive enzymes. (4) Digestive inadequacies, as reflected in excessive amounts of lipids and nitrogenous constituents of the feces, may be due to rapid transit through the intestine rather than to enzyme deficiencies. It is frequently difficult on this basis to distinguish between pancreatic functional impairment and other conditions associated with diarrhea and intestinal malabsorption.

Because pancreatic lipase is the only important lipolytic enzyme of the digestive fluids, severe pancreatic secretory deficiency is usually reflected most strikingly in inadequate fat digestion, particularly on a high fat intake. The stools are characteristically pale and bulky, containing more than 12 g. fat and 3 g. protein (high fat diet for three days). Microscopic examination reveals fat globules and undigested protein fibers. Determination of the relative proportions of neutral fat and fatty acids is of little practical value in distinguishing between pancreatic insufficiency and other causes of steatorrhea (p. 600). The trypsin content of the feces was commonly determined by exposing a gelatin emulsion (x-ray film) to a suspension of feces. Experience with this procedure has not been satisfactory except in young infants with congenital cystic fibrosis.

Many of the fecal abnormalities in pancreatic disease occur also in other conditions accompanied by disturbances of intestinal motility, digestion and absorption. They will be considered in greater detail in the chapter Intestinal Malabsorption Syndromes (p. 600).

References

Adlersberg, D., ed.: The Malabsorption Syndrome. New York, Grune & Stratton, 1957.

Edmondson, H. A., et al.: Calcium, potassium, magnesium, and amylase disturbances in acute pancreatitis. Am. J. Med., *12:*34, 1952.

Grossman, M. I.: Control of pancreatic secretion. *In* Beck, I. T., and Sinclair, D. G., eds.: The Exocrine Pancreas. London, J. & A. Churchill, 1971.

Mallinson, C. N.: Tests of pancreatic function. Brit. J. Hosp. Med., *1:*161, 1968.

Rankin, G. B., and Brown, C. H.: "Routine" studies in suspected pancreatic disease. *In* Brown, C. H. (ed.): Diagnostic Procedures in Gastroenterology. Saint Louis, C. V. Mosby Co., p. 122, 1967.

Volwiler, W.: Gastrointestinal malabsorption syndromes. Am. J. Med., *23:*250, 1957.

Chapter 20

INTESTINAL MALABSORPTION SYNDROMES

Inadequate intestinal absorption of essential nutrients results in (1) escape of excessive quantities of these materials in the feces and (2) systemic manifestations of nutritional defects, including compositional abnormalities in the circulating blood. These clinical and metabolic manifestations, collectively designated the malabsorption syndrome, may result from various causes. They include: (1) critical deficiencies in digestive secretions, e.g., in bile salts or pancreatic enzymes or intestinal enzymes; (2) gross intestinal disease, e.g., jejuno-ileitis, intestinal lipodystrophy, amyloidosis, lymphosarcoma; (3) inadequate contact (extent or time) of nutrient materials with the intestinal absorptive surface, e.g., gastrocolic fistula, extensive intestinal resection, chronic diarrhea; (4) inherent functional defects of the mucosal cells, e.g., sprue and celiac disease; (5) specific defects due to lack of an essential factor, e.g., intrinsic factor deficiency resulting in inadequate vitamin B_{12} absorption (pernicious anemia); (6) hypogammaglobulinemia; (7) a-beta-lipoproteinemia; (8) certain skin disorders; (9) Zollinger-Ellison syndrome; (10) lactase deficiency; (11) certain endocrine disorders and (12) lesions due to drugs.

We are not concerned here with isolated specific absorptive defects, e.g., pernicious anemia (vitamin B_{12} deficiency due to absence of intrinsic factor), but rather with broader defects. These may manifest themselves in a variety of biochemical abnormalities.

Steatorrhea. Decreased intestinal absorption of lipids is the most common single manifestation of the malabsorption syndrome. In normal subjects, with a fat intake that is not excessively high, the lipid content of the feces seldom exceeds 5 per cent (6 g.) of the amount ingested. It is mainly of endogenous and bacterial origin (p. 101). This figure (fecal lipid) may increase to 50 per cent or more in pancreatic disease, 40 per cent in celiac disease and sprue, and 20 per cent in hepatic disease or obstructive jaundice. Although this increased lipid excretion is usually accompanied by diarrhea, excessive amounts of fat may be lost in a single or few semi-formed stools. Obstruction of the mesenteric lymphatics or thoracic duct may result in steatorrhea, as in lymphomatous diseases, lipodystrophy (Whipple's disease), tuberculosis, metastatic carcinoma of the pancreas. In such conditions there may be an associated chylous ascites. As a rule, in the case of gross anatomical abnormalities of the intestine, quantitatively significant absorptive defects seldom occur unless more than 50 per cent of the small bowel is unavailable for absorption of intestinal contents. This may occur in such conditions as extensive small bowel resections (especially distal portion, see p. 659) and gastrocolic fistula. However, steatorrhea may occur occasionally in less extensive disorders, such as diverticulosis, blind loops and intestinal strictures. Its pathogenesis in these conditions is not clear, but it has been attributed to proliferation of intestinal bacteria in the areas of local stasis, the usually accepted basis for the more common manifestations of vitamin B deficiency in these disorders (p. 602). The bacteria may deconjugate bile salts, and so interfere with fat absorption.

The principal underlying cause of steatorrhea in hepatic disease and obstructive jaundice is inadequate digestion of fat resulting from the deficiency in bile acids in the intestine; in pancreatic disease it is inadequate fat digestion resulting from deficiency in pancreatic lipase; in celiac disease (gluten sensitivity) and sprue it is inadequate absorption of lipid resulting from some inherent defect in the intestine, with no abnormality of fat digestion. Although these differences in pathogenesis should theoretically be reflected in differences in the lipid composition of the feces, in practice, the absorptive difficulty in celiac disease and sprue can seldom be distinguished on this basis from the digestive difficulty in hepatobiliary and pancreatic disease. This may be due, in part at least, to the morphologic and functional abnormalities that occur in the pancreas as a result of the disturbances in protein nutrition in celiac disease and sprue (p. 597).

Celiac disease seems to result from an abnormal immunological response. There is a preponderance of IgM in the jejunal plasma cells, where normally there should be IgA. The serum contains IgM type antibodies to fractions of gluten. It is possible that a low IgA response to food antigens in the jejunum might play a role. Failure of production of IgA may play some part in the malabsorption associated with hypogammaglobulinemia. The Zollinger-Ellison syndrome is associated with defective fat absorption because of inactivation of lipase by the high acid content of the stomach entering the duodenum. In a-beta-lipoproteinemia, there is a failure of fat transport from mucosal cell to lymph, because of lack of the necessary apolipoprotein. Acquired lactase deficiency is associated with a number of intestinal disorders, including sprue and celiac disease.

Protein Abnormalities. Inadequate protein digestion (e.g., pancreatic insufficiency or loss from protein-losing gastroenteropathy) and/or amino acid absorption, characterized by excessive nitrogen excretion in the feces and negative nitrogen balance, may lead to clinical manifestations of protein undernutrition. Hypoproteinemia occurs commonly; this is due mainly to decrease in albumin (giving rise to edema), although occasionally the gamma-globulin fraction is also low in severe cases, particularly in celiac disease. Osteoporosis (p. 313), hypochromic anemia, and amenorrhea (deficient gonadotrophin production, p. 698) are additional reflections of protein undernutrition.

Irrespective of the original cause, prolonged protein undernutrition leads to structural and functional abnormalities in the pancreas and intestinal mucosa, setting up a vicious circle. This frequently adds to the difficulties of etiologic diagnosis of the malabsorption syndrome.

Vitamin Deficiencies. Consequences of inadequate absorption of certain vitamins contribute materially to the clinical picture of the malabsorption syndrome. The impairment of fat absorption extends also to fat-soluble vitamins. Vitamin D deficiency occurs commonly in these conditions and, together with the decreased calcium absorption incident to the steatorrhea, may produce characteristic features of rickets during the period of skeletal growth and osteomalacia in adults. Hypocalcemic tetany may occur at any age. These manifestations of vitamin D deficiency are usually most severe in children with celiac disease.

Vitamin K deficiency may be reflected in a decrease in plasma prothrombin, occasionally sufficiently pronounced to produce hemorrhagic manifestations. Although there are seldom recognizable evidences of vitamin A deficiency, the concentrations of carotene and vitamin A in the plasma are often low (p. 606).

Of the water-soluble vitamins, deficiencies in folic acid and/or vitamin B_{12} occur most frequently, producing characteristic macrocytic, megaloblastic types of anemia. This may dominate the clinical picture of malabsorption in cases presenting the intestinal "blind-loop syndrome," due presumably to excessive utilization of vitamin B_{12} by the proliferation of bacteria in the area of intestinal stasis or by the binding of the B_{12}-intrinsic factor complex to bacteria, instead of the lower small intestine acceptor sites. Glossitis, cheilosis, and peripheral neuritis, occurring particularly in celiac disease and sprue, may be due to deficiency in nicotinic acid, riboflavin, and thiamine. There are also reports of ascorbic acid deficiency, although this is unusual.

Anemia. Several factors essential to normal hemoglobin synthesis and erythrocyte maturation may be involved in the absorptive difficulties, including protein, iron, folic acid, and vitamin B_{12}. Anemia, at times profound, occurs frequently in all forms of the malabsorption syndrome. It varies in character, the hematologic picture in a given case depending upon the extent of deficiency in the several erythropoietic factors.

Inadequate absorption of folic acid and/or vitamin B_{12} produces a macrocytic megaloblastic anemia typical of deficiency in these agents (p. 818). This occurs characteristically in sprue, celiac disease, and certain intestinal abnormalities (*Diphyllobothrium latum* infestation, diverticulosis, multiple intestinal strictures, "blind-loop syndrome"). However, the hematologic picture may be complicated by coexisting deficiencies in iron and protein, producing a microcytic hypochromic anemia with an erythroid bone marrow.

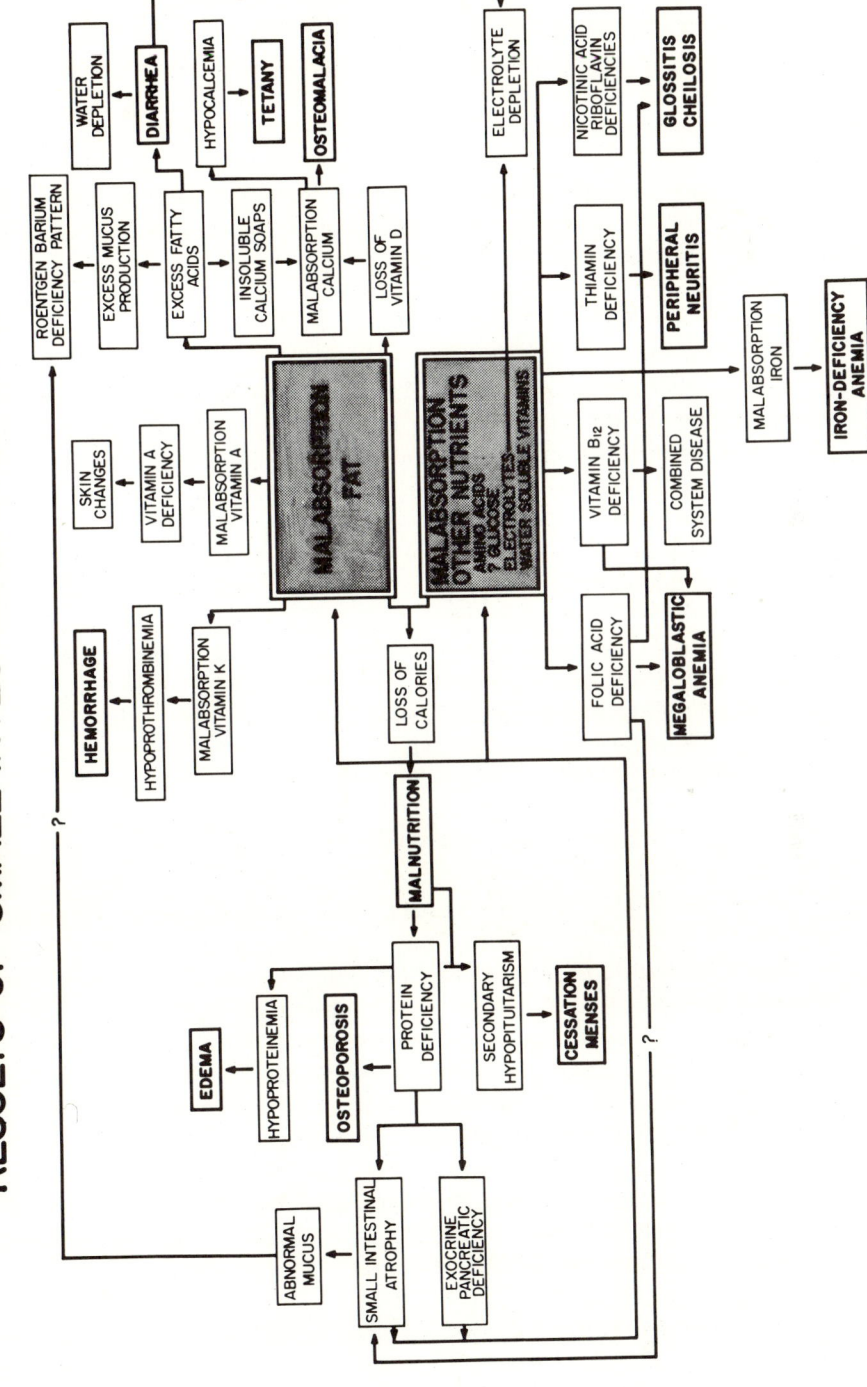

Figure 20-1. Gastrointestinal malabsorption syndromes. (From Volwiler: Am. J. Med., 23:250, 1957.)

This may be the dominant type in the other forms of the malabsorption syndrome.

Skeletal Abnormalities. These vary considerably in nature and extent, being influenced by age and the duration and severity of diarrhea and steatorrhea. Because of the higher requirements for skeletal growth, they are generally more pronounced in children, e.g., celiac disease, fibrocystic disease. Inadequate absorption of vitamin D and calcium (steatorrhea) (p. 302) leads to rickets, osteomalacia and, in severe cases, to hypocalcemic tetany. The skeletal lesion may be aggravated by an associated state of secondary hyperparathyroidism (p. 295). Protein malnutrition may lead to osteoporosis, i.e., defective development of the osteoid matrix (p. 313), producing a clinical picture with respect to skeletal and biochemical abnormalities that may be confusing.

Miscellaneous Abnormalities. The extent of loss of water and sodium depends upon the duration and severity of diarrhea; dehydration may be a serious feature in certain cases. Potassium depletion also occurs under these circumstances. Moreover, in celiac disease and sprue the amount of potassium lost in the feces may be out of all proportion to the water loss.

Increased bacterial fermentation of carbohydrates resulting from their inadequate digestion and absorption may contribute to the production of diarrhea.

One of the commonest and most serious features of the malabsorption syndrome in its advanced form is the state of general undernutrition, including loss of weight, resulting from inadequate provision of energy-producing foodstuffs, i.e., caloric undernutrition.

CLINICAL METHODS OF STUDYING ABSORPTION

The procedures available clinically for evaluating the adequacy of intestinal absorption fall into two general categories:

(1) Modified Balance Studies. These involve oral administration of a standard amount of a test substance and subsequent analysis of the feces to determine the proportion of the substance that has escaped absorption. A number of substances have been employed for this purpose, including proteins, lipids, carbohydrates, vitamin B_{12}, and iron. In principle, procedures of this type provide the most definitive information concerning absorption of the test substances or the adequacy of their conversion to absorbable substances, if this is necessary, e.g., digestion of proteins. However, they present difficulties inherent in all studies involving timed collection of feces in relation to dietary intake. Nevertheless, these are by no means insurmountable.

(2) Tolerance Tests. A standard dose of the test material is administered orally and measurements are made of either (*a*) the increase in concentration of the substance (or its digestion products) in the blood, or (*b*) the amount excreted in the urine. Because of the operation of many variable and uncontrollable factors, results obtained with these procedures are frequently more difficult to interpret than those in the preceding category.

There are, however, important exceptions to this generalization; studies of urinary excretion of radioactive vitamin B_{12} and of D-xylose, under standardized test conditions, are among the more useful clinically available procedures for studying intestinal absorption.

Lipid Absorption. Disturbances in intestinal absorption of lipids usually dominate the clinical picture of the malabsorption syndrome. Several laboratory procedures have been proposed for the demonstration of these defects.

(1) Balance Studies. (a) Fecal Fat after a Standard Fat Meal. This procedure involves quantitative determination of fat excretion in the feces for periods of three to five days on a known fat intake. With a daily fat intake of 100 g., normal subjects excrete in the feces less than 6 g. of fat daily. Apart from certain troublesome technical considerations, this procedure is perhaps the most accurate for demonstration of fat absorption defects. Difficulties in relating periods of collection of feces to timed fat ingestion periods may be minimized by prolonging the study over several days. The procedure may be simplified by eliminating accurate measurement of fat intake. Fecal excretion of more than 6 g. daily may be regarded as excessive on an adequate dietary fat intake.

(b) ^{131}I-labeled Lipids. The substances used most frequently in procedures of this type are triolein and oleic acid. Theoretically, inadequate utilization of triolein may be due to (a) incomplete digestion and/or (b) incomplete absorption; inadequate utilization of oleic acid is attributable to absorptive defects only.

According to a procedure commonly employed, measurements of radioactivity are made in blood samples removed four, five, and six hours following ingestion of the labeled lipid, and in the feces collected over a period of forty-eight hours. In normal subjects, the average activity of the three blood samples is at least 10 per cent of the administered dose and that of the feces less than 5 per cent.

Chemical determination of fecal fat affords a more precise measure of the adequacy of fat absorption than does the ^{131}I-triolein test. However, because of the ease with which radioactivity measurements can be made, the latter procedure can be useful clinically, chemical analysis being resorted to in cases in which equivocal results are obtained. The evening before the test is carried out, the patient must be given 10 drops of Lugol's iodine solution to saturate thyroid uptake of iodide. It has recently been suggested that differently labeled triolein and oleic acid should be given simultaneously.

Normal results with both ^{131}I-oleic acid and ^{131}I-triolein may be obtained at times in subjects with various forms of the malabsorption syndrome. However, a diagnosis of steatorrhea is justified when the blood and feces values are abnormal, i.e., low activity in blood and high activity in feces. Moreover, with occasional exceptions as indicated above, normal values may be regarded as supportive evidence against this diagnosis.

According to some, the combination of abnormal findings with triolein and normal results with oleic acid is helpful in establishing a diagnosis of pancreatic insufficiency, in contrast to other causes of steatorrhea, in which abnormal results are obtained with both procedures. Although this may be helpful, abnormal results with both tests are not uncommon in pancreatic disease. It has been suggested, too, that conversion of an abnormal response to triolein to normal by administration of pancreatic extracts may aid in differentiating pancreatic insufficiency from other types of steatorrhea. With this procedure, as with others, although positive results may be helpful, negative results are not, one important reason being the lack of consistent enzymatic activity in commercially available extracts.

(c) **Blood Lipids after a Standard Fat Meal.** In normal subjects, ingestion of a meal rich in fat is followed by a rise in the concentration of blood lipids, demonstrable by (1) chemical analysis, (2) chylomicron counts (dark-field examination), or (3) serum turbidity measurements. All of these procedures have been employed following a standard fat meal, e.g., butter fat, for purposes of evaluating the adequacy of fat absorption. Their clinical usefulness is limited by the wide variation in normal subjects and by the modifying influences of a number of factors that may be either related to or independent of the intestinal absorptive disturbance, e.g., hormonal influences, carbohydrate nutrition, endocrine function, plasma volume. Moreover, because of technical difficulties, chemical analysis and chylomicron counts do not lend themselves readily to routine use in a general diagnostic laboratory. However, because of its simplicity, measurement of serum turbidity following a standard fat meal is useful as a routine screening procedure; normal findings may be regarded as presumptive evidence against the diagnosis of steatorrhea.

(d) **Vitamin A Tolerance Test.** Determination of the rise in the blood level of vitamin A, a fat-soluble substance, has been employed as a measure of the adequacy of fat absorption. In normal subjects, following administration of a standard dose of vitamin A, its concentration in the serum rises steadily over a period of about six hours (to at least 850 I.U./100 ml.) and then begins to fall. Although a flat curve is strongly suggestive of steatorrhea, it may be obtained occasionally in the absence of this condition. This is useful as a screening procedure in that a normal response makes a diagnosis of steatorrhea unlikely.

(e) **Serum Carotene Concentration.** If the antecedent dietary intake of carotene has been adequate, the fasting serum carotene concentration of normal subjects is 60 to 200 μg./100 ml. This, too, is useful as a screening procedure, a low fasting serum carotene concentration under these conditions being presumptive evidence of inadequate intestinal absorption of fat. However, a normal concentration cannot be assumed to indicate adequate fat absorption.

Carbohydrate Absorption. The only clinically useful procedure in this category is measurement of the urinary excretion of D-xylose after oral administration of a test dose (25 g.). In normal subjects, the blood xylose concentration rises to an average maximum of about 35 mg./100 ml. in one to two hours, falling subsequently to the resting level at about five hours. However, the wide overlap between blood curves in normal subjects and those with absorption defects limits the usefulness of this procedure as an index of adequacy of absorption. Measurement of urinary excretion of xylose under these conditions is of greater value. Normal subjects excrete 6.5 ± 20 per cent in five hours (lower limit of normal is usually taken as 5.0 g.). Distinctly subnormal values suggest an absorptive defect, provided kidney function is normal. Clinical experience with this procedure suggests that its usefulness is not as high as first thought. For children, the dose is 1.1 g./kg. up to a total of 25 g., of which more than 16 per cent should be found in the 5-hour urine specimen.

Protein Absorption. Inadequate assimilation of protein may result in hypoalbuminemia, particularly in children; occasionally the γ-globulin concentration is also lowered. The following studies it was hoped would provide a more direct approach to the demonstration of absorptive defects: (1) determination of fecal nitrogen; (2) measurement of the rise in blood glycine following ingestion of gelatin; (3) measurement of the rise in blood amino acids following a test meal of protein or amino acids; (4) determination of radioactivity in the blood and/or feces following ingestion of test doses of ^{131}I-labeled albumin or casein. There are, however, no routine clinical tests for protein absorption.

Under ordinary dietary conditions, normal subjects excrete 1 to 2 g. nitrogen daily in the feces. Abnormally high values, in the absence of protein-losing gastroenteropathy, are the most definitive indication of inadequate protein assimilation. Except under unusual circumstances, exact determination of the nitrogen intake is not essential, although calculations based on weighed diets are desirable. As with all quantitative studies of feces, difficulty is commonly experienced in timing the stool collections to correspond exactly to twenty-four-hour periods of food intake. This difficulty may be minimized by making determinations on mixed three- to six-day specimens. Usually the defect in protein absorption is less severe than that in fat. There is evidence that, in sprue, in addition to diminished absorption, there is also loss of protein from the intestinal mucosa into the lumen of the bowel (protein-losing enteropathy). Studies of fecal nitrogen do not aid in differentiating pancreatic insufficiency from other causes of intestinal malabsorption.

The existence of a protein-losing enteropathy may be detected by intravenous injection of ^{131}I-labeled polyvinylpyrrolidone. Under the test conditions, normal subjects excrete less than 1.5 per cent of the administered radioactivity in the feces (four-day collection). Abnormally high values are assumed to reflect increased intestinal permeability to this material in the circulating blood and, by implication, also to albumin, which has approximately the same molecular weight. Intravenous ^{131}I-labeled albumin can be used, instead of the polyvinylpyrrolidone; amberlite resin must be taken orally to fix the labeled iodine, since albumin can be broken down in the intestine and the iodine reabsorbed. Alternatively ^{51}Cr-labeled albumin can be used. It is usually sufficient to inject intravenously a small amount of radioactive chromium chloride. This labels the body albumin. Normally, feces should then show very little radioactivity.

Vitamin B_{12} Absorption. Vitamin B_{12} is readily absorbed from the normal intestinal tract in the presence of intrinsic factor (gastric mucosa). The availability of radioactive B_{12} added a useful tool for investigation of the malabsorption syndrome as well as for the diagnosis of pernicious anemia. Evaluation of the adequacy of absorption of orally administered ^{60}Co or ^{58}Co-labeled vitamin B_{12} may be made by one or more of the following procedures: (1) measurement of the radioactivity in the feces (seven days); (2) measurement of the radioactivity in the urine excreted during the subsequent twenty-four hours, the absorbed vitamin B_{12} being flushed out into the urine by subsequent intramuscular injection of 1 mg. unlabeled B_{12}; (3) measurement of radioactivity in the plasma or over the liver, which is a temporary storage site after absorption. The latter procedure is seldom employed in this connection.

Although measurement of radioactivity in the feces affords the most direct approach, because of obvious technical disadvantages the urine procedure is the method of choice and has been employed most widely. Under the test conditions at least 7 per cent of the administered oral dose (0.5–1.0 μCi in 1–2 μg. B_{12}) is excreted in the urine of normal subjects (twenty-four hours). Inadequate absorption is reflected in subnormal urinary excretion values. If this is due to absence of intrinsic factor (pernicious anemia, gastrectomy), repetition of the test with the addition of intrinsic factor will yield a normal response, whereas this has no effect in patients with the malabsorption syndrome. There is evidence, too, that a normal response may be obtained following antibiotic therapy in patients with "intestinal stasis" disorders (blind-loop syndrome, intestinal strictures, multiple diverticulosis), thus serving to distinguish these conditions from other varieties of the malabsorption syndrome.

Miscellaneous Procedures. The variable features of the anemia that occurs frequently in malabsorption syndromes have served to focus attention upon intestinal absorption of factors involved in hemopoiesis. In addition to those already considered (protein, vitamin B_{12}), these include also folic acid and iron. Folic acid absorption has been studied by determining the amount excreted in the urine after ingestion of a test dose. Use of this procedure for this purpose is limited by technical difficulties. Measurement of serum or red cell folate levels is more satisfactory. Low levels indicate the possibility of malabsorption of the vitamin. Alternatively, the level of urinary formiminoglutamic acid (FIGLU) can be measured after histidine loading. In folate deficiency, high values are obtained.

Serum iron levels are frequently low in the malabsorption syndrome.

Biopsy of Small Intestine Mucosa. It is now virtually routine for gastroenterologists to obtain biopsy specimens of mucosa from the small intestine. Frequently in malabsorption syndromes, the mucosa is flattened and virtually devoid of villi. It is possible to demonstrate a variety of enzyme deficiencies by biochemical estimation, as well as by histochemical methods. The former is particularly useful for demonstrating disaccharidase deficiency, especially that of lactase. It is important, however, that the biopsy site should be beyond the duodenal-jejunal junction, and the specimen must be rapidly frozen, as soon as it is obtained by the operator. It has been shown that the enzyme pattern of lactate dehydrogenase, alkaline phosphatase and esterases in the patient with celiac disease differs from that of the normal individual.

References

Adlersberg, D., ed.: The Malabsorption Syndrome. New York, Grune & Stratton, 1957.
Booth, C. C.: Enterocyte in coeliac disease. Brit. Med. J., 3:725, 1970.
Christiansen, P. A., Kirsner, J. B., and Ablaza, J.: D-Xylose and its use in the diagnosis of malabsorptive states. Am. J. Med., 27:443, 1959.
Douglas, A. P.: The value of intestinal biopsy as a diagnostic and research tool. Brit. J. Hosp. Med., 2:1400, 1968.
Farmer, R. G.: Tests for intestinal absorption. In Brown, C. H., ed.: Diagnostic Procedures in Gastroenterology. St. Louis, C. V. Mosby Co., 1967.
Scudamore, H. H., and Green, P. A.: Secondary malabsorption syndromes of intestinal origin. Postgrad. Med., 26:340, 1959.
Tietz, N. W.: Pancreatic function and intestinal absorption. In Tietz, N. W., ed.: Fundamentals of Clinical Chemistry, Philadelphia, W. B. Saunders Company, 1970.
Volwiler, W.: Gastrointestinal malabsorption syndromes. Am. J. Med., 23:250, 1957.

Chapter 21

HEPATIC FUNCTION

The liver is a wedge-shaped organ weighing 1200 to 1600 g. Its situation in relation to the rib cage is shown in Figure 21–1. It is made up of sheets of liver cells (hepatocytes), usually one cell thick. These sheets branch and intermingle in such a way that they form a labyrinth of lacunae (Fig. 21–2), made up mainly by the sinusoids, which are filled with blood and lined by the cells of Kupffer so as to leave a space full of protein-rich fluid (the perisinusoidal space of Disse) between them and the hepatocytes. The solid liver tissue formed in this way is interlaced by two sets of tunnels. These are, respectively, the portal tracts and the hepatic central canals. The former contain the bile ducts, the smaller branches of the portal vein and hepatic artery, lymph vessels and some connective tissue. The hepatic central canals contain the tributaries of the hepatic vein and lymph vessels. The two sets of tunnels, on the whole, run in planes at right angles to each other and are kept completely apart. Their terminal branches are separated from each other by approximately 0.5 mm. The terminal branches of the portal vein and hepatic artery open into the sinusoids, which are themselves connected with the tributaries of the hepatic vein. Both the portal tracts and the hepatic central canals are surrounded by a limiting plate of hepatocytes.

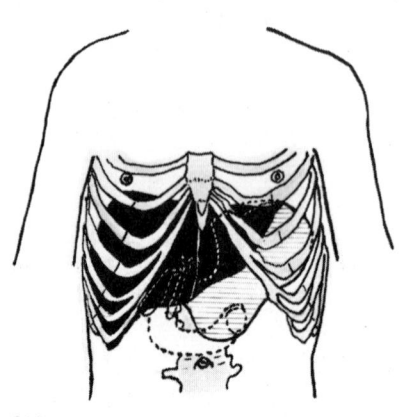

Figure 21-1. Relationship of the liver and stomach to the thoracic cage. The liver is shown in black, the stomach is cross-hatched and the portion behind the liver shown in white dashes, as is the gallbladder and the main bile ducts which open posteriorly into the duodenum (second part), shown as black dashes.

HEPATIC FUNCTION 611

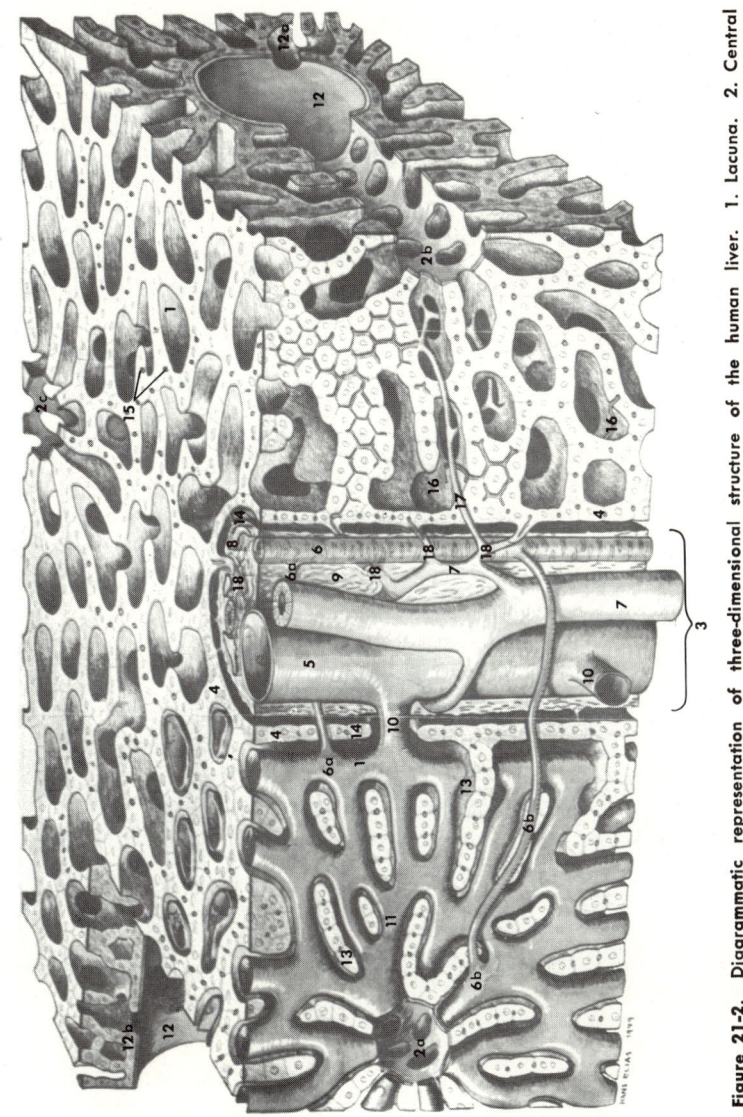

Figure 21-2. Diagrammatic representation of three-dimensional structure of the human liver. 1. Lacuna. 2. Central (hepatic) veins. 3. Portal canal (tract). 4. Limiting plate. 5. Portal vein. 6. Hepatic artery. 6a. Arterial capillary emptying into paraportal sinusoid. 6b. Arterial capillary emptying into intralobular sinusoid. 7. Bile ducts. 8. Lymph vessel. 9. Periportal connective tissue. 10. Inlet venules. 11. Sinusoids. 12. Sublobular vein. 13. Perisinusoidal space of Disse. 14. Periportal space of Mall. 15. Bile canaliculi within liver plates. 16. Bile canaliculi on the surface of liver plates (not infrequent). 17. Intralobular cholangiole. 18. Cholangioles in portal canals. (Reproduced by permission of G. D. Searle & Company.)

The hepatocytes are structured in such a way (Fig. 21-3) that indentations in their walls come together to form the bile canaliculi, which are nevertheless separated from the sinusoids, situated on the sinusoidal surface of the liver cells. The bile canaliculi join together (Fig. 21-2) to form intralobular and then portal cholangioles, which come together in a terminal branch of the portal tract to form small bile ducts. These, in turn, are branches of larger bile ducts, which also join to form the two hepatic ducts in the porta hepatis, and these then join to form the common bile duct, which joins the pancreatic duct as it opens into the second part of the duodenum. The gallbladder is joined to the common bile duct by way of the cystic duct.

It will be appreciated from this structural description that liver function can be affected not only by disturbances in the cells themselves but also by alterations in the biliary as well as in the circulatory system. There is a good deal of interplay among all these factors, which can make functional disturbances difficult to interpret.

Determination of the functional efficiency of the liver offers unusual difficulties for several reasons. One of these is the fact that, whereas the multiplicity of duties performed by the liver may in certain cases be equally involved in the presence of hepatic disease, it is probable that in some cases one or more of these functions may be disturbed relatively more than the others. Then, too, in several aspects of its functional activity the liver is so intimately associated with other organs that, in investigating these functions, it is difficult or impossible to delineate definitely the hepatic factor. Moreover, the liver possesses an enormous functional reserve and remarkable regenerative powers. Within a few weeks after the removal of an entire lobe, the liver of an animal may be as large as before the operation. Furthermore, by means of consecutive operations and by the institution of measures designed to inhibit this regenerative tendency, the total hepatic tissue may be permanently reduced to as low as 15 per cent of the normal amount without remarkable impairment of hepatic functional efficiency if a proper dietary regimen is maintained. Under such circumstances, however, the liver is not able to withstand untoward conditions as well as in its normal state. This may explain the frequent failure to demonstrate functional impairment in the human liver, even in the presence of advanced anatomic changes which may be either primary in the liver or secondary to lesions in the extrahepatic biliary passages. This is particularly true of chronic conditions such as cirrhosis, syphilis, malignancy, etc., which,

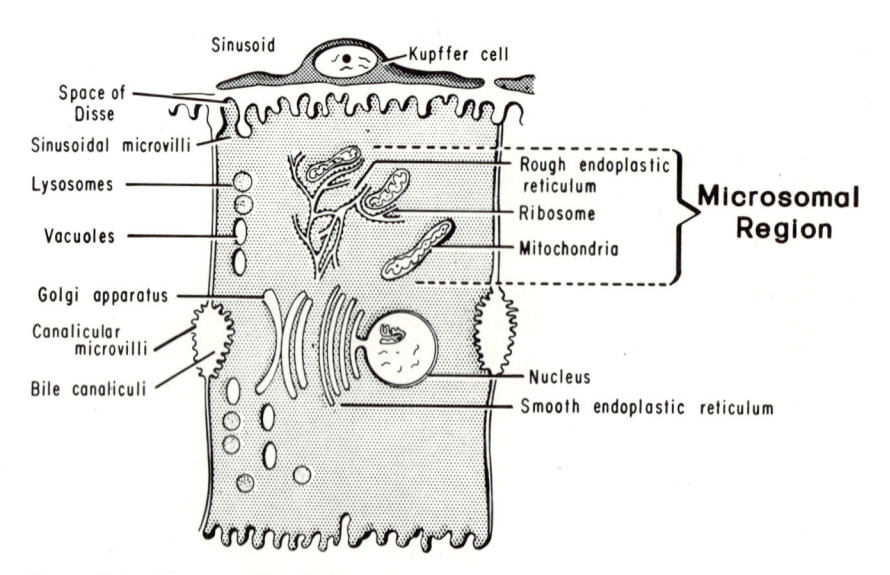

Figure 21-3. Structure and subcellular components of a normal liver cell. (From Tietz: Fundamentals of Clinical Chemistry. W. B. Saunders Company, 1970.)

because of their relatively slowly progressive nature, are associated with concomitant compensatory regeneration of hepatic tissue with little or no impairment of hepatic functional efficiency. In acute widespread hepatic disease, such as occurs in viral and toxic hepatitis and acute hepatic necrosis, the investigation of liver function yields more consistently positive findings.

In spite of the frequently negative findings, valuable information may, at times, be obtained in chronic diseases of the liver and biliary passages by the utilization of certain well established methods of investigating liver function. The routine employment of such procedures is of particular importance as a part of the preoperative study of patients with biliary tract disease, for by the institution of measures designed to increase hepatic functional efficiency, much can be done to diminish postoperative morbidity and mortality in these conditions. The fact is too frequently overlooked that functional tests are designed to indicate the state of functional activity of an organ and not necessarily the presence or extent of morphologic changes in that organ. Obviously, it is possible that an organ such as the liver may be the seat of extensive and even grossly evident chronic disease and still may be capable of carrying on its functions in an essentially normal manner; conversely, it is also conceivable that the functional activity of an organ may be impaired in the absence of demonstrable morphologic changes. If these facts are kept in mind, results obtained by the application of tests of liver function will have considerable clinical significance. On the other hand, any attempt to interpret functional findings alone in terms of disease diagnosis or of the extent of hepatic parenchymal change may lead to serious error. It is also important to point out that such findings not only have application in the diagnostic situation, but also can have an important bearing in relation to therapy and prognosis. Moreover, they may shed light on factors responsible for certain symptoms. It is for these reasons that functional studies with little or no diagnostic importance are included in this chapter.

CARBOHYDRATE METABOLISM

The important part played by the liver in normal carbohydrate metabolism has been considered elsewhere in detail (p. 13). Glucose, preformed or resulting from intestinal digestion of carbohydrates, reaches the liver through the portal circulation. The portion which is not utilized immediately for the maintenance of body activities is stored chiefly in the liver as glycogen which, when the necessity arises, is converted to glucose which passes in the general circulation to the tissues. Liver glycogen is also contributed to from two other sources: (1) glucose resulting from gluconeogenesis from protein, a process which takes place in the liver and is influenced by hormones of the adrenal cortex; (2) lactate, resulting from breakdown of glycogen in the muscles, passing in the blood to the liver, where it undergoes retransformation into glycogen under normal conditions of hepatic function and insulin action. The glycogen in the liver constitutes the reserve for the maintenance of the normal glucose concentration of the blood. This function of hepatic glycogen is not shared by glycogen deposits elsewhere in the body, for, following experimental extirpation of the liver in animals, hypoglycemia occurs which rapidly progresses to a fatal termination although relatively large quantities of glycogen may be present at the time in the muscles.

The influence of the liver in carbohydrate metabolism depends upon the functional activity of the hepatic cells and is not affected by obstruction to the flow of bile per se. Disturbance of hepatic cell function may result in (1) a tendency toward fasting hypoglycemia, (2) diminished glucose, fructose and galactose tolerance, (3) diminished blood sugar response to epinephrine, (4) glycosuria and (5) increased blood lactate concentration. The presence of such abnormalities in a patient with jaundice suggests the existence of impaired liver cell function, presumably due to hepatic cell damage, while their absence suggests that hepatic function is normal (uncomplicated obstructive or hemolytic jaundice). These procedures have, in the past, been employed for the purpose of distinguishing between hepatocellular and other types of jaundice. However, jaundice secondary to biliary obstruction, particularly if due to cholelithiasis, is usually accompanied, sooner or later, by a variable degree of hepatocellular damage. Moreover, hyperglycemia, diminished glucose tolerance and glycosuria may be contributed to by associated pancreatic disease, and the possibility of other, unrelated disturbances of carbohydrate metabolism must be taken into consideration. Because of these facts, the value of studies of carbohydrate metabolism in differentiating biliary obstructive from hepatocellular jaundice is distinctly limited.

Fasting Blood Sugar Level. Because of the extensive functional reserve and enormous regenerative capacity of the liver the fasting blood sugar is usually found to be within normal limits in hepatic disease, even though the hepatic glycogen content may be markedly depleted. This is particularly true of chronic disorders such as portal cirrhosis, biliary cirrhosis, syphilis, carcinoma, etc. Values of 50 to 60 mg. per 100 ml. may, at times, be observed in the terminal stages of cirrhotic processes, particularly in obstructive biliary cirrhosis but occasionally also in portal cirrhosis. Serious and occasionally fatal hypoglycemia may occur after operation for biliary tract disease under general anesthesia, which may exhaust the glycogen reserve already diminished by associated but not readily detectable hepatic disease. Hypoglycemia is more commonly observed in acute, extensive forms of hepatic disease. Values ranging from 25 to 60 mg. per 100 ml. have been reported in cases of phosphorus, cinchophen, chloroform and carbon tetrachloride poisoning and in arsphenamine hepatitis. Acute hepatic necrosis and yellow fever are at times associated with extremely low values (15 to 50 mg.), which may be an important factor in the fatal termination of such cases. It must be realized, however, that the absence of hypoglycemia, even in these acute disorders, is by no means indicative of the absence of hepatic functional impairment. When one realizes the fact that 80 per cent of the total substance of the liver of an animal may be removed without producing hypoglycemia, one can appreciate readily why this is the case.

The phenomenon of spontaneous hypoglycemia, occurring at times paroxysmally, has also been observed in patients with extremely fatty liver, advanced hepatic cirrhosis (especially after alcohol intake) and widespread involvement of the liver with carcinoma. It has been suggested that the occasional occurrence of temporary periods of hypoglycemia in apparently normal individuals subjected to extremely strenuous exercise, such as marathon running, is due perhaps to temporary exhaustion of the hepatic glycogen reserves. As a result, the balance between the utilization of glucose in the muscles and its liberation into the blood stream from the liver is disturbed, and the blood sugar falls unless glucose is supplied from external sources. Whereas the superimposition of hepatic disease usually further diminishes the already lowered glucose tolerance of diabetics and increases their resistance to insulin, the opposite is observed occasionally. Patients with diabetes have been observed in whom the development of extensive parenchymal damage, as in hemochromatosis, was followed by paroxysmal hypoglycemia.

Cases of biliary tract and hepatic disease are encountered occasionally in which the fasting blood sugar concentration is above normal, returning to normal after improvement in the biliary tract condition. This is explained by some on the basis of a disturbance of the mechanism which regulates the rate of breakdown of glycogen in the liver in accordance with the utilization of glucose in the tissues and the supply of glucose to the liver (p. 30). However, great care must be exercised in such cases in eliminating the possibility of the existence of a latent diabetes aggravated by the presence of hepatic disease.

Glucose Tolerance.
When the glycogen-storing function of the liver becomes seriously impaired, as in advanced alcoholic cirrhosis, the blood sugar curve following the ingestion of glucose exhibits an abnormally high rise and relatively delayed fall. If the blood sugar concentration rises above the "renal threshold level," glucose will appear in the urine. The oral glucose tolerance curve can become indistinguishable from that of patients with diabetes mellitus. A similar state of affairs can also occur in the presence of acute hepatitis, as well as in association with obstructive disease of the biliary tract. In patients with cirrhosis, the cortisol glucose tolerance test has given results intermediate between normal and diabetic curves but with no reactive hypoglycemia. In patients with portacaval anastomoses, the fasting glucose level is subnormal, but the remainder of the curve tends to resemble that found in untreated cirrhotic patients. In some cases of toxic jaundice, a high peak level is observed, with a return to normal levels in two hours (lag type curve).

Failure of the glycogen-storing mechanism of the liver may not be the only factor responsible for the production of diminished glucose tolerance in hepatic disease. Under normal conditions a decreased output of sugar from the liver (decreased hepatic glycogenolysis) is an important, although not necessarily the only, factor in determining the characteristic fall in the blood sugar curve after the administration of glucose. In the presence of hepatic functional impairment there may be some interference with the mechanism whereby the liver normally diminishes its supply of glucose to the blood in response to an influx of exogenous sugar. The deleterious influence of diminished hepatic glycogenesis in contributing to the production of diminished glucose tolerance is supplemented by increased or abnormally prolonged hepatic glycogenolysis, the result being an abnormally high and prolonged elevation of blood sugar following the ingestion of glucose.

Although such findings are commonly obtained in patients with hepatic disease, the determination of glucose tolerance has proven to be of little or no practical value in the study of hepatic function. So many factors are involved in the intermediary metabolism of glucose that the interpretation of minor changes in the blood sugar tolerance curve is extremely difficult. Furthermore, because of the activity of extrahepatic factors in the storage and utilization of glucose, essentially normal findings may be obtained occasionally in the presence of advanced hepatic disease.

A peculiar type of disturbance of glucose tolerance has been described in Type I glycogenosis (von Gierke's disease) (p. 80). This is a disturbance of glycogen metabolism which appears in early infancy and is characterized by an abnormal deposit of glycogen in the liver, which becomes engorged with glycogen and enlarged. The peculiarity of this store of glycogen is that it cannot be mobilized readily under ordinary conditions. Because of this abnormal stability (diminished glycogenolysis), the fasting blood sugar concentration may be subnormal, values as low as 20 mg. per 100 ml. having been reported. Due to the operation of the same factors, the alimentary blood sugar response exhibits a more gradual and lesss marked increase than normal and, frequently, a more delayed return to the fasting level. The latter phenomenon is difficult to explain.

Epinephrine Hyperglycemia (see pp. 46 and 74). The increase in blood sugar which follows the administration of epinephrine in normal individuals apparently depends upon the presence of adequate amounts, or availability, of glycogen in the liver. Following the administration of 10 minims of a 1 : 1000 solution of epinephrine hydrochloride by intramuscular injection, the blood sugar concentration normally increases 35 to 45 mg. per 100 ml. in three-fourths to one hour and returns to the resting level in one and three-fourths to two hours. Several observers have noted that in the presence of hepatic disease the degree of rise in the blood sugar concentration is not so great as that observed in normal individuals. In some cases of advanced functional impairment associated with acute hepatic disease practically no rise is obtained. These phenomena are presumably due to the diminution in or depletion of the glycogen reserve in the liver. This test usually yields but little information of value in the early diagnosis of liver disease. Variable results are obtained in chronic liver disease, particularly in portal cirrhosis, although good results have been reported in cases of obstructive biliary cirrhosis. Decrease in or absence of the normal response to epinephrine may frequently be observed in patients with acute diffuse hepatic lesions, such as toxic and viral hepatitis, acute necrosis of the liver, etc. A diminished blood sugar response to epinephrine occurs also in patients with Type I glycogenosis.

Lactate and Pyruvate Levels. Under normal conditions, lactate produced in the muscles (p. 15) is removed from the blood stream mainly by the liver. Normal values in venous blood range from 0.55 to 1.15 mmol./l. Increased concentrations have been reported in severe toxic and infectious hepatitis and portal cirrhosis. However, normal findings are obtained frequently in all types of hepatic disease, and the fact that increased values usually appear only in the terminal stages of these conditions renders the determination of blood lactate of little value from a practical standpoint in the evaluation of hepatic function.

Pyruvate may be increased in the blood and cerebrospinal fluid in advanced hepatic insufficiency, e.g., in decompensated cirrhosis. This has been attributed to inadequate utilization of pyruvate by the liver.

PROTEIN METABOLISM

The liver plays an important part in normal protein metabolism. Its chief functions in this connection, in man, are deamination and transamination of amino acids, urea formation, and formation of certain plasma proteins (p. 173).

Amino Acids (p. 158). The liver is the main site of deamination of amino acids. However, because of the presence of an enormous factor of safety, 90 per cent or more of this organ must be removed before this function is significantly impaired. Complete hepatectomy in animals results in a steadily progressive increase in the concentration of amino acids in the blood. Because of this large functional reserve, the amino acid content of blood and urine is often within normal limits in chronic diseases of the liver. Elevated values may, however, be observed, especially in conditions associated with acute widespread degenerative lesions. This occurs characteristically in acute hepatic necrosis, viral and toxic hepatitis, portal cirrhosis, and some cases of eclampsia. In these conditions the amino acid nitrogen content of the blood may be 10 to 15 mg. per 100 ml., occasionally rising to as high as 30 mg. In one reported case, values above 200 mg. were observed just before death. The increase in amino acids in degenerative lesions of the liver is probably due partly to inefficient deamination of amino acids and partly to extensive autolysis of hepatic tissue. The increased concentration of amino acids in the blood is associated with a corresponding increase in the amino acid content of the urine. Under such circumstances, amino acids may be present in the urine in large quantities. Normally, free amino acids exist in the urine in small amounts (p. 181). Consequently, identification of spontaneously occurring leucine and tyrosine crystals in the urine has been assumed to be strongly indicative of the presence of an extensive hepatic degenerative process such as occurs in acute necrosis of the liver, but this is a rare event.

The abnormal amount of amino acids excreted in the urine of patients with acute hepatic necrosis could to some extent originate from autolyzed liver tissue. In some cases aminoaciduria may occur in the absence of apparent disturbance in the metabolism of amino acids. Continual, massive tyrosinuria (0.9 to 2 g. daily) has been observed only in cases of acute hepatic necrosis with a rapid and fulminating course. Transitory, minimal and moderate tyrosinuria has been observed in cases of subacute necrosis of the liver, in degenerating neoplasm of the liver, in toxic hepatitis and, rarely, in patients with obstructive jaundice of long standing due to stone. Tyrosinuria does not occur in patients with uncomplicated inflammatory lesions of the bile passages but is found occasionally in the presence of extrahepatic autolytic foci, such as degenerating tumors of the lung or extensive sloughs of the skin. During the phase of recovery from degeneration of the liver, tyrosine vanishes from the urine, but with a fresh exacerbation of the hepatocellular lesion tyrosine reappears in the urine. Negative findings may be obtained in the stage of repair, as in nodular cirrhosis, or in the terminal stages of subacute necrosis with a critical reduction in the amount of functioning parenchyma of the liver. Increasing tyrosinuria is assumed to indicate a rapidly progressive degenerative process in the hepatic parenchyma. Although this finding appears to be of distinct prognostic value, the absence of tyrosinuria in such cases does not necessarily warrant an optimistic prognosis.

Excessive aminoaciduria occurs also in patients with cirrhosis of the liver (about 20 per cent of cases) (p. 224), more frequently in those with jaundice. Although the urinary amino acid excretion tends to increase with progression of the disease, and to decrease with clinical improvement, this is not generally of diagnostic or prognostic value. The rather consistent occurrence of aminoaciduria in hepatolenticular degeneration (Wilson's disease) is referred to elsewhere (p. 537). It is not necessarily associated with an increase in blood amino acids in this condition, being due to inadequate reabsorption from the glomerular filtrate by the renal tubular cells, apparently as a result of the toxic effect of their increased copper content (p. 336).

Urea. Urea is formed in the liver from ammonia derived from amino acids during the process of deamination (p. 160) as well as from bacterial metabolism in the large intestine. The liver is the only important site of urea formation. The fact that the production of urea ceases following total hepatectomy in animals would naturally lead to the assumption that hepatic insufficiency should be associated with a decrease in the urea content of blood and urine. In general, the statements made with regard to the relationship of hepatic disease to the metabolism of amino acids apply equally to urea. Deamination and urea formation are closely related phenomena; since the quantity of urea formed by the liver appears to be determined by the efficiency of the mechanism of deamination, it naturally follows that diminution in the concentration of urea in the blood and urine occurs only in those conditions in which the blood and urinary amino acid nitrogen is increased.

In the majority of normal individuals the blood urea nitrogen concentration ranges from 12 to 15 mg. per 100 ml. However, values of 9 to 18 mg. per 100 ml. cannot be considered abnormal. The hepatic lesions which produce subnormal values are those which interfere with the process of deamination and include, as stated above, such conditions as acute necrosis of the liver, phosphorus, arsphenamine, chloroform and carbon tetrachloride poisoning and, at times, eclampsia. In one very severe case of acute toxic necrosis of the liver the blood urea nitrogen was zero shortly before death. Values below 6 mg. per 100 ml. are quite unusual. The level is not infrequently raised because of associated renal failure. In conditions such as cirrhosis, syphilis, malignancy, hepatitis and obstructive jaundice, the blood urea nitrogen is usually within normal limits. Subnormal values occurring in biliary tract disease, particularly in obstructive jaundice, are of serious prognostic significance, since they usually indicate a superimposed lesion of the hepatic parenchyma. The normal ratio of blood urea nitrogen to amino acid nitrogen is approximately 2 to 1. A decrease in this ratio, indicating a decrease in the deaminizing capacity of the liver, is usually significant of advanced grades of hepatic insufficiency. Under such circumstances the total nonprotein nitrogen content of the blood and urine is not necessarily altered. Determination of the nitrogen partition characteristically reveals a decrease in the proportion of nitrogen present in the form of urea and a corresponding increase in amino acid and ammonia nitrogen. Because these changes occur only in far-advanced hepatic disease they are of serious prognostic significance but are of no diagnostic value.

Uric Acid. Although a slight increase in blood uric acid may be observed in some patients with biliary tract disease, cirrhosis and obstructive jaundice, such findings are not common. In fact, low values have been reported in severe cases of acute necrosis of the liver.

Ammonium Ion. The relatively minute amounts of ammonium ion in the blood (less than 0.02 mg. ammonium N/100 ml., p. 179) originate (a) mainly in the intestinal lumen, from action of bacterial and digestive enzymes on protein and urea and, to a minor extent, from (b) deamination of glutamine and other amino acids in the kidney and (c) metabolic reactions in skeletal muscles during exercise. It is removed chiefly in the liver for urea synthesis; smaller amounts may be utilized in various tissues, including brain, for amination of keto acids, e.g., in the reaction α-ketoglutarate $\rightarrow$ glutamic acid $\rightarrow$ glutamine. A small amount may be removed from the body in the expired air; relatively insignificant in normal subjects, this route of excretion may become significant in the presence of severe liver functional impairment.

The blood ammonium ion concentration may be elevated in cases of severe liver functional impairment, particularly in portal cirrhosis, aggravated by the added factor of shunting of blood flow across the liver incident to the portal venous obstruction. According to some, this increased concentration may be an important, although not the only, factor in the pathogenesis of hepatic coma (p. 673). In such cases, increase in the blood, due to increased production in the intestine, may be induced by a high protein diet, administration of ammonium exchange resins or ammonium salts, and following massive or frequent gastrointestinal hemorrhages. An increase has also been observed in association with alkalosis induced by voluntary or involuntary hyperventilation in patients with liver disease (p. 673).

Ammonium tolerance tests may reveal an abnormality in cases in which the resting blood ammonium values may be within normal limits. In one such procedure, blood is withdrawn before and forty-five minutes after oral administration of 3 g. of ammonium chloride. Normal tolerance, i.e., return to resting blood ammonium values at forty-five minutes, is usually observed in patients with jaundice caused by biliary obstruction, metastatic liver cancer, biliary cirrhosis, fatty liver and acute hepatitis. Impaired tolerance, i.e., elevated values at forty-five minutes, occurs in most patients with portal cirrhosis, the incidence and extent of the abnormality being related apparently more to the presence and extent of portal-systemic shunting than to the extent of hepatocellular damage per se. Arterial blood is somewhat more satisfactory than venous blood for demonstration of this abnormality.

Plasma Proteins (p. 182). Certain facts concerning the plasma proteins are of importance in connection with the interpretation of the diagnostic significance of changes that occur in diseases of the liver and bile passages.

(1) Hepatic cells are the sole site of formation of fibrinogen and serum albumin, and the main site of formulation of α- and β-globulins.

(2) Gamma globulins are formed in certain reticuloendothelial cells (plasma cells and lymphocytes) throughout the body. They are, therefore, mainly of extrahepatic origin.

(3) Although various forms of liver disease are associated rather consistently with certain patterns of abnormality of serum protein concentrations, these occur frequently also in a great variety of clinical disorders not necessarily involving the liver. The most consistent abnormalities are a decrease in albumin and an increase in γ-globulins; the β-globulins are often increased; the α-globulins may be decreased.

(a) Fibrinogen. Mild liver injury is associated with normal (200 to 400 mg. per 100 ml.) or slightly increased values. In the presence of severe liver damage the fibrinogen content of the blood plasma may fall to low levels, figures as low as 50 mg. per 100 ml. having been observed in cirrhosis. As a rule, the determination of the plasma fibrinogen content is of little value in the early diagnosis of hepatic disease, but low values in liver disease are usually indicative of extensive damage.

(b) Albumin. The work of the liver cell in protein synthesis is much greater than that of other tissue cells because of the fact that, in addition to construction of cellular proteins, it must produce the bulk of the plasma proteins, mainly albumin. There are about 500 g. of serum albumin in the body of an average normal adult, about one-third of which is in the blood, the remainder in the lymph and tissue fluids. Assuming a half-life of twenty-six days, about 12 g. are formed daily. On this basis, a 50 per cent decrease in production would result in only 20 per cent reduction in serum albumin concentration (e.g., from 5 g. to 4 g. per 100 ml.) in sixteen days. This may account for the fact that the serum albumin concentration is often within normal limits in patients with viral hepatitis. However, studies in human volunteers infected with the virus of infectious hepatitis have revealed a rather consistent fall in albumin within the first few days, reaching a minimum at about eight to ten days, and then usually rising gradually to attain a normal level in five to six weeks. In general, the degree and duration of the hypoalbuminemia parallel the clinical severity of the hepatitis. Recognition of this drop in serum albumin may not be possible in spontaneous forms of this disease because of the facts referred to above. Hypoalbuminemia occurs more consistently in acute and subacute hepatic necrosis (e.g., carbon tetrachloride and chloroform poisoning).

Hypoalbuminemia may occur in chronic liver diseases, such as portal cirrhosis, chronic passive congestion, and carcinoma. It may be quite marked in cirrhosis, contributing to, but not entirely responsible for, the development of ascites and generalized edema in that condition (p. 672). In advanced stages of this disease, hypoalbuminemia, as well as other disturbances of liver function, is contributed to by the progressively increasing shunting of blood in the liver due to the increasing portal pressure and changes in the blood vessels.

Hypoalbuminemia is a rather constant feature of "kwashiorkor," a form of hepatic disease occurring in children in tropical countries, due primarily to protein malnutrition. The low values in cancer of the liver may be due to the frequently present cirrhosis or to the associated inanition. In congestive heart failure, the decrease in albumin, although it may be dependent in part on impaired liver function (p. 669), is contributed to also by other factors, including plasma dilution (p. 189), and gastrointestinal factors (p. 188).

Subnormal albumin concentrations are found in 30 to 50 per cent of patients with extrahepatic biliary obstruction, e.g., carcinoma of the head of the pancreas and common duct stone. This may be due in part to superimposed hepatocellular damage, and in part to imposed or voluntary diet restriction, digestive disturbances (e.g., in associated pancreatic duct obstruction or pancreatitis) or inanition (e.g., in malignancy).

(c) Globulins. The α-globulins, which include mucoproteins (p. 197), may be decreased in viral and toxic hepatitis and portal cirrhosis but are usually within normal limits. They may also be increased in the presence of a wide variety of extrahepatic complications (p. 190). The early decrease in viral hepatitis possibly contributes to the production of abnormal turbidity and flocculation reactions at this stage of the disease. Increase in α_2-globulin may occur in extrahepatic biliary obstruction, metastatic liver cancer, and cholecystitis.

Changes in β-globulins are encountered more consistently than those in α-globulins. They may be increased in viral or toxic hepatitis, portal and biliary cirrhosis, and biliary obstruction. This is frequently related to the abnormality in serum lipids (lipoproteins, p. 191) that occurs in these disorders. In certain instances, paper electrophoresis does not accurately reflect the increase in β-lipoprotein concentration.

The most consistent globulin change in hepatic and biliary tract diseases is an increase in the γ-globulin concentration. This may be due to one or more of several causes. In hepatitis due to infectious agents, e.g., viral hepatitis, it may represent in part the development of anti-virus antibodies. It has been suggested also that the γ-globulin increase may be due to the production of antibodies to altered, i.e., abnormal (foreign), liver-cell proteins resulting from hepatocellular damage induced by infectious and chemical (e.g., carbon tetrachloride, chloroform, arsphenamine, etc.) agents. In the presence of liver damage, the impaired capacity of the liver for utilizing amino acids for synthesis of its cell proteins and plasma albumin may make available increased amounts of these amino acids for synthesis of γ-globulins by extrahepatic tissues. In portal cirrhosis, this is contributed to also by the extensive shunting of blood across the liver. Another possible factor in acute and chronic inflammatory and necrotizing lesions in the liver is the reactive proliferation in that organ of mesenchymal cells, which are concerned with production of γ-globulins.

Studies in human volunteers infected with the virus of infectious hepatitis indicate that the γ-globulins begin to increase in the plasma within the first three days, i.e., before the onset of jaundice, reach a maximum concentration in about eight to ten days, and remain at an elevated level for three to four months in the average case. This is accompanied by hypoalbuminemia, usually of briefer duration, a plasma protein pattern observed frequently in a variety of acute infections. In spontaneously occurring viral hepatitis, abnormally high concentrations of γ-globulin are encountered in about 70 per cent of cases. Although the increase is generally greater in more severe cases, this correlation is not consistent. However, persistence of this abnormality beyond the usual expectation (three and one-half to four and one-half months) may reflect progression of the acute hepatitis. Residual abnormalities in globulin pattern may persist occasionally for one to three years, without any evidence of disease and with no clinical indication of permanent liver damage.

Similar changes, usually more marked, occur in cases of acute and subacute hepatic necrosis, i.e., idiopathic, "kwashiorkor" (dietary), or due to hepatotoxic agents, such as carbon tetrachloride, chloroform, etc. γ-Globulin values over 8 g./100 ml. (average normal, 0.9 g.) have been observed in conditions in this category and also in patients with postnecrotic cirrhosis. In other types of cirrhosis, globulin values up to 6.5 g./100 ml. are not unusual, the increase being largely in the γ-globulin fraction. In biliary cirrhosis, increase in α- and/or β-globulins may be associated with the increase in plasma lipids (lipoproteins) that occurs frequently in this condition (p. 132). The γ-globulins may be increased in 40 to 60 per cent of cases of carcinoma of the liver. Interpretation of the significance of this observation is difficult because of the frequent coexistence of cirrhosis.

In general, increase in γ-globulins is a sensitive reflection of diffuse hepatic disease, inflammatory (hepatitis) or degenerative (necrosis), usually occurring early, and usually persisting after disappearance of subjective and other objective manifestations. Extreme hyperglobulinemia may on occasion indicate an unfavorable prognosis. As indicated above, undue persistence of this abnormality in patients with acute hepatitis or necrosis may indicate progression to chronic disease (active chronic hepatitis, portal cirrhosis). It may be of significance in connection with the plasma protein changes in cirrhosis that liver ischemia induced by surgical restriction of the hepatic circulation has been found to be followed by a decrease in albumin and an increase in β- and γ-globulins. During activity progressing stages of the acute diseases, and also in cirrhosis, the increase in globulins is commonly accompanied by decrease in albumin (p. 190). Whereas these findings are by no means restricted to hepatic disease, occurring also in a variety of acute and chronic infections, the nephrotic syndrome, multiple myeloma, sarcoidosis, etc. (p. 190 ff.), they may aid in differentiating between hepatocellular and uncomplicated extrahepatic obstructive jaundice. Whereas a decrease in plasma albumin, usually of comparatively moderate degree (3.0–3.5 g./100 ml.), is not uncommon in bile duct obstructive types of jaundice, only a small proportion (15 to 25 per cent) of such cases with hypoalbuminemia also exhibit hyperglobulinemia in the absence of other evidence of superimposed hepatocellular damage.

Attention has been drawn to the possible value of specific serum immunoglobin determination in the diagnosis of liver disease. Alcoholic liver disease tends to be associated with elevated levels of serum IgA (46% of cases), primary biliary cirrhosis with elevated IgM (80% of cases) and active chronic hepatitis with elevated IgG (64% of cases). These findings are by no means specific, and raised levels of each of these gamma globulins have been found in association with a variety of liver diseases. For example, increased IgM levels have been found in 20 per cent of patients with extrahepatic obstructive jaundice.

Tissue antibodies have proved to be useful in the differential diagnosis of certain liver diseases. Mitochondrial antibody is virtually always present in the serum of patients with primary biliary cirrhosis, but is not detected in the serum of patients with extrahepatic biliary obstruction. A positive finding is often present in the serum of patients with chronic active hepatitis and occasionally in the serum of those with cryptogenic cirrhosis. Mitochondrial antibody can also be demonstrated in the serum of patients suffering from collagen disorders and has been reported in the serum of those with hepatitis due to halothane as well as in the serum of patients with cholestasis resulting from the administration of chlorpromazine. Up to one per cent of apparently normal individuals also have mitochondrial antibodies present in their serum, but about one-third of these develop demonstrable liver pathology.

Antinuclear factor is frequently demonstrable in the serum of patients with chronic active hepatitis, cryptogenic cirrhosis and primary biliary cirrhosis, as well as those with collagen diseases (e.g., systemic lupus erythematosus).

Smooth muscle antibodies are found in patients with chronic hepatitis, but not in patients with systemic lupus erythematosus. They are found in patients with other collagen disorders (rheumatoid arthritis), as well as in patients with cryptogenic cirrhosis and primary biliary cirrhosis.

Australia antigen is present in the serum of many patients with so-called serum hepatitis (long incubation hepatitis). It can be detected by a variety of means including agar gel diffusion, immunoelectrophoresis and radio-immunoassay among others.

Alpha-l-fetoprotein (which is not an immunoglobulin) is quite frequently but by no means invariably present in the serum of patients with malignant hepatoma. It has also been found to be increased in 45 per cent of cases of Indian childhood cirrhosis.

Carcino-embryonic antigen (CEA) in the plasma may be increased in patients with cirrhosis and possibly with other liver diseases. The original intention was to use this substance as a test for intestinal cancer. This, however, proved to be rather nonspecific, since raised plasma levels can be found in a number of other types of cancer, in various inflammatory disorders and in tissue hyperplasia.

(d) Mucoproteins (p. 196). The serum mucoprotein concentration is frequently decreased in uncomplicated diffuse hepatocellular disease, as in toxic and infectious hepatitis and portal cirrhosis. Normal or elevated values may occur in patients with these conditions in the presence of complicating factors that tend to increase the serum mucoproteins (p. 197), such as primary or metastatic liver cancer or infectious disease, e.g., pneumonia, tuberculosis.

Normal or increased values are usually obtained in various forms of biliary obstruction, often rising as the obstruction increases in degree and duration. The usual increase in cases of hepatomegaly due to extensive metastatic malignancy may aid considerably in differentiation from other causes of hepatomegaly; such procedures as the thymol turbidity, cephalin-cholesterol flocculation, and zinc sulfate turbidity tests may not be helpful in this connection.

(e) Total Protein. Because of the frequent occurrence of simultaneous changes in albumin and globulin in opposite directions in hepatic diseases, as indicated above, the total plasma protein concentration may be within normal limits, subnormal, or elevated, depending upon the relative magnitude of these changes. Normal values may be obtained in the early stages of infectious hepatitis due to an equal decline in albumin and rise in globulins, with elevated values later due to the frequent persistence of hyperglobulinemia for some weeks after the albumin has returned to a normal level. In cirrhosis, values of 5 to 6 grams of globulin and 1 to 3 grams of albumin per 100 ml. are not unusual, the total protein concentration being within normal limits (6–8 g./100 ml.).

It must be remembered that the range of normal variation in total globulin concentration, i.e., 2.1–4.2 g./100 ml., permits a relatively large increase in a single component, e.g., γ-globulin (average normal, 0.9 g./100 ml.), without a demonstrable significant increase in total globulins above the upper limit of normal. Such abnormalities can be revealed by electrophoresis or zinc sulfate turbidity measurements, and by certain of the so-called globulin reactions, e.g., the thymol turbidity and flocculation tests, and the cephalin-cholesterol flocculation test.

The total protein may be quite low in the face of a 100 to 200 per cent increase in γ-globulins, if the albumin concentration is very low (e.g., 1–2 g./100 ml.), as may be the case in cirrhosis or hepatic necrosis. However, rather high total protein values, e.g., over 10 g./100 ml., may be found in these conditions, especially in severe cases, and particularly in subacute necrosis and postnecrotic cirrhosis, in which the decrement in albumin is markedly overcompensated by the large increment in globulins (p. 183).

(f) Globulin Reactions. The various procedures in this category and the mechanisms involved in the production of abnormal reactions are reviewed elsewhere. Those employed most widely in patients with hepatic and biliary tract disease are: (1) the cephalin-cholesterol flocculation test; (2) the thymol turbidity and flocculation tests; (3) the zinc sulfate turbidity test.

It should be emphasized that these are not tests of liver function in the strict sense, at least insofar as the production of abnormal reactions depends upon an increase in γ-globulin. This may be a reflection of morphologic damage to liver cells (in hepatic disease) or of an immunologic response to a causative infectious agent. It may be contributed to also by increased availability of amino acids for extrahepatic synthesis of γ-globulins as a consequence of impaired capacity for synthesis of serum albumin and/or deviation of blood flow from the liver, as in portal cirrhosis. Under these circumstances the liver loses, to a certain extent, its high priority for amino acids in the common pool. Hypoalbuminemia, changes in α- and β-globulins, and hyperlipemia, which may influence certain of these reactions, may result, directly or indirectly, from impairment of liver function. However, none of these phenomena is specifically indicative of liver disease, and abnormalities in these directions do not necessarily parallel the results obtained with true tests of liver function. In fact, their greatest practical value, i.e., in differentiating between hepatocellular and obstructive types of jaundice, arises out of the fact that, unlike true function tests, these reactions appear to reflect the activity and extent of morphological liver damage (in hepatic disease), although they are influenced also by coexisting extrahepatic factors.

Of particular significance in relation to hepatic disease is the fact that results obtained with almost all of these turbidity or flocculation reactions depend upon the state of balance between stabilizing and precipitating factors (p. 205). The test procedures are so adjusted that with normal serum, turbidity or flocculation is at a minimum and within well-defined limits. The stabilizing factors include albumin, α-globulin, and a plasma component that increases in obstructive jaundice. The precipitating factors include γ-globulins, lipids and, perhaps, β-globulins. All globulin reactions are not influenced to the same degree by these factors.

There is a body of opinion which states that these tests are outdated by such procedures as zone electrophoresis and the determination of serum enzymes. It is even suggested that the flocculation tests should be abandoned.

They are, however, simple to perform and have proved to be useful in differentiating jaundice due to hepatocellular disease from that due to extrahepatic obstruction. These tests are dependent, as has been pointed out, on a balance between a variety of serum proteins and will become positive in most conditions associated with increase in gamma-globulin. In other words, false positive results can be obtained in the presence of infection, even of the biliary system. Care must therefore be taken in the interpretation of the results of these tests, if there are such signs as an obvious increase in body temperature in patients under observation, as well as evidence of a leukocytosis or a well-marked increase in erythrocyte sedimentation rate.

Thymol Turbidity and Flocculation (p. 206). Thymol turbidity (protein-thymol-phospholipid complex) is influenced by the following factors:

(1) Decrease in the concentration of the stabilizing albumin.

(2) Decrease in the precipitation-inhibiting properties of the serum albumin, due perhaps to some qualitative change.

(3) Increase in γ-globulin, a precipitating factor.

(4) Increase in β-globulin, a precipitating factor.

(5) Increase in plasma lipid concentration, a precipitating factor. It must be remembered in this connection that an abnormal thymol test may be obtained with any type of hyperlipemic serum, e.g., in diabetes and in normal subjects after a fat meal.

In general, in acute hepatitis, abnormalities of thymol turbidity parallel those in serum lipid concentration, whereas in chronic liver disease, including cirrhosis, they parallel increases in γ-globulin concentration. However, there are many exceptions to these generalizations. For example, marked abnormality of thymol turbidity in viral hepatitis may be related to an increase principally in γ-globulin, in the absence of hyperlipemia.

Clinically, there is generally good qualitative agreement (70 to 90 per cent) between the thymol and cephalin-cholesterol reactions (i.e., both normal or both abnormal) in hepatic and biliary tract disease, although there are frequently differences in the degree of abnormality. Many of the qualitative differences (one test normal, the other abnormal) are of a borderline nature, but they are occasionally striking. The incidence of abnormal thymol turbidity (5 to 12 units, occasionally 20, rarely 30 to 40 units) and flocculation is somewhat lower than that of abnormal cephalin-cholesterol flocculation in acute active stages of viral hepatitis, acute and subacute necrosis and portal cirrhosis. The thymol tests are more often negative in malignant biliary obstruction and in carcinoma of the liver than is the cephalin-cholesterol flocculation test, an observation which may be of some diagnostic value. However, in these conditions, as well as in cirrhosis, instances are encountered of the reverse situation, i.e., positive thymol, negative cephalin-cholesterol tests. During convalescence from hepatitis, abnormal thymol reactions occur more commonly and persist longer than does abnormal cephalin-cholesterol flocculation. Undue persistence of globulin abnormalities in such cases may suggest progression to chronic hepatitis or cirrhosis.

Abnormal thymol turbidity may occur in various extrahepatic diseases other than those accompanied by hyperlipemia. These include: inflammatory diseases, lymphopathia venereum, systemic lupus erythematosus and other "collagen" diseases, rheumatic fever, malaria, infectious mononucleosis, kala-azar, and tuberculosis.

The thymol turbidity test has certain practical advantages over the cephalin-cholesterol test: (1) rapidity (30 minutes vs. twenty-four to forty-eight hours), (2) stability of reagents, and (3) more precise quantitation of results. However, the frequency with which discrepancies between these two reactions occur in common types of hepatic and biliary tract disease makes it undesirable to rely on the thymol tests alone in the study of patients with these disorders. Both should be performed simultaneously, a practice that is entirely feasible because of their technical simplicity and the small amounts of serum required. It should be recalled that, in acute hepatic diseases, significant thymol flocculation (thirty hours) may occur in the absence of abnormally high turbidity values (p. 207). Consequently, the presence of flocculation should be investigated routinely, particularly if the turbidity test yields normal results.

In general, the statements made with regard to the relationship between severity of the liver lesion and changes in plasma globulin (p. 621) at different stages of acute hepatic diseases apply to abnormalities in these globulin reactions. Their prognostic significance is limited during acute phases of hepatic disease (hepatitis and necrosis), but undue persistence of abnormalities may suggest progression to chronic hepatitis.

Cephalin-cholesterol Flocculation (p. 206). Flocculation of the cephalin-cholesterol emulsion with human serum under appropriate conditions appears to depend upon the existence of the following conditions:

(1) Increase in γ-globulin in such quantity that the serum albumin, which stabilizes the emulsion, is unable to inhibit the reaction.

(2) Decrease in serum albumin concentration below that capable of inhibiting the reaction.

(3) Decrease in flocculation-inhibiting properties of the serum albumin fraction due perhaps to some qualitative change.

(4) Decrease in the concentration of the stabilizing α-globulin.

Positive reactions may be obtained with serum of certain normal laboratory animals, including dogs and rabbits, and with normal human serum (a) after standing at refrigerator temperatures for several months, (b) after heating to 56° C. for thirty minutes, or (c) if the reaction mixture is exposed to light during the test period. A 3 or 4 plus flocculation is regarded as a positive (abnormal) result, 2 plus as questionable, in a jaundiced subject.

The clinical value of the cephalin-cholesterol flocculation test in the study of patients with hepatic disease arises out of the fact that the conditions required for the production of an abnormal reaction are present in a large proportion of cases of hepatocellular damage, i.e., decreased serum albumin and increased γ-globulin. This test is positive in: 80 to 85 per cent of cases of viral hepatitis during the active stage; other forms of hepatitis, 20 to 25 per cent; acute and subacute hepatic necrosis, 80 to 90 per cent; portal cirrhosis, 70 to 75 per cent; extrahepatic biliary obstruction, 30 to 35 per cent (mainly calculous obstruction); cancer of the liver, 30 to 60 per cent (usually in patients with hyperbilirubinemia and cirrhosis); occasionally in chronic passive congestion.

An abnormal reaction early in viral hepatitis is due usually to decrease in the stabilizing α-globulin; later it is dependent mainly on the commonly increased concentration of γ-globulin. This reaction is, therefore, frequently abnormal earlier in acute hepatitis than is the zinc sulfate turbidity, which depends mainly on changes in γ-globulin. Hyperlipemia, which increases thymol turbidity, does not influence cephalin-cholesterol flocculation.

Zinc Sulfate Turbidity (p. 205). Although albumin exerts a protective and lipids an aggravating influence on zinc sulfate turbidity, abnormal values are due mainly to increased concentrations of γ-globulins, which are reflected more specifically by this procedure than by the thymol or cephalin-cholesterol tests. Results obtained with this procedure in subjects with hepatocellular damage or biliary tract disease often parallel those obtained with other globulin reactions. However, this parallelism is not observed consistently, since the thymol and cephalin-cholesterol reactions are affected by changes in concentration of serum albumin and lipids to a greater extent than is the zinc sulfate turbidity.

In acute hepatitis, elevated values (above 8 units) may persist longer than abnormalities in other globulin reactions. The incidence of abnormal zinc turbidity in hepatic and biliary tract diseases is approximately as follows; viral hepatitis, 70 to 80 per cent; toxic hepatitis, 50 to 60 per cent; cirrhosis, 80 to 90 per cent; carcinoma of the liver, 60 to 70 per cent; extrahepatic biliary obstruction, 25 to 30 per cent.

The chief clinical value of this procedure is as an aid in distinguishing hepatocellular and extrahepatic obstructive types of jaundice (p. 676 ff.) and, to a lesser degree, in determining progression of acute liver lesions to those which are chronic.

Plasma Prothrombin. Prothrombin, a constituent of the blood plasma essential for normal coagulation of blood, is glycoprotein in nature, being associated with the globulin fraction of the plasma proteins. It is formed in the liver, an adequate supply of vitamin K being necessary for its formation in normal amounts (p. 832).

The hemorrhagic tendency manifested by many patients with obstructive, as well as by some with hepatocellular, jaundice and bile fistula is dependent chiefly upon severe prothrombin deficiency. At least two factors are of significance in this connection, being necessary for the maintenance of a normal plasma prothrombin concentration: (a) adequate hepatocellular function and (b) absorption of an adequate amount of vitamin K from the intestine. Inasmuch as bile salts are necessary for proper absorption of vitamin K (p. 832), inadequate absorption, with consequent hypoprothrombinemia, may occur (1) when bile is absent from the intestine due to bile-duct obstruction or external bile fistula and (2) when there is deficient formation and excretion of bile salts by the liver. Hypoprothrombinemia may also occur in hepatic disease due to inability of the liver to utilize vitamin K satisfactorily for the formation of prothrombin (hepatitis, cirrhosis, acute or subacute hepatic necrosis, chronic passive congestion, fatty liver, extensive hepatic malignancy, infectious hepatitis, etc.). A drop in prothrombin also occurs frequently after operations on the biliary tract.

With the two-stage method for prothrombin determination, hypoprothrombinemia has been demonstrated in a large majority of patients with jaundice of either obstructive or hepatocellular origin. The one-stage method, which is simpler and more commonly employed clinically, reveals this deficiency in a smaller proportion of cases, and it is influenced by deficiencies of factors I, V, VII and X (p. 175). No direct methods are available for accurate clinical determination of prothrombin, the indirect methods in common use representing assays of the thrombin-forming capacity of the plasma under controlled conditions. There is apparently a wide margin of safety in the prothrombin factor, and the coagulation time as measured by usual methods may remain within normal limits until more than 80 per cent of the prothrombin of the blood is lost. When the concentration of prothrombin in the plasma falls below about 30 per cent of normal, the "prothrombin time" becomes longer than twenty seconds (upper limit of normal); with further reduction of the prothrombin concentration to less than 20 per cent of normal, the "prothrombin time" becomes markedly prolonged. When the latter is longer than forty seconds, indicating prothrombin concentration in the plasma of less than 10 per cent of normal, the existence of a hemorrhagic tendency may be expected. Experimentally, plasma prothrombin deficiency has been produced in animals by common-duct ligation, partial hepatectomy, biliary fistula, and the administration of hepatotoxic substances such as chloroform and carbon tetrachloride.

Since hypoprothrombinemia in uncomplicated common-duct obstruction (stone, stricture, tumor, pressure) is due to absence of bile salts in the intestine, with consequent inadequate absorption of vitamin K, a prompt increase should follow parenteral administration of the latter or its oral administration in conjunction with bile salts. On the other hand, hypoprothrombinemia in patients with hepatocellular damage is dependent largely upon inability of the liver cells to form prothrombin, and little or no increase should follow administration of vitamin K. These facts have been utilized in attempts to differentiate between obstructive and hepatocellular types of jaundice, the capacity for responding to vitamin K being interpreted as an index of the adequacy of hepatic cell function. In the presence of hypoprothrombinemia in a patient with jaundice, a significant increase following administration of vitamin K favors the diagnosis of obstructive jaundice, whereas the absence of such increase suggests the presence of impaired hepatocellular function (p. 832). The criteria for a satisfactory response vary considerably. Some regard a failure to obtain a rise of more than 10 per cent in twenty-four hours after intramuscular injection of a vitamin K preparation (2-methyl-1,4-naphthoquinone) as indicative of hepatocellular damage. With the oral administration of 8 mg. of this preparation (menadione) plus 2.5 g. of bile salts or, preferably, parenteral injection of 10 mg. of tetrasodium 2-methyl-1,4-naphthoquinone diphosphoric acid ester (Synkavite), a significant rise of prothrombin may be expected within twenty-four hours in uncomplicated obstructive jaundice and little or no rise in hepatocellular jaundice.

As stated previously, normal values for plasma prothrombin may be obtained (before giving vitamin K) in a considerable number of patients with obstructive or, more frequently, hepatocellular jaundice, especially when the one-stage method is used. Consequently, normal findings may be misleading. Subnormal values are more significant, but it must be remembered that hypoprothrombinemia may occur in conditions other than diseases of the liver and biliary tract (p. 832). Failure of the plasma prothrombin to respond satisfactorily to vitamin K administration is much more significant of impairment of hepatic function. However, the production of a satisfactory rise does not exclude the possibility of the existence of even considerable hepatocellular damage. The usefulness of this and other procedures in differentiating between obstructive and hepatocellular jaundice is discussed elsewhere (p. 676 ff.).

LIPID METABOLISM

The importance of bile in the absorption of fats from the intestine has been recognized for some time and has been considered in detail elsewhere (p. 98). Its action in this connection may be summarized as follows: *(a)* activation of pancreatic lipase; *(b)* the bile salts aid in the emul-

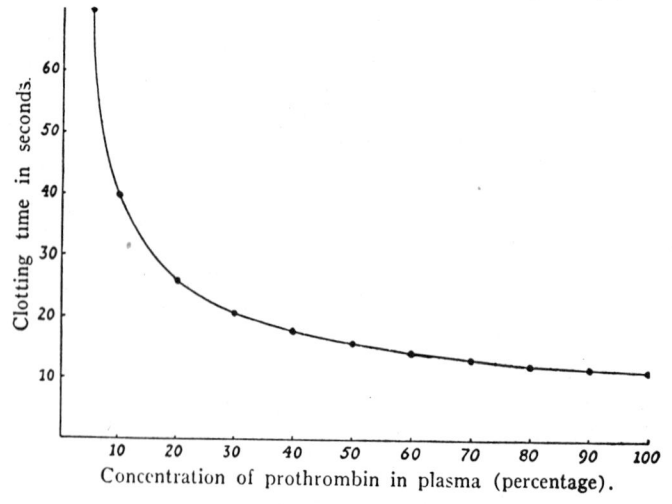

Figure 21-4. Relation of clotting time of recalcified plasma (containing an excess of thromboplastin) to concentration of prothrombin. (After Quick.)

sification and, consequently, the digestion of fat; (c) the bile salts form soluble addition compounds with fat or fat-soluble substances.

Fat in Feces (p. 600). Obstructive jaundice is usually accompanied by the presence of excessive quantities of fat in the feces (total fat, soap fat, free fatty acids). Similar findings may be obtained in infectious hepatitis, cirrhosis of the liver, toxic hepatitis and acute necrosis of the liver. These changes have been attributed generally to decreased absorption of the products of fat digestion, a phenomenon which would naturally be expected to occur when bile is not present in the bowel in adequate amount. However, in view of the fact that fecal lipids consist of fatty material from intestinal mucosal cells as well as unabsorbed portions of ingested lipids (p. 100), this interpretation is open to question. The absence of bile from the intestine causes but little impairment of fat absorption, 65 to 70 per cent of fatty acids of the diet being absorbed in patients and animals with bile fistula (external). It would appear that, as under normal conditions, fecal fat in the absence of bile could, to some extent, be of endogenous origin.

Plasma Cholesterol. The alterations that may occur in plasma cholesterol concentration and partition in patients with hepatic or biliary tract disease have been considered elsewhere (pp. 139, 142). The significance of these changes may be summarized as follows:

(a) Hypercholesterolemia is found in the majority of patients with uncomplicated extrahepatic obstructive jaundice and occasionally in patients with cholelithiasis without obstruction of the common bile duct. The underlying mechanism is not known, although there is evidence which suggests that the increase may be in some way related to and dependent upon the increase in bile acids in the plasma which occurs simultaneously under these circumstances (p. 658). Although both free and ester fractions usually participate in this increase, the former usually rises disproportionately, with consequent decrease in the ester cholesterol : total cholesterol ratio (in about 50 per cent of cases). In general, both total and ester values tend to be lower in neoplastic than in benign (calculous) obstruction.

(b) Hypercholesterolemia occurs in certain forms of hepatic disease in the absence of extrahepatic biliary obstruction, due presumably to obstruction within the intrahepatic duct system. This is observed characteristically in primary biliary cirrhosis (p. 140) and also in certain types of toxic hepatitis, particularly that due to arsphenamine.

Release of biliary obstruction is usually followed promptly by a return of the plasma cholesterol concentration to normal levels. Of importance from a diagnostic and prognostic standpoint is the fact that a similar fall may also occur simultaneously with manifestations of cachexia, terminal cholemia, infection, superimposed hepatocellular damage and other complications, even though the biliary obstruction persists. Under such circumstances, as is indicated below, the fall in plasma cholesterol occurs principally in the ester fraction. Repeated determination of the plasma cholesterol concentration and partition in patients with common duct obstruction is consequently of value in prognosis and treatment.

(c) Hypercholesterolemia occurs occasionally in patients with mild acute hepatitis and cirrhosis, but much less frequently than in obstructive jaundice. When it occurs, the increase is usually mainly in the free cholesterol fraction, the ester cholesterol : free cholesterol ratio being diminished. Normal values for total cholesterol are the rule in infectious hepatitis (except in very severe cases), but the proportion of esters is frequently reduced. Except in the terminal stages, plasma cholesterol concentration and partition are usually normal in uncomplicated portal cirrhosis. Hypocholesterolemia is the rule in advanced forms of this condition, the decrease occurring chiefly in the ester fraction, as indicated below.

(d) Conditions accompanied by severe hepatocellular damage are frequently associated with a diminished proportion of cholesterol esters in the plasma (normally about 60 to 80 per cent of the total), constituting the so-called "Estersturz." This may not be, but usually is, associated with a diminution in total cholesterol concentration. Among the conditions in which this phenomenon has been observed are portal cirrhosis, toxic and infectious hepatitis and yellow fever. In these conditions the plasma cholesterol may be as low as 70 to 100 mg. per 100 ml. (normal 140 to 250 mg.), 0 to 50 per cent being in the ester form (normal 80 to 200 mg. per 100 ml., constituting 60 to 80 per cent of the total). In such cases, the degree of hypocholesterolemia and the fall in the ester fraction are often in proportion to the degree of hyperbilirubinemia. It would appear that the degree of diminution in the cholesterol ester fraction in the plasma may reflect the extent of hepatocellular damage in these conditions. This phenomenon also is present when such damage occurs as a complication in patients with extrahepatic obstructive jaundice, a fall in the ester fraction often occurring in the presence of an initially elevated total cholesterol concentration.

The chief practical significance of this observation lies in the fact that it occasionally offers some assistance in differentiating hepatocellular jaundice from simple obstructive jaundice, in which hypercholesterolemia is the rule. Study of the relationship between the degree of bilirubinemia and the changes in plasma cholesterol may be of value in this connection. In jaundice occurring in acute diseases of the liver, the blood cholesterol does not rise with the bilirubin but usually remains normal or subnormal. This divergence between the degree of blood cholesterol and blood bilirubin elevation in most cases of hepatocellular jaundice contrasts sharply with the parallelism between the hyperbilirubinemia and hypercholesterolemia in obstructive jaundice. The cholesterol ester is usually lowered in acute liver disease and mirrors the severity of the damage. When this phenomenon is observed in patients with common duct obstruction it should be regarded as of serious prognostic significance, particularly if it progresses and if the total cholesterol concentration also falls.

The liver is a major site of formation of the plasma cholesterol as well as the apoprotein of lecithin-cholesterol acyltranferase (p. 113). Accordingly, severe impairment of hepatic function would be expected to result in a fall in both total and ester cholesterol in the plasma. These changes may aid in distinguishing between hepatocellular and extrahepatic obstructive types of jaundice, particularly if interpreted in conjunction with simultaneous changes in serum bilirubin concentration, as indicated above. However, their practical value in this connection is limited by the facts that normal values are not unusual in patients with common duct obstruction, and hypercholesterolemia is not unusual in certain forms of primary intrahepatic disease, e.g., in toxic hepatitis, primary biliary cirrhosis and, occasionally, infectious hepatitis and portal cirrhosis. These facts must be borne in mind whenever data on plasma cholesterol are employed for this purpose.

Other Plasma Lipids. Quantitative studies of blood phospholipids and lipoproteins provide no information of value in the differential diagnosis of jaundice or in the evaluation of liver function that cannot be obtained more readily and more satisfactorily by studies of cholesterol. Changes in serum phospholipid concentration tend to occur in the same direction as those in free cholesterol. Low values are usually seen in severe hepatocellular damage, although a slight increase may occur in acute hepatitis. There may be little or no increase early in biliary obstruction, but chronic bile stasis is usually associated with high phospholipid levels.

The concentration of low-density lipoproteins (mainly β-lipoproteins) is generally increased in obstructive jaundice and biliary cirrhosis; α-lipoproteins are often decreased. In these conditions, when the serum shows lactescence, there is usually a high proportion of the lowest-density components, i.e., $S_f > 12$, including chylomicrons (>400). When there is no lactescence, the increase is mainly in the S_f 0–12 components.

Very high turbidimetric values for total lipids may be obtained in chronic biliary obstruction, both intra- and extrahepatic; moderate increases occasionally occur in acute hepatitis. Changes in concentration of neutral fat in the serum usually parallel those in total lipid, except in cholangiolitic and xanthomatous biliary cirrhosis, in which it may be either low or only slightly elevated.

PIGMENT METABOLISM—JAUNDICE

Bilirubin, the chief pigment of human bile, is derived mainly, but not entirely, from hemoglobin. Cells of the reticuloendothelial system, especially those present in the bone marrow, spleen and liver (Kupffer cells), are concerned with the metabolism and formation of bile pigment. The chief function of the hepatic parenchymal cells in this connection is the removal of bilirubin from the blood stream and its excretion in the bile. The stages of practical importance in the degradation of hemoglobin are outlined in Figure 21-5.

The hemoglobin molecule contains three distinct components: (1) a protein, globin, (2) iron, and (3) a porphyrin, which is a cyclic structure composed of four pyrrole units (p. 261). The iron-porphyrin compound is "heme." When hemoglobin undergoes degradation, the liberated iron and globin are reutilized, but the porphyrin residue is not. Its cyclic structure is broken, the iron is removed, and it is converted to an open-chain green pigment known as biliverdin. It is now believed that biliverdin is directly formed from heme and not from the iron-free protoporphyrin. A microsomal enzyme, heme oxygenase, brings about the cleavage of the porphyrin ring at the α-methene bridge, liberating Fe and carbon monoxide to give the green pigment biliverdin IXα, which is then reduced to the yellow-orange pigment bilirubin IXα by the enzyme, biliverdin reductase. Small amounts of hemoglobin are probably converted to mono- and dipyrrolic compounds. Bilirubin production does not account for all the hemoglobin broken down. The pigment leaves the reticuloendothelial cell and enters the blood plasma, where it is bound to albumin. In this form, it is carried to the liver, where it is separated from its protein carrier and excreted in the bile. It has recently been recognized that the hepatocyte produces a protein known as ligandin, which is involved in the removal of bilirubin from albumin and its transport across the hepatocyte membrane to the cell microsomes.

Bilirubin in the bile differs from that normally present in the blood plasma because it is entirely in conjugated form as the diglucuronide, with smaller amounts of sulfate. Small amounts of bilirubin are also conjugated with glucose, xylose, sugar alcohols and acidic disaccharides. The conjugation of bilirubin with glucuronic acid is effected by an enzyme, glucuronyl transferase, in microsomal membranes, utilizing uridine diphosphate glucuronic acid as a donor substance. Conjugated bilirubin gives the direct van den Bergh reaction, whereas the unconjugated form does not react to anything like the same extent (p. 636). The conjugation process is essential for normal removal of bilirubin from the blood stream and for its excretion in the bile; abnormalities of this mechanism play an important role in the pathogenesis of certain types of jaundice (p. 640). Excretion of conjugated bilirubin into the canaliculus is an active mechanism, disturbances of which can also result in jaundice.

The liver excretes 200 to 370 mg. of bilirubin daily in the bile, up to 30 per cent of which is believed to be derived from sources other than hemoglobin of circulating erythrocytes. These may include erythroid elements in the bone marrow and other heme-proteins (myoglobin, catalase, cytochromes). In the terminal ileum and mainly in the large intestine, bilirubin is reduced to mesobilirubinogen and stercobilinogen by the action of bacteria. These may be loosely collectively referred to as "bilinogens," or "urobilinogen." A fraction of the urobilinogen is reabsorbed into the portal circulation and is carried to the liver (so-called enterohepatic circulation); the major portion of this fraction is removed from the portal blood by the hepatic cells for biliary excretion, the remainder escaping into the general circulation to be excreted in the urine (0.5 to 3.5 mg. daily). The portion which escapes reabsorption from the intestine is oxidized to urobilin and stercobilin (the "fecal urobilins"), which are normal pigments of the feces (40 to 280 mg. daily, usually 100 to 250 mg.). There is some question as to whether any bilirubin is absorbed from the intestine; none is present normally in the feces of adults. It is present in the feces of normal infants, 1 to 25 mg. (av. 8.6 mg.) daily in the first five days of life, 0 to 15 mg. (av. 5.7 mg.) daily during the second five days, decreasing gradually thereafter, disappearing in 8 to 75 days after birth. Fecal excretion of urobilinogen in infants ranges from 0 to 1.0 mg. daily during the first two weeks of life, increasing to an average of 25 to 50 mg. between five and ten years of age. Inasmuch as the molecular weights of the bilinogens and bilins are approximately the same as that of bilirubin, all being about 3.5 per cent of that of hemoglobin, the approximate hemoglobin wastage may be determined indirectly by determining the quantity of these pigments excreted in the urine and feces.

Porphyrins are red pigments with a pyrrole structure, which are important components of hemoglobin, myoglobin, cytochrome and catalase. Their chemistry and metabolism are considered elsewhere (p. 261 ff.), and it is necessary here only to indicate certain features of importance in relation to hepatic function. There are four isomeric etioporphyrins, designated Types I, II, III and IV, but these are of practical importance only as reference types, the naturally occurring porphyrins corresponding to Types I and III. The production of these substances parallels hematopoietic activity and hemoglobin synthesis, Type III being formed in large amounts

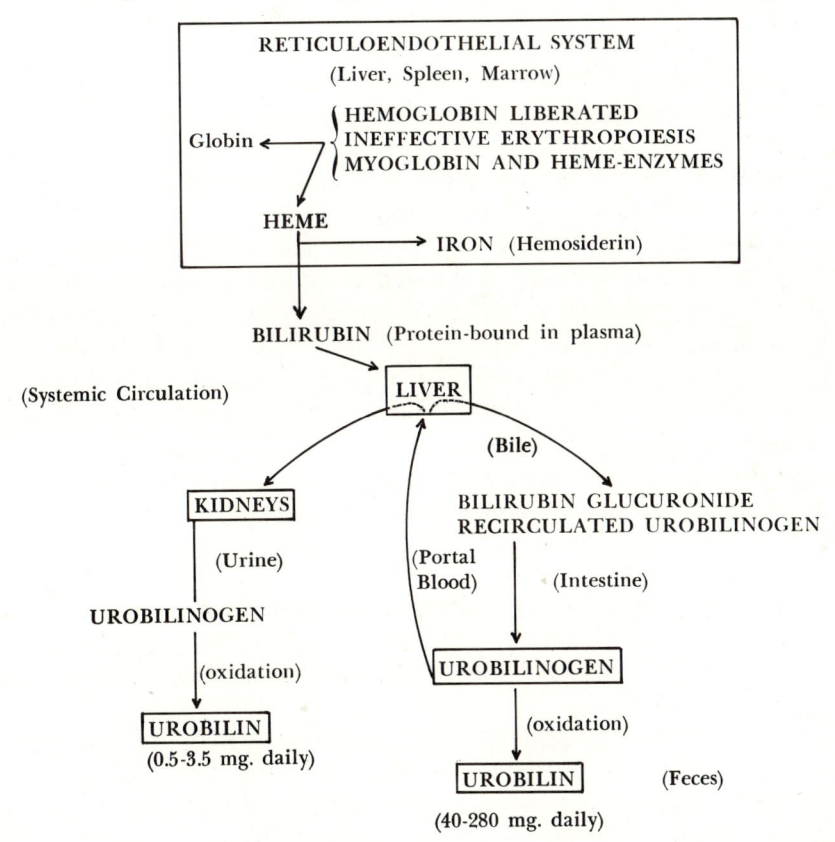

Figure 21-5. Bile pigment metabolism.

and utilized for the formation of hemoglobin and other respiratory factors, which are compounds of Type III porphyrin. Small amounts of coproporphyrin Type I are formed as a by-product in normal hematopoiesis and, with small amounts of coproporphyrin Type III not utilized in hemoglobin synthesis, are excreted, largely by the liver in the bile and eventually in the feces and urine (p. 274).

Clinical investigation of the state of liver and biliary tract function with regard to pigment metabolism includes studies of (a) serum bilirubin, (b) urine bilirubin, (c) excretion of urobilinogen and related products in the urine and feces, and occasionally (d) excretion of porphyrin in the urine and feces.

Serum Bilirubin

The van den Bergh Reaction. The van den Bergh test depends upon the fact that bilirubin couples with diazobenzosulfochloride (Ehrlich's diazo reagent) to form a characteristic red-violet azopigment. Since bilirubin (unconjugated) is insoluble in water and the diazo reagent is in aqueous solution, azotization of free bilirubin is facilitated by the introduction of some substance in which it and the diazo reagent are mutually soluble, e.g., alcohol. Addition of alcohol is not required for coupling of conjugated bilirubin since it is soluble in water. The reaction requiring the addition of alcohol or other catalysts to the serum-reagent mixture is termed the "indirect" reaction and is a test for unconjugated bilirubin. The direct mixture of serum and reagent in aqueous-acid solution constitutes the "direct" reaction. The bilirubin resulting directly from degradation of hemoglobin ("hemolytic" bilirubin), which includes that in normal serum, gives only the indirect reaction. The bilirubin in bile (conjugated) gives both indirect and direct reactions. With the establishment of the basis for these qualitative reactions, the descriptive terms "indirect-reacting bilirubin" and "direct-reacting bilirubin" have been replaced, respectively, by bilirubin and conjugated bilirubin (glucuronide). The production of a positive direct reaction in the blood stream (abnormal) implies the presence of conjugated bilirubin, i.e., usually bilirubin that has re-entered the circulation after having been passed into the bile canaliculi (e.g., biliary obstruction of hepatocellular damage) (p. 644 ff.).

Although normal serum bilirubin does not give a direct qualitative van den Bergh reaction (no visible color development within thirty seconds), it contains a fraction (<0.24 mg./100 ml.) which reacts in aqueous-acid solution (direct reaction) in one minute (one-minute, direct-reacting bilirubin). This represents conjugated pigment reabsorbed from the bile canaliculi.

Serum Bilirubin Concentration. The serum bilirubin concentration is determined by a quantitative van den Bergh procedure. It may also be measured approximately by determination of the icterus index.

Adults. The range of normal serum bilirubin concentration in adults is 0.1 to 1.0 mg. per 100 ml., being below 0.8 mg. in about 75 per cent of cases and below 0.5 mg. in about 50 per cent. Occasionally, values as high as 1.5 mg. are encountered in apparently normal subjects. The serum bilirubin concentration increases in untrained subjects ascending to high altitudes (hypoxia and excessive blood destruction). The liver of the fetus in utero cannot conjugate bilirubin and, therefore, cannot remove it from the circulating blood. This is due to deficiency of two key enzymes: (1) glucuronyl transferase, the conjugating enzyme, and (2) uridine diphosphate glucose dehydrogenase, which catalyzes the formation of the glucuronide donor. Bilirubin in fetal blood readily crosses the placenta into the maternal circulation, and is removed by the maternal liver.

The bilirubin of cord blood serum normally does not exceed 2.8 mg. per 100 ml., averaging 1.8 mg. The normal full-term infant is not jaundiced during the first twenty-four hours of life but becomes so, in about 75 per cent of cases, in two to four days. The serum bilirubin concentration begins to rise immediately after birth, reaching a peak of 2-12 mg./100 ml. (average 7 mg.) at two to four days. Jaundice disappears by the end of the first week, at which time the serum bilirubin concentration is less than 2 mg./100 ml. In the premature infant, there is a more prolonged rise to a higher level, a peak of up to 20 mg./100 ml. being reached at five to seven days. The subsequent fall is correspondingly delayed, depending upon the degree of prematurity. In this physiological jaundice of the newborn, virtually all of the serum bilirubin is unconjugated, giving a negative direct van den Bergh qualitative reaction.

The accumulation of serum bilirubin during the neonatal period is due mainly to two factors: (1) immaturity of the hepatic conjugating mechanism, as indicated above, and (2) an increased rate of destruction of larger numbers of erythrocytes as compared with adults. There is a gradual increase in the activity of glucuronyl transferase and uridine diphosphate glucose dehydrogenase to adult levels during the first few weeks. This is accompanied by excretion of increasing amounts of bilirubin in the bile (conjugated) and diminishing bilirubinemia.

One-Minute, Direct-Reacting Bilirubin. Quantitative determinations of direct-reacting bilirubin are of value, especially in cases in which there is defective conjugation, and aid in the differential diagnosis of jaundice. The recommended time allowed to expire in relation to the determination of direct bilirubin varies from laboratory to laboratory. Some use one minute, some five minutes, and others ten minutes. It has been suggested by some authorities that color development is so rapid that one minute is perhaps too short for accuracy. It has, however, sufficed for most clinical purposes. For the determination of total bilirubin, the technique of Malloy and Evelyn employs 50 per cent methanol as the alcohol component which accelerates the color reaction presumably by solubilization. Color development is measured after 30 minutes (15 minutes for some laboratories). The method has the disadvantage that it can be affected by hemolysis and can occasionally also bring about protein precipitation sufficient to produce some turbidity (minimal with the method of Meites and Hogg). For these reasons, other color accelerators have come into current use. The method of Jendrassik and Grof uses caffeine and sodium benzoate, and a final blue color is developed at a strongly alkaline pH. There is no precipitation of protein. The reaction can be stopped by adding ascorbic acid, and therefore the time of color development can be fixed. It has been recommended that this time should be 10 minutes for both the total and the direct bilirubin reactions. There is no doubt that, no matter what method is employed, unconjugated bilirubin makes some contribution to the direct reaction. Nevertheless, for clinical purposes, the direct reaction appears to suffice as a measure of conjugation.

Icterus Index. Determination of the icterus index consists in comparing the intensity of the yellow color of the blood serum or plasma with that of a standard 1 : 10,000 solution of potassium bichromate (icterus index of the standard = 1). The icterus index of normal serum or plasma is 4 to 6, i.e., the yellow color is 4 to 6 times as intense as that of the standard. This is an entirely nonspecific method but is useful clinically because of its simplicity and because the intensity of the yellow color of the serum is usually dependent upon the bilirubin concentration. However, apart from technical difficulties due to hemolysis and turbidity of the serum, which interfere with the color comparison, difficulties in interpretation may arise from the fact that occasionally substances other than bilirubin may impart a yellow color to the serum. The most important of these are xanthophyll and carotene. In the presence of carotenemia (p. 810) the icterus index is increased and does not, of course, reflect the bilirubin content of the serum. In certain cases of protracted jaundice, particularly of the extrahepatic obstructive variety, a significant amount of biliverdin may appear in the plasma ("biliverdin jaundice"; "verdinicterus"). Inasmuch as this pigment does not give the van den Bergh reaction, icterus index values are relatively higher than the values for bilirubin measured by this procedure.

Pathological Hyperbilirubinemia

The concentration of bilirubin in the blood plasma depends upon (1) the number and rate of destruction of erythrocytes (hemoglobin liberation), (2) the functional activity of reticuloendothelial cells (bilirubin formation), (3) hepatic cell function (removal, conjugation, and excretion of bilirubin), (4) patency of the bile canaliculi and bile ducts (normal flow of bile) and (5) structural integrity of the hepatic cell lining of the bile canaliculi.

Total Serum Bilirubin. Serum bilirubin concentrations above 1.0 mg. per 100 ml. or icterus index values above 6 usually constitute a state of hyperbilirubinemia. The degree of hyperbilirubinemia necessary for the production of clinical icterus varies in different types of jaundice. Discoloration of the sclerae and the skin occur at lower levels of serum bilirubin in obstructive (hepatocellular or extrahepatic) than in hemolytic jaundice. This is explained on the basis that in obstructive jaundice the conjugated, water-soluble bilirubin is more readily diffusible than the free, water-insoluble bilirubin in hemolytic jaundice. It has been found that in obstructive jaundice, when an icterus index value of 16 or a serum bilirubin concentration of 1.6 mg. per 100 ml. is reached and persists for some days, bilirubin diffuses through the capillaries and appears in the tissues. Higher concentrations are usually necessary in hemolytic jaundice for the production of frank icterus. Although generally in obstructive jaundice a relationship exists between the serum bilirubin concentration and the appearance of bilirubin in the urine, this is not invariably so in viral hepatitis, in which bilirubinuria may occur early, at relatively low serum bilirubin levels (p. 649).

Once clinical jaundice has occurred, it usually, but not invariably, persists for some time after the serum bilirubin has fallen below this "threshold" level. During the period of developing jaundice, therefore, serum bilirubin concentrations between 1.0 and 1.6 mg. per 100 ml. and icterus index values between 6 and 15 represent a condition of latent jaundice, usually not detectable by methods other than examination of the blood. At relatively low levels of bilirubinemia, the ratio between the icterus index and serum bilirubin concentration, in milligrams per 100 ml., is roughly 10 : 1, an icterus index of 6 corresponding to a serum bilirubin of 0.6 mg., and one of 15 to a serum bilirubin of 1.5 mg. This relationship has been found not to be so consistent at higher levels of bilirubinemia, but this may have been due to inaccuracies inherent in serum bilirubin determinations by methods employed formerly, or to the presence of biliverdin, which does not give the van den Bergh reaction.

Decrease in the concentration of bilirubin in the blood may occur in aplastic anemia, iron deficiency anemias, and in all "secondary" anemias, particularly those associated with malignancy and with chronic nephritis. A fall may occur in the elevated serum bilirubin concentration of patients with carcinoma of the head of the pancreas even though complete obstruction to the flow of bile persists. This is due perhaps to a decrease in the rate of formation of bilirubin or to the occurrence of biliverdinemia, which is not demonstrable by the van den Bergh test.

With the exception of bilirubinuria in the early stages of infectious hepatitis (p. 649), determination of the serum bilirubin concentration is the only certain means of detecting latent jaundice and gives a much more definite idea of the severity of and variations in hyperbilirubinemia than does examination of the urine or feces or the study of skin pigmentation. This estimation is, therefore, indicated in all cases in which information is desired regarding the possible presence of hepatic functional impairment, biliary stasis or excessive hemolysis. The frequency of latent or frank jaundice in cholecystitis as well as in hepatic disease congestive heart failure and a variety of primarily extrahepatic disease processes emphasizes the importance of the estimation of the degree of bilirubinemia in these conditions. Experience has demonstrated the fallacy of attempting to estimate variations in bilirubinemia in patients with jaundice by observing variations in color of the skin and conjunctivae.

One-Minute, Direct-Reacting Bilirubin. Normal serum contains 0.00-

0.25 mg./100 ml. of bilirubin which reacts with the van den Bergh reagent in aqueous-acid solution (i.e., direct reaction) within one minute (p. 636). This is referred to as "one-minute, direct-reacting bilirubin." Major increases in this fraction (conjugated bilirubin) occur in hepatocellular damage and biliary obstruction. In pernicious anemia and uncomplicated hemolytic jaundice it may comprise up to 15 per cent of the total serum bilirubin concentration. Higher values indicate complicating hepatocellular damage or biliary obstruction. Determination of this factor is of no practical value in differentiating hepatocellular from extrahepatic obstructive jaundice and has little or no advantage over the qualitative van den Bergh reaction in the diagnosis of hemolytic jaundice. Nevertheless it is helpful in assessing whether a jaundice is due to predominant increase of conjugated or unconjugated bilirubin.

PATHOGENESIS OF JAUNDICE

A classification of jaundice on the basis of specific disturbances in the mechanism of bilirubin metabolism and excretion is currently employed. (Table 21-1). The distinguishing criterion in this connection is the serum bilirubin pattern, i.e., the quantitative distribution of unconjugated and conjugated bilirubin.

(1) Predominant Increase in Unconjugated Bilirubin. This is due to: *(a)* increased rate of bilirubin production from circulating erythrocytes (excessive hemolysis) or other sources, e.g., bone marrow (ineffective erythropoiesis), which exceeds the normal functional capacity of the liver for its removal, conjugation and excretion; *(b)* impairment of hepatocyte function, involving the removal of bilirubin from the blood plasma and/or its transfer from the sinusoidal surface of the liver cell to the conjugation site (microsomes) or increased reflux back to the blood; *(c)* impairment of conjugation of bilirubin. In all these causes, there is some degree of reflux of unconjugated bilirubin from the liver cells to the circulation. Predominantly unconjugated hyperbilirubinemia is present when conjugated bilirubin (one-minute, direct-reacting) forms less than one-fifth of the total serum bilirubin. Conjugated bilirubin is usually less than 15 per cent in this condition. There is no bilirubinuria (acholuric jaundice).

(2) Predominant Increase in Conjugated Bilirubin. This is due to: *(a)* impairment of hepatocyte function, involving secretion of conjugated bilirubin into the bile canaliculi; *(b)* disease of the canaliculi, producing obstruction to the flow of bile (cholestasis); *(c)* cholestasis due to disease of the small intrahepatic bile ducts; *(d)* cholestasis due to extrahepatic biliary obstruction (surgical jaundice). In all these cases, jaundice is produced by reflux from the liver cells and/or biliary system into the circulation.

When there is a predominant increase in conjugated bilirubin with hyperbilirubinemia, bilirubin appears in the urine. This is that fraction of conjugated bilirubin which is not bound to albumin. The unbound fraction in plasma is apparently increased when there is an increase in the bile salt content. In urine, conjugated bilirubin appears to be bound to a low molecular weight peptide.

In hepatitis and cirrhosis, there is interference in hepatocyte bilirubin uptake (probably rare), in conjugation and in excretion. The latter is usually the most impaired and results in predominant increase in conjugated bilirubin in the plasma, with bilirubinuria.

Predominant Increase in Unconjugated Serum Bilirubin.

(1) Hemolytic Jaundice. The liver of a normal adult can excrete quite large amounts of bilirubin, but to maintain normal blood levels it can handle up to about 450 mg. bilirubin per day. This would be formed from approximately 13 g. Hb, representing almost a twofold increase in the rate of erythrocyte destruction. Hemolysis in excess of this would lead to bilirubin retention. A number of conditions in which this occurs are enumerated in Table 21-1 on page 643. Because of the fact that bilirubin excretion increases proportionately to the square of its concentration in the plasma, the latter does not exceed 5 mg./100 ml. on the basis of excessive hemolysis alone. In uncomplicated cases the serum gives a negative direct van den Bergh reaction and the one-minute, direct-reacting bilirubin does not exceed 15 per cent of the total.

In patients with hemolytic jaundice, the presence of *(a)* a positive direct van den Bergh reaction, *(b)* a serum bilirubin concentration exceeding 5 mg./100 ml., or *(c)* a one-minute, direct-reacting bilirubin value exceeding 15 per cent of the total indicates associated hepatic functional impairment or biliary obstruction. The former may be due to severe hypoxia (anemia) and the latter to biliary calculi, as in congenital spherocytic jaundice. Erythroblastosis is considered elsewhere (pathological neonatal jaundice, p. 646).

(2) Nonhemolytic Overproduction Jaundice. Mild hyperbilirubinemia, with a negative direct van den Bergh reaction (free bilirubin), occurs frequently in pernicious anemia. In this condition, although there may be a slightly increased rate of destruction of erythrocytes (shorter life span), there is evidence that up to 40 per cent of the bilirubin formed may originate in sources other than the Hb of circulating erythrocytes. It has been suggested that it may arise in the bone marrow from heme or porphyrins not utilized for Hb synthesis or from Hb of immature erythrocytes ("ineffective erythropoiesis"). A similar mechanism may contribute to the hyperbilirubinemia of hemolytic and other disorders associated with erythroid hyperplasia, including congenital spherocytic jaundice.

(3) Familial and Hereditary, Nonhemolytic, Nonobstructive Jaundice. Conditions in this category are characterized by hyperbilirubinemia with a negative direct van den Bergh reaction (predominantly unconjugated bilirubin). They include: *(a)* the Crigler-Najjar syndrome and *(b)* the Lucey-Driscoll syndrome, severe disorders which appear in infancy and early childhood, and *(c)* Gilbert's syndrome, a milder condition, usually first recognized in later life (constitutional hepatic dysfunction).

(a) Crigler-Najjar Syndrome. Intense jaundice is apparent at or soon after birth, the serum bilirubin concentration reaching levels as high as 60 mg./100 ml., almost invariably with kernicterus. This condition is genetically determined and has a familial incidence. Death usually occurs early, but there are occasional survivors to adult life, with persistent jaundice. The basic defect is inability of the liver cells to conjugate bilirubin owing to genetically determined deficiency of glucuronyl transferase. The bile contains no conjugated pigment and free bilirubin accumulates in the blood stream. Except for this specific enzyme deficiency and impaired bilirubin tolerance, there is no demonstrable abnormality of liver function and no detectable morphologic liver damage. There is no evidence of increased hemolysis. Less severe impairment of glucuronide conjugation capacity may be exhibited by asymptomatic parents of children presenting this syndrome.

(b) **Lucey-Driscoll Syndrome.** This is characterized by extreme hyperbilirubinemia (up to 60 mg./100 ml.) in newborn infants, usually more than one, whose mothers are apparently otherwise normal. Kernicterus occurs commonly. In those who survive, jaundice disappears within a few weeks and the serum bilirubin returns to normal levels, usually within the first month. Although its pathogenesis has not been established, it has been suggested that the hyperbilirubinemia may be due to the presence in the blood stream of the child and of the mother (during pregnancy) of a substance, perhaps a steroid, that inhibits conjugation of bilirubin in the liver. There is, however, a view contrary to this hypothesis. A milder type of jaundice, without kernicterus, is also associated with breast feeding.

(c) **Gilbert's Syndrome (Constitutional Hepatic Dysfunction).** This designation is applied to a condition of chronic, fluctuating hyperbilirubinemia, usually not exceeding 6 mg./100 ml., but occasionally as high as 12 mg. The serum gives a negative direct van den Bergh reaction (unconjugated bilirubin) and there is no increase in fecal urobilinogen, i.e., no excess hemolysis. Defective hepatic glucuronide conjugation has been found in some cases, suggesting a mechanism similar to that operating in the Crigler-Najjar syndrome. However, a fundamental difference is indicated by the presence of conjugated bilirubin in the bile, as well as by reports of normal in vivo glucuronide conjugation in certain patients with mild hyperbilirubinemia. It has been suggested that the defect may lie in the intracellular (hepatic cell) transport of bilirubin rather than in the conjugating enzyme. There is good evidence of increased reflux of bilirubin from the hepatocytes back into the circulation.

Possibly this designation has been applied to several conditions of different etiology. The following, with superficially similar manifestations, clinical and biochemical, may have been included: cases of the Crigler-Najjar syndrome who have survived into adult life; persisting hepatic functional impairment following viral hepatitis; impaired hepatocellular function (bilirubin transport) in compensated hemolytic diseases; conditions associated with erythroid hyperplasia in which there may be excessive production of bilirubin from sources other than the hemoglobin of

TABLE 21-1 CLASSIFICATION OF JAUNDICE

1. PREDOMINANT INCREASE OF UNCONJUGATED SERUM BILIRUBIN (ACHOLURIC JAUNDICE)

(a) **Increased bilirubin production (overload jaundice)**
  (i) HEMOLYSIS
   Examples: Hereditary spherocytosis; icterus neonatorum; erythroblastosis fetalis; paroxysmal hemoglobinuria; pernicious anemia; splenic anemia; Cooley's anemia; Marchiafava-Micheli syndrome; extensive burns; sickle-cell anemia; thalassemia major; ruptured ectopic pregnancy or other intraperitoneal hemorrhage; mushroom poisoning; favism; G6PDH deficiency; malaria (black water fever); hemolytic transfusion reactions; increased hemolysis incident to infection with such agents as streptococci, staphylococci, pneumococci, *Clostridium perfringens, Bartonella bacilliformis;* poisoning with snake venoms, arseniuretted hydrogen, toluenediamine, phenylhydrazine, acetanilid, sulfonamides, dinitrobenzol, aniline, benzol and nitrocompounds of phenol, nicotinic acid, etc.; heart surgery

  (ii) INEFFECTIVE ERYTHROPOIESIS
   Examples: Thalassemia minor, pernicious anemia

(b) **Familial and hereditary**
   Examples: Crigler-Najjar syndrome, Lucey-Driscoll syndrome, Gilbert's syndrome

(c) **Due to drugs**
   Examples: Rifomycin, novobiocin, bunamiodyl, flavaspidic acid, sulfonamides, salicylates, etc.

2. PREDOMINANT INCREASE OF CONJUGATED SERUM BILIRUBIN

(a) **Familial and hereditary impaired bilirubin secretion**
   Examples: Dubin-Johnson syndrome, Rotor syndrome

(b) **Cholestasis**
  (i) CANALICULAR CHOLESTASIS
   Idiopathic recurrent intrahepatic cholestasis
   Miscellaneous
   Examples: Pregnancy, oral contraceptives, hepatitis (viral, alcoholic, or drugs), cirrhosis, Hodgkin's disease

  (ii) INTRAHEPATIC BILE DUCT CHOLESTASIS
   Examples: Primary biliary cirrhosis, biliary atresia, certain inflammatory bowel diseases, primary sclerosing cholangitis

  (iii) EXTRAHEPATIC CHOLESTASIS ("SURGICAL JAUNDICE")
   Examples: Calculi, biliary or pancreatic; neoplasm of pancreas, ducts, duodenum or lymph nodes; duodenitis and diverticulum of duodenum; parasites in ducts; stricture of ducts, congenital or acquired; adhesions, pancreatitis; pancreatic cysts; cysts, carcinoma or abscess of the liver; aneurysm of the hepatic artery or renal artery

(c) **Hepatocellular jaundice**
   Examples: Acute and subacute hepatic necrosis; infectious hepatitis; hepatitis due to any infection, such as syphilis, pneumonia, dysentery, *Clostridium perfringens,* tuberculosis, typhus, typhoid, streptococcus, staphylococcus, relapsing fever, Weil's disease; malaria, Oroya fever; infectious mononucleosis, active chronic hepatitis, hepatic cirrhosis; yellow fever; congestive heart failure; hyperthyroidism; diabetes mellitus; sulfonamides, paracetamol, tranquilizing drugs, cytotoxic drugs, etc.; poisoning with *Amanita phalloides,* arsphenamine, mercury, lead, chloroform, carbon tetrachloride, phosphorus, tetrachlorethane, trinitrotoluene, cinchophen, ferrous sulfate; x-ray over liver

circulating erythrocytes, e.g., in the bone marrow. This may account for variations in its reported clinical manifestations. There are contradictory observations on its familial incidence and its tendency to improve with advancing years. There is general agreement that it is asymptomatic, being usually discovered accidentally, that the liver appears normal morphologically, except occasionally for fatty infiltration, and that transient increases in bilirubinemia are often induced by intercurrent infection, fatigue, excessive intake of alcohol and emotional tension.

Predominant Increase of Conjugated Bilirubin in Serum. The various types of this condition can be included in two main categories: (a) impaired hepatic excretion of bile pigment and (b) obstruction (cholestasis), which can result from disease involving the canaliculi, the small intrahepatic bile ducts or the main bile ducts (extrahepatic, or "surgical," jaundice).

(1) Familial and Hereditary Impaired Bilirubin Secretion. There are two conditions in this category due to failure of the hepatocytes to excrete conjugated bilirubin into the canaliculi, and both are rare. They are: (a) the Dubin-Johnson syndrome and (b) the Rotor syndrome.

(a) Dubin-Johnson Syndrome. This condition usually develops before the age of 20 years. There is mild intermittent or persistent jaundice, with bilirubin levels not usually exceeding 10 mg. per 100 ml. of serum. The liver may be slightly enlarged and slightly tender. A brown melanin-like pigment is deposited around the central veins, but biopsy histology is otherwise normal. The Bromsulphalein test (p. 666) results in an initial fall in the blood level of the dye, but then a delayed rise occurs between 60 and 120 minutes after injection. The gallbladder is not visualized on cholecystography, since there is also defective excretion of the x-ray opaque dyes employed in this procedure. The serum alkaline phosphatase and aminotransferase levels are not increased. The disease is inherited as an autosomal recessive. The jaundice may be accentuated by oral contraceptives. Bile salts are excreted normally.

(b) Rotor Syndrome. This is very similar to the Dubin-Johnson syndrome, but there is no pigment deposit in the liver.

(2) Canalicular Cholestasis. There are various types of this condition.

(a) Idiopathic Recurrent Intrahepatic Cholestasis. This condition also is rare. There are intermittent attacks of jaundice with intense pruritus, with bile salt retention. Serum alkaline phosphatase levels are raised. The disease commences in childhood. Biopsy shows bile plugs in dilated canaliculi.

(b) Miscellaneous. Canalicular cholestasis can also occur in pregnancy or result from the use of oral contraceptives, the administration of certain drugs, or from viral or alcoholic hepatitis. Serum alkaline phosphatase levels tend to be raised.

(3) Intrahepatic Bile Duct Cholestasis. Here again, the level of serum alkaline phosphatase tends to be increased. Disease of the small bile ducts is found in primary biliary cirrhosis (p. 643), in biliary atresia, in certain inflammatory bowel diseases and in primary sclerosing cholangitis. Liver biopsy is necessary for final diagnosis.

(4) Extrahepatic (Surgical) Cholestasis. Following obstruction to the flow of bile, bilirubin appears rather promptly in the thoracic duct lymph and, if a major portion of the duct system is involved, increases shortly thereafter in the systemic circulation. The increase is mainly in conjugated bilirubin, due to regurgitation from the bile canaliculi or canals of Hering, or to reversal of the direction of transport within the hepatic cells. There is an increase also of free bilirubin (average about 35 per cent of the total), indicating impairment of hepatocellular function. The level of serum alkaline phosphatase is usually raised.

Occasionally, the direct van den Bergh reaction may become negative following relief of obstruction, after the serum bilirubin has fallen to about 1.5 mg./100 ml. The direct reaction may persist in some cases, however, for some time after the serum bilirubin has fallen to within normal limits.

(5) Hepatocellular Jaundice. The characteristics of the hyperbilirubinemia associated with hepatocellular damage are essentially the same as those in biliary obstruction. Both free and conjugated bilirubin participate in the increase, the latter predominating (average about 65 per cent of the total).

As in the case of obstructive jaundice, increase of free bilirubin (average 35 per cent of total) is attributable to impaired hepatocellular function. The increase of bilirubin diglucuronide has been ascribed to (1) disruption of the continuity of bile canaliculi due to liver cell damage (cloudy swelling), (2) occlusion of their lumen by desquamated and disintegrated cells and bile thrombi, and (3) reversal of the direction of transport of conjugated bilirubin within the hepatic cells. The pathogenesis of this type of jaundice, therefore, as also in biliary obstruction, involves the operation of both retention and regurgitation mechanisms.

In initial stages of mild hepatocellular damage the serum may give a negative direct van den Bergh reaction, but the one-minute, direct-reacting bilirubin is often increased. With increasing damage, the direct van den Bergh reaction becomes positive. Increase in one-minute, direct-reacting bilirubin may be the only demonstrable abnormality of the serum bilirubin in mild cases of infectious hepatitis and also for prolonged periods in patients with portal cirrhosis. Elevated values are also obtained frequently during convalescence from infectious and other types of acute hepatic damage for some time after the total serum bilirubin concentration has fallen to normal levels. This is indicative of persisting liver functional impairment.

Jaundice Due to Drugs

A variety of drugs can give rise to jaundice. This can be predominantly an unconjugated hyperbilirubinemia or predominantly the conjugated type.

(a) Unconjugated Hyperbilirubinemia. This may result from excessive hemolysis (e.g., primaquine and G6PDH deficiency, phenacetin and methyldopa), from competition for binding sites on serum albumin (sulfonamides, salicylates), from inhibition of hepatocyte uptake of serum bilirubin (e.g., bunamiodyl, flavaspidic acid) or from inhibition of the conjugation mechanism (e.g., albamycin).

(b) Conjugated Hyperbilirubinemia. This may result from canalicular cholestasis (e.g., Rifampicin, C_{17} alkyl testosterones, oral contraceptives), from hepatotoxicity (e.g., CCl_4, paracetamol overdosage, intravenous tetracycline, $FeSO_4$) or from hypersensitivity (e.g., isoniazid, methyldopa, phenothiazines, oxyphenisatin, halothane). Hypersensitivity reactions can be very confusing, since they produce the biochemical picture of hepatitis and cholestasis.

It is now necessary, in the treatment of all patients presenting with jaundice, to make careful inquiry about any medicament which they may be receiving.

Pathological Neonatal Jaundice

Definitive criteria for abnormal neonatal hyperbilirubinemia are difficult to define clearly, in certain respects, because of the wide range of normal variation in the upper limit of serum bilirubin concentration in the newborn, particularly the premature (p. 637). However, jaundice may usually be regarded as abnormal under the following circumstances: (*a*) if it appears during the first twenty-four hours of life in full-term infants or the first thirty-six hours in prematures; (*b*) if the maximum serum bilirubin concentration exceeds 12 mg./100 ml. in full-term infants or 20 mg. in prematures; (*c*) if it exceeds 2 mg./100 ml. after the first week of life in full-term infants; (*d*) if the one-minute, direct-reacting bilirubin exceeds 15 per cent of the total; (*e*) if the serum gives a positive direct van den Bergh reaction; (*f*) if the serum bilirubin concentration increases at a rate exceeding 5 mg./100 ml. per day.

(1) Predominant Increase in Free Bilirubin (Unconjugated). Jaundice of this type is due to increased production of bilirubin and/or impaired hepatocellular function (inadequacy of transport or conjugating mechanisms).

(a) Excessive Bilirubin Production. Erythroblastosis (Rh and ABO incompatibilities) is the most important condition in this category. Infants with this condition are jaundiced at birth or within the first twenty-four hours, and the serum bilirubin concentration rises to excessive levels within the first few days. The concentrations are usually higher with Rh than with

TABLE 21-2 NEONATAL JAUNDICE

I. Predominantly Unconjugated Bilirubin
   A. Physiological jaundice.
   B. Excessive bilirubin production.
   C. Hepatocellular dysfunction (transport and/or conjugation).

II. Predominantly Conjugated Bilirubin
   A. Extrahepatic biliary obstruction.
   B. Cholestasis.
   C. Hepatocellular damage.

ABO or minor agglutinin incompatibilities. In the latter, the serum bilirubin concentrations may be within the limits of normal for the newborn.

Certain types of congenital hemolytic disease may manifest themselves at birth or shortly thereafter, e.g., congenital spherocytic anemia (congenital hemolytic jaundice) and nonspherocytic hemolytic anemias. These are rare in the neonatal period, as is hyperbilirubinemia due to thalassemia and sickle-cell anemia.

A hemolytic type of hyperbilirubinemia may result from administration of large doses of vitamin K to newborn infants, especially prematures, or occasionally to the mother. It has occurred also as a result of exposure to naphthalene (in diapers) in infants with congenital absence of glucose-6-phosphate dehydrogenase in the erythrocytes (p. 572).

Jaundice in the newborn due to bacterial, viral and protozoal infections may be due, at least in part, to excessive hemolysis. These include cytomegalic inclusion disease, toxoplasmosis, bacterial sepsis, and congenital syphilis.

Excessive amounts of bilirubin may be produced from large extravasations of blood, particularly subcutaneous.

(b) Hepatocellular Dysfunction (Transport and Conjugation). Conditions in this category are characterized by one or both of the following abnormalities: (1) delayed development of the glucuronide conjugating mechanism, normally absent at birth (p. 637); (2) inadequacy of the mechanism of transport of bilirubin from the sinusoidal surface of the hepatic cell to the site of conjugation (microsomes).

The Crigler-Najjar syndrome is the clearest example of defect in the conjugating mechanism. In this disorder there is a genetically determined absence of glucuronyl transferase, with consequent inability on the part of the liver to excrete bilirubin in the bile, resulting in its accumulation in the blood in very high concentrations (p. 641). A similar mechanism may operate in the Lucey-Driscoll syndrome (p. 642).

Abnormally high levels of serum bilirubin in the neonatal period, in part at least on this basis, occur also in the infants of diabetic mothers, in cretins (athyroidism), and as a result of hypoxia (prolonged, difficult labor; respiratory distress). Hyperbilirubinemia at birth may be aggravated by administration of certain tranquilizing or antihypertensive drugs to the mother during the latter stages of pregnancy.

(2) Predominant Increase in Conjugated Bilirubin. The blood serum gives a positive direct van den Bergh reaction and there is an increase also in free bilirubin (unconjugated). As in adults (p. 644), the mechanisms involved include: (a) biliary obstruction; (b) disturbed hepatocellular function with respect to diglucuronide formation and/or bilirubin excretion. These may be superimposed upon a background of excessive bilirubin formation.

An obstructive mechanism is clearly operative in congenital atresia of the bile ducts. In virtually all other forms of neonatal jaundice in this category, the primary underlying factor is apparently hepatocellular functional impairment or damage; features that are apparently obstructive in nature are probably of secondary rather than primary importance. This is the case in the so-called "inspissated bile syndrome" in erythroblastotic infants. In this condition, bile thrombi, formerly regarded as causative factors (obstructive), are now believed to be consequences of cholestasis incident to hepatocellular functional impairment (decreased excretion rate). The hyperbilirubinemia of galactosemia also belongs in this category. Hepatocellular malfunction and/or damage are also involved, in addition to excessive hemolysis (p. 647), in the production of jaundice incident to bacterial sepsis, cytomegalic disease, herpes simplex infection, toxoplasmosis, congenital syphilis, neonatal hepatitis, and administration of certain drugs (e.g., tranquilizers, antihypertensives) to the mother late in pregnancy.

KERNICTERUS (BILIRUBIN ENCEPHALOPATHY)

This term is applied to a condition of deep-yellow pigmentation of basal ganglia of the brain that occurs in certain deeply jaundiced newborn infants. There are characteristic neurological manifestations, and the condition is always serious, frequently fatal. It occurs most frequently in prematures; in full-term infants, except in the Crigler-Najjar syndrome, it usually develops during the first five days of life. The tissue staining and damage are due to free (unconjugated) bilirubin, which has diffused out of the blood plasma. Kernicterus occurs, therefore, only in those types of jaundice in which the serum bilirubin is predominantly unconjugated (p. 646). Brain pigmentation does not occur as a result of even extremely high concentrations of conjugated bilirubin. This difference has been attributed to an affinity of the brain tissues, with their high lipid content, for the lipid-soluble unconjugated bilirubin and not for the water-soluble conjugated pigment. In vitro, bilirubin, at concentrations of about 20 mg./100 ml., inhibits heme synthesis, uncouples oxidative phosphorylation and depresses oxygen uptake by brain tissue, actions that could conceivably produce the characteristic neurological lesions of kernicterus.

Passage of bilirubin from the blood plasma into the cerebral tissues is influenced by several factors: (1) the concentration of unconjugated bilirubin in the blood; (2) its diffusibility across the capillary walls; (3) the permeability of the blood-brain barrier.

Binding of bilirubin by albumin in the plasma limits its diffusibility across capillary walls and the blood-brain barrier. Consequently, its penetration into the brain substance is related more directly to the extent to which the quantity of bilirubin entering the blood stream exceeds the binding capacity of the plasma than to the serum bilirubin concentration per se. At a pH of 7.4 virtually all of the bilirubin in the plasma is bound to protein, therefore nondiffusible; only 0.1 mg./100 ml. can be maintained in aqueous solution at this pH. The significance of the level of serum bilirubin concentration in relation to the development of kernicterus lies in its indication of the extent of overproduction of bilirubin. A level of 20 mg./100 ml. in the newborn, with a negative direct van den Bergh reaction, is generally regarded as potentially dangerous. However, kernicterus has developed at lower concentrations and has been absent in infants with more marked hyperbilirubinemia. Its occurrence at relatively low serum bilirubin concentrations in some instances may be due to reduction of the binding capacity of the plasma for the pigment. This may result from: (1) hypoalbuminemia; (2) increased hydrogen-ion concentration; (3) the presence of other organic anions (e.g., sulfonamide, sulfisoxazole, salicylate, caffeine sodium benzoate) that compete with bilirubin for binding sites on the albumin molecule, thereby reducing the amount of protein-bound bilirubin and permitting more to diffuse out of the plasma into the brain tissues.

The particular susceptibility of prematures is attributable to: (1) the usually lower serum albumin concentration; (2) immaturity of the blood-brain barrier (increased permeability); (3) immaturity of the hepatic mechanism for conjugating bilirubin, permitting its more rapid accumulation in the blood because of inadequacy of its excretion in the bile.

BILIRUBINURIA

Normally, bilirubin is not present in urine in amounts capable of detection by methods usually employed. In contrast to other types of jaundice, uncomplicated hemolytic jaundice is usually characterized by the absence of bilirubinuria, even at high serum bilirubin concentrations. This difference is perhaps attributable to the insolubility in water of free bilirubin, the predominant pigment in the serum in hemolytic jaundice, and the water-solubility of the glucuronide, the predominant serum pigment in obstructive and hepatocellular jaundice. However, although the urine bilirubin invariably gives a positive direct van den Bergh reaction, the appearance and amount of bilirubinuria are not consistently related to the concentration of direct-reacting serum bilirubin, suggesting that other factors, such as protein binding, may influence renal excretion of the pigment. As a rule, at equal serum bilirubin concentrations, more bilirubin is excreted in the urine in extrahepatic cholestasis than in hepatocellular jaundice. In early stages of viral hepatitis, bilirubinuria may occur with little increase in serum bilirubin, whereas later, during subsidence of the hyperbilirubinemia, it may disappear at direct bilirubin levels in the serum as high as 1.2 mg./100 ml. This difference in behavior may occur also in extrahepatic biliary obstruction. Renal functional impairment results in diminished bilirubin excretion in the urine; in advanced renal failure, the urine may contain no bilirubin, even in obviously jaundiced subjects.

Because of its simplicity, a quantitative adaptation of the methylene blue test for bilirubinuria has been employed in the study of patients with hepatic and biliary tract disease. Positive tests, interpreted as indicating abnormal bilirubinuria, were obtained in a large proportion (40 per cent) of patients with infectious hepatitis one to six days before the onset of clinical icterus and frequently in the presence of normal serum icterus index values. However, there was some danger in interpreting results with this procedure, since the evidence indicated that it was nonspecific, the positive green color obtained in urine containing bilirubin being due chiefly to a blend of yellow and blue colors. The development of a positive reaction was, therefore, influenced by concentration or dilution of the urine and by any yellow constituent. A positive test was obtained with artificially produced bilirubinuria at concentrations of 2 mg. or more of bilirubin per 100 ml. of urine. Despite its nonspecificity, the simplicity of the procedure made it useful in large scale testing during epidemics of infectious hepatitis and where complete laboratory facilities were lacking. More specific, sensitive tests for bilirubinuria (e.g., Harrison spot test, ictotest, ictostix) have been employed to advantage under similar circumstances with better results than the methylene blue test. These procedures have been found to be useful for the following purposes: (1) differential diagnosis of uncomplicated hemolytic jaundice from other types; (2) as a screening procedure for early detection of viral or toxic hepatitis; (3) as a substitute for repeated serum bilirubin determinations in following the course of hepatic or biliary tract disease; (4) as an aid in recognizing the occurrence of renal functional impairment in patients with jaundice.

Bilins and Bilinogens in the Urine and Feces

The relationship between the various derivatives of hemoglobin is illustrated in Figure 21-5, and has been described previously (p. 633). The pertinent facts regarding the metabolism of urobilinogen may be summarized briefly as follows.

Under normal conditions bilirubin entering the intestine in the bile is transformed, through the action of bacteria, into mesobilirubinogen and stercobilinogen (collectively termed bilinogens, or urobilinogen), colorless substances, which are subsequently oxidized to urobilin and stercobilin. The former ("bilinogens") will be collectively referred to as urobilinogen, and the latter ("bilins") as urobilin. Both are present in the feces and, in traces, in the urine of normal subjects. They are usually not present in fresh bile of normal individuals in more than minimal amounts but may appear in the bile in patients with hepatic disease. A portion of the urobilinogen formed in the intestine is reabsorbed into the portal blood stream and carried to the liver where it is re-excreted or metabolized, the products being unknown. This cycle is termed the enterohepatic circulation of bile pigment. The following conclusions seem warranted on the basis of available evidence:

(1) Bacterial infection or putrefaction is necessary for the conversion of bilirubin to urobilin. This may occur in the intestine (normal) or, theoretically at least, in the biliary passages in the presence of infection.

(2) In conditions of moderate hepatic damage, incomplete biliary obstruction, cholangitis and hemolytic jaundice the liver is unable to deal completely with all of the urobilinogen coming to it from the intestine, and some consequently passes into the general circulation and consequently more than the usual small amount is eliminated in the urine. Under such

circumstances urobilinuria is an indication of defective hepatic function. In pure hemolytic icterus, although the function of the liver may be unimpaired, it cannot remove completely the excessive amounts of urobilinogen formed from the increased quantity of bilirubin which passes into the bile and this gives rise to urobilinuria.

Urobilinogen in Feces. The term "urobilinogen" is applied here, and in subsequent sections, to the total bilins and bilinogens, combined, for purposes of convenience. This is permissible because these substances have the same clinical significance. The accurate quantitative estimation of the urobilinogen content of the feces is not employed routinely clinically but may be of value in certain conditions. The normal daily excretion in the feces of adults ranges from 100 to 400 mg. The urobilinogen content of the feces or duodenal contents appears to be an index of the degree of blood destruction in the absence of hepatic or biliary tract disease. Increased quantities have been found in the duodenal bile and feces in hemolytic jaundice, paroxysmal hemoglobinuria and other hemolytic processes and during periods of relapse in pernicious anemia. In the latter condition, during remissions induced by therapy, the urobilin content of the feces drops to within normal limits. The urobilinogen content of the feces of children is low as compared to that of adults, averaging less than 3 mg. daily up to the age of eleven years.

Low values may be obtained in the presence of inanition, inactivity, low-grade infections, hypochromic anemia and certain cases of secondary hyperchromic anemia. Inasmuch as the quantity of urobilinogen in the feces depends primarily upon the quantity of bilirubin entering the intestine, variations in the former in hepatic and biliary tract disease are determined largely by (1) the degree of impairment of bilirubin excretion by the hepatic cells, (2) the presence and degree of obstruction to the flow of bile and (3) the presence and severity of an associated hemolytic process.

This determination may be of value in distinguishing between jaundice due to stone in the common duct and that due to neoplasm of the duct, duodenum or head of the pancreas. Theoretically, in the absence of marked biliary tract infection, complete obstruction of the common duct, with consequent absence of bile pigment from the intestine, should result in a total absence of urobilinogen from the feces and urine. This fact is applied particularly to the differentiation between common duct obstruction due to stone, which is seldom permanently complete, and that due to neoplasm, which usually becomes permanently complete. Absence of urobilinogen from the urine, a quite constant finding in obstructive jaundice due to neoplasm, may occur also at times with obstruction due to stone. However, excretion of urobilinogen in the feces usually differs strikingly in these two groups of patients, being frequently extremely low or absent (0 to 5 mg. daily) in those with malignant obstruction.

In interpreting low values for urobilinogen in the feces or urine in jaundiced subjects, it must be remembered that antibiotic therapy effective in reducing the normal intestinal bacterial flora will thereby diminish urobilinogen formation from bilirubin in the intestine. Under these circumstances, bilirubin appears in the feces and the excretion of urobilinogen in feces and urine will be correspondingly decreased.

In the presence of severe diffuse liver damage, e.g., in infectious hepatitis and acute necrosis, there may be periods, usually relatively brief, in which no bile enters the intestine (acholia). In this condition of total suppression of bile formation, bilins and bilinogens will be absent from the feces (and also urine), as in complete biliary obstruction.

Urobilinogen in Urine. As stated above, absence of urobilinogen from the urine in patients with jaundice is indicative of complete obstruction of the common duct or of complete suppression of bile pigment excretion by the liver, except in those receiving broad-spectrum antibiotic therapy. This finding may also be present, whether or not jaundice is present, in patients with bile fistula in whom the bile is draining either externally or into some viscus other than the intestine. However, some urobilinogen may be present in the urine occasionally in cases of complete obstruction of the common duct. This may occur particularly in cases of calculous obstruction with marked infection of the bile passages or obstructive biliary cirrhosis and, rarely, in patients with marked jaundice in whom some bile pigment may pass into the lumen of the bowel directly from the jaundiced intestinal mucous membrane.

The normal adult eliminates 0.5 to 3.5 mg. urobilinogen daily in the urine. According to the less accurate methods commonly employed (Wallace and Diamond), a positive reaction for urobilinogen may be obtained normally with urine dilutions up to 1 to 20; a positive reaction with dilutions above this figure is considered indicative of the presence of excessive quantities. Slight increases may occur in subjects without hepatic, biliary tract or hemolytic disease in the presence of constipation, allowing an abnormally prolonged period of reabsorption of urobilinogen from the intestine. Although in pathologic conditions the morning urine may contain relatively large quantities of urobilinogen, the rate of elimination varies so much from hour to hour and day to day that all estimations should preferably be made on twelve- or twenty-four-hour specimens and should be repeated for at least five consecutive days before a negative reaction is interpreted as indicating the absence of urobilinuria.

Excessive amounts of urobilinogen and urobilin are found in the urine under the following circumstances:

Excessive Hemolysis. Excessive urobilinogenuria occurs characteristically in congenital and acquired hemolytic jaundice, pernicious anemia, splenic anemia and other conditions associated with hemolytic features or increased bilirubin production, due to the fact that the liver is unable to deal with the excessive amount of urobilinogen formed from the abnormally large quantity of bilirubin in the intestine. A sudden increase in urine urobilinogen has been observed shortly after the occurrence of pulmonary infarction in patients with congestive heart failure or auricular fibrillation, due presumably to hemolysis in the infarcted areas. It has been found that patients with pernicious anemia in whom excessive urobilinuria disappears before the reticulocyte crisis (induced by treatment) exhibit a greater increase of hemoglobin and red blood cells than those in whom excessive urobilinuria persists for even a short time after the reticulocyte crisis. Some believe that excessive urobilinuria in these anemic conditions is due in part to some disturbance of hepatic function, attendant perhaps upon the existing state of hypoxia. Similar findings have been obtained during malarial rigors. The total urobilinogen excretion may be regarded as a quantitative index of the amount of hemoglobin destruction (p. 633).

Hepatic Functional Impairment. (a) **Hepatocellular Jaundice.** Urobilinogenuria occurs regularly in the presence of hepatic functional impairment as long as adequate quantities of bilirubin are entering the bowel. This phenomenon occurs, therefore, in patients with jaundice due to parenchymal hepatic lesions, including hepatitis, acute and subacute hepatic necrosis, toxemia of pregnancy, arsphenamine, chloroform and carbon tetrachloride poisoning, congestive heart failure and liver damage due to a variety of other toxic and infectious agents. In animals poisoned with chloroform, if the hepatic damage is so severe that no bilirubin is excreted by the liver, urobilinogenuria does not occur. This coincides with the common clinical observation of excessive urobilinogenuria in the very early and late stages of acute hepatitis, while it is absent during the stage of complete suppression of bilirubin excretion or obstruction to its excretion due to cholangitis. Excessive urobilinogenuria has been observed in the early stages of acute and subacute hepatic necrosis which disappeared as the condition progressed and the degree of bilirubinemia increased with the development of such marked hepatocellular damage that no bile pigment was eliminated.

Excessive urobilinogenuria is a sensitive index of the presence of liver dysfunction. It frequently persists after all other evidence of hepatic damage has disappeared and may appear before the development of any other manifestation of such damage.

(b) **Hepatic Disease without Jaundice.** Excessive urobilinogenuria occurs commonly in portal cirrhosis even in the early stages of its development. In this condition, it is due in part to hepatic functional impairment and, in addition, to the fact that a portion of the portal blood reaches the general circulation without having passed through the liver. In this condition, excessive urobilinogenuria may occur in the absence of hyperbilirubinemia. Urobilinogenuria has been frequently observed in patients with congestive heart failure, considerable prognostic significance being attributed to it by some authors. In this condition, it is probably dependent upon the presence of hepatic damage, functional or organic, resulting perhaps partly from a state of prolonged hypoxia incident to the existing circulatory disturbance, and partly from injury to the polygonal cells by pressure from the dammed-back blood. It usually precedes the development of hyperbilirubinemia. It may also be observed in the absence of jaundice in certain toxic states, particularly pneumonia, streptococcus infections and malaria, due partly to excessive hemolysis and partly to hepatic parenchymal damage. Because of the nature of the conditions responsible for its occurrence, excessive urobilinogenuria has been regarded as having prognostic significance in pneumonia. This finding has also been obtained in patients with cholecystitis and cholelithiasis in the absence of hyperbilirubinemia, due to an associated mild hepatitis.

(c) **Jaundice Due to Extrahepatic Cholestasis.** Urobilinogenuria rarely if ever occurs in complete obstructive jaundice. In some cases of profound jaundice, a small quantity of urobilinogen may be formed from bile pigment which has passed from the jaundiced intestinal mucosa into the lumen of the bowel. Bilinogens are not present in the urine in complete biliary obstruction even though the liver is simultaneously severely damaged. It has frequently been observed that urobilinogenuria, occurring in patients with cholelithiasis, disappears completely with the development of complete obstruction of the common bile duct. When the bile flow is re-established, urobilinogenuria reappears. This phenomenon is observed not only in cholelithiasis but also in hepatitis with associated cholangitis, e.g., in infectious (viral) hepatitis. It may also be encountered occasionally in patients with complete calculous obstruction complicated by purulent cholangitis. However, and it may be assumed generally, in the majority of cases of obstructive jaundice the presence of urobilinogen in the urine suggests that the obstruction is not permanently complete. Under such circumstances, excessive urobilinogenuria is frequently encountered, dependent upon associated marked hepatic functional impairment with consequent inability of the liver to remove from the blood even relatively small quantities of urobilinogen absorbed from the bowel.

Urobilinogenuria is not pathognomonic of any single disease state, its chief diagnostic significance, apart from its occurrence in hemolytic anemia, pernicious anemia and chronic hemolytic jaundice, being that some bile pigment is entering the intestine and either the functional activity of the liver is very considerably impaired or there is active infection in the bile ducts. When properly interpreted, urobilinogen excretion is of value not only in the study of the state of hepatic function but also in differentiating between complete and incomplete obstruction to the flow of bile. This may be of value in distinguishing between jaundice due to stone in the common duct and that due to neoplasm of the duct, duodenum or head of the pancreas. In the latter connection, simultaneous investigation of the urobilinogen excretion in the feces is of value (p. 655). Occasionally, complications such as severe, protracted diarrhea (decreased absorption of bilinogens from intestine) or renal insufficiency (decreased renal elimination of bilinogens) may interfere with the usual interpretations of changes in urine urobilinogen. The important influence of antibiotic therapy in this connection is referred to elsewhere.

Ingestion of Bicarbonate. A significant increased excretion of urobilinogen has been shown to occur in urine produced during the two hours after ingestion of 10 g. sodium bicarbonate by normal individuals. This is related to the increased urinary pH, and it has been suggested that urinary pH should be taken into account in the clinical interpretation of urobilinogenuria.

Bile Pigments in Feces

Bilirubin is practically completely converted to "bilinogens" and "bilins" in the intestine. Bilirubin is rarely present in the feces, except occasionally in severe diarrhea with extremely rapid passage of material through the intestine, in fistulous communications between the upper and lower bowel, and in patients receiving broad spectrum antibiotic therapy. Bilirubin is also present in the feces of normal infants during the first eight to seventy-five days of life. The persistent absence of pigment ("urobilin") from the feces is indicative of obstruction to the outflow of bile from the liver or of complete suppression of bile pigment excretion by the liver, except in patients receiving antibiotic therapy. Complete obstruction of the bile passages may occur in cholangitis but is most commonly associated with stone in the common duct, stricture of the bile ducts, carcinoma of the head of the pancreas, common duct or duodenum or, rarely, enlargement of the lymph nodes in the region of the common bile duct. The feces are pale, having a so-called "clay-colored" or putty-like appearance, and are usually bulky, extremely offensive and contain excessive amounts of neutral fat and fatty acid (p. 601). If the obstruction is not complete the feces may be fairly normal in color, and even in complete obstruction some urobilin may be present in the feces due to the entrance of small amounts of bile pigment into the intestine from the capillaries of the intestinal mucous membrane.

The significance of accurate quantitative estimations of the "urobilinogen" content of the feces in this connection has been referred to elsewhere (p. 651). It must be remembered, too, that so-called "clay-colored" feces may actually contain considerable amounts of pigment, the presence of which is obscured by large amounts of fat. Before deciding that pigment is absent from the feces under such circumstances, the fat should be removed by extraction. Of greater value in determining whether or not bile is entering the intestine is the procedure of duodenal intubation, the tube being allowed to remain in situ for periods of several hours on successive days.

Porphyrin in Urine and Feces

The pertinent facts regarding normal porphyrin excretion have been outlined elsewhere (p. 274). Under normal conditions the daily excretion of coproporphyrin is about 60 to 180 micrograms in the urine and 12 to 830 micrograms in the feces, the normal urine : feces ratio being about 1 : 4. About 60 to 80 per cent of the urinary coproporphyrin and 70 to 90 per cent of the fecal coproporphyrin is in the form of the Type I isomer, the remainder being Type III.

Increased porphyrinuria has been reported in patients with cirrhosis, hemochromatosis, hepatitis, melanosarcoma of the liver, acute and subacute hepatic necrosis, chronic passive congestion of the liver, obstructive jaundice, infectious hepatitis and metastatic carcinoma of the liver. This is due largely to inability of the liver to excrete the porphyrin in the bile. In the absence of biliary obstruction, an increase in the urine : feces porphyrin ratio may be of more significance in indicating impairment of hepatic function than the absolute amount excreted in the urine. This ratio has been found to be 0.8 to 22.0 in patients with hepatic disease. In these conditions, except those associated with alcoholism, the increase in urine porphyrin is largely in Type I coproporphyrin, i.e., that which predominates normally (Table 5-1, p. 275). Interestingly, in acute alcoholism, and in cases of cirrhosis occurring in alcoholics, the increase is mainly in Type III coproporphyrin, in contrast to the Type I porphyrinuria which occurs in infectious hepatitis and in cirrhosis in nonalcoholic subjects.

Determination of urine coproporphyrins is of no value in differentiating hepatocellular from extrahepatic obstructive jaundice, being rather consistently increased in both. This increase may persist after apparent clinical recovery and after the results of other tests of liver function have returned to normal. Moreover, as indicated above, determination of the type of porphyrin involved may be useful in distinguishing "nonalcoholic" cirrhosis from "alcoholic" cirrhosis and in detecting hepatic disease occurring as a result of heavy metal poisoning.

Apart from these acquired diseases of the liver and biliary tract, there is "porphyria hepatica," which comprises the largest group of the so-called congenital porphyrias (p. 276).

Other types of porphyrinuria are reviewed elsewhere (p. 275). These must be considered in interpreting alterations in urine coproporphyrin in terms of liver-function or hepatic disease.

BILE ACID METABOLISM

The bile acids are derivatives of cholanic acid.

The primary bile acids, formed in the liver from cholesterol, are cholic acid (3,7,12 trihydroxycholanic acid) and chenodeoxycholic acid (3,7 dihydroxycholanic acid). Intestinal bacteria can remove the hydroxyl group at position 7 to form the secondary bile acids, deoxycholic acid (3,12 dihydroxycholanic acid) and lithocholic acid (3 hydroxycholanic acid). All the hydroxyl groups are in the α configuration, which means that they are all on the same side of the molecule; this is also the case with the carboxyl

Cholanic acid

group. Thus, one side of the molecule is polar and the other is nonpolar, and consequently the bile acids are amphipathic, with an affinity for water and for lipids. They are therefore capable of "solubilizing" lipids in the form of micelles. Hence, bile acids play an important part in the intestinal absorption of fat and cholesterol (p. 98), although fatty acids and monoglycerides can be absorbed in the absence of bile salts but in the presence of pancreatic lipase.

Within the hepatocytes, a lysosomal enzyme, which is a transferase, converts the CoA derivatives of the bile acids into conjugates with glycine or taurine (e.g., glycocholic and taurocholic acids). It is believed that the excretion of these conjugates is brought about by the emptying of lysosomes into the canaliculi. Under normal conditions, free bile acids are not found in bile. The ratio of glycine to taurine conjugates is roughly 3 to 1. The conjugates are more soluble than the free acids and are capable of rendering them more soluble. It is likely, therefore, that the conjugation process is highly important in relation to intestinal absorption of lipid material.

Bile has a pH between 6 and 7, and consequently the conjugated bile acids exist as fully ionized salts, since their pKa levels are much lower than 6.0. The same state of affairs exists in the small intestine. Unconjugated chenodeoxycholic and deoxycholic acids have pKa levels above 6.0, and consequently a high proportion of these free bile acids in the lower small and in the large intestine is in the nonionized form. These unconjugated acids occur in the intestine because intestinal bacteria can break down the conjugated forms to give the free acids. The amount of breakdown can be determined by giving ^{14}C glycine-labeled glycocholate by mouth. The labeled glycine liberated in the intestine is broken down by bacteria, and some is reabsorbed and undergoes metabolic breakdown. Both these breakdown processes give rise to labeled CO_2, which appears in the expired air and can be measured. An increase has been found in conditions of stasis and bacterial overgrowth in the small intestine (including blind loop syndrome), as well as after ileal resection. In the latter situation, however, there was also a well-marked increase in fecal ^{14}C.

All types of bile salts can be reabsorbed by one of two mechanisms. The more important is the active transport mechanism in the distal half of the ileum. This is equally effective with both free and conjugated forms. The other type of reabsorption, passive diffusion, can occur anywhere in the small bowel as well as in the colon. It seems to be somewhat more effective with nonionized free acids than with ionized forms. The reabsorbed bile acids pass to the liver, where they are re-excreted and the free forms are conjugated. There is thus an enterohepatic circulation of bile salts. The total bile acid pool in adult man is 3 to 5 g., and the enterohepatic circulation is 20 to 30 g. daily. The amount, which normally escapes absorption and appears in the feces is about 0.8 g. and equals the amount synthesized each day by the normal liver. Very little lithocholic acid is reabsorbed, and consequently only traces appear in the bile. This is just as well, since the substance is hepatotoxic. On an average, bile contains cholate, chenodeoxycholate and deoxycholate conjugates, respectively, in the ratio

1.1:1.0:0.6. The ratio varies considerably with the type of diet. The amount of bile acid reabsorbed inversely governs the amount formed by the liver. Normally, this accounts for some 80 per cent of the breakdown products of cholesterol.

Plasma normally contains 0 to 3.4 mg./l. of total cholic acid and 0 to 1.9 mg./l. of chenodeoxycholic acid together with deoxycholic acid. Adult hepatic bile contains 6.5 to 14.0 g. total bile acids/l., and gallbladder bile 12.6 to 89.0 g./l. In normal Japanese, it has been found that serum contains 0.59 ± 0.31 mg./l. of total bile acids, of which cholic acid was 0.14 ± 0.09 mg./l., chenodeoxycholic acid, 0.23 ± 0.14 mg./l. and deoxycholic acid, 0.22 ± 0.14 mg./l. The ratio of cholic to chenodeoxycholic acid was 0.61 ± 0.30. Total unconjugated serum bile acids were 0.26 ± 0.20 mg./l., and this made up 51.4 ± 22.0 per cent of total bile acids present.

The most important functions attributed to the bile salts are:

1. They facilitate digestion of fats.

2. They aid in absorption of fatty acids, cholesterol, carotene (provitamin A), and the fat-soluble vitamins D and K, by forming complexes more soluble in water ("hydrotropic" action).

3. They are a potent natural stimulus to hepatic bile flow, increasing several-fold the volume flow of bile (choleretic action); this involves both water and electrolytes.

4. They aid in keeping cholesterol in solution in the bile. When the bile acid : cholesterol ratio falls below a critical level, cholesterol may precipitate. This may be a factor in the pathogenesis of biliary calculi.

Because of technical difficulties, the quantitative estimation of bile salts in the blood is not commonly employed in the study of patients with hepatic or biliary tract disease. It has been found, however, in clinical and experimental obstructive jaundice, that bile salts in the blood increase greatly. After prolonged obstruction, the concentration of bile salts in the blood may diminish, due presumably to diminished synthesis of these substances as a result of progressive hepatocellular damage. A similar secondary decrease in the bile acid concentration of the blood in animals with complete biliary obstruction has been observed following the administration of hepatotoxic substances, such as chloroform, carbon tetrachloride and tetrachlorethane. It has been claimed that there is a decreased hepatic output in patients with cirrhosis, because of defective synthesis and abnormal metabolism of bile acids.

The total serum bile acid content is usually increased in patients with hepatobiliary disease. This is well marked in early acute hepatitis and in jaundice due to extrahepatic obstruction. There is usually only a moderate or slight increase in those with active portal cirrhosis or chronic hepatitis. In patients with active portal cirrhosis, there is often an increase in the amount of unconjugated bile acid in the serum. This may reach levels which are ten times the normal level. In patients with obstructive jaundice, on the other hand, the whole increase in serum bile acids is due to the conjugated variety, and the unconjugated bile acids make up a very low percentage (0.2 ± 0.4) of the total serum bile acids.

In the Japanese series, the ratio of serum cholic to chenodeoxycholic acid was increased in individuals with active portal cirrhosis (1.01 ± 0.43) but even more increased (2.74 ± 1.42) in obstructive jaundice.

The importance of certain abnormalities in the composition of bile is discussed elsewhere (p. 678). From a practical standpoint, investigation of the bile acid content of bile may be of value in patients with bile fistula or with a drainage tube in the common duct following operation upon the bile passages. It has been found that bile acids may be virtually absent from the liver bile in cases in which the common bile duct has been obstructed for a week or more. A period of one to four weeks or longer may elapse between the time of release of obstruction and the reappearance of bile acids in the liver bile. This interval appears to be roughly proportional to the degree and duration of obstruction and the degree of cholangitis associated with the obstruction. The persistent absence or very low concentrations of bile acids in the drainage bile under such circumstances has been said to be of poor prognostic significance, indicating severe hepatocellular damage.

Steatorrhea. A variety of conditions associated with bile salt disturbances can give rise to steatorrhea.

(a) Liver Disease. Any liver disease associated with cholestasis will give rise to a deficiency of the amount of bile, and hence bile salts, which enters the small intestine. It has already been pointed out that a fall in bile salt concentration in the upper small intestine leads to malabsorption, particularly of lipids. Steatorrhea also occurs in patients with alcoholic cirrhosis because of reduced output of bile acids (p. 601).

(b) Blind Loop Syndrome. This gives rise to excessive bacterial growth, and consequently excessive deconjugation of bile salts. The free acids are mostly in the nonionized form, and therefore quite readily absorbed into the bloodstream by passive diffusion. Deconjugated bile salts are also readily precipitated. Excessive absorption and precipitation give rise to a deficient amount of bile salts in solution in the jejunum, and hence cause steatorrhea.

(c) Ileal Disease. Since bile salts are absorbed mostly in the lower part of the small intestine, disease of the terminal ileum, or surgical resection, results in diminished absorption of these substances, with excessive amounts in the large intestine and excessive loss in the feces. Although under these conditions the liver synthesis of bile salts is stimulated, maximum synthesis cannot keep up with fecal loss. This gives rise to a diminished concentration of bile salts in the small intestine, and in turn to steatorrhea. The excessive amounts of bile salts in the colon have a marked cathartic action and there is severe diarrhea, usually occurring soon after meals.

After ileal resection, there is an alteration in the bile acids present in bile. There is an increase in the amount of glycine conjugates. Some authors report a fall and others a rise in biliary deoxycholic acids. There is no change in chenodeoxycholate but a marked fall in the cholate fraction. Lithocholic acid is present in increased amount. These changes are believed at least in part to be due to excessive bacterial activity in the intestine, but may also depend on changes in the type of absorption of bile salts.

The large amount of biliary glycine conjugates present in ileal dysfunction give rise to excessive bacterial breakdown of glycine. This can produce glyoxylate, which is absorbed and converted to oxalate in the liver and could be an explanation for the hyperoxaluria known to occur with ileal dysfunction.

Gallstones. The average composition of bile (hepatic and gallbladder) is shown in Table 21-3. The total solids in solution are either inorganic or organic. The former consist largely of bicarbonates, chlorides and phosphates of sodium, potassium, calcium, magnesium, iron and copper. The latter contain a large variety of substances, including vitamin B_{12}, folate, lactate, protein and enzymes, in addition to those shown in the table. Some of the inorganic cations are bound to protein.

In normal bile, primary bile salts form about 80 per cent of the total bile salts present. The cholesterol is "solubilized" by being held in micelles together with conjugated bile salts and phospholipids. The latter cause the cholesterol solubility to be much higher than in bile salts alone. In fact, the solubility appears to be dependent on the ratio of cholesterol to bile salts plus phospholipids. The secretion of phospholipids into the bile is dependent on the availability of the bile salts. If the latter were diminished (decreased bile salt pool), then the phospholipid content of bile would fall, and this would appreciably diminish the solubility of cholesterol in the bile and cause crystallization.

Cholesterol is the major constituent of gallstones in Western communities. A knowledge of the factors which bring about its deposition in the bile is therefore of the greatest importance. Patients with gallstones have increased biliary cholesterol content and a diminished bile salt pool, which can be about 54 per cent of normal. This is probably due to a defect in their enterohepatic circulation. There is in fact an increased incidence of gallstones in patients with diseases of the terminal ileum as well as in patients with cirrhosis. In both cases, there is reduction in the bile salt pool.

It has been suggested that in cholelithiasis the plasma membrane of

TABLE 21-3   AVERAGE COMPOSITION OF BILE

CONSTITUENT	HEPATIC	GALLBLADDER
Water	98%	86%
Inorganic salts	153 mEq./l.	250 mEq./l.
Na	150 mEq./l.	220 mEq./l.
K	5 mEq./l.	14 mEq./l.
HCO_3	30 mEq./l.	19 mEq./l.
Cl	100 mEq./l.	31 mEq./l.
Ca	3.7 mEq./l.	15 mEq./l.
Bilirubin	0.65 g./l.	2.94 g./l.
Bile salts	11 g./l.	32 g./l.
Cholesterol	1.3 g./l.	6.3 g./l.
Phospholipids	2.7 g./l.	34 g./l.
Total fatty acids	2.7 g./l.	24 g./l.
pH	7.5	6.0

the bile canaliculus allows an increased amount of cholesterol to enter the bile. Alternatively, the membrane may be responsible for an increase in the cholesterol-to-phospholipid ratio. These membrane phenomena, together with the diminution in the bile salt pool, would markedly increase the ratio of cholesterol to bile salts plus phospholipids. This would in turn cause cholesterol to be deposited.

Since stones for the most part form in the gallbladder, it is likely that other factors are also involved. A sluggish gallbladder could result in stasis and the retention of seeding factors. The latter could be microspheroliths, desquamated cells or "liposomes." Mucous substances seem to be implicated.

Infection of bile, especially with coliform bacteria, could result in deconjugation of bile acids, with a decrease in their solubility. It could also result in the production of a phospholipase, which could convert lecithin to lysolecithin. This would decrease the stability of the micelles holding cholesterol in solution. Infection, however, does not seem to be important in relation to gallstone production in Western communities, but can probably give rise to calcium-bilirubinate stones, which were frequent in Japan. These result from breakdown of conjugated bilirubin by bacterial β-glucuronidase. The free bilirubin formed then precipitates as calcium bilirubinate.

It is theoretically possible to envisage solubilization of cholesterol-containing gallstones by diminishing the ratio of cholesterol to bile salts plus phospholipid. This could possibly be done by oral administration of lecithin or of a bile salt. Lecithin has proved to be unsuccessful, but some success has been obtained by oral administration of chenodeoxycholate for long periods. It is believed, however, that the major reason for its effectiveness could be that chenodeoxycholate diminishes the rate of secretion of cholesterol into the bile in relation to bile salts and phospholipids, and so affects the appropriate ratio in the right direction. In this way, bile is rendered unsaturated with respect to cholesterol and thus the cholesterol stone can be redissolved. One interesting finding is that administration of chenodeoxycholate also results in a diminution of serum triglyceride levels.

Unfortunately, bacterial action in the intestine converts chenodeoxycholate to lithocholate, which is very hepatotoxic in Rhesus monkeys, producing proliferation of bile ducts and periportal infiltration. Severe changes have not, as a rule, been produced in humans under therapy. There have been increases in serum aspartate aminotransferase and alkaline phosphatase.

DETOXIFICATION-CONJUGATION

The liver plays an important part in protecting the organism from various toxic substances entering through the intestinal tract. This "detoxifying" function involves mainly processes of oxidation and conjugation into relatively nontoxic substances which are subsequently in the bile and urine. For example, indole absorbed from the intestine reaches the liver in the portal circulation, is oxidized to indoxyl, conjugated with sulfuric acid and eliminated in the urine as indoxyl sulfuric acid. Other substances, including salicylic acid, menthol, camphor, phenol, etc., are in this way transformed into conjugate glucuronates and excreted in the urine. The estimation of inorganic and ethereal sulfates in the blood and urine and of conjugate glucuronates or sulfates in the urine following the administration of such substances has been proposed as a method of estimating liver function. These tests have proved to be of little clinical value.

The term "detoxification" is, in a certain sense, a misnomer. It has come to be applied, without regard to the toxicity of the compound involved, to those chemical changes occurring in the body which convert exogenous or endogenous compounds of known chemical structure to derivatives which are excreted in the urine. These include certain hormones, such as the female sex hormones progesterone and estrone. The important role of conjugation mechanisms in the process of hepatic removal of bilirubin and Bromsulphalein from the blood, and their excretion in the bile, is discussed elsewhere (pp. 634 and 667).

Investigation of the ability of the organism to synthesize hippuric acid has been employed for the study of the detoxifying and conjugating activity of the liver.

Hippuric Acid Synthesis. This test, which is now seldom if ever used, is based upon the ability of the organism to conjugate glycine and benzoic acid to form hippuric acid, which is excreted in the urine. In man, the liver is the principal site of formation of this substance. The hourly rate of excretion of hippuric acid in the urine of normal subjects following the ingestion of benzoic acid is remarkably constant, being influenced somewhat by the body surface area. In certain types of hepatic disease the output of hippuric acid is markedly reduced. This reduction is assumed to be due, in part, to a diminished capacity of the liver to synthesize glycine, which is essential for the formation of hippuric acid and, in part, to damage of the enzymatic mechanism which unites benzoic acid with glycine. In man, this conjugating enzyme appears to be present chiefly in the liver, and the small amount which is present in the kidneys is regarded as insufficient to compensate for the hepatic defect in liver disease. The urinary excretion of hippuric acid after the ingestion of benzoic acid is, therefore, regarded as a measure of the capacity of the liver to furnish glycine and also as an index of its conjugating power.

A subject with a normally functioning liver excretes approximately 3 g. of benzoic acid, in the form of hippuric acid (4.41 g.), in the urine in four hours after the ingestion of 6 g. of sodium benzoate. The normal range of variation has been established at 85 to 110 per cent of this figure. In many cases the intravenous route of administration of the sodium benzoate is preferable in order to obviate the possibility of error due to abnormalities of absorption from the intestine. Normal subjects excrete 0.7 to 0.95 g. of benzoic acid (as 1.0 to 1.4 g. hippuric acid) in the urine during the first hour after intravenous injection of 1.77 g. of sodium benzoate. It has been found that the excretion of hippuric acid under these circumstances is influenced by the size of the subjects and the following equation has been suggested for calculating the predicted normal excretion:

Hippuric Acid (grams) = 0.34 + (0.00668 × weight in pounds). Subnormal values have been obtained in patients with a variety of hepatic disorders, including hepatitis, hepatic syphilis, portal cirrhosis, biliary cirrhosis, metastatic carcinoma of the liver, and acute and subacute hepatic necrosis. Normal values have been obtained in uncomplicated obstruction of the common duct, abnormal findings being observed in cases of extremely long duration and in those with superimposed hepatocellular damage. The test has, therefore, been claimed to be of value in differentiating between hepatocellular and obstructive types of jaundice. However, it is necessary to investigate the state of renal function before attaching significance to abnormal findings in patients with hepatic disease. This is due to the fact that diminished or less rapid elimination of hippuric acid may be dependent upon impairment of kidney function, which is present frequently in such cases. It has been found also that synthesis and excretion of hippuric acid are subnormal in conditions other than nephritis and hepatic disease, among these being cachectic states, anemia and congestive heart failure.

THE LIVER AND HORMONE METABOLISM

In view of the fact that liver cells contain enzyme systems capable of catalyzing a large number and wide variety of chemical reactions, it is not surprising that certain hormones undergo metabolic transformations in this organ. Demonstration of abnormalities in this area of hepatic physiology is of little or no practical value in the clinical study of liver function or in differential diagnosis. However, it seems desirable to review briefly certain phases of this subject that appear to have clinical implications.

In general, the action of the liver serves to limit the amount and activity of various hormones in the systemic circulation. This is accomplished by one or more of the following mechanisms: (a) transformation to hormonally inactive or less active substances ("inactivation"); (b) conjugation of these metabolites, usually with sulfate or glucuronate, which renders them more soluble in water and hence more readily excreted in the urine; (c) excretion of the hormone or metabolites in the bile, in some cases with subsequent partial reabsorption from the intestine (enterohepatic circulation).

The liver contains enzymes which inactivate insulin, thyroxine, and certain anterior and posterior pituitary hormones. The liver is the organ chiefly involved in the conversion of progesterone to pregnanediol, a biologically inactive substance, which is in part perhaps excreted in the bile, in part degraded to unknown metabolites, and in part conjugated (glucuronate), passed into the blood stream and excreted in the urine

(p. 779). However, the largest amount of clinically pertinent information in this field is concerned with the role of the liver in the metabolism of estrogens, androgens, and, to a lesser extent, adrenocortical hormones.

Androgens. Testosterone, the major androgen produced in the testes, is metabolized in the liver (in part perhaps in other tissues) to unidentified compounds and to steroids of lower androgenic activity, the most important of which is androsterone, a 17-ketosteroid (p. 794). This substance is in small part excreted in the bile; the largest fraction, however, undergoes conjugation with sulfate, is passed into the blood stream, and is excreted in the urine in this form, together with a relatively small amount of free androsterone. About one-third of the 8 to 20 mg. of neutral 17-ketosteroids excreted daily in the urine of normal men is derived from the testes, the remainder originating in adrenocortical hormones. The latter are the only source of urinary 17-ketosteroids in women (5 to 15 mg. daily) (p. 721 ff.).

Changes in androgen metabolism have been demonstrated in subjects with hepatic disease, e.g., infectious hepatitis, chronic hepatitis, and cirrhosis. These include: *(a)* decrease in urinary 17-ketosteroids; *(b)* increase in free 17-ketosteroids in the urine, a reflection of inadequate conjugation; *(c)* subnormal concentration of 17-ketosteroids in the blood after intravenous injection of testosterone (inadequate metabolism). The rather consistent occurrence in such cases of an increase in urine estrogen and a decrease in gonadotrophin suggests that the lowered 17-ketosteroid output, which is often accompanied by testicular atrophy, may be secondary to an increase in circulating estrogen (depressing gonadotrophin secretion by pituitary). That inadequate metabolic transformation may play a role, however, is suggested by the observation that subjects with hepatic damage excrete a somewhat smaller proportion of exogenous testosterone as 17-ketosteroids than do normal subjects. The main defects in androgen metabolism in patients with severe liver damage, therefore, appear to be: *(a)* diminished secretion of testosterone by the testes, a consequence of decreased gonadotrophin secretion due to an excess of circulating free estrogen; *(b)* perhaps somewhat diminished transformation of testosterone to 17-ketosteroids (mainly androsterone); *(c)* diminished conjugation of androsterone (and other 17-ketosteroids).

Estrogens. The liver plays an important role in the metabolism of estrogens. It contains enzyme systems which can convert estradiol, the primary and most active estrogen, in part to biologically inactive substances and in part to estrogens of lower activity, i.e., estrone and estriol. Conjugation of the estrogens takes place in the liver, with the formation and transfer to the blood stream of water-soluble glucuronates and sulfates, which are ultimately excreted in the urine. Relatively large amounts of free estrogen are present in the bile in certain species, suggesting that biliary excretion may play an important role in the removal of estrogens from the blood stream. Estrogens may subsequently undergo an enterohepatic circulation, being gradually degraded in the liver to compounds of lower activity. There is evidence, too, that estrone is converted in this organ to the more potent estradiol, a process of biological "activation."

The presumed functions of the liver in this connection may be summarized as follows: (1) it effectively removes free estrogens from the systemic circulation by *(a)* biliary excretion and *(b)* metabolic degradation to less active and inactive compounds; (2) it conjugates them with glucuronate and sulfate, converting them to water-soluble forms (inactive until hydrolyzed) which pass into the systemic circulation and are excreted in the urine; (3) it can convert estrone to the more potent estradiol. The urinary estrogens, mainly glucuronates and sulfates of estrone, estriol, and estradiol, are estimated as representing about 5 to 10 per cent of the activity of the estrogens originally secreted.

In subjects with hepatic disease, diminished activity of the liver in removing free estrogens from the blood stream, and in "destroying" and conjugating them, is reflected in the following phenomena: (1) increase in total urine estrogen; (2) increase in free estrogen in the urine; (3) delayed removal of exogenous estrogen from the blood and abnormally high urinary estrogen excretion following injection (decreased estrogen tolerance). Although an estrogen tolerance test has been proposed as a means of studying liver function, there is no close correlation between demonstrable abnormality of estrogen metabolism and the degree of impairment of liver function as measured by more conventional procedures. Moreover, we have found changes in subjects with extrahepatic duct obstruction similar to those obtained in the presence of hepatocellular damage, emphasizing the importance of impaired biliary excretion of estrogens in the production of these phenomena.

The abnormally high concentration of free estrogen in the blood of patients with hepatic disease is regarded as possibly responsible for certain clinical manifestations that occur rather frequently in such cases, particularly in cirrhosis. These include: in women, menstrual disturbances (menorrhagia, metrorrhagia, amenorrhea), infertility, breast changes, and "spider" telangiectasis ("vascular spiders"); in men, testicular atrophy, decreased libido, impotence, infertility, axillary alopecia, vascular spiders, and gynecomastia.

Adrenocortical Hormones. As in the case of the gonadal hormones, the liver is the organ chiefly responsible for transformation of the adrenocortical hormones to biologically inactive, water-soluble metabolites that are returned to the systemic circulation and excreted in the urine (p. 719). The most important reactions involved are preliminary reduction and subsequent conjugation (glucuronides and sulfates). Administration of cortisol to subjects with severe liver damage is followed by a more prolonged increase in free 17-hydroxycorticoids in the blood and a smaller rise in conjugated hormones than in normal subjects. The urinary excretion of 17-ketosteroids is reduced in the presence of liver disease, probably due mainly to diminished activity of the reducing enzymes. As a consequence of this functional defect, therapeutic administration of adrenocortical hormones to patients with liver disease may result in the production of manifestations of overdosage that would not occur in subjects with normal liver function.

However, resting plasma concentrations of free 17-hydroxycorticosteroids are usually within normal limits in patients with hepatic disease. This is perhaps due to the efficient operation of the negative feedback controlling mechanism, whereby any tendency toward an increase in the hormone concentration in the plasma depresses production of ACTH by the adenohypophysis and, therefore, decreases production of adrenocortical hormones. The liver is, therefore, one of the important regulators of adrenocortical function.

The role of decreased inactivation of aldosterone in the pathogenesis of Na and water retention in portal cirrhosis is uncertain. Urinary aldosterone is frequently increased in cirrhotic patients with edema and ascites, but this occurs also in edema due to other causes. The importance of the metabolic activity of the liver in this connection is difficult to evaluate.

ELIMINATION OF DYES

Certain dyes are removed from circulation and eliminated almost entirely by the liver, just as phenosulfonphthalein is excreted almost entirely by the kidneys. Among these dyes are azorubin S., rose bengal, indocyanine green, phenoltetrachlorphthalein, phenoltetraidophthalein and sulfobromophthalein (Bromsulphalein). Several liver function tests have been proposed which have for their basis the estimation of the ability of the liver to eliminate these substances. The procedure in common use at the present time consists in determination of the rate of removal of the dye from the blood stream, the disappearance rate being assumed to be an indication of the state of hepatic function.

Bromsulphalein (BSP) Excretion. Bromsulphalein is the dye most widely employed for this purpose. Its applicability to the study of liver function is based upon the assumption that its removal from the circulation is accomplished virtually exclusively by the liver. Studies in hepatectomized animals indicate that organs other than the liver can remove BSP from the blood stream but, in the presence of a normal liver, probably not more than 5 per cent is removed by extrahepatic tissues. If there is marked retention of the dye in the plasma, as much as 10 per cent may be excreted in the urine, and considerably more in the presence of albuminuria (dye bound to plasma albumin). It has been claimed also that another source of error in interpreting results obtained with this procedure arises out of the fact that some of the BSP excreted in the bile is reabsorbed into the portal circulation, a small amount passing back into the systemic circulation. There seems to be little doubt that, under clinical conditions, the liver is the organ principally involved in this process, the relatively minor factors of extrahepatic removal, enterohepatic circulation, and renal excretion not interfering significantly with the interpretation of results of the BSP test in terms of liver functional efficiency.

Following a single intravenous injection of BSP, about 10 to 15 per cent of the amount in the blood is removed each minute (almost exclusively by the liver cells). In the liver cells, it undergoes conjugation with glutathione and is actively excreted in the bile, in which at least three compounds have been identified in addition to the free dye. Excretion in the bile continues for some time after all dye has been removed from the blood stream in normal subjects. This suggests a biphasic mechanism, i.e., *(a)* primary removal from the plasma with temporary storage in liver cells, and *(b)* subsequent conjugation and secretion into the bile canaliculi. In certain cases, therefore, study of the curve of excretion of BSP in the bile (duodenal intubation) may yield information not given by the usual studies of the rate of its removal from the blood.

Employing the technique of hepatic vein catheterization in subjects receiving a continuous intravenous infusion of BSP, the hepatic blood flow may be estimated from determinations of: (1) the concentration of BSP in the hepatic vein blood and (2) the peripheral blood; (3) the amount of BSP removed by the liver. By this procedure, the hepatic blood flow in normal subjects has been found to be 1085–1845 ml./minute/1.73 sq. m. body surface area, averaging 1500 ml., or approximately 100 ml./100 g. liver. Inasmuch as the removal of BSP from the blood stream is dependent upon the volume flow of blood through the liver, it follows that abnormal retention of the dye may result from decreased hepatic blood flow as well as from impaired liver cell function. Indocyanine green is said to give a better estimate of hepatic blood flow.

Blood BSP Retention Test. As usually performed, this procedure consists in the intravenous injection of BSP in the dosage of 5 mg./per kg. body weight. This gives an initial plasma concentration of 10 mg./100 ml., which is the 100 per cent concentration. Blood samples are withdrawn at desired times subsequently, the serum or plasma separated and alkalinized (to develop purple color), and the concentration of dye measured colorimetrically. Normal values at different time intervals are as follows:

Dose	5 min.	15 min.	30 min.	45 min.
5 mg./kg.	< 85%	< 30%	< 10%	< 6%

The results are, however, affected by gross obesity and by fluid retention. The advantage of the 5 mg. dosage lies in the fact that it imposes a functional burden on the liver cells, which may yield abnormal results in cases of relatively mild functional impairment in which normal results may be obtained with a smaller dosage. Similarly, readings made at five and fifteen minutes may reveal abnormal retention in cases in which later values may be normal. The usual practice is to make a single determination at forty-five minutes. Greater sensitivity is attained by making serial determinations, establishing a curve of dye removal from the blood stream, but this is seldom resorted to in clinical practice. The rate of removal is somewhat more rapid after meals (increased hepatic blood flow) than in the fasting state. Mild impairment of BSP excretion may, therefore, not be demonstrable during periods of active digestion. This test should be performed under uniform conditions, preferably fasting, if results obtained at different times are to be compared.

In the doses employed, Bromsulphalein is practically nontoxic, even in the presence of severe liver damage. The solution used is irritating to the tissues and may cause pain and inflammation if a significant amount is injected outside the vein. It may also produce thrombophlebitis at the site of injection. Occasional toxic reactions have been observed, including fever, nausea, vomiting, vertigo, and syncope. Allergic reactions have been re-reported and also, in a few instances, severe anaphylactoid reactions, with serious cardiorespiratory manifestations and shock. Although these are extremely unusual, care should be exercised in performing this test in subjects with an allergic background. It has been suggested that greater sensitivity is obtained at a dosage of 20 mg./kg. This dosage has not yet been generally applied, partly because of possible toxic reactions.

Significance of Abnormal BSP Retention. Inasmuch as the hepatic excretory mechanisms for bilirubin and BSP are apparently similar, abnormal retention of both would be expected to occur in the same types of conditions, e.g., biliary obstruction and hepatocellular damage. This is, indeed, the case. However, it would appear that the test dose of BSP employed constitutes a greater functional burden on the excretory mechanism than does the amount of bilirubin formed in the course of normal degradation of hemoglobin. Consequently, abnormal BSP retention occurs practically invariably in the presence of hyperbilirubinemia, except, of course, that due to excessive hemolysis. The BSP test, as usually performed, is therefore of no diagnostic value in patients with clinical jaundice (except hemolytic jaundice) or in those with abnormally high total or one-minute, direct-reacting serum bilirubin concentrations. It should be noted, however, that BSP can be measured in jaundiced serum by proper photoelectric colorimetric procedures and that the presence of jaundice or liver damage does not in itself contraindicate the use of this procedure. The BSP test is particularly useful in the study of patients with hepatic or biliary tract disease in whom the serum bilirubin concentrations are within normal limits. However, it has been suggested that quantitative determination of the proportion of retained BSP present in conjugated form may aid in differential diagnosis even in the presence of hyperbilirubinemia.

Extrahepatic Biliary Obstruction. In the presence of obstruction of the common duct, the degree of retention of the dye increases progressively with increasing bilirubinemia until all of the dye injected remains in the blood at the end of the test period. In "surgical" jaundice, therefore, no information is obtained by this method which cannot be obtained by means of the quantitative determination of serum bilirubin. There is evidence from the slopes of the curve of serial estimations that in uncomplicated extrahepatic duct obstruction the percentage of BSP in the serum in conjugated form is often considerably higher than in hepatocellular jaundice, suggesting adequacy of the hepatic mechanisms for uptake and conjugation but impairment of the mechanism for secretion. Following relief of the obstruction, dye retention, although diminishing, frequently persists for a variable period of time after the serum bilirubin concentration has returned to normal. This is due perhaps to the residual liver damage which is present in nearly all patients who have suffered from biliary obstruction for an extended period, especially that due to biliary calculi.

Hepatocellular Damage. Abnormal retention of Bromsulphalein in the blood occurs almost invariably in acute hepatic disease, the degree of retention being roughly indicative of the extent of hepatic functional impairment. This is true of all varieties of hepatitis and acute and subacute hepatic necrosis due to any cause. The degree of dye retention usually, but not invariably, parallels the degree of hyperbilirubinemia in such cases. As indicated above, BSP retention in hepatocellular damage may differ from that in extrahepatic duct obstruction in that, in the former condition, relatively little of the retained dye is in conjugated form. As in the case of obstructive jaundice, abnormal dye retention may occur before and may persist longer than hyperbilirubinemia. This test may be valuable in the early detection of hepatocellular damage in patients receiving hepatotoxic agents for therapeutic purposes.

Varying degrees of dye retention, up to 100 per cent, may be obtained in patients with chronic hepatic lesions. Among these are chronic hepatitis, portal cirrhosis, biliary cirrhosis, malaria, hepatic syphilis, carcinoma of the liver and chronic passive congestion of the liver. In many patients with hepatocellular damage, as in those with extrahepatic obstructive lesions, dye retention occurs only in the presence of hyperbilirubinemia, the degree of retention of dye and pigment being approximately parallel. Frequently, however, this relationship is not maintained. For example, a disproportionately high degree of dye retention may occur in patients with myocardial failure and passive congestion of the liver, as well as in those with hepatic functional impairment complicating thyrotoxicosis, pneumonia and severe anemia. This may be due, in part at least, to the effect of hepatic circulatory disturbance in contributing to decreased removal of BSP from the blood stream (p. 667). In portal cirrhosis, particularly, dye retention may occur in the absence of hyperbilirubinemia. Under such circumstances this procedure is of diagnostic value.

Investigation of the ability of the liver to eliminate dyes, therefore, is of particular value in cases of hepatic dysfunction in which frank hyperbilirubinemia is absent. Among such conditions are certain cases of portal cirrhosis, cholecystitis, cholelithiasis, acute and chronic hepatitis, syphilis and various acute infections, such as pneumonia and typhoid fever. It is often stated that dye retention occurs in cases of gallbladder disease with frank jaundice but not usually in those with latent or no icterus. However, certain patients with cholecystitis, with and without cholelithiasis, may show varying degrees of Bromsulphalein retention, in the absence of, or with very slight degrees of, hyperbilirubinemia.

Other Conditions. Abnormal BSP retention occurs in conditions in which the hepatic blood flow is diminished, with or without hepatocellular damage. These include congestive heart failure and some cases of medical and traumatic shock. However, the hepatic blood flow is often essentially normal in shock and hemorrhage, in the absence of intrinsic liver damage, and the degree of BSP retention does not parallel the severity of shock or the amount of blood lost. In such conditions, impaired liver function may result from hypoxia of liver cells due to vasoconstriction (portal vein and hepatic artery branches).

Normal BSP values are strong evidence against ruptured esophageal varices (i.e., cirrhosis) as the cause of severe gastrointestinal bleeding. It must be recalled in this connection that hyperbilirubinemia and BSP retention may occur in patients with duodenal ulcer.

Slight, transient BSP retention may occur in the presence of fever, in the absence of demonstrable liver damage. It may occur also following operative procedures, particularly those involving the bile passages. This, too, is usually transient in nature. In old age, BSP retention is increased. It is also raised in the last three months of pregnancy. With a low serum albumin level, a falsely low BSP retention can result.

SERUM ENZYMES

The use of serum enzyme studies in liver disease has already been described (p. 575). The serum enzymes which have proved to be of use are alkaline phosphatase, aspartate aminotransferase, lactate dehydrogenase, 5′-nucleotidase, γ-glutamyl transpeptidase (D-glutamyl transferase) and sorbitol dehydrogenase.

In regard to alkaline phosphatase, it is perhaps important to demonstrate by gel electrophoresis that the increase is due to the enzyme produced by the liver.

It is the experience of most that there is no real advantage in undertaking investigation of serum alanine aminotransferase in addition to aspartate aminotransferase. The latter is frequently raised in acute cholecystitis.

An increase in the ratio of serum aspartate aminotransferase to alkaline phosphatase has proved to be useful in differentiating hepatocellular jaundice from "surgical" jaundice.

Certain generalizations may be made concerning the interpretation of serum phosphatase activity in hepatic and biliary tract disease. Determinations of this factor are most useful when repeated at intervals of several days and viewed in relation to the serum bilirubin concentration determined simultaneously.

1. High or rising values, above 215 U/l. (King-Armstrong, 37° C.), in patients with marked or increasing hyperbilirubinemia, suggest uncomplicated extrahepatic biliary obstruction. However, similar findings are common also in cholestasis and other forms of hepatocellular jaundice, in which intrahepatic duct obstruction may be an important feature.

2. Normal or but slightly elevated values, below 140 U/l. (King-

Armstrong, 37° C.), in patients with marked or increasing hyperbilirubinemia, suggest diffuse hepatocellular damage.

3. Values falling from elevated levels in patients with marked or increasing hyperbilirubinemia suggest increasing hepatocellular damage superimposed on biliary obstruction.

4. Elevated values in the presence of normal serum bilirubin concentrations occur in portal cirrhosis, liver abscess, sarcoidosis, metastatic malignancy of the liver, obstruction of single hepatic ducts (e.g., by lymph nodes), histoplasmosis, echinococcus cyst, and pancreatic duct obstruction.

5. Cholangitis and cholangiolitis tend to produce an increase in serum alkaline phosphatase activity. An increase occurs in the presence of active liver regeneration and bile duct proliferation.

6. Although the estimation of serum alkaline phosphatase activity may aid in differential diagnosis in some cases of jaundice, as indicated above, its value in this connection is limited by the considerable overlapping of values in the obstructive and hepatocellular groups, only moderately elevated values occurring in about 15 per cent of patients with extrahepatic duct obstruction, and values between 80 and 160 U/l. in about the same proportion of patients with primary hepatocellular damage.

Determination of serum transaminase activity may aid in the differential diagnosis of neonatal jaundice. Normal values in newborn infants are higher than in normal adults (aspartate aminotransferase, 3-60 U/l. Karmen, at 37° C.). No increase occurs in physiological or in uncomplicated hemolytic jaundice. Occasionally, serum GOT activity may increase in the presence of severe intravascular hemolysis (derived from hemolyzed erythrocytes). A sharp rise occurs in jaundice due to neonatal viral hepatitis and a more gradual increase in the presence of biliary obstruction ("inspissated bile syndrome," congenital atresia of bile ducts, congenital biliary cirrhosis).

Elevations, usually moderate, have been reported in patients with acute attacks of congestive heart failure. This has been related to the extent of liver involvement.

SERUM IRON

The serum iron concentration, normally 50–180 μg./100 ml., is elevated in the majority of patients with acute hepatitis. Although the cause has not been established, this has been attributed to increased hemolysis and/or increased rate of liberation of iron from its storage form in the liver (p. 323). Elevations have been observed in patients with cirrhosis; normal values are the rule in extrahepatic obstructive jaundice. Intravascular hemolysis is associated with high levels of serum iron.

Subjects with hemochromatosis may present either normal or elevated serum iron concentrations, an abnormally large proportion of which is in bound form (transferrin, p. 322). The iron-binding plasma protein is saturated to the extent of about 90 per cent in this disease, as contrasted to the normal 30 to 35 per cent saturation. Another reflection of this abnormality is the fact that administration of iron is followed by a smaller rise of serum iron in patients with hemochromatosis than in normal subjects.

THE LIVER IN WATER AND SALT METABOLISM

In the presence of hepatic disease, disturbances may occur in the regulation of water, sodium and chloride balance. Several mechanisms are undoubtedly involved. The nature of some of these is not well understood, but it seems desirable to review certain observations which have clinical implications.

The tendency of patients with hepatic disease, especially cirrhosis, to develop edema is well known. This is due in part, but by no means entirely, to the hypoalbuminemia which occurs in many of these cases (p. 188). Increase in extracellular fluid volume has been demonstrated in both acute and chronic liver disease (infectious hepatitis, cirrhosis), accompanied by oliguria and decrease in the amount and concentration of sodium and chloride in the urine, sweat, and saliva. Ingestion of a large volume of fluid results in much greater increase in extracellular fluid volume than occurs in normal subjects. The diuretic response to water (water tolerance test) is diminished, and intravenous injection of hypertonic sodium chloride solution does not produce the normal prompt augmentation of urine volume and sodium and chloride excretion, especially in cirrhotic patients with edema and/or ascites. In portal cirrhosis, this abnormal retention of sodium and water is manifested mainly in decompensated stages, associated with a tendency toward the development of edema and ascites. In explanation of this phenomenon, it has been suggested that, although the total is increased, the effective extracellular fluid and plasma volume may be decreased, large portions being sequestered in the peritoneal space and distended hepatic and splanchnic lymph vessels. Accordingly, the homeostatic mechanisms behave as though the organism were in a state of actual sodium deficit (p. 381).

It is now known that decrease in total renal blood flow is quite common in patients with portal cirrhosis, especially if accompanied by ascites. This decrease is not necessarily due to diminution in effective plasma volume, and there is some evidence against this concept. There is, however, evidence for renal vasoconstriction, possibly due to circulating vasoconstrictor substances resulting from defective liver metabolism. There is also a redistribution of renal circulation in cirrhotic patients, with reduction in flow to the outer cortex, which contains most of the glomeruli, and increase of flow to the juxtamedullary nephrons. These have longer loops of Henle and consequently an increased ability for reabsorbing sodium. Vasoconstriction and the alteration in circulation, together with other factors not fully understood, lead to a diminished glomerular filtration rate, which also produces very favorable conditions for reabsorption of sodium in the proximal tubules and loops of Henle. It is believed that this reabsorption is aided by a diminution in the activity of the "third factor" (p. 461). Patients with cirrhosis also have an increased production of aldosterone, and increased amounts are present in their urine. Aldosterone would bring about increased absorption of sodium in the distal tubules. It does not seem to be an important factor in the production of hepatic edema or ascites. It could, at least in part, account for the potassium deficit which accompanies cirrhosis, especially in its terminal stages.

Regarding the production of ascitic fluid in cirrhosis, the fluid loss due to hypoalbuminemia is aggravated by portal hypertension, which would help to push fluid out of the hepatic sinusoids as well as from the capillaries of the splanchnic vessels. Ascitic fluid tends to be relatively low in protein content. Its major contribution must therefore be from the splanchnic circulation, since sinusoidal fluid (spaces of Disse) has a high protein content. Fluid loss into the peritoneal space also is probably aggravated by increased capillary permeability, due to the presence of toxic metabolites arising from, or not inactivated by, the diseased liver.

Another factor of importance in the production of ascites is the lymphatic drainage of the peritoneum and liver. Ascitic fluid can accumulate only after this drainage capacity is saturated. An increased flow of lymph in the thoracic duct has been shown to be present in cirrhosis unaccompanied by demonstrable ascites. As already indicated, functional renal failure also plays a part in the production of ascitic fluid, as with hepatic edema occurring elsewhere.

HEPATIC COMA

This term is applied to an altered state of consciousness in patients with acute and chronic liver disease, accompanied in the early stages by various neurological manifestations which are, however, not specific for this condition. It can occur in the absence of jaundice. Moreover, death may result from hepatic functional insufficiency without evidence of hepatic coma. Because of its gravity and important therapeutic implications, considerable interest has been centered in certain biochemical aspects of its pathogenesis. Attention has been directed to the possible etiological significance of an increased concentration of blood ammonium. Hepatic coma may be induced in subjects with severe liver damage (portal cirrhosis, acute hepatitis) by measures that increase ammonium production and blood ammonium levels, e.g., administration of high protein diets, ammonium salts and ammonium exchange resins, and massive hemorrhage into the gastrointestinal tract. However, it does not always occur in the presence of elevated blood ammonium ion concentrations and may occur with no demonstrable alteration of ammonium metabolism. It may be induced also by administration of morphine and barbiturates, particularly postoperatively, in association with electrolyte disturbances, and following administration of certain diuretics, e.g., chlorothiazide and acetazolamide. Increase of blood ammonium levels may or may not be demonstrable under these circumstances.

Alkalosis may aggravate or precipitate hepatic coma, possibly through the influence of the extracellular H⁺ ion concentration on the transfer of ammonium ion across cell membranes (p. 619). Increase of plasma pH is accompanied by decreased extracellular (blood) and increased intracellular ammonium. The deleterious effects of excessive administration of certain diuretics, e.g., chlorothiazide and acetazolamide, may be due to the development of metabolic alkalosis as a result of the induced potassium deficiency. A complicating factor in this connection is the frequent occurrence of respiratory alkalosis in hepatic coma as a result of hyperventilation, possibly attributable to an action of ammonium ion on the respiratory center.

It would appear that abnormal metabolism of ammonium is a significant etiologic factor in certain but not all cases of hepatic coma. Other reported biochemical abnormalities include increased blood methionine concentration and decreased cerebral uptake of oxygen, the significance of which has not been established. In a considerable proportion of cases, the intestinal bacterial origin of the toxic substance or substances, including ammonium, is indicated by the frequently beneficial effect of oral antibiotic therapy. This also results in disappearance of fetor hepaticus, the cause of which has not been definitely identified, but may, at least in part, be methylmercaptan.

It has been suggested that ammonium, by combination with α-ketoglutarate to form glutamate and then glutamine, interferes with the citric acid cycle, upon which the brain is very dependent. There is, in fact, diminished cerebral oxygen consumption in hepatic coma.

The situation in hepatic coma is, however, far too complicated to be explained in terms of ammonium increase alone. There is even evidence that the brain in hepatic coma can add to the blood ammonium level. Moreover, there is no relationship between blood ammonium levels and coma intensity. Iproniazid, a monoamine oxidase inhibitor, can cause a marked fall in the blood ammonium level without affecting the depth of the coma. Methionine administration can precipitate hepatic coma without a rise in the blood ammonium level. This effect seems to be due to a bacterial metabolite of methionine, since it is prevented by tetracycline administration. L-Tryptophan itself also precipitates coma after oral administration.

A large variety of changes in blood levels has been demonstrated in hepatic coma. There is increase in a number of amino acids (tyrosine, phenylalanine, histidine, cystine, methionine), in citrulline, in short-chain fatty acids (butyrate, valerate, caproate), in α-ketoglutarate, in lactate, pyruvate and citrate, in acetoin and in butylene glycol. In one comatose patient, β-phenylethylamine and isoamylamine were found in the blood. There is frequently alkalemia as well as hypopotassemia. It must be emphasized that hypoglycemia is quite frequently present, even at a level which could account for the coma.

On the whole, it appears that hepatic coma might be produced by the combined action of a variety of toxic metabolites, some of which are produced by intestinal bacteria, which bypass the liver and have a deleterious effect on the mechanisms of consciousness. Some of these toxic substances could be produced by the damaged liver itself. The normal liver also apparently produces substances, necessary for cerebral activity, which may be defective when the liver is damaged. In patients in hepatic coma, an increased breakdown of the neurotransmitter substances dopamine and serotonin has been demonstrated, leading to an increased cerebrospinal fluid concentration of homovanillic acid and 5-hydroxyindole acetic acid, respectively.

There seems little doubt that renal failure also contributes to the production of hepatic coma. It has already been pointed out that alterations in renal circulation occur in patients with advanced cirrhosis. There may even be acute tubular necrosis. The latter condition is comparatively common in fulminant liver failure, as is functional renal failure due to circulatory changes in the kidney. Uremia very frequently accompanies hepatic coma.

Hepatic coma can, of course, result from fulminant liver failure (hepatitis, poisoning, etc.) or from chronic liver disease (cirrhosis). In both instances, the biochemical findings are somewhat similar. It appears that the biochemical mechanism of coma production is probably also closely similar, if not identical, in both fulminant liver failure and chronic liver disease. It is interesting, however, that the neurological manifestations (hepatic encephalopathy, etc.) which precede the onset of coma are frequently quite different in chronic liver disease from those which occur in the acute variety of liver disease.

DRUG INDUCTION OF LIVER ENZYMES

A large number of different drugs are capable of increasing the activity of enzymes located in the endoplasmic reticulum of the hepatocyte. In fact, the smooth reticulum has been shown to hypertrophy. The drugs involved include anticonvulsants, sedatives and some steroids.

Barbiturates (phenobarbitone) seem to be particularly active in this respect. They increase glucuronyltransferase activity, and hence the conjugation of bilirubin. Phenobarbitone is, therefore, used in the treatment of neonatal jaundice in order to prevent kernicterus, as well as in the treatment of Gilbert's syndrome and Crigler-Najjar syndrome. Phenobarbitone also increases the metabolic inactivation of cortisol (and dexamethasone), testosterone and vitamin D. These effects can give rise to symptoms in a number of patients on prolonged anticonvulsant therapy. A low serum calcium level and a raised serum alkaline phosphatase with osteomalacia are by no means infrequent. This is probably due to the effect on vitamin D metabolic breakdown. Low levels of plasma 25-hydroxycholecalciferol have been reported. Administration of the latter effectively reverses the effects of barbiturates on calcium metabolism. It has already been pointed out that barbiturates induce increased synthesis of ALA synthetase (p. 277), and so bring about an attack of certain types of porphyria.

Barbiturates, as well as a number of other therapeutic substances, reduce the effect of other drugs, since enzymes are induced which break down these latter substances. This is of special importance in relation to anticoagulants, which must be given in higher dosage in order to be effective when barbiturates are administered. Uncontrolled withdrawal of the latter without reduction of anticoagulant dosage can result in serious bleeding. Antipyrine, carbamazepine, ethyl alcohol, glutethimide, and griseofulvin have a similar effect on anticoagulant therapy. In a similar manner, a number of drugs (barbiturates, diphenylhydantoin and phenylbutazone) can affect the action of digitoxin. Barbiturate administration can also produce folate deficiency, presumably by enzyme induction.

There are certain hepatotoxins, the toxicity of which is due to metabolites, that can be rendered even more hepatotoxic by drug enzyme induction.

LIVER FUNCTION STUDIES IN DIFFERENTIAL DIAGNOSIS

Difficulties encountered in the clinical employment of so-called tests of liver function have largely arisen from attempts to interpret findings too strictly in terms of disease diagnosis. The fact is too frequently overlooked that tests of function, if they accomplish their purpose, indicate the state of functional activity of an organ and not necessarily the presence or extent of morphologic changes in that organ. Broadly speaking, it has been found empirically that some of the tests tend to be abnormal in the presence of cholestasis, whereas others are abnormal when there has been disturbance of the functional integrity of the hepatocytes. Unfortunately, in the clinical situation, the distinction is not clear-cut, since hepatocellular disease is quite frequently associated with cholestasis and obstruction of the biliary tree can lead to hepatocellular dysfunction. Many drugs used in modern therapy have further clouded the picture, since they can give rise to hepatocyte disturbance or to cholestasis, and not infrequently to both. However, function tests, used properly, may aid materially in establishing a disease diagnosis, complementing information obtained by other methods.

The diagnosis of uncomplicated hemolytic jaundice usually offers little difficulty. Apart from the physical and hematological findings (e.g., reticulocytosis), there are characteristically (a) hyperbilirubinemia (levels above 5 mg./100 ml. are unusual in uncomplicated hemolysis), (b) relatively slight, if any, increase in direct-reacting serum bilirubin, (c) usually no bilirubin in the urine, (d) increase in urobilinogen in the urine and feces, (e) no bile acids in the urine, (f) hypocholesterolemia, (g) increased serum methemalbumin, (h) diminished serum haptoglobin and (i) normal findings by other tests of liver function (dye excretion, serum alkaline phosphatase activity, etc.). In a fairly large proportion of cases of congenital hemolytic jaundice and sickle-cell anemia this characteristic picture is complicated by the occurrence of common duct obstruction by gallstones, with the development of manifestations of obstructive jaundice, which may dominate the clinical picture.

Difficulty is frequently encountered in differentiating between *hepato-*

cellular and *extrahepatic obstructive* jaundice, and it is frequently in this connection that attempts are made to utilize function tests as an aid in disease diagnosis. For most purposes it is advisable to group these tests in two categories: (1) tests of "metabolic functions," which depend entirely upon the functional integrity of the liver cells and are uninfluenced by interference with the flow of bile, unless this produces hepatocellular damage; (2) tests of "excretory function," which depend upon maintenance of a free flow of bile as well as upon hepatocellular function. The most important of the "excretory tests" are the serum bilirubin concentration, serum alkaline phosphatase level and the dye excretion tests. The dye excretion tests are not used, however, in the presence of the two types of jaundice under consideration. The "metabolic tests" include most of the other procedures that have been discussed in connection with the estimation of hepatocellular function. Serum enzyme studies (e.g., aminotransferase) and the so-called "globulin reactions" (thymol turbidity, zinc sulfate turbidity, cephalin-cholesterol flocculation) are often helpful in this connection. In fact, their chief value in differential diagnosis arises out of the fact that they appear to reflect the presence of hepatocellular damage in acute liver disease rather than the state of liver function.

In studying patients with jaundice, the results of the "excretory tests" should be weighed against those of the "metabolic tests." The combination of severe hyperbilirubinemia (predominantly conjugated) with high values for serum alkaline phosphatase, hypercholesterolemia with essentially normal cholesterol : ester ratio, good prothrombin response to vitamin K, negative globulin reactions, and normal or near normal activity of other serum enzymes (e.g., aminotransferase) indicates serious interference with removal of bilirubin and other substances (e.g., bile salts) from the blood but no serious disturbance of the other functions investigated and no apparent active hepatocellular damage. This state of affairs is explicable virtually only on a basis of extrahepatic bile duct obstruction, although similar findings may be obtained in certain cases of toxic hepatitis. However, this hypothetical situation is seldom encountered clinically except in very early stages of obstruction of the common duct due to tumor (pancreas, duodenum, bile duct, lymph nodes). Before very long, even in such cases, there develops a variable degree of hepatocellular damage due to increased intraductal pressure or associated factors, and this damage may be reflected in impairment of the "metabolic" functions of the liver. Except in cases of early, partial obstruction, therefore, there is usually some degree of abnormality of one or more of the tests of hepatocellular function ("metabolic tests"). Under such circumstances, evaluation of the relative degree of impairment of the "metabolic" and "excretory" functions may be helpful in differential diagnosis, the latter often being impaired relatively more than the former in primarily obstructive jaundice. However, moderate or severe impairment in both groups of tests, indicating moderate or severe impairment of hepatocellular function and probably extensive hepatocellular damage, does not by any means exclude the possible presence also of extrahepatic bile duct obstruction. It must be realized that primary hepatocellular damage and that superimposed on extrahepatic duct obstruction may produce essentially the same results with most of the so-called "metabolic tests," insofar as liver damage is reflected in impaired function. From the standpoint of differential diagnosis of the cause of jaundice, therefore, the most definitive findings are obtained in cases of uncomplicated extrahepatic obstruction, e.g., neoplastic. Equivocal results can be obtained in the much more common causes of calculus obstruction, which are not infrequently accompanied by hepatocellular damage.

Duodenal intubation and study of the bile has been claimed as useful in the diagnosis of the nature of a biliary tract lesion (pigmented bile duct and gallbladder epithelium, leukocytes, red blood cells and cholesterol and calcium bilirubinate crystals). In a patient with jaundice, failure to obtain bile after prolonged and repeated duodenal intubation and stimulation is suggestive of complete obstruction of the bile duct. This diagnosis is supported by the absence of urobilinogen from the urine and feces (p. 651).

In the ultimate analysis, accurate diagnosis in lesions of the liver and bile passages depends primarily upon the clinical history and physical examination, which may include x-ray and scintiscan studies, liver biopsy and examination of the bile, supplemented by carefully interpreted biochemical findings. The last must not be accorded undue importance in establishing a disease diagnosis, although they may be very helpful under certain circumstances. Hope for a function test that will clearly differentiate extrahepatic obstructive from hepatocellular jaundice, i.e., one that will make a disease diagnosis, is not founded on reality. A most important contribution to hepatic disease diagnosis has been the increasing use of liver biopsy as a routine procedure. On the other hand, these studies have served to emphasize the frequent lack of correlation between morphologic and functional findings. Whereas significant liver damage may be present in the absence of demonstrable abnormality of function, as measured by available procedures, the reverse may also occur, i.e., evidence of functional disturbance in the absence of significant demonstrable morphologic change. Moreover, the degree of impairment of various functions frequently cannot be predicted on the basis of the nature or extent of the observed morphologic abnormalities.

There is considerable variation in the sensitivity of the many functions of the liver to noxious influences, as well as in the sensitivity of the test procedures available. In the presence of mild functional impairment, significantly abnormal results may be obtained with only one or a few of the tests employed, emphasizing the necessity of applying a number of the available procedures. In view of the fact that the liver is performing a great many different types of chemical reactions, and is involved in excretory as well as purely metabolic processes, its functional activity cannot possibly be measured by a single test. Moreover, functional abnormalities that appear earliest in the course of development of hepatic disease, e.g., viral hepatitis, are not necessarily the last to disappear with subsidence of the disease; i.e., different functions are affected to different degrees at different stages of the same disease.

Serial determinations are much more valuable than single observations. The earlier these are begun in the course of the disease, the more information they provide. Apart from their usefulness in differential diagnosis, serial determinations are invaluable in following the progress of such diseases as calculous obstruction, viral and toxic hepatitis, hepatic necrosis, and cirrhosis.

Selection of Tests. The question is often raised as to which tests are most useful. The fact is too frequently overlooked that selection of the most valuable procedures in a given case is determined largely by the type of information desired. This often results in the indiscriminate routine application of a battery of tests, many of which may be superfluous, in all cases of hepatic and biliary tract disease. For this purpose, clinical situations that are encountered frequently may be considered under four headings: (1) jaundice of undetermined etiology, the primary problem being differential diagnosis; (2) known obstructive jaundice; the problems here are (a) the progress of the obstruction (increasing or decreasing) and (b) detection and evaluation of superimposed hepatocellular damage; (3) known primary hepatic disease; the main problem here is the severity and progress of the disease (improvement or deterioration); (4) suspected hepatic disease without clinical jaundice; the problem here is detection of a focal liver lesion or diffuse hepatocellular damage not severe enough to produce clinical jaundice.

(1) Jaundice of Undetermined Nature. The primary problem is that of establishing the nature of the jaundice, i.e., hemolytic, hepatocellular or cholestatic.

The diagnosis of uncomplicated hemolytic jaundice can usually be made without difficulty (p. 641). Biochemical findings similar in many respects may be obtained in certain forms of nonhemolytic jaundice (p. 641), including the Crigler-Najjar syndrome, Gilbert's syndrome and, rarely, following viral hepatitis. In these conditions, however, there are no hematological findings indicative of excessive hemolysis, nor unduly increased urobilinogen excretion. In the presence of complicating biliary obstruction such as may occur, e.g., in neonatal and spherocytic jaundice, certain of the distinctive features of pure hemolytic icterus disappear: the serum gives a positive direct van den Bergh reaction; there is abnormal BSP retention; and urobilinogen excretion in the feces may not be excessive. However, the clinical and hematological characteristics usually suffice for accurate diagnosis, and quantitative studies of the serum bilirubin reveal an unduly large proportion in the unconjugated state (total bilirubin minus one-minute, direct-reacting bilirubin concentrations).

A problem that is usually most troublesome is that of distinguishing between primary hepatocellular and primary extrahepatic obstructive jaundice. It is usually in this connection that the largest number of test procedures must be employed. A study plan including the following determinations can be most helpful:

(a) Total serum bilirubin concentration. This should be measured as early as possible since it is an important index of the progress of the underlying condition, whether extrahepatic obstructive or hepatocellular. Moreover, the results of certain other tests, e.g., serum alkaline phosphatase (p. 670), are much more informative when interpreted in relation to changes in the serum bilirubin concentration determined simultaneously. In itself, the serum bilirubin concentration is, of course, of no value in differential diagnosis of extrahepatic obstructive and hepatocellular jaundice and, with rare exceptions, the one-minute, direct-reacting bilirubin concentration is not helpful.

(*b*) Plasma prothrombin determinations should be made because of the possibility that surgical intervention may be required. If low, the response to vitamin K administration should be determined. Although this may occasionally yield information of diagnostic value (p. 629), its chief purpose is to anticipate the efficacy of vitamin K therapy, and hence the operative risk, in the event of operation.

(*c*) The serum albumin concentration should be measured. This, too, is done primarily not because of its usefulness in differential diagnosis but because it reflects the state of protein nutrition, a factor of importance, particularly if surgical intervention becomes necessary. If electrophoresis is employed, additional information may be provided by the pattern of serum globulins (p. 621).

(*d*) The test procedures usually most helpful in differentiating between extrahepatic obstructive and hepatocellular jaundice are the globulin reactions (p. 625), electrophoresis of serum proteins, serum alkaline phosphatase and other enzyme activities (pp. 575 and 670), and the serum mucoprotein concentration (p. 624). These are most helpful in this connection when they reflect a degree of hepatocellular involvement that is relatively slight in relation to the degree of hyperbilirubinemia. This supports a diagnosis of biliary obstruction. Similar results may be obtained in certain forms of cholestatic jaundice (p. 644). When the results obtained with these procedures indicate severe hepatocellular involvement, however, the possibility of associated or underlying bile duct obstruction cannot be excluded on this basis alone. The importance should be stressed again of interpreting changes in these factors in relation to simultaneous changes in serum bilirubin concentration. Experience has shown that a serum alkaline phosphatase level above 215 U/l. (King-Armstrong, 37° C.) is usually indicative of extrahepatic obstruction. On the other hand, a serum aspartate aminotransferase level above 250 U/l. (Karmen at 37° C.) is usually indicative of hepatocellular disease. Raised serum levels beow 215 U/l. of alkaline phosphatase usually indicate cholestasis. This may often be intrahepatic. Raised serum levels of aspartate aminotransferase below 250 U/l. can be found in most of the hepatobiliary disorders. Slight rises are quite common in cholecystitis, even in the absence of overt jaundice.

(*e*) In severely jaundiced subjects, determination of urine urobilinogen may be helpful. Its absence implies either complete suppression of bile secretion or complete biliary obstruction. If it persists for more than a few days, the former would involve severe hepatocellular damage and functional impairment, which should be reflected by the tests enumerated above and by the clinical condition of the patient. Persisting absence of urobilinogen from the urine and the presence of bilirubinuria when test procedures suggest only moderate or slight hepatocellular involvement support a diagnosis of complete and permanent obstruction of the common duct. This is due usually to neoplasm or stricture (p. 651). These conclusions in the absence of bilirubinuria are not valid in subjects receiving oral antibiotics.

(f) The detection of α_1-fetoprotein in the serum of a jaundiced adult is strong evidence of the presence of a malignant hepatoma. In Caucasians, this protein is present in only 28 per cent of cases. An increase in serum IgM would help to confirm a diagnosis of primary biliary cirrhosis, but much stronger evidence is the presence of mitochondrial antibody. These findings in this disease occur even before jaundice has developed. The demonstration of Australia antigen indicates the presence of serum hepatitis. This antigen can also be present in nonjaundiced individuals, as well as in chronic carriers, as is the case in some patients with chronic kidney disease.

Liver biopsy frequently gives the best information in relation to the diagnosis of jaundice. This is especially the case in the presence of a predominant increase of conjugated bilirubin.

(2) Known Extrahepatic Duct Obstruction. If the serum bilirubin concentration is falling no other test is required; bile flow is improving. Increasing hyperbilirubinemia may be due to either increasing obstruction or superimposed hepatocellular damage. The former possibility should be investigated by means of the serum alkaline phosphatase activity (p. 670). Studies of serum aspartate aminotransferase (and other enzymes) may provide useful information in regard to the occurrence of hepatocellular failure. Urine urobilinogen should be measured in severely jaundiced patients; its persistent absence, indicating complete and permanent obstruction, has particular diagnostic significance (p. 651).

Determinations should be made repeatedly of plasma prothrombin and, if low, its response to vitamin K administration. This is for the purpose of anticipating possible therapeutic requirements if and when operation is advisable. At that time, too, the serum albumin concentration should be determined.

(3) Known Primary Hepatic Disease. Once the diagnosis of hepatic disease has been established, e.g., hepatitis or cirrhosis, frequently by liver biopsy, the information desired subsequently is concerned with its severity and progress. In patients with initial hyperbilirubinemia, periodic determinations of the total serum bilirubin concentration suffice for this purpose, an increase indicating progression and a decrease improvement. In the latter case, this is repeated until the serum bilirubin has fallen to within normal limits. When normal values are obtained, the BSP test is performed. The estimation of serum aspartate transferase and other serum enzymes is also of value. From a practical standpoint, little is gained by employing more than one of these procedures at a time. In patients with cirrhosis and increasing bilirubinemia, studies of changes in BSP retention and in serum albumin and γ-globulin concentration in relation to the bilirubin concentration may provide information concerning the extent to which blood is being shunted around the liver. In patients with cirrhosis, increased serum mucoprotein concentrations and elevated values for serum lactate dehydrogenase may aid in the diagnosis of complicating primary hepatic malignancy. This may be suggested also by high levels of serum alkaline phosphatase activity, although these may occur occasionally in uncomplicated portal cirrhosis. The demonstration of α_1-fetoprotein may also be of help in this respect.

(4) Suspected Hepatic Disease without Jaundice. In the absence of clinical jaundice, the problem is one of the detection of possible hepatic functional impairment, at times for the purpose of diagnosis of otherwise unrecognizable liver disease, in other instances to follow the course of a known disease, e.g., cirrhosis. Since the presence of latent jaundice can be excluded only by determination of the serum bilirubin concentration, this must be done in all cases. If it is within normal limits, the BSP test can be performed. Urinary urobilinogen should also be measured.

In the absence of evidence of excessive hemolysis, the presence of hepatic (or biliary tract) disease is indicated by increased concentrations of total serum bilirubin or increase in urine urobilinogen, which is, of course, not necessarily diagnostic of liver disease. This diagnosis is supported also by a rising level of total serum bilirubin, even though it is within normal limits (p. 636). Similar significance is attached to abnormal BSP retention in the absence of extrahepatic causes of decreased hepatic blood flow, e.g., the shock syndrome, hemorrhage, congestive heart failure. In patients with cirrhosis, progress of the disease can usually be followed satisfactorily by periodic determinations of total serum bilirubin concentration, serum enzymes and BSP retention. Determinations of serum proteins and electrophoretic procedures are also useful. It should be recalled, too, that bilirubinuria may occur in infectious hepatitis before other functional abnormalities can be demonstrated. A possibly more sensitive indication is the occurrence of serum LD_5.

Measurement of serum aspartate aminotransferase, alkaline phosphatase, and other serum enzyme activities, including isoenzymes, may be useful as an early indication of hepatocellular damage and/or intrahepatic cholestasis in patients receiving hepatotoxic drugs, e.g., certain antibacterial and tranquilizing agents. Increased serum alkaline phosphatase activity, in the absence of other evidence of abnormal liver function, may occur in certain forms of localized or diffuse liver disease, e.g., primary or metastatic carcinoma, sarcoidosis, abscess, echinococcus cyst, minor duct obstruction (e.g., lymphadenopathy).

For the purpose of diagnosis of liver disease without jaundice, liver biopsy often gives most information.

In patients with cirrhosis, progress of the disease can usually be followed satisfactorily by periodic determinations of serum bilirubin, albumin and globulin concentrations, and BSP retention.

In interpreting abnormalities in such factors as the plasma proteins, alkaline phosphatase activity, and abnormalities in thymol and zinc sulfate turbidity and cephalin-cholesterol flocculation, the fact must be kept in mind that changes identical with those present in liver disease may be produced by extrahepatic lesions.

The usefulness of liver function tests in clinical practice may be summarized as follows:

1. The diagnosis of uncomplicated hemolytic jaundice can be established.

2. They aid materially in the diagnosis of uncomplicated extrahepatic duct obstruction.

3. They may reveal the presence of liver damage in patients with known obstructive lesions, e.g., stone, stricture, neoplasm.

4. Serial studies are of great value both in diagnosis and in following the course of acute diffuse liver disease, e.g., infectious and toxic hepatitis and hepatic necrosis. In these conditions, they are particularly useful in revealing residual damage after subsidence of other clinical manifestations and in suggesting possible progression to chronic liver disease.

5. Certain tests may be of quite specific diagnostic value, e.g., Type III porphyrinuria in "alcoholic" cirrhosis, serum α_1-fetoprotein in malignant hepatoma, mitochondrial antibody in primary biliary cirrhosis and Australia antigen in serum hepatitis.

6. They may indicate the presence of unsuspected liver disease, e.g., cirrhosis.

7. They afford the only means of demonstrating the integrity or of evaluating the degree of impairment of various functions of the liver in the presence or absence of evidence of morphologic damage.

References

Arias, I. M.: The chemical basis of kernicterus. Advances in Clinical Chemistry, *3*:36, 1960.
Berliner, D. L., and Dougherty, T. F.: Hepatic and extrahepatic regulation of corticosteroids. Pharmacological Rev., *13*:329, 1961.
Billing, B. H.: Bile pigments in jaundice. Advances in Clinical Chemistry, 2:268, 1959.
Bouchier, I. A. D.: Gallstone formation. Lancet, *1*:711, 1971.
Brauer, R. W., ed.: Liver Function. Washington, D. C., American Institute of Biological Sciences, Pub. No. 4, 1958.
Breen, K. J., and Schenker, S.: Liver function tests. Critical Rev. Clin. Lab Sci., 2:573, 1971.
Brinkhous, K. M.: Plasma prothrombin; vitamin K. Medicine, *19*:329, 1940.
Dam, H.: Vitamin K. Vitamins and Hormones, *6*:27, 1948. New York, Academic Press, Inc.
Dubin, I. N.: Chronic idiopathic jaundice. Am. J. Med., *24*:268, 1958.
Gray, C. H.: Bile Pigments in Health and Disease. Springfield, Ill., Charles C Thomas, 1961.
Greenberg, D. M., and Harper, H. A., eds.: Enzymes in Health and Disease. Springfield, Ill., Charles C Thomas, 1960.
Gutman, A. B.: Serum alkaline phosphatase activity in diseases of the skeletal and hepatobiliary systems. Am. J. Med., *27*:875, 1959.
Gutman, A. B.: The plasma proteins in disease. *In* Advances in Protein Chemistry, edited by M. L. Anson and J. T. Edsal, New York, Academic Press, *4*:155, 1948.
Heirwegh, K. P. M., Meuwissen, J. A. T. P., and Fevery, J.: Critique of the assay and significance of bilirubin conjugation. Advances in Clinical Chemistry, *16*:239, 1973.
Hunter, J.: Enzyme induction and medical treatment. J. Roy. Coll. Phys. London, *8*:163, 1974.
Klatsin, G.: Bile pigment metabolism. Ann. Rev. Med., *12*:211, 1961.

Lathe, G. H.: The degradation of haem by mammals and its excretion as conjugated bilirubin. *In* Campbell, P. N., and Dickens, F., eds.: Essays in Biochemistry (Biochem. Soc.). London, Academic Press, 1972.

Losowsky, M. S., and Scott, B. B.: Ascites and oedema in liver disease. Brit. Med. J., *3*:336, 1973.

Maclagan, N. F., ed.: Symposium on the liver. Brit. Med. Bull., *13*:75, 1957.

Popper, H., and Schaffner, F.: Liver: Structure and Function. New York, McGraw-Hill Book Co., Inc., 1957.

Reinhold, J. G.: Flocculation tests and their application to the study of liver disease. Advances in Clinical Chemistry, *3*:84, 1960.

Scheig, R.: Diseases of lipid metabolism. *In* Bondy, P. K., ed.: Duncan's Diseases of Metabolism. 6th ed. Philadelphia, W. B. Saunders Co., 1969.

Sherlock, S. P. V.: Diseases of the Liver and Biliary System. 4th ed., Oxford, Blackwell Scientific Publications, 1968.

Watson, C. J., and Schwartz, S.: Urinary excretion of porphyrin in liver disease. Josiah Macy, Jr. Foundation Symposium on Liver Injury, New York, 1948.

Wheeler, H. O., Meltzer, J. I., and Bradley, S. E.: Biliary transport and hepatic storage of sulfobromophthalein sodium. J. Clin. Invest., *39*:1131, 1960.

Wolf, P. L., Williams, D., and von der Muehll, E.: Practical Clinical Enzymology and Biochemical Profiling. New York, John Wiley and Sons, 1973.

Wróblewski, F.: The clinical significance of transaminase activities of serum. Am. J. Med., *27*:911, 1959.

Zuelzer, W. W., and Brown, A. K.: Neonatal jaundice. Am. J. Dis. Child., *101*:87, 1961.

Chapter 22

ENDOCRINE FUNCTION

By R. Hall *and* G. M. Besser

The functional state of some of the glands of internal secretion can be determined at the present time only by indirect methods, i.e., by studying certain phases of metabolism which are influenced specifically by certain hormones. For example, the state of parathyroid function is most satisfactorily ascertained clinically by studies of calcium and phosphorus metabolism and serum phosphatase activity. Abnormal function of the pancreatic islet cells is reflected in abnormalities of carbohydrate metabolism, and thyroid function in the basal metabolic rate, cholesterol and carbohydrate metabolism, and creatine metabolism. Water metabolism is affected by the antidiuretic hormone of the posterior lobe of the hypophysis, and carbohydrate metabolism by epinephrine. Carbohydrate, lipid, and protein metabolism are influenced by adrenal cortical and growth hormones and the adrenocorticotrophic hormone of the anterior hypophysis, and protein metabolism also by the male sex hormone. The adrenal cortex exerts a specific effect upon the metabolism of sodium, potassium, chloride, and water. The manner in which these metabolic abnormalities may be used as an index of the state of function of various endocrine glands has been indicated elsewhere in discussing the metabolism of each of the substances in question.

A more direct approach is possible in the case of certain hormones, quantitative determination of their concentration in the blood or urine affording a more or less exact index of the functional activity of the organs in which they originate or of those concerned with their intermediary metabolism. Thus, the concentration of organic iodine in the blood plasma or serum is a measure of the quantity of circulating thyroid hormone (p. 756). Assays can also be made of certain hormones of the anterior hypophysis, the ovary, testis and placenta and the adrenal cortex. Because of the importance of this field of clinical investigation, and because it is fundamentally biochemical in nature, the clinical significance of data obtained by such procedures will be discussed even though the methods employed in certain instances are biological rather than chemical.

CYCLIC ADENOSINE MONOPHOSPHATE

It is now clear that a variety of hormones act by altering the level of intracellular cyclic adenosine monophosphate (cyclic AMP) within their target cells. The primary event in hormone action appears to be a combination of the hormone with a specific receptor on the surface of the cell membrane. This induces an alteration in the activity of a membrane-bound adenyl cyclase system which catalyzes the formation of cyclic AMP from adenosine triphosphate. The cyclic AMP so formed has at least three fates: it can leak out of the cell into the plasma or urine, it can be degraded by an enzyme phosphodiesterase to adenosine 5-phosphate, and it can serve as a second messenger mediating the action of the hormone.

Certain criteria must be fulfilled before it can be concluded that a particular hormone acts by way of cyclic AMP. The hormone should induce a measurable change in the concentration of cyclic AMP in intact cells of its target organ. The alteration in cyclic AMP levels should precede the metabolic responses of the hormone. A dose response relationship of the hormone on cyclic AMP should mirror that of the end-product of hormone action. Active analogues of the hormone should affect cyclic AMP and the product of hormone action similarly. Cyclic AMP or its analogues should mimic the action of the hormone on intact cells. When added to broken cell preparations, the hormone should convert ATP to cyclic AMP. The action of the hormone should be potentiated by phosphodiesterase inhibitors (e.g., methylxanthines) which allow a buildup of cyclic AMP.

Cyclic AMP is thought to act as a kinase activator. Protein kinases catalyze the transfer of phosphate from ATP to a variety of proteins which have a different biological activity in the phosphorylated form. Protein kinases have two components, a catalytic portion and an associated inhibitor of the action of the catalytic portion. Cyclic AMP attaches to the inhibitory component, causing it to become dissociated from the catalytic portion which is then able to phosphorylate protein. In skeletal muscle, the activated protein kinase acts in the presence of ATP and Mg^{2+} to catalyze the phosphorylation and hence activation of phosphorylase b kinase. The activated phosphorylase b kinase then acts in the presence of ATP, Mg^{2+} and Ca^{2+} to catalyze phosphorylation of phosphorylase b to phosphorylase a, the more active form of the enzyme, which can then catalyze the breakdown of glycogen to glucose.

It seems likely that the relative specificity of hormones resides in the specificity of the receptor sites in their target organs. Such specificity for a hormone is not absolute, as is seen in the wide variety of extra-adrenal actions of ACTH. Again, cells in some target organs have a variety of receptors for different hormones, although some hormones can interact at a common receptor site.

Studies of other cyclic nucleotides are under way at present, particularly cyclic guanosine monophosphate (cyclic GMP). Cyclic GMP is present in most tissues at concentrations which are usually about ten times lower than those of cyclic AMP. The formation of cyclic GMP is catalyzed by guanyl cyclase, which is soluble in most instances. Cyclic GMP is degraded by a phosphodiesterase in a manner similar to that of cyclic AMP. The biological role of cyclic GMP has not yet been established, but a variety of hormones have been shown to affect intracellular and extracellular levels of the nucleotide. There is a suggestion that cyclic GMP might oppose some of the actions of cyclic AMP and that it might act as a further regulator of cellular function.

PITUITARY HORMONES

Three anatomical divisions of the hypophysis yield extracts which, on injection, exert well-defined effects in various animal species. These divisions are (1) the neurohypophysis (posterior lobe), (2) the pars intermedia, and (3) the anterior lobe (adenohypophysis). In man and other mammals, the latter is by far the most important inasmuch as it controls the functional activity and structural integrity of other important endocrine glands, e.g., the thyroid, adrenal cortex, and gonads, through its "trophic" (tropic) hormones. In addition, it produces growth hormone (somatotropin), which promotes growth and exerts direct actions in certain fundamental phases of intermediary metabolism, and prolactin which stimulates milk production and controls electrolyte exchanges in some species.

The pars intermedia does not occur in the human pituitary as a distinct entity. Whereas the anterior pituitary hormones are synthesized within the cells of this organ, their secretion is controlled by the hypothalamus. The hypothalamus secretes regulatory hormones, which pass down the portal capillary vessels of the pituitary stalk to act on the anterior pituitary cells (pituitocytes), causing synthesis and release or the inhibition of release of particular anterior pituitary hormones. Posterior pituitary hormones are, however, made in the hypothalamus and pass down the axons of the cells of the hypothalamic neurons, which extend through the pituitary stalk, to be stored in the posterior pituitary. They are directly released from the ends of the axons into the circulation as required. The latter process is called "neurosecretion."

Hypothalamic Regulation of Anterior Lobe Hormones.

The brain controls the secretion of the anterior pituitary hormones by elaborating regulatory hormones within special nuclei of the hypothalamus. The regulation of the production of the hypothalamic hormones depends upon three main types of mechanism: a complex interaction between the effects of the circulating levels of the target organ hormone in the blood (feedback control), intrinsic cyclical mechanisms which operate within the midbrain reticular formation (e.g., controlling monthly or daily hormonal rhythms) and reactions to psychological or physical stimuli (e.g., responses to stress) which involve both higher integrating centers and the reception of afferent sensory messages. The net result is the secretion into the portal capillary vessels of the pituitary stalk of short-chain polypeptide hormones, which either stimulate the synthesis and release of particular anterior pituitary hormones or inhibit their release. Rapid progress is occurring in elucidation of the nature of these substances, and the details are shown in Table 22-1. There are no neural connections between the hypothalamus and the anterior pituitary cells; the latter are bathed in the portal capillary blood in sinusoids which connect with the stalk vessels. Originally, these hypophysiotropic substances were recognized by demonstrating that different extracts of sheep, bovine or pig hypothalami could modify the release of the hormones of the anterior pituitary *in vitro* or *in vivo*, by causing either stimulation or inhibition of hormone secretion. They were therefore called releasing or inhibiting "factors." Recently, however, some of these have been fully identified, their structures ascertained and the compounds synthesized and used in experimental and clinical situations (Table 22-1), and it has become conventional to call them releasing or inhibiting hormones. The term "factor" sometimes is still used when hypothalamic extracts are being used. While these hormones must be secreted into the stalk blood vessels in very small amounts, the presence of some can be detected at times in the general circulation, e.g., CRF after stress in hypophysectomized animals and LH/FSH-RH at the times of the midcycle luteinizing hormone peak in the menstrual cycle.

Thyrotrophin-releasing Hormone (TRH). As little as 10 μg. of this synthetic peptide causes the serum thyroid-stimulating hormone (thyrotrophin—TSH) to rise within two minutes, and 50 μg. and more gives consistent responses. It is absorbed when given by mouth, though only poorly. The tripeptide causes both synthesis and release of TSH by mechanisms which involve the activation of the adenyl cyclase system. The action of TRH on the pituitary cells is blocked by thyroid hormones (triiodothyronine and probably thyroxine), so that if the circulating thyroid hormone levels are too high, TSH release does not occur, and if they are too low, a greater amount of TSH is released. Although the principal site of control of TSH secretion is at the pituitary cell level, it seems likely that TRH production is also altered in a way which is dependent

TABLE 22-1 HYPOTHALAMIC REGULATORY HORMONES

A. RELEASING HORMONES (RH) FOR: **FORMULAE WHERE ESTABLISHED:**

Thyrotrophin—TRH: pyroGLU-HIS-PROamide

Luteinizing hormone and follicle-stimulating hormone } LH/FSH-RH: pyroGLU-HIS-TRP-SER-TYR-GLY-LEU-ARG-PRO-GLYamide

Corticotrophin—CRH

Melanocyte-stimulating hormone—MSH

Prolactin—PRH

Growth hormone—GRH

B. RELEASE-INHIBITING HORMONES (R-IH) FOR:

Growth hormone (GH-RIH)
    H-ALA-GLY-CYS-LYS-ASN-PHE-PHE-TRP-LYS-THR-PHE-THR-SER-CYS-OH

Prolactin— (P-RIH)

Melanocyte-stimulating hormone—M-RIH   PRO-LEU-GLYamide

on circulating thyroid hormone levels. Thus, TRH secretion may be increased in thyroid hormone deficiency states and *vice versa*, although this is not fully established. Therefore, there is "negative feedback" control whereby thyroid hormone production, normally induced by TSH, is reduced when circulating thyroid hormone is too high, and *vice versa*. TRH can be of value in diagnosis of thyroid disease since, when thyroid hormone production is pathologically high (thyrotoxicosis), there will be no serum TSH elevation after a standard intravenous test dose of 200 μg. TRH. In contrast, in thyroid deficiency the serum TSH response is excessive. This TRH test is of great value in confirming a suspected diagnosis of thyroid disease when the diagnosis is difficult by other means. TRH causes prolactin to be released in addition to TSH, but it is distinct from the true physiological prolactin-releasing hormone. It may represent a pharmacological effect of TRH. TRH also releases a small amount of follicle-stimulating hormone, but not luteinizing hormone, in men but not women.

Luteinizing Hormone/Follicle-stimulating Hormone-releasing Hormone (LH/FSH-RH).

It was originally suspected that there were separate releasing hormones for LH and FSH. However, only one hormone, a decapeptide, has been isolated from hypothalami and this causes both LH and FSH release, although LH release is much greater than FSH. It seems likely that the complex changes in LH and FSH levels which occur, for example, during the menstrual cycle, may be produced by an interaction between the different circulating gonadal steroids with this single gonadotrophin-releasing hormone producing differential release of LH or FSH at different phases of the cycle (see following). However, it may be that a predominantly FSH-releasing hormone will be found. It is possible to test the gonadotrophin secretory capacity of the pituitary of patients with hypogonadism using an intravenous test dose of synthetic LH/FSH-RH.

Corticotrophin-releasing Hormone (CRH).

The structure of this hormone is not known. There are three basic control mechanisms which influence the secretion of CRH and in turn pituitary adrenocorticotrophic hormone (ACTH) and adrenocortical hormones:

1. The circadian rhythm, whereby in subjects with a normal sleep-wake cycle the secretion is at its highest at about the time of waking, thereafter falling during the day to reach a low level at about midnight, to begin rising again at about 3 A.M.

2. The negative feedback, whereby secretion diminishes when circulating cortisol is too high and *vice versa*. There is also a negative feedback at the pituitary level since cortisol also seems to block ACTH secretion by an action on pituitary cells.

3. Stress mechanism, whereby ACTH and corticosteroids rise rapidly in response to physical or emotional stress such as pain, a depression or fear. If the stress response is inadequate, the patient may go into circulatory failure or shock.

Vasopressin, the posterior lobe antidiuretic hormone, can act as a CRH in supraphysiological doses, but there is no evidence that it is the physiological CRH.

Melanocyte-stimulating Hormone-releasing Hormone (MRH).

Although melanocyte-stimulating hormone appears to be secreted simultaneously with ACTH, it seems to have a separate releasing hormone. The structure is unknown.

Prolactin-releasing Hormone (PRH). Similarly, experimental evidence strongly suggests the presence of a hypothalamic *prolactin-releasing hormone,* distinct from TRH, which also has prolactin-releasing actions, but its structure is unknown.

Growth Hormone-releasing Hormone (GRH). *Growth hormone-releasing hormone* also exists on the basis of experimental data. At least three peptides have been described with reputed growth hormone-releasing powers, but it is not yet clear which, if any, has physiological meaning.

Growth Hormone-release Inhibiting Hormone (GH-RIH). This hormone has been isolated in a pure form and shown to be a cyclical polypeptide containing 14 aminoacids (Table 22-1). It is very active in man and inhibits the normal growth hormone release in response to stimuli such as hypoglycemia and arginine as well as the excessive secretion found in diseases such as acromegaly. Most interesting, too, is the finding that GH-RIH reduces the TSH but not the prolactin responses to TRH, presumably because there is competition for the same receptors. It does not prevent the LH and FSH response to LH/FSH-RH. The physiological significance of this competition is not clear. When infused into the systemic circulation in large doses, GH-RIH also inhibits gastrin, insulin, and glucagon release from the pancreas. This probably does not represent a physiological action, since GH-RIH is unlikely to reach the peripheral circulation in sufficient quantities, but must indicate similarity between receptors in the pituitary and the gastrointestinal tract.

Prolactin-release Inhibiting Hormone (P-RIH). This hormone has not yet been characterized. However, it is undoubtedly the most important influence controlling the secretion of prolactin, since prolactin secretion is very active if the pituitary is separated from the influence of the hypothalamus *in vitro* or experimentally *in vivo* or in patients with diseases of the hypothalamus. P-RIH secretion is inhibited by many centrally acting drugs, such as the phenothiazines, reserpine, methyldopa, tricyclic antidepressants; these cause excessive prolactin secretion and abnormal lactation.

Melanocyte-stimulating Hormone-release Inhibiting Hormone (M-RIH). Tripeptide and pentapeptide derivatives of oxytocin have been shown to inhibit MSH secretion in some animals, and it has been suggested that oxytocin may act as a prohormone for M-RIH. However, the peptides described do not appear to be active in man and have not been isolated from hypothalamic extracts. Their relevance to human physiology is therefore unclear.

ANTERIOR LOBE (ADENOHYPOPHYSEAL) HORMONES

Following hypophysectomy, the thyroid gland, adrenal cortex and gonads undergo atrophy, and the animal loses weight. In addition, in young animals longitudinal growth is arrested, closure of epiphyseal lines is delayed and the gonads fail to mature, the animals remaining infantile. Certain metabolic defects are apparent. Oxidative processes are depressed; liver and muscle glycogen are depleted; fat oxidation is decreased; ketosis does not appear readily following procedures which usually induce increased ketogenesis; fat turnover is decreased, as reflected in its decreased mobilization from the fat depots. These animals are exceedingly sensitive to insulin.

All of these morphological and metabolic abnormalities can be prevented or corrected by administration of adenohypophyseal hormones. They may be placed in two categories: (1) The so-called "trophic" (tropic) hormones are responsible for the structural integrity and functional activity of certain specific "target" endocrine glands. Their most characteristic effects are due to stimulation of their respective target glands, and, with few exceptions, they exert no important effect in the absence of these organs. They include the following: (*a*) thyrotrophin, (*b*) adrenocorticotrophin, (*c*) two gonadotrophic hormones, follicle-stimulating and leuteinizing, and (*d*) prolactin, a hormone which induces milk production in the prepared breast, but which has other extramammary actions. (2) At least one factor, growth hormone, exerts direct metabolic effects on protein, carbohydrate, and fat metabolism, not mediated by other endocrine glands.

Gonadotrophic Hormones

The anterior pituitary secretes two hormones which exert important effects on the gonads and are, therefore, termed "gonadotrophic." These include: (1) follicle-stimulating hormone (FSH); (2) luteinizing hormone (LH), also called interstitial-cell-stimulating hormone (ICSH) in the male. In the female, they are responsible for the growth, maturation, and expulsion of the ova and the production of the hormones of the ovary. In the male, they stimulate spermatogenesis and production of androgen.

Hypophysectomy is followed by atrophy of the gonads, with abolition of their hormone production and, consequently, of the secondary sex characteristics developed at puberty; impairment of sexual function; and atrophy of the target organs, viz., uterus, vaginal mucosa, prostate, and seminal vesicles. All of these structures and functions are restored by administration of the gonadotrophic hormones. Secretion of these pituitary factors is controlled by the levels of androgen and estrogen in the circulation, an increase in these steroid hormones causing depression and a decreased stimulation of gonadotrophin production. Both LH and FSH appear to be secreted from the basophil cells of the pituitary.

Assays. Pure preparations of the gonadotrophins have been obtained only very recently. In the past, insensitive bioassays have been mainly used for the estimation of gonadotrophins in blood and urine or pituitary extracts, but recently more sensitive radioimmunoassays have been developed for both FSH and LH.

Bioassay for FSH. 1. Ovarian follicle growth in hypophysectomized immature rats. This is a specific test for FSH but is too laborious for clinical use. 2. Augmentation tests. These depend on the increase in ovarian weight of intact immature rats or mice previously treated with excess human chorionic gonadotrophin (HCG). HCG augments the action of FSH on the ovary and eliminates the effects of any LH in the test material, since the action of HCG is similar to that of LH.

Bioassay for LH. 1. The ventral prostatic weight test in hypophysectomized immature male rats is specific for LH but too laborious for routine use. 2. The ovarian ascorbic acid depletion test using intact immature pseudopregnant rats is a satisfactory method for LH. The animals are pretreated with gonadotrophins from pregnant mares' serum, followed by chorionic gonadotrophin. The sensitivity of the test can be increased by measuring depletion of cholesterol.

Nonspecific Bioassays. The mouse uterus test is the most widely used assay. Extracts of test urine are injected into immature mice for several days and the amount required to produce a 100 per cent increase in uterine weight contains one unit. This method measures the combined effects of FSH and LH and provides a clinically useful index of total gonadotrophin activity. The magnitude of response depends on the relative proportions of FSH and LH in the specimens as well as their absolute amounts, so it is difficult to express the results in quantitative terms. The result can be converted to appropriate international reference standards by comparison with control preparations. In most assays, careful extraction of urine is required to remove toxic materials and to concentrate the hormones.

Immunoassays. Immunoassays are now available for LH and FSH in blood and urine. LH can easily be assayed using a specific anti-LH antiserum. A radioimmunological technique is used based on the competition between LH in the sample and ^{125}I-labeled LH for binding sites in the specific anti-LH antiserum. Specific anti-FSH sera now available show little or no cross-reaction with LH. Many of the earlier gonadotrophin antisera were not very specific and showed cross-reaction with each other and with TSH. Radioimmunoassays for FSH and LH now give results that correlate well with those of bioassays, but are much more sensitive.

Standards. The field of gonadotrophin assays has been confused by variations in methodology and standards. Different extraction procedures on urine or blood samples and the different strains of animals used in the assays have made it difficult to compare results from individual laboratories. Again, standard preparations have varied, and results expressed in relation to standards from other species are not always valid in man. Radioimmunoassays require their own standards, and the relationships with the results of bioassays must be worked out for each individual system. Urinary gonadotrophins are assayed against an International Reference Preparation (IRP) of human urinary menopausal gonadotrophin (HMG); for pituitary or blood gonadotrophin assays, a reference standard derived from purified pituitary LH and FSH should be used, since gonadotrophins in the urine have different characteristics from those in blood or the pituitary and presumably represent partially degraded material.

Follicle-stimulating Hormone (FSH). FSH is a glycoprotein with a molecular weight of about 30,000, containing approximately 8 per cent of carbohydrate. Sialic acid is an integral part of the molecule. It is doubtful whether full purification of FSH has yet been achieved. The primary follicle of the ovary develops to the fluid-filled vesicular stage under the action of FSH, which also stimulates granulosa cell proliferation, plays some part in estrogen biosynthesis and increases estrogen production. In the male, FSH increases spermatogenesis and seminiferous tubule development.

Luteinizing Hormone (LH, ICSH). LH is also a glycoprotein with molecular weight in the region of 30,000. It has a high cystine content. The hormone induces ovulation, stimulates estrogen production by the theca interna and initiates and maintains the corpus luteum. In the male, LH stimulates the interstitial or Leydig cells to produce androgens and, therefore, has a secondary role in maturation of spermatocytes and in the development of secondary sexual characteristics.

Human Menopausal Gonadotrophin (HMG). In human blood, the gonadotrophin levels increase after the gonadal failure which results in the menopause. This gonadotrophin is excreted in the urine and may be extracted for therapeutic use. One such preparation—Pergonal—has been used to stimulate ovulation. It contains both LH and FSH activity.

Human Chorionic Gonadotrophin (HCG). HCG is produced by the human placenta and by ovarian, testicular and uterine tumors of trophoblastic origin. It has a close functional and structural relationship to LH and is predominantly luteinizing in function. Its molecular weight is about 30,000, and its carbohydrate content is 30 per cent. A simple reliable pregnancy test has been developed, based on the detection of HCG in the urine of pregnant women by immunological techniques. Both hemagglutination inhibition and complement-fixation tests have been used. Peak levels occur about the ninth week of pregnancy, followed by low levels in midpregnancy and a second smaller peak in the third trimester. In certain cases of threatened abortion, a fall in level of HCG may give a warning of impending placental failure. High levels of HCG are found in certain trophoblastic tumors, and hormone assay can be used as an index of the effectiveness of therapy.

Physiological Considerations. Although unquestionably gonadotrophins are secreted by the pituitary gland from the time of birth or earlier, they circulate in low concentration until about ten to twelve years of age, being subsequently elaborated in increasing amount and producing gonadal stimulation and the phenomena of pubescence. There appear to be no circadian changes in LH or FSH secretion in children or adults of either sex. In the *female*, after the onset of menstruation (the menarche), the pituitary gonadotrophins are secreted in a cyclic fashion (Fig. 22-1). FSH secretion predominates during the first half of the cycle, while LH levels are low. Immediately prior to ovulation there is a very large LH peak which causes the ovum to be released from the ovarian follicle which

THE FEMALE SEX ENDOCRINE CYCLE

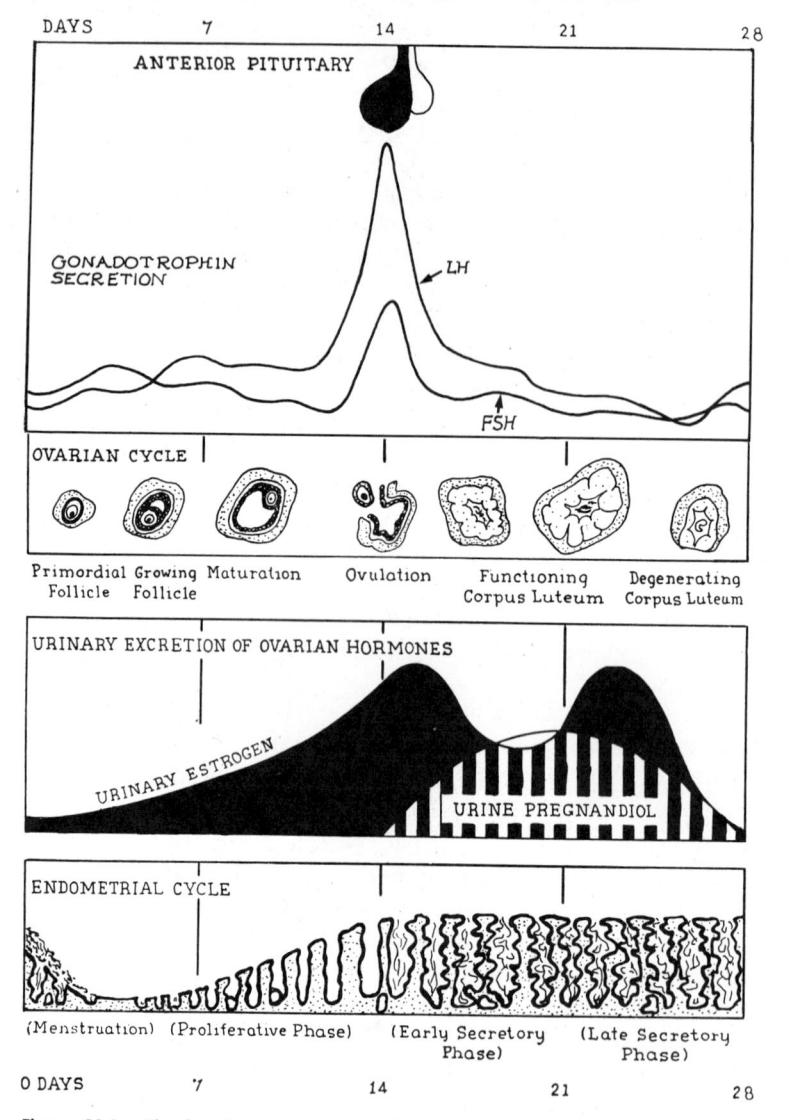

Figure 22-1. The female sex endocrine cycle. (Modified from a figure by Dr. A. E. Rakoff.)

had ripened under the influence of FSH. The LH peak lasts about a day. As a rule, only one of the follicles ripens completely in preparation for ovulation. Apparently a proper synergistic mixture of the follicle-stimulating and luteinizing hormones must be present in order to furnish the necessary stimulus for normal ovulation. The prevalent belief is that ovulation in the human occurs about fourteen days before the next menstrual flow, regardless of the length of the cycle. Following ovulation, the corpus hemorrhagicum is converted to a corpus luteum and during the next ten days it grows and becomes more active functionally. As the gonadotrophic activity diminishes in the premenstrual phase, this structure degenerates and becomes a corpus albicans unless pregnancy has supervened. Thus, under the influence of the gonadotrophic hormones of the hypophysis, the ovary has passed through two endocrine phases: (1) the growing follicle, which produces increasing amounts of estrogen and culminates in ovulation; (2) the corpus luteum, which produces both progesterone and estrogen, preparing the endometrium for pregnancy (See p. 788 ff.).

Menopause. With the approach of the menopause, the ovary becomes progressively less responsive to gonadotrophic stimulation, with a consequent increasing functional demand upon the pituitary for gonadotrophic hormone, chiefly of the follicle-stimulating type. As estrogen production by the ovary continues to diminish, an increasing concentration of follicle-stimulating and luteinizing hormones appears in the blood and urine of the menopausal woman.

Gonadotrophins in the Male. Gonadotrophin secretion increases in males about 11 to 13 years of age, increasing gradually to reach a stable level by 14 years, resulting in increased androgen secretion and puberty. There are no monthly changes in LH or FSH secretion in the male, but there are spontaneous fluctuations in each when blood measurements are made at frequent intervals. These suggest that both LH and FSH are secreted in surges which last one to two hours at a time. A moderate and variable increase in gonadotrophin levels occurs in some males after the age of 50 to 60 years as sexual function declines (the male climacteric).

Normal Gonadotrophin Values. It is not possible to quote, realistically, normal values for LH and FSH levels in blood or urine at different ages in men or women. Bioassays and radioimmunoassays give different answers, since they depend on different properties of the substances being measured. Furthermore, the different techniques make use of different methods for extracting the hormones to be measured, and also have greatly differing sensitivities and cross-react to different degrees with related hormones. Finally, a variety of relatively impure preparations of LH or FSH, derived from urine or pituitary sources, have been used as standards in the assays. It is clear that each laboratory measuring LH and FSH must establish its own normal ranges under a variety of physiological conditions in both sexes. It is often not possible to compare directly results from different laboratories.

Abnormal Gonadotrophin Secretion. The quantity of gonadotrophins in the blood and urine may be influenced by several factors. Increased amounts of hormones may be produced by the pituitary gland in various states of functional overactivity in which gonadotrophins may be produced. Certain nonendocrine tumors may be accompanied by excessive gonadotrophin production, but this is by no means consistent or marked. Tumors of chorionic tissue, however, usually give rise to enormous increases in gonadotrophin of the HCG type. It is questionable whether any other type of tumor can produce gonadotrophic hormone, although this possibility has arisen in the case of other embryonal tumors, particularly of the testes when these contain trophoblastic tissue. Pituitary tumors are often associated with decreased gonadotrophin secretion but are not responsible for excess secretion.

The quantity of hormone in the blood and urine may be diminished in the presence of impaired pituitary activity as a result of (*a*) a tumor involving the anterior hypophysis or neighboring structures, (*b*) states of functional hypopituitarism, or (*c*) suppression of gonadotrophin formation by excessive amounts of gonadal hormones (estrogens, androgens). During pregnancy, there is a fall in gonadotrophins following expulsion of the placenta, in functional abnormalities of the placenta, and in intra-uterine fetal death.

Increase in Gonadotrophic Hormones. (1) Functional Hyperactivity of the Pituitary. The production of increased amounts of gonadotrophic hormones by the pituitary usually results from increased functional demand for gonadal stimulation. At puberty, increased amounts are required to stimulate ovarian or testicular activity, but this increase is promptly checked by the consequent production of estrogens and androgens, except in the presence of a primary ovarian or testicular deficiency, in which case increased gonadotrophic secretion may persist for many years. Such deficiency of the gonads may result from agenesis, orchitis or oophoritis (particularly after mumps), failure of descent of the testes after puberty, spontaneous premature menopause or from gonadal trauma or castration. High gonadotrophin levels are encountered in a syndrome (Turner's syndrome) characterized by congenitally absent ovaries, with infantilism, short stature, and other congenital abnormalities, such as short, webbed neck, and coarctation of the aorta, associated with a deficiency of one of the sex chromosomes (45, XO karyotype).

With the approach of the menopause, increasing amounts of gonadotrophins are produced in an effort to maintain function of the failing ovaries. This increased hypophyseal activity may persist for many years in some women and is believed to be responsible for many of the symptoms of the menopause, particularly the hot flashes and general vasomotor instability. Increased amounts of gonadotrophins may appear within a week after removal of the ovaries or testes. The increase that follows "castration" by x-ray is usually more gradual. In the male, the increase in gonadotrophins at the time of the "climacteric" is not so consistent nor so marked as in the female.

Various menstrual disorders are frequently accompanied by increased secretion of gonadotrophins. In amenorrheic women, this indicates a primary ovarian deficiency. An increase is present in some cases of functional bleeding, particularly at puberty and at the menopause, being usually indicative of primary ovarian dysfunction. The persistent stimulation of abnormal ovaries by increased amounts of gonadotrophin may result in cystic disease of the ovaries. The polycystic ovary syndrome is associated with amenorrhea (usually secondary), infertility and often excessive hair growth of the male type (hirsuties) and often obesity. Circulating LH levels are usually elevated and noncycling, but FSH is low. The disease may be due to a defect in the hypothalamus, causing an imbalance between the mechanisms controlling LH and FSH secretion. The abnormal cystic ovaries produce excessive androgens.

(2) Increased Production by Chorionic Tissue. By far the largest amounts of gonadotrophic hormone are encountered in the presence of growing chorionic tissue. The chorionic gonadotrophin is entirely of the type normally derived from the placenta during pregnancy and differs from pituitary luteinizing hormone.

In normal pregnancy the hormone is demonstrable within ten days after the first missed period, in amounts sufficient to give positive tests with small amounts of blood or urine, increasing enormously within six weeks.

Hydatidiform mole, being a tumor of chorionic tissue, usually produces an even greater amount of gonadotrophic hormone than is observed in normal pregnancy (but usually no estrogen or progesterone).

Very high values are obtained also in choriocarcinoma, which is a condition of malignant degeneration of the chorion. Inasmuch as this condition follows expulsion of a mole, miscarriage, or normal pregnancy, the finding of progressively increasing amounts of gonadotrophin under such circumstances is highly significant, since the hormone titer should normally be diminishing rapidly. It should be emphasized, however, that a positive "pregnancy test" is not in itself diagnostically significant in this connection, since retained pieces of placental tissue or a missed abortion may occasionally produce a positive reaction weeks or even months following delivery. The urinary HCG levels can be used to follow the response of a patient with choriocarcinoma to cytotoxic drugs. Patients are treated until the HCG levels fall to normal and follow-up measurements are made at intervals over many years. This is done because the HCG excretion acts as a "biochemical marker of malignancy," since increased urine levels often precede clinical recurrence of tumor by a long period and early therapy is more likely to be of benefit.

(3) Tumors of the Testicle. Certain malignant tumors of the testicle produce markedly increased amounts of HCG, often enough to give a positive "pregnancy test." In the majority of instances these tumors are believed to be choriocarcinoma, but because of existing difficulties in histologic classification there is some question as to whether other embryonal tumors may not produce gonadotrophic hormones. As a general principle, the largest amounts of hormone are produced by the more embryonal types of tumor. Metastases are often accompanied by a marked increase in hormone concentration, while a decrease often follows removal, irradiation of the testicular tumor or chemotherapy. It must be emphasized that not all malignant tumors of the testis are associated with increased gonadotrophin values.

Certain types of bronchial carcinoma secrete gonadotrophin related to either LH or HCG rather than FSH. Such patients usually have gynecomastia and an elevated estrogen secretion. This type of "ectopic" hormone production by nonendocrine tissues will be dealt with in detail later in this section.

Decrease in Gonadotrophic Hormones. (1) Primary Pituitary Hypofunction. The blood and urine gonadotrophins are diminished or absent in panhypopituitarism, as are all anterior pituitary hormones. They are absent in Simmonds' disease (pituitary cachexia), with consequent atrophy and functional failure of the gonads. Chromophobe tumors of the hypophysis are almost invariably accompanied by a diminution in gonadotrophic hormones. Basophilic (Cushing's syndrome) and eosinophilic (acromegaly) tumors may show no change early in the course of their development, but later are usually associated with diminution in or absence of gonadotrophins.

The fact should be emphasized that many patients with obesity and hypogonadism, classified clinically as "hypopituitarism" or "Fröhlich's syndrome," may have normal or even excessive amounts of gonadotrophin, indicating that these are not true instances of hypopituitarism but should be classified as primary hypogonadism. There is no consistent pattern of gonadotrophin secretion in patients with simple obesity, normal, decreased or increased values having been obtained in these cases.

In the female, diminished or irregular gonadotrophin secretion is commonly observed in association with infertility and with various menstrual disturbances, including amenorrhea, menorrhagia and other types of functional bleeding. Consistent and repeated failure to demonstrate adequate amounts of gonadotrophins in the blood or urine of such patients is the only available method of establishing pituitary insufficiency as the underlying cause of the condition. This is of the utmost importance in arriving at an etiologic diagnosis and in instituting rational therapy.

Diminished or irregular gonadotrophin secretion occurs much less frequently in the male than in the female. Diminished pituitary function is not a common cause of male infertility or impotence, which are much more often due to testicular dysfunction or to systemic causes. However, hypogonadotrophic (primary pituitary) eunuchoidism is by no means rare. Inasmuch as these cases respond at times to gonadotrophin therapy, differentiation between primary pituitary and primary testicular insufficiency as the basis for eunuchoidism is of considerable practical importance. This can usually be accomplished by gonadotrophin assays in conjunction with testicular biopsy.

Malnutrition and various metabolic disorders may be followed by amenorrhea or other evidences of gonadal failure, due in some instances to diminished gonadotrophic activity. In our experience, the same mechanism may operate in cases of amenorrhea due to marked emotional disturbance.

During pregnancy, a decrease in gonadotrophin titer below the normal range for that period of gestation usually indicates placental failure. In cases of intrauterine fetal death or threatened miscarriage, the gonadotrophin content of the blood and urine usually falls progressively.

(2) **Suppression of Gonadotrophin Production or Release.** Excessive amounts of estrogen, androgen and, possibly, progesterone may inhibit the production or release of gonadotrophins by the pituitary. This phenomenon may be observed clinically following administration of large amounts of

TABLE 22-2 VARIATIONS IN GONADOTROPHIN VALUES

CONDITION	GONADOTROPHIN VALUES
Panhypopituitarism Simmonds' disease	Diminished or absent

Pituitary tumors
  Eosinophilic
    Gigantism
    Acromegaly
  Basophilic
    Cushing's disease
  Chromophobe
	Normal early
	Diminished later
	
	Normal early
	Diminished later
	Usually diminished or absent

Hypogonadism
  Primary gonadal failure
    Castration
    Atrophy, etc.
  Secondary (normal gonads)
	
	Increased
	
	Diminished

Menopausal syndrome	Increased
Male climacteric	Usually increased, but not so consistently nor so markedly as in women

Testicular tumors
  Chorionepithelioma
  Other embryonal tumors
  Interstitial cell tumors
Seminomas
	
	Increased, often markedly
	Moderate increase?
	
	Usually normal

Some carcinomas of bronchus with gynecomastia	Increased HCG or LH
Functional menstrual disorders	Variable. Depends on etiology, ovarian function. etc.

Pregnancy
  Normal
 	
  Hydatidiform mole
  Chorionepithelioma
  Intra-uterine fetal death
  Retained placental tissue
  Missed abortion
	
	Enormous increase. Varies with period of gestation
	Usually increased
	Progressive increase
	Progressive decrease
	Delayed fall
	Delayed fall

these hormones and during oral contraceptive medication. It occurs also during normal pregnancy, when the pituitary gonadotrophins are suppressed by the increasing level of estrogenic hormones. The same mechanism may be responsible for the low gonadotrophin values sometimes observed in patients with adrenocortical hyperfunction accompanied by high estrogen or androgen secretion from the adrenal cortex.

Thyroid-stimulating Hormone (Thyrotrophin)

Thyrotrophin (TSH) has been discussed on p. 749.

Assay. A variety of bioassay methods have been used: (1) determination of the number of colloid droplets in the cells of the guinea pig thyroid following injection of preparations containing TSH; (2) measurement of the effect of TSH on the release of radioactive thyroid hormone by mice whose own TSH secretion has been suppressed by thyroxine. The amount of ^{131}I-labeled thyroxine produced depends on the TSH activity injected.

However, the bioassays for TSH are not sensitive enough to measure circulating levels in normal subjects. Radioimmunoassay is therefore used, employing specific antihuman TSH antisera.

Actions. TSH is an important factor in the regulation of thyroid function. Removal of TSH results in atrophy of the thyroid gland and cessation of thyroxine formation. Secretion of this hormone by the anterior pituitary is controlled by the level of circulating thyroxine, increase in the latter depressing, and a decrease stimulating, TSH production.

Injection of TSH is followed by morphological and metabolic evidences of increased thyroid function. However, under ordinary circumstances, an early physiological action is probably to stimulate release of thyroid hormone from the intrafollicular thyroglobulin which is taken by pinocytosis into the follicular cell. Thyroid hormones are released by activation of a proteolytic enzyme which hydrolyzes thyroglobulin, the released hormones being tranported across the follicular cells into the blood stream. Administration of TSH, therefore, produces effects on the organism (except on the thyroid) identical with those produced by thyroxine and triiodothyronine.

Secretion and Metabolism. Secretion of TSH is under the control of thyrotrophin-releasing hormone (TRH) from the hypothalamus and of the circulating thyroid hormones, thyroxine (T_4) and triiodothyronine (T_3). There is good evidence for a circadian rhythm of TSH production, probably mediated by TRH, increased levels occurring approximately two hours or more before the onset of sleep, peaking between 2 and 4 A.M.

The rate of secretion of TSH varies in inverse relation to the concentration of thyroid hormones in the blood stream. This negative feedback mechanism provides for delicate and practically automatic maintenance of a normal level of circulating thyroid hormones in the presence of a normally responsive hypophysis and thyroid. Primary depression of thyroid activity (e.g., thyroidectomy, antithyroid agents) is accompanied by increased secretion of TSH and increase in size and number of pituitary basophils, many showing vacuolation ("thyroidectomy cells"). Conversely, TSH secretion is depressed in thyrotoxicosis.

Abnormal Thyrotrophin (TSH) Secretion. This is dealt with in the section on the thyroid (p. 765).

Prolactin

Prolactin is the lactogenic hormone whose main action is to initiate and sustain lactation. In most species, growth hormone and prolactin can easily be separated, but this has proved difficult in man. There is now, however, no doubt that prolactin is present as a separate protein in human pituitaries and blood.

Assay. Prolactin can induce milk secretion in cultured breast tissue from rabbits, and this has been used as a bioassay for the hormone. It is more sensitive than the older assay, based on the increase in the weight of the pigeon crop. Now, more specific radioimmunoassays for prolactin have been developed, which differentiate it clearly from growth hormone.

Secretion and Action. The secretion of prolactin by the pituitary is normally kept under tonic inhibition by secretion of a prolactin-inhibiting hormone. This is formed in the hypothalamus, stored in the median eminence and secreted into the portal capillaries of the pituitary stalk to reach the anterior pituitary cells. Cessation of secretion of prolactin-inhibiting hormone allows prolactin secretion. Normally, plasma prolactin levels are less than 25 ng./ml. and show a circadian rhythm, the highest levels being found during the night. Prolactin secretion is high in the neonate, which accounts for the common "witch's milk" of the newborn and in the mother during breast feeding, and there is a reflex secretion during stimulation of the breast. It is also secreted during stress.

In animals, prolactin causes milk secretion in the mammary gland that has already been primed by other hormones. Estrogens and progesterone are required for proliferation of the mammary ducts and alveoli. Cortisol must also be present for the ovarian steroids to be fully effective. Milk secretion into the alveoli and terminal ductules results from the action of prolactin, but oxytocin is required to cause contraction of the myo-epithelial cells of the alveoli and ductules, forcing milk into the larger ducts and cisterns. Although prolactin has a definite effect on the corpus luteum in lower species, this action does not appear to be present in man. Corpora lutea following ovulation induced by FSH and HCG in hypophysectomized women have a normal life span.

Prolactin has many metabolic effects on different tissues, similar to those of growth hormone. It causes nitrogen, potassium and phosphate retention and hypercalciuria. A hyperglycemic effect occurs in the hypopituitary subject, and in pituitary dwarfs animal prolactin can cause an increase in linear growth. The action of prolactin at a molecular level is not yet defined but it has been demonstrated to stimulate protein synthesis *in vitro*.

Prolactin secretion is probably maintained by the stimulus of suckling, i.e., afferent nervous pathways from the breast traversing the spinal cord to reach the posterior median eminence.

Overproduction (Galactorrhea). Irritation of the thoracospinal nerve segments after thoracotomy or herpes zoster can initiate persistent lactation, presumably by stimulation of the afferent pathway for prolactin release. Various drugs, such as reserpine, chlorpromazine and other centrally acting drugs, may cause lactation, probably by inhibiting hypothalamic centers which normally secrete prolactin-inhibiting hormone. A tumor in the pituitary-hypothalamic region can also initiate or prolong lactation.

Pathological lactation without suckling may occur spontaneously or follow pregnancy. Such patients are not infrequently found to have a pituitary tumor, and even if there is no enlargement of the sella turcica, careful follow-up is indicated. The amenorrhea that usually occurs is due to a lack of gonadotrophin secretion, which is a common accompaniment of hyperprolactinemia. Owing to a similar mechanism, prolonged lactation may also occur in women who have never been pregnant. Galactorrhea may occur in association with chromophobe, basophil and eosinophil adenomas of the pituitary. It rarely occurs in men. A variety of eponyms have been applied to the various syndromes associated with amenorrhea and galactorrhea. They serve no useful purpose, since one syndrome may develop into the other and all are due to hyperprolactinemia. Acromegaly or Cushing's disease may be found or may develop in patients who present with galactorrhea. Galactorrhea, with or without amenorrhea, may complicate or follow oral contraceptive medication.

Treatment of Inappropriate Lactation. Hyperprolactinemia and galactorrhea disappear if the precipitating drug can be stopped. In the presence of pituitary or hypothalamic disease, radiation, estrogen or clomiphene therapy usually does not work. Recently, treatment with the new ergot alkaloid 2-brom-α-ergocryptine has been shown to reduce elevated prolactin levels, stop the galactorrhea and usually to restore gonadotrophin secretion to normal. Sometimes, L-dopa therapy may improve hyperprolactinemia and inhibit milk production.

Underproduction. This is the cause of failure of lactation in postpartum pituitary necrosis and other causes of panhypopituitarism. In patients with selective failure of gonadotrophin production, e.g., as a result of anorexia nervosa, lactation but not menstruation may follow pregnancy induced by gonadotrophin therapy.

Adrenocorticotrophin (ACTH) and Melanocyte-Stimulating Hormones (α- and β-MSH)

Human ACTH is a single-chain polypeptide of 39 amino acids, with a molecular weight of approximately 4500. It has been completely synthesized, and the amino acid sequences essential for ACTH action have been characterized. The first "amino-terminal" 24 amino acids of ACTH are required for its action on the adrenal cortex, to promote synthesis of corticosteroids—mainly cortisol (hydrocortisone) in man, but also some corticosterone—and to increase the blood flow through the gland. The melanocyte-stimulating hormones, α- and β-MSH, are peptides, closely related to the steroidogenic 1-24 portion of the ACTH molecule. α-MSH has 13 amino acids and the sequence is identical with the first 13 amino acids of ACTH. Specific radioimmunoassay studies suggest that α-MSH is found only in the pituitary gland and does not appear to circulate in the blood. Its functional significance is unknown. Human β-MSH contains 22 amino acids, of which numbers 11 to 17 are the same as 4 to 10 of ACTH and α-MSH. β-MSH circulates in the blood and appears to be secreted under all circumstances together with ACTH, and indeed it is synthesized in the same basophilic cells of the anterior pituitary. β-MSH has little trophic effect on the adrenal cortex, and ACTH has only 3 per cent of the pigmentary activity of β-MSH. The increased pigmentation associated with diseases in which ACTH secretion is excessive, for example, primary adrenocortical insufficiency (Addison's disease), is due to the parallel secretion of β-MSH, which stimulates the melanocytes of the skin, most characteristically in the exposed parts of the body, the genitalia, the pressure points and the buccal mucosa.

The function of the nonsteroidogenic "carboxyl-terminal" portion of the ACTH molecule, amino acids 25 to 39, is unknown. It is within this portion that the differences lie between ACTH of different species, although such differences are small. Several "extra-adrenal" actions of ACTH have been demonstrated *in vitro*, such as increased lipolysis, and some have been ascribed to this C-terminal part of the molecule. However, there is no reliable evidence to suggest that these actions are important in man.

Secretion and Transport. ACTH release from the pituitary is mediated by specific polypeptides produced in the hypothalamus, especially in the median eminence, which reach the anterior pituitary by way of the local portal circulation. These corticotrophin-releasing factors (CRF) have been designated α-CRF and β-CRF. The β-CRF resembles vasopressin and is the more active; α-CRF, which can be further separated into $α_1$- and $α_2$-CRF, has a structure similar to that of α-MSH. It is still uncertain whether they represent the true CRF. The hypothalamic centers are, in turn, controlled by brainstem and suprahypothalamic centers. Distinct hypothalamic sites are responsible for the responses to high and low circulating steroid levels in the negative feedback system and to stressful stimuli, e.g., surgical operations, hypoglycemia or fever. There is also a circadian rhythm of ACTH secretion initiated by the hypothalamus, which leads to changes in cortisol output from the adrenal: the plasma cortisol is at its peak at about the time of waking, at its lowest at about the time of retiring and normally begins to rise at about 3 A.M. Disturbances of the rhythm occur in Cushing's syndrome, heart failure, depressive illness and other forms of stress, but not in Addison's disease.

The concentration of ACTH in the circulation can be determined by bioassay and by radioimmunoassay procedures. In man the normal level in venous blood between 8 and 10 A.M. ranges from 10 to 70 pg./ml. of plasma. ACTH disappears rapidly from the circulation with a half-time of 5 to 20 minutes.

Metabolic Effects. Many theories have been offered to explain the action of ACTH on the adrenal cortex, but none is fully adequate. Synthesis of cortisol and some corticosterone is stimulated and adrenal weight and vascularity increased. It has been suggested that ACTH, via cyclic AMP, affects the mitochondrial membrane, increasing the rate of removal of pregnenolone from the mitochondria. All the subsequent reactions of steroidogenesis except 11-hydroxylation occur outside the mitochondria; hence, the rate of formation of steroids depends on the availability of pregnenolone. Again, pregnenolone inhibits the transformation of cholesterol to 20 α-hydroxycholesterol, the first reaction in its own formation, and removal of pregnenolone from the mitochondrion reduces this inhibition. Certainly, the first step in the action of ACTH is to attach itself to specific membrane receptors on the surface of the adrenocortical cell and activate the cyclic AMP mechanism. Further studies on the mechanism of action of ACTH should indicate whether this proposed scheme is a valid one. Further discussion of corticosteroid synthesis will be found later in this chapter where the adrenal gland is reviewed.

ACTH also has numerous extra-adrenal actions which have been documented in adrenalectomized animals, such as mobilization of fat from adipose tissues, but not as yet in humans. In man, the only effect so far which has proved to be of clinical significance is the melanocyte-stimulating action which causes the pigmentation seen in some patients with Cushing's disease, Addison's disease and congenital adrenal hyperplasia. Increased pigmentation is frequently seen when plasma ACTH levels exceed 500 pg./ml. However, since β-MSH is always secreted in excess along with ACTH, and it is much more pigmenting than ACTH, β-MSH seems to be the hormone principally responsible for the pigmentation. In animals, MSH causes dispersal of melanin granules in the melanocytes, thereby causing darkening of the skin. Such dispersal of granules may occur in man, but MSH may also act by stimulating melanin formation.

Assays for ACTH. Specific bioassays for ACTH are available, which depend on the measurement of the amount of corticosterone produced either in whole, hypophysectomized rats or in rat adrenocortical cells suspended in buffered saline. However, such techniques are difficult and expensive and often require the extraction of large volumes of plasma to measure circulating levels of ACTH. For this reason, radioimmunoassays of ACTH are usually used since they are sensitive, require small volumes of plasma and can be done with speed. However, the results may not always reflect biological activity (i.e., steroidogenic activity), since non-biologically active fragments of ACTH may circulate which can still react with the anti-ACTH antiserum.

Abnormal Adrenocorticotrophin (ACTH) Secretion. Increased ACTH secretion, and consequently glucocorticoid (mainly cortisol) secretion, occurs in normal people if they are stressed. Such stress may be physical, e.g., fever, pain, infection, or psychological, e.g., fear, depression. If such an increase in cortisol production does not occur, as in patients with a pituitary tumor or destroyed adrenal glands, then the subject may go into a state of shock and circulatory failure, with hypotension, tachycardia and heart failure, and this may be fatal. Excess secretion of ACTH may produce Cushing's disease, due to oversecretion of cortisol, although hypercortisolism can also be due to adrenocortical tumor (see later in this section). When the adrenal glands have been destroyed (e.g., by tuberculosis) and cortisol levels in the blood are too low, ACTH and MSH secretion are greatly increased and pigmentation occurs. Similarly, elevated ACTH levels are seen in untreated congenital adrenal hyperperplasia in children. When corticosteroid treatment in high dosage has been given to patients for many months, the adrenals atrophy and ACTH secretion is suppressed and may remain so for some time (weeks or months) after the treatment is withdrawn. The patient is at risk of collapsing if stressed.

Growth Hormone (Somatotropin)

Growth hormone (somatotropin) has been prepared in crystalline form from the hypophyses of several species, including man. All of these are polypeptides, with different molecular weight and exhibiting immunological, structural and physicochemical differences. These differences are reflected in species differences in biological activity. Beef growth hormone is ineffective in man and monkey but is biologically active in rats and fish. Monkey somatotropin, however, is effective in cattle. Human and monkey somatotropins are effective in both of these species as well as in the rat. The nature of the biological activity, when present, is essentially the same in all species. This activity probably resides in an identical core of amino acid sequences, as in the case of other hypophyseal hormones. Growth hormone makes up between 5 and 10 per cent of the dry weight of the human pituitary gland; it is a protein with a molecular weight of 21,000, consisting of 190 amino acids. Clear evidence is now available that prolactin exists separately from growth hormone in man. Owing to the similarities in structure and biological actions and to the fact that prolactin is present in the human pituitary gland in much smaller quantities than growth hormone, their separate identities were doubted for many years. However, with the advent of specific radioimmunoassays for the two peptides, prolactin and growth hormone can be shown to be secreted separately in human subjects and to be controlled by different mechanisms. Nevertheless, growth hormone appears to have some intrinsic lactogenic activity and prolactin, some growth-promoting action, although trace contamination of one with the other cannot be completely excluded in apparently pure preparations.

Human chorionic somatomammotropin (HCS, placental lactogen, or HPL) is the term applied to a protein produced by the placenta that has similarities in structure and action to growth hormone and prolactin. It appears in placental tissue by the ninth week of pregnancy and its concentration in the blood can be high enough to interfere with radioimmunoassays of growth hormone. Its function has not yet been established. Measurement of the excretion of urinary placental lactogen in middle and late pregnancy provides a reliable test of placental function; low levels suggest placental insufficiency.

Secretion and Transport. Sensitive and specific methods are now available for the measurement of growth hormone in blood. Immunoassay, using radioiodine-labeled human growth hormone, is now a standard procedure in most centers. Antiserum to human growth hormone (HGH) is prepared in animals. Labeled HGH is incubated with anti-HGH in the presence of standard amounts of hormone or the serum to be assayed. The amount of HGH in the standard or unknown will vary the amount of labeled HGH bound to antiserum, since competition between them for binding to the antibody will occur. The free and bound HGH can then be separated and a standard curve produced, from which the amount of hormone in the serum specimen can be calculated.

The rate of secretion of growth hormone varies widely throughout the day and responds to many metabolic changes. Normally, growth hormone is secreted in short bursts lasting one to two hours, and these bursts occur mainly during the first half of the night, the period of deep sleep. They are infrequent at other times. Secretion of growth hormone is stimulated by stress, a fall in blood sugar, prolonged fasting, some amino acids, especially arginine, exercise, and sleep. Glucose causes prompt inhibition of growth hormone secretion by the pituitary. Protein administration has a similar but less marked effect. Corticosteroids reduce growth hormone output from the pituitary and also oppose its action at the tissues, which provides a possible explanation for the reduction in growth in children receiving these compounds. The total daily secretion of growth hormone in children and adolescents is greater than that in adults, and many more secretory bursts are found. Serum growth hormone levels in nonfasting adults are variable, usually being less than 10 ng./ml. (1 nanogram is 1×10^{-9} g.). Endogenous growth hormone leaves the circulation rapidly, the half-life being about 20 to 25 minutes. However, the metabolic effects must last much longer, for increased growth in growth hormone-deficient subjects can be maintained with only one or two injections each week.

Growth hormone release from the pituitary can be stimulated by insulin-induced hypoglycemia, by arginine infusions, and by pyrogen. Hypoglycemia and pyrogen also release ACTH but do not affect TSH or the gonadotrophins; arginine also causes a rise in plasma insulin and glucagon by a direct action on the pancreas.

The hypothalamus controls growth hormone synthesis and release by means of two humoral agents, growth hormone-releasing hormone and growth hormone-release inhibiting hormone, which are polypeptides similar to other hypothalamic hormones affecting anterior pituitary hormone formation. The metabolic factors affecting growth hormone release probably act largely at a hypothalamic level.

Growth hormone in blood appears to be transported in bound form, but there is no agreement on the nature of the carrier protein, since both albumin and an α_2-macroglobulin may be involved.

Metabolic Effects. The mechanism by which growth hormone produces its metabolic effects is not yet fully understood but appears to involve the production of an intermediate or large polypeptide, somatomedin or the "sulfation" factor, probably by the liver. This product appears to be related to growth of tissues.

The effect of growth hormone on protein synthesis may, in part, be mediated by an increased transport of certain amino acids into the cell, though the hormone can also stimulate protein synthesis in cell-free preparations. Growth hormone can still exert its effects when messenger-RNA synthesis is blocked by actinomycin and when new protein synthesis is inhibited by puromycin, so it is unlikely that the primary site of action of the hormone is on m-RNA. It may in some way affect the efficiency of the ribosomes, the site of protein synthesis.

Protein Metabolism. Growth hormone is a protein-anabolic principle. It induces a positive nitrogen balance, accompanied by increase in tissue nitrogen and decrease in urea nitrogen, amino nitrogen and total NPN of the blood plasma and urine. Catabolism of amino acids is apparently retarded and their incorporation into body proteins accelerated. This anabolic action of growth hormone is facilitated also by the presence of normal amounts of adrenocortical hormones and thyroxine and is accompanied by simultaneous stimulation of fat and carbohydrate metabolism.

Carbohydrate Metabolism. Hypophysectomy leads to the following changes in carbohydrate metabolism: (1) tendency to hypoglycemia on fasting; (2) decrease in liver and muscle glycogen on fasting; (3) increased sensitivity to insulin; (4) increased utilization of carbohydrate; (5) amelioration of diabetes in depancreatized or alloxanized animals. Certain of these manifestations are due in part to adrenocortical hypofunction resulting from loss of ACTH. However, they are completely reversed only by administration of growth hormone (or crude pituitary extracts) in addition to adrenocortical hormones (or ACTH).

Conversely, administration of growth hormone produces the following changes in carbohydrate metabolism, the effects varying in different species: (1) hyperglycemia (dog, cat) and aggravation of diabetes in depancreatized hypophysectomized animals ("diabetogenic" effect); (2) inhibition of insulin action ("anti-insulin" effect), with decreased utilization of carbohydrate and lowering of the respiratory quotient; (3) increase in muscle glycogen in hypophysectomized animals ("glycostatic" effect). The permanent diabetes produced in certain species (e.g., dog) by growth hormone is due to destruction of the islands of Langerhans. This effect is probably due to the primarily induced hyperglycemia which, in turn, causes hyperactivity and hyperplasia of the islet beta-cells and, subsequently, functional exhaustion and atrophy.

The mechanism of action of growth hormone in producing these effects has not been established definitely. The concept that it inhibits the hexokinase reaction is still controversial. However, uptake of glucose by muscle (diaphragm) is inhibited by growth hormone (and adrenocortical hormones) and increased by insulin. It has been suggested that whereas this action of insulin is exerted on the "glucose transfer mechanism," that of growth hormone may be due to stimulation of production, perhaps in the liver, of a protein which inhibits the glucokinase reaction, possibly somatomedin.

Administration of growth hormone causes increased output of glucose by the liver. Inasmuch as the turnover of glucose in the intact organism is simultaneously increased, any postulated peripheral inhibition of glucose utilization must be overcome by this provision of increased amounts of glucose or by the operation of other components of the homeostatic mechanism. There is suggestive evidence that growth hormone may depress glucose-6-phosphate dehydrogenase activity, with consequent decreased oxidation of glucose via the hexose monophosphate pathway.

Lipid Metabolism. Administration of pituitary extracts stimulates mobilization of fat from the fat depots, with consequent decrease in carcass fat and increase in plasma and liver lipids and increased fatty acid oxidation and ketogenesis. Adrenocortical hormone is necessary for this fat-mobilizing and -catabolizing effect, which is facilitated also by simultaneous depression of carbohydrate utilization. Conversely, hypophysectomy results in retardation of mobilization of depot fat and amelioration of ketosis in diabetic animals.

There is evidence that another fat-mobilizing principle may be elaborated by the hypophysis, which produces a sustained rise in the level of total fatty acids and of free and ester cholesterol in the blood plasma. Release of this substance is presumably stimulated by adrenal corticoids.

Excessive Growth Hormone Secretion. Hyperfunction of the eosinophilic cells of the anterior pituitary (usually eosinophilic adenoma) results in gigantism (longitudinal growth) if it occurs before the epiphyses have closed, and in acromegaly (thickening of bones and soft tissues) if it occurs in older individuals.

Diminished glucose tolerance is demonstrable in the majority of cases of acromegaly at some time during the course of the disease. Frank diabetes, i.e., fasting hyperglycemia and glycosuria, occurs in 15 to 20 per cent of cases. This is the clinical corollary to diabetes produced by growth hormone in certain experimental animals. In late stages, manifestations of hypopituitarism may supervene. In such cases, the blood sugar tends to fall, and the patient is unduly sensitive to insulin.

Increased urinary calcium excretion is common in acromegaly, although the cause is unknown. Less often, the serum calcium is elevated, owing to coincidental hyperparathyroidism. Organic phosphorus levels are often elevated. Bone texture is normal, although the joints become involved in extensive osteoarthrosis. There is marked thickening and overgrowth of bones, especially of the skull vault, sinuses, supraorbital ridges, lower jaw and vertebrae (producing a kyphosis). However, the excessive bone growth is only one feature of the excessive growth of all tissues, so that the skin and subcutaneous tissues and all the viscera are overgrown, including the thyroid, heart, liver, intestines, etc. Despite the goiter, the incidence of hyperthyroidism is not increased.

The pituitary tumor may have local effects, such as pressure on the visual pathways and headache, or it may result eventually in hypopituitarism.

Decreased Growth Hormone Secretion. This is present characteristically in panhypopituitarism, in association with deficiency in adrenocorticotrophic, thyrotrophic, and gonadotrophic hormones. When it occurs in children (usually due to craniopharyngioma, although growth hormone secretion can be deficient in isolation without an obvious organic lesion of the pituitary or hypothalamus), deficiency in this factor is reflected in dwarfism. In adults, it may be due to Sheehan's syndrome (postpartum necrosis of pituitary), chromophobe adenomas of the pituitary, and suprasellar tumors encroaching upon the pituitary, e.g., craniopharyngiomas, gliomas, aneurysms, and meningiomas.

In such cases, as well as in pituitary dwarfism, metabolic abnormalities resulting from growth hormone deficiency (except dwarfism in children) are overshadowed and obscured by the associated states of hypoadrenalism, hypothyroidism, and hypogonadism. The diagnostic approach, from the standpoint of investigation of hormonal abnormalities, is directed to studies of deficiencies in function of the target organs of the trophic hormones rather than to the consequences of growth hormone deficiency. In dwarfed children, however, it is important to demonstrate that the patient cannot secrete growth hormone normally, e.g., in response to a stimulation test such as insulin-induced hypoglycemia, since then human growth hormone therapy will result in resumption of growth. Other forms of dwarfism, with normal growth hormone secretion, do not respond to growth hormone administration.

POSTERIOR LOBE (NEUROHYPOPHYSEAL) HORMONES

Injection of an extract of the posterior lobe of the pituitary (neurohypophysis) produces three well-recognized effects: (1) increase in blood pressure (pressor effect); (2) contraction of the mammalian uterus (oxytocic effect); (3) decreased urine volume in mammals, with increased concentration of urinary solutes (antidiuretic effect). A fourth effect, induced in lower vertebrates, i.e., melanophore expansion or melanophore dispersion, is due to a principle formed in the pars intermedia but present as a contaminant in unfractionated extracts of the posterior lobe. In animals that have no pars intermedia (e.g., whale, chicken, man), this principle occurs in extracts of the anterior lobe only.

The pressor and antidiuretic effects are produced by the same substance, vasopressin, and the oxytocic effect by a second principle, oxytocin, separable from the former. Both are octapeptides, containing eight different amino acids, six of which are identical in the two hormones (Fig. 22-2).

The neurohypophysis consists of three parts, the median eminence, the infundibular stem and the infundibular process or neural lobe. It is composed of nerve cells and their fibers, neuroglia, blood vessels and supporting connective tissue. Nerve fibers reach it from the supraoptic and paraventricular nuclei (SON and PVN), ending in close apposition to capillaries. The cells of the SON and PVN are neurosecretory, synthesizing oxytocin,

CYS-TYR-PHE-GLN-ASN-CYS-PRO-ARG-GLYamide

Arginine vasopressin

CYS-TYR-ISOLEU-GLN-ASN-CYS-PRO-LEU-GLYamide

Oxytocin

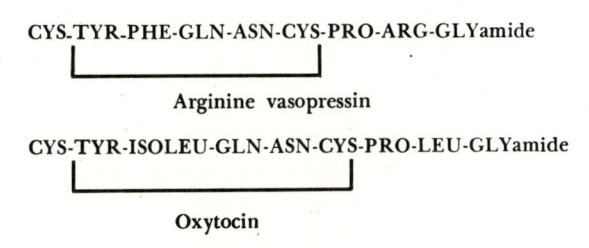

Figure 22-2. Structures of arginine vasopressin and oxytocin.

vasopressin and neurophysin, which pass down their nerve fibers to be stored and released from the posterior lobe. It is likely that the SON is mainly responsible for the synthesis of vasopressin, and the PVN for oxytocin, though it is not yet known whether individual neurons secrete one or both hormones. The SON and PVN, their nerve fibers and the neurohypophysis can be considered as one functional hypothalamoneurohypophyseal system (HNS). The function of neurophysin is unclear. It binds oxytocin and vasopressin and exists in the stores of these hormones in the granules of the neurohypophysis and appears to be released into the circulation with them.

Vasopressin

Vasopressin (antidiuretic hormone) is, like oxytocin, a cyclic octapeptide with a molecular weight of about 1000, consisting of an S-S bonded ring of five amino acids and a tail of three amino acids. Arginine vasopressin is the antidiuretic hormone in man and other mammals, apart from the pig, the peccary and the hippopotamus, where lysine vasopressin is found. Lysine vasopressin is more stable than arginine vasopressin and, since its synthesis was achieved, its main uses have been in the treatment of diabetes insipidus and in testing the integrity of the anterior pituitary by causing release of ACTH. The close chemical similarity of vasopressin and oxytocin explains the overlap of their biological actions. Basic amino acids (lysine or arginine) in position 8 and a phenylalanine group in position 3 increase antidiuretic activity.

The main actions of vasopressin are to promote the renal tubular reabsorption of water and to stimulate smooth muscle contraction, though it is unlikely that vasopressin has any physiological role in the maintenance of blood pressure. In pharmacological doses it causes pallor, coronary vasoconstriction and contraction of smooth muscle in the intestine.

Vasopressin exerts its antidiuretic action on the distal part of the nephron by increasing the osmotic permeability of the distal tubular and collecting duct cells. The hormone acts at the level of the cell membrane where it can be considered to increase the mean pore diameter. It has been suggested that vasopressin may also act on sodium transport from the ascending limb of the loop of Henle, accounting for the buildup of sodium concentration in the renal medulla, but this action has not yet been demonstrated. The disulfide linkage of the peptide may be the means of attachment to the target organ, since blockage of these groups prevents the binding and action of the hormone.

Each day approximately 70 to 100 liters of fluid, iso-osmotic with plasma, are filtered by the glomeruli. Of the filtered water, 85 per cent is reabsorbed passively by the proximal tubules, along with the active reabsorption of solutes, especially sodium and chloride, without the aid of vasopressin. The urine, therefore, remains iso-osmotic with plasma. The long U-shaped loops of Henle form a countercurrent concentrating system in which the ascending limb actively transports sodium to the interstitial fluid of the renal medulla, rendering this region hypertonic and the urine reaching the distal tubule, hypotonic. In the absence of vasopressin, the distal tubules and collecting ducts are impermeable to water, and hypotonic urine is produced in large amount. Vasopressin, by increasing the permeability of these segments of the nephron to water, permits the water to pass from the hypotonic tubular fluid and pass along the concentration gradient. The renal tubules may be unresponsive to vasopressin because of a genetic defect or as a result of potassium depletion, hypercalcemia or amyloidosis.

Release of vasopressin results from a rise in plasma osmolality and from a fall in plasma volume. The osmoreceptors are located somewhere in the vascular bed supplied by the internal carotid artery, possibly in an adjacent but separate area from the SON. Reduction of plasma volume appears to stimulate vasopressin release even if the plasma is hypo-osmolar. Again, the site of the volume receptors has not been accurately determined, but probably they are in the great veins of the thorax. Emotional factors are also potent triggers for vasopressin release, as are nicotine, morphine and ether; alcohol has the opposite effect.

Abnormal Vasopressin Secretion.

Deficiencies of vasopressin, due to disease in the region of the hypothalamus and the pituitary, causes diabetes insipidus. In this disease, large volumes of urine (up to 25 liters daily) and excessive thirst and dehydration occur because the urine cannot be concentrated. If glucocorticoid deficiency is also present due to anterior pituitary diseases then polyuria does not occur since glucocorticoids are required to excrete a water load. Excessive vasopressin secretion may occur from some lung tumors. This ectopic vasopressin secretion is dealt with later.

The diagnosis of diabetes insipidus is suggested by the history of excessive fluid intake and output, but this condition must be differentiated from other causes of polyuria, e.g., diabetes mellitus, renal disease, and psychogenic polyuria and polydipsia. Differentiation from other organic diseases is usually easy, but that from "nephrogenic diabetes insipidus," i.e., congenitally vasopressin-unresponsive kidneys, and from psychogenic polydipsia is at times difficult. In the latter condition, however, if fluids are withheld (which for obvious reasons is difficult to enforce and must be done under strictest supervision), the urine will be concentrated as in normal individuals because there is no deficiency in antidiuretic hormone. Diagnosis is usually made by withholding fluids for up to eight hours and measuring the plasma and urinary osmolality during this time. Normal subjects can concentrate their urine to twice the osmolality of their plasma, which should not rise to more than 300 mOsm./kg. Failure to achieve this suggests diabetes insipidus, and subsequent urine concentration after administration of vasopressin excludes renal disease and confirms the presence of cranial diabetes insipidus. Severe dehydration must be avoided, as it may be dangerous, and the test should be supervised carefully and consideration given to terminating it if the patient loses more than 3 per cent of his body weight. Diabetes insipidus is treated with natural or synthetic vasopressin.

Oxytocin

Oxytocin is an octapeptide, similar in structure to vasopressin with which its biological properties overlap. The actions of oxytocin are largely confined to the uterus and the breast in animals, but it is not clear what role this hormone plays in man.

Uterine Actions. The uterus is more sensitive to oxytocin during the follicular stage of the cycle, and stimulation of uterine contraction may aid the transport of spermatozoa to the fallopian tubes. During human pregnancy and labor, oxytocin is difficult to detect except during very brief bursts of uncertain significance, even using the most sensitive and specific radioimmunoassays; however, the hormone is present in the pituitary and hypothalamus of men and women. During pregnancy, oxytocin levels may be balanced by a rise in the degrading enzyme oxytocinase and progesterone, which may relax uterine muscle. The role of oxytocin in the induction of labor is still controversial but the uterus is certainly very sensitive to oxytocin at the end of pregnancy, particularly if prostaglandin E or $F_2\alpha$ is also given. At term, labor can be induced by administration of these alone or in combination. Oxytocin acts on the myometrial cell membrane, rendering it more permeable to potassium, lowering the membrane potential and increasing excitability. From experiments on rabbits, it has been suggested that mechanical dilatation of the uterus, cervix or vagina reflexly stimulates oxytocin release and, hence, uterine contractions. In women with diabetes insipidus and hypothalamic disease, labor is usually normal, however.

Breast Actions. Ejection of milk from the breast is due to a neurohumoral reflex. Afferent stimuli from the nipple during suckling travel along sensory pathways via the spinal cord to the hypothalamus. Oxytocin is subsequently released from the pituitary and transported to the breast, where it stimulates contraction of myo-epithelial cells around the alveoli, causing milk ejection. The latent period of 30 seconds between the onset of suckling and milk release is largely due to the time involved in oxytocin release and transport. There is some controversy as to whether or not oxytocin itself initiates the release of prolactin necessary for the maintenance of lactation; for although suckling is required to maintain lactation, it is not essential to postulate that this effect is mediated by oxytocin. Drugs such as reserpine and chlorpromazine hydrochloride, which can induce prolactin secretion, actually inhibit the release of oxytocin that normally follows suckling. Secretion of prolactin during lactation is probably maintained by the removal of hypothalamic inhibition.

Oxytocin excess or deficiency has not yet been associated with any disease process, though oxytocin has been isolated from some nonendocrine tumors.

STEROID HORMONES

The term "steroid" is applied to the members of a group of compounds which have in common the cyclopentenophenanthrene structure (Fig. 22-3). This consists of a cyclopentane ring (D), fused to a completely hydrogenated phenanthrene (A-B-C). A carbon atom is present at every junction point in this ring system, its valence requirements (i.e., 4) being satisfied by H atoms. Substituents in the nucleus and on side chains are located by a standard system of numbering the carbon atoms, indicated in Figure 22-3.

The steroid hormones include the adrenal cortical hormones, estrogens, progesterone, and androgens. They are related structurally, and metabolically, to cholesterol.

ADRENAL GLAND

Anatomy and Embryology. The adrenal cortex and medulla have separate embryological origins, the cortex being derived from mesoderm and the medulla from ectoderm. In mammals, the cortex encloses the medulla, but in fish, the cortex and medulla exist as distinct entities. The fetal cortex depends on ACTH stimulation, for it is absent in the anencephalic fetus but can be restored by administration of ACTH. By the third week of life, the adrenal weight is half that at birth, but it is not until the third year of life that the permanent cortex is fully differentiated into the three zones (see below).

The adrenal medulla is, like the sympathetic neurons, derived from ectodermal cells of the neural crest. These chromaffin cells, so called because they stain brown with chromium salts, invade the medial side of the cortex at about the seventh week of intrauterine life, to take up their central position within the gland.

In normal children up to 12 years, each adrenal weighs 1.5 to 3 g. In adults, the average weight of the normal gland is 4 g., with an absolute range in both sexes of 2 to 6 g., but at autopsy the weight varies depending on the cause of death. The lower part of each adrenal contains more medulla than the upper part, the cortex-medullary ratios being 5.1 and 18.1, respectively.

Hormones of the Adrenal Cortex and their Measurement. Although more than 40 steroids have been isolated from the

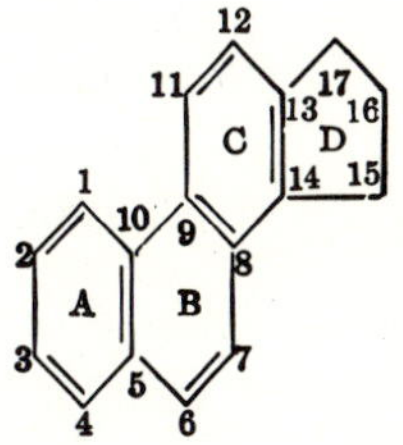

Figure 22-3. Cyclopentenophenanthrene nucleus.

human adrenal cortex, the main ones found in adrenal vein blood are cortisol, corticosterone, aldosterone, dehydroepiandrosterone (DHA), androstenedione and 11-hydroxyandrostenedione. The human adrenal secretes each day about 25 mg. cortisol, 2 mg. corticosterone, 200 µg. of aldosterone and 25 mg. of DHA.

These hormones can be grouped into three types according to their main metabolic activities:

1. Glucocorticoids, prinicpally cortisol, and small amounts of corticosterone.
2. Mineralocorticoids, principally aldosterone, and small amounts of deoxycorticosterone.
3. Androgens, e.g., dehydroepiandrosterone (DHA), androsterone and testosterone.

In addition, small amounts of estrogens and progesterone are produced.

Structure of Steroid Hormones. Natural and synthetic steroids contain the cyclopentenophenanthrene nucleus described earlier and shown in Figure 22-3 with differing side chains. The carbon atoms are numbered as shown in Figure 22-3. The simplest method of classifying natural steroids is by the number of carbon atoms in the molecule. Corticosteroids, a term which includes glucocorticoids and mineralocorticoids, are C_{21} compounds. Androgens are C_{19} compounds, and estrogens, C_{18} derivatives (Fig. 22-4).

Stereoisomerism related to rings A and B is termed "cis" and "trans," rings C and D always being in the "trans" position in biological compounds. Any substitution in rings A and B is related to the methyl group at C-10, which is considered to be above the plane of the ring. Alpha groups lie below the plane and are linked by a dotted line; beta groups lie above the plane and are linked by a solid line. All active corticosteroids have a 17 α-hydroxyl group and a 21 β-alpha ketol grouping

$$\begin{array}{cc} CH_2 & C- \\ | & \parallel \\ OH & O \end{array}$$

at carbon 17. The 11-hydroxyl group of the natural corticosteroids is in the β-position. Desoxycorticosterone, which lacks a hydroxyl group at C-11, is only a very weak glucocorticoid. All active steroids (except DHA) have a double bond between carbon atoms four and five (Δ^4) and an oxygen atom attached to C-3. Aldosterone has an aldehyde group (— CHO) attached to C-18 and is capable of existing in two forms; in one form, the 11-hydroxyl and 18-aldehyde are linked to form the hemiacetal (Figs. 22-4 and 22-5). The natural hormone is d-aldosterone. The androgens, which contain 19 carbon atoms, have an oxygen atom at C-17, hence their designation 17-oxosteroids (also called 17-ketosteroids).

Biosynthesis of Adrenocortical Hormones (Fig. 22-5). Administration of adrenocorticotrophic hormone (ACTH) is promptly followed by increased secretory activity of the adrenal cortex and increased excretion of certain adrenal hormones and their metabolites in the urine. Simultaneously, there is a decrease in the cholesterol and, in most species,

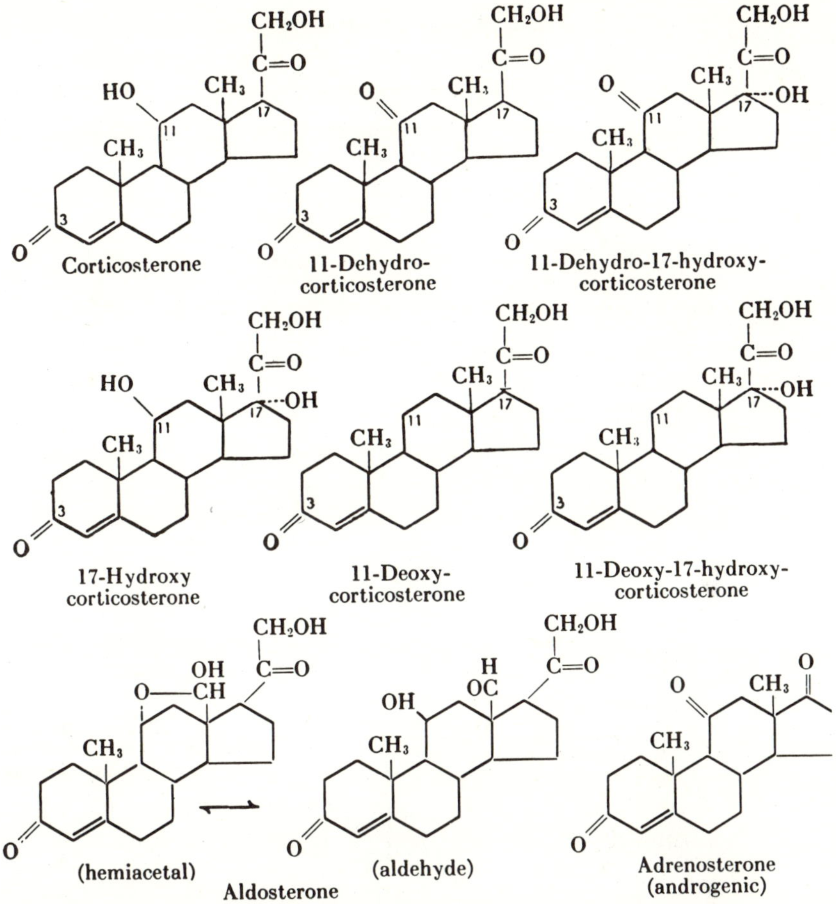

Figure 22-4. Glucocorticoids, mineralocorticoids and androgens derived from the adrenal cortex.

also in the ascorbic acid content of the adrenal cortex. This suggests that cholesterol may be a biological precursor of at least certain of the adrenocortical steroids. Ascorbic acid is not an obligatory participant in this process.

Findings with adrenal perfusion experiments and studies of adrenal-vein blood and of urine indicate that corticosterone, 17-hydroxycorticosterone (cortisol or hydrocortisone), and aldosterone are the most important end-products of corticosteroid biosynthesis in mammalian species, the latter two predominating in man. Present understanding of the pathways of biosynthesis of these hormones is indicated in Figure 22-5 and has contributed greatly to understanding of the pathological physiology of certain clinical forms of adrenal hyperfunction, particularly those accompanied by virilism (p. 735).

The primary compound formed by the adrenal, from either acetate or cholesterol, is apparently pregnenolone, Ring A of which undergoes oxidation, forming progesterone (Fig. 22-6), with the α-,β-unsaturated

ENDOCRINE FUNCTION

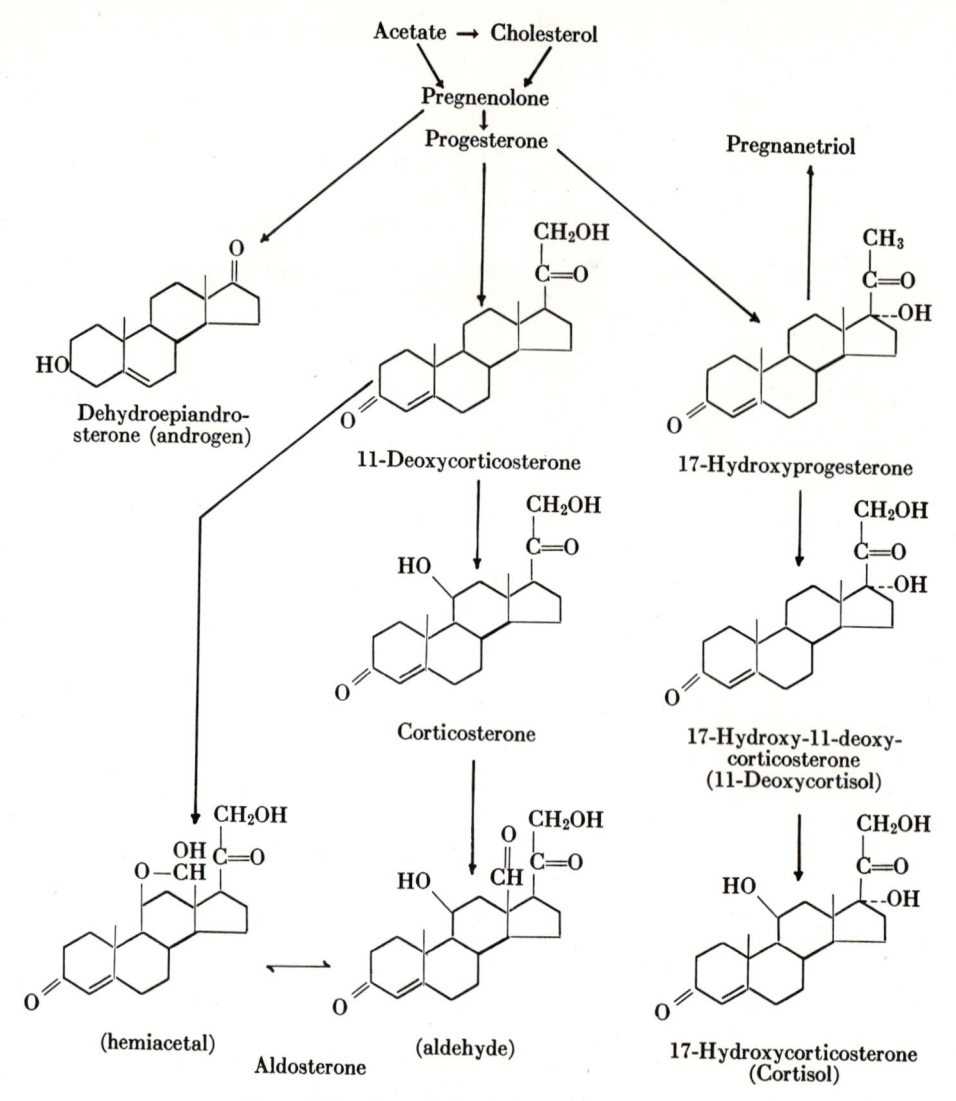

Figure 22-5. Biosynthesis of adrenocortical hormones.

ketonic structure essential for the characteristic biological activities of the corticoids. Two metabolic pathways are taken by progesterone in its subsequent transformations (Fig. 22-5).

(1) The major pathway (in man) involves: (*a*) 17-hydroxylation (i.e., of the 17-C atom) to 17-hydroxyprogesterone; (*b*) 21-hydroxylation of the latter to 11-deoxycortisol; (*c*) 11-hydroxylation of the latter to cortisol.

(2) A minor pathway (in man) involves, successively: (*a*) 21-hydroxylation of progesterone to deoxycorticosterone; (*b*) 11-hydroxylation to corticosterone; (*c*) conversion of the C-18 methyl to an aldehyde, forming aldosterone.

It may be seen that enzymes 17-hydroxylase, 21-hydroxylase and 11-hydroxylase occupy key positions in the biosynthesis of the adrenocortical

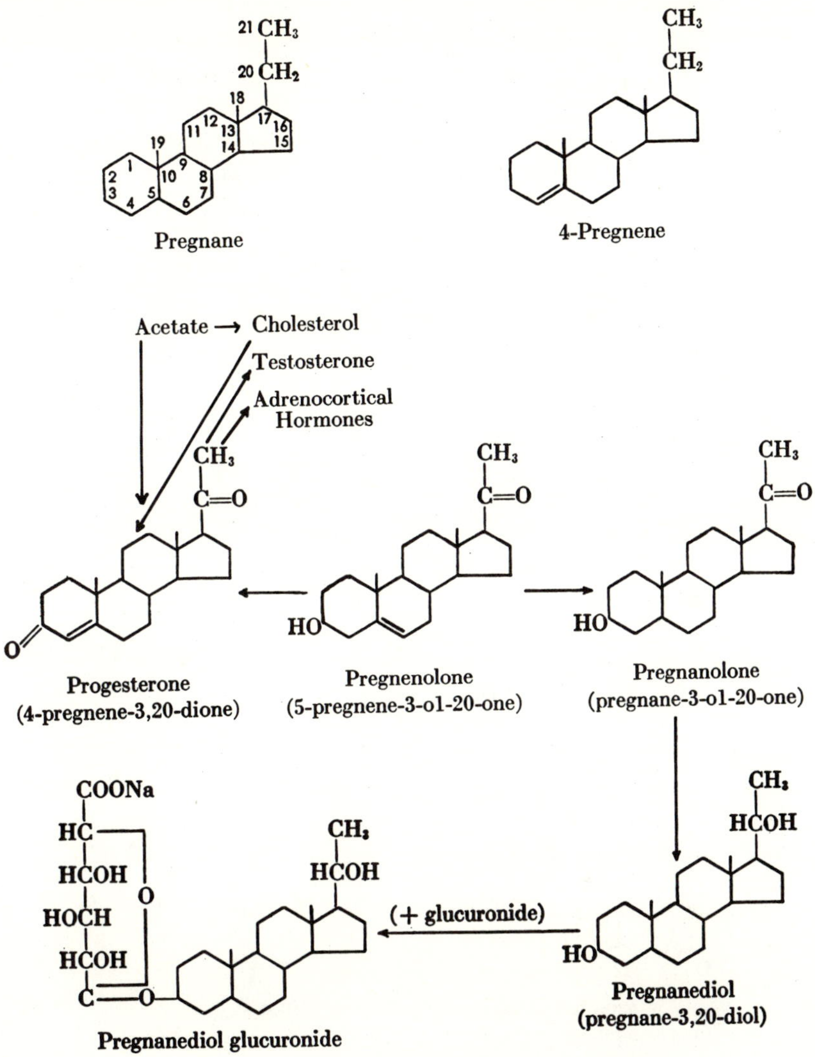

Figure 22-6. Biosynthesis and metabolism of progesterone.

hormones. Deficiencies in the latter two, particularly, form the basis for the metabolic abnormalities and clinical manifestations of certain aberrations of adrenocortical function (p. 735).

Knowledge concerning the pathways of biosynthesis of androgens and estrogens in the adrenal is not so precise. The following are possibilities in the case of androgen: (1) 17-hydroxyprogesterone may be converted to androstenedione and the latter to androsterone and etiocholanolone (Fig. 22-7); (2) pregnenolone may be converted to dehydroepiandrosterone and the latter to androstenedione (Fig. 22-6). Adrenal estrogens are probably derived from androstenedione, a known normal precursor of these hormones.

Transport and Metabolism of Corticosteroids. *A. Protein Binding.* Most of the cortisol entering the bloodstream is bound to pro-

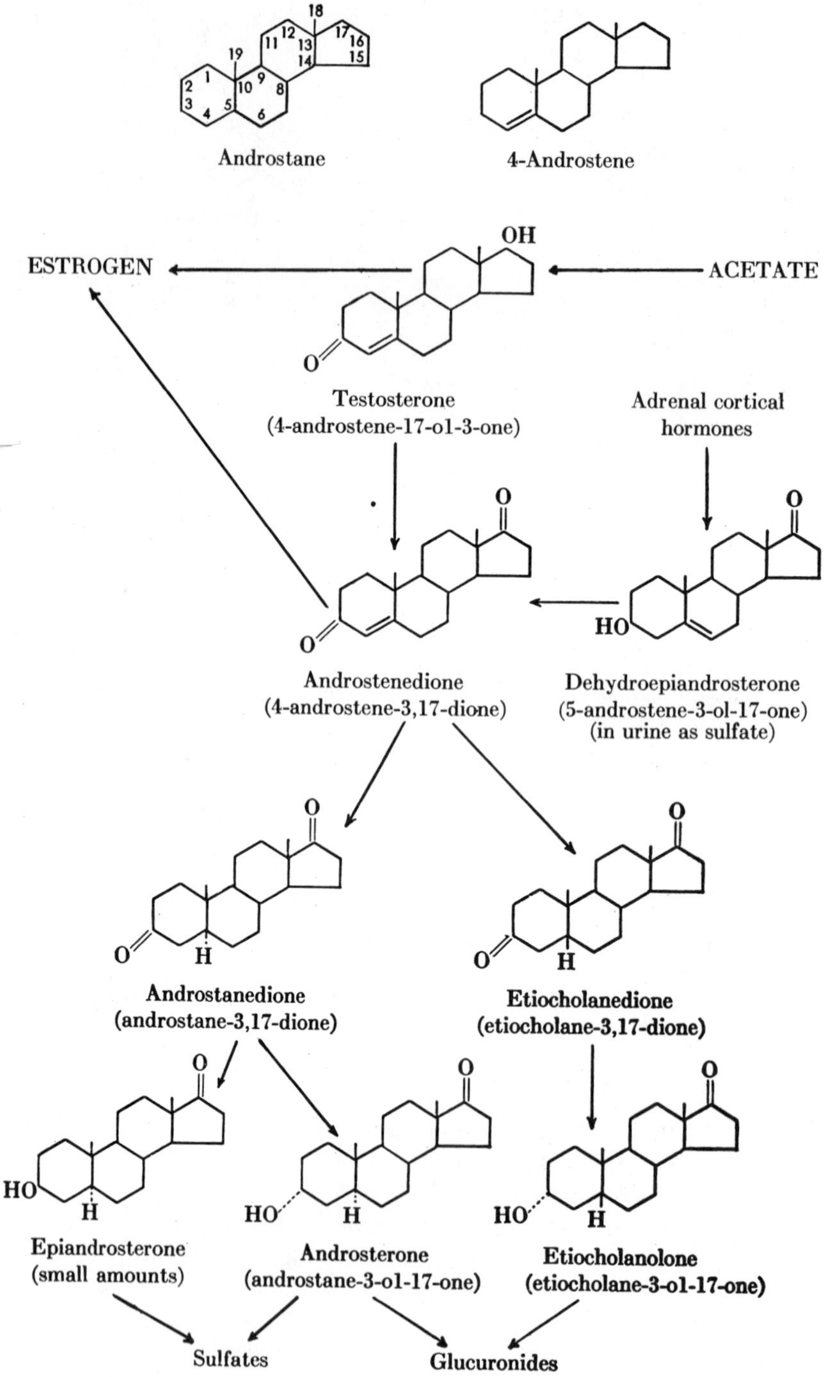

718 ENDOCRINE FUNCTION

Figure 22-7. Metabolism of androgens.

tein, especially to an α-globulin, termed transcortin. This protein has a much higher affinity for cortisol than has albumin, though the total capacity of albumin for cortisol is much greater because it is present in much larger amounts. Usually, more than 95 per cent of the cortisol present in the circulation is bound to protein, this portion being in equilibrium with the free or unbound fraction that is able to enter cells and exert its physiological effect. As the blood levels of cortisol rise, the cortisol-binding capacity of transcortin is exceeded, and the amount of free cortisol rises and may exceed 25 per cent of the total. Only the free hormone can be filtered by the glomerulus, so that the amount of cortisol normally found in urine is very small. During pregnancy or after estrogen administration, circulating levels of transcortin rise, increasing the amounts of bound cortisol. However, since the level of free cortisol is only slightly increased, there are no clinical features of cortisol excess. Because of the increase in bound cortisol the rate of destruction of the compound is reduced during pregnancy. In the presence of cirrhosis and certain dysproteinemias, there is a decrease in cortisol binding power. Normally, about 60 per cent of aldosterone is bound to albumin, but this fraction is reduced when albumin levels fall, e.g., in association with the nephrotic syndrome.

B. Degradation. Because of the high degree of protein binding, cortisol is removed from the circulation fairly slowly, the half-life of injected labeled cortisol being 80 to 110 minutes. The main site of cortisol breakdown is the liver, where the major processes are reduction of the Δ^4, 3-ketone of ring A, with addition of four hydrogen atoms to form tetrahydrocortisol, and the oxidation of the 11-hydroxyl to the 11-ketone to yield cortisone and tetrahydrocortisone. These compounds are rendered more soluble by conjugation with glucuronic acid. Since the glucuronides are poorly bound to protein they are readily excreted in the urine. A small proportion of cortisol is converted to 17-oxosteroids by removal of the side chain.

Aldosterone is converted to its tetrahydro- derivative, and part of this is excreted unchanged and part conjugated with glucuronic acid to be excreted as the glucuronide. About 5 per cent of aldosterone secreted each day is excreted as a 3-oxo- conjugate and only 0.5 per cent in the free form.

DHA is partly converted to its sulfate and is also metabolized to etiocholanolone and androsterone. Androstenedione is similarly converted to these two compounds (Fig. 22-7).

C. Renal Excretion. The kidney excretes steroids and their metabolites; the amounts excreted depend on the extent of protein binding, which largely determines the amount filtered, and the tubular reabsorption. Only 10 per cent of filtered cortisol is excreted because of tubular reabsorption. Less than 80 μg. free unconjugated cortisol are found in the urine each day and about 60 μg. cortisone.

Determination of the Rate of Secretion of Corticosteroids. The rate of secretion of corticosteroids can be determined by measurements of:

1. Gland steroid content
2. Adrenal vein steroid output
3. Blood concentration of steroids
4. Renal excretion of steroids

Only methods (3) and (4) are routinely possible in man.

Blood estimations have the disadvantage of representing only one point in time, and it is known that the adrenal output of steroids is variable. In addition, usually only microgram amounts are measured. Urinary measurements are prone to all the risks of incomplete collection but they do give a measure of steroid output over a longer time; therefore, much larger amounts are involved. Traditionally, assessment of adrenocortical activity in health and disease has depended on estimation of the corticosteroid metabolites in urine, but these only indirectly relate to true cortisol secretion or excretion and may often be misleading. Since the introduction by Mattingly, in 1962, of a rapid and reliable plasma corticosteroid assay, assessment of adrenocortical activity has switched more and more to plasma measurements because the ease of their assay has made frequent sampling during dynamic test procedures a practical proposition. Diagnostic reliability has greatly improved as a result.

Cortisol and Its Metabolites. Compounds which have a 17-hydroxyl group and a "dihydroxyacetone" type of side chain react with the Porter-Silber reagent (phenylhydrazine and strong sulfuric acid) to give a yellow color and are referred to as 17-hydroxycorticosteroids (17-OHCS). These compounds include cortisol, cortisone and their tetrahydrometabolites, as well as other interfering chromogens, but not pregnanetriol. The method is more widely used in the United States than in the United Kingdom and is usually applied to urine corticosteroid estimations. It has in the past also been used for blood corticosteroid measurements, but is tedious to perform and has a low capacity. It has now been replaced by fluorescence or protein-binding techniques, and radioimmunoassay methods are being developed.

Routine estimation of corticosteroids in blood and urine have been greatly facilitated by the method of Mattingly, which measures the total (protein-bound and free) cortisol concentration, together with the small amount of corticosterone present. The technique involves extraction of plasma or urine with methylene chloride and measurement of fluorescence in the presence of concentrated sulfuric acid and ethanol. There is a small amount of nonsteroid fluorescence amounting to about 2 to 4 μg./100 ml. equivalent of cortisol. Plasma fluorogenic corticosteroids (often inaccurately but more simply called "plasma cortisol") vary with the time of day; the 9:30 A.M. values range from 8 to 26 μg./100 ml. in adults. Levels are lowest at midnight and maximal between 6 and 8 A.M. Plasma cortisol can also be measured by "protein-binding" methods, in which radioactively labeled cortisol competes with the unlabeled cortisol extracted from the plasma for the binding sites on transcortin. The source of the transcortin is plasma obtained from estrogen-treated or pregnant patients in whom transcortin levels are high. The more cortisol there is in the original blood sample, the less labeled cortisol will bind, and by comparing the amounts bound with those obtained when standard solutions of cortisol are assayed, the original concentrations can be calculated. This method is no more specific than the fluorimetric assay, since it also measures corticosterone and it is less convenient, but the nonsteroid background measurement is lower.

Both the fluorimetric and protein-binding methods of cortisol assay may be used to measure the amount of free corticosteroids excreted in the urine, and this is a sensitive index of the amount of cortisol excreted per day.

The secretion rate of cortisol can be determined by an isotope dilution method. Isotopically labeled cortisol (usually ^{14}-C-cortisol) is administered, and the secretion rate of naturally produced cortisol calculated from the formula:

$$\frac{\text{administered steroid} \times \text{specific activity of administered steroid}}{\text{specific activity of unique urinary metabolite}}$$

A unique metabolite is one derived solely from the steroid hormone being investigated, e.g., tetrahydrocortisol or tetrahydrocortisone, in the case of cortisol. Cortisol secretion rates in normal subjects range from 5 to 28 mg./day, with a mean of 16 mg./day. The test is time-consuming and requires considerable technical skill, and it is necessary to administer a radioisotope to a patient (though only a small amount of radiation is involved).

During pregnancy, the plasma levels of corticosteroids rise, reaching a peak at the ninth month, normal values being restored by the sixth day after delivery. The rise in cortisol is due to an increase in hepatic production of transcortin, so that most of the circulating corticosteroid is protein bound and metabolically unavailable. The free hormone concentration rises only slightly. By the end of the first week of life, infants have plasma steroid levels similar to those of adults although no circadian rhythm is present for several months.

Aldosterone and Its Metabolites. Aldosterone estimations in urine are not readily available for routine use since methods are tedious and require high technical skill. Satisfactory techniques for determining plasma levels by isotope dilution have only recently been developed, but again are technically very difficult, although radioimmunoassay methods are easier. A mean urinary excretion in normal subjects of 10.5 μg./24 hr. has been reported (range 4.6–18.9 μg.). Output is lowest at night and shows a peak at noon. No difference according to sex has been demonstrated. Potassium administration, sodium restriction or a decrease in extracellular fluid volume can stimulate aldosterone output. During pregnancy, the urinary output of aldosterone increases sixfold.

Plasma aldosterone levels are extremely low—mean 6 μg./100 ml. (range 2–15 μg.). Only 65 per cent of aldosterone in the plasma is bound to protein. Most of the aldosterone is removed from the blood during passage through the liver, which is its major site of degradation. Secretion rates can be determined by isotope dilution methods, the mean figure being about 130 μg./24 hr. (range 50–200 μg.).

Androgens and Their Metabolites. The adrenal androgens are largely excreted as 17-oxosteroids in the urine. In the male, only 24 per cent of testosterone from the testis is converted to 17-oxosteroids, mainly androstenedione, and hence to androsterone and etiocholanolone, which are measured as 17-oxosteroids in the urine. The 5-hydrogen atom of etiocholanolone lies above the plane of the molecule, which is therefore called a 5 β-compound. In androsterone, which has the same formula, the 5-hydrogen atom lies below and the compound is said to have the 5 α configuration. Most of the 17-oxosteroids in the urine are rendered more soluble by conjugation with glucuronic and sulfuric acids. Measurement of 17-oxosteroids depends on the color produced by their reaction with m-dinitrobenzene in the presence of alcoholic potassium hydroxide, the so-called Zimmerman reaction. The effect of other compounds in the urine that produce a color in this reaction can be eliminated by readings at several wavelengths and the application of a correction factor. Some drugs, such as meprobamate, can interfere with the reaction.

DHA can be separated from most of the other oxosteroids because of its 3 β-hydroxyl grouping, which allows its precipitation by digitonin. It is found in large amounts in the urine of some patients with adrenal carcinoma and was formerly thought to be derived only from the adrenal, but it is now known to be produced in the ovary and testis in addition. Much of the DHA present in blood is in the form of sulfate ester, though the role of this ester has not been defined.

Androsterone is derived from DHA, testosterone and androstenedione. Etiocholanolone has a similar origin but can also be produced from 17-hydroxyprogesterone. 11-Oxygenated oxosteroids are mainly formed from cortisol and its metabolites, though this forms only a minor pathway of cortisol degradation. During pregnancy, metabolites of progesterone may be estimated along with 17-oxosteroids, but refined techniques indicate that there is no elevation of true 17-oxosteroid output. The urinary 17-oxosteroid levels are low in children, rising at puberty. Adult levels are somewhat higher in the male, largely because of the greater contribution of the testis than the ovary. The highest values are found in early adult life and decline with age. Since so little testosterone is converted to 17-oxosteroids, and it is a very potent androgen, marked virilization can occur owing to oversecretion of testosterone, yet there may be no change in 17-oxosteroid excretion.

Metabolic Effects. Absence of adrenal cortical hormones (adrenalectomy; Addison's disease) or administration of excessive amounts of these hormones is accompanied by characteristic metabolic changes, which may be grouped broadly under two headings: (1) electrolyte and water metabolism; (2) carbohydrate, protein and fat metabolism. The latter category is influenced principally by cortisol and corticosterone ("glucocorticoids"), whereas the former is influenced predominantly, but by no means exclusively, by aldosterone ("mineralocorticoid").

Electrolytes and Water. Adrenalectomized animals and patients with Addison's disease, if untreated, exhibit the following phenomena: (1) increased urine volume, with (2) disproportionately increased excretion of Na and Cl, leading to (3) decreased concentration of Na and Cl in blood plasma, (4) decrease in body water (dehydration) and (5) decrease in plasma volume. Ultimately there is a fall in blood pressure, with circulatory collapse and impaired renal glomerular filtration, which are aggravated by, but not necessarily entirely dependent upon, the changes in electrolyte and water balance. Simultaneously with the changes in Na, the urinary excretion of K decreases and the concentration of K in the plasma increases.

The kidney is apparently the organ primarily involved in these changes in Na, and perhaps also in K metabolism. In the absence of adrenal hormones, the renal tubular epithelial cells appear to be unable to reabsorb Na (also Cl and water) adequately from the glomerular filtrate in spite of low plasma Na concentration. They also are apparently unable to excrete K adequately (or they reabsorb it excessively) despite the increased plasma K concentration. At the same time there is a disturbance in the equilibrium between the extracellular (plasma and interstitial fluids) and intracellular fluid compartments (extrarenal effect), characterized by an increase in intracellular (muscle) K and a decrease in extracellular Na and Cl. Inadequate $Na^+:H^+$ exchange in the kidney results in decrease in the plasma HCO_3^- concentration (metabolic acidosis). An increase in blood urea nitrogen reflects the diminished glomerular filtration, apparently a primary effect of decrease in corticosterone and aldosterone, aggravated by the superimposed state of dehydration and circulatory collapse.

These abnormalities can be prevented or corrected by administration of adrenal steroids, the Na-retaining potency of aldosterone being twenty to several hundred times that of deoxycorticosterone, and that of the latter two to four times that of the glucocorticoids. There are also qualitative differences in the actions in this connection of these hormones. In the absence of the adrenals, neither water nor Na is excreted adequately when unusually large quantities are given. This difficulty is aggravated by aldosterone, which further increases retention of Na, but is corrected by adrenal cortical extracts or cortisone. Moreover, the elevation of plasma Na resulting from excessive doses of aldosterone or deoxycorticosterone in intact animals can be prevented by simultaneous administration of cortical extract or ACTH.

In addition to the renal effect, the glucocorticoid hormones influence K metabolism by favoring protein catabolism in certain tissues, releasing intracellular K to the extracellular fluids. Then, too, their action in increasing glycogen storage (especially liver) is accompanied by an increase in K in the cells involved in this process. Several aspects of the influence of the adrenocortical hormones on electrolyte and water metabolism are inexplicable at the present time.

The effects of corticoids on water excretion vary considerably, depending upon (1) the amount and nature of the steroid and (2) the conditions under which they are acting, e.g., the extent of the simultaneous water (hydration) or Na load. In animals hydrated with isotonic NaCl solution, aldosterone usually, but not invariably, has little or no effect on water excretion, although Na retention occurs. On the other hand, under these conditions deoxycorticosterone decreases water excretion in proportion to Na retention, whereas the effects of corticosterone, cortisone, and cortisol differ among themselves and from the effects of the other steroids. Obviously, water excretion does not necessarily parallel the rate of Na excretion.

Although adrenocortical insufficiency is characterized by excessive Na and water excretion (and dehydration), it is also associated with inability to excrete a large water load at a normal rate. Normal excretion is restored by cortisol, corticosterone, or adrenocortical extracts, but not by aldosterone or deoxycorticosterone.

Subjects with adrenocortical hyperfunction exhibit (1) decreased urinary excretion of Na (and water) and (2) increased excretion of K (mobilized from intracellular compartment). These lead ultimately to (3) hypernatremia, (4) hypokalemia, and (5) increase in extracellular fluid volume (including blood plasma). As K moves out of the cell, Na moves in, but in smaller amounts; the K is excreted in the urine and, exceeding the amount of Na retained, is accompanied by excessive loss of Cl, with consequent (6) hypochloremia. The epithelial cells of the distal renal tubule share in the K depletion, with consequent (7) increase in the exchange of H^+ for Na^+, and (8) return of increased amounts of HCO_3^- to the blood plasma. The situation is one of hypochloremic alkalosis with increased urinary titratable acidity and ammonia. These electrolyte abnormalities may occur in subjects with generalized adrenocortical hyperfunction (Cushing's syndrome), accompanied by abnormalities in carbohydrate, protein, and lipid metabolism, or in subjects with hyperaldosteronism, without such associated manifestations.

Carbohydrate Metabolism. The fasted, untreated, adrenalectomized animal (or patient with Addison's disease) exhibits the following changes in carbohydrate metabolism: (1) striking decrease in liver glycogen; (2) less marked decrease in muscle glycogen; (3) hypoglycemia; (4) decreased intestinal absorption of glucose (corrected by administration of NaCl). These changes do not occur if a sufficient quantity of carbohydrate is given, and are prevented or corrected by administration of glucocorticoid adrenal hormones. Cortisol is three to five times as potent in this respect as corticosterone or 11-dehydrocorticosterone, and 11-dehydro-17-hydroxycorticosterone (cortisone) is two to three times as potent, but has to be converted to cortisol, predominantly in the liver, before showing this effect. Aldosterone, too, causes glycogen deposition in fasting adrenalectomized animals, its activity being about one-third that of cortisol and two-thirds that of corticosterone. There are other indications of the influence of adrenocortical hormones on carbohydrate metabolism:

1. Adrenalectomized animals exhibit increased sensitivity to insulin.

2. Adrenalectomy results in amelioration of diabetes produced by pancreatectomy or alloxan. Administration of glucocorticoid adrenal hormones or cortical extracts aggravates the diabetes.

3. Administration of cortical extract or glucocorticoid adrenal hormones causes a rise in blood sugar, liver glycogen, and total body carbohydrate, glycogen formation being accelerated and its breakdown to glucose retarded. Glucose tolerance is decreased and, in diabetes, there is

increase in glycosuria and in the insulin requirements; that is, there is "insulin resistance."

The manifestations of adrenocortical deficiency are related mainly to two fundamental phenomena: (1) decreased gluconeogenesis from body protein and (2) decreased output of glucose by the liver. Conversely, cortical hormones active in this sphere increase gluconeogenesis from body protein.

The livers of animals given large amounts of cortisol show selectively increased activity of glucose-6-phosphatase and fructose-1,6-diphosphatase. Inasmuch as these enzymes occupy key positions in reversal of the enzymatic reactions of glycolysis in the liver, such changes could contribute to the hyperglycemic action of the glucocorticoids.

Protein Metabolism. Administration of relatively large amounts of ACTH, adrenal extracts, or glucocorticoids to fasting intact animals is followed by an increase in total urinary nitrogen and free amino acids and a negative nitrogen balance (protein catabolic effect). As indicated above, this is accompanied by an increase in body carbohydrate, and G : N ratios indicate that about 50 to 70 per cent of the extra protein catabolized was converted to glucose. Exogenous protein is apparently not affected in this manner.

There is evidence that more physiological amounts of cortical hormone, on the contrary, exert a protein anabolic effect. Administration of ACTH increases the positive nitrogen balance in children, and cortical extracts, in certain dosages, reduce nitrogen excretion in the urine. It would appear that, as is true of other hormones (eg., thyroid), the quantity secreted and the functional status of the target cells have an important bearing on the character of the metabolic response.

The mechanism of action of these cortical hormones in protein metabolism is not entirely clear. There is some evidence that they increase deamination of amino acids and that adrenalectomized fasted animals do not convert deaminized amino acid residues to glucose so readily as do normal animals. However, protein-fed adrenalectomized animals apparently behave normally in this respect. It is difficult to reconcile these observations. The suggestion has been made that these hormones may exert an "anti-anabolic" rather than, or in addition to, a catabolic action on protein.

Lipid Metabolism. Although they undoubtedly exert a significant influence, the physiological role of glucocorticoids in lipid metabolism is not clearly understood, mainly because their effects vary in different species and under different experimental conditions.

The most pertinent observations may be summarized as follows:

(1) Fat Synthesis and Mobilization. Cortisol increases liver fat in intact (but not in hypophysectomized) animals, and also synthesis of fatty acids and cholesterol in the isolated perfused liver. However, the decreased lipogenesis in the liver of diabetic animals improves following adrenalectomy; this improvement does not occur in adipose tissue. Increase in unsaturated fatty acids in the plasma has been observed following administration of glucocorticoids or ACTH, owing presumably to stimulation of their release from adipose tissue. Moreover, in certain subjects with Cushing's

syndrome (adrenocortical hyperfunction) there is a marked diminution in subcutaneous fat in many regions, accompanied, peculiarly, by an increase in fat deposition in the abdominal and interscapular regions.

(2) Ketogenesis. Glucocorticoids increase ketogenesis in diabetic subjects and depancreatized animals; the converse effect is produced by adrenalectomy. The presence of growth hormone is apparently necessary for this action on ketogenesis.

In normal intact animals ketosis induced by fasting or by cold stress is decreased by administration of cortisol. This may be due to increased production of insulin resulting from increased production of glucose.

(3) Blood Lipids. Observations are contradictory on the effect of glucocorticoids on the level of cholesterol, phospholipids and neutral fat in the blood plasma of normal subjects. A decrease occurs in certain types of hyperlipidemia and hypercholesterolemia.

Anti-inflammatory Action

Cortisol is a potent inhibitor of the inflammatory reaction induced by physical, chemical or bacterial agents. Cortisone is less than 2 per cent as active in this respect unless converted by the tissues, particularly the liver, to cortisol. Corticosterone is virtually inactive, and aldosterone and deoxycorticosterone apparently antagonize this action of cortisol. This anti-inflammatory action includes inhibition of the vascular margination of leukocytes, of the migration of leukocytes from the capillaries, and of the formation of fibrin and accumulation of edema fluid. There is also suppression of the increase in capillary permeability in the inflamed area. Although these effects may in certain respects be detrimental in the case of bacterial infections, they may be desirable in other types of inflammation. Among these are polyarteritis nodosa, lupus erythematosus and rheumatic fever, in which the inflammation of affected tissues often responds dramatically to administration of active corticosteroids. Several synthetic derivatives are available which are highly active in this respect and which exhibit less biological activity in other directions that are therapeutically undesirable.

Effect on Immune Reactions. Certain of the glucocorticoids, including cortisol, exert a pronounced anti-allergic action. Although their main action is to suppress the inflammatory manifestations of the antigen-antibody reaction, they also, to a lesser extent, reduce the production of antibodies following the introduction of an antigen. The mechanism of their action in this connection is not known.

They are partially effective in breaking down genetic barriers to heterotransplantation of tissues, and are used to treat patients receiving organ transplants, such as heterologous kidneys.

Miscellaneous Effects. Administration of glucocorticoids is followed promptly by a decrease in the number of circulating lymphocytes and eosinophils. The lymphopenia is due to active destruction of lymphocytes in lymphatic tissues (intestinal mucosa, lymph nodes, thymus); the mechanism of production of the eosinopenia is not known. Conversely, adrenocortical insufficiency, as in Addison's disease, may be accompanied by increase in circulating eosinophils and lymphocytes and in hypertrophy of lymphoid tissue.

Physiology of Cortisol

The factors which regulate cortisol secretion, via ACTH, are:
1. the circadian rhythm
2. the negative feedback and
3. the stress mechanisms.

The latter is the most powerful of the three.

Plasma cortisol concentrations are not normally constant throughout the day, being highest just before waking, then under quiet conditions gradually falling throughout the day to reach their lowest values at about the time of retiring, only to begin rising again some four hours before waking. This circadian rhythm is maintained by pituitary ACTH secretion and is controlled by an intrinsic "biological clock" mechanism within the midbrain and is mediated by variations in corticotrophin-releasing hormone (CRH) secretion from the median eminence of the hypothalamus. CRH passes down the capillary plexus of the pituitary stalk to act on the basophil cells of the anterior pituitary, causing them to synthesize and release ACTH and also β-MSH. Unlike lower species, there is very little α-MSH in the human pituitary gland, which contains no intermediate lobe cells, and only β-MSH appears to be secreted in man, and then apparently always with ACTH. In addition to the circadian rhythm, there is another control mechanism which operates under basal conditions, the negative feedback, whereby CRH, ACTH and therefore cortisol secretion are reduced or inhibited when plasma corticosteroid levels are too high and increased when corticosteroid levels are inappropriately low. Finally, the third control system is that which operates during stress.

The "Alarm Reaction"; The Adrenal Cortex under Stress

A large number of stimuli, damaging agents or conditions of unusual stress elicit in the body a typical series of events in which adrenal cortical hormones, or rather the hypothalamic-pituitary-adrenocortical "axis," plays an important role. The nature of the alarming stimuli is totally nonspecific; they include almost every conceivable type of damage or change of usual environment, such as muscular exercise, exposure to cold, trauma, burns, drugs and poisons, bacterial infection, exposure to roentgen radiation, psychological stresses, and many others. In response to these stimuli, a shock phase develops which, if not too severe and therefore lethal, is followed by a phase of "countershock." Continued exposure induces a period of resistance. The latter may break down and be followed by the stage of exhaustion. The shock phase will not be discussed here in detail. It is of interest that as an early response to alarming stimuli the output of epinephrine is increased. This is transient, producing a brief rise in blood pressure and blood sugar, followed by the hypotension and hypoglycemia of shock.

Adrenal cortical hormones appear to be necessary for recovery from the changes induced in the shock phase. Adrenalectomized (or hypophysectomized) animals exhibit a considerably decreased resistance to "stress" or alarming stimuli. Exposure to cold, readily tolerated by the intact animal, is fatal in the absence of the adrenals. Subcutaneous injection of relatively small amounts of histamine or formaldehyde kills the adrenalectomized animal, whereas it is tolerated by the intact animal. The intact animal responds to these nonspecific alarming stimuli with an increase in adrenal size. That this actually indicates increased function is suggested by the concomitant reduction in size of the thymus and other lymphatic structures, and by certain metabolic changes, such as negative nitrogen balance and tendency to hyperglycemia. A prompt increase in the level of glucocorticoids in the blood can be demonstrated after application of an alarming stimulus. Resistance to many kinds of stress appears to be dependent upon adrenal cortical hormones, especially the glucocorticoids. The exact nature of their action in overcoming the shock phase of the alarm reaction is unknown.

The response of the adrenal cortex to an alarming stimulus (increased release and increased production of adrenal cortical hormone and eventual hyperplasia) is mediated by the adrenocorticotrophic hormone. In the hypophysectomized animal the adrenal cortex does not respond to such stimuli, which may consequently be fatal. The alarming stimuli cause the same changes in adrenal cholesterol and ascorbic acid as does injection of pure adrenocorticotrophic hormone, and there is a great increase in the secretion rate of cortisol (in man) or corticosterone (in many other mammals). This increased corticosteroid secretion occurs regardless of the time of day or operation of the "feedback." Thus, the stress mechanism is much more powerful than the other control mechanisms, and very high levels of plasma corticoids may be achieved, higher than those ever seen under basal conditions.

The release of adrenocorticotrophic hormone from the anterior pituitary in response to various forms of stress is dependent upon hypothalamic centers located (in the dog) in the paramedian region of the anterior hypothalamus and at the junction of the middle and posterior hypothalamus. Experimental destruction of these nuclei prevents release of ACTH from the anterior pituitary following stress.

The mechanism suggested by recent experiments is that various impulses resulting from stress reach the hypothalamus through nervous pathways. In the hypothalamic nuclei corticotrophin-releasing hormone (CRH) is secreted, which in turn fires the anterior pituitary to release ACTH, and hence corticosteroid secretion increases. If supraphysiological doses of glucocorticoids have been administered to the subject for some time, CRH and ACTH secretion is suppressed by operation of the negative feedback mechanism; in addition, the normal secretory responses of the hypothalamus and pituitary in response to stress do not occur, nor does increased glucocorticoid secretion. Also, the adrenal cortex is atrophic because of the lack of basal ACTH secretion. Growth hormone is secreted in response to stress, and this is also suppressed by long-term glucocorticoid treatment. Epinephrine release from the adrenal medulla is, however, normal under these conditions. It has been shown that mental and emotional stress may act as an "alarming stimulus," causing an increased release of adrenocortical hormones and epinephrine.

Although it is clear that these hormonal responses are required for the organism to cope normally with stress, the mechanism whereby the beneficial effects are produced is unknown. However, if the stress responses fail to occur owing to adrenocortical insufficiency resulting from adrenal gland, pituitary or hypothalamic disease or to glucocorticoid therapy, then the patient is likely to collapse in circulatory failure.

Control of Aldosterone Secretion. The rate of aldosterone production is influenced by many different factors. Increased secretion results from sodium restriction, increased sodium loss caused by diuretics, potassium administration, hemorrhage and dehydration, reduction in plasma volume, injection of angiotensin and assumption of the upright position. Decreased secretion follows potassium depletion, sodium administration and any increase in plasma volume, as well as assumption of the horizontal position.

It is now well established that changes in aldosterone secretion in response to these various stimuli are mediated predominantly by the renin-angiotensin system. Although ACTH infusion induces a transient increase in aldosterone secretion and a pituitary factor is necessary to maintain the adrenal in a condition in which it can respond normally to stimuli to aldosterone production, aldosterone levels are dissociated from cortisol levels in many physiological situations, indicating that ACTH is not the major factor controlling aldosterone. Similarly, hyperkalemia and hyponatremia in the blood perfusing the adrenal are known to stimulate aldosterone secretion but cannot account for changes in aldosterone levels observed under various conditions of diet and posture. By contrast, in normal animals and man changes in plasma aldosterone are almost invariably associated with parallel changes in plasma renin concentration or angiotensin II levels, indicating the overriding importance of renin, a hormone derived from the kidney, in the control of aldosterone (Fig. 22-8 on page 730).

The exact mechanisms by which the various stimuli to aldosterone secretion trigger renin release are not finally established. The most likely main stimulus to aldosterone secretion is a tendency to a reduction in central blood volume brought about by blood or salt and water loss or by pooling of blood in the legs with assumption of the erect posture. This then stimulates the sympathetic nervous system via the carotid sinus, and beta-adrenergic stimuli pass via the renal nerves to the kidney and lead to renin release. This would be brought about either by stimulation of the renin-producing cells directly or possibly by inducing arteriolar con-

730 ENDOCRINE FUNCTION

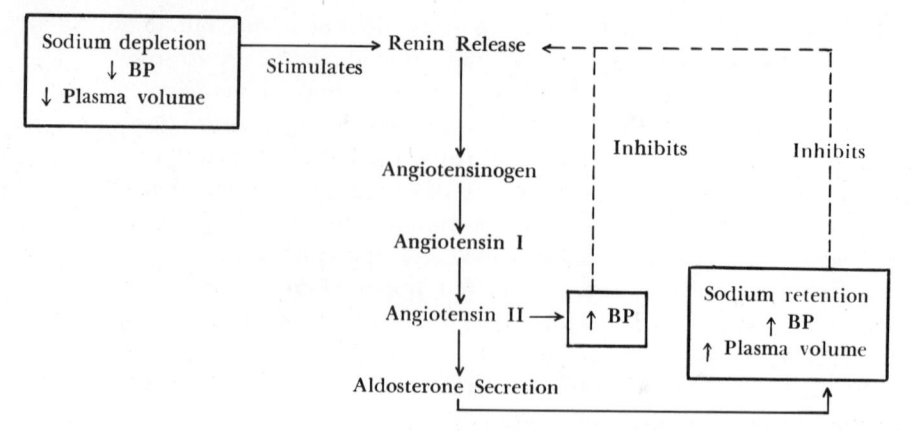

Figure 22-8. The renin-angiotensin-aldosterone mechanism.

striction proximal to baroreceptors in the afferent arterioles, thus lowering the pressure at the receptor. Renin is produced by cells of the juxtaglomerular apparatus in the wall of the afferent arteriole and is released into the circulation, where it acts enzymatically on angiotensinogen derived from the liver to form a decapeptide angiotensin I. This is converted to the octapeptide angiotensin II by "converting enzyme," which is present in the circulation and in high concentration in the lungs. The octapeptide is a potent vasopressor agent as well as the major factor stimulating the secretion of aldosterone (Fig. 22-8).

Adrenocortical Hyperfunction

From the standpoint of hormone secretion, recognized syndromes due to adrenocortical hyperfunction may be classified as follows:

1. Increased secretion of all hormone groups: Cushing's syndrome (panhypercorticalism).
2. Exclusive hypersecretion of androgens: adrenogenital syndrome.
3. Exclusive hypersecretion of estrogens: feminizing syndrome (in males).
4. Exclusive hypersecretion of aldosterone: primary hyperaldosteronism.

There may be instances in which there is overlapping of two or more of these categories. In certain cases of the adrenogenital syndrome, for example, the adrenals, instead of exclusively oversecreting androgens, may "almost exclusively" oversecrete them, producing also one or more of the other hormones somewhat in excess. The latter may give rise to symptoms, thereby causing pictures intermediate between the typical syndromes. In some instances, excessive secretion of one type of hormone may be associated with decreased secretion of others.

Cushing's Syndrome. The term Cushing's syndrome is used to describe the clinical disorder that results from supraphysiological levels of corticosteroids in the circulation, whether this is endogenously produced or administered to the patient. Certain patients with Cushing's syndrome may also show features of androgen excess, especially if the disorder is due to an adrenal adenoma or carcinoma. Like many endocrine diseases, Cushing's syndrome may vary in severity in an individual patient, and spontaneous remissions have occasionally been reported.

Etiology. The syndrome may be divided into two main groups, depending on whether or not the condition derives from exposure to excess ACTH.

ACTH-Dependent Causes

1. Pituitary-dependent bilateral adrenocortical hyperplasia, conventionally called Cushing's disease, due to increased pituitary ACTH secretion
2. The ectopic ACTH syndrome—secretion of ACTH by malignant or benign tumors of nonendocrine origin
3. Iatrogenic—resulting from treatment with ACTH or its synthetic analogues

NonACTH-Dependent Causes

1. Adenomas or carcinomas of the adrenal cortex
2. Iatrogenic—resulting from treatment with supraphysiological doses of corticosteroids

Clearly, the feature distinguishing the two main groups is that there is detectable ACTH in the circulation of those in the first group, whereas in the second group circulating ACTH levels are very low or undetectable. When the iatrogenic and ectopic groups are excluded, pituitary-dependent Cushing's disease comprises 80 per cent of the cases, whereas adenomas and carcinomas each make up about 10 per cent.

Clinical Features. Women are affected with Cushing's syndrome four times more often than men. There may be obesity largely confined to the trunk, hypertension, diabetes mellitus, a plethoric facies, inability to cope with infections, poor healing responses to trauma, bruising, purple striae (skin stretch marks), spontaneous fractures of bone due to thin bones, muscle wasting and weakness, amenorrhea, acne, infertility or impotence and psychiatric disturbances. All these features can be produced by administration of excessive amounts of corticosteroids or ACTH.

Diagnostic Procedures. These include: (1) basal measurements of adrenocortical hormones and their metabolites in the blood and urine and of plasma ACTH; (2) response to stimulation by ACTH; (3) response to pituitary-adrenal suppression by large doses of adrenocortical hormones.

Basal Hormone Values. The concentration of 17-hydroxycorticoids in the blood and their excretion in the urine are frequently, but not invariably increased. Direct evidence of cortisol overproduction can be obtained by measurement of the cortisol production rate by the technique of isotope dilution. This is the absolute test for Cushing's syndrome against which all other tests must be compared, but because the method is difficult less direct estimations must suffice.

Plasma cortisol levels can be used to demonstrate cortisol overproduction. The earlier sign may be loss of the normal circadian rhythm; a raised midnight cortisol level is of considerable value in diagnosis and less often the morning levels are raised. In normal patients, the midnight plasma

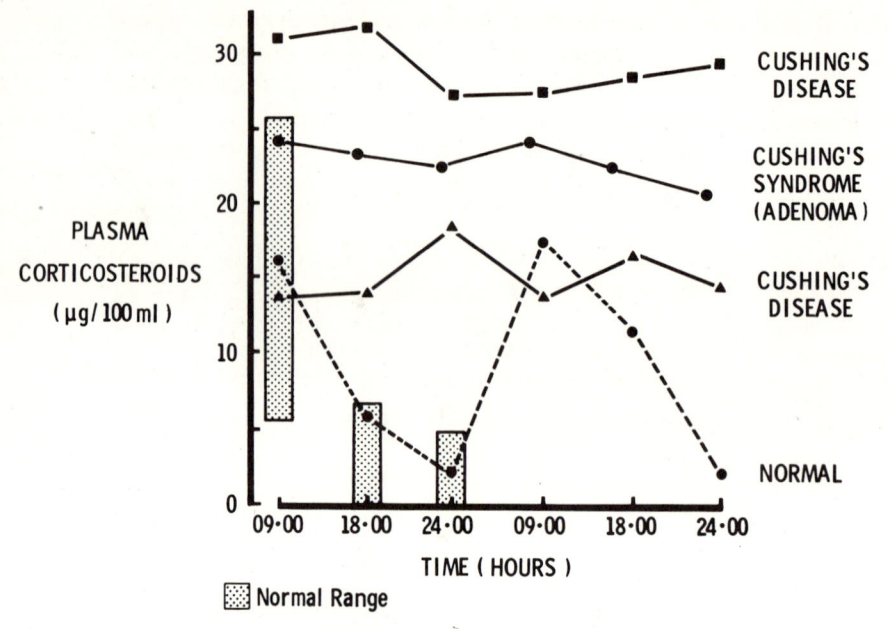

Figure 22-9. Circadian rhythm of plasma fluorogenic corticosteroids in one normal subject, two patients with Cushing's disease (pituitary-dependent adrenal hyperplasia) and one with an adrenal adenoma and Cushing's syndrome. The shaded areas represent the normal ranges at the times indicated. (From Besser and Edwards: Clinics in Endocrinology and Metabolism, 1(2): 451, 1972, Ed. A. S. Mason, W. B. Saunders Co., London.)

cortisol rarely exceeds 8 μg./100 ml. (Fig. 22-9). When taking blood for this estimation, care must be taken not to alarm the patient, because even minimal stress may cause a rise in cortisol production, and it is best taken when the patient has been asleep and as soon as possible after waking.

Estimation of the urinary 17-hydroxycorticosteroids (17-OHCS) is useful in severe cases of Cushing's syndrome but when cortisol overproduction is moderate, urinary tests often show values within the normal range. At higher levels of cortisol production the 24-hour urinary 17-OHCS output is approximately half the cortisol production rate. 17-Oxosteroid (17-OS) output in the urine is even less helpful in the diagnosis of Cushing's syndrome except in those cases due to adrenal tumors where androgen output is often high. Much of the raised 17-oxosteroid output in Cushing's syndrome is due to 11-oxygenated oxosteroid products of cortisol metabolism, but since only a small fraction of cortisol is metabolized by this route, only minimal increases in 17-OS output are to be expected. In some patients, there is also a minor qualitative alteration in the pattern of 17-OS excretion, the reason for which is unknown. By far the most useful urinary excretion test for routine use involves measurement of the urinary free cortisol output by protein binding. This is based on the fact that small elevations of the plasma cortisol increase the percentage of free cortisol in the plasma, and hence its excretion in the urine.

ACTH Stimulation Tests. These were once used in the investigation of suspected Cushing's sydrome, but are valueless because they neither differentiate normal from excessive adrenocortical function nor differentiate the different causes of excessive secretion.

Suppression Tests. These tests are based upon the fact that administration of large doses of corticosteroids to normal subjects causes depression of ACTH secretion by the negative feedback mechanism, and consequent reduction of adrenocortical activity. This is reflected in decreased plasma corticosteroids and excretion of urinary 17-hydroxycorticosteroids. To avoid masking of the changes in endogenous cortisol and its metabolites, synthetic corticosteroids such as dexamethasone are used since they are not measured in significant quantities by the assay techniques used.

In patients with Cushing's syndrome, there is usually no significant decrease in urinary excretion of 17-hydroxycorticosteroids with doses of dexamethasone (0.5 mg. every six hours for 8 doses) adequate to reduce the amount excreted by normal subjects to very low levels. However, when this dosage is doubled, there is frequently suppression of urinary 17-hydroxycorticosteroids in patients with Cushing's syndrome due to pituitary-dependent adrenal hyperplasia (Cushing's disease) but not in those with Cushing's syndrome due to adrenal tumor. However, this distinction cannot always be made reliably on this basis, since many exceptions have been described.

Plasma ACTH Measurements. When available, these are the most reliable means of differentiating the different causes of Cushing's syndrome. If ACTH is detectable, the cause is Cushing's disease (pituitary-dependent adrenal hyperplasia) or the ectopic production of ACTH by nonendocrine tumors (usually ACTH levels are very high in these patients; over 200 pg./ml.). If ACTH is undetectable in the plasma, then the patient has an adrenal adenoma or carcinoma, since the pituitary will be suppressed.

Carbohydrate Metabolism. Diabetes mellitus is a very constant feature of Cushing's syndrome, although not present in all stages. The development of frank diabetes is preceded by a stage of latent or subclinical diabetes, evidenced by diminished glucose tolerance. The diabetes is evidently attributable to the increased secretion of 11-oxysteroids, which increase gluconeogenesis and decrease carbohydrate utilization, thereby leading to hyperglycemia, necessitating increased secretion of insulin, and to secondary changes in the islets of Langerhans. Frank diabetes mellitus ensues when the secretory capacity of the pancreas cannot keep up with the requirements of the body to maintain normal glucose tolerance in the presence of excess glucocorticoid secretion.

Protein Metabolism. Negative nitrogen balance occurs as a result of excessive protein catabolism due to the action of the glucocorticoids. A number of manifestations, including demineralization of bones, thinning of the skin, purplish striae and, in part, diabetes, have been attributed to increased protein catabolism. It is of interest that protein catabolism is increased in Cushing's syndrome in spite of the simultaneous hypersecretion of protein anabolizing androgens. The latter are evidently not increased sufficiently to overcome the protein catabolic action of the glucocorticoids. The fact that therapeutic administration of additional androgens may result in a positive nitrogen balance suggests that the matter is merely one of relative amounts of the two antagonistic hormone groups secreted by the adrenal cortex.

Na, Cl, and K Metabolism. The characteristic changes in Cushing's syndrome are: decreased urinary excretion of Na and Cl and increased excretion of K. Excessive loss of K is accompanied by development of a state of hypokalemic alkalosis.

The electrolyte changes are probably not the result of excess production of aldosterone alone. Glucocorticoids, especially cortisol and desoxycorticosterone, influence sodium, potassium, and chloride metabolism. It would seem that, if present in large amounts, they may cause changes similar to those caused by aldosterone, although they are less potent in this respect than the latter. It is therefore difficult to ascribe the electrolyte changes in Cushing's syndrome entirely to one or the other group of compounds.

Calcium, Phosphorus and Phosphatase. The osteoporosis so frequently present has prompted extensive studies of calcium and phosphorus metabolism. These electrolytes are present in the serum in normal amounts. However, a negative calcium balance has been demonstrated; this loss of calcium may be due to the inability of the matrix to mineralize properly. The serum alkaline phosphatase may be slightly elevated.

Hyperaldosteronism. This may be either primary or secondary. Primary hyperaldosteronism (Conn's syndrome) is a rather distinctive clinical entity due to excessive secretion of aldosterone as a result usually of adrenocortical adenomas, occasionally bilateral hyperplasia. The importance of its early recognition stems from the fact that all clinical manifestations disappear, particularly the hypertension following ablation of the hyperfunctioning tissue if this is accomplished before irreversible kidney changes have occurred. Secondary hyperaldosteronism occurs frequently in patients with congestive heart failure, liver disease and certain types of kidney disease, since in these conditions there may be a fall in plasma volume and renal perfusion with a consequent stimulation of renin secretion, angiotensin I and II production, and hence secondary hyperaldosteronism (Fig. 22-9).

This diagnosis of hyperaldosteronism is not difficult when the characteristic clinical features are accompanied by characteristic laboratory findings: (a) hypokalemia, resistant to potassium therapy; (b) usually normal serum Na concentration, but occasionally hypernatremia; (c) hypochloremic alkalosis; (d) hyposthenuria (defective renal concentrating ability), resistant to antidiuretic hormone therapy; (e) excessive urinary excretion of aldosterone, with normal 17-oxosteroids and 17-hydroxycorticosteroids; if the patient is hypertensive, as in primary hyperaldosteronism and some types of secondary hyperaldosterone, plasma renin concentrations must be measured—elevated renin levels with elevated plasma aldosterone indicates secondary hyperaldosteronism, whereas suppressed renin levels with elevated plasma aldosterone indicates primary hyperaldosteronism (Fig. 22-8).

Adrenogenital Syndrome. This may occur as a result of (1) congenital abnormalities or (2) acquired adrenal disease.

(1) Congenital Adrenal Hyperplasia. Three varieties may be distinguished on the basis of clinical manifestations and patterns of abnormality of adrenocortical hormone metabolism. The normal pathways of biosynthesis of corticoids (Fig. 22-10 on page 736) involve (a) 21-hydroxylation (17-hydroxyprogesterone to 11-deoxycortisol, compound S; progesterone to 11-deoxycorticosterone), and (b) 11-hydroxylation (11-deoxycortisol to cortisol; 11-deoxycorticosterone to corticosterone, the normal pathway to aldosterone). The congenital forms of the adrenogenital syndrome are due to congenital deficiencies in one or other of the enzymes responsible for these reactions, viz., 21-hydroxylase and 11-hydroxylase.

Inadequacy of the 21-hydroxylation reaction results in the following sequence of events: (a) inadequate production of the main normal adrenocortical hormones, i.e., cortisol, corticosterone, and aldosterone; (b) the hypothalamus and anterior pituitary respond to the low levels of circulating cortisol with increased secretion of ACTH and consequent adrenocortical hyperplasia; (c) because of the existing functional block, this stimulation results in accumulation of excessive amounts of 17-hydroxyprogesterone, progesterone, and pregnenolone, with urinary excretion of abnormally large amounts of their metabolites, e.g., pregnanetriol, pregnanediol, dehydroepiandrosterone, and androstenedione with excessive formation of adrenal androgens (see Figs. 22-6 and 22-7).

Inadequacy of the 11-hydroxylation reaction, too, interferes with normal production of cortisol, corticosterone, and aldosterone, with consequent adrenocortical hyperplasia and functional stimulation owing to excessive ACTH production. In this instance, however, the location of the functional block causes accumulation of excessive quantities mainly of 11-deoxycortisol (compound S) and 11-deoxycorticosterone, with smaller accumulations of the compounds principally affected by interference with 21-hydroxylation. This is reflected in urinary excretion of excessive quantities of tetrahydro derivatives of these two hormones.

Three clinical forms of congenital adrenal hyperplasia are recognized: (a) the classical type, characterized by virilization in girls and precocious puberty in boys often manifested at birth, (b) a hypertensive type, in which this feature is present in addition to virilization, (c) a "salt-losing" type, in which, in addition to virilization, there is excessive urinary loss of sodium, with hyponatremia, hyperkalemia and dehydration. Although these clinical manifestations must be consequences of the abnormalities of adrenocortical hormone biosynthesis, there is little precise knowledge linking particular biochemical features with each clinical type. Virilization is perhaps due to excessive secretion of an as yet unidentified androgenic metabolite of pregnenolone. Hypertension is probably dependent upon excessive secretion of deoxycorticosterone. The pathogenesis of the "salt-losing" form is not clear. It has been attributed by some to secretion of a hypothetical "salt-losing" hormone, the presence of which has not been demonstrated.

(2) Adrenogenital Syndrome Developing in Later Life. In children,

736   ENDOCRINE FUNCTION

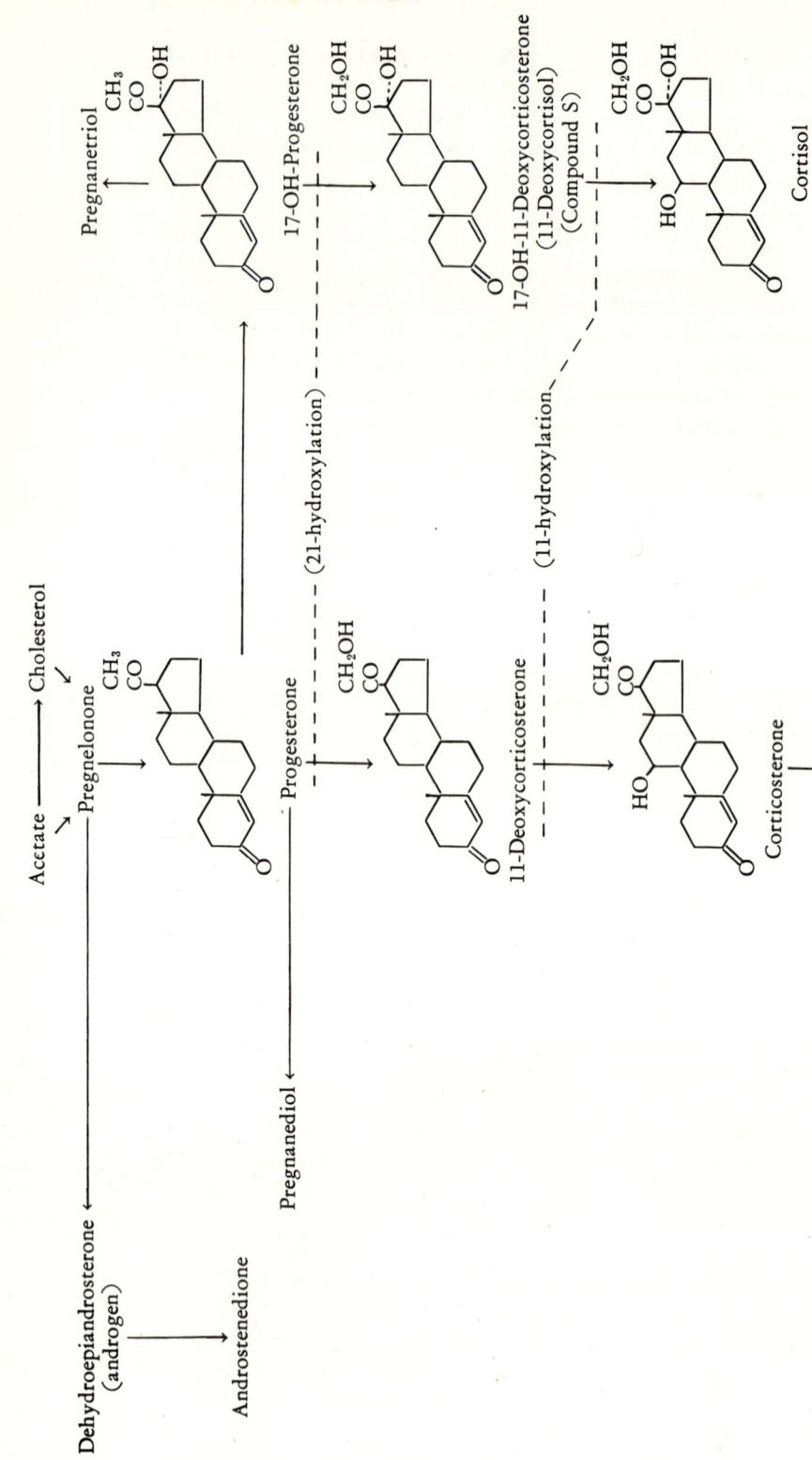

Figure 22-10. Location of metabolic defects in congenital adrenal hyperplasia.

the clinical manifestations are those of precocious pseudopuberty, heterosexual (male) in girls, isosexual (male) in boys. In general, the clinical manifestations and hormonal pattern resemble those in congenital adrenal hyperplasia; the majority of cases are due to unilateral adrenocortical tumors but occasionally to diffuse bilateral hyperplasia. The same applies to the adrenogenital syndrome developing in adult women, characterized by manifestations of virilization.

The endocrinological diagnostic criteria in the adrenogenital syndrome may be summarized as follows:

(1) Urinary 17-Oxosteroids. These substances are usually moderately to markedly increased. The presence of large amounts of dehydroepiandrosterone, which may be measured independently, is suggestive of adrenocortical carcinoma.

(2) Pregnanetriol. Urinary excretion of pregnanetriol, the metabolite of 17-hydroxyprogesterone, present normally in small quantities in adults (< 3 mg./24 hours), usually exceeds 4 mg. in infants with the adrenogenital syndrome and 7 mg. in adults. The values are usually higher in cases of tumor than in hyperplasia.

(3) ACTH in Blood. The concentration of ACTH in blood is elevated in congenital adrenal hyperplasia, although this is of limited diagnostic value.

(4) Urinary 11-Deoxycortisol (Compound S). Increased amounts of this substance (17-hydroxy-11-deoxycorticosterone) are excreted in the urine in cases of the hypertensive form of the adrenogenital syndrome. This does not occur in the other types.

(5) Adrenal Suppression Tests. Characteristically, administration of large doses of synthetic cortisol analogs, e.g., fluoroprednisolone, dexamethasone, is followed by decrease in the urinary excretion of 17-oxosteroids and pregnanetriol in cases of adrenogenital syndrome due to hyperplasia and only occasionally in those due to tumor. This procedure is therefore very useful, but not absolutely reliable, in differentiating these two conditions. Administration of these hormones to patients with the hypertensive form of this syndrome results in decrease also in urinary excretion of compound S and fall in blood pressure.

Adrenocortical Hypofunction

Adrenocortical hypofunction may be caused by a destructive primary lesion of the adrenal glands such as tuberculosis or autoimmune adrenalitis (Addison's disease). It may also be secondary to hypofunction of the hypothalamus or anterior pituitary, in which case lack of adrenocorticotrophic stimulation causes atrophy of the adrenal cortex with depression of function. Adrenocortical insufficiency may be manifest only during severe stress when the patient has limited adrenocortical reserve due to primary or secondary adrenal disease and cannot increase corticosteroid secretion sufficiently when stressed, although functioning adequately under basal conditions. Adrenocortical insufficiency is occasionally induced deliberately by bilateral adrenalectomy, e.g., for treatment of mammary carcinoma. It occurs also in patients who had been receiving large doses of adrenocortical hormones, following cessation of such therapy owing to prolonged suppression of the hypothalamic CRH and pituitary ACTH secretion.

Clinical Features. The patient usually has lost weight, has anorexia, vomiting and hypotension, and is pigmented unless the adrenal insufficiency is due to ACTH deficiency, in which the pigmentation does not occur.

Hormone Studies. The urine steroid studies which were performed in the past are now rarely performed because they are most unreliable as well as inconvenient. Measurement of the plasma corticosteroid response (measured fluorimetrically or by protein-binding methods) to exogenous ACTH is the most useful screening procedure when seeking to confirm a diagnosis of Addison's disease. 250 μg. tetracosactrin (Synacthen or Cortrosyn, a synthetic ACTH analogue, containing the first 24 amino acids of natural ACTH necessary for corticosteroidogenesis) or 40 units of soluble natural ACTH are given intramuscularly at 8:30 A.M., and blood is obtained for plasma cortisol assay before and at 30 and 60 minutes afterwards. The plasma corticosteroids should rise by 7 μg./100 ml. to more than 20 μg./100 ml. An impaired result suggests adrenal insufficiency, and to differentiate between primary adrenal disease (Addison's disease) and secondary adrenal failure (ACTH deficiency) the patient must be given a long-acting ACTH to stimulate a gland which is simply atrophic, as occurs in the prolonged absence of ACTH. Depot-tetracosactrin 2 mg., or ACTH-gel 80 units, is given intramuscularly daily for 3 days and the simple short ACTH stimulation test is repeated. In patients with Addison's disease, there is little or no change in the plasma corticosteroids, but if the glands were merely atrophic the levels rise above 20 μg./100 ml.

Carbohydrate Metabolism. The lack of glucocorticoid secretion results in impairment of carbohydrate metabolism. The most important defects in this connection are (1) impairment or cessation of gluconeogenesis, and (2) increased utilization of glucose by the tissues.

Clinically, the result of these deficiencies is a tendency toward fasting hypoglycemia and increased sensitivity to insulin. There is considerable variation in the extent of this disturbance in different patients. Fasting blood sugar levels (after the conventional overnight fast) tend to be on the low side of normal, or slightly subnormal, but some patients exhaust their carbohydrate stores so rapidly and so completely that even shorter periods of fasting may lead to severe hypoglycemic manifestations. Any condition which further increases carbohydrate utilization will increase the tendency to spontaneous hypoglycemia. Such conditions include severe infectious diseases or exercise.

Oral glucose tolerance tests are characterized by an inadequate rise of blood sugar ("flat curve"), due chiefly to inadequate absorption. However, this type of curve is encountered so frequently in other conditions that it is of no diagnostic value.

Insulin Tolerance Test. Following intravenous injection of a standard dose of insulin (0.1 or 0.15 unit/kg.) in the fasting state, the blood sugar normally falls rapidly, attaining minimum levels of about 30 to 40 mg. per 100 ml. in twenty to thirty minutes and returning to the fasting level in one and a half to two hours. In patients with adrenal insufficiency, the blood sugar usually falls to 20 mg. per cent or less, and remains at hypoglycemic levels for prolonged periods, sometimes showing little or no indication of an attempt to rise from the minimum levels attained. In some cases, severe manifestations of hypoglycemia are induced, including coma, and the test procedure must be interrupted by intravenous injection of glucose. As in the case of glucose-induced hypoglycemia, this phenomenon is due to a lack of adrenal cortical hormones to counteract the effect of the added insulin. This insulin test was formerly used as a diagnostic test for adrenocortical insufficiency but it has largely been replaced by ACTH stimulation tests using plasma corticosteroid assays, as indicated earlier. However, the latter tests will not differentiate mild cases of secondary adrenocortical insufficiency when basal corticosteroid levels are normal, as in the response to ACTH. These patients show inadequate cortisol response to stress. If, therefore, plasma corticosteroids are measured during the insulin-induced hypoglycemia test described above, secondary adrenocortical insufficiency will be indicated if the plasma corticosteroids fail to rise above 20 mg./100 ml. in response to the fall in blood glucose to less than 40 mg./100 ml. A rise to greater than this value excludes adrenal insufficiency enough to require corticosteroid replacement therapy.

Because injection of insulin in fasted patients with adrenal insufficiency may be followed by hypoglycemic coma, this test should be carried out with the utmost care, having glucose solution and syringes at hand to cope with any emergency.

Electrolyte and Water Metabolism. Deficient secretion of mineralocorticoids results in loss of sodium, chloride and water, and retention of potassium. The loss of sodium is somewhat in excess of that of chloride. As a result of this electrolyte disturbance there is a tendency toward hemoconcentration, dehydration, and acidosis; the tendency of the pattern of serum electrolytes is toward low chloride and sodium and elevated potassium values. Patients with adrenocortical insufficiency are unable to excrete a water load normally, since both mineralocorticoids and glucocorticoids are required to maintain the glomerular filtration rate. This was used in the past as a diagnostic test.

However, in chronic stages these changes are often not sufficiently marked to be of diagnostic value. Only extreme changes are recognizable by study of the concentrations of serum electrolytes. These changes are characteristically present in Addisonian crisis, i.e., hypochloremia, low serum sodium, elevated serum potassium, acidosis, and hemoconcentration. It is understandable that many factors precipitating crisis in the Addisonian patient may operate by aggravating the electrolyte imbalance present in the chronic stage, e.g., profuse sweating, diarrhea, pneumonia, etc. Unlike patients with primary adrenal insufficiency, those with pituitary or hypothalamic disease resulting in adrenal failure suffer much less from disturbances of sodium and potassium. This results because aldosterone secretion is only a little dependent on ACTH secretion, so that basal aldosterone levels and the responses to salt depletion and loading remain normal when ACTH is deficient.

Kidney function is seriously impaired in Addisonian crisis, apparently due in part, but not entirely, to the low blood pressure (glomerular filtration pressure inadequate). Even before complete anuria occurs, there is a rise in NPN and blood urea N, and a decrease in creatinine clearance; PSP excretion is reduced.

In periods between crises, with the blood pressure and circulation fairly adequately maintained, kidney function may be normal. However, clearance and concentration tests usually reveal impairment of renal function.

ADRENAL MEDULLARY HORMONES

Two biologically active compounds have been isolated from the adrenal medulla and synthesized, viz., epinephrine (adrenaline) and norepinephrine (noradrenaline), the former being a methylated derivative of the latter. These substances are closely related to the amino acids, tyrosine and phenylalanine.

The term "chromaffin tissue" has been applied to tissues in which the "chromaffin reaction" can be developed, i.e., oxidation of epinephrine and certain related compounds by bichromate or iodate, with the production of a colored compound.

In man, in addition to the adrenal medulla, substances giving this "chromaffin" reaction are present in cells of the paraganglia and various parts of the sympathetic nervous system. The material in these situations is norepinephrine, which is formed in postganglionic sympathetic fibers upon stimulation and is the chemical transmitter of adrenergic nerves.

Storage of Catecholamines. Epinephrine and norepinephrine are localized in the adrenal medulla in storage (chromaffin) granules, which also contain adenosine triphosphate.

The chromaffin granules measure 500–4000 Å in diameter, whereas the granules in sympathetic terminals are only about 500 Å in width. Certain functions may be attributed to the storage granules:

   uptake of dopamine (a precursor of norepinephrine)
   conversion of dopamine to norepinephrine
   storage of norepinephrine and epinephrine
   uptake of norepinephrine and epinephrine from the circulation
   protection of norepinephrine and epinephrine from degradation by mono-amine oxidase
   release of norepinephrine or epinephrine in response to nervous stimulation
   recapture of part of the released norepinephrine

The granules act as a tissue buffer for catecholamines, which can be compared with the tissue or plasma-binding proteins of other hormones.

Release of Catecholamines. Most of the epinephrine in the circulation is derived from the adrenal medulla. Bilateral adrenalectomy causes the urinary epinephrine output to fall by over 80 per cent but norepinephrine excretion remains unchanged. After several years, urinary epinephrine excretion approaches the original level, indicating that extra-adrenal chromaffin tissue has taken over the role of the adrenal medulla in synthesizing this hormone. The adrenal medulla releases catecholamines in response to stress of various sorts, e.g., fear, anger, pain, hypoglycemia, surgical trauma, muscular activity and certain drugs. Usually both catecholamines are released, and it is not known whether separate chromaffin cells are responsible for the release of each hormone.

Norepinephrine is secreted into the circulation both by the adrenal medulla and by sympathetic nerve endings, the latter being the main source. Much of the norepinephrine in the blood is derived from sympathetic nerves in the heart.

Biosynthesis and Metabolism. Tyrosine and phenylalanine are the primary precursors, and conversion of both to norepinephrine and epinephrine has been demonstrated (Fig. 22-11, on page 742). Tyrosine is oxidized (tyrosinase) to 3,4-dihydroxyphenylalanine (dopa). This undergoes decarboxylation (dopa decarboxylase) to 3,4-dihydroxyphenylethylamine (dopamine), which is, in turn, oxidized to norepinephrine. The latter is methylated by the enzyme phenylethanolamine N-methyl transferase to epinephrine.

The three catecholamines (dopamine, norepinephrine, epinephrine) possess hormonal activity. All are present in the adrenal medulla, the synthetic process proceeding to epinephrine as the major end-product under normal conditions. In nervous tissue, biosynthesis stops at the norepinephrine level, but dopamine may act as a local hormone in the lungs, intestine, liver, and brain.

Epinephrine and norepinephrine are present in the blood plasma in both free and conjugated (sulfate, glucuronide) forms; epinephrine is almost completely bound to plasma proteins, chiefly albumin, and norepinephrine to a lesser extent.

Circulating catecholamines either are taken up by tissues to be stored or degraded or are excreted in the urine. Tissue uptake of catecholamines depends on the blood supply of the tissue and the extent of its sympathetic innervation. When the sympathetic innervation is rich, the catecholamines are taken up and bound in the storage granules. Brain tissue takes up very little circulating catecholamines and its content is largely derived from endogenous synthesis. Sympathetic denervation abolishes the ability of a tissue to concentrate catecholamines.

Tissue catecholamines are derived from endogenous synthesis and from the circulation. At least two pools of catecholamines can be detected: an active pool that turns over rapidly and can be released by tyramine or by sympathetic nerve activity when it is destroyed in the circulation or taken up again into nerve endings to be stored or to be O-methylated; and a relatively inactive pool, which acts as a store of catecholamines. Tyramine has little effect on this pool, which is degraded mainly by mono-amine oxidase.

742   ENDOCRINE FUNCTION

The relatively brief duration of action of epinephrine and norepinephrine reflects their active metabolism. They disappear rapidly following injection; less than 5 per cent of the amount administered is excreted as such in the urine during the subsequent 24 hours. The bulk of the remainder appears in the urine in the form of inactive metabolites, the most abundant of which are 3-methoxy-4-hydroxymandelic acid, 3,4-dihydroxymandelic acid, 3-O-methylepinephrine (metanephrine) and 3-O-methylnorepinephrine (normetanephrine) (Fig. 22-11). The most important reactions involved in these transformations are oxidative deamination (amine oxidase) and methylation (catechol O-methyl transferase). There are also sulfate and

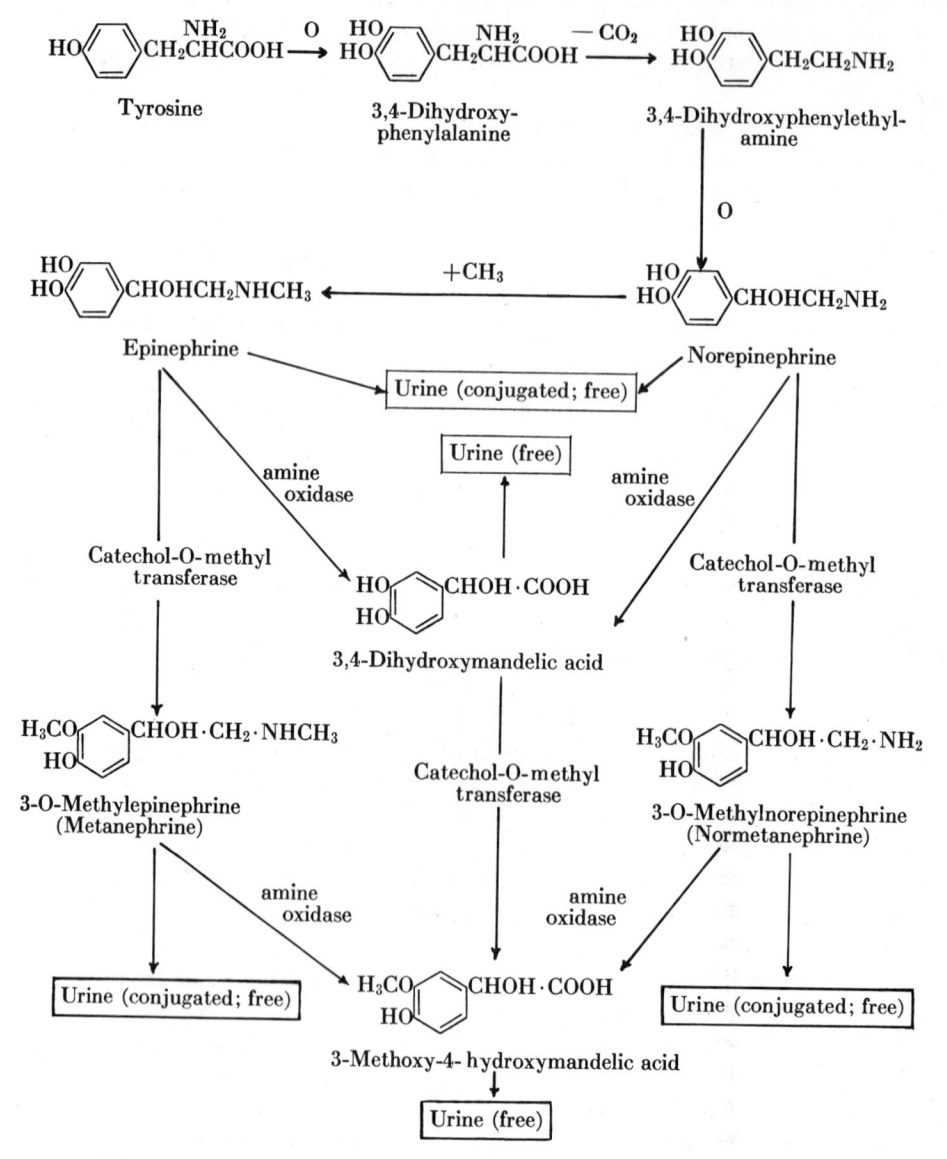

Figure 22-11. Biosynthesis and metabolism of adrenal medullary hormones.

glucuronide conjugates of epinephrine and norepinephrine as well as of certain metabolites.

Small amounts of these hormones are excreted normally in the urine, up to 20 μg. of epinephrine and 10 to 70 μg. of norepinephrine daily. The major metabolite, 3-methoxy-4-hydroxymandelic acid (vanillylmandelic acid, VMA or HMMA), is excreted normally in the urine in amounts up to 8 mg. daily.

Excessive amounts of epinephrine and norepinephrine are produced by certain tumors of chromaffin tissue (pheochromocytomas), with consequent increased urinary excretion of catecholamines and mandelic acids, quantitative determination of which is of diagnostic value in these conditions. Normal values for these hormones in the blood vary with the method employed for their determination.

Metabolic Actions of Catecholamines. The biological effect of catecholamines in the circulation depends on the amount fixed in the tissue and on the sensitivity of the binding sites of the tissue receptors. The tissue receptors have not been isolated but are a convenient theoretical concept. They comprise α-adrenergic receptors, which cause smooth muscle contraction and β-adrenergic receptors, which mediate smooth muscle relaxation. Epinephrine acts on both α- and β-receptors, causing smooth muscle contraction in certain sites and relaxation in others. Epinephrine mobilizes fatty acids by a mechanism mediated through β-adrenergic receptors. The synthetic adrenergic agent isoprenaline acts largely on β-receptors in the heart, smooth muscle of bronchi, skeletal muscle, blood vessels and the gut. Norepinephrine acts mainly on α-receptors and has little action on β-receptors except in the heart. The receptors responsible for some of the metabolic actions of catecholamines cannot be readily fitted in the α-, β-receptor theory. Propranolol and practolol act largely as β-adrenergic blockers, although the action of the latter is more or less restricted to the heart. Thymoxamine, phentolamine and phenoxybenzamine are α-adrenergic inhibitors, although with the exception of thymoxamine the actions of these compounds are not pure.

Mechanism of Action of Catecholamines. The action of adrenaline on carbohydrate metabolism is mediated by stimulation of the formation of cyclic 3′, 5′-adenosine monophosphate from ATP. Cyclic AMP causes activation of phosphorylase which breaks down glycogen to form glucose-1-phosphate; this is then converted to glucose-6-phosphate. Glucose-6-phosphate is available for formation of free glucose or to enter the hexose monophosphate or Embden-Meyerhof pathways.

Effect of Drugs on Catecholamine Metabolism. Many drugs affect catecholamine metabolism, and some of these actions are outlined below. Some drugs have more than one action.

Drugs Interfering with Catecholamine Uptake by Sympathetic Nerve Endings (Cocaine, Imipramine, Ephedrine and Amphetamine). These drugs potentiate catecholamine action, since uptake of the hormones by nervous tissue is the major method by which they are removed from the circulation.

Drugs Causing Release of Bound Catecholamines (Phenylephrine,

Tyramine, Ephedrine, Amphetamine and Reserpine). Some of these agents also elevate the blood pressure by a direct action on α-receptors. Reserpine usually causes a fall in blood pressure because the catecholamines it releases are broken down by mono-amine oxidase, and its central hypotensive action is unopposed.

Drugs Blocking Release of Catecholamines (Bethanidine, Guanethidine, Ganglion-blocking Agents and Inhibitors of Mono-Amine Oxidase). Bethanidine and guanethidine first discharge catecholamines and then block any further release from tissues. Ganglion-blocking drugs which interfere with preganglionic sympathetic nerve impulses prevent the release of catecholamines by the postganglionic terminals. Mono-amine oxidase inhibitors break down catecholamines within the nerve terminals, particularly from the second storage pool.

False Transmitters. These are formed in the presence of α-methyldopa and are taken up by the storage granules in competition with catecholamines. They lower tissue catecholamine content and sympathetic activity.

Drugs Interfering with the Peripheral Action of Catecholamines. These include α-adrenergic blockers (thymoxamine, phenoxybenzamine and phentolamine) and β-adrenergic blockers (propanolol, practolol).

Underproduction of Adrenal Medullary Hormones

Total removal of both adrenal glands can be corrected by administration of adrenocortical hormones alone, and there is no evidence that adrenal medullary deficiency has any clinical sequelae. After adrenalectomy extramedullary chromaffin tissue gradually takes over the role of the adrenal medulla in secreting epinephrine.

Overproduction of Adrenal Medullary Hormones

This is caused by tumors of the adrenal medulla or of accessory chromaffin tissue. Tumors that arise from tissues whose origin is in the neural crest are sympathoblastomas, neuroblastomas, ganglioneuromas, pheochromocytomas, neurofibromas, melanomas, neurilemmomas and argentaffinomas. The first four tumors in this list are derived from sympathetic nervous tissue, the first two of these being embryonic in type and the second two, mature. Recently, some patients with retinoblastomas, which are also of sympathetic origin, have been shown to excrete an excess of catecholamine metabolites in the urine. Very rarely, secretion of catecholamines can be demonstrated by carotid body (glomus jugulare) and melanotic neuroectodermal tumors.

Sympathoblastoma. These highly malignant tumors occur during intrauterine life or during the first year.

Neuroblastoma. Neuroblastomas are highly malignant neoplasms occurring in childhood. Most cases occur before the age of five years. About one-third occur in the adrenal medulla, but it may be difficult to be sure of the site from which a retroperitoneal neuroblastoma has arisen. About

25 per cent of tumors arise in the thorax. Metastases by the blood and lymphatics reach the liver, bones and lymph nodes. Calcification is not infrequent and may be sufficient to be visible on x-ray pictures.

Clinical and Biochemical Features. These are usually the result of local or distant spread and less commonly due to catecholamines which the neoplasm may produce. In most cases, even in those without features of catecholamine excess, catecholamines or their degradation products can be detected in the urine in increased amounts. Dopamine, HVA, epinephrine, norepinephrine and HMMA (VMA) may be found. In practice, HMMA or HVA screening is usually carried out. If this is positive, the diagnosis is confirmed. Occasionally, tests for total catecholamines are positive when HMMA is negative and, rarely, no catecholamines or their metabolites can be found.

Ganglioneuroma. Ganglioneuromas are rare benign or malignant tumors occurring in young adults.

Pheochromocytoma. Pheochromocytomas (PCC) may be familial, and such cases are more often associated with neurofibromatosis or, more rarely, medullary carcinoma of the thyroid. Ninety per cent of all PCC originate in the adrenal medulla, ten per cent being malignant, when they are usually bilateral. Extra-adrenal PCC may develop anywhere along the course of the sympathetic chain from neck to pelvis and even in the bladder wall, but often the paired para-aortic bodies of Zuckerkandl are affected. Children more frequently develop extra-adrenal, multiple PCC.

Clinical Features. These are typically the result of increased catecholamine production by the tumor. The PCC may simulate anxiety neurosis, hyperthyroidism, spontaneous hypoglycemia, diabetes mellitus, renal disease and, most important, essential hypertension. Its clinical features can be grouped under the headings hypertension, hypermetabolism and hyperglycemia.

Diagnosis. The history of intermittent headache and sweating, especially in a hypertensive patient, should always raise the possibility of PCC. All hypertensive patients should be screened for the presence of PCC. Three groups of tests are applied to patients with possible PCC to confirm overproduction of catecholamines:

Direct estimation of catecholamines or their degradation products in blood and urine

Blocking tests with phentolamine

Provocative tests with histamine or tyramine

Direct Estimation of Catecholamines and Metabolites. Plasma levels of total catecholamine are normally less than 1.0 μg. per liter and daily urine excretion contains less than 20 μg. of epinephrine and 70 μg. of norepinephrine. Much higher amounts are found in most patients with pheochromocytomas. It is helpful to know the type of catecholamine which the tumor is secreting, since extra-adrenal PCC secrete largely norepinephrine, as do metastases from adrenal tumors. If epinephrine is secreted in excess, it is likely that the tumor arises in the adrenal medulla. Malignant PCC have a tendency to secrete large amounts of dopamine, but not all tumors secreting dopamine and its metabolites are malignant. Since only a small fraction of the catecholamines secreted each day are excreted unchanged in the urine and since the plasma levels of catecholamines are low, the urinary excretion of their metabolites should be assayed.

The routine procedure of choice in most laboratories is estimation of the daily urine output of HMMA (VMA). Figures vary from different laboratories, but normally less than about 8 mg. of HMMA are excreted each day. To prevent false positive results, the patient should avoid coffee, tea, chocolate, ice cream, bananas and all drugs for 48 hours before the start of the urine collection. Measurement of the excretion of epinephrine and norepinephrine should also be performed not only because the findings may be helpful in determining the type of catecholamine that the tumor is producing, and hence suggesting whether the tumor is intra- or extra-adrenal, but also because some patients with PCC have elevated epinephrine output with normal HMMA excretion. A few patients show normal basal urinary catecholamines with a rise only after an episode of hypertension. Occasionally, PCC secrete hydroxytyramine which has less of a pressor action than norepinephrine.

Blocking Tests. Previously, reliance has been placed on the blood pressure response to the α-adrenergic-blocking drug phentolamine. Unfortunately, false negative and false positive results are common and there now seems to be little place for the phentolamine test.

Provocative Tests. These are dangerous and should no longer be performed.

Localization of Tumor. Once overproduction of catecholamines has been confirmed an attempt is made to determine the site of the tumor. A plain x-ray picture of the abdomen may show displacement of the renal outline or, rarely, calcification. An intravenous pyelogram may show downward displacement of the kidney. Selective adrenal arteriography or aortography is sometimes helpful. Such procedures may provoke hypertensive attacks, and patients should be prepared with adrenal-blocking agents. Selective venous catheterization of the inferior vena cava, taking serial blood samples for catecholamines as the catheter is withdrawn, can sometimes indicate the level and placement of the tumor.

THYROID HORMONES

Chemistry. The thyroid gland contains an iodized glycoprotein, "thyroglobulin," characteristically present in the colloid of the thyroid follicles, which is apparently the "storage form" of the thyroid hormone. It has a molecular weight of about 680,000. Hydrolysis yields several iodine-containing derivatives of tyrosine, viz., mono- and diiodotyrosine (MIT and DIT) and di- and triiodothyronine (two forms) and thyroxine (tetraiodothyronine, T_4) (Fig. 22-12). 3,5,3'-Triiodothyronine (T_3) has four to ten times the biological activity of thyroxine. The naturally occurring forms are levorotatory; synthetic thyroxine shows no optical rotation, being a

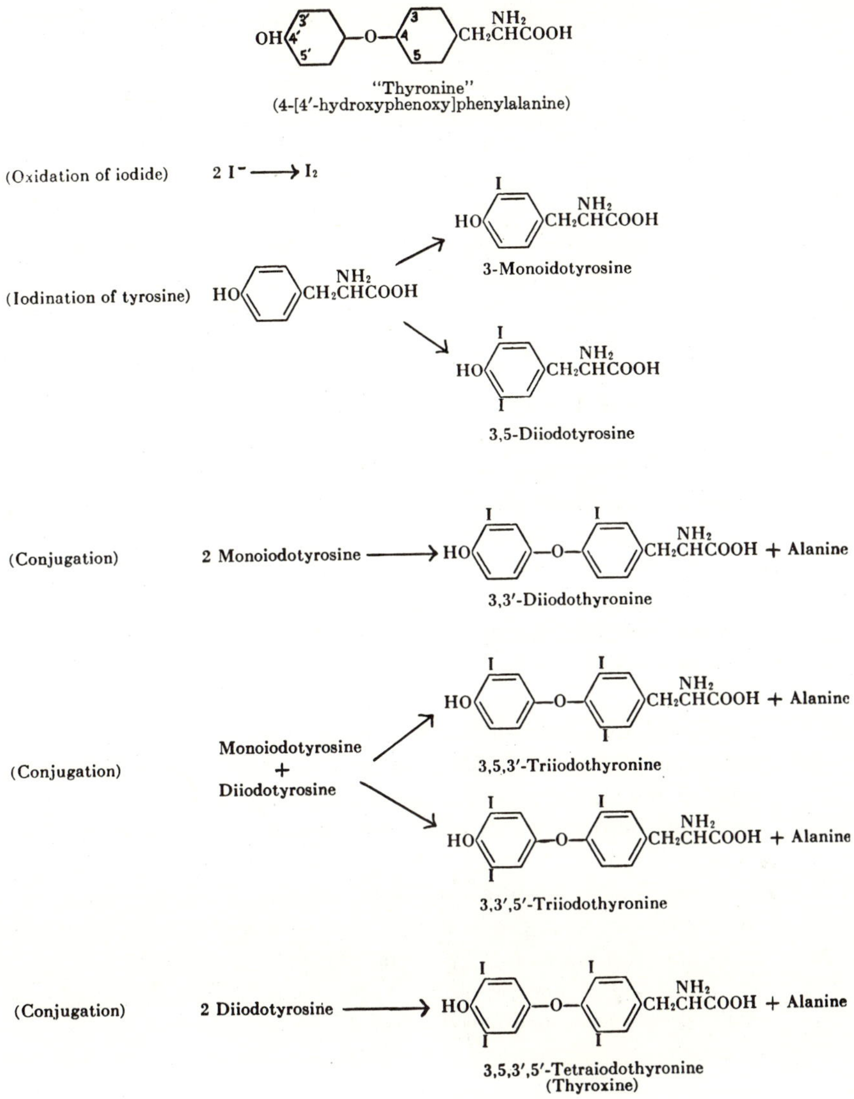

THYROID HORMONES

"Thyronine"
(4-[4'-hydroxyphenoxy]phenylalanine)

(Oxidation of iodide) $2\ I^- \longrightarrow I_2$

(Iodination of tyrosine)   Tyrosine → 3-Monoiodotyrosine → 3,5-Diiodotyrosine

(Conjugation)   2 Monoiodotyrosine ⟶ 3,3'-Diiodothyronine + Alanine

(Conjugation)   Monoiodotyrosine + Diiodotyrosine ⟶ 3,5,3'-Triiodothyronine + Alanine   and   3,3',5'-Triiodothyronine + Alanine

(Conjugation)   2 Diiodotyrosine ⟶ 3,5,3',5'-Tetraiodothyronine (Thyroxine) + Alanine

Figure 22-12. Biosynthesis of thyroid hormones.

mixture of D- and L- forms (racemic), the former being apparently biologically inactive.

Biosynthesis and Secretion. The function of the thyroid in the formation and secretion of thyroid hormone may be divided into six phases: (1) entry of inorganic iodide from the circulation; (2) concentration of inorganic iodide in the gland; (3) synthesis of thyroglobulin in follicular cells; (4) storage of thyroglobulin in colloid; (5) liberation of thyroxine and triiodothyronine from thyroglobulin; (6) passage of these compounds into the circulation.

Accumulation of Iodine in the Thyroid. Inorganic iodide, present in the blood plasma in low concentration (0.04–0.57 μg. per cent), is actively removed by the acinar cells of the thyroid, which are able to "trap" iodine in enormously higher concentration than that in the circulation. Active transport of iodide occurring at the basal cell membrane is an energy-dependent process involving membrane-bound adenosine triphosphatase. The normal human thyroid concentrates iodide from the blood at a rate of about 2 μg. per hour. Accumulation of iodine in the thyroid is accelerated and increased by administration of thyroid-stimulating hormone (TSH) and is markedly reduced by thiocyanate and perchlorate.

The thyroid contains inorganic iodide, monoiodotyrosine, diiodotyrosine, thyroxine, triiodothyronine and other as yet unidentified iodinated compounds. Almost all (about 97%) of the iodine in the gland is firmly bound to protein (thyroglobulin), the organic compounds mentioned above being liberated by hydrolysis. Free thyroxine (possibly loosely linked to protein), though present in small amounts (0.5% of the total I), nevertheless exists in over one-hundred times the concentration of protein-bound thyroxine in the plasma.

Synthesis of Thyroid Hormones. Although all details have not been definitely established, present evidence supports the following sequence of events.

Oxidation of Iodide and Iodination of Tyrosine. Before iodide trapped by the thyroid can be used for hormone synthesis, it is oxidized to some active form by a peroxidase enzyme system utilizing hydrogen peroxide generated by oxidation of TPNH by a flavoprotein, which can be auto-oxidized to produce H_2O_2. Methimazole specifically inhibits the peroxidation of iodide, a reaction responsible for its therapeutic effect. Oxidation of iodide occurs very rapidly, with the result that the thyroid contains very little free iodide. The active form of iodine is able to combine with tyrosine attached to thyroglobulin to form mono- and diiodotyrosine. When iodine intake is normal, approximately equal numbers of mono- and diiodotyrosine residues are present in the thyroglobulin molecule.

Coupling of Iodotyrosines. Appropriately placed molecules of MIT and DIT then couple to form T_4 and T_3. The nature of the coupling reaction is uncertain and no specific enzyme has been identified. It is possible that the reaction is not enzymatic but depends on the steric relationships of adjacent MIT and DIT residues, themselves dependent on the tertiary structure of thyroglobulin. Iodine deficiency is associated with a fall in the number of DIT residues and T_4 per molecule of thyroglobulin and a rise in the MIT/DIT ratio with preferential production of T_3. Hormone-containing molecules of thyroglobulin are then stored in the colloid of the follicular lumen until required.

Release of Thyroid Hormones from Thyroglobulin. The apical surface of the thyroid cell is formed by a series of microvilli which engulf thyroglobulin-containing colloid droplets by pinocytosis. The colloid droplets fuse with lysosomes containing proteolytic enzymes which hydrolyze thyroglobulin and release T_4, T_3, DIT and MIT. The iodotyrosines are rapidly converted to iodide and tyrosine by a dehalogenase enzyme system and the iodide is reutilized for thyroid hormone formation, possibly after entering a separate compartment from iodide trapped from the blood. T_4 and T_3 are released into the circulation along with small amounts of thyroglobulin which escapes hydrolysis and reaches the blood via the thyroid lymph.

Pituitary Hypothalamic Control. The series of reactions involved in the synthesis and release of thyroid hormones is under the control of TSH. A fall in the level of free thyroid hormones in the circulation leads to increased TSH secretion, which stimulates the formation and release of more thyroid hormones. Conversely, increased circulating free thyroid hormone levels lead to suppression of TSH release and reduced thyroid hormone production. Thyroid hormones act directly on the anterior pituitary to influence the release of TSH. Synthesis and release of TSH from the anterior lobe is mediated by the tripeptide pyroglutamylhistidylprolinamide, thyrotrophin-releasing hormone (TRH). Increased circulating levels of thyroid hormones act by some protein intermediate to inhibit TRH-mediated TSH release; hence, serum TSH levels are low in those with hyperthyroidism. TRH is synthesized over a wide area of the hypothalamus and is stored in the median eminence, from which it is released into the portal veins passing down the pituitary stalk to the anterior lobe. Preliminary evidence suggests that thyroxine acting at a hypothalamic level may increase TRH synthesis, but the significance of this is as yet unknown. Animal experiments have shown that inhibition of deiodination of T_4 to T_3 by propylthiouracil prevents the blocking effect of T_4 on TRH-mediated TSH release, implying that T_4 itself is metabolically inert and requires conversion to T_3 before it is effective.

Thyroid-stimulating Hormone (TSH). TSH is a glycoprotein, molecular weight of 28,000, which consists of two nonidentical polypeptide units designated α and β. The α-subunit has a very similar if not identical structure to the α-subunits of the other glycoprotein hormones, LH, FSH and HCG, and is biologically inactive. The β-subunit differs markedly from those of the other glycoproteins, although they have some sequences in common, and has only slight biological activity by itself.

Secretion of TSH is under the control of TRH from the hypothalamus and of circulating T_4 and T_3. There is good evidence for a circadian rhythm of TSH production, probably mediated by TRH, increased levels occurring at about two hours or more after the onset of sleep, peaking between 2 and 4 A.M.

Almost every metabolic process in the thyroid can be stimulated by TSH. Iodide trapping, formation of thyroid hormones and their release are all increased by TSH, as are most of the processes of intermediary metabolism within the gland, e.g., glucose oxidation, ribonucleic acid synthesis and phospholipid formation. The mechanism of action of TSH on the thyroid is still uncertain, but the following sequence is acceptable at present. TSH binds to a specific hormone receptor on the thyroid cell membrane, leading to activation of the enzyme adenyl cyclase, which then converts adenosine triphosphate (ATP) to cyclic adenosine monophosphate (cyclic AMP). The processes by which cyclic AMP mediates the effects of TSH on the thyroid are not yet fully understood, but it seems likely that cyclic AMP activates a protein kinase which then phosphorylates a variety of proteins, thereby causing activation or inactivation of their functions.

Circulating Thyroid Hormones

Thyroid hormones are bound to proteins in the plasma, only a small proportion being present in the free metabolically active state. Approximately 0.024 per cent of T_4 and 0.36 per cent of T_3 are present in the free form. Thyroid hormones are bound to thyroxine-binding globulin (TBG), thyroxine-binding prealbumin (TBPA) and albumin, in decreasing order of affinity. The affinity of T_3 for TBG is two- to sixfold less than that of T_4, whereas the affinity of T_3 for TBPA is similar to its affinity for albumin. Thyroid hormone binding proteins prevent loss of thyroid hormones by the liver and kidney and regulate the delivery of free hormone to its intracellular sites of action, buffering the effects of alterations in thyroid hormone secretion and degradation. TBG forms a stable reservoir of hormone, whereas TBPA provides a labile, rapidly available supply of hormone in stressful situations where its hepatic synthesis is impaired. The lower affinity of T_3 for the binding proteins partly explains its rapidity of action and peripheral turnover.

Abnormal binding capacities of TBG and TBPA are caused by many diseases and drugs as well as genetic defects. Alterations in thyroid hormone binding capacities do not affect the thyroid status of the individual, since a normal free thyroid hormone level is maintained by the feedback mechanism.

Conditions of increased TBG and TBG deficiency are familial with sex-linked inheritance, and three types may be recognized in affected males: TBG deficiency, low TBG capacity and high TBG capacity. Females may be heterozygous, having values intermediate between affected males and those with normal TBG. Overlap in heterozygotes is most common in families with low TBG. Estrogens do not increase TBG in patients with TBG deficiency although heterozygotes do respond with an increase in TBG capacity. This group of patients is euthyroid with a normal level of free T_4 and normal daily degradation of T_4, achieved by an altered turnover rate compensating for the wide variation in the extrathyroidal T_4 pool. It has been suggested that all TBG abnormalities in man described so far are X-chromosome-linked and manifested by quantitative variations in the serum concentration of an apparently structurally unaltered TBG. Mutations at a single locus controlling TBG synthesis could then explain the entire spectrum of genetic TBG abnormalities in man.

Metabolism of Thyroid Hormones. About 10 per cent of thyroxine secreted each day is excreted in the bile, largely as free thyroxine but also conjugated as the glucuronide. Hormone not excreted in this way is deiodinated in the periphery, the iodide released being trapped by the thyroid or excreted in the urine. Although the concentration of T_4 in the tissues is lower than the free level in serum, the reverse is true for T_3, suggesting either differential trapping or conversion of T_4 to T_3 at tissue level. The major sites of T_4 metabolism are liver and muscle where inactivation occurs by deiodination, oxidative deamination and conjugation. In the liver, the main site of metabolism is the smooth endoplasmic reticulum.

There is good evidence for the peripheral conversion of T_4 to T_3, although the sites of this process, its extent, and the factors which control it are poorly understood. Some studies suggest that about one-third of the T_4 metabolized each day is deiodinated to T_3. Assuming an average T_4 degradation of 79 μg. per day, about 22 μg. of T_3 would be produced daily from T_4. If about 33 μg. of T_3 are degraded each day, then only about 11 μg. would be secreted directly from the thyroid. The extent of T_3 secretion and peripheral conversion from T_4 in thyroid and other diseases is not clear. It remains possible that T_3 is the metabolically active thyroid hormone and that T_4 is merely an inactive precursor.

Metabolic Effects. The action of thyroid hormones is characteristically evidenced in the relief of the manifestations of human myxedema. A great variety of effects is involved in the production of this phenomenon. Some of the important body functions influenced by thyroid activity are:

metabolic rate (oxygen consumption)	central nervous system activity
growth and tissue differentiation	reproduction
metabolism of carbohydrate, lipids, protein, electrolytes and water	hematopoiesis
	cardiovascular function
vitamin requirements	muscle activity
temperature sensitivity	resistance to infection
gastrointestinal activity	

Consideration will be given here only to the metabolic effects of the thyroid hormones which are of immediate biochemical interest. The biological activity of triiodothyronine is about five to ten times that of thyroxine, the relative activities differing for different functions and by different assay methods, and also in differing species. It is to be understood that where reference is made, here or elsewhere, to actions of thyroxine, the same statements apply (qualitatively) also to triiodothyronine.

Calorigenic Effect. Thyroid hormones increase the rate of energy exchange and oxygen consumption of all normal tissues except the thyroid gland itself. This is reflected in an increase in the basal metabolic rate (BMR), which is the sum of the effects, of different magnitude, on various tissues.

Protein Metabolism. In physiological doses, thyroxine favors protein anabolism, leading to increased retention of nitrogen (positive nitrogen balance). This explains the growth arrest of young thyroidectomized animals and hypothyroid children and the resumption of growth following administration of thyroid hormone. Large, unphysiological doses stimulate protein catabolism, leading to a negative nitrogen balance. Thyroidectomy (in man) is followed by accumulation of extracellular fluid rich in a hyaluronic acid-containing mucoprotein, apparently similar in composition to fetal mucin. This forms the basis of clinical myxedema. Administration of thyroid hormone is followed by increased catabolism of muscle protein.

Carbohydrate Metabolism. Thyroid hormones increase the rate of absorption of monosaccharides by the intestine. This is the chief cause of the frequently excessive rise in blood sugar concentration following ingestion of glucose by patients with thyrotoxicosis (decreased oral glucose tolerance); this does not occur following intravenous injection of glucose. Hyperthyroidism is a diabetogenic stimulus and may temporarily or, more rarely, permanently unmask latent diabetes. The mechanism of action is not fully resolved through increased turnover, and degradation of insulin is likely to be involved.

Thyroid hormone causes a decrease of glycogen in the liver and, to a lesser extent, in the myocardium and skeletal muscle. This stimulation of glycogenolysis may be due to increased sensitivity of the organism to epinephrine in the presence of excessive amounts of thyroid hormone. In the otherwise normal subject, the hyperglycemic effects of increased hepatic glycogenolysis and gluconeogenesis are usually effectively offset by simultaneously increased utilization of glucose in the tissues generally.

Lipid Metabolism. Fatty acids participate in the increased rate of oxidative processes which follows administration of thyroid hormone. If available stores of carbohydrate are depleted, ketosis develops. Despite the fact that hepatic synthesis of cholesterol and phospholipid is depressed following thyroidectomy and is increased in thyrotoxicosis, the concentration of cholesterol (and to a lesser degree phospholipid) in the plasma is frequently increased in hypothyroidism and decreased in hyperthyroidism. The concentration of plasma lipoproteins of the S_f 10-20 class is frequently increased in hypothyroidism and decreased in thyrotoxicosis or following administration of thyroid hormones to normal subjects. These effects are explicable on the basis that the thyroid hormones increase the rate of both biosynthesis and catabolism of cholesterol (liver and extrahepatic tissues), the catabolic effect predominating over the anabolic.

In the absence of thyroid hormone, the rate of removal of cholesterol and triglycerides from the bloodstream is decreased; this is increased by administration of thyroid hormone and in the presence of hyperthyroidism. Biliary excretion of cholesterol and bile acids is decreased in hypothyroidism; it is increased by administration of thyroid hormone.

Body Water and Solutes. In hypothyroidism, there is an increase in interstitial water and a decrease in plasma volume. Adequate thyroid therapy results in prompt diuresis with increased excretion of Na, Cl, K and N, indicating loss of fluid from both the extracellular and the intracellular compartments. In thyrotoxicosis, the calorigenic and catabolic effects of excessive amounts of thyroid hormone are reflected in increased urinary excretion of K (also N and PO_4) and increased water loss, mainly by vaporization from the skin and lungs (insensible loss); the plasma volume may be increased.

Serum albumin concentrations tend to be low in hyperthyroidism, whereas IgG levels are increased in some patients with autoimmune thyroid disease. Expansion of the blood volume in hyperthyroidism might contribute to the low albumin concentration. A negative nitrogen balance is common in hyperthyroidism, and occasionally muscle wasting is severe.

Serum calcium levels are usually normal in patients with altered thyroid function but rarely hypercalcemia may complicate both hypo- and hyperthyroidism. Usually, no action is required other than correcting thyroid function, when in most instances calcium levels return to normal. If hypercalcemia persists after an adequate period of observation where the patient is euthyroid, associated hyperparathyroidism should be considered. In hyperthyroidism, calcium balance may become negative owing to increased mobilization from the skeleton and increased fecal and urinary excretion. Occasionally, loss of radiological bone density develops in hyperthyroidism, sometimes complicated by vertebral collapse. Histological examination of the bone reveals increased osteoclastic activity. Progress is usually halted by correction of the hyperthyroidism. This sometimes leads to an increase in the serum calcium level.

Vitamins. Administration of large amounts of thyroid hormone increases the requirements for certain members of the vitamin B-complex (thiamine, pyridoxine, pantothenic acid) and for vitamin C. These relationships and the mechanisms involved are obscure but are presumably related to the stimulation of oxidative and catabolic processes. The increased urinary excretion of xanthurenic acid after an oral tryptophan load in those with hyperthyroidism has been interpreted as evidence of pyridoxine deficiency and the increased excretion of formiminoglutamic acid after a histidine load, as lack of folic acid. However, no consistent beneficial effects of administration of vitamins to patients with hyperthyroidism have been proved.

The serum carotene level may be elevated in myxedema, causing yellow discoloration of the skin.

Growth and Development. Absence of the thyroid during the growth period is accompanied by retardation of growth, which is corrected by administration of thyroid hormone. There is evidence that this growth-enhancing and perhaps the protein-anabolic effect of physiological amounts of thyroid hormone may be mediated at least in part by growth hormone of the anterior pituitary.

Thyroid hormone also induces tissue differentiation and maturation. In man, early absence of this hormone not only arrests longitudinal growth, with consequent dwarfing, but also delays the appearance of epiphyseal centers of ossification. This defect is corrected by administration of thyroid hormone.

The effect of this agent in hastening metamorphosis of amphibians has been employed for bioassay purposes. In the absence of thyroid hormone, e.g., in the thyroidectomized tadpole, metamorphosis does not occur; for example, the limbs do not extend and the gills and tail do not regress, although the tadpole continues to grow. Metamorphosis can be induced at any stage by administration of minute amounts of thyroid hormone. Very small tadpoles can be metamorphosed into minute frogs.

This effect in inducing tissue differentiation and development is not related directly to the calorigenic action of thyroxine, since dinitrophenol, which increases this effect, has no calorigenic action. Moreover, certain stages of amphibian metamorphosis, though stimulated by thyroid hormone, are not accompanied by increase in metabolic rate.

Mechanism of Action. The exact manner in which the thyroid hormone acts to produce its characteristic effects is not known. Principally because of the wide variety of these effects, many believe that the hormone exerts a single primary action, probably on energy metabolism, to which all other actions are secondary, depending upon specific local target-organ mechanisms and responsiveness. However, thyroid hormone may have a function in tissue differentiation and development independent of its effect on energy transformations and oxygen consumption. The hormone may be concerned exclusively with energy metabolism and, at the same time, may influence the availability of energy for many specialized processes.

Thyroid hormones alter the number and structure of mitochondria and reduce the efficiency of oxidative phosphorylation. They increase transcription of messenger RNA probably via cyclic AMP, causing an increase in protein synthesis in mitochondria and microsomes and allowing an increase in cell respiration. Many sites in the cell are responsive to thyroid hormones; the nucleus, the endoplasmic reticulum and mitochondria are all affected. A variety of metabolic processes are also accelerated, including protein breakdown in collagen, carbohydrate and lipid turnover and calcium mobilization from bone. Many of the actions of thyroid hormones are compatible with increased sensitivity of β-receptors to catecholamines, and these effects can be suppressed by β-adrenergic blocking agents such as propranolol.

Agents Interfering with Synthesis of Thyroid Hormone

A number of chemical agents can interfere with the formation of thyroid hormone by affecting different phases of the mechanism involved in its synthesis. These may be classified conveniently as follows: (1) thyroid hormone, (2) iodine, (3) thiocyanate, (4) antithyroid agents.

Thyroid Hormone. Administration of thyroid hormone results in regressive changes in the thyroid gland and diminished formation of hormone. This is due to suppression of thyrotrophin (TSH) secretion by the pituitary.

Iodine. Deficiency in iodine intake leads to deficient synthesis of thyroid hormone and compensatory hyperplasia of the thyroid gland (iodine-deficiency goiter). Conversely, administration of iodine to subjects with hyperthyroidism is followed by regressive changes in the gland and diminished output of hormone. High levels of circulating iodide also have a less pronounced and temporary inhibitory effect on thyroid hormone synthesis by the normal thyroid. The mechanism is not clear.

Thiocyanate. Administration of thiocyanate results in enlargement of the thyroid, if the blood iodide concentration is low, and decreased synthesis of thyroid hormone. This effect is due to inhibition of the uptake, concentration and accumulation of iodine by the thyroid, the gland being obliged to form hormone from the iodine which enters by passive diffusion. If the level of circulating iodine is elevated sufficiently by administration of iodine, enough may enter the gland to permit adequate function, inasmuch as thiocyanate does not influence the synthesizing mechanism *per se* (i.e., oxidation, iodination of tyrosine and coupling). The precise mechanism of its action is unknown but it is shared by other compounds which have a similar ionic volume to that of the iodide ion.

Antithyroid Agents. This designation is applied to a number of substances which inhibit synthesis of thyroid hormone, apparently by interfering in some way with the oxidation of inorganic iodide, necessary for iodination of tyrosine.

Most of these compounds fall into two categories: (1) those containing a thiocarbamide grouping (e.g., thiourea, thiouracil and related compounds), and (2) those containing an aminobenzene grouping (aniline derivatives, e.g., sulfonamides). However, other types of compounds are also active in this respect, e.g., 5-vinyl-2-thiooxazolidone (from rapeseed) and 1-methyl-2-mercaptoimidazole.

Administration of these antithyroid compounds leads to hyperplasia of the thyroid (goitrogenic action) as a result of excessive stimulation by TSH, secretion of which by the pituitary is no longer adequately restrained by circulating thyroid hormone.

The mechanism of action of these agents is uncertain. The thiourea-like compounds react rapidly with iodine, reducing free iodine to iodide at neutral pH, a phenomenon which could effectively interfere with iodination of tyrosyl groups. However, other substances which react similarly with iodine, such as glutathione, cystine and thioglycolic acid, do not inhibit thyroid function. Moreover, the aminobenzene derivatives, which apparently inhibit thyroid hormone formation at this same stage, react only very slowly with iodine. Other possible mechanisms include (1) inhibition of the oxidizing enzyme system, (2) competition for hydrogen peroxide or (3) competition for a substrate.

Laboratory Diagnosis of Thyroid Disease

Laboratory tests used in the diagnosis of thyroid disease can be divided into two groups:
1. Those measuring the level of thyroid function and
2. Those indicating the cause of thyroid dysfunction.

A suggested nomenclature for the *in vitro* thyroid function tests (Vivian N. Y. Chan) is shown in Table 22-3. In this classification, both the chemical species being measured and the method employed are clearly stated. If a solvent extraction is used, the solvent name is added in front of the chemical species being measured. Procedures used to assess the degree of saturation of thyroid hormone binding protein (TBP) are referred to as thyroid hormone binding tests (THBT). At present, the concentration of thyroid hormones is expressed as unit weight per 100 ml. (or ml.) of serum. The new International Standard System of units, which will be generally adopted in due course, is not yet being used.

Tests measuring the level of thyroid function:
    a. Thyroid iodide trap tests
    b. Thyroid hormone release tests
    c. Peripheral tissue response tests

a. Thyroid Iodide Trap Tests. These tests are less frequently used now because they involve administration of radioisotope to a patient, a procedure best avoided in infancy and during pregnancy. The isotope ^{131}I was formerly chosen because its half-life of 8 days makes it convenient for clinical use. The dose of radiation may be reduced by using the isotope ^{132}I (half-life 2.3 hours) or Technetium-99m (^{99m}Tc, half-life 6 hours). Because of the short half-life of these isotopes they are best administered intravenously and are only of value for early uptake measurements in the diagnosis of hyperthyroidism (or for thyroid scanning). Measurement of thyroid iodide clearance—the quantity of radioiodine taken up by the thyroid per unit time divided by the mean plasma activity over the time considered—gives a better index of iodide trapping than the iodide uptake to which it has a complex, nonlinear relationship.

b. Thyroid Hormone Release Tests (measuring total circulating levels of thyroid hormone).

Serum protein-bound iodine (PBI)
Serum butanol-extractable iodine (BEI)
Serum thyroxine (T_4)
Serum triiodothyronine (T_3)

Serum Protein-bound Iodine. Estimation of the serum PBI is still a valuable test of thyroid function. It measures the iodine fraction which represents 65.3 per cent of the T_4 molecule and thereby indirectly measures the serum T_4 concentration. The result is more precise than most measurements of thyroxine available at present and the method is capable of automation. Many laboratories use the PBI and a THBT test as the screening thyroid function test and employ T_4 measurements when high values are obtained or if there is any possibility of iodine contamination. Recent improvements in methodology include simple methods of pretreating serum to remove inorganic iodide by anion exchange resins and more rapid and reliable methods of digesting proteins to release bound iodine. In the automated technique (Technicon AutoAnalyzer method), perchloric and sulfuric acids are used and protein digestion to release iodine is accelerated by heating the reaction mixture to 280° C. The liberated iodine then catalyzes the ceric arsenite reaction. The normal range of PBI (mean ± 2 S.D.) is 4 to 8 µg. per 100 ml. of serum.

TABLE 22-3 NOMENCLATURE FOR THYROID FUNCTION TESTS*

NAME OF TEST	ABBREVIATION	CHEMICAL FORM MEASURED	UNITS
A. To assess hypothalamic-pituitary function			
1. Thyroid-stimulating hormone (radioimmunoassay)	TSH (RIA)	TSH	$\mu U/ml.$
B. To determine total circulating levels of thyroid hormone			
1. Protein-bound iodine	PBI	I	$\mu g./100\ ml.$
2. Butanol-extractable iodine	BEI	I	$\mu g./100\ ml.$
3. Thyroxine iodine (column)	$T_4 I$ (C)	I	$\mu g./100\ ml.$
4. Thyroxine (gas liquid chromatography)	T_4 (GLC)	T_4	$\mu g./100\ ml.$
5. Thyroxine (competitive protein-binding assay or radioimmunoassay)	T_4 (CPBA) or T_4 (RIA)	T_4	$\mu g./100\ ml.$
6. Triiodothyronine (competitive protein-binding assay or radioimmunoassay)	T_3 (CPBA) or T_3 (RIA)	T_3	$ng./100\ ml.$
C. To determine circulating levels of free thyroid hormones			
1. Free thyroxine (equilibrium dialysis or ultrafiltration)	FT_4 (D) or FT_4 (UF)	T_4 and $\%FT_4$	$ng./100\ ml.$
D. To assess the degree of saturation of thyroxine-binding sites			
1. Thyroid hormone binding test	THBT	binding of ^{125}I-T_3 by either erythrocytes, resin, charcoal or Sephadex	%
E. To assess, indirectly, circulating levels of free thyroid hormones			
1. Free thyroxine index	FTI	T_4 I or PBI and THBT	
2. Free thyroxine factor	FTF	T_4 I or PBI and THBT	
3. Effective thyroxine ratio	ETR		
4. Urinary thyroxine (competitive protein-binding assay)	T_4 (U, CPBA)	T_4	$\mu g./day$
5. Urinary triiodothyronine (radioimmunoassay)	T_3 (U, RIA)	T_3	$\mu g./day$
F. To assess thyroxine-binding proteins			
1. Thyroxine-binding globulin capacity	TBG cap.	Maximal binding capacity of TBG for T_4	$\mu g./100\ ml.$
2. Thyroxine-binding prealbumin capacity	TBPA cap.	Maximal binding capacity of TBPA for T_4	$\mu g./100\ ml.$
3. Thyroxine-binding globulin (radioimmunoassay)	TBG (RIA)	Concentration of TBG	$\mu g./100\ ml.$
4. Thyroxine-binding globulin (immunoelectrophoresis)	TBG (E)	Concentration of TBG (arbitrary)	

* By permission of Vivian N. Y. Chan, Ph.D. thesis, University of London, 1973.

Serum Butanol-Extractable Iodine (BEI). After acidification of serum, extraction with butanol removes iodinated amino acids, and after evaporation to dryness the iodine content in the residue can be measured after digestion. The method is more laborious than for the PBI and has now largely been replaced by T_4 measurements.

Serum Thyroxine (T_4). Serum thyroxine can be measured by saturation analysis techniques, using the thyroxine-binding protein contained in pooled human serum as the specific binding agent after separation by column chromatography (e.g., Bio-Rad T_4 column test). A variety of kits are available for estimation of serum T_4, e.g., "Thyopac-4" Amersham, which are simple to use but rather more expensive than the PBI and usually less precise. With the development of specific antibodies to T_4, radioimmunoassay of this hormone is now a practical procedure, and it seems likely that radioimmunoassay of T_4 will eventually prove to be the method of choice. Serum T_4 estimations are subject to errors similar to those of the PBI when there are alterations in the levels of TBG, but are not affected by iodine contamination. In hyperthyroid patients, PBI values may be higher than those of T_4 owing to the presence of other iodinated proteins in the circulation.

Serum Triiodothyronine (T_3). Measurements of serum T_3 are difficult because of the very low concentration of this hormone in the circulation and also because of the relatively high concentration of T_4. Radioimmunoassays for T_3 have been made possible by the production of antisera which are highly specific for T_3. Competition between the antibodies and TBG for T_3 have been overcome by the use of agents such as salicylates, diphenylhydantoin or thyroxine, which block the binding of T_3 to serum proteins. In normal subjects, serum T_3 levels are of the order of 1.2 ng. per ml. (range 0.7–1.7).

Iodide depletion is a major cause of an increased T_3/T_4 ratio, and this can be reversed by iodide repletion. Increased T_3/T_4 ratios in iodide deficiency may not be associated with a measurable rise of TSH, although this may be due to lack of assay sensitivity. Alternatively, iodide deficiency might cause an increase in T_3 synthesis by some intrinsic thyroid mechanism. In general, there is a good correlation between the serum concentrations of T_3 and T_4, a notable exception being the syndrome of T_3 toxicosis. Raised T_3 levels in patients with TBG elevation and lowered levels in those with reduction of TBG confirm the role of TBG in the transport of T_3. Similarly, salicylates reduce both T_3 and T_4 levels in the serum by competition for binding sites on TBG. During pregnancy, T_3 levels increase but cord serum shows a markedly reduced T_3 concentration. It is possible that the low T_3 level in the baby at birth is one factor in triggering the marked rise in serum TSH which occurs shortly after birth.

In hypothyroidism, T_3 values are lowered but there is some overlap with the normal range. Some patients with mild hypothyroidism have T_3 levels in the normal range, so a normal T_3 level cannot be used to exclude hypothyroidism. However, a lowered T_3 level is strong evidence for the presence of significant thyroid failure.

In hyperthyroidism, T_3 values are raised and there is remarkably little overlap with the normal range. As described later, occasional patients with hyperthyroidism have raised T_3 levels in the presence of a normal serum T_4. A raised serum T_3 level can also be seen in some apparently euthyroid patients with thyroid nodules or ophthalmic Graves' disease. Serum T_3 measurements provide a sensitive index of acute changes in thyroid hormone secretion—for example, in response to antithyroid therapy.

Thyroid Hormone Binding Tests. These tests are based on the unoccupied binding sites on thyroid hormone binding proteins (TBP). The more thyroxine present, the more saturated the TBP will be and the fewer binding sites will be unoccupied. Conversely, when little thyroxine is present more unoccupied sites will be available. When labeled T_3 or T_4 is added to serum, the amount taken up by the TBP depends on the unoccupied binding sites. Addition of some material which adsorbs thyroid hormone, e.g., a resin or Sephadex, results in uptake of the unbound thyroid hormone by the material which can then be counted to give the resin uptake. In certain kits, e.g., "Thyopac-3" (Amersham), the supernatant is counted rather than the resin and the results are the reciprocal of the resin uptake; low values are observed in hyperthyroidism and high values in hypothyroidism. The test is influenced by both the amount of thyroid hormone and the total amount of TBP. An increase in TBP as a result of pregnancy or estrogen therapy produces hypothyroid values in the thyroid hormone binding test. A decrease in TBP as seen in those with the nephrotic syndrome or as a result of androgen therapy produces "hyperthyroid" results. The thyroid hormone binding test depends on the uptake of labeled T_3 and is sometimes termed the T_3 uptake test. This terminology is best avoided since it leads to confusion of the thyroid hormone binding test with the serum T_3 estimation.

Free Thyroxine. The free thyroxine concentration can be measured directly by equilibrium dialysis methods but these are too cumbersome for routine use. The product of the serum PBI or T_4 and the thyroid hormone binding measured by resin uptake gives a value that is independent of TBP. This product is termed the free thyroxine index and shows a remarkably close correlation with the free thyroxine concentration obtained by equilibrium dialysis or gel filtration. Normal values are obtained for the free thyroxine index in the presence of altered levels of TBP, and it is widely used as the most reliable simple test of thyroid function.

Effective Thyroxine Ratio. The effective thyroxine ratio (ETR) is a new *in vitro* method which combines serum thyroxine and thyroid hormone binding measurements in one kit. The method is rapid and relatively simple, providing similar information to the free thyroxine index.

Urinary Thyroxine and Triiodothyronine. Measurements of urinary T_4 and T_3 might be expected to reflect the circulating content of the free hormones just as the urinary "free cortisol" excretion is related to unbound plasma cortisol levels. Total urinary T_4 measured by competitive protein-binding after extraction with ethyl acetate ranges from 4 to 13 µg. per 24 hours, with a mean value of about 8 µg. in euthyroid subjects; normal values are observed during pregnancy. Raised values are found in hyperthyroid patients and low values in hypothyroid patients. Without extraction, lower values of unconjugated T_4—mean 2.0 µg. per day (range 0.54–3.90)—are found in normal subjects. Similarly, lower values for immunoassayable T_3—0.8 µg. per day (range 0.33–1.91)—are observed in normal subjects. Further investigation is required to determine whether circulating free T_3 and T_4 are better reflected by measurements of unconjugated urinary T_3 and T_4, as would be expected, and to assess the value of urinary measurements in clinical practice.

c. **Peripheral Tissue Response Tests.** Tests based on the peripheral action of thyroid hormone are sometimes useful when routine thyroidal radioiodine uptake and PBI estimations are interfered with by drugs, especially iodides, by various diseases and by pregnancy. Test results also give an indication of the severity of the disease. However, many of the tests are nonspecific and have been poorly validated in clinical practice.

Basal Metabolic Rate. This test is now largely outmoded because of its lack of specificity, its relative insensitivity, the need to secure the collaboration of the patient and the many possibilities for error in the technique.

Tyrosine Tolerance. Fasting levels of plasma tyrosine are elevated in the presence of hyperthyroidism and lowered in the presence of hypothyroidism, and after a tyrosine load plasma levels rise higher than normal in patients with hyperthyroidism. The discrimination between altered levels of thyroid function achieved by the test has been poor and it is not now used routinely.

Red Cell Sodium. In patients with hyperthyroidism, the rate of exchange of sodium inside the red cell for potassium outside is reduced, hence the red cell sodium is increased. About 90 per cent of hyperthyroid patients have results above the normal range. Although hypothyroid patients have low red cell sodium levels, discrimination from normal is poor. The test is somewhat inconvenient for routine use since the cells must be separated from the plasma without delay.

Thyroid-stimulating Hormone Immunoassay

A specific radioimmunoassay for TSH is now widely available. Current assays lack sensitivity and are in general unable to separate low levels from the normal range. Normal TSH levels are probably < 1 µU per ml. of MRC Standard A, which is below the sensitivity of most assays. Most sensitive immunoassays record mean values for TSH of 1 to 2 µU per ml., with an upper limit of normal of about 4 µU per ml. There are no significant differences in TSH levels in men or women, but in children and in the elderly, TSH levels are rather higher and in the neonatal period there is a transient sharp rise of TSH.

Hypothyroidism. The major clinical value of TSH measurements lies in the diagnosis of hypothyroidism due to primary thyroid disease. A raised TSH level does not necessarily imply clinical hypothyroidism, but it does indicate patients at risk for symptomatic hypothyroidism. Patients with nonspecific symptoms compatible with hypothyroidism who have a raised TSH level despite normal routine thyroid function tests warrant a careful therapeutic trial with l-thyroxine. In general, the higher the TSH level, the more likely is the patient to have symptomatic hypothyroidism. A normal TSH level excludes hypothyroidism on the basis of thyroid disease; hence, TSH levels are very useful as a screen for possible thyroid failure, particularly in persons subjected to destructive therapy to the thyroid or with a family or personal history of organ-specific autoimmune disease.

Hyperthyroidism. Only very rarely is hyperthyroidism due to overproduction of TSH from the pituitary. In the majority of subjects with hyperthyroidism, TSH levels are low or undetectable, but because of limitations in the sensitivity of current assays, these values cannot be readily separated from normal. However, the TSH assay can be used to aid in the diagnosis of hyperthyroidism since patients with hyperthyroidism fail to show a rise in TSH in response to TRH (*vide infra*).

Nontoxic Goiter. Most patients with nontoxic goiter have normal TSH levels. The finding of a raised TSH level in such a patient implies some degree of thyroid failure, and the patient is usually suffering from autoimmune thyroid disease or, more rarely, is receiving some goitrogen or has a dyshormonogenetic goiter.

Thyrotrophin-releasing Hormone Test

TRH can be used as a test of thyroid function and as a test of pituitary-hypothalamic function with regard to TSH. The response to TRH can be monitored by measuring serum TSH, T_4 or T_3 levels after intravenous or oral TRH. In the authors' standard TRH test, blood is taken for basal TSH, T_4 and T_3 measurements, and 200 μg. of synthetic TRH is administered rapidly intravenously in 2 ml. of solution. Blood is removed at 20 minutes (peak TSH level), 60 minutes (to detect a delayed TSH response) and, if required, at three hours (peak T_3 level) and at six hours (peak T_4 level). A normal range for TSH response should be established for each sex in each laboratory performing the TSH immunoassay (see Table 22-4). No major side-effects have resulted from TRH but minor side-effects, which are trivial and transient, are common. These include nausea, flushing and a sudden desire to pass urine. They are only observed after administration of the rapid intravenous bolus of TRH and are probably due to stimulation of plain muscle of the gastrointestinal and the genitourinary tracts.

Normal Subjects. Females show a greater response to TRH than males because of the effect of circulating estrogens. Age does not appear to have any marked effect on the response. A variety of drugs alter the TSH response to TRH. Reduction is caused by T_4 and T_3, by pharmacological doses of corticosteroids and by chronic administration of L-DOPA, and an increased response results from estrogen treatment in men, from large doses of theophylline and after overtreatment with antithyroid drugs. Before assessing the TRH response, it is important to withdraw T_4 for three weeks and T_3 for two weeks.

Hypothyroid Subjects. Patients with hypothyroidism have elevated basal levels of TSH and an exaggerated and prolonged rise of TSH in response to TRH but no change in circulating thyroid hormone levels. When the basal TSH level is borderline or only minimally elevated, the enhanced TRH response can provide evidence for some degree of thyroid failure. The TRH test is also useful in separating pituitary from thyroid causes of hypothyroidism. A patient with hypothyroidism due to disease of the pituitary fails to respond to TRH, whereas a patient with hypothyroidism resulting from hypothalamic disease may show a normal response.

Hyperthyroid Subjects. Because of the suppressive action of the increased circulating levels of thyroid hormones on the anterior pituitary, there is suppression of the TSH response to TRH. This is a dose-dependent phenomenon and some patients with mild hyperthyroidism do show a response (usually blunted) to a large intravenous or oral dose of TRH. However, in the authors' experience no patient with proven hyperthyroidism has yet responded to a small (200 μg.) dose of intravenous TRH. The TRH test is of particular value in separating patients with mild hyperthyroidism from those who are euthyroid—a normal response to TRH excluding hyperthyroidism. However, the converse is not true; not all of those who fail to respond to TRH are hyperthyroid. Other causes of an absent or impaired TSH response to TRH include some patients with:

Ophthalmic Graves' disease; Graves' disease—euthyroid after therapy
Autonomous thyroid adenoma
Multinodular goiter
Patients receiving thyroxine or triiodothyronine
Patients with Cushing's syndrome (spontaneous or iatrogenic)
Hypopituitarism
Patients receiving L-DOPA

The TRH test can now replace the T_3 suppression test since it is safer and quicker and does not require the administration of a radioisotope. Patients who fail to suppress with T_3 fail to respond to TRH. Like the T_3 suppression test, the TRH test is helpful in the diagnosis of unilateral exophthalmos, since the majority of patients with ophthalmic Graves' disease show some abnormality of response to TRH.

The following are tests which indicate the cause of thyroid dysfunction:
Triiodothyronine suppression test
TSH stimulation test
Thyroid antibody tests

Triiodothyronine (T_3) Suppression Test. T_3 suppresses TSH output from the pituitary and hence reduces thyroid radioiodine uptake in normal subjects and in patients whose thyroidal iodide trapping mechanism is dependent on TSH. Failure of suppression implies thyroid

TABLE 22-4 TSH RESPONSE TO 200 μg. TRH GIVEN INTRAVENOUSLY

	MEN	WOMEN	SIGNIFICANCE OF DIFFERENCE
Basal	Mean 1.6 Range <0.5-2.8	1.4 <0.5-2.7	Not significant
20 minute	Mean 9.5 Range 3.5-15.6	13.5 6.5-20.5	$p < 0.001$
60 minute	Mean 6.8 Range 2.0-11.5	9.8 4.0-15.6	$p < 0.001$

autonomy as seen in autonomous single or multiple thyroid nodules or abnormal thyroid stimulation by factors other than TSH, as seen in Graves' disease. Suppression of radioiodine uptake can be defined in a variety of ways; however, a fall of uptake to less than 50 per cent of the original value after at least one week of T_3 in divided doses totaling 100 μg. daily is generally accepted as being normal. Normal suppression of thyroidal radioiodine uptake by T_3 is virtually conclusive evidence against a diagnosis of hyperthyroidism, whereas failure of suppression does not necessarily imply hyperthyroidism since it is found in some euthyroid patients with thyroid nodules and in some with the ocular manifestations of Graves' disease (ophthalmic Graves' disease). The T_3 suppression test is now being replaced by the TRH test, which is simpler and safer for the patient.

TSH Stimulation Test. For many years, the TSH stimulation test has been regarded as the absolute criterion for establishing the presence of impaired thyroid reserve. However, critical evaluation of the standard TSH stimulation test has demonstrated a normal response in about 50 per cent of patients with mild hypothyroidism. It seems that a significant increase in thyroidal radioiodine uptake can be produced by the very large and unphysiological dose of TSH used even in the face of mild thyroid failure.

Perchlorate Discharge Test. After an oral dose of radioiodine, administration of potassium perchlorate (600 mg. by mouth) causes no fall in thyroid radioactivity. In patients with defects of the organification of iodine in the thyroid, perchlorate, by competing with iodides in the thyroid trapping mechanism, causes a significant and rapid fall in thyroid radioactivity. Increased perchlorate discharges are found in patients with dyshormonogenetic goiters in which there is an organification defect (e.g., Pendred's syndrome of goiter and nerve deafness) and also in some patients with Hashimoto's and Graves' disease and during therapy with certain antithyroid drugs such as carbimazole.

Thyroid Antibody Tests. Circulating antibodies to a variety of thyroid components are found in patients with the autoimmune thyroid diseases. Thyroglobulin antibodies can be detected by the tanned red cell test, by agar diffusion tests and by immunofluorescence of colloid using fixed sections of thyroid. Antibodies to thyroid microsomes can be detected by complement fixation or by immunofluorescence using unfixed thyroid sections. The long-acting thyroid stimulator, an IgG molecule, occurs only in patients with autoimmune thyroid disease (particularly Graves' disease) or their close relatives. Positive thyroid antibodies occur in a small percentage of the general population, implying the presence of subclinical autoimmune thyroid disease. Detection of a significant titer of antibodies in a patient with a nontoxic goiter implies that this is likely to be due to Hashimoto's disease, in a patient with suspected hypothyroidism that this is likely to be due to autoimmune thyroid disease, and in a patient with hyperthyroidism, that the overactivity is occurring on the basis of Graves' disease.

Application of Tests to Determine the Level of Thyroid Function. Tests to determine the level of thyroid function should always be performed, even if the patient is apparently clinically euthyroid. Minor degrees of thyroid dysfunction may be impossible to recognize clinically; yet their presence may help in diagnosis and their correction benefit the patient. For example, a patient with a goiter who has evidence of thyroid failure can usually be diagnosed as suffering from autoimmune thyroiditis or, more rarely, from an acquired (goitrogens) or congenital defect in thyroid hormone synthesis. Again, the recognition of mild hyperthyroidism in a patient who was apparently euthyroid with a nodular goiter and atrial fibrillation can lead to effective treatment of the complication. Normal conventional tests of thyroid function are not sufficient to rule out minor thyroid dysfunction. Hyperthyroidism can be eliminated by finding a normal TSH response to TRH, normal thyroid suppressibility with T_3 and a normal serum T_3 level. Primary hypothyroidism can easily be eliminated by finding a normal serum TSH level or, if this value is borderline, a normal TSH response to TRH.

NONTOXIC GOITER AND HYPOTHYROIDISM

If the thyroid gland fails to secrete enough thyroid hormone for the body's needs the condition of hypothyroidism develops. Before overt hypothyroidism occurs, there is a period during which TSH stimulation of the gland is sufficient to produce adequate amounts of thyroid hormone, yet thyroid reserve capacity is lost. TSH stimulation of the thyroid produces a goiter, and most goiters can be considered to result from the action of TSH on a thyroid that has failed to produce sufficient thyroid hormone. Sometimes the increased function and increased cell mass that results from TSH stimulation is enough to correct the hormone deficiency leading to a "goiter with compensated thyroid function." If there is marked impairment of thyroid hormone production, compensation is inadequate and goitrous hypothyroidism results. The term nontoxic goiter is used to describe all goiters resulting from interference with thyroid hormone formation. The factors that determine whether or not goiter develops in response to hormone deficiency are poorly understood. It is evident that correction of the hormone deficiency will remove the TSH stimulus to the gland, which should then decrease in size. When the goiter is longstanding and nodular or cystic, however, the situation may be irreversible. Removal of goiters without correction of the hormone deficiency is often followed by recurrence of the thyroid enlargement.

Nontoxic goiter and/or hypothyroidism may result from the following disease processes:

Iodine deficiency
Congenital defects in the enzymes involved in hormone synthesis (dyshormonogenesis)
Acquired defects in enzyme action caused by drugs (goitrogens)
Autoimmune thyroiditis
Riedel's thyroiditis

Hypothyroidism. No reliable estimates are available of the overall incidence of myxedema, but it is by no means a rare disease in temperate countries.

It is now clear that hypothyroidism is a graded phenomenon and it is clinically convenient to define various grades of thyroid failure based on the clinical features, the serum TSH level and the presence of circulating thyroid antibodies.

Overt Hypothyroidism or Myxedema. This is characterized by major symptoms of thyroid hormone deficiency.

Mild Hypothyroidism. This condition presents many diagnostic problems since the symptoms are usually minor and nonspecific.

Compensated Hypothyroidism. This may be defined as an asymptomatic state in which reduction of thyroid activity has been compensated for by an increased TSH output to maintain a euthyroid state.

Autoimmune Thyroid Disease without Disturbance of Thyroid Function. This is quite common. Such patients are asymptomatic and have normal serum TSH levels but circulating thyroid antibodies.

Laboratory Diagnosis of Hypothyroidism

A clinical diagnosis of hypothyroidism should always be confirmed or documented by investigation. Lifelong therapy with thyroid hormone will be necessary, and subsequent physicians seeing a treated patient may, in the absence of adequate documentation, suspect the validity of the diagnosis and withdraw therapy.

Overt Hypothyroidism. This condition rarely presents any diagnostic problem since the clinical diagnosis should be obvious and conventional tests of thyroid function are almost invariably abnormal.

Thyroid Hormone Release Tests. Serum PBI, serum T_4 or thyroid hormone binding tests are abnormal and can be rapidly estimated on a small volume of blood so that the patient's treatment need not be delayed. The serum TSH level is usually grossly elevated. A normal TSH level excludes a diagnosis of hypothyroidism on the basis of primary thyroid disease.

Radioiodine Tests. These may show high, low or normal values in the early stages of goitrous autoimmune thyroiditis, and even in the presence of hypothyroidism a raised level may be found. In those with established nongoitrous hypothyroidism, a 24-hour thyroidal radioiodine uptake value is usually less than 20 per cent of the dose, though the value varies in different laboratories. Tests based on the peripheral action of thyroid hormone are cheap and sometimes helpful. Prolongation of the *Achilles tendon reflex,* abnormalities of the *electrocardiogram* and elevation of the serum *creatine phosphokinase and cholesterol* fall into this category. These tests are relatively nonspecific and there is considerable overlap of values with the normal range.

Mild Hypothyroidism. This is often accompanied by normal routine thyroid function tests, but the diagnosis of primary thyroid failure can be confidently excluded by the finding of a normal serum TSH concentration. A raised TSH level does not necessarily indicate that the patient's symptoms are due to hypothyroidism. However, if the clinical features are suggestive and the TSH level is elevated, a therapeutic trial of thyroid hormone is justified.

Compensated Hypothyroidism. This condition is accompanied by normal routine thyroid function tests but an elevated serum TSH level.

The standard TSH stimulation test cannot be relied upon to identify subjects with mild degrees of thyroid failure.

The presence of thyroid antibodies in the circulation gives evidence of thyroid disease and also indicates its nature.

HYPERTHYROIDISM

The common causes of hyperthyroidism are Graves' disease, toxic multinodular goiter and toxic adenoma, in descending order of frequency.

The eponym Graves' disease is best retained to describe the syndrome comprising goiter, hyperthyroidism, eye signs and rarely localized myxedema and thyroid acropachy. All these clinical features may not occur in an individual patient. The term *ophthalmic Graves' disease* is used to describe patients with the ocular manifestations of Graves' disease in the absence of hyperthyroidism or a past history of hyperthyroidism. *Neonatal Graves' disease* is the term applied to hyperthyroid children born to mothers with Graves' disease and in whom the hyperthyroidism remits spontaneously in a few months. Graves' disease was initially thought to result from increased amounts of TSH, but this theory is now untenable. A long-acting thyroid stimulator (LATS), an immunoglobulin, is found in the circulation of a proportion of patients with Graves' disease but human-specific thyroid-stimulating immunoglobulins are likely to be more important. A comparison of TSH and LATS is shown in Table 22-5. Delayed hypersensitivity mechanisms may also be involved in the pathogenesis of Graves' disease possibly by the action of thymic lymphocytes on the thyroid.

T_3 Toxicosis. Direct measurements of circulating triiodothyronine levels has allowed recognition of the syndrome of T_3 toxicosis. Features of this syndrome are contrasted with those of typical hyperthyroidism in Table 22-6. The frequency of T_3 toxicosis varies in different areas, and true prevalence figures are not yet available.

Diagnosis of Hyperthyroidism. In any patient with suspected hyperthyroidism, it is wise to document the diagnosis with at least two tests of thyroid function. The tests chosen depend on the preference and facilities of the individual physician. The authors routinely use the *PBI* or T_4 and thyroid hormone binding test such as the "Thyopac-3," calculating the free thyroxine index. The effective thyroxine ratio (ETR) can also be used to provide a similar answer. When the initial tests are

TABLE 22-5   COMPARISON OF TSH AND LATS

	LATS	TSH
Peak of action in mouse assay	10-12 hr.	2-3 hr.
Structure and molecular weight	IgG, 150,000	Protein, about 28,000
Inhibition of activity	By anti-IgG serum	By anti-TSH serum
Source	Lymphocytes	Anterior pituitary
Action	Stimulates thyroid	Stimulates thyroid
Circulating level reduced by	High doses of corticosteroids	Thyroid hormone
Overproduced in	Graves' disease	Some pituitary tumors

TABLE 22-6   COMPARISON OF CLINICAL AND BIOCHEMICAL FEATURES OF TYPICAL HYPERTHYROIDISM AND T_3 TOXICOSIS

	TYPICAL HYPERTHYROIDISM	T_3 TOXICOSIS
Cause	Any variety of hyperthyroidism	Any variety of hyperthyroidism
Clinical features	Hyperthyroidism	Hyperthyroidism
Predisposing factors		Iodine deficiency, ? early feature of typical hyperthyroidism and of recurrent hyperthyroidism
Serum triiodothyronine	Raised	Raised
Urine triiodothyronine	Raised	Raised
Serum thyroxine	Raised	Normal
Free thyroxine	Raised	Normal
Thyroid hormone binding tests	Hyperthyroid range	Usually normal
Serum thyroxine-binding globulin	Normal or lowered	Normal
Radioiodine uptake by thyroid	Raised	Raised or normal
T_3 suppression test	Impaired or absent suppression	Impaired or absent suppression
Serum TSH response to TRH	Impaired or absent	Impaired or absent

borderline, it is wise to reassess the clinical picture and to question the patient about drugs that might have modified the clinical features or the tests. Further tests are then employed to confirm the diagnosis. The *triiodothyronine suppression test* is useful in distinguishing raised thyroid radioiodine uptakes caused by iodine deficiency from those of hyperthyroidism. Suppression of the thyroid radioiodine uptake by triiodothyronine excludes a diagnosis of hyperthyroidism, but failure of suppression occurs in Graves' disease without hyperthyroidism and in the presence of one or more autonomous thyroid nodules as well as in hyperthyroidism.

Thyrotrophin-releasing Hormone Test. This test is valuable in the exclusion of hyperthyroidism. A normal TSH response to a small (200 µg.) dose of intravenous TRH excludes hyperthyroidism. An absent or impaired response is often due to hyperthyroidism but may have other explanations, for example in the presence of ophthalmic Graves' disease, Graves' disease in some euthyroid patients in apparent clinical remission, autonomous adenoma, multinodular goiter, pituitary disease and a variety of drugs such as corticosteroids and thyroid hormones. Because of its safety, speed and convenience, the TRH test is likely to replace the T_3 suppression test in clinical practice. There is a good correlation between the results of the two tests; patients with a normal TRH test show normal thyroid suppressibility and those with impaired or absent TRH responses show impaired or absent suppressibility.

CALCITONIN

Calcitonin is a hypocalcemic factor derived from the parafollicular cells of the thyroid and to a lesser extent from the parathyroid and thymus.

The parafollicular cells are distinct from the follicular cells of the thyroid, being derived from the neural crest, from which they migrate to the last branchial pouch. This explains why medullary carcinomas and pheochromocytomas, both of neuroectodermal origin, contain a variety of amines and are sometimes associated in the familial medullary carcinoma of the thyroid syndrome. The C cells are members of the APUD cell series. The term APUD is derived from the initial letters of the three most reliable characteristics *A*mine content, amine-*P*recursor *U*ptake and amino acid *D*ecarboxylase content. The series includes the medullary carcinoma, bronchial and intestinal carcinoids and pheochromocytomas. The com-

```
  ┌─────────────────────────┐
H-Cys-Gly-Asn-Leu-Ser-Thr-Cys-Met-Leu-Gly-
   1   2   3   4   5   6   7   8   9   10
Thr-Tyr-Thr-Gln-Asp-Phe-Asn-Lys-Phe-His-Thr-
 11  12  13  14  15  16  17  18  19  20  21
Phe-Pro-Gln-Thr-Ala-Ile-Gly-Val-Gly-Ala-Pro*-NH₂
 22  23  24  25  26  27  28  29  30  31
```

Figure 22-13. Structure of human calcitonin. (Pro* = prolinamide.)

mon origin explains why C cell tumors can secrete ACTH and carcinoid cells on occasion produce calcitonin.

Structure of Human Calcitonin. Human calcitonin has been purified, sequenced and synthesized. It is a single chain, 32 amino acid polypeptide (Fig. 22-13) with a molecular weight near 3000. There are 18 amino acid substitutions when compared with porcine calcitonin, a surprisingly high number of changes. In addition to the calcitonin monomer (calcitonin M), a dimer (calcitonin D) has also been isolated from human medullary carcinoma tissue. This is composed of two monomers covalently linked in antiparallel fashion by rearrangement of the disulfide bond. The complete 32 amino acid peptide is essential for full biological activity, and integrity of the first 9 amino acids is probably of major importance. Salmon calcitonin, which differs in amino acid sequence from human calcitonin, has a longer half-life and is a more potent calcium-lowering agent on a weight basis.

Factors Affecting Secretion of Calcitonin. The major stimulus for calcitonin secretion is a rise in the serum calcium level, but other factors, including magnesium, may also be involved. Hypercalcemia has a direct action on the C cells, increasing both the synthesis and secretion of calcitonin. Hypermagnesemia also stimulates calcitonin secretion but at levels higher than are seen under physiological or pathological conditions. Glucagon, "pancreozymin" and streptomycin stimulate calcitonin secretion, but the significance of these observations is unknown.

Sites and Mechanisms of Action and Distribution of Calcitonin. Calcitonin inhibits bone resorption, an action reflected by a reduced urinary excretion of hydroxyproline. Unlike parathyroid hormone, calcitonin has no significant actions on the intestine. Urinary excretion of phosphate and sodium is increased but the physiological significance of the renal effects is unknown.

The mechanism of action of calcitonin is also unknown. It acts on bone cells to lower the calcium concentration in the cytosol by increasing calcium efflux from the cell and enhancing calcium uptake into the mitochondrial compartment. Although cyclic AMP concentrations within the cell are increased by the calcitonin-induced fall in cytosol calcium concentration which activates adenyl cyclase, bone resorption is inhibited because of the lowered cytosol calcium which inactivates some of the phosphoprotein products of the cyclic AMP-dependent protein kinases.

Calcitonin is concentrated mainly in the liver and kidney, and the latter is the main site of its degradation. The half-life of human calcitonin is less than 15 minutes.

Assay of Calcitonin. Calcitonin can be assayed by its effect on the serum calcium of rats. The sensitivity of the assay can be greatly increased by the use of a high-phosphate diet or by administration of injections of phosphate at the time when the animals are given calcitonin. Radioimmunoassays for calcitonin capable of detecting the amounts present in the circulation in normal man have now been developed.

Physiological Significance of Calcitonin. The physiological role of calcitonin remains uncertain. It may be a factor involved in the regulation of the serum calcium concentration, acting along with the parathyroid hormone to dampen oscillations in the serum calcium level. Evidence for this action of calcitonin has been obtained in totally thyroidectomized subjects infused with calcium. The rate at which their serum calcium level falls at the end of infusion is slower than normal, suggesting the absence of a hypocalcemic factor. It may protect bone and promote calcium storage in the skeleton, particularly in pregnancy. It may influence calcium concentrations in the cell cytosol, thereby modifying the action of parathyroid hormone on bone and kidney.

Pathological Significance of Calcitonin. No clinical syndromes are as yet associated with calcitonin deficiency, but increased calcitonin secretion has been demonstrated in certain situations. The medullary carcinoma of the thyroid is a tumor of the parafollicular cells and contains high levels of calcitonin. The associations of medullary carcinoma can be divided into two groups, genetic and humoral.

Genetic. The familial occurrence of medullary carcinomas has frequently been reported, and pheochromocytomas may be found in some of the affected individuals. Other inherited features that may be associated are multiple mucosal neuromas, especially of the eyelids and tongue, diffuse thickening of the lips, intestinal ganglioneuromatosis, pes cavus, a high arched palate, a proximal myopathy and skin pigmentation.

Humoral. Medullary carcinomas may secrete a variety of humoral agents in addition to calcitonin. Flushing attacks may be due to secretion of 5-hydroxytryptamine, and diarrhea to release of prostaglandins. Cushing's syndrome may be caused by secretion of ACTH-like peptides from the tumor. Despite the increased circulating levels of calcitonin, serum calcium and phosphate levels are almost invariably normal because extra parathormone is secreted to maintain the calcium level and there are usually no overt skeletal abnormalities. Estimations of plasma calcitonin levels are helpful in the diagnosis of medullary carcinomas and are also useful in predicting complete removal of the tumor or its recurrence. Because of the lack of sensitivity of calcitonin immunoassays, it may be necessary to measure calcitonin levels during a calcium infusion, a procedure which stimulates calcitonin secretion. In view of the familial occurrence of the variety of medullary carcinoma associated with thickening of the lips and Marfanoid habitus, it is wise to screen first-degree relatives of the propositus by calcitonin measurements.

It has recently been shown that there is a high content of the enzyme histaminase in medullary carcinomas, and detection of this enzyme in the circulation is associated with extrathyroidal deposits of the tumor. The enzyme is also present in normal kidney and ileum and can be detected in the serum during pregnancy and after the administration of heparin. Calcitonin can also be demonstrated in some carcinoid tumors, but rarely enters the circulation in this condition.

Clinical Applications of Calcitonin. Calcitonin is of value in the treatment of hypercalcemia due to disseminated malignant disease, parathyroid adenomas, idiopathic hypercalcemia of infancy and vitamin D intoxication. Its major clinical role, however, is in the treatment of Paget's disease of bone. Administration of calcitonin is indicated if bone pain is troublesome and does not respond to simple analgesics, if there is extensive bone deformity or high-output cardiac failure, if there is evidence of compression of nerves of the spinal cord or if there is a fracture through a weight-bearing limb. Bone pain is relieved by calcitonin and there is a progressive fall in the numbers of osteoclasts, in the alkaline phosphatase and in the urinary excretion of hydroxyproline. Calcium balance returns to normal and the bone formed during treatment has a normal lamellar structure. The hormone must be given by injection, initially once or twice daily and then possibly once or twice a week. At present natural porcine calcitonin is available for clinical use in the United Kingdom. Side effects are minor and transient and include a feeling of warmth, flushing of the face and transient nausea and tingling in the pharynx and abdomen. Antibodies have not developed after treatment with the human hormone but could possibly pose problems with animal preparations.

There is as yet no good evidence that calcitonin has a place in the treatment of osteoporosis.

GASTROINTESTINAL HORMONES

Three major gastrointestinal (GI) hormones have been identified so far: gastrin, cholecystokinin (CCK), which is identical to pancreozymin, and secretin. They have a wide spectrum of biological actions and tend to act on the same organs and show interactions with one another producing inhibition or augmentation of action. Despite their wide and often overlapping actions, each hormone has a small group of specific effects.

Gastrin

Gastrin is produced in the "G cells" of the mucosa of the pyloric antrum, which are situated in the neck and mid-zone of the gastric glands. The gastric glands connect with the lumen of the stomach, which explains why the "G cells" respond directly to alterations in concentrations of stomach constituents and to the luminal pH.

Two peptides with gastrin activity have been isolated from human gastric mucosa, designated gastrin I and II. Both are heptadecapeptides (18 amino acids) with similar amino acid sequences. Since biological activity resides in the terminal part of the molecule, short-chain sequences of the gastrin molecule such as pentagastrin are biologically active and are used routinely for tests of gastric acid secretion.

Assay of Gastrin. Bioassays of gastrin depend on measurements of gastric acid output in specially prepared rats, using standard gastrin or histamine as reference preparations. Radioimmunoassays can be used to measure heptadecapeptide gastrins as well as a large more basic immunoreactive form (designated BG), which is present in human plasma. BG is probably composed of the heptadecapeptide covalently linked to a more basic peptide. All three forms of gastrin show very similar immunoreactivity.

Physiology and Regulation of Secretion. Gastrin is released from the antral mucosa in response to vagal reflexes triggered by the intake of food and after chemical and mechanical stimulation of the antrum by food and digestion products. Autoregulation is also involved, since high acid levels inhibit gastrin secretion. Various chemical agents are potent gastrin releasers, e.g., glycine, β-alanine and ethanol, and yet small changes in their structure can inhibit this action.

Gastrin has a variety of other actions and it is uncertain which of these are of physiological significance: release of insulin and secretin, contraction of smooth muscle of the lower esophageal sphincter, stomach, small and large bowel and gallbladder, and relaxation of the sphincter of Oddi; release of histamine from gastric mucosa; stimulation of growth of gastric mucosa (as seen in the Zollinger-Ellison syndrome); stimulation of blood flow in the superior mesenteric artery.

Gastrin interacts with secretin and CCK. CCK and gastrin act at the same receptor site, whereas secretin acts at a separate site which is able to influence the gastrin/CCK site. CCK and secretin inhibit gastrin-stimulated acid secretion.

Syndromes of Gastrin Excess. In the Zollinger-Ellison syndrome a non-beta islet cell tumor of the pancreas secretes gastrin, which causes the production of large amounts of gastric juices with a high hydrochloric acid content. This in turn leads to a fulminant ulcer diathesis. The ulcers cannot be controlled by any medical measures or by surgery short of total gastrectomy. Diarrhea is a frequent association of the syndrome which may form one part of the pluriglandular syndrome. Features of the syndrome include: ulceration in the second and third parts of the duodenum; severe ulcer symptoms in younger persons without a previous ulcer history; recurrent hemorrhage or perforation within a few days of a routine ulcer operation; a family or personal history of parathyroid, pituitary, adrenal or islet cell tumors.

The diagnosis can be confirmed by finding more than one liter of gastric juice, with more than 100 mEq. of free HCl in a 12-hour overnight specimen. A ratio of basal to stimulated acid secretion > 0.6 is also suggestive. Fasting levels of gastrin measured by bioassay or radioimmunoassay are greatly elevated.

In the majority of patients with duodenal ulcers, gastrin levels are lower than normal, whereas in the presence of a gastric ulcer, gastrin levels are higher. This is explained by the lower mean pH associated with duodenal ulcer and the higher value seen in gastric ulceration. High values are also seen in patients with atrophic gastritis and with pernicious anemia.

The action of gastrin on the tone and competence of the lower esophageal sphincter may have clinical relevance. When endogenous gastrin secretion is inhibited by gastric acidification, the resting tone of the esophageal sphincter is reduced. It is well established that the tone of the sphincter is of considerable importance in preventing reflux of gastric contents into the esophagus. It can be speculated that the lower mean gastrin levels in patients with duodenal ulceration might be a factor in causing the esophageal reflux and heartburn which is so common in this condition.

Cholecystokinin

Cholecystokinin (CCK) is produced in the mucosa of the upper gastrointestinal tract and causes gallbladder contraction and stimulation of pancreatic enzyme release. The latter action was attributed to a separate enzyme, pancreozymin, but it is now established that both activities are functions of the CCK molecule. CCK is a polypeptide consisting of 33 amino acids, the C-terminal pentapeptide being common with gastrin.

Assay of CCK. Bioassays depend on measurements of gallbladder contractions (*in situ*) in anesthetized guinea pigs or on assays of pancreatic enzyme production. Radioimmunoassays of CCK have been developed but pure material is in short supply, and because of the similarities between CCK and gastrin it is difficult to produce specific antibodies.

Physiology and Regulation of Secretion. CCK is released from the mucosa of the upper intestine as a result of stimuli such as amino acids, fatty acids and hydrogen ion present in the lumen. Whole protein does not release CCK and it is not known whether small peptides are effective.

CCK has a wide range of actions, some resembling those of gastrin, others competitively inhibiting gastrin and others similar to those of secretin. It causes a striking stimulation of gallbladder contraction, which is unaffected by atropine or vagotomy. Acting on the pancreas, it increases enzyme release and also enhances the secretin-induced water-bicarbonate flow. Its action on the pancreas is independent of cholinergic factors. CCK stimulates gastric acid secretion and, in man, inhibits gastrin-induced acid secretion. It has a variety of other actions whose physiological significance is not established: stimulation of bile secretion, stimulation of pepsin secretion, release of insulin and glucagon from islet cells, stimulation of secretion from Brunner's glands, inhibition of the lower esophageal sphincter and gastric mobility, stimulation of the small and large intestines, relaxation of the sphincter of Oddi, stimulation of growth of the pancreas, and increase in superior mesenteric blood flow.

Syndromes of Excess and Deficiency. At present, no syndromes due to excess CCK have been described. Impaired CCK release has been described in sprue and after vagotomy and pyloroplasty, but the significance of this finding is not known.

Secretin

Secretin is produced by the mucosa of the upper gastrointestinal tract and has been fully characterized and synthesized. It is an open-chain polypeptide composed of 27 amino acids showing structural similarity to glucagon. It is highly basic and carries a strongly positive electric charge which enhances its binding to cell surfaces.

Assay of Secretin. Bioassays of secretin depend on measurements of the bicarbonate content of pancreatic juice in response to intravenous preparations. Radioimmunoassays of the hormone were hampered by the lack of tyrosine in the molecule and by death from diabetes mellitus of animals used to raise antisera. Nevertheless, satisfactory immunoassays are now available for secretin.

Physiology and Regulation of Secretion. Secretin is released from the intestine in response to the presence of hydrogen ions in the lumen. The threshold pH for release of secretin is 4.5. Secretin is a potent stimulator of pancreatic bicarbonate secretion and acts synergistically with gastrin and CCK to increase bicarbonate and water output. It has a variety of other actions whose physiological significance is unknown: stimulation of electrolyte and water secretion by the liver, stimulation of bile secretion, inhibition of gastrin-induced gastric acid secretion, inhibition of gastrin release, stimulation of insulin release, increase in heart rate and stroke volume, increase in blood flow in the superior mesenteric artery.

Syndromes of Excess and Deficiency. Excess secretion of secretin has been implicated in the syndrome of "pancreatic cholera," but firm evidence is still lacking. This syndrome comprises: profuse watery diarrhea without steatorrhea; hypokalemia; lack of peptic ulceration and gastric acid hyposecretion, often with achlorhydria and an impaired response to histamine or pentagastrin; elevation of serum calcium in half the cases reported; a non-beta cell tumor (or hyperplasia) of the islets which is often malignant; and cessation of diarrhea following removal of the tumor. The major features of the syndrome are: watery diarrhea, hypokalemia and achlorhydria.

Deficiency of secretin production, seen particularly in patients with achlorhydria, does not appear to be associated with any clinical sequelae.

OVARIAN HORMONES

Three groups of steroid hormones are produced by the ovaries: estrogens, progesterone, and androgens.

Biosynthesis of the ovarian steroids is controlled by the pituitary gonadotrophins, and is normally cyclical, unless fertilization of an ovum occurs when ovarian secretion becomes continuous. The biosynthetic pathways involved in estrogen production are shown in Fig. 22-14.

Estrogens

Estrogens are secreted by cells of the theca interna and the stratum granulosum. They are characterized by the presence of a benzene ring and a phenolic hydroxyl group (Fig. 22-15). 17β-Estradiol and estrone are

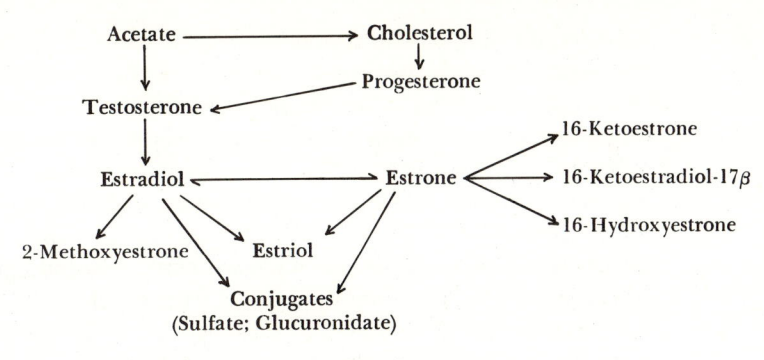

Figure 22-14. Biosynthesis and metabolism of estrogens.

the primary estrogens produced in the ovary, the former being the most potent. They are interconvertible and enzymes responsible for the conversion are found in many tissues. Estriol is a metabolite of estradiol and is much less potent in terms of its effect on the endometrium. It has marked effects on the cervix and vagina, its target organs. All three estrogens circulate in the blood bound to protein and are excreted in the bile and in the urine mainly in the form of conjugates with glucuronic and sulfuric acids. Conjugation of the estrogens occurs in the liver, where the compounds are rendered more soluble.

Actions of Estrogens. Estrogens initiate and maintain maturity of the female genitalia and the secondary sex characteristics. Together with progesterone, they maintain and control menstruation. Estrogens stimulate growth but excessive levels may cause shortness of stature due to premature fusion of the epiphyses. They antagonize the effects of androgens on the skin, hence reducing sebaceous gland activity. They

Figure 22-15. Structure of natural estrogens.

cause deposition of subcutaneous adipose tissue, and are responsible for the characteristic distribution of fat in the mature female. Estrogens inhibit the secretion of FSH. In large doses, they also cause sodium and water retention, causing edema, or, rarely, heart failure in a susceptible individual. In pharmacological amounts, estrogens can cause painful swelling of the breasts, dysfunctional uterine bleeding, nausea and vomiting. Venous thrombosis is a complication of estrogen therapy. Administration of estrogens in early pregnancy may cause vaginal adenosis and adenocarcinoma in female offspring with a latent period of up to two decades.

Assay of Estrogens. Estrogens can be assayed in plasma or in urine by biological, chemical or radioimmunoassay procedures. Bioassays are of historical interest only. Chemical methods for estrone, estriol and 17β-estradiol estimation in urine depending on colorimetry are more practicable than techniques depending on fluorimetry or gas-liquid chromatography. For estimation of the classic estrogens in urine, the methods of Brown are appropriate (Table 22-7). During pregnancy, measurements of urinary estriol by the method of Klopper and Wilson provide adequate information.

Plasma estrogen levels can now be measured by radioimmunoassay and other saturation analysis techniques and also by fluorimetry. Mean levels of estrone and 17β-estradiol have been reported as: 102 and 64 pg. per ml., respectively, in the follicular phase of the cycle; 243 and 172 pg. per ml., respectively, in the luteal phase of the cycle; and 52 and 27 pg. per ml., respectively, in normal adult men. Values are low in the early proliferative phase and just prior to menstruation. Peak levels occur at or just before ovulation and during the luteal phase. The low estrogen levels found in men and postmenopausal women are, at least in part, of adrenal origin.

Indirect estimates of estrogen secretion may also be obtained from examination of vaginal or cervical secretions. Absence of cornification in the vaginal epithelial cells indicates a marked reduction in estrogen secretion. In the fern test, a sample of cervical secretion is spread on a glass slide and dried. A fern-like configuration indicates the presence of adequate estrogen concentrations.

Clinical Applications of Estrogen Assays

Increased Estrogen Values

Increased Production. In true precocious puberty, gonadotrophin and estrogen values similar to those found in normally menstruating women may occur in association with normal manifestations of puberty. Such cases must be differentiated from precocious puberty due to granulosa cell tumor or other neoplasms.

Very few tumors produce increased amounts of estrogen, certainly not quantities comparable with those formed late in pregnancy. *Granulosa cell tumors, thecomas* and *luteomas* of the ovary are usually accompanied by a moderate increase in the estrogen content of the blood and urine.

TABLE 22-7 ESTROGEN LEVELS DURING THE MENSTRUAL CYCLE AND AFTER THE MENOPAUSE

TIME IN CYCLE	URINARY EXCRETION (MG./24 HR.)— MEAN AND RANGE		
	ESTRIOL	ESTRONE	ESTRADIOL
Onset of menstruation	6 (0-15)	5 (4-7)	2 (0-3)
Ovulation	27 (13-54)	20 (11-31)	9 (4-14)
Luteal peak	22 (8-72)	14 (10-23)	7 (4-10)
Postmenopausal peak	3.3 (0.6-8.6)	2.5 (0.8-7.1)	0.6 (0-3.9)

This increase can often be detected in cases occurring in children before puberty and in women after the menopause. It is also evidenced by obvious estrogenic effects in the patient.

Increased amounts of estrogen may be found in some cases of adrenocortical carcinoma. However, this is by no means true of all cases of cortical carcinoma, so that negative findings cannot be regarded as significant in the diagnosis of this condition.

A moderate increase in estrogen excretion has occasionally been observed in cases of *testicular tumors* of chorionepithelioma origin, but normal findings are the rule in such cases.

Functional ovarian disorders are seldom accompanied by marked increase in the blood or 24-hour urine estrogens. It is not unusual, however, to find a persistently high plateau of estrogen excretion, which may result in a state of hyperestrogenism. The monthly output of estrogens under such circumstances may be two to five times the normal. This may be associated with various types of functional bleeding or amenorrhea, depending upon the pattern of estrogen excretion and the responsiveness of the uterus. Not uncommonly, it is associated with amenorrhea followed by prolonged periods of bleeding and a hyperplastic endometrium (metropathia hemorrhagica).

Decreased Destruction. A considerable increase in the free estrogen content of urine, with or without an increase in total estrogen excretion, has been observed in patients with *hepatic cirrhosis*. This has been attributed to diminished destruction or inactivation (or excretion) by the liver. Such patients often exhibit clinical manifestations of hyperestrogenism, such as gynecomastia and testicular atrophy in the male and menstrual disorders in the female.

Decreased Estrogen Values. In women who are in the reproductive period, decreased estrogen values in the blood and urine may result from either (*a*) primary pituitary deficiency, with inadequate gonadotrophic stimulation of the ovaries or (*b*) a primary ovarian defect, despite normal or even excessive gonadotrophic stimulation. The distinction between primary and secondary ovarian (estrogen) deficiency can usually be made with certainty only by gonadotrophin assays.

Decreased estrogen values are found in hypopituitarism and in other conditions associated with diminished gonadotrophin production. Primary ovarian failure which may be idiopathic or associated with organ-specific autoimmune disease or with a variety of chromosomal anomalies, particularly Turner's syndrome, causes estrogen deficiency. This in turn causes primary amenorrhea and failure of development of the secondary sexual characteristics. Relative estrogen deficiency or irregular estrogen production later in the reproductive period may result in a variety of disturbances, including secondary amenorrhea, hypomenorrhea, various types of functional bleeding or sterility. In the mature woman, estrogen deficiency causes reduction in vaginal secretions which causes difficulty during intercourse, libido may be reduced, the amount of body hair is decreased, and the breasts, uterus and vagina decrease in size. These are normal features of the postmenopausal state.

Indications for Estrogen Therapy. Estrogens have been used in the treatment of a wide variety of disorders and only some of the more common uses will be considered here. They are effective in controlling many of the complications of the menopause. In high dosage, they are capable of stopping the prolonged phase of bleeding seen in metropathia hemorrhagica, but oral contraceptive regimens using combined estrogen-progesterone preparations are now preferable. Similarly, combined regimens are effective in the treatment of endometriosis. So long as a uterus is present it is possible to induce withdrawal bleeding with cyclical estrogen regimens, preferably in combination with a progestogen. Estrogens may be used to accelerate epiphyseal fusion in tall girls from tall families but treatment must be started before the age of 12 years. Lactation can be suppressed by a high dosage of estrogens, but this treatment may induce venous thrombosis and is best avoided. Secondary sexual characteristics can be induced or restored with estrogens in females with ovarian failure. Osteoporosis can probably be avoided in women subjected to early oophorectomy if they are treated with estrogens, and estrogen therapy may also prevent the complication of premature coronary atheroma which follows this procedure. In men, estrogens in high doses can cause temporary remission of metastatic prostatic carcinoma. Habitual male sexual offenders may obtain a reduction in libido by periodic estrogen implants. Estrogens may also be used to treat slight elevations of the serum calcium level in postmenopausal women with primary hyperparathyroidism.

Estrogen Preparations. Ethinyl estradiol is the most potent estrogen preparation and is the drug of choice. It is used in a dose of 0.01 mg. once or twice daily to induce withdrawal bleeding or to control menopausal symptoms. Courses of three weeks' duration with one week off the drug are used. For severe menorrhagia, larger doses are required, e.g., ethinyl estradiol 0.05 mg. four times daily for two or three days until the bleeding has stopped and the dose is then gradually reduced over a week. Stilbestrol is not used routinely in females because it tends to cause pigmentation of the nipples. The conjugated equine estrogens such as Premarin are also effective in cyclical regimens and in control of menopausal symptoms, and there is some evidence suggesting that they cause less impairment of carbohydrate tolerance in prediabetic and latent diabetic individuals than do other estrogen preparations. Local applications are of value in the treatment of senile vaginitis, either as creams or as pessaries. Stilbestrol is used for long-term therapy of metastatic prostatic carcinoma, starting with 0.5 mg. twice daily and gradually increasing the dose if the patient tolerates the drug well.

Progesterone

Progesterone, secreted by the corpus luteum, is a steroid with a structure similar to that of the adrenocortical steroids. Progesterone is also synthesized in the adrenal cortex and the testes, where it serves as a precursor in steroid biosynthesis. Progesterone circulates bound to protein. It is degraded in the liver, and a small fraction is excreted in the urine as the inactive conjugate pregnanediol glucuronide.

Actions of Progesterone. Progesterone prepares the endometrium for implantation and the maintenance of pregnancy. It induces secretory changes in the endometrium necessary for implantation of the fertilized ovum and suppresses uterine mobility. During pregnancy, it is at least in part responsible for inhibition of ovulation and for development of the breasts. Its systemic effects are not so clearly defined as those of estradiol. It is pyrogenic, and a daily basal temperature recording which reveals a rise of about 0.5° C. in midcycle maintained until menstruation suggests the presence of a functioning corpus luteum and also that ovulation has occurred. Side-effects of progesterone include acne, breast tenderness, reduced menstrual loss, cholestatic jaundice and virilization of a female fetus.

Assay of Progesterone. Progesterone secretion can be assessed by the estimation of its breakdown product pregnanediol in a 24-hour urine specimen. In the follicular phase of the cycle, the excretion of pregnanediol is less than 1 mg. in 24 hours, which rises in the luteal phase to 2 to 5 mg. in 24 hours. Another indication of progesterone production is the finding of secretory changes in an endometrial biopsy. Plasma progesterone levels can now be measured by radioimmunoassay. During the follicular and midluteal phases of the menstrual cycle, mean reported levels are, respectively, 545 pg. per ml. and 8561 pg. per ml.

Clinical Applications of Progesterone Assays

Increased Progesterone Values. Increased levels of progesterone, usually indicated by urinary pregnanediol excretion, are of little clinical value except to indicate the presence of a corpus luteum and hence demonstrate that ovulation has probably occurred.

Diminished Progesterone Values

Menstrual Disorders and Sterility. Absence of pregnanediol in the urine during the two weeks preceding menstruation is indicative of failure of ovulation and absence of a functionally active corpus luteum. In such instances, the estrogen values are usually also diminished. There is still some question whether or not a normal pregnanediol excretion is conclusive evidence of the occurrence of ovulation. It should be pointed out that not too much significance should be attached to a single low value because of the normally wide daily fluctuations. Furthermore, in studying patients with sterility in whom some defect in ovulation is suspected, it is well to correlate the pregnanediol findings with data obtained by endometrial biopsy, vaginal smears, basal temperature, and so on. Absence of pregnanediol may be associated with anovulatory sterility, amenorrhea and various types of functional bleeding. Patients with dysmenorrhea do not, as a rule, exhibit abnormal pregnanediol values, nor is there any specific relationship between pregnanediol and premenstrual tension, cystic mastitis or various premenstrual disturbances.

Indications for Progesterone Therapy. Progestogens are used in the treatment of dysfunctional uterine bleeding. They are also of value in the treatment of premenstrual swelling and tenderness of the breasts. They have been used in the treatment of habitual abortion but firm evidence for this use is lacking. Some women with metastatic breast cancer improve with progestogens. Progestogens may be used alone or in combination with estrogens as oral contraceptives.

Progesterone Preparations. Progesterone must be given by intramuscular injection in a dose of 20 to 60 mg. daily. Preparations which are active by mouth are now preferred to progesterone, the drug of choice being norethisterone, given in daily doses of 5 to 20 mg.

Androgens

The stroma cells of the ovary synthesize different steroids from those produced by the ovarian follicle and the corpus luteum, mainly androgens. The physiological significance of these ovarian androgens, particularly androstenedione, is uncertain. Androgen secretion by the ovaries is increased in some pathological conditions in which there is abnormal steroidogenesis (as in the polycystic ovary or Stein-Leventhal syndrome) or tumor formation (such as arrhenoblastoma).

The Hormonal Control of the Menstrual Cycle (see Fig. 22-1)

Secretion of estrogens and progesterone is under the control of the pituitary gonadotrophins. During the first half of the cycle, the follicle develops under the influence of FSH and LH. There is gradual increase in the section of 17β-estradiol, which eventually triggers a reflex discharge of the gonadotrophin-releasing hormone (LH/FSH-RH) from the hypothalamus; this in turn causes release of LH and FSH. Estrogens and 17α-hydroxyprogesterone are secreted by the theca interna cells of the developing follicle. Surges of LH rupture the follicle and release an ovum, after a time interval of about 24 hours. The granulosa cells then form the corpus luteum which produces progesterone. Theca interna cells invade the corpus luteum and may cause the rise in 17β-estradiol and 17α-hydroxyprogesterone in the second half of the cycle. After two weeks the corpus luteum regresses, levels of estrogen and progesterone fall, and menstruation ensues.

The endometrium shows proliferative changes during the first half of the cycle in response to estrogens, whereas in the second half of the cycle progesterone induces secretory changes. The factors responsible for corpus luteum regression are unknown.

Clinical tests for ovulation are inadequate. The rise in body temperature resulting from the action of progesterone is a guide, as is measurement of urinary pregnanediol excretion. The increased progesterone production which follows normal ovulation can be measured by radioimmunoassay of plasma progesterone, and this method should soon be generally available.

PUBERTY IN THE FEMALE

At puberty, growth is stimulated by estrogens and androgens, the latter arising mainly from the adrenal cortex in females. The growth spurt often precedes the onset of menstruation. As estrogen secretion increases, growth rate slows, and the epiphyses fuse. Plasma 17β-estradiol levels correlate with bone age and chronological age, and with clinical evaluation of sexual development. It is likely that decreased sensitivity of the hypothalamus to the negative feedback of gonadal hormones is responsible for the onset of puberty. Steroids secreted by the immature gonads become incapable of inhibiting the production of the pituitary gonadotrophins, and puberty begins. Sexual maturation is accompanied by an increase in the secretion of FSH which plateaus after midpuberty, while LH, like 17β-estradiol, slowly increases with advancing sexual maturation.

Menstruation consists of breakdown of the endometrium as a result of withdrawal of estrogens and progesterone. The menstrual loss is composed of endometrial cells, secretions, and blood. It normally occurs in a regular cyclical fashion from the menarche to the menopause. The duration of menstrual loss is three to seven days, and the cycle occurs approximately every four weeks.

THE MENOPAUSE

Since the menopause is a type of primary ovarian insufficiency, the urinary gonadotrophin excretion is found to be high.

Hormonal Changes. In women approaching the menopause, LH levels rise to values seven times higher than those in normal young women, whereas FSH levels are only three times elevated. The increase in urinary LH occurs before the onset of vasomotor symptoms. After the menopause, there is a 15-fold increase in the production rate of FSH and a fivefold increase in LH production and excretion, which is later exceeded by that of FSH. Estrogen secretion by the ovaries falls after the menopause and eventually ceases, when excreted estrogens are derived from peripheral conversion of androgens.

HYPOGONADISM IN THE FEMALE

A classification of hypogonadism in the female is given in Table 22-8.

ORAL CONTRACEPTIVES

Oral contraceptives fall broadly into three groups. Those in the first group contain an estrogen and a progestogen in combination, and the preparation is given for cycles of 20 to 22 days. In the second group, the

TABLE 22-8 HYPOGONADISM IN THE FEMALE

FAILURE OF SEXUAL MATURATION (PRIMARY AMENORRHEA)	FAILURE OF ESTABLISHED SEXUAL FUNCTION (SECONDARY AMENORRHEA)
1. *Ovarian causes* (a) *ovarian failure* congenital hypoplasia "castration" radiation damage chromosomal abnormalities nonmasculinizing tumors (b) *excess androgen production* masculinizing tumors	1. *Ovarian causes* (a) *ovarian failure* menopause "castration" radiation damage chromosomal abnormalities (rare) nonmasculinizing tumors (b) *excess androgen production* masculinizing tumors Stein-Leventhal syndrome
2. *Hypothalamic-pituitary causes* neoplasm vascular disturbance trauma infection granulomas "functional" disturbance	2. *Hypothalamic-pituitary causes* neoplasm vascular disturbance trauma infection granulomas "functional" disturbance including depression, anorexia nervosa and other varieties of psychogenic anorexia iatrogenic causes (including pituitary ablation)
3. *Associated with other endocrine and metabolic disturbances** congenital adrenal hyperplasia Cushing's syndrome Addison's disease obesity undernutrition (including anorexia nervosa) malabsorption syndromes hyperthyroidism, hypothyroidism administration of androgens or anabolic steroids	
4. *Associated with chronic debilitating disease** e.g., renal failure, diabetes, tuberculosis, connective tissue disorders	

* The hypogonadism is assumed in most cases to be mediated through the hypothalamic-pituitary centers.

estrogen is given alone for the first 15 or 16 days after cessation of menstruation, to be followed by a combination tablet containing an estrogen and a progestogen given for a week to complete the course. The estrogenic component is usually either mestranol or ethinyl estradiol. A number of different progestogens are used, and recently a third regimen of oral contraception has been described in which a progestogen is given alone. Here, the progestogen is taken daily without interruption, usually from the first day of the cycle. "Progestogen only" tablets do not always inhibit ovulation and depend for their contraceptive action on their effect on the cervical mucus and the endometrium. They have a small failure rate, rather less than one per cent. They do not inhibit lactation and can be given safely to women who wish to breast feed. Cycle control in the first few months is poor, and even in later months is rarely as good as with combined tablets. Other unpleasant side-effects are few. The tablets are simple to take; however, if tablets are missed, it is more likely to lead to failure. Details of some compounds in common use are given in Table 22-9.

The mode of action of the oral contraceptives has not been fully elucidated, although it is assumed that their effect in preventing ovulation, which is brought about mainly by the estrogenic component, is due to suppression of formation or release of LH/FSH-RH and possibly to a direct action on the anterior pituitary inhibiting release of the gonadotrophins. The combined preparations suppress the release of LH at midcycle, and the sequential regimens cause a decrease in FSH secretion, although there is decreased release of both gonadotrophins if therapy is

TABLE 22-9 ORAL CONTRACEPTIVE PREPARATIONS

PRODUCT	PROGESTOGEN	ESTROGEN
The Combined Preparations		
Anovlar 21	Norethisterone acetate 4 mg.	Ethinyl estradiol 0.05 mg.
Gynovlar 21 (Controvlar)	Norethisterone acetate 3 mg.	Ethinyl estradiol 0.05 mg.
Minovlar	Norethisterone acetate 1 mg.	Ethinyl estradiol 0.05 mg.
Minovlar ED	Norethisterone acetate 1 mg.	Ethinyl estradiol 0.05 mg.
Orlest 28	Norethisterone acetate 1 mg.	Ethinyl estradiol 0.05 mg.
Norlestrin 21	Norethisterone acetate 2.5 mg.	Ethinyl estradiol 0.05 mg.
Demulen 50	Ethynodiol diacetate 0.5 mg.	Ethinyl estradiol 0.05 mg.
Ovulen 50	Ethynodiol diacetate 1 mg.	Ethinyl estradiol 0.05 mg.
Minilyn	Lynestrenol 2.5 mg.	Ethinyl estradiol 0.05 mg.
Volidan 21	Megestrol acetate 4 mg.	Ethinyl estradiol 0.05 mg.
Ortho-Novin 1.50	Norethisterone 1 mg.	Mestranol 0.05 mg.
Norinyl-1	Norethisterone 1 mg.	Mestranol 0.05 mg.
The Sequential Preparations		
Feminor Sequential	15 × Mestranol 0.1 mg.: 5 × Norethynodrel 5 mg. and Mestranol 0.075 mg.	
Sequens	15 × Mestranol 0.08 mg.: 5 × Chlormadinone acetate 2 mg. and Mestranol 0.08 mg.	Strongly estrogenic
C-Quens 21	14 × Mestranol 0.1 mg.: 7 × Chlormadinone acetate 1.5 mg. and Mestranol 0.1 mg.	
Ortho-Norvin S.Q.	14 × Mestranol 0.1 mg.: 7 × Norethisterone 2 mg. and Mestranol 0.1 mg.	
Serial 28	16 × Ethinyl estradiol 0.1 mg.: 5 × Megestrol acetate 1 mg. and Ethinyl estradiol 0.1 mg.	

continued for long periods. Additional peripheral effects on the cervical mucus and the endometrium also occur. There is usually a prompt return of normal pituitary function following withdrawal of the drugs, although amenorrhea may sometimes develop at this time.

Choice of Preparation

Combined (Recommended Dose Containing 0.03-0.05 mg. Estrogen). This group is the most widely used at present. The dose of *estrogen* varies and there may be minor differences in effect between the two types, ethinyl estradiol and mestranol, the latter being a slightly stronger estrogen. *Progestogens* vary in potency and also in dosage. Strong progestogens include norethisterone and its acetate and also ethynodiol diacetate. Lynestrenol, though a strongly potent progestogen, appears in addition to have some "clinically estrogenic" effects. Megestrol acetate and norethynodrel are relatively weaker. The dosage varies from 0.5 mg. to 4 mg. and the aim should be to prescribe the smallest amount of progestogen which will give reasonable cycle control. This varies with each individual, but some assessment can be made by the amount and frequency of menstrual loss—those with heavier and more frequent periods requiring greater amounts of progestogen. Weaker or smaller doses of progestogen are indicated for women with a tendency to obesity or greasy skins.

Combined (Containing More Than 0.05 mg. Estrogen). These preparations are usually prescribed only when hormone therapy is needed for purposes other than contraception. High dose tablets, particularly of progestogen, are indicated in menorrhagia, functional uterine bleeding, endometriosis, and also in cases of dysmenorrhea unresponsive to lower dose products. Higher dose estrogen tablets sometimes benefit certain skin conditions, such as acne.

Sequential. These all contain higher amounts of estrogen and should only be given in special cases. They carry a very small failure rate. They may be a first choice in cases of acne and premenstrual tension. Occasionally, these tablets are the only type of oral contraceptive that a patient can tolerate and may justifiably be prescribed provided that the patient is warned of the "increased health risk."

Side-effects. Those which are commonly encountered can be attributed to one or the other hormone component, or both.

ESTROGEN EFFECTS	PROGESTOGEN EFFECTS
Premenstrual tension	Premenstrual depression
Fluid retention	Leukorrhea
Nausea, vomiting	Dry vagina
Headache	Acne
Mucorrhea	Greasy hair
Cervical erosion	Appetite increase—weight gain
Menorrhagia	Breast discomfort
Excessive tiredness	Reduced menstrual loss
Irritability	Leg and abdominal cramps
Vein complaints	Decrease in libido

Contraindications. The preparations should not be used for patients for those with recent liver disease. They are contraindicated for those with breast or cervical carcinoma, and these lesions should be excluded by appropriate examinations before the drugs are prescribed. Patients who have a history of thromboembolic phenomena should not be given oral contraceptives, and care must be exercised when they are used in the presence of epilepsy, diabetes, hypertension, and advanced cardiac or renal disease with edema.

Uterine fibroids may increase in size under the estrogenic stimulus, although high progestational steroid content pills may produce a reduction in size. There is no evidence of any danger to the fetus associated with oral contraceptives in early pregnancy.

Metabolic Effects of Oral Contraceptives

Abnormalities of carbohydrate and fat metabolism may result from these preparations. Impairment of oral carbohydrate tolerance occurs in up to 20 per cent of patients, and in a similar number an increased maximum pyruvate increment after glucose has been observed. As many as 85 per cent of patients may develop abnormal cortisone-stressed glucose tolerance tests. Overt diabetes mellitus is often aggravated, and diabetes or glycosuria may be precipitated in prediabetic or latent diabetic individuals. The abnormalities noted are similar to those found in steroid diabetes.

The fasting plasma free fatty acid level is often elevated, and the normal fall after glucose administration may be delayed. Thirty per cent of a group of patients receiving cyclical oral contraceptives have shown elevation of the serum triglyceride level, and blood cholesterol tends to be raised.

Many liver function tests are affected, including Bromsulphalein retention and an increase in the serum concentration of a number of enzymes. A variety of changes have been noted in regard to plasma proteins and on occasion jaundice has been produced. Estrogens have an effect on tryptophan metabolism, since the estrogen component enhances the capacity for the conversion of tryptophan to nicotinic acid ribonucleotide. Possibly because this pathway is stimulated, there is depression of the alternative metabolic pathway which produces 5-hydroxytryptamine (serotonin). This could account for the depression occasionally associated with oral contraceptives.

Endocrine Effects of Oral Contraceptives

Hypertension. Hypertension occurs in a variable proportion of normotensive women receiving any variety of oral contraceptive. Prospective studies are of most value in defining the problem and indicate a significant though minor rise in mean systolic but not in diastolic blood pressure. Women with pre-existing hypertension may be more susceptible to a further rise in their blood pressure. The mechanism of the hypertension is not fully understood, but it is probably due to a rise in renin activity which results from an effect of estrogens upon hepatic production of renin substrate. Increased angiotensin activity could then lead to a form of secondary aldosteronism.

Amenorrhea. Amenorrhea may follow withdrawal of any variety of oral contraceptive. Occasionally, it may be due to a premature menopause. The condition is more common in women with a previous history of amenorrhea or menstrual irregularity but may also occur in those with previously regular cycles. Spontaneous remission probably occurs in about one-half of the cases within one year and in another one-fourth as a result of therapy. The mechanism is probably mediated by an action of estrogens on the hypothalamus, which by suppressing the release or formation of LH/FSH-RH cause low LH levels and a reduction in ovarian estrogen production. Hypothalamic dysfunction is supported by the occurrence of galactorrhea, not always recognized by the patient, in about one-third of those with amenorrhea; these patients have hyperprolactinemia.

Pigmentation. Chloasma-like pigmentation similar to that occurring in pregnancy is a frequently seen feature with the contraceptive pill. Recognition of the complication prevents a misdiagnosis of Addison's disease. If the woman is distressed by the pigmentation, withdrawal of the drug is usually followed by a slow regression of the pigmentation. The mechanism of its occurrence is not well understood.

Thyroid and Adrenal Function Tests. Oral contraceptives affect the levels of many plasma proteins largely due to an effect on hepatic protein synthesis. Increased levels of thyroxine-binding globulin (TBG) and of transcortin lead to an elevation of the serum PBI, serum thyroxine and plasma cortisol levels. This may lead to erroneous diagnoses of hyperthyroidism or Cushing's syndrome. The free level of the hormone which is metabolically active is little altered and hence clinical features of hormone excess do not occur. The elevation of the PBI can be allowed for by measurement of some form of thyroid hormone-binding test which will show "hypothyroid" values, to calculate a free-thyroxine index which is normal in pregnant women and in those receiving oral contraceptives.

Thromboembolic Effects of Oral Contraceptives

There is an increased incidence of thrombophlebitis during pregnancy and the puerperium and a rise in the level of certain clotting factors following administered estrogen. During pregnancy, there is an increase of Factors, I, V, VII, VIII and X, and estrogens have been found to raise the level of Factor VIII in women with von Willebrand's disease and in carriers of hemophilia. Estrogens have also been shown to increase platelet adhesiveness by enhancing their sensitivity to ADP. Superficial thrombophlebitis occurs three times more often in women taking oral contraceptives. There is now good evidence that there is a significant increase in venous thrombosis and pulmonary embolism in such women and an increased hazard of cerebral thrombosis, although so far no definite increase in the incidence of coronary thrombosis has been demonstrated. Certain measures can be adopted to minimize the hazard. The tendency to thromboembolism depends upon the estrogen content of the pill, and preparations with an estrogen content of > 50 μg. should not normally be used. Other possible conditions which may predispose to arterial thrombosis are obesity, hypertension, hypercholesterolemia, and possibly worsening migraine.

Gynecological Effects of Oral Contraceptives

Cervical Changes. Changes in appearance of the cervix may be difficult to interpret and florid cervical erosions which bleed on pressure are sometimes found in patients on oral contraceptive therapy. The cytological findings are usually negative in spite of the rather bizarre clinical appearance, although on occasions biopsy may show histological evidence of carcinoma *in situ*. There is no direct evidence linking administration of the pill with the development of carcinoma of the cervix or with malignant lesions in any site. However, cancer of the cervix occurs naturally in women of the reproductive age group, and therefore occasional spontaneous occurrence of malignant change can be expected.

Breast Changes. In some women, cyclical engorgement and fullness occur premenstrually; enlargement of the breasts may also be noted. Some women develop diffuse nodularity in the breasts which subsides when therapy is discontinued. Low dosage oral contraceptive agents have no effect on the maintenance of lactation and can be given safely to lactating women.

Vaginal Candidiasis. Oral contraceptives may precipitate or aggravate candida infection of the vagina, the estrogen component causing changes in the acidity of the vagina which allows "thrush" to become established. This must be distinguished from simple increased vaginal transudate. The patient complains of pruritus vulvae and a discharge, and the appearances of the vagina are usually characteristic.

PLACENTAL HORMONES AND PREGNANCY

The placenta produces progesterone, a variety of estrogens, human chorionic gonadotrophin (HCG), placental lactogen (HPL), human chorionic thyrotrophin (HCT), human molar thyrotrophin (HMT), possibly relaxin, and a variety of prostaglandins. The bulk of steroid hormones elaborated during pregnancy is formed by the complementary activity of the fetus and of the placenta, since both are deficient in certain essential steroidogenic enzyme systems. This fetoplacental unit is capable of synthesizing most if not all of the biologically active steroid hormones required during pregnancy.

There is abundant synthesis of sterols and steroids in the midgestation fetus, whereas little if any sterol or steroid synthesis occurs in the placenta at this time.

The placenta also produces pituitary-like hormones with structural and biological similarities to hormones of the anterior lobe. HPL is similar to both growth hormone and prolactin, HCG is similar to LH and FSH, and HCT and HMT, to TSH.

Estrogens

The true ovarian hormone is 17β-estradiol, which is readily converted to estrone and estriol. By the third month of pregnancy, the placenta has taken over from the ovary as the major source of estrogen formation. Small amounts of estrogens also arise from the maternal adrenal cortex and from the fetus.

The fetal adrenal converts placental progesterone and pregnenolone to dehydroepiandrosterone (DHEA), which is hydroxylated in the fetal liver at C-16. The 16 α-hydroxydehydroepiandrosterone then returns to the placenta to be converted into estriol. This "neutral" pathway is the main route of estriol synthesis. Other routes of estriol formation in the feto-placental unit include: the "phenolic" pathway, by which estrone and estradiol are converted to estriol in the mother and also in the fetal liver; the "ring D steric rearrangement," in which androgens are converted to 16-epiestriol by the mother and to a lesser extent in the placenta; the placental formation of estradiol from cholesterol and the conversion of this estrogen to estriol in the mother.

Estriol is the predominant estrogen excreted in the urine, and the output rises during pregnancy and falls rapidly after delivery. Levels of estriol in maternal blood rise slowly to about 4 μg. per 100 ml. by 32 weeks, after which there is a rapid but variable rise to levels between 8 and 22 μg. per 100 ml. at term. In urine, estriol makes up 75 to 97 per cent of all estrogens, whereas in blood the proportion is lower, from 25 to 80 per cent. Most of the estrone and estradiol in urine is present in the conjugated form as glucosidouronates, whereas in the blood conjugates are present in only trace amounts. Amniotic fluid estriol levels rise steeply in the last few weeks of pregnancy to values as high as 150 to 200 μg. per 100 ml. The main source of amniotic fluid estriol is probably from fetal urine.

The factors that control estrogen production during pregnancy are poorly understood. Estrogen determinations reflect the function of both the fetus and its placenta, in contrast to determinations such as that of progesterone or placental lactogen, which reflect placental activity alone. Isolated observations are of little value, and serial estimations are needed before conclusions can be drawn. Estrogen excretion is reduced in those with moderately severe pre-eclamptic toxemia and diabetic pregnancies and in the presence of hydatidiform moles and chorionepithelioma. Fetal death is associated with a fall in urinary estrogens.

Estrogens cause an increase in RNA, phospholipid and protein synthesis in the uterus. Estradiol is bound to nuclear sites in the uterus and stimulates the activity of the DNA-dependent RNA polymerase. Estrogens control the growth of the decidual lining of the uterus, the myometrium and its blood vessels, acting synergistically with progesterone.

Rising amounts of estrogen during pregnancy finally overcome progesterone inhibition of the uterus, rendering the organ sensitive to oxytocin and initiating labor.

By altering the polymerization of acid mucopolysaccharides, estrogens change the properties of the ground substances between collagen fibers. This may be a factor in allowing stretching of the uterine cervix and increasing the mobility of the pelvic joints, acting synergistically with the hormone relaxin. The generalized water retention in pregnancy is partly due to the increased hygroscopic qualities of mucopolysaccharides induced by estrogens.

Estrogens affect breast development, predominantly of the duct system, and increase the size and mobility of the nipples. High estrogen levels also alter the levels of various circulating clotting factors and hormone-binding proteins.

Progesterone

A few weeks after conception the placenta becomes the main site of progesterone production, and pregnancy can be maintained in the absence of the ovaries. Progesterone is not produced *de novo* by the midterm placenta, which lacks the enzyme systems needed to convert acetate into cholesterol. The midterm fetus produces cholesterol, which is then converted in the placenta to pregnenolone and progesterone and also 17α-hydroxypregnenolone, dehydroepiandrosterone, 17β-estradiol, 20α-dihydroprogesterone and 20β-dihydroprogesterone. The adrenal glands of the mother and fetus also produce a small amount of progesterone. Progesterone acts as a substrate for cortisol and other steroids in the fetal adrenal. About half of the progesterone produced by the placenta reaches the fetus.

Progesterone production rises during pregnancy, leveling off a few weeks from term, when the production rate is above 300 mg./day. Progesterone is bound by albumin and by corticosterone-binding globulin and less than 10 per cent circulates in the free form. Plasma progesterone rises during pregnancy to a level of about 14 μg. per 100 ml. at term.

Progesterone is not bound specifically by any tissue but it does increase the formation of RNA and DNA in the uterus, especially if this has just been primed with estrogen. Progesterone reduces muscle tone in the uterus and in the stomach, colon and ureters. Its role in the treatment of habitual abortion has not been firmly established, and progesterone derivatives are best avoided during pregnancy because of the risk of virilizing the female fetus. A fall in pregnanediol output on serial measurements can provide an indication of placental failure. The rise in body temperature of 0.5° F. during the luteal phase of the menstrual cycle is due to progesterone and its metabolites and continues after conception till about midpregnancy. Progesterone increases alveolar tissue growth in the breast. The increase of depot fat stores during pregnancy may be determined by progesterone. The reduction in alveolar and arterial pCO_2 during pregnancy is due to a central stimulatory action of progesterone.

Human Chorionic Gonadotrophin (HCG)

HCG is a glycoprotein of molecular weight of about 30,000, with structural similarities to LH, FSH and TSH. It comprises α-subunits identical with those in other glycoprotein hormones and hormone-specific β-subunits. Free α-subunits produced by the placenta can be detected in the urine during pregnancy.

HCG is assayed by biological or immunological methods. By definition, 1 IU is the activity contained in 0.1 mg. of the standard preparation. HCG is produced entirely by the trophoblast and is detectable in the plasma as early as 10 days after fertilization. Levels rise rapidly at about the 40th day of pregnancy, peaking at the 60th day and showing a rapid fall by the 80th day to a fairly steady level maintained throughout pregnancy apart from a slight secondary peak between 30 and 36 weeks. Peak values range from 20,000 to 100,000 IU/liter of serum and between 20,000 and 500,000 IU/24 hr. in the urine. HCG is luteotrophic and can prolong the menstrual cycle by maintaining the corpus luteum, leading to a decidual reaction in the endometrium. It also acts on the fetal adrenal cortex, increasing production of dehydroepiandrosterone.

Radioimmunoassay of HCG in the urine is now being used to follow up women with hydatidiform moles as well as to aid in initial diagnosis. In the majority of women delivered of moles, HCG excretion continues for weeks or months, indicating the presence of viable trophoblastic tissue. Serial urinary HCG assays should be performed for two years after evacuation of a mole. Persistence of high levels or a rising titer provides an indication for cytotoxic therapy which is almost always successful if not delayed. Measurement of HCG output is of little value in the diagnosis of placental failure.

Human Placental Lactogen (HPL)
(Human Chorionic Somatomammotrophin, HCS)

HPL is a protein hormone with similarities in structure and immunological and biological properties to growth hormone and prolactin. It is produced in the placenta, the amounts produced being related to placental weight. HPL is present largely in the maternal circulation, very little of it crosses the placenta, and no significant amounts can be detected in the maternal urine. Levels fall rapidly after delivery. The factors which control HPL secretion are poorly understood, but plasma levels may fall in response to glucose and rise after hypoglycemia. Although HPL levels reflect only the function of the placenta, they are still a valuable guide to fetal well-being, since the fetus is very dependent on placental function.

HPL measurements can provide a guide to the outcome of threatened abortion and can be used to screen for fetal distress and neonatal asphyxia. In patients with vaginal bleeding after the eighth week of gestation, low levels of HPL are found in those in whom abortion is completed during their first admission. Women whose pregnancies continue normally or who abort after their first discharge from hospital have normal levels of HPL. In the absence of contraindicating factors such as maternal age or infertility, low HPL levels can provide some guide concerning whether or not to evacuate the uterus in a woman with vaginal bleeding early in pregnancy.

Measurement of HPL levels after the 30th week of pregnancy can help to predict subsequent fetal distress and/or neonatal asphyxia. Three or more levels of HPL less than 4 μg. per ml. between the 35th and 40th weeks of pregnancy indicate high risk of fetal distress in labor or neonatal asphyxia, whereas levels above 5 μg. per ml. are associated with a low frequency of these complications.

The physiological role of HPL has not been fully defined. It may cause: mobilization of free fatty acids to meet fetal requirements, an increase in insulin resistance causing a higher level of circulating insulin, an increase in nitrogen storage, an increase in amino acid transport across the placenta and an increase in breast growth.

Human Chorionic Thyrotrophin (HCT) and Human Molar Thyrotrophin (HMT)

The placenta produces two thyroid-stimulating agents termed HCT and HMT. HMT is produced in large amounts in some patients with hydatidiform moles or choriocarcinoma when it may be responsible for abnormalities of thyroid function tests and occasionally for overt hyperthyroidism. It is possible that HMT may act as a precursor for HCT. A comparison of the two trophoblastic thyrotrophins is shown in Table 22-10.

Prostaglandins (PG)

Prostaglandins are 20-carbon hydroxy fatty acid derivatives of prostanoic acid which are widely distributed in the body. Their role in human reproduction is still uncertain. Amniotic fluid during labor contains high concentrations of $PGF_2\alpha$, which causes myometrial contractions whereas fluid collected earlier contains no $PGF_2\alpha$ and only a little PGE_1. $PGF_2\alpha$ is present in maternal venous blood immediately before uterine

TABLE 22-10 COMPARISON OF THE TROPHOBLASTIC THYROTROPHINS

SOURCE	HCT NORMAL PLACENTA	HMT MOLE AND NORMAL PLACENTA
Reaction with antihuman TSH	Partial	None
Reaction with antibovine TSH	Yes	None
Duration of action	Short	Intermediate
Molecular size	~ 30,000	(> TSH, < LATS) ~ 70,000

contractions in normal spontaneous labor. Both PGE_2 and $PGF_2\alpha$ can be used by oral, intravenous, intravaginal and intra-amniotic routes to induce labor. The placenta is the major source of PGs found in the amniotic fluid and maternal circulations.

The Adrenal Gland and Pregnancy

Plasma cortisol levels rise during pregnancy as a result of the rise in transcortin induced by the high circulating estrogen levels. There is also a rise in plasma-free cortisol indicated by direct measurement and indirectly by measurements of urine-free cortisol and the cortisol production rate. The rate of removal of cortisol from plasma is slowed during pregnancy.

Patients with Addison's disease do not require any adjustment of their replacement therapy during pregnancy unless vomiting or surgical procedures render oral therapy inappropriate.

Aldosterone secretion increases during pregnancy by mechanisms as yet unknown. There is an increase in renin substrate as well as in renin concentration in pregnancy, and a significant positive correlation has been demonstrated between the plasma aldosterone concentration and the product of renin and renin substrate.

The Thyroid Gland and Pregnancy

Enlargement of the thyroid gland is common during pregnancy and the associated hypermetabolic state of pregnancy may lead to an erroneous diagnosis of hyperthyroidism.

Elevation of the serum PBI and thyroxine levels results from an increase in thyroid hormone binding proteins induced by the high estrogen levels. Thyroid hormone binding tests (e.g., Thyopac-3, resin uptake tests, and others) give results in the hypothyroid range. Calculation of the free thyroxine index from the PBI or T_4 and the residual binding capacity gives results similar to the true free thyroxine and allows thyroid status to be assessed during pregnancy. High progesterone levels cause hyperventilation with a fall in the pCO_2 which increases the association of T_4 and T_3 to their binding proteins. This factor may also tend to reduce the level of free thyroid hormones in pregnancy and correlates with the slight elevation of serum TSH in pregnancy demonstrated by very sensitive radioimmunoassays.

Enhanced renal clearance of iodide during pregnancy leads to a fall in the plasma inorganic iodide. This is compensated for by an increase in the absolute iodide uptake, presumably mediated at least in part by the increase in circulating TSH.

The role of HCT and HMT in the control of thyroid function in pregnancy is unknown.

Neonatal hyperthyroidism is a very rare occurrence, affecting children born by mothers with Graves' disease. It is usually possible to detect the long-acting thyroid stimulator (LATS) in the maternal circulation, and transplacental passage of this IgG has been proved by the finding of LATS in the child's blood. The neonatal hyperthyroidism is probably a result of the thyroid stimulatory properties of LATS or other thyroid-stimulating immunoglobulins and the condition remits over a few months as these disappear from the circulation.

The Pituitary Gland and Pregnancy

The pituitary gland increases in size and weight during pregnancy and histologically shows a marked increase in prolactin containing cells of the eosinophil variety. Levels of LH and FSH are low in pregnancy and these hormones are not necessary for maintenance of pregnancy once fertilization and implantation have occurred. Growth hormone levels appear to be normal during pregnancy and respond in the same way to stimuli such as hypoglycemia. Vasopressin levels during pregnancy have not yet been studied by critical immunoassays. During pregnancy, there is a striking fall in plasma osmolality produced by mechanisms as yet unknown. The role of vasopressin in relationship to this fall in osmolality has not been elucidated.

ANDROGENS AND THE TESTIS

The functions of the testis are: (a) the formation of spermatozoa and (b) the synthesis and secretion of testosterone. Spermatogenesis takes place within the seminiferous tubules and the semen is formed by the secretion from the tubules containing the sperm, in addition to secretions from the seminal vesicles (containing high concentrations of fructose) and prostate (containing high concentrations of citrate and acid phosphatase). The production of the seminal fluid depends on the presence of adequate amounts of the male androgenic hormone, testosterone, although spermatogenesis itself depends on follicle-stimulating hormone. Testosterone is secreted by the interstitial cells of Leydig which lie in clumps between the seminiferous tubules. They cannot be distinguished during prepubertal life but develop at puberty under the influence of the increased secretion of interstitial cell-stimulating hormone (ICSH, the same substance as luteinizing hormone; however, this name is usually restricted to females). Testosterone is probably not the hormone actually active at the tissue levels. It appears that it must be converted to dihydrotestosterone first. The negative feedback control of ICSH depends on the circulating levels of testosterone and dihydrotestosterone, whereas that for FSH is dependent on a substance produced during spermatogenesis. This compound, of unknown structure, has provisionally been called "inhibin."

Testosterone Synthesis and Metabolism (Figs. 22-5 and 22-7). Testosterone is the major androgen produced by the Leydig cells of the testis. Small amounts of the hormone are also produced by the adrenal cortex in both sexes and by the ovary. Testosterone is a 17βhydroxylated, C-19 steroid, the formula of which is shown in Fig. 22-5, but it appears to require conversion to an active and much more potent metabolite dihydrotestosterone, and this can occur both in the circulation and in the peripheral target tissues.

Synthesis. Similar enzymatic processes are involved in the synthesis of steroid hormones in the adrenal cortex, ovary and testis. In each of these glands, the side chain of cholesterol is degraded to form pregnenolone, and hence dehydroepiandrosterone. Pregnenolone is converted to progesterone, which in turn is converted to androstenedione and then testosterone.

Transport and Fate. Testosterone and dihydrotestosterone are transported in the blood loosely bound to carrier proteins, mainly a globulin—"sex hormone-binding globulin" (SHBG) or "testosterone-binding globulin" (TeBG)—which also has a high affinity for other 17β-hydroxy-androgens (17-OHA). Testosterone binding capacity of serum (TeBC) is similar in both sexes before puberty and in the adult female range, but is higher in adult females than in adult males. There is increased TeBC in male hypogonadism, hyperthyroidism and cirrhosis of the liver. During pregnancy, TeBC is raised, but at term a low normal level is found. TeBG is made in the liver and production is increased by estrogens and also by high levels of thyroid hormones; progestogens have no effect. Androgenic activity is believed to be a function of the free (unbound) plasma androgen concentration. Testosterone is degraded, particularly in the liver, under the influence of 17β-dehydrogenase enzyme systems and then conjugated with sulfates or glucuronic acid to be excreted in the urine as 17-oxosteroids (androsterone, etiocholanolone and epiandrosterone) (Fig. 22-7). The testis contributes about only 24 per cent of the 17-oxosteroids (ketosteroids) in man, the remainder coming from the adrenal cortex (largely dehydroepiandrosterone, but in smaller amounts androsterone and etiocholanolone). Adrenal 17-oxosteroids derived from cortisol are mainly 11-oxy-17-oxosteroids. Testosterone can also be converted to estrogens, particularly estrone and estradiol, but this is only a minor fate of the hormone. Testosterone is converted into dihydrotestosterone mainly in the blood and at the site of its target tissues. The biological actions of testosterone can be produced by dihydrotestosterone. The plasma concentration of dihydrotestosterone is higher in males than in females, and in both sexes its concentration is far lower than testosterone. In normal men, the production rate of dihydrotestosterone is about 0.4 mg. per day, of which 50 per cent is derived from the transformation of plasma testosterone; in normal females, the production of dihydrotestosterone is only about 0.05 mg. per day, of which only 10 per cent comes from testosterone. In normal males, the production rate of testosterone is 6 to 7 mg. per day, depending on the method used. In normal females, testosterone production rate is much lower at 0.2 mg. per day.

Testosterone Levels in Body Fluids. Testosterone is the most potent naturally occurring androgen but is present in such small amounts in the blood that its estimation is not possible in most routine laboratories. Estimation of the urinary 17-oxosteroids is all that is available to many hospital laboratories. Since the majority of the 17-oxosteroids are metabolic products of percursors other than testosterone, they form a poor tool for the investigation of diseases caused by altered androgen metabolism. Recently, new techniques such as double isotope derivative dilution, competitive protein binding, radioimmunoassay and gas-liquid chromatography have allowed testosterone levels to be measured in blood and urine as a research procedure. In the authors' laboratories the mean concentration of testosterone in peripheral plasma using a competitive protein-binding technique is 5.5 to 12 ng. per ml. In women, the range is 0.2 to 0.8 ng. per ml., with levels slightly higher in the luteal phase of the cycle suggesting an ovarian source. In boys, testosterone levels tend to rise before there is clinical evidence of puberty. Normal levels are seen in many aged men.

Like cortisol, testosterone levels in the blood appear to show circadian rhythmicity, with higher levels observed at 8 A.M. and lower levels between midnight and 4 A.M. A decline in testosterone levels is usually seen with onset of sleep but subsequently there are fluctuations leading to an increase in levels culminating in the 8 A.M. peak. There appears to be an association of individual fluctuations in testosterone levels with periods of REM sleep. The circadian rhythm of testosterone is not suppressed by dexamethasone and does not depend on ACTH. However, a close dependence on LH secretion has not been demonstrated either. Levels of testosterone are raised by administration of human chorionic gonadotropin (HCG), since this has ICSH activity, and lowered by estrogens. Powerful synthetic androgens such as fluoxymesterone and 2α-methyldehydrotestosterone suppress testicular testosterone production, probably by inhibiting ICSH formation, since this effect can be overcome by administration of HCG.

Male sex hormone is concerned with conditioning the mating drive and with maintaining the structure and function of the accessory reproductive organs, particularly the penis, prostate and seminal vesicles. It is also concerned with maintaining certain male characteristics, such as beard growth and masculine hair distribution, deepening of the voice, male type of body development, and so on. Testosterone and other androgens are anabolic and stimulate cell growth and multiplication. They tend to increase body weight and cause nitrogen retention. They also increase sebum secretion and accelerate bone maturation, including fusion of the epiphyses. As with many other hormones, effects on messenger and nuclear RNA have been demonstrated, but the primary site of action of the hormones is still uncertain.

It appears that there is a cytoplasmic steroid receptor for testosterone or dihydrotestosterone which binds the androgen at the target cell and transports it from the cell membrane to the nucleus, where changes in RNA are induced. Androgens are present in the urine of women in amounts up to three-fourths of that in the male. It is possible that in women the hormone is derived from both the ovary and the adrenal cortex, although this is still open to question. The latter is generally believed to be the source of most, if not all, of the androgen in the female.

The characteristic morphological effects of androgens are reflected in the changes induced in the male by castration. In the prepubertal castrate the prostate and seminal vesicles fail to develop; in adults, castration is followed by atrophy of these structures. The penis remains small in the former and regresses in the latter. A normal state is restored by administration of androgens. In prepubertal castrates, beard growth is absent, the body and pubic hair is scant and the voice remains high-pitched. Closure of epiphyseal lines is retarded, resulting in disproportionately long arms and legs ("eunuchoidism").

Abnormally large amounts of androgen in women produce two types of effect, due to (1) "virilization" and (2) suppression of ovarian function. In the human being, the former includes: enlargement of the clitoris, growth of facial and body hair, development of a male forehead hairline, stimulation of secretion and proliferation of the skin sebaceous glands (often with acne), and deepening of the voice. The gonadal effects are due primarily to suppression of pituitary gonadotrophic function. This results in suppression of ovarian follicle maturation and ovulation (i.e., decreased estrogen and progesterone production) followed by atrophy of the uterus, vagina, and often the breasts.

Metabolic Actions. The dominant general metabolic effect of androgens is stimulation of protein anabolism. This is reflected in (1) a decrease in urinary nitrogen without an increase in blood NPN and (2) increase in body weight, due chiefly to an increase in skeletal muscle. In the growing organism, a growth spurt is induced, with increase in bone matrix and skeletal length. In addition to the specific stimulation of growth of the prostate and seminal vesicles, there is a rather selective increase in size and weight of the kidneys (renotropic action). This, together with the changes in renal enzymes and the absence of similar changes in the liver and intestine, has been interpreted as indicating that the protein anabolic effects of androgens are mediated, in part at least, by the kidney. Neither the pituitary growth hormone nor the adrenal cortex is involved in this phenomenon.

Creatine, which is virtually absent from the urine of normal men, increases after castration. This increase is abolished by testosterone, owing to increased storage of creatine in the muscles. After prolonged administration, the quantity of creatinine in the urine may increase, probably reflecting the increase in muscle mass.

Androgen reduces (and estrogen increases) the excretion of citrate in the urine. This is due apparently to increased reabsorption of citrate by the renal tubular epithelium.

The decreased urinary excretion of nitrogen (chiefly urea) that follows administration of androgens is accompanied by a lower urine volume and diminished excretion of Na, Cl, K, SO_4, and PO_4, with no increase in their concentrations in blood plasma. The tissue retention of K, SO_4, and PO_4 is probably related to the increased storage of protein. The retention of Na, Cl, and water is apparently due to an action resembling that of the adrenocortical hormones.

Abnormal Androgen Values. Consistently low values are obtained in male hypogonads whether due to primary gonadal failure or to gonadotrophin deficiency resulting from pituitary or hypothalamic disease. The combination of subnormal androgen and high gonadotrophin values is characteristic of primary testicular failure. This may occur congenitally or as a sequel of orchitis, and from operations interfering with the blood supply to the testes, systemic diseases, and the like. Diminution in or absence of gonadotrophic hormone in men with subnormal androgen values suggests pituitary hypofunction. This may occur in panhypopituitarism and pituitary tumors and from tumors and inflammatory disease of the hypothalamus. Severe nutritional disturbances may also cause a decrease in gonadotrophins and androgens.

Functional disturbances in the pituitary-gonad cycle occur much less commonly in the male than in the female. Moreover, disorders of spermatogenesis, with associated infertility, are often not accompanied by any demonstrable abnormality of male sex hormone excretion.

Excessive excretion of biologically active androgens is encountered in association with masculinizing tumors of the adrenal cortex, arrhenoblastomas of the ovary and interstitial-cell tumors of the testis. In the last condition extremely high values have been obtained. Increased androgen excretion occurs frequently in patients with adrenal cortical hyperplasia. Normal values are usually obtained in patients with simple hirsutism or virilism of constitutional origin.

ECTOPIC HORMONE PRODUCTION BY NONENDOCRINE TUMORS

Many different hormones can be secreted by tumors of tissues other than those normally responsible for their synthesis, the so-called "ectopic" hormone production by tumors. Although these syndromes have been thought to be rare, they are being recognized with increasing frequency. Their development may precede other manifestations of the neoplasm, sometimes by many years, especially when the associated tumor is not malignant, e.g., a bronchial carcinoid. Some syndromes are so unusual that they immediately suggest the possibility of a tumor at a particular site, e.g., the syndrome of inappropriate secretion of vasopressin associated with a bronchogenic carcinoma. Awareness of these syndromes sometimes allows the underlying neoplasm to be diagnosed at an early stage. Improvement of the endocrine manifestation after removal of the neoplasm may be followed by recurrence, along with recurrence of the tumor.

Three theories have been put forward to explain the synthesis of hormones by nonendocrine tumors. The first suggests that hormones are synthesized by chance as a result of random or chaotic protein synthesis characteristic of neoplastic growth; mutations of the DNA of malignant cells would allow coding of peptides with endocrine activity. On the basis of this theory, a tumor might synthesize active peptides with structures similar to the part of the naturally occurring hormone required for biological activity. Since the peptide sequences which constitute an immunoreactive site of a hormone are not necessarily the same as those needed for its biochemical function, the hormone secreted by a tumor might not be immunologically identical with the natural hormone. Thyroid-stimulating hormone and insulin produced by tumors usually appear to be immunologically different from normal TSH and insulin, a finding consistent with the random synthesis theory, but other tumor hormones are both immunologically and biologically similar, e.g., ACTH, MSH and vasopressin. It is not unusual for a tumor to secrete more than one hormone, and this would be rather unlikely if random synthesis of peptides were occurring.

The second possibility is that certain tumors may have a high avidity for hormones, acting as "hormone sponges" which concentrate the hormone from the circulation. Rapid breakdown of malignant cells in an enlarging tumor would allow release of the stored hormone or hormones. However, there does not seem to be any good evidence in support of this theory.

The demonstration that neoplastic tissues can continue to release hormones into the medium when they are maintained in organ culture clearly demonstrates that "ectopic" synthesis as well as release of hormones may occur in tumors and provides strong evidence against the "sponge" theory and strong support for the third "de-repression" theory. Furthermore, the "sponge" concept cannot be applied to the situation in which the ectopically produced hormone would not normally be found in the patient, for example, the ectopic production of placental lactogen in men with bronchial carcinomas. Clearly, in these cases synthesis of the hormone must have occurred *de novo* within the tumor. The "de-repression" theory holds that tumor cells, like all other cells except the gamete cells, inherit an identical complement of DNA and therefore all the coded information requisite for synthesis of all normal proteins. Normal differentiation of cells involves reversible repression of specific segments of the DNA molecule, possibly by combination of DNA with histones, and much of the genetic potential of a normal cell is masked in this way. The malignant cell would revert to synthesis of various peptides, either by inactivation of the histone repressor or by deletion of a regulator gene. This is generally thought to produce the repressor which combines with the operon, normally slowing the manufacture of the messenger RNA molecules. If "de-repression" is involved in tumor hormone synthesis, the hormone produced by the tumor is likely to be identical with the natural hormone, as seems to be the case for ACTH, MSH, vasopressin and parathormone; however, final proof awaits chemical analysis of the hormones produced by tumors.

The explanation for the synthesis of hormones by tumors is therefore still uncertain, current evidence suggesting that some tumors might produce hormones because of random peptide synthesis, whereas most synthesize hormones apparently identical with natural hormones because of "de-repression."

Before it can be accepted that a tumor is responsible for an endocrine syndrome, certain criteria should be fulfilled.

1. The tumor should be shown to have the ability to synthesize the hormone. Usually, this has been assumed by finding a higher tumor content of the hormone, improvement of the syndrome after removal of the tumor and recurrence of the syndrome along with recurrence of the tumor. Incorporation of labeled precursors of the hormone by the tumor *in vivo* or *in vitro* would also be useful evidence. Demonstration of a concentration of the hormone in the venous blood draining the tumor higher than that in the arterial supply is good evidence for release of hormone by the tumor and indirect evidence for its synthesis.
2. The hormonal material should be demonstrable in the circulation and sometimes in the urine.
3. The hormone in the circulation should be capable of producing the endocrine syndrome affecting the patient.
4. Removal of the tumor or its treatment by radiation or other means should be followed by disappearance or fall in the level of the hormone in the circulation and improvement of the endocrine syndrome.

The endocrine syndromes associated with nonendocrine tumors are listed in Table 22-11, together with some examples of the ectopic production of hormones without associated clinical syndromes. Certainly, these conditions are not rare although they are frequently overlooked.

Significance of Ectopic Hormone Production. This far outweighs that suggested by the actual incidence of the clinical syndromes. First, if we could understand the mechanisms whereby the malignant cell produces a hormone apparently quite foreign to the tissues from which the tumor was derived, we might gain some understanding of the nature of the biochemical or genetic changes associated with malignancy. Furthermore, the identification of the hormonal products released by these tumors has allowed us to develop some insight into the reason why tumors actually make people ill even though the lesions are often small and do not involve vital structures. Clearly, they produce toxic chemicals and this concept

TABLE 22-11 HORMONES PRODUCED "ECTOPICALLY" BY NONENDOCRINE TUMORS AND ASSOCIATED CLINICAL SYNDROMES

ECTOPIC HORMONE	CLINICAL SYNDROME
ACTH (with MSH)	Cushing's syndrome
Parathyroid hormone	hypercalcemia
Growth hormone	hypertrophic osteoarthropathy with finger-clubbing and subperiosteal new bone formation
Vasopressin	water retention
Prolactin	galactorrhea
LH, HCG	precocious puberty or gynecomastia
Insulin-like	hypoglycemia
TSH-like	hyperthyroidism
Enteroglucagon	constipation
Erythropoietin-like	polycythemia
Placental lactogen	
Oxytocin	none recognized
Neurophysin	
Glucagon	

of "biochemical malignancy" includes production not only of hormones but also enzymes and other abnormal proteins such as fetal proteins and immunoglobulins. Treatment can be directed at the effects of the toxic products, but in addition, by following the concentration of these "biochemical markers of malignancy" the effects of any therapy directed at the tumor may be assessed. The blood or urine levels of the substance may be related to the mass of active tumor tissue and may indicate whether the tumor has been eradicated or suggest recurrence well before this is clinically obvious.

The concept of endocrine markers of malignancy is already well established in association with nonectopic secretion from functioning malignant tumors of endocrine tissues, in which the blood or urine levels of the hormone are used to follow the progress of the disease. Such tumors include adrenocortical carcinomas (cortisol, androgen or estrogen markers), pheochromocytomas (catecholamines), medullary carcinoma of thyroid (calcitonin), parathyroid carcinoma (parathormone), carcinoid tumors (5-hydroxytryptamine and 5-hydroxytryptophan), islet cell tumors of pancreas (insulin or gastrin), choriocarcinoma of uterus or testis (HCG), interstitial cell tumors of testis (estrogen) and arrhenoblastomas or granulosa cell tumors of the ovary (testosterone or estrogen). The commonest ectopic syndromes will be reviewed in greater detail.

Ectopic ACTH Syndrome. This is most commonly associated with an oat cell carcinoma of the bronchus which secretes the ACTH in association with β-MSH, both of which appear to be very similar or identical with the pituitary hormones. Occasionally, the tumor may be benign, e.g., bronchial carcinoid. Although large amounts of ACTH and cortisol are secreted the patient rarely looks Cushingoid, since such tumors are most often malignant and the patient may not live long enough to develop the classic features normally associated with excess corticoid secretion; indeed, they usually lose weight rapidly. There is, however, a hypokalemic alkalosis and usually frank diabetes mellitus. Plasma corticosteroids usually exceed 40 μg. per 100 ml., and ACTH levels, 200 pg. per ml., without any circadian rhythm. Urinary corticosteroids are also greatly elevated. ACTH, plasma and urinary corticosteroids are resistant to suppression with exogenous corticosteroid administration.

Hypoglycemia. More than 100 patients have been reported with hypoglycemia in association with nonpancreatic neoplasms, usually a large connective-tissue tumor of low-grade malignancy or a primary carcinoma of the liver. Removal of the tumor relieves the symptoms of hypoglycemia —recurrent dizziness, semicoma or coma, convulsions, tachycardia, sweating, and others. It is likely that these tumors elaborate a substance with insulin-like actions, and such material has occasionally been extracted and shown to have activity in the rat epididymal fat pad or rat diaphragm assays. However, radioimmunoassay studies have usually failed to demonstrate material with immunological properties of insulin in these tumors. It is therefore unlikely to be insulin itself.

Hypercalcemia. Hypercalcemia is a common and potentially fatal complication of malignant tumors. In many instances, the hypercalcemia is the result of bony deposits—bone destruction by the tumor, releasing calcium faster than it can be excreted by kidney and gut. However, it may often be found in patients with malignant disease but without bony metastases, and in these, at least, it seems likely that the hypercalcemia is due to a hormone-like substance produced ectopically by the tumor. The most common neoplasms associated with elevated blood calcium levels are squamous carcinoma of the lung, adenocarcinoma of the kidney or ovary and carcinoma of the breast.

A number of these tumors have been shown to contain material with biological activity very similar to parathyroid hormone, and indeed some of them also have immunological activity very like that hormone. Some breast tumors, on the other hand, seem to contain sterols similar to calciferol and other vitamin D-like compounds.

Clinically, the patients present with symptoms and signs of hypercalcemia: thirst, polyuria, lassitude, muscular weakness, nausea, vomiting, cardiac arrhythmias, drowsiness, depression, mental confusion and coma.

Inappropriate Secretion of Vasopressin. This occurs most frequently with oat cell carcinoma of the bronchus. The carcinoma secretes material which is indistinguishable from normal arginine vasopressin, and sometimes in association with oxytocin and neurophysin. Such patients retain water and dilute the body fluids, developing hypotonic plasma. Despite this and owing to the excessive and inappropriate secretion of vasopressin, the urine is concentrated. Thus, the hallmark of this condition is a dilute plasma (osmolality less than 270 mOsmol. per kg.), with a urine which is more concentrated than the plasma. There is often profound hyponatremia and equivalent dilution of the BUN and other electrolytes. When the plasma sodium falls to less than 110 mEq. per liter, neurological abnormalities occur, such as absent reflexes, stupor, convulsions, coma or death. With rather less severe hyponatremia (110-125 mEq. per liter) the early symptoms of water intoxication are seen: depression, lethargy, anorexia, nausea and muscle weakness. Despite the low plasma values, there is little or no over-all electrolyte deficiency, the abnormalities being due simply to water retention and hemodilution. The patients cannot excrete a water load.

Inappropriate secretion of vasopressin may also occur in the absence of malignant disease, although this time the hormone comes from the normal source—the hypothalamus and posterior pituitary gland. The cause of the complication in these conditions is unknown, but it may be found in the presence of any lung infection or brain disease, in hypothyroidism and after head injuries.

Patients with inappropriate water retention due to excess vasopressin secretion do well with simple restriction of water intake, and the condition can be cured if it is possible to treat the underlying condition successfully.

Hyperthyroidism. This condition has been described in tumors of trophoblastic type (such as choriocarcinoma, testicular teratoma or hydatidiform mole) and the material secreted resembles chorionic thyrotrophin rather than pituitary TSH. However, some breast or lung carcinomas have been described in patients with hyperthyroidism and material has been extracted with biological properties similar to pituitary TSH and some immunological similarities too.

Ectopic Growth Hormone Secretion. This usually has no clinical manifestations although the patients have lung tumors and often finger clubbing and subperiosteal new bone formation. There is no evidence that these features are due to the growth hormone.

Ectopic Gonadotrophin Secretion. This occurs from trophoblastic tumors (HCG-like) or carcinoma of the lung (HCG or LH-like). These hormones may produce precocious puberty if they occur in children, or gynecomastia in adults.

References

General Reviews

Hall, R., Anderson, J., Smart, G. A., and Besser, G. M.: Fundamentals of Clinical Endocrinology. Ed. 2. London, Pitman Medical Publishing Co., 1974.
Recent Progress in Hormone Research (published annually).

Cyclic Adenosine Monophosphate

Berson, S. A., and Yalow, R. S., eds.: Methods in Investigative and Diagnostic Endocrinology. New York, American Elsevier Publishing Company, 1973, Part 1, p. 189.
Liddle, G. W., and Hardman, J. G.: Cyclic adenosine monophosphate as a mediator of hormone action. New Eng. J. Med. *285:*560, 1971.
Robison, G. A., Butcher, R. W., and Sutherland, E. W.: Cyclic AMP. New York, Academic Press, Inc., 1971.

Pituitary Hormones

Besser, G. M.: ACTH assays and their clinical application. Clin. Endocr. *2:*175, 1973.
Besser, G. M.: The hypothalamus and the pituitary. *Medicine,* Part 2, 1972, p. 97.
Besser, G. M., and Mortimer, C. H.: Hypothalamic regulatory hormones. Clin. Path., *27:*173, 1974.
Besser, G. M., McNeilly, A. S., Anderson, D. C., et al.: Hormonal responses to synthetic luteinizing hormone and follicle stimulating hormone releasing hormone in man. Br. Med. J. *3:*267, 1972.
Chard, T., and Edwards, C. R. W.: The hypothalamus and the posterior pituitary. Mod. Trends Endocr., *4:*102, 1972.
Faiman, C., and Winter, J. S. D.: Diurnal cycles in plasma FSH, testosterone and cortisol in men. J. Clin. Endocr., *33:*186, 1971.
Goebelsmann, U., Rees-Midgley, Jr., A., Jaffe, R. B.: Regulation of human gonadotrophins VII. Daily individual urinary estrogens, pregnanediol and serum luteinizing and follicle stimulating hormones during the menstrual cycle. J. Clin. Endocr., *29:*1222, 1969.
Jenner, M. R., Kelch, R. P., Kaplan, S. L., and Grumbach, M. M.: Hormonal changes in puberty: iv. Plasma estradiol, LH, and FSH in prepubertal children, pubertal females, and in precocious puberty, premature thelarche, hypogonadism and in a child with a feminizing ovarian tumor. J. Clin. Endocr., *34:*521, 1972.
Raiti, S. M. B., Light, C., and Blizzard, R. M.: Urinary follicle-stimulating hormone secretion in boys and adult males as measured by radioimmunoassay. J. Clin. Endocr., *29:*884, 1969.

Adrenal Hormones

Cope, C. L.: Adrenal Steroids and Disease. Ed. 2. London, Pitman Medical Publishing Co., 1972.
Goodman, L. S., and Gilman, A., eds.: The Pharmacological Basis of Therapeutics. Ed. 4. New York, Macmillan Co., 1970.

Iversen, L.: Catecholamines. Brit. Med. Bull., *29:*1973.
Liddle, G. W., Ilsand, D., and Meador, C. E.: Recent Progr. Horm. Res., *18:*125, 1962.
Mason, A. S., ed.: Adrenal cortex and its diseases. Clinics in Endocrinology & Metabolism, *1*(2):331, 1972.
Varley, H., and Gowenlock, A. H., eds.: The clinical chemistry of monoamines. Proc. Symp. Clin. Chem. Mono-amines, Manchester, 1962.

Thyroid Hormones

Clinical endocrinology. Articles published by Brit. Med. J., British Medical Association, London, 1973.
Werner, S. C., and Ingbar, S. H., eds.: The Thyroid. Ed. 3. New York, Harper & Row, 1971.

Placental Hormones and Pregnancy

Burrow, G. N.: The Thyroid Gland in Pregnancy. Vol. 3 in the series Major Problems in Obstetrics and Gynecology, Philadelphia, W. B. Saunders Co., 1972.
Fuchs, F., and Klopper, A., eds.: Endocrinology of Pregnancy. New York, Harper & Row, 1971.
Hytten, F. E., and Leitch, I.: The Physiology of Human Pregnancy. Ed. 2. Oxford, Blackwell Scientific Publications, 1971.
Hytten, F. E., and Lind, T.: Diagnostic Indices in Pregnancy. Basle, Ciba—Geigy Ltd., 1973.

Androgens and the Testis

Resko, J. A., and Eik-nes, K. B.: Diurnal testosterone levels in peripheral plasma of human male subjects. J. Clin. Endocr., *26:*573, 1966.
Rosenfield, R. L., Eberlein, W. R., and Bongiovanni, A. M.: Measurement of plasma testosterone by means of competitive protein binding analysis. J. Clin. Endocr. *29:*854, 1969.

Ectopic Hormone Production by Nonendocrine Tumors

Rees, L. H., and Ratcliffe, J. G.: Ectopic hormone production by non-endocrine tumours—a review. Clin. Endocr., *3:*263, 1974.

Calcitonin

Calcium metabolism and bone disease, Ed. Iain MacIntyre, Clinics in Endocrinology & Metabolism, Vol. 1, No. 1. W. B. Saunders Co., Philadelphia, 1972.

Ovarian Hormones

Reproductive endocrinology and world population, Ed. J. A. Loraine, Clinics in Endocrinology & Metabolism, Vol. 2, No. 3, W. B. Saunders Co., London, 1973.

Chapter 23

VITAMINS

Although procedures are available for quantitative determination of most of the known vitamins, especially in foods, clinical evaluation of nutritional status with respect to vitamins is often difficult and the available methods unsatisfactory. This is particularly true during the usually long latent period of progressing deficiency which precedes the appearance of frank symptoms and signs. Various types of diagnostic procedures may be employed for this purpose:

1. The concentration of the vitamin or one of its metabolites may be determined in the blood or urine.

2. The curve of concentration in the blood or excretion in the urine may be measured after administration of a standard test dose of the vitamin. This is the so-called "saturation test" procedure, based on the assumption that subsaturation of the tissues with the vitamin will result in a subnormal rise in the blood and subnormal excretion in the urine under conditions of the test.

3. Quantitative determinations may be made of the vitamin content of tissues obtained at biopsy (e.g., liver, muscle).

4. Evidence of certain types of deficiency may be obtained by microscopic studies, e.g., of mucosal scrapings in vitamin A deficiency.

5. Certain consequences of specific deficiencies may be demonstrated by biophysical methods, e.g., impaired dark adaptation in vitamin A deficiency and increased capillary fragility in ascorbic acid deficiency.

6. In the case of certain vitamins, deficiency may manifest itself in some characteristic derangement of metabolism which can be measured quantitatively, e.g., elevation of blood pyruvate in thiamine deficiency, and increase in serum alkaline phosphatase, hypophosphatemia and hypocalcemia in vitamin D deficiency.

7. Of great diagnostic value is the prompt relief of clinical manifestations of suspected deficiency upon administration of adequate amounts of the vitamin in question.

We are concerned here chiefly with quantitative estimations of the vitamins or their metabolic products in the body fluids. Such methods are now available. They still have distinct limitations in their applicability to the appraisal of nutritional status. Plasma levels, for example, may not be lowered until there is gross deficiency. On the other hand, levels may be low, as with ascorbic acid, when deficiency is not present. Inasmuch as the blood is a labile transport medium, its vitamin content in chronic deficiency states may be increased after a comparatively brief period of high vitamin intake without comparable improvement in the morphologic abnormality in the tissues. In the evolution and recession of a state of avitaminosis, changes in concentration of the vitamin in the blood and urine do not occur synchronously with alteration in the tissue state. The same may be said of the so-called "saturation tests," although results obtained by these procedures are perhaps not subject to as rapid fluctuation as the blood values. The extent of the tissue changes, indicating the severity of the deficiency state, can be best determined at times by biomicroscopy. In certain instances, however, rather prolonged deficiency is required for the production of morphologic abnormalities, and blood or urine studies or saturation tests may yield earlier abnormal results. In established chronic deficiency states, which are the type encountered most frequently clinically, blood and urine studies may be misleading, particularly during periods of specific therapy.

There is considerable evidence that the immediately antecedent dietary intake of a vitamin is an important factor in determining the level at which it will be excreted in the urine. It is therefore possible that a high excretion level may result from a recent but transient high intake and a low excretion level from a recent but transient low intake. Such excretion studies might lead to erroneous interpretations of the nutritional state with regard to the factor under investigation. This has been demonstrated to be true in the case of niacin, riboflavin, pyridoxine, pantothenic acid, folic acid, biotin and choline. A lack of correlation was found between the biochemical data and the results of tests of physical (bicycle ergometer) and psychomotor (pursuit meter) performance tests. On the basis of these observations, the conclusion seems justified that biochemical data of this type, although important in nutritional appraisal, are most valuable when used in conjunction with other types of data, e.g., clinical, dietary, physical and psychomotor. Any single type of study, employed alone, may be misleading.

Only those vitamins are considered here which at the present time lend themselves to reasonably accurate quantitative determination in biological fluids as an aid in the detection of nutritional deficiency states, or which influence metabolic processes in such a manner that diagnostically significant biochemical abnormalities are produced as a result of deficiency.

VITAMIN A

There is little substantial evidence that vitamin A deficiency is accompanied by any significant disturbance of protein, fat or carbohydrate metabolism. There have been isolated reports of marked increase in the

Vitamin A

$$\text{Vitamin A structure: } H_2C, H_2C, CH_2, C(CH_3)_2, C-CH_3, -CH=CH-C(CH_3)=CH-CH=CH-C(CH_3)=CH-CH_2OH$$

esterase content of the blood serum in vitamin A deficient rats and of an increase in serum cholesterol following administration of excessive amounts of vitamin A. In the growing dog with vitamin A deficiency there is increased osteoblastic and osteoclastic activity, with proliferation of cancellous bone at the expense of compact bone, the overgrowth of bone causing compression of adjacent nerve fibers and cells, with consequent changes in the central nervous system. This factor may also be necessary for normal development of the teeth. Visual acuity in dim light is dependent upon the presence in the rod cells of the retina of an adequate amount of photosensitive pigment, rhodopsin or visual purple, a dissociable combination of a protein, "opsin," and vitamin A aldehyde (retinene or retinal).

Dietary vitamin A is made up of esters (usually palmitate) of either retinol (vitamin A) or dehydroretinol (vitamin A_2), which is present in the livers of fresh-water fish and has only about one-third of the activity of retinol. A number of carotenoids can give rise to vitamin A in the body. The most effective is β-carotene. This means that dark green leafy vegetables, such as spinach, as well as carrots are good sources of the vitamin. Fish liver oils are very rich in the vitamin itself. The recommended daily intake of retinol from the age of six months to three years is 250 μg. and for adults it is 750 μg.; in pregnancy, 1200 μg. is the appropriate figure.

Vitamin A esters are hydrolyzed in the lumen of the intestine in solution in the mixed lipid micelles (p. 98). Bile salts help this process, and tocopherol is believed to prevent oxidative destruction of the vitamin. Carotene is also dissolved in the micelles. The vitamin and the carotene are now taken up by the intestinal mucosa, where the former is esterified and the latter converted first to retinal and then to retinol (by retinal dehydrogenase), which is then esterified, mostly with palmitic acid. The esters, together with some unchanged retinal and carotenoids (lycopene, lutein, etc.) which are not precursors of vitamin A, are absorbed and enter the intestinal lymphatics, and eventually the circulation, in chylomicrons. In the blood, the vitamin esters are attached to β-lipoprotein and are then taken up by the liver (Kupffer cells), which contains almost all the body store. The vitamin is then released, as retinol, apparently attached to an α_1-globulin (retinol-binding protein) for use elsewhere, for example, in the retinal rods. Plasma carotenoids contain relatively high proportions of lycopene and lutein. Retinol dehydrogenase is present in the liver, and consequently serum retinal can be converted to retinol.

β-carotene

Absorption of vitamin A is impaired in the absence of pancreatic enzymes (impaired hydrolysis of vitamin A esters) and in celiac disease or other conditions in which there is impaired absorption of dietary fat (e.g., obstructive jaundice, chronic pancreatitis).

About 95 per cent of the vitamin A reserves of the body is believed to be held in the liver, a small amount being present in other tissues, e.g., lactating breast, adrenals, lung, intestine. In subjects with liver damage, the capacity for storage and formation of vitamin A is impaired and concentration of this vitamin in the blood is decreased. The hepatic storage capacity is comparatively low in young infants, increasing with age. The quantity stored in the liver varies in different species, but is largely dependent upon the antecedent diet. About 70 per cent of a single large dose of vitamin A may be recovered from the liver of the rat, which can store enough in a few days of adequate intake to satisfy its requirement for months. On the other hand, less than 10 per cent of a similar dose can be recovered from the liver of the guinea pig, which is very sensitive to deficiency in vitamin A. The storage capacity in man is apparently relatively large.

Under conditions of decreased intake the plasma vitamin A concentration is maintained at the expense of the hepatic reserves. Hyperthermia causes depletion of liver vitamin A with a simultaneous decrease in its concentration in the blood. Administration of thyroxine or testosterone accelerates mobilization from the liver, whereas 17-α-hydroxyallopregnane-20-one, an adrenal steroid, produces a decrease in the plasma and a simultaneous increase in the liver. These observations suggest that certain steroid hormones may regulate the equilibria between liver and plasma vitamin A or between the esterified and unesterified forms in the liver.

Reported values for vitamin A and carotenoids in the blood of normal subjects vary widely, owing in part to differences in methods employed, several of which have proved unsatisfactory. The following figures appear to be acceptable for postabsorptive plasma or serum: vitamin A, 20 to 50 μg. per 100 ml.; carotenoids, 80 to 120 μg. per 100 ml.

The plasma vitamin A concentration is usually but not invariably subnormal in subjects maintained for a considerable period on a diet deficient in vitamin A and, although there may be a general parallelism between the intake and the concentration in the blood, the value of blood vitamin A levels in assessing nutritional status in this connection is uncertain. Plasma levels tend to be maintained at or near normal until there is advanced depletion. Values are very low in the presence of eye lesions (see below). Subnormal plasma values have been obtained also in conditions mentioned above in which absorption from the intestine is impaired, in gastrointestinal malignancy and in a number of types of hepatic disease. No definite correlation has been established between the results of biophotometric studies and the vitamin A content of the blood, which apparently bears no constant relation to the adequacy of the vitamin stores in the body. However, the statement has been made, on the basis of an extensive study of this problem, that a high value for vitamin A in the blood is inconsistent with deficiency of this vitamin. Low values for vitamin A and carotene have been obtained in hepatitis and cirrhosis of the liver.

The urine contains no vitamin A or carotene except after administration of excessive amounts. Under normal conditions only very small quantities are excreted in the feces. Administration of mineral oil, especially in young children, may cause excessive loss of carotene in the feces. Appreciable amounts of vitamin A and carotene are present in milk, the concentration being greatest in colostrum and decreasing gradually over the period of lactation. Human colostrum possesses about twice as much vitamin A activity as early milk, the latter, providing about 450 μg. daily, being considerably richer in this factor than cow's milk. The vitamin A content of milk is increased by ingestion of added amounts of vitamin A during pregnancy, but not by similar doses of carotene.

Deficiency in Man. Present understanding of the role of vitamin A in human nutrition is reflected in the following statements:

1. Vitamin A is specific for the cure and prevention of xerophthalmia and certain types of nyctalopia (night blindness), and hemeralopia (day blindness). It alleviates defective dark adaptation due to its deficiency.

2. Vitamin A is essential to the normal structure and behavior of epithelial tissue, e.g., the epithelium forming the lining of the nasal sinuses and respiratory tract, the genitourinary tract, part of the digestive tract, the ducts of the exocrine glands, the conjunctiva and the cornea. In the absence of vitamin A, keratinizing metaplasia occurs. The vitamin probably acts by stabilizing membranes, including those enclosing lysosomes. Interference with the normal structure of these epithelia removes an important barrier to infection.

3. Vitamin A is a growth factor.

Detection of Deficiency in Man. The most important clinical manifestations of vitamin A deficiency in man are: (1) xerophthalmia, i.e., thickening and loss of transparency of the bulbar conjunctiva, with yellowish pigmentation and occasionally Bitot's spots; (2) follicular conjunctivitis; (3) keratomalacia, i.e., softening of the cornea with, in advanced cases, ulceration and necrosis; (4) impairment of dark adaptation, progressing to night blindness (nyctalopia); (5) the evidence for production of follicular hyperkeratosis of the skin is now considered dubious.

The clinical manifestations and their prompt response to administration of adequate amounts of vitamin A constitute the best available means of diagnosis. No reliable procedures are available for detecting subclinical deficiency states. The following objective methods have been investigated: (1) determination of the concentration of vitamin A and carotene in the blood; (2) the dark adaptation test; (3) examination of scrapings of the bulbar conjunctiva and vagina (and other mucous membranes).

Vitamin A Content of Blood. The concentration of vitamin A in the blood (plasma or serum) is not a reliable index of the status of vitamin A nutrition, although extremely low values (below 12 μg./100 ml.) may be regarded as indicative of deficiency. One source of difficulty is lack of agreement as to the limits of normal concentration. Of greater importance, however, is the fact that a normal plasma vitamin A level may be maintained for many months in previously normal subjects receiving diets containing virtually no carotene or vitamin A. There is considerable individual variation in this connection, due undoubtedly to quantitative differences in the hepatic stores of the vitamin which, in well-nourished, normal subjects may be adequate to meet the body requirements for several months. Fever may be accompanied by a sudden drop in plasma (and liver) vitamin A, and a rise follows administration of large doses of the vitamin. There is no satisfactory consistent correlation between the blood vitamin values and the presence or severity of the various clinical deficiency manifestations. The plasma carotene level is of still less diagnostic value. It reflects the immediate past intake of carotenoids, falling promptly after their exclusion from the diet and rising promptly following their administration. A zero value is compatible with normal nutrition, inasmuch as there is no consistent relationship between the level of plasma carotene and the quantity of vitamin A in the organism.

Dark Adaptation Test. The biophotometer has been employed to test the speed of visual adaptation to dim light after a period of exposure to bright light. Although night blindness is an important manifestation of very prolonged vitamin A deficiency, the general experience has been that the dark adaptation test does not satisfactorily reflect subclinical levels of vitamin A deficiency and that biophotometer readings do not parallel the concentration of the vitamin in the blood. The time of development of significant impairment of dark adaptation in subjects on a diet low in vitamin A and carotenoids varies widely, from several days to many months. Moreover, conditions other than vitamin A deficiency may cause this phenomenon.

Examination of Scrapings. In established A-deficiency, keratinized epithelial cells may be demonstrated in scrapings from the bulbar conjunctiva and the vagina. These changes generally occur rather late, being preceded usually by subjective evidence of impaired dark adaptation. This procedure is therefore of little clinical value.

Effects of Excess of Vitamin A. Acute symptoms may follow ingestion of very large amounts of vitamin A. These include drowsiness, sluggishness, severe headache, vomiting, and peeling of the skin about the mouth and elsewhere. This syndrome has been recognized by Eskimos as occurring after eating the livers of polar bears and arctic foxes, which are extremely rich in vitamin A. In infants and young children, there may be a sudden rise of intracranial pressure with bulging fontanelles.

Continued intake of excessive amounts, especially in children, produces roughening of the skin, irritability, coarsening and falling of the hair, anorexia, loss of weight, headache, vertigo, hyperesthesia, occasionally hepatomegaly, splenomegaly, hypoplastic anemia, leukopenia and certain rather characteristic skeletal changes (periosteal thickening of long bones). There may be blurring of vision and diplopia.

The skeletal changes are the result chiefly of acceleration of the normal bone growth sequences, with simultaneous increase in the processes of resorption (osteoclasis) and cortical bone deposition (hyperostosis). Young rats and guinea pigs fed large amounts of vitamin A may show an equivalent of a year's growth in a few weeks, the new bone being inadequately mineralized and fracturing easily. Weanling rats fail to grow, have difficulty in walking and pain in the extremities, with fractures and associated hemorrhages.

Infants may tolerate daily doses of about 60 mg. of vitamin A for a year, but develop symptoms with 150 mg. daily for even brief periods. On this basis the toxic daily dose for adults would be between 300 and 900 mg.

Hypercarotenemia. A marked increase in serum carotenoid levels occurs as a result of prolonged excessive ingestion of green leafy vegetables, or carrots (often as juice), citrus fruits and tomatoes. When the serum carotene level is above 250 μg./100 ml., the pigment is secreted in the sweat and the sebum and reabsorbed by the stratum corneum. The skin develops a yellow discoloration, especially marked in certain areas, including the palms, the soles and the nasolabial folds. The conjunctiva and the buccal mucous membranes are unaffected.

Carotenemia may occur in association with diabetes mellitus, hyperthyroidism and a number of other hyperlipemic states, and is probably due to disturbances of carotene metabolism.

Carotenemia may be confused with hyperbilirubinemia if reliance is placed on the icterus index determination alone (p. 638), which is elevated in this condition. The serum bilirubin concentration is normal in uncomplicated carotenemia and the diagnosis is established by determination of the serum carotenoid (carotene) concentration as well as the absence of yellow discoloration in the conjunctiva and the buccal mucous membrane.

THIAMINE

Thiamine, in the form of the pyrophosphate, the active form of this vitamin, is the cofactor (coenzyme) in reactions involving oxidative decarboxylation of certain important intermediates in carbohydrate metabolism, e.g., pyruvic and α-ketoglutaric acids (p. 11). It is, therefore, referred to as "cocarboxylase." The biochemical and probably also the clinical manifestations of thiamine deficiency are largely related to the consequent defect in this vital phase of intermediary metabolism. The oxidative decarboxylation of pyruvate also requires the action of lipoic acid, NAD and coenzyme A. It gives rise to acetyl-CoA, which forms the link between protein, fat and carbohydrate metabolism. Oxidative decarboxylation of

Thiamine

α-ketoglutarate gives rise to succinyl-CoA, which plays an important part in the biosynthesis of the porphyrin ring of hemoglobin and certain oxidases. Thiamine is also the coenzyme for transketolase activity in the pentose-phosphate cycle. This cycle supplies NADPH required for fatty acid synthesis, etc., and also is the only source of the ribose required for nucleic acid biosynthesis.

Metabolism of Thiamine. Free thiamine is absorbed readily from the intestine, but the pyrophosphate (cocarboxylase) is not. The bulk of the dietary vegetable thiamine (yeast, germ and pericarp of cereals) is in the free form. Meat, fish and poultry contain the pyrophosphate, which must be broken down to the free form before intestinal absorption can occur. The vitamin can be lost by cooking. In carp and some other fish, an enzyme (thiaminase) is present which destroys the vitamin. The enzyme is heat-labile, but in Japan and other countries where large quantities of raw fish are eaten it can bring about thiamine deficiency. The vitamin is very actively phosphorylated to cocarboxylase in the liver and, to a lesser extent, in most other tissues.

It is present in the blood plasma and cerebrospinal fluid in the free form (about 1 μg./100 ml.) largely bound to α- and β-globulins. The largest portion of the blood thiamine, which ranges from 6 to 12 μg./100 ml., is in the blood cells as the pyrophosphate; in protein combination, the red cells contain about 8 μg./100 ml. of cells, and the leukocytes about 70 μg./100 ml. Since free thiamine (base) is readily diffusible and the pyrophosphate is not, the plasma thiamine probably represents the transport form (inactive) of the vitamin, which undergoes phosphorylation (activation) upon entrance into tissue cells.

The capacity of the organism for storing thiamine is apparently limited. It is present, as the pyrophosphate only, in the heart, liver and kidneys and, in lower concentration, in skeletal muscle and brain. Administration of thiamine may result in an increase in the tissues, within certain limits. However, on a thiamine-free diet, the tissue content is depleted within a short time, emphasizing the desirability of providing an adequate daily supply.

If normal amounts of thiamine are ingested (1 to 2 mg. daily; 0.4 mg./1000 calories in diet), about 10 per cent is excreted in the urine. The remainder is apparently partly phosphorylated and utilized for carboxylase action and partly degraded to neutral sulfur compounds and inorganic sulfate. If large amounts are given, the excess is largely excreted in the urine. All of the urinary thiamine is in the free form, but the urine also contains the disulfide, pyramine and other unidentified breakdown products. Normal subjects on an adequate intake excrete at least 50 μg. daily in the urine. That present in the feces is probably largely of bacterial origin (large intestine).

Demonstration of Deficiency. The important biochemical features of human thiamine deficiency include: (*a*) decreased levels of thiamine and thiamine pyrophosphate in the blood (which has not proved very useful) and urine (in beriberi 0–14 μg. in 24 hr.); (b) increased concentrations of pyruvic and lactic acids in the blood; (c) decreased transketolase activity of red cells. Several procedures have been proposed for the laboratory diagnosis of deficiency states, none of which is entirely satisfactory.

Thiamine in Blood. These tests are not very reliable for assessing thiamine status. Some believe that values below 3 micrograms of thiamine per ml. of whole blood are indicative of deficiency in this factor, subnormal levels having been associated with the development of peripheral neuropathy in alcohol addicts. It has been found that the thiamine content of the leukocytes probably reflects states of thiamine deficiency and saturation more satisfactorily than does the thiamine content of whole blood. High thiamine values (three times normal in leukocytes, two times normal in erythrocytes) have been obtained in patients with leukemia, Hodgkin's disease and carcinoma of the gastrointestinal tract. This has been attributed to impaired utilization of thiamine.

Thiamine Excretion. Normal limits of excretion of thiamine are difficult to establish because alterations in intake are reflected promptly in its excretion in the urine. Reported values vary also with the method employed. The daily urinary excretion by normal subjects receiving adequate amounts of thiamine ranges usually from 60 to 500 micrograms; however, values as low as 25 micrograms have been reported, but levels lower than this are usually associated with clinical symptoms.

Blood Pyruvic and Lactic Acids. In the fasting state, normal whole blood contains 0.1 to 1 mg. of pyruvic acid and 5 to 15 mg. of lactic acid per 100 ml. These are increased in thiamine deficiency, but also in other conditions, e.g., congestive heart failure. It has been shown that in normal subjects the ingestion of glucose is followed by a short, steep elevation in blood pyruvic and lactic acids, which reaches a maximum in one hour and returns to the resting level in three hours. The curve is abnormally high and prolonged in subjects with thiamine deficiency. Before a significant increase occurs in the basal pyruvic acid concentration in deficiency states, intravenous injection of 0.4 g. of glucose per kilogram of body weight in 50 per cent solution is followed by an abnormally high rise in blood pyruvic acid and lactic acid. On the basis of this abnormal response to carbohydrate in thiamine deficiency, a "carbohydrate metabolic index" has been evolved which has a value of 5 to 10 for normal subjects and is greater than 10 in those with thiamine deficiency. This phenomenon usually develops some time after "saturation" studies yield abnormal results.

Thiamine Deficiency. Experimental thiamine deficiency in man produced depression, irritability, defective memory and failure to concentrate. There was tenderness of the calf muscles, weakness of the lower limbs, a variety of subjective manifestations (paresthesia, etc.) and diminished reflexes.

Classically, deficiency presents as beriberi, with its "wet" and "dry" forms and neurological manifestations. It is now believed that some degree of thiamine deficiency is not uncommon, especially in the elderly.

RIBOFLAVIN

Riboflavin, in the form of mono- and dinucleotides (riboflavin phosphate and flavine adenine dinucleotide, respectively), is a coenzyme in oxidation-reduction reactions, which are intimately concerned with a number of vital metabolic processes. These enzymes (flavoproteins) serve as bridges over which hydrogen atoms pass from one molecule to another. They play a part in the electron transport system in mitochondria.

The dietary sources of riboflavin are wide. One quart of milk daily provides the recommended intake (0.55 mg./1000 calories). The content is low in cereals, unless they are germinating. The vitamin is phosphorylated in the intestinal mucosa during absorption.

Human blood plasma contains 2.5 to 4.0 μg. of riboflavin per 100 ml., about two-thirds as the dinucleotide, the bulk of the remainder as the mononucleotide. The concentration in erythrocytes has been reported as 15–30 μg./100 grams, and in leukocytes (plus platelets) as about 250 μg./100 grams. These values tend to remain quite constant even in riboflavin deficiency. Determination of riboflavin in the blood is not useful in the clinical evaluation of the state of riboflavin nutrition.

Riboflavin is present in all tissue cells, principally as the nucleotides (coenzymes), a variable proportion of which is bound as flavoprotein. The retina apparently contains free riboflavin. The animal organism does not appear to have a specialized mechanism for storage of riboflavin. The highest concentrations occur in the liver and kidneys, but the tissue content is not increased significantly by administration of large amounts. Certain tissues (e.g., muscle) may retain considerable quantities in the presence of manifestations of riboflavin deficiency. Riboflavin is secreted in the milk, 40 to 80 per cent being in the free state, increasing with increased intake. It is also present in perspiration (10 μg./hour). The riboflavin content of the feces (free and nucleotides) tends to remain quite constant (500 to 750 μg. daily) and is presumably largely of bacterial origin. The urinary excretion (mainly free, but up to 50 per cent nucleotide) varies with the intake. Under ordinary dietary conditions (1 to 2 mg. riboflavin), the daily urinary excretion is about 0.1 to 0.4 mg. (10 to 20 per cent of intake). When large amounts are administered, as much as 50 per cent may be eliminated in the urine. The bulk of the dietary riboflavin is metabolized in the body, largely

Riboflavin

to unknown compounds; a substance of undetermined constitution, uroflavin (aquoflavin), present in the urine, is believed to be a degradation product of riboflavin.

Demonstration of Deficiency. As is true of other B vitamins, demonstration of riboflavin deficiency in man rests mainly on the prompt clinical improvement that follows administration of adequate amounts of the vitamin.

Determination of the blood level or daily urinary excretion of riboflavin has not proved to be of diagnostic value, as indicated above. A urinary excretion of less than 50 µg./24 hr. is usually associated with signs of deficiency. These are soreness and burning of the lips, mouth, tongue and eyes, with photophobia and lacrimation. There are transverse fissures at the angles of the mouth. Dermatitis of the scrotum or vulva is frequently present.

NIACIN (NICOTINIC ACID)

Nicotinic acid occurs principally as the amide (nicotinamide; niacinamide), in which form it enters into the formation of the physiologically active dinucleotides (NAD and NADP). These, like the riboflavin nucleotides, serve as coenzymes in fundamental oxidation-reduction reactions, i.e., as a medium for the transfer of hydrogen from one molecule to another.

Certain tissues and bacteria can synthesize niacin from the amino acid tryptophan. Consequently, the dietary supply of this vitamin is supplemented by its tissue and possibly to some extent by intestinal bacterial synthesis in the presence of adequate provision of proteins rich in tryptophan. The main dietary sources are meat, fish, poultry and wheat. Niacin is present in maize and rice in a form which is not absorbable (niacytin) unless the food is prepared with alkali (tortilla).

Nicotinic acid and its amide are absorbed from the intestine, the concentration in the blood plasma rising promptly after oral administration of large doses (20 mg.). Stated values for human blood are as follows: (1) whole blood, as nicotinic acid, 0.5–0.8 (av. 0.6) mg./100 ml. (almost all as the coenzymes in red cells and leukocytes); (2) erythrocytes, total nicotinic acid activity, 1.3 mg./100 ml. (6.5–9.0 [av. 7.0] mg. as coenzyme); free nicotinic acid, 0.135 mg./100 ml.; (3) plasma, total nicotinic acid activity, 0.025–0.15 (av. 0.075) mg./100 ml. That in the plasma is apparently largely in the free state. The values in the blood are not altered significantly in

Niacin   Niacinamide

severe niacin deficiency (i.e., in pellagra), and their determination is therefore of no value in the detection of clinical deficiency states.

Normal adults, on a normal diet, excrete both nicotinic acid and nicotinamide in the urine (0.25 to 1.25 mg., and 0.5 to 4 mg. daily, respectively). However, the major urinary metabolite is a methylated derivative, N^1-methylnicotinamide, also referred to as F_2 (3.0 to 12.5 mg. daily), which exhibits a bluish-white fluorescence in alkaline butanol in the ultraviolet. There also may be variable amounts of an oxidation product of the latter. Urinary excretion of the latter substance accounts for 40 to 50 per cent and of F_2 for about 10 to 20 per cent of administered nicotinamide. These processes of methylation and oxidation occur in the liver. In rats, administration of large amounts of nicotinic acid or nicotinamide may produce fatty liver, which is prevented by simultaneous administration of methionine, choline or betaine (p. 128). This phenomenon is apparently due to diversion of methyl groups for the formation of N^1-methylnicotinamide.

Traces of nicotinamide are present in the sweat. Small amounts are secreted in human milk, increasing from less than 0.05 mg. on the first day post partum to about 3.0 mg. on the tenth day (intake 16.5 mg. daily). Somewhat larger amounts are present in cow's milk.

Hyperbilirubinemia may result from therapeutic administration of large doses of nicotinic acid.

Demonstration of Deficiency. In its fully developed form (pellagra), the clinical picture is rather characteristic (dermatitis, diarrhea and dementia), and response to administration of adequate doses of nicotinic acid or nicotinamide and a balanced diet is prompt and dramatic. The diagnosis is usually made easily on the basis of clinical features and response to specific therapy.

Available diagnostic laboratory tests, although useful, are not entirely reliable. The most promising are based on the urinary excretion of N^1-methylnicotinamide. The amount excreted is subnormal in niacin deficiency (< 0.8 mg. in 24 hr.). The serum tryptophan level is lowered in individuals with pellagra.

THE VITAMIN B_6 GROUP

There is a group of related compounds with vitamin B_6 activity. It consists of pyridoxol (pyridoxine), pyridoxal and pyridoxamine. The richest dietary sources include meats, fish, vegetables and whole-meal flour, but the vitamin B_6 group is found in almost all foods. The recommended daily intake is 2 mg. in a diet containing 100 g. protein. During pregnancy, and lactation, 2.5 mg. vitamin B_6 is recommended.

The members of the B_6 group are converted by the body to pyridoxal phosphate, which is the coenzyme to a large number of apoenzymes. The enzyme systems involved include the aminotransferases, amino acid decarboxylases, tryptophan synthase, tryptophanase, kynureninase, serine and

Pyridoxine

Pyridoxal

Pyridoxamine

threonine dehydratases, aminolevulinate (ALA) synthase, glycogen phosphorylase and serine hydroxymethyl transferase. This by no means exhausts the list of metabolic reactions involved, but the great importance of the B_6 group is obvious. This especially applies to brain metabolism, since pyridoxal phosphate is necessary for the formation (amino acid decarboxylation) of serotonin, γ-aminobutyric acid and the catecholamines. There is also an important relationship to oxalate metabolism. Hyperoxaluria occurs in deficiency states, probably because of reduced production of glycine from glyoxylate. Treatment with high dosage of pyridoxol has reduced oxalate output in the presence of primary hyperoxaluria. Fortunately, deficiency of B_6 is rare because of its easy availability in most foodstuffs.

Deficiency of vitamin B_6 has been produced experimentally by the administration of a diet low in vitamin B_6, together with 4-deoxypyridoxine. This gave rise to irritability and depression. There were seborrheic manifestations, lymphopenia was common, and peripheral neuropathy occurred in some subjects. There was a high urinary excretion of xanthurenic acid after tryptophan loading. The output of urinary hydroxykynurenine after tryptophan loading is, however, a more sensitive indication of vitamin B_6 deficiency.

Deficiency has occurred in infants on inadequate milk formulas. The major symptom was convulsions, possibly due to depletion of brain γ-aminobutyric acid content.

The drugs isonicotinic acid hydrazide and hydrazaline act as B_6 antagonists, and their administration can result in deficiency symptoms, including hypochromic anemia and peripheral neuropathy.

There are a number of inborn errors of metabolism, which are vitamin B_6-dependency states. They include cystathioninuria, familial xanthurenicaciduria, some pyridoxol-responsive anemias and some infantile seizures. There appears to be some impairment of the coenzyme binding site on the appropriate apoenzyme, although this simple explanation apparently does not always apply.

FOLIC ACID

The term folic acid is applied to any member of a group of substances, of which the parent compound is pteroylglutamic acid. This is the monoglutamic conjugate of pteroic acid, which in turn is a combination of a pterin with p-aminobenzoic acid. The major forms of folic acid in the diet and in the tissues are, however, pteroyltriglutamic and pteroylheptaglutamic acids. It is likely that folate absorbed across the intestinal mucosa is largely in the form of the monoglutamate. Conjugase (γ-glutamyl peptidase) activity, which splits off the additional glutamate residues, exists in the tissues. Some folate is probably absorbed as polyglutamate.

The biosynthesis of folate produces as an end-product the compound 7,8-dihydropteroylglutamic acid (dihydrofolate). The tissues contain an enzyme, dihydrofolate reductase, which brings about further reduction to 5,6,7,8-tetrahydrofolate. It is in this form, and its derivatives, that the folate coenzymes exist as polyglutamates in the tissues. The derivatives are the 10-formyl, 5-formimino, 5,10-methylene, 5-methyl and 5-hydroxymethyl derivatives of the tetrahydrofolate. Enzyme systems exist for attaching these 1-carbon side chains to tetrahydrofolate, and other enzyme systems exist for handing them on during the biosynthesis and degradation of many important compounds, e.g., conversion of serine to glycine, purine synthesis, pyrimidine synthesis, methionine synthesis, choline synthesis and the degradation of histidine.

Folate (as polyglutamates) is present in a wide variety of foods, the richest sources being yeast and liver. The normal daily requirements are probably 0.1 to 0.2 mg. Somewhat more folate is required during pregnancy. Absorption usually occurs in the upper part of the small intestine, but can also occur in the lower portion. Absorption is an active process. Excretion of folate occurs in the feces (largely but not entirely bacterial in origin); a small amount (2–5 μg. daily) appears in the urine. Some is also present in saliva, sweat and bile.

The level of folate in serum and red cells can be estimated microbiologically and more recently by competitive binding techniques, using radioactively labeled folate. Levels vary from laboratory to laboratory. Healthy subjects tend to have a serum range above 3 μg./l. and below 25 μg./l. Hospital patients have a lower serum range, possibly due to diminished intake because of poor health or due to increased requirements of diseased tissues. Red cell levels in normal healthy subjects are seldom below 150 μg./l. Low red cell folate levels are less frequently seen in

Pteroylglutamic acid

hospital patients than are low serum values. A fall in the red cell level is a more reliable indication of folate deficiency than is the serum level.

As has been previously mentioned (p. 608), folate deficiency is associated with an increased output of FIGLU in the urine after histidine loading.

It is possible to assess whether or not folate deficiency is due to defective intestinal absorption by studying blood, fecal and urine levels after oral administration of 1 mg. of the vitamin. It is convenient to use tritium-labeled folate for this purpose.

Folic acid deficiency can result from reduced dietary levels which are seen in the tropics and in geriatric practice, in intestinal malabsorption syndromes and during pregnancy. The deficiency gives rise to a megaloblastic anemia. The nuclei of the neutrophil polymorphonuclear leukocytes contain more than the normal number of lobes. A similar anemia, due to defective folate function, also occurs occasionally after prolonged administration of anticonvulsant drugs (phenytoin sodium and primidone).

VITAMIN B_{12}

This vitamin was first isolated in the form of dark red crystals of the compound cyanocobalamin. Its molecule is built on the foundation of a planar corrin ring and a nucleotide-like portion nearly at right angles to it. The ring has a structure very similar to that of a porphyrin (p. 261), but has only three methene bridges and is closed by direct linkage between two adjacent pyrrole rings. In vitamin B_{12}, a cobalt atom is attached to the four pyrrole nitrogen atoms. This, together with the appropriate side chains including ribosyl phosphate, forms a structure known as cobamide. Since the cobalt atom is also attached to a cyanide group, the vitamin is a derivative of cobamide cyanide. It is, in fact, α-(5,6-dimethylbenzimidazolyl) cobamide cyanide. The nucleotide-like portion is the dimethylbenzimidazole attached to ribosyl phosphate. There is also a linkage between an imidazole nitrogen and the cobalt atom. The portion of the vitamin molecule without the cyanide group is known as cobalamin, and the vitamin is referred to as cyanocobalamin. The CN can be replaced by CH_3 to give methylcobalamin or by a 5′deoxyadenosyl group to give a cobamide (coenzyme B_{12}), which is the major form in which the vitamin exists in the body, mainly in the liver. It also exists to some extent as methyl- and hydroxycobalamin. The cyano- derivative originally isolated is probably an artifact of preparation but is readily converted by the body into the naturally occurring forms.

Vitamin B_{12}, as the coenzyme, is involved with tetrahydrofolate in the synthesis of labile methyl groups, which can be transferred to homocysteine to form methionine, possibly with methylcobalamin as an intermediate. This is not sufficient to supply the body with its total methionine requirement, and the amino acid is therefore a member of the essential group. The vitamin is also a coenzyme for the mutase, which converts methylmalonyl-CoA into succinyl-CoA; hence, methylmalonate accumulates in the urine in deficiency states. The vitamin is also involved in the mainte-

Cyanocobalamin

nance of sulfhydryl groups in the reduced state. A deficiency of B_{12} results in a reduction of the amount of glutathione in the blood. Vitamin B_{12} is probably also involved in the production of deoxyribose during the biosynthesis of DNA.

The original source of vitamin B_{12} is probably bacterial, from which it finds its way into animals. In man's diet, the richest sources of the vitamin are liver and kidney. It is also present in meat, fish, poultry and dairy products. The daily mean intake varies from 2.7 µg. on a poor diet to 31.6 µg. on a high-cost diet. Only 2 to 3 µg. daily are absorbed across the intestinal mucosa. Absorption of B_{12} requires the presence of gastric intrinsic factor, which is a mucopolysaccharide of molecular weight 119,000 and is probably a dimer. Intrinsic factor is microheterogenous, because of varying content of sialic acid. The factor binds vitamin B_{12}, and the complex then becomes attached to acceptor sites in the lower small intestine for absorption, which probably requires calcium ions. There is evidence that bound cyanocobala-

Cobamide coenzyme

min is converted to coenzyme B_{12} during absorption. The absorbed vitamin is bound to plasma proteins (transcobalamin I and transcobalamin II). Most is eventually stored in the liver. There is a high content of the vitamin in bile, but most of this is reabsorbed.

Vitamin B_{12} may be assayed microbiologically (*Euglena gracilis, Lactobacillus leichmannii*) or by saturation analysis using radioactive vitamin B_{12}. The mean content of normal human liver is 910 ng./g. That of normal human serum is 450 pg./ml., with a range of 160 to 1000 pg./ml. Levels below 100 pg./ml. are definitely abnormal. Raised serum levels occur in patients with certain types of liver disease and in some leukemias.

In man, deficiency of vitamin B_{12} may result from poor dietary intake, which occurs in the tropics; it may be found in strict vegetarians (vegans) and occasionally in the elderly. It may result from a deficiency of intrinsic factor (Addisonian pernicious anemia; total or partial gastrectomy) or from interference with the function of intrinsic factor (fish tapeworm, blind loop syndrome, etc.) and from intestinal malabsorption (p. 602). Deficiency has also resulted from the prolonged use of anticonvulsant drugs and of para-aminosalicylic acid, which is used in the treatment of tuberculosis. In deficiency states, the serum level is low, and there is a megaloblastic macrocytic anemia with glossitis. This is very similar to the state of affairs in folate deficiency. With vitamin B_{12} deficiency, there can also be severe disease of the nervous system, both central (subacute combined degeneration of the cord) and peripheral. There may also be an amblyopia. Tobacco amblyopia is closely related to vitamin B_{12} deficiency. Psychiatric symptoms are not uncommon in deficiency of this vitamin.

Involvement of the nervous system is possibly due to disturbed propionic acid metabolism resulting from failure of conversion of malonyl-CoA to succinyl-CoA.

Congenital Methylmalonic Aciduria. A severe form of acidosis has been reported in children up to the age of one year. It is associated with excessive production of methylmalonate. There are two types, one of which is due to a congenital defect in the biosynthesis of 5′deoxyadenosylcobalamin. These children do not have a megalobastic anemia.

ASCORBIC ACID

The fundamental role played by ascorbic acid in metabolic processes is not known. The fact that it is very sensitive to reversible oxidation (ascorbic acid ⇌ dehydroascorbic acid) suggests that it may be involved in cellular oxidation-reduction reactions. However, there is no direct evidence that this is the case. There is evidence that ascorbic acid may be involved in the intermediary metabolism of tyrosine and phenylalanine, in the conversion of folic acid to the citrovorum factor (folinic acid), and in the formation of noradrenaline. It seems to be involved in hydroxylation of steroids in the adrenal cortex.

Ascorbic acid is essential for the normal regulation of the colloidal

L-Ascorbic acid   L-Dehydroascorbic acid
(ascorbone)

condition of intercellular substances, including the fibrils and collagen of connective tissue, osteoid tissue, dentin, and perhaps the intercellular "cement substance" of the capillaries. Ascorbic acid is concerned in the hydroxylation of proline, and hydroxyproline is an important constituent of collagen. Hydroxylation actually involves the prolyl residues of the collagen precursor. Many of the important clinical manifestations of vitamin C deficiency are directly dependent upon abnormal development and maintenance of these structures. The capillary defect in scurvy may be related to an inhibitory effect of ascorbic acid on the hyaluronidase-hyaluronic acid system.

Ascorbic acid may be involved in the conversion of pteroylglutamic (folic) acid to the active formyltetrahydrofolic derivative; as evidence of this relationship, severe deficiency in ascorbic acid in infants may be accompanied by a megaloblastic type of anemia, relieved by folinic acid or large amounts of folic acid.

Only man, other primates and guinea pigs, of the many species investigated, are unable to synthesize ascorbic acid, the entire human requirement for which must consequently be supplied by the diet. Ascorbic acid is absorbed readily from the small intestine, peritoneum, and subcutaneous tissues. It is widely distributed throughout the body, in local concentrations roughly paralleling the metabolic activity of the tissue. Ascorbic acid is supplied to the fetus from the maternal circulation, passing the placental barrier readily. The concentration in umbilical cord blood is higher than in the maternal blood, suggesting that the placenta may be able to concentrate the vitamin.

There is no evidence that any particular organ or tissue serves as a storage reservoir. That the body does contain limited mobilizable reserve stores of ascorbic acid is indicated, however, by the fact that clinical manifestations of scurvy develop in man only after several months on an ascorbic acid-free diet.

Normal human blood plasma contains about 0.6 to 1.5 mg. ascorbic acid per 100 ml. Under adequate dietary conditions the concentration in the erythrocytes is one to two and one-half times, and in the white blood cells and platelets ("white layer") twenty to forty times that in the plasma.

The vitamin exists in the body largely in the reduced form (ascorbic acid), in reversible equilibrium with a relatively small amount of dehydroascorbic acid (oxidized form). Under conditions of normal dietary intake (60 mg., recommended daily intake), about 50 to 75 per cent of ingested ascorbic acid undergoes metabolic conversion to inactive compounds. The remainder is excreted, as such, in the urine. It is also present in milk in active form. The main catabolic pathway is degradation to CO_2. The other demonstrated important metabolic end-product is oxalate, which is eliminated in the urine.

Detection of Deficiency in Man. Advanced scurvy is readily diagnosed but is rarely seen. Detection of subclinical deficiency is difficult, as is demonstration of ascorbic acid deficiency as a cause of various clinical conditions. As in the case of vitamin deficiencies generally, prompt improvement following administration of ascorbic acid constitutes the most reliable evidence that deficiency in this substance had contributed to the development of the condition in question. Certain objective methods have been proposed for the demonstration of subclinical deficiency states. They include: (1) the concentration of ascorbic acid in the blood; (2) urinary excretion of ascorbic acid; (3) urinary excretion following administration of the vitamin (saturation test); (4) intradermal test for ascorbic acid; (5) capillary fragility test.

Blood Ascorbic Acid. The plasma ascorbic acid concentration falls relatively promptly after removal of the vitamin from the diet, approaching zero in forty to eighty days. Usually, when the plasma values fall below 0.3 mg./100 ml. the concentration in the leukocytes and platelets begins to decrease, and when the former reaches 0.1 mg./100 ml. the latter may be assumed to be seriously reduced. As a rule, recognizable clinical manifestations of scurvy appear only after this degree of depletion has been reached. Consequently, decrease in ascorbic acid of the "white blood cell-platelet layer" or, less significantly, of whole blood, is a more reliable index of the scorbutic nature of clinical signs than is a fall in the plasma ascorbic acid concentration.

According to some observers, plasma values above 1 mg. per 100 ml. may be regarded as indicating normal vitamin C saturation, and values below 0.6 mg. undersaturation. However, values below 0.4 mg. are obtained so frequently in apparently healthy subjects that a diagnosis of subclinical vitamin C deficiency based upon this finding alone is questionable. The plasma ascorbic acid concentration reacts almost immediately to variations in intake and, therefore, much more reliance is to be placed upon the concentration in whole blood or the white cell–platelet layer. In a normal subject given a diet deficient in vitamin C, ascorbic acid fell to zero in the plasma after 41 days, and in the whole blood and "white layer" after 124 days, the first clinical signs of scurvy appearing on the 134th day. Thus, the plasma is the first and the "white layer" the last to be depleted, the latter being apparently the most sensitive index of the state of vitamin C nutrition. It has been suggested that a diagnosis of vitamin C deficiency is warranted when the plasma ascorbic acid concentration is zero, and a diagnosis of a prescorbutic state when the whole blood ascorbic acid is zero.

Urine Ascorbic Acid. Urinary excretion of ascorbic acid decreases rather promptly following lowering of the intake and does not, therefore, reflect the nutritional status of the organism in this regard. Test procedures which embody the "tissue saturation" principle are perhaps more useful clinically. These saturation tests are based upon the principle that if the tissues contain an adequate amount of ascorbic acid, i.e., are "saturated" with the vitamin, a certain proportion of an administered test dose will be excreted in the urine. If the tissues are "undersaturated," they will retain an abnormally large proportion of the test dose, and a subnormal amount will be excreted in the urine. A number of different procedures and dosages have been employed, with rather equivocal results. The chief difficulty has been in the establishment of normal values because of the wide range of individual variation. Such saturation tests may be valuable under carefully controlled conditions.

Normal adults receiving an adequate amount of vitamin C usually excrete 20 mg. or more daily in the urine. In clinical scurvy there is no ascorbic acid in the urine, the traces reported occasionally probably consisting of other reducing substances. The twenty-four hour excretion of ascorbic acid is a poor index of vitamin C deficiency, being dependent upon the degree of saturation of the tissues with regard to this factor. This is due to the fact that in the "unsaturated" subject about 99 per cent of the ascorbic acid presented to the kidneys is reabsorbed in the tubules, whereas in the "saturated" subject all the excess of the vitamin ingested above that which can be utilized is excreted.

A more satisfactory index of the state of vitamin C nutrition is afforded by the use of the saturation test, involving measurement of the urinary excretion of ascorbic acid following administration of a test dose. The interpretation of this test is based upon the hypothesis that if the vitamin C content of the tissues is subnormal an abnormally large proportion of the administered dose will be retained in the body and, consequently, a smaller amount will be eliminated in the urine than if the vitamin C content of the tissues is normal. A variety of methods have been proposed, including varying amounts of the vitamin, administered orally and intravenously, and a test period of varying duration. The intravenous route of administration and a short test period seem preferable. Following the injection (intravenous) of 100 mg. of ascorbic acid, at least 40 mg. (40 per cent) should be excreted in the urine within three hours in normal subjects. Following the intravenous injection of 1000 mg., at least 400 mg. should be excreted in the urine in five hours. The excretion of quantities less than these is interpreted as possibly indicative of subnormal saturation of the tissues with vitamin C.

There is considerable question as to the significance of results obtained by this procedure. It has been suggested that in some cases abnormal findings may be due to greater utilization of ascorbic acid rather than to subsaturation of the tissues with this factor. Moreover, there is no definite proof of a direct relation between a state of apparent "subsaturation" and the disease conditions in which it may be present. Among these are tuberculosis, rheumatic fever, rheumatoid arthritis, osteomyelitis, congestive heart failure, renal and hepatic disease, gastrointestinal disturbances, purpura, metabolic and endocrine disorders, and malignancy. An increase in ascorbic acid excretion in the urine has been observed after ether anesthesia and, in children, after administration of acetylsalicylic acid. It would appear that the requirement of the organism for vitamin C and the excretion of this factor in the urine can be influenced by a great variety of factors, and that some degree of vitamin C deficiency may contribute to the symptomatology of a number of clinical disorders.

Capillary Resistance (Fragility) Test. This is designed to disclose, by the application of mechanical stress, increased fragility of the capillary walls. This is generally ascribed to defective formation of the intercellular capillary cement substance, but increased activity of the hyaluronidase-hyaluronic acid system may be involved. A positive reaction may be obtained before the defect manifests itself spontaneously by petechial or other hemorrhages. The mechanical stress is applied in the form of either negative (cupping) or positive (tourniquet) pressures.

These procedures are of value in aiding the diagnosis of ascorbic acid deficiency but, unfortunately, capillary fragility may be increased in other conditions. They are useful chiefly as screening procedures, a normal (negative) response constituting evidence against serious ascorbic acid deficiency.

VITAMIN D

The inactive natural precursors of the D vitamins, the corresponding provitamins, are steroids (p. 96). At least ten such substances are known, only two of which have been found in nature: (1) ergosterol (provitamin D_2) and (2) 7-dehydrocholesterol (provitamin D_3), the former in plants (fungi, yeasts), the latter in animals and human skin. Transformation from the inactive to the active forms is accomplished physiologically by the ultraviolet rays present in sunlight, or artificially by ultraviolet irradiation. The only D vitamins of importance in human nutrition are activated ergosterol (vitamin D_2; calciferol) and activated 7-dehydrocholesterol (vitamin D_3; cholecalciferol), present in animal tissues, fish liver oils and irradiated milk. Prolonged irradiation of ergosterol yields tachysterol, which gives dihydrotachysterol on reduction.

The international unit is equal to 0.025 μg. crystalline cholecalciferol. Provided that there is adequate exposure to sunlight, the body can usually synthesize enough vitamin D for its own needs. In the absence of sufficient exposure and especially during the period of active skeletal growth, and during pregnancy and lactation, supplements may be required. The daily oral administration of 400 I.U. is considered adequate under these circumstances.

The D vitamins are readily absorbed in the small intestines. Being fat-soluble, their absorption from the bowel is enhanced by factors which favor fat absorption, including an adequate quantity of bile salts. After absorption, the vitamin is stored, largely in the liver (20–200 I.U./100 g.), kidneys, intestines, adrenals and bones. A small amount may be excreted in the bile but is at least partly reabsorbed in the intestine. None is eliminated in the urine. The vitamin D content of milk reflects the intake; e.g., ingestion of a single large dose (600,000 I.U.) increases the concentration in the milk (human) from about 10 I.U./liter to 1000 I.U./liter in one day, the level being still elevated (100 I.U./liter) after one month.

In adults, the serum level of vitamin D is 700 to 3100 I.U./l. In children, it is 860 to 2100 I.U./l. The assay can be biological (tibia line) or chemical. The latter involves preliminary separation from extracts followed by gas chromatography. Vitamin D is transported in the blood

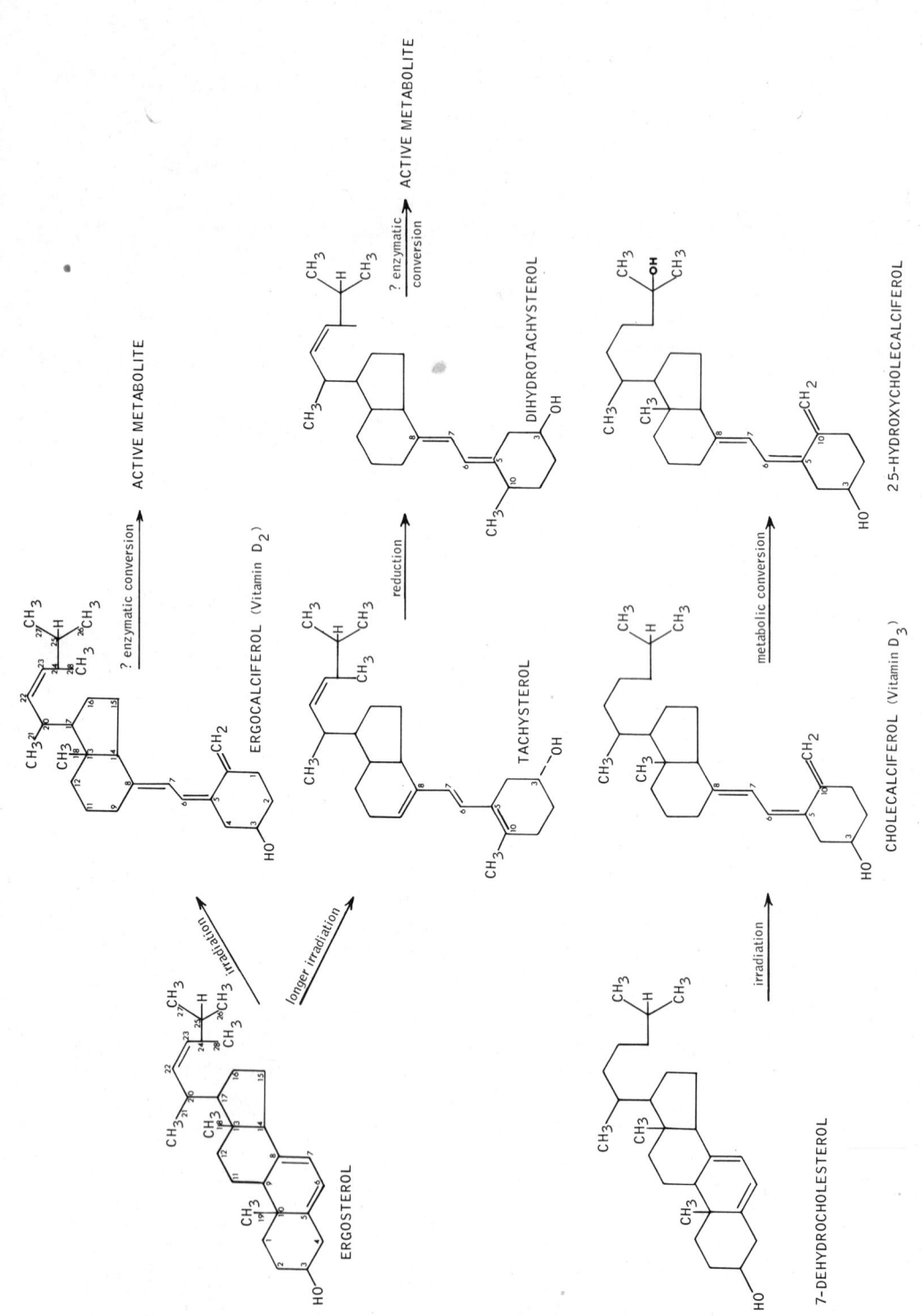

Figure 23-1. Schematic illustrating structure, origin and active metabolic products of vitamins D₂ and D₃. The formation in vivo of 25-hydroxycholecalciferol, the active metabolite of vitamin D₃, has been established. It is likely, but has yet to be proved, that 25-hydroxylation may be a common mechanism of metabolic conversion for the antirachitic compounds in vivo; hence, this is shown as probable for dihydrotachysterol and ergocalciferol. (Data provided through the courtesy of Dr. H. F. DeLuca.) (From Bondy: Duncan's Diseases of Metabolism, 6th ed. W. B. Saunders Company, 1969.)

bound to α_2-globulins and albumin. Cholecalciferol is now known to be converted to the more active 25-hydroxycholecalciferol in the liver (p. 281) and the even more active metabolite 1,25-dihydroxycholecalciferol in the kidney, especially on a low-calcium diet. The kidneys of animals on a high-calcium diet or under the influence of parathormone produce 24,25-dihydroxycholecalciferol, which is a metabolite with low activity. Recent evidence seems to indicate that the 25,26-dihydroxy- derivative is formed in the human being. In actual fact, it is believed that the determining factor for the production of the active or inactive dihydroxy-metabolite is the intracellular concentration of calcium ions. This is, of course, affected by parathormone.

It appears that the major function of vitamin D is to stimulate transcription of the mRNA for a calcium transport protein (p. 281). In this way, it is concerned with intestinal absorption, renal excretion and bone metabolism of calcium. It is now recognized that its ultimate effect is intimately related to the concurrent activity of parathormone and calcitonin. The complex interrelationship has not been completely elucidated.

Vitamin D directly stimulates intestinal absorption of calcium and, indirectly, that of phosphate. In individuals with hypoparathyroidism, these effects are still produced, but larger doses are necessary. The renal clearances of calcium and phosphate are decreased by the vitamin in the intact animal, but the effect is completely reversed after parathyroidectomy. With regard to bone, it appears that vitamin D gives rise to the intracellular conditions necessary for the action of parathormone, and possibly calcitonin. Administration of vitamin D in very large dosages can lead to bone dissolution. It is not clear whether or not this is really an increase of sensitivity to the effect of parathormone. It has been shown that in animals rendered chronically hypocalcemic by complete deprivation of vitamin D, parathormone will not mobilize bone calcium. It is interesting, however, that under these conditions calcitonin will lead to even lower levels of serum calcium.

Deficiency Manifestations. Vitamin D deficiency during the period of skeletal growth results in rickets (rachitis), with a characteristic defect in endochondral bone growth and mineralization of the "zone of provisional calcification" of long bones (junction or epiphysis and diaphysis) and corresponding areas of flat bones. This lesion cannot occur in fully grown bones, the characteristic skeletal manifestation of deficiency in adults being a type of defective mineralization of osteoid tissue termed "osteomalacia."

Metabolic Manifestations. Vitamin D deficiency results in decreased net retention of calcium and inorganic phosphate in the organism. The mineral content of the bones is diminished, the water and organic matter being correspondingly increased. The bones have a lower calcium, phosphate, and carbonate content and a higher magnesium content. Urinary excretion of calcium and phosphate falls and fecal excretion increases. The serum inorganic phosphate concentration usually falls, whereas the serum calcium usually remains within normal limits for some time; in advanced deficiency it, too, decreases, often to markedly hypocalcemic levels (infantile tetany). The serum alkaline phosphatase activity increases, usually in proportion to the severity of the skeletal defect.

The sequence of events is believed to be as follows:

1. There is decreased intestinal absorption of calcium, resulting in tendency toward decrease in serum calcium and decrease in urine calcium.

2. As soon as the serum calcium tends to fall the parathyroids are stimulated to increased activity (also hyperplasia), maintaining the serum calcium level, now at the expense of the skeleton (p. 302).

3. Increased parathyroid activity results in increased urinary excretion of inorganic phosphate. This may be due also to a direct effect of vitamin D deficiency on renal tubular absorption of phosphate.

4. Increase in fecal phosphate may result, in part at least, from diminished absorption due to the excess of unabsorbed calcium in the bowel (unfavorable Ca : PO_4 ratio).

5. As deficiency progresses, there is insufficient vitamin necessary for the action of parathormone. Hypocalcemia then develops and, if of sufficient degree, is manifested clinically by neuromuscular hyperexcitability (infantile or osteomalacic tetany).

6. Decreased mineralization of the bones, both generalized and also in growing long bones, in the zone of provisional calcification, results in softening and deformities. This strain constitutes a stimulus to proliferation of osteoblasts, which form alkaline phosphatase. The local increase in this enzyme is reflected in increased alkaline phosphatase activity in the blood.

All of the above manifestations are reversed promptly after administration of adequate amounts of vitamin D, calcium, and phosphate. The increased alkaline phosphatase activity is the last abnormality to disappear.

The occurrence of hyperchloremic acidosis and aminoaciduria has already been discussed (p. 314).

Vitamin D deficiency can be the result of a number of causes. The cause may be dietary or insufficient exposure to sunlight. This can occur in children and adults. It has been reported in men working underground, in night workers and in the elderly. The deficiency may be associated with gastrointestinal disorders such as malabsorption syndromes and chronic obstructive jaundice, or it may be a result of surgery (partial gastrectomy; resection of terminal ileum). A variety of conditions, frequently renal in origin, are associated with vitamin D resistance. Prolonged treatment with anticonvulsant drugs can lead to increased inactivation of the vitamin.

Demonstration of Vitamin D Deficiency. This depends on the presence of (1) suggestive clinical manifestations, (2) characteristic roentgenographic abnormalities, and (3) metabolic aberrations. The most important of the latter, which are discussed in detail elsewhere, are: (1) hypophosphatemia (p. 310); (2) increased serum alkaline phosphatase activity (p. 563); (3) normal or decreased serum calcium (p. 302); (4) decreased urinary and increased fecal excretion of calcium and phosphate (p. 308). Prompt response to specific therapy, except in cases of so-called "vitamin D-resistant rickets," is of great diagnostic importance.

Effects of Excess Vitamin D. Vitamin D is well tolerated in doses many times the normal requirement. However, seriously deleterious effects may be produced by extremely large amounts (500 to 1000 times the normal requirement). The early symptoms are due chiefly perhaps to the induced hypercalcemia (increased intestinal absorption of calcium and possibly increased skeletal mobilization). These include anorexia, thirst, lassitude, constipation, and polyuria, followed later by nausea, vomiting, and diarrhea. Hyperphosphatemia may occur.

Increased urinary excretion of calcium and phosphate may lead to urinary lithiasis, and the hypercalcemia and hyperphosphatemia may lead to metastatic calcification. The kidneys, arteries, bronchi, pulmonary alveoli, muscles, and gastric mucosa are principally involved. Renal failure may develop, leading to death. In growing children there may be excessive mineralization of the zone of provisional calcification at the expense of the diaphysis, which undergoes demineralization (with extremely large doses).

VITAMIN E (TOCOPHEROLS)

The tocopherols are largely methyl derivatives of the parent compound tocol. α-Tocopherol is 5,7,8-trimethyltocol; β-tocopherol is 5,8-dimethyltocol, and so on. Some tocopherols are derivatives of tocotrienol, which contains three double bonds in the aliphatic side chain and is therefore terpenoid in structure. The tocopherols are yellow oily substances, which are fat-soluble. Intestinal absorption requires the presence of bile salts. Estimations of the daily intake of vitamin E (vegetable oil, lettuce, milk products, egg yolks, etc.) vary from 7.4 to 14 mg. α-tocopherol in a total tocopherol intake of about 24 mg. It has been suggested that 10 to 30 mg. is the daily requirement of the human adult—the greater the intake of polyunsaturated fats, the greater the tocopherol requirement.

The vitamin is stored in the liver (mitochondria; microsomes) and fatty tissues. It is present in relatively high concentration in the adrenals, the pituitary, the uterus and the testes. The normal concentration of vitamin E in adult serum is stated to be approximately 10 mg./l., and about half this in the newborn. It is now realized that blood concentrations

Tocol

have no real meaning unless the concentration of the blood lipids is also known.

The tocopherols probably play a part in the stabilization of cellular membranes. In this respect, they are believed to be associated with selenium, and together with the appropriate protein to play some part in electron transport. Much has been made of their antioxidant properties regarding the prevention of the formation of peroxides of unsaturated fatty acids. It is doubtful, however, whether such peroxides occur *in vivo*. The tocopherols do prevent the oxidation of vitamin A, and, in association with Se, probably of thiol groups. Vitamin E seems to be involved in heme synthesis.

Experimentally produced deficiency of vitamin E in human adult males led to increased susceptibility of erythrocytes to hemolysis by hydrogen peroxide and decrease in the erythrocyte life span. Infants fed formulas containing unsaturated oils have developed a vitamin E deficiency, which gave rise to anemia, edema and skin changes. The danger is greatest in premature infants. Decreased erythrocyte life span, hemolysis, creatinuria and ceroid deposition in malabsorption syndromes in children are believed to be due to vitamin E deficiency.

The curative effect of vitamin E in large doses in a number of disorders, including myocardial ischemia, is still highly controversial but should not be dismissed lightly.

VITAMIN K

This is a fat-soluble substance required for the formation of prothrombin which is an essential factor in the mechanism of coagulation of blood. The great practical importance of this vitamin arose out of the observation that certain common hemorrhagic disorders in man are dependent upon hypoprothrombinemia (e.g., in obstructive and hepatocellular jaundice; in the newborn; in certain intestinal disorders). Administration of vitamin K was found to be strikingly effective in controlling the hemorrhagic diathesis in many of these conditions.

The several substances, natural and synthetic, with vitamin K (antihemorrhagic) activity are naphthoquinones. It has been suggested that the relative effectiveness of these compounds in this connection may be related to their capacity for forming 2-methyl-1,4-napthoquinone ("menadione") in the body; this is the most potent known substance exerting this effect.

Two naturally occurring vitamins K have been identified. Vitamin K_1, isolated originally from alfalfa, and vitamin K_2, isolated originally from putrid fish meal. Phthiocol is a constituent of tubercle bacilli possessing slight vitamin K activity. Menadione, because of its high activity, is the most important of the synthetic vitamins K, a number of which have been prepared. Vitamin K_1 has also been synthesized.

The natural vitamins K are "fat-soluble" vitamins, i.e., they are insoluble in water and are quite soluble in most fat solvents. The synthetic

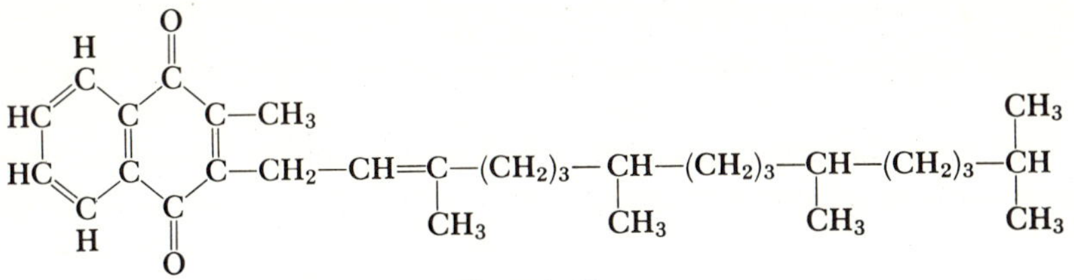

Vitamin K₁

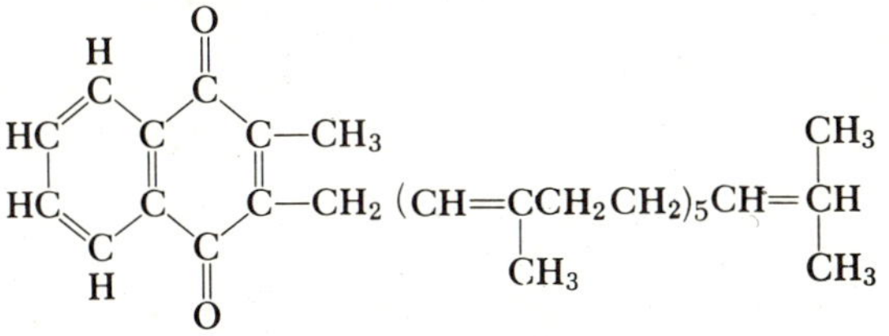

Vitamin K₂

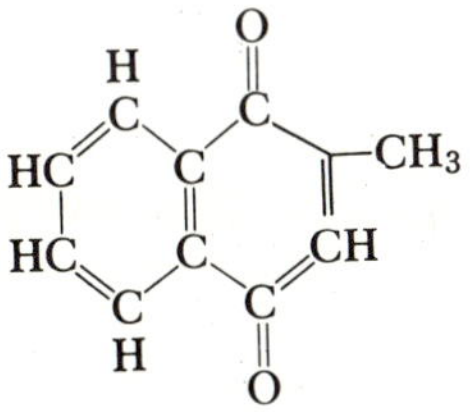

Menadione

forms, lacking the long hydrocarbon chain, are somewhat soluble in water but not sufficiently so for practical purposes. However, the hydroquinones of the vitamins K form esters (disulfate, diphosphate, diacetate) which are more soluble (water), less irritating, and more stable than the parent compounds, although less active. Certain of these are available and satisfactory for parenteral administration, e.g., 2-methyl-1,4-naphthohydroquinone-3-sodium sulfonate; 4-amino-2-methyl-l-naphthol hydrochloride.

A normal diet contains an abundance of vitamin K, which is also formed by intestinal bacteria (chiefly *Escherichia coli*) in amounts adequate to meet normal requirements. Being fat-soluble, its absorption, which occurs predominantly in the jejunum, by way of the lymphatics, is influenced by factors which affect the absorption of lipids. Adequate amounts of bile salts must be present for optimal absorption, a fact of great importance in relation to the hemorrhagic manifestations of biliary obstruction. Absorption is diminished by large amounts of liquid petrolatum. It is doubtful whether the intestinal bacterial synthesis has any importance in human nutrition.

Vitamin K has not been found consistently in the blood stream in significant amounts. During pregnancy, it apparently passes readily from the mother to the fetus. The capacity of the organism for storing this vitamin is extremely limited, at least in experimental animals, and evidence suggests that this is true also of man. Both vitamin K and prothrombin are utilized or metabolized rather rapidly, so that deficiency manifestations appear after a relatively brief period of deprivation (twenty-four to forty-eight hours in rats).

Vitamin K is apparently not excreted in the urine or bile. It has been found in the milk of women receiving adequate amounts of the vitamin. Rather large quantities may be present in the feces; whereas this may represent actual excretion by the intestinal mucosa, a more probable source is the intestinal bacterial flora.

Vitamin K is apparently essential for the formation of prothrombin, proconvertin (Factor VII), Christmas factor (Factor IX) and Stuart factor (Factor X) by the hepatic polygonal cells. The mechanism of its action is unknown. It has been suggested that it acts as a coenzyme in the synthesis of these blood coagulation factors or as an oxidation-reduction catalyst in this process.

Deficiency. Vitamin K deficiency can be induced by dietary restriction in several species (chick and other birds, rat, mouse, rabbit). In man, deficiency rarely results from dietary inadequacy. The low plasma prothrombin which occurs consistently in the newborn, during the first few days of life, is attributed to vitamin K deficiency, possibly due in part to inadequacy of the intestinal bacterial flora, and in part perhaps to inadequate bile flow and intestinal hypermotility with consequent poor absorption.

Deficiency manifestations result usually from conditions which interfere with (1) absorption of the vitamin from the intestine or (2) its utilization in formation of the blood coagulation factors.

(1) Impaired Absorption. Bile acids appear to act as carrying agents for the passage of vitamin K across the intestinal wall, a function similar to that which they perform for vitamin D and carotene. Absorption from the intestine is impaired in the absence of adequate amounts of bile (bile acids) in the intestine, as in obstructive jaundice and external bile fistula, with a consequent fall in plasma prothrombin and development of a hemorrhagic tendency. Adequate absorption under such circumstances follows administration of bile salts, the most effective of which appears to be sodium deoxycholate. Inadequate absorption of vitamin K may occur in conditions of severely disturbed intestinal function (malabsorption syndrome, p. 602), as in patients with intestinal obstruction, intestinal fistula, enterostomy, gastrocolic fistula, chronic pancreatitis, sprue, celiac disease, chronic ulcerative colitis, prolonged diarrhea, and anorexia nervosa. Absorption is also interfered with by large amounts of liquid petrolatum.

(2) Hepatic Disease. The liver is the chief if not the only site of formation of plasma prothrombin and the other coagulation factors involved and an adequately functioning liver is essential for the proper utilization of vitamin K in this process. The plasma prothrombin may be diminished experimentally by partial hepatectomy and administration of hepatotoxic agents, such as chloroform, carbon tetrachloride, and phos-

phorus. The same phenomenon occurs in many patients with hepatocellular damage due to any cause (p. 629). In contrast to the beneficial effect of vitamin K administration in patients with uncomplicated obstructive jaundice, those with extensive hepatocellular damage characteristically exhibit little or no elevation of prothrombin following injection of vitamin K. This failure to respond to this agent has been suggested as a test of liver function (p. 629).

(3) Miscellaneous. The plasma prothrombin is low in the newborn infant (physiological hypoprothrombinemia) and is one cause of hemorrhagic disease of the newborn. A condition of idiopathic hypoprothrombinemia in adults has been described. Large doses of salicylates and barbiturates may be followed by decrease in plasma prothrombin. Hypoprothrombinemia has been observed in pulmonary tuberculosis, pneumonia, following hemorrhage, and after operations, particularly on the biliary tract. In the majority of these conditions, if not accompanied by severe hepatic damage, the prothrombin level increases after administration of adequate doses of vitamin K. Extremely large doses are required to combat the hypoprothrombinemia induced by administration of dicumarol and similar anticoagulants. Hypoprothrombinemia may be induced also by oral administration of certain sulfonamides and antibiotics for the purpose of inhibiting growth of intestinal bacteria, e.g., prior to operations on the gastrointestinal tract. This can be overcome by adequate vitamin K therapy.

Large doses of vitamin K, employed therapeutically, may produce hyperbilirubinemia, the mechanism being unknown (in part excessive hemolysis, in part hepatocellular damage). The increase is mainly in unconjugated bilirubin. This is of particular concern during the neonatal period because of its implications with respect to the development of kernicterus (p. 648).

Demonstration of Vitamin K Deficiency. The methods employed for the demonstration of vitamin K deficiency consist in determination of the quantity of prothrombin and the three other factors in the plasma. These methods are indirect in nature. The one-stage method (Quick) commonly employed consists in the determination of the clotting time of oxalated plasma at 37.5° C. after addition of an excess of thromboplastin and a fixed amount of calcium. The two-stage method is more satisfactory but more troublesome. In the first, or prothrombin conversion stage, prothrombin is converted completely to thrombin with an optimal amount of calcium and an excess of thromboplastin. In the second, or clotting stage, the amount of thrombin formed is estimated by the time required for clotting of a standard fibrinogen solution.

Decrease in prothrombin is evidenced by prolongation of the clotting time ("prothrombin time") under these circumstances (see p. 629). Experience has shown that bleeding occurs commonly in patients with obstructive jaundice, bile fistula and other conditions mentioned above when the plasma prothrombin concentration, as measured by these tests, falls below 30 to 40 per cent of normal. Values of 40 to 70 per cent of normal are potentially dangerous.

OTHER VITAMINS

Biotin and pantothenic acid have not been detailed here, since deficiency in man is extremely rare. Both vitamin deficiencies have been produced experimentally.

Biotin deficiency has been reported as a result of a diet containing an excess of raw eggs. It also occurs occasionally in young children (Leiner's disease). The main manifestations are seborrheic dermatitis and desquamative erythroderma.

Pantothenic acid, suitably modified, forms part of the molecule of coenzyme A. Biotin is the prosthetic group (coenzyme) of propionyl CoA carboxylase and plays an important part in many other reactions.

References

Burns, J. J., ed.: Vitamin C. Ann. New York Acad. Sci., *92*:1, 1961.
Chanarin, L.: The Megaloblastic Anaemias. Oxford, Blackwell Scientific Publications, 1969.
Dam, H.: Vitamin K. Vitamins and Hormones, New York, Academic Press, Inc., *6*:28, 1948.
Dann, W. J.: The appraisal of nutritional status in humans. Physiol. Rev., *25*:326, 1945.
Follis, R. H., Jr.: Deficiency Disease. Springfield, Ill., Charles C Thomas, 1958.
Isler, O., and Wiss, O.: Chemistry and biochemistry of the K vitamins. Vitamins and Hormones, *17*:54, 1959..
Marks, J.: The Vitamins in Health and Disease. London, J. A. Churchill, 1968.
McLaren, D. S.: The vitamins. *In* Bondy, P. K., ed.: Duncan's Diseases of Metabolism. Ed. 6, Philadelphia, W. B. Saunders Co., 1969.
Ungley, C. C.: The chemotherapeutic action of vitamin B_{12}. Vitamins and Hormones, *13*:139, 1955.
Williams, R. J., et al.: The Biochemistry of B Vitamins. New York, Reinhold Publishing Corporation, 1950.

Chapter 24

CEREBROSPINAL FLUID

PROTEIN

The protein content of cerebrospinal fluid is lower than that of any other normal body fluid with the exception of the aqueous humor of the eye, which closely resembles the cerebrospinal fluid in chemical composition. The amount of protein in normal fluid obtained by lumbar puncture is 20 to 45 mg. per 100 ml., cisternal fluid containing 10 to 15 mg. per 100 ml. The appearance of increased amounts of protein in the fluid is dependent upon the same factors which govern the presence of protein in all body fluids, normal and pathologic. Chief among these is the permeability of the barrier interposed between the plasma and the subarachnoid space, which under normal circumstances may be considered to be the choroid plexuses. Normally these vessels, as well as those of the meninges, are practically impermeable to proteins; in inflammatory conditions of the brain, cord and meninges the capillary walls become more permeable and, depending upon the degree of inflammation, increasing amounts of albumin, globulins and fibrinogen pass into the subarachnoid space, the last appearing only in severe inflammatory processes. Investigation by modern methods, including isoelectric focusing in a gel followed by gel electrophoresis, has shown that a surprising number of serum proteins occur in normal cerebrospinal fluid. The major component is albumin, followed by pre-albumin. The immunoglobulin IgG is invariably present, but its proportion to albumin is much less than in serum. Transferrin is easily demonstrated, as well as the so-called tau protein. Haptoglobin type 1-1 is frequently present, since its molecular size is small enough for it to pass the blood brain barrier in small amounts. When the protein content of the fluid is high, as in inflammatory lesions of the central nervous system, IgA as well as the other types of haptoglobin are able to pass into the cerebrospinal fluid. In monoclonal gammopathies, especially those involving IgG, the M protein can be found in the fluid. Immunological techniques also demonstrate the presence of a number of proteins in cerebrospinal fluid.

Increase in the protein content may be found in conditions other than inflammations in which capillary and cell permeability are increased. These include: toxic states such as uremia, pneumonia and typhoid fever; convulsive states, as epilepsy and spasmophilia; conditions causing generalized or localized elevation of subarachnoid tension, as brain and spinal cord tumor.

Since in most pathologic fluids containing an increased quantity of protein, globulin is present in abnormally high concentration, globulin tests were commonly used as a means of roughly estimating the protein content of cerebrospinal fluid. The most popular of these tests were the Noguchi, Nonne-Apelt, Ross-Jones and Pandy reactions; the results were reported as normal, 1 plus, 2 plus, 3 plus, and 4 plus, depending upon the amount of globulin precipitated. More exact quantitative determinations are now made (electrophoresis; zinc precipitation), the results being recorded in milligrams per 100 ml. of fluid.

Meningitis. Inflammatory exudates in any situation contain comparatively large amounts of protein. In the various forms of suppurative meningitis the total protein content of the cerebrospinal fluid is high (125 to 3000 mg. per 100 ml.), consisting of globulin, albumin and, at times, small amounts of fibrinogen and many other proteins. High values are also the rule in tuberculous meningitis (200 to 2000 mg.) and in acute luetic meningitis (180 to 540 mg.). In true meningitis the increase in protein usually parallels the increase in cellular content. It must be realized that in some cases of early meningitis the fluid protein may be normal.

Serous Meningitis. This term is applied to a condition of meningeal irritation of a noninfective nature which may be due to toxic states such as pneumonia, influenza, typhoid fever, and the like, or uremia.

The cerebrospinal fluid in serous meningitis is, in most cases, entirely normal except for some increase in amount and pressure. However, in

TABLE 24-1 PROTEIN IN CEREBROSPINAL FLUID

CONDITION	SPINAL FLUID*	CISTERNAL FLUID*
Normal	20– 45	10– 15
Tabes	40– 200	
Paresis	30– 190	
Cerebrospinal lues	20– 160	
Pneumococcal meningitis	150–2000	
Meningococcal meningitis	180–3000	
Streptococcal meningitis	160–2000	
Tuberculous meningitis	200–2000	
Brain tumor	30– 500	
Brain abscess	30– 500	
Froin syndrome	80–3000	25–180
Epidemic encephalitis	50– 300	
Cerebral hemorrhage	20– 220	
Cerebral thrombosis	20– 140	
Epilepsy	15– 80	
Serous meningitis	20– 70	
Myxedema	60– 200	

* Expressed in milligrams per 100 ml.

isolated cases, abnormally high concentrations of protein may be found (up to 70 mg.), other chemical and cytologic findings being normal.

Aseptic Meningitis. The syndrome of aseptic meningitis is frequently due to viral infection or associated with allergic encephalomyelitis. The cerebrospinal fluid glucose concentration is normal, and protein levels are usually normal or slightly elevated. There is an increase in mononuclear cells in the fluid.

A similar picture is sometimes found in bacterial infections lying adjacent to the meninges (paranasal sinusitis, mastoiditis) and in early tubercular or acute syphilitic meningitis.

In leukemic and carcinomatous infiltration of the meninges ("meningitis"), the sugar content can be very low indeed and the protein level is raised.

Convulsive States. During and shortly after epileptic seizures there may be an increase in cerebrospinal fluid protein. Several investigators have reported increased albumin in the fluid in this condition. In the interval between convulsions the fluid is essentially normal. The same is true of the convulsions of spasmophilia of children. The protein concentration rises also during uremic convulsions.

Organic Disease of Brain and Cord. Protein is increased in many organic diseases of the brain and cord, with or without associated pathologic change in the meninges. In the luetic affections, paresis, tabes and cerebrospinal syphilis, values of 20 to 200 mg. per 100 ml. are found, usually accompanied by an increase in cells. In epidemic encephalitis the protein ranges from 50 to 300 mg. per 100 ml. In cerebral hemorrhage, thrombosis and embolism the protein content may be normal or increased (20 to 220 mg.). In brain abscess without meningitis and in brain tumor with increased intracranial pressure the protein varies from 30 to 100 mg. per 100 ml. with slight or no cellular increase. The highest values for protein are found in fluids presenting the so-called "Froin syndrome" (xanthochromia, greatly increased protein, spontaneous coagulation and mononuclear pleocytosis). This syndrome is characteristic of spinal cord compression, due usually to tumor, associated with retention of spinal fluid in a cul-de-sac. Values of over 2000 mg. per 100 ml. have been reported in this condition. The protein content of the fluid above and below the point of obstruction may show marked differences in protein concentration. In one series of cases the protein content of the fluid removed by cisternal puncture was 24 to 180 mg. per 100 ml. while that obtained by lumbar puncture contained 80 to 3000 mg. per 100 ml. Such findings are of considerable diagnostic importance. The gamma-globulin content of cerebrospinal fluid has now become of importance in relation to the diagnosis of demyelinating diseases. In about 60 to 70 per cent of patients with multiple sclerosis, gamma-globulin comprises one-fifth or more of the total protein.

Myxedema. The protein content of cerebrospinal fluid is rather consistently increased in myxedema. Values as high as 100 mg. per 100 ml. are observed commonly and up to 200 mg. occasionally.

Myelography. Intrathecal injection of radiopaque materials for diagnostic purposes may produce an inflammatory reaction in the meninges with increase in cerebrospinal fluid protein concentration and leukocyte content. This occurs particularly following cisternal puncture for cervical myelography. Values as high as 150–200 mg./100 ml. have been observed for periods as long as four months following this procedure, usually without symptoms of meningeal irritation. This possibility should be considered in interpreting abnormally high spinal fluid protein values under such circumstances.

GLUCOSE

The concentration of glucose in the cerebrospinal fluid is dependent upon: (1) the blood sugar concentration; (2) the permeability of the protective barrier, represented by the lining of the choroid plexuses and possibly of the capillary terminations of the cerebrospinal vessels which are surrounded by prolongations of the subarachnoid space; (3) the rate of glycolysis within the fluid. It has been fairly well established that, in the absence of disease of the brain, cord or meninges, the concentration of sugar in the cerebrospinal fluid is approximately 60 to 70 per cent of that in the blood within a wide range of values extending from moderate hypoglycemia (40 mg.) to moderate hyperglycemia (400 mg.). The normal range has been found to be from 40 to 70 mg. per 100 ml. in the fasting adult. The normal values for children up to ten years of age have been found to be 70 to 90 mg. per 100 ml. It has been shown that the fluid must be withdrawn under fasting conditions, and, in order to be properly interpreted, must be compared with the sugar content of the blood withdrawn at the same time. There is a delayed postprandial rise in spinal fluid sugar following that which occurs in the blood. Because of the lack of sufficient information regarding the blood and cerebrospinal fluid sugar curve relationship following the ingestion of food, no significance should be attached to variations in the ratio of spinal fluid sugar to blood sugar except in the

TABLE 24-2 GLUCOSE IN CEREBROSPINAL FLUID

CONDITION	SUGAR*
Normal fasting adult	40– 70
Normal fasting child (to ten years)	70– 90
Functional mental disease	70– 95
Lues of central nervous system	30–110
Epidemic encephalitis	70–110
Suppurative meningitis	0– 25
Tuberculous meningitis	18– 36
Brain abscess	70–110
Brain tumor	70–110

* Expressed in milligrams per 100 ml.

fasting state. The sugar content of ventricular and cisternal fluid is usually slightly higher than that of lumbar spinal fluid.

Practically all injuries to semipermeable membranes result in an increase in their permeability. Hence, it is naturally inferred that, as with protein, any disease associated with injury to the cerebrospinal vascular system, particularly the choroid plexuses and cerebrospinal capillaries, may result in an increase in the sugar content of the spinal fluid, a condition termed hyperglycorrhachia. A decrease in spinal fluid sugar, hypoglycorrhachia, is almost invariably dependent upon an increased rate of glycolysis, this being probably the only agency by which the sugar concentration can be lowered independently of a diminution in the sugar content of the blood. Increased glycolysis occurs most strikingly in the suppurative meningitides and in leukemic meningitis; it occurs to a lesser degree in tuberculous meningitis and in some cases of luetic meningitis. It is due to the glycolytic action of either the leukocytes or the organisms present in the fluid.

Hyperglycorrhachia. *Epidemic Encephalitis.* A frequent finding in the cerebrospinal fluid of patients with acute epidemic encephalitis is an increased glucose concentration. The figures range from 70 to 110 mg. per 100 ml. in adults. The presence of a subnormal spinal fluid sugar concentration militates against the diagnosis of acute epidemic encephalitis, a fact of importance in view of the varied symptomatology of this condition and the difficulty occasionally encountered in differentiating it from other disorders, particularly tuberculous meningitis.

Syphilis of the Central Nervous System. Increased sugar concentrations are found in some cases of syphilis, chiefly in that form affecting the cerebrospinal vessels, particularly the vascular form of cerebrospinal syphilis with little or no meningeal involvement. The figures range from 70 to 110 mg. per 100 ml. In cerebrospinal syphilis of the parenchymatous type, or paresis, with slight meningeal lesion, the sugar content is normal, as it is in most cases of tabes. In acute luetic meningitis, low normal or slightly subnormal values are obtained.

Increased Intracranial Pressure. High sugar values (70 to 110 mg.) are usually found in conditions causing marked increase in intracranial tension. These include brain tumor, convulsive disorders, and brain abscess not associated with meningitis. Increased sugar concentrations are present in some cases of serous meningitis, particularly in that type associated with uremia.

Functional Mental Disorders. Certain psychiatric disorders, particularly dementia praecox, may be accompanied by hyperglycorrhachia (70 to 95 mg.). This finding is not constant and is of no diagnostic value in these conditions.

Diabetes Mellitus. As stated above, high spinal fluid sugar values may be due to hyperglycemia. Recognition of this fact is essential for the proper interpretation of spinal fluid sugar determinations.

Hypoglycorrhachia. *Suppurative Meningitis.* In all cases of acute meningitis associated with polymorphonuclear pleocytosis, sugar is either entirely absent or present only in very small amounts (0 to 25 mg. per 100 ml.). Improvement in the condition of patients with suppurative meningitis following the administration of antibiotics is accompanied by a rise in the spinal fluid sugar concentration.

Tuberculous Meningitis. Hypoglycorrhachia is a feature of the fluid of tuberculous meningitis (18 to 36 mg. per 100 ml.). Sugar is rarely entirely absent, as in the suppurative meningitides. This finding is of importance from the standpoint of differential diagnosis, particularly, as stated above, from acute epidemic encephalitis, which may in other respects simulate it very closely.

Luetic Meningitis. In a comparatively small number of cases of syphilis of the central nervous system, particularly those involving primarily the meninges and associated with a marked pleocytosis, the sugar content of the fluid may be slightly decreased (30 to 40 mg. per 100 ml.).

Leukemic Meningitis. It has already been pointed out that in leukemic or carcinomatous infiltration of the meninges, the sugar content of the fluid may be very low and occasionally it is not detectable.

NONPROTEIN NITROGENOUS CONSTITUENTS

Practically all of the known nonprotein nitrogenous constituents of the blood plasma are represented in the cerebrospinal fluid. The concentration of these substances in the fluid differs from that in the blood, however, depending upon their degree of diffusibility and other factors such as transport mechanisms and contribution from cerebral tissues. The total nonprotein nitrogen varies from 12.5 to 30 mg. per 100 ml. Urea, being one of the most diffusible constituents of the blood, exists in all body fluids in practically the same concentration. The urea content of cerebrospinal fluid, expressed as nitrogen, is 6 to 15 mg. per 100 ml. The creatinine varies between 0.45 and 1.5 mg., amino acids from 1 to 4 mg., and uric acid from 0.25 to 1.0 mg. per 100 ml., the last being slightly higher in children (0.3 to 1.5 mg.). The residual or undetermined nitrogen normally is approximately 50 per cent of that of the blood, ranging from 2 to 6 mg. per 100 ml. The concentration of an amino acid in the cerebrospinal fluid is usually lower than that found in plasma. In other words, the plasma : CSF ratio is greater than unity. In certain aminoacidopathies, in which the enzyme defect is present in the brain, the amino acids accumulate and enter the cerebrospinal fluid. This causes a lowering of the plasma : CSF ratio, which might become less than unity. Such a state of affairs is found in disorders of the urea cycle (p. 230).

The determination of nonprotein nitrogenous constituents of the spinal fluid is, in most instances, of little practical benefit either diagnostically or prognostically. With the possible exception of uric acid, they are altered in few conditions other than nephritis and uremia. Uric acid has been reported in increased concentration in all forms of meningitis. In most cases of nephritis and uremia the ratio between the total nonprotein nitrogen of the spinal fluid and blood is maintained, values as high as 375 mg. per 100 ml. having been reported. As these values increase, the urea nitrogen fraction constitutes an increasingly large proportion of the total. Thus, in one case, with a total nonprotein nitrogen of 110 mg. per 100 ml., the urea nitrogen was found to be 90 mg. per 100 ml. Creatinine, uric acid and amino acids usually increase in proportion to their concentration in the blood. In some cases of uremia, strangely enough in the asthenic type, without convulsions, the undetermined nitrogen fraction is found to be increased out of all proportion to its level in the blood. In a few instances it constituted 70 to 75 per cent of the total nonprotein nitrogen, which ranged between 240 and 375 mg. per 100 ml.

As stated above, at the present time little or no practical importance can be attached to the determination of these constituents of the spinal fluid. However, the findings in nephritis and uremia are of considerable interest in view of the insight which they afford into the concentration of the various nonprotein nitrogen fractions in the tissues and body fluids as contrasted with that in the blood.

CHLORIDE

The chloride content of the cerebrospinal fluid is from 720 to 750 mg. per 100 ml., expressed as sodium chloride (123 to 128 mmol./l.). Values in infants exhibit slightly more variation, ranging from 650 to 750 mg. per 100 ml. (111 to 128 mmol./l.). This increased concentration in the spinal fluid as compared with that in the plasma (570 to 620 mg.) is explained in part on the basis of the Donnan equilibrium governing the concentrations of ions on either side of a semipermeable membrane (the capillary walls of the choroid plexuses) when the fluid on one side (plasma) contains molecules which are not diffusible (protein) (see p. 353). The spinal fluid chloride has been found to be increased in some cases of nephritis but is of neither diagnostic nor prognostic importance in that condition.

In most forms of meningitis the chloride is decreased, the diminution being most marked in tuberculous meningitis, but not in the early and best treatable stage of this disease. In the suppurative meningitides the values range between 600 and 700 mg. per 100 ml. In established tuberculous meningitis the figures are quite consistently below 600 mg., usually being from 450 to 580 mg. per 100 ml. In luetic infections of the central nervous system, in epidemic encephalitis, poliomyelitis, and in practically all other diseases of the brain and cord the chloride content of the cerebrospinal fluid is but little altered. The decrease in cerebrospinal fluid chloride which occurs in inflammatory conditions of the meninges is probably due in a large measure to the increased protein content of the fluid in those conditions, the more equal distribution of protein between the blood plasma and cerebrospinal fluid resulting in a more equal distribution of chloride in those fluids (see p. 353). The decrease in plasma chloride concentration, which occurs at times in meningitis, also plays an important part in the production of this phenomenon.

Variations in the chloride content of the blood plasma are reflected in corresponding changes in the chloride content of the cerebrospinal fluid. Consequently, in such conditions as lobar pneumonia and upper intestinal or pyloric obstruction, which are commonly associated with hypochloremia, the concentration of chloride in the cerebrospinal fluid is correspondingly decreased. The plasma chloride concentration must, therefore, be considered in interpreting subnormal values for cerebrospinal fluid chloride, particularly in acute infectious diseases such as pneumonia, in which the development of symptoms of meningeal irritation may arouse suspicion of the presence of meningitis.

INORGANIC PHOSPHATE

The inorganic phosphate content of cerebrospinal fluid averages approximately 30 to 50 per cent of that of blood serum, ranging between 1 and 2 mg. per 100 ml. in adults and from 1.5 to 3.5 mg. in children. It is increased in all forms of meningitis, in nephritis and uremia with phosphate retention, and has been found to be increased in conditions associated with degenerative processes in the brain and cord, such as tumor, tabes and paresis. The determination of the cerebrospinal fluid phosphorus concentration has no practical value.

CHOLESTEROL

Most investigators have found that the normal cerebrospinal fluid contains either no cholesterol or minute traces only. An increase occurs in many diseases of the central nervous system, including meningitis (trace to 12 mg.), brain tumor and abscess (5 to 15 mg.), and cerebral hemorrhage (5 to 20 mg.). In the majority of cases of syphilis of the central nervous system the cholesterol content of the fluid is not appreciably increased. In various mental disorders, figures ranging from 0.2 to 0.7 mg. per 100 ml. have been reported.

LACTATE

The cerebrospinal fluid normally contains no lactate. If allowed to stand at room temperature, lactate is formed in the process of glycolysis of the glucose present in the fluid. In all conditions in which the sugar content of the fluid is decreased, lactate will be found to be correspondingly increased. It exists in highest concentration, therefore, in the suppurative meningitides and, to a lesser extent, in tuberculous meningitis. It has also been found in patients with uremia.

HYDROGEN ION CONCENTRATION

The hydrogen ion concentration of cerebrospinal fluid examined immediately after withdrawal is almost the same as that of the blood, the pH varying normally from 7.4 to 7.6. Levinson has found that upon standing, exposed to air, the pH increases, due to the loss of CO_2 from the fluid. The CO_2 combining power is also practically identical with that of blood plasma (55 to 75 volumes per cent).

The blood-brain barrier to some extent hinders the passage of HCO_3^- ions. In disturbances of hydrogen ion regulation, the pH changes in the cerebrospinal fluid do not immediately match those in the circulation. There is a tendency for the disturbance in the cerebrospinal fluid to lag behind that in the blood. The same state of affairs occurs in relation to corrective processes, whether homeostatic or therapeutic.

The pH of the fluid in tuberculous meningitis is usually within normal limits (7.4 to 7.6), the pH increasing upon standing, as in normal fluid. In the suppurative meningitides the fluid is slightly more acid (pH 7.2 to 7.5), the increase being due, in all probability, to the presence of lactate. Upon standing, the acidity shows much less tendency to decrease than in the case of normal fluids, actually increasing in some instances (Levinson). This is perhaps due either to an increased rate of glycolysis or fermentation or to increased production of CO_2 by the cells present in the fluid, balancing the loss of CO_2 which occurs upon standing exposed to air.

SODIUM, POTASSIUM, CALCIUM AND MAGNESIUM

Determination of the sodium and potassium content of cerebrospinal fluid is, as far as is known, of no practical value. Figures for sodium range from 129 to 153 mmol./l. and those for potassium from 2.06 to 3.86 mmol./l., being essentially the same as the values for the concentrations of these elements in the blood plasma under normal and abnormal conditions.

The calcium content of normal cerebrospinal fluid varies from 0.98 to 1.30 mmol./l. It is increased in practically all cases of meningitis and epidemic encephalitis. This is due to the fact that in the presence of protein in tissue fluids their calcium content is increased by an amount proportional to the quantity of protein present, the added calcium being nondiffusible in nature and the diffusible fraction being essentially unaltered. The terms "diffusible" and "nondiffusible," as here employed, refer to the ability to pass through a semipermeable artificial membrane or the capillary wall. The normal cerebrospinal fluid calcium is all in ionized form and, under normal conditions, is quantitatively virtually identical with the diffusible fraction of serum calcium. This correspondence between values for cerebrospinal fluid calcium and diffusible calcium of the blood serum as determined by artificial membrane methods is, however, lost under conditions of abnormal serum calcium concentration. For example, marked decrease or increase in serum calcium produced by parathyroidectomy or administration of parathyroid hormone is accompanied by little or no change in the cerebrospinal fluid calcium concentration. Instances of hypoparathyroidism have been reported in which the serum and spinal fluid calcium concentrations were identical (4.5 mg. per 100 ml.). Low values for spinal fluid calcium may be obtained in patients with renal failure, the decrease being roughly proportional to the increase in spinal fluid phosphate concentration.

Reported values for magnesium concentration of the spinal fluid are slightly higher than those for normal blood serum, averaging about 1.12 mmol./l. (range 0.23-2.01). It would appear that changes in the spinal fluid magnesium concentration are not completely dependent upon variations in its concentration in the blood. Increased, normal and decreased values for spinal fluid magnesium have been reported in various forms of meningitis.

XANTHOCHROMIA

Xanthochromia is a term applied to a condition in which the spinal fluid exhibits a clear yellow color. In most cases the pigment responsible for this discoloration is bilirubin. The intensity of color may be expressed in terms of milligrams of bilirubin or, more simply, by estimation of the icterus index.

Xanthochromia may be due to hemorrhage into the subarachnoid space from any cause. If blood is present in the subarachnoid space for more than a few hours, the cerebrospinal fluid, following centrifugation, will exhibit a clear yellow color above the layer of packed red cells. This observation is at times of value in differentiating between pathologic hemorrhage into the subarachnoid space and the introduction of blood into the cerebrospinal fluid by injury to the subdural venous plexus at the time of puncture, provided there is no significant contribution by the bilirubin in the blood. The development of the yellow color is due to the occurrence of hemolysis with subsequent transformation of hemoglobin to a pigment closely resembling, if not identical with, bilirubin. If the blood has been present for more than a week the yellow color becomes much more intense and then gradually fades, due probably to reabsorption of pigment, finally disappearing entirely in from one to two weeks, depending upon the extent and duration of the hemorrhage.

Xanthochromia may occur in compression of the cord by intramedullary tumors or by extradural compression, as in the case of tuberculosis of the vertebrae; it is particularly frequently observed in tumors in the region of the cauda equina. Under such circumstances a meningeal pocket is formed and venous stasis occurs with consequent capillary hemorrhage into the cerebrospinal fluid. A similar condition occurs in some cases of acute myelitis and in chronic meningitis with the formation of adhesions. In such cases, particularly in cord tumors, there is, in addition, interruption of the cerebrospinal fluid circulation and transudation of serum in the area of venous and capillary stasis. The cerebrospinal fluid under such circumstances contains large quantities of protein. The term "Froin syndrome" is applied to xanthochromia in a fluid of high protein content which may coagulate spontaneously. This syndrome, because of its frequent occurrence in cases of tumor of the spinal cord, is of considerable diagnostic import.

Xanthochromia of varying degree occurs as a physiologic phenomenon in practically all premature infants during the first two months of life. About 60 per cent of such fluids respond to the indirect van den Bergh reaction, the icterus index being about 1 during the first twenty-four hours of life, increasing to reach a maximum of about 4 during the second week, and then gradually decreasing to disappear in about eight weeks or earlier, depending upon the degree of immaturity of the infant. In cases unassociated with hemorrhage into the subarachnoid space, the xanthochromia of premature infants appears to be related to the increased bilirubin content of the blood during that period (icterus neonatorum) and to increased permeability of the blood plasma-cerebrospinal fluid barrier in premature infants and in mature infants during the first few days of life.

In adults, hepatocellular or obstructive jaundice, even of severe degree, may or may not be accompanied by visible xanthochromic discoloration of the cerebrospinal fluid. Nevertheless, except in mild cases, the presence of bilirubin can be demonstrated almost invariably by means of the quantitative van den Bergh procedure. Even in severely jaundiced subjects, its concentration in the cerebrospinal fluid seldom exceeds 1.0 mg./100 ml., being usually below 0.5 mg. The relative amounts of conjugated and free bilirubin reflect those in the blood plasma; i.e., conjugated bilirubin (p. 640) predominates in the cerebrospinal fluid of patients with hepatocellular and extrahepatic obstructive jaundice.

ENZYMES

Several enzymes present in the blood plasma occur also in the cerebrospinal fluid. These include aminotransferases, lactate dehydrogenase and glucose phosphate isomerase.

Isomerase activity (normally < 38 U/l. at $37°$ C.) is increased in about 60 per cent of subjects with malignant brain tumors, primary or secondary, and seldom in those with benign tumors. This may occasionally be useful in differential diagnosis. Elevations occur also, but less consistently, in meningitis and cerebral thrombosis.

Cerebral vascular accidents are frequently accompanied by increase in cerebrospinal fluid lactate dehydrogenase (normally < 30 U/l. at $25°$ C.)

and aspartate aminotransferase (normally < 10 U/l. at $25°$ C.). The former is occasionally increased in malignant brain tumors, the latter seldom.

In patients with convulsive disorders, increased levels of creatine kinase, aspartate aminotransferase and lactate dehydrogenase have been found in cerebrospinal fluid and in blood.

In those with carcinomatous neuropathies, frequently there are increased levels of aspartate aminotransferase in the fluid.

In the lipoidoses, blood and cerebrospinal fluid levels of lactate dehydrogenase and aspartate aminotransferase are elevated in Tay-Sachs disease, and only the aminotransferase shows increases in Niemann-Pick disease. Cerebrospinal fluid aldolase levels are frequently raised in all central nervous system lipoidoses.

The changes occurring in cerebrovascular accidents, bacterial meningitis, Korsakoff's psychosis and Wernicke's encephalopathy have been described elsewhere (p. 577).

References

Davies-Jones, G. A. B.: Cerebrospinal fluid enzyme abnormalities. *In* Studies in Clinical Enzymology. London, William Heinemann Ltd., 1969.
Davson, H.: Physiology of the Cerebrospinal Fluid. London, J. & A. Churchill, 1967.
Fossard, C., Dale, G., and Latner, A. L.: Separation of the proteins of cerebrospinal fluid using gel electrofocusing followed by electrophoresis. J. Clin. Path., *23*:586, 1970.
Lord Brain, and Walton, J. N.: Diseases of the Nervous System. Ed. 7, London, Oxford University Press, 1971.
Merritt, H. H., and Fremont-Smith, F.: The Cerebrospinal Fluid. Philadelphia, W. B. Saunders Co., 1937.
Wolf, P. L., Williams, D., and Von der Muehll, E.: Practical Clinical Enzymology and Biochemical Profiling, New York, John Wiley and Sons, 1973.

Chapter 25

HISTORY AND DEVELOPMENT OF CLINICAL BIOCHEMISTRY

The crude examination of body fluids, including urine, probably dates back to the time of Hippocrates (c. 460–c. 375 B.C.). At a much later date, the iatro-chemists made more intensive studies of body fluids, and in the second half of the 17th Century, Thomas Willis wrote a dissertation on urine and used the presence or absence of a sweet taste to differentiate between the two varieties of diabetes. In 1684, Robert Boyle listed headings under which blood should be examined; these included observations on blood coagulation. He also listed a number of headings for the examination of urine.

Boerhaave of Leyden (1688–1738), who has had lasting influence on medicine, taught quite emphatically his belief that medical phenomena could be interpreted in terms of chemistry. This undoubtedly influenced greatly the whole contemporary medical outlook. Physicians had long been interested in urinary calculi, but another hundred years elapsed before Wollaston described five chemically distinct types. In 1810, he added a sixth variety, composed of what he called cystic oxide, now known as cystine. By 1825, much of our present-day knowledge in relation to the chemical composition of such calculi already existed. In 1836, Richard Bright demonstrated the association of albuminous urine with kidney disease. This he did by heating urine in a spoon and demonstrating that cloudiness developed just before boiling.

In the 19th Century, the work and writings of Justus Liebig profoundly influenced the application of chemistry to biological problems, including those of medicine. His students included Henry Bence Jones, who lectured in chemistry at St. George's Hospital, London, and whose name is associated with the "protein," and J. W. L. Thudichum, who became lecturer in chemical pathology at St. Thomas' Hospital, London. The latter, in addition to working with brain lipids, also did a good deal of work with urinary pigments and hematin. Just over 100 years ago, Alfred Garrod devised one of the first quantitative analytical methods applied to blood. He found that blood uric acid crystallized quantitatively on threads of huckaback placed in sera in watch glasses on his consulting-room mantelpiece. A number of well-known qualitative tests for urine originated at about this time. These included Fehling's test for sugar, Jaffe's tests for indican and creatinine, Gerhardt's test for ketone bodies, Millon's test for protein, Gmelin's test for bile pigments and Hay's test for bile salts. In fact, by the end of the 19th Century much of modern urine testing had already been devised.

It was undoubtedly at the beginning of the 20th Century that "Clinical Chemistry" really began to develop into the discipline which we know today. During the first two decades, laboratory examination was mainly confined to gastric contents and urine. The former was examined for free hydrochloric and lactic acids and also for free and total acidity, which were often expressed as degrees of acidity. With urine, quantitative tests were carried out in regard to sugar and urea, and a series of qualitative tests for acetone, albumin, bile and sugar were performed. Osazones were occasionally prepared and urinary deposits were examined for crystals. Calculi and cyst fluids were also occasionally investigated. Additional quantitative methods developed for urine were the Wohlgemuth technique for diastase in 1908, Benedict's qualitative and quantitative solutions for urinary sugar in 1909 and 1911 and the alkaline picrate method for creatinine, published by Folin in 1904.

Techniques now began to appear, mainly in the United States, for quantitative estimations applied to blood. In 1912, Folin and Denis described the phosphotungstic acid reagent for uric acid; in 1913, Bang described his micro-method for blood sugar, and van den Bergh used Ehrlich's diazo-reagent for the determination of bilirubin. Further micro-methods for sugar estimation soon appeared. These included that of MacLean in 1915 and that of Shaffer and Hartman in 1921. In 1914, Van Slyke and Cullen published the urease-aeration method for the estimation of blood urea. In 1914, methods appeared for the estimation of fat in feces, either wet or dry. In 1915, MacLean and Van Slyke published their iodometric method for chloride, and a year later van den Bergh distinguished between the direct and indirect bilirubin reactions. In the same year, the Liebermann-Burchard reaction was applied to the determination of cholesterol in blood; digitonin was used to separate free from ester cholesterol. Van Slyke's volumetric apparatus for carbon dioxide combining power was published in 1917 and was used for the study of oxygen capacity in 1918 and for the determination of carbon monoxide in 1919. Folin and Wu published their *System of Blood Analysis* in 1918. This included the determination, in their well-known tungstic acid filtrate, of sugar, creatinine and creatine, nonprotein nitrogen, urea and uric acid. In 1918, there appeared the cobaltinitrite method for potassium and in 1920 a gravimetric technique for the estimation of sodium. The iodometric titration variant of the latter appeared in 1924. Using hydroquinone as reducing agent, inorganic phosphorus was determined in 1920 by a colorimetric method similar to those in use today. Kramer and Tisdall's permanganate titration method for blood calcium appeared in 1922. The investigation of cerebrospinal fluid received a great stimulus by the use of the Lange gold curve in 1912. There is no doubt that the newly introduced clinical application of insulin to the treatment of diabetes greatly stimulated urine and blood sugar estimations as well as those of electrolytes. As a result of the latter, much was added to therapy in relation to diabetic coma.

In the years between the two world wars, the range of quantitative tests performed in the clinical chemistry laboratory increased remarkably. It included estimation of the blood sugar, reducing substances in urine, carbon dioxide combining power of blood, the glucose tolerance test, proteins in urine, blood urea, serum calcium, cholesterol, chloride, creatinine and creatine, uric acid and phosphate. The B.M.R. (basal metabolic rate) was also estimated as well as urea concentration tests, the phenol red excretion test, the fractional test meal and the test for occult blood. Pancreatic disease was studied by means of urinary diastase, fecal fat and muscle fibers in the feces. Clinical study of liver disease involved the van den Bergh reaction, the urine bilirubin and urobilin, the levulose tolerance test and a dye removal test using phenoltetrachlorophthalein. In relation to neurological disease, cerebrospinal fluid was fairly well investigated. Various kinds of calculi were analyzed and pigments in urine identified. Not unnaturally, this period was marked by the appearance of a number of texts dealing with practical clinical chemistry. These included Harrison's *Chemical Methods in Clinical Medicine* and Stewart and Dunlop's *Clinical Chemistry in Practical Medicine,* which appeared in 1930. A year later, the two volumes of Peters and Van Slyke were published. These three publications undoubtedly formed the foundation for the numerous textbooks on the subject in circulation today.

During the interwar period, color development was measured by visual colorimeters. In the late thirties in the United States and in the early forties in the United Kingdom, there began to appear various types of photoelectric colorimeters, which resulted in great speeding up of routine clinical chemical work, as well as an undoubted increase in accuracy. These photo-electric colorimeters were undoubtedly the forerunners of automation. It would be wrong at this stage to omit the name of E. J. King, who played so great a part in the development of these instruments as well as of numerous microchemical estimations which depended on them. The clinical chemistry laboratory was now equipped to deal with far larger numbers of specimens than in the past. Another great advance, in the middle and late forties, was the appearance of the flame photometer, which enabled the estimation of sodium and potassium in body fluids to be done quite quickly and remarkably accurately. The speed of estimation meant that clinicians could be quickly provided with electrolyte results applicable to patients whom they were treating, and this had an undoubtedly profound influence on the lowering of the mortality rate from such diseases as diabetic coma.

In the 1920's and 1930's, Somogyi undertook his important work on blood glucose estimations, which played a large part in the diagnosis and therapy of diabetes mellitus.

In 1938, Callow published the use of the Zimmerman reaction to determine 17-ketosteroids. This was the first of the steroid determinations applied for routine purposes. The application of the Norymberski techniques for corticosteroids occurred in the fifties. These were rapidly followed by the development of reliable techniques for pregnanediol, pregnanetriol and estrogens. Satisfactory methods were developed for the estimation of catecholamines and their metabolites, particularly in urine. Urinary hormone assay had now become a recognized portion of the work of the clinical biochemist. The more recent development of radioimmunological assay and protein-binding techniques has led to the estimation of the hormones themselves, both in blood and other biological fluids. These methods are now used for the estimation of therapeutic substances in the plasma and other tissue fluids.

In the early thirties, Bodansky as well as King and Armstrong introduced their techniques for phosphatase estimation, and in 1938 the Gutmans modified the latter for use in the estimation of acid phosphatase. Specific inhibition techniques then led to the estimation of prostatic acid phosphatase itself, which played a great part in the diagnosis and control of therapy of prostatic carcinoma. The range of enzymes estimated in the blood rapidly increased and began to include such assays as those for aminotransferases, lactate dehydrogenase and a number of other intracellular enzymes. This led to a great advance in the diagnosis of myocardial infarction and certain types of hepatic disease. Isoenzyme determinations added to this. Increasing experience with enzymes also led to more specific methods of assay of substances in biological fluids, for example, the enzymatic determination of blood glucose.

In the early forties, the use of dithizone came into vogue. This enabled accurate estimations of lead, and at a later date, of mercury, in urine. Techniques also became available for the estimation of elements such as iron in the blood. Interest in the trace elements increased greatly with the advent of the atomic absorption spectrophotometer, which quantitates trace elements rapidly and accurately for routine purposes.

Free boundary electrophoresis was first described in 1892. The techniques of Tiselius published in 1937 enormously increased our knowledge of proteins in serum and other biological fluids. We could now get more insight into antibodies, as well as into protein disorders such as myelomatosis. The advent of zone electrophoresis occurred in 1886, but it did not become widely popular until after 1950. This saw the introduction of paper electrophoresis, which was soon followed by electrophoresis in cellulose acetate and various gels. Immunoelectrophoresis was put on a firm basis from 1953 by Grabar and his associates. The paraproteinemias could now be studied in detail and, in relation to other disorders, individual serum proteins could be investigated, not only by immunoelectrophoresis, but also by other immunological techniques, including radioimmunoassay.

Absorption chromatography was described by Tswett in 1906, ion-exchange chromatography by Adams and Holmes in 1936, partition chromatography by Martin and Synge in 1941, paper chromatography by Consden, Gordon and Martin in 1944, thin-layer chromatography by Mottier and his colleagues in the early fifties, and gas-liquid chromatography by James and Martin in 1952. All these techniques are in current use by clinical chemists and have greatly increased the range of substances which can be investigated in relation to disease. The Stein and Moore column ion-exchange technique for amino acids, 1954, has led to the development of automatic apparatus for amino acid analysis in biological fluids for the clinical study of the aminoacidurias, metabolic changes after surgery and a host of other problems.

In the fifties and sixties, the use of the Astrup technique improved the control of respiratory disorders, open-heart surgery and renal dialysis. Microbiological methods of assay were used to estimate vitamins, such as vitamin B_{12}, and more recently to estimate amino acids for the study and detection of hereditary metabolic disorders.

All these developments have led to an enormous increase in the number of estimations undertaken in clinical chemistry laboratories. The advent of the AutoAnalyzer and a number of other types of automatic analytical equipment has therefore been of considerable importance. The increase in work load now merits the use of computers for data processing and laboratory control.

Once clinical chemistry had achieved the status of a recognized discipline in its own right, national societies devoted to the subject were formed in various countries. The first of these was that formed in the Netherlands in 1947 and was soon followed by those in the United States of America and in the United Kingdom. Societies of clinical chemistry have now grown up all over the world. The United States gave birth to the first journal, *Clinical Chemistry,* in 1955. This was soon followed by *Clinica Chimica Acta,* and there are now numerous journals published in many parts of the world. In addition, there are a number of excellent books dealing with details of analytical methodology.

Once national societies had been formed, it was realized that problems related to clinical chemistry should be considered at an international level. This has led to the formation of the Clinical Chemical Section of the International Union of Pure and Applied Chemistry (IUPAC) and the International Federation of Clinical Chemistry (IFCC). These organizations have a number of joint committees. Much of the account of the history of clinical biochemistry given in this chapter was originally produced for the Joint Commission on Education of these two bodies.

The discipline, of course, continues to advance. High resolution analysis is developing rapidly. This employs pressure chromatography and techniques such as mass spectroscopy and electron spin resonance. Sophisticated techniques of this kind will enable the clinical chemist to estimate relatively large numbers of metabolites in relatively small amounts of blood, tissue fluid, tissue culture and biopsy samples. Such development must certainly lead to greater understanding of disease processes as well as to more accurate diagnostic procedures and therapeutic control, since they will provide metabolic profiles far more important than the somewhat arbitrary biochemical profiles carried out on serum, whole blood or plasma. The latter have, however, played an important part in regard to our attitude to what used to be referred to as "the normal range." To perform these tests in sufficient number, automatic equipment is necessary. This has greatly increased accuracy and enabled laboratories to perform very large numbers of analyses. This in turn has provided sufficient statistical data amply to demonstrate that the levels of blood constituents are dependent on a variety of factors. So much is this the case that it is now becoming almost mandatory to abolish the old-fashioned concept of "normal range" and replace it by "reference values."

THE "NORMAL RANGE"

So-called normal ranges were at first determined on specimens obtained from healthy medical students or laboratory staff. It soon became apparent, however, that relatively small groups had been used and, in any case, factors such as bedrest could affect certain important biochemical values. A little more than 20 years ago, the author's laboratory, and no doubt other laboratories, began to use ranges determined on hospital patients. For example, in order to determine the normal range of blood urea, samples were obtained from a large group of patients not suffering from conditions in which the blood urea was expected to be abnormal. Others demonstrated that it was really necessary to separate many in-patient values from those for out-patients, since these were not at bedrest. Statistical analysis related to Gaussian distribution was employed and a normal range expressed as a mean value plus or minus two standard deviations, to give 95 per cent confidence limits. Groups of at least 300 patients were used. Where distributions did not appear to fit a Gaussian distribution, it could be demonstrated that plotting the logarithm of the concentration gave a curve closely approximating the correct shape, and 95 per cent confidence limits were expressed in terms of logarithmic values. Whether or not this was justified is open to serious question, since the apparent correction in shape of the curve depended on the fact that logarithms of small numbers are more or less proportional to those numbers, whereas larger numbers have logarithms with relatively smaller differences.

With hospital patients, the cumulative frequency can be plotted on probability paper, and only the straight line portion used to calculate the Gaussian distribution. Even this manipulation is open to criticism, and in any case, values (ordinary or logarithmic) for certain normal blood constituents do not give a Gaussian distribution. It seems safer just to collect a large number of values and determine 95 per cent confidence limits by cutting off the 2.5 per cent at the lowest and highest levels. Here again there is difficulty, since it is not agreed by all that this type of cut-off can be fully justified and is especially difficult when a high proportion of values is in the region approaching zero, the differences being well within experimental error.

Whatever method is used to determine ranges, it soon becomes obvious that these values are dependent on a whole variety of factors, including method of collection and handling of the sample for analysis, time of collection, seasonal changes, laboratory analytical method employed, laboratory accuracy, patient's age, sex, ethnic group, social class, diet, physiological factors such as pregnancy or environment, and so on. There is also the effect of therapeutic agents on the analytical method employed as well as on the actual blood levels of certain constituents. Many of these variables can be standardized, and then it is possible to obtain so-called reference populations. These must be defined in terms of the other factors, such as sex, age, ethnic group, social class, etc. It should be pointed out here that in this respect biochemical values resemble many other physical signs obtained at the bedside or by ancillary methods of investigation.

It will take a considerable period of time, and much computer study, to determine adequately the appropriate number of reference values. It is for this reason that so-called normal ranges have been quoted throughout this book. There is no doubt that revised values will have to be considered in subsequent editions and that eventually the concept of "normal range" will give way to "reference values" or "reference intervals."

There is at least one other important factor which is often not considered by those without adequate clinical experience. This is the effect of disease itself on biochemical values, which in themselves are not diagnostic of that disease. For example, in renal disease, in addition to an increase of blood urea there is frequently an increase in blood urate. There is a rough clinical correlation. The clinician learns to associate certain blood urate levels with appropriate increases in blood urea. If the level of the former is too high in relation to the latter, then the possibility of gout is considered. The latter disease can also be associated with an increase in the blood urea, as well as of blood urate, and a similar reasoning process must be employed.

All this discussion indicates that, in relation to diagnosis, biochemical values are merely additional, and frequently very important, physical signs. Biochemical findings, as with most physical signs, have significance only in relation to the history of the patient's illness and the physician's findings, at the bedside as well as from ancillary methods of investigation.

Special difficulty is encountered in attempting to obtain normal reference values in the elderly. In this group, disease is so frequently present that it is extremely difficult to define the "normal" state. In pediatrics, possibly because of difficulties of sample collection and emotional obstacles, we are only just beginning to obtain meaningful values.

In all groups of patients, in order to obtain meaningful "reference intervals" it will be necessary to standardize analytical methodology. This does not necessarily mean that laboratories throughout the world will be expected to use identical analytical methods. It does mean that it will be necessary to set up reliable reference methods, by which other methods can be standardized. This is particularly the case in relation to enzyme determinations. It is also important to report laboratory findings in terms of standard units. It has been decided, for this purpose, to adopt SI units.

SI UNITS

In 1960, the Conférence Générale des Poids et Mesures confirmed the Système International d'Unites (SI), originally adopted in 1954. This system is now based on eight base units related to eight basic kinds of quantity.

In spite of the fact that the katal has been included in Table 25-1, it should be pointed out that a number of authorities believe that the concept catalytic amount is not really viable and that even if this were so, its unit would really be derived rather than base. It is probably a consensus of opinion that we should talk about catalytic activity and express this in terms of mole/s. One could then introduce the concept catalytic activity per liter (derived coherent unit), which would give a figure which would be the same as if the katal were employed.

The *mole* (p. 347) and the *katal* (p. 548) have been defined previously. In order to define the *kelvin,* it is merely necessary to state that 273.16 K represents the temperature interval between the absolute zero and the triple point of water. Temperature is more commonly expressed in terms of Celsius temperature, which is the thermodynamic temperature minus 273.15 K, since the Celsius temperature at the triple point of water is 0.01 degree Celsius (0.01° C). The former term "centigrade" should no longer be used. The steam point Celsius temperature is 100° C above the Celsius triple point of water.

Other units can be derived from the base units.

Derived Coherent Units. These are units constructed exclusively from base units, for example, kilogram per cubic meter or mole per kilogram. In this category, unit area (unit length × unit length) or unit velocity (unit length divided by unit time) can also be included. The coherent unit of volume is the cubic meter. For the purposes of clinical

TABLE 25-1 BASE UNITS OF SI SYSTEM

KIND OF QUANTITY	SI BASE UNIT	SYMBOL FOR UNIT
Length	meter	m
Mass	kilogram	kg
Time	second	s
Electric current	ampere	A
Thermodynamic temperature	kelvin	K
Luminous intensity	candela	cd
Amount of substance	mole	mol
Catalytic amount	katal	kat

biochemistry, the liter has been retained as the unit and redefined as exactly one-thousandth part of a cubic meter, or in other words as a cubic decimeter.

Derived Noncoherent Units. These are units constructed from base units and numerical factors, for example, milligram per deciliter. In other words the numerical factors are multiples or submultiples, which can be denoted by prefixes as shown in Table 25-2.

Certain derived coherent units are used in clinical biochemistry. The unit of force is the *newton* (N) and $1 N = 1$ kg m/s^2. The unit of energy is the *joule* (J) and $1 J = 1$ Nm: 4.2 kj corresponds to 1 kcal (kilocalorie), and the latter unit, which is the Medical Calorie, will be phased out. The unit of power is the *watt* (W) and $1 W = 1$ J/s. The electrical units are the *volt* (V), the *farad* (F) and the *ohm* (Ω): $1 V = 1$ W/A; $1 F = 1$ A/s/V; $1 \Omega = 1$ V/A. The unit of pressure is the *pascal* (Pa), which has already been defined (p. 434).

Concentration can be expressed in two types of unit. *Mass concentration* is the mass of the component divided by the volume of the mixture (kg/l, g/l, mg/l, μg/l, etc.) *Substance concentration* is the amount of substance of the component divided by the volume of the mixture (mol/l, mmol/l, μmol/l, etc.). The latter form of expression is preferable if the molecular weight of the substance is known. It can also be used for formula units (p. 347). Hemoglobin, vitamin B$_{12}$ and folate are expressed as mass concentrations because of uncertainty in relation to the exact elementary entity of hemoglobin, which should be used for calculation purposes, and because of doubt in relation to the molecular weight of the biologically active forms of the two vitamins.

Substance concentration has not been expressed in SI units in a number of chapters of this book, since expressions of concentration in common clinical usage have been employed (mg./100 ml., mEq./liter, etc.). To

TABLE 25-2 PREFIXES DENOTING DECIMAL FACTORS

FACTOR	NAME	SYMBOL
10^{12}	tera	T
10^{9}	giga	G
10^{6}	mega	M
10^{3}	kilo	k
10^{2}	hecto	h
10^{1}	deca	da
10^{-1}	deci	d
10^{-2}	centi	c
10^{-3}	milli	m
10^{-6}	micro	μ
10^{-9}	nano	n
10^{-12}	pico	p
10^{-15}	femto	f
10^{-18}	atto	a

convert these values into substance concentration in SI units (mmol/l. μmol/l, etc.), it is necessary to multiply by certain factors, some of which are shown in Table 25-3.

It is now correct, as will be seen from Table 25-1 and the paragraph defining certain derived coherent units, to omit the period punctuation mark from the symbol for any unit. For various reasons, the period has been included in the main part of the text of this edition. It will be omitted in the future.

The multiplication factor is easily derived from the former ways of expressing mass concentration, since it equals $\frac{n}{V \times MW}$, where n is the numerical value in a stated volume V (measured in liters) and MW is the molecular weight of the substance concerned. When concentration has been expressed as mEq/l, it is merely necessary to divide by the valency of the component being determined.

Certain standard abbreviations are also recommended, as shown below.

Whole blood	B
Arterial blood	aB
Venous blood	vB
Day	d (preferably 24 h)
Fasting measurement	f Pt
Feces	F
Night	n
Plasma	P
Serum	S
Spinal fluid	Sp
Urine	U

TABLE 25-3 MULTIPLICATION FACTORS FOR CONVERSION TO SUBSTANCE CONCENTRATION IN PLASMA

SUBSTANCE	CURRENT CONCENTRATION EXPRESSION	SUBSTANCE CONCENTRATION	MULTIPLICATION FACTOR FOR CONVERSION TO SUBSTANCE CONCENTRATION
Amino acid N	mg/100 ml	mmol/l	0.714
Ammonium	μg/100 ml	μmol/l	0.587
Ascorbate	mg/100 ml	μmol/l	56.8
Bilirubin	mg/100 ml	μmol/l	17.1
Calcium ion	mg/100 ml	mmol/l	0.250
	mEq/l	mmol/l	0.5
Cholesterol	mg/100 ml	mmol/l	0.0259
Copper	μg/100 ml	μmol/l	0.157
Creatinine	mg/100 ml	μmol/l	88.4
Glucose	mg/100 ml	mmol/l	0.0555
Iron	μg/100 ml	μmol/l	0.179
Lead	μg/100 ml	μmol/l	0.0483
Magnesium	mg/100 ml	mmol/l	0.411
	mEq/l	mmol/l	0.5
Phosphate (inorganic P)	mg/100 ml	mmol/l	0.323
Protein-bound iodine	μg/100 ml	mmol/l	78.8
Sodium ion	mEq/l	mmol/l	1.0
Urate	mg/100 ml	mmol/l	0.0595
Urea	mg/100 ml	mmol/l	0.166

SYSTEM	COMPONENT	KIND OF QUANTITY	NUMERICAL VALUE	UNIT
serum	calcium (total)	substance concentration	2.5	mmol/l

A laboratory report of a measurement should ideally always indicate (1) the system, (2) the component, (3) the kind of quantity, (4) the numerical value and (5) the unit.

It is doubtful whether in actual fact the words "substance concentration" will be included in the average laboratory report.

References

Caraway, W. T.: The scientific development of clinical chemistry to 1948. Clin. Chem. *19*:373, 1973.
Henry, R. J.: Clinical Chemistry Principles and Techniques. Ed. 2. New York, Harper & Row, 1973.
IUPAC & IFCC Information Bulletin: Quantities and Units in Clinical Chemistry, 1972.
Robinson, R.: Clinical Chemistry and Automation. London, Chas. Griffin & Co. Ltd., 1971.
Tietz, N. W. (ed.): Fundamentals of Clinical Chemistry. Philadelphia, W. B. Saunders Co., 1970.
Young, D. S., Thomas, D. W., Friedman, and Pestaner, L. C.: Bibliography: Drug Interferences with Clinical Laboratory Tests. Clin. Chem., *18*:1041, 1972.

Index

Page numbers in *italics* refer to figures; page numbers followed by (t) refer to tables.

Abbreviations, standard for laboratory reports, 856
Abdominal tumors. See *Tumors, abdominal.*
Abortion, missed, plasma fibrinogen and, 184
Acanthocyte(s), a-beta-lipoproteinemia and, 136
Acetate, carbohydrate metabolism and, 11-12, 17
    fatty acids and, 101-103
    lipid metabolism and, *102*
    protein digestion and, 151
Acetazolamide, diuresis and, 469
Acetic acid, carbohydrate metabolism and, 12
    serum cholinesterase and, 552
Acetoacetate, carbohydrate metabolism and, 17, *18*, 21
    diabetes mellitus and, 137-138
    ketosis and, 109
    lipid metabolism and, *102*
    metabolic interrelations and, 163(t)
Acetoacetic acid, ketolysis and, 109
    ketosis and, 109
    pancreatectomy and, 34
Acetoacetyl, fatty acid synthesis and, 103
Acetoacetyl-CoA, carbohydrate metabolism and, 21-22
    cholesterol biosynthesis and, 96
    diabetes mellitus and, 137
Acetoin, hepatic coma and, 674
Acetone, 21, 109
Acetphenetidin, methemoglobinemia and, 272
Acetyl–CoA, carbohydrate metabolism and, 12, 17, 21-22, 125
    cholesterol biosynthesis and, 96
    diabetes mellitus and, 137
    fatty acid synthesis and, 103
    lipogenesis and, 123
    tricarboxylic acid cycle and, 107
Acetylcholine, serum cholinesterase and, 552
N-Acetylgalactosamine, 92
N-Acetylgalactosaminyl, 92
N-Acetylneuraminidogalactosyl-glucosyl ceramide, 92
Acetylthiocholine, serum cholinesterase and, 552

Achlorhydria, gastric investigations and, 584
    gastric response in, 585
    glucose response and, 3, 70
    hydrochloric acid and, 585-586
    iron absorption and, 321
    pepsin activity and, 587
Achylia, 584, 587
Acid(s), catalytic, 15
    deficit of, alkalosis and, 422-424
    excess of, 416-417
    excretion of, nonrespiratory acidosis and, 420
    gastric secretion and, 582
    hydrogen ions and, 400, 415(t)
    nonvolatile, buffer systems for, 405-406
    nucleic. See *Nucleic acids.*
    production of, gastric function and, 580-582
    renal excretion of, 407-408
    urinary, acid-base balance and, 430
Acid-base balance, 399-453. See also *Hydrogen ions, concentration of.*
    abnormal, 413-415, 415(t)
    acidosis and, 415-421
    alkalosis and, 421-425
    ammonia and, 411
    bone and, 406
    buffer systems and, 402-412
    calcium concentration and, 291
    chloride and, 405
    disturbances of, 425-426
    hemoglobin and, 402-403
    methods of study, 426-432
    renal regulation and, 407
    respiratory regulation and, 406
Acid lipase, deficiency of, 130(t)
Acid mucopolysaccharides, urea cycle disorders and, 232-234
Acid phosphatase, 566-569
Acidemia, blood sugar and, 53
    calcium and, 307
    defined, 399
    diabetic, plasma albumin and, 185
    glucose tolerance and, 59
    ketoacidosis and, 78-79
    ketosis and, 109
    pancreatectomy and, 34
    urinary, ammonia and, 229

Acidosis, 415-421
  alveolar air CO_2 tension in, 430
  ammonia formation and, 221, 229, 411, 430
  biochemical characteristics of, 417, 418(t)
  blood sugar and, 53
  blood volume and, 392
  chloride and, 530
  compensated, 430
  cystinosis and, 538
  defined, 399
  diabetic, ketoacidosis and, 78-79
    potassium and, 387
  dilutional, nonrespiratory acidosis and, 421
  equation for, 415
  Fanconi syndrome and, 311
  glucose and, 53
  glucose tolerance and, 59
  glycogen and, 53
  glycosuria and, 87
  hydrogen ion regulation and, 415-421
  hyperphosphatemia and, 310, 529
  ketosis and, 109, 431
  kidney disease and, 417, 529, 538, 539
  lactic, nonrespiratory acidosis and, 415(t), 419
  Lowe syndrome and, 538
  nitrogen partition and, 221, 229
  nonrespiratory, base excess and, 417-421
    base reduction and, 421
    causes of, 415(t)
    compensatory mechanisms in, 420
    hydrogen ion regulation and, 415(t)
  pancreatectomy and, 34
  potassium and, 383, 412
  renal failure and, 529-530
  renal tubular, 538-539
    aminoaciduria and, 224
    electrolytes and, 376
    hypophosphatemia and, 311
    magnesium and, 317
    nonrespiratory acidosis and, 421
  respiratory, 416-417
    hydrogen ion regulation and, 415(t)
Aciduria, congenital methylmalonic, vitamin B_{12} and, 821
  orotic, nucleic acids and, 259
Acinar cells, islands of Langerhans and, 590
  secretion of, 589
Acromegaly, glucose and, 51, 64
  hydroxyproline and, 224
  insulin resistance and, 71
  ketonuria and, 110
ACTH. See *Adrenocorticotrophic hormone.*
Actin, 16
Actinomycin D, choriocarcinoma and, 250
  protein synthesis and, 247
  Wilms' tumor and, 250
Acyl-CoA, fatty acid synthesis and, 103
  lipid deposition and, 123
Addison's disease, adenohypophyseal hypofunction and, 69

Addison's disease *(Continued)*
  adrenocortical hypofunction and, 738, 740
  adrenocortical insufficiency and, 67
  carbohydrate metabolism and, 24, 57, 162
  electrolytes and, 379
  epinephrine tolerance test in, 74
  glomerular filtration and, 457
  hypoglycemia and, 70
  plasma albumin and, 185
  potassium and, 387
Adenine, nucleoside triphosphates and, 250
Adenohypophysis. See also *Pituitary, anterior.*
  carbohydrate metabolism and, 7, 33
  growth hormone and, 705-709
  hormones of, 690-709
    adrenocorticotrophin and, 702-705
    experimental diabetes mellitus and, 33
    gonadotrophic, 691-699
    hypothalamic regulation of, 687
    melanocyte-stimulating, 702-705
    prolactin and, 700-702
    thyroid-stimulating, 699-700
  hypofunction of, glucose tolerance and, 68-69
  importance of, 697
  protein metabolism and, 164-165
  somatotropin and, 705-709
Adenoma, organic hyperinsulinism and, 54
  toxic, hyperthyroidism and, 766
Adenopathy, heavy-chain disease and, 202
Adenosine, nucleoside triphosphates and, 250
Adenosine monophosphate, carbohydrate metabolism and, 10, 27. See also *Adenylic acid.*
  cyclic, 686-687
    insulin and, 124
  triglyceride conversion and, 105
Adenosine phosphates, nucleoside triphosphates and, 250
Adenosinediphosphate, 16, *16*
Adenosinetriphosphate, adenosine monophosphate and, 10
  anaerobic metabolism and, 7
  cyclic adenosine monophosphate and, 686
  muscle contraction and, 16, *16*
Adenoviruses, nucleic acids and, 240-241
Adenylcyclase, 10, 27
Adenylic acid, carbohydrate metabolism and, 16, *16*
  muscle contraction and, 16, *16*
  nucleic acid biosynthesis and, 237
ADH. See *Antidiuretic hormone.*
Adipose tissue, nutrient metabolism and, 122-129
ADP. See *Adenosinediphosphate.*
Adrenal cortex. See also *Adrenal gland.*
  abnormal function of, 33
  atrophy of, 57
  function of, deficiency, 25
    excess, 17-18
  hormones of, 713-714

Adrenal cortex *(Continued)*
  hormones of, biosynthesis of, 714, *716*
    carbohydrate metabolism and, 7, 21, 24-25, 724-725
    cholesterol and, 96
    diabetes and, 35
    experimental diabetes mellitus and, 33
    fructose metabolism and, 43
    globulin concentration and, 190
    gluconeogenesis and, 14
    glucose utilization and, 32
    insulin deficiency and, 47
    lipid metabolism and, 725
    liver and, 665
    oxygenated, 24
    postoperative states and, 389
    potassium and, 387
    protein metabolism and, 725
    steroids and, 96
    triglyceride conversion and, 106
  insulin deficiency and, 48
  preponderance of, 22
Adrenal cortical hormones. See *Adrenal cortex, hormones of.*
Adrenal cortical hyperfunction, 730-737. See also *Cushing's syndrome.*
  adrenogenital syndrome and, 735-737
  Cushing's syndrome and, 730-735
  excess electrolytes and, 382
  glucose tolerance and, 59
  glycosuria and, 86
  insulin and, 63
  insulin tolerance test and, 71
Adrenal cortical hypofunction, 737-740
  electrolytes and, 379, 739-740
  glucose tolerance and, 67
  hypoglycemia and, 57
  mucoproteins and, 197
  potassium and, 387
Adrenal diabetes. See *Diabetes, adrenal.*
Adrenal estrogens. See *Estrogen(s), adrenal.*
Adrenal function tests, oral contraceptives and, 786
Adrenal gland(s), alarm reaction and, 727
  anatomy and embryology of, 713
  cholesterol synthesis and, 112
  cortisol and, 727
  diabetes mellitus and, 33
  phospholipids and, 112
  pregnancy and, 792
Adrenal glucocorticoids. See *Glucocorticoids, adrenal.*
Adrenal hormones. See *Adrenal cortex, hormones of, Epineprine and Norepinephrine.*
Adrenal 11-hydroxysteroids, 17
Adrenal hyperplasia, hypoglycemia and, 51, 57
Adrenal medulla, blood sugar and, 51
  hormones of, 740-746
    biosynthesis of, 741, *742*
    catecholamines and, 740-741
    metabolism of, 741-744

Adrenal medulla *(Continued)*
  hormones of, overproduction of, 744-746
    underproduction of, 744
  hyperfunction of, 744, 745
  hyperglycemia and, 51
  tumor of, 86
Adrenalectomy, 24-25
  bilateral, electrolytes and, 389
  effects of, 722
Adrenalin. See *Epinephrine.*
β-Adrenergic nerves, 24
Adrenergic receptors, catecholamines and, 743
Adrenocortical hormones. See *renal cortex, hormones of.*
Adrenocortical insufficiency. See *Adrenal cortical hypofunction.*
Adrenocorticotrophic hormone, 702-705
  abnormal secretion of, 704
  Addison's disease and, 702-703
  adenohypophysis and, 702-705
  adrenocortical hypofunction and, 737
  adrenogenital syndrome and, 735-737
  alarm reaction and, 727-729
  aldosterone secretion and, 729
  anterior pituitary insufficiency and, 57
  assays, 704
  carbohydrate metabolism and, *9*
  circadian rhythm and, 727
  cortisol and, 727
  Cushing's syndrome, 731-734
  lipid metabolism and, in adipose tissue, 123
  lipolytic mechanisms and, 124
  metabolic effects, 703, 722
  secretion of, 703
  transport of, 702-703
  unesterified fatty acid concentration and, 119
Adrenocorticotrophic hormone stimulation tests, 732
Adrenogenital syndrome, 735-737. See also *Adrenal cortical hyperfunction.*
  adrenocortical hyperfunction and, 735-737
  blood sugar and 51, 64
  diagnostic criteria for, 737
  glucose tolerance and, 64
Aerobic catabolism, 15
Aerobic metabolism, 6-13
Aerobic pathway, 12-13
Afferent arteriole, kidney anatomy and, 455
Afibrinogenemia, congenital, pregnancy and, 184
Agammaglobulinemia, globulin concentration and, 190
  immunoglobulin synthesis and, 196
  serum protein abnormalities and, 203-204
Age, calcium absorption and, 281
  glucose tolerance and, 40
  iron absorption and, 320
  lipid metabolism and, 121-122
Agonal phase, lactic acidosis and, 77
Ahaptoglobinemia, 266

INDEX

Air, pressure ratios in, 434(t)
Akee nuts, hypoglycemia and, 58
Alanine, amino acids and, 161-162
    carbohydrate metabolism and, 12, 17
    protein digestion and, 151
β-Alanine, nucleotide synthesis and, 252
Alanine pool, protein digestion and, 151
β-Alaninuria, aminoaciduria and, 228
Alarm reaction, 727-729
Albinism, phenalanine metabolism and, 226
Albumin, abnormal levels of, related disorders and, 185-190
    blood plasma and, 187
    calcium and, 288
    cerebrospinal fluid and, 835
    colloidal osmotic pressure and, 185
    decreased. See *Hypoalbuminemia.*
    diseases and, 185-190
    edema and, 394-395
    free fatty acids and, 119
    hepatitis and, 621
    hypoalbuminemia and, 186-189
    jaundice diagnosis and, 680
    liver and, 173
    liver function and, 620
    plasma fibrinogen and, 184
    plasma protein abnormalities and, 185-190
    plasma proteins and, 166-177
    properties of, 169(t)
    protein concentrations and, 168(t)
    protein nutrition and, 174
    serum, nephrotic syndrome and, 138
        renal albuminuria and, 218
    study methods and, 171(t)
    synthesis of, genetic factors and, 189
    test for, 170
    thymol turbidity and, 626
    thyroxine and, 332
    urine and, 179. See also *Albuminuria.*
Albumin-globulin ratio. See *Albumin, plasma proteins and* and *Albumin, in urine.*
Albuminuria, 214-218
    alimentary, 214
    benign, abnormal urinary nitrogen and, 214-215
        adolescent, 214
        cyclic, 214
        essential albuminuria and, 214
        intermittent, 214
        of pregnancy, 215
        orthostatic, 214
        postural, 214
        premenstrual, 214
    febrile, 216
    organic, abnormal urinary nitrogen and, 215-218
        postrenal, 216-217
        prerenal, 215-216
        proteinuria and, 214
        renal, 217-218
    related disorders and, 214-218

Albuminuria *(Continued)*
    renal tubular defects and, 534
    urinary protein studies and, 218-219
Albustix strips, albuminuria and, 214
Alcohol, gastric secretion and, 582
    iron absorption and, 322
    lipuria and, 111
    nonrespiratory acidosis and, 419
Alcoholism, anoxia and, 444
    blood sugar tolerance curve and, 63
    lactic acidosis and, 77
    magnesium and, 317
    porphyrinuria and, 656
Aldolase, 558, 846
Aldosterone. See also *Adrenal cortex, hormones of.*
    adrenal cortex and, 714
    antagonists of, diuresis and, 469
    cirrhosis and, 673
    electrolytes and, 382
    ionic regulation and, 358
    metabolites of, 721
    secretion control of, 727
    urinary, related disorders and, 382
Alimentary albuminuria, 214
Alimentary fructosuria, 87
Alimentary galactosuria, 89
Alimentary glycosuria, 84
Alimentary pentosuria, 88
Alimentary response, abnormal, carbohydrate metabolism and, 58
    epinephrine tolerance test and, 74
    normal carbohydrate metabolism and, 36-43
Alkalemia, ammonia and, 229
    defined, 399
    hypocalcemia and, 306
    ketonuria and, 109
    ketosis and, 109
Alkali, deficit of. See *Acidosis.*
    excess of. See *Alkalosis.*
    intoxication with, hypercalcemia and, 299-300
    iron absorption and, 321
    nonrespiratory alkalosis and, 424
    reserve of. See *Acidosis, Alkalosis,* and *Carbon dioxide.*
Alkaline phosphatase, 560-566
    biliary tract disease and, 565-566
    cobalt and, 337
    diseases and, *562*
    heavy-chain disease and, 202
    isoenzymes and, 549
    liver disease and, 565-566
    skeletal diseases and, 562-565
Alkalosis, 421-425. See also *Acid-base balance.*
    ammonia and, 229
    bicarbonate radical and, 410
    biochemical characteristics of, 418(t)
    defined, 399
    hepatic coma and, 673-674

INDEX

Alkalosis *(Continued)*
  hydrogen ion regulation and, 415(t), 421-425
  hypochloremic, postoperative, 389
    salt solution and, 389
  ketonuria and, 109
  ketosis and, 109
  nonrespiratory, acid deficit and, 423-424
    base excess and, 424
      compensatory mechanisms in, 424
    causes of, 415(t)
    hydrogen ion regulation and, 415(t)
  potassium and, 386
  related disorders and, 421-425
  respiratory, acid deficit and, 422-423
    compensatory mechanisms in, 423
    hydrogen ion regulation and, 415(t)
Alkaptonuria, phenylalanine metabolism and, 225
Alkylating agents, cancer chemotherapy and, 249
Allantoin, normal urine and, 472
  uric acid and, 252
Alloxan, diabetes and, 34-35
Alpha cells. See *Langerhans, islands of*.
Alpha–1–fetoprotein, hepatoma and, 624
Alpha-lipoproteinemia, hyperlipoproteinemia type five and, 136
Altitude, blood bilirubin and, 637
  hypoxia and, 442
Aluminum, phosphate absorption and, 278
  rachitogenic action of, 278
Alveoli, acidosis and, 418(t), 419
  air composition and, 434-435
  alkalosis and, 418(t), 419
  plasma—air exchange and, 353
Amenorrhea, hypogonadism and, 782(t)
  oral contraceptives and, 785
  protein abnormalities and, 602
Amino acid(s), blood nonprotein nitrogen and, abnormal, 211
  blood urea and, 160
  carbohydrate synthesis and, 4
  catabolism of, 159
  circulating, protein metabolism and, 158
  deamination of, 159
  formation of, 158-159
  glucogenic, defined, 5-6
    fatty acid synthesis and, 104
    metabolic interrelations and, 163(t)
    phosphoenolpyruvic acid and, 161
    pyruvate and, 13
  glucose and, 6
  hepatic coma and, 674
  in blood, related disorders and, 211
  ketogenic, lipid synthesis and, 107
    metabolic interrelations and, 163(t)
  lipid metabolism and, distribution of, *102*
  lipid synthesis and, 107
  liver function and, 617
  muscle and, 25

Amino acid(s) *(Continued)*
  nitrogen excretion and, 181
  nonprotein nitrogen and, 178-179
  nonprotein nitrogenous substances and, 840
  plasma concentration conversion and, 856(t)
  protein digestion and, 149
  protein metabolism and, 154-156
  rat growth and, 154
  sparing systems and, 154-155
  transamination of, 159
  transport, processes of, 149
  urinary. See also *Aminoaciduria*.
  urinary nonprotein nitrogen and, 474
Amino acid nitrogen, blood nonprotein nitrogen and, 210-211
β-Amino-isobutyric acid, nucleotide synthesis and, 252
Aminoaciduria, abnormal, 223-229
  galactosemia and, 89
  glycosuria and, 84
  hepatic function and, 617-618
  hypophosphatemia and, 311-312
  renal tubular defects and, 534-535
  renal tubular injury and, 89
γ-Aminobutyric acid, 161
p-Aminohippurate clearance, renal blood flow and, 487
  renal tubules and, 465
Aminopeptidases, protein digestion and, 148
Aminopterin, cancer chemotherapy and, 249
p-Aminosalicylic acid, renal tubules and, 465
Aminotransferases, 553-554
  cerebrospinal fluid and, 845
  plasma proteins and, 177
Ammonia, acid-base balance and, 462-463
  acid and, 410-411, 430, 472-473
  acidosis and, 221, 229, 411, 421, 430
  alkalosis and, 221, 229, 418(t)
  kidney and, 411, 430, 472
  blood and, abnormal, 211, 619
    normal, 179
  blood nonprotein nitrogen and, abnormal, 211
  excretion of, 410, 462
  formation of, acid-base balance and, 221, 229, 410-411, 421, 430, 462-463, 372-373
    acid ingestion and, 410-411, 430, 472-473
    acidosis and, 221, 229, 411, 421, 430
    alkalosis and, 221, 229, 418(t), 430
    amino acids and, 160, 181, 229
    base stabilization and, 410-411, 430, 472-473
    glutamine and, 160, 181, 229
    kidneys and, 410, 430, 472-473
    respiratory acidosis and, 417
  hepatic coma and, 674
  hyperargininemia and, 231
  liver function and, 619
  nitrogen excretion and, 182
  nonprotein nitrogen and, 179

Ammonia *(Continued)*
  nonrespiratory acidosis and, 420
  plasma concentration conversion and, 856
  tolerance, liver disease and, 619
  urinary, 229, 411, 430
    acid-base balance and, 430
    diabetes and, 229, 421, 430
    Fanconi syndrome and, 311
    formation of, 160, 410, 430, 472
    nephritis and, 529
    nitrogen disposal and, 160
    nonprotein nitrogen and, 229, 472
    normal, 181, 462-463
Ammonium chloride, diuresis and, 469
  hyperchloremia and, 383
  nonrespiratory acidosis and, 419
AMP. See *Adenosine monophosphate.*
Amphetamine, blood sugar level and, 54
  catecholamines and, 743
Amylase, as digestive enzyme, 550-551
  carbohydrate metabolism and, 1
  cholecystokinin-pancreozymin and, 593
  insulin and, 590
  isoenzymes, pancreatic function and, 596
  Lundh test meal and, 593
  pancreatic, 589
  secretin test and, 592
  serum, pancreatic function and, 594-595
  urinary, pancreatic function and, 596-597
Amylo-1,6-glucosidase, 10
Amyloid nephrosis, hypoalbuminemia and, 201
Amyloidosis, congo red test for, urea cycle disorders and, 231-232
  fecal fat and, 101
  hyperglobulinemia and, 202
  macroglobulins and, 199
Amylolytic enzyme, pancreas and, 589
Amylopectinosis, glycogen storage disease and, 80
Anabolic reactions, 14
Anabolism, lipids and, 105
  protein, hormones and, 164-166
Anaerobic metabolism, 6-13, 7
Anaerobiosis, relative, 6
Analbuminemia, genetic defects and, 189
Anaphylatoxin, plasma proteins and, 175
Androgens, 793-797
  abnormal values of, 797
    related disorders, 793-797
  adrenal cortex and, *715*
  amino acids and, 179, 211
  lactose stimulation and, 30
  lipoproteins and, 121
  liver and, 664
  metabolic effects of, 722
  metabolism of, *718*, 796
  metabolites of, 721
  ovaries and, 780
  protein metabolism and, 165
  steroids and, 96
  testosterone and, 793-796
Androstenedione, adrenal cortex and, 714

Androstenedione *(Continued)*
  protein metabolism and, 165
Androsterone, protein metabolism and, 165
Anemia, albuminuria and, 216, 217
  amino acids and, 211
  anoxia, and 443
  aplastic, 275(t), 276, 633
  bilirubin and, 639
  blood albumin and, 188
  blood volume in, 391-392
  ceruloplasmin and, 335
  chlorotic, 633
  cholesterol acyltransferase and, 137
  cholesterol and, 141, 142
  chronic renal failure and, 533
  copper deficiency and, 335-336
  fecal urobilinogen and, 639
  glucose tolerance and, 64
  hemolytic, enzyme defects and, 270
    hematinemia and, 273
    hemoglobinemia and, 270-271
    hemosiderin and, 325-326
    hexokinase and, 559
    porphyrinuria and, 275(t), 276
  hemorrhagic, cholesterol and, 141
  hypochromic, cholesterol and, 141
    hypocholesterolemia and, 142
    iron and, 325-326
  hypoproteinemia and, 188
  hypoxia and, 443
  iron-deficiency, achlorhydria and, 585
  lactic acidosis and, 77
  macrocytic megaloblastic, folic acid and, 602
  malabsorption syndrome and, 601-602, 608
  megaloblastic, intestinal malabsorption and, *603*
  microcytic hypochromic, bone marrow and, 602
  neonatal jaundice and, 647
  pernicious, achlorhydria and, 585-586
    achylia and, 587
    Addisonian, vitamin B_{12} deficiency and, 821
    albuminuria and, 216, 217
    blood bilirubin and, 635
      iron in, 325-326
      uric acid and, 255
      volume in, 391-392
    cholesterol and, 142
    cobalt and, 337
    direct action bilirubin in, 633
    fecal urobilinogen and, 639
    gastric investigations and, 584
    hematinemia and, 273
    hyperbilirubinemia and, 635
    hypocholesterolemia and, 142
    icterus index and, 635
    iron and, 325-326
    methemalbuminemia and, 273
    plasma cholesterol and, 142
    porphyrinuria and, 275(t), 276
    uric acid metabolism and, 255

INDEX

Anemia *(Continued)*
  pernicious, urobilinogen and, 639, 641
    urobilinuria and, 641
    van den Bergh reaction and, 635
  plasma cholesterol and, 141
  protein and, 188
  secondary, achylia and, 587
    blood bilirubin and, 633
      iron in, 325-326
    hypoacidity and, 586
    hypobilirubinemia and, 633
    hypocholesterolemia and, 142
    hyposecretion and, 586
    icterus index and, 633
  sickle-cell, 500, 636-637
  splenic, 391-392, 636-637
    blood volume and, 391
  tissue iron and, 325-326
  urea clearance and, 504-505
Anesthesia, acidosis and, 419
  alkali reserve and, 419
  blood cholesterol and, 138, 392
  blood sugar level and, 51-52
  glycogen and, 51
  glycosuria and, 87
  hypercholesterolemia and, 138
  hypoglycemia and, 56
  plasma volume and, 392
Angiotensin, ionic regulation and, 360
Aniline, methemoglobinemia and, 272
Anion-cation balance. See *Acid-base balance, Acidosis, Alkalosis.*
Anorexia nervosa, adenohypophyseal hypofunction and, 69
  basal metabolism and, 451
  blood sugar and, 72
  glucose tolerance and, 70
  insulin tolerance test and, 71
Anoxemia. See also *Hypoxia.*
  hypoxic, respiratory alkalosis and, 422
Anoxia. See also *Hypoxia.*
  kidney function and, 456
  lactate and, 76
  liver glycogen and, 52
Anterior lobe hormones. See *Adenohypophysis.*
Anterior pituitary. See *Pituitary, anterior.*
Anterior pituitary-like hormone. See *Gonadotrophic hormones, chorionic.*
Anti-insulin effect, 25
Antibiotics, hypoprothrombinemia after, 833
  urobilinogen and, 655
Antibodies, liver function and, 623
  plasma proteins and, 175
Anticonvulsants, hypocalcemia and, 305
  liver enzymes and, 675
Antidiuretic hormone, 358-359. See also *Vasopressin.*
Antigen(s), Australia, liver function and, 624
Antihemophilic globulin, coagulation and, 175
Antilipotropism, lipotropism and, 128

Antimetabolites, cancer chemotherapy and, 249
Antipyrine, methemoglobinemia after, 272
Antithyroid agents, 755
Antitoxin, hemoglobinemia and, 271
a_1-Antitrypsin, properties of, 169(t)
Antral mucosa, gastric function and, 580
Anuria, hyperkalemia and, 389
  kidney function and, 492-494
Aortic stenosis, hyperlipoproteinemia type two and, 134
Apoferritin, iron absorption and, 320
Apoproteins, plasma lipoproteins and, 115
Appendicitis, hyposecretion and, 586
Arcus cornea, cholesterol acyltransferase and, 137
  hyperlipoproteinemia type two and, 134
Argentaffinomas, adrenal medulla and, 744
Arginase, urea formation and, 160
Arginine, creatine and, 180
  nitrogen equilibrium and, 154
  urea formation and, 160
Argininosuccinate, urea formation and, 160
Argininosuccinicaciduria, aminoaciduria and, 227
  hyperammonemia and, 231
  urea cycle disorders and, 230
Armstrong technique, 850
Arsenic poisoning, 49. See also *Liver, diseases of, Necrosis, hepatic,* and *Jaundice, hepatocellular.*
Arsphenamines, blood sugar tolerance curve and, 63
Arterenol. See *Norepinephrine.*
Arterial-venous blood glucose difference. See *Glucose.*
Arterioles, kidney anatomy and, 455, 456
Arteriosclerosis. See also *Atherosclerosis.*
  blood cholesterol and, 132
  glucose tolerance and, 66
  lactic acidosis and, 77
Artery disease, lactate and, 76
Arthritis, basal metabolism and, 450-451
  blood cholesterol and, 144
  glucose tolerance and, 65-66
  hypertrophic, magnesium and, 317
  hypocholesterolemia and, 144
Ascites, cirrhosis and, 672
  hypochloremia and, 380-382
  prerenal albuminuria and, 216
  protein and, 187
Ascorbate, plasma concentration conversion and, 856
Ascorbic acid, 821-825
  calcitonin and, 284
  deficiency of, 822-825
  in blood, 823
  urine and, 823
Asparaginase, tumors and, 250
Asparagine, tumors and, 250
Aspartate, amino acids and, 162
  metabolism and, *18*
  nucleic acid metabolism and, *241*

Aspartate aminotransferase, cerebrospinal fluid and, 846
Aspartic acid, amino acids and, 161
  carbohydrate metabolism and, 12, 17
Asphyxia, acidosis and, 416
  blood sugar and, 52
  glycosuria and, 87
  mechanical, hydrogen ion regulation and, 415(t)
Asthma, cryoglobulins and, 198
Astrup microtechnique, 851
  acid-base balance and, 429
Atherosclerosis, cholesterol synthesis and, 113
  hyperlipidemia and, 133
  hyperlipoproteinemia type three and, 135
  hyperlipoproteinimeia type two and, 134
  lipoproteins and, 93, 121
  magnesium and, 317
  plasma cholesterol and, 120
  plasma lipoprotein abnormalities and, 131-134
  plasma lipoprotein and, 141
Athrocytosis, 213
Athyroidism, neonatal jaundice and, 647
ATP. See *Adenosinetriphosphate.*
Atrophic gastritis. See *Gastritis, atrophic.*
Atrophy, adenohypophyseal hypofunction and, 69
Atropine, gastric secretion and, 582
  vagus nerves and, 24
AutoAnalyzer, blood sugar and, 38
  diabetes testing and, 61
Autonomic imbalance, albuminuria and, 215
  basal metabolism and, 451
  glucose tolerance and, 56, 67
Avitaminosis. See *Vitamins, deficiency of.*
Avocado pear, 23
Azide, 2
Azotemia. See *Urea, in blood, increased.*
Azotorrhea. See *Protein, in feces.*
Azure A, gastric analysis and, 587

Bacteremia, bilirubin and, 636-637
  hemoglobin and, 271
  plasma fibrinogen and, 184
  urobilinuria and, 642
  van den Bergh reaction and, 636-637
Bacteria, deoxyribonucleic acid and, 248
  hemoglobin disorders and, 271
  nucleic acids and, 243
Balance studies, lipid absorption and, 605
Bang technique, 848
Barbiturates, albuminuria and, 216
  blood glucose and, 52
  glycosuria and, 87
  liver enzymes and, 675
  porphyrinuria and, 277-278
  vitamin K deficiency and, 833
Baroreceptors, extracellular fluids and, 358
Basal ganglia, kernicterus and, 648

Basal metabolism, clinical significance of, 453
  defined, 450
  hyperthyroidism and, 144
  respiratory exchange and, 433-453
  variations in, 450
Base. See also *Alkalosis.*
  excess of, alkalosis and, 424
    hydrogen ion regulation and, 415
  hydrogen ions and, 400
  phosphatidyl compounds and, 91
  reduction of, hydrogen ion regulation and, 415(t)
    nonrespiratory acidosis and, 421
Basophilism, pituitary, 51
Bauer test, globulin reactions and, 205
Bedford diabetes survey, 61
BEI. See *Butanol extractable iodine.*
Bence Jones, Henry, 847
Bence Jones proteins, 199-200
  gamma-globulins and, 192
  plasma protein abnormalities and, 199-200
Bence Jones proteinuria, abnormal globulin and, 202
  Waldenström's marcoglobulinemia and, 203
Benedict's solutions, 848
Benzene, lipids and, 93
Benzidine test, gastric residuum and, 583
Benzodioxane, 51
Benzoic acid, hippuric acid and, 662
Benzothiadiazines, 24
Benzoylcholine, serum cholinesterase and, 552
Benzoylglycine, normal urine and, 476
Beriberi, 12, See also *Thiamine, deficiency of.*
Beta cells. See *Langerhans, islands of.*
Betahydroxybutyric acid. *Ketosis* and *Ketonuria.*
Bethanidine, catecholamines and, 744
Bicarbonate, kidney tubules and, 463
  pancreatectomy and, 34
  respiratory acidosis and, 417
  secretin test and, 591
  urobilinogen and, 654
Bicarbonate radical, alkalosis and, 410
  buffer systems and, *404*
  electrolytes and, 372
  renal tubular acidosis and, 376
Bile, calcium excretion and, 292
  composition of, 660(t)
  iodine and, 332
  obstruction of, abnormal bromsulphalein retention and, 668
  pigment metabolism of, *635*
  secretin test and, 591
  serum enzymes and, 670
Bile acid, cholesterol catabolism and, 113-114
  cholic acid and, 96
  iron absorption and, 320
  liver and, 124

Bile acid *(Continued)*
　metabolism of, 112-114, 656-661
　　gallstones and, 660-661
　　steatorrhea and, 659
　steroids and, 96
Bile ducts, atresia of, bilirubin and, 647
Bile pigment. See also *Bilirubin.*
　in bile, *635*, 636
　in feces, 655
Bile salts. See also *Bile acids.*
　importance of, 658
　lipid digestion and, 98
　vitamin D absorption and, 100
　vitamin K absorption and, 100
Biliary obstruction, alkaline phosphatase in, 670-671
　bile acids in blood and, 656-659
　bromsulphalein test and, 669
　cephalin-cholesterol flocculation and, 627
　function tests and, 676-683
　hyperbilirubinemia and, 640, 643(t), 644
　hypocholesterolemia and, 139, 631
　porphyrinuria and, 277-278
　steatorrhea and, 101, 631
　thymol turbidity test and, 626
　vitamin K absorption and, 629
　xanthomatosis and, 141
　zinc sulfate turbidity and, 628-629, 676-683
Biliary tract, ceruloplasmin and, 335
　disease of, 63-64
　　alkaline phosphatase and, 565-566
　　aminotransferases and, 553
　　glucose tolerance and, 63-64
　　　diminished, 59
　　hypercholesterolemia and, 139-140
　　hypoglycemia and, 56
　lesions to, diagnosis of, 678
Bilinogens. See also *Urobilinogen.*
　feces and, 650-654
　hemoglobin catabolism and, 265
　urine and, 650-654
Bilins. See also *Urobilinogen.*
　feces and, 650-654
　hemoglobin catabolism and, 265
　urine and, 650-654
Bilirubin, 636-648. See also *Hypobilirubinemia,* and *Hyperbilirubinemia, Jaundice, Icterus Index* and *van den Bergh reaction.*
　albumin and, 175
　atresia of bile ducts and, 647
　conjugated, increase in, neonatal jaundice and, 647
　　jaundice and, 644-648
　extrahepatic duct obstruction and, 681
　hyperbilirubinemia and, 638-640
　　one-minute, direct reacting in, 639-640
　increased. See *Hyperbilirubinemia.*
　liver function and, 633
　plasma concentration conversion and, 856
　secretion of, impaired, jaundice and, *643,* 644

Bilirubin *(Continued)*
　serum, concentration of, 636-637
　　one-minute direct reacting bilirubin and, 637-638
　　icterus index and, 638
　　jaundice diagnosis and, 679-680
　　liver function and, 636-638
　　van den Bergh reaction and, 636
　unconjugated, hemolytic jaundice and, 641
　　jaundice and, 640, *643*
　　nonhemolytic overproduction jaundice and, 641
　　nonobstructive jaundice and, 641-644
　xanthochromia and, 845
Bilirubin encephalopathy. See *Kernicterus.*
Bilirubinemia, hypercholesterolemia and, 139
Bilirubinuria, liver function and, 649-656
Biliverdinemia, liver function and, 639
Biochemistry, history of, 847-857
Biotin, carbohydrate metabolism and, 15
　fatty acid synthesis and, 103
　vitamins and, 834
Bisalbuminemia, genetic factors and, 189
Bisphosphatidyl glycerol, lipid metabolism and, 94
Black water fever, 272, 273
Bladder, urinary, postrenal albuminuria and, 216
Bladder disease. See *Cystitis.*
Blind loop syndrome, steatorrhea and, 659
Blood, arterial sugar concentration in, 38
　ascorbic acid and, 823
　capillary, glucose concentration in, 38
　circulation of, lactic acidosis and, 77
　colloidal osmotic pressure of, albumin and, 185
　diseases of, prerenal albuminuria and, 216
　disorders of, hyperuricemia and, 255
　fibrinolytic activity of, 185
　gastric juice and, 583
　in urine, albuminuria and, 214
　iodine and, 331
　kidney function and, 456
　　clearance tests and, 487-488
　magnesium and, 317-318
　nitrogenous constituents of, protein metabolism and, 166-179
　pH determination of, 429-430
　plasma phospholipids and, 118
　stomach contents and, 583
　sulfur and, 327
　thiamine in, 812
　uric acid and, 253
　vitamin A in, 808-809
　volume of, abnormal, 391-393
　　carbon monoxide method, 343
　　decreased, 391-393
　　dye methods, 343
　　I-labeled proteins and, 344
　　increased, 391-393

Blood *(Continued)*
  volume of, radioactive iron method, 343
Blood bromsulphalein retention test, 667-668
Blood carbon dioxide tension, 429
Blood glucose. See *Glucose, blood.*
Blood lactate, relation to pyruvate, 13
Blood plasma, albumin and, 187
  glucose in, 28
  kidney and, *458*
  phosphate in, 46
  zinc and, 337
Blood pressure, systemic, glomerular filtration and, 457
Blood protein. See *Protein.*
Blood sugar. See *Glucose.*
Blood urea. See *Urea.*
Blood urea clearance. See *Urea, clearance of.*
Blue diaper syndrome, aminoaciduria and, 227
BMR. See *Basal metabolism.*
Bodansky technique, 850
Body fluid. See *Blood plasma, Fluids, body, Interstitial, Seminal, Synovial, Water.*
Boerhaave of Leyden, 847
Bohr effect, hypoxia and, 442
Bone. See also *Marrow* and *Skeleton.*
  calcium absorption and, 285
  cholesterol synthesis and, 112
  development of, disturbances in, 313-315
  disease of, aminoaciduria and, 224
  formation of, 281-284
  hydrogen ions and, 406
  minerals and, 281-285
Bone resorption, nonrespiratory alkalosis and, 424
Bowel, dehydration and, 370(t)
Bowman's capsule, kidney anatomy and, 455
Boyle, Robert, 847
Brain, abscess of, cerebrospinal fluid and, 836(t), 837
    cholesterol and, 841
  diseases of, cerebrospinal fluid and, 837
  disorders of, electrolytes and, 379
  phospholipids and, 112
  tumor of, cerebrospinal fluid and, 836(t), 837
    cholesterol and, 841
Branching enzyme, 9
Branching-point linkages, 9
Breasts, oral contraceptives and, 787
Breast-feeding, iron loss and, 320
Breathing, rapid shallow, 442
Bright, Richard, 847
Bromide, intoxication and, hypochloremia and, 383
Bromsulphalein, abnormal retention of, 668-670
  excretion of, liver and, 666
Bronchography, iodine and, 331
Brucellosis, cryoglobulins and, 198
BSP. See *Bromsulphalein.*

Buffer systems, 402-412
  carbonic acid and, 403
  effect of, 406
  hemoglobin and, 402
  hydrogen ions and, 402
  nonvolatile acids and, 405-406
Burkitt's lymphoma, chemotherapy for, 250
Burns, gamma-globulins and, 191
  glomerular filtration and, 458
  hypofibrinogenemia and, 184
  negative nitrogen balance and, 229
  plasma albumin and, 185
Butanol extractable iodine, blood iodine and, 331
  serum, thyroid hormone release tests and, 758
Butylene glycol, hepatic coma and, 674
Butyric acid, stomach contents and, 583
Butyryl, fatty acid synthesis and, 103

Cachexia, pituitary, 57
Cadmium poisoning, 84
Caffeine, blood sugar level and, 54
  diuresis and, 469
Caffeine sodium benzoate, gastric analysis and, 587
Calcification, secretin test and, 592
Calcinosis universalis, urine calcium and, 308
Calcitonin, assay of, 769
  bone mineral deposition and, 284-285
  calcium concentration and, 290
  clinical applications of, 770-771
  endocrine function and, 768-771
  significance of, 769-770
  structure of, *768*
Calcium, absorption of, 280-281
  blood and, 287-291. See also *Hypocalcemia* and *Hypercalcemia.*
  body fluids and, 294
  cerebrospinal fluid and, 843-844
  concentration of, 288, *288*
    parathyroid hormone and, 289
    Vitamin D and, 290
  Cushing's syndrome and, 734
  excretion of, 292-293
  fecal, abnormal, 309
  function of, 279
  insulin release and, 24
  intake of, 280
  magnesium and, 316-317
  metabolism of, 279-315
    bone mineral deposition and, 281
    citrate metabolism and, 285
  normal urine and, 478
  plasma concentration conversion and, 856
  plasma proteins and, 290-291
  potassium and, 384
  requirement of, 293-294
  serum, abnormal, 294-307
    hypercalcemia and, 294-300

INDEX 869

Calcium *(Continued)*
  serum, abnormal, hypocalcemia and, 300-307
    alterations in, conditions, 306(t)
    calcium salts and, 291
    hypermagnesemia and, 317
    magnesium salts and, 291
    regulation of, 285-287
  source of, 280
  urinary, abnormal, 307-309
    acidemia and, 307
    decreased, 308-309
    hyperparathyroidism and, 307
    hyperthyroidism and, 307
    hypervitaminosis D and, 307
    hypothyroidism and, 308
    idiopathic hypercalciuria and, 308
    increased, 307-308
  vitamin D deficiency and, 308
Calcium chloride, diuresis and, 469
Calcium salts, serum calcium and, 291
Calcium soaps, insoluble, intestinal malabsorption and, *603*
Callow, Zimmerman reaction and, 850
Caloric equivalents, 446(t)
Calorie(s), defined, 444
  requirement, 452-453, 452(t), 453(t)
  thyroid hormone effects on, 751
Calorimetry, 448-450
  direct, 449
  indirect, 449
Cancer. See *Malignancy*.
Capillary resistance test, ascorbic acid and, 824
Carbamino compounds, 440
Carbohydrates, absorbed, 4-5
  absorption of, clinical study of, 607
  caloric comparisons, 446(t)
  conjugated proteins and, 196-197
  digestion and absorption of, 1-5
  functions of, 1
  glycoproteins and, 196-197
  ingestion of, type two fatty liver and, 127
  lipids and, metabolic interrelations with, 93, 106-110, 107(t)
  metabolism of, 1-90
    abnormal alimentary response in, 59
    acetoacetate and, 17, *18*, 21
    adipose tissue fat stores and, 124
    adrenocortical hormones and, 24-25, 722
    adrenocortical hypofunction and, 738
    aerobic phase in, 11-12
    alimentary reaction in, 36-43
    amino acids and, 159
    anaerobic phase, 7-11
    anterior pituitary and, 25
    blood-glucose concentration in, 30
    blood lactic acid and, 76-78
    blood pyruvic acid and, 76-78
    blood sugar, normal postabsorptive, 27-29
    carbon dioxide assimilation in, 14-15
    Cushing's syndrome and, 733

Carbohydrates *(Continued)*
  metabolism of, diabetes mellitus in, 33-36
    diabetic ketoacidosis and, 78-79
    digestion and absorption in, 1-5
    disaccharidase deficiency in, 3
    endocrine influences in, 18-27
    enzymes, 554-559
    general processes in, 6-13
    glucagon in, 26
    gluconeogenesis in, 14-15
    glucose in non-blood body fluid and, 28-29
    glucose utilization in, 5-6
    glycogen storage disease and, 79-82
    growth hormone and, 707
    insulin and, 18-24
    lipid metabolism and, 17-18
    liver and, 13-15, 613-617
      in lipid metabolism, 124-125
    muscle in, 15-17
    non-glucose sugar in, 29-30, 43-47
    nucleic acids and, 241
    pentose shunt in, 12-13
    postabsorptive blood sugar, abnormalities in, 47-58
    protein metabolism and, 17-18, 162-164
    pyruvic acid and, 76-78
    sugar secretion in urine and, 82-83
    sugar tolerance in, 36-43
    thyroid hormone and, 26, 751
  mucoproteins and, 196-197
  normal urine and, 480
  respiratory quotient and, 446, 448(t)
  sphingolipids and, synthesis of, 92
  tolerance. See *Glucose* and *Galactose Tolerance*.
  utilization of, ketonuria and, 109
    ketosis and, 109
Carbohydrate nutrition, blood sugar and, 41-42, *42*
Carbohydrate tolerance tests, 36-43
Carbon dioxide. See also *Carbonic acid*.
  aerobic metabolism and, 12
  anabolic reactions and, 14
  anaerobic metabolism and, 10-11
  arterial blood and, 440, 441(t)
  assimilation of, 14
  blood content and, 437(t)
  blood plasma and, 427-428
  buffer systems for, 403-405
  buffering of, hemoglobin and, 404
  entrance to tissues and, 440, 441(t)
  fatty acid synthesis and, 103-104
  pentose shunt and, 13, *13*
  pressure in blood, 437(t)
  respiratory regulation and, 407
  transport of, 440-442
  tricarboxylic acid cycle and, 15
Carbon dioxide saturation reaction, globulin reactions and, 205
Carbon dioxide tension, hemoglobin oxygenation and, *439*

Carbon dioxide tension *(Continued)*
  influence of, 436
2-Carbon fragment, 17
Carbon monoxide, carboxyhemoglobin and, 443
  hyperglycemia and, 52
  hypoxia and, 443
Carbon tetrachloride, blood sugar tolerance curve and, 63
  hypercholesterolemia and, 142
  poisoning from. See *Liver, diseases of.*
Carbonic acid. See also *Carbon dioxide.*
  buffer systems and, 403, 406
Carbonic anhydrase, cobalt and, 337
Carboxyhemoglobin, 273
Carboxypeptidase, pancreatic, cobalt and, 337
  protein digestion and, 149
Carcino-embryonic antigen, liver function and, 624
Carcinogenesis, nucleic acids and, 243-244
Carcinogens, 96
Carcinoma. See also *Malignancy.*
  alimentary response and, 63
  bronchogenic, hyponatremia and, 379
  hypouricemia and, 259
  gastric, achlorhydria and, 585-586
  metastatic, of bone, alkaline phosphatase and, 564
  organic hyperinsulinism and, 54
  pancreatic, glucose tolerance and, 597
  secretin test and, 592
  proteinuria and, 219
  stomach, gastric investigations and, 584
  hyposecretion and, 586
Cardiac decompensation, creatine and, 210
Cardiolipins, lipid metabolism and, 94
Carinamide, kidney tubules and, 464, 466
Carnitine, fatty acid synthesis and, 103
Carnosinuria, aminoaciduria and, 228
Carotene, icterus index and, 638
  lipid absorption and, 100
  serum, concentration of, 606
Carotenemia, vitamin A deficiency and, 810
Carotenoids, phospholipids and, 97
Casein hydrolysates, 55
Catabolic reactions, carbohydrate metabolism and, 6-13
  hormonal imbalance and, 22
  lipids and, 105
  protein, hormones and, 163-164
Catalase, synthesis of, iron and, 324
Cataract, senile, hypercholesterolemia and, 141
Catecholamines, 740-744
Cation concentration, 3, 349
CCK. See *Cholecystokinin.*
CCK-PZ. See *Cholecystokinin-pancreozymin.*
CEA. See *Carcino-embryonic antigen.*
Celiac disease. See also *Steatorrhea* and *Intestines, malabsorption of.*
  a-beta-lipoproteinemia and, 136
  calcium and, 302

Celiac disease *(Continued)*
  fecal fat and, 101
  globulin concentration and, 190
  glucose tolerance and, 69
  hyperplasia and, 296
  hypocholesterolemia and, 145
  hypogammaglobulinemia and, 601, 604
  hypophosphatemia and, 310
  hypoproteinemia and, 189, 191
  steatorrhea and, 302, 601
  tetany and, 301, 305
  vitamin A deficiency and, 807
  vitamin D deficiency and, 301, 305, 602
  vitamin K absorption and, 832
Cell(s), gastric function and, 580. See also under specific names, as *Acinar cell(s).*
  growth, cyclic guanosine monophosphate and, 251
Cells of Kupffer, 610
Central nervous system. See *Nervous system, central.*
Cephalic phase. See *Psychic phase.*
Cephalin. See also *Phosphatidyl ethanolamine* and *Phospholipids.*
  blood plasma and, 117(t)
Cephalin-cholesterol flocculation. See also *Globulin reactions.*
  liver function and, 627-628
Cephalin-cholesterol flocculation test, globulins and, 190
  globulin reactions and, 206
  vs. thymol turbidity test, 627
Ceramides, sphingosine and, 91-92
Cerebral. See *Brain.*
Cerebrosides. See also *Phospholipids.*
  carbohydrate to lipid conversion and, 106
Cerebroside sulfatase, deficiency of, 130(t)
Cerebrospinal fluid, metabolism and, 835-846(t)
Cerebrospinal lues, glucose and, 838(t)
  protein and, 836(t), 837
Cerebrotendinous xanthomatosis, 130(t)
Ceroid disposition, vitamin E and, 830
Ceroid storage disease, 129
Ceruloplasmin, copper and, 335
  plasma protein abnormalities and, 197
  properties of, 169(t)
Cervix, oral contraceptives and, 787
Cheilosis, intestinal malabsorption and, *603*
  vitamin deficiency and, 602
Chemoreceptors, extracellular fluids and, 358
Chenodeoxycholic acid, bile acids and, 658
Chloride, abnormalities of, 368-398
  absorption of, 361-364
    kidney tubules and, 461
  acid-base balance and, 405
  blood and, 368-383. See also *Hyperchloremia* and *Hypochloremia.*
  blood value, normal, 364-365
  buffer systems and, *404*
  cerebrospinal fluid and, 841-842
  deficits of, 368-380
    dehydration and, 368-369

Chloride *(Continued)*
  deficits of, gastrointestinal, 369
    routes of loss, 369-370
    skin and, 368
    urine and, 369
  depletion of, chronic renal failure and, 526
  diabetes mellitus and, 376
  electrolyte redistribution and, 372
  excess of, 380-383
    adrenocortical hyperfunction and, 382
    congestive heart failure and, 381
    kidney disease and, 380
  excretion of, digestive fluids and, 363
    skin and, 363
    urine and, 363
  extracellular fluid and, 345
  exudates and, 395
  importance of, 360
  metabolism and, 365-367
  metabolism of, Cushing's syndrome and, 734
  normal urine and, 478
  perspiration and, 341
  physiological considerations, 339-367
  postoperative response, 382
  transudates and, 395
Chloride-bicarbonate shift, *404*, 405
Chloroform, blood sugar tolerance curve and, 63
  hypocholesterolemia and, 142
  lipids and, 93
  poisoning from. See *Liver, diseases of; Necrosis, hepatic;* and *Jaundice, hepatocellular.*
Chlorpromazine, lactation and, 701
Chlorpropamide, 23
Cholangiolitis, blood sugar tolerance curve and, 63
  serum enzymes and, 670
Cholangitis. See also *Biliary tract, diseases of; Cholelithiasis;* and *Jaundice, obstructive.*
  hyperbilirubinemia and, 643(t)
  urobilinuria and, 654
Cholanic acids, bile acids and, 656
Cholecystitis. See also *Biliary tract, diseases of.*
  plasma cholesterol and, 140
Cholecystography, iodine and, 331
Cholecystokinin, 773
Cholecystokinin-pancreozymin, amylase and, 593
  enzymes and, 589
  intestinal phase and, 581
Cholelithiasis, bilirubin and, 643, 644(t), 645
  cholesterol and, 140, 631-633
  liver function and, 631
  pancreas and, 592
  secretion test and, 592
Cholera, plasma albumin and, 185
Cholestanol, cholesterol catabolism and, 114
  lipid absorption and, 99
Cholestasis, canalicular, jaundice and, 644

Cholestasis *(Continued)*
  extrahepatic, jaundice and, 645
  idiopathic recurrent intrahepatic, 644
  intrahepatic bile duct, 644-645
  jaundice and, *643*
    with urobilinogenuria, 654
  serum enzymes and, 670
Cholesterol, animal metabolism and, 112
  bile acid and, 113
  biosynthesis of, 96-97
  blood plasma and, 117(t)
  brain tissue and, 93
  catabolism of, 113
  cerebrospinal fluid and, 842
  chylomicrons and, 99
  diabetes and, 21
  diabetic acidosis and, 138
  diseases and, 121
  distribution of, *102*
  esterification of, 113
    plasma lipids and, 118
  excretion of, 113
  exudates and, 396
  free, hypercholesterolemia and, 139
    plasma lipids and, 118
  gallstones and, 660
  hormones and, 96
  lipid absorption and, 99
  liver and, 125
    in lipid metabolism, 124
  metabolism of, *114*
  phospholipids and, 96, 121
  plasma. See also *Hypercholesterolemia.*
    age and, 121
    anemia and, 141
    anesthesia and, 138
    cholecystitis, 140
    cholelithiasis, 140
    diabetes mellitus and, 137
    goiter and, 143
    hemolytic jaundice and, 142
    hemorrhage and, 141
    jaundice and, 139
  liver and, 118
  liver function and, 631
  nephrotic syndrome and, 138
  pernicious anemia and, 142
  plasma concentration conversion and, 856
  plasma lipids and, 118
  plasma lipoproteins and, 115
  polyunsaturated fatty acids and, 100
  racial differences in, 120
  serum, dietary fat and, 120
  steroids and, 96
Cholesterol acyltransferase, deficiency of, 137
Cholesterol ester storage disease, hepatic, 129, 130(t)
Cholesterol fatty liver, lipotropism and, 128
  type one, 127
Cholesterolemia, hypocholesterolemia and, 142
Cholic acid, cholesterol and, 96
Choline, amino acids and, 162
  anabolism and catabolism of, 92

Choline *(Continued)*
  fatty liver and, 128
  lipid synthesis and, 107
  lipotropism and, 128-129
  phosphatidyl compounds and, 91
  type two fatty liver and, 127
Choline phosphoglycerides, lipid metabolism and, 94
Cholinesterase, properties of, 169(t)
  serum, 552
Chondroitin, acid mucopolysaccharides and, 233
Chondromucoprotein, acid mucopolysaccharides and, 232
Choriocarcinoma, chemotherapy and, 250
Chorionic tissue, gonadotrophin and, 697
Choroid plexus, cerebrospinal fluid and, 835
Christmas factor, coagulation and, 175
  vitamin K and, 832
Chromaffin tissue, defined, 740
Chromaffinoma. See *Pheochromocytoma.*
Chromatography, history of, 851
Chromium, trace elements and, 338
Chromophobe tumor, adenohypophyseal hypofunction and, 69
Chromosomes, nucleic acids and, 240
Chronic renal failure. See *Glomerulonephritis.*
Chvostek sign, tetany and, 317
Chyliform effusions, 396
Chylomicron(s), adipose tissue and, 123, 124
  cholesterol and, 99
  fecal fat and, 101
  lipid digestion and, 98-99
  lipid transport and, 115-116
  lipoprotein electrophoresis and, 116
  lipoproteins and, 98
  plasma lipid and, 120
Chylous effusions, exudates and, 396
Chyluria, 111
Chymotrypsin, protein digestion and, 149
Chymotrypsinogen, pancreas and, 589
Cinchophen, hypocholesterolemia and, 142
  poisoning from. See *Liver, diseases of; Necrosis, hepatic;* and *Jaundice, hepatocellular.*
Circadian rhythm, adrenocorticotrophic hormone and, 727
  corticosteroids and, *732*
Cirrhosis. See *Liver, cirrhosis of.*
  alimentary response and, 63
  biliary, hypercholesterolemia and, 140
  biliary, xanthomata and, 141
  blood sugar tolerance curve and, 63
  cellular fluid and, 672
  cephalin-cholesterol test and, 628
  dye retention and, 669
  epinephrine tolerance test and, 74
  hepatic, gamma-globulins and, 191
    plasma protein abnormalities and, 183
  hypoglycemia and, 56
  iron absorption and, 322
  liver function and, 49

Cirrhosis *(Continued)*
  macroglobulins and, 199
  plasma fibrinogen and, 184
  porphyrinuria and, 655-656
  portal, blood volume and, 391
    cryoglobulins and, 198
    hippuric acid and, 663
    hypokalemia and, 387
    magnesium and, 317
    mucoproteins and, 197
    serum enzymes and, 670
    urobilinogenuria and, 653-654
  zinc and, 338
Citrate, 12, *18*
  hepatic coma and, 674
  metabolism of, calcium and, 285
Citric acid, fatty acid synthesis and, 104
  normal urine and, 476
Citric acid cycle, 10-11
Citrulline, hepatic coma and, 674
  urea formation and, 160
Citrullinemia, urea cycle disorders and, 231
Clearance tests, filtration fraction and, 489
  kidney function and, 481-491
  normal values, 486(t)
  renal blood flow and, 487
  tubular excretory capacity and, 488
  tubular reabsorption and, 489-490
Clinistix, diabetes testing and, 61
Closed-circuit system, indirect calorimetry and, 449
Clotting, relation to plasma, *630*
Coagulation, plasma proteins and, 175
Cobalt, trace elements and, 336-337
Cobamide coenzyme, *820*
Cocaine, catecholamines and, 743
  acetate transformation and, 77
Cocarboxylase. See also *Thiamine.*
  acetate transformation and, 77
Coenzyme(s), free nucleotides and, 250-251
Coenzyme A, 12
  free nucleotides and, 251
  pantothenic acid and, 834
Cohn fractions, plasma proteins and, 168
Cold exposure, albuminuria and, 214
Colitis, electrolytes and, 375
  glucose and, 70
  vitamin K and, 832
Collagen, bone development and, 282, *282*
Collagen disorders, gamma-globulins and, 191
Colloid osmotic pressure, 354
Colloidal gold curve, globulin reaction and, 205
Coma, hepatic, 673-675
  lactic acidosis and, 78
Compound A. See *Adrenal cortex, hormones of.*
Compound E. See *Adrenal cortex, hormones of.*
Compound F. See *Adrenal cortex, hormones of.*
Compound S. See *Adrenal cortex, hormones of.*

INDEX 873

Concentration, defined, 855
Concentration tests. See, *Kidneys, function of.*
Concussion, hyperglycemia and, 50
Congestion, passive, alimentary response and, 63
Congo red test, amyloidosis and, 231-232
Constipation, hyposecretion and, 586
Contraceptives, blood lipids and, 121
   iodine and, 331
   oral, 782-787
      adrenal function tests and, 786
      amenorrhea and, 786
      endocrine effects of, 785
      endocrine function and, 782
      gynecological effects of, 787
      hypertension and, 785
      metabolic effects of, 785
      pigmentation and, 786
      preparations, 783-784
      thromboembolic effects of, 786
      thyroid function tests and, 786
Convulsions, albuminuria and, 215
   cerebrospinal fluid and, 837
   hypoglycemia and, 53, 56
Convulsive disorders, prerenal albuminuria and, 216
C.O.P. See *Colloid osmotic pressure.*
Copper, body function and, 334-336
   ceruloplasmin and, 197-198
   plasma concentration conversion and, 856
Coproporphyria, erythropoietic, 276
Coproporphyrins. See also *Porphyrins.*
   defined, 261
   porphyrinurias and, 275(t)
   urine and, 274
Coprosterol, cholesterol catabolism and, 114
   lipid absorption and, 99
Coronary artery occlusion, glomerular filtration and, 481-483, 493
   glucose and, 52
   glycosuria and, 87
   hyperglycemia and, 52
   lactic acid and, 76
   lipoproteins and, 132
Corpuscles, blood volume determination and, 391
Cortex, adrenal. See *Adrenal cortex.*
Corticosteroid(s), secretion of, rate, 719
   See also *Adrenal cortex, hormones of.*
   transport of, 717
Corticosteroid therapy, hypoglycemia and, 57
Corticosterone, adrenal cortex and, 714
   carbohydrate metabolism and, 24
Corticotrophin releasing factors, adrenocorticotrophic hormones and, 703
Corticotrophin-releasing hormone, 689
   circadian rhythm and, 727
Cortisol. See also *Adrenal cortex, hormones of.*
   adrenal cortex and, 714
   anti-inflammatory action of, 726
   carbohydrate metabolism and, 24

Cortisol *(Continued)*
   metabolites and, 720
   physiology of, 727
Cortisol-glucose tolerance test, in diabetes testing, 62
Cortisone, anti-inflammatory action of, 726
   lipid mobilization and, 105
   unesterified fatty acid concentration and, 119
Cortisone acetate, 45
Cortisone-glucose tolerance test, 45
   abnormal, latent diabetes and, 72
Cortisone-prednisone tolerance test, 45
C-reactive proteins, 198
Cranium, pressure in, cerebrospinal fluid and, 839
Creatine, body fluid concentration and, 180
   castration and, 796
   metabolic interrelations and, 163(t)
   methyltestosterone and, 165
   nitrogen excretion and, 180-181
   urinary. See also *Creatinuria.*
      nonprotein nitrogen and, 222-223, 473
Creatine kinase, 569-570
   cerebrospinal fluid and, 846
Creatine phosphate, *16*
Creatine tolerance, 223
Creatinine, blood and, 178
   blood nonprotein nitrogen and, 209
      nephritis and, 210
   endogenous, urinary nonprotein nitrogen and, 506-507
   excretion of, 153
   nitrogen excretion and, 180-181
   nonprotein nitrogenous substances and, 840
   plasma concentration conversion and, 856
   protein and, 161
   urinary nonprotein nitrogen and, 473
Creatinine coefficient, nitrogen excretion and, 181
Creatinuria, diseases with, 222
   normal, 180-181, 473-474
   vitamin E and, 830
Creatorrhea. See *Proteins, feces and.*
Cretinism. See also *Hypothyroidism.*
   glucose tolerance and, 69
   hypoglycemia and, 58
   neonatal jaundice and, 647
CRH. See *Corticotrophin-releasing hormone.*
Crigler-Najjar syndrome, 641-642
   neonatal jaundice and, 647
   phenobarbitone and, 675
Crotonyl, fatty acid synthesis and, 103
Cryoglobulin(s), multiple myeloma and, 201
   plasma protein abnormalities and, 183, 198
   study methods for, 171(t), 172
Cryoglobulinemia, macroglobulinemia and, 198
Cullen method, 848
Cushing's syndrome. See also *Adrenocortical hyperfunction.*

Cushing's syndrome *(Continued)*
  adrenocortical hyperfunction and, 51, 730-735
  adrenocortical hyperplasia and, 87
  globulin concentration and, 190
  glucose tolerance and, 64
  hyperpituitarism and, 51
  insulin tolerance test and, 71
  ketonuria and, 110
  potassium and, 387
Cyanide, 2, anoxia and, 444
Cyanocobalamin, vitamin B_{12} and, 818, *819*
Cyanosis, methemoglobin and, 272
Cyclopentenophenanthrene nucleus, steroid hormones and, 714
Cyclophosphamide, cancer chemotherapy and, 250
Cyst(s), echinococcus, serum enzymes and, 670
  of pancreas, secretin test and, 592
Cystathioninuria, aminoaciduria and, 227
Cysteine, diabetes and, 35
  gamma-globulins and, 191-192
  sulfur and, 327
Cysteinyl residues, proinsulin structure and, *23*
Cystic fibrosis, enzyme abnormalities and, 579
  secretin test and, 592
Cystine, fatty liver and, 122
  metabolism of, 328
  protein metabolism and, 152
  urine and, 224
Cystine aminopeptidase, 571-572
Cystinosis, aminoaciduria and, 224
  renal tubular defects and, 538
Cystinuria, aminoaciduria and, 223-224
  lithiasis and, 542
  renal tubular defects and, 535-536
Cystitis, albuminuria and, 216
  nucleoproteins and, 219
Cytidine monophosphate, ganglioside biosynthesis and, 92
Cytochrome(s), synthesis of, iron and, 324
Cytochrome C, copper and, 335
Cytomegaly, neonatal jaundice and, 648
Cytosine, nucleic acids and, 235
  nucleoside triphosphates and, 250

Dark adaptation test, Vitamin A and, 809
Deamination, liver and 159, 617-618
Dehydration, ammonia and, 229
  biochemical consequences of, 374-375
  blood components and, 374
  electrolytes and, 368-370
  glomerular filtration and, 457
  kidney function, 492
  pancreatectomy and, 34
  plasma albumin and, 185
  prerenal deviation of water and, 508
  renal function and, 374

Dehydration *(Continued)*
  urinary nonprotein nitrogen and, 509
Dehydroandrosterone. See *Androgens.*
11-Dehydrocorticosterone, 24
Dehydroepiandrosterone, adrenal cortex and, 714
  protein metabolism and, 165
Dehydrogenase, cobalt and, 337
11-Dehydro-17-hydroxycorticosterone, 24
Deiodinase, iodotyrosines and, 332
Dementia praecox, cerebrospinal fluid and, 839
  glucose tolerance and, 64
Denis technique, 848
Dental caries, fluorine and, 338
Deoxycholic acid, bile acids and, 658
Deoxycorticosterone. See *Adrenal cortex, hormones of.*
Deoxycorticosterone acetate, glycemic response and, 67
Deoxyglucose, carbohydrate absorption and, 2
2-Deoxyglucose, insulin release and, 23
Deoxyribonucleic acid, biosynthesis of, 238
  distribution of, 240
  genes and, 241-243
  protein synthesis and, 244
Deoxyribonucleoproteins, chromosomes and, 241
  viruses and, 240
Deoxyribose, glucose conversion and, 6
  nucleic acid synthesis and, 6
Deposition, lipid metabolism and, 93
Dermatan sulfate, acid mucopolysaccharides and, 233
Dermatitis, hypercholesterolemia and, 141
  seborrheic, biotin and, 834
Desferrioxamine test, 326
Desoxyribonuclease, pancreas and, 589
Detoxification-conjugation, hippuric acid and, 662
  liver function and, 661-663
Detoxification reactions, glucose conversion and, 6
Dextrose. See *Glucose.*
Dextrose tolerance curve, 42
Diabetes, adrenal, 70, 433, 436. See also *Adrenal Cortical Hyperfunction.*
Diabetes, adrenal vs. pancreatic, 35
  adrenocortical hormones and, 21
  alloxan, 34-35
  ammonia and, 229
  frank, 41
  latent, cortisone-glucose tolerance test and, 72
  metathyroid, 335. See also *Glycosuria, renal.*
  pancreatic, blood lactate and pyruvate and, 77
  renal, 83-84
  steroid, 51
  glucose tolerance and, 64
Diabetes innocens, 83-84

INDEX 875

Diabetes insipidus, blood volume and, 391-392
  nephrogenic, 380, 539-540
Diabetes mellitus, 47-49, 60-63
  alimentary glycosuria and, 62
  anterior pituitary hormones and, 21
  autoimmunity and, 62
  blood volume and, 392
  creatinuria and, 222
  diminished glucose tolerance and, 59
  electrolytes and, 376
  etiology of, 62
  experimental, 33-36
  hemoconcentration and, 183
  hypercholesterolemia and, 21, 137
  hyperglycemic glycosuria and, 85
  hyperlipemia and, 21
  hyperlipoproteinemia type four and, 135
  hypoalbuminemia and, 188
  insulin and, 62
  insulin antagonist synalbumin, 62
  insulin resistance vs. sensitivity, 71
  ketogenesis and, 21
  ketone bodies and, 78-79
  ketosis and, 109
  lactic acidosis and, 77
  latent, 62
  lipolysis and, 21
  liver fat in, 21
  magnesium and, 317
  mucoproteins and, 197
  nonrespiratory acidosis and, 415(t), 417
  obesity and, 73
  one-hour, two dose test and, 62
  overt, 62
  pancreatectomic, 35
  plasma lipoprotein abnormalities and, 132
  renal glycosuria and, 62-63
  screening of, 61-62
  secretin test and, 592
  stages of, 62
  subclinical, 62
  tolbutamide tolerance test and, 72
  urine specific gravity and, 86
  xanthomata and, 141
Diabetic coma, glomerular filtration and, 458
  plasma dilution and, 189
Diabetic piqûre, 50
Diabetogenic effect, 25
Dialysis, chronic, hemodialysis and, 544
Diarrhea, 3
  a-beta-lipoproteinemia and, 136
  ammonia and, 229
  electrolytes and, 375
  fatty. See *Steatorrhea*.
  glomerular filtration and, 457
  hemoconcentration and, 183
  intestinal malabsorption and, *603*
  plasma albumin and, 185
  plasma dilution and, 189
  potassium and, 386
Diastase, 550-551. See also *Amylase*.

Diazoxide, effect on plasma insulin, 24
Diet, amino acid imbalance and, 154, 155
  blood cholesterol and, 119-121
  blood lipoproteins and, 119-121
  blood proteins and, 156
  blood urea and, 178
  calcium absorption and, 281
  fat mobilization and, 105, 119
  glucose tolerance and, 41-42, 66
  hemoglobin and, 156
  liver fat and, 125-129
  nitrogen balance and, 153-154, 162-164
  protein requirement and, 155
  rachitogenic, 310
  renal blood flow and, 487
  rickets and, 310
  urea clearance and, 504
  uric acid and, 253
  urinary urea and, 180
  urine ammonia and, 182
  urine sulfate and, 328
Digestion. See also under metabolism of each nutrient.
  calcium excretion and, 292
  fluids of, ionic excretion and, 363(t)
  protein metabolism and, 148-152
Digitalis, potassium and, 384
Digitalis glycosides, lipid absorption and, 99
Diglycerides, conversion of, 105
Dihydrocholesterol, cholesterol absorption and, 100
  cholesterol catabolism and, 114
Dihydroxyacetonephosphate, glycerophosphate and, 122
Dihydroxyphenylalanine, melanuria and, 228
Diiodotyrosine, iodine absorption and, 330
  thyroglobulin and, 332
Dilution test, kidney function and, 494
Dimercaptopropanol, diabetes mellitus and, 35
Dinitrophenol, metabolism and, 2
Diodrast, renal tubules and, 465
Diphtheria, glucose tolerance and, 64
  hypoglycemia in, 56
  proteinuria and, 219
Disaccharides, 1
Disaccharidase, 1
Disaccharidase activity, carbohydrate digestion and absorption and, 2
Disaccharidase deficiency, 3
Diseases, hereditary, enzyme abnormalities and, 579
  infectious, ceruloplasmin and, 335
    hypocholesterolemia and, 143
  lymphomatous, gamma-globulins and, 191
  neoplastic, macroglobulinemia and, 199
  parasitic, plasma protein abnormalities and, 201
  protozoal, gamma-globulins and, 191
    plasma protein abnormalities and, 183, 201
  serum protein and, 200

Diseases, hereditary *(Continued)*
  viral, gamma-globulins and, 191
    plasma protein abnormalities and, 201
Disequilibrium syndrome, hemodialysis and, 543-544
Distal convolution, water absorption and, 461
Distention, 3
  abdominal, a-beta-lipoproteinemia and, 136
Disulfide bridge, in proinsulin structure, 23
Diuresis, blood volume and, 392
  drug induced, potassium and, 387-388
  nonrespiratory alkalosis and, 424
  renal function and, *468*
Diuretics, renal function and, 468-470
Donnan equilibrium, cerebrospinal fluid and, 841
  plasma electrolytes and, 364-365
Dopa, catecholamines and, 741
Dopadecarboxylase, catecholamines and, 741
Dopamine, catecholamines and, 741
  hepatic coma and, 675
Down's syndrome, lactic acidosis and, 77
Drugs, albuminuria and, 216
  liver enzymes and, 675-676
  respiratory alkalosis and, 422
Dubin-Johnson syndrome, jaundice and, *643*, 644
Dubois standards, 451(t)
Duodenum, copper and, 335
  glomerular filtration and, 457
  hyperacidity and, 584
  iron absorption and, 320
  ulcer of, hypersecretion and, 586
  zinc and, 338
Dyes, elimination of, liver and, 666-670. See also *Bromsulphalein*.
Dynamic state, protein digestion and, 150
Dysgammaglobulinemia, serum protein abnormalities and, 204

ECG. See *Electrocardiogram*.
Eclampsia. See also *Pregnancy, toxemias of*.
  amino acid nitrogen and, 210
  hyperglycemia in, 51
  lipuria and, 111
  plasma fibrinogen and, 184
  plasma phospholipids and, 141
  renal albuminuria and, 218
Ectopic adrenocorticotrophic hormone syndrome, 800
Ectopic gonadotrophin secretion, 802
Ectopic growth hormone, secretion of, 802
Ectopic hormones, 799((t)
  endocrine function and, 797-801
  hypercalcemia and, 800-801
  hyperthyroidism and, 801-802
  hypoglycemia and, 800
  production of, 799

Ectopic hormones *(Continued)*
  tumors and, 797-802
  vasopressin secretion and, 801
Edema, albumin and, 394-395
  electrolytes and, 393-397
  famine, 185, 187
  fluid. See also *Transudates* and *Exudates*.
  inflammatory. See *Exudates*.
  intestinal malabsorption and, *603*
  kidney function and, 493
  noninflammatory. See *Transudates*.
  palatal, heavy-chain disease and, 202
  pancreatic, secretin test and, 592
  prerenal deviation of water and, 508
  protein and, 394
  specific gravity of, 393
Edetate, zinc excretion and, 338
EDTA. See *Edetate*.
Efferent arteriole, kidney anatomy and, 456
Effusions, chyliform, 396. See also *Exudates* and *Transudates*.
Elastase, protein digestion and, 149
Electrocardiogram, postoperative, potassium and, 389
Electrolytes. See also under individual ions.
  abnormalities of, fluid compartments and, 370-371
  acute glomerulonephritis and, 377
  Addison's disease and, 379
  adrenal cortex and, 722
  body fluids and, *364*
  cerebral disorders and, 379
  chronic glomerulonephritis and, 376
  dehydration and, 370(t), 374-375
  depletion of, intestinal malabsorption and, *603*
  diabetes mellitus and, 376
  digestive fluids and, *363*
  excesses of, 380-383
  exudates and, 393-397
  fever and, 373
  gastrointestinal disorders and, 375
  heart failure and, congestive, 378
  hyperlipemia and, 374
  ketosis and, 374
  kidney disease and, 376
  metabolism of, thyroid hormone and, 752-753
  nephritis and, 377
  osmolality and, 372-373
  perspiration and, excessive, 378
  pH and, 372-373
  plasma pattern and, *377*
  postoperative states and, 388
  pulmonary neoplasms and, 379
  redistribution of, 372
  renal tubular acidosis and, 376
  starvation and, 374
  transudates and, 393-397
  urea retention and, 373
Electrophoresis, jaundice diagnosis and, 680
  plasma protein abnormalities and, 183

Electrophoresis *(Continued)*
　plasma protein study and, 171
Electrophoretic mobility, myeloma proteins and, 199
Emiocytosis, 23
Emotional states, blood sugar and, 41
　epinephrine in, 50
　glucose tolerance and, 64
　glycosuria and, 87
　hypoglycemia in, 56
　renal blood flow and, 487
Empyema, indican and, 329
Enamel, mottled, 338
Encephalitis, cerebrospinal fluid and, 836-839, 836(t), 838(t)
　respiratory alkalosis and, 422
Endergonic, defined, 444
Endocarditis, cryoglobulins and, 198
Endocrine syndromes, tumors and, criteria for, 799
Endocrine system, calcium absorption and, 281
　carbohydrate metabolism and, 18-27
　disorders of, hyperuricemia and, 258
　function of, 685-803
　　adenohypophyseal hormones and, 690-709
　　adrenal glands and, 713-740
　　adrenal medullary hormones and, 740-746
　　androgens and, 793-797
　　anterior lobe hormones and, 690-709
　　calcitonin and, 768-771
　　cyclic adenosine monophosphate and, 686-687
　　ectopic hormone production and, 797-801
　　female hypogonadism and, 782
　　female puberty and, 781
　　gastrointestinal hormones and, 771-774
　　hyperthyroidism and, 766-768
　　hypothyroidism and, 764-766
　　menopause and, 781-782
　　nontoxic goiter and, 764-766
　　neurohypophyseal hormones and, 709-713
　　oral contraceptives and, 782
　　ovarian hormones and, 774-781
　　pituitary hormones and, 687-690
　　placental hormones and, 787-793
　　posterior lobe hormones and, 709-713
　　pregnancy and, 787-793
　　steroid hormones and, 713
　　testis and, 793-797
　　thyroid hormones and, 746-764
　imbalance of mucoproteins and, 197
Endogenous creatinine clearance, urinary nonprotein nitrogen and, 506-507
Endogenous sources of glucose, 4
Endometrium, prostaglandins in, 95
Endopeptidases, protein digestion and, 148, 149

Energy, metabolism of, 444-448
Enteritis, electrolytes and, 375
　regional, 3
　tuberculous, glucose tolerance and, 70
Enterogastrone, intestinal phase and, 582
Enterogenous cyanosis, 272
Enteroglucagon, 26
Enzymes, 547-579. See also under specific enzymes.
　anaerobic metabolism and, 7
　brush border membrane and, 1-2
　calcium absorption and, 281
　carbohydrate digestion and absorption and, 1-3
　carbohydrate metabolism and, 554-559
　cerebrospinal fluid and, 574, 845-846
　coenzyme A, 12
　debranching system and, 10
　deficiency of, glycogen and, 79
　diagnostic abnormalities and, 574-579
　digestive, 549-552
　gastric activity and, 580
　glycolytic, plasma proteins and, 177
　hemoglobin biosynthesis and, 263
　international units for, 548
　kinases and, 7
　nucleic acid biosynthesis and, 237-238
　of liver, drug induction of, 675-676
　pancreatic, 589
　peritoneal effusions and, 573-574
　plasma proteins and, 177
　pleural effusions and, 573-574
　porphyrin biosynthesis and, 263
　serum, liver function and, 670-671
　standard values for, 547
Eosinopenia, postoperative states and, 389
Eosinophilia, cryoglobulins and, 198
Eosinophilic tumor, acromegaly and, 64
Ephedrine, catecholamines and, 743
Epilepsy, albuminuria and, 216
　cerebrospinal fluid and, 836, 836(t), 837
　hyperglycemia and, 51
　hypocholesterolemia and, 144
　Jacksonian, 51
　magnesium and, 317
Epinephrine, adrenal medulla and, 740-746
　amino acid concentration and, 179
　carbohydrate metabolism and, 26
　emotional glycosuria and, 86-87
　glycogen synthesis and, 10
　glycosuria and, 86-87
　hepatic glycogenesis and, 20
　hepatic glycogenolysis and, 31
　hyperglycemia and, 26, 50, 74, 616
　insulin production and, 59
　insulin secretion and, 23, 24
　lipid metabolism and, in adipose tissue, 123
　lipid mobilization and, 105
　lipolytic mechanism and, 124
　metabolic interrelations and, 163(t)
　proinsulin structure and, *9*

Epinephrine *(Continued)*
  secretion of, 50
    diminished glucose tolerance and, 59
    triglyceride conversion and, 105
    unesterified fatty acid concentration and, 119
Epinephrine hydrochloride, 46, 50
Epinephrine tolerance test, 46, 74
Epithelium, degeneration of, lipuria and, 111
  secretion of, 589
  seminal vesicular, 30
  vitamin A deficiency and, 808
Ergosterol, lipid absorption and, 99
  plant metabolism and, 112
Erogtamine, 51
Ergothioneine, carbohydrate metabolism and, 27
  protein metabolism and, 152
Erysipelas, hypochloremia and, 383
Erythema, heavy-chain disease and, 202
Erythremia. See *Polycythemia*.
Erythroblastosis, bilirubin and,
Erythrocuprein, copper and, 335
Erythrocytes, buffer system of, 402-404
  copper and, 335
  glucose distribution and, 27
  iodide ions and, 332
  lifespan of, vitamin E and, 830
  pentose shunt and, 12
  plasma proteins and, 175
Erythrocyte enzymes, defects of, 572-573
Erythrocyte sedimentation, plasma fibrinogen and, 184
Erythroderma, desquamative, biotin and, 834
Erythropoiesis, copper and, 335
  ineffective, jaundice, *643*
  iron absorption and, 322
Erythropoietic porphyrias. See *Porphyrias, erythropoietic*.
Erythropoietin, iron absorption and, 321
Escherichia coli, vitamin K and, 831
Esterase, serum lipase and, 596
Estersturz, 632
Estradiol. See *Estrogens*.
Estrogens, 774-779
  actions of, 775-776
  assay of, 776
  biosynthesis of, *775*
  contraceptives and, effects of, 784
  copper and, 335
  lipoproteins and, 121
  liver and, 664
  metabolism of, *775*
  placental hormones and, 788
  preparations of, 778-779
  steroids and, 96
  structure of, *775*
  values for, decreased, 777-778
    increased, 776
    menstrual cycle and, 777(t)
Estrogen therapy, indications for, 778

Estrone. See *Estrogens*.
Ethacrynic acid, diuretics and, 469
Ethanol, lactic acidosis and, 78
  precipitation of, 170
Ethanolamine, amino acids and, 162
  anabolism and catabolism of, 92
  lipid synthesis and, 107
  phosphatidyl compounds and, 91
  phosphoglycerides, 94
Ether. See also *Anesthesia*.
  hyperglycemia and, 138
  lipids and, 93
ETR. See *Effective thyroxine ratio*.
Euglobulin, serum protein concentration and, 168(t)
  study methods for, 171(t)
Euglycemic diabetic ketoacidosis, 79
Eukaryotes, specialized ribosomes and, 246
Exercise, benign albuminuria and, 214
Exergonic, defined, 444
Exertion, hyperuricemia and, 255
Exocrine, pancreatic, excretion of, 589
Exopeptidases, protein digestion and, 148, 149
Exophthalmic goiter, 49
Extracellular fluid. See *Transudates, Exudates, Dehydration* and *Edema*.
Extrarenal azotemia, increased urea nitrogen and, 207-208
Exudates, 393-397
  enzymes and, 574
  sulfur and, 329

F_{ab} fragments, gamma-globulins and, 191-192
F_c fragments, gamma-globulins and, 191-192, 196
  heavy-chain disease and, 202
F_d fragments, gamma-globulins and, 191-192
Fabry's disease, 129, 130(t)
Factor VII, vitamin K and, 832
Factor IX, vitamin K and, 832
Factor X, vitamin K and, 832
Famine edema. See *Edema, famine*.
Fanconi syndrome, aminoaciduria and, 224
  hypophosphatemia and, 311
  ketosis and, 110
  renal tubular defects and, 537
  Type VI glycogenosis and, 81
  Wilson's disease and, 84
Farad, defined, 855
Fat(s). See *Fatty acids* and *Lipids*.
  caloric comparisons, 446(t)
  fecal, lipid digestion and, 100-101
    standard fat meal and, 606
  gastric secretion and, 582
  glucose and, 5
  lipid storage and, 93-94
  metabolism of, 101-106
  neutral, blood plasma and, 117(t)
    distribution of, *102*

Fat(s) *(Continued)*
  neutral, fatty acid esters and, 91
    fecal fat and, 110
    hyperlipemia and, 119
    phospholipids and, 121
    plasma and, 114
    plasma lipoproteins and, 115
  respiratory quotients and, 446, 448(t)
  urine and, 111
Fatty acid(s). See also *Cholesterol* and *Lipids*.
  anabolism of, 102-105
  catabolism of, 102-105
  distribution of, *102*
  essential, lipotropism and, 128
  esters, of alcohol, 91
    of glycerol, 91
  excess of, intestinal malabsorption and, 603
  free. See *Free fatty acids*.
  lipid digestion and, 98-99
  lipids and, 91
  lipoid metabolism and, 93
  lipoprotein lipase and, in lipid deposition, 124
  liver and, 124, 125
  non-esterified, lipid transport and, *123*
    plasma lipids and, 119
  oxidation of, pancreatectomy and, 34
  plasma and, 114
  polyunsaturated, cholesterol and, 100
    serum cholesterol and, 120
    type four fatty liver and, 128
  role of, 91
  saturated, cholesterol and, 100
  short-chain, hepatic coma and, 674
  stomach contents and, 583
  synthesis of, carbohydrates and, 103
    insulin and, 21
    NADP and, 12
  synthetase complex of, 103
  unesterified, blood plasma and, 117(t)
    plasma lipids and, 118-119
    plasma lipoproteins and, 115
  unsaturated, cholesterol and, 100
Fatty liver. See *Liver, fatty*.
Favism, hemoglobinemia and, 271
Feces, bile pigments in, 655
  bilinogens and, 650-654
  bilins and, 650-654
  examination of, pancreatic function and, 598, 603, 606
  fat in, 110-111, 600-601
  fatty acids and, 110, 600, 605
  iron in, 320
  nitrogen and, 152
  phosphate excretion and, 292
  porphyrin and, 655-656
  protein and, 598, 603, 607
  protein metabolism and, 152
  urobilinogen and, 633, 651
  water output and, 340
Fehling's sugar test, 848

Females, endocrine cycle in, *694*
Feminizing syndrome. See *Adrenal cortical hyperfunction*.
Ferritin, iron absorption and, 320
  iron storage and, 323
Ferrous iron, absorption of, 320
Fetal hemoglobin. See *Hemoglobin, fetal*.
α_1-Fetoprotein, jaundice diagnosis and, 681
Fetus, dead, plasma fibrinogen and, 184
  immunoglobulins and, 196
Fever, albuminuria and, 216, 217
  alkalosis and, respiratory, 422
  aminoaciduria and, 224
  basal metabolism and, 450
  blood urea and, 208
  creatinuria and, 222
  dehydration and, 368-369
  electrolytes and, 373
  nonprotein nitrogen and, 208
  renal blood flow and, 488
  urea clearance and, 504
  urinary creatine and, 222
FFA. See *Free fatty acids*.
Fibrinogen, coagulation and, 175
  liver and, 173, 620
  plasma protein abnormalities and, 184
  pneumococcus exudates and, 395
  properties of, 169(t)
  serum protein concentration and, 168(t)
  study methods for, 171(t)
  test for, 170
Fibrinogenopenia, congenital. See *Afibrinogenemia*.
Fibrinolysin, plasminogen and, 185
Fibrocystic disease, phenylalanine metabolism and, 226
Fibrous dysplasia, hydroxyproline and, 224
FIGLU. See *Formiminoglutamic acid*.
Filariasis, hyperglobulinemia and, 201
Fistula(s), biliary, blood volume and, 392
  electrolytes and, 375
  intestinal, blood volume and, 392
    plasma albumin and, 185
  pancreatic, blood volume and, 392
  thoracic duct, albumin and, 188
Flavin-adenine dinucleotide, free nucleotides and, 251
Flocculation reactions, plasma protein abnormalities and, 183
Flotation. See *Lipoproteins*.
Fluids. See also under specific fluids and components.
  amniotic, prostaglandins in, 95
  ascitic, cirrhosis and, 673
  blister, protein and, 394
  body, blood volume, 343-344
    calcium content of, 294
    compartments, composition of, 346
      electrolyte abnormalities and, 370
      exchanges between, 352-357
    volume, 342-348
    extracellular, 348-351
      total volume of, 344-345

INDEX

Fluids (Continued)
  body, hydrogen ion concentration of, 399-432
    osmolality and, 347
    plasma volume, 343-344
    testosterone level in, 794-795
  cellular, increased, liver function and, 672-673
  cerebrospinal, enzymes of, 574
    insulin and, 24
    protein and, 393
  disturbances of, hydrogen ion concentration and, 414
  exchanges of, Gibbs-Donnan equilibrium and, 353
  extracellular, composition of, *350*
    osmolar equality of, *349*
    pH of, 399
  input vs. output, kidney function and, 491
  interstitial, glucose in, 28
    intracellular fluid and, 353-357
    pH and, 400
  intracellular, 351-352
    composition of, *350*
    osmolar equality of, *349*
  menstrual, prostaglandins in, 95
  molar concentration of, 347
  seminal, prostaglandins in, 95
  synovial, glucose in, 28
Fluoride, osteosclerosis and, 315
Fluorine, body function and, 338
  trace elements and, 338
Fluoroacetate, 2
5-Fluoro-2'-deoxyuridine, cancer chemotherapy and, 249
5-Fluorouracil, cancer chemotherapy and, 249
Folic acid, 817-818
  anemia and, 602
  ascorbic acid and, 821
  cancer chemotherapy and, 249
  intestinal malabsorption and, *603*
  malabsorption syndromes and, 608
Folinic acid. See also *Pteroylglutamic acid*.
Folin's method, 848
Follicle-stimulating hormone, 693
Food(s), caloric value of, 444
  nitrogen of, protein metabolism and, 152
  preparation of, protein and, 149
Formate, nucleic acid metabolism and, *241*
Formiminoglutamic acid, malabsorption syndromes and, 608
Formol-gel reaction. See also *Globulin reactions*.
  globulin reactions and, 205
Fractures, negative nitrogen balance and, 229
Free fatty acids, acetyl-CoA and, 22
  adipose tissue and, 25
  albumin and, 119
  carbohydrate metabolism and, 125
  fecal fat and, 110

Free fatty acids (Continued)
  glycolysis and, 22
  insulin and, 21
  lipoproteins and, 119
  plasma lipids and, 119
Free nucleotides. See *Nucleotides, free*.
Friedreich's ataxia, a-beta-lipoproteinemia and, 136
Fröhlich's syndrome. See *Hypopituitarism*.
Froin syndrome, cerebrospinal fluid and, 836(t), 837
  xanthochromia and, 845
Fructokinase reaction, carbohydrate absorption and, 3
Fructose, absorption of, 2
  alimentary response to, 74-75
  anaerobic metabolism and, 7, *7*
  body fluids and, 29
  fructosuria and, 82
  metabolism of, 43-44
  urine and. See *Fructosuria*.
Fructose-1,6-diphosphate, anaerobic metabolism and, *7*
  cortisol and, 25
Fructose-6-phosphate, anaerobic metabolism and, *7*
  pentose shunt and, 13
Fructose tolerance test, liver function and, 44
Fructosuria, alimentary, 87
  detection in body fluids, 29
  essential, 87-88
  fructose level and, 82
  fructose tolerance and, 75
  hypoglycemia and, 58
FSH. See *Follicle-stimulating hormone*.
Functional albuminuria. See *Albuminuria, functional*.
Functional hyperinsulinism, 55
Furosemide, diuretics and, 469
Fusidic acid, protein synthesis and, 247

G cells, gastric function and, 580
G_{M1} Galactosidase, 130(t)
G_{M1} Gangliosidoses, 130, 130(t)
Galactocerebroside β-galactosidase, deficiency of, 130(t)
  glycosphingolipids and, 92
Galactokinase, deficiency of, galactosemia and, 89
Galactolipids, 29
Galactorrhea, prolactin production and, 701
Galactosamine, 6
Galactose, alimentary response to, 75-76
  anaerobic metabolism and, 7, *7*
Galactose,
  body fluids and, 29
  carbohydrate to lipid conversion and, 106
  endogenous sources of, 4
  galactosuria and, 82
  glycolipids and, 6

Galactose *(Continued)*
  lactose and, 6
  urinary. See *Galactosuria, Galactosemia.*
Galactose-1-phosphate, 7
  galactosemia and, 89
Galactose-3-sulfate, sphingomyelins and, 92
Galactose tolerance, 44
Galactosemia, aminoaciduria and, 224
  enzyme abnormalities and, 579
  galactose tolerance and, 76
  hyperbilirubinemia and, 648
  renal tubular defects and, 536
  sugar secretion and, 89
Galactosidase, bacteria and, 248
  carbohydrate metabolism and, 2
β-Galactosidase, deficiency of, 130(t)
Galactoside, bacteria and, 248
Galactosuria. See also *Galactose tolerance.*
  alimentary, 89
  galactose level and, 82
Galactosyl, glycosphingolipids and, 92
α-Galactosyl hydrolase, deficiency of, 130(t)
Galactosylceramide, glycosphingolipids and, 92
Gallbladder. See *Biliary tract.*
Gallstones, bile acids and, 660
Gamma-glutamyl transpeptidase, 570-571
Ganglioneuromas, adrenal medulla and, 744, 745
Gangliosides, carbohydrate to lipid conversion and, 106
  classification of, 92
  of ceramide, sphingomyelins and, 92
Gangliosidosis Type I, 130(t)
Garrod, Alfred, 848
Gastrectomy, fecal fat and, 101
  hypoglycemia in, 55
  vitamin B_{12} deficiency and, 821
Gastric. See also *Stomach.*
Gastric antrum, gastric function and, 580
Gastric lipase, triglycerides and, 98
Gastric phase, gastric function and, 580, 581
Gastric residuum, gastric contents and, 582
Gastric retention, stomach carcinoma and, 584
Gastrin, 771-773
  gastric function and, 580
  pancreas and, 589
  serum, gastric investigation and, 585, 587
Gastritis, atrophic, hypoacidity and, 584
  electrolytes and, 375
Gastroenteritis, achylia and, 587
Gastroferrin, iron absorption and, 321
Gastrointestinal diseases. See under specific diseases.
Gastrointestinal mucosa, prostaglandins in, 95
Gaucher's disease, 129, 130(t)
Genes, nucleic acids and, 240
Genes, regulatory, bacteria and, 128
Gerhardt tests, 848
GH-RIH. See *Growth hormone-release inhibiting hormone.*

Gibbs-Donnan equilibrium, 353
Gilbert's syndrome, 641-642
  phenobarbitone and, 675
Glands. See specific gland.
Globulin(s), abnormal, incidence of, 202
  alpha, 190
    properties of, 169(t)
    serum protein abnormalities and, 190
  Bence Jones proteins and, 199-200
  beta, 191
    lipoprotein electrophoresis and, 116
    properties of, 169(t)
    serum protein abnormalities and, 190
  blood coagulation and, 175
  C-reactive proteins and, 198
  ceruloplasmin and, 197-198
  cryoglobulins and, 198
  disorders and, 190-192
  gamma, 191-196
    properties of, 169(t)
    serum protein abnormalities and, 190, 191-196
  glycoproteins and, 196-197
  liver function and, 621-626
  macroglobulins and, 198-199
  mucoproteins and, 196-197
  myeloma proteins and, 199
  phospholipids and, 118
  plasma protein and, 190
  properties of, 169(t)
  pyroglobulins and, 198
  serum protein concentrations and, 168(t)
  study methods and, 171(t)
  test for, 170
  thymol turbidity and, 626
  thyroxine binding, properties of, 169(t)
Globulin reactions, 205-206
  jaundice diagnosis and, 680
  liver function and, 625
Glomerular filtration, 457-459, 482-487
  arterial hypotension and, 493
  fraction, 489
  hemoconcentration and, 208
  nephritis and, 518-522
  nephrosclerosis and, 527
Glomerulonephritis. See also *Nephritis.*
  acute, defect localization and, 518-520
    electrolytes and, 377
    nonprotein nitrogen and, 509-510
  biochemical manifestations of, 526-533
  chronic, defect localization and, 520
    electrolytes and, 376
    nonprotein nitrogen and, 510-512
  chronic active, nephrotic syndrome and, 138
  glucose tolerance and, 64
  glycosuria and, 85
  hemoconcentration and, 183
  hyperkalemia and, 389
  hyperphosphatemia and, 309
  hypoalbuminemia and, 187, 201
  hypocalcemia and, 303-304
  increased urea nitrogen and, 208

Glomerulonephritis *(Continued)*
  magnesium and, 317
  nonrespiratory acidosis and, 419
  protein quantity in urine and, 218
  renal albuminuria and, 217
Glomerulosclerosis, 209
Glossitis, intestinal malabsorption and, *603*
  vitamin deficiency and, 602
Glucagon, alpha cells and, 589
  carbohydrate metabolism and, 26
  glycogen metabolism and, 9, 10
  insulin release and, 23
  lipid mobilization and, 105
  pancreatic, 26
  pancreatic secretion and, 590
  protein metabolism and, 166
  triglyceride conversion and, 105-106
α-1,4-Glucan, 9
α-1,4-Glucan 6-glycosyltransferase, 9
Glucocerebroside glucosylceramide and, 92
Glucocorticoids. See also *Adrenal cortex, hormones of.*
  adrenal, protein metabolism and, 165
  adrenal cortex and, *715*
  anti-inflammatory action of, *726*
  carbohydrate metabolism change and, 24
Glucogenesis, glucose conversion and, 6
  non-nitrogenous residue and, 161
Glucogenic amino acids. See *Amino acids, glucogenic.*
Glucokinase, anaerobic metabolism and, 7
  fructose metabolism and, 43
  hepatic, carbohydrates and, 125
Gluconeogenesis, absorbed carbohydrates and, 4
  Addison's disease and, 57
  adrenocortical deficiency and, 25
  aerobic metabolism and, 12
  blood glucose concentration and, 31
  blood glucose drop and, 40
  diabetes mellitus and, 47, 61
  diabetic ketoacidosis and, 79
  hepatic, 59-60, 64
  Houssay animals and, 36
  insulin and, 22
  liver and, 14, 162
  pancreatectomy and, 34
  pyruvate kinase and, 125
  transamination and, 161
Gluconeogenic precursors, release of, 25
Glucosamine, 6
Glucose, absorption of, 3
  adrenalectomy and, 24
  glucose tolerance and, 59
  hyperthyroidism and, 59
  kidney tubules and, 460
  administration of, electrolyte changes in, 22
  adrenocortical deficiency and, 25
  aerobic metabolism of, 11-13
  amino acids and, 6
  amino acid concentration and, 179
  anaerobic metabolism of, 7

Glucose *(Continued)*
  blood, absorption of, 31
    cerebrospinal fluid and, 838-840
    concentration of, 30-36
      diminished glucose tolerance and, 59, *60*
      normalcy, 28
      postabsorptive, abnormal, 47-58
      postabsorptive, normal, 27-30
    fall in, 39-40
    fasting, 614-615
    glycogenesis and, 13
    glycogenolysis and, 30
    hypoglycemia and, 72
    lactate and, 77
    regulation of, 30-36
    sources of, 7
    tolerance curve of, 63
  carbohydrate conversion and, 6
  carbohydrate synthesis and, 6
  diabetes mellitus and, 48
  diabetic acidemia and, 53
  endogenous sources of, 4
  exudates and, 395
  fat and, 5
  glucogenic amino acids and, 161
  glycogen metabolism and, *9*
  glycosuria and, 82
  hypochloremia and, 383
  hyponatremia and, 383
  hypophosphatemia and, 47
  insulin hypoglycemia and, 43
  interrelation of nutrient metabolism and, *18*
  intravenous, hypophosphatemia and, 312
  lipid metabolism and, distribution of, *102*
  lipid to carbohydrate conversion and, 107
  nonblood body fluids and, 28
  noncarbohydrate sources and, 4
  oxidation of, 5
  pancreatectomy and, 34
  plasma concentration conversion and, 856
  potassium and, 385-386
  pregnancy and, 62
  reabsorption of, 83
  removal from blood, 31
  storage of, 5
  transudates and, 395
  unesterified fatty acid concentration and, 119
  utilization of, 5-6
    regulatory mechanism in, 32
Glucose-galactose intolerance, 85
Glucose-6-phosphatase, 559
Glucose-6-phosphatase reaction, 13
Glucose-1-phosphate, anaerobic reaction and, 7
  galactosemia and, 89
  glycogen metabolism and, 9, *9*
  glycogenolysis and, 13-14, *14*
  phosphorylase activity and, 27
Glucose-6-phosphate, anaerobic metabolism and, 7, *7*

Glucose-6-phosphate *(Continued)*
   cortisol and, 25
   dehydrogenase, pentose shunt and, 12
   glycogen metabolism and, *9*
   glycogenolysis and, 13-14, *14*
   pentose shunt and, 12, 13(t)
Glucose phosphate isomerase, 559
   cerebrospinal fluid and, 845
Glucose reaction, alimentary, 46-47
Glucose tolerance, biliary tract disorders and, 63-64
   decreased response, 66-70
   diabetes mellitus and, 60
   diminished, 58
   hyperthyroidism and, 63
   liver disorders and, 63-64
   liver function and, 614
   obese diabetics and, 73
   pancreatic carcinoma and, 597
   pancreatitis and, 597
   pituitary hypofunction and, 69
   pregnancy and, 64
Glucose tolerance curve, blood glucose drop and, 40
   normal, 39
Glucose tolerance test, functional hypoglycemia and, 55
   intravenous, 38-39, *39*
   one-hour—two dose, 42-43
   oral, 36-38
      diabetes mellitus and, 61
Glucosidases, carbohydrate digestion and, 1-2
Glucoside phloridzin, glycosuria and, 83
Glucosuria. See *Glycosuria*.
Glucosyl, 9
Glucosylceramide, neutral glycosphingolipids and, 92
Glucosylceramide hydrolase, deficiency of, 130(t)
Glucuronic acid, 6
   normal urine and, 476
Glucuronyltransferase activity, barbiturates and, 675
Glutamate, *18*
   amino acids and, 162
Glutamic acid, 12, 17
   amino acids and, 161
   central nervous system and, 161
Glutamic oxalacetic transaminase, 553
Glutamic pyruvic transaminase, 553
Glutamine, nitrogen surplus and, 159
   nucleic acid metabolism and, *241*
Glutathione, 27, 35
   protein metabolism and, 152
   sulfur and, 327
Glutathione-insulin transhydrogenase, 24
Glycemia, alimentary, curve of, 63
   Staub-Traugott effect and, 42
Glyceraldehyde-3-phosphate, 7, 13, *13*
Glycerokinase, adipose tissue and, 123
Glycerol, free, lipid digestion and, 99
   insulin and, 21

Glycerol, free *(Continued)*
   lipid storage and, 93-94
   lipid to carbohydrate synthesis and, 106
   pyruvate and, 13
Glycerol moiety, 4, 5
   of fats, degradation of, 92
α-Glycerol phosphate, fatty acid synthesis and, 103
   lipid digestion and, 99
   triglyceride conversion and, 106
   triglyceride synthesis and, 108
Glycerophosphate, fat deposition and, 122-123
Glycerophosphatides, carbohydrate to lipid conversion and, 106
Glycine, bone development and, 282
   creatine and, 180
   creatinuria and, 222
   nucleic acid metabolism and, *241*
Glycinuria, aminoaciduria and, 224, 227
   renal tubular defects and, 536
Glycogen, adrenalectomy and, 24
   anaerobic metabolism and, 7
   biliary tract surgery and, 56
   formation of, anaerobic metabolism and, 7, 8, *7, 8*
   glucose drop and, 39-40
   glucose storage and, 5
   glycogenolysis and, 14
   Houssay animals and, 36
   lipid metabolism and, distribution of, *102*
   liver and, 613-614
      anesthesia and, 52
      anoxia and, 52
      blood glucose and, 49
      epinephrine tolerance test and, 46
      glucose drop and, 40
      storage of, 59
   metabolism of, pathways in, *9*
   muscle contraction and, *16*
   myocardial, 20
   storage of, diminished glucose tolerance and, 59
Glycogen α-4-glucosyltransferase, 9
Glycogen storage diseases, 10, 14, 82
   aminoaciduria and, 224
   carbohydrate metabolism and, 79-82
   epinephrine tolerance test and, 74
   glycogenosis and, 80
   hyperlipoproteinemia type four and, 135
   ketosis and, 109
   Type I, hyperuricemia and, 257-258
      lactic acidosis and, 77
   xanthomas and, 141
Glycogen synthetase, 9, 10
Glycogen synthetase kinase, 10
Glycogenesis, absorbed carbohydrates and, 4
   anaerobic metabolism and, 8, *8*
   blood glucose and, 13
   glucose storage and, 5
   hepatic, diabetes mellitus and, 60
      rate of, 30
      regulatory mechanism in, 32

Glycogenesis *(Continued)*
  hepatic, venous blood sugar and, 37
  liver and, 13
  muscle in, 15
  pyruvate in, 13
Glycogenolysis, absorbed carbohydrates and, 4
  anaerobic metabolism and, 9
  epinephrine and, 26
  glucose tolerance and, 69
  hepatic, diabetes mellitus and, 47
    diminished glucose tolerance and, 59
    fasting hyperglycemia and, 47
    glucose drop and, 40
    glucose tolerance and, 64
    hyperthyroidism and, 63
    increase in, 59
    insulin secretion and, 71
    rate of, 30
    thyroxine in, 26
  hyperthyroidism and, 49
  liver and, 13-14
  venous blood sugar and, 37
Glycogenolytic mechanism, 47
Glycogenosis, 14
  type I, 10
  types of, 10, 80, 616
Glycolipids, plasma and, 114
Glycolysis, anaerobic metabolism and, 7-11, 7
  blood lactic acid and, 76
  blood pyruvic acid and, 76
  carbohydrate to lipid conversion and, 106
  defined, 6
  glucose oxidation and, 5
  insulin and, 22
  lactic acidosis and, 77
  muscle and, 15, 16
Glycolytic reaction, 13, 14
Glycolytic sequence, 6
Glyconeogenesis, hepatic, glucose tolerance and, 64
Glycoprotein(s), plasma proteins and, 196-197
α_1-Glycoprotein, properties of, 169(t)
Glycoprotein synthesis, 6
Glycosidases, in carbohydrate digestion and absorption, 1
Glycosphingolipid(s), 92
Glycosphingolipidosis, 92
Glycostatic effect, 25
Glycosuria, acidosis and, 87
  alimentary, 84-85
    criteria for, 62
  asphyxia and, 87
  benign, 83-84
  coronary artery occlusion and, 52
  defined, 83
  emotional, 87
  glomerulonephritis and, 85
  Houssay animals and, 36
  hyperglycemic, 83, 85
  mechanisms of, sugar excretion and, 82

Glycosuria *(Continued)*
  nephrosis and, 85
  nonhyperglycemic, 83
  pancreatectomy and, 34
  paroxysmal hypertension and, 86
  renal, 58, 83-84
    criteria for, 62-63
    glucose tolerance and, 69
    ketonuria and, 110
  renal tubular defects and, 534
  vs. glucosuria, 82
Gmelin test, 848
Goiter, nontoxic, 764-766
  thyroid stimulating hormone immunoassay and, 760
  plasma cholesterol and, 143
  toxic multinodular, hyperthyroidism and, 766
Gonadotrophic hormones, adenohypophysis and, 691-699
  assays for, 691-692
  decrease in, 698-699
  follicle stimulating hormones and, 693
  human chorionic, 693
  human menopausal, 693
  increase in, 696-698
  luteinizing hormone and, 691-693
  physiologic considerations and, 693-695
Gonadotrophin, normal values for, 695
  pituitary gland and, 696
  secretion of, abnormal, 695-696
  suppression of, 698-699
  value variation of, 699(t)
Gonads, cholesterol synthesis and, 113
  dysfunction of, galactose tolerance and, 76
GOT. See *Glutamic oxalacetic transaminase.*
Gout, cryoglobulins and, 198
  hyperlipoproteinemia type four and, 135
  hyperuricemia and, 255
GPT. See *Glutamic pyruvic transaminase.*
Grabar, immunoelectrophoresis and, 850
Granulomatous proliferation, gamma-globulins and, 191
Grave's disease, hyperthyroidism and, 766
GRH. See *Growth hormone-releasing hormone.*
Growth, thyroid hormone and, 753
  vitamin A deficiency and, 808
Growth hormones. See *Hormones, growth.*
Growth hormone-release inhibiting hormone, 690
Growth hormone-releasing hormone, 690
GTP. See *Guanosine triphosphate.*
Guanethidine, catecholamines and, 744
Guanidine, 532
Guanine, nucleoside triphosphates and, 250
Guanosine monophosphate, cyclic, 251
  adenosine monophosphate and, 686
Guanosine triphosphate (GTP), protein synthesis and, 247
Guanylic acid, nucleic acid biosynthesis and, 237

Haff disease, myoglobinuria and, 272
Hageman factor, coagulation and, 175
Hair, iodine and, 332
Hamman-Hirschman effect, 42
Hand-Schüller-Christian disease, 130
  xanthomatosis and, 141
Haptoglobin(s), hemoglobin and, 197, 265
  hemoglobin transport and, 175
  properties of, 169(t)
Haptoglobin type I-I, cerebrospinal fluid and, 835
Harrison spot test, bilirubinuria and, 649-656
Hartman method, 848
Hartnup syndrome, aminoaciduria and, 223-224
  renal tubular defects and, 536
Harvard fractions, plasma proteins and, 168
Hay test, 848
HCG. See *Human chorionic gonadotrophin.*
HCS. See *Human chorionic somatomammotropin.*
HCT. See *Human chorionic thyrotrophin.*
Heart, cholesterol synthesis and, 112
  diseases of. See also *Coronary artery occlusion* and *Heart Failure.*
    aminotransferases and, 553
    arterial, cryoglobulins and, 198
    congenital, hypoxia and, 442-443
    congestive, lactate and, 76
    creatine kinase and, 570
    enzyme abnormalities and, 576
    Framingham studies and, 133
    hyperlipoproteinemia type four and, 135
    hyperlipoproteinemia type two and, 134
    prerenal albuminuria and, 215
    uric acid and, 256
  glycogen and, 79
Heart failure. See also *Myocardial infarction.*
  congestive, blood volume and, 391
    electrolytes and, 378
    excess electrolytes and, 381
    hypokalemia and, 387
    kidney function and, 526
  glomerular filtration and, 457
Heat, production of, 445
Heavy-chain disease, serum protein abnormalities and, 202
Heinz bodies, hemoglobin and, 269
Hematinemia, hemoglobin and, 274
Hematocrit dehydration and, 374
Hematuria, vs. hemoglobinuria, 219
Heme, iron absorption and, 321
Hemochromatosis, glucose tolerance and, 64
  glycosuria and, 87
  insulin resistance and, 57
  iron metabolism and, 325-326
  porphyrinuria and, 655
  secretin test and, 592

Hemoconcentration, diabetes mellitus and, 392
  electrolytes and, 382
  excretion and, 183
  glomerular filtration and, 457
  hyperparathyroidism and, 208, 295
  oliguria and, 493
  plasma protein abnormalities and, 183
  urea nitrogen and, 208
Hemodialysis, kidney function and, 545
Hemoglobin, abnormal, 267-269
  acid-base balance and, 402-403
  biosynthesis of, 262-265
    regulation of, *264*
  buffer action of, 402-406
  carbon dioxide buffering and, 404
  carboxyhemoglobin and, 273
  catabolism of, 265
  defined, 260
  dehydration and, 374
  derivatives of, abnormal, 272
  ethnic groups and, 268
  fetal, 266-267
    hereditary persistence of, 267
  haptoglobins and, 265
  identification of, procedures for, 270
  iron and, 320
  kidneys and, 219
  metabolism of, 260-278, 633-636
  methemalbuminemia and, 273-274
  methemoglobin and, 272
  myoglobinuria and, 271-272
  oxygenation of, oxygen/carbon dioxide tension and, *439*
  sulfhemoglobin and, 273
  synthesis of, iron and, 324
  transport of, plasma proteins and, 175
  varieties of, human, 266-267
Hemoglobin A, sickle cell anemia and, 268
Hemoglobin F, abnormal hemoglobin and, 268
Hemoglobin H, thalassemia and, 270
Hemoglobin M, abnormal hemoglobin and, 269-270
Hemoglobin S, sickle cell anemia and, 268
Hemoglobinemia, 270-271
Hemoglobinopathy, hemoglobin identification and, 270
Hemoglobinuria, 270-271
  paroxysmal, urobilinogen levels and, 651
  proteinuria and, 219
Hemolysis, enzyme defects and, 12
  glucose-6-phosphate dehydrogenase and, 13
  hematinemia and, 274
  jaundice and, *643*
  pentose shunt and, 12
  porphyrinuria and, 274, 275(t)
  urobilinogenuria and, 652-653
  vitamin E and, 830
Hemolytic jaundice. See *Jaundice, hemolytic.*

INDEX

Hemopexin, heme and, 197
  properties of, 169(t)
Hemopoiesis, malabsorption syndrome and, 608
Hemorrhage, adenohypophyseal hypofunction, 69
  bilirubin and, 643(t)
  cerebral, cerebrospinal fluid and, 836(t), 837
    cholesterol and, 841
    glomerular filtration and, 457
    glucose tolerance and, 64
    hemoglobinemia and, 270-271
    hemoglobinuria and, 270-271
    intestinal malabsorption and, *603*
    plasma cholesterol and, 141
    subarachnoid, hypoglycemia in, 58
Hemosiderin, iron storage and, 323
Hemosiderosis, excess iron and, 326
Henderson-Hasselbalch equation, 414
  respiratory regulation and, 407
Heparan sulfate, acid mucopolysaccharides and, 233
Heparin, acid mucopolysaccharides and, 233
  lipemia and, 133
  plasma and, 117
Heparinase, hyperlipemia and, 117
Hepatic. See also *Liver*.
Hepatic coma, hypoglycemia in, 56
Hepatic disease. See *Liver, disorders of*.
Hepatic insufficiency. See *Liver, disorders of*.
Hepatic tricarboxylic acid cycle, 105
Hepatitis. See also *Liver, diseases of*.
  albumin and, 621
  bilirubin and, 643(t), 644-648
  blood sugar tolerance curve and, 63
  cephalin-cholesterol test and, 628
  cholesterol and, 631-633
  cryoglobulins and, 198
  dye retention and, 669
  epinephrine tolerance test and, 74
  function tests and, 676-683
  globulins and, 200-205, 625-626
  hippuric acid and, 662
  hypercholesterolemia and, 140
  hypocholesterolemia and, 142
  hypoglycemia in, 56
  macroglobulinemia and, 199
  mucoproteins and, 197
  neonatal, jaundice and, 648
  plasma fibrinogen and, 184
  plasma protein abnormalities and, 201
  porphyrinuria and, 655
  proteins and, 617-630
Hepatocellular dysfunction, neonatal jaundice and, 647
Hepatolenticular degeneration, aminoaciduria and, 224
  ceruloplasmin and, 336
  renal tubular defects and, 537
Hepatoma, alpha-1-fetoprotein and, 624
Hepatorenal syndrome, 208, 525

Hepatosplenomegaly, cobalt and, 337
  heavy chain disease and, 202
  hyperlipoproteinemia type five and, 136
  hypogammaglobulinemia and, 204
  hypolipoproteinemia and, 137
  Waldenström's macroglobulinemia and, 203
Hereditary angioedema, plasma proteins and, 175
Heredity, disorders of, hyperuricemia and, 257-258
Herpes simplex, neonatal jaundice and, 648
Hers' disease, glycogen storage disease and, 830
Hexokinase, 7, 7
  anaerobic metabolism and, 7, 7
  glycolytic reaction equilibria and, *14*
  hemolytic anemia and, 559
  in liver and muscle, 43
Hexokinase reaction,
  carbohydrate absorption and, 2
Hexosaminidase-A, deficiency of, 130(t)
Hexose, absorption of, thyroxine and, 26
  in carbohydrate metabolism, 4, 6
Hexose monophosphate shunt, 6, 12-13
Hexose phosphates, *16*, 46
Hippuric acid, normal urine and, 476
  synthesis of, 662
Hiroshima diabetes survey, 61
Histamine, gastric analysis and, 587
  plasma proteins and, 175
Histidine, hemoglobin buffer system and, 402
  iron absorption and, 321
  malabsorption syndromes and, 608
  nitrogen equilibrium and, 154
  pregnancy and, 182
Histidinuria, aminoaciduria and, 227
Histoplasmosis, hyperglobulinemia and, 201
  serum enzymes and, 670
HMG. See *Human menopausal gonadotrophin*.
HMT. See *Human molar thyrotrophin*.
HnRNA. See *Ribonucleic acid, heteronuclear*.
Homeostatic mechanism, blood sugar and, 28
  diabetes mellitus and, 60
Homeostatic response, hypoglycemia and, 71
Homocystinuria, aminoaciduria and, 227
Homovanillic acid, hepatic coma and, 675
Hormones. See also *Endocrine system*.
  adenohypophyseal. See *Adenohypophysis, hormones of*.
  adrenal medullary. See *Adrenal medulla, hormones of*.
  adrenocortical. See *Adrenal cortex, hormones of*.
  adrenocorticotrophic. See *Adrenocorticotrophic hormone*.
  calcium metabolism and, 291-292
  catecholamines and, 741

Hormones *(Continued)*
  ectopic. See *Ectopic hormones.*
  endocrine, protein metabolism and, 164-166
  gastrointestinal, endocrine function and, 771-774
  glucocorticoid, 9, 25
  growth, adenohypophysis and, 705-709
    amino acid concentration and, 179
    fructose metabolism and, 43
    hypoglycemia and, 125
    ketonuria and, 110
    lipase biosynthesis and, 106
    lipolytic mechanism and, 124
    magnesium and, 317
    metabolic effects of, 706-708
    pituitary and, insulin balance, 17
    protein metabolism and, 164-165
    secretion of, 705-706, 708
    transport of, 705-706
    unesterified fatty acid concentration and, 119
  hypothalamic regulatory, 688(t)
  imbalances of,
    adrenocortical preponderance and, 22
  ionic regulation and, 358
  lipid mobilization and, 105-106
  menopause and, 781
  menstrual cycle and, 780-781
  metabolism of, liver and, 663-666
  11-oxygenated adrenocortical, gluconeogenesis and, 162
    ketonuria and, 110
  parathyroid, magnesium and, 316
  pituitary, glucose utilization and, 32
  potassium excretion and, 363
  protein anabolism and, 164
  thyroid, metathyroid diabetes and, 35
Houssay animals, 35-36
Houssay preparation, 32
Howe fractionation procedure, 190
HPL. See *Human placental lactogen.*
Hüfner factor, 437
Human chorionic gonadotrophin, 693, 790
Human chorionic somatomammotropin, 705, 790-791
Human chorionic thyrotrophin, pregnancy and, 791
Human menopausal gonadotrophin, 693
Human molar thyrotrophin, pregnancy and, 791
Human placental lactogen, 705, 790-791
Hunter's syndrome, mucopolysaccharidoses and, 233
Hurler's syndrome, mucopolysaccharidoses and, 233
Hyaluronic acid, acid mucopolysaccharides and, 232-233
Hydatidiform mole, plasma fibrinogen and, 184
Hydrochloric acid, achlorhydria and, 585-586

Hydrochloric acid *(Continued)*
  gastric activity and, 580
  iron absorption and, 320
  loss of, nonrespiratory alkalosis and, 423
  pepsin and, 587
Hydrogen, fatty acid synthesis and, 103
Hydrogen ions, acid-base balance and, methods of study, 426-432
  acidosis and, 415-421
  alkalosis and, 421-425
  bone and, 406
  buffer systems and, 402
  concentration of, abnormalities, 413
    cerebrospinal fluid and, 843
    disturbances and, 414
    factors determining, 413-414
  fluid concentration of, 399-432
  pH relations and, 412-413
  regulation of, disturbances in, 415(t)
    plasma proteins and, 174
  renal regulation and, 407
  respiratory regulation and, 406-407
Hydrolysis, enzyme, insulin and, 21
  muscle contraction and, 16
11-Hydroxyandrostenedione, adrenal cortex and, 714
Hydroxybutyrate, 21-22
β-Hydroxybutyrate, ketosis and, 109
β-Hydroxybutyric acids, ketosis and, 109
  pancreatectomy and, 34
β-Hydroxybutyryl, fatty acid synthesis and, 103
Hydroxylysine, bone development and, 282
3-Hydroxy-methylglutaryl CoA, cholesterol biosynthesis and, 96
Hydroxyphenylalanine, phenylalanine and, 225
p-Hydroxyphenylpyruvic acid, hydroxyphenylalanine and, 225
Hydroxyproline, acromegaly and, 224
  bone development and, 282
  calcitonin and, 284
  urinary, bone disease and, 224
Hydroxyprolinuria, aminoaciduria and, 228
Hyperacidity, gastric investigations and, 584-586
Hyperadrenalism, diminished glucose tolerance and, 59
Hyperaldosteronism, adrenocortical hyperfunction and, 734
  magnesium and, 317
  potassium and, 387
  potassium-losing nephritis and, 390
Hyperammonemia, urea cycle disorders and, 231
Hyperargininemia, urea cycle disorders and, 231
Hyperbilirubinemia, conjugated, drug-induced jaundice and, 646
  drugs and, 646
  galactosemia and, 89
  hypercholesterolemia and, 139, 142

Hyperbilirubinemia *(Continued)*
  niacin and, 815
  pancreatic, 597
  pathological, 638-640
  serum enzymes and, 670
  unconjugated, drugs and, 645-646
  vitamin K and, 833
Hypercalcemia, alkali intoxication and, 299-300
  ectopic hormones and, 800
  hyperglobulinemia and, 202
  hyperlipoproteinemia type four and, 135
  hyperparathyroidism and, familial, 296
    primary, 294-295
    secondary, 295-296
  hypersecretion and, 587
  hypervitaminosis and, 297
  idiopathic infantile, 299
  immobilization and, 298
  magnesium and, 317
  milk intoxication and, 299-300
  multiple myeloma and, 297
  neoplastic disease and, 298
  sarcoidosis and, 298-299
Hypercalciuria, chronic renal failure and, 528-529
  idiopathic, renal tubular defects and, 538
  urinary calcium and, 308
  lithiasis and, 540-541
  magnesium and, 317
Hypercapnia, hydrogen ion regulation and, 415(t)
Hypercarotenemia, vitamin A and, 810
Hyperchloremia, 380-383
  gastrointestinal disorders and, 375
  hypokalemia and, 390
Hyperchloremic acidosis, galactosemia and, 89
Hypercholesterolemia, biliary tract disease and, 139-140
  diabetes and, 21
  lipid metabolism and, 137-142
  lipid utilization and, 94
  liver disorders and, 139-140, 631
  pancreatectomy and, 34
  plasma cholesterol and, 120
  plasma lipoprotein abnormalities and, 131-134
Hyperestrogenism, mucoproteins and, 197
Hypergammaglobulinemia, hepatitis and, 201
Hyperglobulinemia, diseases and, 201
  hypercalcemia and, 202
  liver function and, 621-626
  plasma protein abnormalities and, 183, 190-205
Hyperglycemia, adrenal medullary tumor and, 86
  alimentary response and, 63
  carbohydrate nutrition and, 41-42
  diabetic ketoacidosis and, 78-79
  diminished glucose tolerance and, 59

Hyperglycemia *(Continued)*
  diuresis and, 469
  epinephrine, liver function and, 616
  ether and, 138
  fasting, 47-54
    diabetes mellitus and, 60
    tolbutamide tolerance test and, 72
  glucose absorption in, 59
  glucose tolerance and, 64
  Houssay animals and, 36
  insulin deficiency and, 22
  paroxysmal, 51
  serum gastrin and, 588
  Staub-Traugott effect and, 42
  steroid induced, 45
  without glycosuria, 86
Hyperglycemic factor, 26
Hyperglycemic phase, venous blood sugar and, 37, *37*
Hyperglycorrhachia, cerebrospinal fluid and, 839
Hyperinsulinism, epinephrine tolerance test and, 74
  fasting hypoglycemia and, 54-56
  glucose tolerance, diminished response, 66-67
  hypoglycemia and, 70
  ketosis and, 109
  organic, 54
Hyperkalemia, 388-389
Hyperketonemia. See *Ketosis*.
Hyperlactatemia, carbohydrate metabolism and, 77
Hyperlipemia, adipose tissue and, 124
  diabetes and, 21
  diabetic ketoacidosis and, 79
  electrolytes and, 374
  hypoalbuminemia and, 138
  lipid utilization and, 94
  pancreatitis and, 597
  plasma lipoprotein abnormalities and, 131-134
  post-alimentary, 117
  postprandial, food intake and, 119
  starvation and, 121
  unesterified fatty acids and, 119
Hyperlipidemia, 133
Hyperlipoproteinemia, 134-136, 141
Hypermagnesemia, chronic renal failure and, 528
  serum calcium and, 317
Hypernatremia, electrolyte concentration and, 371
  gastrointestinal disorders and, 375
  hypertonicity and, 373
  primary water deficit and, 380
  sodium excess and, 380
Hyperosmolality, diabetic ketoacidosis and, 78-79
Hyperoxaluria, lithiasis and, 542
Hyperparathyroidism, diagnosis of, 296
  electrolytes and, 382

Hyperparathyroidism *(Continued)*
  familial, hypercalcemia and, 296
  hemoconcentration and, 208, 295
  hydroxyproline and, 224
  hypercalciuria and, 317
  hypomagnesemia and, 317-318
  hypophosphatemia and, 311
  primary, hypercalcemia and, 294-295
  secondary, hypercalcemia and, 295-296
  skeletal diseases and, 563
  urinary calcium and, increased, 307
Hyperphosphatemia, 309-310
Hyperpituitarism, 51
  galactose tolerance and, 76
  glycosuria and, 86
Hyperplasia, adrenal, congenital, 735, *736*
  adrenocortical, intrathoracic neoplasms and, 379
Hyperpotassemia, 383-391
Hyperprolactinemia, lactation and, 702
Hyperproteinemia, 183, 184, 190-200
Hypersecretion, gastric investigation and, 584, 586-587
Hypertension, albuminuria and, 217
  blood volume and, 391
  essential, 523. See also *Nephrosclerosis, benign.*
  glucose and, 51, 53
  glucose tolerance and, 64
  hyperglycemia and, 53
  magnesium and, 317
  oral contraceptives and, 785
  paroxysmal, 51
    glycosuria and, 86
  renal blood flow and, 489, 493
  renal failure and, 512, 523
  urea clearance and, 504, 523
Hypertensive encephalopathy, 51
Hyperthyroidism, alimentary response and, 63
  blood volume and, 391
  ceruloplasmin and, 335
  creatinuria and, 222
  diagnosis of, 767
  diminished glucose tolerance and, 59
  ectopic hormones and, 801-802
  endocrine function and, 766-768
  fasting hypoglycemia and, 59
  galactose tolerance and, 76
  glucose absorption and, 3
  glucose tolerance and, 64
  glycogen conversion and, 49
  glycosuria and, 86
  hydroxyproline and, 224
  hypocholesterolemia and, 143-144
  hyposecretion and, 586
  iodine and, 332
  mucoproteins and, 197
  prerenal albuminuria and, 216
  thyroid stimulating hormone immunoassay and, 760
  thyrotrophin-releasing hormone test, 762

Hyperthyroidism *(Continued)*
  urinary calcium and, increased, 307
  vs. T_3 toxicosis, 767(t)
Hypertonic contraction, electrolyte concentration and, 371
Hypertonicity, hypernatremia and, 373
Hypertrophic hypersecretory gastropathy syndrome, 586
Hyperuricemia. See also *Uric acid, blood and.*
  nucleic acids and, 255-257
  uric acid and, 255
Hyperuricuria, lithiasis and, 543
Hypervitaminosis, hypercalcemia and, 297
  hyperphosphatemia and, 309
Hypervitaminosis D, 307, 829
Hypoacidity, gastric investigation and, 584
Hypoalbuminemia, albumin loss and, 186-187
  genetic defects and, 189
  hyperlipemia and, 138
  lactation and, 189
  lipoprotein metabolism and, 139
  liver disorders and, 621
  multiple myeloma and, 201
  plasma dilution and, 189
  plasma protein abnormalities and, 183
  plasmapheresis and, 139
  pregnancy and, 189
  protein absorption and, 607
  proteins and, 188-189
  renal albuminuria and, 217
  sarcoidosis and, 201
Hypobeta-lipoproteinemia, fecal fat and, 101
  hypolipoproteinemia and, 136
Hypocalcemia, abnormal serum calcium and, 300-307
  chronic renal failure and, 303-304, 528-529
  citrate tetany and, 305
  intestinal malabsorption and, *603*
  pancreatitis and, 305, 597
  potassium and, in diarrhea, 387
  vitamin D deficiency and, 302, 828
Hypochloremia, 368-380
  acute glomerulonephritis and, 377
  bromide intoxication and, 383
  diabetic ketoacidosis and, 78
  erysipelas and, 382
  gastrointestinal disorders and, 375
  glomerulonephritis and, 376
  glucose and, 383
  hemoconcentration and, 382
  hyperparathyroidism and, 382
  meningococcal meningitis and, 382
  oliguria and, 382
  pneumonia and, 382
  primary potassium deficiency and, 382
  renal failure and, 382
  rheumatic fever and, 382
  tuberculosis and, 383
  typhoid fever and, 383

Hypochloremia *(Continued)*
  urinary chloride and, 383
Hypochlorhydria. See also *Hypoacidity*.
  stomach carcinoma and, 584
Hypocholesterolemia. See also *Cholesterol, plasma*.
  anemia and, 142
  hyperthyroidism and, 143-144
  infectious diseases and, 142
  lipid metabolism and, 142-145
  liver disorders and, 142-143
Hypochromic anemia, protein abnormalities and, 602
Hypocupremia, copper and, 335
Hypofibrinogenemia, plasma fibrinogen and, 184
Hypogammaglobulinemia, hypoalbuminemia and, 188
  immunoglobulin synthesis and, 196
  serum protein abnormalities and, 203-204
  steatorrhea and, 601
Hypoglycemia, adrenalectomy and, 24
  ectopic hormones and, 800
  epinephrine secretion and, 50
  fasting, 54-58
  gastric investigation and, 584
  growth hormone and, 125
  Houssay animals and, 36
  hyperinsulinism and, 67
  insulin-glucose and fructose in, 43
  insulin-glucose tolerance test and, 71
  insulin induced, 20
  insulinoma and, 45
  leucine in, 23
  liver function and, 614-615
  poisoning and, 614
  serum gastrin and, 588
  spontaneous, 46
  tolbutamide tolerance test and, 72
  unresponsiveness of, 70
Hypoglycemic phase, venous blood sugar curve and, 37, 38
Hypoglycorrhachia, 839-840
Hypogonadism, cobalt and, 337
  female, 782, 782(t)
Hypoinsulinism. See also *Diabetes mellitus*.
  glucose tolerance, diminished response to, 67
Hypokalemia, 383-390
  diabetic ketoacidosis and, 78-79
  galactosemia and, 89
  hepatic coma and, 674
Hypolipemia. See also *Fat, Fatty acid, Phospholipids, Cholesterol, plasma, Hypocholesterolemia*.
  fatty liver and, type four, 127
Hypolipoproteinemia, lipid metabolism and, 136-137
Hypomagnesemia, diabetic ketoacidosis and, 78
  hyperparathyroidism and, 317-318
Hyponatremia, 368-380

Hyponatremia *(Continued)*
  acute glomerulonephritis and, 377
  diabetic ketoacidosis and, 78-79
  electrolyte concentration and, 371
  gastrointestinal disorders and, 375
  glomerulonephritis and, 376
  glucose and, 383
  hemoconcentration and, 382
  hyperparathyroidism and, 382
  oliguria and, 382
  pulmonary neoplasms and, 379
  renal failure, 382
  renal insufficiency and, 527
Hypoparathyroidism, hyperphosphatemia and, 309
  hypocalcemia and, 300-301
  urinary calcium and, 308
Hypophosphatasia, 563-564
  bone development and, 314
  phosphoethanolamine and, 224
Hypophosphatemia, 310-312
  glucose utilization and, 47
Hypophysectomy, carbohydrate metabolism and, 25
  gonads and, 691
  sterol metabolism and, 113
Hypophysis, anterior, hyperfunction of, 63
  insulin deficiency and, 48
  divisions of, importance, 697
Hypopituitarism, basal metabolism and, 451
  insulin and, 57, 68-69, 73-74
  intestinal malabsorption and, *603*
Hypopotassemia. See *Hypokalemia*.
Hypoproteinemia, albuminemia and, 188
  ceruloplasmin and, 336
  chronic renal failure and, 530-531
  hypocalcemia and, 303
  idiopathic, globulin concentration and, 190
  intestinal malabsorption and, *603*
  protein abnormalities and, 602
Hypoprothrombinemia, intestinal malabsorption and, *603*
  liver disease and, 628
  vitamin K deficiency and, 833
Hyposecretion, gastric investigation and, 586
Hyposthenuria, hypokalemia and, 390
  significance of, kidney function and, 498-501
Hypotension, 493, 523
Hypothalamus, anterior lobe regulation and, 687-688, 688(t)
  extracellular fluid and, 359
  hypogonadism and, 782(t)
  importance of, 687
  lesions to, glucose tolerance and, 64
    glycosuria and, 87
  pituitary control of, 687, 749
  psychic phase and, 581
Hypothermia, potassium and, 390
Hypothyroidism, compensated, 766

Hypothyroidism *(Continued)*
  endocrine system function and, 764-766
  glucose absorption and, 3
  glucose tolerance, diminished response to, 69
  hypercholesterolemia and, 140
  hyperlipoproteinemia type four and, 135
  hypoglycemia and, 58
  iodine and, 332
  laboratory diagnosis of, 765
  mild, 766
  mucoproteins and, 197
  overt, 765-766
  plasma lipoprotein abnormalities and, 132
  primary, insulin tolerance test and, 71
  thyroid stimulating hormone immunoassay and, 760
  thyrotrophin-releasing hormone test, 762
  urinary calcium and, 308
  xanthomas and, 141
Hypotonic contraction, electrolyte concentration and, 371
Hypouricemia, nucleic acids and, 258
Hypoxanthine, nucleotide, synthesis and, 252
Hypoxia, anemic, 443
  basal metabolism and, 442-444
  blood sugar level and, 52
  histotoxic, 444
  hypoxic, 442
  lactic acidosis and, 77
  neonatal jaundice and, 647
  relative, 444
  stagnant, 443-444
Hysteria, respiratory alkalosis and, 422

ICD. See *Isocitrate dehydrogenase*.
ICSH. See *Luteinizing hormone*.
Icterus index, 638. See also under specific disorder.
Ictostix, bilirubinuria and, 649-656
Ictotest, bilirubinuria and, 649-656
Ig. See *Immunoglobulins*.
Ileum, diseases of, steatorrhea and, 659
Iminoglycinuria, familial, renal tubular defects and, 536
Imipramine, catecholamines and, 743
  negative nitrogen balance and, 229
Immobilization, hypercalcemia and, 298
Immunity, plasma proteins and, 175
  properdin system and, 177
Immunoglobulin(s), gamma-globulins and, 190-196
  groups of, 193-196
Immunoglobulin IgG, cerebrospinal fluid and, 835
Immunologic technique, plasma protein abnormalities and, 183
Inanition, hypocholesterolemia and, 144

Inanition *(Continued)*
  protein and, 188
  urobilinogen and, 651
Indican, urine and, 329, 532
Infantilism, pituitary. See *Pituitary infantilism*.
Infants, premature, xanthochromia and, 844
Infections. See also *Diseases, infectious*.
  bacterial, gamma-globulins and, 191
    lactic acidosis and, 77
  C-reactive proteins and, 198
  glycosuria and, 87
  negative nitrogen balance and, 229
  neonatal jaundice and, 647
  plasma fibrinogen and, 184
  pyogenic, glucose tolerance and, 64
  serum protein abnormalities and, 200
  urinary nonprotein nitrogen and, 508
Inosine, iron absorption and, 322
Inositol, phosphatidyl compounds and, 94
  type two fatty liver and, 127
  type four fatty liver and, 128
Inositol phosphoglycerides, lipid metabolism and, 94
Inspissated bile syndrome, neonatal jaundice and, 648
Insulin. See also *Proinsulin, Pancreas* and *Langerhans, islands of*.
  actions of, 19-20
  adipose tissue and, 124
  administration of, 19
    electrolyte changes and, 22
  adrenal 11-hydroxysteroids and, 17
  amino acid concentration and, 179
  anaerobic metabolism and, 10
  "big," 46
  biosynthesis of, origin of, 23
  carbohydrate metabolism and, 18-24
  deficiency of, 17-21
    diabetes mellitus and, 47-49
  disposal of, 24
  endogenous, obese diabetics and, 73
  fatty acid synthesis and, 21, 104
  fructose metabolism and, 43
  gastric secretion and, 582
  glucose utilization and, 31
  glycolysis and, 22
  hepatic glycogenolysis and, 30
  lipid mobilization and, 105
  obese diabetics and, 73
  pituitary growth hormone and, 17
  plasma, chronic renal failure and, 531
    diazoxide and, 24
    levels of, 46
    obesity and, 72
  production of, factors inhibiting, 59
  protamine zinc, 54
  protein metabolism and, 165
  resistance to, 57
    diabetic ketoacidosis and, 78
  resistance vs. sensitivity, 71
  secretion of, 23-24

Insulin *(Continued)*
  sensitivity to, Houssay animals and, 36
    steroid diabetes and, 64
  sulfur and, 327
  triglyceride synthesis and, 104
  unesterified fatty acid concentration and, 119
  urine and, 24
  vagus effect and, 590
Insulin-glucose tolerance test, 44-45
  hypoglycemia and, 71
Insulin metabolism, 24
Insulin tolerance, abnormal, 70-72
Insulin tolerance test, 44
  adrenocortical hypofunction and, 739
  functional hypoglycemia and, 55
Insulinases,
  insulin disposal and, 24
Insulinoma(s), 24
  glucose concentration and, 72
  obesity and, 74
  tolbutamide tolerance test and, 45-46
Intestinal fistulas, 369, 375, 392
  plasma protein and, 185, 188, 189
Intestinal glucagon, 26
Intestinal malabsorption syndromes. See *Intestines, malabsorption syndromes of*.
Intestinal mucosa, carbohydrate metabolism and, 1
  cholesterol and, 112
Intestinal phase, gastric function and, 580, 581-582
Intestine(s), cholesterol esterification and, 113
  disorders of, electrolytes and, 375
    potassium and, 386
  magnesium and, 316
  malabsorption syndromes of, 600-609
  obstruction of, albuminuria and, 216
    creatine and, 210
    electrolytes and, 375
    plasma albumin and, 185
  pH of, calcium absorption and, 281
  phospholipids and, 112
  small, atrophy of, *603*
    heavy-chain disease and, 202
    malabsorption syndromes of, *603*
    protein digestion and, 148
Intracellular water. See *Water, intracellular*.
Intravenous glucose tolerance test. See *Glucose tolerance tests*.
Inulin clearance. See *Glomerular Filtration*.
Iodide, plasma inorganic, blood iodine and, 331
Iodine, metabolism of, 330-333
  plasma concentration conversion and, 856
  thyroid and, 747-748
  thyroid hormone interference and, 754
5-Iodo-2'-deoxyuridine, cancer chemotherapy and, 249
Iodotyrosines, coupling of, 748

Iodotyrosines *(Continued)*
  thyroglobulin and, 332
Ions. See also specific ions.
  concentration of, in plasma, 349
  potassium, insulin release and, 24
  regulation of, Renin-Angiotensin and, 360
Iproniazide, hepatic coma and, 674
Iron, hemoglobin and, 156, 263-264
  metabolism of, 319-326, *324*
    abnormal, 325-326
    absorption, 320-322
    deficiency, 320, 322, 325-326
    distribution, *323*, 324
    excess, 325-326
    excretion, 320-322
    function, 319
    storage, 323-325
    transport, 322-323
    utilization, 323-325
  plasma concentration conversion and, 856
  porphyrin biosynthesis and, 263-264
  serum, 671-672
  transport of, plasma proteins and, 175
Irritable bowel syndrome, hyposecretion and, 586
Ischemia, kidney function and, 456
Iso-amylamine, hepatic coma and, 674
Isocitrate dehydrogenase, 558
Isocitric acid, fatty acid synthesis and, 104
Isoenzymes, 548-549
  alkaline phosphatase and, 561-562, *561*
  amylase, pancreatic function and, 596
  aspartate aminotransferase and, 554
  lactate dehydrogenase and, *555, 556, 557*
  leucine aminopeptidases and, 571
Isoleucine, ketogenic pathway and, 162
Isomaltose, carbohydrate metabolism and, 1-2
Isomerase, cerebrospinal fluid and, 845
Isoproterenol, insulin release and, 24
Isosthenuria. See *Hyposthenuria*.
Isotonic contraction, electrolyte concentration and, 371
Isotope(s), dilution of, iron storage and, 326
  glucose conversion to fat and, 5
Isovaleric acid, 55
Isozymes. See *Isoenzymes*.

Jaffe tests, 848
Jaundice, albuminuria and, 216
  classification of, *643(t)*
  diagnosis of, 676
  drug induced, 645-646
  familial, serum bilirubin and, 641
  galactose tolerance and, 75
  hemolytic, plasma cholesterol and, 142
    unconjugated serum bilirubin and, 641
    urobilinogen levels and, 651
  hepatocellular, 645

Jaundice *(Continued)*
  hepatocellular, hypercholesterolemia and, 140
    serum enzymes and, 670
  hepatocellular vs. extrahepatic obstructive, 676-677
  hereditary, serum bilirubin and, 641
  kernicterus and, 648
  liver function and, 49
    neonatal, phenobarbitone and, 675
  nonhemolytic, serum bilirubin and, 641
    serum bilirubin and, 641
  nonobstructive, serum bilirubin and, 641
  obstructive, blood sugar tolerance curve and, 63
    plasma lipoprotein abnormalities and, 132
    porphyrinuria and, 655
    prolonged, steatorrhea and, 302
    steatorrhea and, 601
  pathogenesis of, 640-648
  pathological neonatal, 646-648, 646(t)
  pigment metabolism and, 633-640
  plasma cholesterol and, 139, 140
  spirochetal, hypocholesterolemia and, 142
  undetermined, diagnosis of, 679-681
  xanthochromia and, 845
Jejunum, pH of, iron absorption and, 320
Joule, defined, 855
Juice, gastric. See *Gastric juice.*
Juvenile G_{M1} ganglosidosis, 130(t)

Kala-azar, calcium and, 303
  cryoglobulins and, 198
  macroglobulins and, 199
  plasma protein abnormalities and, 201
  thymol turbidity test and, 627
Katal, defined, 548
Keratan sulfate, acid mucopolysaccharides and, 233
Keratin, protein digestion and, 149
Keratomalacia, vitamin A and, 808
Kernicterus, liver function and, 648-649
α-Keto acid(s), 17
  non-nitrogenous residue and, 161
Ketoacidemia, diabetic ketoacidosis and, 79
Ketoacidosis, diabetic, acidemia and, 78-79
  carbohydrate metabolism and, 78-79
Ketogenesis. See also *Ketosis.*
  adrenal cortical hormones and, 726
  diabetes and, 21
  Houssay animals and, 36
  ketosis and, 109
  lipid metabolism and, *102*
  nutrient metabolism interrelation and, 17, *18*
Ketogenic amino acids, fatty acid synthesis and, 104
Ketogenic pathway, protein metabolism and, 162

α-Ketoglutarate, *18*
  hepatic coma and, 674
  metabolic interrelations and, 163(t)
  nucleic acid metabolism and, *241*
  protein digestion and, 151
  proteins and, 163
α-Ketoglutaric acid, amino acids and, 161
  nutrient metabolism interrelation, 17
  tricarboxylic acid cycle and, 161
Ketolysis, acetoacetic acid and, 109
  lipid metabolism and, *102*
Ketone bodies. See also *Ketosis, Ketogenesis* and *Ketonuria.*
  acid-base balance and, 431
  blood plasma, 117(t)
  liver and, 125
  normal urine and, 477
Ketonemia, acetoacetate and, 105
Ketonuria, acetoacetate and, 105
  defined, 109-110
  Fanconi syndrome and, 110, 311
  pancreatectomy and, 34
Ketosis, diabetes and, 78-79
  electrolytes and, 374
  Houssay animals and, 36
  lipid metabolism and, 109
  lipid utilization and, 94
  pancreatectomy and, 34
17-Ketosteroids. See *Androsterone* and *Estrone.*
Kidney(s). See also *Renal.*
  acid-base balance and, 407, 409-411
  Addison's disease and, 740
  blood plasma and, *458*
  blood supply to, 456
  cholesterol synthesis and, 113
  creatine and, 178
  damage to, benign albuminuria and, 214
    disaccharidase deficiency and, 3
  defects of, hypouricemia and, 259
  disease of, electrolytes and, 376
    excess electrolytes and, 380-381
    glucose reabsorption and, 82
    increased urea nitrogen and, 208
    plasma protein and, 201
    potassium and, 389-390
    procedures for, 519(t)
    water elimination and, 493-494
  disturbances of, hydrogen ion concentration and, 414
  excretion of, acid and, 407
  failure of, acute, defect localization and, 522
    hemodialysis and, 544
  function of, 454-546
    chronic renal failure and, biochemical manifestations, 526-533
    clearance tests and, 481-491
    clinical study, 481
    concentration test, 496-501
    defect localization and, 518
    disturbances of, *527*

894 INDEX

Kidney(s) *(Continued)*
  function of, diuresis and, *468*
    diuretics and, 468-470
    foreign substances and, 515-526
    glomerular filtration and, 457-459
    hemodialysis and, 543-545
    homotransplantation and, 545-546
    impaired, urinary nonprotein nitrogen and, 507
    nonprotein nitrogenous substances and, 501-515
    normal urine characteristics and, 470-481
    normal values, 486(t)
    phenolsulfonphthalein test and, 515
    regulation of, hydrogen ions and, 407-412
    solids elimination and, 495-501
    urinary lithiasis and, 540-543
    urine specific gravity and, 495-501
    water elimination and, 491-495
    water homeostasis and, 466-468
  functional impairment of, urinary nonprotein nitrogen and, 507
  glycogen and, 79
  importance of, 454
  insulin disposal and, 24
  ion excretion and, 363
  ionic regulation and, 358
  lesions to, glomerular filtration and, 459
    renal albuminuria and, 217
    urinary nonprotein nitrogen and, 512-513
  malfunction of, potassium and, 389-390
  morphology of, 455-457, *455*
  phospholipids and, 112
  prostaglandins in, 95
  protein production and, 157
Kidney tubules, defects of, 533-540
  dysfunction of, 534(t)
  epithelial damage of, 498-499
  excretion and, 465, 488
  function of, 459-466
  proteinuria and, 89
  reabsorption of, 459-465, 489
    capacity, 490
  synthesis and, 465
  transport mechanism competition and, 465-466
  uricosuric action and, 464
Kilocalorie, defined, 444
Kinases, 7, 10
King, E. J., 849
King technique, 850
Kinin, immunity and, plasma proteins and, 175
Korsakoff's psychosis, cerebrospinal fluid and, 846
Krabbe's disease, 130, 130(t)
Krebs cycle, 10, 12
  in aerobic metabolism, 11

Kupffer's cells, 610
  Vitamin A and, 806
Kwashiorkor, hypoalbuminemia and, 201, 621

Labor, tumultous, plasma fibrinogen and, 184
Lactase, deficiency of, 3-4
Lactate, cerebrospinal fluid and, 842-843
  hepatic coma and, 674
  liver function and, 617
  muscles and, 76
Lactate dehydrogenase, carbohydrate metabolism and, 555
  cerebrospinal fluid and, 845
  related disorders and, *555, 556, 557*
Lactation, chlorpromazine and, 701
  hypoalbuminemia and, 189
  hypoglycemia in, 58
  reserpine and, 701
Lactic acid, blood, carbohydrate metabolism and, 76-78
  carbohydrate metabolism and, 5-10, *7*
  muscle contraction and, 16, *16*
  stomach contents and, 583
  thiamine and, 812
Lactic acidosis, carbohydrate metabolism and, 77
Lactose, body fluids and, 30, 480
  intolerance to, congenital, 3
  urea and, 82
Lactosuria, lactose and, 82, 480
  sugar secretion and, 88
Langerhans, islands of. See also *Pancreas.*
  abnormal function in, 33
  acinar cells and, 590
  alpha cells and, 34
  beta cells and, 25, 34, 35
    insulin biosynthesis and, 23
  cell exhaustion in, 62
  diabetes mellitus and, 47-49
  functional hyperinsulinism and, 55
  glucose tolerance and, 64
  pancreatic function and, 589-599
Lanosterol, cholesterol biosynthesis and, 96
LATS. See *Long-acting thyroid stimulator.*
Lead, plasma concentration conversion and, 856
Lead poisoning, 84
  aminoaciduria and, 224, 538
  bilirubin and, 643
  hyperuricemia and, 255
  porphyrinuria and, 278
Lecithin. See also *Phosphatidyl choline.*
  blood plasma and, 117(t)
  cholesterol acyltransferase deficiency and, 137
  fatty liver and, 128
  phospholipids and, 118

INDEX

Lecithin *(Continued)*
  plasma and, 114
  protein and, in exudates, 396
Lecithin-cholesterol acyltransferase, cholesterol esterification and, 113
  deficiency of, 137
Lecithinases, lipid absorption and, 100
  phospholipid catabolism and, 112
Leiner's disease, biotin and, 834
Lesch-Nyhan syndrome, enzyme abnormalities and, 579
  hyperuricemia and, 258
Lesions, C-reactive proteins and, 198
Letterer-Siwe disease, 130
Leucine, *18*, 23
  fatty acid synthesis and, 104
  ketogenic pathway and, 162
L-Leucine, 55
Leucine aminopeptidases, 571-572
Leukemia, ceruloplasmin and, 335
  chemotherapy and, 249
  cryoglobulins and, 198
  gamma-globulins and, 191
  hypogammaglobulinemia and, 204
  lactic acidosis and, 77
  zinc and, 337
Leukocyte(s), degeneration of, lipuria and, 111
Leukocyte acid phosphatase, 569
LH. See *Luteinizing hormone.*
LH/FSH-RH. See *Luteinizing hormone/Follicle-stimulating and releasing hormone.*
Liebermann-Burchard method, 848
Liebig, Justus, 847
Limit dextrinosis, glycogen storage disease and, 80
Lincomycin, protein synthesis and, 247
Linoleic acid, unsaturated fatty acids and, 100
Linolenic acid, fatty acid synthesis and, 103
Lipase(s), acid, deficiency of, 130(t)
  fat deposition and, in adipose tissue, 123
  lipoprotein, lipid deposition and, 124
  Lundh test meal and, 593
  pancreatic, 589
    fecal examination and, 598
    lipid digestion and, 98
  plasma proteins and, 177
  secretin test and, 592
  serum, 551
    pancreatic function and, 595-596
  triglyceride conversion and, 105
Lipemia, diabetes mellitus and, lipuria and, 111
  heparin and, 117
  idiopathic, familial, 132, 141
  kidney disorders and, 132, 134
  xanthomatosis and, 130, 140
Lipemia retinalis, hyperlipoproteinemia type five and, 136

Lipid(s). See also *Fats and Fatty Acids.*
  absorption of, clinical study of, 605
    phosphoglycerides and, 100
  anabolism and catabolism of, 105
  blood, adrenal cortical hormones and, 726
    plasma and, 117(t)
    standard fat meal and, 606
  carbohydrates and, metabolic interrelations with, 106-110, 107(t)
  definition of, 91
  deposition of, 122-124
  dietary, immediate fate of, 101-102
  digestion and absorption of, 98-101
    fecal fat and, 100-101
    sterols and, 99-100
    triglycerides and, 98-99
  exudates and, 395-396
  ^{131}I-labeled, absorption and, 605
    lipid absorption and, 605
  metabolism of, 91-146, *102*
    acetoacetate and, *102*
    adrenal cortical hormones and, 722
    age and, 121-122
    carbohydrate metabolism and, 17-18, 93
    characteristics, 93
    deposition of, 122-129
    fecal fat and, 110-111
    growth hormone and, 707
    hypercholesterolemia and, 137-142
    hyperlipoproteinemia and, 134-136
    hypocholesterolemia and, 142-145
    hypolipoproteinemia and, 136-137
    liver function and, 630-633
    nucleic acids and, *241*
    plasma lipid abnormalities and, 131
    plasma lipoprotein abnormalities and, 131-134
    protein and, 93
      metabolism and, 162-164
    thyroid hormone and, 751
    transport and deposition in, 93
  mobilization of, 105
    adrenal cortical hormones and, 722
    pancreatectomy and, 34
  plasma, abnormalities of, 131
    absorbed lipid and, 101
    body fat and, 120-121
    chronic renal failure and, 531
    concentration of, 117-119
    food and, 119-120
    postabsorptive concentration of, 117-119
  proteins and metabolic interrelations with, 106-110, 107(t)
  role of, 91
  serum albumin and, 139
  storage of, 122-129
    diseases, 129-130, 130(t)
    fats and, 93-94
    fatty liver and, 125-130
    lipotropism and, 125-130

Lipid(s) *(Continued)*
  synthesis of, adrenal cortical hormones and, 722
    amino acids and, 107
  transport of, 114-122
    food and, 119-121
    plasma proteins and, 175
  transudates and, 395-396
Lipid storage diseases, 130(t)
  mucopolysaccharidoses and, 234
Lipidosis, lipid metabolism and, 129-130
  xanthomatosis and, 141
Lipodystrophy, steatorrhea and, 601
Lipogenesis, 22
  lipid metabolism and, *102*
Lipogenic effect, 22
Lipoic acid, 12
Lipoidoses, cerebrospinal fluid and, 846
Lipolysis, 21
  adipose tissue and, 124
  enzymes of, pancreas and, 589
  fecal fat and, 101
  growth hormones and, in protein anabolism, 125
  lipid digestion and, 98
  triglycerides and, 98
Lipomucopolysaccharidosis, 234
Lipoprotein(s), adipose tissue and, 123, 124
  atherosclerosis and, 93
  beta, lipoprotein electrophoresis and, 116
  chemical composition of, 115(t)
  concentration of, male vs. female, 121-122
  density of, 115
  free fatty acids and, 119
  high-density, 116
  lipid metabolism and, 97-98, 105
  lipid transport and, 115, *123*
  liver and, 124
  low-density, 116
  phospholipids and, 94, 118
  plasma, 115
    abnormalities of, 131-134
    atherosclerosis and, 141
    electrophoresis of, 116
  prebeta, electrophoresis and, 116
  very high-density, 116
  very low-density, 116
α_1-Lipoprotein, properties of, 169(t)
β-Lipoprotein, properties of, 169(t)
Lipoprotein lipase. See also *Lipase, lipoprotein.*
  hyperlipemia and, 117
  lipid digestion and, 99
β-Lipoproteinemia, fecal fat and, 101
Lipotropism, lipid storage and, 125-130
Lipovitellin, lipoproteins and, 97
Lipuria, nephrotic syndrome and, 111
Lithiasis, idiopathic renal, 541
  kidney function and, 540-543
Liver. See also *Hepatic.*
  abscess of, serum enzymes and, 670
  acute diffuse necrosis of, 49

Liver *(Continued)*
  adrenocortical deficiency and, 25
  albumin formation and, 173
  alkaline phosphatase and, 561
  amino acids and, 159
  carbohydrate metabolism and, 4, 13, 613-617
    fatty acids and, 125
  carcinoma of, dye retention and, 669
  cell structure of, 612
  cholesterol synthesis and, 112
  chronic passive obstruction of, porphyrinuria and, 655
  cirrhosis of, secretin test and, 592
  congestion of, dye retention and, 669
  damage to, bromsulphalein retention and, 668
    diagnosis of, 677-678
    mucoproteins and, 197
  differential diagnosis and, 676-683
    test selection and, 679
  diseases of, 63-64
    albuminuria and, 216
    aldolase and, 558
    alkaline phosphatase and, 565-566
    aminoaciduria and, 224
    aminotransferases and, 553
    ceruloplasmin and, 335
    excess electrolyte and, 381
    hemorrhage and, 628
    insufficiency, amino acid nitrogen and, 210
      decreased urea nitrogen and, 209
    isocitrate dehydrogenase and, 558
    lactate dehydrogenase and, 557
    lactic acidosis and, 77
    nonrespiratory acidosis and, 415, 419
    plasma protein and, 201
    primary, diagnosis of, 681-682
    steatorrhea and, 601, 659
    urea nitrogen and, 208
    without jaundice, diagnosis of, 682-683
  disorders of
    diminished glucose tolerance and, 59
    disaccharidase deficiency and, 3
    glucose tolerance and, 63-64
    hypercholesterolemia and, 139-140
    hyperglycemia and, 53
    hyperuricemia, 255
    hypocholesterolemia and, 142-143
    hypoglycemia, 56
  ducts of, obstruction of, 670
  enzyme abnormalities and, 575-576
  enzymes of, drug induction of, 675-676
  experimental difficulty with, 612
  failure of, plasma fibrinogen and, 184
  fatty, blood sugar tolerance curve and, 63
    epinephrine tolerance test and, 74
    lipid storage and, 125-130
    pathological, type four, 127
    physiological, type three, 127
    type four, lipotropism and, 128-129

Liver *(Continued)*
    fatty, types of, 125-131
    fibrinogen formation and, 173
    fructose and, 74
    fructose tolerance test for, 44
    function of, 610-684
        adrenocortical hormones and, 665
        albumin and, 620
        amino acids and, 617
        ammonium ion and, 619
        antibodies and, 623
        bile acid metabolism and, 656-661
        bilirubinuria and, 649-656
        detoxification-conjugation and, 661-663
        elimination of dyes and, 666-670
        epinephrine hyperglycemia and, 616
        fecal fat and, 631
        fibrinogen and, 620
        fructose tolerance and, 75
        globulin reactions and, 625
        globulins and, 621-626
        glucose tolerance and, 614
        hormone metabolism and, 663
        hyperbilirubinemia and, 638-640
        hypoglycemia and, 614-615
        jaundice and, 640-648
        kernicterus and, 648-649
        lactate and, 617
        lipids and, 630-633
        mucoproteins and, 624
        pigment metabolism and, 633-640
        plasma cholesterol and, 631
        plasma proteins and, 620
        porphyrins and, 635-636
        protein metabolism and, 617
        protein values and, *624*
        prothrombin and, 628
        pyruvate and, 617
        salt metabolism and, 672-673
        serum bilirubin and, 636-638
        serum enzymes and, 670-671
        tyrosinuria and, 618
        urea and, 618
        uric acid and, 619
        water metabolism and, 672-673
        zinc sulfate turbidity and, 628
    functional impairment of, urobilinogenuria and, 652-653
    galactose tolerance and, 44
    gluconeogenesis and, 5, 162
    glucose oxidation and, 5
    glucose-6-phosphatase reaction and, 13
    glycogen and, 79
    glycogenesis and, 13
    glycogenic function of, 49
    glycogenolysis and, 13-14
    homeostatic mechanism of, 47
    insulin disposal and, 24
    iron storage and, 324
    lactate and, 77
    lipid metabolism and, 124-125
    lipid transport and, *123*

Liver *(Continued)*
    lymphatic drainage of, cirrhosis and, 673
    metastatic carcinoma of, hippuric acid and, 663
    metastatic malignancy of, serum enzymes and, 670
    morphology of, 610-613
    phospholipids and, 111-112
    plasma cholesterol and, 118
    plasma phospholipids and, 118
    position of, *610*
    protein production and, 157
    prothrombin formation and, 173
    structure of, *611*
    thymol turbidity and, 626-627
    unesterified fatty acids and, 119
    vitamin A and, 805-810
Liver-depot axis, lipid storage and, 125
Liver glycogen. See *Glycogen, liver.*
Liver phosphorylase. See *Phosphorylase, liver.*
Long-acting thyroid stimulator, 766
    vs. thyroid stimulating hormone, 767(t)
Loop of Henle, water absorption and, 460
Lowe's syndrome, aminoaciduria and, 224
    renal tubular defects and, 538
Lucey-Driscoll syndrome, 641-642
    neonatal jaundice and, 647
Lues, cerebrospinal, glucose and, 838(t)
        protein and, 836(t), 837
    of liver, alimentary response and, 63
Lundh test meal, pancreas and, 593-594
Lungs, cholesterol synthesis and, 113
    congestion of, 442
    phospholipids and, 112
    prostaglandins in, 95
Lupus erythematosus, cryoglobulins and, 198
    systemic, hyperlipoproteinemia and, 134
        thymol turbidity test and, 627
Luteinizing hormone, gonadotrophic hormones and, 691-693
    pituitary gland and, 689
Luteinizing hormone / follicle-stimulating and releasing hormone, 689
Lymph glands, diseases of, hypogammaglobulinemia and, 204
Lymphadenopathy, hypolipoproteinemia and, 137
    Waldenström's macroglobulinemia and, 203
Lymphatics, intestinal, hyperlipemia and, 119
    lipid transport and, *123*
Lymphatic drainage, 24
Lymphatic system, absorbed lipid and, 101
Lymphoma(s), hyperlipoproteinemia and, 134
    macroglobulinemia and, 199
Lymphopathia venereum, plasma fibrinogen and, 184
    plasma protein abnormalities and, 201

Lymphopathia venereum *(Continued)*
  thymol turbidity test and, 627
Lymphosarcoma, Bence Jones proteins and, 199-200
Lymphuria, 111
Lysine, iron absorption and, 321
  protein digestion and, 151
Lysocephalins, lipid absorption and, 100
Lysol, tubular damage and, 84
Lysolecithins, lipid absorption and, 100
  lipid digestion and, 98
Lysosomal acid phosphatase, deficiency of, 569
Lysosomes, 2

M protein, abnormal globulin and, 202
  gamma-globulins and, 192
  Waldenström's macroglobulinemia and, 203
MacLean method, 848
Macroamylasemia, 551
Macroglobulin(s), multiple myeloma and, 201
  plasma protein abnormalities and, 198-199
α_2-Macroglobulin, properties of, 169(t)
Macroglobulinemia, Bence Jones proteins and, 199-200
  cryoglobulins and, 198
  hyperlipoproteinemia type one and, 134
  hyperlipoproteinemia type four and, 135
  plasma protein and, 199
Magnesium, absorption of, 316-317
  blood and, 317-318
  cerebrospinal fluid and, 843-844
  diabetic ketoacidosis and, 78-79
  excretion of, 316-317
  hypocalcemia and, 305
  metabolism of, 316-318
  normal urine and, 478
  plasma concentration conversion and, 856
  renal tubular acidosis and, 317
  serum, abnormal, 317-318
Magnesium salts, serum calcium and, 291
Magnesium sulfate, intestinal phase and, 581
Malaria, albuminuria and, 216
  bilirubinuria and, 643(t)
  bromsulphalein test and, 669
  cryoglobulins and, 198
  dye retention and, 669
  hemoglobinuria and, 271
  hyperglobulinemia and, 201
  methemalbuminemia and, 274
  thymol turbidity test and, 627
  urobilinogen and, 643, 654
Malate, *18*
Malate dehydrogenase, 557
Malignancy, aldolase and, 559
  Bence Jones proteins and, 199-200
  C-reactive proteins and, 198

Malignancy *(Continued)*
  chemotherapy of, 249-250
    alkylating agents and, 249
    antimetabolites and, 249
  enzyme abnormalities and, 577
  gastric, electrolytes and, 375
    hypoacidity and, 584
  glycosuria and, 87
  lactate dehydrogenase and, 557
  macroglobulins and, 199
  protein and, 188
Malnutrition, globulin concentration and, 190
  intestinal malabsorption and, *603*
  negative nitrogen balance and, 229
  plasma albumin and, 185
  potassium and, 385-386
Malonate, 2
Malonic acid, urea and, 253
Malonyl-CoA, fatty acid synthesis and, 103-104
  insulin and, 21
Maltase, carbohydrate metabolism and, 2
Maltodextrin, 9
Maltose, carbohydrate metabolism and, 1
Maltosuria, sugar secretion and, 89
Manganese, trace elements and, 338
Manic-depression, glucose tolerance and, 64
Mannoheptulose, sugar, 23
Mannose, carbohydrate metabolism and, 2, 6
  endogenous sources of, 4
Maple syrup urine disease, aminoaciduria and, 223-224, 226-227
Marasmus, glucose tolerance and, 70
Marfan's disease, hydroxyproline and, 224
Maroteaux-Lamy syndrome, mucopolysaccharidoses and, 233
Marrow. See also *Bone* and *Skeleton*.
  cobalt and, 337
  fracture of, lipuria and, 111
  heavy-chain disease and, 202
Mass concentration, 855
McArdle's disease, glycogen storage disease and, 80
Meat, elemental composition of, 447(t)
Mecholyl, gastric secretion and, 582
Mecholyl chloride, vagus effect and, 590
Melanocyte-stimulating hormone, adenohypophysis and, 702-705
Melanocyte-stimulating hormone-release inhibiting hormone, 690
Melanocyte-stimulating hormone releasing hormone, 689
Melanomas, adrenal medulla and, 744
Melanosarcoma, of liver, porphyrinuria and, 655
Melanuria, aminoaciduria and, 228
Melituria, sugar secretion and, 82, 83-89
Menadione, vitamin K and, 830
Meningitis, aseptic, cerebrospinal fluid and, 837

Meningitis *(Continued)*
    bacterial, cerebrospinal fluid and, 846
    cerebrospinal fluid and, 836
    chloride and, 841
    cholesterol and, 841
    leukemic, hypoglycorrhachia and, 840
    luetic, cerebrospinal fluid and, 839
    meningococcal, cerebrospinal fluid and, 836(t), 837
        hypochloremia and, 382-383
    phosphate and, 841
    pneumococcal, cerebrospinal fluid and, 836(t), 837
    serous, cerebrospinal fluid and, 836-837, 836(t)
    streptococcal, cerebrospinal fluid and, 836(t), 837
    suppurative, cerebrospinal fluid and, 838(t)
        hypoglycorrhachia and, 839
        lactate and, 842-843
    tuberculous, cerebrospinal fluid and, 836(t), 837, 838(t)
        hydrogen ions and, 843
        lactate and, 843
Menopause, endocrine function and, 781-782
    galactose tolerance and, 76
    gonadotrophins and, 695
    hormonal change and, 781
    osteoporosis and, 285
Menses, cessation of, intestinal malabsorption and, *603*
Menstruation, galactose tolerance and, 76
    hormonal control and, 780-781
    human chrorionic gonadotrophin and, 790
    iron loss and, 320
    plasma fibrinogen and, 184
Mental states. See *Emotional states*.
6-Mercaptopurine, leukemia and, 249
Mercuhydrin, diuresis and, 469
Mercury poisoning, albuminuria and, 216
    bilirubin and, 636
Mesobilirubinogen, 650. See also *Urobilinogen*.
Metabolism. See under specific substances.
Metabolism, basal. See *Basal metabolism*.
Metachromatic leukodystrophy, 129, 130, 130(t)
    sulfur and, 329
Metastases, C-reactive proteins and, 198
Metcuhydrin, diuresis and, 469
Methemalbuminemia, 273-274
Methemoglobin, hemoglobin and, 272
Methemoglobinemia, causes of, 272-273
    hypoxia and, 443
Methionine, creatine and, 180
    hepatic coma and, 674
    malabsorption of, renal tubular defects and, 536
    protein metabolism and, 152
    protein synthesis and, 246

Methionine *(Continued)*
    sulfur and, 327
Methionine malabsorption syndrome, aminoaciduria and, 227
Methotrexate, cancer chemotherapy and, 249
Methyl salicylate poisoning, 54
Methylmalonic acid, urea and, 253
Methylmalonyl-CoA, fatty acid synthesis and, 105
Methyltestosterone, protein metabolism and, 165
Mevalonic acid, cholesterol biosynthesis and, 96
Microsomes, lipids in, 97
Milk, hypercalcemia and, 299-300
Milk-alkali syndrome, nonrespiratory alkalosis and, 424
Milliequivalents, body fluid and, 346
Millimole, definition of, 347
Millon test, 848
Mineralocorticoids, adrenal cortex and, *715*
Miscelles, lipid digestion and, 98
Mitochondria,
    ketogenic pathway in, 21
    lipids in, 97
Modified balance studies, intestinal malabsorption and, 604
Mole(s), definition of, 347
    hydatidiform, human chorionic gonadotrophin and, 790
Molybdenum, copper absorption and, 335
    trace elements and, 338
Mongolism, lactic acidosis and, 77
Monoclonal gammapathies, cerebrospinal fluid and, 835
Monoglycerides, lipid digestion and, 98
Mononucleosis, thymol turbidity test and, 627
Mononucleotides, nucleotide synthesis and, 252
Monosaccharides, carbohydrate metabolism and, 1
    sphingomyelins and, 92
Monosialo-gangliosides, 92
Morphine, 52, 87, 416
Morquio's syndrome, mucopolysaccharidoses and, 233
Mosenthal test, 496
M-RIH. See Melanocyte-stimulating hormone-release inhibiting hormone, 690
MRH. See *Melanocyte-stimulating hormone releasing hormone*.
mRNA. See *Ribonucleic acid, messenger*.
Mucin, exudates and, 395
    urine and, 219
Mucopolysaccharide(s), glucose conversion and, 6
    acid, urea cycle and, 232-234
    protein metabolism and, 232-234
    sulfur and, 329
    urea cycle disorders and, 232-234

Mucopolysaccharide synthesis, glucose and, 6
Mucopolysaccharidoses, acid mucopolysaccharides and, 233
  enzyme abnormalities and, 579
Mucoproteins, haptoglobins and, 265
  jaundice diagnosis and, 680
  liver function and, 624
  plasma proteins and, 196-197
  study methods for, 171(t)
  test for, 170
Mucous colitis, glucose tolerance and, 70
Mucosa, gastric, achlorhydria and, 585-586
  intestinal, biopsy of, 608
    iron storage and, 324
    protein production and, 157
Mucus, abnormal, small intestine malabsorption and, 603
  gastric activity and, 580
Multiple myeloma, aminoaciduria and, 224
  Bence Jones proteins and, 199-200
  cryoglobulins and, 198
  hypercalcemia and, 297
  hypogammaglobulinemia and, 204
  macroglobulins and, 199
  mucoproteins and, 197
  plasma fibrinogen and, 184
  plasma protein abnormalities and, 183
  protein abnormality and, 201
  pyroglobulins and, 198
Muscle, atrophy of, creatinuria and, 222
  carbohydrate metabolism and, 15-17
  cholesterol synthesis and, 113
  contraction of, chemistry of, 16, *16*
  dystrophies of, creatinuria and, 222
  glycogen and, 79
    adrenalectomy and, 24
    as carbohydrate source, 4
  hypertrophy of, creatinuria and, 222
  iodine and, 332
  ketolysis and, 109
  lactate and, 76
  phosphatidyl compounds in, 94
  phospholipid : cholesterol ratio and, 96
  phospholipids and, 112
  skeletal, aldolase and, 559
    disease of, creatine kinase and, 569-570
Muscular dystrophy, hypoglycemia in, 58
Mustard gas, cancer chemotherapy and, 249
Mutase, glycolytic reaction equilibria and, 14, *14*
Mutation, nucleic acids and, 243-244
Myasthenia gravis, creatinuria and, 222
Myelography, cerebrospinal fluid and, 838
Myeloma, multiple. See *Multiple myeloma*.
Myeloma globulin, plasma protein abnormalities and, 183
Myeloma protein(s), in serum, 201
  plasma protein abnormalities and, 187, 199
Myelomatosis, abnormal globulin and, 202
  chemotherapy for, 250
  hyperglobulinemia and, 202

Myelomatosis *(Continued)*
  hyperlipoproteinemia type one and, 134
  hyperlipoproteinemia type four and, 135
  paresis and, 202
  Waldenström's macroglobulinemia and, 203
Myleran, cancer chemotherapy and, 250
Myocardial glycogen, 20
Myocardial infarction, aldolase and, 558
  C-reactive proteins and, 198
  ceruloplasmin and, 336
  isocitrate dehydrogenase and, 558
  lactate dehydrogenase and, 555
  uric acid and, 256
Myoglobin, synthesis of, iron and, 324
Myoglobinuria, 271-272
  proteinuria and, 219
  urinary protein and, 219
Myosin, 16
Myositis, creatinuria and, 222
Myotonia congenita, creatinuria and, 222
Myxedema, cerebrospinal fluid and, 836(t), 837
  glucose tolerance and, 69
  hypoglycemia and, 58

NADP. See *Nicotinamide-adenine dinucleotide phosphate*.
NADPH. See *Nicotinamide-adenine dinucleotide phosphate hydrogen*.
Naphthalene, neonatal jaundice and, 647
Narcotics, glycosuria and, 87
Natulan, cancer chemotherapy and, 250
Necrosis, acute, plasma fibrinogen and, 184
  secretin test and, 592
  alimentary response and, 63
  C-reactive proteins and, 198
  cephalin-cholesterol test and, 628
  hepatic, gamma-globulins and, 191
    porphyrinuria and, 655
  hypocholesterolemia and, 142
Necrotizing arteriolitis, defect localization and, 523
NEFA. See *Fatty acids, non-esterified*.
Neoplasms, intrathoracic, adrenocortical hyperplasia and, 379
  pulmonary, electrolytes and, 379
Nephritis. See also *Kidney, diseases of*.
  acidosis and, 417, 529
  acute, edema protein and, 394
    pathological, 518
  alkali deficit and, 417, 426, 529
  basal metabolism and, 451
  blood albumin and, 186-187
  chloride and, 841
  creatine and, 209
  hypercalcemia and, 300
  hypoalbuminemia and, 186
  nonprotein nitrogen and, 207-212, 509-514, 518-520
  potassium-losing, hyperaldosteronism and, 390

INDEX 901

Nephritis *(Continued)*
    potassium-losing, renal tubular defects and, 539
    protein and, 185-189, 190-195, 530
    renal tubular acidosis and, 377
    salt-losing, 377
        renal tubular defects and, 539
    water-losing, renal tubular defects and, 539-540
Nephrocalcinosis, renal tubular acidosis and, 377
Nephrogenic diabetes insipidus, 380
Nephrons, decreased number of, hyposthenuria and, 498
    diuresis and, *468*
    morphology of, 455, *455*
Nephrosclerosis, benign, defect localization and, 523
    glucose tolerance and, 64
    malignant, defect localization and, 523
    renal albuminuria and, 217
    urinary nonprotein nitrogen and, 512
Nephrosis. See also *Nephrotic syndrome.*
    amyloid, 138
    glycosuria and, 85
Nephrotic syndrome, blood volume and, 391
    ceruloplasmin and, 336
    defect localization and, 523-524
    globulin concentration and, 190
    hypercholesterolemia and, 138
    hyperlipoproteinemia type four and, 135
    hypoalbuminemia and, 186-187, 201
    hypocalcemia and, 303
    hypokalemia and, 387
    lipuria and, 111
    macroglobulins and, 199
    mucoproteins and, 197
    pathogenesis of, *525*
    plasma fibrinogen and, 184
    plasma lipoprotein abnormalities and, 132
    protein quantity in urine and, 218
    renal albuminuria and, 217
    zinc and, 338
Nervous system, central, diseases of, enzyme abnormalities and, 577
    glutamic acid and, 161
    diseases of, ceruloplasmin and, 336
        lactic acidosis and, 77
        hypoglycemia in, 58
    insulin release and, 24
    lipids and, 91
    phosphatidyl compounds and, 94
Neurilemmomas, adrenal medulla and, 744
Neuritis, peripheral, intestinal malabsorption and, *603*
    vitamin deficiency and, 602
Neuroblastomas, adrenal medulla and, 744-745
Neurocirculatory asthenia, hypoglycemia in, 58
Neurofibromas, adrenal medulla and, 744
Neurohypophysis, hormones of, 709-713
    oxytocin and, 712

Neurohypophysis *(Continued)*
    hormones of, vasopressin and, 710-712
Neurologic. See *Nervous system.*
Neurosecretion, defined, 687
Neuroses, gastric, hyposecretion and, 586
Newton, defined, 855
Niacin, 814-815
Nicotinamide, coenzymes and, 251
Nicotinamide-adenine dinucleotide, free nucleotides and, 251
Nicotinamide-adenine dinucleotide phosphate, 12-13, 13(t)
    free nucleotides and, 251
Nicotinamide-adenine dinucleotide phosphate hydrogen, lipogenesis and, 123
    pentose shunt and, 12-13
Nicotine, epinephrine-stimulating effect of, 54
    gastric secretion and, 582
Nicotinic acid, 814-815
    intestinal malabsorption and, *603*
    vitamin deficiency and, 602
Niemann-Pick disease, 129, 130(t)
    cerebrospinal fluid and, 846
Night blindness, vitamin A deficiency and, 808
Nitrobenzene, tubular damage and, 84
Nitrogen. See also *Nonprotein nitrogen.*
    amino acid and, blood nonprotein nitrogen and, 210-211
    amino acid synthesis and, 159
    balance of, diets and, 163-164
        negative, 229-230
        postoperative states and, 388
        protein metabolism and, 153
    excretion of, protein metabolism and, 152, 179-182
    protein. See *Proteins, plasma.*
    protein metabolism and, 152
    undetermined, nonprotein nitrogen and, 179
    urea, blood nonprotein nitrogen and, 207
        decreased, disorders and, 209
        glomerulonephritis and, 208
        increased, disorders and, 207-209
        kidney diseases and, 208
    urinary, protein metabolism and, 212-230
        proteinuria and, 212-220
Nitrogen mustard, cancer chemotherapy and, 250
Nitrogenous nonproteins. See *Nonprotein nitrogen.*
Nitrogenous substances, synthesis of, 158-159
Noguchi test, cerebrospinal fluid and, 836
Non-esterified fatty acids. See *Fatty acids, non-esterified.*
Nonne-Apelt test, cerebrospinal fluid and, 836
Non-nitrogenous residue, protein metabolism and, 161
Nonocclusive coronary insufficiency, C-reactive proteins and, 198

Nonprotein nitrogen. See also *Amino acids, Creatine, Ammonia, Urea* and *Uric acid.*
  abnormal, amino acid nitrogen and, 210-211
  blood, abnormal, ammonia and, 211
    creatine and, 209
    urea and, 207-209
    uric acid and, 209
  blood urea and, 178
  cerebrospinal fluid and, 840-841
  elimination of, renal function and, 501-515
  intermediary protein metabolism and, 177-179
  normal urine and, 472-475, 473(t)
  protein metabolism and, 207-212
  total, blood nonprotein nitrogen and, 211
  urine and, 472
  urea and, 472
  uric acid and, 471-475
  urinary, 220-230
    amino acids and, 474
    ammonium and, 472
    creatine and, 473
    creatinine and, 473
    kidney function and, 501-515
      blood nitrogen studies and, 507-514
      body fluids and, 514-515
      simultaneous study of blood and urine, 501-507
      urinary studies, 501
Noradrenaline. See also *Norepinephrine.*
Norepinephrine, 9, 23
  adrenal medulla and, 740-746
  lipolytic mechanism and, 124
  triglyceride conversion and, 105
  unesterified fatty acid concentration and, 119
Normal range, 852-854
Norymberski technique, 850
Nuclei, extracellular fluid and, 359
  lipids in, 97
Nucleic acids, bacteria and, transforming substance of, 243
  biological significance of, 240-250
  biosynthesis of, 237-239
  cancer chemotherapy and, 249
  carcinogenesis and, 243-244
  importance of, 235
  metabolism of, 235-259
    digestion and absorption, 239
    hyperuricemia and, 255-258
    hypouricemia and, 259
    intermediary, 251-259
    interrelations with other foodstuffs, 239-240, *241*
  mutation and, 243-244
  protein synthesis and, 244-248
  viruses and, 240-241
Nucleoproteins, in urine, 219
  metabolism of, 235-259
  types of, 235-236

Nucleoside(s), monophosphates, cyclic, 251
Nucleoside(s), nucleic acids and, 236
Nucleoside triphosphates, free nucleotides and, 250
Nucleotide(s), cancer chemotherapy and, 249
  free, nucleic acid metabolism and, 250-251
  in pentose shunt, 13
Nucleotide units, nucleic acids and, 236
Nutrition, lipid vs. carbohydrate, 91
  protein and, 156

Oasthouse syndrome, renal tubular defects and, 536
Obesity, diabetes mellitus and, 73
  glycosuria and, 87
  hyperglycemia in, 53-54
  insulinoma and, 74
Ohm, defined, 855
Oleic acid, cholesterol and, 99
  I-labeled lipids and, 605
Oligosaccharides, sphingomyelins and, 92
Oliguria, electrolytes and, 382
  glomerulonephritis and, 509
  hyperkalemia and, 389
  kidney function and, 492
Omentum, cholesterol synthesis and, 112
One-hour—two dose glucose tolerance test, 42-43, 62
Open-circuit system, indirect calorimetry and, 449
Oral glucose tolerance test. See *Glucose tolerance tests, oral.*
Ornithine, urea formation and, 160
Ornithine-citrulline-arginine cycle, urea formation and, 160
Ornithine transcarbamylase, hyperammonemia and, 231
Orosomucoid, prealbumin and, 189-190
  properties of, 169(t)
Orotic acid, nucleic acid biosynthesis and, 237
Orotic aciduria. See *Aciduria, orotic.*
Orotidylic acid, nucleic acid biosynthesis and, 237
Osmolality, electrolytes and, 372
Osmole, definition of, 347
Osmoreceptors, extracellular fluids and, 358
Osmotic diuresis, diabetic ketoacidosis and, 78-79
Osteitis deformans, alkaline phosphatase and, 563
  glucose tolerance and, 64
Osteitis fibrosa, hyperparathyroidism and, 315
Osteoarthritis, hypertrophic, hypercholesterolemia and, 141
Osteoblasts, bone development and, 282
Osteodystrophy, chronic renal failure and, 528-529
Osteomalacia, alkaline phosphatase and, 564
  bone development and, 314, 315(t)

Osteomalacia *(Continued)*
  hypocalcemia and, 302
  hypophosphatemia and, 310
  intestinal malabsorption and, *603*
  proteinuria and, 219
  skeletal abnormalities and, 604
  starvation and, 302
Osteoporosis, calcified tissue and, 313, 314(t)
  intestinal malabsorption and, *603*
  menopause and, 285
  protein abnormalities and, 602
  skeletal abnormalities and, 604
Osteosclerosis, bone development and, 315
Ovaries, failure of, hypogonadism and, 782(t)
  hormones of, 774-781
    androgens and, 780
    estrogens, 774-779
    progesterone and, 779-780
  iodine and, 332
Oxalate, hypocalcemia and, 305
Oxalic acid, normal urine and, 475
  poisoning from, magnesium and, 317
Oxaloacetate, aerobic metabolism and, 12
  metabolic interrelations and, 163(t)
  nucleic acid metabolism and, 241
  nutrient metabolism interrelation and, 17
  phosphoenolpyruvate and, 27, 107
  protein digestion and, 151, 163
Oxaloacetic acid, amino acids and, 161
  pyruvic acid and, 15
  tricarboxylic acid cycle and, 107, 161
Oxidative deamination, 12
Oxidative decarboxylation, of 6-phosphogluconic acid, 13
Oxidative shunt, 6, 12-13
Oxidative step, 7
Oxidative transamination, 12
Oxygen, absorption of, mechanical interference with, 442
  blood content and, 437(t)
  carbohydrate metabolism and, 6
  pressure in blood, 437(t)
  pressure of, formula for, 434
  transport of, 436-440
Oxygen tension, hemoglobin oxygenation and, *439*
  influence of, 436
Oxyhemoglobin, 272
  dissociation of, 437, *438*
Oxysteroids. See also *Adrenal cortex, hormones of.*
  adrenal, amino acid transamination and, 161
Oxytocin, 712
Oxytocinase, 571-572

Paget's disease, alkaline phosphatase and, 563
  glucose tolerance and, 64
  hydroxyproline and, 224

PAH. See *p-Aminohippurate*.
Palmitic acid, fatty acid synthesis and, 103
  saturated fatty acids and, 100
Palmityl, fatty acid synthesis and, 103
Palmityl-CoA, fatty acid synthesis and, 103
Pancreas. See also *Langerhans, islands of.*
  carcinoma of, glucose tolerance and, 597
  cyst of, secretin test and, 592
  cystic fibrosis of, iron absorption of, 322
  diminished glucose tolerance and, 59
  diseases of, steatorrhea and, 601
  disorders of, steatorrhea and, 302
  edema of, secretin test and, 592
  function of, 589-599
    chemical examination of, 590-593
    fecal examination and, 598
    Lundh test meal and, 593
    secretin test and, 591
    serum enzymes and, 594-596
    tests for, 590
  hypofibrinogenemia and, 184
  iron absorption and, 321
  metabolic significance of, 589
  phospholipids and, 112
Pancrease, exocrine deficiency of, *603*
  gastrin producing tumor of, 586
Pancreatectomy, phenomena following, 19, 33-34, 109-110
Pancreatic amylase, 1
Pancreatic duct, obstruction of, serum enzymes and, 670
Pancreatic islets. See *Langerhans, islands of.*
Pancreatic islet insufficiency, 17-18
Pancreatic vein, 24
Pancreatitis, chronic, secretin test and, 592
  chronic relapsing, xanthomata and, 141
  glomerular filtration and, 457
  glucose tolerance and, 597
  glycosuria and, 87
  hyperbilirubinemia and, 597
  hypocalcemia and, 305, 597
  pancreatic function and, 590-598
Pancreozymin, serum amylase and, 595
Pandy test, cerebrospinal fluid and, 836
Panhypopituitarism, growth hormone and, 708
  mucoproteins and, 197
Pantothenic acid, 3, 12, 834
  coenzymes and, 251
  lipotropism and, 128
  type five fatty liver and, 129
Paralysis, familial hyperkalemic, potassium and, 388
  familial periodic, potassium and, 390
Paralytic ileus, sulfur and, 329
Paraprotein, abnormal globulin and, 202
Parathormone, bone development and, 283
  calcium absorption and, 281
  vs. calcitonin, 284
Parathyroid, secretory control of, 290

Parathryoid hormones. See *Hormones, parathyroid*.
   calcium concentration and, 289
   injection of, effects, 289
   urinary phosphate and, 312
Parathyroidectomy, calcium concentration and, 289
   magnesium and, 317
Paraventricular nuclei, extracellular fluid and, 359
Paresis, cerebrospinal fluid and, 836(t), 837
   hypoglycemia in, 58
Parietal cells, secretion of, 582
Parkinsonian disease, creatinuria and, 222
Parotid saliva, insulin in, 24
Pars intermedia, 697
Pascal, defined, 434, 855
PBI. See *Protein-bound iodine*.
pCO_2, formula for, 434
   hemoglobin and, 439
   oxyhemoglobin and, *438*
   ratio in air, 434(t)
Pelvis, renal, postrenal albuminuria and, 216
Penicillin, renal tubules and, 465
Pentagastrin, hydrochloric acid and, 585-586
   stomach evacuation and, 584
   Zollinger-Ellison syndrome and, 587
Pentose(s),
   body fluids and, 29
   carbohydrate metabolism and, 6
   endogenous sources of, 4
   pentosuria and, 82
Pentose-P, *18*
Pentose phosphate cycle, 21
   fatty acid synthesis and, 104
Pentose shunt, 6, 12-13, *13*
Pentosuria, alimentary, 88
   body fluids and, 29
   essential, 88
   pentose and, 82
   sugar secretion and, 88
Pepsin, activity of, 587
   protein digestion and, 148
Pepsinogen, intestinal phase and, 582
   protein digestion and, 148
Peptidases,
   insulin disposal and, 24
Peptides, intestinal phase and, 581
Peptidyl transferase, protein synthesis and, 247
Peptones proteinuria, and, 219
Perchlorate discharge test, 763
Perhydrocyclopentano-phenanthrene, nucleus of, steroids and, 96
Periarteritis nodosa, cryoglobulins and, 198
   macroglobulins and, 199
Peripheral neuritis. See *Neuritis, peripheral*.
Peristalsis, lipid digestion and, 98
Peritoneal effusions, enzymes and, 573-574
Peritoneum, lymphatic drainage of, cirrhosis and, 673
Peritonitis, electrolytes and, **375**

Peritonitis *(Continued)*
   glomerular filtration and, 457
   hypoalbuminemia and, 188
   sulfur and, 329
   urea and, 208
Pernicious anemia. See *Anemia, pernicious*.
Peroxidase, synthesis of, iron and, 324
Perspiration, blood volume and, 392
   dehydration and, 370(t)
   excessive, electrolytes and, 378
   insensible, 341
   pancreatic function and, 597-598
   water loss and, 341
PG. See *Prostaglandins*.
pH, electrolytes and, 372
   extracellular fluid and, 399
   hemoglobin dissociation and, 439
   interstitial fluid and, 400
   intracellular, 400
   intracellular-extracellular relations, 412-413
   of blood, determination of, 429-430
   of plasma, determination of, 429-430
   of serum, determination of, 429-430
   of urine, 407
   oxyhemoglobin and, *438*
   urinary acidity and, 471
Phagocytosis, plasma proteins and, 175
Pheneturide, osteomalacia and, 305
Phenformin, lactic acidosis and, 77
Phenobarbital, hypocholesterolemia and, 142
Phenobarbitone, osteomalacia and, 305
Phenol, chronic renal failure and, 531
Phenol oxidases, copper and, 335
Phenolsulfonphthalein, renal tubules and, 465
Phenolsulfonphthalein test, foreign substances and, 515
Phenoxybenzamine, catecholamines and, 743
Phentolamine, catecholamines and, 743
Phenylalanine, *18*
   ascorbic acid and, 821
   catecholamines and, 741
   fatty acid synthesis and, 104
   ketogenic pathway and, 162
   metabolism of, abnormal, 225
   protein digestion and, 148
Phenylephrine, catecholamines and, 743
β-Phenylethylamine, hepatic coma and, 674
Phenylketonuria, aminoaciduria and, 223-224
   phenylalanine metabolism and, 226
Phenytoin, osteomalacia and, 305
Pheochromocytoma(s), adrenal medulla and, 744, 745
   glucose tolerance and, 64
   hyperglycemia and, 86
Phlebotomy, iron storage and, 326
Phloridzin, glucose reabsorption and, 83
Phosphatases, 559-569
   acid, 566-569

Phosphatases *(Continued)*
  alkaline, 560-566, *562*
    isoenzymes and, 561-562
    liver diseases and, 565-566
    related disorders and, 566
    skeletal diseases and, 562-565
  Cushing's syndrome and, 734
  plasma proteins and, 177
Phosphate. See also under specific phosphate compound.
  absorption of, 280-281
  blood calcium and, 287-291
  cerebrospinal fluid and, 842
  concentration of, alterations in, conditions, 306(t)
  electrolyte changes and, 22
  excretion of, 292
  glycogen metabolism and, 9
  glycolytic reaction equilibria and, *14*
  hexose, 46
  iron absorption and, 320-321
  metabolism of, 279
  normal urine and, 478
  nucleoside triphosphates and, 250
  parathyroid hormone and, 289-290
  plasma, calcium and, 291
  plasma concentration conversion and, 856
  serum, alimentary glucose reaction and, 47
    diabetes mellitus and, 61
  urinary, abnormal, 312-313
Phosphate ions, of bone, buffering and, 406
Phosphatides. See also *Phospholipids*.
  plasma lipids and, 118
Phosphatidic acid, phosphoglycerides and, 94
Phosphatidyl choline. See also *Lecithin*.
  lipid metabolism and, 94
Phosphatidyl compounds, fatty acids and, 91
Phosphatidyl derivatives, lipid absorption and, 100
  lipid metabolism and, 94-95
Phosphatidyl ethanolamine, lipid metabolism and, 94
Phosphatidyl inositol, lipid metabolism and, 94
Phosphatidyl serine, lipid metabolism and, 94
Phosphocreatine, creatine and, 180
  muscle contraction and, 16
Phosphoenolpyruvate, 27
  amino acids and, 161
  oxaloacetate and, 107
Phosphoenolpyruvate carboxylase, cortisol and, 25
Phosphoenolpyruvic acid, carbohydrate to lipid conversion and, 107
  glucogenesis and, 161
Phosphoethanolamine, hypophosphatasia and, 224
Phosphoethanolaminuria, aminoaciduria and, 227
Phosphoglyceric acid, 7

Phosphoglycerides, lipid absorption and, 100
  lipid metabolism and, 94-95
Phosphoglucomutase, 9
6-Phosphogluconic acid,
  pentose shunt and, 12, 13
Phospholipase A, lipid digestion and, 98
Phospholipids, blood plasma and, 117(t)
  catabolism of, 112
  diabetic acidosis and, 138
  diseases and, 121
  distribution of, *102*
  function of, 94-95
  lipid digestion and, 98
  lipovitellin and, 97
  liver and, 124, 125
  metabolic interrelations and, 163(t)
  metabolism of, 111-112
  plasma and, 114
    age and, 121
    hypothyroidism and, 141
  plasma lipids and, 118
  plasma lipoproteins and, 115
  ratio to cholesterol, 96
  tissue and, 93
  turnover of, 112
Phosphoprotein, urinary, 219
Phosphopyruvic acid, 7
5'-phosphoribosyl-1'-pyrophosphate, nucleic acid biosynthesis and, 237
Phosphoric acid,
  phosphatidyl compounds and, 91
Phosphorus. See also *Phosphate*.
  blood sugar tolerance curve and, 63
  Cushing's syndrome and, 734
  function of, 279
  hypocholesterolemia and, 142
  kidney tubules and, 464
  lipuria and, 111
  requirement of, 293-294
Phosphorus poisoning,
  liver function and, 49
Phosphorylase, glycolytic reaction and, *14*
  liver, glucagon and, 27
Phosphorylase *a*, 10
Phosphorylase activity, glucose-1-phosphate and, 27
  glycogenolysis and, 26
Phosphorylase *b*, 10
Phosphorylation, anaerobic metabolism and, 6
Phosphorylcholine, sphingomyelins and, 92
Phthiocol, vitamin K and, 830
Phytanic acid oxidase, deficiency of, 130(t)
Phytic acid, calcium absorption and, 280
Phytosterols, lipid absorption and, 99
Pick's disease, albumin and, 188
Pigment(s), bile, in feces, 655
  metabolism of, liver function and, 633-640
  respiratory, copper and, 335
Pigmentation, oral contraceptives and, 785
Pilocarpine, gastric secretion and, 582
  vagus effect and, 590
Pinocytosis, lipid digestion and, 98

Pitocin, plasma fibrinogen and, 184
Pituitary, anterior, 687-709
  carbohydrate metabolism and, 25, 35
  diabetes and, 21, 35
  excess, 17-18
  experimental diabetes mellitus and, 33
  hormones of, 327, 687-709
  hyperfunction of, 71. See also *Acromegaly*.
  hypofunction of, hypoglycemia and, 70
  insulin tolerance test and, 71
  insufficiency, 57
Pituitary, posterior, hormones of. See *Neurohypophysis*.
Pituitary cachexia, adenohypophyseal hypofunction and, 69
  epinephrine tolerance test and, 74
Pituitary gland. See also *Hyperpituitarism* and *Hypopituitarism*.
  hormones of, 687-712
  hypofunction of, glucose tolerance and, 69
  gonadotrophin and, 698
  hypogonadism and, 782(t)
  hypothalamic control and, 749
  iodine and, 332
  pregnancy and, 793
Pituitary infantilism, glucose tolerance and, 69
  hypoglycemia and, 70
Pituitrin, 54. See *Pitocin* and *Vasopressin*.
Placenta, enzyme abnormalities and, 576
  hormones of, 787-793
  IgG and, 196
  separated, plasma fibrinogen and, 184
Plasma, blood, air exchange with alveoli and, 353
  buffer system of, 402
  calcium and, 287
  carbon dioxide and, in buffer systems, 404
  carbon dioxide and, combining power with, 428
  carbon dioxide content of, 427-428
  carbonic acid buffer systems and, 403
  cholesterol level and, 113
  clotting and, 630
  conversion table for, 856(t)
  dilution of, hypoalbuminemia and, 189
  electrolyte pattern and, 377
  erythrocyte exchange and, 353
  freezing point of, 347
  hypercalcemia and, 298
  interstitial fluid exchanges and, 353-355
  iodine and, 331
  ionic concentration in, 349
  lipoproteins and, 97
  pH determination of, 429-430
  phosphate in, 46
  thyroxine and, 332
  volume of, 391-393
Plasma membrane, carbohydrate metabolism and, 2

Plasma thromboplastin antecedent, coagulation and, 175
Plasma urate level, 253
Plasmacytoma, Bence Jones proteins and, 199-200
  heavy-chain disease and, 202
Plasmapheresis, hypoalbuminemia and, 139
Plasminogen, coagulation and, 175
  fibrinolysin and, 185
  properties of, 169(t)
Plasmogens, fatty acids and, 91
Platelets, plasma proteins and, 175
Pleural effusions, enzymes and, 573
pN_2, formula for, 434
  ratio in air, 434(t)
Pneumonia, cerebrospinal fluid and, 836
  cryoglobulins and, 198
  glomerular filtration and, 457
  glucose tolerance and, 64
  hypocholesterolemia and, 142
  plasma fibrinogen and, 184
  pneumococcal, hypochloremia and, 382
  proteinuria and, 219
pO_2, formula for, 434
  oxyhemoglobin dissociation and, 437
  ratio in air, 434(t)
Poisoning, albuminuria and, 216
  animal, hemoglobin disorders and, 271
  chemical, hemoglobin disorders and, 271
  hypoglycemia and, 614
  metallic, urinary phosphate and, 312
  nephrotoxic agents and, 224
  oxalic acid, magnesium and, 317
  renal tubular defects and, 537-538
  salicylate, acid-base balance and, 425
  vegetable, hemoglobin disorders and, 271
Poliomyelitis, cerebrospinal fluid and, 841
  porphyrin and, 275(t), 278
Polycistron, deoxyribonucleic acid, 242
Polycythemia vera, blood volume and, 391
Polymorphonuclear leukocytes, plasma proteins and, 175
Polypeptides, protein metabolism and, 152
Polysaccharides, composition of, 1
Polyserositis, albumin and, 188
Polyuria, Houssay animals and, 36
  pancreatectomy and, 34
  urine specific gravity and, 86
Pompe's disease, enzyme abnormalities and, 579
  glycogen storage disease and, 80
Porphobilinogen, porphyria and, 274
  structure of, 263
Porphyria, 274-278
  erythropoietic, congenital, 276
  hepatic, 276-278
    acute intermittent, 276-277
    coproporphyria, 277
    cutanea tarda, 277-278
    variegata, 277
Porphyrin(s), biosynthesis of, 262-265
  defined, 261
  in feces, 655-656

Porphyrin(s) *(Continued)*
  in urine, 655-656
  metabolic interrelations and, 163(t)
  metabolism of, 260-278
  nitrogen surplus and, 159
  pigment metabolism and, 635-636
  structure of, *261*
Porphyrinogen, porphyria and, 274
Porphyrinuria, 272-278, 275(t), 655
Porter-Silber reagent, cortisol and, 720
Posterior lobe hormones. See *Neurohypophysis.*
Potassium, abnormalities of, 383-391
    chronic renal failure and, 528
  absorption of, 361-364
  adrenocortical hormones and, 387
  blood value, normal, 364-365
  cerebrospinal fluid and, 843-844
  Cushing's syndrome and, 734
  deficits of, 368-380
  depletion of, nonrespiratory alkalosis and, 423
  diabetic ketoacidosis and, 78-79
  diarrhea and, 386
  diuretic drugs and, 387
  electrolyte changes and, 22
  electrolyte redistribution and, 372
  excretion of, 361-364
  familial hyperkalemic paralysis, 388
  familial periodic paralysis and, 390
  gastrointestinal disorders and, 386
  glucose and, 385-386
  hyperkalemia and, 388
  hypothermia and, 390
  importance of, 360
  kidney tubules and, 463
  loss of, pancreatectomy and, 34
  malnutrition and, 385-386
  metabolism and, 365-367
  normal urine and, 478
  pH relations and, 412-413
  physiological considerations, 360-367
  postoperative response, 382
  primary deficiency, hypochloremia and, 382
  renal disease and, 389-390
  salt and, 385-386
  serum, alimentary glucose reaction and, 47
  starvation and, 385-386
  urine and, 384
  vomiting and, 386
Practolol, catecholamines and, 743
Prealbumin(s), cerebrospinal fluid and, 835
  plasma proteins and, 189-190
  properties of, 169(t)
  thyroxine and, 332
Prediabetes, 62
Prednisone-glucose tolerance test, abnormal, latent diabetes and, 72
Pregnancy, abnormal alimentary response and, 64
  acid-base balance and, 415(t), 423, 425

Pregnancy *(Continued)*
  adrenal glands and, 792
  calcium and, 294
  caloric requirements and, 452, 452(t)
  carbohydrate metabolism and, 58, 61, 64
  cholesterol synthesis and, 113, 121
  decreased urea nitrogen and, 209
  diabetes mellitus and, 61
  glucose testing and, 62
  glycosuria of, 87
  histidine and, 182
  hormones and, 787-793
  human chorionic gonadotrophin and, 790
  human chorionic somatomammotrophin and, 790
  human chorionic thyrotrophin and, 791
  human molar thyrotrophin and, 791
  human placental lactogen and, 790
  hypoalbuminemia and, 189
  hypoglycemia in, 58
  hyposecretion and, 586
  iron and, 325
  lipids and, 121
  magnesium and, 317
  phosphorus and, 294
  pituitary gland and, 793
  plasma fibrinogen and, 184
  prostaglandins and, 791-792
  respiratory alkalosis and, 423
  sterol metabolism and, 113
  thyroid gland and, 792-793
  toxemia of, excess electrolytes and, 381-382
    kidney function and, 525-526
    liver function and, 49
  urea clearance and, 504
  vitamin K and, 831
Pregnanediol. See also *Progesterone.*
  liver and, 663
Pregnanetriol, adrenogenital syndrome and, 737
Pregnenolone, corticotrophin and, 703
Pressure, formulas for, 434
  relationships in, *435*
PRH. See *Prolactin-releasing hormone.*
Primaquine, pentose shunt and, 12
Primidone, osteomalacia and, 305
P-RIH. See *Prolactin-release inhibiting hormone.*
Priodax, albuminuria and, 214
Proaccelerin, hypofibrinogenemia and, 185
Probenecid, kidney tubules and, 464, 466
Procarboxypeptidase, pancreas and, 589
Proconvertin, vitamin K and, 832
Progesterone, 779-780
  biosynthesis of, *717*
  cholesterol and, 96
  liver and, 663
  placental hormones and, 789
  preparations for, 780
  steroids and, 96
  values for, 779-780
Progesterone therapy, indications for, 780

Progestogen, contraceptives and, effects of, 784
Proinsulin, obesity and, 74
  plasma insulin level and, 46
  primary structure of, 23, *23*
Prolactin, adenohypophysis and, 700-702
Prolactin-release inhibiting hormone, 690
Prolactin-releasing hormone, 690
Prolinuria, aminoaciduria and, 228
  renal tubular defects and, 536
Properdin system, plasma proteins and, 177
Propionyl CoA, fatty acid synthesis and, 105
  oxaloacetate and, 107
Propranolol, catecholamines and, 743
  insulin release and, 24
Prostaglandins, lipid metabolism and, 95-96
  lipolysis and, 124
  pregnancy and, 791-792
Prostigmin, vagus effect and, 590
Prostanoic acid, prostaglandins and, 95
Prostate, cancer of, hypofibrinogenemia and, 184
  postrenal albuminuria and, 216
Protamine, hyperlipemia and, 117
Protamine zinc insulin, 54
Protein(s), abnormalities of, intestinal malabsorption syndromes and, 601-602
  absorption of, clinical study of, 607
  breakdown of, aminoaciduria and, 224
  C-reactive, plasma protein abnormalities and, 198
  caloric comparisons, 446(t)
  catabolism of, hypoalbuminemia and, 188-189
    excessive, urinary nonprotein nitrogen and, 508
    urea nitrogen and, 208
  cerebrospinal fluid and, 835-838
  concentration, liver function and, 624
  conjugated, carbohydrates and, 196-197
  deficiency of, intestinal malabsorption and, *603*
  defined, 147
  deprivation of, 157
  dietary requirement of, 155, 156
  extracellular fluid and, *350, 351*
  hypoalbuminemia and, 188
  intermediary metabolism of, pathways, 158, *159*
  intestinal malabsorption syndromes and, 601
  lipids and, metabolic interrelations with, 93, 106-110, 107(t)
  meat, elemental composition of, 447(t)
  metabolism of, 147-234
    abnormal urinary nitrogen and, 212-230
    acid mucopolysaccharides and, 232-234
    adrenal cortical hormones and, 722
    androgens and, 796
    blood nonprotein nitrogen and, 207

Protein(s) *(Continued)*
  metabolism of, carbohydrate metabolism and, 17-18, 162-164
    Carbohydrate-fat metabolism and, 159
    Cushing's syndrome and, 733
    digestion and, 158-162
    endocrine influence and, 164-166
    growth hormone and, 707
    intermediary, 158-162
    lipid metabolism and, 162-164
    liver function and, 617
    nitrogen excretion and, 179-182
    nitrogenous constituents of the blood and, 166-179
    nonprotein nitrogen and, 207-212
    nucleic acids and, *241*
    over-all, 152-158
    plasma protein abnormalities and, 182-200
    serum protein abnormalities and, 200-207
    thyroid hormone and, 751
    urea and, 178
    urea cycle and, 230-232
    urinary nitrogen and, 212-230
  myeloma, 201
  nitrogen and, urinary, 161
  plasma, abnormalities of, 200
    fibrinogen and, 198
    globulins and, 189-196
    protein metabolism and, 182-200
    albumin and, 185-190
    blood plasma and, 166
    calcium and, 290-291
    diabetic acidosis and, 138
    enzymes and, 177
    function of, 173-177
    globulins and, 190-200
    immunity and, 175
    liver function and, 620
    properdin system and, 177
    properties of, 169(t)
    protein metabolism and, 179-200
    techniques for study, 200
    tests for, 168-173, 171(t)
    total, abnormalities of, 183
  respiratory quotients and, 446
  serum, abnormalities of, dysgammaglobulinemia and, 205
    globulin reactions and, 205-207
    heavy-chain disease and, 202
    multiple myeloma and, 201-202
    sarcoidosis and, 201
    Waldenström's macroglobulinemia, 203-204
  concentration of, 168(t)
  disease and, 200-207
  protein metabolism and, 200-207
sparing effect of, 157
special, synthesis of, 156

Protein(s) *(Continued)*
   transudates and, 393
   urinary, variety of, 219
   value of, 156
Protein-bound iodine, blood iodine and, 331
   serum, thyroid hormone release tests and, 756
Protein sulfur, sulfur absorption and, 327
Protein synthesis, cancer chemotherapy and, 249
   hypoalbuminemia and, 188
   nucleic acids and, 244-248
Proteinuria, abnormal urinary nitrogen and, 212-220
   cholesterol acyltransferase and, 137
   multiple myeloma and, 201
   protein and, 218-219
Proteoses, urinary, 219
Prothrombin, coagulation and, 175
   glycoproteins and, 197
   hypofibrinogenemia and, 184
   liver and, 173, 628
   plasma, jaundice diagnosis and, 680
   plasma, vitamin K and, 602
   properties of, 169(t)
   vitamin K and, 832
Protocollagen, bone development and, 282
Protoporphyria, erythropoietic, 276
Protoporphyrin, defined, 261
Protozoa, hemoglobin disorders and, 271
Provitamin D, steroids and, 96
Proximal convolution, kidney tubules and, 460
Proximal tubule, insulin disposal and, 24
PRPP, nucleic acid biosynthesis and, 237
Pseudoglobulin, serum protein concentrations and, 168(t)
   study methods and, 171(t)
Pseudo-Hurler's disease, 130(t)
Pseudohypoparathyroidism, hypocalcemia and, 301
   renal tubular defects and, 538
Pseudo-pseudohypoparathyroidism, hypocalcemia and, 301-302
Psoriasis, hypercholesterolemia and, 141
Psychic hyperglycemia, 50
Psychic phase, gastric function and, 580-581
Psychogenic vomiting, blood sugar and, 72
Psychoses, hypoglycemia in, 58
Pteridines, diuretics and, 469
Pteroylglutamic acid, folic acid and, 817
Puberty, female, endocrine function and, 781
Pulmonary tuberculosis, proteinuria and, 219
Purine(s), metabolic interrelations and, 163(t)
   metabolism of, uric acid and, 178
   nitrogen surplus and, 159
   nucleic acid metabolism and, *241*

Purine(s) *(Continued)*
   nucleotide synthesis and, 236, 252-253
   protein metabolism and, 152
   synthesis of, cancer chemotherapy and, 249
   excessive, 256
Puromycin, protein synthesis and, 247
Purpura cryoglobulinemia, 198
Pyelonephritis, defect localization and, 522
   lipuria and, 111
   magnesium and, 317
Pyloric cells, gastric antrum and, 580
Pyloric mechanism,
   carbohydrate metabolism and, 55
Pyogenic infections, glucose tolerance and, 64
Pyridoxal, vitamin B_6 and, 815
Pyridoxamine, vitamin B_6 and, 815
Pyridoxine, 3
   lipotropism and, 128
   vitamin B_6 and, 815
Pyridoxal phosphate, amino acid transamination and, 159, 161
Pyridoxol, vitamin B_6 and, 815
Pyrimidines, metabolic interrelations and, 163(t)
   nitrogen surplus and, 159
   nucleic acid synthesis and, 252
   protein metabolism and, 152
Pyrimidine nucleotides, nucleic acids and, 236
Pyroglobulins, plasma protein abnormalities and, 183, 198
   study methods for, 171(t), 172
Pyrophosphatase, calcitonin and, 284
Pyrophosphate. See *Thiamine*.
Pyruvate, glycogenesis and, 13
   hepatic coma and, 674
   lipid metabolism and, distribution of, *102*
   liver function and, 617
   metabolic interrelations and, 163(t)
   phosphoenolpyruvate and, 27
   protein digestion and, 151, 163
Pyruvate carboxylase, 25
   Acetyl-CoA and, 22
Pyruvate kinase, liver and, in carbohydrate metabolism, 125
Pyruvic acid, amino acids and, 161
   blood, carbohydrate metabolism and, 76-78
      thiamine and, 812
   carbohydrate metabolism and, 6, 7
   fatty acid synthesis and, 103
   oxaloacetic acid and, 15
   oxidation of, 11
Pyruvic acid cycle, 10
Pyruvic carboxylase, 15

Quinine, blood sugar and, 54

Race, cholesterol difference and, 120
Radio-immunoassay, insulin metabolism research and, 24
  plasma insulin levels and, 46
Radioiodine tests, 765-766
Raynaud's disease, blood volume and, 391
Raynaud's phenomena, Waldenström's macroglobulinemia and, 203
Rebreathing, hydrogen ion regulation and, 415(t)
Refsum's disease, 129, 130(t)
Regurgitation, gastric residuum and, 583
Renal. See *Kidney.*
Renal glycosuria. See *Glycosuria, renal.*
Renin-angiotensin-aldosterone mechanism, 730
Renin-angiotensin system, ionic regulation and, 360
Repressor protein, bacteria and, 128
Reserpine, catecholamines and, 743
  lactation and, 701
Respiration, chemical control of, 435
  disturbances of, hydrogen ions and, 414
  inhibiting agents in, 2
  regulation of, hydrogen ions and, 406-407
Respiratory exchange, basal metabolism and, 433-453
Respiratory quotient, 445-448
  alimentary glucose reaction and, 47
  carbohydrates and, 448(t)
  diabetes mellitus and, 61
  fat and, 448(t)
  hepatic disease and, 63
  significance of, 448
  Houssay animals and, 36
Rest nitrogen. See *Nitrogen, undetermined.*
Reticuloendothelial system, hypogammaglobulinemia and, 204
  iron storage and, 322
Reticuloses, chemotherapy for, 250
Retinitis pigmentosa, a-beta-lipoproteinemia and, 136
Reverse transcriptase, deoxyribonucleic acid and, 242
Rheumatic fever, cryoglobulins and, 198
  hypochloremia and, 382
  thymol turbidity test and, 627
Rheumatoid arthritis, C-reactive proteins and, 198
  cryoglobulins and, 198
  glucose tolerance and, 64
  macroglobulins and, 199
Riboflavin, 813-814
  coenzymes and, 251
  intestinal malabsorption and, 603
  vitamin deficiency and, 602
Ribonuclease, pancreas and, 589
Ribonucleic acid, biosynthesis of, 238
  distribution of, 240
  heteronuclear, 248
  messenger, protein synthesis and, 244-248

Ribonucleic acid *(Continued)*
  protein synthesis and, 244
  transfer, protein synthesis and, 244-248
Ribonucleoproteins, viruses and, 240
Ribose, glucose conversion and, 6
  nucleic acid synthesis and, 6
  nucleoside triphosphates and, 250
  nucleotide synthesis and, 252
Ribose-5-phosphate, pentose shunt and, 13, *13*
Ribulose-5-phosphate, pentose shunt and, 13, *13*
Rickets, alkaline phosphatase and, 563
  hypocalcemia and, 302
  hypophosphatemia, 310
  magnesium and, 317
  renal, 296, 303, 564
  vitamin D-resistant, hypophosphatemia and, 311-312
  renal tubular defects and, 538
RNA. See *Ribonucleic acid.*
Ross-Jones test, cerebrospinal fluid and, 836
Rotor syndrome, jaundice and, *643*, 644
Rouleaux formation, plasma fibrinogen and, 184
R.Q. See *Respiratory quotient.*

Salicylate, alkalosis and, 422
  poisoning, acid-base balance and, 425
  hyperglycemia and, 54
  vitamin K deficiency and, 833
Saliva, dehydration and, 370(t)
  iodine and, 331
  urea and, 514
  volume of, 340(t)
Salivary glands, iodine and, 332
Salt(s), metabolism of, liver and, 672
  perspiration and, excessive, 378
  potassium and, 385-386
  serum calcium and, 291
Salyrgan, diuresis and, 469
Sanfilippo's syndrome, mucopolysaccharidoses and, 233
Saponin, intestinal phase and, 581
Sarcoidosis, hypercalcemia and, 298-299
  hyperglobulinemia and, 201
  serum enzymes and, 670
Scarlet fever, glucose tolerance and, 64
Scheie's syndrome, mucopolysaccharidoses and, 233
Schistosomiasis, hyperglobulinemia and, 201
SDA. See *Specific dynamic action.*
Secretin, 773-774
  assay of, 774
  insulin release and, 23-24
  intestinal phase and, 581-582
  pancreas and, 589, 590-593
  serum amylase and, 595
  syndromes and, 774

Sedatives, liver enzymes and, 675
Sedoheptulose-7-phosphate, 13
Sepsis, bacterial, neonatal jaundice and, 648
Septicemias, hematinemia and, 213
    hyperbilirubinemia and, 642-645
    plasma fibrinogen and, 184
Serine, amino acids and, 162
    anabolism and catabolism of, 92
    lipid synthesis and, 107
    nucleic acid metabolism and, *241*
    phosphatidyl compounds and, 91
Serine phosphoglycerides, lipid metabolism and, 94
Serotonin, *9*
    hepatic coma and, 675
    metabolites of, in urine, 228
Serous meningitis, cerebrospinal fluid and, 836(t), 837
Serum prothrombin conversion accelerator, 175
Shaffer method, 848
Sheehan's syndrome, growth hormone and, 708
Shock, anoxia and, 444
    anuria and, 493, 494
    basal metabolism and, 451
    glomerular filtration and, 458
    glucose and, 52
    hypofibrinogenemia and, 184
    kidney function and, 484, 487, 503, 513, 522
    lactic acidosis and, 77
    nephrosis and, 513, 522
    nonprotein nitrogen and, 208, 513
    oxygen and, 444
    plasma albumin and, 185
    urea clearance and, 504
    urea nitrogen and, 208
SI units, 854-857
    base, 854(t)
    derived, coherent, 854-855
    noncoherent, 855-857
Sialic acid, gangliosides and, 92
    oligosaccharides and, 92
Sickle cell anemia, hemoglobin S and, 268
Sideropenia, iron storage and, 326
Siderophilin, iron transport and, 175
Siderosis, excessive iron and, 326
Simmond's disease, 57
    adenohypophyseal hypofunction and, 69
    hypoglycemia and, 70
    insulin tolerance test and, 71
Sitosterol, cholesterol absorption and, 100
    lipid absorption and, 99
Skeleton, calcium interchange and, 286
    diseases of, alkaline phosphatase and, 562-565
    intestinal malabsorption syndromes and, 604
    osteoblastic metastases of, hypocalcemia and, 306

Skin, cholesterol synthesis and, 113
    ionic excretion and, 363
Soap, intestinal phase and, 581
Soap fat, fecal fat and, 110
Sodium, abnormalities of, 368-398
    absorption of, 361-364
    kidney tubules and, 461
    blood value, normal, 364-365
    calcium absorption and, 281
    cerebrospinal fluid and, 843-844
    concentration of, effects, 371
    Cushing's syndrome and, 734
    deficits of, 368-380
    dehydration and, 374, 375
    depletion of, chronic renal failure and, 526
    diabetes mellitus and, 376
    electrolyte redistribution and, 372
    excess of, 380-383
        adrenocortical hyperfunction and, 382
        congestive heart failure and, 381
        kidney disease and, 380
        toxemia of pregnancy and, 381-382
    excretion of, 361-364
    extracellular fluid and, 345
    importance of, 360
    liver disease and, 381
    loss of, 34
    metabolism and, 365-367
    normal urine and, 478
    perspiration and, 341
    pH relations and, 412-413
    physiological considerations, 360-367
    plasma concentration conversion and, 856
    postoperative response, 382
    urinary excretion and, 379
Solids, elimination of, renal function and, 495-501
Somatotropin, adenohypophysis and, 705-709
    bone development and, 284
    metabolic effects, 706-708
    secretion of, 705-706
Somogyi, diabetes mellitus and, 850
Sorbitol dehydrogenase, 571
Spaces of Disse, 610
Sparsomycin, protein synthesis and, 247
Spasmophilia, cerebrospinal fluid and, 836
Specific dynamic action, of foods, 451-452
Specific gravity, normal urine and, 470
    of urine, renal function and, 495-501
4-Sphingenine, sphingosine and, 91-92
Sphingolipids, blood plasma and, 117(t)
    brain tissue and, 93
    carbohydrates of, synthesis of, 92
    lipidosis and, 129
    sphingosine and, 91-92
Sphingomyelin(s), 92
    blood plasma and, 117(t)
Sphingomyelinase, deficiency of, 130(t)
Sphingosine, amino acids and, 162

Sphingosine *(Continued)*
   lipid synthesis and, 107
   metabolism and, 93
   sphingolipids and, 91-92
Spinal cord, diseases of, cerebrospinal fluid and, 837
Spironolactone, diuresis and, 469
Spleen, cholesterol synthesis and, 112
   iron storage and, 324
   prostaglandins in, 95
Sprue, calcium and, 302
   feces and, 600, 601
   globulins and, 190
   glucose tolerance and, 69
   hypocholesterolemia and, 145
   hypophosphatemia and, 310
   parathyroid hyperplasia and, 295
   steatorrhea and, 302, 601
   vitamin deficiency and, 602
Squalene, cholesterol biosynthesis and, 96
Standard units. See *SI units*.
Starvation, ammonia and, 229
   creatinuria, 222
   electrolytes and, 374
   gamma-globulins and, 191
   hypocalcemia and, 302
   lactic acidosis and, 77
   nonrespiratory acidosis and, 415(t), 419
   potassium and, 385-386
   Staub-Traugott effect, 42
Stearic acid, saturated fatty acids and, 100
Steatorrhea. See also *Fat, fecal*.
   a-beta-lipoproteinemia and, 136
   bile acid metabolism and, 659
   hypocalcemia and, 302
   idiopathic, hypophosphatemia and, 311
   intestinal absorption and, 600-601
   lipid absorption and, 100
Stercobilinogen, 650
Steroids, lipid metabolism and, 95-97
   liver enzymes and, 675
   phospholipids and, 97
Steroid hormones, 713-714
Sterols, lipid absorption and, 99-100
   metabolism of, 112-114
   natural, cholesterol catabolism and, 114
Stomach. See also *Gastric function*.
   contents of, 580
   disorders of, electrolytes and, 375
      potassium and, 386
   evacuation of, danger, 584
   function of, chemical investigation of, 580-588
   hypofibrinogenemia and, 184
   protein digestion and, 148
   secretion of, investigation of, 584-588
Stomach juices, normal, gastric function and, 582-584
   pepsins and, 148
Stones. See *Lithiasis*.
Streptomycin, protein synthesis and, 247
Stuart factor, vitamin K and, 832

Stuart-Prower factor, coagulation and, 175
Substance concentration, 855
Succinate, free, 22
   iron absorption and, 322
Succinyl-CoA, 22
   fatty acid synthesis and, 105
   oxaloacetate and, 107
Sucrase, carbohydrate metabolism and, 2
Sucrose, body fluids and, 29
   carbohydrate metabolism and, 1
   sucrosuria and, 82
Sucrosuria, sucrose and, 82
Sudbury diabetes survey, 61
Sugar tolerance, 36-43
Sugar tolerance curve, diabetes mellitus and, 60
Sulfate, inorganic copper and, 335
   urine and, 328-329
Sulfatides, of ceramide, sphingomyelins and, 92
   sulfur absorption and, 327
Sulfhemoglobinemia, hemoglobin and, 273
   hypoxia and, 443
Sulfolipids, sulfur absorption and, 327
Sulfonamides, blood sugar tolerance curve and, 63
   hemoglobinemia and, 272
   pentose shunt and, 12
Sulfonylgalactosyl-ceramide, sulfur and, 329
Sulfonylureas,
   glucagon and, 27
   insulin release and, 24
Sulfur, absorption of, 327
   blood and, 327
   function of, 327
   intermediary metabolism of, 327-328
   metabolism of, 327-329
   normal urine and, 477
   urine and, 328-329
Supraoptic nuclei, extracellular fluid and, 359
Surgery, negative nitrogen balance and, 229
Svedberg flotation units, 97
Sweat. See *Perspiration*.
Sympathoblastomas, adrenal medulla and, 744
Syndrome. See under specific syndrome.
Synthetase system, 14
Syphilis, cerebrospinal fluid and, 837
   congenital, macroglobulins and, 199
      neonatal jaundice and, 648
   dye retention and, 669
   hepatic, hippuric acid and, 663
   hyperglobulinemia and, 201
   secretin test and, 592
Systemic lupus, macroglobulins and, 199

Tabes, cerebrospinal fluid and, 836(t), 837
Takata-ara test, globulin reactions and, 205
Tangier disease, hypolipoproteinemia and, 136

INDEX 913

Tau protein, cerebrospinal fluid and, 835
Taurocholic acid, sulfur and, 327
Tay Sachs disease, 92, 129-130, 129(t)
    cerebrospinal fluid and, 846
TBG. See *Thyroxine-binding globulin.*
TBP. See *Thyroid hormone binding proteins.*
TBPA. See *Thyroxine-binding prealbumin.*
Techniques, history of, 847-857
Telepaque, albuminuria and, 214
Testes, function of, 793-797
    tumors of, gonadotrophin and, 697
Testosterone. See also *Androgens.*
    amino acid concentration and, 179
    cholesterol and, 96
    creatinuria and, 222
    effects of, 795
    metabolism of, 793-796
    protein metabolism and, 165
    synthesis of, 793-796
Tetanus, albuminuria and, 216
    hyperglycemia in, 51
Tetany, citrate, hypocalcemia and, 305
    intestinal malabsorption and, *603*
    magnesium and, 317
    maternal, hypocalcemia and, 305
    vitamin D and, 828
Thalassemia, abnormal hemoglobin and, 270
    neonatal jaundice and, 647
Theobromine, diuresis and, 469
Theophylline, diuresis and, 469
Thiamine, 3, 810-815
    blood and, 812
    deficiency of, 12, 811-812
        intestinal malabsorption and, *603*
    excretion of, 812
    metabolism of, 811
    sulfur and, 327
    vitamin deficiency and, 602
Thiamine pyrophosphate, 12
    acetate transformation and, 77
Thiazide, diuretics and, 469
Thiocyanate, thyroid hormone interference and, 755
Thioglycollic acid,
    diabetes and, 35
Thioguanine, chronic myeloid leukemia and, 249
Thioguanosine, chronic myeloid leukemia and, 249
Thoracic duct, lymph of, 24
Threonine, nucleic acid metabolism and, *241*
    protein digestion and, 151
Thromboangitis obliterans, blood volume and, 391
Thromboplastic protein, lipoproteins and, 97
Thrombosis, cerebral, cerebrospinal fluid and, 836(t), 837
Thudichum, J.W.L., 847
Thymicolymphaticus, hypoglycemia in, 58

Thymine, nucleic acid and, 235
Thymol, turbidity of, liver function and, 626-627
Thymol turbidity test, globulins and, 190, 206
    vs. cephalin-cholesterol test, 627
Thymoxamine, catecholamines and, 743
Thyroglobulin(s), iodine and, 332
    iodotyrosines and, 332
    thyroid hormones and, 748
Thyroid antibody tests, 763-764
Thyroid function tests, 757(t)
    oral contraceptives and, 786
Thyroid gland, disease of, diagnosis, 755-760
    function tests, 757(t)
    gamma-globulins and, 191
    hormones of, 746-764. See also *Thyroxine.*
        biosynthesis of, 747-750
        carbohydrate metabolism and, 26
        chemistry of, 746
        circulating, 750-754
        iodine and, 330
        interfering agents of, 754-755
        metabolism and, 750
        protein metabolism and, 166
        secretion of, 747-750
        synthesis of, 748
        unesterified fatty acids and, 119
    iodine and, 332
    plasma cholesterol level and, 113
    pregnancy and, 792-793
    prostaglandins in, 95
    stimulating hormone for (TSH), 749
    thyrotrophin and, 699-700
Thyroid hormones. See *Thyroid gland, hormones of.*
Thyroid hormone binding tests, 759
Thyroid hormone release tests, 756, 765
Thyroid iodide trap tests, 756
Thyroid stimulating hormone, 749
    vs. long-acting thyroid stimulator, 767
Thyroid stimulating hormone immunoassay, 760-761
Thyroid stimulating hormone stimulation test, 763
Thyroidectomy, glucose and, 58
    glucose tolerance and, 69
Thyrotoxicosis. See also *Hyperthyroidism.*
    hypoalbuminemia and, 188
    ketosis and, 110
Thyrotrophin, adenohypophysis and, 699-700
Thyrotrophin, trophoblastic, comparison of, 791(t)
Thyrotrophin-releasing hormone, 688-689
Thyrotrophin-releasing hormone test, 761-764, 763(t)
    hyperthyroidism and, 768
Thyroxine, free, thyroid hormone release tests and, 759
    glycogenolytic action of, 49

Thyroxine *(Continued)*
  hepatic glycogenolysis and, 26, 31
  iodine absorption and, 330
  lipid mobilization and, 105
  metabolic interrelations and, 163(t)
  plasma and, 332
  prealbumins and, 190
  serum, thyroid hormone release tests and, 758
  synthesis of, 332
  thyroglobulins and, 332
  triglyceride conversion and, 106
  urinary, thyroid hormone release tests and, 759
Thyroxine-binding globulin, 750
  blood iodine and, 331
Thyroxine-binding prealbumin, 750
Thyroxine ratio, effective, thyroid hormone release tests and, 759
TIBC. See *Total iron-binding capacity*.
Tissue, connective, macroglobulins and, 199
  fat content of, 93
  utilization of, diminished glucose tolerance and, 59
Tocopherols. See *Vitamin E*.
Tolbutamide, insulin release and, 24
Tolbutamide tolerance test, 45-46
  abnormal, diabetes and, 72
Tolerance curve. See under specific types.
Tolerance tests, intestinal malabsorption and, 604
Tonsils, heavy-chain disease and, 202
Total iron-binding capacity, iron transport and, 322
Toxemia, insulin production and, 59
  of pregnancy, acid-base balance and, 425
  electrolytes and, 375
Toxic adenoma, 49
T_3 Toxicosis, hyperthyroidism and, 766
  vs. hyperthyroidism, 767(t)
Toxoplasmosis, macroglobulins and, 199
  neonatal jaundice and, 648
Trace elements, 334-338
Tranquilizers, bilirubin and, 647
Transaminases, 553-554
Transamination, gluconeogenesis and, 161
  protein digestion and, 151
Transcortin, plasma cortisol and, 197
  properties of, 169(t)
Transduction, bacterial, 243
Transferase enzyme system, 9
  glycolytic reaction equilibria and, *14*
Transferrin, cerebrospinal fluid and, 835
  iron transport and, 175, 322
  iron utilization and, 323
  plasma iron and, 197
  properties of, 169(t)
Translocase, protein synthesis and, 247
Transudates, chloride and, 395
  electrolytes and, 393-397
  glucose and, 395
  lipids and, 395-396
Trauma, C-reactive proteins and, 198

Trauma *(Continued)*
  hypoalbuminemia and, 189
  negative nitrogen balance and, 229
Trehalase, activity of, glucosidases, 2
Trehalose, carbohydrate metabolism and, 2
TRH. See *Thyrotrophin-releasing hormone*.
Triamterene, diuretics and, 469
Tricarboxylic acid cycle,
  aerobic metabolism and, 11
  anaerobic metabolism and, 10
  carbohydrate to lipid conversion and, 107
  carbon dioxide and, 15
  α-ketoglutaric acid and, 161
  ketolysis and, 109
  lipid metabolism and, *102*
  muscle contraction and, 16
  pentose shunt and, 13
Tricarboxylic acid oxidative pathway, 105
Triglycerides. See also *Fats*.
  carbohydrate to lipid conversion and, 106
  conversion of, 105
  fatty acid esters and, 91
  glucose storage and, 5
  α-glycerol-phosphate and, 108
  hyperlipoproteinemia, type four and, 135
  insulin and, 21
  lipid digestion and, 98-99
  lipid storage and, 93-94
  lipid transport and, *123*
  plasma lipids and, 118
  role of, 91
Triiodothyronine, serum, thyroid hormone release tests and, 758
  thyroglobulin and, 332
  thyroid hormone release tests and, 759
Triiodothyronine suppression test, 762-763
  hyperthyroidism and, 768
Triiodothyronine toxicosis, 766
Triolein, I-labeled lipids and, 605
Trioses, 7
Tropocollagen, bone development and, 282
Trypanosomiasis, hyperglobulinemia and, 201
Trypsin, fecal examination and, 598
  Lundh test meal and, 593
  protein digestion and, 149
  regurgitation and, 587
  secretin test and, 592
  serum, serum lipase and, 596
Trypsinogen, insulin and, 590
  pancreas and, 589
  protein digestion and, 149
Tryptophan, prealbumin and, 189-190
  protein digestion and, 148
Tuberculosis, albumin and, 185-186
  basal metabolism, 450
  cryoglobulins and, 198
  globulins and, 383
  hypochloremia and, 383
  hypocholesterolemia and, 143
  hypoprothrombinemia and, 833
  hyposecretion and, 586
  glucose tolerance and, 64

INDEX

Tuberculosis *(Continued)*
  protein and, 188
  thymol turbidity test and, 627
Tuberculosis enteritis, glucose tolerance and, 70
Tubular necrosis, defect localization and, 522
  urinary nonprotein nitrogen and, 513
Tubules. See *Kidney tubules*.
Tumors, abdominal, prerenal albuminuria and, 216
  endocrine syndromes and, criteria, 799
  hypoglycemia and, 58
Turbidity reactions, plasma protein abnormalities and, 183
Turner's syndrome, gonadotrophin and, 696-698
Two-hour specific gravity test, 496-497
Typhoid fever, cerebrospinal fluid and, 836
  hypochloremia and, 383
  sulfur and, 329
Typhus, plasma protein abnormalities and, 201
Tyramine, catecholamines and, 741, 743
Tyrosinase, copper and, 335
Tyrosine, *18*
  ascorbic acid and, 821
  catecholamines and, 741
  fatty acid synthesis and, 104
  iodination of, 748
  ketogenic pathway and, 162
  metabolism of, abnormal, 225
  protein digestion and, 148
  tolerance to, thyroid disorders and, 760
Tyrosinemia, hereditary, phenylalanine and, 225
  neonatal, phenylalanine metabolism and, 225
Tyrosinosis, phenylalanine metabolism and, 225
Tyrosinuria, liver function and, 618
  phenylalanine metabolism and, 226

U. See *International units*.
UDPG. See *Uridine diphosphoglucose*.
UFA. See *Fatty acids, unesterified*.
Ulcer(s). See also under specific organ.
  glomerular filtration and, 458
  peptic, electrolytes and, 375
    gastric investigations and, 584
    proteinuria and, 219
Ultracentrifugation, plasma protein abnormalities and, 183
  plasma protein study and, 171, 171(t)
Undernutrition, cholesterol and, 118, 119
  glucose and, 66, 68
  hypoglycemia in, 58
  protein and, 230
Undetermined nitrogen. See *Nitrogen, undetermined*.
Unesterified fatty acids. See *Fatty acids, unesterified*.

Uracil, nucleic acids and, 235
  nucleoside triphosphates and, 250
Uranium poisoning, 84
Urate, concentration in urine, 254
  plasma concentration conversion and, 856
Urea, blood amino acids and, 160
  blood and, 509-514
  body fluids and, 514-515
  clearance, 501-506
  formation of, nitrogen disposal and, 160
  kidney tubules and, 464
  liver function and, 618
  nitrogen excretion and, 180
  nonprotein nitrogen and, 178
  plasma concentration conversion and, 856
  retention of, electrolytes and, 373
  urinary, nonprotein nitrogen and, 220-222, 472
Urea cycle, disorders of, amyloidosis and, 231-232
  arginosuccinic aciduria and, 230
  protein metabolism and, 230-232
Urea nitrogen. See *Nitrogen, urea*.
Uremia, blood volume and, 392
  cerebrospinal fluid and, 835
  diseases of, electrolytes and, 375
  lactic acidosis and, 77
  magnesium and, 317
  nonprotein nitrogen and, 207-212, 507-515, 518-526
  postrenal albuminuria and, 216
Ureterosigmoidostomy, hyperchloremia and, 383
  nonrespiratory acidosis and, 419
Uric acid, concentration in urine, 254
  diminished, 256
  hyperuricemia and, 255
  in blood, nucleotide synthesis and, 253
  liver function and, 619
  meningitis and, 840
  nitrogen excretion and, 180
  nonprotein nitrogen and, 178
  nucleotide synthesis and, 253
  purine catabolism and, 252
  urinary nonprotein nitrogen and, 222, 474-475
  urine and, purine and, 253
Uricase, uric acid and, 252
Uricosuric effect, kidney tubules and, 464
Uridine diphosphate, lipid metabolism and, 9
  ganglioside biosynthesis and, 92
Uridine diphosphoglucose, *9*
Uridine triphosphate, 9, *9*
Uridylic acid, nucleic acid biosynthesis and, 238
Uridyltransferase, *9*
Urinary 11-deoxycortisol, adrenogenital syndrome and, 737
Urinary 17-oxosteroids, adrenogenital syndrome and, 737
Urine, abnormal sugar and, 83-89
  acidification of, 408-409, *409*

916 INDEX

Urine *(Continued)*
  acidification of, kidney tubules and, 462
  acidity of, respiratory acidosis and, 417
  albumin and, 179
  amino acids in, nonprotein nitrogen and, 223-229
  ammonia and, 229
    acid-base balance and, 430
    excretion of, 410
  amylase and, pancreatic function and, 596-597
  ascorbic acid and, 823
  bicarbonate content of, 410
  bilinogens and, 650-654
  bilins and, 650-654
  calcium excretion and, 285, 292
  chloride and, 361-363
  creatine in, nonprotein nitrogen and, 222-223
  creatinine in, nonprotein nitrogen and, 222-223
  dehydration and, 370
  excretion of, amino acid nitrogen and, 210
    protein metabolism and, 158
  fat in, 111
  gamma-globulins and, 192
  histidine and, 182
  insulin in, 24
  iron excretion and, 320
  magnesium and, 316
  nonprotein nitrogen and, 220-222
  normal, characteristics of, 470-481
    acidity, 471-472
    allantoin and, 475
    calcium and, 479
    carbohydrates and, 480
    chloride and, 478-479
    citric acid and, 476-477
    glucuronic acid and, 479
    hippuric acid and, 479
    ketone bodies and, 477
    magnesium and, 479-480
    nonprotein nitrogenous constituents of, 472-475
    organic acids and, 477
    oxalic acid and, 475
    phosphate and, 478
    potassium and, 479
    sodium and, 479
    specific gravity, 470-471
    sulfur compounds and, 477
    volume, 470
  obstruction of, creatine and, 210
    nonprotein nitrogen and, 514
    water elimination and, 494
  pH of, 407
  phosphate excretion and, 292
  porphyrin and, 655-656
  postoperative states and, 388
  potassium and, 361-363, 384

Urine *(Continued)*
  protein metabolism and, 153
  proteins in, variety of, 219
  proteinuria and, 218-219
  sodium and, 361-363
  sodium excretion and, 379
  specific gravity of, kidney function and, 495-501
    polyuria and, 86
  sugar secretion and, 82-83
  sulfur and, 328-329
  suppression of, creatine and, 210
  titratable acids in, acid-base balance and, 430
  urate ion concentration in, 254
  uric acid in, nonprotein nitrogen and, 222
    purine and, 253
  uropepsin and, 587
  water loss and, 341
  zinc and, 338
Urine concentration test, kidney function and, 497
Uriniferous tubule. See *Nephrons*.
Urobilinogen, in feces, 651
  in urine, 652-654
  jaundice diagnosis and, 680
  liver function and, 634-635
Urogastrone, intestinal phase and, 582
Urogenital system, diseases of, enzyme abnormalities and, 577
Urography, iodine and, 331
Uropepsin, 551
  pepsin and, 587
Uroporphyrins, defined, 261
  porphyrinurias and, 275(t)
Uterus, prostaglandins and, 95
UTP. See *Uridine triphosphate*.

Vaginal candidiasis, oral contraceptives and, 787
Vagotonia, hypoglycemia in, 58
Vagus center, gastric activity and, 585
Vagus nerves, insulin release and, 24
Valinuria, aminoaciduria and, 228
Values, normal range of, 852
van den Bergh reaction, serum bilirubin and, 636
  xanthochromia and, 845
van den Bergh technique, 848
Van Slyke method, 848
Vasa recta, diuresis and, *468*
Vascular disease, hyperuricemia and, 258
Vasopressin. See also *Antidiuretic hormone*.
  abnormal secretion of, 711-712, 801
  neurohypophysis and, 710-712
Venous plexus, kidney anatomy and, 456
Ventilation, pulmonary, nonrespiratory acidosis and, 420
  respiratory acidosis and, 416

Vesicular glands, prostaglandins in, 95
Vincaleukoblastine, choriocarcinoma and, 250
Viral disease. See *Diseases, viral.*
Virus(i), gamma-globulins and, 191
  nucleic acids and, 240-241
Vitamin(s), 804-834. See also under specific vitamins.
  deficiency of, aminoaciduria and, 224
    renal tubular defects and, 537
  fat-soluble, lipid absorption and, 100
  intestinal malabsorption syndromes and, 602
  lipid-soluble, phospholipids and, 97
  metabolism of, thyroid hormone and, 753
Vitamin A, 805-810
  blood concentration of, 808-809
  bone development and, 284
  dark adaptation test and, 809
  deficiency of, 808
    carotenemia and, 810
    detection of, 808
    intestinal malabsorption and, *603*
  excess of, 809-810
  hypercarotenemia and, 810
  lipid absorption and, 100
Vitamin A tolerance test, 606
Vitamin B, coenzymes and, 251
  imbalance of, lipotropism and, 128
  type two fatty liver and, 127
Vitamin B_6, 815-816
Vitamin B_{12}, 818-821
  absorption of, clinical study of, 607-608
  cobalt and, 337
  deficiency of, 820-821
    intestinal malabsorption and, *603*
  macrocytic megaloblastic, folic acid and, 602
Vitamin D, 825-829
  bile salts and, 100
  calcium absorption and, 281
  calcium concentration and, 290
  deficiency of, demonstration of, 828-829
    hypocalcemia and, 302
    metabolic manifestations, 827
    urinary calcium and, 308
  excess of, 829
  hypercalcemia and, 297, 299
  loss of, intestinal malabsorption and, *603*
  metabolic products of, *826*
  urine and, 316
Vitamin E, 829-830
Vitamin K, 830-833
  bile salts and, 100
  deficiency of, 832-833
    demonstration of, 833
    hepatic disease and, 832-833
    hypoprothrombinemia and, 833
    impaired absorption and, 832
  intestinal malabsorption and, *603*
  neonatal jaundice and, 647

Volt, defined, 855
Vomiting, ammonia and, 229
  blood volume and, 392
  electrolytes and, 375
  hemoconcentration and, 183
  plasma albumin and, 185
  plasma dilution and, 189
  potassium and, 386
  psychogenic, blood sugar and, 72
von Gierke's disease, carbohydrate metabolism and, 14
  epinephrine tolerance test and, 74
  glucose tolerance and, 616
  glycogen storage disease and, 80

Waldenström's macroglobulinemia, serum protein and, 203
Water. See also *Fluids.*
  abnormalities of, 368-398
  absorption of, kidney tubules and, 460
  adrenocortical hormones and, 722
  adrenocortical hypofunction and, 739
  aerobic metabolism and, 12
  anaerobic metabolism and, 10-11
  body compartments and, *340*, 342-348
  deficits of, 368-380
    dehydration and, 368-369
    gastrointestinal, 369
    routes of loss, 369-370
    skin and, 369
    urine and, 369
  dehydration and, consequences of, 374-375
  depletion of, intestinal malabsorption and, *603*
  distribution of, plasma proteins and, 174
  elimination of, kidney function and, 491-495
    anuria and, 492-494
    input vs. output, 491-492
    oliguria and, 492-494
  equilibrium requirements for, 341
  excess of, 380-383
    congestive heart failure and, 381
    toxemia of pregnancy and, 381-382
  homeostasis of, renal function and, 466-468
  intake of, 339
  intracellular, glucose in, 28
  intracellular fluid and, 351
  metabolism of, liver and, 672
    thyroid hormone and, 752
  normal turnover of, 340(t)
  output of, 340-342
    feces and, 341
    perspiration and, 341
    urine and, 341
  physiological considerations, 339-367
  prerenal deviation of, 507-508
  primary deficit of, 380

Water *(Continued)*
  regulatory mechanisms and, 357-360
  restriction of, blood volume and, 392
  sources of, *340*
  total in body, 345-346
Water function test, kidney function and, 494
Watt, defined, 855
Weight, loss of, heavy-chain disease and, 202
Weltmann test, globulin reactions and, 205
Werner's syndrome, hyperlipoproteinemia type four and, 135
Wernicke's encephalopathy, cerebrospinal fluid and, 846
Whipple's disease, fecal fat and, 101
  steatorrhea and, 601
Willis, Thomas, 847
Wilm's tumor, actinomycin D and, 250
Wilson's disease, amino aciduria and, 224
  ceruloplasmin and, 198, 336
  Fanconi syndrome and, 84
  renal tubular defects and, 537
Wohlgemuth technique, 848
Wolman's disease, 129, 130(t)

X-ray irradiation, plasma fibrinogen and, 184
Xanthelasma, hyperlipoproteinemia type two and, 134
Xanthine, nucleotide synthesis and, 252
Xanthinuria, hereditary, hypouricemia and, 259
  lithiasis and, 543

Xanthochromia, cerebrospinal fluid and, 844-845
Xanthomas, hyperlipoproteinemia type three and, 135
  hyperlipoproteinemia type five and, 136
Xanthomatosis, hypercholesterolemia and, 141
  lipid metabolism and, 129
Xanthophyll, icterus index and, 638
L-Xylitol, pentosuria and, 88
L-Xyloketose, pentosuria and, 88
L-Xylulose, pentosuria and, 88
Xylulose-5-phosphate, pentose shunt and, 13

Yellow fever, bilirubin and,
  hypocholesterolemia and, 142

Zimmerman reaction, 850
Zinc, copper absorption and, 335
  metabolism and, 337-338
  trace elements and, 337
Zinc sulfate turbidity, globulin reactions and, 205-206
  liver function and, 628
Zinc sulfate turbidity test, globulins and, 190
Zollinger-Ellison syndrome, gastrin and, 772
  hypersecretion and, 586-587
  pancreatic function and, 597
  steatorrhea and, 601
Zymogens, pancreas and, 589